T0181646

Solid Mechanics and Its Applications

Volume 225

Series editors

J.R. Barber, Ann Arbor, USA
Anders Klarbring, Linköping, Sweden

Founding editor

G.M.L. Gladwell, Waterloo, ON, Canada

Aims and Scope of the Series

The fundamental questions arising in mechanics are: *Why?, How?, and How much?* The aim of this series is to provide lucid accounts written by authoritative researchers giving vision and insight in answering these questions on the subject of mechanics as it relates to solids.

The scope of the series covers the entire spectrum of solid mechanics. Thus it includes the foundation of mechanics; variational formulations; computational mechanics; statics, kinematics and dynamics of rigid and elastic bodies; vibrations of solids and structures; dynamical systems and chaos; the theories of elasticity, plasticity and viscoelasticity; composite materials; rods, beams, shells and membranes; structural control and stability; soils, rocks and geomechanics; fracture; tribology; experimental mechanics; biomechanics and machine design.

The median level of presentation is to the first year graduate student. Some texts are monographs defining the current state of the field; others are accessible to final year undergraduates; but essentially the emphasis is on readability and clarity.

More information about this series at http://www.springer.com/series/6557

Zdeněk P. Bažant · Milan Jirásek

Creep and Hygrothermal Effects in Concrete Structures

 Springer

Zdeněk P. Bažant
Department of Civil and Environmental
  Engineering
Northwestern University
Evanston, IL
USA

Milan Jirásek
Department of Mechanics, Faculty of Civil
  Engineering
Czech Technical University in Prague
Prague
Czech Republic

ISSN 0925-0042   ISSN 2214-7764   (electronic)
Solid Mechanics and Its Applications
ISBN 978-94-024-1496-7   ISBN 978-94-024-1138-6   (eBook)
https://doi.org/10.1007/978-94-024-1138-6

© Springer Science+Business Media B.V. 2018
Softcover re-print of the Hardcover 1st edition 2018
This work is subject to copyright. All rights are reserved by the Publisher, whether the whole or part of the material is concerned, specifically the rights of translation, reprinting, reuse of illustrations, recitation, broadcasting, reproduction on microfilms or in any other physical way, and transmission or information storage and retrieval, electronic adaptation, computer software, or by similar or dissimilar methodology now known or hereafter developed.
The use of general descriptive names, registered names, trademarks, service marks, etc. in this publication does not imply, even in the absence of a specific statement, that such names are exempt from the relevant protective laws and regulations and therefore free for general use.
The publisher, the authors and the editors are safe to assume that the advice and information in this book are believed to be true and accurate at the date of publication. Neither the publisher nor the authors or the editors give a warranty, express or implied, with respect to the material contained herein or for any errors or omissions that may have been made. The publisher remains neutral with regard to jurisdictional claims in published maps and institutional affiliations.

Printed on acid-free paper

This Springer imprint is published by Springer Nature
The registered company is Springer Science+Business Media B.V.
The registered company address is: Van Godewijckstraat 30, 3311 GX Dordrecht, The Netherlands

To our wives Iva and Vlasta without whose loving support this book could not have come to fruition, and to the memory of late Treval C. Powers, a giant of 20th century cement chemistry and physics underpinning important parts of this book.

To the memory of
Treval Clifford Powers
1900–1997

widely regarded as the leading cement and concrete physicist of the middle part of the twentieth century, who made major lasting contributions to the understanding of the structure of fresh and hardened cement paste and achieved fundamental results on concrete rheology, workability, consistency, durability, shrinkage and swelling, creep, and resistance of concrete to frost, sulfates, and abrasion.

Born on February 8, 1900, in Palouse, WA, Powers studied chemistry at Willamette University in Salem, Oregon, a small private college founded in 1846. From 1930 until his retirement in 1965, he conducted research at the famous Portland Cement Association (PCA) Laboratories, located in Chicago and later in Skokie, Illinois. For many years until his retirement in 1965, he served as the PCA Director. His stellar research achievement earned him top honors. The American Concrete Institute (ACI) bestowed on Powers its highest award, the Wason Medal for Materials Research; in fact, it did so three times, in 1933, 1940, and 1948, which is a singular case in ACI history testifying to his fundamental achievements. In 1957, Powers received the S.E. Thompson Award from the American Society for Testing and Materials (ASTM) and, in 1976, the Arthur R. Anderson Award from the ACI. In 1961, he became an Honorary Member of ACI and, also in 1961, he was awarded an honorary Doctor of Science degree by the University of Toledo. During 1967–68, he lectured as Visiting Professor at the University of Toronto. He died on June 30, 1997, in Green Valley, AZ. The founding conference of IA-ConCreep at M.I.T. in 2001 was dedicated to Powers, to honor his memory.

Guest Preface

Concrete, the solid that forms at room temperature from mixing Portland cement with water, sand, and aggregates, suffers from time-dependent deformation under load. This creep occurs at a rate that degrades the durability and truncates the life span of concrete structures, and it can lead in some (fortunately rare) cases to catastrophic failures unless a rational approach is put in place that blends the underlying mechanics and physics of concrete creep and shrinkage with engineering ingenuity and mathematical eloquence to gain predictive impact at the scale of engineering operations.

It is precisely this challenging mission that Zdeněk Bažant and Milan Jirásek embraced and accomplished in this generational masterpiece; the definite book on creep and shrinkage of concrete the community of engineers and scientists has been waiting for.

For generations of scientists and engineers, concrete creep has been a daunting task: Creep rates are intrinsically low, thus requiring typically long time scales for laboratory experimentation under highly controlled hygrothermal conditions. The load-induced deformation must be separated from other sources of deformation related to an ever evolving microstructure and other chemo-physical aging and out-of-equilibrium phenomena that define the very nature of concrete's life cycle. Moreover, creep of concrete is by its very nature dissipative. This means that the work provided to the material or structural system in form of load is not recovered, but irreversibly lost in the creation of deformation in excess of elastic reversible deformation. An engineer in charge of a structural design will thus aim at monitoring via modeling and simulations a controlled energy dissipation, so as to avoid that this energy would be dissipated in an uncontrolled way in, e.g., fracture creation. With limited experimental creep data thus available, engineers must rely on models to predict, over extended periods of time, the impact of creep deformation on structural functionality, integrity and stability, force and moment distribution (in statically indeterminate structures), to ultimately minimize, by inverse design of materials and structures, the impact of structural creep on performance.

With a life worth of experience in Structural Engineering and Design, the core proposition of Zdeněk Bažant and Milan Jirásek is the need for reliable engineering models of concrete creep that permit engineers to meet the tasks ahead of designing high-performance concrete structures with high confidence levels. These models are calibrated against a large database of creep and shrinkage painstakingly collected and developed by Zdeněk Bažant since the early 1960s. In their most advanced version, these models permit a rapid recalibration from short-term tests. Enabled by the physics of the phenomena at stake, these models become an integral part of the innovation pathway for sustainable concrete and concrete structures.

This is science-enabled engineering at its best! Decoded by two engineering scientists of eminent status and encyclopedic knowledge of the mechanics and physics of concrete creep, this book is a must-read for any structural engineer and engineering scientist in search of concrete innovation.

Franz-Josef Ulm
Massachusetts Institute of Technology,
Former Vice-President of the Engineering
Mechanics Institute of the
American Society of Civil Engineers,
Director, Concrete Sustainability Hub at MIT

Preface

The literature on concrete creep and shrinkage is vast and scattered over many media. A host of books have already been devoted to this subject. Some are valuable compilations of diverse properties and experimental results, but lack a coherent system. A few champion formalistic beauty of sophisticated mathematical treatment based, however, on oversimplified unrealistic hypotheses. Others are focused on simple methods of analysis for practical design (nothing is, of course, wrong with simple methods except when they are simplistic).

In this book, we present a different kind of exposition of the subject. We attempt to balance a sound, theoretically justified, mathematical modeling at the level of current knowledge with careful attention to laboratory test results, measurements on structures, structural design applications, analysis of design standards or recommendations, and numerical algorithms.

To make our book useful to different kinds of readers, we divide it in two parts: Part I deals with the essentials required for designing structures, and Part II deals with advanced subjects concerned with the effects of moisture, solidification, aging, cracking, and temperature, the consideration of which is necessary for more accurate predictions. We mark by asterisks the titles of the sections that elaborate on various highly theoretical aspects and can be skipped by a reader interested mainly in practical design. Throughout the book, we emphasize the randomness of creep and shrinkage effects and their probabilistic treatment.

Both parts together give an all-encompassing presentation, although with some exceptions. We must admit that our exposition of the nanoscale mechanism of creep and shrinkage may soon be regarded incomplete because the current research, driven by increased concern with sustainability of infrastructure and facilitated by advanced micro- and nanoscale measurements as well as computer simulations, is, at the time of writing, advancing rapidly (especially in the research group of Franz-Josef Ulm at M.I.T., with collaborators worldwide). We must also admit that our coverage of the consequences of the large autogenous shrinkage and self-desiccation in modern concretes is only superficial because the long-term data and the understanding of nanomechanics are still too limited. We must admit, too, that our discussions of time-dependent growth of cracking damage and fracture, of deformation rate effects under impact and explosions, and of high-temperature effects in fire or hypothetical nuclear accidents, could be more thorough if the scope of the book permitted.

Our book is intended for university researchers and educators, for members of committees formulating design recommendations, and for practicing engineers designing or evaluating creep sensitive structures such as large-span prestressed bridges, supertall buildings, large roof shells and nuclear containments, airport pavements, underground excavation linings, and ocean oil platforms. Parts of the book can be used for teaching graduate level courses. Supplementary materials will be made available at http://mech.fsv.cvut.cz/ConcreteCreep.

The book grew out of various courses taught by each of the authors, including: sections of the first author's short intensive courses on Material Modeling of Concrete (including creep and shrinkage) taught at Swedish Cement & Concrete Institute in Stockholm in 1976, Chalmers University in 1977, University of Mexico in 1977 and École nationale des ponts et chausées in Paris in 1978, his short course on Concrete Creep and Shrinkage at Politecnico di Milano in 1982, sections of his short courses on Inelastic Materials and Structures at EPF de Lausanne in 1983, 1988, and 1991 and at Luleå University in 1994, and sections of his course on Material Modeling taught at Northwestern University in the 1980s; and sections of the second author's course on Deformation and Failure of Materials taught at CTU Prague. The first author also deeply values the three-year experience in creep analysis of large bridges that he gained as a bridge engineer in Dopravoprojekt, Prague (1961–63). Valuable was also his experience as a staff consultant at Sargent & Lundy Engineers, Chicago, during 1974. He also benefited from the experience with creep and hygro-thermal effect that he gained while serving during 1974–94 as a staff consultant to the Reactor Analysis Division of Argonne National Laboratory. He feels particularly grateful for 48 years of almost continuous funding of numerous research projects, concerned fully or partly with creep, shrinkage, and durability, by the US National Science Foundation, Department of Transportation, Department of Energy, Electric Power Research Institute and W.R. Grace Co.

We wish to express our deep thanks to many respected colleagues for valuable discussions on the subject. The first author was introduced to the subject in 1959 by his undergraduate advisor Jan Klimeš at CTU Prague. During 1967–69, the first author was lucky to have visionary mentors in Robert L'Hermite at CEBTP Paris, Boris Bresler at UC Berkeley, and especially Treval C. Powers of PCA Skokie, a giant of cement physics who inspired the first author's research direction while both held visiting appointments at the University of Toronto during 1967–68. The first author wishes to acknowledge the stimulating interactions and collaborations with Gianluca Cusatis at Northwestern University, Qiang Yu of Pittsburgh University, Franz-Josef Ulm and Roland Pelenq of M.I.T., Jialiang Le at University of Minnesota, Kaspar Willam, Yunping Xi, and Mija Hubler at UC Boulder, Roman Wendner at BOKU Vienna, Folker H. Wittmann, Christian Huet, and Thomas Zimmermann at EPF Lausanne, Ignacio Carol at UPC Barcelona, Matthieu Vandamme of Université Paris Est, Vladimír Křístek, Vít Šmilauer, Zdeněk Bittnar, and Petr Havlásek at CTU Prague, and Joško Ožbolt at Stuttgart University. Thanks for valuable collaborations on the subject are also due to many of the first author's former doctoral students at Northwestern University,[1] as well as postdoctoral associates and visiting scholars.[2]

The second author is grateful for stimulating discussions with his colleagues at the Czech Technical University in Prague, in particular with Jan Zeman, Vít Šmilauer, Zdeněk Bittnar, Bořek Patzák, Radek Štefan, Vladimír Křístek, Jan Vítek, Lukáš Vráblík, Tomáš Vogel, and Pavel Demo, as well as with many international experts, including Christian Huet, Gilles Pijaudier-Cabot, Ignacio Carol, Joško Ožbolt, Peter Grassl, Franz-Josef Ulm, Mija Hubler, Dariusz Gawin, Francesco Pesavento, Luca Sorelli, and Jean-Michel Torrenti. Special thanks are due to Petr Havlásek, who collaborated with the second author on the development of modeling techniques and numerical algorithms for creep, shrinkage, moisture transport, and heat transfer; this research was funded by the Czech Science Foundation (projects 103/09/H078 and P105/10/2400) and by the European Social Fund (project CZ.1.07/2.3.00/30.0034). Assistance with the preparation of some of the figures was provided by students of the Czech Technical University.[3]

Last but not least, as expressed in our dedication, we wish to thank our beloved wives Iva and Vlasta for their sustained support of our professional activities, which made the arduous work on this book possible.

Evanston, USA   Zdeněk P. Bažant
Prague, Czech Republic   Milan Jirásek
July 2017

---

[1] They included Leonard J. Najjar, Spencer T. Wu, Ali A. Asghari, Elmamoun Abdalla Osman, Werapol Thonguthai, Sang-Sik Kim, Liisa Panula, Tatsuya Tsubaki, Jenn-Chuan Chern, Santosh Prasannan, Joong-Koo Kim, Yunping Xi, Ravindra Gettu, Sandeep Baweja, Goangseup Zi, Qiang Yu, Guang-Hua Li, and Mija Helena Hubler.

[2] They included Laurent Granger, Anders Boe Hauggaard, Franz-Josef Ulm, Alexander Steffens, Geir Horrigmoe, Henrik O. Madsen, Tong-Sheng Wang, Joško Ožbolt, M. Elisabeth Karr, Jaroslav Navrátil, Larissa Molina, Zhishen Wu, Milan Holický, Daniele Ferretti, Vít Šmilauer, Goangseup Zi, Abdullah Dönmez, Enrico Masoero, and Mohammad J.A. Qomi.

[3] They included Marek Vinkler, Hana Hasníková, Pavel Fišar, Dominika Majerová, Aneta Bulíčková, and Michal Šmejkal.

Contents

7.14.2   Curvature and Residual Stresses Due to Cyclic Creep 255

7.14.3   Appraisal of the Magnitude of Cyclic Creep Effects in Structures 258

7.14.4   Recapitulation 265

7.15   Conclusions for Method of Analysis and Design 266

Part II   Advanced Topics

8   Moisture Transport in Concrete 271

8.1   Water in Concrete 272

8.2   Pore Fluids at Thermodynamic Equilibrium 276

8.2.1   Multiphase Porous Medium 276

8.2.2   State Equations 278

8.2.3   Capillary Pressure and Relative Humidity 280

8.2.4   Kelvin Equation 284

8.2.5   Sorption Isotherm 290

8.2.6   Free and Hindered Surface Adsorption, Disjoining Pressure, and Its Continuum Thermodynamics* 294

8.3   Moisture Transport 302

8.3.1   Transport Mechanisms 302

8.3.2   Darcy's Law 303

8.3.3   Mass Balance Equation 309

8.3.4   Differential Equations for Moisture Transport 312

8.3.5   Scaling Properties 320

8.3.6   Effect of Distributed Cracking on the Rate of Drying 321

8.4   One-Dimensional Moisture Transport 323

8.4.1   One-Dimensional Diffusion Equation 323

8.4.2   Numerical Solution by Finite Differences 325

8.4.3   Drying of a Slab 345

8.4.4   Initial Drying and Analysis of Infinite Half-Space 353

8.4.5   Evolution of Total Water Loss from a Specimen 361

8.4.6   Effects of Variable Environmental Humidity 373

8.5   Spreading of Hydraulic Pressure Front Into Unsaturated Concrete 381

8.6   Shrinkage and Stresses Due to Nonuniform Drying 386

8.7   Effects of Self-Desiccation and Autogenous Shrinkage in Drying or Swelling Specimens—A Problem Requiring Further Research 402

8.7.1   Recent Paradigm-Changing Observations 402

8.7.2   Improved Aging Characterization via a Model for Hydration 404

8.8   Creep and Diffusion as Processes Controlling Alkali–Silica Reaction 406

Symbols

Since this book covers a wide range of phenomena, theories, and models, it is next to impossible to keep the notation systematic and simple and at the same time consistent with the standard notation in each particular area of research. For the readers' convenience, the following list provides an overview of most symbols used throughout the book, with brief definitions and occasional comments explaining in which equation, section, or chapter the symbol is introduced. Completely generic symbols, e.g., c used as some constant or f used as some function, are not included.

The list shows that certain duplicities arise and are tolerated, in particular in cases when a symbol has two or more completely different meanings that can hardly be confused. For instance, w denotes the deflection of a beam, but also the water content in a concrete mix. On the other hand, symbols that could be confused must be kept unique. For this reason, we systematically denote the total water content in drying concrete as w_t and the evaporable water content as w_e, even though w is often used in the literature.

Symbol	Meaning
A	Area, cross-sectional area
A	Affinity (Chap. 13)
A	Parameter used by Hansen isotherm (Appendix I.1)
A	Water absorption coefficient (Appendix J.16)
A_T	Temperature factor (Chaps. 4 and 7)
A_T	Parameter used in Eqs. (12.41)–(12.42)
A_c	Area of concrete part of a section
A_c	Coefficient used in Eq. (13.89)
A_g	Exponent used in Eq. (J.17)
A_n	Coefficient used in Eq. (13.74)
A_p	Area of prestressing steel part of a section (Chap. 4)
A_p	Parameter used in Eqs. (12.41)–(12.42)
A_s	Area of steel part of a section
A_w	Exponent used in Eq. (J.16)

A_λ	Parameter used in Eq. (13.88)
A_ω	Parameter of Bary formula (12.37)
$A_0, \ldots A_5$	Parameters in Eq. (13.71)
A_0, A_1	Parameters in Eq. (D.54)
a	Aggregate content in concrete mix
a	Permeability (Chap. 13)
a	Crack size
a	Parameter used by ACI model in Eq. (E.23)
a	Moisture ratio corresponding to a complete monolayer (Appendix I.1)
a	Scale function of the gamma process (Appendix K.8)
a_T	Parameter used in Eqs. (13.99) and (13.103)
a_{Tw}, a_{TT}	Parameters used in Eq. (13.77)
a_g^*	Auxiliary coefficient defined in Eq. (J.25)
a_l^*	Auxiliary coefficient defined in Eq. (J.23)
a_{ww}, a_{wT}	Parameters used in Eq. (13.76)
a_0	Initial crack size
a_0	Permeability at saturation (Chap. 13)
$a_0, \ldots a_5$	Parameters in Eq. (13.71)
a_1, a_2, a_3	Coefficients of cubic function in Eq. (13.63)
B	Auxiliary variable defined in Eq. (J.29)
B_T	Surface heat transfer coefficient (Chap. 13)
B_v	Surface vapor transfer coefficient
B_1, B_2	Parameters in Eq. (13.32)
$\boldsymbol{B}$	Strain-displacement matrix
b	Width of rectangular section
b	Parameter used by ACI model in Eq. (E.23)
b	Parameter used by Langmuir isotherm (Appendix I.1)
b	Parameter used by Künzel isotherm (Appendix I.1)
b	Shape function of the gamma process (Appendix K.8)
b_1, b_2	Coefficients in Eq. (J.36)
b_3, b_4	Coefficients in Eq. (J.37)
$\boldsymbol{b}$	Body force vector
$\bar{\boldsymbol{b}}$	Column matrix of body forces
C	Moisture diffusivity
C_A	Axial sectional compliance
C_I	Bending sectional compliance
C_S	Coupling sectional compliance (Sect. 4.3)
C_S	Parameter used by microprestress theory (Chap. 10)
C_T	Function used by BSB model (Appendix I.1)
$\bar{C}_f$	Incremental flow compliance
C_l	Diffusivity of liquid water in saturated concrete
C_p	Effective specific heat capacity of concrete (Chap. 13)
C_p	Isobaric specific heat capacity (Sect. 13.5)

C_{pa}	Specific heat capacity of dry air (Chap. 13)
C_{pag}	Specific heat capacity of aggregates (Chap. 13)
C_{pc}	Specific heat capacity of cement (Chap. 13)
C_{pg}	Specific heat capacity of pore gas (Chap. 13)
C_{pl}	Specific heat capacity of liquid water (Chap. 13)
C_{ps}	Specific heat capacity of solid skeleton (Chap. 13)
C_{ps0}	Specific heat capacity of solid skeleton at room temperature (Chap. 13)
C_{pv}	Specific heat capacity of water vapor (Chap. 13)
C_{pw}	Specific heat capacity of moisture (Chap. 13)
C_{p0}	Specific heat capacity of fresh concrete (Chap. 13)
C_v	Molar concentration of vapor
C_w	Factor defined in Eq. (8.249)
C_γ	Parameter used in Eq. (10.1)
$\boldsymbol{C}_v$	Dimensionless elastic compliance matrix
C_ω	Parameter of Bary formula (12.38) (Sect. 12.7)
C_0	Moisture diffusivity at zero humidity (Appendix I.4.1)
C_0	Parameter of the BET isotherm (8.54)
$C_0, \ldots C_5$	Parameters in Eq. (13.72)
C_1	Parameter in cyclic creep law (Sect. 7.13)
C_1	Moisture diffusivity at saturation (Chap. 8)
$C_{1,28}$	Moisture diffusivity at saturation and at the age of 28 days (Appendix I.4.2)
c	Cement content in concrete mix
c, c_0, c_1	Parameters used by microprestress theory (Chap. 10)
c_f	Material length (Chap. 10)
c_h	Parameter used by GL2000 model (Appendix E.4)
c_k	Auxiliary constants (Appendix F)
c_p	Moisture permeability
c_{pl}	Moisture permeability at saturation
c_1, c_2	Parameters used by Lykow isotherm (Appendix I.1)
D	Effective cross-sectional thickness
D	Characteristic dimension (Chap. 12)
D	Beam depth (Chap. 12)
D	Modified age-dependent modulus (Sect. A.4.2)
D	Dissipated energy (Appendix G)
$\mathscr{D}_M$	Mechanical dissipation (Appendix G)
D_T	Thermal diffusivity (Chap. 13)
D_{av}	Free air-vapor diffusion coefficient
$\boldsymbol{D}_e$	Elastic material stiffness matrix
D_h	Liquid conduction coefficient (Appendix J.6)
$\boldsymbol{D}_k$	Incremental material stiffness matrix
$D_t\ldots$	Material time derivative of ...
D_w	Capillary transport coefficient (Appendix J.6)

D_μ	Parameter of μth aging Kelvin unit
$\mathbf{D}_\nu$	Dimensionless elastic stiffness matrix
D_0	Transitional size
D_0	Free air-vapor diffusion coefficient at reference conditions (Sect. J.1)
d	Parameter used by ACI model in Eq. (E.25)
d	Parameter used by Posnow isotherm (Appendix I.1)
$\mathbf{d}$	Column matrix of displacement parameters
d_a	Maximum aggregate size
E	Elastic (Young's) modulus, spring stiffness
$E\{\ldots\}$	Expectation of ... (Appendix K.8)
E_N	Normal microplane stiffness (Sect. 12.8)
E_{N0}	Initial normal microplane stiffness (Sect. 12.8)
E_T	Shear microplane stiffness (Sect. 12.8)
E_{T0}	Initial shear microplane stiffness (Sect. 12.8)
E_{as}	Asymptotic modulus (age-dependent)
E_c	Elastic modulus of concrete
E_{cyc}	Cyclic effective modulus (Sect. 7.14)
E_{dyn}	Dynamic modulus
E_{ef}	Effective modulus
$\bar{E}_k$	Incremental modulus in step number k
E_{ref}	Reference value of elastic modulus
E_s	Elastic modulus of steel
E_t	Tangent modulus of prestressing steel
E_μ	Stiffness of μth rheologic unit
$E^{(\infty)}$	Final value of elastic stiffness at full solidification
E_∞	Final value of elastic modulus (Appendix D)
E''	Age-adjusted effective modulus
E_0	Asymptotic modulus (age-independent)
E_1	Short-term elastic modulus at age t_1
E_{28}	Conventional elastic modulus
$E_{28}^{(c)}$	Estimate of conventional elastic modulus based on creep compliance
$E_{28}^{(s)}$	Estimate of conventional elastic modulus based on strength
e	Eccentricity
e	Auxiliary strain-like variable (Chap. 9)
e_r	Dimensionless strain rate
$\mathbf{e}_{r\mu}$	Column matrix of dimensionless strain rates in unit μ
$\mathbf{e}_z$	Unit vector in direction opposite to gravity acceleration
$e_{\beta,\alpha}$	Rate of energy transfer from phase β to phase α, per unit volume (Sect. 13.5.6)
F	Concentrated force
$\mathscr{F}$	Isothermal strain energy density (Appendix G)
$\mathscr{F}^*$	Isothermal complementary energy density (Appendix G)

F_b	Interatomic force (Sect. 12.6)
F_p	Function describing the stress–strain law for prestressing steel (Sect. 4.3.4)
f	Filler content in concrete mix
$\mathbf{f}$	Column matrix of equivalent nodal forces
f_c	Compression strength
f_c	Allowable compressive stress (Sect. 7.14)
f_c^0	Reference strength (Sect. 11.4.2)
f_{ck}	Characteristic compression strength (Appendix E)
f_{cm}	Mean compression strength (Appendix E)
$\bar{f}_c$	Standard compression strength
f_c'	Reduced (characteristic, specified) compression strength
$\tilde{f}_c$	Current (age-dependent) compression strength
f_{cr}	Function describing a cohesive cracking law (Chap. 12)
f_{cr}'	Required mean compression strength (Appendix E)
f_k	Auxiliary functions (Appendix F)
$\bar{f}_t$	Mean tensile strength
f_t	Allowable tensile stress (Sect. 7.14)
$f, \bar{f}$	Distributed load
f	Joint probability density
f	Microwave frequency (Sect. 13.3.4)
f_N	Normal loading function (Sect. 12.8)
f_{ref}	Reference strength (Appendix C)
f_T	Shear loading function (Sect. 12.8)
f_{VD}	Volumetric-deviatoric loading function (Sect. 12.8)
f_X	Marginal probability density of random variable X
f_{YX}	Conditional probability density of random variable Y for given X
f_Y'	Prior probability density of long-time response
f_Y''	Posterior probability density of long-time response
$\bar{f}_y$	Mean yield stress of prestressing steel
f_p	Ultimate strength of prestressing steel
f_{py}	Specified yield strength of prestressing steel
f_{pu}	Specified tensile strength of prestressing steel
f_s	Stoichiometric coefficient in Eq. (13.63)
f_y'	Marginal probability density of short-time response
f_ω	Function used by damage models (Sect. 12.7)
f_1, f_2, f_3	Functions defined in Eqs. (13.80)–(13.82)
G	Shear modulus of elasticity
G	Generic nonlinear transformation (Sect. 6.2)
G	Gibbs free energy, free enthalpy
$\mathscr{G}$	Energy release rate
G_f	Fracture energy

g	Gravity acceleration
g	Dimensionless energy release function (Chap. 10)
g	Auxiliary function defined in Eq. (3.23)
μ	Specific Gibbs free energy (Sect. 13.5 and Appendix G)
g_{sh}	Function defined in Eq. (13.105)
$g_\omega, \tilde{g}_\omega$	Functions used by damage models (Sect. 12.7)
H	Heaviside step function
H	Height of a water column (Sect. 8.3.2)
H_c	Released hydration heat (Chap. 13)
H_{max}	Potential hydration heat (Chap. 13)
H_p	Pressure head, hydraulic head
H_t	Total head
H_{inf}	Ultimate hydration heat (Chap. 13)
H_Δ	Redistribution function
h	Pore relative humidity
h	Depth of rectangular section
h	Planck constant (Sect. 12.6)
$\bar{h}$	Average pore relative humidity
$\bar{\bar{h}}$	Pore relative humidity averaged in space and time (Sect. 8.4.6)
h^*	Specific enthalpy (Sect. 13.5)
h_c	Parameter of Bažant–Najjar model
h_c	Process zone thickness (Chap. 12)
h_{env}	Environmental relative humidity
$\bar{h}_{env}$	Average ambient relative humidity (Sect. 8.4.6)
$\hat{h}_{env}$	Peak amplitude of ambient relative humidity (Sect. 8.4.6)
h_g^w	Specific enthalpy of water vapor (Chap. 13)
h_{in}	Initial relative humidity (Chap. 13)
h_l^w	Specific enthalpy of liquid water (Chap. 13)
h_s	Thickness of computational crack band (Chap. 12)
h_s^*	Humidity rate due to self-desiccation
h_s^w	Specific enthalpy of water in hydrates (Chap. 13)
h_0	Initial relative humidity
I	Moment of inertia
$\mathscr{I}$	Identity operator
I_b	Moment of inertia of a beam (Sect. 4.1)
I_c	Moment of inertia of a column (Sect. 4.1)
I_c	Moment of inertia of concrete part of a section
I_k	Auxiliary constants (Appendix F)
I_s	Moment of inertia of steel part of a section
I_4, I_5, I_6	Higher-order moments of sectional area (Sect. 7.14)
i_s	Radius of inertia of steel part of a section
J	Compliance function for aging material
J	Jacobian (Sect. 13.5)
$\bar{J}$	Average compliance

$\mathscr{J}$	Compliance operator (acting on variable t)
$\mathscr{J}'$	Compliance operator acting on variable t'
J_G	Shear compliance function
J_K	Bulk compliance function
J_b	Basic creep compliance function
J_d	Drying creep compliance function
J_d^∞	Final value of drying creep compliance (Appendix D)
J_f	Flow compliance function
J_{tot}	Total material compliance in presence of cyclic loading
J_v	Viscoelastic compliance function
J_Δ	Function defined in Eq. (4.22)
J_0	Compliance function for nonaging material
$\boldsymbol{j}$	Mass flux vector
$\boldsymbol{j}_a$	Mass flux of dry air
$\boldsymbol{j}_a^A$	Advective mass flux of dry air
$\boldsymbol{j}_a^D$	Diffusive mass flux of dry air
$\boldsymbol{j}_g$	Mass flux of gas
$\boldsymbol{j}_g^A$	Advective mass flux of gas
$\boldsymbol{j}_l$	Mass flux of liquid water
j_r	Radial mass flux
$\boldsymbol{j}_s$	Mass flux of solid phase
$\boldsymbol{j}_v$	Mass flux of water vapor
$\boldsymbol{j}_v^A$	Advective mass flux of water vapor
$\boldsymbol{j}_v^D$	Diffusive mass flux of water vapor
j_w	One-dimensional mass flux of moisture (all phases of water)
$\boldsymbol{j}_w$	Mass flux of moisture (all phases of water)
j_x, j_y, j_z	Components of mass flux vector
K	Bulk modulus of elasticity
K	Stress intensity factor
$\boldsymbol{K}$	Structural stiffness matrix
$\tilde{\boldsymbol{K}}$	Structural stiffness matrix for unit elastic modulus
K_I	Mode-I stress intensity factor
K_c	Critical stress intensity factor, fracture toughness
K_h	Hydraulic permeability, hydraulic conductivity, filtration coefficient
K_l	Bulk modulus of liquid water
K_{max}	Maximum stress intensity factor in cycles
K_{min}	Minimum stress intensity factor in cycles
K_0	Intrinsic permeability
$K_{0,ref}$	Reference value of intrinsic permeability
K_1	Permeability coefficient (Appendix J.6)
k	Reciprocal moisture capacity (Chap. 8)
$\bar{k}$	Average reciprocal moisture capacity (Chap. 8)
k_B	Boltzmann constant

k_T	Thermal conductivity (Chap. 13)
k_T	Function used by BSB model (Appendix I.1)
$k_T^{(dry)}$	Thermal conductivity of dry concrete (Chap. 13)
$k_T^{(dry,0)}$	Thermal conductivity of dry concrete at room temperature (Chap. 13)
k_{T0}	Thermal conductivity of fresh concrete (Chap. 13)
k_h	Parameter used by B3
k_r	Relative permeability
$k_{r,g}$	Relative permeability to gas
$k_{r,l}$	Relative permeability to liquid water
k_s	Parameter used by B3 and B4
k_{sh}	Shrinkage coefficient
k_{sh}^*	Shrinkage ratio
$k_{sh,ij}^*$	Tensorial components of shrinkage ratio (Sect. 13.3.3.3.2)
k_t	Parameter used by B3
k_{tw}	Proportionality factor used in Eq. (8.225)
k_{tw0}	Proportionality factor used in Eq. (8.209)
$k_{\varepsilon a}$	Parameter used by model B4 (Appendix D)
$k_{\tau a}$	Parameter used by B4
k_1	Reciprocal moisture capacity at saturation (Chap. 8)
k_1	Parameter used by microprestress theory (Chap. 10)
L	Span
L	Stress impulse memory function for aging material
L	Continuous retardation spectrum
L^*	Auxiliary function derived from retardation spectrum (Appendix F)
$\mathscr{L}$	Likelihood function (Appendix K.6)
L_c	Height of a column
L_f	Continuous retardation spectrum (age-independent)
L_{ij}	Coefficients used for evaluation of shear strain on microplane (Sect. 12.8)
L_k	kth order approximation of continuous retardation spectrum
L_0^*	Stress impulse memory function for nonaging material
l	Tangential vector on microplane (Sect. 12.8)
l_a	Size of atomic lattice block (Sect. 12.6)
l_d	Thickness of hindered adsorbed layer
l_i	Components of tangential vector on microplane (Sect. 12.8)
l_0	Material characteristic length
M	Number of rheologic units in a chain (without isolated spring)
M	Bending moment
M	Total mass of a body (Sect. 13.5)
M_C	Moment resultant of stresses in concrete (Sect. 7.14)
M_{CEB}	Mean deviation (Appendix K.4)
M_D	Moment due to dead load (Sect. 7.14)

M_{DLP}	Moment due to dead load, live load, and prestress (Sect. 7.14)
M_L	Moment due to live load (Sect. 12.8)
M_a	Molar mass of dry air
M_{ij}	Coefficients used for evaluation of shear strain on microplane (Sect. 7.13)
M_{sh}	Bending moment due to shrinkage
M_w	Molar mass of water
m	Exponent in creep laws, for B3 model set to 0.5
m	Exponent in Paris law (Sect. 7.13)
m	Parameter of van Genuchten law (8.16)
m	Parameter in isotherm formula (13.66)
m	Exponent used by ACI model in Eq. (E.30)
m	Measure defined in (K.14)
$\boldsymbol{m}$	Tangential vector on microplane (Sect. 12.8)
m_a	Apparent density of dry air
m_{deh}	Mass of water released by dehydration per unit volume (Chap. 13)
m_{ev}	Mass of evaporated water per unit volume (Chap. 13)
m_i	Components of tangential vector on microplane (Sect. 12.8)
m_l	Apparent density of liquid water
m_s	Apparent density of solid skeleton
m_v	Apparent density of water vapor
$\dot{m}_{\beta,\alpha}$	Rate of phase change from phase β to phase α (Sect. 13.5)
N	Normal force, axial force
N	Number of strata (Sect. 6.2)
$\boldsymbol{N}$	Displacement interpolation matrix
N_C	Force resultant of stresses in concrete (Sect. 7.14)
N_{cyc}	Number of cycles
N_{ij}	Coefficients used for evaluation of normal strain on microplane (Sect. 12.8)
N_m	Number of microplanes (Sect. 12.8)
N_{sh}	Normal force due to shrinkage
n	Exponent in creep laws, for B3 model set to 0.1
n	Parameter used by Hansen isotherm (Appendix I.1)
$\boldsymbol{n}$	Unit outward normal vector (on a boundary)
$\boldsymbol{n}$	Normal vector on microplane (Sect. 12.8)
n_f	Parameter of Kachanov law (12.33)
n_i	Components of normal vector on microplane (Sect. 12.8)
n_k	Auxiliary constants (Appendix F)
n_p	Porosity
n_{p0}	Capillary porosity at reference temperature (Chap. 13)
P	Axially applied compressive force
$\mathscr{P}$	Probability

P_D	Dead load
P_D^*	Factored dead load
P_L	Live load
P_L^*	Factored live load
P_{max}	Maximum load, ultimate load
P_{cr}	Critical load, buckling load
p	Pressure
p	Number of random parameters (Sect. 6.2)
p	Parameter used by microprestress theory (Chap. 10)
p_a	Air pressure
p_{ad}	Average longitudinal stress in hindered adsorbed layer
$p_{ad,f}$	Longitudinal stress when a nanopore just gets filled
p_{atm}	Atmospheric pressure
p_c	Capillary pressure
p_d	Disjoining pressure
p_{entry}	Entry pressure
p_f	Pressure at the moment a nanopore gets filled
p_g	Gas pressure
p_{gel}	Average gel pressure (Sect. 8.8)
p_h	Hydrostatic pressure
p_l	Pressure in liquid (capillary) water
p_{l0}	Reference pressure
p_v	Partial vapor pressure in capillary pores
$p_{v,env}$	Ambient vapor pressure
$p_{v,f}$	Vapor pressure when a nanopore just gets filled
p_{sat}	Saturation vapor pressure of water
p_1, p_2	Update parameters used in Eq. (3.34) or Eq. (11.4)
p_3	Update parameter used in Eq. (H.6)
$p_1, p_2, p_3, p_4, p_5,$	Parameters used by model B4 (Appendix D)
p_{5H}	
$p_{\tau c}, p_{\tau w}, p_{\varepsilon w}$	Exponents used by model B4 (Appendix D)
Q	Auxiliary function used by models B3 and B4
Q	Activation energy
Q_T	Activation energy used by *fib* model (Appendix E.22)
Q_a	Latent heat of adsorption minus latent heat of condensation
Q_e	Activation energy of hydration
Q_f	Auxiliary function used by B3 model (Appendix C)
Q_p	Activation energy of flow of prestressing steel
Q_r	Activation energy of viscous processes
Q_s	Activation energy of microprestress relaxation
Q_w	Activation energy of water migration at low temperature (Chap. 13)
Q_0	Activation energy of separation of interatomic bonds (Sect. 12.5)

q	Quotient in geometric progression
q	One-dimensional conductive heat flux (Chap. 13)
$\boldsymbol{q}$	Conductive heat flux vector (Chap. 13)
q_{conv}	Rate of energy loss by convective heat flux, per unit surface area (Chap. 13)
$\boldsymbol{q}_g$	Conductive heat flux vector in gaseous phase (Sect. 13.5.6)
$\boldsymbol{q}_l$	Conductive heat flux vector in liquid phase (Sect. 13.5.6)
q_n	Normal heat flux
$\boldsymbol{q}_r$	Radiation heat flux vector (Chap. 13)
q_s	Parameter used by log-double-power creep law
$\boldsymbol{q}_s$	Conductive heat flux vector in solid phase (Sect. 13.5.6)
q_1, q_2, q_3, q_4, q_5	Parameters of models B3 and B4
q_5^*	Auxiliary parameter defined in Eq. (D.72)
R	Relaxation function for aging material
R	Universal gas constant
$\mathscr{R}$	Relaxation operator (acting on variable t)
$\mathscr{R}_{t}$	Relaxation operator acting on variable t'
$\mathscr{R}_A$	Operator defined in Eq. (4.82)
$\mathscr{R}_I$	Operator defined in Eq. (4.78)
$\mathscr{R}_S$	Operator defined in Eq. (4.77)
$\mathscr{R}_c$	Relaxation operator for concrete
$\mathscr{R}_s$	Relaxation operator for steel
R_0	Relaxation function for nonaging material
r	Radial coordinate
r	Radius of capillary meniscus
r	Parameter of Bažant–Najjar model
r	Distributed heat source per unit mass (Chap. 13)
r	Auxiliary function used by B3 model (Appendix C)
r, r'	Parameters used in Eq. (13.103)
r_g	Distributed heat source in gaseous phase (Sect. 13.5.6)
r_l	Distributed heat source in liquid phase (Sect. 13.5.6)
r_s	Distributed heat source in solid phase (Sect. 13.5.6)
r_α	Parameter used by model B4 (Appendix D)
$r_{\varepsilon w}$	Exponent used by model B4 (Appendix D)
r_1, r_2	Principal radii of curvature
S	Strain impulse memory function for aging material
S	Shrinkage function defined in Eq. (3.16)
S	Generic static quantity
S	Static moment of a section (Sect. 4.3)
S	Microprestress (Chap. 10)
S	Boundary of a spatial domain representing a body of interest
S	Standard deviation of compression strength tests (Appendix E)
S	Auxiliary function (Appendix F.4)

S_{CEB}	Relative error (Appendix K.3)
S_{cr}	Critical saturation degree (Appendix J.2)
S_e	Exposed surface area of a concrete part
S_{ir}	Irreducible saturation degree (Appendix J.2)
S_j	Points separating individual strata (Sect. 6.2)
S_l	Saturation degree (by liquid)
S_{spec}	Specific area (of pores)
S_t	Unsupported (free) part of boundary of a body
S_0^*	Strain impulse memory function for nonaging material
s	Standard deviation
s	Distance between neighboring atoms (Sect. 8.2)
s	Crack spacing
s	Weighted standard error (Sect. 11.4)
s	Specific entropy (Sect. 13.5 and Appendix G)
s	Parameter used by CEB, *fib*, and GL2000 models (Appendix E)
s^*	Specific entropy production (Sect. 13.5 and Appendix G)
$\dot{s}$	Rate of instantaneous microprestress (Chap. 10)
s_Y	Standard deviation of random variable Y
s_Y'	Prior approximation of standard deviation of Y
s_Y''	Posterior approximation of standard deviation of Y
s_c	Spacing of dominant cracks
$s_{c,max}$	Maximum possible crack spacing
$s_{\varepsilon f}$	Exponent used by model B4s (Appendix D)
$s_{\tau f}$	Exponent used by model B4s (Appendix D)
s_2, s_5	Parameters used by model B4s (Appendix D)
T	Temperature (absolute)
$T, T_1, \ldots T_4$	Auxiliary functions (Appendix F.4)
T_C	Temperature expressed in °C
T_{cr}	Critical temperature of water
T_{env}	Ambient temperature (Chap. 13)
T_g	Characteristic time of aging process (Sect. 9.8.2)
T_h	Period of humidity cycles (Sect. 8.4.6)
T_{init}	Initial temperature
T_{tr}	Transition temperature in Eq. (13.79)
T_0	Room temperature (usually 293 K)
t	Time (general)
t	Current age
$\mathbf{t}$	Surface force vector
$\bar{t}$	Column matrix of prescribed surface forces
t'	Age at loading in a creep test
$\hat{t}, \hat{t}_d$	Time of drying
t_T'	Temperature-adjusted age (Appendix E.2.2)
t_{adj}'	Adjusted age at loading (Appendix E.2.2)

t_a	Average age at self-weight application (Sect. 7.9)
t_c	Age at construction end (Chap. 7)
t_c	Characteristic time of hydration (Chap. 13)
t_d	Delay time for the onset of hydration (Chap. 13)
t_e	Equivalent age
t_{e0}	Equivalent time at the onset of drying (Appendix D)
t_f	Age at failure or time to failure
t_k	Time of kth stress jump or kth time instant in a numerical solution
t_m	Age of 1000 days or 1500 days (Sect. 7.9)
t_p	Time at peak humidity (Sect. 8.4.6)
t_r	Reduced time
t_{ref}	Reference time (Sect. 9.4 and Eq. I.4.2)
t_s	Reduced microprestress time
t_0	Age at the onset of drying (end of curing)
t_1	Age at first loading
t_{1c}	Age at which a bridge was open for traffic (Sect. 7.14)
u	Moisture ratio
u	Specific internal energy (Sect. 13.5)
u	Sectional perimeter in contact with the atmosphere (Appendix E.2.1)
u_a	Axial displacement
u_f	Displacement due to fracture (Sec 7.13.2)
u_h	Maximum hygroscopic moisture ratio (Appendix I.1)
u_p	Displacement of prestressing steel
V	Volume (of a concrete part, of a body)
V	Spatial domain representing a body of interest
V_a	Activation volume (Sect. 12.6)
V_g	Volume occupied by gas
V_l	Volume occupied by liquid water
V_m	Monolayer capacity (Appendix I.1)
V_p	Pore volume
V_s	Volume of solid skeleton
V_{tot}	Total volume
v	Volume growth function used in solidification theory
v	Specific volume
$\boldsymbol{v}$	Velocity vector
$\boldsymbol{v}_a$	Activation volume tensor (Sect. 12.6)
v_c	Subcritical crack growth velocity (Sect. 12.5)
v_l	Filtration velocity, Darcy velocity
$\boldsymbol{v}_l$	Filtration velocity vector, volumetric water flux
v_n	Normal velocity
$\boldsymbol{v}_v$	Volume flux of vapor

$v^{(\infty)}$	Final value of volume growth function used in solidification theory
W_L	Total water loss per unit surface area
w	Water content in concrete mix
w	Deflection
$\tilde{w}$	Volumetric water content (Appendix J.3)
w_c	Crack opening
w_d	Water content released by dehydration (Chap. 13)
w_e	Evaporable water content
w_f	Critical crack opening (at zero stress)
w_f	Evaporable water content at free saturation (Appendix J.6)
w_h	Water deficiency due to hydration
$w_{h,\infty}$	Terminal water deficiency
w_{mid}	Midspan deflection
w_n	Non-evaporable water content
w_{sh}	Deflection due to shrinkage
w_t	Total water content
w_μ	Integration weight of microplane number μ (Sect. 12.8)
w_1	Evaporable water content in saturated concrete at 25°C (Chap. 13)
X	Generic random variable
$\mathbf{X}$	Set of random variables
$\mathbf{X}$	Position vector in initial (reference) configuration (Sect. 13.5)
X_1	Redundant force
x, y, z	Spatial coordinates
$\mathbf{x}$	Position vector in current configuration (Sect. 13.5)
x_d	Penetration depth
x_s	Position of saturation front
Y	Generic response variable
$\mathbf{Y}$	Set of response variables
$\bar{Y}$	Mean value of variable Y
$\bar{Y}'$	Prior mean of Y
$\bar{Y}''$	Posterior mean of Y
$Y_{95\%}$	95% confidence limit
Z	Auxiliary function used by B3 model (Appendix C)
Z_{sh}, Z_w	Sums of squared deviations (Appendix H)
z_0	Initial deviation of column axis from straight line
α	Parameter of generalized trapezoidal and midpoint rules
α	Parameter used by creep models (double-power and log-double-power)
α	Relative crack length (Sect. 12.6)
α	Temperature-dependent parameter defined in Eq. (13.83)
α	Exponent used by model B4 (Appendix D)
α	Exponent used by *fib* model (Appendix E.2.2)

α	Collection of internal variables (Sect. 13.5)
α_E	Parameter used by CEB and *fib* models (Appendix E.2)
α_T	(linear) coefficient of thermal expansion
$\alpha_{T,ij}$	Tensorial coefficients of thermal expansion (Sect. 13.3.3.3.2)
α_e, α_r, α_s	Parameters describing influence of humidity on transformed times t_e, t_r, t_s
α_h	Parameter used in Eq. (I.17)
α_{vT}	Volumetric coefficient of thermal expansion
α_μ^A, α_μ^C, α_μ^D, α_μ^f	Correction factors (Appendix F)
α_0	Parameter of Bažant–Najjar model
α_1, α_2	Parameters used by B3
β	Scaling factor
β	Brittleness number (Sect. 12.5)
β	Parameter used in Eq. (13.106)
β	Dimensionless shape factor (Appendix I.4.1)
β_D	Dissipative thermodynamic forces (Sect. 13.5)
β_H	Parameter used by CEB and *fib* models (Appendix E.2)
β_N	Stiffness reduction factor (Sect. 12.7)
β_Q	Quasi-conservative thermodynamic forces (Sect. 13.5)
β_T	Function used by model B4 (Appendix D)
β_T^{cur}, β_T^{dl}	Parameters used by model B4 (Appendix D)
β_c	Convective mass transfer coefficient (Sect. J.7)
β_{eT}, β_{rT}, β_{sT}	Factors describing influence of temperature on transformed times t_e, t_r, t_s
β_{eh}, β_{rh}, β_{sh}	Factors describing influence of humidity on transformed times t_e, t_r, t_s
β_f, β_T	Parameters used by CEB model (Appendix E.2.1)
β_h	Parameter used in Eq. (I.17)
β_k	Auxiliary constant used by exponential algorithm
Γ_a	Surface water concentration
Γ_1	Mass of full mononuclear water layer per unit area
γ	Parameter used in Sect. 4.3.4
γ	Surface tension at solid–liquid interface
γ	Normalized permeability (Sect. 8.4)
γ	Exponent used by *fib* model (Appendix E.2.2)
γ, γ_1, $\ldots\gamma_6$	Parameters used by ACI model (Appendix E.3)
γ_a	Surface tension at interface between solid and free adsorbed water
γ_{a1}	Parameter used in Eq. (10.1)
γ_e	Surface heat emissivity (Chap. 13)
γ_{gl}	Surface tension at gas–liquid interface
γ_h	Parameter used in Eq. (I.17)
γ_ω	Exponent used in Eq. (12.40)
γ_0	Solid surface tension

$\gamma_0,\ \gamma_1,\ \gamma_2$	Dimensionless geometry factors (Sect. 7.13.2)
$\Delta \ldots$	Increment of ...
ΔC_v	Additional moisture diffusivity due to cracking (Sect. 8.36)
ΔJ_N^cyc	Cyclic creep compliance
ΔT_max	Ultimate temperature increase (Chap. 13)
ΔW	Water loss from a wall per unit area of its mid-surface
ΔW_∞	Final water loss from a wall per unit area of its mid-surface
$\Delta h_\mathrm{l,g}^\mathrm{w}$	Specific enthalpy of vaporization (Chap. 13)
$\Delta h_\mathrm{s,l}^\mathrm{w}$	Specific enthalpy of dehydration (Chap. 13)
Δt_crit^*	Estimate of critical time step
Δt_s	Conventional delay
Δw	Water loss
Δw_∞	Final water loss per unit volume
$\Delta \varepsilon_N^\mathrm{cyc}$	Strain increment due to cyclic loading
$\Delta \varepsilon_{\mathrm{f},k}''$	Flow strain increment under constant stress in step number k
$\Delta \varepsilon_k''$	Strain increment due to creep in step number k
$\Delta \kappa^\mathrm{cyc}$	Curvature increment due to cyclic loading
δ	Generic kinematic quantity
δ	Crack width (Sect. 8.3.6)
δ	Dirac distribution
δ	Water vapor diffusion coefficient (Appendix J.6)
δ_a	Thickness of free adsorbed layer
δ_a	Initial distance between atoms (Sect. 12.6)
δ_c	Depth of layer affected by humidity variations (Sect. 8.4.6)
δ_{ij}	Kronecker delta
δ_p	Water vapor permeability (Appendix J.6)
ε	Strain
$\boldsymbol{\varepsilon}$	Column matrix of strain components
$\boldsymbol{\varepsilon}$	Strain tensor (Sect. 13.5)
$\hat{\varepsilon}$	Strain level in a relaxation test
$\tilde{\varepsilon}$	Eigenstrain
$\tilde{\varepsilon}$	Equivalent strain (Sect. 12.7)
$\varepsilon_\mathrm{L},\ \varepsilon_\mathrm{M}$	Shear strains on microplane (Sect. 12.8)
$\varepsilon_\mathrm{Le},\ \varepsilon_\mathrm{Me}$	Elastic shear strains on microplane (Sect. 12.8)
$\varepsilon_\mathrm{LITS}$	Load-induced thermal strain
$\varepsilon_{\mathrm{L}\sigma},\ \varepsilon_{\mathrm{M}\sigma}$	Mechanical shear strains on microplane (Sect. 12.8)
ε_N	Normal strain on microplane (Sect. 12.8)
$\varepsilon_{\mathrm{N}e}$	Elastic normal strain on microplane (Sect. 12.8)
$\varepsilon_{\mathrm{N}\sigma}$	Mechanical normal strain on microplane (Sect. 12.8)
$\varepsilon_\mathrm{Nmax},\ \varepsilon_\mathrm{N\sigma max},$ $\varepsilon_\mathrm{Nmin},\ \varepsilon_\mathrm{N\sigma min}$	Internal variables used by microplane model (Sect. 12.8)
ε_T	Thermal strain
$\varepsilon_\mathrm{Tx},\ \varepsilon_\mathrm{Ty},\ \varepsilon_\mathrm{Tz}$	Normal strains due to thermal expansion
ε_T0	Thermal strain in the absence of loading

ε_V	Volumetric strain used by microplane model (Sect. 12.8)
$\varepsilon_{V\sigma}$	Mechanical volumetric strain used by microplane model (Sect. 12.8)
ε_V	Relative change of volume (Sect. 13.2.3)
ε_a	Axial strain
ε_{au}	Autogenous shrinkage strain
$\varepsilon_{au,cem}$	Parameter used by model B4 (Appendix D)
ε_{au}^∞	Final autogenous shrinkage strain
ε_c	Creep strain
ε_{cc}	Strain in concrete (Sect. 4.3)
ε_{cem}	Parameter used by model B4 (Appendix D)
ε_{cyc}	Total cyclic creep strain
$\varepsilon_{e,ij}$	Components of elastic strain tensor
ε_e	Elastic strain
ε_f	Viscous flow strain
ε_{ij}	Tensorial strain components
ε_{lve}	Strain predicted by linear viscoelasticity (Sect. 12.7)
ε_p	Strain in prestressing steel (Sect. 4.3)
ε_r	Residual strain
ε_s^∞	Final shrinkage strain (at given humidity)
ε_{sc}	Smeared cracking strain
$\varepsilon_{s,cem}$	Parameter used by model B4s (Appendix D)
ε_{sh}	Shrinkage strain
$\bar{\varepsilon}_{sh}$	Predicted shrinkage strain (Appendix H)
ε_{sh}^∞	Final shrinkage strain (at zero humidity)
$\varepsilon_{sh,tot}$	Total shrinkage strain
ε_{sh}^*	Updated prediction of shrinkage strain (Appendix H)
ε_{stat}	Static creep strain
ε_v	Viscous strain, viscoelastic strain
$\varepsilon_x, \varepsilon_y, \varepsilon_z$	Normal strain components
ε_σ	Mechanical strain
ε_0	Initial strain
ε_1	Initial strain at time t_1
$\varepsilon_1, \varepsilon_2, \varepsilon_3$	Principal strains
ζ, ζ_σ	Internal variable used by microplane model (Sect. 12.7)
η	Viscosity
η	Parameter in Eq. (13.32)
η	Auxiliary variable (Appendix F)
$\tilde{\eta}$	Modified viscosity defined in Eq. (10.75)
η_a	Dynamic shear viscosity of dry air
η_e, η_e^*	Surface emissivity
η_f	Viscosity associated with viscous flow term
η_f	Dynamic shear viscosity of a fluid (Chap. 8 and 13)
η_g	Dynamic shear viscosity of a gas

η_g	Volume fraction of gaseous phase (Sect. 13.5.6)
η_l	Dynamic shear viscosity of liquid phase
η_l	Volume fraction of liquid (Sect. 13.5.6)
η_s	Volume fraction of solid phase (Sect. 13.5.6)
η_v	Dynamic shear viscosity of water vapor
η_α	Volume fraction of phase α (Sect. 13.5.6)
η_μ	Viscosity of μth rheologic unit
$\eta^{(\infty)}$	Final value of viscosity at full solidification
θ	Dimensionless diffusivity (Sect. 8.4)
θ	Relative water content (Appendix H)
κ	Curvature
κ	Disjoining ratio (Sect. 8.26)
κ	Normalized reciprocal moisture capacity (Sect. 8.4)
κ_c	Curvature due to creep
κ_e	Elastic curvature
κ_h	Parameter defined in Eq. (8.233)
κ_{pr}	Curvature of a prestressed section (Sect. 4.3.6)
κ_{sh}	Curvature due to shrinkage
Λ	Compressibility of pore water (Sect. 12.8)
$\dot{\lambda}$	Auxiliary multiplier used by microplane model (Sect. 12.8)
λ_k	Auxiliary constant used by exponential algorithm
λ_0	Parameter of B3 model, typically set to one day
λ_0, λ_1	Parameters used in Sect. 4.3.4
μ	Short-term safety factor (Sect. sec:creepbuck)
μ	Water vapor diffusion resistance factor (Appendix J.6)
μ_L	Dimensionless factor in Sect. 7.14.2
μ_S	Parameter used by microprestress theory (Chap. 10)
μ_{ad}	Specific Gibbs free energy of adsorbed water
μ_l	Specific Gibbs free energy of liquid water
μ_v	Specific Gibbs free energy of water vapor
ν	Poisson ratio
ν_T	Characteristic attempt frequency (Sect. 12.6)
ν_c	Number of traffic load cycles per unit time
ξ	Dimensionless coordinate
ξ	Dimensionless time (Appendix F)
ξ	Hydration degree (Chap. 13)
$\boldsymbol{\xi}$	Vector function specifying the spatial position of a material particle
ξ_{inf}	Ultimate hydration degree (Chap. 13)
Π	Free energy of atomic lattice block (Sect. 12.6)
Π^*	Complementary energy
Π_f^*	Complementary energy due to fracture
Π_1	Local potential (Sect. 12.6)
π	Spreading pressure

π_a	Spreading pressure in free adsorbed layer
π'_a	Total spreading pressure in free adsorbed layer
π_d	Spreading pressure in hindered adsorbed layer
π'_d	Total spreading pressure in hindered adsorbed layer
π_0	Parameter of van Genuchten law (Chap. 8)
ρ	Mass density
ρ	Relaxation ratio
ρ	Correlation coefficient (Sects. 11.4–11.5)
ρ	Trost's relaxation coefficient (Appendix B)
ρ, ρ'	Parameters used in Eq. (13.99)
ρ_{100}, ρ_{1000}	Parameters used in Sect. 4.3.4
ρ_a	Density of dry air
ρ_{ad}	Density of adsorbed water
ρ_c	Density of concrete
ρ_{cp}	Density of cement paste
ρ_d	Density of dry concrete
ρ_g	Density of gas (wet air)
ρ_l	Density of liquid water
ρ_{l0}	Reference density of liquid water (at standard conditions)
ρ_s	Density of steel
ρ_s	Density of solid skeleton (Chap. 8 and 13)
ρ_v	Density of water vapor
σ	Stress
$\boldsymbol{\sigma}$	Column matrix of stress components
$\boldsymbol{\sigma}$	Stress tensor
$\hat{\sigma}$	Stress level in a creep test
$\bar{\sigma}$	Effective stress (Sect. 12.7)
$\tilde{\sigma}$	Dimensionless stress-to-strength ratio (Sect. 12.7)
σ_D	Stress in concrete caused by dead load
$\boldsymbol{\sigma}_D$	Deviatoric stress tensor
$\boldsymbol{\sigma}_D$	Dissipative stress tensor (Sect. 13.5)
σ_{Db}	Deviatoric bounding stress (Sect. 12.8)
σ_{DP}	Stress in concrete caused by dead load and prestressing
σ_{DLP}	Stress in concrete caused by dead and live load and prestressing
σ_L	Stress in concrete caused by live load
σ_L, σ_M	Shear stresses on microplane (Sect. 12.8)
$\sigma_L^{(tr)}, \sigma_M^{(tr)}$	Trial shear stresses on microplane (Sect. 12.8)
σ_N	Nominal strength
σ_N	Normal stress on microplane (Sect. 12.8)
$\sigma_N'', \sigma_N''^{+}, \sigma_N''^{-}$	Inelastic normal stresses on microplane (Sect. 12.8)
σ_{Nb}	Normal bounding stress (Sect. 12.8)
$\sigma_N^{(tr)}$	Trial normal stress on microplane (Sect. 12.8)
σ_P	Stress in concrete due to prestressing

σ_Q	Quasi-conservative stress tensor (Sect. 13.5)
σ_{SB}	Stefan–Boltzmann constant (Chap. 13)
σ_T	Magnitude of shear stress on microplane (Sect. 12.8)
$\sigma_T^{(tr)}$	Magnitude of trial shear stress on microplane (Sect. 12.8)
σ_{Tb}	Shear bounding stress (Sect. 12.8)
σ_V	Volumetric stress used by microplane model (Sect. 12.8)
σ_{Vb}	Volumetric bounding stress (Sect. 12.8)
σ_c	Stress in concrete (Sect. 4.3)
σ_e	Stress in an elastic spring
σ_{ij}	Tensorial stress components
σ_m	Mean stress
σ_{max}	Maximum stress
σ_{min}	Minimum stress in periodic cycles
σ_p	Stress in prestressing steel (Sect. 4.3)
σ_{p0}	Initial prestress (Sect. 4.3)
σ_r	Residual stress
σ_{rt}	Residual tensile stress
σ_v	Viscous stress
σ_0	Asymptotic value of nominal strength
τ	Relaxation or retardation time, characteristic time
τ	Stress parameter (Sect. 12.6)
τ	Tortuosity (Appendix J)
τ_{au}	Parameter used by model B4 (Appendix D)
$\tau_{au,cem}$	Parameter used by model B4 (Appendix D)
τ_{cem}	Parameter used by model B4 (Appendix D)
τ_f	Parameter of Kachanov law (12.33)
τ_p	Characteristic time of gel pressure relaxation (Sect. 8.8)
$\tau_{s,cem}$	Parameter used by model B4s (Appendix D)
τ_{sh}	Shrinkage halftime
$\bar{\tau}_{sh}$	Shrinkage halftimes estimated from water loss data (Appendix H)
τ_w	Halftime of drying (water loss)
τ_{w0}	Characteristic time of drying process defined in Eq. (8.199)
$\tau_{w\infty}$	Characteristic time of drying process defined in Eq. (8.204)
τ_μ	Relaxation or retardation time of μth rheologic unit
τ_ω	Characteristic time of damage development (Sect. 12.7)
τ_0	Parameter used by B4
Φ	Compliance function of cement gel (nonaging constituent)
Φ	Interatomic pair potential (Sect. 8.2)
Φ	Parameter used by GL2000 model (Appendix E.4)
$\Phi_A, \ \Phi_C, \ \Phi_D, \ \Phi_f, \ \Phi_J$	Normalized compliance functions for various specific models
Φ_{hT}	Function linking saturation degree to humidity and temperature

Φ_{pT}	Function linking saturation degree to pressure and temperature
ϕ	Function linking evaporable water content to humidity and temperature (Chap. 13 and Appendix J)
ϕ	Dimensionless coefficient in Eq. (12.2)
ϕ	Strength reduction factor (Sect. sec:creepbuck)
ϕ_K	Stiffness reduction factor (Sect. sec:creepbuck)
ϕ_{RH}	Parameter used by CEB and *fib* models (Appendix E.2)
ϕ_T	Parameter used by *fib* model (Appendix E.2.2)
ϕ_k	Auxiliary functions (Appendix F)
ϕ_1	Parameter used by double-power creep law
φ	Creep coefficient
φ_{bc}	Basic creep coefficient (Appendix E.2.2)
φ_{dc}	Drying creep coefficient (Appendix E.2.2)
φ_u	Ultimate creep coefficient (Appendix E.2.3)
φ_{28}	Creep coefficient in its alternative definition
χ	Aging coefficient
Ψ	Relaxation function of cement gel (nonaging constituent)
ψ	Parameter used by log-double-power creep law
ψ	Correction factor (Sect. 11.6)
ψ	Empirical function used in Eq. (B.1)
ψ	Exponent used by ACI model in Eq. (E.25)
ψ	Auxiliary variable (Appendix H)
ψ	Specific Helmholtz free energy (Appendix G)
ψ_e, ψ_r, ψ_s	Factors describing influence of temperature and humidity on transformed times t_e, t_r, t_s
ψ_k	Auxiliary functions (Appendix F)
Ω	Unit hemisphere (Sect. 12.8)
ω	Coefficient of variation, C.o.V
ω	Damage variable (Sect. 12.7 and Eq. (J.7))
ω_{CEB}	Coefficient of variation of prediction errors (Appendix K.2)
ω_G	Overall coefficient of variation of prediction errors (Appendix K.1)
ω_R	Coefficient of variation of data/prediction ratios (Appendix K.5)
ω_T	Thermal damage (Sect. J.7)
ω_Y	Coefficient of variation of random variable Y
ω_m	Mechanical damage (Sect. J.7)
ω_0	Parameter used in Eq. (12.39)
ω_1	Initial damage (Sect. 12.7)
ω_∞	Terminal damage (Sect. 12.7)
$\mathbf{1}$	Unit second-order tensor (Sect. 13.5)

Acronyms

AAEM	Age-adjusted effective modulus (method)
AASHTO	American Association of State Highway and Transportation Officials
ACI	American Concrete Institute
AEA	Air entraining agent
ASME	American Society of Mechanical Engineers
ASR	Alkali–silica reaction
ASTM	American Society for Testing and Materials
BDDT	Brunauer-Deming-Deming-Teller (isotherm)
BET	Brunauer-Emmett-Teller (isotherm)
BEu	Backward Euler (method)
BP	Bažant–Panula (model)
BP-KX	Bažant–Panula-Kim (model)
BSB	Brunauer-Skalny-Bodor (isotherm)
BT	Bažant–Thonguthai (model)
CEB	Euro-International Committee for Concrete [Comité euro-international du béton]
CEE	Civil and Environmental Engineering
CN	Crank–Nicolson (method)
CoV	Coefficient of variation
C-S-H	Calcium silicate hydrate
DPL	Double-power law
FEu	Forward Euler (method)
FFT	Fast Fourier transform
fib	International Federation for Structural Concrete [Fédération internationale du béton]
FIP	International Federation for Prestressing [Fédération internationale de la précontrainte]
FPZ	Fracture process zone
GL	Gardner–Lockman (model)
GTR	Generalized trapezoidal rule

GZ	Gardner–Zhao (model)
IAPWS	International Association for the Properties of Water and Steam
ITZ	Interfacial transition zone
JICA	Japan International Cooperation Agency
JRA	Japan Road Association
JSCE	Japan Society of Civil Engineers
KB	Koror–Babeldaob (bridge)
LDPL	Log-double-power law
LDPM	Lattice discrete particle model
LEFM	Linear elastic fracture mechanics
LHS	Latin hypercube sampling
LITS	Load-induced thermal strain
MC	Model code
MD	Molecular dynamics
MIT	Massachusetts Institute of Technology
MPS	Microprestress-solidification (theory)
MR	Midpoint rule
NIST	National Institute of Standards and Technology
NU	Northwestern University (database)
NU-ITI	Northwestern University—Infrastructure Technology Institute (database)
RH	Relative humidity
RILEM	International Union of Laboratories and Experts in Construction Materials, Systems and Structures [Réunion Internationale des Laboratoires et Experts des Matériaux, systèmes de construction et ouvrages]
RS	Rapid hardening (cement)
RVE	Representative volume element
SI	International System of Units
SL	Slow hardening (cement)
SMiRT	Structural Mechanics in Reactor Technology
STR	Standard trapezoidal rule
TR	Trapezoidal rule
WR	Water reducer

Part I
Fundamentals

Chapter 1
Introduction: How the Theory Evolved and How It Impacts Practice

Abstract Complex physical phenomena usually have a long history, must draw on many fields of science, and have a multifaceted practical impact. The creep, shrinkage, moisture diffusion, and thermal effects in concrete are an excellent example. This brief chapter highlights the main historical advances, beginning in 1887. It introduces the reader to the general problematics and breadth of the present complex phenomenon and points out diverse creep, moisture, and thermal effects on practical concrete structures, not only negative but also positive. Additionally, recognizing that various concrete structures have very different levels of sensitivity to these effects, the present chapter discerns five different structure types requiring different levels of sophistication in design calculations, depending on the practical impact. They range from simple and crude quasi-elastic estimates to computations based on experimentally verified theory of greater, though inevitable, complexity.

The shrinkage of hardened Portland cement paste was identified by chemist Henri Louis Le Chatelier in Paris in 1887 [568], and the creep of concrete was discovered by William Kendrick Hatt at Purdue University in 1907 [470]. Yet, despite the long history of research, the understanding of these phenomena is still far less than complete. Advances in research came in spurts, mostly in response to new needs and problems of construction.

The first major impetus for research came in the 1930s and 1940s, first due to design of long-span arches, whose long-term stability could not have been ensured without creep buckling analysis, and then, mainly, due to the invention of prestressed concrete, which would have been impossible without getting grasp of the prestress losses caused by creep and shrinkage. The names of Eugène Freyssinet in Paris, Gustaaf Paul Robert Magnel in Ghent, and Franz Dischinger in Munich may be mentioned among the pioneers from that period. Simultaneously, another major impetus came during the 1930s due to construction of very large dams, in which the relaxation by creep of the stresses caused by hydration heat and shrinkage was the main factor limiting the speed of construction and played an important role in ensuring the integrity of the dam. Important test data were obtained by many experimental researchers, among whom the contributions of George Earl Troxell and Raymond E. Davis in Berkeley and of the Bureau of Reclamation in Denver were prominent.

© Springer Science+Business Media B.V. 2018

Z.P. Bažant and M. Jirásek, *Creep and Hygrothermal Effects in Concrete Structures*, Solid Mechanics and Its Applications 225, https://doi.org/10.1007/978-94-024-1138-6_1

3

The next impetus was provided during the 1950s by the development of double cantilever method of segmental erection of long-span prestressed box girder bridges, which was the brainchild of Ulrich Finsterwalder in Munich. The greatest spurt in research was engendered by the flourishing of nuclear power during the 1960s and 1970s, particularly by the problems of integrity of large containment shells and reactor vessels, for which creep at elevated temperature is of great concern. This research was extended during the 1970s and 1980s to the problems of preventing or mitigating hypothetical high-temperature nuclear accidents, with spinoff to explosive spalling of tunnel linings in fire. The biannual SMiRT conferences (Structural Mechanics in Reactor Technology) provided at that time were the main forum for research on creep and shrinkage.

Problems of sustainability of cement production and use currently motivate extensive researches in the group of Franz–Josef Ulm at M.I.T. based on molecular dynamics simulations of deformation and the role of water movements on the level of nanopores in hydrated cement. These researches are having various fundamental implications for the nanoscale mechanism of concrete creep and shrinkage and its consequences for macroscale modeling.

Another current surge of interest in creep was spurred in 2008 by the release of previously sealed technical data on the 1996 collapse of Koror-Babeldaob (KB) Bridge in Palau. Although most large bridges are supposed to be designed for at least a 100-year lifetime, this bridge of a span of 241 m (a world record among prestressed box girders) deflected within 18 years by 1.61 m (compared to the design camber). The analysis at Northwestern University (cf. Chap. 7) blamed the observed deflection mainly on obsoleteness of the standard recommendations of engineering societies. In consequence, a concerted effort was launched (under the auspices of RILEM[1]) to gather information on other similar bridges. Despite widespread reluctance and pernicious legal obstacles to releasing data, the result has been a collection of deflection data of 71 large-span box girders, most of which developed within 20–30 years excessive deflections that required costly retrofit or bridge closing.

This latest experience documents one historical impediment to progress—the structural damage due to creep often takes two to three decades to develop. Retroactively, it thus becomes very difficult to pin the blame precisely, and the legal litigation of the damage often leads to an incorrect settlement, e.g., blaming the excessive deflections on shoddy construction. Not being too concerned about distant future, many practicing engineers and consultants, even those sitting on code-making committees, do not seem to be very worried about creep.

Pervasive problems have been the reluctance of most structural firms to reveal data from failures and damages, and particularly the legal sealing of data from the litigation of these damages in courts. Progress in structural engineering has traditionally depended on the analysis of failures. If the data on failures are not released, progress is impeded, with enormous costs to the society. This may, for example, help to explain why an obsolete, simplistic, and misleading creep and shrinkage

---

[1]RILEM is the French acronym of the International Union of Laboratories and Experts in Construction Materials, Systems and Structures.

model adopted in 1971 is, after more than four decades, still surviving as a standard recommendation.

Another obstacle to the development of a rational theory of creep and shrinkage has been the enormous effect of concrete composition and diverse admixtures on the creep and shrinkage properties. The worldwide database is vast but is complicated by large differences in creep and shrinkage properties of various concretes, which are very hard to describe and filter out mathematically, yet obscure the trends with respect to time, age, sectional thickness, environmental humidity, etc. This is one reason for the proliferation of diverse creep and shrinkage models calibrated by only limited selections of test data.

In practice-oriented studies, there has been a tendency to deduce all the concrete design equations by curve fitting of test data. This empirical approach has worked adequately in many cases. But for creep, the purely empirical approach is insufficient and a solidly founded theory is indispensable. The reason is that, in the available database of thousands of creep and shrinkage tests, most of creep tests (about 95%) have a duration of less than 6 years and only a few exceed 12 years, while the design lifetimes of many creep-sensitive structures are supposed to be at least a century. To fill this huge time gap, a well-founded theory is inevitable. To validate the theory, data on multidecade deformations of large structures, such as bridges and tall buildings, should be collected and their inverse analysis should be conducted to validate the theory. Only meager efforts of this kind have so far been made.

The theory and modeling of creep and shrinkage has recently been unsettled by the realization that the self-desiccation and autogenous shrinkage of modern concretes with very low water-cement ratios or large amounts of admixtures, or both, are phenomena of decades long duration, reaching much larger magnitudes than in old concretes. The self-desiccation can cause the relative humidity in the pores to drop as low as to 70%. The autogenous shrinkage can even exceed the drying shrinkage. They both evolve logarithmically for at least a decade, and probably centuries. Unfortunately, at the time of writing, the experimental information on the long-term self-desiccation and autogenous shrinkage is meager and, in the existing database, the multiyear autogenous shrinkage has been neither recorded nor separated from creep and drying shrinkage data. Consequently, these phenomena cannot be modeled here as thoroughly as they deserve (and their full treatment must be relegated to a future second edition).

Perhaps surprisingly, the creep of concrete has little in common with the creep of other materials, particularly metals, polymers, and clays. The physical origin is completely different, and the mathematical models, too. The problem of concrete creep and shrinkage intersects with many fields of engineering and science—structural design, experimental methods, design of experiments, aging viscoelasticity, mechanics of materials and structures, mathematical modeling by differential and integral equations, materials science, microscopy, silicate and colloid chemistry, poromechanics, thermodynamics, theory of constitutive relations, water diffusion and sorption in concrete, nanomechanics, molecular dynamics simulations at the atomistic level, modeling of the associated cracking and fracture, probabilistic modeling, optimization, statistical evaluation and prognosis. In prestressed structures, the creep of concrete is intertwined with the viscoplasticity of prestressing steel, in retrofitted structures with the creep of polymeric laminates, and in structures undergoing uneven long-time settlement with the consolidation and creep of clay.

The effects of creep and shrinkage are mainly of interest for the long-term serviceability and durability of structures and have strong implications for sustainability of civil engineering infrastructure. Errors in the prediction of creep and shrinkage effects can lead to intolerable excessive deflections of bridges, differential column shortening in super-tall buildings, shrinkage cracking with consequent ingress of water and corrosion of reinforcement, stress redistributions which also lead to cracking, endangerment of leak-tightness of nuclear containments, etc. Creep can also compromise structural safety since it reduces the long-time critical load in buckling of columns, thin roof shells, tunnel linings, and tunnels or buildings in fire. For instance, the tragic collapse of the KB Bridge three months after retrofit was likely triggered by creep buckling of a previously delaminated top slab, overloaded by additional prestressing tendons.

Accuracy Levels Recommended for Practice

The degree of sensitivity of various structures to creep and shrinkage varies widely. Accurate and laborious analysis of creep and shrinkage is necessary only for certain special types of structures. The following approximate classification of *sensitivity levels of structures* may be made on the basis of general experience [107]:

Level 1.   Reinforced concrete beams, frames, and slabs with spans under 20 m (65 ft.) and heights of up to 30 m (100 ft.), plain concrete footings, retaining walls.

Level 2.   Prestressed beams or slabs of spans up to 20 m (65 ft.), high-rise building frames up to 100 m (325 ft.) high.

Level 3.   Medium-span box girder, cable-stayed or arch bridges with spans of up to 80 m (260 ft.), ordinary tanks, silos, pavements, tunnel linings.

Level 4.   Long-span prestressed box girders, cable-stayed or arch bridges; large bridges built sequentially in stages by joining parts; large gravity, arch or buttress dams; cooling towers; large roof shells; very tall buildings.

Level 5.   Record-span bridges, nuclear containments and vessels, large offshore structures, large cooling towers, record-span thin roof shells, record-span slender arch bridges, super-tall buildings.

Level 5 requires the most realistic and accurate analysis based on a model such as B3 or B4—typically performed using step-by-step numerical integration of a rate-type constitutive law, coupled with the solution of the differential equations for drying and heat conduction, statistical estimation of confidence limits and with updating based on short-time tests of given concrete. The designers usually prefer a simpler creep and shrinkage model. But it makes little sense to run a detailed finite element analysis, sometimes with statistical estimates and experimental updates, while at the same time using a poor material model introducing much larger errors. Anyway, the cost of proper level 5 analysis is minuscule compared to the cost of large structures of extreme designs. The error in maximum deflections, stresses, and cracking predictions caused by replacing a realistic creep and shrinkage model with a simplistic estimation of creep and shrinkage effects is often larger than the gain from replacing old fashioned frame analysis by pencil with finite element analysis by computer.

Examples of level 5 structures are long-span prestressed box girders, especially when segmentally erected, cable-stayed bridges and arches, large gravity arch or buttress dams, cooling towers, large roof shells, super-tall buildings, nuclear containments and vessels, large offshore structures, large-span roof shells.

An accurate analysis is also necessary for level 4 and is recommended, though not requisite, for level 3. This includes, e.g., medium-span box girders, cable-stayed or arch bridges, silos, pavements, prestressed beams with spans over 65–165 ft. (20–50 m), and high-rise buildings frames up to 325 ft. (100 m) tall.

Although most of this book is focused on levels 4 and 5, a simplified method, such as the age-adjusted effective modulus method (Chap. 4, [76]), endorsed by the American Concrete Institute (ACI) and by *fib*,[2] is recommended for levels 3 and lower. This method is nevertheless also useful for preliminary design estimates at levels 4 and 5. The effective modulus method (Chap. 4) suffices for level 2. For level 1, creep and shrinkage analysis of the structure is not required, although a crude empirically based estimate using the effective modulus is desirable to check whether level 1 is indeed applicable.

Since creep and shrinkage deformations inevitably exhibit large statistical scatter, a statistical analysis (Chap. 6) with estimation of 95% confidence limits should be mandatory for level 5 and is recommended for level 4. If high temperatures occur, their analysis ought to be detailed for level 5 and approximate for level 4. For level 3, their analysis is unnecessary though advisable and can be ignored for levels 1 and 2 (except for the effects of hydration heat).

In many situations, the neglect or simplistic estimation of creep and shrinkage may not only compromise durability but may also lead to overdesign. For example, creep is very beneficial in mitigating the damage and cracking caused by long-term expansive processes in concrete, such as the alkali–silica reaction or reinforcement corrosion. It greatly reduces the stresses caused by drying shrinkage or swelling, by nonuniform autogenous shrinkage, by gradual long-time differential settlement of structures on consolidating clayey foundations, or by changes of structural system, which are typical of modern prestressed concrete construction procedures.

Some research subjects, such as the elastic analysis of structural frames, become closed after several decades. Not the science field of creep and hygrothermal effects. After 130 years of research, this field has expanded to enormous breadth and the end is not yet in sight. Thus, the present coverage of the cutting-edge subjects will surely have to be updated in less than a decade, although many parts of the present coverage are probably definitive.

The present field now includes, or impinges on, a number of scientific disciplines—aside from deformations and strength of structures and traditional laboratory testing,

---

[2]The acronym "*fib*" (officially written in lower case italics) stands for "fédération internationale du béton," in English the International Federation for Structural Concrete.

also modern computational mechanics, fracture mechanics, chemo-mechanics, thermomechanics and thermodynamics, diffusion theory, nanomechanics, scale bridging, probabilistic modeling of materials and structures, stochastic processes, stability of structures and crack systems, statistical interpretation and fitting optimization of big data gathered in enormous worldwide databases, intelligent use of sensors to collect data from structures, optimum inverse analysis of sensor data collected from structures or laboratory specimens, effects of chemical composition and of the material micro or nanostructure, and nanoscale experimental methods. This book aims to interpret these scientific disciplines, some in detail, some only tangentially, in one coherent exposition.

Finally, let us point out that what would greatly reduce uncertainties in the development of creep and shrinkage theories would be to agree internationally, among all laboratories, to accompany every creep and shrinkage testing program by similar tests performed on one and the same standard concrete, with a predefined composition, curing procedures, and moisture conditions. But, at present, this is just a dream.

Chapter 2
Fundamentals of Linear Viscoelasticity

Abstract Despite complexities such as aging due to cement hydration or strong moisture sensitivity, the creep of concrete at service stress levels belongs to the broad realm of linear viscoelasticity, a theory that has been extensively studied beginning with the works of Maxwell, Kelvin, and Boltzmann in the late nineteenth century. Thanks to linearity, we introduce the compliance function as the basic material characteristic for constant sustained stress. Then we use the principle of superposition to characterize the creep at variable stress, calculate the relaxation function for constant imposed strain, and generalize the creep model to multiaxial stress. Finally we introduce an operator notation, which simplifies the mathematical exposition and illuminates the main concepts.

2.1 Characterization of Creep by Compliance Function

Laboratory tests and measurements on real structures indicate that, for many materials, strain tends to grow when the stress is kept at a constant level. This phenomenon is usually referred to as *creep*. For materials that exhibit creep, stress tends to decrease when the strain is kept constant, which is referred to as *relaxation*. Creep and relaxation are intimately linked and have a common origin in viscous deformation processes in the material microstructure. Such time-dependent behavior can lead to undesirable effects on the structural level, e.g., to dramatic growth of bridge deflections or to loss of prestress in cables, and so it needs to be taken into account by using appropriate constitutive models. One possible approach consists in replacing the dependence between the current values of stress and strain by the dependence of the current value of stress (or strain) on the entire previous history of strain (or stress).

In general, the stress–strain relation can be nonlinear, but it is convenient to start from a relatively simple linear theory and later generalize it as needed, e.g., by adding the strain induced by cracking or plastic yielding, and also the stress-independent part of deformation due to shrinkage and thermal effects. Classical *linear viscoelasticity* is based on the *superposition principle*, which states that the responses to individual loading histories can be superimposed. In mathematical terms, this means that the

© Springer Science+Business Media B.V. 2018
Z.P. Bažant and M. Jirásek, *Creep and Hygrothermal Effects
in Concrete Structures*, Solid Mechanics and Its Applications 225,
https://doi.org/10.1007/978-94-024-1138-6_2

mapping of the stress history onto the corresponding strain history (or vice versa) is described by a linear operator.

Boltzmann Superposition Principle: (2.1)

If, for a given material,
stress history $\sigma_a(t)$ corresponds to strain history $\varepsilon_a(t)$
and stress history $\sigma_b(t)$ corresponds to strain history $\varepsilon_b(t)$
then,
for arbitrary real constants c_a and c_b,
$\sigma(t) = c_a\sigma_a(t) + c_b\sigma_b(t)$ corresponds to $\varepsilon(t) = c_a\varepsilon_a(t) + c_b\varepsilon_b(t)$.

This principle, first proposed by Boltzmann [243] for nonaging phenomena and later generalized by Volterra [840] for aging phenomena, is an enhanced form of the superposition principle used in linear elasticity, which works with the current values of stress and strain. Linear viscoelasticity takes into account the entire history of stress and strain, described by functions of time t.

The superposition principle is not, of course, a fundamental law of physics. It is a convenient hypothesis, valid only approximately and not under all conditions. For example, for concrete the superposition is a realistic hypothesis only if all the principal stresses remain in the service stress range (i.e., below 40–50% of uniaxial strength). At higher stresses, a damage law must be appended. For basic creep (i.e., creep of a sealed specimen, see Sect. 3.1) within the service stress range, the assumption of linearity agrees with test results very well. But in the case of drying or variable temperature, the superposition principle is accurate only if it is coupled with a realistic constitutive law for cracking damage. It has sometimes been suggested that the superposition did not apply to unloading during strain reversal, but these were cases of creep with drying in which the irreversibility of cracking was not accurately accounted for by a constitutive law applied at each material point. However, the use of an effective creep model for the whole cross section of a member exposed to drying (the so-called sectional approach, see Sects. 3.5 and 3.6), as defined in codes or society recommendations and also by models B3 and B4, may lead, even within the service stress range, to significant deviations from the principle of superposition.

Once accepted, the principle of superposition serves as a handy tool for constitutive modeling. Based on superposition, the uniaxial stress–strain relation of a given viscoelastic material can be constructed from one single function $J(t, t')$, called the *compliance function*, or, alternatively, from a closely related function $R(t, t')$, called the *relaxation function*.

Consider a special experiment—the *creep test*, in which a previously stress-free material sample is subjected to a given uniaxial stress $\hat{\sigma}$ at time t', and the stress level is then kept constant at all later times. Mathematically, the prescribed stress history is characterized by the function

$$\sigma(t) = \hat{\sigma}\, H(t - t') \qquad\qquad (2.2)$$

where H is the *Heaviside step function*, defined by the rule[1]

$$H(s) = \begin{cases} 0 & \text{for } s < 0 \\ 1 & \text{for } s \geq 0 \end{cases} \tag{2.3}$$

Up to time t', the material is at zero stress and therefore does not deform.[2] At time t', stress jumps to a given level $\hat{\sigma}$, and this induces a certain instantaneous[3] strain. For a rate-independent material model (not necessarily elastic, but possibly elastoplastic, damage model, fracture model, etc.), the strain would afterward remain constant as long as the stress is kept constant. In reality, delayed deformation processes taking place in the material microstructure lead to creep, which manifests itself by a gradual strain increase, described by a function $\varepsilon(t)$. The superposition principle implies that if we divide the resulting strain history $\varepsilon(t)$ by the stress level $\hat{\sigma}$ at which the creep test was performed, we obtain a function that is independent of the applied stress level and characterizes the material. Therefore, the evolution of strain induced by the special prescribed stress history (2.2) is given by

$$\varepsilon(t) = \hat{\sigma} \, J(t, t') \tag{2.4}$$

where J is the *compliance function* of the viscoelastic material; see Fig. 2.1a. The value of J at (t, t') can be interpreted as the strain at time t induced by a unit stress acting from time t'. Of course, if $t < t'$, the value of J must be zero.

For a linear elastic material characterized by Young's modulus E, the mechanical strain measured during the creep test is equal to zero for all times t preceding the time t' when the stress is applied and equal to $\hat{\sigma}/E$ for all times t after the stress application. Such a strain history is described by the function $\varepsilon(t) = (\hat{\sigma}/E)H(t-t')$, and so the compliance function of an elastic material is

$$J(t, t') = \frac{1}{E}H(t - t') \tag{2.5}$$

The compliance function of a linear viscoelastic material has a similar meaning to the inverse value of the elastic modulus in time-independent linear elasticity, and its units are 1/Pa. If the material properties do not evolve in time, the compliance function depends only on the time lag $t - t'$, i.e., on the duration of loading (elapsed time), and not on times t and t' separately. In that case we can write

---

[1]The value of Heaviside function at $s = 0$ is often defined as $1/2$, for symmetry reasons. For the present purpose, it is preferable to set $H(0) = 1$.

[2]Here, we consider only the mechanical strain, i.e., the stress-induced part of deformation. Of course, in general one needs to account for other sources of deformation, such as thermal changes or shrinkage, which will be incorporated in Sect. 2.5.

[3]What exactly is meant by "instantaneous" strain depends on the time scale at which we work. It is hard to make a clear distinction between the truly instantaneous strain and the part of creep that takes place very shortly after loading.

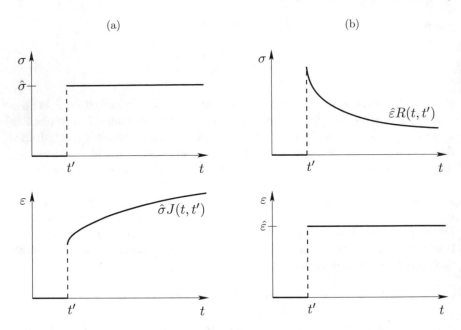

Fig. 2.1 Histories of stress and strain in a (a) creep test, (b) relaxation test

$$J(t, t') = J_0(t - t') \tag{2.6}$$

where J_0 is also a compliance function, but this time is considered as a function of only one variable, the duration of loading. Compliance functions in the form (2.6) are characteristic of *nonaging materials*, such as polymers. In contrast to that, concrete is a typical example of a material with evolving microstructure due to chemical processes such as cement hydration. Chemical changes in the microstructure as well as relaxation of self-equilibrated stresses in the nanostructure result into changes of viscoelastic properties, and so the compliance function depends not only on the load duration but also on the age of the sample. The age is usually measured from the initial setting time, i.e., from the time when the hardening fresh concrete first becomes a solid (typically several hours after mixing). Only for sufficiently old concrete, further changes of mechanical properties might be negligible, in which case the compliance function can be considered in the simple form (2.6). Realistic analysis of concrete structure must be based on the general form of compliance function, characteristic of *aging materials*.

For real materials, the superposition principle is applicable with sufficient accuracy only within a certain range of stresses. This is illustrated in Fig. 2.2a, b, which presents the results of creep tests performed by Komendant, Polivka and Pirtz [551] on sealed concrete cylinders. From age $t' = 90$ days, the specimens were loaded by

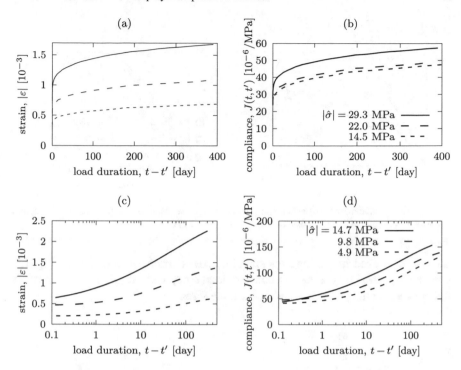

Fig. 2.2 Creep tests at various stress levels: (a, c) strain evolution, (b, d) compliance function

uniaxial compression at three different stress levels. Figure 2.2a shows the strain[4] as a function of the load duration, $t - t'$, up to more than 1 year. Higher stresses of course induce larger strains, but the evolution of compliance (defined as strain divided by stress) is almost the same at low and medium stresses ($|\hat{\sigma}| = 14.5$ MPa and 22.0 MPa), as documented in Fig. 2.2b. Therefore, proportionality is verified at least up to the stress level of 22 MPa, which is about one half of the compression strength of this particular concrete, $\bar{f}_c = 45.4$ MPa. At stress $|\hat{\sigma}| = 29.3$ MPa, the compliance is visibly higher, which indicates a deviation from proportionality. To complement the picture, Fig. 2.2c, d shows, this time in a semi-logarithmic scale, the results of creep experiments reported by Mamillan [598]. The tested concrete had a lower 28-day compressive strength, $\bar{f}_c = 33.9$ MPa, and was loaded at an early age ($t' = 7$ days). Consequently, the deviation from linearity occurs already at lower stress levels than in Fig. 2.2a, b.

The results plotted in Fig. 2.2 confirm the empirical rule that creep becomes non-linear for stresses exceeding about one half of the strength limit. The cause of non-linearity is the time-dependent growth of microcracks. The nonlinear creep behavior

---

[4]Concrete creep tests are typically performed under compression, and the resulting creep strains are thus negative. For simplicity, the negative sign is sometimes omitted in graphical representations of the results, e.g., in Fig. 2.2a, but the sign must be included in all calculations. Of course, the compliance is always positive, no matter whether it is determined from a compressive or a tensile test.

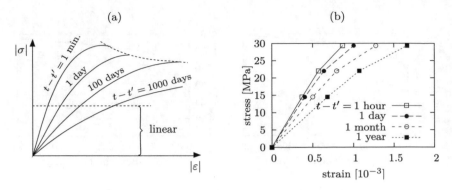

Fig. 2.3 Creep isochrones: (a) schematic plot covering the range up to peak stress, (b) actual isochrones constructed from the data of Komendant et al. [551]

may be visualized by plotting the so-called stress-strain *isochrones*; see Fig. 2.3. They are obtained by conducting constant load creep tests at different stress levels and connecting the points reached at the same time.

Fig. 2.4 Basic rheologic units: (a) elastic spring, (b) viscous dashpot

The compliance function of a specific material can be determined by the creep test. The experimentally measured creep curve could be described by a sequence of points, but for theoretical derivations and practical applications it is useful to approximate it by a suitable analytical function. The general form of such function can be motivated by closed-form solutions obtained for rheologic models, which consist of elementary units coupled in series or in parallel. In viscoelasticity, the basic units are a spring (Fig. 2.4a), representing an idealized elastic response, and a dashpot (Fig. 2.4b), representing an idealized viscous behavior. In linear viscoelasticity, both types of units are governed by linear laws. The force transmitted by a linear spring is assumed to be proportional to the elongation (change of length with respect to the initial stress-free configuration), and the force transmitted by a linear dashpot is assumed to be proportional to the elongation rate. Since the rheologic model represents the behavior of an infinitesimal material volume (sometimes called the material point), forces actually represent stresses and elongation represents strain. The spring stiffness E then has the meaning of an elastic modulus, measured in Pa, and the material constant η characterizing the dashpot is the *viscosity*, measured in Pa $\cdot$ s (i.e., in kg $\cdot$ m^{-1}s^{-1}). The inverse of viscosity, $1/\eta$, is called the *fluidity*.

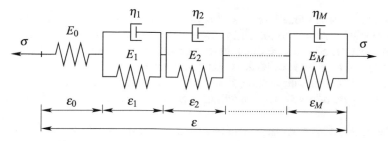

Fig. 2.5 Kelvin chain

The compliance functions of various rheologic models are derived in Appendix A. For our purpose, the most important case is the Kelvin chain, consisting of a finite number of Kelvin units coupled in series (Fig. 2.5). Each Kelvin unit is formed by a spring and a dashpot coupled in parallel. One of the units in the Kelvin chain is usually taken as a simple spring, which is a special case of a Kelvin unit with zero viscosity of the dashpot (so that this dashpot carries no stress and can be removed). The compliance function of a Kelvin chain composed of $M + 1$ Kelvin units with elastic moduli $E_\mu, \mu = 0, 1, 2, \ldots M$, and viscosities $\eta_0 = 0$ and $\eta_\mu, \mu = 1, 2, \ldots M$, is given by

$$J_0(t) = \left[\frac{1}{E_0} + \sum_{\mu=1}^{M} \frac{1}{E_\mu} \left(1 - e^{-t/\tau_\mu} \right) \right] H(t) \qquad (2.7)$$

where $\tau_\mu = \eta_\mu / E_\mu$ are the *retardation times*. From the mathematical point of view, the time-dependent part of (2.7) is the *Dirichlet series* (or *Prony series*); see Hardy and Riesz [459].

For concrete, as an aging viscoelastic material, an appropriate generalization is needed. Since the time is measured from the set of concrete and the creep test starts at a certain age $t' > 0$, the variable t on the right-hand side of (2.7) must be replaced by the load duration, $t - t'$. In tests started at different ages, different compliance curves are obtained, and so the model parameters should be considered as functions of t'. It turns out that the retardation times can be fixed, and only the moduli need to be taken as age-dependent. The generalized form of (2.7) thus reads

$$J(t, t') = \left[\frac{1}{D_0(t')} + \sum_{\mu=1}^{M} \frac{1}{D_\mu(t')} \left(1 - e^{-(t-t')/\tau_\mu} \right) \right] H(t - t') \qquad (2.8)$$

The age-dependent moduli $D_\mu(t')$ cannot be directly interpreted as current stiffnesses of aging elastic springs, as explained in detail in Appendix A. This is why they are denoted by symbols different from the age-independent spring stiffnesses E_μ from (2.7).

Example 2.1. Approximation of a measured compliance function by Dirichlet series

Real data measured in creep tests are usually scattered, and a piecewise linear interpolation between the measured values may lead to irregularities, such as a negative creep rate in certain time intervals. It is much better to use a global approximation by a smooth function, and the Dirichlet series is perfectly suitable for this purpose.

Consider the compliance function extracted from the creep test of Komendant et al. [551] at the age of $t' = 90$ days and at the low stress level of $\hat{\sigma} = 14.5$ MPa. In this illustrative example, variable t' will be considered as fixed, and instead of age-dependent parameters $D_\mu(t')$ we will deal with their values at age 90 days, denoted simply as D_μ.

The simplest approximation can be based on the so-called standard linear solid, i.e., on a Kelvin chain consisting of an elastic spring and a Kelvin unit (which corresponds to $M = 1$ in (2.8)), with the compliance function given by

$$J(t, t') = J_0(t - t') = \frac{1}{D_0} + \frac{1}{D_1}\left(1 - e^{-(t-t')/\tau_1}\right), \qquad t \geq t' = 90 \text{ days} \quad (2.9)$$

This model possesses only one characteristic time, τ_1. For load durations $t - t'$ much larger than τ_1, the compliance function is almost constant. Since the data in Fig. 2.2 indicate that the compliance still grows at several hundred days, let us set $\tau_1 = 100$ days. Moduli D_0 and D_1 can then be determined by optimal fitting, based, e.g., on the least-square method. The best agreement is obtained for $D_0 = 33.30$ GPa and $D_1 = 62.74$ GPa (see also Table 2.1a), and the resulting approximation is shown in Fig. 2.6a.

At a first glance, the fitting seems to be perfect, but a closer examination reveals that the analytical approximation deviates from the measured data for times of loading shorter than about 10 days and also for those longer than about 300 days. This is best seen if the results are replotted with the load duration in logarithmic scale; see Fig. 2.6b. Such a plot reveals that the exponential function in (2.9) is almost constant for times of loading much shorter or much longer than the characteristic time τ_1. By changing τ_1, we can shift the range in which the measured data can be fitted but we

Table 2.1 Parameters of Dirichlet series used in Example 2.1

(a) M $= 1$		
μ	D_μ [GPa]	τ_μ [day]
0	33.30	-
1	62.74	100
(b) M $= 4$		
μ	D_μ [GPa]	τ_μ [day]
0	40.43	-
1	324.63	0.5
2	421.47	5
3	145.80	50
4	52.04	500

can never extend that range. Covering a wider range is possible only by adding more exponential terms with different characteristic times, i.e., by increasing the number of Kelvin units in the chain.

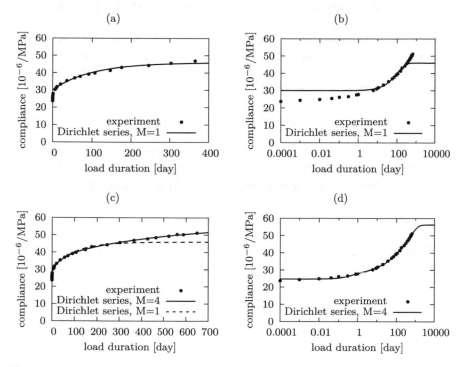

Fig. 2.6 Fitting of a measured compliance function by Dirichlet series

Figure 2.6d shows the best fit obtained for a chain consisting of an elastic spring and four Kelvin units, with retardation times ranging from 0.5 to 500 days in a geometric progression with quotient 10. The model parameters are listed in Table 2.1b. The graph of the compliance function in semi-logarithmic scale (logarithmic scale for the load duration, linear scale for the compliance) clearly indicates that the measured compliance keeps increasing even after long load durations, which is not so obvious if the horizontal axis is kept in the linear scale; see Fig. 2.6c. The experimental data used in this example have been recorded only up to 2 years, but the fact that the creep process never really stops has been confirmed by other tests running up to 10 or even 30 years, and by bridge deflection over even longer periods; see Chap. 7. However, every Kelvin chain with a finite number of units can capture the growth of compliance only up to times that are comparable to the maximum retardation time. The number of units (i.e., the number of terms in the Dirichlet series) and their retardation times must be adjusted to the specific application, having in mind the shortest and the longest loading times for which the viscoelastic response needs to be represented accurately. ∎

2.2   Integral Stress–Strain Relation

Once the compliance function of a viscoelastic material is known, it is possible to evaluate the strain induced by any given stress history. All that is needed is the principle of superposition.

Consider first a simple loading program with stress increasing by a jump from zero to σ_1 at time t_1 and then remaining constant up to time t_2, when it increases by a jump to σ_2 and then remains constant again. Such a stress history is the sum of two jump functions and can be described as

$$\sigma(t) = \Delta\sigma_1 \, H(t - t_1) + \Delta\sigma_2 \, H(t - t_2) \tag{2.10}$$

where $\Delta\sigma_1 = \sigma_1$ and $\Delta\sigma_2 = \sigma_2 - \sigma_1$. Since the stress history $\Delta\sigma_1 \, H(t - t_1)$ would lead to strain history $\Delta\sigma_1 \, J(t, t_1)$ and stress history $\Delta\sigma_2 \, H(t - t_2)$ to strain history $\Delta\sigma_2 \, J(t, t_2)$, the prescribed stress history (2.10) leads to strain history

$$\varepsilon(t) = \Delta\sigma_1 \, J(t, t_1) + \Delta\sigma_2 \, J(t, t_2) \tag{2.11}$$

This argument can be extended by induction to an arbitrary finite series, and so the stress history

$$\sigma(t) = \sum_{k=1}^{n} \Delta\sigma_k \, H(t - t_k) \tag{2.12}$$

leads to strain history

$$\varepsilon(t) = \sum_{k=1}^{n} \Delta\sigma_k \, J(t, t_k) \tag{2.13}$$

This is graphically shown in Fig. 2.7a.

Expressions (2.12)–(2.13) correspond to loading programs for which the stress changes by jumps at a finite number of time instants and between the jumps remains constant. By reducing the length of intervals between the jumps and simultaneously reducing the size of jumps, we can proceed in the limit to the description of more general stress histories, as illustrated in Fig. 2.7b. The sum in (2.13) turns into an integral, and the strain evaluation formula can be written as[5]

$$\varepsilon(t) = \int_{0}^{t} J(t, t') \, d\sigma(t') \tag{2.14}$$

---

[5]In integral expressions similar to (2.14), t is called the *current time*, t' is the *historic time*, and $t - t'$ is the *elapsed time*. In mathematical terms, the compliance function J is called the *kernel* of the integral transform mapping the stress rate onto the strain. For nonaging materials, $J(t, t')$ is replaced by $J_0(t - t')$, called the *convolution kernel*, and the integral in (2.14) represents the *convolution* of the compliance function and the stress rate.

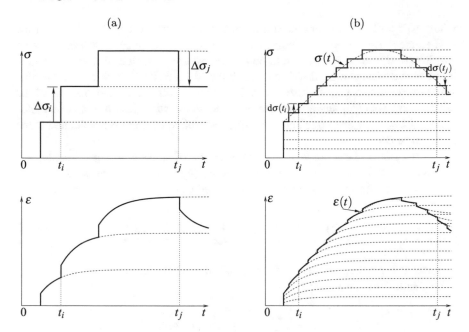

Fig. 2.7 Application of the superposition principle: (a) piecewise constant stress history with a finite number of jumps, (b) limit transition to a continuous stress history

The isolated time instants t_k are replaced by the integration variable t' that sweeps through the history from time zero to the current time t, and the finite increments $\Delta\sigma_k$ at times t_k are replaced by increments $d\sigma(t')$. To cover the general case of stress histories which can be piecewise continuous with finite jumps at certain isolated time instants, the right-hand side of (2.14) must be interpreted as a generalized type of integral called the *Stieltjes integral*. A finite stress jump $\Delta\sigma_k$ at time t_k contributes to the value of the integral by $J(t, t_k)\,\Delta\sigma_k$.

Equation (2.14) is not only a consequence of the principle of superposition but its equivalent alternative statement. Indeed, (2.1) follows from (2.14). Using superposition, one may further generalize Eq. (2.14) for the case of multiaxial stress and strain; see Sect. 2.4.

In terms of the classical Riemann integral, formula (2.14) can be rewritten in a somewhat clumsy form as

$$\varepsilon(t) = \sum_{k=1}^{n} J(t, t_k)\,\Delta\sigma_k + \int_{0}^{t_1^-} J(t, t')\,\dot{\sigma}(t')\,dt'$$

$$+ \sum_{k=2}^{n} \int_{t_{k-1}^+}^{t_k^-} J(t, t')\,\dot{\sigma}(t')\,dt' + \int_{t_n^+}^{t} J(t, t')\,\dot{\sigma}(t')\,dt' \qquad (2.15)$$

where t_k, $k = 1, 2, \ldots n$, are those time instants between zero and the current time t when the stress changes by jumps $\Delta\sigma_k$, $k = 1, 2, \ldots n$. In each interval between two jumps, the stress is assumed to be a differentiable function of time, and so the infinitesimal increments $d\sigma(t')$ can be replaced by $\dot\sigma(t')\,dt'$ where $\dot\sigma$ is the time derivative of stress, i.e., the stress rate. The superscripts "plus" and "minus" at the integration limits t_{k-1}^+ and t_k^- indicate that integration starts just after t_{k-1} and ends just before t_k, because at t_{k-1} and t_k the time derivative of stress is not defined in the classical sense.

For the frequent case in which concrete is free of stress until some loading age t_1 at which a finite stress σ_1 is applied suddenly (by a jump), and afterward $\sigma(t)$ is continuous and differentiable, formula (2.15) reduces to

$$\varepsilon(t) = J(t, t_1)\,\sigma_1 + \int_{t_1^+}^t J(t, t')\,\dot\sigma(t')\,dt' \tag{2.16}$$

Often it is practical to work with the strain rate. Differentiating (2.14) according to the Leibniz rule, we have

$$\dot\varepsilon(t) = \frac{\dot\sigma(t)}{E(t)} + \int_0^t \dot J(t, t')\,d\sigma(t') \tag{2.17}$$

where we tacitly assume that the current time t at which we compute the strain rate is not one of the times t_k at which σ has a jump; $E(t)$ denotes the (instantaneous) elastic modulus at time t, defined as $E(t) = 1/J(t, t)$, and the dot over J denotes partial derivative with respect to t (not t'), i.e., $\dot J(t, t') = \partial J(t, t')/\partial t$. For a smooth stress history starting with a jump, Eq. (2.17) may again be written in terms of the Riemann integral similar to (2.16):

$$\dot\varepsilon(t) = \dot J(t, t_1)\,\sigma_1 + \frac{\dot\sigma(t)}{E(t)} + \int_{t_1^+}^t \dot J(t, t')\,\dot\sigma(t')\,dt' \tag{2.18}$$

The integral stress–strain relation (2.14) is based on the superposition of infinitely many stress histories, each of which exhibits a stress jump by $d\sigma(t')$ at time t' and afterward remains constant; see Fig. 2.7b. Alternatively, one could use the superposition of infinitely many stress impulses, each of which acts at a finite level $\sigma(t')$ during an infinitesimal time interval starting at time t'. The corresponding formula can be derived from Eq. (2.16) if the integral on the right-hand side is integrated by parts. The resulting expression contains the partial derivative of the compliance function J with respect to its second argument, t'. For convenience, we introduce the so-called *stress impulse memory function*

$$L(t, t') = -\frac{\partial J(t, t')}{\partial t'} \tag{2.19}$$

which will later be shown to have a specific physical meaning. Since $J(t, t')$ has a jump at $t = t'$, it is not differentiable at this point in the classical sense. One could use the theory of distributions and deal with generalized derivatives, but it is perhaps simpler to consider the variable t' in the integral in (2.16) as running to an upper limit t^-, just before time t. Integration by parts then yields

$$\int_{t_1^+}^{t^-} J(t, t') \dot{\sigma}(t') \, dt' = J(t, t^-)\sigma(t^-) - J(t, t_1^+)\sigma(t_1^+) - \int_{t_1^+}^{t^-} \frac{\partial J(t, t')}{\partial t'}\sigma(t') \, dt' =$$

$$= \frac{\sigma(t)}{E(t)} - J(t, t_1)\sigma_1 + \int_{t_1}^{t} L(t, t')\sigma(t') \, dt' \tag{2.20}$$

where $E(t) = 1/J(t, t^-)$ is the instantaneous modulus at age t. Substituting (2.20) into (2.16), we obtain the modified integral stress–strain relation

$$\varepsilon(t) = \frac{\sigma(t)}{E(t)} + \int_{t_1}^{t} L(t, t')\sigma(t') \, dt' \tag{2.21}$$

which was for concrete introduced by Maslov [609] and used in some simplified analytical solutions by Arutyunian [39].

The first term on the right-hand side of (2.21) reflects the instantaneous response to the stress acting at the current time t. The stress impulse memory function L indicates how the strain at the current time t is affected by a unit stress impulse acting at a previous time t'. Function L is closely linked to the compliance function J. In practice, the compliance function is preferred, for two reasons: (i) The compliance function can be directly determined from the creep test, and (ii) for realistic compliance functions, the impulse function $L(t, t')$ tends to infinity as $t \to t'$ and the numerical evaluation of the integral in (2.21) becomes difficult. Of course, from the mathematical point of view, the stress–strain relations (2.16) and (2.21) are equivalent.

2.3 Relaxation Function

If the strain history $\varepsilon(t)$ is given, then (2.16) or (2.21) represents a *Volterra integral equation* for the unknown stress history.[6] For realistic forms of $J(t, t')$, this integral equation cannot be solved analytically.[7] But accurate numerical solutions, in which the integral is approximated by a discrete sum, are easy. In this way, one can for

---

[6]It is interesting to note that (2.16) is a Volterra equation of the first kind, with the stress rate as the unknown function and the compliance function as the kernel of the integral operator, while (2.21) is a Volterra equation of the second kind, with the stress as the unknown function and the stress impulse memory function as the kernel.

[7]Analytical solutions exist for nonaging materials with compliance functions corresponding to rheologic chains, but for chains consisting of many units they are usually very complicated.

example find the *relaxation function* $R(t, t')$, which is defined as the stress history $\sigma(t)$ caused by unit constant strain imposed at age t'.

As already mentioned, relaxation means spontaneous stress decrease under constant strain. A relaxation test is a direct counterpart of the creep test: A certain strain $\hat{\varepsilon}$ is suddenly imposed at time t' and afterward kept constant, and the corresponding stress history is measured; see Fig. 2.1b. But practically, a relaxation test is more difficult to perform because either the strain must be electronically or manually adjusted in short intervals to a constant value, or the load frame must be much stiffer than the specimen, which is next to impossible.

By virtue of the principle of superposition, we can write the stress history as

$$\sigma(t) = \hat{\varepsilon} R(t, t') \tag{2.22}$$

where $R(t, t')$ is the relaxation function. Following the same line of reasoning as in the preceding section, the superposition principle can equivalently be stated in its inverse form

$$\sigma(t) = \int_0^t R(t, t') \, d\varepsilon(t') \tag{2.23}$$

which expresses the stress history in terms of the strain history. This compact expression based on the Stieltjes integral could be rewritten in a form analogous to (2.15) using the Riemann integral. For strain histories that are discontinuous only at the onset of loading at time t_1 and afterward remain differentiable, we can write, in analogy to (2.16),

$$\sigma(t) = R(t, t_1) \varepsilon_1 + \int_{t_1^+}^t R(t, t') \dot{\varepsilon}(t') \, dt' \tag{2.24}$$

Formulae (2.23)–(2.24) for evaluation of the stress history from a given strain history represent the inversion of formulae (2.14) and (2.16) for evaluation of the strain history from a given stress history. So the compliance function and the relaxation function are not independent. If the compliance function is known, the corresponding relaxation function can be obtained from Eq. (2.16) written for the stress history $\sigma(t) = \hat{\varepsilon} R(t, t_1)$ and strain history $\varepsilon(t) = \hat{\varepsilon} H(t - t_1)$ which describe the relaxation test. Setting $\sigma_1 = \hat{\varepsilon} R(t_1, t_1)$, we obtain after easy manipulations the integral equation

$$J(t, t_1) R(t_1, t_1) + \int_{t_1^+}^t J(t, t') \dot{R}(t', t_1) \, dt' = 1 \quad \text{for all } t \geq t_1 \tag{2.25}$$

with $R(t, t_1)$ as the unknown function of t (with fixed t_1). Here, $\dot{R}$ is the partial derivative of the relaxation function with respect to its first argument, t. In the special case $t = t_1$, the integral vanishes and the equation reduces to an algebraic one, from which $R(t_1, t_1) = 1/J(t_1, t_1) = E(t_1)$. So the instantaneous material stiffness is the reciprocal value of the instantaneous material compliance, but in general the value of the relaxation function is not equal to the reciprocal value of the compliance function.

Equation (2.25) can be solved numerically, with the integral approximated by a finite sum. In this way, the values of the relaxation function $R(t_k, t_1)$ for a series of time instants t_k, $k = 1, 2, 3 \ldots N$, can be constructed. This will be explained in detail in Chap. 5; see Eqs. (5.22)–(5.23). One should note that if the compliance function $J(t, t_1)$ of an aging viscoelastic material is known only for one specific value of t_1 (e.g., determined from a creep test started at age t_1), it is not possible to determine the corresponding relaxation function $R(t, t_1)$, because evaluation of the integral in (2.25) requires evaluation of the compliance function $J(t, t')$ for general values of t' between t_1 and t.

The foregoing solution of relaxation curves based on the principle of superposition agrees very well with experiments for concrete, but only in absence of severe drying; see Fig. 3.8b.

The integral stress–strain equation can also be presented in terms of rates. In analogy to (2.17) or (2.18), we can write

$$\dot{\sigma}(t) = E(t)\dot{\varepsilon}(t) + \int_0^t \dot{R}(t, t')\, d\varepsilon(t') \tag{2.26}$$

or

$$\dot{\sigma}(t) = \dot{R}(t, t_1)\,\varepsilon_1 + E(t)\dot{\varepsilon}(t) + \int_{t_1^+}^t \dot{R}(t, t')\,\dot{\varepsilon}(t')\, dt' \tag{2.27}$$

where $E(t) = R(t, t) =$ instantaneous elastic modulus at age t, and the dot over R denotes partial derivative with respect to the first variable, t, i.e., $\dot{R}(t, t') = \partial R(t, t')/\partial t$.

Finally, in analogy to (2.21), we can write

$$\sigma(t) = E(t)\varepsilon(t) + \int_{t_1}^t S(t, t')\varepsilon(t')\, dt' \tag{2.28}$$

where

$$S(t, t') = -\frac{\partial R(t, t')}{\partial t'} \tag{2.29}$$

is the so-called *strain impulse memory function*.

In general, neither $\sigma(t)$ nor $\varepsilon(t)$ is prescribed. Then (2.14) or (2.23) represents a uniaxial stress–strain relation for linear aging creep. The fact that, for concrete, $J(t, t')$ and $R(t, t')$ do not depend merely on the time lag (elapsed time) $t - t'$, but on t and t' separately, is a consequence of aging (Figs. 3.7 and 3.8a). The theory defined by constitutive equation (2.14) or (2.23) is called the *aging linear viscoelasticity*. The aging, unfortunately, prevents transplanting from nonaging viscoelasticity the Laplace transform methods for solving structural creep problems. Thus, the aging is a major obstacle to analytical solutions. It requires that accurate solutions be obtained numerically, by step-by-step integration in time; see Chap. 5.

In the special case of nonaging *classical linear viscoelasticity*, widely used for polymers, the compliance function depends only on the elapsed time $t - t'$, i.e., $J(t, t') = J_0(t - t')$, and then the relaxation function has a similar property and can be expressed as $R(t, t') = R_0(t - t')$. One can also define the corresponding stress impulse memory function L_0^* and strain impulse memory function S_0^* as the time derivatives of J_0 and R_0, respectively. If these derivatives are considered in the sense of distributions and contain a singular Dirac-like component, the right-hand sides of Eqs. (2.21) and (2.28) can be written as convolutions and after transformation into the Laplace space they reduce to simple products between the Laplace images. In classical viscoelasticity, the Laplace transform of the stress impulse memory function L_0^* is called the *retardance* and the Laplace transform of the strain impulse memory function S_0^* is called the *relaxance*.

As proven by Roscoe [733], the compliance function $J_0(t)$ of any nonaging viscoelastic material can be approximated with arbitrary accuracy by the series (2.7), representing the compliance function of a Kelvin chain. The relaxation function $R_0(t)$ of a nonaging viscoelastic material can be approximated by the relaxation function of a *Maxwell chain* (also called a *Wienert model*), consisting of a sufficient number of Maxwell units coupled in parallel (Fig. 2.8). Each Maxwell unit is formed by a spring and a dashpot coupled in series. In Appendix A, it is shown that the relaxation function of a Maxwell chain with M units characterized by stiffnesses E_μ and viscosities η_μ, $\mu = 1, 2, \ldots M$, is given by

$$R_0(t) = \left(\sum_{\mu=1}^{M} E_\mu e^{-t/\tau_\mu} \right) H(t) \tag{2.30}$$

where the parameters $\tau_\mu = \eta_\mu / E_\mu$ are in this context called the *relaxation times*. The sum in (2.30) is a special case of Dirichlet series.

For nonaging chains, it is possible to prove that every Maxwell chain is exactly equivalent to a certain conjugate Kelvin chain and vice versa. The rules for constructing the conjugate of a given chain were formulated by Alfrey and Doty [32].

Approximation of the relaxation function for aging materials could be based on an *aging Maxwell chain*, for which the stiffnesses E_μ and viscosities η_μ are considered as age-dependent. The corresponding relaxation function

$$R(t, t') = \left(E_0(t') + \sum_{\mu=1}^{M} E_\mu(t') e^{-(t-t')/\tau_\mu} \right) H(t - t') \tag{2.31}$$

is a straightforward generalization of the Dirichlet series (2.30). Treatment of rheologic models with age-dependent properties is discussed in detail in Appendix A, but for practical application it is more convenient to use nonaging chains in combination with the solidification theory; see Chap. 9.

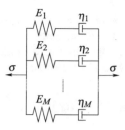

Fig. 2.8 Maxwell chain

2.4 Viscoelasticity Under Multiaxial Stress

So far we have considered stress and strain as scalars, tacitly assuming that the material is subjected to uniaxial stress. This is sufficient for the description of trusses, and also of beams and frames, if the effects of shear and torsion are neglected. For walls, slabs, shells, and massive structures, such as dams or nuclear containments, two- and three-dimensional constitutive laws are needed.

Recall that the generalized Hooke's law, describing a linear elastic isotropic material, works with two independent elastic constants, e.g., with Young's modulus E and Poisson's ratio ν, or with the bulk modulus K and shear modulus G. The extension to viscoelasticity should, in general, consider two independent compliance functions, e.g., $J_K(t, t')$ for the bulk compliance (related to changes of volume) and $J_G(t, t')$ for the shear compliance (related to changes of shape). Although the nanoscale creep mechanism consists presumably in sliding, the macroscale deformation involves both deviatoric and volumetric changes, which is a consequence of porosity. Since truly three-dimensional creep data are still scarce, in practical applications it is usually assumed that all compliance functions can be obtained by appropriate scaling of the uniaxial compliance function $J(t, t')$, in other words, that Poisson's ratio is not affected by creep and can be considered as a constant.[8] When this assumption is accepted, the three-dimensional generalization of the uniaxial viscoelastic strain–stress law (2.14) is written as

$$\boldsymbol{\varepsilon}(t) = \int_0^t J(t, t')\mathbf{C}_\nu \, \mathrm{d}\boldsymbol{\sigma}(t') = \mathbf{C}_\nu \int_0^t J(t, t') \, \mathrm{d}\boldsymbol{\sigma}(t') \tag{2.32}$$

Here,

$$\mathbf{C}_\nu = \begin{bmatrix} 1 & -\nu & -\nu & 0 & 0 & 0 \\ -\nu & 1 & -\nu & 0 & 0 & 0 \\ -\nu & -\nu & 1 & 0 & 0 & 0 \\ 0 & 0 & 0 & 2(1+\nu) & 0 & 0 \\ 0 & 0 & 0 & 0 & 2(1+\nu) & 0 \\ 0 & 0 & 0 & 0 & 0 & 2(1+\nu) \end{bmatrix} \tag{2.33}$$

---

[8]This assumption would not be realistic at young ages, when the material evolves from a liquid (fresh concrete mix) to a solid (hardened concrete) and Poisson's ratio decreases from initial values near 0.5 to final values near 0.2.

is the dimensionless elastic compliance matrix corresponding to a unit value of Young's modulus, $\boldsymbol{\varepsilon}$ is the 6×1 column matrix of strain components, and $\boldsymbol{\sigma}$ is the 6×1 column matrix of stress components. In a similar spirit, the three-dimensional generalization of the uniaxial viscoelastic stress–strain law (2.23) is written as

$$\boldsymbol{\sigma}(t) = \int_0^t R(t, t') \mathbf{D}_v \, \mathrm{d}\boldsymbol{\varepsilon}(t') = \mathbf{D}_v \int_0^t R(t, t') \, \mathrm{d}\boldsymbol{\varepsilon}(t') \tag{2.34}$$

where

$$\mathbf{D}_v = \mathbf{C}_v^{-1} = \frac{1}{(1+v)(1-2v)}\begin{bmatrix} 1-v & v & v & 0 & 0 & 0 \\ v & 1-v & v & 0 & 0 & 0 \\ v & v & 1-v & 0 & 0 & 0 \\ 0 & 0 & 0 & 0.5-v & 0 & 0 \\ 0 & 0 & 0 & 0 & 0.5-v & 0 \\ 0 & 0 & 0 & 0 & 0 & 0.5-v \end{bmatrix} \tag{2.35}$$

is the dimensionless elastic stiffness matrix, corresponding to a unit value of Young's modulus. Two-dimensional versions of the stress–strain law valid under plane–stress or plane–strain conditions are easily obtained by an appropriate modification of matrix $\mathbf{D}_v$.

2.5 Operator Notation

Defining a Volterra integral operator for creep on the basis of (2.14), one can approach the structural creep analysis in a powerful, general, and elegant manner. Recall that an operator assigns to each function from a suitable domain of definition and another function as its image. A simple example is a differential operator that assigns to each differentiable function its derivative. In viscoelasticity, we introduce the *compliance operator* (or *creep operator*) $\mathscr{J}$ that maps the stress history onto the corresponding strain history, and the *relaxation operator* $\mathscr{R}$ that maps the strain history onto the corresponding stress history. From this definition, it is clear that these operators are mutually inverse—their composition produces the identity operator $\mathscr{I}$, which maps an arbitrary function onto itself.

The operator notation is more compact and also more general than the explicit integral notation. For instance, the strain–stress relations (2.14), (2.15), (2.16), and (2.21) are all transcribed as

$$\varepsilon(t) = \mathscr{J}\{\sigma(t)\} \tag{2.36}$$

where the braces emphasize that $\mathscr{J}$ is an operator applied on the entire function $\sigma(t)$ (and not just a function of the value of σ at a fixed point t). In a similar spirit, the stress–strain relations (2.23), (2.24), and (2.28) are transcribed as

$$\sigma(t) = \mathscr{R}\{\varepsilon(t)\} \tag{2.37}$$

Obviously, operators $\mathscr{J}$ and $\mathscr{R}$ are mutually inverse:

$$\mathscr{J}^{-1} = \mathscr{R}, \qquad \mathscr{R}^{-1} = \mathscr{J} \tag{2.38}$$

This relation is useful when equations involving the rheologic operators $\mathscr{J}$ and $\mathscr{R}$ need to be solved. Another useful property is the linearity of these operators, formally expressed by the formulae

$$\mathscr{J}\{c_1\sigma_1(t) + c_2\sigma_2(t)\} = c_1\mathscr{J}\{\sigma_1(t)\} + c_2\mathscr{J}\{\sigma_2(t)\} \tag{2.39}$$

$$\mathscr{R}\{c_1\varepsilon_1(t) + c_2\varepsilon_2(t)\} = c_1\mathscr{R}\{\varepsilon_1(t)\} + c_2\mathscr{R}\{\varepsilon_2(t)\} \tag{2.40}$$

valid for all real numbers c_1 and c_2 and all integrable functions σ_1, σ_2, ε_1, and ε_2. Equations (2.39) and (2.40) are yet another way of writing the superposition principle.

Finally, it is good to know that the creep operator maps the Heaviside function on the compliance function and the relaxation operator maps the Heaviside function on the relaxation function[9]:

$$\mathscr{J}\{H(t - t')\} = J(t, t') \tag{2.41}$$

$$\mathscr{R}\{H(t - t')\} = R(t, t') \tag{2.42}$$

Inverting these relations, we get

$$H(t - t') = \mathscr{R}\{J(t, t')\} = \mathscr{J}\{R(t, t')\} \tag{2.43}$$

For a linear elastic material characterized by Young's modulus E, the stress–strain relations $\varepsilon(t) = \sigma(t)/E$ and $\sigma(t) = E\varepsilon(t)$ can formally be presented as (2.36) and (2.37) with the rheologic operators $\mathscr{J} = \mathscr{I}/E$ and $\mathscr{R} = E\mathscr{I}$ set to multiples of the identity operator $\mathscr{I}$. In view of (2.41) and (2.42), the corresponding compliance and relaxation functions are $J(t, t') = H(t - t')/E$ and $R(t, t') = EH(t - t')$.

So far, we have considered only the part of strain directly induced by stress. In general, strain is also affected by temperature and humidity changes. Usually, it is assumed that the total strain can be additively decomposed into the mechanical strain (induced by stress), thermal strain, and shrinkage strain. Equations (2.36)–(2.37) are then generalized to

---

[9]In this book, whenever an operator is applied on a function of several variables, it is supposed that it acts on variable t, and the other variables are considered as parameters. In some rare cases, it will be necessary to let an operator act on variable t', and this will be emphasized by denoting the operator as $\mathscr{J}'$ or $\mathscr{R}'$ instead of $\mathscr{J}$ or $\mathscr{R}$.

$$\varepsilon(t) = \mathscr{J}\{\sigma(t)\} + \varepsilon_{\text{sh}}(t) + \varepsilon_T(t) \tag{2.44}$$

$$\sigma(t) = \mathscr{R}\{\varepsilon(t) - \varepsilon_{\text{sh}}(t) - \varepsilon_T(t)\} \tag{2.45}$$

where ε_{sh} is the shrinkage strain (see Sect. 3.5) and ε_T is the thermal strain (see Sect. 10.6.2).

Chapter 3
Basic Properties of Concrete Creep, Shrinkage, and Drying

Abstract Clear and unambiguous definition and characterization of material properties is the essential basis of analysis, although it has not been achieved in much of the literature. We begin by discussing the dependence of elastic modulus on the rate or duration of short-time loading and introduce the notion of asymptotic modulus for infinitely fast loading. Then, we define the basic creep as the creep at constant moisture content, introduce the creep coefficient as the ratio of creep strain to properly defined elastic strain, and proceed to discuss shrinkage and creep of cross sections at drying exposure. Our attention is then focused on common misconceptions in measuring, defining, and reporting creep and shrinkage data, such as initial strains incompatible with the elastic modulus. We warn about false extrapolations caused by plots in linear time scale and point out problems due to autogenous shrinkage in modern concretes. Finally, we emphasize the importance of updating long-term predictions on the basis of short-time measurements on structures.

3.1 Sources and Characterization of Time-Dependent Deformations

In structural metals, time-dependent deformation at constant stress is observed only under elevated temperatures or under very high stresses. By contrast, the strains in concrete increase with time even if the applied stress is much smaller than the material strength. Although, with a few exceptions such as creep buckling [115, Chapter 9], the time-dependent deformations of concrete normally have little effect on the safety against collapse, they play an important role in serviceability and durability of structures, and their economic impact is enormous

Two components of the time-dependent strains of concrete can be distinguished:

1. *hygro-thermal strain*, which is independent of stress σ and at constant temperature represents the *shrinkage strain*, ε_{sh} (negative), and
2. *additional (delayed) mechanical strain* produced by stress, called in general *creep*.

© Springer Science+Business Media B.V. 2018
Z.P. Bažant and M. Jirásek, *Creep and Hygrothermal Effects in Concrete Structures*, Solid Mechanics and Its Applications 225, https://doi.org/10.1007/978-94-024-1138-6_3

The time-dependent shear strain is purely due to creep because thermal expansion and shrinkage are volumetric and do not produce shear strains.

The existence of *shrinkage* has been vaguely known since the invention of modern concrete in the early 1800s, and was first clearly documented by Le Chatelier in 1887 [568]. The existence of creep was discovered in 1907 by Hatt [470]. For service stress levels (up to 40% of strength) and at constant moisture content, creep depends on applied stress σ linearly and can be described by linear viscoelasticity using the general framework outlined in Chap. 2. During drying or at high stress levels, concrete creep is nonlinear, but the term viscoplasticity, used for nonlinear creep of metals at high temperature, is inappropriate because concrete does not exhibit plasticity (except under enormous confining pressure). The nonlinear dependence of creep on stress is caused essentially by cracking or microcracking damage.

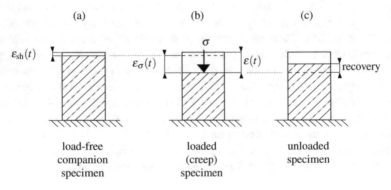

Fig. 3.1 (a) Shrinkage deformation, (b) mechanical deformation (elastic + creep) produced by stress, and (c) recovery after unloading

When exposed to a dry atmosphere, a concrete specimen gradually shrinks (Fig. 3.1a). The magnitude of shrinkage strain ε_{sh} is an increasing function of time t which approaches a finite bound at a gradually decreasing rate (Figs. 3.2a and 3.3). In normal concretes, most of shrinkage represents the *drying shrinkage*, which is caused mainly by increase of the capillary tension of pore water and the solid surface tension of pore walls, as well as thinning of multimolecular hindered adsorbed water layers in cement gel micropores. This kind of shrinkage is engendered by diffusion of water out of pores. Specimens immersed in water exhibit *swelling* (positive ε_{sh}), which is normally an order of magnitude smaller than drying shrinkage, and thus often negligible. The cause of swelling is incorporation of additional water into the porous nanostructure as well as an increase of the water content of hindered adsorbed layers only a few molecules thick, acting as part of the solid microstructure.

A part of shrinkage, called the *autogenous shrinkage*, is caused by volume changes exhibited by the chemical reactions of cement hydration under sealed conditions.[1] It

---

[1] The term "autogenous shrinkage" refers to the macroscopic (bulk) changes, while a related term "chemical shrinkage" refers to the internal volume changes due to chemical reactions. Autogenous shrinkage is smaller than the chemical shrinkage, due to the voids generated by hydration.

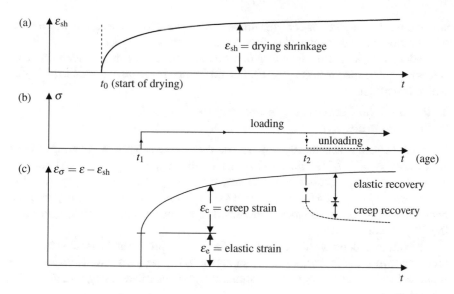

Fig. 3.2 Curves of shrinkage, creep, and recovery after unloading

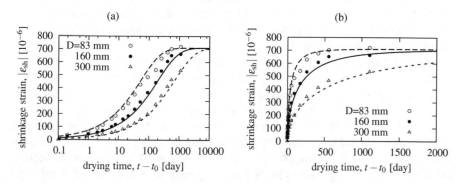

Fig. 3.3 Shrinkage measured by Wittmann, Bažant, Alou, and Kim [878] on cylinders with different diameters D (data points) compared to theoretical curves: (a) logarithmic time scale, (b) linear time scale

is the only type of shrinkage taking place in sealed specimens. For normal-strength concrete of high water-cement ratio (above 0.55) and no admixtures, it represents only a small fraction of the drying shrinkage and is usually neglected. However, for the modern high-strength concretes cast with a very low water-cement ratio (0.4 or less) and various admixtures, the autogenous shrinkage is comparable in magnitude to the drying shrinkage and must be taken into account. After the relative humidity[2] of

---

[2]The *pore relative humidity* is defined as $h(p_v, T) = p_v/p_{sat}(T)$ where p_v is the partial vapor pressure in the capillary pores and $p_{sat}(T)$ is the saturation vapor pressure of water at temperature T.

water vapor in the pores of concrete drops below about 65%, the chemical reactions of hydration virtually stop, and so does the autogenous shrinkage.

Under stress, a concrete specimen deforms with time more than an identical load-free companion specimen (Fig. 3.1b). The difference is the *mechanical strain* $\varepsilon_\sigma(t)$, which consists of

1. the *elastic (or instantaneous, short-time) strain*, ε_e and
2. the *creep strain*, ε_c.

If uniaxial stress σ is suddenly imposed on a concrete sample at age t_1, the elastic strain is $\varepsilon_e = \sigma/E(t_1)$ where $E(t_1)$ is the elastic modulus at age t_1. At constant stress, creep strain increases at a gradually decreasing rate (Fig. 3.2b, c). As far as it is known, no bound on the creep strain exists, but this is not an alarming property because a 500-year compliance is projected to be only about 20–25% higher than the 50-year compliance.[3]

Unloading of concrete at some age t_2 (Fig. 3.1c) results in an instantaneous strain recovery corresponding to the elastic modulus $E(t_2)$ at age t_2, followed by a long-time monotonic partial recovery of creep strain at a gradually decreasing rate (Fig. 3.2c, dashed curve).

The creep that occurs at constant moisture content of concrete is called the *basic creep*. It is caused by breakage and reformation of atomic bonds at various highly stressed sites within the colloidal microstructure of the calcium silicate hydrate gels in the hardened cement paste.

Simultaneous drying causes additional creep, called the *drying creep* (or *Pickett effect*, or *stress-induced shrinkage*). The drying creep evolves in time similar to shrinkage and exhibits a similar dependence on cross-sectional thickness D (Fig. 3.3), while the basic creep is independent of D. Figure 3.4 shows the classical tests of Pickett [690], which were the first to demonstrate the drying creep and revealed that it occurs both for drying and wetting. The specimens were plain concrete beams of square cross section (with side 50.8 mm, span 813 mm, and midspan load 222 N); data point series LC corresponds to loading accompanied by humidity cycles (drying–wetting–drying–wetting, etc.), series LD to loading with drying, series L to loading without drying, and series D to drying without loading.

The drying creep has complex physical causes. One is that drying elevates the local stress peaks within the microstructure of calcium silicate hydrates and thus increases the rate of bond breakages [132]. Another cause is apparent, due to the fact that a large part of the observed drying creep in compression has its origin in cracking, and is treated as creep only for convenience [198, 874]. The reason is that the drying of concrete specimens produces nonuniform distribution of pore humidity and local shrinkage, which creates self-equilibrated stresses that may cause extensive microcracking. The microcracking is expansive and thus it diminishes the observed drying shrinkage of the companion compressed specimen, compared to what would

---

[3]Like all materials, the primary creep of concrete, proceeding under constant stress at decaying rate, must eventually transit to the so-called secondary creep, which proceeds at constant rate. The transition time, called the Maxwell time, occurs in rocks at 100 years, as approximately inferred from geologic processes. Doubtless the same can be expected for concrete.

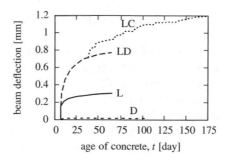

Fig. 3.4 Test data by Pickett [690]

be expected for no microcracking. Therefore, the shrinkage customarily measured on load-free specimens is in fact an apparent shrinkage, which is significantly less than the true shrinkage in the absence of microcracking. The true shrinkage can occur only under sufficient compressive stress, simultaneously with compressive creep. A deeper discussion of the physical sources of drying creep will be presented in Sect. 10.1, and a method for their separate assessment will be described in Sect. 12.4.

A salient property of concrete creep is *aging*. What is meant by aging is very different from the aging effects in other materials (e.g., the gradual degradation of strength of fiber composites). In concrete, the aging causes the strength as well as the elastic modulus E to increase with age t, with the rate of increase gradually diminishing in time. One manifestation is that the elastic recovery is smaller than the initial elastic strain (Fig. 3.2c). Another is that concrete specimens loaded at a high age creep much less than those loaded at low age (Figs. 3.5, 3.7 and 3.8a).

An important, though not the sole, cause of aging in creep is the chemical process of *hydration*. The hydration products, chiefly the tricalcium silicate hydrate gel (essentially identical to the mineral tobermorite), gradually fill the pores of hardened cement paste; hence, the total volume and the mean size of capillary pores decrease, which gradually stiffens and strengthens the microstructure. However, since the aging process continues (at pore humidities above 75%) for many years after the chemical process of hydration becomes very slow, there must be another cause. It is the gradual relaxation of stress peaks of self-equilibrated microprestress on the nanoscale in cement gel microstructure [132]. This provides one motivation for the microprestress-solidification theory, described in Chap. 10.

Typically, the ratios of the creep rates (at the same load duration) of concretes loaded at the ages of 3 days, 1 year, and 10 years to the creep rate of concrete loaded at the age of 28 days are about 7, 0.3, and 0.1. In view of this paramount role of aging, **the time is always measured from the initial set of concrete**, i.e., from the instant when concrete first becomes a solid. Thus, in concrete creep calculations, the time variable t always corresponds to the age of concrete.

During a certain initial period, concrete remains in the formwork and thus cannot dry. At surfaces without formwork, drying should be (and normally is) prevented by moist covers (e.g., burlap), sealing membranes, or sprays, to achieve proper curing. For shrinkage calculations, it is important to know the approximate age t_0 at the start

of drying, which roughly corresponds to the age at the removal of formwork or the
end of other drying protection.

The mathematical description of time-dependent behavior of concrete can be
developed within the framework of linear aging viscoelasticity (see Chap. 2), with
certain adjustments and generalizations. To capture all the important phenomena,
it is necessary to account for the additional strain due to cracking (governed by
an appropriate smeared crack model or damage model), and for the nonmechanical
(hygro-thermal) strain induced by thermal expansion and shrinkage. All these effects
will be lumped into the so-called *eigenstrain*, $\tilde{\varepsilon}$. In most of what follows, we consider
only the shrinkage strain, setting $\tilde{\varepsilon}(t) = \varepsilon_{\mathrm{sh}}(t)$. The strain evaluation formula (2.14)
needs to be enriched by the eigenstrain and generalized to

$$\varepsilon(t) = \int_0^t J(t, t') \, \mathrm{d}\sigma(t') + \tilde{\varepsilon}(t) \tag{3.1}$$

A similar adjustment is required for the stress evaluation formula, in which only the
mechanical strain (without its part caused by cracking) enters as the cause of stress,
and so (2.23) is rewritten as

$$\sigma(t) = \int_0^t R(t, t') \, \mathrm{d}\left[\varepsilon(t') - \tilde{\varepsilon}(t')\right] \tag{3.2}$$

Development of a realistic model for predicting the compliance function and
shrinkage function of a given concrete is a difficult problem that has engendered
intense polemics for a long time. Since many structures are designed for lifetime
over 100 years while laboratory data for multidecade creep are very scant (and for
more than 30 years nonexistent), the combination of laboratory data with multidecade
observations of creep deformations on structures has recently become a problem of
great interest. The problem calls for careful analysis of extensive data from long-
time tests conducted on different concretes and in different environments, generally
exhibiting large statistical scatter. When short-time measurements on a given concrete
are to be extrapolated to long times, it is necessary to use a compliance function that
is correct for both long and short times.

It is important to note that the compliance function should account not only for
basic creep but also for drying creep, which depends on the variations of humidity
and cannot be considered as a function of the current time t and the age at loading
t' only. For the sake of simplicity, we will still write the compliance function with
two arguments t and t', but this function will be considered as the sum of the basic
compliance function and the additional compliance due to drying creep, denoted as
J_{d} and dependent on factors that influence the drying process; see Sect. 3.6.

A realistic prediction tool for creep and shrinkage of concrete is *model B3*
[104, 107], which represents a RILEM standard recommendation (1995). Its recent
improvement, *model B4*, has also been accepted as a RILEM recommendation, developed by the RILEM Technical Committee TC-242-MDC [136]. Model B3 and, to
some extent, model B4 will be presented in the following sections of this chapter.

Both of these models decompose the compliance function into three additive parts which have a different physical origin:

$$J(t, t') = \frac{1}{E_0} + J_b(t, t') + J_d(t, t') \tag{3.3}$$

Here, E_0 is the time-independent *asymptotic modulus*, J_b is the basic creep compliance, and J_d is the drying creep compliance. The individual terms will be discussed separately in Sects. 3.2, 3.3, and 3.6. All the times such as t or t' will be expressed in days and measured from the set of concrete.

3.2 Asymptotic Modulus

Concrete happens to exhibit non-negligible creep even for extremely short load durations, and even the dynamic modulus $E_{dyn}(t)$, which is obtained by sound velocity measurements and corresponds to the durations of sound vibration periods, is age-dependent and non-negligibly affected by creep [104, 107].

The asymptotic modulus, i.e., parameter E_0 in (3.3), corresponds to extrapolating the creep curve to loading durations many orders of magnitude shorter than ever measured. Bažant and Osman [173] and Bažant and Baweja [105, 107] demonstrated the age-independence of E_0 by considering the compliances for load durations $t - t'$ ranging from about 10 seconds to several days. They fitted the compliances by a function of the type $J = 1/E_0 + c(t - t')^n$ and obtained the parameters by optimizing the fit of data for various ages t' at loading. The E_0 values for various t' were nearly the same, and the coefficient of variation of errors of the fit did not increase significantly when E_0 was forced to be exactly the same for all t'.

Figure 3.5a shows such fits constructed for creep curves corresponding to the same concrete tested at different ages $t' = 28, 90$, and 270 days, using $n = 0.1$. Note that the variable on the horizontal axis is the load duration raised to power 0.1. The extrapolation to zero load duration gives for all these tests almost the same value, which we will consider as a constant (age-independent) material parameter E_0. This is also consistent with the data for loading times as short as 0.001 s, deduced from the complex modulus corresponding to vibration tests of Radjy and Richards [712]. The advantage of introducing E_0 corresponding to such extrapolation is that, unlike the short-term (or static) elastic modulus $E(t')$, the asymptotic modulus E_0 is unaffected by creep. Of course, the assumption that E_0 is age-independent is empirical and holds with reasonable accuracy only for concrete older than at least 3 days (Fig. 3.5b); a more refined model would be needed for creep at very young ages.

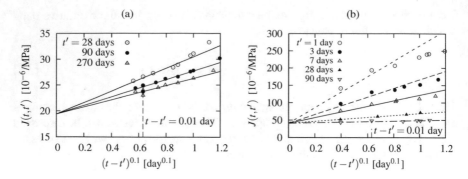

Fig. 3.5 Extrapolations of short-time creep data to zero load duration: (a) data of Komendant et al. [551], (b) data of Pirtz [696] measured on Dworshak Dam concrete

Even though the asymptotic modulus of very young concrete is probably not constant (i.e., not age-independent), the creep strain for short load durations still grows proportionally to $(t - t')^{0.1}$. This is confirmed by the experimental data of Boulay presented by Acker and Ulm [21] and replotted in Fig. 3.6. Boulay's measurements, performed on concrete at the age of 1 day, cover extremely short load durations starting from 8.5 microseconds.

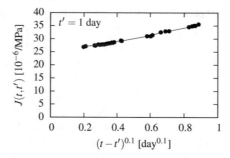

Fig. 3.6 Creep compliance as a function of the load duration raised to the power of 0.1, based on the experimental data of Boulay presented by Acker and Ulm [21]

The *short-term* (or static) *modulus of elasticity*, including its age dependence, is defined as

$$E(t') = \frac{1}{J(t' + \Delta t_s, t')}$$ (3.4)

where Δt_s is a load duration selected such that the strain delayed by less than Δt_s is from the practical point of view considered as instantaneous. The choice of $\Delta t_s \approx 0.01$ day ≈ 15 min gives (for $t' = 28$ days) good agreement with the *conventional elastic modulus* E_{28} defined in the ACI Building Code 318-05. The *dynamic modulus*

E_{dyn}, including its age dependence, is obtained if Δt_s is replaced by $\Delta t_d \approx 10^{-7}$ day ≈ 0.01 s. Equation (3.4) with J computed according to model B3 captures the age dependence of E better than the American Concrete Institute (ACI) recommendation of E being proportional to the square root of the compression strength at age t [107]; see formulae (D.54)–(D.56) in Appendix D.

As a crude empirical estimate, the asymptotic modulus can be taken as $E_0 = E_{28}/0.6$. If the conventional elastic modulus at age 28 days, E_{28}, is not measured directly, it can be estimated from the *standard compression strength* $\bar{f}_c$ (mean uniaxial compression strength determined on standard cylinders at age 28 days) using the empirical formula recommended by ACI (1999),

$$E_{28} = 57 \text{ ksi} \cdot \sqrt{\frac{\bar{f}_c}{1 \text{ psi}}} \tag{3.5}$$

or, in SI units,

$$E_{28} = 4.733 \text{ GPa} \cdot \sqrt{\frac{\bar{f}_c}{1 \text{ MPa}}} \tag{3.6}$$

In practice, E_{28} is often estimated from the *reduced strength*, f_c', which is the strength value required for design and is about 30% smaller than the mean strength $\bar{f}_c$. But to estimate the mean value of E_{28}, the mean value of strength $\bar{f}_c$ must, of course, be used in (3.5) because that is how this empirical formula was calibrated [671].

The *fib* Model Code for Concrete Structures 2010 recommends a somewhat different empirical formula for evaluation of the conventional elastic modulus,

$$E_{28} = 21.5 \text{ GPa} \cdot \alpha_E \cdot \left(\frac{\bar{f}_c}{10 \text{ MPa}}\right)^{1/3} \tag{3.7}$$

which corresponds to the load duration in the order of 10 s. Here, α_E is a dimensionless coefficient that depends on the aggregate type and is equal to 1.0 for quartzite aggregates, 0.9 for limestone aggregates, 0.7 for sandstone aggregates, and 1.2 for basalt and dense limestone aggregates.

3.3 Basic Creep

Figure 3.7 exhibits the typical compliance curves of sealed concrete specimens (basic creep) for unit stresses applied at various ages at loading, t'. If the first reading is taken 15 minutes after loading and the corresponding deformation is regarded as elastic (instantaneous), as has often been done, then the initial compliance $J(t' + \Delta t_s, t') = 1/E(t')$ decreases with the age at loading t' at a gradually decreasing rate (see the dotted curve in Fig. 3.7a). Since concrete exhibits non-negligible creep already for load durations as short as a fraction of a second, the age effect on $E(t')$ can be regarded as a special case of the age effect on the creep curves. Note that for loading of young concrete, the effect of increasing the loading age t' is mainly to shift the creep curve vertically downwards, while for the loading of old concrete, the effect is mainly to shift the creep curve horizontally to the right (Fig. 3.7a).

Since the range of significant creep stretches over at least ten orders of magnitude of load duration, it is insufficient and potentially misleading to plot the creep curves in the actual (linear) time scale (Fig. 3.7a). Such a scale can display the creep trend for only one to two orders of magnitude of t and t'. The plots in the linear time scale obscure either long times or short times, or both. Therefore, experimental validations of creep models based on plots in the linear time scale must be distrusted. Proper validation requires the creep curves to be plotted in the logarithmic time scale.

A typical creep compliance plot in the logarithmic scale of load duration $t - t'$ is shown in Fig. 3.7b. Now, we see the reason for the term asymptotic modulus—as $t - t'$ tends to 0, all the creep curves approach a horizontal asymptote at compliance level $1/E_0$. This is confirmed by test data, with the exception of creep curves obtained for very young ages at loading, below 3 days [107]. The slope of the creep curves in the log-time initially increases, but later each of the curves approaches a straight line, which has about the same slope for all the ages at loading. For a higher age t' at loading, this transition occurs at a longer duration $t - t'$ but at smaller strain. Experimental evidence of the very large effect of the age at loading, t', on both the compliance function and the relaxation function is exemplified by Figs. 3.8 and 3.9.

Figures 3.10 and 3.11 demonstrate further simple properties of basic creep. In the semilogarithmic plots in Fig. 3.10, the graphs of compliance functions for various ages at loading approach parallel straight lines. This implies that, after some initial period, the compliance function becomes logarithmic. The graphs of the compliance rate ($\dot{J} \equiv \partial J/\partial t$) versus time in Fig. 3.11, which are plotted as logarithmic on both axes, show that after some lapse of time the compliance rates for various ages at loading approach a single straight line of slope -1. This means that $\dot{J}(t, t') \propto -1/t$, the integration of which confirms that the long-time basic creep is logarithmic. These

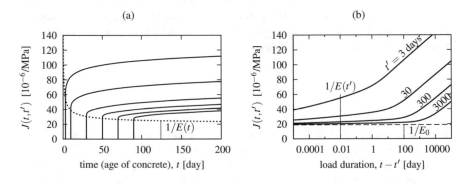

Fig. 3.7 Compliance functions for basic creep at various ages t' at loading: (a) linear time scale, t' ranging from 3 to 90 days, (b) logarithmic time scale, t' ranging from 3 to 3000 days

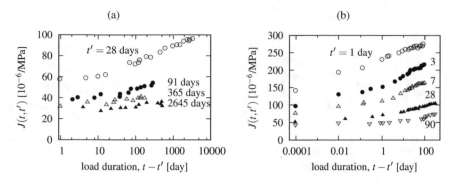

Fig. 3.8 Effect of aging on compliance function measured on (a) Shasta Dam concrete (after [126]), (b) Dworschak Dam concrete [696]

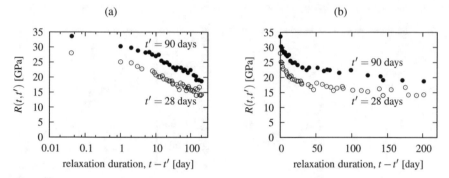

Fig. 3.9 Effect of aging on relaxation function measured on Ross Dam concrete [455, 458] in (a) semilogarithmic scale, (b) linear scale

basic properties are what motivated the formulation of the compliance function of the B3 and B4 models. The theoretical curves obtained with the B3 model are plotted in Fig. 3.12.

A very simple form of the basic creep compliance function is provided by the *double-power law* [173],

$$J_b(t, t') = \frac{\phi_1}{E_0} \left(t'^{-m} + \alpha \right) \left(t - t' \right)^n \tag{3.8}$$

which gives acceptable approximation only for, roughly, load duration $t - t' \leq 365$ days and age at loading t' between 14 and 365 days, with the parameter values $n = 0.1$, $m \approx 1/3$, $\alpha \approx 0.05$, and ϕ_1 ranging from 2 to 6, depending on the specific concrete, and with t and t' substituted in days. If parameters m and α are taken by their default values, the asymptotic modulus should be set to $E_0 = (1 + 0.239\phi_1)E_{28}$ where E_{28} is the conventional elastic modulus of concrete at age 28 days.

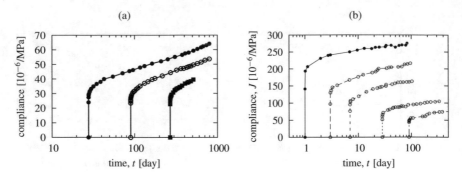

Fig. 3.10 Compliance functions J for basic creep at various ages at loading: (a) data of Komendant et al. [551], (b) data of Pirtz [696]

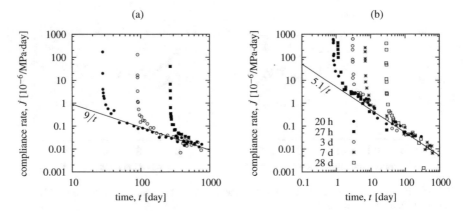

Fig. 3.11 Compliance rates $\dot{J} \equiv \partial J/\partial t$ for basic creep at various ages at loading: (a) data of Komendant et al. [551], (b) data of Ulm and Acker [822] and Acker and Ulm [21]

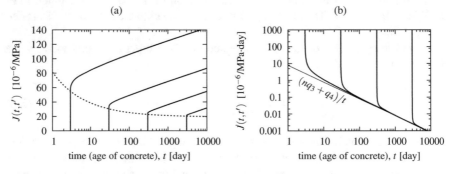

Fig. 3.12 Compliance functions J and their rates $\dot{J} \equiv \partial J/\partial t$ for basic creep at ages at loading ranging from 3 to 3000 days, according to the B3 model

Improved predictions can be obtained with the *log-double-power law*

$$J_b(t, t') = q_s \ln \left[1 + \psi \left(t'^{-m} + \alpha \right) (t - t')^n \right] \tag{3.9}$$

first proposed by Bažant and Chern [117] and used in the *short form of B3 model* [108]. The typical values of the model parameters are $n = 0.1$, $m = 0.5$, $\alpha = 0.001$, $\psi = 0.3$, and $q_s = 11.4/E_{28}$. For J_b evaluated according to this model, the asymptotic modulus in (3.3) should be set to $E_0 = E_{28}/0.6$.

The *full form of model B3* [104, 107] has been based on a systematic theoretical formulation of the basic physical phenomena involved, and on statistical optimization with regard to most of the test data that exist in the literature. Since the model is based on the solidification theory (to be explained in detail in Chap. 9), the basic creep compliance is more conveniently defined by its time rate than its accumulated value:

$$\frac{\partial J_b(t, t')}{\partial t} = \frac{n(q_2 t^{-m} + q_3)}{(t - t') + (t - t')^{1-n}} + \frac{q_4}{t} \tag{3.10}$$

in which $n = 0.1$ and $m = 0.5$ are empirical parameters whose values can be taken the same for all normal concretes, and q_2, q_3, and q_4 are empirical constitutive parameters whose prediction is discussed in Appendix C. The total basic creep compliance is obtained by integrating (3.10) with the initial condition $J_b(t', t') = 0$ (because the instantaneous compliance is fully reflected by the constant term $1/E_0 \equiv q_1$ in (3.3)). The terms containing q_3 and q_4 can be integrated in closed form, but the term containing q_2 leads to a binomial integral which cannot be expressed analytically. So the basic creep compliance function has the form

$$J_b(t, t') = q_2 Q(t, t') + q_3 \ln[1 + (t - t')^n] + q_4 \ln \left(\frac{t}{t'} \right) \tag{3.11}$$

where

$$Q(t, t') = \int_{t'}^{t} \frac{ns^{-m}}{(s - t') + (s - t')^{1-n}} \, ds \tag{3.12}$$

is a function that can be obtained by numerical integration or by interpolation from a table computed in Bažant and Baweja [104, 107]. The values of $Q(t, t')$ can also be calculated from the approximate explicit formula (C.2) given in Appendix C.

The new B4 model [136] uses the same form of compliance function (3.11) as the B3 model but recommends different empirical formulae for the estimation of parameters q_1 to q_4; see Appendix D.1.1.

A characteristic property of the basic creep curves for different t' is that they do not diverge, i.e., always approach each other with increasing t, which is mathematically expressed by the inequality

$$\partial^2 J(t, t')/\partial t \partial t' \geq 0 \tag{3.13}$$

This so-called *nondivergence condition* [152] is satisfied by the compliance function (3.11) used by the full B3 model and by the B4 model, but is violated by the short form of B3 (3.9) as well as by the double-power law (3.8) and by other forms endorsed by ACI, CEB-*fib* and JSCE; see Sect. 9.6. When a creep formula with divergent creep curves is used in the superposition principle to predict creep recovery after unloading, the recovery curve is nonmonotonic. This feature is thermodynamically inadmissible for the Kelvin chain model [139] and may inhibit the convergence of a computer simulation of a nonlinear problem that involves creep. Therefore, the short form of B3 is not suitable for large-scale computer analyses; exclusively the full B3 model or, even better, the new B4 model, should be used for that purpose.

3.4   Creep Coefficient

In creep analysis of structures, it is convenient to characterize creep by a dimensionless *creep coefficient*, which is defined as[4]

$$
\varphi(t, t') = \frac{\text{creep strain at time t for stress } \hat{\sigma} \text{ applied at age } t'}{\text{"initial elastic" strain at age } t'} =
$$
$$
= \frac{\hat{\sigma} J(t, t') - \hat{\sigma}/E(t')}{\hat{\sigma}/E(t')} = E(t')J(t, t') - 1 \tag{3.14}
$$

Most authors (and most design codes) have in the past defined creep by the creep coefficient φ instead of the compliance function J, the latter being a sum of the initial elastic compliance and the creep compliance. However, in structural design one is interested only in the response for load durations at least an order of magnitude longer than the load duration corresponding to the definition of the (static, short-term) elastic modulus E. Then, it is only the sum (i.e., the total compliance J) that matters for the results of analysis, the subdivision between the creep and elastic parts being irrelevant. The definition of creep coefficient runs into the problem that the range of freedom in the definition of the "initial" deformation is considerable—e.g., the ratio of the J-values for the load durations of 1 hour and 1 second can for young concrete exceed 1.5. This introduces a danger.

The danger is that the code makers specify only the creep coefficient φ but not the corresponding value of E that gives the correct compliance J, in agreement with the measurement used to calibrate the code. The designer may then combine this creep coefficient with different values of elastic modulus, for example, the value specified in the design code. This value is incompatible because it has been calibrated by tests involving load cycles and different load application times, and may be very different from the moduli corresponding to the initial deformations in the creep tests.

---

[4]Formula (3.14) represents the standard definition of creep coefficient. Some models and codes use a modified definition, with $E(t')$ replaced by $E(28)$; see Appendix E for details.

Likewise, many experimenters reported only the measured values of the creep coefficient or creep strain without specifying the measured initial elastic strain. Such reports of test data are virtually useless. Therefore, the designer should always first establish the values of the compliance function J and then use them to determine φ according to (3.14). If that is done, the choice of the particular definition of elastic modulus has no appreciable effect on the calculated long-time response at variable stress (and no effect at all at constant stress).

3.5 Mean Cross-Sectional Shrinkage

Shrinkage evolves differently at different points of the cross section of a concrete beam or plate. Therefore, its effects should properly be analyzed two- or three-dimensionally, e.g., by subdividing the cross section or the structure into many finite elements. Such analysis necessitates a local constitutive relation for shrinkage (see Sect. 8.6), in which the free (unhindered) shrinkage strain rate $\dot{\varepsilon}_{sh}$ at a generic point of structure may be given as a function of the rate of pore relative humidity $\dot{h}$.

To keep structural analysis simple, a uniform "average" or "effective" shrinkage of the whole cross section is usually assumed. In view of deflection measurements on prestressed box girder bridges that came to light during the last decade [558], this classical simplifying assumption often causes major errors in predicting long-time deflections and cracking (see Chap. 7). Nevertheless, this assumption remains acceptable for structures of low creep and shrinkage sensitivity and still forms the basis of current design practice. While it greatly simplifies long-time structural analysis, it makes the expressions for the shrinkage function considerably more complex because the cross section is typically in a highly nonuniform state of pore humidity and residual stress.

According to models B3 [104, 107] and B4 [136], the average longitudinal shrinkage of a cross section of a long beam or plate may be approximately calculated as

$$\varepsilon_{sh}(t) = -\varepsilon_{sh}^{\infty} k_h \, S(t - t_0) \tag{3.15}$$

where t is the current age of concrete, t_0 is the age at the start of drying, $\varepsilon_{sh}^{\infty}$ is the theoretical magnitude of the final shrinkage strain at zero ambient humidity (typically $0.0003 - 0.0011$), k_h is a coefficient depending on the average environmental humidity h_{env} (relative vapor pressure), and $S(\hat{t})$ is an increasing function of the duration of drying, $\hat{t} = t - t_0$. This function describes the evolution of normalized shrinkage strain $|\varepsilon_{sh}|/\varepsilon_{sh}^{\infty}$ in a perfectly dry environment, starting from its initial value 0 at $\hat{t} = 0$ and approaching asymptotically 1 as $\hat{t} \to \infty$. A suitable formula, theoretically justified in Sect. 8.4.5.1, is

$$S(\hat{t}) = \tanh \sqrt{\frac{\hat{t}}{\tau_{sh}}} \tag{3.16}$$

where τ_{sh} is traditionally called the *shrinkage halftime*, because it roughly indicates the time at which ε_{sh} reaches one half of its final value.[5]

The shrinkage halftime in model B3 can be estimated as

$$\tau_{sh} = k_t (k_s D)^2 \tag{3.17}$$

in which k_t is a factor dependent on concrete diffusivity, k_s is a cross-sectional shape factor, and D is the effective cross-sectional thickness. Factor k_t is inversely proportional to the diffusivity of pore water in concrete.[6] It can be estimated using the empirical formula in line 5 of Table C.2 in Appendix C, based on the compression strength and age of concrete at the onset of drying. The values of k_s for different specimen shapes given in Table 3.1 are based on solutions of the nonlinear diffusion equation for drying of concrete [166]. The definition of D is such that, for an infinite slab, it represents the actual thickness. For a general concrete member, it can be estimated as $D = 2V/S_e$ where V and S_e are the volume and the exposed surface area of the concrete part.

Table 3.1 Values of shape factor k_s

Specimen shape	k_s
Infinite slab	1.00
Infinite cylinder	1.15
Infinite square prism	1.25
Sphere	1.30
Cube	1.55

Model B4 estimates the shrinkage halftime using a slightly modified formula

$$\tau_{sh} = \tau_0 k_{\tau a} \left(k_s \frac{D}{1\,\text{mm}} \right)^2 \tag{3.18}$$

where k_s is the same shape factor as in model B3, $k_{\tau a}$ is a factor dependent on the aggregate type, with default value 1, and parameter τ_0 takes into account the composition of the concrete mix; see Appendix D for details. Recently, Donmez and Bažant [356] have pointed out that, for a nonlinear diffusion model, the optimal value of the shape factor depends on the ambient humidity. The refinement is minor and will be described in Sect. 8.4.5.2.

---

[5]In fact, one half of the final shrinkage is reached already at time $t = 0.3\tau_{sh}$, while at time $t = \tau_{sh}$ the shrinkage function $S(\hat{t})$ is at 76% of its final value. Despite that, we will stick to the traditional terminology commonly used in the diffusion theory.

[6]The diffusivity of concrete is extremely low, as manifested, for instance, by the fact that the core of a standard 6-inch (150 mm) cylinder may take up to 20 years to dry to the environmental humidity.

The typical shapes of the shrinkage curves, proportional to function $S(\hat{t})$, are portrayed in Fig. 3.3. The proportionality of τ_{sh} to the square of thickness is a salient property of all kinds of diffusion processes, linear as well as nonlinear. This property is well verified by many data (see, e.g. Fig. 3.3). Other data show some deviations occurring at longer times (Fig. 3.13). This can be explained by two causes: the simultaneous aging and the microcracking. They usually have opposite effects at long times. Often they nearly cancel each other, with the result that the scaling $\tau_{sh} \propto D^2$ still works reasonably well.

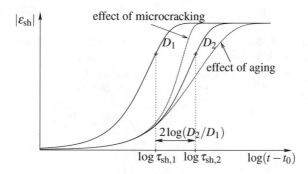

Fig. 3.13 Horizontal shift of the shrinkage curve in logarithmic time scale by distance $2\log(D_2/D_1)$ caused by a change of diameter from D_1 to D_2, and the effects of cracking and aging

Another consequence of the diffusion origin of drying shrinkage is that the initial part of the shrinkage curve must evolve as $\sqrt{\hat{t}}$.[7] Function $S(\hat{t})$ in (3.16) is used because it satisfies this property asymptotically for small $\hat{t}$ and because it also exhibits a reasonable asymptotic form for long times $\hat{t}$. The proportionality of shrinkage strain to the square root of drying time at the early stages of the drying process is confirmed by Fig. 3.14a, which shows that the plot of the logarithm of shrinkage versus the logarithm of drying time is initially a straight line of slope $1/2$. The small deviations for very short times are due to neglecting finiteness of surface emissivity. The deviations for long times reveal the limited reach of the square-root function, which ought to be, and is, shorter for thinner specimens. The expected form of the initial asymptotics is further confirmed by Fig. 3.14b, in which the shrinkage strain is plotted as a function of the square root of drying time and the initial part of the graph is seen to be very close to a straight line. The concept of asymptotic matching is illustrated in Fig. 3.15.

The final shrinkage strain $\varepsilon_{sh}^{\infty}$ corresponds to a perfectly dry environment ($h_{env} = 0$) and, for a concrete of given composition, can be roughly estimated from the empirical formulae [106] given in Appendix C. In an environment of relative humidity $h_{env} > 0$, it must be reduced by the factor

---

[7] An exception is a cement paste specimen less than a few millimeters thick, in which the finiteness of surface emissivity modifies the initial shrinkage curve. The emissivity is roughly equivalent to adding on the surface a layer about 1 mm thick.

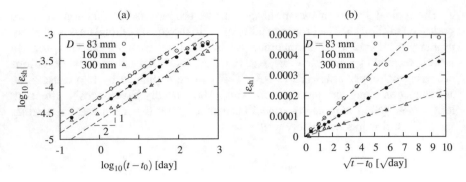

Fig. 3.14 Test data of Wittmann et al. [878] (a) in logarithmic scale, (b) with shrinkage plotted against the square root of drying time

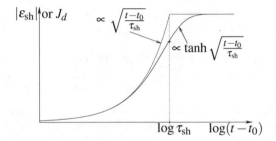

Fig. 3.15 Shrinkage curve viewed as asymptotic matching of the theoretically determined asymptotic behaviors for short and long drying times

$$k_h = 1 - h_{\text{env}}^3 \tag{3.19}$$

This formula can be used only up to $h_{\text{env}} = 98\%$, which is the value recommended as a suitable approximation of relative environmental humidity of sealed normal-strength concrete. Of course, the initial relative humidity in the pores of concrete is 100%, but under sealed conditions, this value decreases in time due to *self-desiccation* caused by cement hydration.[8] With $k_h = 1 - 0.98^3 = 0.0588$, formula (3.15) gives, for normal concretes (but not high-strength concretes), reasonable approximate values for the autogenous shrinkage. Fully saturated conditions with $h_{\text{env}} = 100\%$ are achieved only if the concrete surface is kept wet, which happens, e.g., for concrete immersed in water. Under such conditions, concrete exhibits swelling, i.e., an expansion of volume, which is the opposite of shrinkage. This phenomenon can be approximately modeled by formula (3.15) with $k_h = -0.2$.

---

[8]For instance, for ordinary concrete with water-cement ratio $w/c = 0.46$ hydrating under sealed conditions, Baroghel-Bouny et al. [57] reported that the pore relative humidity decreased to 97% after 1 month, 95% after 6 months, and 93% after 2 years. In contrast to that, for high-strength concrete with $w/c = 0.26$, containing silica fume and superplasticizer, they found much lower values: 77% after 1 month, 72% after 6 months, and 64% after 2 years.

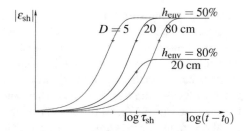

Fig. 3.16 Thickness and environmental humidity effects on shrinkage curves

Figure 3.16 compares the curves of shrinkage ε_{sh} for various relative environmental humidities h_{env} and for various effective thicknesses D. A change of h_{env} is manifested as a vertical scaling of the shrinkage curve. On the other hand, a change of D causes a horizontal shift of the curve in the logarithmic time scale. In some older models [11, 14], a change of D is manifested as vertical scaling, which is incorrect.

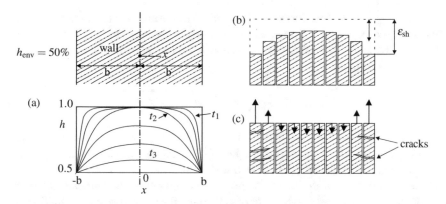

Fig. 3.17 (a) Pore humidity distributions during drying, (b) free shrinkage of slices imagined to be cut out, and (c) shrinkage stresses

The character of the pore water diffusion process, which is governed by the diffusion equation (Sect. 8.3), is illustrated by Fig. 3.17a, which shows the typical profiles at subsequent times of the pore relative humidity h over the cross section. The free shrinkage of a material point is roughly proportional to the loss of water from concrete, which in turn is roughly proportional to the relative humidity h in the pores. Due to nonuniformity of pore humidity profiles, shrinkage produces long-time self-equilibrated stresses, called shrinkage stresses. Although they are partly relaxed due to creep, they are large enough to cause micro- or macrocracking, which in turn further reduces these stresses.

The origin of shrinkage stress is explained by Fig. 3.17b, which shows what would happen if the cross section were cut in slices without affecting the pore humidity profiles. As shown, each slice would shrink independently, according to its current humidity. This would cause the slices to develop different shortenings. In a long concrete beam or slab strip, however, the cross section must remain plane (or else arbitrarily large shear stress would develop between the slices). As a consequence, stresses required to achieve equal lengths of all the slices must develop. This obviously produces longitudinal tensile stresses near the surface and longitudinal compressive stresses in the core; see Fig. 3.17c. At the end of drying, though, this picture gets reversed, because of creep, microcracking, and nonuniform aging.

3.6 Mean Drying Creep in the Cross Section

The additional mean cross-sectional compliance caused by simultaneous drying, expressing a coupling between creep and shrinkage, can be estimated from the formula [104]

$$J_d(t, t') = q_5\sqrt{e^{-g(t-t_0)} - e^{-g(t'-t_0)}} \tag{3.20}$$

valid for $t \geq t' \geq t_0$. The formula can be used as a rough approximation (for long times only) even if $t_0 > t'$, in which case t' must be replaced by t_0 [107]. Of course, if $t < t'$, the compliance function vanishes. All these adjustments can be captured by a general formula

$$J_d(t, t') = q_5\sqrt{\langle e^{-g(t-t_0)} - e^{-g(\langle t'-t_0\rangle)}\rangle} \tag{3.21}$$

in which the angular brackets (called Macauley brackets) denote an operator extracting the positive part of the argument, i.e., $\langle x \rangle \equiv \max(x, 0)$. This means that $\langle t' - t_0\rangle = t' - t_0$ if $t' \geq t_0$, and $\langle t' - t_0\rangle = 0$ if $t' < t_0$.

Like shrinkage, and unlike basic creep, the drying creep is bounded. It depends on humidity and cross-sectional thickness through the shrinkage shape function $S(\hat{t})$ defined in (3.16). Recall that the drying creep compliance J_d is added to the asymptotic compliance $1/E_0$ and to the basic compliance J_b describing the basic creep; see (3.3). Since the formulae for basic creep compliance and also for the asymptotic modulus in the full version and the short version of model B3 are different, the optimal agreement of the total compliance with the experimental data is obtained for a different function $g(\hat{t})$ and a different value of parameter q_5.

For the short form of model B3 with basic compliance function given by the log-double-power law (3.9), it is recommended to use

$$g(\hat{t}) = 3\left[1 - (1 - h_{env})S(\hat{t})\right] \tag{3.22}$$

and, as an empirical estimate, $q_5 = 0.0006/\bar{f}_c$.

For the full model B3 with basic compliance function given by (3.11), the recommended formulae are

$$g(\hat{t}) = 8\left[1 - (1 - h_{\text{env}})S(\hat{t})\right] \tag{3.23}$$

$$q_5 = \frac{7.57 \times 10^5}{\bar{f_c}(\varepsilon_{\text{sh}}^\infty)^{0.6}} \tag{3.24}$$

Example 3.1. Compliance curves predicted by model B3

The relative contributions of instantaneous compliance, basic creep, and drying creep according to the full form of model B3 are illustrated by a specific example.

Consider a concrete mix with water content $w = 170$ kg/m^3, type-I cement content c kg/m^3, and aggregate content $a = 1800$ kg/m^3. This is very close to the composition used by Komendant et al. [551] for their second mix, series 14–26. The standard compression strength measured on companion specimens was $\bar{f_c} = 45.4$ MPa. Suppose that the mix is used to produce a concrete slab of thickness $D = 200$ mm, cured in air with initial protection against drying until the age $t_0 = 7$ days. Subsequently, the slab is exposed to an average environmental humidity $h_{\text{env}} = 70\%$.

The parameters of the full B3 model are estimated according to the recommendations described in Table C.2 in Appendix C and in formula (3.17) as[9]

$$q_1 \equiv 1/E_0 = 126.77\,\bar{f_c}^{-0.5} = 18.81 \quad [\times 10^{-6}/\text{MPa}] \tag{3.25}$$

$$q_2 = 185.4\,c^{0.5}\,\bar{f_c}^{-0.9} = 126.9 \quad [\times 10^{-6}/\text{MPa}] \tag{3.26}$$

$$q_3 = 0.29(w/c)^4 q_2 = 0.7494 \quad [\times 10^{-6}/\text{MPa}] \tag{3.27}$$

$$q_4 = 20.3(a/c)^{-0.7} = 7.692 \quad [\times 10^{-6}/\text{MPa}] \tag{3.28}$$

$$k_t = 0.085\,t_0^{-0.08}\,\bar{f_c}^{-0.25} = 0.02803 \quad [\text{day/mm}^2] \tag{3.29}$$

$$\tau_{\text{sh}} = k_t(k_s D)^2 = 1121 \quad [\text{day}] \tag{3.30}$$

$$\varepsilon_s^\infty = \alpha_1\alpha_2\left(0.019\,w^{2.1}\,\bar{f_c}^{-0.28} + 270\right) = 702.4 \quad [\times 10^{-6}] \tag{3.31}$$

$$\varepsilon_{\text{sh}}^\infty = \varepsilon_s^\infty \times 0.57514\sqrt{3 + 14/(t_0 + \tau_{\text{sh}})} = 701.1 \quad [\times 10^{-6}] \tag{3.32}$$

$$q_5 = 7.57 \times 10^5\,\bar{f_c}^{-1}\left(\varepsilon_{\text{sh}}^\infty\right)^{-0.6} = 327.0 \quad [\times 10^{-6}/\text{MPa}] \tag{3.33}$$

In (3.30), we have used $k_s = 1$ for an infinite slab, and in (3.31) $\alpha_1 = 1$ for type-I cement and $\alpha_2 = 1.2$ for curing in air with initial protection against drying. Note that the value of $\varepsilon_{\text{sh}}^\infty$ to be substituted into formula (3.33) is 701.1 and not 701.1×10^{-6}. The environmental humidity is not needed for evaluation of parameters (3.25)–(3.33), but it is incorporated into the definition (3.23) of function $g(\hat{t})$ that influences the drying creep compliance J_d according to formula (3.20).

The total compliance $J(t, t')$ is plotted in Fig. 3.18a as a function of the current age t and in Fig. 3.18b in semilogarithmic scale as a function of the load duration $t - t'$ for four different ages at loading, $t' = 1$ week, 1 month, 3 months, and 1 year. The dashed horizontal line indicates the asymptotic elastic compliance $q_1 = 1/E_0$ (considered

---

[9]We evaluate all the parameters to four significant digits, even though the intrinsic error of the empirical formulae is certainly much larger. However, highly accurate values calculated here can be useful for checking the correct implementation of the formulae into various programs and design tools.

by model B3 as age-independent), and the dotted descending curve in Fig. 3.18a shows the evolution of the static elastic compliance $1/E(t) = 1/J(t + \Delta t_s, t)$ (the reciprocal value of the short-term elastic modulus defined in (3.4)) as a function of the concrete age t. In Fig. 3.18b, the short-term elastic compliance would be found at the intersection of the compliance curve with a vertical line at $t - t' = \Delta t_s = 0.01$ day. The decrease of this compliance with age reflects the growth of the short-term modulus due to aging.

Figure 3.18a is plotted with time in the linear scale, and it shows that the compliance increases dramatically just after loading, while later its evolution slows down. To provide a better idea about the behavior for very short or very long load durations, Fig. 3.18b shows the compliance as a function of the load duration in logarithmic scale on the horizontal axis. The graph covers a wide range of load durations from 10^{-5} day (roughly 1 second) to 10^5 days (roughly 300 years). It is interesting to note that the actual compliances after just 1 second of loading are way above the asymptotic compliance $1/E_0$. In fact, even for a load duration as short as 1 microsecond (about 10^{-11} day), there would be a non-negligible difference between the compliance and its asymptotic limit for $t - t' \to 0^+$. The reason is that the initial part of the compliance curve is dominated by a term proportional to $(t - t')^{0.1}$, and the load duration must be decreased by ten orders of magnitude in order to reduce this term by one order of magnitude.

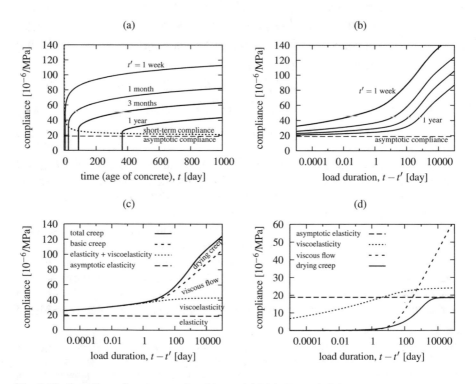

Fig. 3.18 Compliance functions predicted by model B3 in Example 3.1

Another interesting point revealed by Fig. 3.18b is that the compliance curves for very large load durations approach straight lines (in semilogarithmic scale), the slope of which is independent of the age at loading, but the maximum slope is attained before that, for load durations in the order of 1000 days. This is caused by the contribution of drying creep, which approaches a finite limit and remains almost constant for load durations much larger than the shrinkage halftime τ_{sh}, which is in the present example equal to 1121 days. If only the basic creep is considered, the compliance curve is convex (in the semilogarithmic scale), as indicated by the dashed curve in Fig. 3.18c. This figure presents a decomposition of the total compliance curve for loading at age $t' = 28$ days into individual contributions, which are then plotted separately in Fig. 3.18d. The first component of the compliance is the asymptotic elastic compliance $q_1 = 1/E_0$, which is represented by the horizontal dashed line in both parts of the figure. The second component, attributed to viscoelastic processes and represented in model B3 by the terms with parameters q_2 and q_3 (see the first two terms on the right-hand side of (3.11)), is plotted by the dotted curve and dominates the short-term creep, in the present example up to a hundred days (here, we use the adjective "short-term" in a different meaning than when the "short-term modulus" was discussed). The long-term creep is dominated by the logarithmic expression with parameter q_4 (last term on the right-hand side of (3.11)), which is physically interpreted as a viscous flow (with increasing age-dependent viscosity) and is responsible for the asymptotic behavior as the load duration approaches infinity. Up to about 10 days of loading, this term has a negligible contribution. Finally, the drying creep compliance is controlled by parameter q_5 and in Fig. 3.18d is plotted by the solid curve. As already discussed, this part of creep is bounded, as is shrinkage. The fact that the final value of the drying creep compliance in Fig. 3.18d seems to be close to the asymptotic compliance is just a coincidence, because this value depends on environmental humidity and for another choice of humidity would be different.

The graphs constructed in this example illustrate the basic trends, but the relative importance of individual contributions to the total creep compliance depends on many factors. For instance, drying creep may become very important for concrete that starts drying early and is exposed to low environmental humidity.

Estimation of the model parameters using empirical formulae of the new model B4 is, for the same input data, presented in Appendix D; see Example D.1.

∎

3.7 Common Misconceptions in Measuring and Defining Creep

3.7.1 Incompatible Initial Strain

For linear creep analysis of structures, the only required material property is the compliance function $J(t, t')$. The alternative description of creep by two functions, the creep coefficient $\varphi(t, t')$ and the elastic modulus $E(t')$, is only a matter of convenience. Different combinations of $\varphi(t, t')$ and $E(t')$ that give the same compliance function $J(t, t') = [1 + \varphi(t, t')]/E(t')$ give also identical results in structural creep analysis.

However, the fact that the definition of initial elastic deformation $1/E(t')$ is, to a large extent, ambiguous has been a source of trouble. Many investigators in the past reported the compliance increase $\Delta J(t, t') = J(t, t') - J(t' + \Delta t_s, t')$, but not the total compliance $J(t, t')$. Instead, they reported the value of elastic modulus $E(t')$. That would be correct only if concrete exhibited no short-time creep. But it does.

The sustained load has often been initially applied in the loading frame under manual control through springs or hydraulic jacks, in which case the load application lasted between $\Delta t_s = 1$ min. to 1 h. Pistons loaded by compressed air released by a fast valve from a bottle have also been used, and in that case, the load application may take as little as $\Delta t_d = 0.001$ s. The studies that omit the initial deformation report the value of elastic modulus $E(t')$. However, this value is measured by the standard code procedure for elastic modulus tests, which is very different from the actual process of applying the load on the creep specimens and involves load cycling. Thus, the value of such $1/E(t')$ can differ very much from the unreported initial compliance $J(t' + \Delta t_s, t')$ in the creep test.

To realize the range of ambiguity, note that for the concrete from Example 3.1 loaded at the age of 7 days, the ratios of the strains caused by load durations 1 s, 1 min, and 1 hour to the strain after 0.001 s are respectively 1.24, 1.46, and 1.74 (for loading at the age of 28 days, these ratios are 1.14, 1.27, and 1.44); see Fig. 3.18b.

Consequently, experimental studies in which the total strain in the creep apparatus has not been reported are virtually worthless. The creep coefficient and elastic modulus to be used in creep analysis must always be evaluated from the same $J(t, t')$. Whether Δt_s in (3.4) is chosen as 0.001 day or 0.1 day makes no difference for long-time creep predictions.

3.7.2   Plotting Creep Curves in Actual, Rather than Logarithmic, Time Scale

It has unfortunately been prevalent to plot the creep test results (as well as creep deflections of structures) in the actual (linear) scale. Figure 3.19a–c shows such plots of $J(t, t')$ versus $t - t'$ for typical parameters of model B3, and for the creep durations of 30, 500, and 10,000 days. Also inserted are data points giving a realistic impression of the scatter of measurements. Figure 3.19d gives the corresponding plots of $J(t, t')$ versus $\log(t - t')$. This graph gives clear information on creep evolution through the entire period from 0.01 day to 30 years, but the linear scale plots visualize creep only for one order of magnitude. Moreover, due to inevitable scatter, one is tempted to conclude, from the terminal slope of the linear scale plots ending at 30, 500, or 10,000 days, that the creep rise is leveling off at the end of the plot and that the creep growth will be over at 100, 1,500, or 30,000 days, respectively.

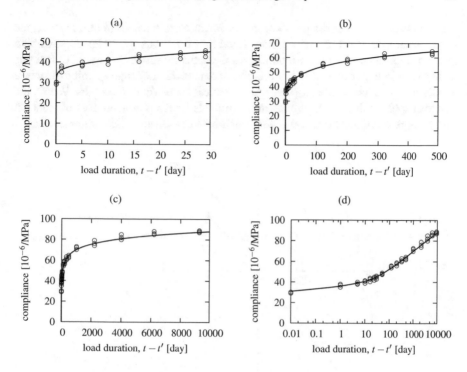

Fig. 3.19   Compliance functions in different ranges of load duration

3.7.3   Creep "Inflation"

Plotting creep test data in the actual (linear) time scale explains the historical phenomenon called the creep "inflation." During the 1930 and 1940s, when most available tests had the duration of only about 1 year, the codes or standard practice recommendations featured creep prediction formulae with a rather small "final" value. During the 1960s, longer creep tests became abundant, and the "final" value was raised, or "inflated." Only recently, thanks to plotting in the logarithmic time scale, it has been accepted widely (though not yet universally) that there is no evidence for a "final" creep value.

The absence of an upper bound on creep is, of course, no problem for design. Indeed, according to model B3, the 500-year compliance is predicted to be only about 20–25% higher than the 50-year compliance.

3.7.4   Is Tensile Creep Different from Compression Creep?

For more than half-century, this apparently plausible point has repeatedly been falsely exploited by critics to dismiss mathematical creep models and research proposals. Simply saying "tension," with no qualifier, means that a different (greater) creep

occurs as soon as the stress passes from compression to tension. In other words, the
creep isochrones (Sect. 2.1) for various fixed time durations would have to exhibit a
sudden slope change at the zero point, as schematically shown in Fig. 3.20. But this
is not seen in experiments. If it were, then some sudden change of microstructure
would have to occur while crossing the zero stress state. Yet none occurs. Within the
range of applicability of linear constitutive models, which is nonzero for both tension
and compression, there can be no difference between tension and compression.

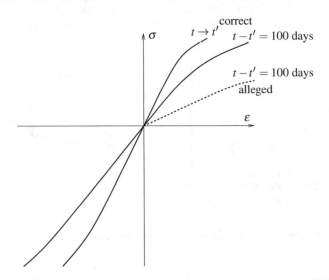

Fig. 3.20 Realistic creep isochrones (solid) and a fictitious isochrone (dashed) with a nonphysical
change of slope at the origin

Further note that a zero macroscopic stress does not correspond to zero stresses
in the microstructure; it only corresponds to a zero resultant of the microstresses.
The microstresses, in general, are scattered and do not transit from compression to
tension at the moment the macrostress crosses zero. Some transit earlier, some later.
So, even if the compression–tension transition caused a sudden change of creep at
the microscale, there would still be a smooth continuity of the macroscale creep law
when passing from compression to tension.

There is a difference, of course, in the transition to nonlinear creep at higher
stress. This transition occurs at about the same percentage of the strength limit under
uniaxial stress, about 40 to 50%. But in uniaxial tension it may occur at a much
smaller stress magnitude than in uniaxial compression.

The early studies prior to 1960 tried to make the viscosity coefficient in the rate-
of-creep model (Appendix B) stress-dependent, or to modify the compliance by a
nonlinear function of uniaxial tensile stress, and then use that compliance in memory
integrals, sometimes with further nonlinearity introduced through compound memory integrals. But that was simplistic, while the compound integrals were artificially
complicated. These models were not consistently formulated for general triaxial

stress states and ignored the necessity of yield surfaces in the tensorial constitutive model for creep. This subject, however, is better discussed in the context of the theories of damage mechanics and plasticity.

The nonlinearity of creep is caused by damage, in tension mainly by microcracking damage. That is properly captured by using a general triaxial damage model, such as the microplane model (Sect. 12.8), or a plastic-fracturing model based on yield surfaces, to calculate the additional nonlinear strain (for the same stress) that must be added to the linearly calculated creep strain. The damage model should also include the rate effect, i.e., the dependence of the damage strain rate on the stress.

There is one constitutive law that includes a sudden stiffness change at zero stress. It is the classical "no-tension" material model, which is the limit of multiaxial plasticity for yield strength approaching zero. According to this model, the material can carry no tensile stress in all directions. The extension to creep is obvious.

3.7.5 Autogenous Shrinkage

Autogenous shrinkage is the shrinkage caused by the hydration of concrete independently of external drying. It used to be considered negligible and its growth short-lived, but ongoing research shows that, for modern concretes with low water-cement ratios and a variety of admixtures, it is significant and long-lasting. It can even exceed the drying shrinkage. After an initial transient period, it grows logarithmically for at least ten years, and probably a century. Unlike drying shrinkage, it is the strongest in sealed specimens and is insensitive to specimen size, except indirectly

The autogenous shrinkage will require changes in the evaluation of creep and shrinkage experiments. It must be measured for the whole duration of creep and shrinkage tests. It must be subtracted from the basic creep data measured on sealed specimens (or else one gets a false impression of nonlinear dependence of stress). In drying shrinkage and drying creep specimens, the autogenous shrinkage proceeds in the specimen core until the drying front arrives from the exposed surface (thus, its effect may last for decades and is large in thick specimens but is negligible in thin ones). It decreases with decreasing relative humidity h in the pores and does not proceed if h drops below about 0.7. For more detail, see Sect. 8.7.

3.8 Updating Long-Time Creep and Shrinkage Predictions from Short-Time Measurements

Problems with durability and long-term serviceability of concrete structures attest to the uncertainty in creep and shrinkage predictions. Three kinds of uncertainty may be distinguished:

1. The intrinsic uncertainty, due to the fact that concrete creep and shrinkage are random processes even if all the influencing parameters are fixed.
2. The uncertainty stemming from the random variability of concrete composition and curing, which can have two origins:

   a. basic variability due to inevitable imperfections of normally affordable quality control, and
   b. additional variability due to deliberate vagueness of design specifications, permitting a significant range of compositions.

3. The uncertainty due to randomness of environmental conditions, particularly the relative humidity and temperature.

The intrinsic uncertainty is mathematically more complicated since it calls for random process treatment (e.g., [300]). It is relatively small compared to the second and third kinds, which are the sole focus in this section.

The greatest is the second kind of uncertainty. It is aggravated by the sophistication of modern concretes, especially high-strength concretes, because diverse admixtures, high-range water reducers, and pozzolanic ingredients have an appreciable, yet poorly understood, effect on creep and shrinkage [263, 264].

Unless precise specifications are made and precise quality control is ensured, the scatter of the databases of creep and shrinkage tests around the world [107, 160, 175, 488, 727, 871] is symptomatic of the scatter to be expected in design. The scatter seen in these databases is mainly of the second kind and is very large (in fact, so large that prediction models giving realistic and unrealistic time curves have nearly the same overall error in comparison with an unfiltered database).

Since materials science of cement and concrete has not advanced enough to capture the effects of concrete composition, it is necessary to update the prediction model on the basis of short-time tests on the given concrete. If the tests, their extrapolation, and model updating are conducted properly, model B3 permits a drastic improvement in long-time predictions [104, 107]. So does model B4.

To assess the expected behavior of the concrete considered for a design and construction project, only a limited time such as 1 to 3 months is usually available. The updating is most effective only if the prediction model has a form amenable to linear regression and if it gives realistic shrinkage and creep curves from the shortest durations up. Models B3 and B4 satisfy these conditions.

The updating of creep predictions from short-time tests should be mandatory for highly creep-sensitive structures, characterized in Chap. 1 as level 5; it is advisable for level 4, and often useful for level 3, but not needed for levels 1 and 2.

3.8.1 Updating Creep Predictions

The simplest update procedure could consist in proportional scaling of the entire compliance function by a single factor. However, better results can be achieved by separate scaling of the constant part of the compliance function (corresponding to instantaneous elasticity) and the variable part of the compliance function (corresponding to creep), using two update parameters, p_1 and p_2. The improved prediction of the compliance function is constructed in the form

$$J(t, t') = p_1 + p_2 F(t, t') \tag{3.34}$$

where function

$$F(t, t') = J_b(t, t') + J_d(t, t') \tag{3.35}$$

is evaluated according to model B3 using the empirical formulae for the effect of composition parameters and strength; see Table C.2 in Appendix C. If the data agreed with the form of model B3 exactly, the plot of the actually measured compliance values J_i versus the predicted creep compliance values $F_i = F(t_i, t'_i)$ would be a single straight line for all the measurements labeled by subscript i, with $p_1 = 1/E_0$ and $p_2 = 1$. The vertical deviations of the data points from this straight line represent errors which are to be minimized by linear least-square regression.

So we consider the plot of the known (measured) short-time values J_i ($i = 1, 2, \ldots n$) (say, up to 28 days of creep duration) versus the corresponding values of F_i, calculated from model B3, and pass through these points a regression line; see Fig. 3.22. The Y-intercept and the slope of this line give the values of p_1 and p_2 that are optimum in the sense of the least-square method. According to the well-known normal equations of least-square linear regression (e.g., [330]),

$$p_2 = \frac{\displaystyle\sum_{i=1}^{n} F_i J_i - n \bar{F} \bar{J}}{\displaystyle\sum_{i=1}^{n} F_i^2 - n \bar{F}^2}, \quad p_1 = \bar{J} - p_2 \bar{F} \tag{3.36}$$

where n is the number of measured values, $\bar{J} = \sum_i J_i/n = $ mean of all the measured compliance values, and $\bar{F} = \sum_i F_i/n = $ mean of all the corresponding predicted creep compliance values.

Example 3.2. Updating basic creep prediction

Let us illustrate the procedure by considering, as an example, the data for basic creep by L'Hermite, Mamillan, and Lefèvre [580] (see also L'Hermite and Mamillan [579]). The composition of concrete used in their tests was characterized by water content $w = 171.5$ kg/m^3, cement content $c = 350$ kg/m^3, and aggregate content $a = 1685$ kg/m^3, and the measured mean (cylindrical) compressive strength was $\bar{f}_c = 33.9$ MPa. The empirical formulae from Table C.2 in Appendix C give the following parameter estimates: $q_1 = 21.77 \times 10^{-6}$/MPa, $q_2 = 145.5 \times 10^{-6}$/MPa, $q_3 = 2.433 \times 10^{-6}$/MPa, and $q_4 = 6.757 \times 10^{-6}$/MPa. The corresponding predic-

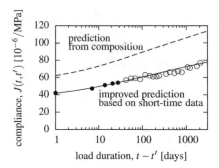

Fig. 3.21 Example of improving the prediction of basic creep by the use of short-time test data

tion of the compliance function for loading at age $t' = 7$ days is not very good, as is apparent from Fig. 3.21 (dashed curve).

We now pretend we know only the first 5 data points for the first 28 days of creep duration, which are indicated in Fig. 3.21 by solid circles. The values of measured compliances J_i and of predicted creep compliances F_i, $i = 1, 2, \ldots 5$, are listed in Table 3.2 and graphically presented in Fig. 3.22. According to formula (3.36), the optimal values of update parameters are

$$p_2 = \frac{\sum_{i=1}^{n} F_i J_i - n \bar{F} \bar{J}}{\sum_{i=1}^{n} F_i^2 - n \bar{F}^2} = \frac{12825 - 5 \times 51.37 \times 49.4}{13389 - 5 \times 51.37^2} = 0.7016 \qquad (3.37)$$

$$p_1 = \bar{J} - p_2 \bar{F} = 49.4 - 0.7016 \times 51.37 = 13.36 \qquad [10^{-6}/\text{MPa}] \quad (3.38)$$

After the correction, the updated compliance function (3.34) gives not only an excellent fit within the range of load duration up to 28 days, but also a very good prediction up to the last measured value at 2070 days; see the solid curve in Fig. 3.21.

Updating of the compliance function according to (3.34) is equivalent to replacing the values q_1, q_2, q_3, and q_4 calculated from the formulae in Table C.2 by the values $q_1^* = p_1 = 13.36 \times 10^{-6}/\text{MPa}$, $q_2^* = p_2 q_2 = 102.1 \times 10^{-6}/\text{MPa}$, $q_3^* = p_2 q_3 = 1.707 \times 10^{-6}/\text{MPa}$, and $q_4^* = p_2 q_4 = 4.740 \times 10^{-6}/\text{MPa}$. Note that this equivalence would not hold if parameters q_i were not involved linearly in the expression for the compliance function. Based on the updated parameters, the compliance function $J(t, t')$ can be evaluated for real structures, using ages at loading other than the t'-value used for the test specimens. The effective thickness, environmental humidity, etc., may also differ. For drying creep, parameter q_5 would be replaced by $q_5^* = p_2 q_5$. ∎

Table 3.2 Evaluation of an improved prediction of basic creep

i	$t_i - t'$ [day]	J_i [10^{-6}/MPa]	F_i [10^{-6}/MPa]	$F_i J_i$ [10^{-12}/(MPa)^2]	F_i^2 [10^{-12}/(MPa)^2]
1	1	42	40.40	1697	1632
2	7	47	49.06	2306	2407
3	14	51	53.30	2718	2841
4	21	53	56.03	2970	3139
5	28	54	58.05	3135	3369
Sum		247	256.84	12825	13389
Average		49.40	51.37		

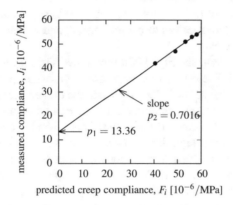

Fig. 3.22 Predicted versus measured compliance values and the regression line from which parameters p_1 and p_2 can be determined

As shown in Fig. 3.21, a major improvement of long-time prediction is achieved by updating based on short-time measurements. The well-known formulae of linear regression [330] also yield the coefficients of variation of p_1 and p_2, which in turn provide the coefficient of variation of $J(t, t')$ for any given t and t'. However, better estimates of the coefficients of variation may be obtained by the Bayesian statistical approach, to be discussed in Sect. 6.4.

To maximize the benefit of updating, the measurements should start immediately after applying the load, because it helps to anchor the overall slope of the creep curve in the logarithmic time scale. Short-time data spanning from load durations $t - t' = 10$ s to 3 days give a much better update than those from 2 hours to 3 days (provided that, of course, the model has a realistic time curve in that range). In our example, the available readings span from 1 day to 28 days, but an equally good update could probably be achieved if they spanned from 1 hour to 3 days. To minimize statistical bias (see Chap. 11), the readings used for updating should be spaced at approximately equal intervals in the logarithmic time scale (i.e., should form a geometric progression in time; cf. Chaps. 5 and 11).

To improve the prediction of the drying creep part, an update based on water loss measurements on companion specimens is very helpful, as discussed in Sect. H.1.

3.8.2 Difficulties in Updating Shrinkage Predictions

For shrinkage, the problem of extrapolating short-time data is much harder than it is for creep. The reason is that, if the time range of shrinkage measurements does not extend into the final stage at which the shrinkage curve of ε_{sh} versus $\log(t - t_0)$ levels off, the problem of fitting the shrinkage formulae (3.15)–(3.16) to the measured strain values is what is known in mathematics as an ill-conditioned problem, meaning that the parameters of the optimum fit are very sensitive to small changes in the data. In other words, very different values of parameters ε_{sh}^∞ and τ_{sh} can give almost equally good fits of shrinkage data.

To clarify the problem, see Fig. 3.23a where two shrinkage curves according to model B3 are plotted. The solid curve corresponds to shrinkage halftime $\tau_{sh} = 1000$ days and final shrinkage $\varepsilon_{sh}^\infty = 1.5 \times 10^{-3}$, and the dashed curve to $\tau_{sh} = 400$ days and $\varepsilon_{sh}^\infty = 1.0 \times 10^{-3}$. Despite the large difference in parameter values, the curves nearly coincide up to 200 days. If the data do not reach beyond the time at which the curves begin to diverge significantly (which may be unattainable for normal size specimens), there is no way to determine the model parameters unambiguously. This is true not only for models B3 and B4 but also for other shrinkage formulae.

Possible pitfalls are further illustrated by Fig. 3.23b which compares shrinkage measurements on different concretes whose durations are not long enough. The solid curve corresponds to a relatively porous concrete A that dries quickly and reaches moisture equilibrium soon but has a low final shrinkage, while the dashed curve corresponds to a dense concrete B which dries very slowly but has a large final shrinkage. However, a short-time shrinkage test, terminating at the points marked, would be misleading, suggesting that concrete A has a higher final shrinkage than concrete B, whereas the opposite is true.

Such plots, combined with simple estimates of the shrinkage halftime, reveal that a reliable determination of the final value of shrinkage would require, for 6-inch (15 cm)-diameter cylinders, measurements of at least 5 years in duration, which is much too long for a designer. Even with a 3-inch (7.5 cm)-diameter cylinder, this would exceed 15 months. A significant acceleration of the shrinkage process would require using specimens of diameter 0.5 or 1 in. (1.27 or 2.54 cm). Such specimens, smaller than the aggregate size, would need to cut by a saw. A correction, probably small, would then have to be made for the wall effect stemming from different composite interaction of cement mortar and aggregate pieces near the surface.

Increasing the temperature of the shrinkage tests to about 50°C would not shorten the drying times drastically and would introduce further uncertainties about the effects of temperature, such as the effect of thermally accelerated hydration on permeability, the thermal effect on shrinkage stress relaxation and shrinkage microcracking,

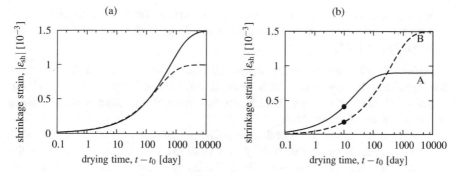

Fig. 3.23 (a) Example of shrinkage evolutions with nearly the same initial shrinkage but very different final values, (b) possible shrinkage evolutions for geometrically identical specimens of different concretes, exposed to the same environment

etc. A greater increase of temperature would raise the uncertainty of inferences for the room temperature.

To overcome the difficulties with shrinkage updating, the following two refined extrapolation methods have been proposed:

1. **Weight Loss Method**: It has been widely accepted that drying shrinkage strains are approximately proportional to the water loss, Δw. Indeed, the water loss curve is similar to the shrinkage curve. But there is one important difference: Unlike shrinkage, even if the water loss is measured (by weighting) only during the first 1 to 3 months, the final water loss can be predicted reasonably well, in two ways: (a) By drying the specimen in the oven (at 105 °C), one can determine the water loss for zero relative humidity and then use an approximate desorption isotherm to interpolate the water loss to the given relative humidity; or (b) knowing w/c ratio in the concrete mix, one can, at least in theory, estimate the water content evaporable at zero humidity and then interpolate. Once the final value of water loss is known, one can obtain the halftime of the water loss curve τ_w (assuming that the water loss follows the same type of function as shrinkage). Considering the shrinkage halftime τ_{sh} to be nearly the same, one can calibrate the formula for the entire shrinkage curve. The details of this procedure, proposed by Bažant and Baweja [104], are described in Appendix H.

2. **Diffusion Size Effect Method**: Assuming all the shrinkage in drying environment to be caused by diffusion of water from the specimen (i.e., autogenous shrinkage to be absent), and shrinkage and specific water loss increments at each point of specimen to be roughly proportional, one could, in theory, exploit the diffusion size effect. The test of a cast 6-inch-diameter cylinder ($D = 2V/S_e = 3$ in.) can be accompanied by a test of 1 in. square prism cut out from concrete ($D = 0.5$ in.), which dries (and supposedly shrinks) about 36 times faster (ignoring a small correction for the wall effect). Thus, the shrinkage curve of the small specimen supposedly approaches its final value within a short enough time. By simultaneous optimization of the fit of this curve and of the short-time data for the

large specimen, τ_{sh} can thus be identified. It is then simple to calculate the final shrinkage for the 6-inch-diameter cylinder; see the detailed procedure in Bažant and Donmez [124].

Initially, from a few examples, it seemed that the weight loss method worked. But significant discrepancies have been observed later [471]; see, e.g., Fig. H.3 in Appendix H. Likewise, the method of diffusion size effect seemed to work in some examples but in others was poor.

Therefore, while the use of both extrapolation methods is surely better than the alternative of intuitive extrapolation by a curve drawn by eye, none of them is truly predictive and reliable in practice. Why? The likely cause is that the role of autogenous shrinkage is much bigger than expected, especially in modern concretes, as proposed in Bažant, Donmez, Masoero, and Rahimi Aghdam [125]. This will be discussed in Sect. 8.7.

Chapter 4
Structural Effects of Creep and Age-Adjusted Effective Modulus Method

Abstract We begin with creep analysis of structures having homogeneous material properties, which is a feature requiring uniformity of concrete age and, in the case of drying exposure, also uniformity of concrete wall thickness. We explain the Boltzmann–Volterra elastic–viscoelastic analogy for creep with a linear stress dependence and discuss the effects of changes in structural system, which occur commonly during modern construction. Then, we derive and expound in detail the age-adjusted effective modulus method (AAEM), the simplest method to obtain good approximate estimates of the creep effects in the presence of aging—a method that is featured by now in most, if not all, design codes and recommendations. Our discussion then proceeds to calculating the stress redistributions in structures, and cross sections in which the creep properties are not homogeneous, or in which the nonuniformity is introduced by drying. Careful attention is given to stress relaxation in prestressed members, which is caused not only by the creep and shrinkage of concrete but also by the relaxation of prestressing steel at variable stress and temperature. We present a recently developed viscoplastic model for prestressing steel, justify it mathematically, emphasize that the tendon relaxation should be computed as part of creep structural analysis, and point out that daily cycles of tendon heating can greatly increase the prestress loss. We finish our discussion with the creep buckling of columns and shells, accurate assessment of which is particularly important for structural safety.

4.1 Homogeneous Structures

Perfectly homogeneous structures are rarely encountered in practice, but a solid understanding of their behavior and analysis constitutes a good starting point for deeper discussion of more general problems. A structure is considered as homogeneous, if all its parts are made of the same material with the same properties; for concrete structures, this means that the concrete used in all parts should have not only the same composition but also the same age and curing. Strictly speaking, drying should also develop in the same way everywhere in the structure, because it affects the drying creep compliance. This last condition is, of course, impossible to satisfy pointwise, but as an approximation one can introduce the effective cross-

© Springer Science+Business Media B.V. 2018

Z.P. Bažant and M. Jirásek, *Creep and Hygrothermal Effects in Concrete Structures*, Solid Mechanics and Its Applications 225, https://doi.org/10.1007/978-94-024-1138-6_4

sectional compliance that reflects the average properties over the cross section, and in this sense, the structure is homogeneous, e.g., if all cross sections are the same and subjected to the same environmental humidity.

4.1.1 Elastic–Viscoelastic Analogy

For homogeneous structures with no change of structural system (supports, internal hinges, etc.), it is possible to prove that all equations derived under the assumption of a linear elastic behavior remain valid for a linear viscoelastic behavior, after the following adjustments, equivalent to the theorems of McHenry [619]:

- multiplication by the elastic modulus is replaced by application of the relaxation operator, $\mathscr{R}$,
- division by the elastic modulus is replaced by application of the compliance (creep) operator, $\mathscr{J}$.

To illustrate this general rule, consider the well-known relation between the bending moment, M, and the curvature of the beam axis, κ. In elasticity, the moment–curvature relation is derived by combining the integral formula for the bending moment expressed as the moment resultant of normal stresses,

$$M = \int_A z\sigma(z)\,\mathrm{d}A \tag{4.1}$$

with the elastic stress–strain law, $\sigma = E\varepsilon$, and with the formula

$$\varepsilon(z) = \varepsilon_a + \kappa z \tag{4.2}$$

describing the linear strain distribution across the section (which is a consequence of the assumption that the sections remain planar even after deformation). After easy manipulations, the above equations yield the moment–curvature relation

$$M = \int_A z\sigma(z)\,\mathrm{d}A = \int_A zE\varepsilon(z)\,\mathrm{d}A = \int_A zE(\varepsilon_a + \kappa z)\,\mathrm{d}A =$$
$$= E\varepsilon_a \int_A z\,\mathrm{d}A + E\kappa \int_A z^2\,\mathrm{d}A = EI\kappa \tag{4.3}$$

in which $I = \int_A z^2\,\mathrm{d}A$ is the sectional moment of inertia. Note that, according to (4.2), the strain, in general, depends not only on the curvature, κ, but also on the axial strain, ε_a. However, if the coordinate z is measured from a centroidal axis y, then $\int_A z\,\mathrm{d}A$ vanishes and there is no cross-coupling between the axial stretching and bending. The resulting formula (4.3) shows that the bending moment M is proportional to the beam curvature κ, with the bending stiffness EI as the proportionality coefficient.

The foregoing derivation can be extended to a viscoelastic beam. Equations (4.1)–(4.2) remain valid and only the stress–strain law needs an adjustment. Instead of the simple Hooke's law $\sigma = E\varepsilon$, we must consider the viscoelastic stress–strain law, most easily formulated in the operator notation as $\sigma = \mathscr{R}\{\varepsilon\}$.[1] Since the relaxation operator $\mathscr{R}$ is linear, we can write

$$M = \int_A z\sigma(z)\,\mathrm{d}A = \int_A z\mathscr{R}\{\varepsilon(z)\}\,\mathrm{d}A = \int_A z\mathscr{R}\{\varepsilon_a + \kappa z\}\,\mathrm{d}A =$$
$$= \mathscr{R}\{\varepsilon_a\}\int_A z\,\mathrm{d}A + \mathscr{R}\{\kappa\}\int_A z^2\,\mathrm{d}A = I\mathscr{R}\{\kappa\} \tag{4.4}$$

Of course, the resulting formula can also be presented as $M = \mathscr{R}\{I\kappa\}$, because the sectional moment of inertia, I, is not a function of time and therefore is perceived by the relaxation operator as a constant (even though it may depend on the coordinate x measured along the beam axis). For simplicity, we have not marked explicitly which quantities may depend on time, since this is obvious.

Using the operator formalism, the moment–curvature relation (4.4) can easily be converted into its inverse form,

$$\kappa = \mathscr{J}\left\{\frac{M}{I}\right\} = \frac{\mathscr{J}\{M\}}{I} \tag{4.5}$$

This is the viscoelastic counterpart of the elastic relation $\kappa = M/EI$. As already alluded to, division by the elastic modulus E (i.e., multiplication by the elastic compliance $1/E$) is replaced by application of the compliance operator $\mathscr{J}$.

The replacement rules based on the *elastic–viscoelastic analogy* [619] for aging materials greatly facilitate the analysis of homogeneous aging viscoelastic structures. For instance, it is known from basic structural analysis that if an elastic beam clamped at both ends is loaded by a uniformly distributed load of intensity $\bar{f}$ and by vertical settlement $\bar{w}_b$ of its right support, then the moment reaction at the left support is calculated as $M_a = \bar{f}L^2/12 + 6EI\bar{w}_b/L^2$ and the deflection at midspan as $w_m = \bar{f}L^4/384EI + \bar{w}_b/2$, where L is the beam span and EI is the sectional bending stiffness. If the material of the beam is linear viscoelastic, the foregoing formulae need to be rewritten as follows:

$$M_a = \frac{\bar{f}L^2}{12} + \mathscr{R}\left\{\frac{6I\bar{w}_b}{L^2}\right\} = \frac{\bar{f}L^2}{12} + \frac{6I\mathscr{R}\{\bar{w}_b\}}{L^2} \tag{4.6}$$

$$w_m = \mathscr{J}\left\{\frac{\bar{f}L^4}{384I}\right\} + \frac{\bar{w}_b}{2} = \frac{\mathscr{J}\{\bar{f}\}L^4}{384I} + \frac{\bar{w}_b}{2} \tag{4.7}$$

Note that the terms which do not depend on the elastic modulus are unaffected. So, for instance, the reactions caused by the distributed load remain the same, no matter whether the material is elastic or viscoelastic, and this is true not only for statically

---

[1]For simplicity, the effects of shrinkage and thermal strain are neglected here. They could be easily added by using the general form of the stress–strain law (2.45).

determinate (isostatic) structures, for which the reactions follow directly from the equilibrium equations, but also for statically indeterminate (or redundant) structures, as long as they are homogeneous.

By contrast, the reactions caused by nonuniform support settlement or by temperature changes are proportional to the elastic modulus and if the material is viscoelastic, they tend to relax. If the support settlement takes place abruptly at time t_1, its history is described by the function $\bar{w}_b(t) = \hat{w}_b H(t - t_1)$ where $\hat{w}_b$ is a constant, and the corresponding moment reaction $M_a(t) = (6I/L^2)\mathscr{R}\left\{\hat{w}_b H(t - t_1)\right\} = (6I\hat{w}_b/L^2)\mathscr{R}\{H(t - t_1)\} = (6I\hat{w}_b/L^2)R(t, t_1)$ evolves proportionally to the relaxation function $R(t, t_1)$. For general loading histories, $\mathscr{R}\{\bar{w}_b(t)\}$ needs to be computed numerically, but this is the only extra effort required by the viscoelastic solution as compared to the elastic one. The simplicity of this procedure is what makes the assumption of structural homogeneity so convenient and popular, but the error induced by such a simplification should always be examined with great care.

The elastic–viscoelastic analogy is applicable not only to simple formulae derived by pencil calculations or found in design manuals but also to the sets of equations used by the slope-deflection method for analysis of redundant beam structures, or by the finite element method for analysis of arbitrarily complicated structural models (plates, shells, three-dimensional models, etc.). All such methods applied to linear elastic problems with small strains and displacements lead to a linear system of equations $Kd = f$ where K is the structural stiffness matrix, d is the column matrix of unknown displacement parameters (e.g., of displacements and rotations of the joints), and f is the column matrix of the equivalent external forces reflecting the applied load. All the stiffness coefficients in K are proportional to the elastic modulus, and so we can formally write $K = E\tilde{K}$ where $\tilde{K}$ is the stiffness matrix evaluated with a unit value of the elastic modulus. The set of equations describing the viscoelastic structure is then

$$\tilde{K}\mathscr{R}\{d(t)\} = f(t) \tag{4.8}$$

provided that the equivalent external forces $f(t)$ do not depend on the elastic modulus (which is the case if they come only from applied loads, but not if they express the effects of temperature changes or support settlement). The solution of (4.8) can formally be written as

$$d(t) = \mathscr{J}\left\{\tilde{K}^{-1}f(t)\right\} = \tilde{K}^{-1}\mathscr{J}\{f(t)\} \tag{4.9}$$

where, of course, the inverse stiffness matrix $\tilde{K}^{-1}$ is never constructed explicitly (multiplication by $\tilde{K}^{-1}$ is based on a suitable decomposition of $\tilde{K}$, e.g., on the Cholesky or LDL^T decomposition). The right-hand side of (4.9) indicates that it is sufficient to transform the load history using the compliance operator and then solve the elastic problem for a structure with unit elastic modulus, subjected to transformed loading $\mathscr{J}\{f(t)\}$. In the particular case of external loading applied abruptly at time t_1 and then remaining constant, all the displacement parameters $d(t)$ evolve proportionally to the compliance function $J(t, t_1)$, as discussed before for a simpler beam problem.

The general rules describing the behavior of a homogeneous viscoelastic structure under constant external solicitation (by dead loads, temperature changes, and support settlement) acting from time t_1 can be summarized as follows:

- The **static quantities** (such as reactions, internal forces, or stresses) induced by constant **external forces** remain **constant**.
- The **static quantities** induced by constant **temperature changes and support settlement** decrease in proportion to the **relaxation function** $R(t, t_1)$.
- The **kinematic quantities** (such as deflections, rotations, curvatures, or strains) induced by constant **external forces** increase in proportion to the **compliance function** $J(t, t_1)$.
- The **kinematic quantities** induced by constant **temperature changes and support settlement** remain **constant**.

Under time variable loads, the governing equations are still set up quite easily, but their solution may require application of the relaxation or creep operator to certain functions of time. The corresponding integrals can be computed numerically or approximated in the spirit of the AAEM method, to be developed in Sect. 4.2.

Example 4.1. Shrinkage of a restrained bar

In statically indeterminate structures, thermal or shrinkage strains cannot be fully accommodated by deformation of the structure, and compatibility constraints lead to the development of internal forces.

As the simplest case, consider a restrained bar, e.g., a pavement slab, subjected to drying shrinkage. In reality, the drying process is nonuniform across the section and stresses would develop even in a statically determinate bar with zero resultant internal forces (see Fig. 3.17). This phenomenon requires a solution of the diffusion equation that governs the evolution of relative pore humidity and will be discussed in Sect. 8.6; see Fig. 8.48. Here, we present a simplified approach, which deals with the average value of shrinkage strain over the section, estimated using model B3.

In a restrained bar, the total strain must be zero, and thus the mechanical strain (related to stress by a linear viscoelastic model) is equal to minus the shrinkage strain. In other words, in the general stress–strain relation (2.45), we set $\varepsilon(t) = 0$ and $\varepsilon_T(t) = 0$ and express the stress as

$$\sigma(t) = -\mathscr{R}\{\varepsilon_{\text{sh}}(t)\} \tag{4.10}$$

The stress history is thus obtained by applying the relaxation operator to the given function $\varepsilon_{\text{sh}}(t)$ that describes the history of shrinkage strain (and changing the sign). Since the shrinkage strain is negative (contractive), positive (tensile) stresses can be expected. For an elastic material, such stresses would be proportional to the shrinkage strain. For a viscoelastic material, they are partially relaxed by viscous effects.

The competition between stress build-up due to growing shrinkage and stress relaxation due to viscoelasticity results into a nonmonotonic stress evolution with a peak at a time comparable to the shrinkage halftime. This is documented in Fig. 4.1, which shows the stress history corresponding to concrete with the same properties

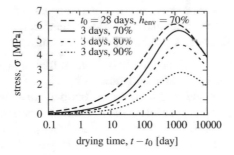

Fig. 4.1 History of stress induced by restrained shrinkage

as in Example 3.1. The solid curve represents the stress history in concrete exposed to the environmental humidity $h_{env} = 70\%$ at age $t_0 = 3$ days. The stress evaluated according to (4.10) grows up to its peak value 5.68 MPa, which is attained approximately at 1300 days. Of course, such a high tensile stress could hardly be transmitted by concrete with compressive strength 45.4 MPa, for which the tensile strength can be estimated as 3.4 MPa. Restrained shrinkage would lead to cracking, and the stresses would be reduced by tensile softening of the material.

The maximum stress calculated by a viscoelastic model (with cracking neglected) is even higher if the drying process starts later, e.g., at $t_0 = 28$ days, as indicated by the top dashed curve in Fig. 4.1. The reason is that the shrinkage strains remain almost the same, but the material becomes stiffer due to aging. Only if the concrete is exposed to a higher environmental humidity, shrinkage is reduced and the induced stresses as well. The dotted curve in Fig. 4.1 corresponds to $h_{env} = 90\%$ and has its peak at stress level 2.85 MPa and drying time approximately 1500 days. ∎

As demonstrated in the previous example, the shrinkage-induced stresses in a fully restrained slab are likely to exceed the tensile strength, and cracking would need to be taken into account in order to get more realistic results. However, the stress histories computed according to Eq. (4.10), i.e., with cracking neglected, can be useful when solving problems with partially restrained deformation. For instance, the problem of shrinkage in a statically indeterminate frame has the same mathematical structure as shrinkage in a restrained bar, but the resulting stress is lower and depends on the flexibility of the structure.

Example 4.2. Restrained shrinkage of a frame

Consider a two-hinge portal frame in Fig. 4.2a, which is statically indeterminate of degree 1, and so it can be analyzed using a single unknown redundant force, e.g., the horizontal reaction X_1 introduced according to Fig. 4.2b. If the axial flexibility of the bars is neglected (as compared to the bending flexibility) and the material is considered as linear elastic, the governing equation of the force (or compliance) method reads

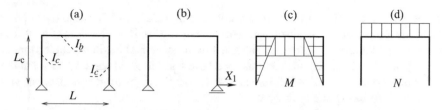

Fig. 4.2 Two-hinge portal frame: (a) frame geometry, (b) statically redundant reaction, (c) bending moment diagram, (d) normal force diagram

$$\frac{1}{E}\left(\frac{2L_c^3}{3I_c} + \frac{LL_c^2}{I_b}\right) X_1 + L\varepsilon_{sh} = 0 \tag{4.11}$$

where L and L_c are the lengths of the beam and column, I_b and I_c are their sectional moments of inertia, and E is the elastic modulus. For a viscoelastic material, multiplication by the elastic compliance $1/E$ must be replaced by application of the compliance operator $\mathscr{J}$. Using the relaxation operator $\mathscr{R}$, the equation can be formally solved and the solution can be written as

$$X_1(t) = -\left(\frac{2L_c^3}{3LI_c} + \frac{L_c^2}{I_b}\right)^{-1} \mathscr{R}\{\varepsilon_{sh}(t)\} \tag{4.12}$$

Same as in the case of a restrained bar, we need to apply the relaxation operator to the history of shrinkage strain. The resulting redundant force X_1 thus evolves in proportion to the stress that would develop in a fully restrained bar, with a proportionality factor dependent on the frame geometry. The same holds (with different proportionality factors) for other static quantities, such as bending moments or normal forces. The time at which the maximum stress is attained is the same as for the restrained bar, but, of course, the value of that stress is different, affected by the frame geometry. The horizontal member of the frame is subjected to normal force $N(t) = X_1(t)$ and bending moment $M(t) = L_c X_1(t)$ (both constant along the member); see Fig. 4.2c,d. According to the beam theory, the corresponding maximum tensile stress in the bottom fibers is

$$\sigma_{max}(t) = \frac{N(t)}{A_b} + \frac{M(t)}{W_b} = \left(\frac{1}{A_b} + \frac{L_c}{W_b}\right) X_1(t) =$$

$$= -\left(\frac{1}{A_b} + \frac{L_c}{W_b}\right)\left(\frac{2L_c^3}{3LI_c} + \frac{L_c^2}{I_b}\right)^{-1} \mathscr{R}\{\varepsilon_{sh}(t)\} \tag{4.13}$$

where A_b is the cross-sectional area of the member and W_b is its elastic sectional modulus (for a rectangular section of width b and depth h, we have $A_b = bh$ and $W_b = bh^2/6$). The proportionality factors in (4.12 and 4.13) decrease with increasing L_c, which is quite natural, since higher columns (with all the other dimensions unchanged) are more flexible. However, it may be somewhat surprising that for

$L_c \to 0$ the factors tend to infinity and not to finite values which would correspond to the response of a restrained bar. The reason is that if L_c becomes small, the contributions of axial deformation and shear to the structural flexibility are no longer negligible. For simplicity, the formula used here was based on bending only, but the other terms need to be added if frames with $L_c \ll L$ and the limit case of a bar with $L_c = 0$ are to be captured properly.

For a typical case with $L_c = L = 10h$ and equal rectangular sections of the beam and columns, the geometrical factor multiplying $-\mathscr{R}\{\varepsilon_{sh}(t)\}$ in (4.13) is about 0.03. This means that the shrinkage-induced stresses are more than 30 times lower than in a fully restrained bar, and their peak values certainly do not exceed the tensile strength. Of course, these stresses need to be combined with those caused by the self-weight and other loads.   ∎

4.1.2   Change of Structural System

An interesting type of problem arises if the structural system is changed during the life of the structure (typically during the construction phase). This will be illustrated by the following example.

Example 4.3.   Cantilever end placed on support

In cantilever construction of prestressed concrete box girder bridges,[2] the balancing of the pier is made easier if the end span is cast from the pier toward the abutment as a cantilever (Fig. 4.3). To simplify the analysis, we assume the whole cantilever to have the same age t, representing the average of the actual ages of the casting segments of the cantilever, and the dead load to be applied when this average age is t_1. At age $t_2 > t_1$, a bearing is placed under the end of the cantilever and is made to fit snugly (which may induce an initial vertical reaction at the support). This changes the structural system from statically determinate (system I, for $t < t_2$) to statically indeterminate, or redundant (system II, for $t \geq t_2$). Assuming, for the sake of simplicity, the forces acting at the end of the opposite, counterbalancing cantilever erected from the same pier to have a negligible effect on the vertical reaction $F_b(t)$ at the abutment, $F_b(t)$ is a single statically indeterminate force (Fig. 4.3).

Elementary analysis of an elastic cantilever provides the formulae for deflection at the free end. For simplicity, we assume a constant cross section with moment of inertia I, but it would be easy to adapt the analysis to the case of a variable cross

---

[2]One important cause of underestimation of long-time creep deflections of prestressed concrete box girders by beam-type analysis has been the neglect of the so-called shear lag, a phenomenon characterized by out-of-plane warping of cross sections in zones of high shear force, and by nonlinear stress distribution, particularly the development of stress peaks in the top slab near the webs. This effect is classical knowledge in elasticity but has often been overlooked in creep analysis. For shear lag correction of beam-type creep analysis, see Křístek and Bažant [557]. If three-dimensional or shell-type finite elements are used for creep analysis, the shear lag is, of course, accounted for automatically.

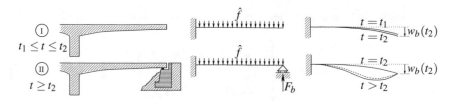

Fig. 4.3 Effect of placing the end of a cantilever on a support

section. The deflection due to uniformly distributed load f is $fL^4/8EI$ and the deflection due to concentrated load F acting at the free end is $FL^3/3EI$. This is all we need to know in order to write the general expression for the viscoelastic cantilever loaded by a time-dependent uniform load $f(t)$ and time-dependent reaction force $F_b(t)$:

$$w_b(t) = \mathscr{J}\left\{\frac{f(t)L^4}{8I} - \frac{F_b(t)L^3}{3I}\right\} = \frac{L^4}{8I}\mathscr{J}\{f(t)\} - \frac{L^3}{3I}\mathscr{J}\{F_b(t)\} \qquad (4.14)$$

The negative sign at the second term is due to the fact that the reaction F_b is considered as positive if it acts upward while the deflection is positive if it is downward; see Fig. 4.3.

In our particular problem, the uniform load is applied abruptly at time t_1 and afterward held fixed. So it is described by $f(t) = \hat{f}H(t - t_1)$ where $\hat{f}$ is a given constant (dead weight per unit length). The concentrated force $F_b(t)$ represents the vertical reaction at the right support and is yet to be determined. Before the change of the structural system, i.e., at times $t < t_2$, the support does not exist, and force $F_b(t)$ is known to be zero while the deflection $w_b(t)$ is unknown. After the change of the structural system, i.e., at times $t > t_2$, the reaction force $F_b(t)$ is unknown and the end deflection $w_b(t)$ must remain constant (unless the support experiences some settlement). Equation (4.14) is valid throughout the entire history, and it serves for the evaluation of deflection $w_b(t)$ at times $t \leq t_2$ and of reaction $F_b(t)$ at $t > t_2$.

Assuming further that there is no jacking up nor pulling down of the cantilever end before the bearing at the abutment is installed, we require continuity of deflection $w_b(t)$ at time $t = t_2$, which implies continuity of the reaction $F_b(t)$. Substituting $\mathscr{J}\{f(t)\} = \mathscr{J}\{\hat{f}H(t - t_1)\} = \hat{f}J(t, t_1)$ into (4.14), we get

$$w_b(t) = \frac{\hat{f}L^4}{8I}J(t, t_1) - \frac{L^3}{3I}\mathscr{J}\{F_b(t)\} \qquad (4.15)$$

For $t \leq t_2$, the second term on the right-hand side vanishes and the deflection increases proportionally to the compliance function, as expected. The deflection at time $t = t_2$ is

$$w_b(t_2) = \frac{\hat{f}L^4}{8I}J(t_2, t_1) \qquad (4.16)$$

and subsequently remains constant. Therefore, the entire history of deflection can be described as

$$
w_b(t) = \begin{cases} \dfrac{\hat{f}L^4}{8I} J(t, t_1) & \text{for } t \le t_2 \\[3mm] \dfrac{\hat{f}L^4}{8I} J(t_2, t_1) & \text{for } t \ge t_2 \end{cases} \tag{4.17}
$$

The history of the reaction force can now be calculated by inverting (4.15):

$$
F_b(t) = \mathscr{R}\left\{ \frac{3I}{L^3}\left(\frac{\hat{f}L^4}{8I} J(t, t_1) - w_b(t) \right) \right\} = \frac{3\hat{f}L}{8} H(t - t_1) - \frac{3I}{L^3}\mathscr{R}\{w_b(t)\} \tag{4.18}
$$

Function $w_b(t)$ is already known, but in (4.17) it is described by different expressions in two intervals. However, to get $\mathscr{R}\{w_b(t)\}$, the relaxation operator must be applied to the entire history—it would be incorrect to evaluate $\mathscr{R}\{w_b(t)\}$ for $t \ge t_2$ by applying $\mathscr{R}$ on the constant function in the second line of (4.17). Therefore, it is useful to transform the piecewise description of $w_b(t)$ into a single expression, valid at all times t:

$$
w_b(t) = \frac{\hat{f}L^4}{8I} J(t, t_1)\,[1 - H(t - t_2)] + \frac{\hat{f}L^4}{8I} J(t_2, t_1) H(t - t_2) \tag{4.19}
$$

If this is substituted into (4.18), the resulting expression for the history of the reaction force can be written, after some rearrangements, as

$$
F_b(t) = \frac{3\hat{f}L}{8} H_\Delta(t, t_2, t_1) \tag{4.20}
$$

where

$$
H_\Delta(t, t_2, t_1) = \mathscr{R}\{J_\Delta(t, t_2, t_1)\} \tag{4.21}
$$
$$
J_\Delta(t, t_2, t_1) = [J(t, t_1) - J(t_2, t_1)]\, H(t - t_2) \tag{4.22}
$$

■

Note that the factor $3\hat{f}L/8$ in (4.20) corresponds to the reaction force F_b that would be induced if the given load were applied on the structure **after** the change of the structural system, i.e., on a beam clamped at its left end and simply supported at its right end (system II in Fig. 4.3). The function $H_\Delta(t, t_2, t_1)$ multiplying this coefficient, defined in (4.21)–(4.22), looks complicated but has quite a simple physical meaning. Note that the function $J_\Delta(t, t_2, t_1)$ defined in (4.22) vanishes for $t < t_2$ (because then $H(t - t_2) = 0$) and equals $J(t, t_1) - J(t_2, t_1)$ for $t \ge t_2$. Therefore, $J_\Delta(t, t_2, t_1)$ represents the difference between the strain history generated in a creep test at a unit stress level started at time t_1, and the strain history in a mixed creep-relaxation test, in which creep at constant unit stress takes place from time t_1 till time t_2, and subsequently, the strain level is fixed and stress is relaxed; see Fig. 4.4a. Applying the relaxation operator $\mathscr{R}$ to $J_\Delta(t, t_2, t_1)$, we obtain the dimen-

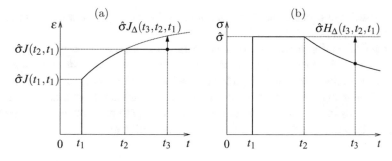

Fig. 4.4 Illustration of functions J_Δ and H_Δ

sionless function $H_\Delta(t, t_2, t_1)$ describing the decrease of stress with respect to its initial unit level in such a mixed test; see Fig. 4.4b. Since $H_\Delta(t, t_2, t_1)$ characterizes the stress redistribution after the change from creep to relaxation, it may be called the *redistribution function* [310, 311]. For fixed t_1 and t_2, function J_Δ is a continuous function of t that vanishes up to $t = t_2 \geq t_1$ and then gradually increases, without a jump. Moreover, its derivative with respect to t is, for $t > t_2$, equal to $\dot{J}(t, t_1)$. The redistribution function H_Δ, formally defined in (4.21)–(4.22), can thus be evaluated as

$$H_\Delta(t, t_2, t_1) = \int_{t_2}^t R(t, t') \dot{J}(t', t_1) \, dt' \qquad \text{for } t \geq t_2 \geq t_1 \qquad (4.23)$$

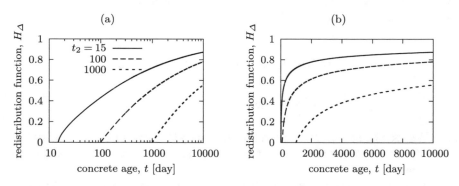

Fig. 4.5 Redistribution function $H_\Delta(t, t_2, t_1)$ plotted as a function of variable t for $t_1 = 14$ days and $t_2 = 15$, 100 and 1000 days, in (a) logarithmic scale, (b) linear scale

For illustration, the redistribution function is plotted in Fig. 4.5 for the concrete from Example 3.1, with the first loading at age $t_1 = 14$ days and the change of structural scheme, respectively, at $t_2 = 15$ days, 100 days, and 1000 days. The solid curves represent the "exact" redistribution functions, obtained by an accurate numerical integration, as described in Chap. 5. It can be observed that if the redistribution starts early after loading, the redistribution function rises steeply and approaches values close to 1 (which would correspond to full redistribution). On the other hand, if the redistribution starts late, the redistribution function grows more gradually and even after 30 years is still far below 1.

The redistribution function serves as a handy tool for the description of evolution of reactions and internal forces after a change of structural system. Consider for instance the moment reaction M_a at the left support of the cantilever whose right end is placed on a support at time t_2; see Example 4.7. Since we have already derived formula (4.20) for the history of statically indeterminate reaction F_b, we can easily compute the value of M_a at any time, only from equilibrium:

$$M_a(t) = \frac{1}{2} f(t)L^2 - F_b(t)l. = \frac{1}{2} \hat{f} L^2 H(t - t_1) - \frac{3\hat{f} L^2}{8} H_\Delta(t, t_2, t_1) \qquad (4.24)$$

Here, $\hat{f} L^2/2$ is the moment reaction on the cantilever (system I), i.e., **before** the change of structural system. It is worth noting that for $H_\Delta = 1$, we would obtain $M_a = \hat{f} L^2/8$, which is the moment reaction that would arise if the load were applied on system II, i.e., **after** the change of structural system. So the factor $-3\hat{f} L^2/8$ multiplying H_Δ in (4.24) represents the difference between the moment reactions computed for the given load on an elastic structure after and before the change of the structural system. The redistribution function $H_\Delta(t, t_2, t_1)$ specifies the relative amount of redistribution at time t if the load is applied at time t_1 and the structural system is changed at time t_2. The same general description of the redistribution process applies to any static quantity such as a reaction or an internal force. For instance, if we want to express the force reaction at the left support, F_a, it is sufficient to compute its value on the cantilever (system I), $\hat{f} L$, and on the statically indeterminate beam with left end clamped and right end simply supported (system II), $5\hat{f} L/8$, and then write

$$F_a(t) = \hat{f} L H(t - t_1) + \left(\frac{5\hat{f} L}{8} - \hat{f} L \right) H_\Delta(t, t_2, t_1) \qquad (4.25)$$

$$= \hat{f} L H(t - t_1) - \frac{3\hat{f} L}{8} H_\Delta(t, t_2, t_1)$$

The Heaviside function $H(t - t_1)$ is used here only for formal reasons, to make the formula valid even for $t < t_1$, and it can be omitted if we specify that only times $t \geq t_1$ are considered.

The general rule for redistribution of any **static quantity** S upon a change of structural system from I to II is formally expressed by the formula

$$S(t) = \tilde{S} H(t - t_1) + \Delta\tilde{S} H_\Delta(t, t_2, t_1) \qquad (4.26)$$

in which $\tilde{S}$ is the value of this static quantity (e.g., internal force) computed on the initial structural system I, and $\Delta\tilde{S}$ is the difference between the values computed on systems II and I. The general symbol S can be replaced by a reaction, bending moment, shear force, stress, etc.

For **kinematic quantities** δ, such as deflections, rotations, curvatures, etc., the following rule can be derived:

$$\delta(t) = \tilde{\delta} J(t, t_1) + \Delta\tilde{\delta} J_\Delta(t, t_2, t_1) \tag{4.27}$$

Recall that $J_\Delta = \mathscr{J}\{H_\Delta\}$ is defined in (4.22). Symbol $\tilde{\delta}$ denotes the value of the kinematic quantity of interest computed on system I using a unit elastic modulus, and $\Delta\tilde{\delta}$ is the difference between the values computed on systems II and I using (in both cases) the unit elastic modulus.

Application of these general rules will be illustrated by the next example.

Example 4.4. Simply supported beams made continuous

Continuous bridge beams are often assembled from precast prestressed concrete beams, which are transported from a plant to the site, raised on the supports and later joined at time t_2 to create a continuous beam (this is more efficient than simply supported beams and has a better seismic performance); see Fig. 4.6a. The joining can, for instance, be achieved by placing a continuous reinforced concrete slab on top of the beams, by the welding of steel bars, or by installing additional prestressing tendons running from one span to the next.

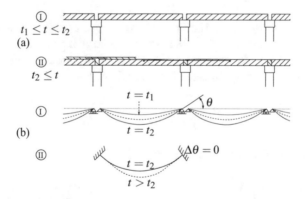

Fig. 4.6 Simply supported beams made continuous

Let the age of concrete at which the beams are placed on the supporting piers be denoted as t_1. Initially, each span acts as a simply supported beam and is statically determinate. The deflections due to self-weight, acting since time t_1, increase freely, which causes the beam ends to gradually undergo rotations $\theta(t)$ (slope increase, Fig. 4.6b). This continues until age t_2 at which the beam ends are joined.

The joining at age t_2 changes the structural system from statically determinate (system I) to statically indeterminate (system II). After that, the additional end rotations of the adjacent beams are forced to be the same. Since the continuing deflection is opposed by this restraint, negative moments above the supports gradually build up.

For the sake of simplicity, we will consider a continuous beam of infinitely many identical spans. Then, by virtue of symmetry conditions, the additional rotations $\Delta\theta$ above the support must be zero (Fig. 4.6b), and each beam after the change of structural system behaves as clamped at both ends. The moments above the supports computed on the simply supported beams (system I) are zero, and the moments computed on the clamped beams (system II) are $-\hat{f}L^2/12$. The negative sign indicates that tensile stresses arise in the top part of the section. The actual evolution of the moments M_s above the supports (extreme negative moments) in the viscoelastic structure with a change of structural system is, according to the general rule (4.26), given by

$$M_s(t) = -\frac{\hat{f}L^2}{12} H_\Delta(t, t_2, t_1) \qquad (4.28)$$

The extreme positive moments M_m at midspan of each beam would be on a simply supported beam (system I) equal to $\hat{f}L^2/8$ and on a clamped beam (system II) to $\hat{f}L^2/24$. Their redistribution due to the change of structural system is described by

$$M_m(t) = \frac{\hat{f}L^2}{8} H(t - t_1) + \left(\frac{\hat{f}L^2}{24} - \frac{\hat{f}L^2}{8}\right) H_\Delta(t, t_2, t_1) =$$

$$= \frac{\hat{f}L^2}{8} H(t - t_1) - \frac{\hat{f}L^2}{12} H_\Delta(t, t_2, t_1) \qquad (4.29)$$

The deflection at midspan would be $5\hat{f}L^4/384I$ on an elastic simply supported beam with unit elastic modulus, and $\hat{f}L^4/384I$ on an elastic clamped beam with unit elastic modulus. According to the general rule (4.27), the evolution of the deflection on the viscoelastic structure with a change of structural system is given by

$$w_m(t) = \frac{5\hat{f}L^4}{384I} J(t, t_1) - \frac{4\hat{f}L^4}{384I} J_\Delta(t, t_2, t_1) \qquad (4.30)$$

Substituting for J_Δ from (4.22), we can rewrite the result as

$$w_m(t) = \begin{cases} \dfrac{5\hat{f}L^4}{384I} J(t, t_1) & \text{if } t_1 \le t \le t_2 \\[2mm] \dfrac{\hat{f}L^4}{384I} J(t, t_1) + \dfrac{4\hat{f}L^4}{384I} J(t_2, t_1) & \text{if } t_2 \le t \end{cases} \qquad (4.31)$$

For illustration, the history of extreme bending moments and of deflection at midspan is plotted in Fig. 4.7 for the following specific case: beam span $L = 25$ m, cross section of area $A = 0.4$ m^2, and moment of inertia $I = 0.07$ m^4, effective depth $D = 0.2$ m and shape factor $k_s = 1.25$ (estimated), concrete properties, curing and environmental humidity the same as in Example 3.1, basic and drying creep described by model B3, effect of shrinkage on bending neglected, loading by self-weight of the simply supported beam applied at age $t_1 = 14$ days, and structural system changed to clamped beam at age $t_2 = 28$ days. For specific weight $\rho g = 24$ kN/m^3, we obtain $\hat{f} = \rho g A = 9.6$ kN/m and $\hat{f}L^4/384I = 139.5$ MN/m. For the given concrete, the conventional modulus of elasticity is $E_{28} = 31.9$ GPa, and the maximum deflection caused by the self-weight on a simply supported beam with elastic modulus E_{28} would be $5\hat{f}L^4/384E_{28}I = 21.9$ mm.

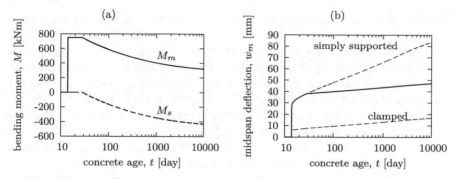

Fig. 4.7 Simply supported beams loaded at $t_1 = 14$ days and made continuous at $t_2 = 28$ days: history of (a) extreme bending moments and (b) deflection at midspan

As shown in Fig. 4.7a, the bending moment at midspan jumps to $\hat{f}L^2/8 = 750$ kN at the age of 14 days and remains constant until the change of structural system at the age of 28 days. After the change, it gradually decreases due to the development of a negative moment at the support. The difference between the moments at midspan and at the support remains constant, as dictated by equilibrium. At the age of 10,000 days, the moment at midspan is reduced to 320 kN and the moment at the support is -430 kN. If the beam was clamped from the beginning, the moments would be 250 kN and -500 kN. The corresponding value of the redistribution function $H_\Delta(10000, 28, 14)$ is $(750 - 320)/(750 - 250) = 430/500 = 0.86$. The bending moment distribution at the age of 10,000 days is thus "composed" of 86% of the moment distribution corresponding to system II (clamped beam) and 14% of the moment distribution corresponding to system I (simply supported beam). The results presented here are based on the "exact" redistribution function, evaluated by accurate numerical integration according to linear viscoelasticity.

Figure 4.7b shows the history of midspan deflection (solid curve). For comparison, the dashed curves indicate the deflections that would evolve on a simply supported beam (top curve) and on a clamped beam (bottom curve), both loaded by self-weight at the age of 14 days. ∎

4.2 Age-Adjusted Effective Modulus Method

4.2.1 Background

For a realistic compliance function such as model B3, and even for relatively simple functions such as the log-double-power law (3.9) or the typical design creep formulations, concrete creep problems cannot be solved analytically, even within the framework of aging linear viscoelasticity. They require a step-by-step numerical integration, which is unnecessarily tedious for preliminary design or simple structures.

A simple approximate solution method is, therefore, needed for design practice. Such a method is adequate if the nonlinear effects of drying, with the inherent cracking, and the large statistical uncertainties of creep and shrinkage, are either ignored or handled too simplistically. The inherent errors of creep structural analysis are often large enough to dwarf the errors caused by simplifications of aging viscoelastic analysis, making an accurate linear viscoelastic solution almost useless. In any case, the simplified approach presented in this chapter is generally insufficient for the actual design of large creep-sensitive structures. Nevertheless, even then this simplified approach is instructive, to develop understanding of the qualitative nature of creep and shrinkage effects, and thus has a place in preliminary design studies.

The history of simplified approaches to creep analysis is reviewed in Appendix B. The earliest of such methods was the *effective modulus method* [623], in which the creep solution for time t is obtained by elastic structural analysis based on the so-called *effective modulus*

$$E_{\text{ef}} = \frac{1}{J(t, t_1)} = \frac{E(t_1)}{1 + \varphi(t, t_1)} \tag{4.32}$$

where t_1 is the time at which the load was applied. Evidently, this method is exact only if the loads and stresses in a structure have a single-step history (i.e., are constant since the instant of first loading, t_1); see Fig. 4.8a. The corresponding strain or displacement history must be a multiple of the compliance function. This condition is far from reality if one deals with a statically indeterminate system afflicted by significant long-time stress redistributions caused by nonuniformity of creep properties or by drying, or if the structural system is changed during the construction, or if the permanent loads are applied gradually. The error magnitude depends on the degree of deviation from a single-step history of load and stresses. The error can be very large, especially for long-time response of structures loaded at a young age [167].

Two more sophisticated methods, which attempt to facilitate the creep analysis of structures by introducing a **simplified form** of the compliance function $J(t, t')$, are historically important:

1. One such method was the *Glanville–Dischinger method*, or briefly Dischinger method [353, 427], also called the rate-of-creep method (and in Russia the theory of aging; [820]), and its later refinement known as the improved Dischinger method [746] or the rate-of-flow method [369]. In all the variants of this method, the compliance function was simplified to a form that allows reducing structural creep problems to ordinary linear differential equations in a transformed time, with constant coefficients, one first-order equation for each static or kinematic unknown.

2. Another such method was the *Maslov–Arutyunyan method* [39, 609], which used another (purportedly better) simplification of $J(t, t')$ permitting the creep problem to be reduced to ordinary linear first-order differential equations with time-dependent coefficients, again one equation for each static or kinematic unknown. In the case of one unknown, this equation can be solved analytically

in terms of an incomplete gamma function. This method, which became popular with mathematically inclined Soviet researchers, is obviously much more complicated than the first, while the first is still substantially more complicated than the effective modulus method.

After more extensive test data and data of long duration became available and were systematically analyzed [80, 175], it turned out that the aforementioned two methods leading to first-order differential equations are, on the average, not more accurate than the simple effective modulus method, which leads to algebraic linear equations with time as a parameter. Thus, the experimental evidence rendered the two aforementioned methods pointless. None of them is sufficiently accurate compared to the computer solutions for a realistic (unsimplified) compliance function based directly on long-time measurements spanning broad ranges of load durations and ages at loading.

A remedy that is sufficiently accurate for linear aging viscoelasticity in most basic situations and accepts an arbitrary form of compliance function was found in the *age-adjusted effective modulus method* (AAEM). It was formulated and mathematically proven by Bažant [76], as a modification and rigorous refinement of the earlier *relaxation method* proposed by Trost [816], who used perspicacious though intuitive semiempirical arguments and ignored the effects of aging on the elastic modulus and relaxation coefficient.

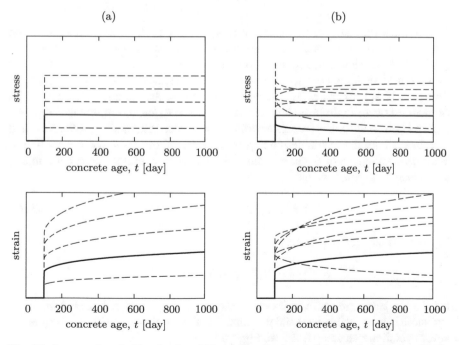

Fig. 4.8 Stress and strain histories for which (a) the effective modulus method is exact, (b) the age-adjusted effective modulus method is exact

The AAEM is formulated for a one-step loading history: the load is applied suddenly at age t_1 and then is either constant until the current time t or varies monotonically at a gradually decreasing rate; see Fig. 4.8b. The corresponding strain (or displacement) history is an arbitrary linear combination of a (shifted) Heaviside function and the compliance function. The response to multistep load histories can, of course, be obtained by superposing the solutions for several one-step histories. While the effective modulus method takes one step from the unstressed state of structure at time t_1^- **just prior** to the first loading to the current state at time t, the AAEM takes one step from the initial stressed state at time t_1^+ **just after** application of the load to the current state at time t. So, the initial state just after loading, which plays no role in the effective modulus method, must be calculated separately, in advance. This is accomplished by standard elastic analysis of the structure based on modulus $E(t_1) = R(t_1^+, t_1) = 1/J(t_1^+, t_1)$. In practical applications, time t_1^+ is usually considered as $t_1 + \Delta t_c$ where Δt_c is the conventional delay, about 15 min. Therefore, $E(t_1)$ is not the asymptotic modulus but rather the static (short-term) modulus of elasticity, and the relation $R(t_1^+, t_1) = 1/J(t_1^+, t_1)$ holds only approximately.

The reader who is not interested in the theoretical background can skip the details of the derivation and proceed directly to the final formulae (4.40) and (4.52)–(4.53).

4.2.2   Fundamental Equation of AAEM

Suppose that the material was under zero stress and strain until time t_1, at which the stress and strain suddenly increased by a jump and then evolved smoothly, at a gradually decreasing rate, until the current time t. If the initial strain $\varepsilon_1 = \varepsilon(t_1^+)$ and the current strain increment $\Delta\varepsilon(t) = \varepsilon(t) - \varepsilon(t_1^+)$ are known, the strain history up to time t can be approximated by a linear combination of the jump function $H(t' - t_1)$ (which describes the strain evolution in a relaxation test) and the compliance function $J(t', t_1)$ (which describes the strain evolution in a creep test); see Fig. 4.9a. Note that t' is an auxiliary time variable running from the initial time t_1 to the current time t. With the coefficients of linear combination denoted as α and β, the (approximate) strain history is expressed as

$$\varepsilon(t') = \alpha + \beta J(t', t_1), \qquad t' \geq t_1 \tag{4.33}$$

The corresponding stress history

$$\sigma(t') = \alpha R(t', t_1) + \beta, \qquad t' \geq t_1 \tag{4.34}$$

is a linear combination of the relaxation function $R(t', t_1)$ (which describes the stress evolution in a relaxation test) and the jump function $H(t' - t_1)$ (which describes the stress evolution in a creep test); see Fig. 4.9b.

Constants α and β can be determined from the given initial strain ε_1 and the strain increment $\Delta\varepsilon(t)$. Recalling that $J(t_1^+, t_1) = 1/E(t_1)$, we can rewrite conditions $\varepsilon(t_1^+) = \varepsilon_1$ and $\varepsilon(t) - \varepsilon(t_1^+) = \Delta\varepsilon(t)$ as

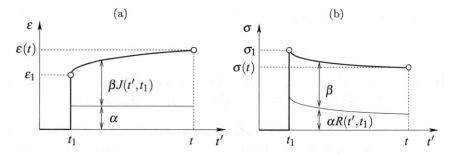

Fig. 4.9 (a) AAEM approximation of the strain history and (b) the corresponding stress history

$$\alpha + \frac{\beta}{E(t_1)} = \varepsilon_1 \tag{4.35}$$

$$\beta \left[J(t, t_1) - \frac{1}{E(t_1)} \right] = \Delta\varepsilon(t) \tag{4.36}$$

and solve for the constants

$$\beta = \frac{\Delta\varepsilon(t)}{J(t, t_1) - \frac{1}{E(t_1)}} = \frac{E(t_1)\Delta\varepsilon(t)}{E(t_1)J(t, t_1) - 1} = \frac{E(t_1)\Delta\varepsilon(t)}{\varphi(t, t_1)} \tag{4.37}$$

$$\alpha = \varepsilon_1 - \frac{\beta}{E(t_1)} = \varepsilon_1 - \frac{\Delta\varepsilon(t)}{\varphi(t, t_1)} \tag{4.38}$$

In the last step of (4.37), we have exploited the relation $E(t_1)J(t, t_1) - 1 = \varphi(t, t_1)$, which follows from the definition of the creep coefficient (3.14).

Substituting the derived expressions for α and β into (4.34), we obtain the approximation of the stress history in terms of the initial strain and the strain increment between times t_1 and t:

$$\sigma(t') = \left[\varepsilon_1 - \frac{\Delta\varepsilon(t)}{\varphi(t, t_1)} \right] R(t', t_1) + \frac{E(t_1)\Delta\varepsilon(t)}{\varphi(t, t_1)} = R(t', t_1)\varepsilon_1 + \frac{E(t_1) - R(t', t_1)}{\varphi(t, t_1)} \Delta\varepsilon(t) \tag{4.39}$$

The key point is that this stress approximation has been constructed from a strain approximation that exactly matches the values of strain at times t_1 and t. Therefore, even though (4.39) is defined for an arbitrary time t', it is actually accepted as the stress approximation only at time t. If t changes, the values of "constants" α and β are adjusted accordingly, as described by formulae (4.37)–(4.38). The resulting approximation of the stress history for an arbitrary t is obtained simply by substituting t for t' in (4.39). Introducing the *age-adjusted effective modulus* [76]

$$E''(t, t_1) = \frac{E(t_1) - R(t, t_1)}{\varphi(t, t_1)} = \frac{E(t_1) - R(t, t_1)}{E(t_1)J(t, t_1) - 1}, \qquad t > t_1 \tag{4.40}$$

we can write the final formula as

$$\sigma(t) = R(t, t_1)\varepsilon_1 + E''(t, t_1)\Delta\varepsilon(t) \tag{4.41}$$

This is one form of the *fundamental equation of the AAEM*, which replaces the "exact" integral relation between the strain and stress histories (2.24) by a much simpler algebraic relation.

The attentive reader has certainly noticed that the numerator and the denominator in (4.40) become very small for short elapsed times $t - t_1$. With $E(t_1)$ considered as the asymptotic modulus, E_0, the fraction would tend to E_0 as $t - t_1 \to 0^+$. However, in practical applications of the AAEM method, $E(t_1)$ is considered as the static (short-term) modulus of elasticity, corresponding to the conventional delay $\Delta t_s = 0.01$ day. The fraction then tends to infinity as $t - t_1 \to \Delta t_s^+$ (because the denominator tends to zero but the numerator does not), and the formula is not applicable to elapsed times $t - t_1$ close to Δt_s. This will be explained in more detail in Example 4.5, and a remedy will be proposed.

4.2.3 Alternative Derivation of AAEM*

The first term on the right-hand side of (4.41), $R(t, t_1)\varepsilon_1$, describes stress relaxation at constant strain and perfectly agrees with the term that appears in the exact formula (2.24),

$$\sigma(t) = R(t, t_1)\varepsilon_1 + \int_{t_1^+}^{t} R(t, t')\dot{\varepsilon}(t')\,dt' \tag{4.42}$$

The second term on the right-hand side of (4.41) represents the influence of strain changes between time t_1^+ (just after the initial load application) and current time t, and it can be considered as an approximation of the integral in (4.42):

$$\int_{t_1^+}^{t} R(t, t')\dot{\varepsilon}(t')\,dt' \approx E''(t, t_1) \int_{t_1^+}^{t} \dot{\varepsilon}(t')\,dt' = E''(t, t_1)\Delta\varepsilon(t) \tag{4.43}$$

So the age-adjusted effective modulus $E''(t, t_1)$ represents the relaxation function $R(t, t')$ averaged (in some generalized sense) over all t' between t_1 and t.

If the effective value of $R(t, t')$ were computed by simple averaging, the approximation (4.43) would be exact for histories with constant strain rate $\dot{\varepsilon}(t')$, i.e., for strain evolving as a linear function of time. This could be a good approximation of the real history in certain special cases (see, e.g., the effects of cyclic loading described in Sect. 7.13), but a typical response of a viscoelastic structure is characterized by a gradual decay of the strain rate. Therefore, the AAEM is based on the condition that the result be exact for any linear combination of relaxation (at constant strain) and creep (at constant stress). In a relaxation test, the strain is constant, its rate vanishes and the integral in (4.42) vanishes as well, and so its approximation (4.43) is in this case exact since $\Delta\varepsilon(t) = 0$. For a creep test at a unit stress level, we can set $\sigma(t) = H(t - t_1)$ and $\varepsilon(t) = J(t, t_1)$. Substituting this into (4.42) and considering $t > t_1$, we get

$$1 = R(t, t_1) J(t_1^+, t_1) + \int_{t_1^+}^{t} R(t, t') \dot{J}(t', t_1) \, dt' \tag{4.44}$$

where $\dot{J}$ denotes the derivative of the compliance function with respect to its first argument. Thus, the exact value of the integral on the left-hand side of (4.43) is

$$\int_{t_1^+}^{t} R(t, t') \dot{J}(t', t_1) \, dt' = 1 - R(t, t_1) J(t_1^+, t_1) = 1 - \frac{R(t, t_1)}{E(t_1)} \tag{4.45}$$

and the approximation on the right-hand side of (4.43) is exact if

$$1 - \frac{R(t, t_1)}{E(t_1)} = E''(t, t_1)\left[J(t, t_1) - \frac{1}{E(t_1)}\right] \tag{4.46}$$

From this condition, we can determine the age-adjusted effective modulus

$$E''(t, t_1) = \frac{1 - R(t, t_1)J(t_1^+, t_1)}{J(t, t_1) - J(t_1^+, t_1)} = \frac{E(t_1) - R(t, t_1)}{E(t_1)J(t, t_1) - 1} = \frac{E(t_1) - R(t, t_1)}{\varphi(t, t_1)} \tag{4.47}$$

Equations (4.42)–(4.47) in effect represent an alternative derivation of the age-adjusted effective modulus, and confirm formula (4.40). Yet another derivation is given in Bažant [76, 87, 126] and Jirásek and Bažant [521].

4.2.4 Ramifications of AAEM

The stress–strain relation (4.41) is applicable at all times $t \geq t_1$. In the special case of $t = t_1$, it reduces to the simple elastic stress–strain relation describing the instantaneous response,

$$\sigma_1 = E(t_1)\varepsilon_1 \tag{4.48}$$

where $\sigma_1 = \sigma(t_1^+) =$ initial stress just after the application of strain ε_1.

Equation (4.41) can easily be inverted to obtain an expression for the strain increment in terms of the initial strain and the stress increment:

$$\Delta\varepsilon(t) = \frac{\sigma_1 + \Delta\sigma(t) - R(t, t_1)\varepsilon_1}{E''(t, t_1)} = \frac{E(t_1) - R(t, t_1)}{E''(t, t_1)}\varepsilon_1 + \frac{\Delta\sigma(t)}{E''(t, t_1)} =$$

$$= \varphi(t, t_1)\varepsilon_1 + \frac{\Delta\sigma(t)}{E''(t, t_1)} \tag{4.49}$$

This is the traditional form of the *fundamental equation of AAEM*, derived by Bažant [76]. Adding the initial strain ε_1 and using the relation $\varepsilon_1 = \sigma_1/E(t_1)$, we get the total strain at time t,

$$\varepsilon(t) = \varepsilon_1 + \Delta\varepsilon(t) = [1 + \varphi(t, t_1)]\frac{\sigma_1}{E(t_1)} + \frac{\Delta\sigma(t)}{E''(t, t_1)} = J(t, t_1)\sigma_1 + \frac{\Delta\sigma(t)}{E''(t, t_1)} \tag{4.50}$$

This inverted form of the stress–strain relation has exactly the same structure as the original formula (4.41), with the strain replaced by the stress, the relaxation function by the compliance function, and the age-adjusted effective modulus by its reciprocal value. Formula (4.50) can be considered as an approximation of the exact integral formula

$$\varepsilon(t) = J(t, t_1)\sigma_1 + \int_{t_1^+}^{t} J(t, t')\dot\sigma(t')\,dt' \tag{4.51}$$

and the reciprocal value of $E''(t, t_1)$ plays the role of a weighted average of the compliance $J(t, t')$ over all t' between t_1 and t, determined such that (4.50) gives the exact result for the stress history corresponding to the relaxation test.

In summary, the age-adjusted effective modulus method consists in the following approximations of the relaxation and creep operators:

$$\mathcal{R}\{\varepsilon(t)\} \approx R(t, t_1)\varepsilon(t_1^+) + E''(t, t_1)[\varepsilon(t) - \varepsilon(t_1^+)] \tag{4.52}$$

$$\mathcal{J}\{\sigma(t)\} \approx J(t, t_1)\sigma(t_1^+) + \frac{1}{E''(t, t_1)}[\sigma(t) - \sigma(t_1^+)] \tag{4.53}$$

Such approximations can be applied not only on the level of the stress–strain relation but also on the structural level, whenever the relation between two quantities (e.g., between load and deflection) is described by the relaxation or creep operator.

Since, for $t \geq t_1$, we have $\varphi(t, t_1) = E(t_1)J(t, t_1) - 1$ and $\mathcal{R}\{\varphi(t, t_1)\} = E(t_1) - R(t, t_1)$, the definition of the age-adjusted effective modulus (4.40) can be transformed into the following equivalent expressions:

$$E''(t, t_1) = \frac{\mathcal{R}\{\varphi(t, t_1)\}}{\varphi(t, t_1)} = \frac{\mathcal{R}\{\Delta J(t, t_1)\}}{\Delta J(t, t_1)} = \frac{\Delta R(t, t_1)}{\mathcal{J}\{\Delta R(t, t_1)\}} \tag{4.54}$$

where $\Delta J(t, t_1) = J(t, t_1) - J(t_1^+, t_1) = J(t, t_1) - 1/E(t_1) = \varphi(t, t_1)/E(t_1)$ and $\Delta R(t, t_1) = R(t, t_1) - R(t_1^+, t_1) = R(t, t_1) - E(t_1)$ are the increments of the compliance function or of the relaxation function over the interval $(t_1, t]$.

To keep the previous derivations simple, we have neglected the effects of shrinkage and temperature changes. Such effects can be incorporated by replacing $\varepsilon(t)$ with the mechanical strain, $\varepsilon_\sigma(t) = \varepsilon(t) - \varepsilon_{\text{sh}}(t) - \varepsilon_T(t)$; see Eqs. (2.44)–(2.45).

For convenience, the age-adjusted effective modulus, whose primary definition is (4.40), is normally expressed in the form

$$E''(t, t_1) = \frac{E(t_1)}{1 + \chi(t, t_1)\varphi(t, t_1)} \tag{4.55}$$

where

$$\chi(t, t_1) = \frac{E(t_1)}{E(t_1) - R(t, t_1)} - \frac{1}{\varphi(t, t_1)} \tag{4.56}$$

is the so-called *aging coefficient*. For $\chi(t, t_1) = 1$, the effective modulus (4.32) would be obtained as a special case. It turns out that χ varies relatively little, usually from 0.5 to 1.0, with 0.8 as the most typical value. The exact evaluation of the age-adjusted effective modulus according to (4.40) requires the evaluation of the relaxation function, which is not readily available because most creep models specify the compliance function or the creep coefficient as the primary material characteristics and the corresponding relaxation function would need to be solved numerically from (2.25) or similar equations. Instead of that, the age-adjusted effective modulus can be estimated from (4.55), with the aging coefficient χ roughly approximated by a constant. Tables of χ, computed for certain compliance functions, have been included in Bažant [76] and ACI Committee 209 design recommendations [12, 13].

The basic Eqs. (4.52)–(4.53) of AAEM are exact for all the strain histories representing a linear transformation of the creep coefficient curve, i.e.,

$$\varepsilon(t) - \varepsilon_{sh}(t) = a + b\varphi(t, t_1), \qquad t \geq t_1 \tag{4.57}$$

where a and b are arbitrary constants. This includes a broad range of strain histories illustrated in Fig. 4.8b. The actual strain histories in structures under permanent load are usually well approximated by (4.57). This is the reason for the good accuracy of AAEM in a broad range of problems. By contrast, the effective modulus method is exact only for a more limited set of histories, namely for creep at constant stress (Fig. 4.8a).

The AAEM has been generalized in a matrix form for the vectors of force and displacement components [93], and also for bending creep of composite cross sections [566, 567] and for loads varying in time monotonically at a decreasing rate [541].

4.2.5   Approximation of Relaxation Function

Formula (4.40) defining the age-adjusted effective modulus contains not only the compliance function, which is directly prescribed by the creep model, but also the relaxation function, which needs to be evaluated by inverting the integral relation between the strain history and the stress history. To dispense with a computer solution of the relaxation function $R(t, t')$ for a given compliance function $J(t, t')$, one may use the following semiempirical approximate formula developed by Bažant, Hubler, and Jirásek [135]:

$$R(t, t') = \frac{1}{J(t, t')} \left[1 + \frac{c_1(t') J(t, t')}{10 \, J(t, t - \Delta t)} \left(\frac{J(t_m, t')}{J(t, t_m)} - 1 \right) \right]^{-10} \tag{4.58}$$

where $\Delta t = 1$ day, $t_m = (t + t')/2$, and

$$c_1(t') = 0.08 + 0.0119 \ln t' \tag{4.59}$$

Formula (4.58) is a refinement of a previously used formula, developed by [153], which had been formulated for a predecessor of model B3 and is not suitable for the full version of model B3 (especially not for multidecade relaxation of young concrete).

Example 4.5. Approximation of relaxation function

For comparison, approximations of the relaxation function based on the explicit formula (4.58) are plotted in Fig. 4.10 (dashed curves, marked as "R approx"), along with a highly accurate numerical solution (solid curves, marked as "R exact") and with the reciprocal value of the compliance function (dotted curves, marked as "1/J"). The input parameters (concrete composition and strength, curing, environmental humidity) are the same as in Example 3.1.

Figure 4.10a shows the relaxation curves for basic creep and Fig. 4.10b for drying creep, in both cases for three ages at loading, $t' = 10, 100,$ and 1000 days (from top to bottom). For short elapsed times $t - t'$, the relaxation function $R(t, t')$ is very close to the reciprocal compliance function $1/J(t, t')$, while for long times the values of R are substantially smaller than $1/J$. The accuracy of the analytical approximation of R by formula (4.58) can be considered as sufficient for quick practical estimates.

Figure 4.11 shows the age-adjusted effective modulus $E''(t, t_1)$ for three ages at loading in the same plot (from top to bottom, the three families of curves correspond to $t_1 = 1000, 100,$ and 10 days). Again, exact values are indicated by the solid curves, approximate values based on (4.58) by the dashed curves and, for comparison, the reciprocal compliance values (i.e., the effective modulus (4.32)) by the dotted curves. In practical applications, the AAEM method is useful for long-time predictions of structural behavior, and so it is the accuracy for large elapsed times $t - t_1$ that matters. However, it is interesting to note that the standard AAEM formula (4.40) with $E(t_1)$ considered as the static modulus of elasticity, $1/J(t_1 + \Delta t_s, t_1)$, has a singularity at $t - t_1 = \Delta t_s = 0.01$ day. The reason is that the relation $R(t_1 + \Delta t_s, t_1) = 1/J(t_1 + \Delta t_s, t_1)$ holds only approximately. For $t \to t_1 + \Delta t_s$, the denominator in (4.40) tends to zero but the numerator does not, and the fraction blows up. This would not happen if $E(t_1)$ was considered as the asymptotic modulus. However, the overall accuracy of the AAEM method is usually better if the initial response includes creep up to elapsed time Δt_s, i.e., if its evaluation is based on the conventional static modulus rather than on the asymptotic one.

The singularity of E'' at $t - t_1 = \Delta t_s$ is harmless if the AAEM predictions are constructed for elapsed times substantially larger than the conventional delay Δt_s. If one intends to plot the complete evolution of the response, starting from elapsed times near or even below Δt_s, it can be useful to adopt a modified definition of AAEM, which provides a continuous transition from the formula based on the asymptotic modulus to the standard formula based on the static modulus. The key idea is that $E(t_1)$ is considered as the static modulus only if the elapsed time exceeds 10 times the conventional delay (i.e., if $t - t_1 > 0.1$ day); otherwise, it is replaced by a modulus corresponding to the delay $(t - t_1)/10$ instead of Δt_s. Formula (4.40) is then rewritten as

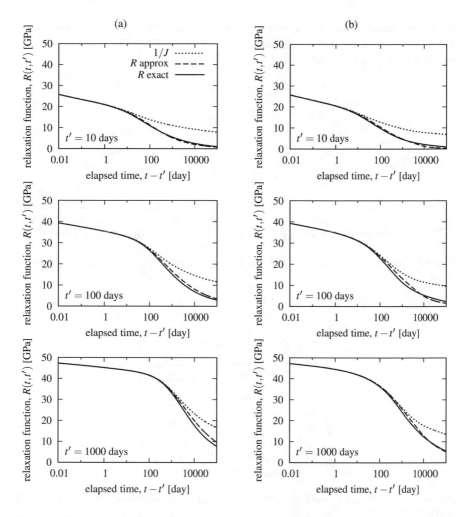

Fig. 4.10 Comparison of the explicit approximation of relaxation function by formula (4.58) and a highly accurate numerical solution: (a) basic creep, (b) drying creep

$$E''(t, t_1) = \frac{1 - R(t, t_1) J(t_1^*, t_1)}{J(t, t_1) - J(t_1^*, t_1)} \tag{4.60}$$

where

$$t_1^* = \begin{cases} 0.9t_1 + 0.1t & \text{if } t_1 < t < t_1 + 10\Delta t_s \\ t_1 + \Delta t_s & \text{if } t_1 + 10\Delta t_s \leq t \end{cases} \tag{4.61}$$

This modified definition of AAEM is graphically compared to the standard one in Fig. 4.12 for elapsed times $t - t_1$ between 10^{-4} day and 1 day. For $t_1 - t \geq 10\Delta t_s = 0.1$ day, both definitions coincide.

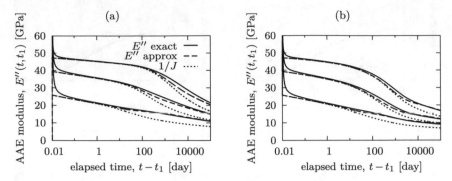

Fig. 4.11 Age-adjusted effective modulus and its approximation for (a) basic creep, (b) drying creep

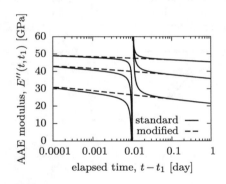

Fig. 4.12 Age-adjusted effective modulus according to the standard formula (4.40) and its modification (4.60)

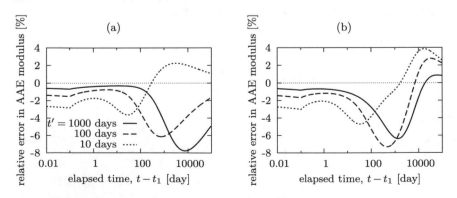

Fig. 4.13 Relative error of the approximation of age-adjusted effective modulus for (a) basic creep, (b) drying creep

The relative error of the approximate evaluation of AAEM based on the approximation of the relaxation function by formula (4.58) is shown in Fig. 4.13. In the present case, this error ranges from -7% to $+4\%$. The relative error is taken here with respect to the actual value of AAEM. If it was taken with respect to the static modulus, the relative errors for long-term loading would be much smaller. Note that evaluation of AAEM is based on the modified definition (4.60); otherwise, the magnitude of the error would blow up for $t - t_1$ approaching 0.01 day.

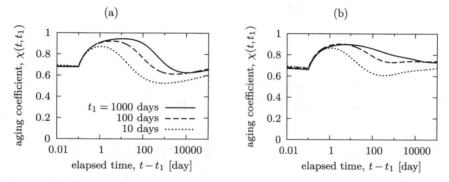

Fig. 4.14 Aging coefficient for (a) basic creep, (b) drying creep

Finally, Fig. 4.14 shows the aging coefficient, again for three ages at loading and for basic and drying creep. The aging coefficient varies approximately between 0.6 and 0.9. In very crude approximations, it can be taken as a constant, equal to 0.8. ∎

4.2.6 Simple Applications of AAEM

Example 4.6. Stresses due to restrained shrinkage: AAEM approximation

In Examples 4.1 and 4.2, the history of the shrinkage strain ε_{sh} had to be transformed by the relaxation operator, which was achieved by numerical integration. A quick estimate can be constructed by the AAEM method. The initial time t_1 in the sense of AAEM must be considered as the time t_0 at the onset of drying. Since the initial value $\varepsilon_{sh}(t_0)$ is zero, the first term in formula (4.52) vanishes and the AAEM approximation of (4.10) can be written simply as

$$\sigma(t) = -\mathscr{R}\{\varepsilon_{sh}(t)\} \approx -E''(t, t_0)\varepsilon_{sh}(t) \tag{4.62}$$

The age-adjusted effective modulus E'' can be computed from its definition (4.60), or estimated according to (4.55) with a constant value of the aging coefficient. Definition (4.60) contains the relaxation function, which can be accurately evaluated by numerical solution of Eq. (2.25), or estimated according to formula (4.58).

In Fig. 4.15a, the solid curve corresponds to the exact history of shrinkage-induced stress (computed by an accurate numerical evaluation of the integral defining the relaxation operator) for concrete cured until time $t_0 = 28$ days and then exposed to an environment of relative humidity $h_{env} = 70\%$. The dashed and dotted curves are AAEM approximations computed in four different ways:

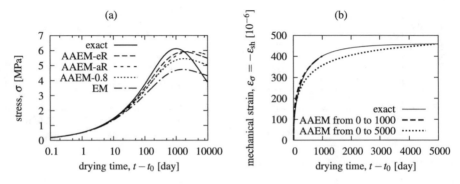

Fig. 4.15 (a) History of stress induced by restrained shrinkage, (b) evolution of mechanical strain (minus the shrinkage strain) and its AAEM approximations for two different drying times, 1000 and 5000 days

- AAEM-eR ... from (4.60) using an "exact" value of relaxation function, obtained numerically,
- AAEM-aR ... from (4.60) using an approximate value of relaxation function, obtained from formula (4.58),
- AAEM-0.8 ... from (4.55) using a constant value of the aging coefficient, $\chi = 0.8$,
- EM ... from (4.55) using aging coefficient $\chi = 1$, which means that the AAEM is replaced by the effective modulus (4.32).

All these methods correctly predict that the evolution of stress is not monotonic. Up to 1000 days, the best accuracy is achieved by AAEM-eR, followed by AAEM-aR. The other two methods underestimate the peak stress. In the range of drying times above 1000 days, when the stress decreases, none of the methods is reliable. All of them substantially underestimate the rate at which stress is relaxed. The reason is that the AAEM method approximates the actual evolution of mechanical strain (in the present case, of minus the shrinkage strain) as a constant plus a multiple of the compliance function. The initial strain is in the present case almost zero (it corresponds to the shrinkage strain after $\Delta t_s = 0.01$ day of drying, which is extremely small), and so the approximation is in fact a multiple of the function defining the evolution of the creep coefficient. As long as the shrinkage strain keeps growing sufficiently fast, such an approximation is quite realistic; see the dashed curve in Fig. 4.15b, which is used as the AAEM approximation when estimating the stress at drying time $t - t_0 = 1000$ days. For times at which the drying process is almost complete and the shrinkage strain curve approaches a horizontal asymptote, the AAEM approach leads to a strain evolution that initially lags behind the actual one and only later catches up; see the dotted curve in Fig. 4.15b, which is used as the AAEM approximation when estimating the stress at drying time $t - t_0 = 5000$ days. Since the strain increments are imposed later than the actual ones, there remains less time for relaxation and the resulting approximate stress is higher than the correct one according to linear viscoelasticity. ∎

Example 4.7. Redistribution function

The redistribution function H_Δ, formally defined in (4.21)–(4.22), can be evaluated by numerical integration according to (4.23). For quick estimates, it is more convenient to use an approximation of the integral operator by an algebraic formula based on the age-adjusted effective modulus. Here, it is essential to realize that the redistribution function (4.21) is obtained by applying the relaxation operator to a special function $J_\Delta(t, t_2, t_1)$, which is zero up to time $t = t_2$ (greater than t_1) and only then starts gradually increasing. Therefore, the initial time of the AAEM approximation, in the original derivation denoted as t_1, should now be taken as t_2, and the appropriate age-adjusted effective modulus will be $E''(t, t_2)$. Since $J_\Delta(t, t_2, t_1)$ does not change by a jump at $t = t_2$, the AAEM approximation (4.52) of the relaxation operator simplifies to

$$H_\Delta(t, t_2, t_1) = \mathscr{R}\{J_\Delta(t, t_2, t_1)\} \approx E''(t, t_2) J_\Delta(t, t_2, t_1) =$$
$$= E''(t, t_2)[J(t, t_1) - J(t_2, t_1)] H(t - t_2) \qquad (4.63)$$

In Fig. 4.16, the dotted curves, marked as "AAEM-eR," are the AAEM approximations of the redistribution function computed using an "exact" value of the relaxation function, obtained numerically, and the dashed curves, marked as "AAEM-aR," are the AAEM approximations computed using an approximate value of the relaxation function according to formula (4.58). Of course, from the practical point of view, the approach denoted as AAEM-eR makes little sense. If the designer is ready to invest the effort into an accurate numerical evaluation of the relaxation function, it is better to evaluate directly the redistribution function (rather than its AAEM approximation), because the numerical effort is comparable. If a quick estimate is needed, the relaxation function entering the definition of the AAEM should be approximated by the closed-form formula (4.58), which corresponds to the AAEM-aR approach. However, plotting the curves that would be obtained with the AAEM-eR approach is useful for comparison of the relative contribution of two sources of error, coming from the AAEM approximation of the actual history and from the approximation of the relaxation function. As illustrated by the results plotted in Fig. 4.16, the relative importance of these two sources of error varies depending on the specific case. For redistribution starting at age $t_2 = 15$ days, the approximation of R is quite accurate and the main source of error is in the AAEM approximation. For $t_2 = 100$ days, both sources contribute, and for $t_2 = 1000$ days, the main source of error is in the approximation of R. Of course, this discussion refers only to the present illustrative example and the observations cannot be considered as general.   ■

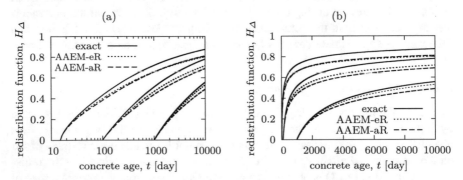

Fig. 4.16 Redistribution function $H_\Delta(t, t_2, t_1)$ for $t_1 = 14$ days and $t_2 = 15$, 100 and 1000 days, plotted in (a) logarithmic scale, (b) linear scale

4.3 Nonhomogeneous Structures

In practice, most concrete structures need to be treated as nonhomogeneous. The sources of heterogeneity are multiple: differences in age of individual segments, presence of passive reinforcement or prestressed tendons, nonuniformity of the drying process and the related effect of wall thickness differences, etc.

The stress redistributions caused by creep in nonhomogeneous, statically indeterminate structures generally consist of a gradual transfer of stress from the parts that creep more (e.g., a younger concrete, a lower-strength concrete, a thinner cross section, a drying member) to the parts that creep less. Steel parts in composite steel–concrete structures do not creep (except possibly in fire), and so the effect of creep is, in general, a gradual transfer of stress into the steel parts.

Unprestressed steel reinforcement, likewise, does not creep (except in fire). Because of their very high stress, the prestressing tendons do creep, but very differently from concrete. Their behavior is viscoplastic and causes prestress relaxation.

The reinforcement thus introduces an inhomogeneity into the creep properties of the structure and leads to stress redistributions between steel and concrete. A rigorous approach is to consider the reinforcement and the concrete as separate parts of the structure.

In prestressed concrete structures, though, the effect of reinforcement stiffness on the creep deformations of beams is generally small, for two reasons: (1) the ratio of the cross-sectional area of reinforcement to the total cross-sectional area is small, compared to unprestressed structure, and (2) the cross sections do not undergo cracking. In all the preceding examples and in Example 4.8, this effect is therefore neglected.

The restraining effect of reinforcement on creep becomes significant in structures that develop distributed cracking, which includes all unprestressed reinforced concrete beams and plates. This effect is nonlinear and its treatment will be discussed in Chap. 12.

4.3.1 Stress Redistributions Due to Differences in Age of Concrete

Example 4.8. Joining two cantilevers of different age

Since concrete loaded at young age creeps more than concrete loaded at old age, large differences in age lead to significant redistributions of internal forces in structures. As an example, consider again a box girder bridge cast by the cantilever method. In the interest of economy, the pair of steel trusses at each pier that support new segments during erection is used repeatedly to erect one cantilever pair after another. Consider, therefore, that in Fig. 4.17a the left cantilever is older by Δt than the right cantilever (a difference Δt of 3–6 months is not unusual).

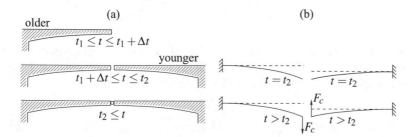

Fig. 4.17 Effect of creep differences due to differences in age

Let the general (global) time variable t be measured from the set of the left (older) cantilever, and let t_1 be the age of each cantilever at which the self-weight is assumed to be applied.[3] This means that the older cantilever is loaded at time t_1 and the younger one at time $t_1 + \Delta t$ (when its own age is t_1). As long as the opposite cantilevers of the pair in one span are separated, they deflect independently of each other. At time $t = t_2$, they are joined. For the sake of simplicity, assume that their joining is done by installing a horizontally sliding hinge, which provides only one

---

[3]This is of course a simplification; in reality, the bending moments due to self-weight grow gradually as the cantilever is being erected.

statically indeterminate internal force $F_c(t)$ in each span (full continuity at midspan is a better, deflection mitigating, design, but it provides three statically indeterminate forces at midspan).

If the cantilevers were allowed to deflect independently, the left (older) one would deflect after time t_2 less than the right (younger) one because more of creep has occurred before t_2 in the left cantilever. Therefore, if the differences in deflection are prevented, the pair of vertical forces representing F_c will act on the left cantilever downward, and on the right cantilever upward (Fig. 4.17b). In other words, F_c is the shear force transmitted by the sliding hinge c.

The age difference makes the structure nonhomogeneous, but in the sense of our simplifying assumptions, each cantilever is homogeneous. Similar to the examples in the preceding section, we decompose the problem into two loading cases, one corresponding to the permanent load and the other to the statically indeterminate force. The deflection at the right end of the left cantilever can easily be expressed by superposition of these two loading cases as

$$w_{ca}(t) = \mathscr{J}\left\{\frac{f(t)L^4}{8I}\right\} + \mathscr{J}\left\{\frac{F_c(t)L^3}{3I}\right\} \tag{4.64}$$

For simplicity, we assume that both cantilevers have the same dimensions, but, of course, it would be no problem to adapt the solution to a more general case. The self-weight is constant from time t_1, so we can set $f(t) = \hat{f}H(t - t_1)$ and rewrite (4.64) as

$$w_{ca}(t) = \frac{\hat{f}L^4}{8I}J(t, t_1) + \frac{L^3}{3I}\mathscr{J}\{F_c(t)\} \tag{4.65}$$

A similar formula holds for the left end of the right cantilever, but we must take into account that (i) the age of this cantilever at time t is $t - \Delta t$, and (ii) the positive force F_c acts on this cantilever upward. So (4.65) must be modified to

$$w_{cb}(t) = \frac{\hat{f}L^4}{8I}J(t - \Delta t, t_1) - \frac{L^3}{3I}\mathscr{J}_{t-\Delta t}\{F_c(t)\} \tag{4.66}$$

Here, $\mathscr{J}_{t-\Delta t}$ is a time-shifted compliance operator that uses compliance function $J(t - \Delta t, t' - \Delta t)$ instead of $J(t, t')$, which reflects the fact that a load acting from global time t' till global time t is actually acting from age (of the younger cantilever) $t' - \Delta t$ until age $t - \Delta t$.

Until time t_2, the deflections $w_{ca}(t)$ and $w_{cb}(t)$ evolve independently and the force $F_c(t)$ is zero. After the change of structural system at time t_2, further increments of deflections at the joined cantilever ends must be the same, and the history of statically indeterminate force $F_c(t)$ can be computed from the compatibility condition

$$w_{ca}(t) = w_{cb}(t) + \Delta w, \qquad t \geq t_2 \tag{4.67}$$

where Δw is a constant that corresponds to the difference between w_{ca} and w_{cb} just after the joining. If the construction procedure used to join the cantilever ends at time t_2 does not introduce any initial force F_c, then Δw can be directly computed as the difference between $w_{ca}(t_2)$ and $w_{cb}(t_2)$, which are evaluated from (4.65)–(4.66) with the terms reflecting the influence of F_c still equal to zero. However, since, in general, a nonzero initial force F_c may be introduced, we consider Δw for a while as an additional unknown that will later be related to the initial value of F_c.

Substituting (4.65) and (4.66) into (4.67), we obtain an equation from which the history of F_c can be solved. Since this equation has an integral character, it is good to present it in a form valid at all times t, not just at times $t \geq t_2$ as the original compatibility condition (4.67). This can be formally achieved if both sides of (4.67) are first multiplied by $H(t - t_2)$. The point is that for $t < t_2$ the value of $H(t - t_2)$ is zero and the resulting equation reduces to the identity $0 = 0$. After the aforementioned substitutions and subsequent simple rearrangement, the compatibility condition (4.67) is transformed to

$$\mathcal{J}\{F_c(t)\} + \mathcal{J}_{t-\Delta t}\{F_c(t)\} = \frac{3\hat{f}L}{8}\{J(t - \Delta t, t_1) - J(t, t_1)\}\,H(t - t_2) +$$
$$+ \frac{3I}{L^3}\Delta w\,H(t - t_2) \tag{4.68}$$

Equation (4.68) is valid at all times t. Writing it specifically for $t = t_2^+ = $ time just after the change of structural system, we obtain a relation between the initial force $F_c(t_2^+)$ and the yet unknown displacement difference Δw:

$$\frac{F_c(t_2^+)}{E(t_2)} + \frac{F_c(t_2^+)}{E(t_2 - \Delta t)} = \frac{3\hat{f}L}{8}\{J(t_2 - \Delta t, t_1) - J(t_2, t_1)\} + \frac{3I}{L^3}\Delta w \tag{4.69}$$

This is a universal relation that covers also the specific case with $F_c(t_2^+) = 0$. The constant Δw can now be expressed in terms of $F_c(t_2^+)$ and eliminated from (4.68). Interestingly, after this substitution, the difference of compliance functions $J(t, t_1) - J(t_2, t_1)$ multiplied by $H(t - t_2)$ appears on the right-hand side, and this is in fact the previously introduced function $J_A(t, t_2, t_1)$ defined in (4.22). Another similar expression, $[J(t - \Delta t, t_1) - J(t_2 - \Delta t, t_1)]H(t - t_2)$, can be identified as $J_A(t - \Delta t, t_2 - \Delta t, t_1)$. This brings the resulting equation for the history of the statically indeterminate force into the relatively compact form

$$\mathcal{J}\{F_c(t)\} + \mathcal{J}_{t-\Delta t}\{F_c(t)\} = \frac{3\hat{f}L}{8}[J_A(t - \Delta t, t_2 - \Delta t, t_1) - J_A(t, t_2, t_1)] +$$
$$+ \left[\frac{1}{E(t_2)} + \frac{1}{E(t_2 - \Delta t)}\right]F_c(t_2^+)H(t - t_2) \tag{4.70}$$

Such an integral equation cannot be solved by a simple application of the usual relaxation operator, because the operator acting on the unknown function $F_c(t)$ is the sum of the compliance operator $\mathcal{J}$ and the shifted compliance operator $\mathcal{J}_{t-\Delta t}$. The solution can be obtained numerically, replacing the integrals by finite sums. Once the history of the internal reaction F_c has been determined, the deflection history can be obtained from (4.65)–(4.66).

For illustration, an accurate numerical solution computed for concrete with the same properties as in Example 3.1 is plotted in Fig. 4.18. In this specific example, the older cantilever is cast at time $t = 0$ and loaded at age $t = t_1 = 30$ days, while the younger cantilever is cast at time $t = \Delta t = 60$ days and loaded at time $t = t_1 + \Delta t = 90$ days, when its age is 30 days. Drying creep in an environment of 70% relative humidity is considered. The internal reaction just after the change of structural system, $F_c(t_2^+)$, is set to zero. The solid curves in Fig. 4.18a correspond to the deflections that would evolve on two independent cantilevers. They have the same shape and are shifted in time by $\Delta t = 60$ days. These curves reflect the first terms in Eqs. (4.65) and (4.66), respectively, and until time $t = t_2 = 120$ days they represent the actual deflection. The subsequent evolution of deflections of the joined cantilevers is marked by the dashed curves, whose vertical distance remains constant. Deflection of the older cantilever is augmented and deflection of the younger one is reduced. The time variable t on the horizontal axis is taken as the age of the older cantilever. The deflections on the vertical axis are normalized by the "instantaneous" deflection, calculated with the conventional modulus of elasticity.

Figure 4.18b shows the internal reaction F_c as a function of the time $t - t_2$ elapsed after joining the cantilevers. The horizontal axis corresponds to the time elapsed after joining the cantilevers, in logarithmic scale. Force F_c on the vertical axis is normalized by the total weight of one cantilever, $\hat{f}L$. It is interesting to note that the force transmitted by the hinge first increases, but approximately after 360 days, it attains its maximum and then decreases. ∎

(a)

(b)

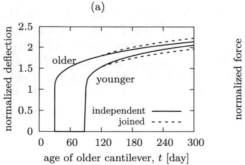
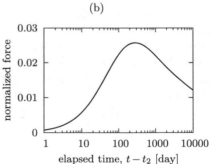

Fig. 4.18 Joined cantilevers of different age: evolution of (a) deflection, (b) force transmitted by the hinge

Example 4.9. *Joining two cantilevers of different age: AAEM approximation*

An approximate solution of Eq. (4.70) from the previous example can easily be constructed using the AAEM method, with t_2 considered as the starting time. For $t > t_2$, the terms on the left-hand side of (4.70) are approximated as

$$\mathscr{J}\{F_c(t)\} \approx J(t, t_2)F_c(t_2^+) + \frac{\Delta F_c(t)}{E''(t, t_2)} \tag{4.71}$$

$$\mathscr{J}_{t-\Delta t}\{F_c(t)\} \approx J(t - \Delta t, t_2 - \Delta t)F_c(t_2^+) + \frac{\Delta F_c(t)}{E''(t - \Delta t, t_2 - \Delta t)} \tag{4.72}$$

where $\Delta F_c(t) = F_c(t) - F_c(t_2^+)$ is the continuous increment of the joining force F_c from the value $F_c(t_2^+)$ just after the change of the structural system to the current value $F_c(t)$. Now it is easy to substitute the approximations (4.71)–(4.72) into (4.70) and solve for the increment

$$\Delta F_c(t) = \frac{J_\Delta^* \dfrac{3\hat{f}L}{8} - \varphi^* F_c(t_2^+)}{\dfrac{1}{E''(t, t_2)} + \dfrac{1}{E''(t - \Delta t, t_2 - \Delta t)}} \tag{4.73}$$

For brevity, we have denoted

$$J_\Delta^* = J_\Delta(t - \Delta t, t_2 - \Delta t, t_1) - J_\Delta(t, t_2, t_1) \tag{4.74}$$

$$\varphi^* = \frac{\varphi(t, t_2)}{E(t_2)} + \frac{\varphi(t - \Delta t, t_2 - \Delta t)}{E(t_2 - \Delta t)} \tag{4.75}$$

Accuracy of the AAEM solution is illustrated in Fig. 4.19. The solid curve represents the accurate numerical solution, already shown in Fig. 4.18b. The internal reaction just after the change of structural system, $F_c(t_2^+)$, is set to zero. Various dashed and dotted curves in Fig. 4.18b show the approximations based on (4.73), with the age-adjusted effective moduli $E''(t, t_2)$ and $E''(t - \Delta t, t_2 - \Delta t)$ evaluated in four different ways: (i) AAEM-eR ... using an "exact" value of relaxation function, obtained numerically, (ii) AAEM-aR ... using an approximate value of relaxation function, obtained from formula (4.58), (iii) AAEM-0.8 ... using a constant value of the aging coefficient, $\chi = 0.8$, (iv) EM ... using aging coefficient $\chi = 1$, which means that the AAEM is replaced by the effective modulus.

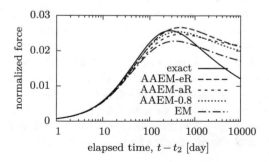

Fig. 4.19 Joined cantilevers of different age: comparison of approximate solutions

The performance of individual methods is qualitatively very similar to Example 4.6, in which the force (stress) evolution was also nonmonotonic. The early response is captured by the AAEM approximation fairly well. However, the accuracy deteriorates at later stages, after the peak. This can be explained by the fact that the histories of stress or strain (and thus also of force or deflection) that are captured by AAEM exactly are, in their continuous part (i.e., with the exception of the initial jump), all monotonic; see Fig. 4.9. If the actual stress (force) history is not monotonic, as is the case here, the approximation becomes too crude and the error increases. This has been illustrated by Fig. 4.15b in Example 4.6. In the range where the force decreases, none of the methods is reliable. The smallest error is obtained with the effective modulus, but this happens only by chance, because the EM method obviously underestimates the changes of the force, both positive and negative.   ■

4.3.2 Stress Redistributions in Beams of Composite Cross Section

In steel–concrete composite beams, as well as composite beams with cross sections consisting of parts of very different age or quality, creep causes stress redistributions within the cross sections, which generally transfer stress to the part creeping less, that is, to the steel part or to the concrete part of higher age or higher strength [80, 87, 110].

Consider a general composite beam consisting of several parts, each of which exhibits different creep and shrinkage properties (a steel part is a special case with zero creep and shrinkage). For simplicity, we assume that the cross section is symmetric with respect to its vertical axis z, but not necessarily symmetric with respect to the horizontal axis y about which the beam bends (the symmetry conditions must, of course, be satisfied not only by the geometric shape of the section but also by the distribution of material properties). Our goal is to derive general relations between the internal forces (normal force N and bending moment M) and the variables characterizing the deformation of an infinitesimal beam segment (axial strain ε_a and curvature κ). For a homogeneous cross section, the moment–curvature relation has already been derived in (4.4). However, for a nonhomogeneous cross section the relaxation operator $\mathscr{R}$ is not independent of the cross-sectional coordinates y and z (because the relaxation functions of different parts are not the same), and so we cannot bring $\mathscr{R}$ in front of the integrals over the cross section. We also consider the influence of shrinkage, and so the stress–strain relation is expressed as $\sigma = \mathscr{R}\{\varepsilon - \varepsilon_{sh}\}$. The shrinkage strain ε_{sh} can vary across the section, i.e., it is a function of the cross-sectional coordinates y and z. The derivation in (4.4) is then modified as follows (for brevity, we do not mark the dependence on time t explicitly):

$$M = \int_A z\sigma(z)\,\mathrm{d}A = \int_A z\mathscr{R}\{\varepsilon(z) - \varepsilon_{\mathrm{sh}}(y,z)\}\,\mathrm{d}A = \int_A z\mathscr{R}\{\varepsilon_{\mathrm{a}} + \kappa z - \varepsilon_{\mathrm{sh}}(y,z)\}\,\mathrm{d}A =$$

$$= \int_A z\mathscr{R}\,\mathrm{d}A\{\varepsilon_{\mathrm{a}}\} + \int_A z^2\mathscr{R}\,\mathrm{d}A\{\kappa\} - \int_A z\mathscr{R}\{\varepsilon_{\mathrm{sh}}(y,z)\}\,\mathrm{d}A = \mathscr{R}_S\{\varepsilon_{\mathrm{a}}\} + \mathscr{R}_I\{\kappa\} + M_{\mathrm{sh}} \quad (4.76)$$

where we have introduced new operators $\mathscr{R}_S$ and $\mathscr{R}_I$, formally defined as

$$\mathscr{R}_S = \int_A z\mathscr{R}\,\mathrm{d}A \quad (4.77)$$

$$\mathscr{R}_I = \int_A z^2\mathscr{R}\,\mathrm{d}A \quad (4.78)$$

This is a compact way of writing that operators $\mathscr{R}_S$ and $\mathscr{R}_I$ are defined by the same general formula (2.23) or (2.24) as the relaxation operator $\mathscr{R}$, but the relaxation function $R(t, t')$ is replaced by $R_S(t, t') = \int_A zR(t, t'; y, z)\,\mathrm{d}A$ and $R_I(t, t') = \int_A z^2 R(t, t'; y, z)\,\mathrm{d}A$, resp. By listing the sectional coordinates y and z explicitly among the arguments of R, we have emphasized that the relaxation function at individual points, in general, depends on their coordinates (because different parts of the section are made of different materials). For a homogeneous cross section, $\mathscr{R}_I$ is equal to $I\mathscr{R}$ where I is the sectional moment of inertia (with respect to coordinate axis y), and if the material is elastic, application of $I\mathscr{R}$ is replaced by multiplication by the bending stiffness EI. The operator $\mathscr{R}_S$ for a homogeneous section equals to $S\mathscr{R}$ where S is the static moment of the section, which vanishes if the coordinate axis y passes through the centroid. For a nonhomogeneous section, the operator $\mathscr{R}_S$ vanishes only in special cases, e.g., if the geometric and material properties of the section are symmetric with respect to the horizontal axis y. If $\mathscr{R}_S$ vanishes, Eq. (4.76) reduces to the moment–curvature relation

$$M = \mathscr{R}_I\{\kappa\} + M_{\mathrm{sh}} \quad (4.79)$$

and the bending effects get decoupled from the axial effects (described by the relation between the normal force and the axial strain, to be derived later).

Symbol M_{sh} that appears in (4.76) and (4.79) stands for the bending moment due to shrinkage,

$$M_{\mathrm{sh}} = -\int_A z\mathscr{R}\{\varepsilon_{\mathrm{sh}}(y,z)\}\,\mathrm{d}A \quad (4.80)$$

This is the moment that would build up from stresses due to restrained shrinkage if the beam segment were forced to remain in the undeformed state (with zero axial strain and zero curvature). If the cross section and the drying process are symmetric with respect to axis y, the moment due to shrinkage vanishes.

As already mentioned, unless the operator $\mathscr{R}_S$ defined in (4.77) happens to vanish, bending and axial deformation of the beam are coupled. It is therefore useful to derive the relation between the normal force N and the parameters that characterize the deformation of the beam segment, ε_{a} and κ. The derivation is fully analogous to (4.76):

$$N = \int_A \sigma(z)\, dA = \int_A \mathscr{R}\{\varepsilon(z) - \varepsilon_{sh}(y, z)\}\, dA = \int_A \mathscr{R}\{\varepsilon_a + \kappa z - \varepsilon_{sh}(y, z)\}\, dA =$$
$$= \int_A \mathscr{R}\, dA\{\varepsilon_a\} + \int_A z\mathscr{R}\, dA\{\kappa\} - \int_A \mathscr{R}\{\varepsilon_{sh}(y, z)\}\, dA = \mathscr{R}_A\{\varepsilon_a\} + \mathscr{R}_S\{\kappa\} + N_{sh} \qquad (4.81)$$

where

$$\mathscr{R}_A = \int_A \mathscr{R}\, dA \qquad (4.82)$$

is yet another operator describing the sectional stiffness, which reduces, for a homogeneous viscoelastic section, to $A\mathscr{R}$, and for an elastic section to multiplication by the normal stiffness EA, and

$$N_{sh} = -\int_A \mathscr{R}\{\varepsilon_{sh}(y, z)\}\, dA \qquad (4.83)$$

is the normal force due to shrinkage.

In summary, the relation between the internal forces in a composite viscoelastic cross section and the deformation parameters is described by (4.81) and (4.76), i.e.,

$$N = \mathscr{R}_A\{\varepsilon_a\} + \mathscr{R}_S\{\kappa\} + N_{sh} \qquad (4.84)$$
$$M = \mathscr{R}_S\{\varepsilon_a\} + \mathscr{R}_I\{\kappa\} + M_{sh} \qquad (4.85)$$

If the operator $\mathscr{R}_S$ vanishes, e.g., due to symmetry, these equations describe separately the axial deformation and bending effects. In a general case, the equations are coupled. In principle, one could decouple them by shifting the y axis such that $\mathscr{R}_S$ vanishes, in the spirit of the method of transformed cross section, but since this particular position of the axis would in general vary in time, the calculation would be quite tedious [73, 167]. Instead of that, it is always possible to solve (4.84)–(4.85) as a set of two equations. For instance, if the cross section is subjected to a bending moment only (with zero normal force), the history of ε_a can be expressed from (4.84) in terms of the history of κ and then substituted into (4.85) to obtain a pure moment–curvature relation. This procedure will be demonstrated by the following example.

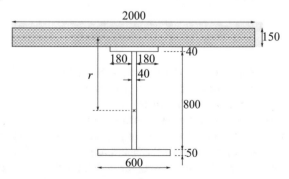

Fig. 4.20 Concrete slab on a steel beam (dimensions in mm)

Example 4.10. Restrained shrinkage of top slab in composite beam

Consider the effect of shrinkage of a top slab of a simply supported steel–concrete composite beam (Fig. 4.20) in which the bending moments from permanent load are large enough to prevent shrinkage cracking of the slab. In that case, the full drying shrinkage and drying creep (or stress-induced shrinkage) takes place. The slab is assumed to be so thin that the bending moment in the slab is negligible. For the sake of simplicity, we assume that the drying and loading start at the same age $t_0 = t_1$.

The problem is a special case of the general composite cross section consisting of two parts (steel and concrete), one of which (steel) does not creep. For this particular problem, the basic Eqs. (4.84)–(4.85) turn out to have the simplest possible form if the origin of the sectional coordinates y and z is placed at the centroid of the concrete slab section. Since the normal force is zero, this does not affect the value of the bending moment (which is normally expressed with respect to the centroidal axis of the entire cross section). Let us introduce the following notation (see also Fig. 4.20): $A_c =$ sectional area of the concrete slab, $A_s =$ sectional area of the steel beam, $I_s =$ moment of inertia of the steel beam with respect to its own centroidal axis, $i_s = \sqrt{I_s/A_s} =$ radius of inertia of the steel beam, $r =$ vertical distance between the centroids of the concrete slab and of the steel beam, $E_s =$ elastic modulus of steel. We also assume that the moment of inertia of the concrete slab with respect to its centroidal axis (which is used here as the coordinate axis y) is negligible. The "sectional" operators defined in (4.82) and (4.77)–(4.78) can now be evaluated as follows:

$$\mathscr{R}_A = \int_A \mathscr{R} \, dA = A_s E_s \mathscr{I} + A_c \mathscr{R}_c \tag{4.86}$$

$$\mathscr{R}_S = \int_A z \mathscr{R} \, dA = r A_s E_s \mathscr{I} \tag{4.87}$$

$$\mathscr{R}_I = \int_A z^2 \mathscr{R} \, dA = (I_s + r^2 A_s) E_s \mathscr{I} = (i_s^2 + r^2) A_s E_s \mathscr{I} \tag{4.88}$$

Here, we have denoted the relaxation operator of concrete as $\mathscr{R}_c$, to formally distinguish it from the position-dependent relaxation operator $\mathscr{R}$, which is equal to $\mathscr{R}_c$ in the concrete part of the section but is equal to $\mathscr{R}_s$ in the steel part of the section. Since the steel is assumed to be elastic, its relaxation operator is $\mathscr{R}_s = E_s \mathscr{I}$ where $\mathscr{I}$ denotes the identity operator (mapping each function onto itself), because then the stress–strain law for steel reduces to Hooke's law: $\sigma = \mathscr{R}_s\{\varepsilon\} = E_s \mathscr{I}\{\varepsilon\} = E_s \varepsilon$.

Furthermore, we can evaluate the normal force and bending moment due to shrinkage according to Eqs. (4.83) and (4.80). The normal force due to shrinkage turns out to be $N_{sh} = -A_c \mathscr{R}_c\{\varepsilon_{sh}\}$ where ε_{sh} is the shrinkage strain in concrete (calculated as the average shrinkage strain for the given thickness of the slab) and the moment due to shrinkage M_{sh} vanishes due to our particular choice of the coordinate system. Now, we can write the basic equations (4.84)–(4.85) with ε_a replaced by $\varepsilon_{cc} =$ strain in the concrete slab[4]:

---

[4]We denote the strain in the concrete slab as ε_{cc} and not simply as ε_c, because ε_c is reserved for the creep strain while ε_{cc} denotes the total strain in concrete.

$$A_s E_s \varepsilon_{cc} + A_c \mathscr{R}_c \{\varepsilon_{cc}\} + r A_s E_s \kappa - A_c \mathscr{R}_c \{\varepsilon_{sh}\} = 0 \tag{4.89}$$

$$r A_s E_s \varepsilon_{cc} + \left(I_s + r^2 A_s\right) E_s \kappa = M \tag{4.90}$$

Here, the normal force N has been set to zero and the bending moment M is considered as given, since for a statically determinate structure it follows from the equilibrium equations. Equation (4.90) has an algebraic character, and it can be used to eliminate the curvature from (4.89). After simple manipulations, we obtain

$$\frac{E_s I_s}{i_s^2 + r^2} \varepsilon_{cc} + A_c \mathscr{R}_c \{\varepsilon_{cc}\} = A_c \mathscr{R}_c \{\varepsilon_{sh}\} - \frac{r}{i_s^2 + r^2} M \tag{4.91}$$

This is an integral equation for the unknown ε_{cc}, which describes the strain history at the beam "axis," but due to our special choice of the coordinate system it is in fact the (average) strain in the concrete slab. To clearly show the nature of this equation, we rewrite it from the compact operator notation into the more explicit notation using integrals:

$$\frac{E_s I_s}{i_s^2 + r^2} \varepsilon_{cc}(t) + A_c R(t, t_1)\varepsilon_{cc}(t_1^+) + A_c \int_{t_1^+}^t R(t, t')\dot{\varepsilon}_c(t') \, dt' =$$

$$= A_c \int_{t_1}^t R(t, t')\dot{\varepsilon}_{sh}(t') \, dt' - \frac{r M(t)}{i_s^2 + r^2} \tag{4.92}$$

When processing the term $\mathscr{R}_c \{\varepsilon_{sh}\}$ on the right-hand side, we have taken into account that the drying is assumed to start at time $t_0 = t_1$. Thus, the shrinkage strain has no jump at time t_1 and the term $R(t, t_1)\varepsilon_{sh}(t_1)$ can be omitted.

Equation (4.92) could be solved numerically, if the integrals are approximated by finite sums. For a quick estimate, it is possible to use the AAEM method and approximate (4.91) by

$$\frac{E_s I_s}{i_s^2 + r^2} \varepsilon_{cc}(t) + A_c R(t, t_1)\varepsilon_{cc}(t_1^+) + A_c E''(t, t_1) \left[\varepsilon_{cc}(t) - \varepsilon_{cc}(t_1^+)\right] =$$

$$= A_c E''(t, t_1)\varepsilon_{sh}(t) - \frac{r M(t)}{i_s^2 + r^2} \tag{4.93}$$

This algebraic equation needs to be solved first at the initial time $t = t_1^+$, just after application of the load, and then, with $\varepsilon_{cc}(t_1^+)$ already known, it can be solved at an arbitrary time $t > t_1$ to get $\varepsilon_{cc}(t)$. The initial strain

$$\varepsilon_{cc}(t_1^+) = -\frac{r}{E_s I_s + (i_s^2 + r^2)A_c E_c(t_1)} M(t_1^+) \tag{4.94}$$

must of course agree with the elastic solution based on the static elasticity modulus of concrete, $E_c(t_1)$. At a general time $t > t_1$, we obtain

$$\Delta\varepsilon_{cc}(t) = \cfrac{\varphi(t, t_1)\varepsilon_{cc}(t_1^+) + \varepsilon_{sh}(t)}{1 + \cfrac{E_s I_s}{(i_s^2 + r^2)A_c E''(t, t_1)}} - \frac{r\,\Delta M(t)}{E_s I_s + (i_s^2 + r^2)A_c E''(t, t_1)} \qquad (4.95)$$

with $\Delta M(t) = M(t) - M(t_1^+)$ denoting the increment of bending moment, which can be set to zero if the applied load remains constant and the structure is statically determinate. Formula (4.95) describes the time evolution of the strain increment in concrete, and the corresponding stress in concrete can be obtained in the spirit of the AAEM method as

$$\sigma(t) = \mathscr{R}_c\{\varepsilon_{cc}(t) - \varepsilon_{sh}(t)\} \approx R(t, t_1)\varepsilon_{cc}(t_1^+) + E''(t, t_1)\,[\Delta\varepsilon_{cc}(t) - \varepsilon_{sh}(t)] \qquad (4.96)$$

It is also easy to express the curvature $\kappa(t)$, e.g., from (4.90), and to evaluate the strain and stress evolution at an arbitrary point of the steel section. Finally, from the distribution of curvature along the beam axis (with moment M depending on the position x), one can compute the deflections.

Note that if the steel area is much larger than the concrete area, we have $A_c/A_s \approx 0$ and the stress change in concrete is the largest possible, i.e., the same as if the concrete slab were fully restrained. On the other hand, if the steel area is much smaller than the concrete area, we have $A_s/A_c \approx 0$, and then there is no stress change in concrete, the same as in the case of free shrinkage.

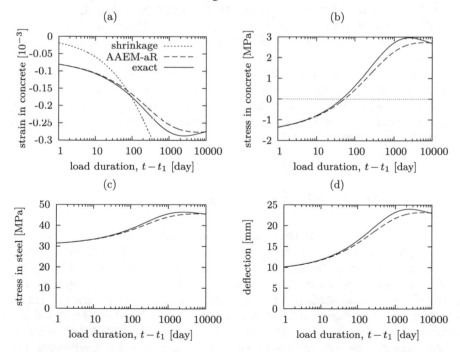

Fig. 4.21 Concrete slab on a steel beam: history of (a) strain in concrete, (b) stress in concrete, (c) maximum stress in steel, (d) midspan deflection

For illustration, Fig. 4.21 shows the history of strain in concrete, stress in concrete, maximum tensile stress in steel, and midspan deflection for the following specific case: sectional dimensions according to Fig. 4.20, concrete properties the same as in Example 3.1, onset of drying and loading by self-weight at age $t_0 = t_1 = 14$ days, environmental humidity $h_{env} = 70\%$, simply supported beam of span $L = 20$ m. The distance between the centroids of the steel beam and concrete slab is $r = 442.3$ mm, and the sectional characteristics are $A_c = 0.3\,\text{m}^2$, $A_s = 0.078\,\text{m}^2$, $I_s = 9.49 \times 10^{-3}\,\text{m}^4$ and $i_s = 348.8$ mm. Young's modulus of steel is considered as $E_s = 210$ GPa and the densities of steel and concrete as $\rho_s = 7850$ and $\rho_c = 2420$ kg/m^3. The self-weight $f = g\rho_c A_c + g\rho_s A_s = 13.13$ kN/m leads to the bending moment at midspan $M_m = fL^2/8 = 656.4$ kNm.

The solid curves in all parts of Fig. 4.21 have been computed by an accurate numerical method, the dashed curves by the AAEM method with the relaxation function approximated by formula (4.58). Just after loading by self-weight, the stress in concrete is compressive (Fig. 4.21b). Even without drying, a partial redistribution of stress from concrete to steel would take place, because concrete creeps but steel does not. Drying shrinkage accelerates this redistribution, and the stress in concrete changes sign after approximately 80 days of loading and drying. Tensile stresses in concrete gradually build up but, once the shrinkage process slows down, they are reduced by relaxation. The maximum tensile stress of 3 MPa (i.e., slightly below the estimated tensile strength, $\bar{f}_t = 3.4$ MPa) is attained approximately at 2500 days. Due to the redistribution, the maximum tensile stress in steel increases from its initial value of 30 MPa to its peak value of 46.4 MPa (Fig. 4.21c). Note that the total strain in concrete is negative (Fig. 4.21a) but, except for the very early stage, is smaller in magnitude than the shrinkage strain, and so the mechanical strain is positive. For comparison, the shrinkage strain is indicated in Fig. 4.21a by the dotted curve. In general, the AAEM approximation provides quite a good estimate of all the quantities of interest. The peak tensile stress in concrete is slightly underestimated, and its relaxation for load durations above 10,000 days (not covered by the figure) would be underestimated as well. ∎

4.3.3   Effects of Nonuniform Drying

The average cross-sectional compliance and shrinkage functions predicted by model B3, or any other model serving similar purposes, have a more limited usefulness than traditionally assumed. The reason is that the drying process, which drives shrinkage and drying creep and causes nonhomogeneity of shrinkage and creep properties throughout the cross section, is usually nonsymmetric and thus leads to a curvature change of the beam axis. Recently, it became clear [558] that neglect of this nonsymmetry has often been one major cause of gross mispredictions of long-time deflections of structures, particularly large-span prestressed concrete bridges. The importance of this problem had not been realized until the late 1990s, after long-time deflection measurements on long-span prestressed box girders became available and systematic occurrence of excessive deflections was documented [560, 836].

Although it has been the standard practice for decades to take into account the nonhomogeneity (or nonuniformity) of shrinkage and creep properties in composite cross sections combining steel and concrete or different concretes, the nonhomogeneity caused by drying has typically been ignored. Only the average shrinkage and creep properties of the concrete parts of the cross section have been considered in calculations. However, these average properties are defined under the assumption of symmetric drying. They cannot capture the coupling between axial deformation and bending of a beam—the main consequence of drying nonsymmetry. This coupling is particularly large in box girders built by the cantilever method, in which the top and bottom slabs often have very different thicknesses, causing order-of-magnitude differences in drying halftimes. The effects of nonuniform drying on creep and shrinkage are referred to as *differential creep* and *differential shrinkage*.

An accurate analysis of the effects of drying requires solving the differential equation for water transport through the pores of concrete. This will be discussed in Chap. 8. However, for cross sections consisting of flanges (slabs, plates), such as box or T cross sections, a simple estimation of the effects of nonuniform drying can usually be made on the basis of model B3, provided that each flange is considered as a distinct body, and that model B3 is applied to each flange **separately**, to determine its average long-time deformation [558]. Let us explain it by a series of simple examples.

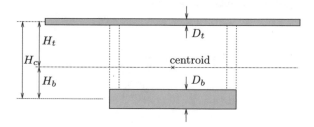

Fig. 4.22 Idealized box section (webs neglected)

Example 4.11. Idealized webless box: differential shrinkage

Figure 4.22 shows a segmented cantilever, having an idealized cross section consisting only of top and bottom slabs, with massless webs. Applying model B3 to each slab separately, and assuming the same environmental conditions for each, the longitudinal shrinkage strains of the top and bottom slabs evolve as

$$\varepsilon_t(\hat{t}) = -\varepsilon_\infty \tanh \sqrt{\hat{t}/\tau_t}, \quad \varepsilon_b(\hat{t}) = -\varepsilon_\infty \tanh \sqrt{\hat{t}/\tau_b} \tag{4.97}$$

respectively, where $\hat{t}$ is the drying duration, ε_∞ is the magnitude of the final shrinkage, and τ_t and τ_b are the drying halftimes of the top and bottom slabs. The final shrinkage is given by $\varepsilon_\infty = k_h \varepsilon_{sh}^\infty$ where k_h depends on the ambient relative humidity, and ε_{sh}^∞ depends on the concrete composition and curing; see Eqs. (3.15) and (3.19) and Table C.2. According to formula (3.17), based on the diffusion theory, we have

$$\tau_t = k_t (k_s D_t)^2, \quad \tau_b = k_t (k_s D_b)^2 \tag{4.98}$$

where k_t is a material parameter, k_s is the shape factor, for slabs approximately equal to 1, and D_t and D_b are the thicknesses of the slabs. Consequently,

$$\tau_b = \left(\frac{D_b}{D_t}\right)^2 \tau_t \tag{4.99}$$

Strictly speaking, one should consider $\varepsilon_{\mathrm{sh}}^\infty$ as dependent on the shrinkage halftime and thus on the slab thickness; see the empirical formula in line 7 of Table C.2. This is only a minor effect which, in the present crude approximation, can be neglected (this assumption will be verified in Example 4.15).

If $D_b > D_t$, the top slab is drying faster and shrinkage produces a positive girder curvature

$$\kappa_{\mathrm{sh}}(\hat{t}) = \frac{\varepsilon_b(\hat{t}) - \varepsilon_t(\hat{t})}{H_c} = \frac{\varepsilon_\infty}{H_c} \left(\tanh \sqrt{\hat{t}/\tau_t} - \tanh \sqrt{\hat{t}/\tau_b} \right) \tag{4.100}$$

where H_c is the distance between the centroids of the top and bottom slabs; see Fig. 4.22. The bending stiffness of these slabs is assumed to be much smaller than the bending stiffness of the whole cross section.

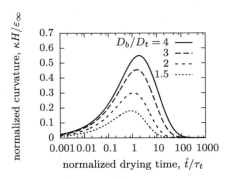

Fig. 4.23 Evolution of curvature due to differential shrinkage

From (4.99), we note that if, for instance, $D_b/D_t = 3$ (which is quite typical), the second, negative, term in (4.100) has a halftime $9\times$ longer than the first. Therefore, drying causes initially a positive curvature, which later decreases. If the segment is a part of a cantilever, positive curvature contributes to lifting of the cantilever end, and the decrease of curvature at later stages contributes to sinking of the cantilever end. For several ratios D_b/D_t, the evolution of curvature according to (4.100) is plotted in Fig. 4.23. ∎

Consider a cantilever of length L, with its left end at $x = 0$ clamped. The shrinkage deflection at the right end at $x = L$ is

$$w_{sh}(\hat{t}) = -\int_0^L (L-x)\kappa_{sh}(x,\hat{t})\,dx \qquad (4.101)$$

The negative sign corresponds to upward deflection (lifting). Since the sectional dimensions are often variable along the beam axis, the curvature depends on the axial coordinate x. For given functions $D_t(x)$, $D_b(x)$ and $H_c(x)$, the integral in (4.101) can be evaluated numerically. The complete formula obtained by combining (4.98), (4.100), and (4.101) reads

$$w_{sh}(\hat{t}) = -\varepsilon_\infty \int_0^L \frac{L-x}{H_c(x)} \left(\tanh \sqrt{\frac{\hat{t}}{k_t k_s^2 D_t^2(x)}} - \tanh \sqrt{\frac{\hat{t}}{k_t k_s^2 D_b^2(x)}} \right) dx \quad (4.102)$$

One should also consider that different segments of the cantilever can be exposed to drying at different times. Introducing a global time variable t, defined for instance as the age of the oldest segment, and denoting the (global) time at the onset of drying of the segment at position x as $t_0(x)$, we can rewrite (4.102) in a still more general form,

$$w_{sh}(t) = -\varepsilon_\infty \int_0^L \frac{L-x}{H_c(x)} \left(\tanh \sqrt{\frac{t-t_0(x)}{k_t k_s^2 D_t^2(x)}} - \tanh \sqrt{\frac{t-t_0(x)}{k_t k_s^2 D_b^2(x)}} \right) dx \quad (4.103)$$

Nonuniform drying leads not only to changes of curvature due to nonuniform shrinkage but also to differences in drying creep compliance of individual slabs. The growth of drying creep compliance of the thicker slab is delayed with respect to the thinner slab. This has an effect on the evolution of the bending compliance of the cross section.

Example 4.12. Idealized webless box: differential creep

For illustration, consider again the idealized cross section from Fig. 4.22, with areas of the top and bottom slabs denoted as A_t and A_b and the total area denoted as $A = A_t + A_c$. Suppose that the section is subjected to a bending moment M and normal force N (the internal forces are considered here as the resultants of stresses in concrete only, including the stresses in concrete generated by prestressing). The stresses in each slab can be considered approximately as uniform (neglecting again the bending stiffnesses of the slabs) and are given by

$$\sigma_t = \frac{N}{A} - \frac{M}{H_c A_t}, \qquad \sigma_b = \frac{N}{A} + \frac{M}{H_c A_b} \qquad (4.104)$$

The growth of the corresponding average strains ε_t and ε_b in the slabs is determined by compliance functions J_t and J_b, which can differ due to the influence of the thickness on the drying halftime and thus on the drying creep (note that J_b in this example refers to the total compliance function of the **bottom** slab, and not to the **basic** compliance function).

If the internal forces arise at age t_1 and remain constant afterward, the evolution of strains is given by

$$\varepsilon_t(t) = \sigma_t J_t(t, t_1), \qquad \varepsilon_b(t) = \sigma_b J_b(t, t_1) \qquad (4.105)$$

and the corresponding axial strain a curvature can be evaluated as

$$\varepsilon_a(t) = \frac{A_b \varepsilon_b(t) + A_t \varepsilon_t(t)}{A} = \frac{A_b J_b(t, t_1) + A_t J_t(t, t_1)}{A^2} N + \frac{J_b(t, t_1) - J_t(t, t_1)}{A H_c} M \qquad (4.106)$$

$$\kappa(t) = \frac{\varepsilon_b(t) - \varepsilon_t(t)}{H_c} = \frac{J_b(t, t_1) - J_t(t, t_1)}{A H_c} N + \frac{1}{H_c^2} \left(\frac{J_b(t, t_1)}{A_b} + \frac{J_t(t, t_1)}{A_t} \right) M$$

$$(4.107)$$

The factor multiplying N on the right-hand side of (4.106) represents the axial compliance of the section, C_A, and the factor multiplying M on the right-hand side of (4.107) represents the bending compliance of the section, C_I. The factor multiplying M in (4.106) is the same as the factor multiplying N in (4.107), and it represents a coupling compliance, C_S, which characterizes the interaction between axial deformation and bending. If the creep compliance functions of both slabs are the same, $J_t = J_b = J$, the sectional compliances are $C_A = J(t, t_1)/A$, $C_I = J(t, t_1)/I$ and $C_S = 0$, where $I = H_c^2 A_t A_b / A$ is the moment of inertia of the section with respect to its horizontal centroidal axis. In this case, bending is uncoupled from the axial deformation, same as in elasticity, and the axial and bending compliances grow proportionally to the (unique) creep compliance function. Differences between the drying creep compliances of the top and bottom slabs lead to a somewhat different evolution of the compliances and to coupling between axial deformation and bending. Let us now examine the errors induced by neglecting the differential creep effects, i.e., by setting both J_t and J_b to the creep compliance function J that corresponds to the overall effective thickness.

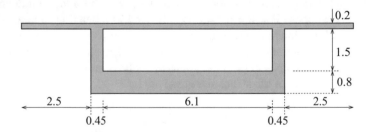

Fig. 4.24 Pier cross section of a typical box girder (dimensions in meters)

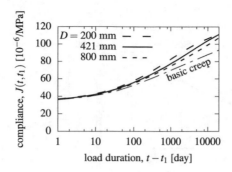

Fig. 4.25 Creep compliance functions for three different slab thicknesses

For the purpose of quantitative comparison, consider a typical pier section of a box girder in Fig. 4.24 with $D_t = 0.2$ m, $D_b = 0.8$ m, $A_t = 2.4$ m^2, $A_b = 5.6$ m^2, $A = 8$ m^2, and $H_c = 2.0$ m. The overall effective thickness of the section (with webs neglected) is evaluated from the condition $A/D = A_t/D_t + A_b/D_b$ as $D = 0.421$ m. The concrete properties, curing, and humidity are taken the same as in Example 3.1, and the age at loading is set to $t_1 = 28$ days. Figure 4.25 shows the creep compliance functions corresponding to the top slab thickness, average thickness, and bottom slab thickness. The drying halftimes are $\tau_t = 1, 121$ days, $\tau_{aver} = 4, 968$ days and $\tau_b = 17, 936$ days. Figure 4.26 shows the sectional compliances evaluated from formulae (4.106)–(4.107) (solid curves), or estimated as $J(t, t_1)/A$ and $J(t, t_1)/I$ from the compliance function J that corresponds to the overall effective thickness (dashed curve). It turns out that the maximum relative error is only 2.8% for the axial compliance and 3.5% for the bending compliance. A similar analysis has been done for a real case—the main pier cross section of the Koror–Babeldaob Bridge in Palau, for which $D_t = 280$ mm and $D_b = 1150$ mm; see Fig. 7.3 in Chap. 7. The maximum relative error turns out to be 3.2% for the axial compliance and 4.1% for the bending compliance.

To complete the analysis, one should also check the error arising from the coupling compliance, C_S. Here, it does not make sense to evaluate the relative error because the approximation leads to $C_S = 0$. The coupling compliance can be expected to be small, but what should it be compared to? What matters is whether the curvature due to the normal force can be comparable to the curvature caused by other effects (bending moment and differential shrinkage). This will be addressed in the next example. ∎

Example 4.13. Idealized webless box: differential creep and shrinkage combined

To assess the relative importance of differential creep and differential shrinkage, let us consider both effects simultaneously. Of course, the behavior depends on the actual combination of internal forces, which should correspond to a reasonable stress distribution. We will examine two extreme cases:

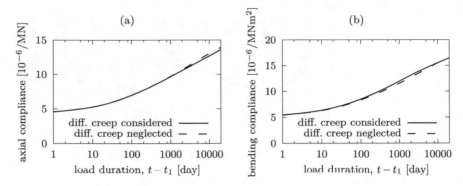

Fig. 4.26 (a) Axial and (b) sectional bending compliances evaluated either from the compliance function corresponding to the average thickness, or using more accurate expressions from formulae (4.106)–(4.107)

1. the girder is prestressed such that the stress in concrete resulting from a combination of prestressing and dead load is uniform and equal to $-0.4 f_c'$;
2. the girder is prestressed such that the stress in concrete vanishes at the centroid of the top slab and equals $-0.4 f_c'$ at the centroid of the bottom slab.

In **case 1**, the bending moment vanishes and the normal force is $N = -0.4 f_c' A$. The resulting curvature can be expressed as

$$\kappa(t) = \frac{\varepsilon_\infty}{H_c} \left(\tanh \sqrt{\frac{t - t_0)}{\tau_t}} - \tanh \sqrt{\frac{t - t_0}{\tau_b}} \right) - \frac{0.4 f_c'}{H_c} \left(J_b(t, t_1) - J_t(t, t_1) \right)$$

$$(4.108)$$

where the first term is the contribution of differential shrinkage according to (4.100) and the second term is the contribution of differential creep according to (4.107). As usual, t_0 is the age at the onset of drying and t_1 is the age at loading. The curvature is seen to be inversely proportional to the distance H_c between the centroids of the top and bottom slabs. The relative error due to various simplifications is properly reflected by the dimensionless curvature, $H_c\kappa$, which is in fact equal to the difference of strains in bottom and top slabs, $\varepsilon_b - \varepsilon_t$.

For the concrete from Example 3.1 (with characteristic strength $f_c' = \bar{f}_c - 8$ MPa $= 37.4$ MPa) and box cross section from Fig. 4.24, environmental humidity $h_{env} = 70\%$, and times $t_0 = 7$ days and $t_1 = 28$ days, the dimensionless curvature evaluated according to the "exact" formula (4.108) is plotted in Fig. 4.27a by the solid curve, marked as DC+DS, which means that both differential creep and differential shrinkage are included. The dashed curve (DS) and dotted curve (DC) show separately the contribution of differential shrinkage and differential creep. It is seen that the effect of differential creep on the curvature is less pronounced than, but still comparable to, the effect of differential shrinkage. Both of them contribute to positive curvatures (convex shape of the deflection curve). This is logical, since the top slab is thinner, dries out faster, and thus the negative strains due to creep and shrinkage are larger in magnitude than those in the bottom slab.

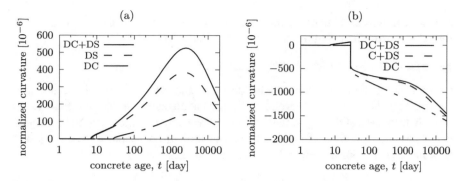

Fig. 4.27 Evolution of dimensionless curvature and the effect of neglecting differential shrinkage and differential creep: (a) for uniform compressive stress, (b) for zero stress in top slab and compressive stress in bottom slab (note the different vertical scales)

In **case 2**, the strain in the top slab is just the shrinkage strain (evaluated for effective thickness D_t) and the strain in the bottom slab is a sum of the shrinkage strain (for effective thickness D_b) and creep strain, including drying creep. The corresponding time evolution of curvature is given by

$$\kappa(t) = \frac{\varepsilon_\infty}{H_c} \left(\tanh \sqrt{\frac{t - t_0}{\tau_t}} - \tanh \sqrt{\frac{t - t_0}{\tau_b}} \right) - \frac{0.4 f_c'}{H_c} J_b(t, t_1) \qquad (4.109)$$

For the same conditions as in case 1, the dimensionless curvature evaluated according to (4.109) is plotted in Fig. 4.27b by the solid curve (DC+DS). The dashed curve (C+DS) shows the result that would be obtained with differential shrinkage taken into account but creep evaluated from the compliance function that corresponds to the average thickness (of course under bending, it does make sense to neglect creep completely). Finally, the dotted curve (DC) shows what would be obtained if differential creep is considered but differential shrinkage neglected.

The curvatures are negative (concave deflection curve), except for the period between the onset of drying and the onset of loading, when there is no creep and the differential shrinkage produces the same effect as in case 1. Thus, the initial part of the solid curve is the same as in Fig. 4.27a. Note that the two graphs displayed in Fig. 4.27 are plotted in a different scale on the vertical axis. After the onset of loading by a combination of compressive normal force and negative bending moment (case 2, Fig. 4.27b), large negative curvatures occur. Their magnitude is alleviated mainly by the effects of differential shrinkage while differential creep plays only a minor role. Neglecting differential creep leads in this case to a relative error (in curvature) not exceeding 7%. On the other hand, neglecting differential shrinkage would lead to a relative error of up to 47%. ∎

The example has shown that the effect of differential shrinkage on the curvature of prestressed box girders can be important while the effect of differential creep (on curvature) is less pronounced. However, differential creep may also have an influence on the axial strain and may lead to warping of the floors in tall buildings with massive exterior columns and relatively thin interior core walls, even if they are exposed to the same environmental humidity. This will be illustrated next.

Example 4.14. Tall building: differential creep and shrinkage

Consider a tall building with exterior columns and interior walls that have been designed for the same stress level. For simplification, suppose that the stress varies linearly from zero at the top to the maximum level $\sigma < 0$ at the base but the section is constant (designed for the base). Then, the vertical displacement at the top of a column or wall of initial height H can be evaluated as

$$w(t) = -H \left(\frac{\sigma}{2} J(t, t_1) + \varepsilon_{\text{sh}}(t - t_0) \right) \tag{4.110}$$

and the difference in vertical displacements of the wall and column is

$$\Delta w(t) = w_w(t) - w_c(t) = -H \left(\frac{\sigma}{2} (J_w(t, t_1) - J_c(t, t_1)) + \varepsilon_{\text{sh},w}(t - t_0) - \varepsilon_{\text{sh},c}(t - t_0) \right) =$$

$$= -\frac{H\sigma}{2} (J_w(t, t_1) - J_c(t, t_1)) + H\varepsilon_\infty \left(\tanh \sqrt{\frac{t - t_0}{\tau_w}} - \tanh \sqrt{\frac{t - t_0}{\tau_c}} \right) \tag{4.111}$$

For the concrete from Example 3.1, stress $\sigma = -0.4 f'_c = -15$ MPa, building height $H = 200$ m, effective thicknesses $D_c = 1$ m and $D_w = 0.15$ m, ambient humidity $h_{\text{env}} = 60\%$, and times $t_0 = 7$ days and $t_1 = 28$ days, the evolution of Δw is plotted in Fig. 4.28a. The displacement difference culminates after 5–6 years and attains about 125 mm.

For comparison, Fig. 4.28b shows analogous results for high-strength concrete characterized by $\bar{f}_c = 95$ MPa, $w/c = 0.31$, $a/c = 3.11$ and $c = 530$ kg/m³. The stress level is set to $\sigma = -35$ MPa, and the building height is considered as $H = 400$ m while all the other parameters remain the same. The shape of the curves is similar to the previous case, but the maximum displacement difference is about 237 mm. In actual design, the cross-sectional area will decrease with the height to

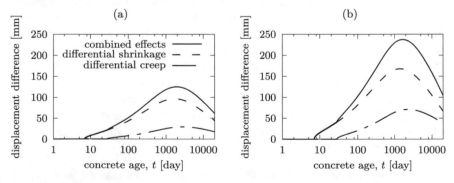

Fig. 4.28 Evolution of displacement difference between exterior columns and interior walls in a tall building, caused by differential creep (DC) and differential shrinkage (DS): (a) normal-strength concrete and building height of 200 m, (b) high-strength concrete and building height of 400 m

keep the stress nearly uniform, and then the displacement difference will, of course, be larger, though not much larger because the shrinkage contributes to the displacement difference much more than does the creep. ∎

Examples 4.11–4.13 presented a simplified analysis, illustrating the basic trends. In real situations, the cross section of a box girder needs to be subdivided into at least 3 parts: top slab, bottom slab, and webs. Because each has a different effective thickness D_i, model B3 yields different shrinkage and creep for each. Obviously, one may also take into account different average environmental humidities and temperatures (during exposure to, or shielding from, the sun; sealing by asphalt pavement; kind of ventilation inside the box; etc.), and possible differences in concrete compositions and ages.

Rigorous analysis of the problem can be based on general equations for a composite cross section, summarized in (4.84)–(4.85). Each part of the section is now characterized by a certain compliance function $J_i(t, t')$ and shrinkage strain evolution $\varepsilon_{sh,i}(\hat{t})$, both dependent on the thickness D_i. Let us denote the area of part number i as A_i, its static moment with respect to the centroidal axis (of the entire section) as S_i, and its moment of inertia with respect to the centroidal axis as I_i. For each part of the section, the compliance function $J_i(t, t')$ uniquely determines the corresponding relaxation function $R_i(t, t')$ and relaxation operator $\mathscr{R}_i$. Operators used in (4.84)–(4.85) and defined in (4.82) and (4.77)–(4.78) are then expressed as

$$\mathscr{R}_A = \int_A \mathscr{R} \, dA = \sum_i A_i \mathscr{R}_i \tag{4.112}$$

$$\mathscr{R}_S = \int_A z \mathscr{R} \, dA = \sum_i S_i \mathscr{R}_i \tag{4.113}$$

$$\mathscr{R}_I = \int_A z^2 \mathscr{R} \, dA = \sum_i I_i \mathscr{R}_i \tag{4.114}$$

where the sum is taken over all the parts of the cross section.

If operators $\mathscr{R}_i$ are indeed different, Eqs. (4.84)–(4.85) for the normal force and bending moment are coupled (this ocurred in Example 4.12, see Eqs. (4.106)–(4.107)). Their solution would require step-by-step numerical procedures or approximation by the AAEM method. However, as illustrated by Examples 4.12 and 4.13 (see Figs. 4.26 and 4.27), the error induced by neglecting differential creep is often acceptable (if not negligible). If all parts of the section have the same properties and age, and if differential creep is neglected, all operators $\mathscr{R}_i$ become identical and equal to the average operator $\mathscr{R}$, which corresponds to the creep compliance function determined for the overall effective thickness of the entire section. The big advantage is that operator

$$\mathscr{R}_S = \int_A z \mathscr{R} \, dA = \sum_i S_i \mathscr{R}_i = \left(\sum_i S_i \right) \mathscr{R} = S \mathscr{R} = 0 \tag{4.115}$$

in this case vanishes, because $S = \sum_i S_i$ is the static moment of the entire section with respect to its centroidal axis, and Eqs. (4.84)–(4.85) get uncoupled. Operators $\mathscr{R}_A$ and $\mathscr{R}_I$ reduce to operator $\mathscr{R}$ multiplied by the cross-sectional area A and moment of inertia I, respectively, and Eq. (4.85) reduces to the simple moment–curvature relation (4.79). The effects of differential shrinkage are reflected in the definition of internal forces due to shrinkage, Eqs. (4.80) and (4.83), which can now be rewritten as

$$N_{\text{sh}} = -\int_A \mathscr{R}\{\varepsilon_{\text{sh}}(y,z)\}\, dA = -\mathscr{R}\left\{\int_A \varepsilon_{\text{sh}}(y,z)\, dA\right\} = -\mathscr{R}\left\{\sum_i A_i \varepsilon_{\text{sh},i}\right\} \quad (4.116)$$

$$M_{\text{sh}} = -\int_A z\mathscr{R}\{\varepsilon_{\text{sh}}(y,z)\}\, dA = -\mathscr{R}\left\{\int_A z\varepsilon_{\text{sh}}(y,z)\, dA\right\} = -\mathscr{R}\left\{\sum_i S_i \varepsilon_{\text{sh},i}\right\} \quad (4.117)$$

The relations between internal forces and deformation variables are then given by

$$N = \mathscr{R}_A\{\varepsilon_a\} + N_{\text{sh}} = A\mathscr{R}\{\varepsilon_a\} - \mathscr{R}\left\{\sum_i A_i \varepsilon_{\text{sh},i}\right\} = A\mathscr{R}\left\{\varepsilon_a - \varepsilon_{a,\text{sh}}\right\} \quad (4.118)$$

$$M = \mathscr{R}_I\{\kappa\} + M_{\text{sh}} = I\mathscr{R}\{\kappa\} - \mathscr{R}\left\{\sum_i S_i \varepsilon_{\text{sh},i}\right\} = I\mathscr{R}\{\kappa - \kappa_{\text{sh}}\} \quad (4.119)$$

where

$$\varepsilon_{a,\text{sh}} = \frac{1}{A}\sum_i A_i \varepsilon_{\text{sh},i} \quad (4.120)$$

is the axial strain induced by differential shrinkage and

$$\kappa_{\text{sh}} = \frac{1}{I}\sum_i S_i \varepsilon_{\text{sh},i} \quad (4.121)$$

is the curvature induced by differential shrinkage.

Note that Eq. (4.100) exploited in Example 4.11 is a special case of (4.121), with $S_1 = -A_t H_t$, $S_2 = A_b H_b = -S_1$, $\varepsilon_{\text{sh},1} = \varepsilon_t$, $\varepsilon_{\text{sh},2} = \varepsilon_b$ and I approximated by $A_t H_t^2 + A_b H_b^2$, where H_t and H_b are the distances of the centroids of the top and bottom slabs from the centroid of the entire section, and $H_t + H_b = H_c$. Let us check how inclusion of the web affects the results of Example 4.11.

Example 4.15. Complete box girder cross section with nonuniform drying

Consider a typical section of a box girder in Fig. 4.24, divided into three parts: top slab, bottom slab, and webs, denoted by subscripts 1, 2, and 3. The centroid of the section is found to be at the distance of 1.421 m from the top fibers, and the centroidal moment of inertia is $I = 7.629$ m^4. The areas, static moments, and effective thicknesses of individual parts are summarized in Table 4.1. The table also contains the estimated shrinkage halftimes and final shrinkage values (referring to a completely dry environment) for the same concrete and the same curing conditions as in Example 3.1. Despite the dramatic differences in shrinkage halftimes, the final shrinkage values differ by a fraction of a percent, which confirms that the simplifying assumption of a unique value of final shrinkage in Example 4.11 was, for practical purposes, fully justified. Recall that $\varepsilon_{\text{sh}}^{\infty}$ denotes the theoretical magnitude of final shrinkage in a completely dry environment, and it needs to be multiplied by a humidity-dependent factor to get the final shrinkage at the actual humidity.

Table 4.1 Geometric properties and shrinkage characteristics for individual parts of the box girder cross section from Fig. 4.24

part	i	A_i [m^2]	S_i [m^3]	D_i [m]	$\tau_{\text{sh},i}$ [day]	$\varepsilon_{\text{sh},i}^{\infty}$ [10^{-6}]
Top slab	1	2.4	−3.1694	0.2	1121	701.11
Bottom slab	2	5.6	3.8047	0.8	17936	699.76
Webs	3	1.35	−0.6353	0.45	5675	699.96

The evolution of axial strain and curvature caused by shrinkage can now be evaluated from an expanded version of formulae (4.120)–(4.121),

$$\varepsilon_{\text{a,sh}}(\hat{t}) = -\frac{k_h}{A} \sum_{i=1}^{3} A_i \varepsilon_{\text{sh},i}^{\infty} \tanh \sqrt{\frac{\hat{t}}{\tau_{\text{sh},i}}} \qquad (4.122)$$

$$\kappa_{\text{sh}}(\hat{t}) = -\frac{k_h}{I} \sum_{i=1}^{3} S_i \varepsilon_{\text{sh},i}^{\infty} \tanh \sqrt{\frac{\hat{t}}{\tau_{\text{sh},i}}} \qquad (4.123)$$

in which $k_h = 1 - h_{\text{env}}^3 = 1 - 0.7^3 = 0.657$ is the factor that takes into account the ambient relative humidity of 70%; see formula (3.19). The dependence of axial strain and curvature on the drying time is represented by the solid curves in Fig. 4.29. For comparison, the dashed curves show the approximations that would be obtained for an idealized webless section. For curvature, the corresponding formula (4.100) has been derived in Example 4.11; it could also be obtained from (4.123) by setting $A = A_1 + A_2$, $H_1 = H_c A_2/A$, $H_2 = H_c A_1/A$, $S_1 = -A_1 H_1$, $S_2 = A_2 H_2 = -S_1$, $S_3 = 0$, and $I = A_1 H_1^2 + A_2 H_2^2$ (section consisting of top and bottom slabs only). In a similar fashion, an approximate formula for axial strain can be derived from (4.122) by setting $A_3 = 0$ and $A = A_1 + A_2$.

It can be expected that the simplified method neglecting the webs overestimates the actual curvature. Free shrinkage of the webs would lead to their uniform contraction, but if the top slab shrinks more than the bottom slab, the webs are forced into bending and their flexural stiffness reduces the resulting curvature (as compared to the case without webs). Indeed, in the present example, the simplified method overestimates the maximum curvature by about 11%; see Fig. 4.29b. For the axial strain, the error induced by neglecting the web is very small; see Fig. 4.29a.

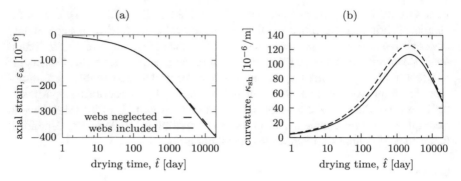

Fig. 4.29 Evolution of (a) axial strain, (b) curvature caused by differential shrinkage; approximation based on (4.100), with webs neglected, and more accurate results based on (4.123), with webs accounted for

Since the shrinkage is now constrained by the condition that the section remain planar (which did not represent a constraint for the webless section), self-equilibrated stresses are induced in the section. They can be calculated by combining the general viscoelastic stress–strain Eq. (2.45) with formula (4.2) for linear distribution of normal strain across the section, in which the axial strain ε_a and curvature κ are set to their values caused by shrinkage:

$$\sigma(z, t) = \mathscr{R}\{\varepsilon(z, t) - \varepsilon_{sh}(z, t)\} = \mathscr{R}\{\varepsilon_{a,sh}(t) + \kappa_{sh}(t)z - \varepsilon_{sh}(z, t)\} =$$

$$= -k_h \sum_{i=1}^{3} \left(\frac{A_i}{A} + \frac{zS_i}{I} \right) \varepsilon_{sh,i}^{\infty} \mathscr{R}\left\{ \tanh\sqrt{\frac{t - t_0}{\tau_{sh,i}}} \right\} + k_h \varepsilon_{sh,k(z)}^{\infty} \mathscr{R}\left\{ \tanh\sqrt{\frac{t - t_0}{\tau_{sh,k(z)}}} \right\}$$

(4.124)

Here, $k(z)$ is an integer-valued function that provides the number of the section part to which the point with coordinate z belongs (in general, k could also depend on the y coordinate, but in the present case, the individual parts are bounded by horizontal lines, and thus, the vertical coordinate uniquely determines the corresponding part).

As usual, application of the relaxation operator can be approximated by the AAEM method, which leads to

$$\sigma(z, t) \approx E''(t, t_0)\left[\varepsilon_{a,sh}(t) + \kappa_{sh}(t)z - \varepsilon_{sh}(z, t)\right] =$$

$$= k_h E''(t, t_0)\left[-\sum_{i=1}^{3} \left(\frac{A_i}{A} + \frac{zS_i}{I} \right) \varepsilon_{sh,i}^{\infty} \tanh\sqrt{\frac{t - t_0}{\tau_{sh,i}}} + \varepsilon_{sh,k(z)}^{\infty} \tanh\sqrt{\frac{t - t_0}{\tau_{sh,k(z)}}} \right]$$

(4.125)

For the specific concrete, section, and conditions considered in this example, the evolution of stresses due to differential shrinkage at four characteristic points of the section is plotted in Fig. 4.30. The selected points are the centroid of the top slab

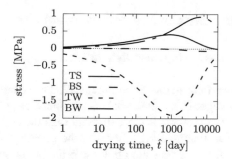

Fig. 4.30 Evolution of stresses caused by differential shrinkage

(TS), centroid of the bottom slab (BS), top of the webs (TW), and bottom of the webs (BW). As seen in Fig. 4.30, shrinkage-induced stresses in the top slab and at the bottom of the webs are tensile while stresses at the top of the web and in the bottom slab are compressive. This is logical, since free shrinkage would lead to the fastest growth of strain magnitude in the top slab, slower growth in the webs, and the slowest growth in the bottom slab (see the drying halftimes in Table 4.1). To ensure planarity of the section, the top slab must be stretched by tensile stresses and the bottom slab must be shortened by compressive stresses, and the resulting negative bending moment is compensated by a positive moment in the webs, leading to tension in the bottom part of the web and compression in the top part. The top slab dries out relatively fast, and the maximum tensile stress in the top slab develops approximately after 1000 days. Simultaneously, the compressive stress in the top part of the webs attains its maximum. The development of tensile stress in the bottom part of the web is delayed because it is mainly driven by the difference between shrinkage in the web and in the bottom slab. It is also interesting to note that the compressive stress in the bottom slab remains very small. This is because the bottom slab is massive (it accounts for 60% of the total area) and dries out extremely slowly, with a halftime of almost 50 years.

Finally, it should be noted that the self-equilibrated stresses due to shrinkage have been evaluated here using the assumption that shrinkage is uniform through the thickness of each of the section parts (top slab, bottom slab, webs). Such an assumption is implicitly contained in the adopted sectional approach. In reality, the drying process is faster near the surface and slower in the core, which leads to additional self-equilibrated stresses, tensile near the surface, and compressive in the core. Their evaluation requires more detailed modeling of the drying process and will be discussed in Sect. 8.6. ∎

The foregoing results refer to typical concretes and environmental humidities. There exist concretes with a significantly higher drying creep and structures exposed to lower humidities. For these, the differential effects would be stronger.

4.3.4   Stress Relaxation in Prestressed Members

4.3.4.1   Model of Prestressing Steel Relaxation

Over time, the prestress force in a steel tendon installed in a concrete member relaxes due to creep and shrinkage of concrete as well as viscoplasticity of steel, manifested as prestress relaxation. The problem is more difficult than the problem of a composite cross section treated in Example 4.10, because the prestressing tendons exhibit a time-dependent behavior that does not obey the principle of superposition. Before analyzing the complex processes in a prestressed concrete member, it is necessary to postulate an appropriate constitutive law governing the behavior of the tendons.

For prestress relaxation at constant strain and temperature, Sect. 2.3.4.5 of the CEB Model Code 1990 [322] specifies the formula

$$\frac{\sigma_{p0} - \sigma_p(t)}{\sigma_{p0}} = \rho_{1000} \left(\frac{t}{\lambda_1}\right)^k \tag{4.126}$$

where σ_p is the stress in the steel tendon; $\sigma_{p0} = \sigma_p(0)$ is the initial prestress, i.e., the stress in tendon when the prestressing force is transferred to the anchor (which is usually also the maximum stress ever experienced by the tendon); t is the time elapsed since the transfer of prestress force onto concrete; $\lambda_1 = 1000$ h; and ρ_{1000} and k are parameters. Note that the difference $\sigma_{p0} - \sigma_p(t)$ represents the prestress loss at time t, and the fraction on the left-hand side of (4.126) is the relative prestress loss, also called the *relaxation ratio*, $\rho(t)$.

Parameter ρ_{1000} represents the relaxation ratio after $t = 1000$ h. Its recommended values are different for three classes of prestressing steel: normal wires and strands (class 1), improved ones (class 2), and bars (class 3); see Table 4.2. The value of ρ_{1000} is also affected by the initial prestress level, i.e., by the ratio σ_{p0}/f_p where f_p is the *ultimate strength* of prestressing steel. Parameter k can be estimated from relaxation tests as $k \approx \log_{10}(\rho_{1000}/\rho_{100})$ where ρ_{100} is the value of relaxation ratio measured after $t = 100$ h of relaxation. It is recommended to set $k = 0.12$ for class 1 and $k = 0.19$ for class 2. An example of the evolution of prestress (normalized by the ultimate strength f_p) for class 1 of prestressing steel and for different values of initial prestress is plotted in Fig. 4.31a. Note that the time scale is logarithmic and the time variable t is given in days.

Table 4.2 Values of parameter ρ_{1000} for three classes of prestressing steel and various levels of initial prestress (according to the CEB Model Code)

Prestressing steel	Class	σ_{p0}/f_p		
		0.6	0.7	0.8
Normal wires and strands	1	4%	8%	12%
Improved wires and strands	2	1%	2%	5%
Bars	3	2%	4%	7%

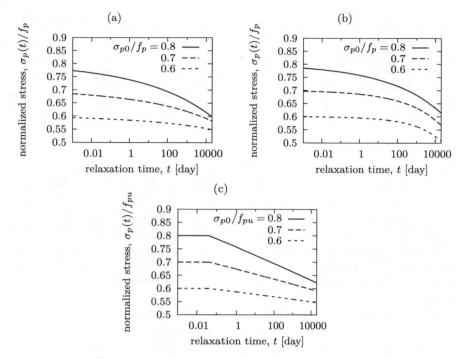

Fig. 4.31 Prestress relaxation according to (a) CEB formula (4.126), for class 1 ($k = 0.12$), (b) Eurocode formulae (4.126)–(4.128), for class 1 ($A = 43.12 \times 10^{-5}$, $B = 6.7$), (c) formula (4.129) used in American practice, for stress-relieved tendons ($f_{py}/f_{pu} = 0.85$)

The Eurocode [373] uses the same power law (4.126) as the CEB Model Code, but the recommended values of parameters ρ_{1000} and k are different. The relaxation ratio after 1000 h is given by

$$\rho_{1000} = Ae^{B\sigma_{p0}/f_p} \tag{4.127}$$

where parameters A and B depend on the class and are listed in Table 4.3. Exponent k is considered as dependent on the initial prestress ratio and is given by

$$k = 0.75\left(1 - \frac{\sigma_{p0}}{f_p}\right) \tag{4.128}$$

In American practice, the manufacturers' data on prestress loss due to steel relaxation at constant strain and temperature are often approximated by another formula [591] [647, Sect. 3.3], which can be written in the present notation as follows:

$$\frac{\sigma_{p0} - \sigma_p(t)}{\sigma_{p0}} = \frac{\langle \sigma_{p0} - s_0 \rangle}{s_y}\left\langle \ln\left(\frac{t}{\lambda_0}\right)\right\rangle \tag{4.129}$$

Table 4.3 Values of parameters A and B for three classes of prestressing steel (according to the Eurocode)

Prestressing steel	Class	A	B
Normal wires and strands	1	43.12×10^{-5}	6.7
Improved wires and strands	2	1.65×10^{-5}	9.1
Bars	3	7.92×10^{-5}	8.0

Parameters s_0 and s_y depend on the *specified yield strength* of prestressing steel, f_{py} and are given by $s_0 = 0.55 f_{py}$ and $s_y = (10 \ln 10) f_{py} = 23.03 f_{py}$. Parameter λ_0 is equal to 1 h. The Macauley brackets $\langle \ \rangle$ extract the positive part, defined as $\langle x \rangle = \max(x, 0)$. While formula (4.129) gives generally a slower evolution of prestress loss than the CEB formula (4.126) and is meaningful only for $t \gg \lambda_0$, it is more realistic in that it has a bound, which is set at $0.55 f_{py}$. The typical values of f_{py} are $0.80 f_{pu}$ for prestressing bars, $0.85 f_{pu}$ for stress-relieved tendons, and $0.90 f_{pu}$ for low relaxation tendons, where f_{pu} is the *specified tensile strength* of prestressing steel.

If plotted in the semilogarithmic scale, the ACI empirical formula (4.129) is represented by a straight line; see Fig. 4.31b. Therefore, it cannot capture the short-time relaxation accurately, although its long-time estimate is acceptable.

As exemplified by the KB Bridge in Palau, to be analyzed in detail in Chap. 7, the strain variation in steel bonded to concrete need not be negligible when dealing with creep-sensitive structures. Also, the sun exposure of the pavement can cause appreciable heating of the tendons embedded in the top slab [210]. The heating can be very important not only for bridges but also for prestressed nuclear containments or vessels. Plenty of test data on prestressing steel relaxation exist for constant strain and constant room temperature [207]. The most extensive data are given in Magura et al. [591]. The data for variable strain and variable temperature are much more limited but those that exist [281, 737] suffice for calibration.

Relaxation of prestressing steel can be described by a constitutive model that considers the total strain as a sum of the elastic, plastic, and viscous strains, with the viscous strain ε_v defined by the rate equation

$$\dot{\varepsilon}_v = A_T(T) \, f(\varepsilon, \sigma_p) \tag{4.130}$$

Here, σ_p is the stress in prestressing steel, ε is the strain in steel, T is the absolute temperature, and A_T is a temperature factor, which equals 1 at temperature $T_0 = 293$ K ($20°$C). Realistic expressions for functions f and A_T, derived and experimentally calibrated in Bažant and Yu [207], have the form

$$A_T(T) = \exp\left(\frac{Q_p}{k_B T_0} - \frac{Q_p}{k_B T}\right) \tag{4.131}$$

$$f(\varepsilon, \sigma_p) = \begin{cases} kc^{1-1/k} \rho_0^{1/k} \dfrac{\langle F_{\mathrm{p}}(\varepsilon) - \gamma \bar{f}_y \rangle}{E_t(\varepsilon)} \dfrac{[\zeta^{1/c}(\varepsilon, \sigma_p) - 1]^{1-1/k}}{\lambda_1 \zeta^{1+1/c}(\varepsilon, \sigma_p)} & \text{for } \sigma_p > \gamma \bar{f}_y \\ 0 & \text{for } \sigma_p \le \gamma \bar{f}_y \end{cases}$$
$$\tag{4.132}$$

where

$$\zeta(\varepsilon, \sigma_p) = \frac{F_{\mathrm{p}}(\varepsilon) - \gamma \bar{f}_y}{\sigma_p - \gamma \bar{f}_y} \tag{4.133}$$

Expression (4.131) is based on the rate process theory, with Q_p denoting the *activation energy of flow of prestressing steel* and $k_B = 1.38 \cdot 10^{-23}$ J/K the *Boltzmann constant*.[5] For practical applications, it is convenient to consider the ratio Q_p/k_B as a single parameter, with the dimension of temperature. Its values determined by fitting are $Q_p/k_B = 14,600$ K for the data of Shinko Wire Company, Ltd. (Amagasaki, Japan), and $Q_p/k_B = 7000$ K for the data of Rostásy et al. [737]. The value of $Q_p/k_B = 12,000$ K gives a reasonable agreement with the Eurocode [373] formula $A_T(T) = 1.14^{T-T_0}$ in the range between $10\,°\mathrm{C}$ and $45\,°\mathrm{C}$. Note that the activation energy can vary enormously depending on the type of steel. Therefore, tests are recommended whenever the temperature effects play an important role.

In formulae (4.132)–(4.133), $\lambda_1 = 1000\,\mathrm{h}$ and k, c, ρ_0 and γ are positive empirical constants for the given steel. Examples of their values determined by fitting of experimental data for various types of prestressing steel are given in Table 4.4. In Bažant and Yu [207], parameter ρ_0 was considered as dependent on the strain level, but this dependence is rather weak and makes only a little difference.

Furthermore, F_{p} is the function describing the stress–strain law $\sigma_p = F_{\mathrm{p}}(\varepsilon)$ for short-time loading. At constant strain, $F_{\mathrm{p}}(\varepsilon) = \sigma_{p0} = $ initial prestress. Function $E_t(\varepsilon) = \mathrm{d}F_{\mathrm{p}}(\varepsilon)/\mathrm{d}\varepsilon = $ describes the tangent modulus of steel for loading ($\dot{\varepsilon} > 0$), which is equal to Young's modulus $E = 200$ GPa if the prestress is not above the linear range and for unloading ($\dot{\varepsilon} < 0$); $\bar{f}_y = $ mean yield strength of prestressing steel (defined by 1% offset); $\gamma \bar{f}_y = $ threshold below which there is no relaxation (parameter γ is taken safely as $\gamma = 0.45$, although the American practice uses $\gamma = 0.55$). Constant $c \approx 2$ controls the transition from short-time relaxation to the long-time asymptotic value, and exponent k characterizes the initial relaxation curve $\sigma_{p0} - \sigma_p \propto t^k$. According to optimum fitting of published data, Bažant and Yu [207] recommended the value of $k \approx 0.08$, which differs from the values used in the CEB Model Code.

For the special case of constant ε and T, Eqs. (4.130)–(4.133) integrate to a formula which, for high stress, is close to the CEB formula (4.126), and for lower stress approaches a threshold, $\gamma \bar{f}_y$, in agreement with the American practice formula (4.129). At variable ε or T, numerical integration is needed.

---

[5]In expressions similar to (4.131), the universal gas constant R is sometimes used instead of the Boltzmann constant k_B. Both approaches are equivalent, and they differ only by the meaning of activation energy Q, which is taken per mole if R is used, or per elementary entity (atom or molecule) if k_B is used. In any case, for practical applications it is more convenient to specify directly the fraction Q/R (or Q/k_B), which has the dimension of temperature.

The viscous strain defined by the rate Eq. (4.130) is added to the elastic and plastic strains and to the thermal strain. The resulting stress–strain equation is conveniently written in the rate form

$$\dot{\varepsilon} = \frac{\dot{\sigma}_p}{E_t} + \dot{\varepsilon}_v + \alpha_T \dot{T} \tag{4.134}$$

where the first term on the right-hand side represents the sum of elastic and plastic strain rates, and α_T is the *thermal expansion coefficient* of prestressing steel. Of course, under variable temperature, the strain that enters the viscous law (4.130) as the first argument of function f should not be the total strain but the mechanical strain,

$$\varepsilon^* = \varepsilon - \alpha_T (T - T_{init}) \tag{4.135}$$

obtained from the total strain by subtracting the thermal strain. Here, T_{init} denotes the initial temperature. Combining (4.134) with (4.130) in which ε is replaced by ε^*, we can express the rate of stress in the prestressed tendon as

$$\dot{\sigma}_p = E_t \left[\dot{\varepsilon}^* - A_T(T) \, f(\varepsilon^*, \sigma_p) \right] \tag{4.136}$$

For the purpose of general finite element analysis, the rate Eq. (4.136) must be converted to an incremental form. Approximating the rates within a typical time step Δt by finite differences, we can express the prestress increment (typically negative) as

$$\Delta\sigma_p = E_t(\varepsilon^*) \left[\Delta\varepsilon^* - A_T(T) \, \Delta t \, f(\varepsilon^*, \sigma_p) \right] \tag{4.137}$$

The term $E_t(\varepsilon^*)\Delta\varepsilon^*$ on the right-hand side of (4.137) represents the stress change calculated according to the short-time stress–strain law. In most cases, the tangent modulus E_t is equal to the Young modulus of prestressing steel. Only if the yield stress is exceeded and the strain increment is positive (i.e., plastic yielding takes place), E_t should be understood as the elastoplastic tangent modulus. In that case, a more accurate result is obtained if $E_t(\varepsilon^*)\Delta\varepsilon^*$ is replaced by the increment $F_p(\varepsilon^* + \Delta\varepsilon^*) - F_p(\varepsilon^*)$.

The right-hand side of (4.137) depends on both stress and strain. For a given strain and temperature evolution, the mechanical strain ε^* can be substituted, e.g., by its midpoint value. In an explicit approach, the stress σ_p at the beginning of the step can be used. Higher accuracy can be achieved by an implicit approach, with (4.137) considered as a nonlinear equation from which the stress σ_p at midstep or at the end of the step can be computed by a local iterative procedure. At the structural level, this numerical scheme has to be combined with an appropriate implementation of a creep and shrinkage model for concrete.

Figures 4.32, 4.33, and 4.34 compare the curves computed from Eq. (4.137) with typical test data from the literature. The corresponding parameters are listed in Table 4.4. The mean yield strength $\bar{f}_y$ is a physical parameter taken according to the original sources. The other parameters have been obtained by fitting of the experimental relaxation curves [207] and slightly modified for the present purpose.

Figure 4.32 shows the evolution of stress in standard relaxation tests performed by Magura et al. [591] at room temperature, with different levels of initial stress. Labels NR and OT refer to two different types of prestressing steel. The effect of elevated temperature is illustrated in Fig. 4.33 for the temperature range from the room temperature to 80 °C in the tests of Shinko Wire Company (Fig. 4.33a), and to 175 °C in the tests of Rostásy et al. [737] (Fig. 4.33b). The measured data clearly demonstrate that elevated temperatures due to sun exposure can enormously accelerate the stress relaxation, especially in hot countries. Finally, in the relaxation tests of Buckler and Scribner [281] plotted in Fig. 4.34, the strain was kept constant for 24 h, then suddenly reduced and subsequently kept constant again. In most cases presented here, the fits of the test data based on the model of Bažant and Yu [207] are excellent, with the exception of high temperatures above 150 °C, which occur in fire. For such temperatures, the basic trend is still captured but the shape of the simulated relaxation curves deviates from the experimental one; see the lowest curves in Fig. 4.33b.

Table 4.4 Parameters used for fitting of relaxation tests

Parameter	Magura-NR	Magura-OT	Buckler	Shinko	Rostásy
$\tilde{f}_y$ [MPa]	1565	1334	1806	1550	1670
k	0.082	0.083	0.298	0.223	0.26
ρ_0	0.32	0.33	0.22	0.17	0.063
γ	0.45	0.5	0.53	0.52	0.55
c	3	2	2	3	2
Q_p/k_B [K]	–	–	–	14600	7000

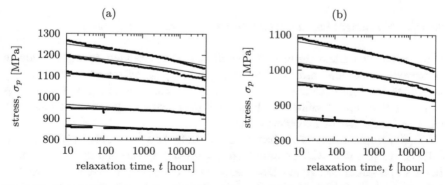

Fig. 4.32 Fitting of steel relaxation tests [591] by the model based on (4.130)–(4.133): (a) series NR101-105, (b) series OT101-104

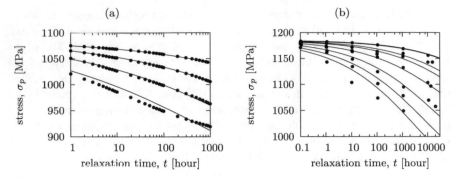

Fig. 4.33 Fitting of steel relaxation tests at different constant temperatures by the model based on (4.130)–(4.133): (a) tests of Shinko Wire Company, temperature levels 20 °C, 40 °C, 60 °C, and 80 °C (from top to bottom), (b) Rostásy, Thienel, and Schütt [737], temperature levels 20 °C, 55 °C, 70 °C, 110 °C, 130 °C, 155 °C, and 175 °C (from top to bottom)

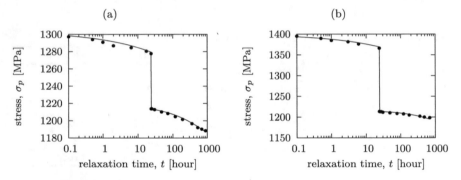

Fig. 4.34 Fitting of steel tests under varied strain [281] by the model based on (4.130)–(4.133): (a) tests SR8-5, (b) test SR14-10

4.3.4.2 Numerical Solution of a Centrically Prestressed Strut

In the preceding section, the history of strain in a prestressed tendon was considered as prescribed and the corresponding stress history was evaluated by numerically solving differential equation (4.136), in the numerical approximation rewritten as (4.137). Let us now examine what happens if the prestress relaxation interacts with creep and shrinkage of a concrete member.

For illustration of the basic phenomena, consider first the simple case of a centrically prestressed strut with a constant cross section. Suppose that the concrete strut freely shrinks until time t_1, at which the steel tendons are prestressed and anchored. Let us denote the initial prestress, just after installation of the tendons, as σ_{p1}. The loss of prestress due to slip is already incorporated in the value of σ_{p1}, and the loss of prestress due to wobble friction is neglected, so that we can consider the prestress as uniform along the strut.

As long as the total normal force transmitted by the prestressed strut is zero, the stress in concrete, σ_c, and the stress in prestressing tendons, σ_p, are linked by the condition

$$A_c\sigma_c(t) + A_p\sigma_p(t) = 0 \qquad (4.138)$$

with A_c and A_p, respectively, denoting the sectional areas of concrete and of prestressed tendons.

During the prestressing procedure, the strain in steel jumps from zero to ε_{p1} and the strain in concrete from $\varepsilon_{sh,1}$ to ε_{cc1}. Here, $\varepsilon_{sh,1} = \varepsilon_{sh}(t_1)$ denotes the shrinkage strain (in the sense of sectional average) at time t_1. If the prestress is applied sufficiently fast, the strain in concrete just after prestressing,

$$\varepsilon_{cc1} = \varepsilon_{sh,1} + \frac{\sigma_{c1}}{E_{c1}} \qquad (4.139)$$

can be evaluated using the short-term elastic modulus of concrete at age t_1, $E_{c1} = E_c(t_1)$. Subsequently, the strains in concrete and prestressing steel evolve, but their difference remains constant. We can therefore write

$$\varepsilon_{cc}(t) = \varepsilon_{cc1} + \varepsilon_p(t) - \varepsilon_{p1} \qquad (4.140)$$

The corresponding evolution of stresses can formally be described by relations

$$\sigma_c(t) = \mathscr{R}_c\{\varepsilon_{cc}(t) - \varepsilon_{sh}(t)\} \qquad (4.141)$$
$$\sigma_p(t) = \mathscr{R}_p\{\varepsilon_p(t)\} \qquad (4.142)$$

in which $\mathscr{R}_c$ denotes the linear relaxation operator for concrete and $\mathscr{R}_p$ is the nonlinear relaxation operator that formally describes the stress–strain relation for prestressing steel.

Substituting (4.140) into (4.141) and then substituting (4.141)–(4.142) into (4.138), we obtain an equation

$$A_c\mathscr{R}_c\{\varepsilon_p(t) - \varepsilon_{p1} + \varepsilon_{cc1} - \varepsilon_{sh}(t)\} + A_p\mathscr{R}_p\{\varepsilon_p(t)\} = 0 \qquad (4.143)$$

from which the unknown function $\varepsilon_p(t)$ can be computed. Note that the evolution of shrinkage strain, $\varepsilon_{sh}(t)$, is considered as known. The initial strain in steel, $\varepsilon_{p1} = \varepsilon_p(t_1)$, is evaluated from the given initial prestress σ_{p1} by inverting the short-term stress–strain law for steel, $\sigma_{p1} = F_p(\varepsilon_{p1})$, and the initial stress and strain in concrete, σ_{c1} and ε_{cc1}, then follow from (4.138)–(4.139).

Equation (4.143) is highly nonlinear and needs to be solved numerically. Its approximate solution will be constructed step by step, starting from the prescribed value ε_{p1} at time t_1 and proceeding to strains $\varepsilon_{pk} = \varepsilon_p(t_k)$ at time instants t_k, $k = 2, 3, \ldots n$. In a typical computational step number k, the value of ε_{pk} is already known and our task is to compute $\varepsilon_{p,k+1} = \varepsilon_{pk} + \Delta\varepsilon_{pk}$. We could directly manipulate Eq. (4.143) and construct its numerical counterpart, but the derivation will be easier to follow if we get back to the original set of Eqs. (4.138) and (4.140)–(4.142), discretize them separately and then eliminate all unknowns except for $\Delta\varepsilon_{pk}$.

For the state at the end of step number k, condition (4.138) is written as

$$A_c\sigma_{c,k+1} + A_p\sigma_{p,k+1} = 0 \qquad (4.144)$$

where $\sigma_{c,k+1}$ and $\sigma_{p,k+1}$, respectively, denote the stresses in concrete and steel at the end of the step, which have to be expressed in terms of the unknown increment $\Delta\varepsilon_{pk}$ based on constitutive Eqs. (4.141)–(4.142) combined with the compatibility condition (4.140).

Application of the concrete relaxation operator in (4.141) can be approximated in the spirit of the AAEM method by the formula

$$\sigma_{c,k+1} \approx R_{k+1}(\varepsilon_{cc1} - \varepsilon_{sh,1}) + E''_{k+1}(\varepsilon_{cc,k+1} - \varepsilon_{cc1} - \varepsilon_{sh,k+1} + \varepsilon_{sh,1}) \qquad (4.145)$$

in which $R_{k+1} = R(t_{k+1}, t_1)$ is the relaxation function and $E''_{k+1} = E''(t_{k+1}, t_1)$ is the age-adjusted effective modulus, both evaluated for the time interval from t_1 to t_{k+1}. Based on (4.140), we can write

$$\varepsilon_{cc,k+1} - \varepsilon_{cc1} = \varepsilon_{p,k+1} - \varepsilon_{p1} = \varepsilon_{pk} + \Delta\varepsilon_{pk} - \varepsilon_{p1} \qquad (4.146)$$

and thus the right-hand side of (4.145) can be converted to a linear expression in terms of the basic unknown, $\Delta\varepsilon_{pk}$. Introducing an auxiliary constant

$$\tilde{\sigma}_{c,k+1} = R_{k+1}(\varepsilon_{cc1} - \varepsilon_{sh,1}) + E''_{k+1}(\varepsilon_{p,k} - \varepsilon_{p1} - \varepsilon_{sh,k+1} + \varepsilon_{sh,1}) \qquad (4.147)$$

we can rewrite (4.145) in the simple form

$$\sigma_{c,k+1} = \tilde{\sigma}_{c,k+1} + E''_{k+1}\Delta\varepsilon_{pk} \qquad (4.148)$$

Note that $\tilde{\sigma}_{c,k+1}$ has a direct physical meaning—it represents the stress that would arise in concrete at the end of the step if the strain remained constant during step number k.

The prestress relaxation operator $\mathscr{R}_p$ in (4.142) describes in an abstract way the constitutive law for the tendon, which is actually specified by differential Eq. (4.136), in the numerical approximation rewritten as (4.137). In the present case, the relation $\sigma_p = \mathscr{R}_p\{\varepsilon_p\}$ is within step number k approximated by

$$\Delta\sigma_{pk} = E_t\left[\Delta\varepsilon_{pk} - \Delta t_k f(\varepsilon_{pk} + \Delta\varepsilon_{pk}, \sigma_{pk} + \Delta\sigma_{pk})\right] \qquad (4.149)$$

For simplicity, we assume that the temperature is kept at the reference level and thus the factor A_T can be omitted. Equation (4.149) defines an implicit link between the stress and strain increments. Formally, we can define function s_k which assigns to each strain increment $\Delta\varepsilon_{pk}$ the corresponding stress increment

$$\Delta\sigma_{pk} = s_k(\Delta\varepsilon_{pk}) \tag{4.150}$$

obtained as the solution of (4.149).

Substituting (4.148) and (4.150) combined with $\sigma_{p,k+1} = \sigma_{pk} + \Delta\sigma_{pk}$ into (4.144) and rearranging the terms, we get a nonlinear equation

$$A_c E''_{k+1} \Delta\varepsilon_{pk} + A_p s_k(\Delta\varepsilon_{pk}) = -A_c\tilde\sigma_{c,k+1} - A_p\sigma_{pk} \tag{4.151}$$

for the unknown strain increment $\Delta\varepsilon_{pk}$. The solution is computed by iteration. The right-hand side of (4.151) is a constant that can be directly evaluated. It corresponds to minus the normal force calculated under the assumption that the strain during the step remains constant. Starting from the initial guess $\Delta\varepsilon_{pk}^{(0)} = 0$, the left-hand side is then successively linearized around the current (iterated) solution $\Delta\varepsilon_{pk}^{(i)}, i = 0, 1, 2, \ldots$, and the solution is updated until the left-hand side of (4.151) becomes sufficiently close to the right-hand side.

Note that evaluation of function s_k, defined implicitly as the solution of (4.149), also requires iteration, which is embedded in the iterative solution of (4.151). The need for two nested iterative loops can be avoided by using an alternative approach, in which (4.149) is considered as the fundamental equation. From linear Eq. (4.144) and (4.148), the unknown stress in steel at the end of the step can easily be expressed in terms of the strain increment $\Delta\varepsilon_{pk}$ as

$$\sigma_{p,k+1} = -\frac{A_c}{A_p}\sigma_{c,k+1} = -\frac{A_c}{A_p}\left(\tilde\sigma_{c,k+1} + E''_{k+1}\Delta\varepsilon_{pk}\right) = \tilde\sigma_{p,k+1} - E^*_{k+1}\Delta\varepsilon_{pk} \tag{4.152}$$

where

$$\tilde\sigma_{p,k+1} = -\frac{A_c}{A_p}\tilde\sigma_{c,k+1}, \qquad E^*_{k+1} = \frac{A_c}{A_p}E''_{k+1} \tag{4.153}$$

are auxiliary constants, introduced to simplify the notation. Substituting (4.152) into (4.149), we obtain a single nonlinear equation

$$(E_t + E^*_{k+1})\Delta\varepsilon_{pk} - E_t\Delta t_k f(\varepsilon_{pk} + \Delta\varepsilon_{pk}, \tilde\sigma_{p,k+1} - E^*_{k+1}\Delta\varepsilon_{pk}) = \tilde\sigma_{p,k+1} - \sigma_{pk} \tag{4.154}$$

which needs to be solved iteratively for the unknown $\Delta\varepsilon_{pk}$. A slight disadvantage could be that such an approach is not applicable in the limit case when $A_p = 0$. However, this is the trivial case with no prestressing tendons, which does not require any computations.

When solving (4.154), one needs to be careful, because the value of $f(\varepsilon_p, \sigma_p)$ tends to infinity for σ_p approaching $F_p(\varepsilon_p)$ from below and is undefined for $\sigma_p \geq F_p(\varepsilon_p) > \gamma\bar f_y$ (note that $\zeta < 1$ for $\sigma_p > F_p(\varepsilon_p) > \gamma\bar f_y$, and $\zeta^{1/c} - 1$ is then negative, which leads to an undefined power expression in the numerator of the last fraction in (4.132)). Physically, this is reasonable, because the actual stress σ_p can never exceed the stress computed from the short-term stress–strain law, with relaxation effects neglected. However, numerical iteration could in some cases require the evaluation of function f from formula (4.132) for certain nonconverged combinations of stress and strain which are physically inadmissible, and this must be avoided. It is therefore advisable to determine in advance a range of values of the strain increment $\Delta\varepsilon_{pk}$ within which the value of f is defined and the difference between both sides of Eq. (4.154) changes sign. A lower bound on the solution $\Delta\varepsilon_{pk}$ is provided by the condition $F_p(\varepsilon_{p,k+1}) > \sigma_{p,k+1}$ which, in the range of elastic short-term steel behavior, can be rewritten as $E_t\left(\varepsilon_{pk} + \Delta\varepsilon_{pk}\right) > \tilde{\sigma}_{p,k+1} - E^*_{k+1}\Delta\varepsilon_{pk}$ and leads to the constraint

$$\Delta\varepsilon_{pk} > \frac{\tilde{\sigma}_{p,k+1} - E_t\varepsilon_{pk}}{E_t + E^*_{k+1}} \tag{4.155}$$

For values of $\Delta\varepsilon_{pk}$ close to this lower bound, f is very large and the left-hand side of (4.154) is negative, very large in magnitude, and therefore algebraically smaller than the right-hand side.

An upper bound is obtained if we find an increment $\Delta\varepsilon_{pk}$ for which $f = 0$ and simultaneously $(E_t + E^*_{k+1})\Delta\varepsilon_{pk} \geq \tilde{\sigma}_{p,k+1} - \sigma_{pk}$. According to (4.132), we have $f(\varepsilon_{p,k+1}, \sigma_{p,k+1}) = 0$ for $\sigma_{p,k+1} \leq \gamma\bar{f}_y$, which in our case means $\tilde{\sigma}_{p,k+1} - E^*_{k+1}\Delta\varepsilon_{pk} \leq \gamma\bar{f}_y$. If both aforementioned conditions are satisfied as strict inequalities, the left-hand side of (4.154) is larger than the right-hand side and $\Delta\varepsilon_{pk}$ cannot be a solution. From these considerations, we obtain the constraint

$$\Delta\varepsilon_{pk} \leq \max\left(\frac{\tilde{\sigma}_{p,k+1} - \gamma\bar{f}_y}{E^*_{k+1}}, \frac{\tilde{\sigma}_{p,k+1} - \sigma_{pk}}{E_t + E^*_{k+1}}\right) \tag{4.156}$$

Once the interval in which the solution must be located has been identified, Eq. (4.154) can be solved for instance by the Newton method, making sure that the initial guess and all subsequent iterated approximations remain within this interval. Should the iterative process lead to a value outside the interval, it is better to switch to a slower but more robust technique, such as the secant method, or even the bisection method.

Example 4.16. Interaction of prestress relaxation with creep and shrinkage in a centrically prestressed strut

To get an idea about the relative importance of individual phenomena, let us compute the evolution of strain and stress for a specific case. The concrete is supposed to have the same properties as in Example 3.1, and the properties of prestressing tendons are taken from the column of Table 4.4 that corresponds to the experiments of Buckler and Scribner [281]. The reinforcement ratio is set to $A_p/A_c = 0.01$. Concrete is cured until 7 days and then left stress-free until the age of 28 days, when the tendons are prestressed to $\sigma_{p1} = 1300$ MPa. The stress in concrete is computed from condition (4.138) and is, for the given reinforcement ratio of 1%, initially equal to $\sigma_{c1} = -13$ MPa.

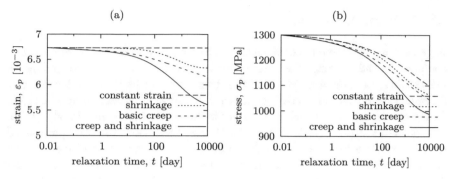

Fig. 4.35 Relaxation of prestress in a tendon and its interaction with creep and shrinkage: (a) strain evolution, (b) stress evolution

For simplicity, we do not consider any external loading of the strut (otherwise, we would need to replace zero on the right-hand side of (4.144) by the normal force). Figure 4.35 shows the evolution of strain and stress in the tendon computed by the numerical algorithm with an initially very small and gradually increasing time step. The strain in concrete differs from the strain in prestressing steel by a constant, and the compressive stress in concrete is 100 times smaller in magnitude than the tensile stress in the tendon. The top curves (dashed) in both parts of Fig. 4.35 correspond to the case of prestress relaxation under constant strain (note that the scale on the time axis is in days, not in hours as in Figs. 4.32, 4.33, and 4.34). It is seen that, in a standard relaxation test, the prestress would decrease from its initial value 1300 MPa to 1223 MPa after 100 days and to 1099 MPa after 10,000 days. But the strain in the prestressed concrete strut would actually remain constant only if the concrete did not shrink and remained elastic (with no creep). For shrinkage alone (still with no creep), the strain in the tendon would decrease as indicated by the dotted curve in Fig. 4.35a, and the prestress would decrease to 1209 MPa after 100 days and to 1063 MPa after 10,000 days. Of course, this is a fictitious case, but the results illustrate the relative importance of shrinkage. On the other hand, if there is no shrinkage (e.g., under sealed conditions) but the concrete exhibits basic creep, the strain and stress evolutions correspond to the short-dashed curves, and the prestress decreases to 1187 MPa after 100 days and to 1045 MPa after 10,000 days. Finally, if creep and shrinkage in an environment of 70% relative humidity are taken into account (including drying creep), the strain and stress evolve as shown by the solid curves, and the prestress decreases to 1165 MPa after 100 days and to 988 MPa after 10,000 days.

In summary, the relative loss of prestress after 10,000 days due to steel relaxation (at constant strain) is, in this particular example, about 15%, and the effects of basic creep under sealed conditions would add another 4%, while the effects of creep and shrinkage under 70% ambient relative humidity would add almost 9% (to the original 15%). Of course, all these results refer to the specific combination of material properties and reinforcement ratio used in the present example and should not be considered as generally valid. ∎

4.3.4.3 Prestressed Beams

In general, a prestressed concrete beam can be subjected to a combination of axial compression and bending. The deformation of each section is characterized by the axial strain ε_a (measured at the centroid) and curvature κ, which may both vary along the beam axis. For a reinforced concrete beam, as a special case of a composite beam, the internal forces would be linked to the deformation variables by Eqs. (4.84)–(4.85), with operators $\mathscr{R}_A$, $\mathscr{R}_S$, and $\mathscr{R}_I$ given by (4.86)–(4.88). However, for a prestressed beam we can use formulae (4.84)–(4.85) only for the contribution of the concrete part of the cross section. For the prestressed steel part, we must take into account three aspects that call for appropriate modifications of the mathematical model:

1. the stress–strain law is nonlinear and time-dependent,
2. the initial strain in prestressed steel differs from the initial strain in concrete, and
3. the difference between the strains in steel and in adjacent concrete may even evolve in time.

The first aspect implies that the operator $\mathscr{R}_p$ describing the special constitutive law for prestressing steel is nonlinear. If this were the only difference compared to the case of reinforced concrete, it would be sufficient to replace in (4.86)–(4.88) the product $E_s \mathscr{I}$ (i.e., the product of the steel elastic modulus and the identity operator) by $\mathscr{R}_p$, and the reinforcing steel area A_s by the prestressing steel area A_p. However, due to the second aspect listed above, such a simple replacement is not possible because $\mathscr{R}_p$ must be applied on the strain in prestressing steel, ε_p, which is not the same as the strain in the adjacent concrete. When the prestress is applied, the tendons are stretched and, if the beam is posttensioned, the adjacent concrete is compressed. The strain in concrete may also contain another negative component caused by previous shrinkage. It is therefore better to consider the strain in prestressing steel, ε_p, as an additional (in principle independent) deformation variable and evaluate the internal forces as the sum of contributions from concrete and from prestressing steel.

The internal forces in an eccentrically prestressed beam are expressed as

$$N = \mathscr{R}_A\{\varepsilon_a\} + \mathscr{R}_S\{\kappa\} + A_p\mathscr{R}_p\{\varepsilon_p\} + N_{sh} \tag{4.157}$$

$$M = \mathscr{R}_S\{\varepsilon_a\} + \mathscr{R}_I\{\kappa\} + eA_p\mathscr{R}_p\{\varepsilon_p\} + M_{sh} \tag{4.158}$$

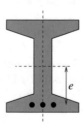

Fig. 4.36 Cross section of concrete beam with prestressed tendons

where operators $\mathscr{R}_A$, $\mathscr{R}_S$, and $\mathscr{R}_I$ are given by formulae analogous to (4.86)–(4.88) but with the steel part omitted, and e denotes the distance of the tendon centroid (or of the centroid of a group of tendons) from the horizontal centroidal axis; see Fig. 4.36. If the compliance function of concrete is the same across the whole section (i.e., if differential creep is neglected), the above equations reduce to

$$N = A_c \mathscr{R}_c \{\varepsilon_a\} + A_p \mathscr{R}_p \{\varepsilon_p\} + N_{sh} \tag{4.159}$$

$$M = I_c \mathscr{R}_c \{\kappa\} + e A_p \mathscr{R}_p \{\varepsilon_p\} + M_{sh} \tag{4.160}$$

where I_c is the moment of inertia of the concrete part of the section. Although both of these equations are nonlinear (due to nonlinearity of $\mathscr{R}_p$), they can be combined such that the resulting equation becomes linear. Indeed, multiplying (4.159) by e and subtracting it from (4.160), we obtain

$$M - Ne = \mathscr{R}_c \{I_c \kappa - e A_c \varepsilon_a\} + M_{sh} - e N_{sh} \tag{4.161}$$

Physically, the right-hand side of (4.161) corresponds to the moment resultant of stresses, expressed with respect to the horizontal axis that passes through the tendon centroid (i.e., is shifted by e with respect to the centroidal axis of the section). The force in the tendon does not contribute to this moment, and this is why ε_p does not appear in the condition. Equation (4.161) is linear but still needs to be combined with one of the original nonlinear equations, (4.159) or (4.160). The resulting set of two equations contains two unknown functions. To proceed further, we have to take into account how the strain in prestressing steel, ε_p, is related to the deformation variables ε_a and κ, which correspond to the axial strain in concrete and curvature of the concrete segment.

A direct relation between the strains in prestressing steel and in concrete can be established only if the two materials are **perfectly bonded**. This is the case for pretensioned members, and approximately also for posttensioned members if they are injected with grout. Even then, the initial strains in the two materials are not the same, but their difference remains constant in time. Thus, one can write

$$\varepsilon_p(t) = \varepsilon_a(t) + e\kappa(t) + \Delta\varepsilon_{pc1}, \qquad \text{for } t \geq t_1 \tag{4.162}$$

where $\Delta\varepsilon_{pc1}$ is the difference between the steel and concrete strains at time t_1 when the bond is formed. Based on (4.162), ε_p can be eliminated from the set of equations describing the problem.

If the structure is **statically determinate**, the internal forces can be precomputed from equilibrium with external forces (independently of the constitutive and kinematic equations) and afterward treated as known functions. Inverting Eq. (4.161) and taking into account definitions (4.80) and (4.83) of N_{sh} and M_{sh}, we can express the curvature in terms of the axial strain. The result can be presented in the form

$$\kappa = \frac{e A_c}{I_c} \varepsilon_a + \kappa^* \tag{4.163}$$

where

$$\kappa^* = \frac{1}{I_c} \left(\mathscr{J}_c\{M\} - e \, \mathscr{J}_c\{N\} + \int_A (z - e)\varepsilon_{\text{sh}} \, dA \right) \qquad (4.164)$$

is a function that can be computed from the history of internal forces and shrinkage strain. Based on (4.162)–(4.163), it is easy to convert (4.159) into an equation

$$A_c\mathscr{R}_c\{\varepsilon_a\} + A_p\mathscr{R}_p \left\{ \left(1 + \frac{e^2 A_c}{I_c}\right)\varepsilon_a + e\kappa^* + \Delta\varepsilon_{pc1} \right\} = N - N_{\text{sh}} \qquad (4.165)$$

with ε_a as the only unknown function.

Due to nonlinearity of $\mathscr{R}_p$, Eq. (4.165) requires a numerical solution. Its formal structure is similar to Eq. (4.143) solved in Sect. 4.3.4.2, and thus the numerical procedure developed in that section can be reused after straightforward modifications. In fact, the problem addressed in Sect. 4.3.4.2 (an axially prestressed strut with zero internal forces and uniform strain) can be interpreted as a special case of the present, more general formulation. For $r = 0$ and $N = 0$, Eq. (4.165) is equivalent to (4.143), with ε_a replaced by $\varepsilon_p - \Delta\varepsilon_{pc1}$ where $\Delta\varepsilon_{pc1} = \varepsilon_{p1} - \varepsilon_{cc1}$, and with N_{sh} rewritten as $-A_c\mathscr{R}_c\{\varepsilon_{\text{sh}}\}$. For the special case of uniform strain, the assumption that strains in steel and concrete differ by a constant is justified even in the absence of a bond.

If the structure is **statically indeterminate**, Eqs. (4.159)–(4.160) and (4.162) must be combined with the equilibrium equations and with kinematic equations $\varepsilon_a = u'_a$ and $\kappa = -w''$ that link the deformation variables to the axial displacement u_a and deflection w. Due to nonlinearity of the prestress relaxation operator $\mathscr{R}_p$, the problem always requires a numerical solution.

The situation becomes even more complicated if a perfect **bond** between steel tendons and concrete is **not ensured**. The strain in the tendon then cannot be expressed directly from the axial strain (in concrete) and curvature. The tendon can slip, and its axial displacement must be introduced as an additional unknown function, u_p. The strain in the tendon (or group of tendons subjected to the same strain) is expressed as $\varepsilon_p = u'_p$, and the difference between the increments of u_p and $u_a - w'z$ is the slip displacement, which is related by an additional constitutive law to the shear flux between the tendon and concrete. The shear flux then enters an additional equilibrium equation for the tendon in axial direction, which, combined with the standard two equilibrium equations for the whole beam, provides three equations for three unknown functions, u_a, w and u_p. These equations have a differential character with respect to the spatial variable x, but they also contain the constitutive operators $\mathscr{R}_c$ and $\mathscr{R}_p$ describing the dependence of stress on the strain history. If the beam contains several groups of tendons with a different strain in each group, one unknown displacement function is introduced for each group and one additional equilibrium equation is set up for each group.

4.3.5 Creep Buckling

Creep and shrinkage cause excessive deflections and stress redistributions with cracking, which compromise durability but not much structural safety. From the safety viewpoint, it is important to take into account the reduction of the long-time buckling strength of slender columns, thin plates and shells; see Bažant and Cedolin [115]. As a prototype buckling problem, consider a beam-column with cross section characterized by the (possibly variable) moment of inertia I and with a slight initial deviation from the perfectly straight shape described by function z_0 (distance of the centroids of individual cross sections from the straight line connecting the centroids of the end sections). The beam-column is loaded by a lateral load of intensity f (possibly variable along the axis) and by an axially applied force $P > 0$ which induces a compressive normal force $N = -P$ (constant along the axis). If the material behavior is linearly elastic, with Young's modulus E, the general equation for the deflection[6] w may be written as

$$(EIw'')'' + P(w + z_0)'' = f \tag{4.166}$$

As a special case, for a beam with constant cross section and with no lateral load, the governing equation can be simplified to

$$EIw^{IV} + P(w + z_0)'' = 0 \tag{4.167}$$

This fourth-order differential equation must be combined with two boundary conditions at each end, which depend on the type of supports. For a pin-ended column of length L, they read

$$w(0) = 0, \qquad w''(0) = 0, \qquad w(L) = 0, \qquad w''(L) = 0 \tag{4.168}$$

If the initial imperfection is approximated by a harmonic function

$$z_0(x) = \bar{z}_0 \sin \frac{\pi x}{L} \tag{4.169}$$

then equation (4.167) with boundary conditions (4.168) has the analytical solution

$$w_0(x) = \bar{w} \sin \frac{\pi x}{L} \tag{4.170}$$

with maximum deflection

$$\bar{w} = \bar{z}_0 \frac{P}{P_{cr} - P} \tag{4.171}$$

---

[6]By "deflection" w we mean the additional lateral displacement with respect to the initial unstressed state. The initial deviation of the column axis from a straight line is characterized by function z_0, and the sum $z = z_0 + w$ will be called the "deflection ordinate."

where

$$P_{cr} = \frac{EI\pi^2}{L^2} \tag{4.172}$$

is the *critical load* (buckling load). It is also useful to express the maximum deflection ordinate (i.e., the total deviation from the straight shape),

$$\bar{z} = \bar{z}_0 + \bar{w} = \bar{z}_0 \left(1 + \frac{P}{P_{cr} - P}\right) = \bar{z}_0 \frac{1}{1 - P/P_{cr}} \tag{4.173}$$

With growing load P, the deflection grows and tends to infinity as P approaches P_{cr}. For a perfectly straight column with $\bar{z}_0 = 0$, the deflection remains zero as long as $P < P_{cr}$, and it can become arbitrarily large for $P = P_{cr}$. Of course, this is only an idealized case. A real column always has an initial imperfection, and axial compression is accompanied by bending. Since the material has a finite strength, the combination of internal forces (axial force and bending moment) becomes inadmissible before the critical load computed for an idealized linear elastic material is attained.

If the material exhibits creep, the deflections due to the dead load grow in time, which amplifies the bending moment and contributes to failure. In Eq. (4.166), it is necessary to replace the multiplication by elastic modulus by application of the relaxation operator. For reinforced concrete, this equation is generalized to

$$(E_s I_s w'' + I_c \mathcal{R}\{w''\})'' + P(w + z_0)'' = f \tag{4.174}$$

where E_s is the elastic modulus of steel, I_s and I_c are the moments of inertia of the reinforcing bars and of the concrete part of the section (with respect to the centroidal axis of the whole section), and $\mathcal{R}$ is the relaxation operator characterizing the viscoelastic behavior of concrete. The deflection w now depends not only on the spatial coordinate x but also on the time t.

In general, Eq. (4.174) would need to be solved numerically in space and time. To illustrate the basic phenomena caused by the time-dependent behavior of the material, let us consider the simple case of a column with a constant cross section and no lateral loading, and with an initial imperfection described by the harmonic function (4.169). The governing equation then takes the form

$$E_s I_s w^{IV} + I_c \mathcal{R}\{w^{IV}\} + Pw'' = P\frac{\pi^2}{L^2}\bar{z}_0 \sin\frac{\pi x}{L} \tag{4.175}$$

This equation with boundary conditions (4.168) has a solution of the form (4.170) with the constant $\bar{w}$ replaced by a function of time satisfying the integral equation

$$\frac{\pi^4}{L^4}(E_s I_s \bar{w} + I_c \mathcal{R}\{\bar{w}\}) - \frac{\pi^2}{L^2}P\bar{w} = P\frac{\pi^2}{L^2}\bar{z}_0 \tag{4.176}$$

The solution can be constructed by a step-by-step procedure, or approximated using the AAEM method; see Sect. 4.2.

Suppose that the force P is applied at time t_1 and afterward remains constant. In the spirit of the AAEM method, we can replace $\mathscr{R}\{\bar{w}(t)\}$ by $R(t, t_1)\bar{w}_1 + E''(t, t_1)[\bar{w}(t) - \bar{w}_1]$ where $\bar{w}_1 = \bar{w}(t_1^+)$ is the maximum deflection just after loading. To keep the derivation easier to follow, we omit arguments t and t_1 of R and E'', and we rewrite (4.176) as

$$\frac{\pi^2}{L^2}\left(E_s I_s \bar{w} + I_c\left[R\bar{w}_1 + E''(\bar{w} - \bar{w}_1)\right]\right) - P\bar{w} = P\bar{z}_0 \qquad (4.177)$$

At time $t = t_1^+$, we have $\bar{w} = \bar{w}_1$ and $R = E'' = E_1 =$ short-term elastic modulus at age t_1. The initial (short-term, elastic) value of the maximum deflection calculated from (4.177) can be expressed in the form

$$\bar{w}_1 = \bar{z}_0 \frac{P}{P_{cr,1} - P} \qquad (4.178)$$

where

$$P_{cr,1} = \frac{(E_s I_s + E_1 I_c)\pi^2}{L^2} \qquad (4.179)$$

is the critical load evaluated according to formula (4.172) valid for elasticity, with the elastic bending stiffness of the section EI set equal to $E_s I_s + E_1 I_c$, i.e., calculated using the short-term elastic modulus of concrete. Once $\bar{w}_1$ is known, we can proceed to a general time t and express the corresponding maximum deflection $\bar{w}$ as

$$\bar{w} = \frac{P\bar{z}_0 + \frac{\pi^2}{L^2}(E'' - R)I_c\bar{w}_1}{P_{cr}'' - P} = \frac{P\bar{z}_0}{P_{cr}'' - P}\left(1 + \frac{\frac{\pi^2}{L^2}(E'' - R)I_c}{P_{cr,1} - P}\right) \qquad (4.180)$$

in which

$$P_{cr}'' = \frac{(E_s I_s + E'' I_c)\pi^2}{L^2} \qquad (4.181)$$

is the critical load evaluated again from formula (4.172), but this time with the elastic modulus of concrete replaced by the age-adjusted effective modulus E''. Formula (4.180) can be recast as

$$\bar{w} = \bar{z}_0 \frac{P(P_{cr}^* - P)}{(P_{cr}'' - P)(P_{cr,1} - P)} \qquad (4.182)$$

where

$$P_{cr}^* = P_{cr,1} + \frac{\pi^2}{L^2}(E'' - R)I_c = \frac{(E_s I_s + E^* I_c)\pi^2}{L^2} \qquad (4.183)$$

is yet another critical load, calculated with the modulus of concrete set to

$$E^* = E_1 + E'' - R = E'' E_1 J \tag{4.184}$$

The last identity in (4.184) follows from the definition of the age-adjusted effective modulus (4.47).

The derived formula (4.182) for the maximum deflection contains three auxiliary critical loads. The short-term critical load $P_{cr,1}$ is constant while P_{cr}'' and P_{cr}^* evolve in time. Initially, at $t = t_1^+$, all these critical loads coincide because $E''(t_1^+, t_1) = E_1$ and $J(t_1^+, t_1) = 1/E_1$. If the load P applied at time t_1 attains or exceeds the short-term critical load $P_{cr,1}$, stability is lost immediately. If $P < P_{cr,1}$, the maximum deflection starts from the initial value $\bar{w}_1$ given by (4.178) and its subsequent growth at constant load P is described by (4.182) with time-dependent values of P_{cr}'' and P_{cr}^*. For times $t > t_1$, we have $P_{cr}'' < P_{cr,1} < P_{cr}^*$ because $E'' < E_1 < E^*$ (the last inequality is equivalent to $E''J < 1$, which follows from $RJ \leq 1$ and from the definition of the age-adjusted effective modulus (4.47). Due to the decrease of the age-adjusted effective modulus E'', the critical load P_{cr}'' decreases in time, and as it approaches P from above, the deflection growth is accelerated and the column fails. The theoretical time to failure $t_f - t_1$ can be calculated from the condition $P_{cr}''(t_f, t_1) = P$. Of course, in reality the column fails somewhat earlier because at large deflections the material behavior ceases to be linear viscoelastic.

Formula (4.182) can be rewritten in the dimensionless form as

$$\frac{\bar{w}}{\bar{z}_0} = \frac{\frac{P_{cr}^*}{P} - 1}{\left(\frac{P_{cr}''}{P} - 1\right)\left(\frac{P_{cr,1}}{P} - 1\right)} = \frac{\mu\pi^* - 1}{(\mu\pi'' - 1)(\mu - 1)} \tag{4.185}$$

where

$$\mu = \frac{P_{cr,1}}{P}, \qquad \pi^* = \frac{P_{cr}^*}{P_{cr,1}}, \qquad \pi'' = \frac{P_{cr}''}{P_{cr,1}} \tag{4.186}$$

Parameter μ plays the role of a short-term safety factor (ratio between the short-term critical load and the actually applied load). Factors π^* and π'' represent the time-dependent critical loads P_{cr}^* and P_{cr}'' normalized by the short-term critical load $P_{cr,1}$. They can be evaluated as

$$\pi^* = \frac{E_s I_s + E^* I_c}{E_s I_s + E_1 I_c}, \qquad \pi'' = \frac{E_s I_s + E'' I_c}{E_s I_s + E_1 I_c} \tag{4.187}$$

and if the contribution of reinforcement to the bending stiffness is neglected, they reduce to normalized moduli:

$$\pi^* \approx \frac{E^*}{E_1} = E'' J, \qquad \pi'' \approx \frac{E''}{E_1} \tag{4.188}$$

For illustration, Fig. 4.37a shows the dependence of the normalized critical loads π^* and π'' on the load duration $t - t_1$ for concrete described by the B3 model with parameters from Example 3.1. The age at loading is set to $t_1 = 100$ days and drying creep is taken into account, with $t_0 = 7$ days and $h_{env} = 70\%$. As expected, factor π^* increases in time while factor π'' decreases. In Fig. 4.37b, the growth of the normalized deflection $\bar{w}/\bar{z}_0$ in time is plotted for different values of the short-term safety factor μ. For instance, $\mu = 1.25$ means that the applied load is at 80% of the short-term critical load ($P/P_{cr,1} = 1/\mu = 1/1.25 = 0.8$). The short-term deflection (meaning the deflection at 0.01 day after loading) is then $\bar{w}_1 = 4\bar{z}_0$, as follows from (4.178) with $P_{cr,1}/P = \mu = 1.25$. At 1 day after loading, the deflection is already 9.3 times the initial imperfection $\bar{z}_0$, at 10 days it is 115 times $\bar{z}_0$, and it blows up within a few additional days. For $\mu = 2$, i.e., for an applied load at 50% of the short-term critical load, the short-term deflection is equal to the initial imperfection $\bar{z}_0$ and the deflection after 1 day is still only $1.3\,\bar{z}_0$. However, after 100 days it grows to $3.5\,\bar{z}_0$, and it blows up at about 2 years. The actual shape of this curve in linear scale is better seen in Fig. 4.37c.

Figure 4.37d shows the time to failure $t - t_f$ as a function of the short-term safety factor μ. The solid curve corresponds to the theoretical time at which $P''_{cr} = P$, and the deflection becomes infinite. From the practical point of view, the deflections should be limited by a finite value. The dashed curve shows the time at which the deflection attains 20 times the initial imperfection, and the dotted curve corresponds to 5 times the initial imperfection.

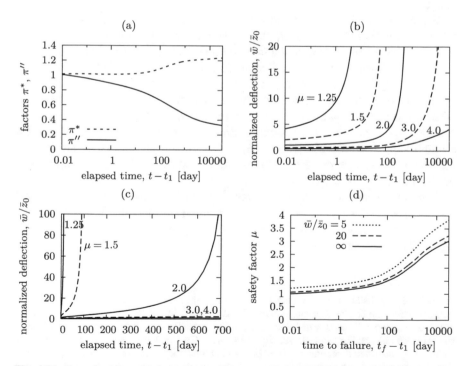

Fig. 4.37 Creep buckling: (a) dependence of the normalized critical loads on the load duration, (b)–(c) dependence of the normalized deflection on the load duration, (d) dependence between the short-term safety factor and the time to failure

The **design for creep buckling** is based on the evaluation of the maximum bending moment due to dead load combined with live load. The effects of creep are taken into account for the dead load only, while the response to the superimposed live load is considered as elastic. The combination of the resulting maximum bending moment and normal force must not fall outside the cross-sectional interaction diagram. To take into account the statistical uncertainties of the loads, the design values of the dead and live loads, P_D and P_L, are replaced by the factored loads $P_D^* = \psi_{Du} P_D$ and $P_L^* = \psi_{Lu} P_L$ where, according to the ACI Standard [17], the factors guarding against overload uncertainty are $\psi_{Du} = 1.2$ and $\psi_{Lu} = 1.6$ when both dead and live load act jointly, and $\psi_{Du} = 1.4$ when the dead load is considered separately.

The statistical uncertainties of the column properties are reflected by two other factors. The stiffness reduction factor, ϕ_K, accounts for the variability in the sectional bending stiffness EI and the moment magnification analysis and is applied on the critical load. According to ACI Committee 318 [17], $\phi_K = 0.75$ for an isolated column and 0.875 for a multistory frame. The strength reduction factor, ϕ, accounts for variations in material strength and dimensions and for inaccuracies in the design equations and reflects the degree of ductility of the member under the load effects being considered and the importance of the member. For compression-controlled failure, $\phi = 0.75$ for members with spiral reinforcement and 0.65 for other reinforced members.

For sustained loading by the factored dead load P_D^*, we can calculate (e.g., by the method indicated above, or by step-by-step computer analysis with a rate-type creep law described in Chap. 5) the maximum deflection ordinate $\bar{z}(t) = \bar{z}_0 + \bar{w}(t)$ at time t corresponding to the specified lifetime. Consider now that at time t the axial load is rapidly increased from P_D^* to $P_D^* + P_L^*$ where P_L^* is the factored live load. For this overload, the short-term response can be evaluated using an elastic analysis, with the elastic modulus taken as the short-term modulus $E(t)$ corresponding to the current age t. Let us denote the corresponding short-term critical load as $P_{cr}(t)$ (for a simply supported column, $P_{cr}(t) = (E_s I_s + E(t) I_c) \pi^2 / L^2$).[7] To take into account the uncertainties in the properties of the column, this critical load is later multiplied by the stiffness reduction factor ϕ_K.

Before the overload by P_L^*, the column is loaded by axial load P_D^* and its total deflection is $\bar{z}(t) = \bar{z}_0 f(t)$. For short-term loading (or unloading), it behaves like an elastic column with an initial imperfection $\bar{z}_0$, which can be evaluated from Eq. (4.173), rewritten as

---

[7]In fact, the ACI Standard [17] recommends to estimate the sectional bending stiffness as $E_s I_s + 0.2 E_c I_c$ where the concrete elastic modulus E_c is reduced by factor 0.2 (it is also possible to use a more accurate formula, which incorporates the dependence on the specific combination of normal force and bending moment; see Eq. (10-8) in [17]). This huge reduction of E_c is in code commentary attributed to the effects of cracking and nonlinearity in the stress–strain response. The consideration of cracking apparently refers to combinations of axial force with significant bending moments, which can lead to tensile strains. Similar provisions occur in Eurocode [373], with 0.2 replaced by a factor that depends on concrete grade, member slenderness ratio, and the relative axial force (ratio of axial force to maximum allowable force).

$$\bar{z}(t) = \tilde{z}_0(t) \frac{1}{1 - \dfrac{P_D^*}{\phi_K P_{cr}(t)}} \qquad (4.189)$$

Physically, $\tilde{z}_0$ can be interpreted as the deflection ordinate which would be found right after a sudden removal of the axial load. For a viscoelastic column, $\tilde{z}_0$ is larger than the initial imperfection $\bar{z}_0$ because of the accumulated deformation due to creep and because of aging.

From (4.189), it is easy to express

$$\tilde{z}_0(t) = \bar{z}(t) \left(1 - \frac{P_D^*}{\phi_K P_{cr}(t)} \right) \qquad (4.190)$$

and then the total deflection ordinate for a column axially loaded by $P_D^* + P_L^*$ is calculated as

$$\bar{z}_{DL}(t) = \tilde{z}_0(t) \frac{\phi_K P_{cr}(t)}{\phi_K P_{cr}(t) - P_D^* - P_L^*} = \bar{z}(t) \frac{\phi_K P_{cr}(t) - P_D^*}{\phi_K P_{cr}(t) - P_D^* - P_L^*} \qquad (4.191)$$

The magnified bending moment for which a column under sustained load and sudden overload must be designed is

$$M_{DL}^*(t) = (P_D^* + P_L^*) \bar{z}_{DL}(t) \qquad (4.192)$$

The design requirement is that, in the limit interaction envelope of bending moment M versus axial load P, the point $(M_{DL}^*/\phi, (P_D^* + P_L^*)/\phi)$ must not lie outside the envelope. Here, P_D^* and P_L^* are the factored loads and ϕ is the strength reduction factor.

4.3.6 Reduction of Flexural Creep Due to Cracking in Unprestressed Reinforced Concrete

In unprestressed reinforced concrete (RC) beams, the tensile cracking causes that the effective creep coefficient for flexure is much smaller than in prestressed concrete beams, in which cracking is suppressed. The situation is explained in Fig. 4.38, where a cross section is subjected to a sustained bending moment M which causes distributed cracking up to the neutral axis (n.a.) of zero strain. The middle diagram shows the typical stress distribution $\sigma(z)$ under service load, the resultant C of the compressive stresses in concrete, and the tensile force T in the steel bar. The diagram on the right depicts the corresponding elastic strain distribution, for which the strain at the level of C is $\varepsilon_e^C = \sigma^C/E$ (segment 23 in the figure). The elastic bending curvature is given by the strain diagram slope $\kappa_e = d\varepsilon/dz$, which is the slope of line 103 in the figure.

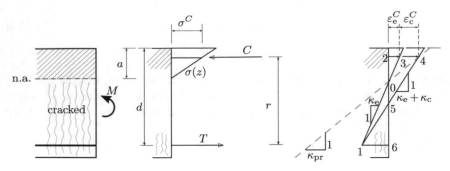

Fig. 4.38 Reinforced concrete cross section subjected to bending

The key point to note is that the tensile strain at the level of steel bar remains, under constant moment M, constant, because the steel bar does not creep and is under approximately constant stress. So, point 1 does not move. On the other hand, the strain at the level of compression resultant C will grow by creep increment $\varepsilon_c^C = \varphi \sigma^C / E$ (segment 34 in the figure) where φ is the creep coefficient. The curvature increment due to concrete creep is then $\kappa_c = \varepsilon_c^C / r$ where r is the arm of internal forces ($C = T$). The sum of the initial elastic curvature and the additional curvature due to creep, $\kappa_e + \kappa_c$, is thus equal to the slope of the line connecting points 1 and 4. The curvature change at no strain change in steel will of course require a downward shift of the neutral axis from point 0 to point 5, which would cause closing of the distributed cracks over portion 05 of cross-sectional depth.

For comparison note that, in the absence of cracking, which is the case of pre-stressed beam, the beam curvature after creep, κ_{pr}, would be given by the slope of the dashed line 04, which can be 2 to 5 times greater than the cracked beam curvature that has just been calculated (compare slopes of lines 451 and 40 in the figure). Furthermore, prestressed beams are in reality much more slender and, consequently, they can have 3 to 10 times greater creep curvatures and deflections than reinforced concrete beams over the same span.

This observation shows that the creep analysis of unprestressed reinforced concrete beams and frames is much less important than it is for prestressed beams and frames. For unprestressed beams and frames, the estimates of creep deflections can be very crude or even skipped. Aside from the prestress loss, this was another reason why the introduction of prestressed concrete in the 1940s required (and depended on) taking creep into account.

Chapter 5
Numerical Analysis of Creep Problems

Abstract Introduction of computers several decades ago was a quantum jump in what is feasible in creep structural analysis. It made possible efficient step-by-step evaluation of history integrals or integration of differential equations for both frame-type and finite element creep analysis of structures. Closer to the first principles is the numerical analysis based on history integrals. We discuss it first and point out its limitations due to excessive computer demands. Then, we present the rate-type conversion of creep analysis with aging, which represents a generalization of the Kelvin chain model of classical viscoelasticity, leads to far more efficient calculations, and makes it easy to take into account the effects of drying, variable environment, and cracking. We examine the accuracy and numerical stability of various numerical integration schemes and emphasize the exponential algorithm, which is unconditionally stable, allowing arbitrarily increasing time steps as the stress changes fade out. The algorithm is first presented for a nonaging Kelvin chain and then extended to a solidifying chain and to a chain with general aging.

5.1 Numerical Analysis of Structural Creep Problems Based on History Integrals

In linear viscoelasticity, the relations between stress and strain histories are given by the integral expressions (2.14) and (2.23). Analytical evaluation of the integrals is possible only for simple models and simple histories of the prescribed variable (stress or strain). For general applications, numerical integration schemes are needed. They can rely on standard quadrature rules known from numerical mathematics, such as the trapezoidal rule or the Simpson rule. For highly nonlinear compliance functions, the accuracy can be increased by special quadrature rules.

Approximations of the impulse memory integral by sums were proposed, e.g., by Prokopovich [709, 710]. Early numerical analyses of structural problems based on approximating the integral $\int_0^t J(t, t') \, d\sigma(t')$ with a sum were presented by Bresler and Selna [255], Ghali, Neville and Jha [425], Selna [771], Tadros, Ghali and Dilger [798, 799], Cederberg and David [302], and Rashid [718]. The last two were the

© Springer Science+Business Media B.V. 2018
Z.P. Bažant and M. Jirásek, *Creep and Hygrothermal Effects in Concrete Structures*, Solid Mechanics and Its Applications 225, https://doi.org/10.1007/978-94-024-1138-6_5

first finite element studies of creep in nuclear reactor vessels. Composite redundant steel-concrete beams were solved numerically by Bažant [73].

Suppose that the stress history $\sigma(t)$ in a certain time interval $[t_1, t_{max}]$ is prescribed (with no stress acting before time t_1), and the corresponding strain history $\varepsilon(t)$ is to be computed. The evolution of strain can be characterized by values $\varepsilon_k = \varepsilon(t_k)$ at discrete time instants $t_1, t_2 = t_1 + \Delta t_1, t_3 = t_2 + \Delta t_2, \ldots t_{n-1} = t_{n-2} + \Delta t_{n-2}, t_n = t_{n-1} + \Delta t_{n-1} = t_{max}$. In typical structural applications, it is appropriate to use increasing time steps Δt_k, because the creep process slows down in time. Optimally (for constant load), the time increments Δt_k may be selected as constant in the logarithmic scale, which means that they form a geometric progression. Several steps per decade usually suffice.

The initial value of strain at time t_1^-, just before the onset of loading, is prescribed, usually as $\varepsilon_1 = 0$. The strain at times t_k, $k = 2, 3, \ldots n$, is approximated according to the numerical quadrature rule by[1]

$$\varepsilon_k = \int_{t_1^-}^{t_k} J(t_k, t') \, d\sigma(t') \approx \sum_{i=1}^{k-1} J_{k,i} \, \Delta\sigma_i \tag{5.1}$$

where

$$\Delta\sigma_i = \sigma(t_{i+1}) - \sigma(t_i) \tag{5.2}$$

is the stress increment in time step number i, and coefficients $J_{k,i}$ are evaluated as

$$J_{k,i} = J\left(t_k, \frac{t_{i+1} + t_i}{2}\right) \tag{5.3}$$

for the *midpoint* (or rectangular) *rule*, or as

$$J_{k,i} = \frac{J(t_k, t_{i+1}) + J(t_k, t_i)}{2} \tag{5.4}$$

for the *trapezoidal rule*. For convenience, we denote

$$t_{i+1/2} = \frac{t_{i+1} + t_i}{2} = t_i + \frac{1}{2}\Delta t_i \tag{5.5}$$

---

[1] The superscript "minus" at the lower bound t_1^- of the integral in (5.1) emphasizes that integration in the Stieltjes sense starts just before time t_1, so that the potential initial change of stress by a jump is captured correctly. In fact, if the stress history contains jumps, numerical accuracy is increased by treating the corresponding terms separately, in the spirit of formula (2.15). This is equivalent to using zero-duration time steps associated with the stress jumps at certain discrete times. More specifically, if the stress has a jump at time t_{jump}, we select the time steps of the numerical scheme such that $t_k = t_{jump}^-$ and $t_{k+1} = t_{jump}^+$ for a certain time step number k. The duration of that step, Δt_k, is then zero, but the stress increment, $\Delta\sigma_k = \sigma(t_{jump}^+) - \sigma(t_{jump}^-)$, is nonzero and corresponds to the stress jump.

and rewrite (5.3) as

$$J_{k,i} = J\left(t_k, t_{i+1/2}\right) \tag{5.6}$$

Note that, for simplicity, we use J with subscripts k, i, even though a more logical (but clumsier) notation would be $J_{k,i+1/2}$.

The increment of strain in step number k is

$$\Delta\varepsilon_k = \varepsilon_{k+1} - \varepsilon_k = J_{k+1,k}\Delta\sigma_k + \sum_{i=1}^{k-1} \Delta J_{k,i}\,\Delta\sigma_i \tag{5.7}$$

where

$$\Delta J_{k,i} = J_{k+1,i} - J_{k,i} \tag{5.8}$$

Equation (5.7) can be interpreted as the numerical counterpart of the analytical formula (2.17), and it shows that the incremental stress–strain relation is linear. To emphasize that, we rewrite (5.7) as

$$\Delta\varepsilon_k = \frac{\Delta\sigma_k}{\bar{E}_k} + \Delta\varepsilon_k'' \tag{5.9}$$

where

$$\bar{E}_k = \frac{1}{J_{k+1,k}} \tag{5.10}$$

is the *incremental modulus* in step number k and

$$\Delta\varepsilon_k'' = \sum_{i=1}^{k-1} \Delta J_{k,i}\,\Delta\sigma_i \tag{5.11}$$

is the strain increment caused by creep at constant stress (in step number k). In this way, the strain increment is presented as the sum of two terms, one of which is proportional to the stress increment and represents the "almost" instantaneous deformation while the other is independent of the stress increment and represents the delayed effects.

The asymptotic rate of convergence is the same for both rules (midpoint and trapezoidal). It depends on the type of stress history and may be affected by the special form of the compliance function, but in the optimal case is quadratic [75].

Example 5.1. Accuracy and convergence rate of integration schemes*

Let us apply the midpoint rule (MR) and the trapezoidal rule (TR) to a simple calculation of the strain history corresponding to the prescribed stress history shown in Fig. 5.1a. The material is loaded at age $t = 30$ days by suddenly applied stress 5 MPa, which remains constant until $t = 60$ days. Subsequently, the stress increases quadratically up to the level of 15 MPa, attained at $t = 90$ days, and afterward is kept constant.

First, consider a very simple nonaging viscoelastic model with compliance function given by

$$J(t, t') = \frac{1}{E_0} + \frac{1}{E_1} \exp\left(-\frac{t - t'}{\tau_1}\right) \tag{5.12}$$

which corresponds to the so-called standard linear solid (elastic spring of stiffness E_0 coupled in series with a Kelvin unit characterized by stiffness E_1 and retardation time τ_1). The specific values of model parameters are taken as $1/E_0 = 20 \times 10^{-6}/$MPa, $1/E_1 = 24 \times 10^{-6}/$MPa, and $\tau_1 = 15$ days. Owing to the simple form of compliance function and stress history, the integral expression for the strain evolution can be evaluated analytically. The exact strain history is shown in Fig. 5.1b by the solid curve. The isolated points represent the values obtained with a uniform time step $\Delta t = 15$ days using the MR (filled circles) or the TR (crosses). Since the stress has a jump at time $t = 30$ days, a zero-size step is inserted before the regular steps. Up to $t = 60$ days, the numerical solution is exact, because creep takes place at constant stress. Later the numerical solution slightly deviates from the exact one, but the accuracy is seen to be good, even with a relatively large step. In this particular case, the MR leads to a larger error than the TR. The convergence rate can be assessed by plotting the error against the step size in logarithmic scale. In Fig. 5.1c, the relative error of strain at time $t = 90$ days is used for this purpose. For comparison, the dashed line indicates slope 2:1. The actual slope of the error plot is very close to this ideal slope, which means that the error is for both rules (almost) proportional to the square of the step size. In other words, the asymptotic convergence rate is quadratic. The accuracy of the TR is found to be higher than the accuracy of the MR, but this conclusion cannot be considered as a general rule. For instance, if the convex quadratic evolution of stress between times 60 days and 90 days is replaced by a linear one, or by a concave quadratic one, the strain at 90 days numerically evaluated by the MR would be more accurate than if the TR (with the same step size) is used. The asymptotic convergence rate would still remain quadratic.

It can be proven that quadratic convergence rate is guaranteed if the compliance function and the stress history have bounded second derivatives. It is sufficient if this condition is satisfied "in the piecewise sense", i.e., the stress history may exhibit kinks (discontinuities of the first derivative) or jumps at a finite number of time instants. At those time instants at which the stress history is discontinuous, zero-size computational steps should be inserted.

As explained in Chap. 3, the actual creep compliance function of concrete is, for very short load durations, proportional to a power function with exponent n, usually taken as 0.1. The derivative of such a function is unbounded (and the second derivative as well), which may spoil the quadratic convergence rate. For instance, consider that the material is described by the log-double-power law (3.9) (short form of B3 model), with conventional elastic modulus $E_{28} = 30$ GPa and with standard values of all other parameters. The corresponding strain history is plotted in Fig. 5.1d. The solid curve shows the converged results (computed with a very short time step), while the isolated points represent the values obtained with a uniform time step $\Delta t = 15$ days using the MR (filled circles) or the TR (crosses). To capture

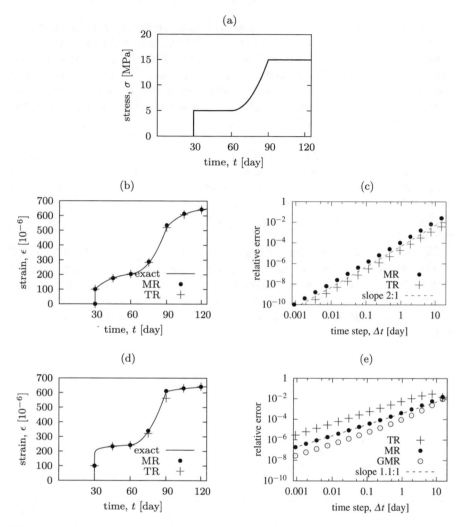

Fig. 5.1 (a) Prescribed stress history considered in Example 5.1, (b) the corresponding strain history for the standard linear solid (simple case of Kelvin chain), (c) relative error of numerically evaluated strain at time $t = 90$ days for the standard linear solid, (d) strain history for the short form of model B3, (e) relative error of numerically evaluated strain at time $t = 90$ days for the short form of model B3

properly the stress jump at $t = 30$ days, the first integration step has zero size and the corresponding stress increment is $\Delta\sigma_1 = 5$ MPa. In this first step, the strain jumps to $J(t_1^+, t_1)\Delta\sigma_1 = q_1\Delta\sigma_1 = \Delta\sigma_1/E_0 = 5\,\text{MPa}/50\,\text{GPa} = 100 \times 10^{-6}$, where $E_0 = E_{28}/0.6 = 50$ GPa is the asymptotic modulus and $q_1 = 1/E_0$ is the instantaneous compliance. In Fig. 5.1d, the first cross and circle seem to be located in the middle of a vertical segment representing the exact solution, but actually only the lower part of that segment up to the first point is exactly vertical, while the upper part follows the shape of the compliance function and slightly bends to the right, which is not visible in the present scale of the time axis covering many days. In fact, the numerical solution is exact up to time $t = 60$ days, independently of the size of subsequent steps. Indeed, if all the stress increments except the first one vanish, the sum in (5.1) reduces to $J_{k,1}\Delta\sigma_1$, and since $t_1 = t_2$, formulae (5.6) and (5.4) both yield the same $J_{k,1} = J(t_k, t_1)$. Note that if the stress jump was smeared over a nonzero time step Δt_1, some error would be induced. Of course, this error would tend to zero as $\Delta t_1 \to 0^+$.

The error[2] plot in Fig. 5.1e indicates that the asymptotic convergence rate is no longer quadratic. Detailed convergence analysis leads to a theoretical convergence rate slightly faster than linear, with the error proportional to the step size raised to the power of $1 + n$, where $n = 0.1$ is the exponent in the power law that dominates the initial part of the creep compliance function. The actual slope of the error plot in Fig. 5.1e is indeed very close to the theoretical slope 1.1:1, which is indicated by the dashed line.

Fig. 5.1e refers to the error in strain at time 90 days. Interestingly, if a similar plot is constructed for the error in strain at time 120 days (or at any other time larger than 90 days), quadratic convergence rate is detected. This is because the stress rate is zero starting from 90 days, and thus the product $J(t, t')\dot{\sigma}(t')$ vanishes for $t' > 90$ days. However, for any nonconstant stress history, the quadratic convergence rate would be spoiled.

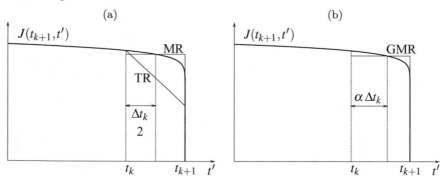

Fig. 5.2 Illustration of the accuracy of integration schemes for a highly nonlinear compliance function: (a) standard midpoint and trapezoidal rules, (b) generalized midpoint rule with optimal position of the integration point

For typical, highly nonlinear compliance functions with an unbounded derivative, the MR can be expected to provide better results that the TR. The reason is illustrated in Fig. 5.2a: The actual area under the graph of the (highly nonlinear part of)

---

[2]Since the exact value of the strain at time 90 days is not available (for the log-double-power law, the integral expression for strain cannot be evaluated analytically), it is not possible to evaluate the true error. Instead of the difference between the numerical solution and the exact one, the error is considered here as the difference between the numerical solution for the given step size and a more accurate numerical solution obtained with the step size divided by two. Such a pseudo-error exhibits the same asymptotic behavior as the actual error and thus can serve for examination of the convergence rate.

compliance function is much better approximated by the rectangle corresponding to the MR than by the trapezoid corresponding to the TR.

Optimal accuracy can be achieved with the *generalized midpoint rule* (GMR), if the sampling point is shifted to a special position, determined such that a power function with exponent n is integrated exactly. Consider a function f having the form $f(t, t') = a + b(t - t')^n$, where n is given and a and b are arbitrary constants. The following integral can be evaluated analytically:

$$I = \int_{t_k}^{t_{k+1}} f(t_{k+1}, t') \, dt' = a \, \Delta t_k + b \int_0^{\Delta t} s^n \, ds = a \, \Delta t_k + \frac{b}{n+1} (\Delta t_k)^{n+1} \quad (5.13)$$

The generalized midpoint rule uses one sampling point placed at $t^* = (1 - \alpha)t_k + \alpha t_{k+1}$ where α is an adjustable parameter between 0 and 1 (note that $\alpha = 0.5$ would correspond to the standard midpoint rule). Using such a rule, the integral is approximated as

$$I \approx \Delta t_k f(t_{k+1}, t^*) = \Delta t_k \left[a + b(t_{k+1} - t^*)^n \right] = a \, \Delta t_k + b \, \Delta t_k \left[(1 - \alpha) \Delta t_k \right]^n$$
$$(5.14)$$

Comparing this with the exact value in (5.13), we find the optimal value of $\alpha = 1 - (1 + n)^{-1/n}$. For a linear function f, exponent n is 1 and the standard midpoint rule with $\alpha = 0.5$ turns out to be optimal, as expected. However, for exponent $n = 0.1$ characteristic of creep compliance functions, optimal accuracy is achieved with $\alpha = 1 - 1.1^{-10} \approx 0.614457$. This is graphically illustrated in Fig. 5.2b.

In application to the evaluation of the integral in (5.1), the generalized midpoint rule with $\alpha = 1 - (1 + n)^{-1/n}$ has to be applied to the last interval only (i.e., for $i = k - 1$), because this is the only interval in which the integrated function has the assumed form (constant plus a power function of the distance from the end point of the interval). In other intervals, the optimal position of the sampling point would need to be adjusted depending on the distance from the point at which the derivative of the compliance function is infinite. In this way, a higher convergence rate could be achieved. If, for simplicity, the standard midpoint rule with $\alpha = 0.5$ is applied in all intervals except the last one, the accuracy is increased, but the asymptotic convergence rate remains the same (of order 1.1); see the convergence plot labeled as GMR in Fig. 5.1e. ∎

Formula (5.8) is suitable for models which specify a closed-form expression for the compliance function. For the solidification theory (Chap. 9), and in particular the B3 model (Chap. 3), the compliance function is defined not directly but in terms of its rate. For the standard midpoint rule, the coefficient $J_{k,i}$ is given by (5.6) and its increment corresponding to an increase of t_k by Δt_k can be approximated as

$$\Delta J_{k,i} = J_{k+1,i} - J_{k,i} = J(t_{k+1}, t_{i+1/2}) - J(t_k, t_{i+1/2}) \approx \dot{J} \left(t_{k+1/2}, t_{i+1/2} \right) \Delta t_k$$
$$(5.15)$$

where $\dot{J}$ is the derivative of the compliance function with respect to its first argument. In this way, the factors $\Delta J_{k,i}$ are computed directly, without the need to evaluate the compliance function (which is not available in a closed form for this class of models). However, the compliance value $J_{k+1,k}$ is still needed because it appears in definition (5.10) of the incremental modulus. Its accurate evaluation requires special attention, as follows.

Recall that, according to the midpoint rule (5.6), $J_{k+1,k} = J\left(t_{k+1}, t_{k+1/2}\right)$ corresponds to a load duration of half a step, $\Delta t_k/2$, which can be very short. For short load durations, the compliance function is highly nonlinear in time, and its rate tends to infinity as the load duration approaches zero. The term in the compliance function $J(t, t')$ responsible for such behavior is proportional to $(t - t')^n$, and its time derivative is proportional to $(t - t')^{n-1}$. The situation is thus similar to Example 5.1, but now the power function to be integrated has exponent $n - 1$ instead of n, and the integration variable is t instead of t'. The optimal value of parameter α to be used in the generalized midpoint rule is then $\alpha = (1 + n - 1)^{-1/(n-1)} = n^{1/(1-n)}$. For the typical value $n = 0.1$, we get $\alpha = 0.1^{10/9} \approx 0.077426$. Since the integration is performed over an interval of size $\Delta t_k/2$, the integration point is at the distance of $\alpha \Delta t_k/2 = 0.038713 \Delta t_k$ from its left boundary, and the optimal formula for the compliance reciprocal to the incremental modulus reads

$$J_{k+1,k} = J\left(t_{k+1}, t_{k+1/2}\right) = J(t_{k+1/2}, t_{k+1/2}) + \int_{t_{k+1/2}}^{t_{k+1}} \dot{J}(t, t_{k+1/2})\, dt \approx$$

$$\approx J(t_{k+1/2}, t_{k+1/2}) + \dot{J}(t_{k+1/2} + 0.038713 \Delta t_k, t_{k+1/2})\, \frac{\Delta t_k}{2} \qquad (5.16)$$

The instantaneous (asymptotic) compliance $J(t_{k+1/2}, t_{k+1/2})$ is obtained easily; e.g., for the B3 model, it is equal to parameter q_1, which is in fact the reciprocal value of the asymptotic modulus.

The foregoing equations have been written for the uniaxial case, but they are easily generalized to multiaxial stress. Stress and strain are then characterized by column matrices, and the stress is premultiplied by the unit elastic compliance matrix C_ν defined in (2.33), which depends only on the Poisson ratio, assumed here to be unaffected by creep. For instance, the incremental relation (5.9) is for multiaxial stress rewritten as

$$\Delta \boldsymbol{\varepsilon}_k = \frac{1}{\bar{E}_k} C_\nu \Delta \boldsymbol{\sigma}_k + \Delta \boldsymbol{\varepsilon}_k'' \qquad (5.17)$$

where the incremental modulus $\bar{E}_k$ is still given by (5.10) and the expression for creep strain increment,

$$\Delta \boldsymbol{\varepsilon}_k'' = C_\nu \sum_{i=1}^{k-1} \Delta J_{k,i}\, \Delta \boldsymbol{\sigma}_i \qquad (5.18)$$

is a straightforward generalization of (5.11).

Equation (5.9) or (5.17) is easy to invert, and so it can be adapted to the case when the strain history is given and the stress history needs to be computed. The numerically approximated incremental stress–strain relation reads

$$\Delta\sigma_k = \bar{E}_k \left(\Delta\varepsilon_k - \Delta\varepsilon_k'' \right) \tag{5.19}$$

for uniaxial stress and

$$\Delta\sigma_k = \bar{E}_k \boldsymbol{D}_v \left(\Delta\boldsymbol{\varepsilon}_k - \Delta\boldsymbol{\varepsilon}_k'' \right) \tag{5.20}$$

for multiaxial stress. Recall that $\boldsymbol{D}_v = \boldsymbol{C}_v^{-1}$ is the unit elastic stiffness matrix defined in (2.35).

In the context of a finite element simulation of a viscoelastic structure, Eq. (5.20) is written at each Gauss integration point of the finite element model and the stress is substituted into the standard expression for equivalent nodal forces. The resulting equations at the structural level are incrementally linear (and thus can be solved exactly, without the need for equilibrium iteration), but the structural stiffness matrix needs to be recomputed in each time step, because the incremental modulus $\bar{E}_k$ changes in time.

In each time step, $\Delta\boldsymbol{\varepsilon}_k''$ can be evaluated from the previous stress history, and thus is known. Therefore, (5.20) is equivalent to an elastic stress–strain relation with eigenstrains due to creep, and the response of the structure can be analyzed using the following algorithm [75, 80].

Algorithm 5.1

1. For a given time step number k, assemble the structural stiffness matrix from element contributions based on the material stiffness matrix $\bar{E}_k \boldsymbol{D}_v$, with the incremental modulus $\bar{E}_k$ given by (5.10).
2. Assemble the increment of the equivalent load vector corresponding to stress decrements due to relaxation,

$$\bar{E}_k \boldsymbol{D}_v \Delta\boldsymbol{\varepsilon}_k'' = \bar{E}_k \sum_{i=1}^{k-1} \Delta J_{k,i} \, \Delta\boldsymbol{\sigma}_i \tag{5.21}$$

This is the contribution to the right-hand side of the incremental equilibrium equations, to be added to the contribution of the external load increments and of the nonmechanical effects, such as temperature changes and shrinkage, during step number k.
3. Solve the set of linear equilibrium equations at the structural level to obtain the displacement increments.
4. Update the displacements and stresses, increment the step counter k by 1, and proceed to the next time step (go to 1).

One possible application of Eq. (5.19) is the numerical evaluation of the relaxation function corresponding to a given compliance function. In the first time step of zero duration, a unit strain increment is prescribed, and in the subsequent time steps the strain increments are set to zero. The computed stresses then represent the approximate values of the relaxation function. This is formally described by the recursive formula

$$R_1 = 0, \qquad R_2 = \bar{E}_1 = \frac{1}{J(t_1, t_1)} \tag{5.22}$$

$$R_{k+1} = R_k - \frac{1}{J_{k+1,k}} \sum_{i=1}^{k-1} \Delta J_{k,i} \left(R_{i+1} - R_i\right), \qquad k = 2, 3, \ldots n - 1 \tag{5.23}$$

The computed values R_k are approximations of $R(t_k, t_1)$. For an aging material, the computation needs to be repeated with different values of the initial time t_1, in order to obtain a complete description of the relaxation function. Since the first time step is chosen to have zero duration, we have $t_2 = t_1$ and, according to (5.10) and (5.22), $R_2 = \bar{E}_1 = 1/J_{2,1} = 1/J(t_1, t_1)$. Thus, the theoretical initial value of the relaxation function is captured exactly.[3] For instance, for the B3 model, we obtain $R_2 = E_0 =$ asymptotic modulus.

As will be shown in the next example, formulae (5.22)–(5.23) lead to an underestimation of the early part of the relaxation curve. In particular, the approximation R_3 is always below the exact value $R(t_3, t_1)$. This is caused by the highly nonlinear early stress evolution in a relaxation test. The numerical scheme approximates the stress within one time step as a linear function of time, but the early part of the relaxation function is very close to the reciprocal compliance function, and its graph (plotted in the linear scale) is convex and strongly curved; see Fig. 5.3b. With R_3 equal to the exact value, the assumed linear approximation connecting the points at t_1 and t_3 would be above the actual curve (inclined dotted line in Fig. 5.3b) and the corresponding strain at t_3 would be larger than the actual one. However, the strain in the relaxation test is prescribed (by a unit value), and so the value of R_3 used by the numerical scheme with linear stress interpolation must be below the exact value $R(t_3, t_1)$ to make sure that the numerically evaluated strain is equal to the actual one.

It turns out that the relaxation value R_3 obtained according to (5.22)–(5.23) is, at least for a reasonably sized initial step Δt_2, less accurate than a simple estimate based on the reciprocal value of compliance function (i.e., on the effective modulus). This observation leads to a modified recursive procedure:

$$R_1 = 0, \qquad R_2 = R_3 = \frac{1}{J(t_3, t_1)} \tag{5.24}$$

$$R_{k+1} = R_k - \frac{1}{J_{k+1,k}} \sum_{i=1}^{k-1} \Delta J_{k,i} \left(R_{i+1} - R_i\right), \qquad k = 3, 4, \ldots n - 1 \tag{5.25}$$

Note that not only R_3 but also R_2 has been set equal to the effective modulus, $1/J(t_3, t_1)$. Although we know that the exact value of R_2 is $1/J(t_1, t_1)$, for the evaluation of R_4, R_5 etc. it is essential to use a reduced value of R_2, otherwise the problem with underestimated relaxation values would reappear. Theoretical justification is

---

[3]Note that $R_1 = 0$ represents the value $R(t_1^-, t_1)$ just before the application of strain, and $R_2 = \bar{E}_1$ represents the value $R(t_1^+, t_1)$ just after the application of strain, i.e., the asymptotic modulus.

provided by the fact that the assumption of constant stress $1/J(t_3, t_1)$ during the first nonzero step from $t_1 = t_2$ to t_3 (horizontal dashed line in Fig. 5.3b) is closer to the actual, highly nonlinear stress evolution (solid curve) than the assumption of a linear stress variation between $1/J(t_1, t_1)$ at $t_1 = t_2$ and $1/J(t_3, t_1)$ at t_3 (inclined dotted line). Of course, the value of R_2 specified in (5.24) needs to be used in the recursive evaluation of (5.25) but in the output can be replaced by the correct value according to (5.22).

Example 5.2. Numerical evaluation of relaxation function

Consider concrete with the same properties as in Example 3.1, subjected to relaxation without drying. The compliance function is approximated by model B3 with parameters q_1, q_2, q_3, and q_4 determined in (3.25)–(3.28). Figure 5.3a shows the corresponding relaxation function over a very wide range of load durations, $t - t'$, with the age at the onset of relaxation set to $t' = 28$ days. It is clear that the relaxation process is initially very fast, because of the big difference between the asymptotic and conventional moduli. For comparison, the horizontal dashed line shows the asymptotic modulus, which would be approached for $t - t' \to 0^+$. Already after 1 second of relaxation (approximately 10^{-5} day), the relaxation function drops to 73% of the asymptotic modulus, and after 1 day it drops to 52%; see also the solid curve in Fig. 5.3b.

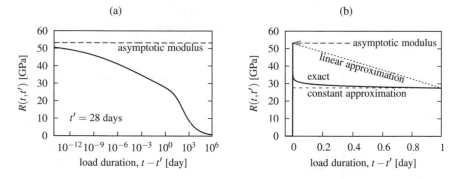

Fig. 5.3 Accurately evaluated relaxation function (a) in semilogarithmic scale over a wide range of load durations, (b) in linear scale up to 1 day

The curve plotted in Fig. 5.3a has been constructed using a highly accurate numerical solution, with the first nonzero time step $\Delta t_2 = 10^{-15}$ day and with subsequent time steps forming a geometric progression with quotient $q = 1.1$. The influence of the choice of the time step on the accuracy of numerical solution is illustrated in Fig. 5.4a. The solid curve corresponds to the (almost exact) reference solution, and the isolated points to values obtained according to formulae (5.22)–(5.23) with initial step $\Delta t_2 = 1$ day (circles) and $\Delta t_2 = 10$ days (crosses), in both cases using $q = 2$. The results confirm that the relaxation value after the first nonzero step is underestimated, as explained in the discussion preceding the present example. For $\Delta t_2 = 1$ day, the error is relatively small and in subsequent steps decreases and becomes almost negligible. On the other hand, for $\Delta t_2 = 10$ days, the error is large and is only partially reduced in subsequent steps. In principle, the basic procedure described by (5.22)–(5.23) could be used, but for good accuracy it would be advisable to select the initial step size at least an order of magnitude smaller than the shortest time of interest, and not larger than 1 day (for concrete loaded at a young age, one would need to be even more careful).

A modified procedure described by formulae (5.24)–(5.25) leads to more accurate results, as documented in Fig. 5.4b. The choice of the quotient and of the initial time steps is the same as in Fig. 5.4a, but the value of relaxation function after the first nonzero step is set to the reciprocal value of the compliance function. This leads to a slight overestimation, but the error is much lower than in the previous case, and the accuracy is preserved in subsequent steps as well. For comparison, the dashed curve indicates the reciprocal of the compliance function. For initial steps larger than 10 days, the initial error would be much more pronounced, but up to 1 day the curves of R and $1/J$ almost coincide, and up to 10 days the difference remains quite small.

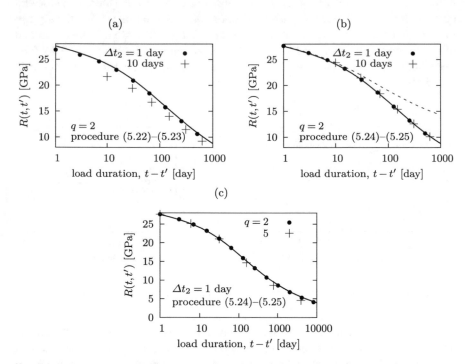

Fig. 5.4 Values of relaxation function computed numerically, with different initial time steps Δt_2 and quotients q

Finally, Fig. 5.4c demonstrates the influence of the quotient q that characterizes the growth of the time step. The solution obtained with initial step $\Delta t_2 = 1$ day and quotient $q = 2$ (marked by circles) remains sufficiently accurate up to 10,000 days (and even later). For the same initial step but quotient $q = 5$ (marked by crosses), some error would build up gradually. Graphically, this error may still look negligible, but at elapsed time 3906 days the numerical approximation underestimates the exact value of relaxation function by more than 17%, and at elapsed time 19,531 days (not shown in the figure) by 25%. A dramatic increase of the step size is thus not recommended. The choice of $q = 2$ (doubling the step size after each step) seems to be a very reasonable compromise between accuracy and efficiency. For instance, in the present example the recommended procedure covers the range from 1 day to 90 years in just 15 steps, with relative error not exceeding 5% (with respect to the current value of relaxation function). ∎

The integral approach presented in this section has been based on numerical approximation of formula (2.14) for the computation of the strain history from the stress history, using repeated evaluation of the compliance function. A dual procedure could be developed, which would use a given relaxation function and approximate the integral in (2.23). However, since most of the available experimental data correspond to creep tests (and not to relaxation tests), the compliance function is considered as the primary characteristic of the viscoelastic behavior of concrete. Thus, the dual approach dealing with the relaxation function is not really practical.

For the evaluation of the creep strain increment according to (5.11), the values of stress increments in all previous time steps must be known. In very large structural systems, the storage of the entire stress history at each integration point of each finite element and the evaluation of history integrals according to (5.1) can become prohibitively expensive, even with the most powerful computers. Note that the storage requirements and the number of operations needed to assemble *one* structural stiffness matrix and *one* right-hand side are proportional to the number of preceding time steps. If the total number of time steps in the analysis, N_{step}, is large, the total CPU time becomes proportional to N_{step}^2 (or even worse, if the memory capacity is exceeded and external storage must be used). For example, the total number of assembly-related operations in an analysis with 500 time steps is 100 times larger than in an analysis with 50 time steps (on the same finite element mesh), and with 5000 time steps it would be 10,000 times larger.

The computational complexity of the problem can be substantially reduced if the integral stress–strain relation is replaced by a differential one, dealing with certain history (internal) variables. This leads to more efficient computational schemes, described in the next section.

5.2   Efficient Rate-Type Creep Analysis

Approaches based on the differential (rate-type) formulation are computationally more efficient. Moreover, they are inevitable for incorporating the effects of variable humidity and temperature (treated in Sects. 10.6.1–10.8), and for generalization to nonlinear effects such as distributed cracking and damage (Chap. 12). The key idea of the rate-type approach is that a general compliance function (nonaging or aging) can be approximated by a Dirichlet series (A.25) or (A.40) corresponding to a Kelvin chain. Each unit of the chain is described by a differential equation that can be integrated in a step-by-step manner. In contrast to the integral approach, it is not necessary to store the entire previous history but only a limited (and fixed) number of history variables that are updated after each step. The number of numerical operations per step is constant, independent of the total number of steps.

The solution proceeds again in a sequence of time steps Δt_k, $k = 1, 2, \ldots n - 1$, starting from time t_1 at which the load is first applied. At the beginning of each step, the values of history variables are known from the previous step (or, for the first step, from the initial conditions). In typical long-time creep analysis of structures under constant loads, the response initially varies fast and Δt_k needs to be very short, e.g., 0.01 day, while at the end the response varies slowly, and Δt_k can be very long, e.g., 100 days. So, the time step must be greatly increased for efficiency of computation. As will be demonstrated in Sect. 5.2.1, the most common integration algorithms for ordinary differential equations lose numerical stability or at least accuracy if Δt_k becomes much larger than the shortest retardation time of the Kelvin chain. This is overcome by the *exponential algorithm*, originally proposed for nonaging materials by Zienkiewicz, Watson and King [897] and Taylor, Pister and Goudreau [803] and extended to aging materials by Bažant [74]. The exponential algorithm will be developed for a nonaging Kelvin model in Sects. 5.2.2–5.2.3, extended to a Kelvin chain in Sect. 5.2.4, and adapted to solidification and general aging in Sects. 5.2.5–5.2.8.

5.2.1 Generalized Trapezoidal Rule*

A simple yet important rheologic model, which can be used as a building block of more general ones, is the nonaging Kelvin model in Fig. 5.5, described by the first-order differential equation

$$E\varepsilon(t) + \eta\dot{\varepsilon}(t) = \sigma(t) \tag{5.26}$$

where E and η are model parameters characterizing, respectively, the elastic and viscous behavior (for more details see Appendix A). The ratio $\tau = \eta/E$ is the retardation time that sets the time scale at which the viscoelastic processes captured by this model take place.

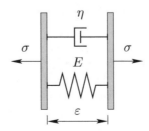

Fig. 5.5 Kelvin model

Suppose that the stress history $\sigma(t)$ is given and the strain history $\varepsilon(t)$ is to be evaluated. As already mentioned, the numerical solution proceeds in a sequence of time steps. The increment of strain over a time step can be obtained by integrating the strain rate. To this end, Eq. (5.26) is rewritten as

$$\dot{\varepsilon}(t) = \frac{1}{\tau}\left(\frac{\sigma(t)}{E} - \varepsilon(t)\right) \tag{5.27}$$

and formally integrated over time step number k, from time t_k to time t_{k+1}:

$$\varepsilon_{k+1} - \varepsilon_k = \frac{1}{\tau}\int_{t_k}^{t_{k+1}}\left(\frac{\sigma(t)}{E} - \varepsilon(t)\right)\,dt \tag{5.28}$$

The stress history $\sigma(t)$ is given, but the strain history $\varepsilon(t)$, which also appears in the integrand, is characterized only by the discrete values $\varepsilon_1, \varepsilon_2, \ldots \varepsilon_n$ at times $t_1, t_2, \ldots t_n$. It is thus natural to approximate the integral using the trapezoidal rule. The *standard trapezoidal rule* would give equal weights to the values at both ends of the interval. However, to allow for more flexibility of the resulting scheme, it is useful to consider a *generalized trapezoidal rule* (GTR) with weight factors $1 - \alpha$ and α assigned to times t_k and t_{k+1}, where α is an adjustable parameter between 0 and 1.[4] Approximating the integral in (5.28) by the GTR yields the algebraic equation

$$\varepsilon_{k+1} - \varepsilon_k = \frac{\Delta t_k}{\tau}\left(\frac{(1-\alpha)\sigma_k + \alpha\sigma_{k+1}}{E} - (1-\alpha)\varepsilon_k - \alpha\varepsilon_{k+1}\right) \tag{5.29}$$

from which the strain at the end of the step is easily expressed as

$$\varepsilon_{k+1} = \frac{\tau - (1-\alpha)\Delta t_k}{\tau + \alpha\Delta t_k}\varepsilon_k + \frac{(1-\alpha)\Delta t_k}{E(\tau + \alpha\Delta t_k)}\sigma_k + \frac{\alpha\Delta t_k}{E(\tau + \alpha\Delta t_k)}\sigma_{k+1} \tag{5.30}$$

This formula provides the final strain in step number k, based on the initial strain and on the initial and final stress. By its recursive application, an approximation of the strain history can be constructed. The factors multiplying ε_k, σ_k and σ_{k+1} depend on the model parameters E and τ, on the step size Δt_k and on parameter α of the numerical scheme. For $\alpha = 0$, the GTR reduces to the forward Euler (FEu) method, with a very simple update formula

$$\varepsilon_{k+1} = \varepsilon_k + \frac{\Delta t_k}{\tau}\left(\frac{\sigma_k}{E} - \varepsilon_k\right) \tag{5.31}$$

---

[4]In fact, $\alpha = 0$ means that the strain rate is evaluated at the beginning of the step and then assumed to be constant, which corresponds to the *forward Euler method*. Similarly, $\alpha = 1$ means that the strain rate (considered as constant during the step) is evaluated at the end of the step, which corresponds to the *backward Euler method*. The standard trapezoidal rule, which deals with the average of the initial and final rates, is obtained for $\alpha = 0.5$.

For $\alpha = 1$, the backward Euler (BEu) method is obtained, and $\alpha = 0.5$ gives the standard trapezoidal rule (STR).

Numerical stability of the scheme depends on the factor multiplying ε_k in (5.30), which determines whether an error $\delta\varepsilon$ in the strain value is amplified or damped by the numerical scheme. If this factor is larger than 1 in magnitude, the error grows and after some time dominates over the actual solution. Since the denominator $\tau + \alpha\Delta t_k$ is always positive, the condition of numerical stability can be written as

$$|\tau - (1 - \alpha)\Delta t_k| \le \tau + \alpha\Delta t_k \tag{5.32}$$

which is equivalent to $(1 - 2\alpha)\Delta t_k \le 2\tau$. The method is thus unconditionally stable for $\alpha \ge 0.5$, while for $\alpha < 0.5$ it is only conditionally stable and the critical time step is $\Delta t_{\text{crit}} = 2\tau/(1 - 2\alpha)$. In particular, the BEu method and the STR are unconditionally stable, while the FEu method is conditionally stable with critical time step $\Delta t_{\text{crit}} = 2\tau$.

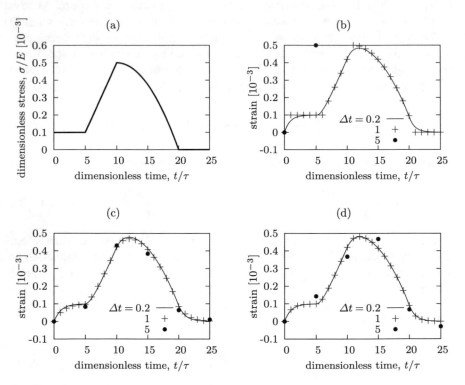

Fig. 5.6 (a) Prescribed history of stress acting on a Kelvin model; (b)-(d) histories of strain computed by the (b) forward Euler method, (c) backward Euler method, (d) standard trapezoidal rule

Example 5.3. Solution of Kelvin model by generalized trapezoidal rule

For illustration, the graphs in Fig. 5.6b-d show the numerical solutions of the strain history induced in a Kelvin model by a prescribed stress history specified in Fig. 5.6a.

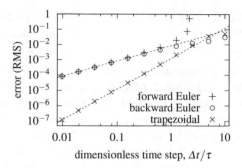

Fig. 5.7 Convergence diagram for different versions of the GTR, showing the dependence of the error on the step size

The stress scale is normalized by the spring stiffness E and the time scale by the retardation time τ. The stress jumps to $10^{-4}E$ at time 0 and remains constant until time 5τ. Then, it increases linearly to value $5 \times 10^{-4}E$ attained at time 10τ and subsequently decreases quadratically and reaches zero at time 20τ. From that moment on, the stress remains at zero level.

The strain histories numerically computed by the three special versions of the GTR are presented in Fig. 5.6b-d. For each integration scheme, three different time steps (in this example kept constant during the entire solution) have been used. For the shortest time step, $\Delta t = 0.2\tau$, the results (plotted as solid curves) are almost the same for all the three methods and are visually indistinguishable from the exact solution. Some differences can be observed for larger time steps. The forward Euler method (Fig. 5.6b) is inaccurate for $\Delta t = \tau$ and becomes unstable for $\Delta t = 5\tau > 2\tau = \Delta t_{crit}$. The backward Euler method (Fig. 5.6c) is somewhat more accurate for $\Delta t = \tau$, and the error remains reasonable even for $\Delta t = 5\tau$. The standard trapezoidal rule leads to a good accuracy for $\Delta t = \tau$ (Fig. 5.6d), but large deviations of an oscillatory character appear for $\Delta t = 5\tau$.

The accuracy of the methods can be compared quantitatively if a specific measure of the error is defined and evaluated. In this example, we use the root-mean-square of the differences between the numerical solution and the exact one at all times t_k between 0 and 30τ. For a piecewise polynomial stress history, Eq. (5.26) admits an analytical solution, and so the error of the numerical scheme can be evaluated precisely. The dependence of the error on the step size is shown in logarithmic scale in the convergence diagram in Fig. 5.7. Individual points correspond to errors of the numerical schemes for different time steps, ranging from 0.01τ to 10τ. The dashed straight lines do not directly connect the points; they just indicate slopes 1:1 and 2:1 and help to identify the asymptotic convergence rates. As the step size tends to zero, the error of the forward as well as backward Euler methods is proportional to the step size, which means that in the logarithmic plot the points lie on a straight line of slope 1. For the standard trapezoidal rule, the error is proportional to the square of the step size and the points lie on a straight line of slope 2. This is in agreement with the theoretical analysis of the numerical scheme[5], according to which $\alpha = 0.5$ (corresponding to the STR) is the only value that leads to a quadratic convergence rate, and for all the other values the convergence rate is linear. ■

The asymptotic convergence rate determines the error evolution as the step size tends to zero and thus is related to accuracy for very short time steps. From this point of view, the STR is clearly superior to the forward or backward Euler methods. However, it is also interesting to look at the accuracy for medium or even large step sizes. As seen in Fig. 5.7, the error of the forward Euler method blows up for steps larger than $t_{\text{crit}} = 2\tau$, due to the loss of numerical stability. Already for step sizes below t_{crit} but comparable to it, the error is substantially larger than for the other methods; see also the strain values for $\Delta t = \tau$ in Fig. 5.6b. The STR gives the best accuracy for step sizes below 2.5τ, but for larger steps the backward Euler method seems to be superior. This can be confirmed by a computation with step size 10τ continued over a longer time interval (still using the prescribed stress history from Fig. 5.6a, with stress after time 20τ identically equal to zero). Of course, one cannot expect a high accuracy with such a long step, but the numerical results should at least reflect the main features of the solution, and in particular they should quickly approach zero after time 20τ. This is indeed the case for the backward Euler scheme but not for the STR. As shown in Fig. 5.8a, the strain values computed by the STR oscillate even after complete removal of the stress, and the amplitude of oscillations decreases only slowly. The pollution of the results by such oscillations becomes especially strong if the step size is progressively increased Fig. 5.8b shows the strain values obtained when the initial step size $\Delta t_1 = 0.1\tau$ is doubled in each subsequent step. For the STR, a large error is still observed at times one or two orders of magnitude larger than the duration of the loading impulse. Again, the backward Euler method gives acceptable results.

The dramatic differences in the performance of the STR and the backward Euler method with large steps can be explained by looking at the factor multiplying ε_k in (5.30), which approaches $1 - 1/\alpha$ as $\Delta t_k \to \infty$. For the STR, $\alpha = 0.5$ and the factor approaches -1. This means that if no stress is applied and a large time step is used, ε_{k+1} has the opposite sign than and almost the same magnitude as ε_k. Similar oscillations, albeit less dramatic, can be expected for all values of α between 0.5 and 1, as soon as the step size becomes sufficiently large. For $\alpha < 0.5$, the magnitude of the oscillations would grow, which corresponds to numerical instability. Only for $\alpha = 1$, i.e., for the backward Euler method, the factor multiplying ε_k remains positive for all step sizes, and the oscillations cannot appear.

The foregoing analysis indicates that no version of the GTR is perfectly suited for applications to creep problems. The STR would give the highest accuracy for

---

[5]The numerical approximation of the integral in (5.28) based on the STR is exact for a linear function, and so the error of integration from t_k to $t_k + \Delta t$ is dominated by the quadratic part of the integrand and is proportional to $(\Delta t)^3$. The total number of time steps over an interval of fixed length is inversely proportional to Δt, and so the cumulative error is proportional to $(\Delta t)^2$. For all other versions of the GTR, with $\alpha \neq 0.5$, the integration is exact for a constant function only, and the error is due to the linear part of the integrand, thus being proportional to $(\Delta t)^2$ in one time step and to Δt after accumulation over a fixed interval.

short steps but generates spurious oscillations if the steps are long. The backward Euler method is free of such oscillations, but its convergence rate is inferior. In this discussion, "long" and "short" steps are taken relative to the retardation time of the model, τ. In real applications, Kelvin chains consisting of units with very different retardation times are used, and the step size is usually increased during the analysis by orders of magnitude. Therefore, it cannot be avoided that the steps are "short" for some of the units and "long" for others. A method exhibiting a quadratic convergence rate and at the same time providing a reasonable response for long steps is needed. Such a method will be developed in the next sections.

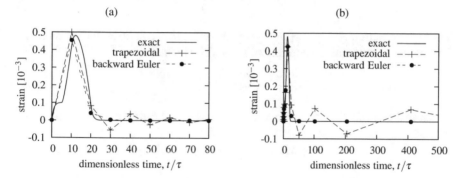

Fig. 5.8 Comparison of strain histories computed by the standard trapezoidal rule and the backward Euler method using (a) constant step size 10τ, (b) initial step size 0.1τ increased in geometric progression with quotient 2

5.2.2   First-Order Exponential Algorithm*

The key idea of the exponential algorithm is that if the stress on the right-hand side of (5.26) is constant, this equation has an exact analytical solution of an exponential type. If the stress is variable, we can replace it in each step of numerical integration by a constant, e.g., by weighted average of the initial and final value, and then construct the analytical solution and evaluate the strain increment over the time step. The value of strain at the end of the step is then used as initial condition for the analytical solution in the next step.

For convenience, we divide Eq. (5.26) by E and take into account that $\eta/E = \tau = $ retardation time of the model. Within time step number k, the stress can be approximated by a constant, $\sigma_{k+\alpha} = (1 - \alpha)\sigma_k + \alpha\sigma_{k+1}$, where α is an adjustable parameter between 0 and 1, with default value $\alpha = 0.5$. The resulting differential equation with a constant right-hand side,

$$\varepsilon(t) + \tau\dot{\varepsilon}(t) = \frac{\sigma_{k+\alpha}}{E} \tag{5.33}$$

has the general solution

$$\varepsilon(t) = \frac{\sigma_{k+\alpha}}{E} + C\,\mathrm{e}^{-t/\tau} \tag{5.34}$$

where C is an arbitrary integration constant.

The solution is valid for $t_k \leq t \leq t_{k+1}$ and must satisfy the initial condition $\varepsilon(t_k) = \varepsilon_k = $ strain at the end of the previous step (or, in the first step, zero). Substituting the general solution (5.34) into the initial condition, we evaluate constant C and construct the particular solution

$$\varepsilon(t) = \frac{\sigma_{k+\alpha}}{E} + \left(\varepsilon_k - \frac{\sigma_{k+\alpha}}{E}\right)\mathrm{e}^{-(t-t_k)/\tau} \tag{5.35}$$

The value of strain at $t = t_{k+1}$ then provides the initial condition for the next step. To simplify the notation, we introduce an auxiliary constant

$$\beta_k = \mathrm{e}^{-\Delta t_k/\tau} \tag{5.36}$$

which makes it possible to express the strain at the end of the step by the simple formula

$$\varepsilon_{k+1} = \varepsilon(t_{k+1}) = \beta_k \varepsilon_k + (1 - \beta_k)\frac{\sigma_{k+\alpha}}{E} \tag{5.37}$$

Subtracting ε_k, we obtain the strain increment

$$\Delta\varepsilon_k = \varepsilon_{k+1} - \varepsilon_k = (1 - \beta_k)\left(\frac{\sigma_{k+\alpha}}{E} - \varepsilon_k\right) \tag{5.38}$$

Formula (5.38) has a clear physical interpretation. At the beginning of the increment, the elastic spring transmits stress $E\varepsilon_k$. The total stress applied during the increment, $\sigma_{k+\alpha}$, is assumed to be constant. The difference between the total stress and the elastic stress is the viscous stress transmitted by the dashpot, σ_v. If the initial viscous stress $\sigma_v(t_k) = \sigma_{k+\alpha} - E\varepsilon_k$ vanishes, the system is in equilibrium at a zero strain rate and the strain remains constant during the entire step. The expression in the last parentheses in (5.38) represents the initial viscous stress divided by the spring stiffness, and if it vanishes, the numerically evaluated strain increment vanishes as well. In general, the strain rate is proportional to the viscous stress, but since an increase of strain leads to an increase of elastic stress while the total stress remains constant, the viscous stress decreases and the strain rate as well. This is why the strain increment according to (5.38) is not simply proportional to the increment of time. The dependence on the time increment Δt_k is captured by the factor $1 - \beta_k = 1 - \mathrm{e}^{-\Delta t_k/\tau}$, which is for $\Delta t_k \ll \tau$ approximately equal to $\Delta t_k/\tau$ and for very large time increments tends to 1. The expression in the last parentheses in (5.38) is in fact the strain increment that would be needed to transmit the given total stress exclusively by the elastic spring, with no contribution from the dashpot.

Let us emphasize that if the total stress indeed remains constant, formula (5.38) gives the exact solution for arbitrarily large time increments. This is the big advantage of the exponential algorithm as compared to the most common integration schemes which usually approximate the solution by polynomials. The exponential algorithm is unconditionally stable, because the factor β_k multiplying ε_k in (5.37) is always smaller in magnitude than 1. Moreover, it is always positive, and so spurious oscillations for large steps are not expected.

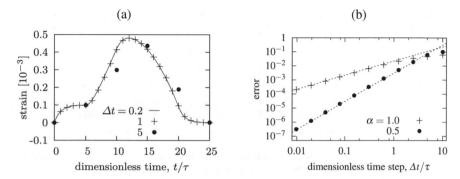

Fig. 5.9 (a) Strain histories computed by the first-order exponential algorithm using different time increments, (b) convergence diagram showing the dependence of the error on the step size

Example 5.4. Solution of Kelvin model by first-order exponential algorithm

For illustration, the numerically evaluated strain histories corresponding to the stress history from Fig. 5.6a are computed by the exponential algorithm using three different time steps and plotted in Fig. 5.9a. Factor α is set to its default value 0.5. As expected, the initial loading stage at constant stress is reproduced exactly, independently of the time step. In the second and third stage with linear and quadratic stress history, the solution with time step $\Delta t = \tau$ remains quite accurate, but relatively large errors are observed for $\Delta t = 5\tau$. The last stage at constant (zero) stress would be reproduced exactly if the initial conditions were exact. Due to the error accumulated during the preceding stages, the numerical solutions are not really exact, but with increasing time all of them correctly tend to zero strain.

The error (defined again as the root-mean-square of the deviations from the exact solution at all times t_k between 0 and 30τ) is shown in Fig. 5.9b as a function of the step size Δt. For comparison, the error has also been evaluated for the nonstandard value $\alpha = 1$. From the slope of the convergence diagram, it can be seen that the convergence rate is quadratic for the default value $\alpha = 0.5$ but only linear for $\alpha = 1$. The default value is thus optimal for short steps. However, closer examination of Fig. 5.9b reveals that if the step size is 5τ or larger, $\alpha = 1$ provides higher accuracy. The reason for that will be explained in the next section.

Finally, the graphs in Fig. 5.10 show that the exponential algorithm does not lead to any spurious oscillations if a large time step is used (Fig. 5.10a), nor if the initially short time step is progressively increased (Fig. 5.10b). This is true for both $\alpha = 0.5$

and $\alpha = 1$, but the latter choice leads to higher accuracy, which confirms the previous conclusion based on Fig. 5.9b.   ∎

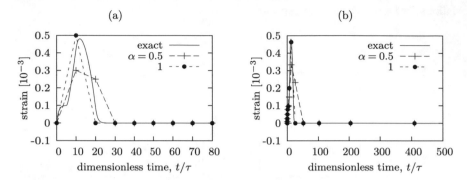

Fig. 5.10 Strain histories computed by the first-order exponential algorithm using (a) constant step size 10τ, (b) initial step size 0.1τ increased in geometric progression with quotient 2

5.2.3   Second-Order Exponential Algorithm

Accuracy of the exponential algorithm can be increased at almost no additional expense by using a piecewise linear approximation of the stress history instead of a piecewise constant one. The derivation could be based on (5.26) with the right-hand side replaced by $\sigma_k + (t - t_k)\Delta\sigma_k/\Delta t_k$, where $\Delta\sigma_k = \sigma_{k+1} - \sigma_k$ is the stress increment in step k. The same final formula for the strain increment is obtained by an alternative derivation that is preferred here because it permits an easy generalization to the case of aging, to be treated in Sects. 5.2.5–5.2.8.

Equation (5.26) governing the behavior of a nonaging Kelvin model is derived in Appendix A by substituting the equations describing an elastic spring and a viscous dashpot,

$$\sigma_e(t) = E\varepsilon(t) \tag{5.39}$$

$$\sigma_v(t) = \eta\dot{\varepsilon}(t) \tag{5.40}$$

into the stress equivalence (internal equilibrium) condition

$$\sigma_e(t) + \sigma_v(t) = \sigma(t) \tag{5.41}$$

Differentiating (5.39) with respect to time and combining it with (5.40), we find a link between the elastic stress rate and the viscous stress:

$$\dot{\sigma}_e(t) = E\dot{\varepsilon}(t) = E\frac{\sigma_v(t)}{\eta} = \frac{\sigma_v(t)}{\tau} \tag{5.42}$$

where $\tau = \eta/E$ is, as usual, the retardation time. Substituting (5.42) into the rate form of (5.41),

$$\dot{\sigma}_e(t) + \dot{\sigma}_v(t) = \dot{\sigma}(t) \tag{5.43}$$

we obtain a differential equation for the viscous stress,

$$\frac{\sigma_v(t)}{\tau} + \dot{\sigma}_v(t) = \dot{\sigma}(t) \tag{5.44}$$

with the stress rate on the right-hand side. Within time step number k, the stress rate can be approximated by a constant, and (5.44) becomes

$$\frac{\sigma_v(t)}{\tau} + \dot{\sigma}_v(t) = \frac{\Delta\sigma_k}{\Delta t_k} \tag{5.45}$$

The approximation is exact if the stress rate $\dot{\sigma}(t)$ is constant, i.e., if the stress history is linear (within the time step). Equation (5.45) is again a first-order linear differential equation with constant coefficients and constant right-hand side, similar to (5.33), and its particular solution satisfying the initial condition $\sigma_v(t_k) = \sigma_{vk}$ is

$$\sigma_v(t) = \tau \frac{\Delta\sigma_k}{\Delta t_k} + \left(\sigma_{vk} - \tau\frac{\Delta\sigma_k}{\Delta t_k}\right) e^{-(t-t_k)/\tau} \tag{5.46}$$

The increment of the viscous stress is then easily evaluated as

$$\Delta\sigma_{vk} = \sigma_v(t_{k+1}) - \sigma_v(t_k) = (1 - \beta_k)\left(\tau\frac{\Delta\sigma_k}{\Delta t_k} - \sigma_{vk}\right) = \lambda_k \Delta\sigma_k - (1 - \beta_k)\sigma_{vk} \tag{5.47}$$

where β_k is the factor defined in (5.36) and

$$\lambda_k = (1 - \beta_k)\frac{\tau}{\Delta t_k} = \left(1 - e^{-\Delta t_k/\tau}\right)\frac{\tau}{\Delta t_k} \tag{5.48}$$

is another auxiliary factor introduced for convenience.

Note that both β_k and λ_k monotonically decrease from 1 and asymptotically approach 0 as the time step Δt_k is varied from 0 to ∞. When $\Delta t_k/\tau$ is too large, the exponential expression in (5.36) may cause in computations an underflow, which should be avoided by setting $\beta_k = 0$. On the other hand, when $\Delta t_k/\tau$ is much smaller than 1, β_k approaches 1 and the evaluation of λ_k from (5.48) is inaccurate; it should be replaced by the first few terms of the Taylor series,

$$\lambda_k = 1 - \frac{1}{2}(\Delta t_k/\tau) + \frac{1}{6}(\Delta t_k/\tau)^2 - \ldots \tag{5.49}$$

So far, the solution (5.47) is formally the same as that for the constant stress approximation (see Eq. (5.38)), but it is written in terms of the viscous stress instead of the strain. To evaluate the strain, we need to express the strain rate as $\dot{\varepsilon}(t) = \sigma_v(t)/\eta$

and integrate.[6] Using the initial condition $\varepsilon(t_k) = \varepsilon_k$ and integrating the right-hand side of (5.46) divided by the viscosity η (which is equal to $E\tau$), we get

$$\varepsilon(t) = \varepsilon(t_k) + \int_{t_k}^{t} \frac{\sigma_v(t')}{\eta}\, dt' = \varepsilon_k + \frac{\Delta\sigma_k}{E}\frac{t - t_k}{\Delta t_k} - \frac{1}{E}\left(\sigma_{vk} - \tau\frac{\Delta\sigma_k}{\Delta t_k}\right)\left(e^{-(t-t_k)/\tau} - 1\right)$$
(5.50)

and the strain increment over the time step then is

$$\Delta\varepsilon_k = \varepsilon(t_{k+1}) - \varepsilon(t_k) = (1 - \lambda_k)\frac{\Delta\sigma_k}{E} + (1 - \beta_k)\frac{\sigma_{vk}}{E}$$
(5.51)

Note that this is again an incrementally linear stress–strain relation, which could be presented in the form (5.9) with $\bar{E}_k = E/(1 - \lambda_k)$ and $\Delta\varepsilon_k'' = (1 - \beta_k)\sigma_{vk}/E$. The big advantage compared to the integral approach is that, in contrast to (5.11), the evaluation of $\Delta\varepsilon_k''$ is very simple and does not require the knowledge of the entire previous history but only the value of the viscous stress σ_v at the beginning of the step. So, in principle, the viscous stress should be stored as an internal variable, updated by increments evaluated from (5.47). Another variable that needs to be stored is of course the strain, ε. Before the first step, both σ_v and ε are set to zero.

Incidentally, for the present simple model (nonaging Kelvin unit), variables σ_v and ε are not independent—they are linked by the stress equivalence condition $E\varepsilon + \sigma_v = \sigma$, which follows from (5.39) and (5.41). So the viscous stress can always be evaluated from the values of total stress and strain as

$$\sigma_v(t) = \sigma(t) - E\varepsilon(t)$$
(5.52)

and it does not really need to be stored, if the strain ε is stored. This will no longer be true for the solidifying Kelvin model, to be treated in Sect. 5.2.5.

Evaluation of the viscous stress from (5.52) brings an additional benefit: No special treatment is needed if the stress history is discontinuous, provided that the time instants at which the stress changes by a jump are included among the times t_k used by the numerical scheme. If, at time t_k, the stress happens to change by a jump from σ_k^- to σ_k^+, the stress increment $\Delta\sigma_{k-1}$ corresponding to the time interval (t_{k-1}, t_k) is calculated with σ_k^- as the final value, the viscous stress at the beginning of the next step is set to $\sigma_{vk}^+ = \sigma_k^+ - E\varepsilon_k$, and the next stress increment $\Delta\sigma_k$ is calculated using σ_k^+ as the initial value. This is more accurate than approximating the stress history by a continuous function with the value $(\sigma_k^+ + \sigma_k^-)/2$ at time t_k. In fact, a proper treatment of the stress jump is obtained simply by using a zero-size step from t_k^- to $t_k^+ \equiv t_{k+1}$, because for $\Delta t_k = 0$ we have $\beta_k = 1$ and $\lambda_k = 1$, and from (5.47) and (5.51) we obtain $\Delta\sigma_{vk} = \Delta\sigma_k = \sigma_k^+ - \sigma_k^-$ and $\Delta\varepsilon_k = 0$.

---

[6]Continuity of strain follows from boundedness of the strain rate, which in turn follows from the boundedness of the viscous stress (equal to the product of a finite viscosity and the strain rate). Note that we work here on the level of one Kelvin unit, which will later become a part of a Kelvin chain. The chain usually contains a spring that reflects the instantaneous elastic response. So the total strain of the chain can change by a jump, but such a jump is fully accommodated by the elastic spring, and the partial strains in Kelvin units remain continuous.

If (5.52) is taken into account, it is no longer necessary to use formula (5.47) for the update of the viscous stress, and formula (5.51) for the strain increment can be rewritten as

$$\Delta\varepsilon_k = (1 - \lambda_k)\frac{\Delta\sigma_k}{E} + (1 - \beta_k)\left(\frac{\sigma_k}{E} - \varepsilon_k\right) \tag{5.53}$$

By adding ε_k, one obtains a formula for ε_{k+1} in which ε_k is multiplied by β_k. This is the factor that determines numerical stability of the algorithm, and it happens to have the same value as for the first-order version; cf. (5.37). Consequently, the second-order exponential algorithm is also unconditionally stable and does not generate spurious oscillations for very large time steps.

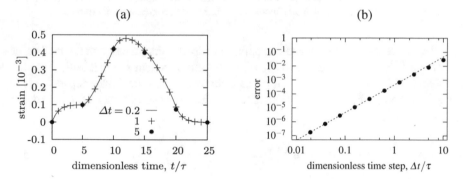

Fig. 5.11 (a) Strain histories computed by the second-order exponential algorithm using different time increments, (b) convergence diagram showing the dependence of the error on the step size

Example 5.5. Solution of Kelvin model by second-order exponential algorithm

The accuracy of the improved (second-order) formula of exponential algorithm (5.53) is superior to the basic (first-order) formula (5.38). The result is exact not only for constant stress histories but also for linear ones. This is illustrated by the numerical solution of the strain history corresponding to the prescribed stress history from Fig. 5.6a. Independently of the size of the time increment, the response during the first two stages of loading is captured exactly. In the last two stages, the deviations from the exact solution are quite small; see Fig. 5.11a. The error diagram in Fig. 5.11b indicates that the convergence rate is quadratic. So the convergence rate is the same as for the first-order version (with default value $\alpha = 0.5$), but the absolute accuracy of the second-order version is higher. Note that the straight line in Fig. 5.11b is shifted down with respect to the straight line of the same slope in Fig. 5.9b. In this particular example, the error of the second-order exponential algorithm turns out to be seven times smaller than the error of the first-order exponential algorithm (for the same step size). It can also be verified that no oscillations appear if the second-order exponential algorithm is used with a large or a progressively increasing time step; see Fig. 5.12. ∎

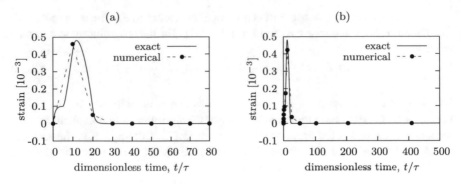

Fig. 5.12 Strain histories computed by the second-order exponential algorithm using (a) constant step size 10τ, (b) initial step size 0.1τ increased in geometric progression with quotient 2

The second-order formula (5.53) permits a direct comparison with the first-order formula (5.38). Recalling that $\sigma_{k+\alpha} = \sigma_k + \alpha \Delta\sigma_k$, we can see that both formulae coincide if $1 - \lambda_k = \alpha(1 - \beta_k)$. It is instructive to look at the Taylor expansions for short time steps. For the sake of simplicity, we leave out subscripts k at β, λ, and Δt:

$$1 - \beta = 1 - e^{-\Delta t/\tau} = \frac{\Delta t}{\tau} - \frac{1}{2}\left(\frac{\Delta t}{\tau}\right)^2 + \dots \tag{5.54}$$

$$1 - \lambda = 1 - (1 - \beta)\frac{\tau}{\Delta t} = \frac{1}{2}\frac{\Delta t}{\tau} - \frac{1}{6}\left(\frac{\Delta t}{\tau}\right)^2 + \dots \tag{5.55}$$

Both $1 - \beta$ and $1 - \lambda$ are equal to zero for $\Delta t = 0$ and tend to 1 for $\Delta t \to \infty$, but their values for finite time steps are different. The identity $1 - \lambda = \alpha(1 - \beta)$ cannot hold for arbitrary step sizes with a fixed value of α. For $\alpha = 0.5$, the first-order terms in the expansions of $\alpha(1 - \beta)$ and of $1 - \lambda$ coincide, which justifies $\alpha = 0.5$ as the default choice for the first-order method. This choice is optimal for short steps (compared to the retardation time τ), and it leads to the same (quadratic) rate of convergence as the second-order method. On the other hand, for long time steps the accuracy is higher with α close to 1, because the ratio $(1 - \lambda)/(1 - \beta)$ approaches 1 as Δt tends to infinity. This explains why the error of the first-order algorithm with time step 5τ or larger was smaller when α was set to 1 than it was when the default value 0.5 was used (see Figs. 5.9b and 5.10).

For all the integration rules discussed so far, the strain at the end of the step, ε_{k+1}, can be expressed as a linear combination of the strain at the beginning of the step, ε_k, and the stresses at the beginning and at the end of the step, σ_k and σ_{k+1}. The coefficients in this linear combination depend on the physical parameters E and τ and on the step size Δt_k. They are summarized in Table 5.1. Numerical stability is guaranteed if the coefficient at ε_k, does not exceed 1 in magnitude. Spurious oscillations appear if this coefficient is negative.

5.2.4   Nonaging Kelvin Chain

So far, the numerical solution based on the rate-type approach has been developed for a single Kelvin unit. As explained in Example 2.1, such a simple model can reflect viscoelastic properties of concrete over a limited range of load durations only. This range can be enhanced by using a Kelvin chain, consisting of several Kelvin units with different retardation times, coupled in series. Recall that the compliance function of such a chain is given by the Dirichlet series (2.7).

Table 5.1 Coefficients in strain update formula according to different integration methods

Method		ε_k	σ_k	σ_{k+1}
Generalized trapezoidal rule	GTR	$\dfrac{\tau - (1-\alpha)\Delta t_k}{\tau + \alpha\Delta t_k}$	$\dfrac{(1-\alpha)\Delta t_k}{E(\tau + \alpha\Delta t_k)}$	$\dfrac{\alpha\Delta t_k}{E(\tau + \alpha\Delta t_k)}$
Forward Euler	FEu	$\dfrac{\tau - \Delta t_k}{\tau}$	$\dfrac{\Delta t_k}{E\tau}$	0
Backward Euler	BEu	$\dfrac{\tau}{\tau + \Delta t_k}$	0	$\dfrac{\Delta t_k}{E(\tau + \alpha\Delta t_k)}$
Standard trapezoidal rule	STR	$\dfrac{2\tau - \Delta t_k}{2\tau + \Delta t_k}$	$\dfrac{\Delta t_k}{E(2\tau + \Delta t_k)}$	$\dfrac{\Delta t_k}{E(2\tau + \Delta t_k)}$
First-order exponential algorithm	EXP-1	β_k	$\dfrac{1 - \beta_k}{2E}$	$\dfrac{1 - \beta_k}{2E}$
Second-order exponential algorithm	EXP-2	β_k	$\dfrac{\lambda_k - \beta_k}{E}$	$\dfrac{1 - \lambda_k}{E}$

Generalization of the exponential algorithm to a nonaging Kelvin chain is easy, because all the units in the chain are under the same stress σ, and the total strain ε is the sum of partial strains ε_μ, $\mu = 0, 1, 2 \ldots M$. Therefore, Eq. (5.53) can be written separately for each unit of the chain, with all variables except for the stress and stress increment marked by an additional subscript μ that refers to the number of unit. The overall strain increment is then obtained by summing the contributions of individual units:

$$\Delta\varepsilon_k = \sum_{\mu=0}^{M} \Delta\varepsilon_{\mu k} = \left(\frac{1}{E_0} + \sum_{\mu=1}^{M} \frac{1 - \lambda_{\mu k}}{E_\mu} \right) \Delta\sigma_k + \sum_{\mu=1}^{M} (1 - \beta_{\mu k}) \left(\frac{\sigma_k}{E_\mu} - \varepsilon_{\mu k} \right)$$

(5.56)

Note that we have included the "zeroth" unit, which is an elastic spring of stiffness E_0 (without any dashpot). The spring could be considered as a Kelvin unit with zero viscosity and therefore zero (or negligibly small) retardation time, but it is more instructive to treat it separately.

The first term on the right-hand side of (5.56) can be interpreted as the strain increment due to a change of the applied stress. In this term, the stress increment is multiplied by the incremental compliance

$$\frac{1}{\bar{E}_k} = \frac{1}{E_0} + \sum_{\mu=1}^{M} \frac{1 - \lambda_{\mu k}}{E_\mu} \tag{5.57}$$

Note that $\lambda_{\mu k}$ is very close to 1 (larger than 0.99) if $\Delta t_k / \tau_\mu < 0.02$ and very close to 0 (smaller than 0.01) if $\Delta t_k / \tau_\mu > 100$. In the former case, the μ-th Kelvin unit acts as almost rigid, while in the latter case it acts as almost elastic (i.e., almost as a spring with no viscous damping).

The second term on the right-hand side of (5.56) is the strain increment due to creep at constant stress,

$$\Delta \varepsilon_k'' = \sum_{\mu=1}^{M} (1 - \beta_{\mu k}) \left(\frac{\sigma_k}{E_\mu} - \varepsilon_{\mu k} \right) \tag{5.58}$$

With this notation, (5.56) can be written in the compact form

$$\Delta \varepsilon_k = \frac{\Delta \sigma_k}{\bar{E}_k} + \Delta \varepsilon_k'' \tag{5.59}$$

This is formally the same incrementally linear strain–stress equation as equation (5.9) used by the integral-type approach. The difference is that in the integral-type approach the incremental modulus $\bar{E}_k$ and the strain increment due to creep $\Delta \varepsilon_k''$ are evaluated using formulae (5.10)–(5.11), while the differential approach leads to formulae (5.57)–(5.58). The integral approach requires storing the entire history of stress, and the number of terms in formula (5.11) increases with increasing step number. By contrast, the differential approach requires storing a limited and fixed number of internal variables (partial strains ε_μ) that are updated after each step, and the number of terms in formula (5.58) does not depend on the step number.

In a strain-driven material subroutine of a finite element code, the strain increment is prescribed and the stress increment has to be computed. Equation (5.59) is easily inverted, and the stress increment is evaluated as

$$\Delta \sigma_k = \bar{E}_k (\Delta \varepsilon_k - \Delta \varepsilon_k'') \tag{5.60}$$

The derivation of the exponential algorithm has been presented for stress and strain as scalars (uniaxial case), but it is easily extensible to the general multiaxial case. The only important modification is that the matrix $\boldsymbol{D}_\nu$ or $\boldsymbol{C}_\nu$ must be inserted into the relation between stresses and strains. Recall that $\boldsymbol{D}_\nu$ is the unit elastic stiffness matrix, given by (2.35), and $\boldsymbol{C}_\nu = \boldsymbol{D}_\nu^{-1}$ is the unit elastic compliance matrix, given by (2.33). Equations (5.58)–(5.59) are thus replaced by

$$\Delta\varepsilon_k'' = \sum_{\mu=1}^{M}(1-\beta_{\mu k})\left(\frac{1}{E_\mu}C_\nu\sigma_k - \varepsilon_{\mu k}\right)$$

(5.61)

$$\Delta\varepsilon_k = \frac{1}{\bar{E}_k}C_\nu\Delta\sigma_k + \Delta\varepsilon_k''$$

(5.62)

Algorithm 5.2   Second-order exponential algorithm for nonaging Kelvin chain

1. At the beginning of the simulation (i.e., at time $t = t_1$), set all partial strains ε_μ, $\mu = 1, 2, \ldots M$, to zero.
2. For a given time step k from t_k to $t_{k+1} = t_k + \Delta t_k$, compute the factors

$$\left.\begin{aligned}\beta_{\mu k} &= e^{-\Delta t_k/\tau_\mu} \\ \lambda_{\mu k} &= (1-\beta_{\mu k})\tfrac{\tau_\mu}{\Delta t_k}\end{aligned}\right\} \quad \mu = 1, 2, \ldots M$$

(5.63)

the incremental modulus

$$\bar{E}_k = \left(\frac{1}{E_0} + \sum_{\mu=1}^{M}\frac{1-\lambda_{\mu k}}{E_\mu}\right)^{-1}$$

(5.64)

and the auxiliary compliance

$$\bar{C}_k = \sum_{\mu=1}^{M}\frac{1-\beta_{\mu k}}{E_\mu}$$

(5.65)

3. Compute the creep strain increment

$$\Delta\varepsilon_k'' = \bar{C}_k C_\nu\sigma_k - \sum_{\mu=1}^{M}(1-\beta_{\mu k})\varepsilon_{\mu k}$$

(5.66)

4. For a given strain increment $\Delta\varepsilon_k$, compute the stress increment

$$\Delta\sigma_k = \bar{E}_k D_\nu(\Delta\varepsilon_k - \Delta\varepsilon_k'')$$

(5.67)

5. Update the partial strains

$$\varepsilon_{\mu,k+1} = \beta_{\mu k}\varepsilon_{\mu k} + \frac{1-\beta_{\mu k}}{E_\mu}C_\nu\sigma_k + \frac{1-\lambda_{\mu k}}{E_\mu}C_\nu\Delta\sigma_k, \quad \mu = 1, 2, \ldots M$$

(5.68)

6. Increment the step counter k and proceed to the next step (go to 2).

Note that step 1 is executed only once, before the start of the time incrementation loop. Steps 2 and 3 are executed once per time step, and step 4 must be repeated in every equilibrium iteration. If the entire structure is linear viscoelastic and the incremental material stiffness $\bar{E}_k \mathbf{D}_v$ is used in setting up the structural stiffness matrix, equilibrium is restored immediately after the first iteration. However, if certain parts of the model are nonlinear (e.g., due to cracking), additional iterations may be needed. Also note that step 2 of the algorithm does not need to be repeated if the time increment Δt_k remains the same as in the preceding time step.

5.2.5 Solidifying Kelvin Unit

Derivation of the second-order exponential algorithm for a nonaging Kelvin model started from Eq. (5.43), which is the rate form of the stress equivalence condition. The rate of stress in the elastic spring, $\dot{\sigma}_e$, was eliminated using relations (5.39)–(5.40), which govern the behavior of individual rheologic units (elastic spring and viscous dashpot). In the case of aging, the elastic law must be formulated in the rate form; see Appendix A.4 for a detailed justification. Equations (5.39)–(5.40) are then generalized to

$$\dot{\sigma}_e(t) = E(t)\dot{\varepsilon}(t) \tag{5.69}$$

$$\sigma_v(t) = \eta(t)\dot{\varepsilon}(t) \tag{5.70}$$

Functions $E(t)$ and $\eta(t)$, describing the evolution of elastic stiffness and viscosity due to aging, could be in principle independent.

The solidification theory, to be explained in detail in Chap. 9, attributes the changes of material properties to gradual deposition of a solidifying constituent with age-independent properties. Therefore, both $E(t)$ and $\eta(t)$ are assumed to be proportional to the same function $v(t)$ that describes the growth of the relative volume of the solidifying constituent. The ratio $\eta(t)/E(t) = \tau$ thus remains constant, and the relation $\dot{\sigma}_e(t) = E(t)\sigma_v(t)/\eta(t) = \sigma_v(t)/\tau$, similar to (5.42), still holds. Consequently, Eqs. (5.44)–(5.48) remain valid for a solidifying Kelvin unit. Only Eq. (5.50) needs an adjustment, because the viscosity (which appears in the denominator of the integrand) is now variable. For simple forms of function $v(t)$, one could try to integrate the strain rate exactly, but in general it is more convenient to approximate the viscosity within the current time step by a constant, $\eta_{k+1/2} = \tau E_{k+1/2}$. Equation (5.51) then remains valid if E is replaced by $E_{k+1/2}$:

$$\Delta\varepsilon_k = (1 - \lambda_k)\frac{\Delta\sigma_k}{E_{k+1/2}} + (1 - \beta_k)\frac{\sigma_{vk}}{E_{k+1/2}} \tag{5.71}$$

A substantial difference as compared to the nonaging case is that Eq. (5.52) is no longer applicable (a similar relation could be written in terms of rates but not in terms of the total values). Therefore, the viscous stress cannot be recovered from the values of stress and strain. It must be stored as an internal variable and updated according to formula (5.47). Equation (5.53) is also inapplicable, because its derivation was based on (5.52). The strain increment must be evaluated from (5.71).

5.2.6 Solidifying Kelvin Chain

The main assumption of the solidification theory is that the changes of apparent rheologic properties are caused by deposition of a nonaging constituent. The growth of the relative volume of the solidified material is characterized by a function $v(t)$ that starts from $v(0) = 0$ and asymptotically tends to a certain finite limit, v_∞. The viscoelastic behavior of the constituent can be described by a nonaging Kelvin chain with constant partial moduli E_μ^∞ and constant viscosities $\eta_\mu^\infty = \tau_\mu E_\mu^\infty$, which determine the compliance function[7]

$$\Phi(t) = \sum_{\mu=1}^{M} \frac{1 - e^{-t/\tau_\mu}}{E_\mu^\infty} \tag{5.72}$$

The aging material is then characterized by a solidifying Kelvin chain with partial moduli

$$E_\mu(t) = E_\mu^\infty v(t), \qquad \mu = 1, 2, \ldots M \tag{5.73}$$

and partial viscosities

$$\eta_\mu(t) = \eta_\mu^\infty v(t) = \tau_\mu E_\mu(t), \qquad \mu = 1, 2, \ldots M \tag{5.74}$$

which are all proportional to one and the same function $v(t)$.

Regarding the algorithmic treatment, we can proceed from a solidifying Kelvin unit to a solidifying Kelvin chain along similar lines as in the nonaging case; see Sect. 5.2.4. The total strain ε is written as the sum of partial strains $\varepsilon_{\mu k}$ in individual units, with the same stress increment $\Delta\sigma$ applied to all the units. Exploiting formula (5.71) for each unit, we obtain the incremental strain–stress law

$$\Delta\varepsilon_k = \sum_{\mu=1}^{M} \Delta\varepsilon_{\mu k} = \frac{\Delta\sigma_k}{v_{k+1/2}} \sum_{\mu=1}^{M} \frac{1 - \lambda_{\mu k}}{E_\mu^\infty} + \frac{1}{v_{k+1/2}} \sum_{\mu=1}^{M} (1 - \beta_{\mu k}) \frac{\sigma_{v\mu k}}{E_\mu^\infty} = \frac{1}{\bar{E}_k} \Delta\sigma_k + \Delta\varepsilon_k'' \tag{5.75}$$

where

$$\bar{E}_k = v_{k+1/2} \left(\sum_{\mu=1}^{M} \frac{1 - \lambda_{\mu k}}{E_\mu^\infty} \right)^{-1} \tag{5.76}$$

and

$$\Delta\varepsilon_k'' = \frac{1}{v_{k+1/2}} \sum_{\mu=1}^{M} \frac{1 - \beta_{\mu k}}{E_\mu^\infty} \sigma_{v\mu k} \tag{5.77}$$

---

[7]Note that the constant term $1/E_0$, which reflects the instantaneous elastic compliance, does not appear in (5.72). The reason for leaving it out will be explained after Algorithm 5.3.

The updating formula for viscous stresses in individual units (playing the role of internal variables) easily follows from (5.47):

$$\sigma_{v\mu,k+1} = \lambda_{\mu k} \Delta\sigma_k + \beta_{\mu k}\sigma_{v\mu k} \tag{5.78}$$

Note that there is no need to evaluate or store the partial strains $\varepsilon_{\mu k}$ because they do not explicitly appear in Eqs. (5.76)–(5.78). Only the variables $\sigma_{v\mu k}$, $\mu = 1, 2, \ldots M$, need to be updated in each computational step. So the memory requirements are the same as for the nonaging chain, for which the partial strains are stored.

Generalization to multiaxial stress is also straightforward, and the resulting computational procedure is described by the following algorithm. The internal variables have the meaning of viscous stress and thus have the same number of components as the stress (i.e., 6 in the three-dimensional setting). They are arranged in column matrices σ_v.

Algorithm 5.3 Exponential algorithm for solidifying Kelvin chain

1. At the beginning of the simulation (time $t = t_1$), set all components of internal variables $\sigma_{v\mu}$, $\mu = 1, 2 \ldots M$, to zero. Set the step counter k to 1.
2. For a given time step from t_k to $t_{k+1} = t_k + \Delta t_k$, compute the mid-step (or average) relative volume $v_{k+1/2}$, the factors $\beta_{\mu k}$ and $\lambda_{\mu k}$ given by (5.63), and the incremental modulus $\bar{E}_k$ given by (5.76).
3. Compute the strain increment due to creep,

$$\Delta\varepsilon_k'' = \frac{C_v}{v_{k+1/2}} \sum_{\mu=1}^{M} \frac{1 - \beta_{\mu k}}{E_\mu^\infty} \sigma_{v\mu k} \tag{5.79}$$

4. For a given strain increment $\Delta\varepsilon_k$, compute the stress increment

$$\Delta\sigma_k = \bar{E}_k D_v (\Delta\varepsilon_k - \Delta\varepsilon_k'') \tag{5.80}$$

5. Update the internal variables using the formula

$$\sigma_{v\mu,k+1} = \lambda_{\mu k} \Delta\sigma_k + \beta_{\mu k}\sigma_{v\mu k} \tag{5.81}$$

6. Increment the step counter k by 1 and proceed to the next step (go to 2).

Note that if the time step remains constant, the factors $\beta_{\mu k}$ and $\lambda_{\mu k}$ do not need to be reevaluated, but the incremental modulus $\bar{E}_k$ does, because of the aging effect.

Algorithm 5.3 describes the numerical treatment of a solidifying Kelvin chain. Note that it does not explicitly mention a term that would correspond to an isolated spring, without a parallel dashpot, which is necessary for the description of instantaneous elasticity. The reason is that the solidifying Kelvin chain is just a building block of a complete creep model. In model B3, it is coupled in series not only with an elastic spring, but also with an aging dashpot that reflects the logarithmic nature of long-term creep. Yet another contribution to strain is added if the effects of drying on creep are incorporated using the sectional approach.

For the moment, let us restrict attention to basic creep. Truly instantaneous elastic effects are, according to model B3, captured by a nonaging spring of stiffness E_0 equal to the asymptotic modulus. The corresponding contribution to the strain increment due to a stress increment $\Delta\sigma_k$ is simply $C_v \Delta\sigma_k/E_0$. The aging dashpot that reflects long-term viscous flow has a variable viscosity $\eta_f(t) = t/q_4$, which is not proportional to function $v(t)$ used by the solidification theory, because the aging process is attributed to a different physical mechanism (relaxation of microprestress—see Chap. 10). The increment of viscous flow strain can be approximated as

$$\Delta\varepsilon_{fk} = \int_{t_k}^{t_{k+1}} \frac{C_v \sigma(t)}{\eta_f(t)} \, dt \approx \frac{\Delta t_k}{\eta_{f,k+1/2}} C_v \left(\sigma_k + \frac{1}{2}\Delta\sigma_k \right) \quad (5.82)$$

In summary, after adding a nonaging spring and an aging dashpot, Eqs. (5.76) and (5.79) have to be modified as follows:

$$\bar{E}_k = \left(\frac{1}{E_0} + \frac{1}{v_{k+1/2}} \sum_{\mu=1}^{M} \frac{1-\lambda_{\mu k}}{E_\mu^\infty} + \frac{\Delta t_k}{2\eta_{f,k+1/2}} \right)^{-1} \quad (5.83)$$

$$\Delta\varepsilon_k'' = \frac{C_v}{v_{k+1/2}} \sum_{\mu=1}^{M} \frac{1-\beta_{\mu k}}{E_\mu^\infty} \sigma_{v\mu k} + \frac{\Delta t_k}{\eta_{f,k+1/2}} C_v \sigma_k \quad (5.84)$$

After these modifications, Algorithm 5.3 can be used for the numerical treatment of basic creep described by model B3.

The drying creep compliance function (3.20) used by model B3 does not have the format implied by the solidification theory and needs to be treated separately, in the context of general aging viscoelasticity. The same is true for typical compliance functions recommended by codes such as CEB or ACI, which do not even separate basic and drying creep.

5.2.7 Aging Kelvin Unit

As explained in detail in Appendix A, a general compliance function of an aging viscoelastic material (not necessarily derived from the solidification theory) can be approximated by the Dirichlet series

$$J(t, t') = \left[\frac{1}{E_0} + \sum_{\mu=1}^{M} \frac{1-e^{-(t-t')/\tau_\mu}}{D_\mu(t')} \right] H(t - t') \quad (5.85)$$

representing an aging Kelvin chain. Note that variables D_μ are different from the moduli E_μ of individual Kelvin units; see Sect. A.4.2 for a detailed explanation. The underlying equation that governs the behavior of one aging Kelvin unit is the second-order differential equation

$$\dot{\sigma}(t) = D(t)\dot{\varepsilon}(t) + \eta(t)\ddot{\varepsilon}(t) \tag{5.86}$$

in which $\eta(t) = \tau D(t)$, according to the assumption that is used in Appendix A to construct the compliance function (5.85). Equation (5.86) can thus be rewritten as

$$\dot{\varepsilon}(t) + \tau\ddot{\varepsilon}(t) = \frac{\dot{\sigma}(t)}{D(t)} \tag{5.87}$$

This is a second-order differential equation, but it can be converted to a first-order equation in terms of the strain rate. It is convenient (albeit not necessary) to use the dimensionless strain rate $e_r(t) = \tau\dot{\varepsilon}(t)$ as the unknown function. Equation (5.87) then becomes

$$\frac{e_r(t)}{\tau} + \dot{e}_r(t) = \frac{\dot{\sigma}(t)}{D(t)} \tag{5.88}$$

which has the same format as (5.44), only with the unknown function $\sigma_v(t)$ replaced by $e_r(t)$ and the right-hand side $\dot{\sigma}(t)$ by $\dot{\sigma}(t)/D(t)$. Following the same approach as in Sect. 5.2.3, we replace the right-hand side within the current computational step by a constant, $\Delta\sigma_k/\Delta t_k D_{k+1/2}$, and impose the initial condition $e_r(t_k) = e_{rk} = $ value of e_r at the end of the previous step. In analogy to (5.46), the solution is then

$$e_r(t) = \frac{\tau \Delta\sigma_k}{\Delta t_k D_{k+1/2}} + \left(e_{rk} - \frac{\tau \Delta\sigma_k}{\Delta t_k D_{k+1/2}}\right)e^{-(t-t_k)/\tau} \tag{5.89}$$

and the increment of e_r over the k-th step is evaluated as

$$\Delta e_{rk} = \frac{\tau \Delta\sigma_k}{\Delta t_k D_{k+1/2}}(1 - \beta_k) - (1 - \beta_k)e_{rk} = \lambda_k\frac{\Delta\sigma_k}{D_{k+1/2}} - (1 - \beta_k)e_{rk} \tag{5.90}$$

Integrating the strain rate, we obtain the strain increment

$$\Delta\varepsilon_k = \frac{1}{\tau}\int_{t_k}^{t_{k+1}} e_r(t)\,\mathrm{d}t = (1 - \lambda_k)\frac{\Delta\sigma_k}{D_{k+1/2}} + (1 - \beta_k)e_{rk} \tag{5.91}$$

The final formula linking the strain increment to the stress increment is very similar to (5.51), except that the constant modulus E is replaced by the age-dependent value $D_{k+1/2}$ and σ_{vk}/E is replaced by e_{rk} (recall that, for a nonaging material, $\sigma_{vk} = \eta\dot{\varepsilon}_k = E\tau\dot{\varepsilon}_k = Ee_{rk}$). The dimensionless strain rate e_r plays here the role of an internal variable.

5.2.8   *Aging Kelvin Chain*

From an aging Kelvin unit, we can proceed to an aging Kelvin chain along similar lines as in the previous cases of nonaging and solidifying Kelvin models; see Sects. 5.2.4 and 5.2.6. The resulting relations can be summarized in the following algorithm, written in the multidimensional context. The internal variables have the meaning of dimensionless strain rates and thus have the same number of components as the strain (i.e., 6 in the three-dimensional setting). They are arranged in column matrices $e_{r\mu}$, $\mu = 1, 2, \ldots M$. Note that the memory requirements are the same as for a nonaging or solidifying chain.

Algorithm 5.4   Exponential algorithm for aging Kelvin chain

1. At the beginning of the simulation (time $t = t_1$), set all components of internal variables $e_{r\mu}$, $\mu = 1, 2 \ldots M$, to zero. Set the step counter k to 1.
2. For a given time step from t_k to $t_{k+1} = t_k + \Delta t_k$, compute the mid-step (or average) stiffness moduli $D_{\mu,k+1/2}$, the factors $\beta_{\mu k}$ and $\lambda_{\mu k}$ given by (5.63), and the incremental modulus

$$\bar{E}_k = \left(\frac{1}{E_0} + \sum_{\mu=1}^{M} \frac{1 - \lambda_{\mu k}}{D_{\mu,k+1/2}} \right)^{-1} \tag{5.92}$$

3. Compute the strain increment due to creep,

$$\Delta\varepsilon_k'' = \sum_{\mu=1}^{M} (1 - \beta_{\mu k}) e_{r\mu k} \tag{5.93}$$

4. For a given strain increment $\Delta\varepsilon_k$, compute the stress increment

$$\Delta\sigma_k = \bar{E}_k D_v (\Delta\varepsilon_k - \Delta\varepsilon_k'') \tag{5.94}$$

5. Update the internal variables using the formula

$$e_{r\mu,k+1} = \frac{\lambda_{\mu k}}{D_{\mu,k+1/2}} C_v \Delta\sigma_k + \beta_{\mu k} e_{r\mu k} \tag{5.95}$$

6. Increment the step counter k by 1 and proceed to the next step (go to 2).

If the time step remains constant, the factors $\beta_{\mu k}$ and $\lambda_{\mu k}$ do not need to be reevaluated, but the incremental modulus $\bar{E}_k$ does, because of the aging effect.

Chapter 6
Uncertainty Due to Parameter Randomness via Sampling of Deterministic Solutions

Abstract Concrete creep and shrinkage are notorious for high random scatter of input parameters and high uncertainty of long-time predictions. Obviously, the structural design should not be based merely on mean predictions. In this chapter, we show how to estimate realistic confidence limits on the long-time creep and shrinkage predictions. Except for creep buckling of columns and shells, these limits are, fortunately, far less stringent than those required for failure prevention, but nevertheless important to make premature structural repair or demolition rare enough. Because sustainable design calls for structural lifetimes in excess of a century, we focus on long-term predictions of structural performance in the light of the randomness of material and environmental parameters. We emphasize the numerical sampling approach based on repeated runs of deterministic analysis of creep and shrinkage effects for judiciously selected random samples of input data. Finally, our discussion is focused on improving long-term predictions by means of the Bayesian updating in which the uncertainty due to the experimental data and prediction model is reduced by prior data on short-time creep of the given concrete and on the observed initial deformations of the given structure.

Although, under precise laboratory conditions, shrinkage and creep of one and the same concrete exhibit relatively small scatter, similar to that of elastic modulus [878], the random variability affecting concrete structures is very large. The reason is the variability of environment, curing and composition, and unaffordability of perfect quality control. The consequence is a high random scatter in long-time structural behavior, impairing durability. When the type of concrete is undecided at the time of design, then the scatter due to differences among various types of concrete that could be used is the largest of all.

A quantitative assessment of long-time random scatter has not been a standard practice. However, for creep sensitive structures, it should be, and it can be performed relatively easily. An effective method to do that is to apply probabilistic sampling to a deterministic creep and shrinkage prediction model. A model such as B3, giving a realistic picture of the mean long-time behavior, must be used for the scatter estimates to be meaningful.

© Springer Science+Business Media B.V. 2018
Z.P. Bažant and M. Jirásek, *Creep and Hygrothermal Effects in Concrete Structures*, Solid Mechanics and Its Applications 225,
https://doi.org/10.1007/978-94-024-1138-6_6

6.1   Random Parameters in Creep and Shrinkage Model

While creep and shrinkage are stochastic processes [300], their values and structural effects can approximately be treated as functions of a set of *random parameters*. In principle, each parameter of the selected creep and shrinkage model could be considered as a random variable, but no information on their separate statistical properties can be extracted from the existing databases, and so it is more convenient and sufficiently realistic to consider a strong correlation among some of the parameters and assume that several model parameters are controlled by the same random variable. Further simplification is achieved by assuming that all these random variables have the normal (Gaussian) distribution and are statistically independent, although some correlations surely exist and have been considered by Xi and Bažant [882], along with expanding the number of random variables.

For instance, in the case of model B3, it is assumed that parameters q_i, $i = 1, 2, \ldots 5$, are perfectly correlated and depend on one single random variable $X^{(1)}$ in the following way:

$$q_i = (1 + \omega_1 X^{(1)})\bar{q}_i, \qquad i = 1, 2, \ldots 5 \tag{6.1}$$

Here, the bar over q denotes the *mean value*, which corresponds to the deterministic version of the model and is estimated from the concrete composition using the formulae from Table C.2 in Appendix C. Symbols q_i without the bar denote (in this chapter) the parameters considered as random variables. Random variable $X^{(1)}$ has the standard normal distribution (i.e., has a zero mean and a unit standard deviation), and ω_1 is the *coefficient of variation* of random parameters q_i, which characterizes the scatter of elastic and creep compliances. Based on statistical analysis of the data in the RILEM database, Bažant and Baweja [107] recommended to use $\omega_1 = 0.23 = 23\%$.

The final value of shrinkage (in a perfectly dry environment) is considered to be statistically independent of the compliance parameters and is described as

$$\varepsilon_{\text{sh}}^{\infty} = (1 + \omega_2 X^{(2)})\bar{\varepsilon}_{\text{sh}}^{\infty} \tag{6.2}$$

where $\omega_2 = 0.34$ is the coefficient of variation and $X^{(2)}$ is a random variable having again the standard normal distribution. Thus, random variables $X^{(1)}$ and $X^{(2)}$ have the same type of statistical distribution, but they characterize different phenomena and are considered as statistically independent. Two additional random variables with the standard normal distribution, $X^{(3)}$ and $X^{(4)}$, are introduced to characterize the variability of environmental humidity and compression strength:

$$h_{\text{env}} = (1 + \omega_3 X^{(3)})\bar{h}_{\text{env}} \tag{6.3}$$
$$f_c = (1 + \omega_4 X^{(4)})\bar{f}_c \tag{6.4}$$

The values of the coefficients of variation recommended by Bažant and Baweja [107] are $\omega_3 = 0.20$ and $\omega_4 = 0.15$.

Bažant and Liu [163] considered a total of eight random parameters, but here we will use the formulation with four parameters only, as in Bažant and Baweja [107].

Since the empirical formulae for the estimation of parameters q_1, q_2, q_3, q_5, and ε_{sh}^∞ contain the compression strength, the random value f_c of compression strength must be computed before Eqs. (6.1)–(6.2) are used, and this random value must be substituted in the formulae in Table C.2 instead of the mean compression strength $\bar{f_c}$. Of course, if this is done, ω_1 is not the true coefficient of variation of random variables q_i ($i = 1, 2, 3, 5$) and ω_2 is not the true coefficient of variation of random variable ε_{sh}^∞, but this is the price to pay for simplicity.

As already mentioned, Bažant and Baweja [107] recommended to set the coefficient of variation of environmental humidity to $\omega_3 = 0.2$. This should be considered as a general recommendation to be used in the absence of detailed information on the specific climatic conditions at the site of the structure. If such information is available, it is often possible to use a reduced value of ω_3. First, it is important to understand the precise meaning of the model parameter denoted as h_{env}, which represents the **average** value of environmental humidity during a certain period of interest. The environmental humidity measured at a given location exhibits large fluctuations due to complex atmospheric processes, but short-term oscillations have a very limited influence on the evolution of humidity inside the concrete structure, which is the actual driving force behind shrinkage and drying creep (a detailed analysis of the penetration depth of such oscillations and its relation to their frequency will be presented in Sect. 8.4.6). Therefore, the coefficient of variation to be used in the description of random environmental effects on creep and shrinkage does not characterize the deviations of **instantaneous** values of humidity from their overall mean. It is rather related to the uncertainty caused by the random character of the average humidity. The key point here is the specification of the period of interest over which the humidity is averaged. What exactly this means can be best explained by an example.

Suppose that we are concerned about potential cracking due to restrained shrinkage, and we would like to estimate the stresses arising in a concrete frame one week after the end of curing. In such a case, the calculation should be based on the average value of environmental humidity during that week. If we know that the frame will be exposed to the environment e.g., in Fresno, California, we can use the available data recorded in the past by the local weather station (downloadable from http://ipm.ucanr.edu/weather). Each day of the period 2000–2014 is represented in Fig. 6.1a by a dot, with horizontal coordinate corresponding to the date (ignoring the year) on which the measurement was taken and vertical coordinate to the daily average of relative humidity[1]. The statistical ensemble of all these daily averages has the mean value of 61.2% (which is the average humidity over the 15 years considered) and the standard deviation of 13.5%, which gives a coefficient of variation equal to $13.5/61.2 = 0.221$. However, we are interested in the average humidity during a (randomly selected) week, so we cluster the daily values into groups of 7 consecutive days and for each of them calculate the weekly average. The resulting statistical ensemble has of course still the same mean value, but the standard devia-

---

[1] Since the available data sets contain only the maximum and minimum relative humidities measured on each day, the daily average is estimated as the average of these two reported values. This is sufficient for the present purpose. For real calculations, it would be better to get access to the actual daily averages evaluated from more detailed records.

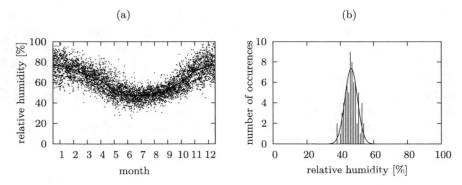

Fig. 6.1 (a) Daily average humidities recorded in Fresno, California, in 2000–2014 (indicated by dots), mean humidities in various months (solid) and the range of mean plus and minus standard deviation of daily averages (dashed), (b) histogram of weekly average humidities in July (years 2000–2014 combined) and its approximation by normal distribution

tion is slightly reduced to 12.4%, which corresponds to a coefficient of variation of 0.203. If we process in a similar way the monthly averages, we find another slight reduction to a standard deviation of 11.6% and a coefficient of variation of 0.19. A dramatic reduction is observed for annual averages: The standard deviation drops to 2.2%, which corresponds to a coefficient of variation of 0.036.

The origin of the high variability of daily, weekly, and monthly average humidities is of course in the seasonal variations, which are clearly manifested in Fig. 6.1a. The solid line connects the mean values of relative humidities calculated for each month separately. The dashed lines mark the mean plus and minus the standard deviation of the ensemble of daily average humidities for each month. For the given set of data, the highest mean relative humidity of 76.8% is found in January and the lowest mean relative humidity of 46.6% is found in July. Averaging over the whole year wipes off such seasonal variations; it is irrelevant when exactly the one-year period starts because it always covers all seasons equally.

For shorter periods of interest, the variability can also be reduced if we specify, at least approximately, at which time of the year the period will start. For instance, if a concrete frame is to be exposed to the environment in July and the calculation aims at the response after one week of exposure, one should consider the ensemble of weekly averages evaluated for the weeks of July only. Figure 6.1b shows the corresponding histogram and its approximation by the normal distribution. The mean value of humidity, $\bar{h}_{env}$, is then set to the mean humidity in July, i.e., to 46.6%, and the standard deviation to 3.7%, which corresponds to a coefficient of variation $\omega_3 = 0.08$.

The statistical characteristics of daily, weekly, and monthly averages for all months of the year (still based on the data from Fresno, 2000–2014) are summarized in Table 6.1. In each row, the name of the month is followed by the mean value of relative humidity and by the coefficient of variation of daily, weekly, and monthly averages. This is complemented by the coefficients of variation of quarterly averages, which are specified for the four quarters in the last column of the table. The last row contains the characteristics that refer to all the data recorded (in any month). It should be emphasized that the table is in no way universal; it refers to the data collected

Table 6.1 Statistical characteristics of average humidities based on data recorded in Fresno, California, between 2000 and 2014

	Mean	Day	Week	Month	Quarter
Jan	76.8	0.129	0.098	0.087	
Feb	72.6	0.108	0.080	0.061	0.059
Mar	67.4	0.112	0.080	0.055	
Apr	61.6	0.140	0.104	0.085	
May	52.8	0.161	0.123	0.074	0.061
Jun	48.0	0.144	0.098	0.071	
Jul	46.6	0.115	0.080	0.058	
Aug	49.0	0.107	0.066	0.037	0.040
Sep	52.3	0.115	0.079	0.049	
Oct	60.5	0.146	0.118	0.070	
Nov	70.9	0.125	0.091	0.061	0.056
Dec	76.5	0.125	0.095	0.074	
All	61.2	0.221	0.203	0.190	0.169

at the Fresno weather station between 2000 and 2014. The table cannot be directly used for calculations of structures at other locations and is not meant to become a design tool. It has been set up to illustrate a suitable methodology and basic trends.

The presented case study confirms that the generally recommended coefficient of variation of 0.2 is quite realistic, provided that the period of interest is relatively short (between a day and several months) and that, at the same time, it is not known in which season of the year the drying process will take place. For one-year or multiyear calculations, the uncertainty caused by humidity variation in time is much lower and the coefficient of variation can be reduced. Some reduction is also possible for shorter periods of interest if they refer to a specific date range within the year (e.g., to a specific month) and parameter $\bar{h}_{env}$ is set to the mean humidity evaluated for that range. However, when introducing such reductions, one needs to account for the influence of spatial variability. The actual on-site conditions differ from the conditions at the nearest weather station, and thus the mean humidity is not known precisely, even if long-term records from the station are available. Depending on local conditions, the coefficient of variation determined by statistical analysis of the recorded data may need to be increased.

Instead of being treated as a random parameter, the random environmental humidity can be more realistically modeled as a *random process* in time. This refined approach, however, makes sense only if it is based on the diffusion equation for pore water transfer. An effective way of dealing with random processes is the *spectral approach* [92, 192, 193]. Unfortunately, it is applicable only to the linear diffusion equation, which means that some effective value of diffusivity, independent of pore humidity, must be used. Since the diffusivity decreases about 20 times during drying, the error due to considering it as constant can offset the advantages of the spectral approach. Therefore, we will restrict our attention to the description based on random variables.

6.2   Latin Hypercube Sampling of Parameters of Creep and Shrinkage Model

For given values of random variables $X^{(i)}$, one can calculate deterministically the time-dependent response $Y(X^{(i)}, t)$. The response of interest can, for example, be the stress or strain predicted by model B3, or a structural response such as the bending moment at a given cross section, the reaction at a given support, or the maximum deflection, delivered by a computer program or pencil calculation. To obtain the statistics of response, the idea is to generate random parameter samples of equal probability, calculate for each sample the deterministic response, and then obtain the statistics of the collection of all responses, especially the mean and variance.

A rigorous fundamental approach is the *Monte Carlo simulation*, which is, however, computationally demanding and inefficient. At the opposite extreme of simplicity is the *method of two-point estimates*, applied to creep by Madsen and Bažant [589], in which each random variable $X^{(i)}$ is sampled only at ± 1, which means that the corresponding model parameters are sampled at their mean plus or minus the standard deviation. A more accurate and more effective approach is the *stratified sampling*, in which the range [0, 1] of the cumulative distribution function (cdf) of each random variable $X^{(i)}$ is subdivided into N strata (or layers) of equal probability $1/N$, and the random variables used to generate the input parameters for the deterministic calculations are sampled only at one point of each stratum (see Fig. 6.2, for $N = 8$).

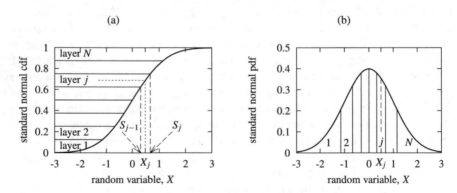

Fig. 6.2 Illustration of stratified sampling: (a) standard Gaussian cumulative distribution function and (b) the corresponding probability density function

To justify the optimal choice of the sampling points, let us show how the stratified sampling is exploited for the evaluation of the basic stochastic characteristics. For simplicity, we consider one random variable Y whose dependence on another random variable X is described by a deterministic rule $Y = G(X)$, G being a given function. Variable X is a random model parameter, function G represents the deterministic model, and Y is the random response variable of interest. Random variable X is characterized by the cumulative distribution function (cdf) $F(X)$, or by its derivative, the probability density function (pdf) $f(X) = \mathrm{d}F(X)/\mathrm{d}X$. If the dependence of Y

on X is nonlinear, the exact cdf or pdf of random variable Y is hard to determine, but we can compute the basic stochastic characteristics, such as the *mean* (mathematical expectation)

$$\bar{Y} = \int_{-\infty}^{\infty} Y f(X) \, dX = \int_{-\infty}^{\infty} G(X) f(X) \, dX \qquad (6.5)$$

or the *standard deviation* s_Y, which is the square root of the *variance*

$$s_Y^2 = \int_{-\infty}^{\infty} (Y - \bar{Y})^2 f(X) \, dX = \int_{-\infty}^{\infty} Y^2 f(X) \, dX - \bar{Y}^2 = \int_{-\infty}^{\infty} [G(X)]^2 f(X) \, dX - \bar{Y}^2$$

$$(6.6)$$

The integrals can be evaluated analytically only in special cases. In general, the mean and variance need to be computed numerically, usually as

$$\bar{Y} \approx \sum_{j=1}^{N} Y_j \, \Delta F_j \qquad (6.7)$$

$$s_Y^2 \approx \sum_{j=1}^{N} Y_j^2 \Delta F_j - \bar{Y}^2 \qquad (6.8)$$

where $Y_j = G(X_j)$, $j = 1, 2, \ldots N$, are the response values determined at suitably selected sampling points X_j, $j = 1, 2, \ldots N$, and ΔF_j, $j = 1, 2, \ldots N$, are the probabilities associated with those sampling points (they are the finite counterpart of the infinitesimal increments $dF(X) = f(X) dX$).

In the crude Monte Carlo method, we generate a sequence of N numbers Φ_j between 0 and 1 using a quasi-random generator with uniform probability density, and for each of them, we determine the sampling value X_j from the condition $F(X_j) = \Phi_j$ and assign the weight $\Delta F_j = 1/N$. However, the spacing of generated values Φ_j in the interval $[0, 1]$ is not really uniform, and convergence of the approximations to the exact values with increasing N is relatively slow.

An improvement can be achieved by a *stratification method*, which divides the interval $[0, 1]$ into N subintervals (strata) of equal length and uses one sampling point per subinterval, again with equal weights $\Delta F_j = 1/N$. A natural choice would be to take $\Phi_j = (2j - 1)/2N$ as the midpoint of the jth stratum. However, a somewhat better accuracy is achieved if the sampling points are chosen such that, for the special case of $Y = X$, the mean value be computed exactly.

Let us define points S_j, $j = 0, 1, \ldots N$, as the points in the domain of the random variable X whose images $F(S_j)$ separate the individual strata in the interval $[0, 1]$; see Fig. 6.2a. Mathematically, these points can be defined as the solutions of equations $F(S_j) = j/N$, $j = 0, 1, \ldots N$, admitting that $S_0 \to -\infty$ and $S_N \to \infty$ are possible. The integral defining the mean value of X can then be split into a sum of integrals over intervals $[S_{j-1}, S_j]$, and each of these contributions to the total integral should be represented by the corresponding term in the sum (6.7) exactly. In mathematical writing, the approximation

$$\sum_{j=1}^{N} \int_{S_{j-1}}^{S_j} X f(X) \, dX \approx \frac{1}{N} \sum_{j=1}^{N} X_j \tag{6.9}$$

is exact if

$$X_j = N \int_{S_{j-1}}^{S_j} X f(X) \, dX, \qquad j = 1, 2, \ldots N \tag{6.10}$$

This result has a clear geometrical interpretation. If we divide the area under the graph of the probability density function by vertical cuts at points S_j into N slices of equal areas (Fig. 6.2b), then the integral in (6.10) is the first-order moment (static moment) of the jth slice with respect to the vertical line $X = 0$, and the area of that slice is $1/N$. The right-hand side of (6.10) is thus the horizontal coordinate of the centroid of the jth slice.

It is known that if we concentrate the mass of each slice into the centroid of that slice, the overall centroid (whose horizontal coordinate corresponds to the mean value $\bar{X}$) does not move, and this is why the numerical approximation of the mean is exact if the sampling points are selected according to (6.10). Based on a similar geometrical interpretation it can be shown that, with this choice of the sampling points, the variance S_X^2, which corresponds to the centroidal moment of inertia, will always be underestimated (though negligibly if N is large).

With sampling points X_j defined in (6.10), the approximation of the mean response $\bar{Y}$ is exact not only for the identity mapping $Y = X$, but also for any linear transformation $Y = aX + b$ where a and b are arbitrary constants. This follows from the fact that in such cases $\bar{Y} = a\bar{X} + b$. A good, though not exact, approximation of the mean can be expected for general nonlinear transformations $Y = G(X)$.

As an example, Table 6.2 gives the coordinates of sampling points X_j for the cases of $N = 8$, 16, or 32 strata of the standard Gaussian cdf. The sum of these N coordinates is always zero, because $\sum_j X_j / N$ must give the mean value $\bar{X}$, which is zero for the standard Gaussian distribution. In fact, for the Gaussian distribution and for any other symmetric distribution, the approximation of the mean value would be exact even for other (symmetric) choices of the sampling points, e.g., $X_j = F^{-1}((j - \frac{1}{2})/N)$, but this is only because the error in stratum number j is compensated by the same error with the opposite sign in the opposite stratum, number $N + 1 - j$ (this would not be the case for nonsymmetric distributions).

The sum of squares of the sampling point coordinates divided by N represents the approximation of the variance of X and ideally should be equal to one (again for the standard Gaussian distribution). The actual values for $N = 8$, $N = 16$, and $N = 32$ are 0.9450, 0.9778, and 0.9908, respectively. As expected, the variance is slightly underestimated, but the error tends to zero as the number of strata is increased. If the sampling points were chosen so as to correspond to the cdf values at the midheight of each stratum, the approximated value of the variance would be 0.8510 for $N = 8$, 0.9237 for $N = 16$, and 0.9611 for $N = 32$, and so the error would be bigger. Interestingly, if the estimate of the variance is computed using $N - 1$ instead of N in the denominator, the foregoing values change to 0.9726, 0.9852, and 0.9922, respectively, and they get even closer to the exact value 1 than for sampling points at centroids.

Table 6.2 Coordinates of sampling points to be used for a random variable with the standard Gaussian distribution in stratified sampling with (a) 8 strata, (b) 16 strata, (c) 32 strata (to save space, points 1–16 with negative coordinates have been omitted for $N = 32$)

j	X_j (centroid)	X_j (Midpoint)
(a) $N = 8$		
1	−1.646828	−1.534112
2	−0.895384	−0.887146
3	−0.491349	−0.488776
4	−0.157976	−0.157311
5	0.157976	0.157311
6	0.491349	0.488776
7	0.895384	0.887146
8	1.646828	1.534112
(b) $N = 16$		
1	−1.967743	−1.862675
2	−1.325913	−1.318009
3	−1.012887	−1.009990
4	−0.777881	−0.776422
5	−0.579964	−0.579132
6	−0.402735	−0.402250
7	−0.237459	−0.237202
8	−0.078493	−0.078412
9	0.078493	0.078412
10	0.237459	0.237202
11	0.402735	0.402250
12	0.579964	0.579132
13	0.777881	0.776422
14	1.012887	1.009990
15	1.325913	1.318009
16	1.967743	1.862675
(c) $N = 32$		
17	0.039186	0.039176
18	0.117800	0.117770
19	0.197151	0.197099
20	0.277767	0.277690
21	0.360235	0.360130
22	0.445235	0.445097
23	0.533591	0.533410
24	0.626336	0.626099
25	0.724828	0.724514
26	0.830935	0.830511
27	0.947376	0.946782
28	1.078398	1.077515
29	1.231294	1.229857
30	1.420532	1.417793
31	1.683275	1.675920
32	2.252211	2.153875

If the random response Y depends on p random model parameters, the stratification can be applied to each parameter separately, and the sampling values $X_j^{(i)}$, $j = 1, 2, \ldots N$, can be determined for each parameter $X^{(i)}$, $i = 1, 2, \ldots p$. But if we consider all possible combinations of N values of $X^{(1)}$ with N values of $X^{(2)}$, etc., we end up with N^p sampling points, which can be a huge number.

The number of sampling points is greatly reduced by the *Latin hypercube sampling* (LHS) method [505, 620]. Its main idea can best be explained if only two random parameters are considered. The potential sampling points can be represented by a table with N rows (corresponding to the strata of parameter $X^{(1)}$) and N columns (corresponding to the strata of parameter $X^{(2)}$). To cover all these combinations of random parameters, we would need to perform N^2 deterministic calculations (e.g., computer runs or formula evaluations) of response values $Y_{jk} = G(X_j^{(1)}, X_k^{(2)})$ and then estimate the mean and variance of Y using a generalized version of formulae (6.7)–(6.8) (with double sums over all j and k). Of course, a preliminary estimate based on a subset of the sampling points could be constructed even before all those N^2 calculations are finished. The key idea of LHS is that the sampling points are not processed in a random order, or in a regular order (e.g., in a double loop with j and k running from 1 to N), but in a special order constructed so that, in the first N calculations, each stratum of $X^{(1)}$ gets sampled once and only once and each stratum of $X^{(2)}$ gets sampled once and only once, and the same holds also for the second series of N calculations, for the third series, etc. Of course, if we finish all N series of N calculations, all N^2 sampling points are covered and the final result is exactly the same as it is with any other ordering. However, the intermediate estimates constructed after the first series, after the second, etc., in general provide better approximations than if the sampling points are processed in a random order, and much better than if the points are processed in a regular order. In practice, only one series is used, so that the number of calculations is reduced from N^2 to N.

The condition that each stratum of each variable be processed once and only once in the first series of N calculations represents a constraint but does not imply a unique choice of this subset of sampling points. In terms of the table with N rows and N columns, the condition means that if we mark by a certain symbol those points that will be processed in the first series, this symbol should appear once and only once in each row and in each column. The same condition should be satisfied for the second series, marked by another symbol.

If all the N series are described by N different symbols, the table gets filled completely and represents what is in mathematics known as the *Latin square*; see the examples in Fig 6.3. The adjective "Latin" refers to the fact that when Euler studied this type of tables, he was filling them by letters of the Latin alphabet, as in Fig 6.3a. Of course, in the application to sampling it is more natural to use integers from 1 to N, corresponding to the number of series (Fig 6.3b, c). A Latin square describes the sampling technique for $p = 2$ parameters, but the concept can be extended to a Latin cube for $p = 3$, and further to a p-dimensional Latin hypercube for a general case. This explains the name of the LHS method.

	(a)		(b)	(c)

(a)	(b)	(c)

B	D	A	C
C	A	D	B
A	B	C	D
D	C	B	A

2	4	1	3
3	1	4	2
1	2	3	4
4	3	2	1

1	2	3	4
4	1	2	3
3	4	1	2
2	3	4	1

Fig. 6.3 Examples of Latin squares

As already mentioned, the advantage of the LHS method is that good estimates of the basic response characteristics can be obtained with a reduced number of sampling points. In fact, already after the first series of N response calculations, the first estimates can be constructed, and this is how the method is normally used. The quality of sampling can further be improved if the basic condition (each stratum of each parameter sampled once) is supplemented by additional constraints that ensure a reasonably "uniform" coverage of the entire space of random parameters.

For instance, constructing the rows by cyclic permutation leads to the Latin square shown in Fig 6.3c, which formally satisfies the basic condition but the sampling points that would be used in the first series are located on the diagonal, as if the parameters $X^{(1)}$ and $X^{(2)}$ were strongly correlated. So the first series provides no information on the response for low values of $X^{(1)}$ combined with high values of $X^{(2)}$, or for high values of $X^{(1)}$ combined with low values of $X^{(2)}$. Using such series can bias the results, and since we have a choice, it is better to work with sampling series that are "better balanced." Formula (6.10) for the coordinates of sampling points for one random parameter X has been derived from the condition that the mean value of X be reproduced exactly, and this property is preserved for each random parameter $X^{(i)}$ by each series of samples based on the LHS technique. The variances of each $X^{(i)}$ are also captured with a good accuracy.

However, nothing is known a priori about the covariances, which may or may not be reproduced well, depending on the specific sampling series. If parameters $X^{(i)}$ are assumed to be stochastically independent, their covariances should be zero, and the deviation from these theoretical values serves as a measure of bias introduced by the specific series of sampling points.

Based on this idea, powerful tools for optimization of LHS tables have been developed [843]. They can generate good series of sampling points not only for uncorrelated parameters $X^{(i)}$ with normal distribution but also for parameters with many other distributions and with an arbitrary covariance matrix specified by the user.

To relieve an engineer from having to generate Latin hypercubes and check the covariance matrices, Table 6.3 describes a possible choice of sampling points for $N = 8$, 16, and 32 strata and $p = 4$ (almost) uncorrelated random parameters with normal distribution. The absolute values of all sample Pearson correlation coefficients are below 0.04 for $N = 8$ and $N = 16$ and below 0.02 for $N = 32$. Each integer m_{ik} represents the number of the stratum to be sampled for the input variable $X^{(i)}$ in the kth response calculation. In other words, the sampling point in the kth response calculation has coordinates $(X^{(1)}_{m_{1k}}, X^{(2)}_{m_{2k}}, \ldots X^{(p)}_{m_{pk}})$. Computational experience indicates that one series of eight response calculations usually gives acceptable engineering accuracy. Repeating the calculations with other series, a picture of accuracy of statistical estimates can be obtained.

Table 6.3 Uncorrelated LHS sampling

run k	m_{1k}	m_{2k}	m_{3k}	m_{4k}
(a) $N = 8$				
1	6	7	4	1
2	7	5	8	3
3	8	2	1	5
4	1	3	2	2
5	4	8	3	8
6	3	1	7	7
7	5	4	6	6
8	2	6	5	4
(b) $N = 16$				
1	6	13	1	15
2	15	14	8	8
3	2	16	10	2
4	9	9	5	4
5	14	12	4	10
6	4	8	16	16
7	1	2	2	7
8	3	7	6	13
9	13	10	3	9
10	16	4	9	6
11	7	15	14	3
12	11	5	7	12
13	12	6	15	11
14	5	3	12	1
15	10	1	11	5
16	8	11	13	14
(c) $N = 32$				
1	15	16	31	12
2	17	30	10	24
3	27	20	8	10
4	18	7	7	14

(continued)

Table 6.3 (continued)

run k	m_{1k}	m_{2k}	m_{3k}	m_{4k}
5	24	6	2	3
6	12	24	19	26
7	2	8	12	8
8	1	3	29	25
9	28	19	22	7
10	32	1	24	17
11	26	11	6	28
12	11	9	4	30
13	19	17	20	18
14	21	23	1	27
15	16	27	32	4
16	22	25	16	23
17	9	28	3	11
18	3	2	25	15
19	8	18	15	19
20	29	31	30	31
21	5	21	21	6
22	25	22	28	32
23	7	14	17	21
24	20	15	26	13
25	23	5	5	5
26	14	10	14	9
27	30	26	27	1
28	13	4	23	20
29	31	12	13	22
30	4	32	9	16
31	6	13	18	29
32	10	29	11	2

6.3 Histograms and Statistics of Response, and Confidence Limits for Design

The response values $Y_1, Y_2, \ldots Y_N$ for all the deterministic calculations (or computers runs), made for N different combinations of random parameters, may be arranged by increasing values. Then they may be plotted as the histograms of Y versus the percentile of responses that are smaller than Y.

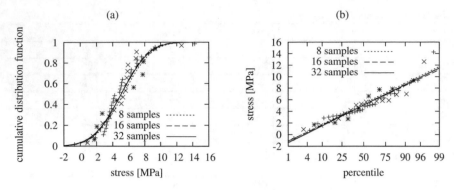

Fig. 6.4 Cumulative histograms plotted (a) in linear scales, (b) in the normal probability paper

Such histograms (calculated for the problem of shrinkage-induced stress, to be discussed in Example 6.1) are exemplified in Fig. 6.4. The random variable whose distribution is examined is the stress in a restrained bar after 220 days of drying. As an alternative to the usual representation in Fig. 6.4a, it is instructive to plot the cumulative histograms in the normal probability paper, which has scales such that the normal distribution appears as a straight line (Fig. 6.4b). The slope of that line is proportional to the standard deviation and the value at 50% is the median, which, for the normal distribution, coincides with the mean. Typically, the calculated histograms are approximately straight, which confirms that the distribution of responses Y_j is almost normal (Gaussian). Figure 6.4 demonstrates that the histograms for 8, 16, and 32 response calculations are not significantly different. Generally, 8 are acceptable and 16 good enough for engineering purposes. For the same number of deterministic calculations (e.g., computer runs) of response, Latin hypercube sampling has been shown to give better statistical estimators than other methods [620, 621].

The estimates of the mean and the variance of response are evaluated using formulae (6.7)–(6.8) with uniform weights $\Delta F_j = 1/N$. The square root of the variance is then the standard deviation. For easy reference, the resulting formulae for the estimates of the mean, standard deviation, and coefficient of variation are rewritten here as[2]

$$\bar{Y} = \frac{1}{N}\sum_{j=1}^{N} Y_j, \quad s_Y = \sqrt{\frac{1}{N}\sum_{j=1}^{N}(Y_j - \bar{Y})^2}, \quad \omega_Y = \frac{s_Y}{\bar{Y}} \qquad (6.11)$$

---

[2]The estimate of the standard deviation in (6.11) is the so-called *maximum likelihood estimate*, which would be unbiased in the statistical sense (i.e., free of systematic error) only if the exact mean value were known in advance, which is not the case here. An *unbiased estimate* is obtained if the factor $1/N$ is replaced by $1/(N-1)$. Obviously, for large N, the biased and unbiased estimates differ negligibly. For more details see Bulmer [284], p. 130, Mandel [604], p. 134, Song [783], p. 262, Freund [399].

After determining these basic characteristics of response, one can easily estimate the confidence limits with a desired percentage cutoff. For this purpose, the type of pdf must be rationally chosen.

This choice is a difficult question for the safety of structures under extreme loads of short duration, because the margins of distribution matter. However, for concrete creep and shrinkage problems, such temporary overloads do not matter, except when a sudden overload comes after large creep buckling deflections (Sects. 9.5 and 9.6 in Bažant and Cedolin [115]). In typical creep and shrinkage problems, only the central range of pdf matters. In such cases, the assumption of normal (Gaussian) distribution generally appears to be adequate.

According to the table of Gaussian pdf, the values $\bar{Y} \pm s_Y$ or $(1 \pm \omega_Y)\bar{Y}$ represent the 16 and 84% probability cutoffs. This means that if these cutoffs were used for bridge design, then about 1 in 6 concrete structures would not have the design lifetime, i.e., would have to be closed or repaired prematurely. On the other hand, emulating for creep deflections the standard safety requirement that not more than 1 in a million of structures may collapse, would be excessive by far and economically wasteful.

As a reasonable and affordable requirement, concrete structures such as bridges should be designed so that no more than 1 in 20 structures would require closure or repairs (for nuclear concrete structures, of course, more stringent limits are necessary). This means that the design of concrete structures should be based on 95% probability that the tolerable limits (e.g., maximum deflection) are not exceeded. This will henceforth be referred to as the one-sided 95% *confidence limit* (for a symmetric distribution, this is equal to the usual two-sided 90% confidence limit). If the response variable Y used in the design criterion has a normal distribution, the one-sided 95% confidence limit that must not exceed the tolerable value is obtained as

$$Y_{95\%} = \bar{Y} + 1.65\,s_Y = (1 + 1.65\,\omega_Y)\,\bar{Y} \qquad (6.12)$$

Example 6.1. Calculating the statistics of creep and shrinkage prediction and structural effects

The simplest application of the Latin hypercube sampling just described is to determine the mean and coefficient of variation of the predictions of model B3. Figure 6.5 gives an example of the curves that represent the evolution of the mean value and of the mean $\pm$ standard deviation (i.e., for the normal distribution, the two-sided 68% confidence limits) for the strain in a creep test of drying concrete with the same properties as in Example 3.1, loaded from time $t_1 = 7$ days by constant compressive stress $\sigma = -5$ MPa and simultaneously exposed to environmental humidity $h_{env} = 70\%$. Note the large spread between the top and bottom curves, which represent 16 and 84% probability cutoffs. About 1/6 of the responses must be expected to lie above, and 1/6 below, these curves. The mean curve lies precisely in the middle. Note that simulations with different numbers of strata, ranging from 8 to 32, give in this case almost the same results.

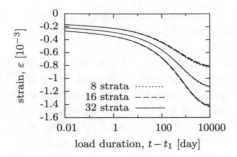

Fig. 6.5 Mean value and two-sided 68% confidence limits (i.e., curves of mean ± standard deviation) for the strain in a creep test of drying concrete

Figures 6.6–6.9 show the curves of $\bar{Y}$ and of $\bar{Y} \pm s_Y$ obtained in the numerical examples of typical structural responses analyzed in Chap. 4. All of them have been constructed for concrete with the same properties as in Example 3.1, using model B3. The cdf range was partitioned into 8, 16, and 32 strata. The solid curves in Figs. 6.5–6.9 correspond to 32 strata, the dashed curves to 16 strata, and the dotted curves to 8 strata. Comparisons indicate that 8 strata give acceptable engineering accuracy, and 16 strata give good accuracy.

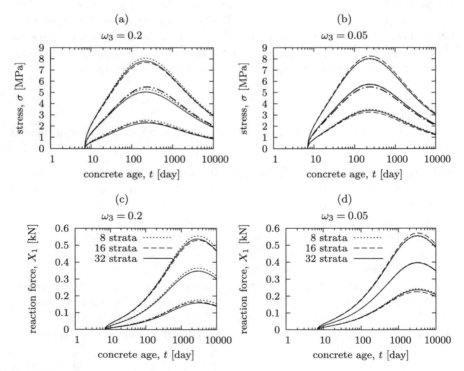

Fig. 6.6 Mean value and two-sided 68% confidence limits for (a)–(b) stress due to shrinkage in a restrained bar, (c)–(d) horizontal reaction due to shrinkage in a two-hinge portal frame

Figure 6.6a shows the stress history in a restrained bar (e.g., a pavement slab) of thickness $D = 50$ mm exposed to the environmental humidity $h_{env} = 70\%$ at age $t_0 = 7$ days, evaluated according to Eq. (4.10) from Example 4.1. The curve characterizing the mean response is for 16 strata almost the same as it is for 32 strata and is not too far from the thick dash-dotted curve curve computed with one simulation using the mean values of all parameters. The histograms for this case are plotted in Fig. 6.4, and for the other examples the histograms look similar.

The graphs in Fig. 6.6a have been obtained with the standard values of coefficients of variation recommended by Bažant and Baweja [107], i.e., $\omega_1 = 0.23$, $\omega_2 = 0.34$, $\omega_3 = 0.2$, and $\omega_4 = 0.15$. For comparison, Fig. 6.6b shows analogous graphs that would be obtained with ω_3 reduced to 0.05 and the other coefficients of variation kept unchanged. Recall that ω_3 describes the uncertainty associated with the ambient humidity, discussed in detail in Sect. 6.1. It is interesting that even though the present example deals with stresses induced by drying shrinkage, a reduction of the coefficient of variation of the average ambient humidity leads to only a moderate reduction of the band between the confidence limits. At the same time, the mean response is shifted slightly upwards, which is caused by the nonlinear dependence of the ultimate shrinkage on the ambient humidity.

The graphs in Fig. 6.6c,d show the history of the horizontal reaction force in a two-hinge portal frame in Fig. 4.2 exposed to environmental humidity $h_{env} = 70\%$ at age $t_0 = 7$ days. Again, the results have been computed for the standard value of $\omega_3 = 0.2$ (Fig. 6.6c) and for the reduced value of $\omega_3 = 0.05$ (Fig. 6.6d). This example has been solved using the approach from Example 4.2, and the specific frame geometry has been taken the same as in Madsen and Bažant [589]: frame dimensions $L = 10$ m and $B = 5$ m, moment of inertia of all members $I_b = I_c = 2.13 \times 10^{-3}$ m^4, effective thickness $D = 200$ mm, and shape factor $k_s = 1.25$.

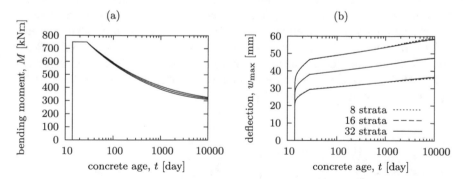

Fig. 6.7 Simply supported beams made continuous: mean value and two-sided 68% confidence limits for the (a) bending moment and (b) deflection at midspan

Figure 6.7a shows the evolution of the bending moment at midspan after a change of structural system, in which simply supported beams are made continuous; see Example 4.4. In contrast to Fig. 6.6, the spread of the curves is extremely small. Only the results obtained with 8 strata are plotted here because the results with 16 or 32 strata are very similar and the curves would overlap. The reason for the low sensitivity of the bending moment to the random parameters is that the bending

moment before the change of structural system, 750 kNm, is independent of material
properties (because the initial structural system is statically determinate, the internal
forces being fully determined by equilibrium), and the evolution of bending moment
after the change to a continuous beam is controlled by the redistribution function and
is not affected by shrinkage. The redistribution function is constructed by applying
the relaxation operator on function J_Δ, which depends on the compliance function.
If, for a specific combination of random parameters, the concrete exhibits higher
creep, it also exhibits faster relaxation and both these effects partially cancel. On the
other hand, the deflection at midspan depends on the compliance function and is more
sensitive to the randomness of creep than the moment redistribution; see Fig. 6.7b.

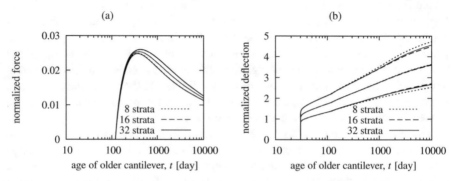

Fig. 6.8 Two opposite cantilevers of different age joined at midspan: mean value and two-sided
68% confidence limits for the (a) normalized internal reaction force and (b) normalized deflection
of older cantilever

Figure 6.8 refers to the problem of interaction between two opposite cantilevers
of different age joined at midspan, already discussed in Example 4.8. The evolution
of the normalized reaction force in Fig. 6.8a exhibits quite a low scatter, only slightly
higher than the scatter of bending moment in Fig. 6.7a, while the evolution of normalized deflection in Fig. 6.8b exhibits higher scatter, similar to the deflection in
Fig. 6.7b. As in Example 4.8, the force is normalized by the weight of one cantilever
and the deflection by the deflection of an elastic cantilever with Young's modulus
equal to the conventional modulus E_{28}.

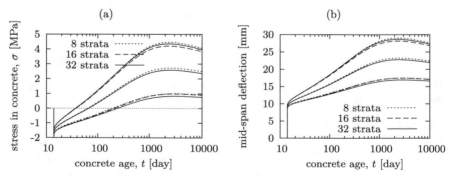

Fig. 6.9 Steel–concrete composite beam: mean value and two-sided 68% confidence limits for the
(a) stress in concrete and (b) midspan deflection

Finally, Fig. 6.9 shows the evolution of the stress in concrete and of the midspan deflection of the steel–concrete composite beam already analyzed in Example 4.10. The effect of randomness is very strong, because both the stress and the deflection depend on shrinkage, and the randomness of shrinkage is combined here with the randomness of creep. Note that, by contrast, shrinkage had no influence on the results plotted in Figs. 6.7 and 6.8, because of the assumed symmetry of the cross sections with respect to the horizontal axis. The composite beam does not possess such symmetry and shrinkage of the concrete part leads to a change of curvature and thus of deflection. The mean stress in c dded, the tensile strength would be exceeded. This indicates that cracking can be expected for some of the composite beams with properties and loading considered in Example 4.10.

In summary, the examples revealed the strongest influence of randomness in the cases where both the creep and shrinkage have an effect on the result, followed by the cases where only the creep (or relaxation) plays a role. Interestingly, the redistribution of internal forces and stresses is much less affected by the random variation of creep properties. Of course, these are only observations of the basic trends, which should not be understood as general rules. All the problems considered in this example were treated by Bažant and Liu [163] using a predecessor of model B3. The results presented here have been obtained with model B3, using an accurate numerical solution (not an AAEM approximation). ∎

The wide spread between the curves in Figs. 6.5, 6.6, 6.7b, 6.8b, and 6.9 documents the uncertainty in predicting the creep and shrinkage and their structural effects. This gives ample justification to the point made in the preceding chapter, namely that updating based on short-time measurements should be conducted for every sensitive structure. It also underscores the importance of designing not for the mean but for the one-sided 95% confidence limits (i.e., values not exceeded with 95% probability).

Furthermore, the wide spread of statistical results indicates that designs based only on accurate and tedious deterministic analysis of creep and shrinkage and their structural effects are futile. They make sense only if the uncertainties are also taken into account. Otherwise, the simple AAEM is just as good for design as are accurate numerical solutions.

6.4   Bayesian Improvement of Statistical Prediction of Creep and Shrinkage Effects*

Updated predictions of the long-term structural response, which help to minimize the uncertainties associated with the model parameters, should ideally be based on short-time measurements taken on the specific structure of interest or on laboratory specimens of the same composition, casting, and curing as the concrete used in the structure. If such measurements are so limited that their extrapolation is unreliable, or if they do not exist at all, the updating can exploit other sources of information, e.g., available measurements on a similar concrete, extracted from the database, or deflections of another similar bridge in the same geographical area.

The updating can be carried out on the basis of the *Bayes theorem* of the theory of probability [36, 229, 446, 716, 783]. The Bayesian approach can be used to update either the long-time strain predictions of the creep and shrinkage model such as B3, or the predictions obtained by a computer program for long-time creep and shrinkage effects on a given structure, such as the deflections. The former update can be reduced analytically to numerical evaluation of a certain integral [116], while the latter normally requires an entirely numerical discrete approach such as the Latin hypercube sampling [145, 557].

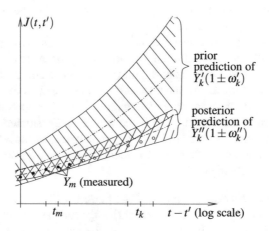

Fig. 6.10 Updating of creep predictions

The problem of updating may be explained by means of Fig. 6.10. Without any test data for the given concrete or a similar concrete, we can predict the statistical characteristics of the response using the general information on the random varia- tion of material properties for all concretes of the same or similar composition, as described in Sect. 6.1. The mean values of model parameters are estimated from a few basic material characteristics (such as the strength and the mass fractions of water, cement, and aggregates), which cannot define the resulting concrete uniquely. The actual material properties are affected by many other factors, e.g., by the type and shape of aggregates, mineral type, granulometry, admixtures, and the procedures of mixing, casting and curing. In addition to that, modern concretes often contain components other than just water, cement, gravel, and sand, e.g., superplasticizers, water reducers, silica fume, limestone fillers, fly ash, and blast-furnace slag. An exact characterization of the concrete mix composition and its treatment is thus next to impossible, and if we estimate the response from the basic composition character- istics only, we cannot expect high accuracy. This was documented by the examples in Sect. 6.3 and is reflected by the wide spread of the *prior prediction* in Fig. 6.10.

As shown in Sect. 3.8, the predictions of long-time behavior can be greatly improved if short-time test data are available. The updating procedures presented in that chapter focused on the mean response only. Although the statistics of regression could have been calculated also, they might not have been accurate. Now we will develop tools for a better prediction of the statistical characteristics of such updates, which lead to much closer confidence limits than the original predictions; see the *posterior prediction* in Fig. 6.10, which has been constructed based on the short-time measurements denoted as Y_m.

*6.4.1 Background on Bayesian Statistics**

Some mathematical background needs to be reviewed first. If $\mathscr{P}(A)$ and $\mathscr{P}(B)$ are the probabilities of two random events A and B, the probability $\mathscr{P}(A \cap B)$ that events A and B occur simultaneously is equal to the product $\mathscr{P}(A)\mathscr{P}(B)$ only if the events are stochastically independent. In general one can write

$$\mathscr{P}(A \cap B) = \mathscr{P}(A)\mathscr{P}(B|A) \tag{6.13}$$

where $\mathscr{P}(B|A)$ is the *conditional probability* that event B occurs provided that event A has occurred. In fact, relation (6.13) serves as the definition of conditional probability

$$\mathscr{P}(B|A) = \frac{\mathscr{P}(A \cap B)}{\mathscr{P}(A)} \tag{6.14}$$

which can be interpreted as the "number of cases in which events A and B occur simultaneously" divided by the "number of cases in which event A occurs." Since the conditional probability $\mathscr{P}(A|B)$ is defined as $\mathscr{P}(A \cap B)/\mathscr{P}(B)$, it follows that

$$\mathscr{P}(B|A) = \frac{\mathscr{P}(B)}{\mathscr{P}(A)}\mathscr{P}(A|B) \tag{6.15}$$

This relation is known as the *Bayes theorem*, dating back to the eighteenth century.

All these considerations can be extended to random variables, say X and Y, characterized by the *joint probability density* function $f(X, Y)$. Recall that $f(\tilde{X}, \tilde{Y})\,dX\,dY$ gives the probability that the value of X lies between X and $X + dX$ and at the same time the value of Y lies between Y and $Y + dY$. Integrating over the domain $\mathscr{Y}$ of all possible values of Y, we obtain the *marginal probability density* of variable X,

$$f_X(X) = \int_{\mathscr{Y}} f(X, Y)\,dY \tag{6.16}$$

Typically, $\mathscr{Y}$ is the interval $(-\infty, \infty)$ or $[0, \infty)$. The *conditional probability density* of Y for a given value of X can be defined, in analogy to (6.14), as

$$f_{YX}(Y|X) = \frac{f(X, Y)}{f_X(X)} = \frac{f(X, Y)}{\int_{\mathscr{Y}} f(X, Y') \, dY'} \tag{6.17}$$

and the Bayes theorem is written in analogy to (6.15) as

$$f_{YX}(Y|X) = \frac{f_Y(Y)}{f_X(X)} f_{XY}(X|Y) \tag{6.18}$$

Equation (6.18) represents a link between the conditional probabilities $f_{YX}(Y|X)$ and $f_{XY}(X|Y)$ and provides one of them if the other is known. Now let us see how this theory can be applied to the updating of creep and shrinkage predictions.

6.4.2   Method of Bayesian Analysis*

We will denote by X the random variables that influence the material properties, by y the short-time response, and by Y the long-time response. To give a simple example, we can imagine that y is just a scalar that corresponds to the strain measured in the creep test after one day of loading and Y is a scalar that corresponds to the strain measured in the creep test after one year of loading. In general, y will be a column matrix collecting the information from several measurements, e.g., strains after 1, 7, and 28 days of loading, and Y may also contain the values referring to several long-time predictions, e.g., strains after 1, 5, and 30 years. For all concretes sharing the same basic composition characteristics (but differing in many other aspects, as discussed before), the random variables X have a certain distribution with a relatively large scatter. Let us characterize this distribution by the marginal probability density function $f_X(X)$ where $X \in \mathscr{X}$.

For a given value of X, the material properties are known and the corresponding responses y and Y can, in principle, be calculated. The dependence of y and Y on X is still not deterministic, because even if we consider precisely the same concrete mix subjected to precisely the same treatment and environmental conditions, the response will vary from sample to sample, or from structure to structure, as known from experiments. However, the scatter of the response under such controlled conditions would no doubt be much smaller than the scatter of the responses of all concretes with the same values of the basic composition parameters only. Let us denote by $f_{yX}(y|X)$ and $f_{YX}(Y|X)$ the conditional probability density functions of random variables y and Y for fixed material properties X.

Put in simple terms, for one response variable y and for given properties $X = \tilde{X}$, function $f_{yX}(y|\tilde{X})$ can be constructed from the histogram of measurements of strain after 1 day in the same type of creep test repeated under the same environmental conditions on many samples made of the same concrete batch, prepared and cured following the same procedure. The coefficient of variation of this random distribution can be expected to be about 5–8%.

If the marginal probability density of X and the conditional probability density of y for all possible values of X are known (in practice they will usually be approximated by functions that correspond to the normal distribution), the joint probability density

$$f(X, y) = f_X(X)f_{yX}(y|X) \tag{6.19}$$

can be constructed according to (6.17), and then the marginal probability density of the short-time response y can be computed as

$$f_y'(y) = \int_{\mathscr{X}} f(X, y)\, dX = \int_{\mathscr{X}} f_{yX}(y|X)f_X(X)\, dX \tag{6.20}$$

The prime in f_y' indicates that this is the so-called *prior*, i.e., a prediction that does not make use of any previous knowledge other than the basic composition characteristics. At this stage there is no fundamental difference in the treatment of y and Y yet. The *prior probability density* of the long-time response is evaluated using an analogous formula

$$f_Y'(Y) = \int_{\mathscr{X}} f_{YX}(Y|X)f_X(X)\, dX \tag{6.21}$$

Note the difference between the prior probability density $f_Y'(Y)$ and the conditional probability density $f_{YX}(Y|X)$. The former characterizes the scatter in response of many kinds of concrete in general while the latter characterizes the statistical scatter for one particular concrete.

Now the crucial point is that if, in a certain specific test, we measure the actual value y_m of the short-time response y, we can improve the prediction of the long-time response Y in this same test. From the measured y, we cannot determine the precise value of X because the dependence of y on X is not deterministic, and even if it were, it would not be uniquely invertible. Still, the measured information is very useful, because it can be used to "narrow down the range in which X may lie." In mathematical terms, the prior probability density function of X can be modified based on the newly acquired information, and the integral in (6.21) can be re-evaluated with f_X replaced by the modified (posterior) probability density function f_X'', to get the *posterior probability density* function of the long-time response,

$$f_Y''(Y) = \int_{\mathscr{X}} f_{YX}(Y|X)f_X''(X)\, dX \tag{6.22}$$

The key step of the updating procedure is the determination of the posterior probability density function f_X''. This function is in fact the conditional probability of X provided that y is equal to the measured value y_m. So, using the Bayes theorem (6.18), we can write

$$f_X''(X) = f_{Xy}(X|y_m) = \frac{f_X(X)}{f_y'(y_m)} f_{yX}(y_m|X) \tag{6.23}$$

and, substituting this in (6.22), we get the final formula

$$f_Y''(Y) = \int_{\mathscr{X}} f_{YX}(Y|X) \frac{f_X(X)}{f_y'(y_m)} f_{yX}(y_m|X) \, dX = \frac{\int_{\mathscr{X}} f_{YX}(Y|X) f_{yX}(y_m|X) f_X(X) \, dX}{\int_{\mathscr{X}} f_{yX}(y_m|X) f_X(X) \, dX} \quad (6.24)$$

Before formula (6.24) can be applied to a real case, the prior probability density f_X and the conditional probability densities f_{yX} and f_{YX} must be known or reasonably estimated. Probability density f_X describes the variability of material properties among all concretes of the same basic composition. This was already discussed in Sect. 6.3, where the variability of parameters of model B3 was described by four random variables $X^{(1)}$, $X^{(2)}$, $X^{(3)}$, and $X^{(4)}$, which were assumed to be normally distributed and mutually independent (uncorrelated). In that section, we used variables with standard normal distribution and their relation to the actual model parameters was described by linear transformations (6.1)–(6.4). Each of these transformations contained two constants defining the mean value and the coefficient of variation of the respective model parameter. The mean values are estimated from the composition using the empirical formulae that represent the deterministic version of model B3, and the coefficients of variation associated with variables $X^{(1)}$, $X^{(2)}$, $X^{(3)}$, and $X^{(4)}$ were taken according to the recommendations of Bažant and Baweja [107] as 23, 34, 20 and 15%, respectively.

It remains to specify the conditional probability densities f_{yX} and f_{YX}, which describe the variability of measured short-time and long-time responses for one and the same concrete. Again, the simplest assumption is that, for fixed properties X, the individual components of y are normally distributed and uncorrelated. The assumption of no correlation is of course somewhat questionable, because closely spaced measurements in the same test are certainly not independent. However, for reasonably spaced measurements, the correlation is weak and can be neglected. Each component of y represents the value measured at a specific time, e.g., after one day of loading, and its mean value is expected to be close to the prediction obtained, for that time, using the deterministic version of model B3 with parameters determined from the given value of X.

So it is sufficient to estimate the coefficient of variation of measurements from many tests on the same concrete. It can be expected that this coefficient of variation does not strongly depend on X (i.e., is about the same for different concretes) and for long-time measurements is larger than for short-time ones. We will further use the symbol ω_{yX} for the coefficient of variation corresponding to the conditional probability density f_{yX} and ω_{YX} for that corresponding to f_{YX}, knowing that for different components of y and Y it may be different (especially if the long-time predictions cover a wide range of times).

Having outlined the assumptions regarding the random distribution of parameters X and the conditional probabilities of the short-time and long-time responses, we proceed to the evaluation of the prior and posterior response predictions. From the practical point of view, the fundamental characteristics of the response are the mean and the standard deviation. Other important stochastic characteristics, such as the confidence limits, can be estimated from the mean and standard deviation if a certain type of distribution (usually normal) is assumed.

Let us first discuss the evaluation of the prior characteristics, denoted by single primes and derived from the prior distribution. For instance, to characterize the prior prediction of the long-time response, we need to evaluate the mean and variance of random variable Y with the prior pdf f_Y' given by (6.21). For simplicity, we will develop the formulae related to one single component of Y denoted as Y, knowing that each component can be treated separately. Using the specific form (6.21) of the pdf, we can express the *prior mean* of Y as

$$\bar{Y}' = \int_{\mathcal{Y}} Y f_Y'(Y) \, dY = \int_{\mathcal{X}} \left[\int_{\mathcal{Y}} Y f_{YX}(Y|X) \, dY \right] f_X(X) \, dX \qquad (6.25)$$

where $\mathcal{Y}$ denotes the domain of the random variable Y, usually the set of all real numbers. The integral in the brackets represents the mean response Y for fixed parameters X, which is supposed to be equal to the prediction of the deterministic model with the given parameters. Therefore, there is no need to evaluate the integral in brackets numerically, and we can replace it by the symbol $G(X)$ where function G describes the deterministic model. The right-hand side of (6.25) then reduces to an integral over the parameter space $\mathcal{X}$, which is a generalized form of (6.5). In analogy to (6.7), we can construct the numerical approximation

$$\bar{Y}' = \int_{\mathcal{X}} G(X) f_X(X) \, dX \approx \sum_{j=1}^{N} G(X_j) \Delta F_j = \frac{1}{N} \sum_{j=1}^{N} G(X_j) \qquad (6.26)$$

where X_j are sampling points with equal prior probabilities $\Delta F_j = 1/N$. Note the slight difference between (6.5)–(6.8), where Y was considered as a deterministic function of X, and the present situation, in which even for fixed X the value of Y is random and the deterministic function G describes the dependence of the **mean** response Y on parameters X. This had no effect on the resulting formula (6.26) for the mean, but the expression for the variance will be more involved than (6.8). To derive it, we start from the definition of *prior variance*

$$s_Y'^2 = \int_{\mathcal{Y}} Y^2 f_Y'(Y) \, dY - \bar{Y}^2 = \int_{\mathcal{X}} \left[\int_{\mathcal{Y}} Y^2 f_{YX}(Y|X) \, dY \right] f_X(X) \, dX - \bar{Y}^2 \quad (6.27)$$

Now the point is that the integral in the brackets is not just the square of $G(X)$ (as it would be if Y were a deterministic function of X), but it is equal to $G^2(X) + s_Y^2(X) = (1 + \omega_Y^2) G^2(X)$ where $s_Y^2(X) = \omega_Y^2 G^2(X)$ is the variance of Y under fixed X and ω_Y is the coefficient of variation of Y for fixed X, assumed to be independent of the parameters X. Therefore, the prior approximation of the response variance is computed as

$$s_Y'^2 \approx \sum_{j=1}^{N} \left[G^2(X_j) + \omega_Y^2 \right] \Delta F_j - \bar{Y}^2 = \frac{1}{N} \sum_{j=1}^{N} G^2(X_j) + \omega_Y^2 - \bar{Y}^2 \qquad (6.28)$$

For simplicity, we have shown the derivation for one long-time response variable Y, but the same formulae would apply to all components of Y and also to the components of the short-time response y (with the appropriate G and ω_Y for each component).

The main objective in this section is to show how the posterior estimate of the long-time response can be evaluated. The posterior mean and variance are defined by formulae analogous to (6.25) and (6.27), but with the prior pdf $f_X(X)$ replaced by the posterior pdf $f_X''(X)$, given by (6.23). So the *posterior mean* is given by

$$\bar{Y}'' = \int_{\mathcal{X}} G(X) f_X''(X) \, dX = \frac{1}{f_y'(y_m)} \int_{\mathcal{X}} G(X) f_{yX}(y_m|X) f_X(X) \, dX \qquad (6.29)$$

Introducing, for convenience, weight factors

$$p_j = f_{yX}(y_m|X_j), \qquad j = 1, 2, \dots N \qquad (6.30)$$

and normalized weight factors

$$\Delta F_j'' = \frac{p_j}{\sum_{k=1}^{N} p_k}, \qquad j = 1, 2, \dots N \qquad (6.31)$$

we can approximate (6.29) by

$$\bar{Y}'' \approx \frac{\sum_{j=1}^{N} G(X_j) p_j}{\sum_{j=1}^{N} p_j} = \sum_{j=1}^{N} G(X_j) \Delta F_j'' \qquad (6.32)$$

This is very similar to the prior estimate (6.26), but the uniform weights $\Delta F_j = 1/N$ are replaced by the nonuniform weights $\Delta F_j''$ given by (6.31). The contribution of individual sampling points is in the posterior weighted by factors which express the relative likelihood that the material with properties corresponding to the sampling parameter values would exhibit the actually measured short-time response y_m. The *posterior* estimate of the *variance* is constructed in an analogous way, starting from (6.28) and replacing the uniform weights by the updated ones:

$$s_Y''^2 \approx \sum_{j=1}^{N} G^2(X_j) \Delta F_j'' + \omega_Y^2 - \left(\bar{Y}'' \right)^2 \qquad (6.33)$$

It suffices to assume that all the distributions are normal, and then the foregoing values for the mean and standard deviation (square root of variance) permit fixing the posterior probability distributions and determining the confidence limits.

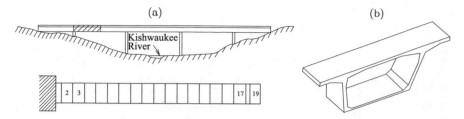

Fig. 6.11 Kishwaukee River Bridge (Illinois): (a) longitudinal section and close-up of cantilever (half-span) analyzed, (b) box-girder segment; reprinted from [145] with permission from ASCE

Example 6.2. Updating prediction of bridge deflections

Bayesian analysis in combination with the Latin hypercube sampling of the random parameter sets was applied to the analysis of long-time bridge deflections by Bažant and Kim [145]. The Kishwaukee River Bridge, used in this example, is a segmental box-girder prestressed concrete bridge constructed by the cantilever erection method (Fig. 6.11a top). The analysis focused on one half-span cantilever (Fig. 6.11a bottom) consisting of 19 segments of different ages (Fig. 6.11b). The creep and shrinkage of concrete were described by a predecessor of model B3 and the model parameters were controlled by 8 independent random variables.

Typical cumulative histograms of the prior and posterior distribution of deflection of the 17th segment after 45 years are plotted on the normal probability paper in Fig. 6.12. The histogram in Fig. 6.12a was obtained with one series of 16 simulations, using 16 strata for each random variable. The histogram in Fig. 6.12b combines the results obtained in 4 different series, each consisting of 16 simulations. Analogous histograms for 1 or 4 series consisting of 32 simulations using 32 strata are shown in Fig. 6.12c, d. The closeness of the histograms to a straight line fit confirms that the normal distribution is a good assumption. As expected, the straight line approximation has a smaller slope for the posterior distribution than for the prior one, because the updating reduces the standard deviation.

Figure 6.13 gives an example of the deflection histories corresponding to the mean response and to 68% confidence limits (mean ± standard deviation) for the prior prediction and for the posterior prediction updated on the basis of the short-time data points shown, which documents that the uncertainty of prediction has been drastically curtailed by the Bayesian sampling approach. The predictions in Fig. 6.13a are based on one series of 16 simulations, and those in Fig. 6.13b on one series of 32 simulations. For details of these calculations, see Bažant and Kim [145]. Treatment of certain anomalies in the measured deflections is described on p. 2543 of that paper. ∎

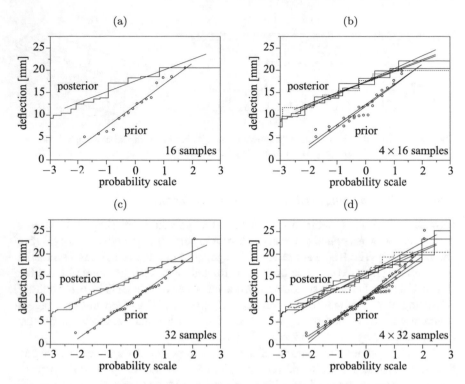

Fig. 6.12 Cumulative histograms of the prior and posterior distribution of bridge deflection at segment number 17 after 45 years

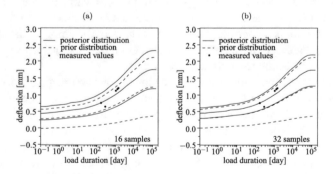

Fig. 6.13 Prior and posterior predictions of the history of bridge deflection (mean values and 68% confidence limits) at segment number 17

Chapter 7
Paradigms of Application, Phenomena Affecting Creep Deformations, and Comparisons to Measurements on Structures

Abstract In this chapter, we explain by examples how to apply in practice the methods presented in the preceding chapters. We compare the calculation results to the measurements on structures and assess the errors of various methods. The most revealing examples aiding progress are those of major serviceability loss. First, we discuss the La Lutrive Bridge, which was an early documented example of excessive creep deflections. Then, we present a striking paradigm of excessive creep deflections and prestress loss, offered by the Koror–Babeldaob (KB) Bridge, which has recently been analyzed in detail at Northwestern University. In this context, we outline the way to adapt commercial finite element software to concrete creep analysis and comment on algorithmic aspects. We also discuss the effects of wall thickness, cracking, and temperature (including the effect of solar heating of prestressing steel embedded in concrete) and explain the method of determining the model parameters. Evidence of excessive deflections from 69 other bridges is summarized and interpreted. Finally, we give a préci of recent results on cyclic creep, with application to prestressed box girder bridges, and show that its effects on deflections are negligible, especially for large spans, while its effects on stress redistributions and cracking may be appreciable for medium spans.

In this chapter, whose main part closely follows a series of papers by Bažant et al. [209–211], predictions of various models will be compared. The models to be considered include the B3 model, the ACI-209 model, the CEB-*fib* model, the GL 2000 model, and the JSCE model. For their description, see Appendices C and E. In Sect. 7.8, we will also refer to the design recommendation of Japan Road Association (JRA) [528], which is a slightly updated version of the JSCE model.

© Springer Science+Business Media B.V. 2018
Z.P. Bažant and M. Jirásek, *Creep and Hygrothermal Effects in Concrete Structures*, Solid Mechanics and Its Applications 225,
https://doi.org/10.1007/978-94-024-1138-6_7

7.1 Drying Effects in Viaduct La Lutrive

Unrealistic representation of drying and its effect on shrinkage and creep is often an important, though not dominant, factor in excessive long-time deflections. This has been early on documented by the viaduct La Lutrive (near Montreux, Switzerland) (Fig. 7.1). As has been typical, the thin top slab dried and shrank first, also exhibiting drying creep. This caused lifting. Only after the top slab almost finished drying, the bottom slab started to dry and shrink, with some additional drying creep, which was pulling the bridge down faster than predicted by the traditional analysis based on uniform (average) shrinkage and creep properties of the whole box.

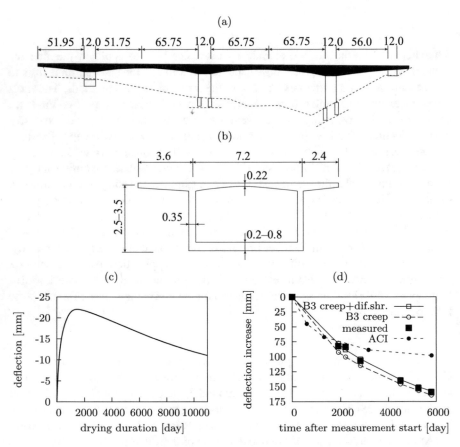

Fig. 7.1 Lutrive bridge: (a) elevation, (b) cross section (dimensions in meters), (c) deflection component due to differential shrinkage, (d) deflection measured or predicted by various models; figure originally published in ACI Concrete International, 2006

The curve in Fig. 7.1c shows (in actual time scale) the model B3 prediction of the upward vertical deflection [558]. After this curve is superposed on the model B3 prediction of creep (with drying creep), the deflection curve closely matches the measurements of maximum deflection (shown by data points in Fig. 7.1d). Note the large difference from the deflection curve predicted according to the ACI recommendation [11], reapproved in 2008; see Appendix E.3. This difference must be attributed not only to the neglect of differential shrinkage and creep, but also to the fact that the early creep models (which still languish in various standard recommendations) greatly underestimate the long-time creep.

The unexpectedly small downward deflections which are often observed during the first few years lead to euphoric self-satisfaction. Then, as the bridge ominously begins deflecting downward at a faster rate than expected, gloom sets in. The picture is aggravated if, as has been typical, the design was based on an outdated prediction model such as that of ACI [11], which greatly underestimates long-time creep. When the deflections become much larger than predicted in design, panic may develop and provoke risky remedial measures. The records for the pioneering German bridges in the 1950s (Worms, Koblenz, Bendorf) are either nonexistent or sealed from the public.

A long record of excessive deflections also exists for the Zvíkov Bridge in Bohemia, built in 1962 [558].

The foregoing discussion highlights the importance of taking into account the effects of nonuniformity and nonsymmetry of drying. It also shows that approximate estimates of these effects can be achieved quite easily by applying model B3 separately to different flanges in the cross section.

7.2 Description of the KB Bridge in Palau and Input Data

Recent studies [558–560, 646, 836] reported measurements documenting grossly excessive long-time deflections on a number of long-span prestressed box girder bridges [558]. The most blatant example of excessive deflections is provided by the *KB Bridge*, which crossed the Toegel Channel between the islands of Koror and Babeldaob in the Republic of Palau in the tropical Western Pacific (Fig. 7.2). When completed in 1977, its main span of 241 m (791 ft.) set the world record for prestressed concrete box girders [886]. The final deflection, measured as a difference from the design camber of −0.3 m (or −12 in.), was expected to terminate at 0.76 to 0.88 m (30.0 to 34.6 in.), as predicted in design [3, 775] based on the original CEB-FIP recommendations (1970-1972). According to the 1971 ACI model [11], reapproved in 2008 [14], the deflection (measured from the design camber) would have been predicted as 0.71 m (28 in.), according to [615], and 0.74 m (29 in.), according to the present calculations with the ACI model.

After 18 years, the deflection measured since the end of the construction reached 1.39 m (54.7 in.) and kept growing [3, 367].

(a) (b) (c) (d) (e)

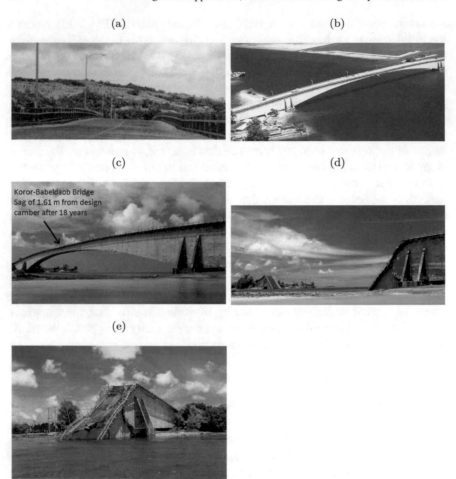

Fig. 7.2 Koror–Babeldaob Bridge in Palau: (a) deck view at midspan hinge before retrofit, (b)–(c) side view before retrofit, (d)–(e) after collapse (in 1996); images courtesy ACI; DYWIDAG Systems International; Wiss, Janney, Elstner Associates, Inc.; image (a) originally published on the cover of ACI SP-194, 2002; images (b)–(d) originally published in ACI Concrete International, June 2010; image (e) originally published in ACI Structural Journal, Nov-Dec 2011

The design camber of 0.305 m (12 in.) was not met [775], because an additional creep deflection of 0.229 m (9 in.) was accumulated during the segmental construction, making the actual camber only 0.076 m (3 in.) when the cantilevers were joined. Thus, the total 18-year deflection at midspan was 1.61 m (5.3 ft.). Remedial prestressing was undertaken, but caused collapse 3 months after its completion (on September 26, 1996), with two fatalities and many injuries [286, 615, 670, 694, 695, 785]; see Fig. 7.2d-e.

As a result of legal litigation, the technical data collected on this major disaster by the investigating agencies have been unavailable to the engineering public for many years. In view of this fact, a resolution (proposed by Bažant, in the name of a worldwide group of 47 experts listed in [209]) was introduced at the 3rd Structural Engineers World Congress in Bangalore. The resolution called, on the grounds of engineering ethics, for the release of all the technical data necessary for analyzing major structural collapses, including the bridge in Palau. It passed on November 6, 2007, and was circulated widely. In January 2008, the Attorney General of the Republic of Palau permitted the release of the necessary technical data.

The main span of 241 m (791 ft.) consisted of two symmetric concrete cantilevers connected at midspan by a horizontally sliding hinge. Each cantilever was made of 25 cast-in-place segments of depths varying from 14.17 m (46.5 ft.) at the main piers to 3.66 m (12 ft.) at midspan. The main span was flanked by 72.2 m (237 ft.) long end spans, in which the box girder was partially filled with rock ballast to balance the moment at the main pier. The total length of the bridge was 386 m (1266 ft.). The thickness of the top slab ranged from 432 mm (17 in.) at the main piers to 280 mm (11 in.) at the midspan. The thickness of the bottom slab varied from 1153 mm (45.4 in.) at the main piers to 178 mm (7 in.) at the midspan. Compared to the depth of girder, the webs had a relatively small thickness of 356 mm (14 in.), constant through the whole main span. The elevation and two cross sections are shown in Fig. 7.3.

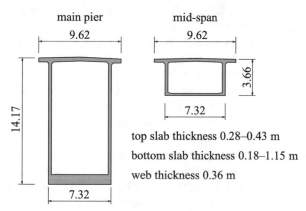

Fig. 7.3 Cross sections of box girder at main pier and midspan of main span (dimensions in meters)

Type-I Portland cement was used for the superstructure [775]. The mass density of concrete was $\rho = 2325$ kg/m^3 (145 lb/ft.^{3}). The top slab was covered by concrete pavement of average thickness 76 mm (3 in.) and density 2233 kg/m^3 (139 lb/ft^3). The aggregate was crushed basalt rock of the maximum aggregate size of about 19 mm (3/4 in.), supplied from a quarry on the island of Malakal. Beach sand from Palau, washed to keep the chloride content well within allowable limits, was used as the fine aggregate [368].

Although no original measurements of the elastic modulus E of concrete are known, some information was obtained in 1990 by core sample tests [510], which yielded the static modulus of elasticity $E = 22.1$ GPa (3200 ksi). In 1995, further core sample tests [367] made just before the retrofit revealed the porosity to be high and E to be about 21.7 GPa (3150 ksi). Both investigations showed values about

23% lower than the value $E = 28.3$ GPa (4110 ksi) estimated from the compression strength[1] according to the ACI empirical formula (3.5). In JICA's on-site investigation just before the retrofit, truck load tests were conducted, and matching the deflection measured at midspan by finite element elastic analysis one gets, after correction for concrete age according to the ACI formula, the average conventional modulus E_{28} of about 22.0 GPa, or 3190 ksi [510]. This is the value adopted for analysis since the load test gives the average elastic modulus in the box girder while the core samples are subjected to local random scatter.

The prestress was generated by Dywidag threaded alloy bars (tendons) of yield strength 1034 MPa (150 ksi) and diameter 31.8 mm (1.25 in.), extended by couplers, anchored by nuts, and grouted in ducts of diameter of 47.6 mm, or 1.9 in. [3, 359]. Some tendons were stressed from one end, and some from both [615, 886]. The jacking force of each tendon was 0.60 MN, or 135 kips [359]. There were 316 tendons above the main pier, densely packed in four layers within the top slab. Their combined initial prestressing force was about 190 MN, or 42600 kips [615, 694, 886]. The same threaded bars were used to provide vertical prestress in the webs and horizontal transverse prestress in the top slab. The tendon spacing in the webs ranged from 0.3 to 3 m, or 1 to 10 ft. [775]. The horizontal transverse tendons in the top slab were spaced at 0.56 m, or 22 in. [3, 615].

The alloy steel of the tendons had yield strength 1034 MPa (150 ksi) and ultimate strength 1054 MPa, or 153 ksi [359]. Its Young modulus was assumed as 200 GPa (29000 ksi) and Poisson ratio as 0.3. There was also unprestressed steel reinforcement [3], which was taken into account in calculations. In post-collapse examination, neither the prestressed nor the unprestressed steel showed any signs of serious corrosion, despite the tropical marine environment (some of the ducts showed mild corrosion, irrelevant for deflections).

The construction of each segment took slightly more than 1 week [796]. When the concrete strength in the segment just cast attained 17.2 MPa (2500 psi), 6 to 12 tendons were stressed to 50% of their final jacking force [796], and when the concrete strength reached 24.1 MPa (3500 psi), all the tendons terminating in this segment were stressed fully. The segmental erection of the opposite symmetric cantilevers was almost simultaneous and took about 6 to 7 months [886].

Despite close monitoring, the camber planned to offset the anticipated long-time deflections was not met. The creep and shrinkage during the segmental erection caused an unintended initial sag of 229 mm (9 in.) at midspan, which could not be corrected during the erection because it would have required abrupt large changes of slope [775]. The initial sag before installation of the midspan hinge is included neither in the reported deflection measurements nor in the deflection curves in figures.

The initial deflections for the first two years were benign. However, the longer term deflections came as a surprise. In 1990, the midspan deflection reached 1.22 m,

---

[1]The mean 28-day compressive strength $\bar{f}_c = 35.9$ MPa (5200 psi) is based on the results of cylinder tests during construction, reported in a private communication by Zelinski [890], the former Resident Engineer at the KB Bridge construction, after he read Bažant et al. [211]. For further corrections by Zelinski, see [209].

or 48 in. [510], which caused ride discomfort, vibrations after each vehicle passage, and excessive deterioration of road surface. By 1993, the deflection reached 1.32 m, or 52 in. [3]. In 1995, just before the removal of roadway pavement (of average thickness of 76 mm, or 3 in.), the midspan deflection reached 1.39 m (54.7 in.) and was still growing [367].

7.3   Creep Structural Analysis Utilizing Commercial General-Purpose Finite Element Program

As justified in the preceding chapters, concrete under service conditions and nonde-creasing strains can be assumed to follow aging linear viscoelasticity with corrections for tensile cracking, humidity and temperature variations, and drying creep (or Pickett effect). The compliance function $J(t, t')$ of concrete is deduced from the chosen creep and shrinkage prediction model. The ACI, CEB-FIP, JSCE, GL, and B3 models are considered [14, 104, 107, 390, 405, 407, 529]; see Chap. 3 and Appendices C and E. The same finite element program with the same step-by-step time integration based on Kelvin chain is used for all these models.

As explained in Chap. 5 and further discussed in Sect. 7.4, step-by-step analysis in short time steps reduces the structural creep problem to a sequence of elastic problems with eigenstrains, one such problem for each time step [68]. Each such analysis can be carried out with an adapted elastic finite element program. So one merely needs to find a suitable commercial finite element program which has the requisite geometric and material modeling features. The software ABAQUS/Standard (Simulia, Providence, Rhode Island) has been chosen, and various supplemental computer subroutines have been developed to introduce the incremental effects of creep and shrinkage based on different models.

The plates (slabs and walls) of the box girder are subdivided into eight-node hexa-hedral isoparametric finite elements (Fig. 7.4). Since the stresses caused by load do not vary significantly through the thickness and plate bending is not important when traffic loads are not considered, the finite elements are chosen to extend through the whole thickness, except that the top slab thickness is subdivided into two elements. Although accurate modeling of drying creep may generally require the wall thickness to be subdivided into at least six finite elements, model B3 makes it possible to avoid thickness subdivision because it is based on an approximate solution of drying profile evolution depending on plate thickness and on analytically established scaling properties of the nonlinear diffusion equation. Because the creep specimens in the database are loaded centrically and are drying symmetrically, this approach would not be accurate for cross sections subjected to flexure or highly eccentric compression. However, for the present case of box cross sections with relatively thin walls, this approach is accurate because the resultant of normal stresses across the wall is everywhere nearly centric, same as in standard creep tests.

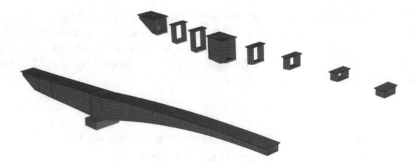

Fig. 7.4 Three-dimensional finite element mesh of the KB Bridge; figure originally published in ACI Concrete International, June 2010

In view of symmetry, only one half of the bridge is analyzed. Together with the pier, it is subdivided into 5036 hexahedral elements; see the mesh in Fig. 7.4. The prestressing tendons and the nonprestressed steel bars are subdivided into further 6764 bar elements connected rigidly (with no slip) to the nodes of the three-dimensional mesh (for reinforcement description, see [3]). Sufficiency of mesh fineness has been validated by checking that a finer mesh with 20,144 hexahedral elements would yield only a negligible correction of the computed elastic deflections.

The differences in the ages of various segments, the sequential prestressing at various times, and the stepwise load increase in individual segments during construction (including the extra weight of 1068 kN, or 240 kips, introduced during the construction by the form carrying traveler) are taken into account according to the actual cantilever erection procedure. To simulate this segmental erection procedure, the elements are deactivated at first and then progressively reactivated according to the construction sequence.

The initial time steps after the hinge installation at midspan are 100×0.1 days and then 10×10 days. After that, the time step is kept constant at 100 days up to 19 years. For the deflection prediction up to 150 years, all the subsequent time steps are 1000 days.

The individual prestressing bars, the number of which is 316 above the main pier, are modeled as two-node line elements, attached to concrete at the nodes. The individual bars of unprestressed steel reinforcement are modeled similarly. Although the tendons were not prestressed fully at one time [796], each tendon is assumed to get fully prestressed 7 days after its anchoring segment had been cast. Capturing the time schedule accurately matters for the initial deflection history, but not for multidecade deflections.

Since the tendons are straight, the curvature friction is nil and only the wobble friction needs to be taken into account. To do that, the initial prestress is diminished according to the length of each individual tendon with an approximate wobble coefficient $0.0003/\text{ft}$ or $0.00098/\text{m}$ [359, 660].

The prestress losses caused by creep and shrinkage and by sequential prestressing of tendons are automatically reproduced by the finite element simulation, provided that the behavior of concrete is described by an appropriate creep and shrinkage model. Modeling of the loss due to prestressing steel relaxation is based on the approach proposed by Bažant and Yu [207] and presented in Sect. 4.3.4.1.

Because of significant strain change in the tendons, caused by axial contraction and flexure of the box girder due to creep and shrinkage, Eqs. (4.130)–(4.133) yield for the KB Bridge considerably smaller steel relaxation than the CEB Code formula which ignores decreasing strain. However, this difference is more than offset by the effect of periodic daily heating of the tendons in the top slab [207]. For the KB Bridge, the tropical sunlight is considered to heat the pavement to 55°C at least, and a simple calculation of heat conduction indicated that the top layer of tendons must have reached at least 30°C within 2.5 hours after pavement heating, and for about 6 hours daily. Based on Eq. (4.131) and on the value $Q_p/k_B \approx 14,600$ K, deduced from the data of Shinko Wire Co. (see Fig. 4.33a), it is estimated that, as a time-averaged value, parameter $A_T \approx 2$ for these tendons. Parameters of the prestress relaxation model proposed by Bažant and Yu [207] and described in Sect. 4.3.4 are set to $k = 0.12$ and $\rho_0 = 6.5\%$.

7.4 Numerical Implementation and Algorithmic Aspects

In the material library of ABAQUS, no material model can capture the characteristics of the creep compliance functions $J(t, t')$ of the prediction models under consideration. Therefore, an individual material constitutive law corresponding to every different creep and shrinkage prediction model had to be developed for the user subroutine UMAT. The three-dimensional generalization is obtained assuming material isotropy, characterized by a time-independent Poisson ratio ν [80, 87, 126]; see Sect. 2.4. Based on core sample tests [367], $\nu = 0.21$ is used in simulations.

For the sake of efficiency of large-scale computer analysis, it is advantageous to avoid computation of history integrals. This is made possible by converting the stress–strain equations to an equivalent rate-type form, based on the Kelvin chain model; see Sect. 5.2.

In the case of model B3 (which represents RILEM Recommendation 1995), conversion of the compliance function of basic creep to a rate-type creep law is particularly easy, as explained in Sect. 5.2.6. The Kelvin chain model is in this case applied to the nonaging solidifying constituent, and thus, its moduli are age-independent, but only for basic creep, and except the aging viscous flow term. As explained in Sect. 5.2.6, the flow term can be treated separately, because it has no heredity, i.e., no memory, no history integral; see the modified version of Algorithm 5.3 based on Eqs. (5.83) and (5.84).

Besides the basic creep, there is in model B3 a separate drying creep term, which captures the Pickett effect. It is approximated by compliance function J_d given by (3.20), which does not have the structure corresponding to the solidification theory. Such a compliance function can be approximated by general Dirichlet series (2.8), which corresponds to an aging Kelvin chain. The age-dependent moduli D_μ can be evaluated using an approximation of the retardation spectrum, as explained in Sect. F.4. Alternatively, the spectra of the basic and drying creep compliance can be combined and the compliance function can be approximated by one aging Kelvin chain, which is the most efficient approach.

The microprestress-solidification theory, which will be explained in Chap. 10, allows a more accurate representation of drying creep, aging, and thermal effects and thus would have been more realistic. However, it would have required calculating the distributions of pore relative humidity across the thickness of each slab, which would have necessitated not one but at least five finite elements over the slab thickness.

For empirical models such as those of ACI, CEB, JSCE, and GL, a nonaging constituent in the sense of the solidification theory cannot be identified. Therefore, approximations by aging Kelvin chains must be used, with retardation spectra determined by the formulae from Sect. F.3.

As will be discussed in Sect. 11.2, the models other than B3 exhibit some serious deficiencies, theoretical as well as practical. One deficiency, especially for the ACI model, is that the long-time creep is strongly underestimated. Another deficiency of the ACI model, and to a lesser extent the CEB model, is that the drying creep, which is very sensitive to the cross-sectional thickness, is not separated from the basic creep and that the effects of thickness on shrinkage and drying creep are described by a scaling factor on the strain at fixed time, rather than by scaling of the time rate of strain [98].

Algorithm 7.1 Finite element analysis using a creep and shrinkage model

1. For each integration point of each finite element that represents concrete: Initialize (by zero values) the column matrix of internal variables $e_{r\mu,1}$, $\mu = 1, 2, \ldots M$. Set the step counter k to 1.
2. Start the step number k from time t_k to $t_{k+1} = t_k + \Delta t_k$.
3. For each integration point of each finite element that represents concrete, compute

    a. the midstep stiffness moduli $D_{\mu,k+1/2}$, using suitable formulae based on the retardation spectrum, e.g., formulae from Sect. F.3 for the ACI or CEB model,
    b. the factors $\beta_{\mu k}$ and $\lambda_{\mu k}$ given by (5.63),
    c. the incremental modulus

$$\bar{E}_k = \left(\frac{1}{E_0} + \sum_{\mu=1}^{M} \frac{1 - \lambda_{\mu k}}{D_{\mu,k+1/2}} \right)^{-1} \tag{7.1}$$

    d. the column matrix of strain increments due to creep

$$\Delta \varepsilon_k'' = \sum_{\mu=1}^{M} (1 - \beta_{\mu k}) e_{r\mu k} \tag{7.2}$$

e. the increment of shrinkage strain $\Delta \varepsilon_{\mathrm{sh},k} = \varepsilon_{\mathrm{sh}}(t_{k+1}) - \varepsilon_{\mathrm{sh}}(t_k)$, based on the effective thickness t_b and environmental humidity h_{env},

f. the increment of thermal strain $\Delta \boldsymbol{\varepsilon}_{\mathrm{T},k} = \alpha_T [T(t_{k+1}) - T(t_k)]$, which may be considered to take into account seasonal changes but is not needed here because the temperature in the tropics is constant, and

g. the increment of eigenstrain

$$\Delta \tilde{\boldsymbol{\varepsilon}}_k = (\Delta \varepsilon_{\mathrm{sh},k} + \Delta \varepsilon_{\mathrm{T},k})\boldsymbol{\delta} + \Delta \boldsymbol{\varepsilon}_k'' \tag{7.3}$$

where $\boldsymbol{\delta}$ is the column matrix corresponding to unit normal strains and zero shear strains (in the three-dimensional setting, it has three unit entries and three zero entries).

4. For each integration point of each finite element that represents steel (passive reinforcement or prestressed cables), compute the incremental stiffness and the increment of eigenstrain using an appropriate material model (e.g., an elastoplastic model for passive reinforcement and an elastoviscoplastic model described by (4.136) for prestressing steel).

5. Assemble the incremental structural stiffness matrix $\boldsymbol{K}_k$, using the standard finite element procedure based on numerical evaluation of the integral

$$\boldsymbol{K}_k = \int_V \boldsymbol{B}^T \boldsymbol{D}_k \boldsymbol{B} \, \mathrm{d}V \tag{7.4}$$

where V is the spatial domain occupied by the investigated structure, $\boldsymbol{B}$ is the standard strain-displacement matrix containing the derivatives of the shape functions, and $\boldsymbol{D}_k$ is the incremental material stiffness matrix. At integration points of elements that represent concrete, $\boldsymbol{D}_k$ is equal to $\bar{E}_k \boldsymbol{D}_v$, where $\bar{E}_k$ is the incremental modulus given by (7.1) and $\boldsymbol{D}_v$ is the dimensionless elastic material stiffness matrix (2.35), corresponding to unit Young's modulus. At integration points of elements that represent steel, $\boldsymbol{D}_k$ is the tangent material stiffness matrix of the appropriate material model.

6. Assemble the incremental equivalent force vector $\Delta \boldsymbol{f}_k$ (right-hand side of the discretized equilibrium equations), using the standard finite element procedure based on numerical evaluation of the integral

$$\Delta \boldsymbol{f}_k = \int_V \boldsymbol{N}^T \Delta \bar{\boldsymbol{b}}_k \, \mathrm{d}V + \int_{S_t} \boldsymbol{N}^T \Delta \bar{\boldsymbol{t}}_k \, \mathrm{d}S + \int_V \boldsymbol{B}^T \boldsymbol{D}_k \Delta \tilde{\boldsymbol{\varepsilon}}_k \, \mathrm{d}V \tag{7.5}$$

where $\boldsymbol{N}$ is the displacement interpolation matrix containing the shape functions, $\Delta \bar{\boldsymbol{b}}_k$ is the vector of increments of prescribed body forces, $\Delta \bar{\boldsymbol{t}}_k$ is the vector of increments of prescribed surface forces on the unsupported part of boundary, S_t, and $\Delta \tilde{\boldsymbol{\varepsilon}}_k$ is the increment of eigenstrains evaluated according to (7.3).

7. Solve the incremental discretized equilibrium equations[2]

$$K_k \Delta d_k = \Delta f_k \tag{7.6}$$

and determine the column matrix of incremental nodal displacements Δd_k.

8. For each integration point of each finite element:

    a. evaluate the strain increment $\Delta \varepsilon_k = B \Delta d_k$,

    b. compute the stress increment using the formula

$$\Delta \sigma_k = \bar{E}_k D_v (\Delta \varepsilon_k - \Delta \tilde{\varepsilon}_k) \tag{7.7}$$

    for concrete and an appropriate stress–strain equation for steel (e.g., an elastic-perfectly plastic model for passive reinforcement and formula (4.137) for prestressed cables), and update the stress $\sigma_{k+1} = \sigma_k + \Delta \sigma_k$,

    c. update the internal variables for concrete using the formula

$$e_{r\mu,k+1} = \frac{\lambda_{\mu k}}{D_{\mu,k+1/2}} C_v \Delta \sigma_k + \beta_{\mu k} e_{r\mu k} \tag{7.8}$$

9. Increment the step counter k by 1 and proceed to the next step (go to 2).

7.5   Effects of Slab Thickness, Temperature, and Cracking

Model B3 combined with the sectional approach predicts separately the basic creep of the material (i.e., the part of creep unaffected by moisture content variation) and the additional effects of drying. It gives the average shrinkage and average drying creep (or stress-induced shrinkage) in a centrically loaded and symmetrically drying cross section.

The shrinkage is modeled by eigenstrain increments. In each of the 25 segments of the central half-span, the plate thicknesses and concrete ages are different, resulting in a different shrinkage function and different compliance function for each plate of each segment.

Křístek et al. [558] revealed extreme sensitivity of the early box girder deflections to the differences in the rates of shrinkage and drying creep between the top and bottom slabs, as illustrated by Examples 4.11 and 4.15 and the example of La Lutrive Bridge in Sect. 7.1. Thanks to the fact that model B3 is physically based, the differences in its parameters between the top and bottom can be assessed realistically, based on the known drying rates.

---

[2]If the structural model contains some nonlinear components (e.g., nonlinear equation for prestress relaxation, plastic yielding of steel reinforcing bars, or the effect of concrete cracking), this step requires iterations.

These rates are characterized by the shrinkage halftimes, which can be estimated as $\tau_{sh} = k_t(k_s D)^2$; see formula (3.7). The shape factor k_s is for plates equal to 1; see Table 3.1. Parameter k_t can be estimated from the empirical formula given in line 5 of Table C.2 in Appendix C. For the onset of drying at $t_0 = 7$ days and mean compressive strength $\bar{f_c} = 35.9$ MPa (5200 psi), its approximate value is 2.98 day/cm^2 (19.2 day/in.^{2}).

The mean temperature of the bottom slab and webs was probably 25°C, but the top slab with the roadway layer, exposed on top to intense tropical sunlight, was probably some 20°C warmer during the day. According to the curves for the effect of temperature on permeability in Fig. 10.3b,c of Bažant and Kaplan [142], this likely caused a 10-fold decrease of τ_{sh} for the top slab. Furthermore, while no cracking could have occurred in the compressed bottom slab, the top slab near supports may have developed hairline cracks since it was under tension, albeit low, due to excessive prestress loss. Based on the experiments reported by Bažant, Şener, and Kim [122], cracks of width 0.15 mm increase the drying rate about three times. The same may be assumed for the top slab, and so the value of $k_t = 19.2$ days/in.^{2} was used for the bottom slab and the webs, and the value of $k_t = 19.2/(3 \times 10) = 0.64$ days/in.^{2} for the top slab. In calculations of deflections, the stiffness of the pavement layer is entirely neglected, since it is unreinforced and in tension. However, since the pavement tends to decelerate the drying and heating of top slab, its thickness is taken into account in shrinkage and temperature calculations.

The shrinkage halftime τ_{sh} depends on the effective member thickness $D = 2V/S_e$ with V = volume and S_e = surface of the member. It used to be commonplace to consider one V/S_e value as a characteristic of the whole cross section, i.e., to take V and S_e as the characteristics of the whole box. In that case, D was a property of the whole cross section, resulting in a supposedly uniform shrinkage and supposedly uniform creep properties. Recently, however, it has been shown by Křístek, Bažant, Zich, and Kohoutková [558] that, to avoid serious errors (which usually lead to overoptimistic interpretation of early deflections), differences in the drying rate due to different thicknesses of the top slab, the bottom slab, and the webs must be taken into account.

A simple way to do that, demonstrated in [558], is to apply a model such as B3 separately to each part of the cross section. Since the drying halftimes are proportional to slab-thickness square, the thickness differences then yield different shrinkage and drying creep compliance in each different plate.

According to the ACI model, an increase in thickness reduces creep and shrinkage through a certain constant multiplicative factor and scales down the alleged final value for infinite time. However, in reality (except for a small multiplicative reduction due to a higher degree of hydration reached in thicker slabs), a thickness increase causes a delay, properly modeled as deceleration and characterized as an increase of the shrinkage halftime, which is proportional to the thickness square (e.g., if the ultimate shrinkage for a slab 0.10 m thick is reached in 5 years, for a slab 1 m thick it is reached in 500 years, i.e., virtually never).

Because of excessive prestress loss, the top slab is found to get into tension after the first year. Although no large tensile cracks were observed [3, 510], sparse fine cracks in the first six segments from the midspan were reported. The mean tensile strength is estimated according to ACI as $\bar{f}_t = 6$ psi $\times \sqrt{\bar{f}_c/\text{psi}} = 433$ psi (3.0 MPa), where $\bar{f}_c = 5200$ psi = mean compression strength. Calculations show that if the tensile strength limit $\bar{f}_t$ is ignored, the tensile stresses in the top slab would in later years reach the level of about $2\bar{f}_t$. The most realistic model would be the cohesive crack model with rate-dependent softening, applied to growing parallel cracks of uniform spacing, with the material between the cracks considered as viscoelastic [162]. However, to implement this model with ABAQUS has been found to be virtually impossible.

After trying with ABAQUS various simplifications, the computations were eventually run under the simplifying assumption that the effective incremental modulus $\bar{E}_k$ for the current time step (which includes the effect of creep and is evaluated in step 3c of Algorithm 7.1) gets reduced to $\bar{E}_k/4$ when the tensile stress exceeds $0.7\bar{f}_t$. With this simplification, the maximum computed tensile stress in the top slab is about 3.0 MPa, and the corresponding strain is $1.83\times$ larger than the actual strain at peak tensile stress. With hardening due to positive $\bar{E}_k/4$ compensated by the strength reduction to 70%, the tensile stress resultant happens to be about the same as that obtained with a more realistic model consisting of a bilinear softening stress–strain relation with an unreduced tensile strength limit and the softening modulus of about $-\bar{E}_k/3$. The error compared to this more realistic model is estimated as $< 1\%$ of the deflection. For comparison, if unlimited tensile strength were assumed, the computed deflections would have been about 4% smaller.

Combined with the steel stiffness, the softening of concrete would have resulted in overall tension stiffening, which would have been easy to implement had all the computations been programmed. But, in the algorithm with ABAQUS, the tensile softening turned out to be intractable because it would have interfered with the programming of exponential algorithm for creep. This is why a positive modulus $\bar{E}_k/4$ had to be adopted.

7.6   Determination of Model Parameters

The input parameters of the creep and shrinkage prediction models may be divided into extrinsic and intrinsic. For all models, the *extrinsic parameters*, which include the environmental factors, are:

1. the age at the start of drying, taken here as $t_0 = 7$ days, which is the mean period of the segmental erection cycle ranging from 5 to 10 days [775, 796];
2. the average environmental humidity $h_{\text{env}} = 0.70$;
3. the effective thickness of cross section $D = 2V/S_e$, which is different for the web, top slab, and bottom slab and varies along the span;
4. for the extended model B3 capturing the thermal effects on creep, also the temperature.

The *intrinsic input parameters*, which reflect the composition of concrete, vary from model to model. The formulation of ACI, CEB, and GL models was driven by simplicity, as desired by many engineers. Accordingly, the main intrinsic parameter in these models is the mean standard 28-day compression strength $\bar{f}_c$. The other major influencing parameters such as the cement content and the water-cement and aggregate-cement ratios are taken into account by factors some of which act merely as vertical scaling even when the shape of the curve should change, too (as in drying effect in ACI model).

Model B3, which became a RILEM Recommendation [727], is special in that the freely adjustable parameters of the compliance function $(q_1, q_2, \ldots q_5)$ are more than one. They can be adjusted according to concrete composition. If unknown, they can, of course, be set equal to their recommended default values. If the composition is unknown or its effect represented poorly, they have the advantage that one can explore the reasonable ranges of the unknown concrete mix parameters, run the computation of structural response for various plausible sets of values of these parameters, and thus get a picture of the possible range of structural responses to expect. Two sets of input parameters have been considered in the present computations.

Set 1 (Pure Prediction):

For simple prediction on the basis of composition, the following input is used:

1. The mean compressive strength at 28 days, $\bar{f}_c = 35.9$ MPa (5200 psi), reported by Zelinski [890]. Note that ABAM Engineers [3] interpret the value of 5200 psi as the so-called *specified* compression strength (in terminology of the ACI code), which is lower than $\bar{f}_c$ by 1.34× standard deviation, and this is how the strength value is considered by Bažant [211]. However, the records of Raymond Zelinski, the Resident Engineer at the KB Bridge construction, are deemed more reliable, and so the results of Bažant [211] had to be recalculated [209, 210].
2. The conventional (28-day) elastic modulus E_{28} was neither specified in design nor measured during construction. The E_{28} value identified from the truck load test just before the retrofit is appropriate for set 2 but must not be used for set 1, which is intended to check purely the prediction capability of the model. The only way E_{28} could have been estimated at the time of design was from the approximate ACI formula (3.5) (already in use at that time), which gives $E_{28} = 28.3$ GPa (4110 ksi).
3. According to Zelinski's records, the specific cement content was $c = 535$ kg/m^3 (33.4 lb/ft^3), the aggregate-cement ratio was $a/c = 2.9$, and the water-cement ratio was $w/c = 0.40$ (a superplasticizer was used).

The resulting parameters determined from concrete composition using the B3 empirical formulae (Appendix C) are shown in Table 7.1 in the columns denoted as "set 1". To account for the plasticizer effects observed by Brooks [266], the values of q_2, q_3, and q_4 have been increased by 20%.

Set 2 (Updated):

For a better estimate, only the values of q_2, q_5, ε_s^∞, and k_t, governing mainly the response for the first few years, were estimated from the composition, and the estimates of the remaining parameters were improved. Parameter q_1 was adjusted according to the elastic modulus obtained in the truck load test, and parameters q_3 and q_4 were identified by a trial-and-error procedure, conducted with two objectives in mind:

1. Obtain the closest possible fit of the measured deflection of the KB Bridge.
2. Check that the parameter values give compliance curves within a realistic range for structural concretes, matching some existing tests.

The resulting parameter values are given in Table 7.1 in the columns marked as "set 2". They give compliance curves similar to those observed in the 30-year tests of Brooks [267], although the composition was different from Brooks' tests.

Table 7.1 Two sets of parameters of model B3 used in the analysis of Palau Bridge

Parameter	Set 1	Set 2	Unit	Set 1	Set 2	Unit
q_1	0.146	0.188	10^{-6}/psi	21.2	27.3	10^{-6}/MPa
q_2	1.420	1.420	10^{-6}/psi	206.0	206.0	10^{-6}/MPa
q_3	0.011	0.262	10^{-6}/psi	1.6	38.0	10^{-6}/MPa
q_4	0.080	0.140	10^{-6}/psi	11.6	20.3	10^{-6}/MPa
q_5	2.33	2.33	10^{-6}/psi	337.9	337.9	10^{-6}/MPa
ε_s^∞	0.981	0.981	10^{-3}	0.981	0.981	10^{-3}
k_t	19.2	19.2	days/in.^{2}	2.97	2.97	days/cm^2
k_t (top slab)	0.64	0.64	days/in.^{2}	0.10	0.10	days/cm^2

It is easy to check that the terminal asymptotic slope of the compliance curves of model B3 is proportional to $nq_3 + q_4$; see, e.g., the integrated form of Eq. (9.31). Vice versa, to determine parameter q_3 and especially q_4, multidecade creep tests are necessary (note that $n = 0.1$, and thus, q_3 has only a weak influence on the terminal slope). Such information is scant in the existing worldwide database of creep tests. The database is totally dominated by the test data for load durations < 6 years, which are relatively insensitive to q_4. Consequently, parameter q_4 cannot be uniquely identified from the database. For this reason, and because multidecade laboratory data cannot be acquired soon, the best way to obtain an improved empirical estimate of parameter q_4 from concrete composition would be an inverse finite element creep analysis of measured multidecade deflections of many existing box girder bridges [211].

To calculate and compare the predictions of various models, all the properties of concrete and environmental histories of the KB Bridge concrete would have to be known. But they are not. So, comparing the predictions of various models either mutually or with the observations is not fully informative. Nevertheless, what can be compared is whether the observed deflections are within the realistic range of possible predictions of each model. It has been shown [209, 210] that they are within the realistic range for model B3, but not at all for other models, particularly the ACI, CEB, JSCE, and GL models.

For model B3, the predictions are not fixed because there exist input parameters that are unavailable for the KB Bridge and are thus free to set within their realistic range. But the predictions of the other models are fixed by the reported value of concrete strength, with no flexibility of adjustment (a partial exception is the JSCE model which takes into account the water content w and cement content c). The data available for the KB Bridge, as presented here, do not suffice to obtain for this bridge a unique compliance function (unless the default parameter values are used). But they do suffice to obtain unique compliance functions for the ACI, CEB, JSCE, and GL models, although at the cost of ignoring many important influences.

Some engineers view it as advantageous if the model predicts creep and shrinkage from as few parameters as possible, particularly from the concrete design strength only [14]. Slightly more convenient though it may be, realistic it is not. If the additional parameters of model B3 for the given concrete are known, a better prediction can be made. If they are unknown, they can be assigned their typical, or default, values, and thus, predictions can still be made even if only the strength is known. Furthermore, varying the main influencing parameters of model B3 through their realistic range, one can explore the realistic range of the responses to expect, and design the structure for the most unfavorable realistic combination. With the other models, among the intrinsic influences only the effect of strength variation can be explored.

7.7 Results of Simulations and Comparisons to Measurements

7.7.1 Calculated Deflections

As the first check of the computational model, comparison was made with the bridge stiffness, which was measured in January 1990 in a load test by Japan International Cooperation Agency [510]. An average downward deflection of 30.5 mm (0.10 ft.) was recorded at midspan when two 12.5 ton trucks were parked side by side on each side of the midspan hinge (one previous paper erroneously assumed that only one truck was parked on each side). The front wheels of two trucks on each side are assumed to have been 3 m away from the midspan. The rear wheels, 12 m behind the front wheels, are assumed to carry 60% of the truck weight. The finite element code gives the deflection of 30 mm (0.098 ft.) under the load of 245 kN (55.1 kips). The deflection was assumed to be measured 0.1 day after load application, which means that the calculation included the 0.1-day creep. Given the uncertainty about the actual test duration and rate of loading, the agreement with the measurement is good enough.

The results of long-term creep calculations are shown in Figs. 7.5 and 7.7, both in the linear and logarithmic time scales ($t - t_c$ denotes the time measured from the end of construction, and t_c is the time when the midspan hinge is installed). The data points show the measured values. The circles represent the data reported by the firm investigating the excessive deflection [367], and the square boxes the data accepted from a secondary source [510]. For comparison, the figures show the results obtained with model B3 and the ACI, CEB, JSCE, and GL models. All these responses have been computed with the same finite element program and the same step-by-step time integration algorithm. For model B3, it was possible to consider the effect of the differences in thickness of the slabs and webs on their drying rates.

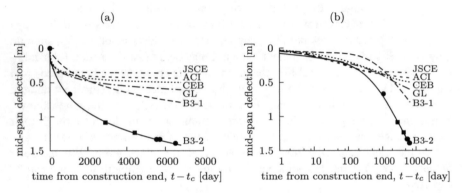

Fig. 7.5 Midspan deflections calculated by various creep and shrinkage models in normal and logarithmic scales, compared with measurements (data points)

Figure 7.5 shows the deflection curves up to the moment of retrofit at about 19 years of age. Since well-designed bridges (such as the Brooklyn Bridge) have lasted

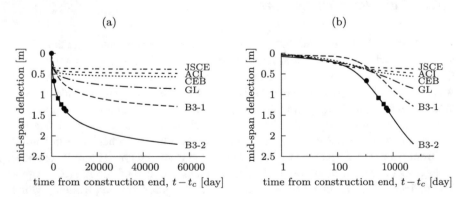

Fig. 7.6 Calculated midspan deflections extended up to 150 years, compared with measurements (data points)

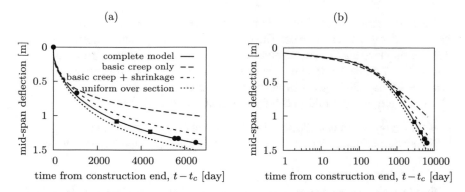

Fig. 7.7 Midspan deflections obtained with B3 model using different assumptions on drying creep and shrinkage, compared with measurements (data points)

much longer, Fig. 7.6 shows the same curves extended up to 150 years under the assumption that there has been no retrofit (and thus no collapse).

According to the ACI, CEB, and JSCE models, the compliance curves and the deflection curves terminate with a horizontal asymptote. But according to model B3, the long-time compliance curves are logarithmic. Thus, model B3 predicts the deflection curve to terminate in the logarithmic time scale with an asymptote that is a straight line of a finite slope; see Fig. 7.5b, which shows that the final slope obtained by model B3 with set-2 parameters agrees very well with the observations. For set-1 parameters, the deflection predicted at 18 years is grossly underestimated (but not as grossly as for the other models).

The measured deflection at 18 years since span closing, which was 1.39 m, is closely matched by the deflection calculated from model B3 with set-2 parameters. This measured deflection is roughly 3 times larger than that calculated for the ACI or CEB model (which is 0.47 m or 0.53 m), and about the double of that calculated for the GL model (which is 0.65 m); see Fig. 7.5. Besides, the ACI, CEB, JSCE, and GL deflection curves have shapes rather different from those of model B3 as well as the observed deflection history. They all give relatively high deflection growth during the first year, but far too low from 3 years on, especially for the ACI, CEB, and JSCE models.

For times longer than about 3 years, the deflections are seen to evolve almost linearly in the logarithmic time scale, which is to be expected for theoretical reasons [98]. So the terminal deflections can be extrapolated to longer times graphically; see Fig. 7.6. The graphical straight-line extrapolation is seen to agree almost exactly with model B3 calculations up to 150 years. It is virtually certain that if the bridge were left standing without any retrofit, the bridge would not have collapsed and the 150-year deflection would have reached 2.2 m (7.3 ft.), well beyond the limit of serviceability.

Capturing correctly the initial deflection history is essential for correct extrapolation, to foresee later troubles. The differential shrinkage and differential drying creep compliance due to nonuniform drying is important in this respect [558, 561]. Note in Fig. 7.5 that, for the first measured deflection at about 1000 days, model B3 (with set-2 parameters) gives by far the closest prediction.

Fig. 7.7 presents the midspan deflection that is obtained

1. if the drying creep is neglected (labeled "basic creep + shrinkage"),
2. if both the shrinkage and drying creep are neglected (labeled "basic creep only"), and
3. if the shrinkage and the drying creep compliance are considered to be uniform over the cross section, being deduced from the overall effective thickness $D = 2V/S_e$ of the whole cross section (labeled "uniform over section").

The solid curve (labeled "complete model") represents the reference solution using the complete model B3 with set-2 parameters. Note that the use of uniform creep and shrinkage properties throughout the cross section neglects the curvature growth due to differential shrinkage and differential drying creep and gives results dominated by the unusually thin webs. Also note that the effect of mean drying can be very different from the mean of the effects of drying in the individual slabs [155].

The full three-dimensional analysis reveals that the classical engineering theory of bending (in which the cross sections are assumed to remain plane and normal) is too simplified to capture the true deformation of box girders. Its main deficiency is that it lacks the effects of shear lags, which are different for the self-weight and for the prestress loads from tendon anchors. In Fig. 7.8, the distribution of normal and shear stress is shown in the cross section located at 14.63 m (48 ft.) from the main pier under self-weight, and in the cross section at 60.35 m (198 ft.) from the main pier under prestress. The shear stress distribution looks unusual; the reason is the arch form of the box girder and the fact that the distribution is plotted for a section that is vertical and thus not orthogonal to the centroidal axis of the curved arch-like beam. Significant shear stresses are seen to exist in the top and bottom slabs. The computations show that the shear lag occurs in four different ways— in the transmission of vertical shear force due to vertical reaction at the pier, in the

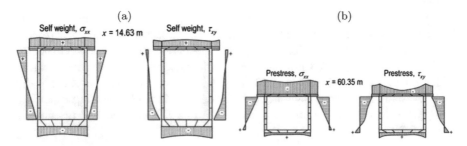

Fig. 7.8 Normal and shear stress distributions caused by (a) self-weight, (b) prestress; reprinted from [210] with permission from ASCE

transmission of the concentrated forces from tendon anchors, and for each of these in both horizontal slabs and in the vertical walls. Only full three-dimensional analysis can capture this behavior. It always yields larger deflections and larger prestress losses.

7.7.2 Calculation of Prestress Losses Due to Creep, Shrinkage, Cyclic Creep, and Steel Relaxation

The significant discrepancy between the full three-dimensional analysis and the analysis based on the engineering theory of bending is demonstrated in Fig. 7.9. The deflections and prestress losses[3] for both cases are evaluated using the B3 model with set-2 parameters. Compared to the three-dimensional analysis, the analysis using the classical bending theory is found to underpredict the deflections by 20% and the prestress loss by 10%. This is despite the relatively high span-to-box-width ratio of this two-lane bridge. For multilane bridges with a wider box, these errors would be larger. For the ACI model, the shear lag effect on deflection and prestress loss is less pronounced (and both phenomena are grossly underestimated). Fig. 7.9a also includes an example of prediction of a widely used commercial program for creep of prestressed bridges, named SOFiSTiK [782], in which the CEB model from early 1970s is embedded as a black box. The creep effects are calculated either by beam elements with memory integrals, or by two- or three-dimensional finite elements using the Trost method [816], which is a predecessor of the AAEM method where the aging of elastic modulus and the time variability of aging coefficient (called in this case relaxation coefficient) is neglected. The figure documents that the predictions of this commercial code, still in wide use, are inferior to all the other methods studied; see in detail Yu, Bažant and Wendner [889].

An approximate correction for the shear lag has actually been considered in the KB Bridge design [775]. But it was considered only due to self-weight and only in the top slab. It was done by introducing into a beam-type analysis the classical effective width [6, 232, 352, 596, 722], which is not very accurate and still misses the shear lags in the webs and bottom slabs and those due to prestress forces from anchors, which add up to a major error.

Accuracy in calculating the prestress loss is crucial because the bridge deflection is a small difference of two large but uncertain numbers—the downward deflection

---

[3]The simulation of prestress losses must take into account the construction sequence. The cantilever consists of 25 segments, which are added sequentially, with 7 days spent by construction of each segment and 1.1 day by its prestressing. The plots in Figs. 7.9b and 7.10 show the evolution of stress in tendons used for prestressing of segment number 7, and the elapsed time is measured from the end of construction of this segment, t_{c7}, which precedes the end of construction of the whole cantilever, t_c, by 146.9 days. The jumps apparent in Fig. 7.10b are caused by prestressing of additional segments number 8 to 25.

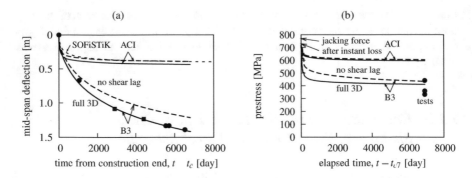

Fig. 7.9 Comparison of (a) deflections and (b) prestress loss obtained by full three-dimensional analysis (solid curves) with those according to the bending theory with cross sections remaining planar (dashed curves), using the B3 model with set-2 parameters and ACI model

due to self-weight, and the upward deflection due to prestress. The shear lag plays a relatively more important role in the former. The error due to neglecting the shear lags is thus magnified. Calculations show that, compared to the classical theory of bending, all the shear lags combined will increase the elastic downward deflection due to self-weight by 18% and the elastic upward deflection due to initial prestress by 14%, which jointly produce the aforementioned total shear lag effect of 20%.

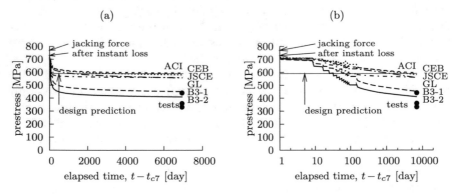

Fig. 7.10 Prestress loss in tendons at main pier, predicted using B3, ACI, CEB, JSCE, and GL models (the elapsed time is measured from the end of construction of the seventh segment)

It is also important to consider that the steel relaxation is a viscoplastic phenomenon which occurs at variable rather than constant strain and is strongly influenced by elevated temperature, as described by the viscoplastic constitutive law proposed in Bažant and Yu [207] and presented in Sect. 4.3.4. That law builds on the formulae for constant strain relaxation used in both the European code [322, 381] and the

American practice [591, 647], formulates the strain change effects as validated by the experiments of Buckler and Scribner [281], and introduces the temperature effect through the activation energy and the Arrhenius factor [326] calibrated by the test data of Rostásy et al. [737], roughly also matching the data from *fib* [381]. Calculations by Bažant et al. [210] showed that the heating of the top slab by tropical sun must have significantly intensified the steel stress relaxation, which in turn must have increased the deflections.

As transpired from the simulation of the bridge in Palau, the excessive deflection is accompanied by longitudinal creep and shrinkage contraction of the box girder. Although this contraction tends to cause a reduction of the subsequent prestress loss due to steel relaxation, the temperature increase due to solar heating of the top slab, together with the continuing longitudinal creep and shrinkage contraction of concrete, more than compensates for this reduction, thus causing the excessive prestress loss to continue and eventually reach values much higher than those for constant strain of steel and constant temperature.

The environmental fluctuations of both temperature and humidity, of course, affect also the concrete. However, the daily temperature fluctuations have little overall effect on the concrete, and the seasonal ones are negligible in the tropics. Even at higher latitudes, the humidity fluctuations due to weather and the seasonal changes are not too important for multidecade deflections because the moisture diffusivity is about 1000 times lower than the thermal diffusivity [150]; see Chaps. 8 and 13. Anyway, the effect of seasonal changes cannot be extracted from the measured deflection histories of 69 bridges presented in [138].

An important point to note is that the predicted 19-year prestress loss is only 22% and 24% when the ACI and CEB models are used in the present finite element code, but about 46% when the B3 model with parameter set 2 is used (Fig. 7.10). The correctness of prestress loss predicted by model B3 is confirmed by nine stress relief tests which were made by ABAM on three tendons just before the retrofit [367]. Normally, only nondestructive methods are permitted. This makes it next to impossible to measure the stresses in grouted tendons. But for the KB Bridge, the cutting of tendons was not a big sacrifice because additional tendons were to be installed anyway. Thus, the decision to retrofit furnished a unique opportunity to learn about the actual prestress losses. Sections of three tendons were bared, and strain gauge were glued at three different locations on each of the three tendons. Each of these tendons was then cut, and the stress was calculated from the shortening measured by the gauge next to the cut; see Table 7.2.

The average stress obtained from nine measurements on the tendons was 377 MPa (54.7 ksi). This means that the average prestress loss over 19 years was 50%. The coefficient of variation was 12.3%. Similar tests were also conducted by another investigating company (Wiss, Janney and Elstner, Highland Park, Illinois), and the average measured prestress loss was almost the same. This confirms that the prestress loss predicted by finite element simulations based on model B3 is realistic, while the ACI and CEB models grossly underestimate the actual values.

Table 7.2 Summary of strain relief tests of prestressed tendons of KB Bridge in Palau [367]

Tendon	Location	Measured strains [10^{-6}]			Mean strain [10^{-6}]	Mean stress [MPa]
	1	1640	1640	1630	1637	327.3
1	2	1650	1640	1650	1647	329.3
	3	1680	1700	1710	1697	339.3
	Average					332.0
	4	1810	1820	1790	1807	361.3
2	5	1810	1800	1790	1800	360.0
	6	1780	1790	1790	1787	357.3
	Average					359.6
	7	2250	2230	2220	2233	446.7
3	8	2220	2220	2210	2217	443.3
	9	2170	2150	2170	2163	432.7
	Average					440.9

In the mid-1970s, the prestress loss used to be calculated not by finite elements but by simple formulae based on the beam theory [660]. A lump estimate of the final prestress loss was generally used, and, according to Shawwaf [775], it was used for the KB Bridge. According to the lump estimate, the prestress loss would have been 22%, which is marked in Fig. 7.10 by a thin horizontal line labeled as "design prediction." Compared to the measurements, the errors of this estimate are enormous, and so are the errors compared to the present calculation based on model B3. These errors are one reason why the long-time deflections were so badly underestimated in design. One must conclude that, for large box girders, the standard textbook formulae for prestress loss are inadequate and dangerously misleading.

7.8 Excessive Long-Term Deflections of Other Box Girders

It is deplorable that the data on excessive deflections usually go unpublished. Nevertheless, Dr. Yasumitsu Watanabe, the Chief Engineer of Shimizu Corp., Tokyo, graciously made available the data on some of the excessively deflecting large Japanese bridges which epitomize the experience in many other countries. These deflection data are plotted in Fig. 7.11, where the data points represent the measured deflections, and the dashed curves show the prediction based on the design recommendation of Japan Road Association (JRA) [528]. The solid curves give the predictions of model B3 calculated in the same way as for the KB Bridge, after adjusting the composition parameters similarly to set 2, as mentioned before. The deflection of one of the four Japanese bridges, Koshirazu, is not excessive yet, but the trends portend trouble for the future.

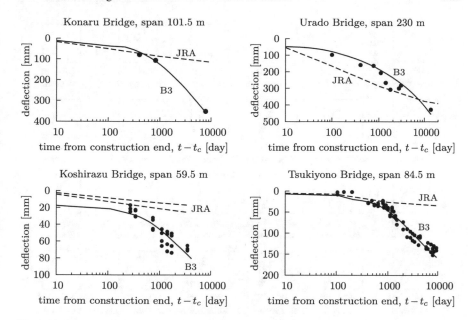

Fig. 7.11 Excessive deflections observed in four Japanese bridges

The results of the analysis of the KB Bridge in Palau and four Japanese bridges have shown that the excessive deflections and prestress loss can be explained and even closely matched with realistic material parameters. Are such excessive deflections rare?

They are not. A subsequent search of various papers, society reports, and company reports [138, 208] under the auspices of the RILEM Committee TC-MDC (Multi-Decade Creep, founded in 2010) led to a veritable **wake-up call**—see Figs. 7.12 and 7.13 documenting the deflection histories of 69 large bridge spans [251, 387, 608, 683, 836], many of them excessive. All, except for one arch, are segmental box girders, and many of them exceed the maximum acceptable deflection, 1/800 of the span according to the AASHTO standard [2]. Hard to obtain though such examples are, hundreds more probably exist. In the original reports, the data were plotted in a linear time scale, which obscured the systematic long-term logarithmic trend.

Of course, segmental bridges that have not deflected excessively (such as the Pine Valley Creek Bridge in California built in 1975) exist, too. Note that even if a poor creep model is used, the deflections can be low if one adopts some or all of the six precautionary measures listed in Sect. 7.11 [210, 211]. Most of them, though, are span-limiting or costly, or both, and render esthetically pleasing slenderness harder to achieve.

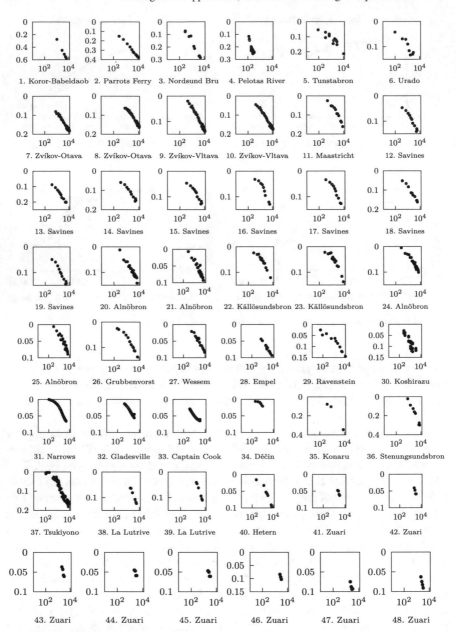

Fig. 7.12 Deflections of 69 bridge spans (part 1), many of them excessive within lifetime (horizontal axis: time from construction end, $t - t_c$ [day]; vertical axis: deflection/span [%])

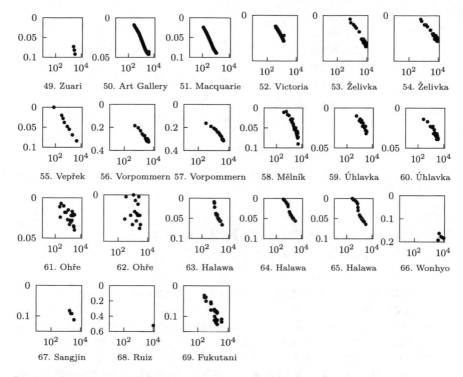

Fig. 7.13 Deflections of 69 bridge spans (part 2), many of them excessive within lifetime (horizontal axis: time from construction end, $t - t_c$ [day]; vertical axis: deflection/span [%])

The absence of a midspan hinge has been known to reduce deflections. However, it is not a panacea. Even bridges without a midspan hinge, designed by the code, can suffer excessive deflections. This is, for example, documented by the data on the Děčín Bridge over Elbe in North Bohemia; see Fig. 7.14. About 10 of the 69 bridges in Figs. 7.12–7.13 have no midspan hinge.

Contrary to concerns expressed by many researchers, the cyclic creep does not contribute significantly to the deflection of large-span prestressed bridges. However, in medium-span bridges (40-80 m), the cyclic creep can lead to nonnegligible additional inelastic strains causing cracking damage; see Sect. 7.13.

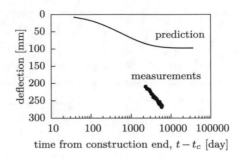

Fig. 7.14 Excessive deflections observed on the continuous bridge over the Elbe River in Děčín, Czech Republic

7.9   Approximate Multidecade Extrapolation of Medium-Term Deflection

The logarithmic time plots in Figs. 7.12 and 7.13 give no hint of an approach to an asymptotic bound. Such a bound is a feature incorrectly implied by all the society recommendations except model B3 (and its predecessors since 1978) and the new *fib* model. They document that the long-term creep is a logarithmic curve, as observed on the basis of laboratory data in Bažant, Carreira and Walser [114].

Why have the recommendations on creep been misleading, for decades? Aside from disregard of the theoretical basis and questionable interpretation of the laboratory test data, the cause also lies in inevitable statistical bias of the worldwide laboratory database [160]. In the latest and largest database, assembled at Northwestern University during 2010–2013 [488], only 5% of creep test curves exceed 6 years in duration, and less than 2% attain 18 years. Most of them were obtained on old types of concretes, very different from modern concretes. Moreover, the data readings are heavily biased for short times and ages.

Also, when the database values of J or ε_{sh} are plotted as a function of $t - t'$, the averages of the values of h_{env}, t' or D in subsequent time intervals are far from being equal in all the intervals, which brings in another source of bias. Therefore, even if the statistical bias is filtered by proper weighting [161], the database does not suffice to bring the multidecade trend to light.

Consequently, inverse interpretation of the bridge deflection histories in Figs. 7.12 and 7.13 appears to be the salvation. Ideally, one should conduct statistical inverse three-dimensional finite element creep analysis of these bridges. But it has appeared impossible to obtain data that would suffice for finite element analysis of these 69 bridge spans, except for six of them.

The time at which the compliance curve becomes a straight line in the logarithmic time scale depends on many factors, but on the average, it is about three years. It is interesting to note in Figs. 7.12 and 7.13 that, beginning at about 1000 days since the span closing, the bridge deflection curves, too, become straight lines in the logarithmic time scale. This means that the complex transient processes, particularly the gradual filling of the capillary pores by cement hydration products, drying shrinkage and drying creep, and the steel relaxation rate greatly attenuate within a few years, and that the relaxation of microprestress becomes a stationary process preserving the logarithmic creep curve.

This observation suggests that deflection w_m measured roughly at $t_m = 1000$ days after the span closing could simply be extrapolated to long times by assuming similarity to $J(t, t')$. For some bridges, $t_m = 1500$ days is better. To keep the extrapolation simple, two severe simplifications of the regime prior to span closing need to be introduced [138]:

1. The age differences among the box girder segments must be ignored, and the age of concrete must be characterized by one common effective (or average) age t_c at the span closing.
2. Instead of the gradual increase of the bending moment in the cantilever segments during the erection, one must consider one common average age t_a at which the moments due to self-weight are applied to the erected cantilever.

Because of these simplifications, the long-time deflections do not grow in proportion to the total compliance $J(t, t_a)$. Nevertheless, for the additional deflection w that develops after the span closing time t_c, the errors in approximating the early loading history must decay with time and eventually become negligible when $t \gg t_c$, i.e., after the lapse of a sufficient time, t_m. It appears that the smallest such time is roughly $t_m = 1000$ days (measured from span closing).

Before the span closing and for a few years afterward, the drying and aging processes, of course, make the box girder response very complicated. But later, after time t_m, these effects nearly die out and the box girder begins to behave as a nearly homogeneous structure (Sect. 4.1), for which the growth of deflection w should be roughly proportional to the increment of the compliance function that has occurred since the closing time t_c; i.e., $w(t) = C[J(t, t_a) - J(t_c, t_a)]$, where C is a certain constant (of course different for each bridge). The value of C or w_m can vary widely and its calculation would require a detailed finite element analysis considering creep with drying and the construction sequence. However, C may be calibrated experimentally from w_m. To get $w(t_m) = w_m$, we need to set $C = w_m/[J(t_m, t_a) - J(t_c, t_a)]$. This leads to the approximate extrapolation formula [138]

$$w(t) = w_m \frac{J(t, t_a) - J(t_c, t_a)}{J(t_m, t_a) - J(t_c, t_a)} \tag{7.9}$$

To analyze the data in Figs. 7.12 and 7.13, the values $t_m = 1000$ days, $t_c = 120$ days and $t_a = 60$ days were assumed for all the bridges [138]. In recent studies, $t_m = 1500$ days seems preferable.

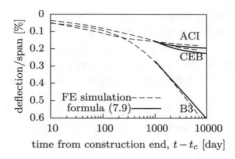

Fig. 7.15 Verification of extrapolation formula (7.9) by comparisons with accurate finite element creep solutions for the KB Bridge in Palau

To check how good this formula is, we may take advantage of the creep deflection curves accurately calculated by finite elements for the KB Bridge in Palau using the B3, ACI, and CEB material models [209–211]. For each curve, we use (7.9) to extrapolate the computed deflection w_m using the same compliance function $J(t, t')$ as that used in finite element computations. The resulting extrapolations are shown in Fig. 7.15. It is remarkable how close to the computed curve each extrapolation is, especially in the case of model B3. This validates the usefulness of formula (7.9).

To apply formula (7.9) to the bridges in Figs. 7.12 and 7.13, one must specify the mean concrete strength $\bar{f}_c$, and for model B3 also w/c, c, and ρ. Unfortunately, these parameters are known for only six among the 69 bridges. So, individual comparisons for each bridge are impossible. Nevertheless, we can make a useful comparison at least in the mean relative sense for all the bridges combined.

We will assume that the specified design (or characteristic) strength of concrete in these older bridges was on the average 31 MPa (4500 psi), which implies (according to CEB-*fib*, with the characteristic strength set to 31 MPa) the mean strength to be at least $\bar{f}_c = 39$ MPa (5660 psi); see formula (E.1). Furthermore, we will assume the average effective cross-sectional thickness of $D = 0.25$ m (10 in.), and the environmental humidity of 70% for the Scandinavian bridges (Norsund Bru, Tunstabron, Alnöbron), and 65% for other bridges. For the other parameters, we assume $w/c = 0.5$, $c = 400$ kg/m^3 (25 lb./ft.^{3}), and $\rho = 2300$ kg/m^3 (143 lb./ft.^{3}). Of course, the deflection curve extrapolated in this way from w_m will likely be incorrect for each particular bridge. But the mean of the extrapolations for all the bridges should still be approximately equal to the mean of the correct extrapolated slopes that would be obtained if the properties of each individual concrete were known.

Fourteen of the 69 bridge spans in Figs. 7.12 and 7.13 are omitted from the extrapolation exercise. The reason is that, for La Lutrive, Zuari, Wonhyo, Sangjin, and Ruiz, the record of deflection values starts at times much longer than 1000 days, and so it would be very difficult to get a reasonable estimate of w_m.

The measured deflections corresponding to 55 bridge spans are represented by the dots in Fig.7.16. To make the presentation of data for different bridges in the

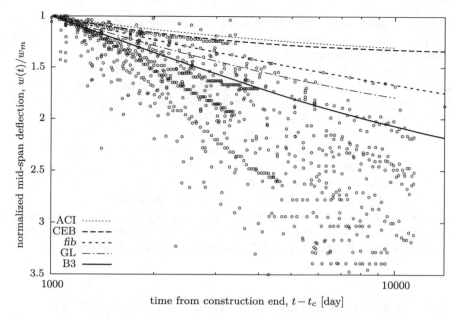

Fig. 7.16 Extrapolations by formula (7.9) from measured 1000-day deflection w_m for 55 sets of bridge deflection records for average concrete properties, obtained using models ACI-209, CEB Model Code 1990, *fib* Model Code 2010, GL2000 and B3

same graph meaningful, the deflections for each bridge are normalized by the value at 1000 days from construction end, interpolated from the nearest measured values. The advantage of the normalized plot is that the prediction constructed according to (7.9) is the same for all bridges and depends only on the compliance function, i.e., on the creep model adopted. The extrapolations obtained with the ACI, CEB, *fib*, GL, and B3 models are shown in Fig. 7.16 as smooth curves. None of these models is found to be satisfactory since they all systematically and significantly underestimate the measured long-time deflections. Nevertheless, model B3 does not perform as poorly as the others. It is seen to have two important advantages:

1. The long-time form of model B3 is a logarithmic curve (seen as a straight line in the semilogarithmic scale), which agrees with the long-time shape of the data, while the long-time curves for the ACI and CEB models (as well as JSCE and JRA models, not plotted here) level off as they approach a horizontal asymptote. The GL model would also level off, but much later. The *fib* Model Code 2010 terminates with an unbounded logarithmic curve similar to B3.
2. Model B3 is the only one that can be updated without compromising the short-time performance, because the slope of the straight long-time asymptote can be separately controlled.

These advantages are shared by model B4, which gives for this specific concrete composition somewhat slower deflection growth after 1000 days than B3. The same holds true for the new *fib* model.

From Fig. 7.16, it can be observed that the extrapolations from 1000 days to a high age t such as 30 to 100 years could be improved by multiplying the deflection increment from 1000 days to time t by the factor of about 1.6. Then, the extrapolation would be roughly in the middle of the range of data in Fig. 7.16.

In segmental bridges that are made continuous through the midspan (and thus have no midspan hinge), the internal forces redistribute so as to approach the elastic moment distribution for a continuous bridge. The redistribution can be estimated as explained in Sect. 4.1.2, and a correction to (7.9) could be devised. However, it is interesting that this simple formula gives good extrapolations even for these bridges, e.g., for the Děčín and Vepřek bridges. The explanation is that the degree of redistribution after 1000 days must have been very small, which could be explained by the relative shallowness and high flexibility of the cross section at midspan.

7.10   Uncertainty of Deflection Predictions and Calculation of Confidence Limits

Creep and shrinkage are notorious for their relatively high random scatter. For this reason, it has been argued during the last two decades [145, 147] that the design should be made not for the mean deflections, but for some suitable confidence limits such as 95% [163]. Adopting the Latin hypercube sampling of the input parameters explained in Sect. 6.2, one can easily obtain such confidence limits by repeating the deterministic computer analysis of the bridge according to model B3 8 times, one run for each of eight different randomly generated samples of eight input parameters.

As explained in Chap. 6, the range of the cumulative distribution of each random input variable (assumed to be Gaussian) is partitioned into $N = 8$ intervals of equal probability. The parameter values corresponding to the centroids of these intervals are selected according to randomly generated Latin hypercube tables, such as Table 6.4a. The values are then used as the input parameters for eight deterministic computer runs of creep and shrinkage analysis.

One random input variable is the environmental relative humidity h_{env}, whose mean and coefficient of variation are estimated as 0.7 and 20%. The others are B3 model parameters q_1, q_2, q_3, q_4, q_5, k_t, and ε_s^∞. Their means are taken from set 2 (see Table 7.1), except for parameter $q_2 = 1.04 \times 10^{-6}$/MPa (these are preliminary parameter values available when the stochastic analysis was made, and differ slightly from the final values used to calculate Fig. 7.5). The estimated coefficient of variation is 23% for creep parameters q_1, q_2, q_5, and 30% for q_3, q_4 (these have a higher uncertainty as they relate to long-term creep, for which the data are scarce). For shrinkage parameters k_t and ε_s^∞, it is 34% [104, 107, 175].

The responses from each deterministic computer run for model B3, particularly the midspan deflections at specified times, are collected in one histogram of eight values, whose mean $\bar{w}$ and coefficient of variation ω_w are the desired statistics. Knowing these, and assuming the Gaussian (or normal) distribution, one can get the one-sided 95% confidence limit as $w_{95} = (1 + 1.65\omega_w)\,\bar{w}$. This is the limit that is exceeded with the probability of 5%; in other words, the limit would be exceeded by one out of 20 identical bridges, which seems to give optimal balance between risk and cost.

(a)   (b)

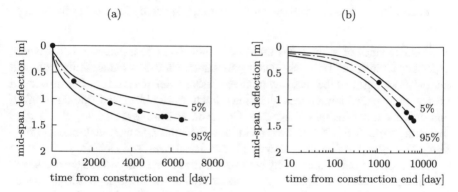

Fig. 7.17 Mean values and 5% and 95% confidence limits of deflections predicted by model B3 in normal and logarithmic scales

The curves of the mean, and of the one-sided 95% and 5% confidence limits for the KB Bridge in Palau, are shown in Fig. 7.17. Note that the curves of the present three-dimensional finite element creep calculations according to the ACI and CEB models lie way outside the statistical confidence band obtained with model B3.

Using the NU database [488] as the prior information, one could further improve the statistics of long-term deflection by means of Bayesian statistical analysis [557], as described in Sect. 6.4.

Note that the probabilistic problem of deflections is, fortunately, much easier than that of structural safety. For the latter, the extreme value statistical theory must be used since the tolerable probability of failure is $< 10^{-6}$, which is far less than the value of 0.05 that is acceptable for deflections. Unlike here, the distribution tail becomes a major problem.

7.11 Precautionary Deflection-Minimizing Design and Tendon Layout

In design, it is prudent to minimize deflections and prestress losses by the following measures, most of which have been known though often not followed:

1. choose a concrete with a low long-time creep;
2. give concrete more time to gain strength before prestressing;
3. install empty ducts for possible later installation of additional tendons;
4. use a tendon layout that minimizes deflections [561];
5. use excess prestress causing upward deflection;
6. avoid midspan hinge and make the girder stiff at midspan;
7. use stiffer (deeper) girders, especially at midspan;
8. minimize steel relaxation by using low relaxation steel and by reducing the ratio of initial prestress to tendon strength;
9. locate tendons so as to minimize their heating (especially for bridges in the tropics).

The avoidance of a hinge has often been thought to be a panacea, but it is not, especially if the flexural stiffness at midspan is not high. It does reduce deflections and has the advantage that the roadway of an excessively deflected bridge has no sudden change of slope, which reduces vibrations due to traffic loads and ride discomfort, and makes deflections less noticeable. However, at least seven among the bridges deflecting more than 0.001 of the span in Figs. 7.12 and 7.13 have no hinges (Parrots Ferry, Grubbenvorst, Wessem, Empel, Hetern, Ravenstein and Děčín).

Measures 4–8 may explain why some old bridges have not deflected excessively even if a poor creep prediction model was used in design. One such bridge may be the first large US prestressed segmental box girder, the Pine Valley Creek Bridge (later named N. I. Greer Memorial Bridge) on Interstate 8 in the San Diego County, California, designed in 1972 by Man-Chung Tang and built in 1974, with the main span of 137 m. Although the deflections of this bridge have not been reported, there are no signs of problems. This multispan girder has a constant depth and a high stiffness at midspan. The main span has no hinge, but the adjacent shorter spans do. To avoid the risk of errors in calculation of moment redistribution due to creep, jacking was used at span closure to produce the same moment distribution as in a bridge built on a falsework, i.e., as in a continuous beam. So there was no moment redistribution due to creep. The stress at the pier due to self-weight was reduced approximately to that for a continuous beam, and the stress profile was changed roughly to that in Fig. 7.18c. Since the segment of positive moments created by jacking was short, the stress due to self-weight did not dominate, producing at midspan a self-weight stress distribution like that in Fig. 7.18c. Also, a significant camber was planned and actually achieved. It must be warned, however, that avoidance of a hinge helps significantly only if the midspan is nearly as stiff in flexure as the ends of the span and if jacking is used.

To explain the effect of excess prestress, consider the stress profiles in a negative moment cross section of a segmental girder sketched in Fig. 7.18, in which σ_P, σ_D, and σ_L are the stresses caused by the prestress, by the dead load (self-weight), and by the live load (traffic). The dashed lines mark the compressive and tensile strength limits that cannot be exceeded.

For a small enough span, the live load dominates, $|\sigma_L| \gg |\sigma_D|$, as shown in Fig. 7.18a. To satisfy the tensile and compressive stress limits, the bending moment

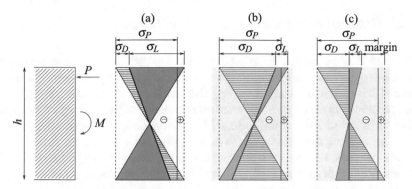

Fig. 7.18 Stress distributions in a prestressed cross section under bending: (a) small span, (b) large span, economic design, (c) large span, precautionary design

under prestress and dead load alone must be positive (the corresponding stress distribution is marked by the thick solid line in Fig. 7.18a). This means that the creep will cause an upward deflection and one need not worry about creep (no cases of excessive upward deflections are known).

For a large span, $|\sigma_L| \ll |\sigma_D|$. If the girder is designed most efficiently from the strength viewpoint, with a full use of the tensile and compressive stress limits (Fig. 7.18b), the dead load stress profile corresponds to a large negative bending moment. This might or might not lead to unacceptable creep deflections, but in any case, a realistic and accurate creep analysis is required.

Alternatively, the large-span girder may be designed so that the dead load and the prestress cause an approximately uniform stress profile (Fig. 7.18c), which is inefficient from the strength viewpoint. As a result of such a design, the creep will cause mainly shortening of the girder but no significant vertical deflection compared to the design in Fig. 7.18b, and errors in creep prediction will not matter. Of course, even after superposing the live load stress profile σ_L, there will be large margins against the tensile stress limit on top and against the compressive strength limit at bottom (Fig. 7.18c). A deeper and more massive cross section, and a higher level of prestress, will have to be used. This will force the design to be heavier, less slender, and more expensive, and the increased self-weight may even prevent bridging the desired span. Hence, such a precautionary but inefficient design is often impossible.

Modern tall buildings and nuclear containments are highly sensitive to creep and shrinkage. It is again important to use a realistic prediction model and calibrate it by short-time tests (Sect. 3.8, Appendix H). Differential shortening of concrete columns in a very tall tower is a potential problem, even if they are made of high-strength concrete. To minimize it, it is wise to make the cross sections of all columns identical and their environment the same (this measure was used by W. Baker of SOM, Chicago, in the design of Burj Khalifa tower in Dubai).

7.12 Deflection-Mitigating Layout of Tendons

Compared to statically determinate beams, the statically indeterminate beams provide more opportunity to minimize the deflections. Moving the tendon anchorages
between the segments of positive and negative bending moments, or changing the tendon eccentricity in the segments of transition between positive and negative moments
(in which the prestress is not fully utilized from the stress level viewpoint), can greatly
affect the midspan deflection. For the sake of simplicity, we consider the structure
as homogeneous, i.e., we neglect the stress redistributions due to creep. We also
consider other deflection sources, e.g., the differential shrinkage, as unimportant for
multidecade deflections. With these simplifications, the problem of optimum layout
of the prestressing tendon can be limited to elastic deflections.

Consider a prestressed tendon anchored at sections a and b of a beam of total
length L, with eccentricity[4] described by a function $e(x)$. Prestressing of the tendon
by a force $P > 0$ (considered for simplicity as uniform) generates in concrete primary
bending moments

$$M_1(x) = \begin{cases} 0 & \text{for } 0 \le x < a \\ -Pe(x) & \text{for } a \le x \le b \\ 0 & \text{for } b < x \le L \end{cases} \tag{7.10}$$

If the beam is statically indeterminate (e.g., a continuous beam), additional self-
equilibrated moments $M_2(x)$ arise. In an elastic calculation, the corresponding total
curvature is expressed simply as $\kappa(x) = [M_1(x) + M_2(x)]/EI(x)$, where E is the
elastic modulus and I is the sectional moment of inertia. The deflection at a specific
section, e.g., at midspan, can be evaluated using the principle of virtual forces. Let
us denote by $\bar{M}(x)$ the distribution of bending moments caused on the same elastic
beam by a unit vertical force applied at midspan. The midspan deflection is then
given by

$$w_{mid} = \int_0^L \bar{M}(x)\kappa(x)\,\mathrm{d}x = \int_0^L \frac{\bar{M}(x)[M_1(x) + M_2(x)]}{EI(x)}\,\mathrm{d}x \tag{7.11}$$

Taking into account that $\bar{M}(x)/EI(x) = \bar{\kappa}(x) =$ curvature that would be caused by
the unit force, we can rewrite (7.11) as

$$w_{mid} = \int_0^L \bar{\kappa}(x)[M_1(x) + M_2(x)]\,\mathrm{d}x = -P\int_a^b \bar{\kappa}(x)e(x)\,\mathrm{d}x \tag{7.12}$$

Note that the integral $\int_0^L M_2(x)\bar{\kappa}(x)\,\mathrm{d}x$ vanishes because moments M_2 are self-
equilibrated and curvatures $\bar{\kappa}$ are compatible. This means that the secondary moments
M_2 do not need to be calculated if only the deflection w_{mid} is of interest. If the integral

---

[4]The eccentricity is considered as positive if the centroid of the prestressed tendons is below the
centroid of the section; see Fig. 4.36.

$\int_0^L \bar{\kappa}(x)e(x)\,\mathrm{d}x$ is positive, the deflection due to prestress is negative and the midspan lifts up, which is advantageous for the reduction of the resulting total deflections after adding the self-weight and traffic loads. Since the sign of curvature $\bar{\kappa}$ is the same as the sign of moment $\bar{M}$, it is good to use maximum positive eccentricity (tendons below the beam axis) in the region where moments due to the unit force are positive (tension in bottom fibers) and maximum negative eccentricity in the region where moments due to the unit force are negative. Some of the best designers have used such wisdom intuitively. A practical example based on the bridge over the Labe (Elbe) River in Mělník is available in Křístek et al. [561].

For a simply supported beam, $\bar{\kappa}$ is positive along the whole span, which means that tendons below the beam axis cause lifting of the beam, as expected. For a continuous beam, an example of the distribution of moments $\bar{M}$ is shown in Fig. 7.19. To reduce the midspan deflection in the main span, the tendons should be placed, as much as possible, below the beam axis in the central part of the main span, and above the beam axis in the outer parts of the main span and in the side spans. For a symmetric three-span beam with uniform cross-sectional stiffness EI, the relative size of the zone with positive moments $\bar{M}$ is characterized by

$$\alpha = \frac{1}{2} + \frac{L_1}{2L_1 + 3L_2} \tag{7.13}$$

with the meaning of L_1, L_2, and α specified in Fig. 7.19. For $L_1 = 0$, we obtain $\alpha = 0.5$, which corresponds to a beam clamped at both ends. The limit $L_1 \to \infty$ corresponds to a simply supported beam with $\alpha = 1$.

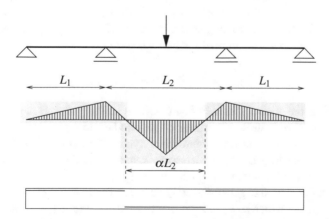

Fig. 7.19 Continuous beam with moments due to a unit force, and examples of tendons contributing to a negative deflection at midspan

In statically determinate structures, the opportunity to minimize deflection by tendon layout is lesser, but some exists. In a cantilever of a box girder with midspan hinge, the prestress force, cross-sectional height, and wall thicknesses are normally decreased with distance from the support. However, the deflections can get mitigated if these design variables are decreased less than allowed by stress considerations. This makes it possible to use a higher prestress than necessary, which will pull the midspan up.

7.13   Effect of Cyclic Stress Variations on Creep Compliance

7.13.1   History of Cyclic Creep Models

In discussions at conferences, it has often been claimed that the cyclic creep caused by repeated traffic loads explains the excessive long-time deflections of box girder bridges. But this is not true. Following a recent in-depth study of Bažant and Hubler [134], let us examine why.

The *cyclic creep* of concrete, also called the *fatigue creep* (or vibro-creep, as a literal translation of the Russian term vibropolzuchest') is the long-time deformation produced by cyclic load. This phenomenon was experimentally detected by Féret in 1906 and was also observed by Probst in 1925, Mehmel and Heim in 1926, and Ban in 1933 (cf. [223]). More systematic experiments permitting quantitative characterization had to wait until the works of Gaede [402] and of Mehmel and Kern [625]. After World Word II, many researchers studied this phenomenon experimentally and proposed various approximate and mutually contradictory empirical formulas [59, 69, 70, 149, 176, 231, 268, 287, 409, 443, 483, 486, 574, 575, 597, 644, 654, 662, 668, 863, 872]. After conducting extensive tests of cyclic creep in compression, Gaede [402] proposed a formula of the general form

$$\Delta \varepsilon_N^{cyc} = c \sigma_{max} N_{cyc}^r \tag{7.14}$$

where $\Delta \varepsilon_N^{cyc}$ is the strain increment due to cyclic loading after N_{cyc} cycles, σ_{max} is the maximum uniaxial stress in periodic cycles, and r and c are empirical constants. Equation (7.14) was based on compression cycles from 0.14 $\bar{f_c}$ to 0.75 $\bar{f_c}$. Such cycles reached way beyond the service stress range allowed in bridge design, which is limited by 40% of standard compression strength.

Wittmann [872] tried to generalize his power law for a (static) creep curve, $\varepsilon(t) = at^n \sinh(b\sigma / \bar{f_c})$ in which a, b, and n are empirical constants. He ignored the aging and used a hyperbolic sine function based on his assumption of thermally activated transitions. Considering cyclic stress $\sigma(t) = \sigma_m + \Delta\sigma \sin \omega t$ where ω is the circular frequency and σ_m and $\Delta\sigma$ are the mean value and half-amplitude of cyclic stress, Wittmann empirically generalized this power law by replacing the constant exponent n with the variable $n = n_0 + c(\Delta\sigma / \bar{f_c})^d$, where n_0, c, and d were constants calibrated by Gaede's data.

The most comprehensive diverse tests were conducted by Neville et al. [483, 654, 863]. Neville and Hirst [654] proposed that the cyclic creep is an inelastic deformation caused by microcracking, but made no attempt to model the microcracking per se. In view of the hardening effect under low stress cyclic creep observed in some experiments [59, 231, 483, 654, 863], they suggested that the microcracking occurs at the aggregate interfaces. Garrett [409] speculated that these microcracks could expose unhydrated cement to further hydration which in turn might cause further deformation. Hirst and Neville [483], and later also Brooks and Forsyth [268], assumed that the total cyclic creep strain $\varepsilon_{cyc} = \varepsilon_{stat} A(\ln t)^B$, where ε_{stat} is the static creep strain and A and B are calibration parameters.

Later on, Pandolfi and Taliercio [668] suggested a more complicated formula for cyclic creep of concrete based on numerical simulations. They emphasized two concepts: The time is only implicitly related to the number of cycles, N_{cyc}, i.e., the tests should be interpreted in terms of N_{cyc}, and the loading frequency is indirectly related to the loading rate [486]. Damage evolution models based on failure surfaces in the stress space have also been suggested. However, prior to 2014, no model based on fatigue growth of individual microcracks under cyclic loading has been published.

The phenomenological formulations treated cyclic creep in two ways: either as a deformation $\Delta \varepsilon_N^{cyc}$ that is additional to the static creep [70], or as an acceleration of the static creep [149, 176, 210]. Both were able to provide acceptable fit of the main data, doubtless because of their limited duration—mostly less than 10 days, and 28 days as the maximum. This is much less than the desirable lifetimes of large-span bridges, which are 100 to 150 years. For such extrapolation to be realistic, the cyclic creep model must based on a sound theory rooted in micromechanics. Although the micromechanics has been discussed intuitively in qualitative terms [409, 654, e.g.] or in terms of damage mechanics [404, 590], no micromechanics-based and experimentally validated constitutive model seems to have been published prior to 2014.

7.13.2 Macroscopic Strain Due to Small Growth of Microcracks*

Unlike metals and fine-grained ceramics, the microstructure of concrete is disordered and full of microcracks on all the scales—from the nanoscale to the macroscale of a representative volume element (RVE), whose size is typically 0.1 m (assuming normal size aggregates). The growth of cracks larger than the RVE reduces strength and stiffness and is countered by reinforcing bars. Cracks much smaller than the RVE do not appreciably affect strength and do not reduce material stiffness. Under fatigue loading, such cracks must be expected to grow, which causes additional deformation referred to as cyclic creep, but do not appreciably reduce the unloading stiffness or elastic modulus. Indeed, the experimental results for cyclic compressive loading of concrete within the service stress range (i.e., for stresses less than 40%

of the strength) show no degradation of material strength for subsequent short-time loading up to failure and indicate only a slight decrease of concrete stiffness [475].

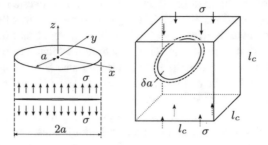

Fig. 7.20 Mode-I penny-shaped crack and a generic microcrack in three dimensions

Consider a generic three-dimensional planar crack of size a, for instance a penny-shaped crack of radius a (Fig. 7.20). In the case of compression loading, we imagine a shear crack with combined modes II and III, and in the case of tensile loading, a pure mode-I crack. For the sake of simplicity, the crack is assumed to grow in a self-similar way, expanding in its plane in proportion to a. The *energy release rate* due to three-dimensional self-similar growth of the crack may generally be expressed as

$$\mathscr{G}(\sigma, a) = \frac{\gamma_1}{a} \frac{\partial \Pi_f^*(\sigma, a)}{\partial a} \tag{7.15}$$

where Π_f^* is the *complementary energy* (or Gibbs' free energy) due to fracture alone, per microcrack, σ is the applied remote stress (precisely, stress at infinity), and γ_1 is a dimensionless geometry factor which, e.g., would be equal to $1/2\pi$ in the case of a penny-shaped crack in mode I. Even though the stress intensity factors must vary along the crack edge in three dimensions, one can define an effective *stress intensity factor* at the crack edge,

$$K(\sigma, a) = \sqrt{E\mathscr{G}(\sigma, a)} \tag{7.16}$$

on the basis of the average energy release rate of the microcrack,

$$\mathscr{G}(\sigma, a) = \gamma_2 a \sigma^2 / E \tag{7.17}$$

Here, E is Young's modulus and γ_2 is a dimensionless geometry factor. For the simple case of a mode-I penny-shaped crack, $K = K_I = 2\sigma \sqrt{a/\pi}$ [178, 797], which gives $\gamma_2 = 4/\pi$. According to Eqs. (7.15) and (7.17), the total energy release rate per crack, due to fracture alone, is

$$\frac{\partial \Pi_f^*(\sigma, a)}{\partial a} = \frac{a}{\gamma_1} \mathscr{G}(\sigma, a) = \frac{\gamma_2 \sigma^2}{\gamma_1 E} a^2 \tag{7.18}$$

Integration at constant σ furnishes the complementary energy due to fracture alone,

$$\Pi_f^*(\sigma, a) = \frac{\gamma_2 \sigma^2}{3\gamma_1 E} a^3 \tag{7.19}$$

Let the volume per microcrack be l_c^3 and consider, for the sake of simplicity, all the microcracks to be orthogonal to the direction of applied stress σ. According to Castigliano's theorem [298], we may calculate the displacement u_f due to fracture (at infinity, per crack) as follows:

$$u_f = \frac{\partial \Pi_f^*(P, a)}{\partial P} = \frac{1}{l_c^2} \frac{\partial \Pi_f^*(\sigma, a)}{\partial \sigma} = \frac{\gamma_0}{E l_c^2} \sigma a^3 \tag{7.20}$$

Here, $P = \sigma l_c^2$ is the remotely applied force per crack, and $\gamma_0 = 2\gamma_2/3\gamma_1$ is yet another dimensionless constant characterizing the geometry. The macroscopic strain caused by the formation of microcracks of size a under remotely applied stress σ is then

$$\varepsilon_{sc} = \frac{u_f}{l_c} = \frac{\gamma_0}{E l_c^3} \sigma a^3 = \gamma_0 \frac{a^3}{l_c^3} \frac{\sigma}{E} \tag{7.21}$$

To obtain the total compliance (or deformation), one would have to add the part of compliance of the structure before crack formation. But this part is included in the viscoelastic compliance (or creep deformation) as is not part of the cyclic creep.

The total microcrack size increment over N_{cyc} cycles is $\Delta a_N = a_N - a_0$, where a_N is the crack size after N_{cyc} cycles and a_0 is the initial crack size before cyclic loading. If Δa_N were not small compared to a_0, the cyclic loading of concrete in the service stress range would have to cause a significant loss of structural stiffness. If the loss of stiffness is small, it will be offset (or even overcompensated) by the increase of strength due to aging and to long-time compression. This can explain why no loss of short-time unload–reload stiffness due to cyclic creep has been reported. According to (7.21), the strain increment due to cyclic loading may now be expressed as

$$\Delta \varepsilon_N^{cyc} = \frac{\gamma_0}{E l_c^3} \sigma (a_N^3 - a_0^3) = \frac{\gamma_0}{E l_c^3} \sigma a_0^3 \left[\left(1 + \frac{\Delta a_N}{a_0} \right)^3 - 1 \right] \tag{7.22}$$

In failure analysis, large Δa_N need to be considered. But since the creep strains in service are always small, we may assume that $\Delta a_N/a_0 \ll 1$. Then, upon noting that $(1+x)^3 \approx 1 + 3x$ when $x \ll 1$, we may linearize (7.22) as follows:

$$\Delta \varepsilon_N^{cyc} = 3\gamma_0 \frac{\sigma}{E} \left(\frac{a_0}{l_c} \right)^3 \frac{\Delta a_N}{a_0} \tag{7.23}$$

7.13.3   Strain According to Paris Law for Subcritical Microcrack Growth*

Consider cyclic loading of amplitude $\Delta\sigma = \sigma_{max} - \sigma_{min}$ (Fig. 7.21a). [669] showed that, except for very large stress amplitudes and very high σ_{max} occurring in failure analysis, the fatigue growth of a crack depends only on the amplitude ΔK of the stress intensity factor K but not on its maximum and minimum individually. The dependence of the crack length increment per cycle on the amplitude can be approximated by the *Paris law*

$$\frac{\Delta a}{\Delta N_{\text{cyc}}} = \lambda \left(\frac{\Delta K}{K_c} \right)^m \tag{7.24}$$

in which K_c is the critical stress intensity factor for monotonic loading and prefactor λ and exponent m are empirical constants.

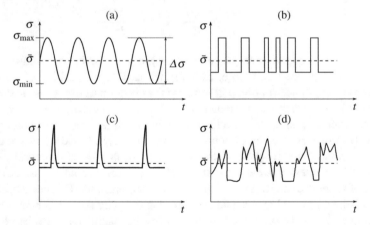

Fig. 7.21 Typical cyclic stress histories: (a) harmonic loading, (b) infrequent cycles, (c) rare cycles, (d) real loading

The Paris law is a good approximation for the intermediate range of fatigue crack growth, which is relevant for creep deflections of structures in service. For very large or small ΔK, the crack growth rate deviates from the slope m, producing S-shaped deviations when K_{max} and ΔK exceed certain thresholds. While exceeding these thresholds is important for failure analysis, it is not for deformations in the service stress range (i.e., for stresses less than 40% of the strength limit). When K_{max} and ΔK vary broadly, it is further important to take into account the dependence of prefactor λ on the ratio $K_{\text{max}}/K_{\text{min}}$, but again this is not important within the service stress range. Recently, it has been shown that on the atomic scale, exponent m must be equal to 2 and that m must increase when moving up to higher scales [156, 569].

Thus, for microcracks much smaller than the RVE, exponent m must be expected to be much smaller than its value for macrocracks in concrete, which is 10.

The amplitude ΔK is controlled by the remotely applied stress amplitude $\Delta \sigma$, and so one can write

$$\Delta K = c \sqrt{a} \, \Delta \sigma \tag{7.25}$$

where c is a dimensionless geometry constant. For example, for mode-I penny-shaped cracks, $c = 2/\sqrt{\pi}$. Substituting (7.25) into (7.24), we would get variable a on both sides of the equation. Although the resulting differential equation could be easily integrated by parts, we may consider that, in the case of creep under service loads, $\Delta a \ll a_0$. So, $\Delta K = c \sqrt{a_0} \Delta \sigma$. Integration at constant $\Delta \sigma$ then delivers, for small crack extensions:

$$\Delta a_N = a_N - a_0 = \lambda \left(\frac{c \sqrt{a_0} \, \Delta \sigma}{K_c} \right)^m N_{\text{cyc}} \tag{7.26}$$

Substituting (7.26) into (7.23) and rearranging, we obtain for the strain increment due to cyclic creep after N_{cyc} cycles the formula

$$\Delta \varepsilon_N^{\text{cyc}} = C_1 \sigma \left(\frac{\Delta \sigma}{\bar{f}_c} \right)^m N_{\text{cyc}} \tag{7.27}$$

where

$$C_1 = \frac{3 \gamma_0}{E} \frac{\lambda}{a_0} \left(\frac{a_0}{l_c} \right)^3 \left(\frac{\bar{f}_c c \sqrt{a_0}}{K_c} \right)^m \tag{7.28}$$

Here, $\bar{f}_c$ is the standard compression strength of concrete, introduced merely for convenience of dimensionality.

It is noteworthy that $\Delta \varepsilon_N^{\text{cyc}}$ is predicted to depend on both σ and N_{cyc} linearly. This agrees with the available cyclic creep measurements and is convenient for structural analysis. Of course, parameters C_1 and m must be calibrated using experimental data.

7.13.4 Compressive Cycles via Dimensional Analysis and Similitude*

Although the analysis in (7.15)–(7.22) was exemplified by a mode-I crack, it is equally applicable to cyclic creep under compressive and shear loadings, which are of main practical interest for prestressed structures in which almost no concrete is under tension. Under compression, five types of cracks producing inelastic compressive strain can be distinguished:

1. crushing band propagating transversely to compression [366, 794, 795], shown in Fig. 7.22a;

2. wedge-splitting cracks [204] at hard inclusions, parallel to compression (Fig. 7.22b);
3. interface cracks at inclusions (Fig. 7.22c);
4. pore-opening cracks parallel to compression [379, 539, 751], shown in Fig. 7.22d; and
5. wing-tip frictional cracks inclined to compression direction [42, 485, 506, 531, 768], shown in Fig. 7.22e.

The last is observed only rarely and is the only type that can be conceived to develop in a homogeneous medium that contains only preexisting microcracks or weak planes but neither inclusions nor pores [768]. The first type, i.e., Suresh's crack growth, differs from the others in that only a finite crack extension is possible, but in the present case, it does not matter since very small crack length growth is assumed at the outset.

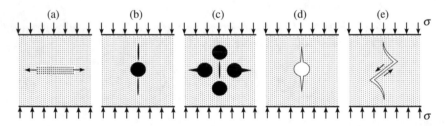

Fig. 7.22 Five types of cracks producing axial inelastic strain: (a) transversely propagating crushing band, (b) axial wedge-splitting cracks at hard inclusions in hardened cement paste, (c) interface cracks at inclusions, (d) pore-opening axial cracks, and (e) inclined wing-tipped frictional cracks

Detailed mathematical analysis of these compressive cracks would be rather complicated. We will, therefore, resort to dimensional analysis and similitude considerations [51, 53]. According to Buckingham's Π-theorem [51, 280] (which should in fairness be called the *Vashy-Buckingham theorem* [835]), the description of a physical system in terms of N_t parameters can be reduced to a description in terms of $N_i = N_t - N_d$ independent dimensionless parameters, where N_d is the number of independent physical dimensions among the parameters.

In Eq. (7.17) for the energy release rate, from which everything follows, the original parameters are the crack size a [dimension m], applied remote stress σ [dimension N/m²], elastic modulus E [dimension N/m²], and the energy release rate per unit area of crack $\mathscr{G}$ [dimension N/m], which means that $N_t = 4$. These parameters have only $N_d = 2$ independent dimensions, length and force. So the number of dimensionless independent parameters is $N_i = 4 - 2 = 2$. They can be selected as

$$\Pi_1 = \frac{E\mathscr{G}}{\sigma^2 a}, \quad \Pi_2 = \frac{\sigma}{E} \tag{7.29}$$

where $K^2 = E\mathcal{G}$. The governing relation must have the form $\Phi(\Pi_1, \Pi_2) = 0$. In the special case of a mode-I penny-shaped crack, the governing relation must coincide with (7.18), and so it must have the form $\Pi_1 - \mu_0\Pi_2 = 0$, where μ_0 is a constant. Indeed, this gives

$$\mathcal{G} = \mu_0 \frac{\sigma^2}{E} a^2 \tag{7.30}$$

which is identical to (7.18) (with $\mu_0 = \gamma_2/\gamma_1$). The subsequent derivation of the cyclic creep strain proceeds the same way as before, i.e., the integration of (7.30) at constant load yields the complementary energy, whose differentiation at constant length then yields (according to Castigliano's theorem) the displacement. This is then combined with the Paris law and leads to the same form of the cyclic creep law. Although the Paris law seems not to have been tested for cyclic compression (except for transverse crushing band [794]), it is expected to apply.

7.13.5 Cyclic Creep Compliance and Multiaxial Generalization

Because σ appears in (7.27) linearly, it is possible to define the cyclic creep compliance

$$\Delta J_N^{\text{cyc}}(t) = C_1 \left(\frac{\Delta\sigma}{\bar{f}_c}\right)^m N_{\text{cyc}}(t) \tag{7.31}$$

The total material compliance in presence of cyclic loading component is

$$J_{tot}(t, t') = J(t, t') + \Delta J_N^{\text{cyc}}(t) \tag{7.32}$$

where $J(t, t')$ is the standard compliance function (which reflects the elastic deformation and creep under sustained load), and $N_{\text{cyc}}(t)$ is the number of stress cycles imposed on the material up to time t. The standard engineering theory of bending cannot be used, since the distribution of $\Delta\sigma$ over the cross section is nonlinear and thus makes the distribution of $\Delta\varepsilon_N^{\text{cyc}}$ over the cross section nonlinear [70].

No experimental information seems to exist for cyclic creep under multiaxial stress or tensile loading, nor cyclic creep with cycles of varying amplitudes. Nevertheless, it appears reasonable to neglect transverse strains and to expect that under tension the cyclic creep per unit stress is at least as large as it is under compression. Also note that, for the special case of a crack normal to the maximum principal stress direction, an analysis similar to (7.15)–(7.21) gives a zero additional lateral strain due to crack opening.

A three-dimensional finite element program for creep and cyclic creep in a bridge has been formulated by [210] under the assumption that the cyclic creep eigenstrains corresponding to individual principal stresses can be superposed.

It would be convenient to model the cyclic creep as an acceleration of static creep. This alternative [149, 176], attractive for simplicity of computations, was tried but later rejected. It works only within one order of magnitude of the number of cycles and gives great error when stretched to several orders of magnitude. The reason that in previous publications the existing test data could be fitted using this alternative is that the data had very limited time span (only about 10 days). So the cyclic creep must be considered to be an eigenstrain, as written in (7.32), which is in fact the way the cyclic creep was empirically treated in the earliest mathematical model [69, 70].

7.13.6   Calibration by Existing Test Data

Experimental verification presented by Bažant and Hubler [134] relied on well-documented laboratory data from the literature. Because Bažant and Hubler's cyclic creep model described here is applicable only within the service stress range (i.e., for stresses up to 40% of concrete strength), the tests in which this limit was exceeded, or those in which fatigue failure occurred, were excluded from calibration. For each curve, the B3 creep model was first calibrated to fit the static creep curve (i.e., the curve for zero stress amplitude), which must be subtracted from the data to separate the effect of cyclic creep. The B3 parameters thus calibrated were then used in the joint least-square optimization of the fits of all cyclic test curves to determine the optimum values of cyclic creep parameters C_1 and m in (7.31). The fits of the test curves are shown in Fig. 7.23. They include both cyclic creep and static creep, the latter under either sealed conditions (basic creep) or drying conditions (drying creep).

The data set of Whaley and Neville [863] is the most comprehensive set available, containing cyclic tests done under both sealed and drying conditions. The initial strains, unfortunately, were not measured, and so they are estimated by extrapolation of the model B3 curve fitted to the initial portion of the data. Whaley and Neville measured the cyclic creep at various amplitudes and mean stress levels, at a frequency of 585 cycles/min. One of their data sets, studying the effect of mean stress, could not be used because the maximum stresses in the cycles exceeded the fatigue limit.

Also considered for calibration were the tests of Mehmel and Kern [625], which reported the effect of stress cycling on the drying creep at various average stresses and various amplitudes of the cyclic load, at two different frequencies, 380 cycles/min. and 3000 cycles/min. Moreover, Hirst and Neville [483] reported tests at 3000 cycles/min. and various stress amplitudes. Further cyclic test data do exist, but were not used because they either lacked the reference static creep test (at zero amplitude) or used a concrete with unusual mix proportions.

The optimum fit of (7.31) to the test data is presented in Fig. 7.23. With the optimized parameter values $C_1 = 46 \cdot 10^{-6}/$ MPa and $m = 4$, the fit of the data is quite close, with a coefficient of variation of only 5% (root-mean-square error divided by data mean).

The fact that the experiments verify the exponent value $m = 4$ is interesting. According to the activation-energy-based probabilistic theory of quasibrittle fracture

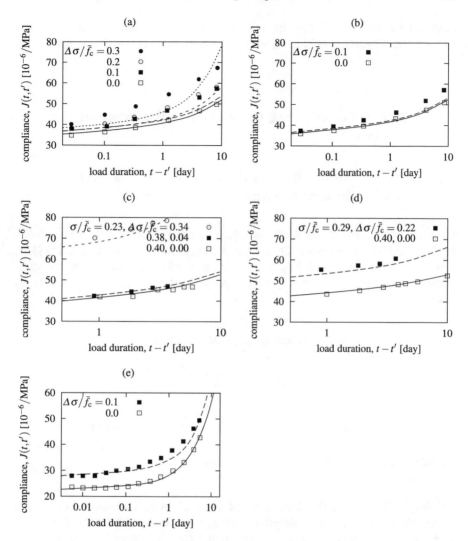

Fig. 7.23 Optimal fitting of test data by (a)–(b) Whaley and Neville [863], (c)–(d) Mehmel and Kern [625], and (e) Hirst and Neville [483]

[156, 569], exponent m of the Paris law should be equal to 2 on the nanoscale. Moving up through the scales causes the exponent m of the crack growth law to increase by approximately 2 for each order of scale magnitude. This matches the fact that for metals and fine-grained ceramics, in which the RVE is of micrometer size, $m \approx 4$. In concrete, one has to cross several more scales to reach the scale of an RVE, which is of the size of 0.1 m. Thus, it is no wonder that the exponent of Paris law for macroscale crack propagation in concrete is about 10 [206]. In this light, it is not at all surprising that the cyclic creep exponent of concrete is equal to 4. In view of the theory of Bažant and Le [156], this value of m implies that the relevant cracks that cause the cyclic creep should be micrometer-scale cracks, far smaller than the RVE size.

7.14   Effects of Cyclic Creep on Bridge Deflections and Cracking

7.14.1   Stress Distribution in a Prestressed Cross Section Under Variable Loading

Consider now prestressed segmental box girders, many of which were shown to suffer from excessive long-time deflections and often develop excessive cracking [137, 138]. Whenever excessive bridge deflections or cracking are discussed at conferences, usually someone blames cyclic creep, despite the absence of any supportive calculations.

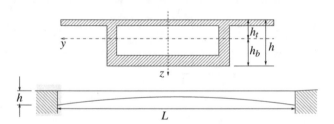

Fig. 7.24 Typical bridge cross section at the pier

Figure 7.24 shows a simplified form of a typical cross section at pier of a box girder bridge. For the sake of simplicity, we use the engineering theory of bending, with the cross sections remaining plane. If the material strength is utilized fully (which is for record-span bridges inevitable), cross section dimensions and prestress level are designed so that the longitudinal normal stresses σ reach their allowable limits, $\sigma = -f_c$ ($f_c > 0$) at the bottom face of coordinate $z = h_b$ and $\sigma = f_t$ at the top face of coordinate $z = -h_t$, where coordinate z is measured from the horizontal centroidal axis y and is positive downward, $h_b > 0$ and $h_t > 0$ are the distances of the bottom and top fibers from this axis, with $h_b + h_t = h = $ cross-sectional depth. After prestress losses, the allowable stresses in highway bridges are, according to AASHTO [1], $f_c = 0.4\, f_c'$ in compression and $f_t = 6\ \mathrm{psi}\sqrt{f_c'/\mathrm{psi}}$ in tension, where f_c' is the specified compressive strength. For structures other than bridges, ACI [18] sets a higher limit for compression, $f_c = 0.45\, f_c'$, while the limit for tension remains the same as according to AASHTO. The calculations to be presented here will be done with the limits $f_t = 0$ and $f_c = 0.4\, \bar{f_c}$ where $\bar{f_c}$ is the mean compressive strength, which can be estimated from the specified compressive strength as $\bar{f}_c \approx 0.4(f'_c + 1200\mathrm{psi})$; see the ACI formula (E.4). However, these differences in allowable stress limits do not affect the general conclusions significantly.

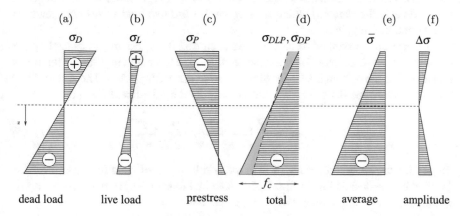

Fig. 7.25 Bridge cross section at the pier, designed to reach allowable limits: profiles of the stresses due to (a) dead load, (b) cyclic live load and (c) prestress, and profiles of the (d) total stress, (e) average stress, and (f) stress amplitude

If the cross sections remain planar and the stress–strain equation for concrete is linear (elastic or viscoelastic), the distribution of normal stress in concrete across the depth, due to bending moment M_C and normal force N_C, is described by the well-known formula

$$\sigma(z) = \frac{M_C}{I}z + \frac{N_C}{A} \tag{7.33}$$

where A and I are the sectional area and sectional moment of inertia with respect to the horizontal centroidal axis. Here, M_C and N_C are understood as the resultants of stresses in concrete only (note that M and N in Eqs. (4.157)–(4.158) were the total internal forces, which included the contribution of stresses in steel). For a fully loaded bridge, we substitute $N_C = -P$ and $M_C = M_{DLP} \equiv M_D + M_L - Pe$ where $P > 0$ is the prestress force and e is its eccentricity[5], and M_D and M_L are, respectively, the bending moments[6] due to dead load (including self-weight) and due to live load, representing the traffic load (for the sake of simplicity, we assume the statically indeterminate moments due to prestressing, denoted as M_2 in Sect. 7.12, to

---

[5]The eccentricity is considered as positive if the centroid of the prestressed tendons is below the centroid of the section; see Fig. 4.36. In pier cross sections, which transmit negative bending moments, the tendons are near the top fibers and the eccentricity is thus negative.

[6]The bending moment is considered as positive if it causes tension in the bottom fibers. At the pier section, the moments due to dead load and live load are negative, while the moment due to prestress, $-Pe$, is positive, because $e < 0$.

be negligible, as they often are in large box girders with a flexible midspan; otherwise they would have to be added to $-Pe$). The corresponding linear stress distributions are plotted in Fig. 7.25a–c separately for the effects of dead load, live load, and prestress. The total stress distribution is then shown in Fig. 7.25d, where the shaded area indicates the range in which stresses vary if the live load varies between zero and its maximum value.

The section is assumed to be designed such that the stress σ_{DLP}, caused by the combination of dead load, full live load, and prestress, is equal to f_t in the top fibers and to $-f_c$ in the bottom fibers; see the thick solid line in Fig. 7.25d. These conditions uniquely determine the stress distribution described by the linear function

$$\sigma_{DLP}(z) = \frac{f_t(h_b - z) - f_c(h_t + z)}{h} \tag{7.34}$$

To get the stress σ_{DP} corresponding to the combination of dead load and prestress (plotted by the dashed line in Fig. 7.25d), we need to subtract from σ_{DLP} the contribution of the bending moment M_L due to live load:

$$\sigma_{DP}(z) = \sigma_{DLP}(z) - \frac{M_L}{I}z = \frac{f_t(h_b - z) - f_c(h_t + z)}{h} - \frac{M_L}{I}z \tag{7.35}$$

In service conditions, the actual stress varies between σ_{DP} and σ_{DLP}, with mean value (Fig. 7.25e)

$$\bar{\sigma}(z) = \frac{\sigma_{DLP}(z) + \sigma_{DP}(z)}{2} = \frac{f_t(h_b - z) - f_c(h_t + z)}{h} - \frac{M_L}{2I}z \tag{7.36}$$

and amplitude (Fig. 7.25f)

$$\Delta\sigma(z) = \frac{|\sigma_{DLP}(z) - \sigma_{DP}(z)|}{2} = \frac{|M_L z|}{I} \tag{7.37}$$

The use of an arithmetic average in (7.36) means that the periods of positive and negative deviations from the average are considered to have equal durations and similar time profiles, which is what characterizes the experiments used in calibration (Fig. 7.21a). When, however, the load cycles are asymmetric as shown in Fig. 7.21b,c, the proper value of $\bar{\sigma}$ is debatable since no such tests have been reported. If linear superposition of the load effects were applicable, then $\bar{\sigma}$ would represent the time average of stress over the cycle (shown by the horizontal dashed lines in Fig. 7.21). But since the dependence of cyclic creep on the stress deviation from the mean is highly nonlinear, the time shape of cycle is probably unimportant and mainly the extremes matter. Then, the average value $\bar{\sigma} = (\sigma_{DP} + \sigma_{DLP})/2 = \sigma_D + \sigma_L/2 + \sigma_P$ may be used for all cycle profiles.

7.14.2 Curvature and Residual Stresses Due to Cyclic Creep

The distributions of the mean stress $\bar{\sigma}$ and stress amplitude $\Delta\sigma$ across the depth, already shown in Fig. 7.25e,f, are for easier reference replotted in Fig. 7.26a,b. Due to the dependence on the fourth power of $\Delta\sigma$, the additional compliance caused by cyclic creep,

$$\Delta J_N^{\mathrm{cyc}}(z,t) = C_1 N_{\mathrm{cyc}}(t)\left(\frac{\Delta\sigma(z)}{\bar{f_c}}\right)^4 = C_1 N_{\mathrm{cyc}}(t)\left(\frac{|M_L\,z|}{I\bar{f_c}}\right)^4 \qquad (7.38)$$

evaluated from (7.31) with $m = 4$, is very small near the centroidal axis and much higher near the top and bottom fibers; see Fig. 7.26c. For simplicity, we have set exponent m to 4, which is the value obtained by optimum fitting of experimental data [134]. Of course, the subsequent derivation could be generalized to an arbitrary value of m.

The product of the additional compliance $\Delta J_N^{\mathrm{cyc}}$ and the mean stress, which is here denoted as $\bar{\sigma}$, gives the inelastic strains (or eigenstrains) due to cyclic creep,

$$\Delta\varepsilon_N^{\mathrm{cyc}}(z,t) = \Delta J_N^{\mathrm{cyc}}(z,t)\bar{\sigma}(z) = C_1 N_{\mathrm{cyc}}(t)\left(\frac{|M_L\,z|}{I\bar{f_c}}\right)^4\left(\frac{f_t(h_b-z)-f_c(h_t+z)}{h}-\frac{M_L}{2I}z\right) \qquad (7.39)$$

which are plotted in Fig. 7.26d. To simplify the notation, let us introduce a dimensionless factor

$$\mu_L = -\frac{M_L h}{2I\bar{f_c}} \qquad (7.40)$$

where the minus sign is used in order to get positive μ_L for negative M_L (at the pier section, the moment due to the traffic load is negative). Formula (7.39) can then be rewritten as

$$\Delta\varepsilon_N^{\mathrm{cyc}}(z,t) = \frac{16C_1 N_{\mathrm{cyc}}(t)\mu_L^4}{h^5}\left[(f_t h_b - f_c h_t)\,z^4 + (\mu_L\bar{f_c}-f_t-f_c)z^5\right] \qquad (7.41)$$

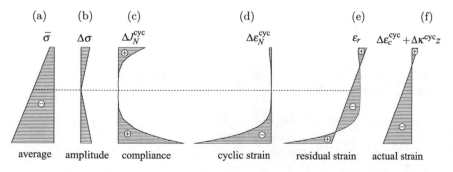

(a) $\bar{\sigma}$ (b) $\Delta\sigma$ (c) $\Delta J_N^{\mathrm{cyc}}$ (d) $\Delta\varepsilon_N^{\mathrm{cyc}}$ (e) ε_r (f) $\Delta\varepsilon_c^{\mathrm{cyc}}+\Delta\kappa^{\mathrm{cyc}}z$

average   amplitude   compliance   cyclic strain   residual strain   actual strain

Fig. 7.26 Bridge cross section at the pier, for which the prestressing eccentricity and bridge span has been designed to reach allowable limits: (a) average stress, (b) amplitude of cyclic stress variation, (c) cyclic creep compliance, (d) free cyclic strain, (e) residual strain, (f) actual linearly distributed strain due to cyclic creep effects on the section

The dependence of the inelastic strains $\Delta\varepsilon_N^{\mathrm{cyc}}$ on the vertical coordinate z is highly nonlinear and the condition that the cross sections must remain planar would be violated. Therefore, residual strains ε_r shown in Fig. 7.26e must develop to enforce the planarity of cross sections, in the same way as it happens for shrinkage or thermal strains. The resulting linearly distributed strains caused by the effects of cyclic creep on the section, shown in Fig. 7.26f, can be expressed as

$$\Delta\varepsilon_N^{\mathrm{cyc}}(z,t) + \varepsilon_r(z,t) = \Delta\varepsilon_a^{\mathrm{cyc}}(t) + \Delta\kappa^{\mathrm{cyc}}(t)\,z \qquad (7.42)$$

where $\Delta\varepsilon_a^{\mathrm{cyc}}$ and $\Delta\kappa^{\mathrm{cyc}}$ are deformations due to cyclic creep at the sectional level (additional axial strain at the centroid and additional curvature).

The residual strains ε_r give rise to residual stresses σ_r, which are linked to ε_r by the viscoelastic law

$$\sigma_r(z,t) = \mathscr{R}\{\varepsilon_r(z,t)\} = \mathscr{R}\left\{\Delta\varepsilon_a^{\mathrm{cyc}}(t) + \Delta\kappa^{\mathrm{cyc}}(t)\,z - \Delta\varepsilon_N^{\mathrm{cyc}}(z,t)\right\} \qquad (7.43)$$

in which $\mathscr{R}$ is the relaxation operator; see Sect. 2.5. The residual stresses are self-equilibrated, because the normal force and bending moment may be considered to be unaffected by cyclic creep (this is a simplifying assumption in the case of statically indeterminate structures, in which cyclic creep causes some redistribution of internal forces). The conditions that self-equilibrated stresses σ_r have zero resultants (vanishing normal force and bending moment) are mathematically written as

$$\int_A \sigma_r(z,t)\,\mathrm{d}A = 0 \qquad (7.44)$$

$$\int_A z\,\sigma_r(z,t)\,\mathrm{d}A = 0 \qquad (7.45)$$

Substituting (7.43) into (7.44)–(7.45) and applying the compliance operator $\mathscr{J}$ on both sides of the equations, we get

$$\int_A \left[\Delta\varepsilon_a^{\mathrm{cyc}}(t) + \Delta\kappa^{\mathrm{cyc}}(t)\,z - \Delta\varepsilon_N^{\mathrm{cyc}}(z,t)\right]\mathrm{d}A = 0 \qquad (7.46)$$

$$\int_A z\left[\Delta\varepsilon_a^{\mathrm{cyc}}(t) + \Delta\kappa^{\mathrm{cyc}}(t)\,z - \Delta\varepsilon_N^{\mathrm{cyc}}(z,t)\right]\mathrm{d}A = 0 \qquad (7.47)$$

Since $\int_A \mathrm{d}A = A = $ sectional area, $\int_A z\,\mathrm{d}A = 0$, and $\int_A z^2\,\mathrm{d}A = I = $ sectional moment of inertia, conditions (7.46)–(7.47) combined with (7.41) directly lead to

$$\Delta\varepsilon_a^{\mathrm{cyc}}(t) = \frac{1}{A}\int_A \Delta\varepsilon_N^{\mathrm{cyc}}(z,t)\,\mathrm{d}A = \frac{16C_1 N_{\mathrm{cyc}}(t)\mu_L^4}{Ah^5}\left[(f_t h_b - f_c h_t)\,I_4 + (\mu_L\bar{f}_c - f_t - f_c)I_5\right] \quad (7.48)$$

$$\Delta\kappa^{\mathrm{cyc}}(t) = \frac{1}{I}\int_A z\,\Delta\varepsilon_N^{\mathrm{cyc}}(z,t)\,\mathrm{d}A = \frac{16C_1 N_{\mathrm{cyc}}(t)\mu_L^4}{Ih^5}\left[(f_t h_b - f_c h_t)\,I_5 + (\mu_L\bar{f}_c - f_t - f_c)I_6\right] \quad (7.49)$$

where $I_k = \int_A z^k\,\mathrm{d}A$, $k = 4, 5, 6$, are higher-order moments of the sectional area.

To assess the relative importance of the additional curvature due to cyclic creep, it is useful to introduce the dimensionless ratio $\Delta \kappa^{\text{cyc}}(t)/\kappa_{\text{ref}}$ where $\kappa_{\text{ref}} = M_{DLP}/EI = -(f_c + f_t)/Eh$ is the elastic curvature caused by the maximum moment

$$M_{DLP} = \frac{I}{h}\left(\sigma_{DLP}(h_b) - \sigma_{DLP}(-h_t)\right) = -\frac{I}{h}\left(f_c + f_t\right) \qquad (7.50)$$

Formula (7.49) can be transformed into

$$\frac{\Delta \kappa^{\text{cyc}}(t)}{\kappa_{\text{ref}}} = EC_1 N_{\text{cyc}}(t)\left(\beta_{4\kappa}\mu_L^4 + \beta_{5\kappa}\mu_L^5\right) \qquad (7.51)$$

where

$$\beta_{4\kappa} = \frac{16}{Ih^4}\left(\frac{f_c h_t - f_t h_b}{f_c + f_t}I_5 + I_6\right) \qquad (7.52)$$

$$\beta_{5\kappa} = -\frac{16 I_6 \bar{f}_c}{Ih^4(f_c + f_t)} \qquad (7.53)$$

are dimensionless coefficients dependent on the geometry of the section and on the ratios between allowable stresses f_c and f_t and the mean compressive strength $\bar{f}_c$.

Once the sectional deformation measures $\Delta\varepsilon_a^{\text{cyc}}$ and $\Delta\kappa^{\text{cyc}}$ have been determined, it is easy to evaluate the residual stresses from (7.43). A typical distribution of these stresses across the depth of the section has the same shape as the profile of residual strains in Fig. 7.26e. The residual stresses grow proportionally to the number of cycles and are positive near the top and bottom fibers and negative near the centroid. Near the bottom fibers, the additional tensile residual stresses that develop due to cyclic creep effects are superposed onto large compressive stresses induced by prestress and sustained loading, and the resulting stresses certainly remain for a long time compressive. The most critical situation arises near the top fibers, where the stresses due to prestress and dead and live loads are small and compressive, or even tensile (if the allowable stress f_t is considered as positive). Therefore, let us examine the residual stress at the top fibers,

$$\sigma_r(-h_t, t) = \mathcal{R}\left\{\Delta\varepsilon_a^{\text{cyc}}(t) - \Delta\kappa^{\text{cyc}}(t)\,h_t - \Delta\varepsilon_N^{\text{cyc}}(-h_t, t)\right\} \qquad (7.54)$$

Based on (7.41) and (7.48)–(7.49), the residual stress can be expressed as

$$\sigma_{rt}(t) \equiv \sigma_r(-h_t, t) = C_1 \bar{f}_c \left(\beta_{4\sigma}\mu_L^4 + \beta_{5\sigma}\mu_L^5\right)\mathcal{R}\{N_{\text{cyc}}(t)\} \qquad (7.55)$$

where

$$\beta_{5\sigma} = \frac{16}{h^5}\left(\frac{I_5}{A} - \frac{I_6 h_t}{I} + h_t^5\right) \qquad (7.56)$$

$$\beta_{4\sigma} = 16\frac{f_t h_b - f_c h_t}{\bar{f}_c h^5}\left(\frac{I_4}{A} - \frac{I_5 h_t}{I} - h_t^4\right) - \frac{f_t + f_c}{f_c}\beta_{5\sigma} \qquad (7.57)$$

are dimensionless coefficients that depend on the cross section and on the ratios $f_c/\bar{f}_c$ and $f_t/\bar{f}_c$.

7.14.3 Appraisal of the Magnitude of Cyclic Creep Effects in Structures

In the preceding section, we have derived formulae (7.51) and (7.55), which characterize the additional curvature due to cyclic creep and the residual stress in the top fibers. The additional curvature leads to an increase of deflection, which can be roughly estimated as the elastically computed deflection (due to the full load) multiplied by the ratio $\Delta\kappa^{\mathrm{cyc}}/\kappa_{\mathrm{ref}}$ from (7.51). If this ratio is much smaller than 1, the effect of cyclic creep on deflections is negligible.

Since we assume that the section has been designed with allowable tensile stress $f_t = 0$, the residual stress σ_{rt} given by (7.55) is at the same time the maximum (tensile) stress in the whole section—it is the total stress in the top fibers when the full load (including the live load) is applied. Cracking can be expected to occur when this maximum stress reaches the mean tensile strength of the material, $\bar{f}_t$. Therefore, if the ratio $\sigma_{rt}/\bar{f}_t$ is much smaller than 1, there is no danger of cracking induced by cyclic creep.

Example 7.1. Rectangular section

To get a rough idea about the order of magnitude of the consequences of cyclic creep, consider first a rectangular section of width b and depth h, for which $h_t = h_b = h/2$, $A = bh$, $I = bh^3/12$, $I_4 = bh^5/80$, $I_5 = 0$, and $I_6 = bh^7/448$. The allowable stresses are taken as $f_c = 0.4\bar{f}_c$ and $f_t = 0$. Formulae (7.52)–(7.53) then yield $\beta_{4\kappa} = 3/7$ and $\beta_{5\kappa} = -15/14$, and formulae (7.56)–(7.57) yield $\beta_{5\sigma} = 2/7$ and $\beta_{4\sigma} = 8/175$. Using the value $C_1 = 46 \times 10^{-6}/\mathrm{MPa}$ identified by [134] and assuming typical values $E = 24$ GPa and $\bar{f}_t = 0.1\bar{f}_c$, we can rewrite (7.51) and (7.55) as

$$\frac{\Delta\kappa^{\mathrm{cyc}}(t)}{\kappa_{\mathrm{ref}}} = 0.473\mu_L^4 \left(1 - 2.5\mu_L\right) N_{\mathrm{cyc}}(t) \tag{7.58}$$

$$\frac{\sigma_{rt}(t)}{\bar{f}_t} = 0.505\mu_L^4 \left(1 + 6.25\mu_L\right) \frac{\mathscr{R}\{N_{\mathrm{cyc}}(t)\}}{E} \tag{7.59}$$

Recall that the extreme (negative) moment due to the dead load, live load, and prestress is $M_{DLP} = -(f_c + f_t)I/h$; see (7.50). Combining this with (7.40) and setting $f_c = 0.4\bar{f}_c$ and $f_t = 0$, we get

$$\frac{M_L}{M_{DLP}} = -\frac{M_L h}{(f_c + f_t)I} = \mu_L \frac{2\bar{f}_c}{f_c + f_t} = 5\mu_L \tag{7.60}$$

which means that

$$\mu_L = \frac{M_L}{5M_{DLP}} \tag{7.61}$$

If, for instance, the moment due to live load is 10% of the total moment, then $\mu_L = 0.02$, and formulae (7.58)–(7.59) combined with the rough estimate $\mathscr{R}\{N_{\mathrm{cyc}}(t)\} \approx E N_{\mathrm{cyc}}(t)$ give

$$\frac{\Delta\kappa^{\mathrm{cyc}}(t)}{\kappa_{\mathrm{ref}}} = 7.2 \times 10^{-8} N_{\mathrm{cyc}}(t) \tag{7.62}$$

$$\frac{\sigma_{rt}(t)}{\bar{f}_t} = 9.1 \times 10^{-8} \frac{\mathscr{R}\{N_{\mathrm{cyc}}(t)\}}{E} \approx 9.1 \times 10^{-8} N_{\mathrm{cyc}}(t) \tag{7.63}$$

This means that, after 1 million cycles, the contribution of cyclic creep to the deflection would be about 7% of the elastic deflection, and the tensile residual stress in the top fibers would be about 9% of the tensile strength.

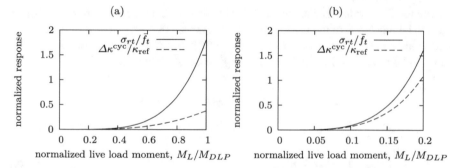

Fig. 7.27 Dependence of normalized residual stress and normalized curvature caused by cyclic creep on normalized moment due to live load for a rectangular cross section: (a) after 10^3 cycles, (b) after 10^6 cycles

Fig. 7.27 shows the normalized residual stress $\sigma_{rt}/\bar{f}_t$ (for $\bar{f}_t = 0.1\bar{f}_c$) and the normalized curvature $\Delta\kappa^{\mathrm{cyc}}/\kappa_{\mathrm{ref}}$ as a function of the normalized moment due to live load, M_L/M_{DLP}. After 10^3 cycles (Fig. 7.27a), the effects of cyclic creep are negligible for moments due to live load up to one half of the maximum moment. However, after 10^6 cycles (Fig. 7.27b), these effects are negligible only if the moments due to live load are not more than 10% of the maximum moment. ∎

The foregoing estimates (7.58)–(7.59) are just simple approximations, for two reasons: (i) the section has been assumed to be rectangular, which leads to different residual strains near the top fibers than for a box girder, and (ii) the relation between residual strain and residual stress has been assumed to be elastic (because $\mathscr{R}\{N_{\mathrm{cyc}}(t)\}$ has been replaced by $E N_{\mathrm{cyc}}(t)$) while in reality the stress would be affected by relaxation and aging. Nevertheless, if such a strongly simplified analysis indicates that the effect of cyclic creep is negligible, it is fully sufficient. If, on the other hand, the crude estimate predicts that considerable stresses might be generated by the effects of cyclic creep, the analysis needs to be refined.

Let us first explore the effect of the cross-sectional geometry. For a given section, the geometrical characteristics such as A, I, I_4, I_5, and I_6 can be evaluated numerically.

Example 7.2. Koror–Babeldaob Bridge

As a real example, consider the KB Bridge described in Sect. 7.2. The section at the pier (Fig. 7.3) consisted of the top slab of thickness 0.432 m and width 9.62 m, bottom slab of thickness 1.153 m and width 7.32 m, and two webs, each of thickness 0.356 m. The total height of the section was 14.17 m. From these dimensions, all necessary geometrical characteristics can be calculated: $h_t = 8.16$ m, $h_b = 6.01$ m, $A = 21.6\,\mathrm{m}^2$, $I = 649\,\mathrm{m}^4$, $I_4 = 28.4 \times 10^3\,\mathrm{m}^6$, $I_5 = -114 \times 10^3\,\mathrm{m}^7$, and $I_6 = 1.45 \times 10^6\,\mathrm{m}^8$. The corresponding dimensionless coefficients evaluated from (7.52)–(7.53) and (7.56)–(7.57) are $\beta_{4\kappa} = 0.319$, $\beta_{5\kappa} = -2.22$, $\beta_{5\sigma} = 0.354$, and $\beta_{4\sigma} = 0.0122$.

For the KB Bridge, Bažant and Hubler [134] estimated the moment due to traffic load to be just 2.5% of the total moment. This means that $M_L/M_{DLP} = 0.025$, and, according to (7.61), $\mu_L = M_L/5M_{DLP} = 0.005$. Formulae (7.51) and (7.55) with $E = 24$ GPa, $C_1 = 46 \times 10^{-6}/\mathrm{MPa}$, and $\bar{f}_t = 0.1\bar{f}_c$ then yield

$$\frac{\Delta\kappa^{\mathrm{cyc}}(t)}{\kappa_{\mathrm{ref}}} = 0.352\mu_L^4 \left(1 - 6.96\mu_L\right) N_{\mathrm{cyc}}(t) = 2.2 \times 10^{-10} N_{\mathrm{cyc}}(t) \qquad (7.64)$$

$$\frac{\sigma_{rt}(t)}{\bar{f}_t} = 0.135\mu_L^4 \left(1 + 29.0\mu_L\right) \frac{\mathscr{R}\{N_{\mathrm{cyc}}(t)\}}{E} = 9.6 \times 10^{-11} \frac{\mathscr{R}\{N_{\mathrm{cyc}}(t)\}}{E}$$

$$\approx 9.6 \times 10^{-11} N_{\mathrm{cyc}}(t) \qquad (7.65)$$

These are negligible effects even for $N_{\mathrm{cyc}} = 10^9$ cycles, while a realistic number is $N_{\mathrm{cyc}} = 10^7$. ∎

To get more insight into the role played by the shape of the section, it is useful to construct analytical expressions for an idealized box or I-shaped section, characterized by the total area, A, and areas of the top and bottom flanges, A_t and A_b. In calculations of second- and higher-order moments, the thickness of the flanges is neglected, i.e., each flange is represented by a "concentrated area" A_t or A_b at distance h_t or h_b from the centroidal axis. The web (or, in a box section, both webs combined) is represented by a rectangle of width $b = (A - A_t - A_b)/h$ and depth h. Introducing dimensionless parameters $\alpha_t = A_t/A$, $\alpha_b = A_b/A$, $\alpha_f = \alpha_t + \alpha_b$, and $\alpha_{tb} = \alpha_t - \alpha_b$, the geometrical characteristics can be expressed as

$$b = \frac{A - A_t - A_b}{h} = \frac{A}{h}(1 - \alpha_f) \qquad (7.66)$$

$$h_t = \frac{(A - A_t - A_b)h/2 + A_b h}{A} = \frac{h}{2}(1 - \alpha_t + \alpha_b) = \frac{h}{2}(1 - \alpha_{tb}) \qquad (7.67)$$

$$h_b = h - h_t = \frac{h}{2}(1 + \alpha_t - \alpha_b) = \frac{h}{2}(1 + \alpha_{tb}) \qquad (7.68)$$

$$I_m = \frac{b}{m+1}\left(h_b^{m+1} - (-h_t)^{m+1}\right) + A_t\,(-h_t)^m + A_b h_b^m = \qquad (7.69)$$

$$= \frac{Ah^m}{2^m}\left\{\frac{1-\alpha_f}{2(m+1)}\left[(\alpha_{tb}+1)^{m+1} - (\alpha_{tb}-1)^{m+1}\right] + \alpha_b(\alpha_{tb}+1)^m + \alpha_t(\alpha_{tb}-1)^m\right\}$$

Based on this, it can be shown that the dimensionless coefficients $\beta_{4\kappa}$, $\beta_{5\kappa}$, $\beta_{4\sigma}$, and $\beta_{5\sigma}$, defined in (7.52)–(7.53) and (7.56)–(7.57), depend exclusively on α_t and α_b or, alternatively, on α_f and α_{tb} (and, of course, on the ratios between the allowable stresses and the mean compressive strength, which are considered here as $f_c/\bar{f}_c = 0.4$ and $f_t/\bar{f}_c = 0$). Note that $\alpha_f = (A_t + A_b)/A$ is the relative area of flanges and $\alpha_{tb} = (A_t - A_b)/A$ reflects the asymmetry of the section. The dependence of coefficients β on the relative flange area α_f is plotted in Fig. 7.28 for different values of the asymmetry coefficient α_{tb}. The case of $\alpha_f = 0$ corresponds to a rectangular section with no flanges, and the values of coefficients β are then those from Example 7.1. The thick master curves are valid for sections with a horizontal axis of symmetry, for which $\alpha_{tb} = 0$. The other curves cover the range of α_{tb} between -0.2 and 0.2. Their leftmost points correspond to sections with one flange only, i.e., a T-shaped section for positive α_{tb} and an inverted T section for negative α_{tb}. For the section of the KB Bridge plotted in Fig. 7.3 left and analyzed in Example 7.2, the value of the asymmetry coefficient would be $\alpha_{tb} \approx -0.2$ (negative because the bottom flange is very massive).

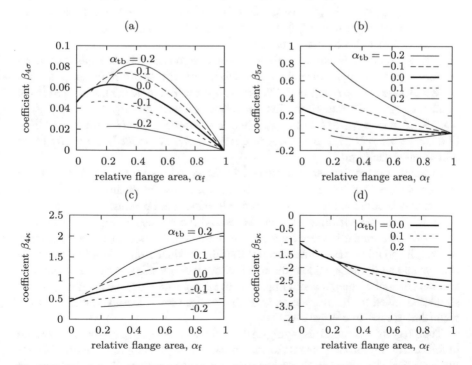

Fig. 7.28 Dependence of coefficients (a) $\beta_{4\sigma}$, (b) $\beta_{5\sigma}$, (c) $\beta_{4\kappa}$, and (d) $\beta_{5\kappa}$ on the relative flange area, $\alpha_f = (A_t + A_b)/A$, for different values of the asymmetry coefficient, $\alpha_{tb} = (A_t - A_b)/A$

Even though formulae (7.67)–(7.69) are just approximate, the graphs in Fig. 7.28 illustrate the basic trends and reveal the influence of various parameters. For large-span bridges with small ratios between the traffic load and the self-weight, parameter μ_L is small and the terms with μ_L^4 dominate, which means that the response is controlled by coefficients $\beta_{4\sigma}$ and $\beta_{4\kappa}$. As shown in Fig. 7.28a, the maximum value of $\beta_{4\sigma}$ for a symmetric section is about 0.06 and is attained if the flanges represent about 20 to 25% of the total area. For larger relative flange areas and for asymmetric sections with the bottom flange more massive than the top one (i.e., for $\alpha_{tb} < 0$), coefficient $\beta_{4\sigma}$ decreases. Thus, knowing that the cyclic creep coefficient C_1 is close to $1/E$, we can construct a very crude estimate of the maximum residual stress, $\sigma_{rt}(t)/\bar{f}_c \approx 0.06\mu_L^4 N_{\text{cyc}}(t)$. Here, we neglect the effects of relaxation and aging on the residual stress, and at the same time, we ignore the term with μ_L^5, which may become important for moderate ratios between the traffic load and self-weight, i.e., for medium- and small-span bridges. Fig. 7.28b shows that coefficient $\beta_{5\sigma}$ multiplying μ_L^5 in (7.56) is never larger than 0.29 for section with equal flanges, but for sections with a more massive bottom flange, it may become larger (recall that its value was 0.354 for the KB Bridge).

The additional curvature due to cyclic creep is described by formula (7.51), with coefficients $\beta_{4\kappa}$ and $\beta_{5\kappa}$. Since the latter coefficient is always negative (see Fig. 7.28d), the term with μ_L^5 can safely be neglected. For symmetric sections and sections with the bottom flange more massive than the top one, coefficient $\beta_{4\kappa}$ never exceeds 1 (see Fig. 7.28c). Therefore, a conservative estimate of the additional curvature would be $\Delta\kappa_{\text{cyc}}(t)/\kappa_{\text{ref}} \approx \mu_L^4 N_{\text{cyc}}(t)$. Based on this, it can be expected that if the number of cycles is not larger than $1/\mu_L^4$, the additional deflection due to cyclic creep will not exceed the short-term elastic deflection.

The foregoing crude estimates should be understood as tools for a quick decision whether a deeper analysis of cyclic creep effects is worth the effort. For instance, for the KB Bridge, we have $\mu_L = 0.005$ (see [134] and Example 7.2), and even if the number of cycles is set to 10 million, the crude estimates give $\sigma_{rt} \approx 4 \times 10^{-4}\bar{f}_c \approx 4 \times 10^{-3}\bar{f}_t$ and $\Delta\kappa_{\text{cyc}} \approx 6 \times 10^{-3}\kappa_{\text{ref}}$ (while the more accurate estimates (7.64)–(7.65) give $\sigma_{rt} \approx 10^{-3}\bar{f}_t$ and $\Delta\kappa_{\text{cyc}} \approx 2 \times 10^{-3}\kappa_{\text{ref}}$). So, indeed, the cyclic creep effects are virtually nil.

In small- and medium-span prestressed bridges, with very low moments due to self-weight, the relative stress amplitude is much larger, which means that cyclic creep effects may become important. To appraise it, one should perform a more accurate analysis. First, the sectional characteristics and the corresponding coefficients β should be determined by exact integration, based on the real sectional geometry. Second, it should be taken into account that the residual strains develop gradually over years and decades, and so the residual stresses get relaxed by creep but also feel the effect of aging. At a still higher level of sophistication, the residual stresses should be added to the average stresses due to dead load, live load, and prestress when the residual strains are evaluated.

The effects of relaxation and aging on the residual stresses are already incorporated in formula (7.55) through the relaxation operator $\mathscr{R}$ applied on the function $N_{\text{cyc}}(t)$ that describes the increase of the cumulative number of cycles in time. As a first approximation, the intensity of traffic can be considered as constant, and then the number of cycles

$$N_{\mathrm{cyc}}(t) = v_c \langle t - t_{1c} \rangle \tag{7.70}$$

increases proportionally to the time elapsed from the time t_{1c} at which the bridge was open for the traffic, with v_c denoting the frequency (number of traffic load cycles per unit of time). Based on the definition of the relaxation operator (see (2.37) and (2.24)), we can write

$$\mathscr{R}\{N_{\mathrm{cyc}}(t)\} = \int_{t_{1c}}^{t} R(t,t') \dot{N}_{\mathrm{cyc}}(t')\,\mathrm{d}t' = v_c \int_{t_{1c}}^{t} R(t,t')\,\mathrm{d}t' \tag{7.71}$$

In previous simple estimates, which ignored the relaxation effects, the term $\mathscr{R}\{N_{\mathrm{cyc}}(t)\}$ was replaced by $E N_{\mathrm{cyc}}(t)$, with E denoting the elastic modulus. The improved estimate accounting for relaxation and aging should use

$$\mathscr{R}\{N_{\mathrm{cyc}}(t)\} = E_{\mathrm{cyc}}(t, t_{1c}) N_{\mathrm{cyc}}(t) \tag{7.72}$$

where

$$E_{\mathrm{cyc}}(t, t_{1c}) = \frac{\mathscr{R}\{N_{\mathrm{cyc}}(t)\}}{N_{\mathrm{cyc}}(t)} = \frac{v_c}{v_c(t - t_{1c})} \int_{t_{1c}}^{t} R(t,t')\,\mathrm{d}t' = \frac{1}{t - t_{1c}} \int_{t_{1c}}^{t} R(t,t')\,\mathrm{d}t' \tag{7.73}$$

is the *cyclic effective modulus*—a special type of effective modulus obtained by averaging the relaxation function $R(t,t')$ over t' ranging from t_{1c} to t.

Example 7.3. Cyclic effective modulus for the B3 model

For illustration, let us evaluate the cyclic effective modulus E_{cyc} defined in (7.73) for the B3 model with parameters that were used by Bažant, Yu and Li [209] in their analysis of the KB Bridge.

The objective is to compute $E_{\mathrm{cyc}}(t, t_{1c})$ for fixed t_{1c} and variable t. Instead of a direct evaluation of the integral in (7.73), which would require multiple costly computations of the relaxation function (repeated for each t), it is more efficient to consider this integral as the stress at time t generated by strain history $\varepsilon(t) = \langle t - t_{1c} \rangle$. Evaluation of the stress history generated by such a prescribed strain evolution can be done, e.g., using the numerical scheme described by Eqs. (5.8)–(5.11) in Sect. 5.1. The algorithm starts from $\sigma_1 = 0$ at time $t_1 = t_{1c}$, and then, in each time step Δt_k, $k = 2, 3, \ldots$, the strain increment $\Delta \varepsilon_k$ is set to Δt_k and the stress increment $\Delta \sigma_k$ is obtained from (5.9). The cumulated stress increments provide the stress σ_k, and dividing it by $t_k - t_{1c}$, we obtain the cyclic effective modulus $E_{\mathrm{cyc}}(t_k, t_{1c})$.

Using this numerical scheme, the cyclic effective modulus has been evaluated for the B3 model with parameters taken from set 1 in Table 7.1 and with the same conditions as in the analysis of the KB Bridge: start of drying at $t_0 = 7$ days, ambient humidity $h_{\mathrm{env}} = 70\%$, and effective thickness $D = 432$ mm corresponding to the top slab of the KB Bridge. Parameter k_t has been considered by its standard value 2.97

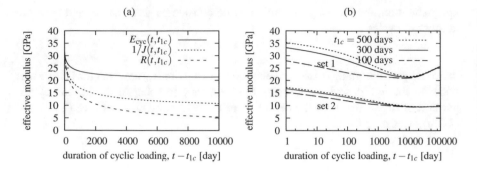

Fig. 7.29 (a) Cyclic effective modulus for model B3 with set-1 parameters, with cyclic loading from age $t_{1c} = 300$ days, (b) influence of model parameters and age on cyclic effective modulus

days/cm², not by the reduced value from the last row of Table 7.1, which is somewhat speculative.

In Fig. 7.29a, the dependence of the cyclic effective modulus $E_{cyc}(t, t_{1c})$ on the duration of cyclic loading is compared to the relaxation function $R(t, t_{1c})$ and to the reciprocal value of the compliance function, $1/J(t, t_{1c})$. These graphs have been plotted for $t_{1c} = 300$ days. Interestingly, the beneficial relaxation effects are felt only during the first few years and the decrease of E_{cyc} is much slower than the decrease of the relaxation function. After a few years, the decrease becomes extremely slow and later the trend can even get reverted and the cyclic effective modulus increases. This is better seen from the graphs in semilogarithmic scale, plotted in Fig. 7.29b. Here, the evolution of the cyclic effective modulus is shown for two sets of model parameters (set 1 and set 2 from Table 7.1) and for three different ages at the start of cyclic loading. After about 20 years, the cyclic effective modulus becomes almost independent of the age at which the cyclic loading started and remains almost constant (for set 2) or even increases (for set 1).

The numerical results show that an approximation of $\mathcal{R}\{N_{cyc}(t)\}$ by $N_{cyc}(t)/J(t, t_{1c})$ (i.e., multiplication of $N_{cyc}(t)$ by the standard effective modulus) would be unsafe and would underestimate the residual stresses. In fact, a simple replacement of $\mathcal{R}\{N_{cyc}(t)\}$ by $EN_{cyc}(t)$ with E taken as constant (e.g., equal to the conventional modulus of elasticity) turns out to be more realistic. The reason is that the beneficial effect of relaxation (i.e., reduction of the residual stresses) is partially or fully compensated, or even overcompensated, by the effect of aging. Due to aging, the instantaneous increment of residual stress generated by a given increment of residual strain is higher for older concrete, and also the subsequent relaxation of this stress is much less pronounced than for young concrete. Since the residual strains are gradually increased over the whole life of the bridge, the residual stresses at age of say 30 years are much more affected by the properties of "old" concrete (say from 5 to 30 years of age) than by the properties of relatively young concrete. ∎

There are probably other effects which cannot be analyzed because of the lack of test data. For example, the prefactor C_1 in (7.27) is likely to decrease with age.

To appraise the cyclic creep effects in large prestressed segmental box girders, Bažant and Hubler [134] analyzed six large-span bridges—the ill-fated KB Bridge in Palau, bridges Koshirazu, Tsukiyono, Konaru, and Urado in Japan, and the North Halawa Valley Viaduct in Hawaii (for details of these bridges see [209] and [210]). To make comparisons easier, they assumed that, in all these bridges, the P and M_{DL} values were the maximum allowable for each cross section. Based on the dimensions of each bridge, the live and dead loads (including the self-weight) were determined, and the bending moments at the pier were calculated assuming the girder to be rather flexible at midspan, in which case the moment at the pier is almost as large as it would be for a hinge at midspan. The results were presented in terms of (i) the inelastic residual strain ratio, defined as the maximum residual strain produced in the pier cross section (at either top or bottom face) by stress cycling, ε_r, divided by the elastic strain, f_t/E, and (ii) the inelastic residual curvature ratio, defined as the additional curvature due to load cycling, $\Delta\kappa^{cyc}$, divided by the elastic curvature, $(f_c + f_t)/Eh$. The inelastic residual curvature ratio after 2 million cycles was found to be below 2.3% in absolute value for all bridges analyzed. The inelastic residual strain ratio was extremely low for two bridges with the largest span—the KB Bridge in Palau (about 2×10^{-6}) and the Urado Bridge (about 1%). For the other bridges, with spans around 100 m, this ratio was between 5% and 56%.

It is interesting to note that whereas the static long-time creep deflections grow in time approximately logarithmically, the cyclic creep deflections grow linearly, provided that the traffic load frequency and amplitude remain constant. This property is verified by experiments and, theoretically, is a consequence of Paris law (7.24), which states that the crack extension is proportional to the number of cycles, N_{cyc}. Consequently, even if the cyclic creep effects are insignificant within the first 10 years of service, they may become significant, compared to creep, after 50 years.

For the KB Bridge in Palau, the calculated cyclic creep effects were virtually nil. Why? Partly because heavy trucks were rare, but mainly because this record-span bridge was totally dominated by self-weight. The bending moment due to traffic represented only about 2.5% of the total bending moment. Consequently, the ratio $\Delta\sigma/\bar{\sigma}$ was unusually small, only about 0.025. Since the exponent, equal to 4, is high, one has $(\Delta\sigma/\bar{\sigma})^4 \approx 4 \times 10^{-7}$. This explains why the cyclic creep contribution to curvature, deflection, and cracking must have been, in this bridge, virtually zero. However, the high value, 4, of the exponent causes that a 20-fold increase of the ratio of traffic load to self-weight will increase the cyclic creep contribution about 10^5 times.

The foregoing analysis also indicates that, for heavily travelled bridges (unlike the KB Bridge), the time average of the traffic load should be included in creep analysis.

7.14.4 Recapitulation

A realistic hypothesis for the mechanism of cyclic creep is that it is caused by growth of preexisting micrometer-scale subcritical cracks governed by the Paris law. The relative increase of crack lengths must be considered small, because no significant decrease of stiffness of bridge girders has been reported. The cyclic creep strain is obtained from the sum of openings of the microcracks calculated from the amplitude of the stress intensity factor, which itself is assumed to be proportional to the amplitude of the applied cyclic load. Detailed derivation of the form of the cyclic creep law is possible for mode-I cracks, either tensile or (in the sense of crushing band) compressive. A general applicability of this law for other types of cracks can be justified by dimensional analysis and similitude arguments.

The derived cyclic creep law can fit the test data available in the literature quite well and has been calibrated by them. The data show that the Paris law exponent must be about 4. This observation is consistent with the recent finding that the Paris law exponent on the atomic scale must be 2 and that the exponent should roughly double during each scale transition, in this case from the nanoscale to the micrometer scale. The exponent of 4 is high enough to make cyclic creep deflections enormously sensitive to the relative amplitude of the applied cyclic stress. While the static creep deflections grow in time approximately logarithmically, the cyclic creep deflections grow linearly. This is a consequence of the Paris law. Consequently, even when the cyclic creep is unimportant for ten years of service, it may become important for hundred years of service.

Calculation examples indicate that, contrary to often voiced opinions, the cyclic creep effects are absolutely negligible for large-span prestressed segmental box girders, which are totally dominated by self-weight. In particular, the cyclic creep could have played no role in the excessive deflections of the ill-fated world-record KB Bridge in Palau and other large-span bridges. For small spans (up to 40 m), having a small proportion of self-weight, the ratio of cyclic creep deflections to the span is not negligible but it does not matter since, due to prestress needed to resist the live load, the static creep in such bridges causes upward deflections.

Cyclic creep can produce major tensile strains in small- and medium-span prestressed bridges and thus may contribute significantly to their cracking and corrosion. Other structures, such as those supporting large turbines or electric generators, can also suffer from surface tensile cracking caused by cyclic creep.

7.15   Conclusions for Method of Analysis and Design

The excessive deflections of the KB Bridge in Palau are perfectly explicable. In the order of decreasing importance, the main causes of underestimation of deflections and prestress loss are as follows:

1. Poor material model for creep and shrinkage.
2. Beam-type analysis instead of a full three-dimensional analysis.
3. Differences in the rates of shrinkage and drying creep due to different thicknesses of slabs in the box cross section.
4. Lack of statistical estimation of the range of possible responses.

The analysis in this chapter leads to the following conclusions:

1. As a purely predictive tool, none of the available material models for predicting creep and shrinkage is satisfactory.
2. The 1971 ACI model (reapproved in 2008) and, to a somewhat lesser extent, the CEB and JSCE models severely underestimate multidecade deflections as well as the prestress losses, and give an unrealistic shape of deflection growth curves. The more recent GL model gives better predictions, but not sufficiently better. Unlike the parameters in the ACI, fib, GL, and JSCE models, the basic parameters $q_1, \dots q_5$ of model B3 can be updated by linear regression of compliance data.
3. Model B3, which is to a large extent theoretically based and has been calibrated by filtering out the database bias for short durations and ages, gives significantly better multidecade predictions of deflection history and its shape, and of deflection growth rate.
4. Even model B3 is unsatisfactory when its input parameters are estimated from the composition of concrete or taken at their default values. Nevertheless, thanks to its free parameters, model B3 can be made to fit the measurements perfectly with parameter values that are within their realistic range. Thus, the form of model B3 appears to be correct, and the problem is with the empirical formulae predicting the input parameters from the composition of concrete. Obviously, these formulae must be improved.
5. The box girders are thick-walled shells for which the beam-type analysis is inadequate. Three-dimensional analysis must be used. Its main purpose is to capture the shear lag effects, which are rather different for self-weight and for the loads from prestressing tendons, and occur not only in the top slab but also in the webs and the bottom slab. At the piers, the self-weight produces large vertical shear forces in the web, while the loads from tendon anchors produce shear lags mainly in the top slab. The shear lag for the self-weight is stronger than it is for the prestress. Since the total deflection is a small difference of two large numbers, one for the downward deflection due to self-weight and the other for the upward deflection due to prestress, small percentage errors in each will result in a far larger percentage error in the total deflection.
6. For box girders with a larger width-to-span ratio than the KB Bridge (which had only two lanes for a span of 241 m), the difference between the beam-type analysis of creep and shrinkage and the three-dimensional analysis which captures the shear lag effects must be expected to be larger.
7. The effect of thickness differences among the webs and the top and bottom slabs on shrinkage and drying creep must be taken into account. This leads to nonuniform creep and shrinkage properties throughout the cross section, manifested as differential drying creep compliances and differential shrinkage.
8. In the creep and shrinkage prediction model, the drying creep should be separated from the basic creep, because the former is thickness-dependent and approaches a finite terminal value while the latter is thickness-independent and unbounded. Only model B3 and the new fib model have this feature. The thickness-induced differences in the compliance functions for drying creep are often more important than those in shrinkage.

9. The prestress loss in box girders can be 2 to 3 times higher than that predicted by simple textbook formulae or lump estimates. It can also be much higher than that calculated by the theory of beam bending, in which the cross sections are assumed to remain plane. The prestress loss should be calculated as part of the three-dimensional finite element creep analysis, rather than estimated in advance.

10. When dealing with large creep-sensitive structures, the creep and shrinkage prediction model must be updated by means of short-time tests of the creep and shrinkage of the given concrete, as discussed in Sect. 3.8 and Appendix H. The updating is effective only if the curves of creep and shrinkage growth have correct shapes for short times. To this end, differential shrinkage and differential drying creep must be estimated realistically in conformity with the diffusion theory, which is a feature of model B3.

11. The shrinkage tests should be accompanied by simultaneous measurements of water loss due to drying (Sect. H.1) or else the extrapolations can have errors of the order of 100%. As either an alternative, or an additional measure to improve extrapolation, shrinkage tests of standard cylinders (of 7.5 to 15 cm diameter) can be supplemented by tests of much smaller prisms (which can be sawed out of larger blocks even if the prism thickness is less than the aggregate size). Such prisms shrink much faster and approach the terminal phase of shrinkage curve much earlier. This allows using the diffusion scaling to improve the extrapolation [124]. B3 and B4 are models that have been specifically formulated so as to allow easy updating by linear regression, while for other models the updating problem is nonlinear.

12. Large bridges should be designed not for the mean but for the 95% confidence limit on the predicted deflection. The necessary statistical analysis is easy. It suffices to repeat a deterministic computer run of structural response about 10 times, using random samples of the input parameters, generated according to the Latin hypercube sampling. Then, one merely needs to estimate the mean and variance of the calculated response values. Since the distribution of structural response can be assumed to be normal (within such confidence limits), the mean and variance suffice to obtain the confidence limit for the response.

13. As observed in Křístek et al. [558], the deflection evolution of large box girders is often counterintuitive. The deflections at first grow slowly or are even negative, which may lead to unwarranted optimism, and after a few years, a rapid and excessive deflection growth sets in.

14. In design, it is prudent to minimize deflections and prestress losses by the afore-mentioned nine measures, listed in Sect. 7.11.

15. Cyclic creep is not a significant cause of excessive deflections, although it can produce large residual tensile strains at the bottom and top faces.

Part II
Advanced Topics

Chapter 8
Moisture Transport in Concrete

Abstract After presenting the fundamentals in the previous chapters constituting Part I, we begin here Part II dealing with advanced topics. In this chapter, we study concrete as porous material, the mechanical behavior of which is strongly affected by the presence and migration of moisture (i.e., various phases of water) through the pore space. We present the basic concepts and equations characterizing the moisture transport under isothermal conditions. We discuss the thermodynamic aspects and briefly describe various transport mechanisms. Then, we focus attention on relatively simple models with a limited number of parameters, particularly on the classical Bažant–Najjar model, which can be effectively used in practical applications and is recommended in design codes. After deriving a nonlinear moisture diffusion equation, we study various problems of practical interest by combining analytical and numerical techniques. The cases we cover include drying of a slab or half-space (under constant or variable ambient humidity), steady flux of moisture through a wall, and spreading of a hydraulic pressure front into unsaturated or self-desiccated concrete. The link between moisture transport and shrinkage is also discussed. Finally, we briefly comment on the changes required to take into account the effects of self-desiccation and autogenous shrinkage and outline the diffusion processes affecting the alkali–silica reactions (ASR).

The transport of moisture in porous materials and its interplay with the mechanical behavior is in general described by the theory of partially saturated porous media [221, 222, 328, 424, 572, 766], with the theoretical support of thermodynamics of multiphase systems [435, 467, 468]. In general, the hygromechanical phenomena should be considered simultaneously with heat transport and with the chemical processes of hydration. Sophisticated thermo-hygro-chemo-mechanical models have been proposed in the literature [419, 420]. Unfortunately, the aforementioned theories are hard to apply to concrete because the delineation of the solid and fluid phases is blurred and variable in time, and the parameters to account for it are not known and would be hard to determine. For this reason, simpler models with a limited number of parameters, such as Bažant and Najjar's [166], turn out to be more effective.

Much of the moisture in concrete consists of the hindered adsorbed water in nanopores $0.3 \sim 3$ nm thick. Although evaporable (by definition), its movement

© Springer Science+Business Media B.V. 2018

Z.P. Bažant and M. Jirásek, *Creep and Hygrothermal Effects in Concrete Structures*, Solid Mechanics and Its Applications 225, https://doi.org/10.1007/978-94-024-1138-6_8

is controlled by surface forces of adsorption. Depending on many factors, it may act as part of the solid skeleton or the fluid system. Therefore, the aforementioned theories of multiphase media are not easily applicable to concrete (except perhaps near saturation).

In this chapter, we restrict attention to a simplified description of moisture transport under isothermal conditions. The influence of mechanical fields on the transport properties is neglected, although in the case of extensive damage it is not negligible. The effect of hydration is incorporated only approximately. More general approaches will be discussed in Chaps. 10 and 13.

8.1 Water in Concrete

The hardened cement pastes, and thus also cement mortars and concrete, are porous and strongly hydrophilic materials. Aside from the water that is *chemically bound* in the hydration products (mainly calcium silicate hydrates and calcium hydroxide) and can get liberated only upon heating to more than 550 °C, these materials contain a large amount of *evaporable water*, filling the pores. In water-saturated hardened cement paste, the evaporable water typically represents 15 to 40% of volume. In saturated concrete, the evaporable water content depends on the cement fraction. At most about 90% of volume can be occupied by mineral aggregates, and then, if the aggregate porosity is negligible, the content of initially available evaporable water can be as low as 2 to 3% of volume and even slightly less when air-entraining admixtures and entrapped air are present.

Three phases of evaporable water may be distinguished (Fig. 8.1):

1. *capillary water*, i.e., liquid water residing in capillary pores, which typically have micrometer dimensions, and bounded by capillary menisci (pressure[1] p_l in liquid capillary water is often negative, which means that the water is under tension);
2. *water vapor* in capillary pores, of partial pressure $p_v > 0$; and
3. *adsorbed water*, which is of two kinds:

    a. the *free adsorbed* layers, which can be up to 5 water molecules in thickness ($\approx 5 \times 0.27$ nm), are held at pore walls and have contact with water vapor, and
    b. *the hindered adsorbed* water layers in cement gel pores (nanopores) of width smaller than 10 water molecules (2.7 nm), in which the maximum thickness of adsorbed water layers cannot develop, causing transverse pressure across the layer called the *disjoining pressure*, p_d. The hindered adsorbed water communicates with vapor and capillary water only by diffusion along the nanopore.

---

[1] Variables referred to as pressures characterize hydrostatic stress states, but with an opposite sign convention—positive pressure corresponds to compressive (negative) hydrostatic stress.

When convenient, water in both liquid and vapor forms will be referred to as the *moisture*.

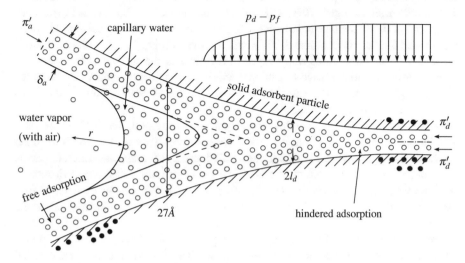

Fig. 8.1 Idealized sketch of a hindered adsorbed water layer with transition into the capillary pore; π'_a is the total spreading pressure in the free adsorbed layer, π'_d is the total spreading pressure in the hindered adsorbed layer, p_d is the disjoining pressure, p_f is the pressure at the moment the nanopore gets filled, r is the radius of the capillary meniscus, δ_a is the thickness of the free adsorbed layer, and l_d is the thickness of the hindered adsorbed layer

In this book, the *evaporable water content* (mass of evaporable water per unit volume of concrete) is denoted as w_e, and the *nonevaporable water content* (chemically bound water, sometimes including what cement chemists call the *interlayer water*, which is contained between the layers of C-S-H) is denoted as w_n. Of course, both w_e and w_n in general vary in time. Their initial values are $w_e(0) = w$ and $w_n(0) = 0$, where w is the initial water content in the mix, which is one of the fundamental parameters defining the concrete mix composition. The sum $w_t = w_e + w_n$ is the *total water content*, which remains constant (and equal to w) under sealed conditions but decreases when drying takes place.

Because of very low permeability of concrete, the evaporable (mobile) water content w_e always varies slowly enough to justify the assumption that all the water phases in contact within a capillary pore of concrete are in thermodynamic equilibrium at any time.

As will be shown later, the evaporable water content is closely related to the relative humidity of water vapor in the pore space. The *pore relative humidity*

$$h = \frac{p_v}{p_{sat}} \tag{8.1}$$

(for brevity sometimes called just the "humidity") is defined as the partial pressure of water vapor p_v divided by the saturated vapor pressure p_{sat} (which is a function of the temperature, see Eq. (8.17) for the precise definition). The pore relative humidity in general varies throughout the specimen or structure and will be used as the primary unknown of the diffusion equation describing the moisture transport. It should not be confused with the environmental (ambient) relative humidity, h_{env}, which is derived from the partial pressure of water vapor in the air near the surface of the specimen or structure and which is used as a prescribed quantity in the boundary condition for the diffusion equation. The capillary water dominates at relative humidities above 85% and almost disappears at 50%.

At the moment of set, concrete starts as perfectly saturated, with 100% pore relative humidity ($h = 1$). If the surface of the concrete specimen or structure is exposed to the environment, which is usually at lower humidity (unless the concrete part is submerged in water), moisture migrates from concrete pores into the environment and the total water content in concrete decreases. This drying process leads to a decrease of the pore relative humidity h, which causes a part of shrinkage, called the *drying shrinkage*.

However, it turns out that shrinkage also occurs even in perfectly sealed specimens, in which the total water content remains constant. This is the part of shrinkage called the *autogenous shrinkage* which is driven chemically, by a slight volume decrease during the hydration reactions and by chemically caused self-desiccation. The autogenous shrinkage is large in concretes with low water-cement ratios ($w/c < 0.35$), which is the case of modern high-strength concretes. It is sometimes nonnegligible even for medium w/c around 0.5. In the B3 model, it is lumped with drying shrinkage (which is insufficient for low, and some cases of medium, w/c), while in the new B4 model, it is described by a separate formula. In normal-strength concretes, autogenous shrinkage represents only about 5% of the total shrinkage, but in modern high-strength concretes, it can lead to strains reaching $350 \sim 700 \times 10^{-6}$ in magnitude [484, 570, 679, 857], in some extreme cases even higher. For instance, Holt [484] measured autogenous shrinkage strain of 1200×10^{-6} in magnitude on concrete with water/cement ratio 0.30 (rapidly hardening cement). On cement paste with silica fume, autogenous shrinkage strains exceeding 2000×10^{-6} in magnitude can be observed [515, 586].

The autogenous shrinkage is consequence of *self-desiccation* [700]. The chemical reaction of water with cement creates solid hydration products that contain chemically bound water. The amount of evaporable water occupying the pore space therefore decreases, but the pore volume decreases as well, due to the deposition of hydration products on the pore walls. If the volume of water and cement consumed by the chemical reaction were exactly equal to the volume of products resulting from that reaction, the pores would remain fully saturated by water. Although the reaction products on the nanoscale have a slightly lower volume than the initial reactants, the expansion is reversed to compression by viscoelastic deformation of the solid skeleton caused by the decrease of pore humidity, due either to self-desiccation or to external drying, except in water immersion in which the expansion is manifested as swelling; see also [111, 125].

The shrinkage that would occur without the resistance of the solid skeleton is called the *chemical shrinkage*. It can be assessed by mixing cement with an excessive amount of water (to avoid formation of a solid skeleton) and measuring the volume decrease during hydration. The solid skeleton that develops in a hardening cement paste restricts the shrinkage process and reduces the autogenous shrinkage (the actual relative size change of a sealed specimen) as compared to the chemical shrinkage. The fact that the reduced volume of the hydration products is only partially accommodated by autogenous shrinkage leads to water deficiency in the pore space and thus to a decrease of pore humidity even at constant total water content.

The self-desiccation in sealed normal-strength concretes without admixtures is relatively limited, but, in high-strength concretes which are made with a very low water-cement ratio, it can reduce the pore relative humidity in sealed specimens to levels comparable with typical environmental relative humidities [282, 516, 676]. For instance, Baroghel-Bouny et al. [57] reported values of $h = 93\%$ for ordinary concrete (water-cement ratio 0.48) after 2 years in sealed conditions, and much lower values for high-strength concrete (water-cement ratio 0.26, with silica fume and superplasticizer), with h decreasing to 72% after 6 months, 69% after 1 year, and 64% after 2 years in sealed conditions.

Any rewetting causes *swelling*, which used to be explained by expansion of hindered adsorbed water layers in the nanopores. But the ongoing research at Northwestern University led to a new paradigm of deformations during hydration, which gives a more accurate picture of the swelling mechanism [125].

The fact that the volume of cement hydration products is slightly smaller than the original volume of cement and water, known since 1887, is valid only on the nanoscale. It does not mean that the hydration reaction causes overall contraction of the porous cement paste and concrete. As first suggested in 2015 [125], the opposite is true for porous cement paste as a whole. The growth of C-S-H shells around anhydrous cement grains pushes the adjacent contacting shells apart and thus causes *volume expansion* of the porous cement paste on the material scale, while the nanoscale volume contraction of hydration products contributes to porosity. The growth of ettringite and portlandite crystals may also cause additional expansion, but is never dominant. On the material scale, the expansion always dominates over the contraction, i.e., the hydration per se is, in the bulk, always expansive, while the source of all of the observed shrinkage, whether autogenous or due to external drying, is the compressive elastic or viscoelastic strain in the solid caused by a decrease of chemical potential of pore water, with the corresponding decrease in pore humidity, increase of solid surface tension and, mainly, decrease of disjoining pressure. The expansiveness of hydration reaction is the only way to explain swelling of thin specimens under water. However, there must be a large size effect of diffusion type, such that thicker walls shrink rather than swell even under water immersion. Thanks to expansiveness of hydration, the swelling and both the drying and autogenous shrinkage can all be predicted from one and the same unified model. Comparisons with the existing experimental evidence confirm that.

Furthermore, despite current lack of experimental evidence, the model implies that there must be a large size effect on the diffusion of water that gradually fills the self-desiccated pores to 100% humidity. This diffusion is much slower than drying.

As shown in Bažant [79] (see also Sect. 8.5), the amount of water that needs to be delivered to the propagating front of wetting is large because the humidity at the self-desiccated front must be raised to 100% before the wetting front can advance farther. Thus the interface condition at the advancing wetting front gives a large interface sink of water, which greatly slows down the advance of the diffusion front of wetting (note that a similar negative sink is absent at the front of drying which, therefore, propagates relatively much faster). The result is that structural members thicker than about 0.2 m must be expected to always shrink (autogenously) under water for at least a century, due to self-desiccation of most of their interior volume. The swelling of the (relatively thin) surface layer is cancelled by inducing a compressive stress parallel to the surface, which is beneficial as it prevents the formation of surface cracks.

At room temperature, the process of concrete drying is very slow, due to extremely low permeability of concrete. A slab of thickness $D = 15$ cm exposed to an environment of 50% relative humidity may take more than 10 years to dry to a nearly uniform specific water content; see Fig. 8.27a for an example. The drying times increase with the square of thickness (and so the wall of a nuclear containment, about 1 m thick, would take about 900 years to dry, if there were no cracking). Therefore, the nonuniform self-equilibrated stresses caused by drying shrinkage persist long enough to be strongly reduced by creep.

By contrast, the diffusion process of heat conduction in concrete is about 1000 times faster, which has the fortunate consequence that, in most situations, the heating and drying problems may be analyzed as uncoupled [150]. Exceptions are the rapid drying at high temperature, as in a fire or nuclear accident [142, 188], or the peeling of contaminated surface layers of concrete by microwave blast [213]; see Chap. 13. Another exception is moisture movement in young mass concrete, in which the chemical processes of hydration can significantly raise the temperature.

8.2 Pore Fluids at Thermodynamic Equilibrium

8.2.1 Multiphase Porous Medium

Although the multiphase porous medium is not a sufficient model for concrete, because of nanopore water that behaves as both the solid and fluid, it is nevertheless a useful simple basis for further refinements. Transport properties of porous materials strongly depend on the morphology of the pore space. If the material microstructure is known, the effective macroscopic properties can, in theory [357], be determined by appropriate upscaling techniques (this would probably be quite complicated if the surface adsorption forces were taken into account).

To keep the presentation simple, we will use a phenomenological approach and start directly from a macroscopic description. Instead of treating the pore space as a three-dimensional subdomain of a very complicated shape, we consider a smeared description, which deals with the solid skeleton and the pores as two overlaid continua. The pores are characterized by the relative volume they occupy, i.e., by the *porosity*, n_p, which may vary across the body as a function of the macroscopic coordinates. In the fully saturated case, all pores are filled with liquid water. However, in general, the pores also contain wet air, i.e., air mixed with water vapor. The relative volume of liquid water with respect to the pore volume is called the *saturation degree*, S_l (strictly speaking, one should distinguish between the *capillary degree of saturation* and the *vacuum degree of saturation*). Formally,

$$n_p = \frac{V_p}{V_{tot}} = \frac{V_l + V_g}{V_s + V_l + V_g} \tag{8.2}$$

$$S_l = \frac{V_l}{V_p} = \frac{V_l}{V_l + V_g} \tag{8.3}$$

where $V_{tot} = V_s + V_l + V_g$ is the total volume, V_s is the volume of the solid skeleton, $V_p = V_l + V_g$ is the pore volume, V_l is the volume occupied by liquid water, and V_g is the volume occupied by gas (wet air).

In general, all the volumes mentioned above may vary not only in space but also in time, due to deformation processes, cracking, phase changes, and chemical reactions (such as hydration). If we consider a chemically inert material with small deformations only (and ignore hindered adsorbed water), the changes of porosity n_p can be neglected because both V_{tot} and V_s remain constant, up to negligible correction terms. However, the subdivision of V_p into V_l and V_g may evolve, and thus, the saturation degree S_l is a time-dependent variable.

In a partially saturated porous medium, the following phases can be distinguished:

1. solid skeleton, which contains chemically bound (nonevaporable) water;
2. liquid pore water, consisting of capillary water;
3. adsorbed water, which is neither liquid nor solid and consists of

    a. free adsorbed water,
    b. hindered adsorbed water;

4. pore gas (wet air), which is a binary mixture of two ideal gases, namely

    a. dry air,
    b. water vapor.

However, the existing multiphase models do not distinguish the adsorbed and liquid water, and so we will consider the adsorbed water as part of liquid water. Variables referring to one of the phases will be denoted by subscripts s (solid), l (liquid), g (gas), v (vapor), and a (air). Subscript w will refer to water as chemical species (in the form of liquid or adsorbed water or water vapor). For instance, ρ with a subscript will denote the true (bulk) mass density of the corresponding phase (per unit volume of the space occupied by this phase), while m with a subscript will be the apparent mass density (per unit volume of the porous medium). Since mass is additive, it is easy to establish the relations

$$m = m_s + m_l + m_a + m_v = \frac{\rho_s V_s + \rho_l V_l + \rho_a V_g + \rho_v V_g}{V_{tot}} =$$

$$= \rho_s(1 - n_p) + \rho_l n_p S_l + (\rho_a + \rho_v)n_p(1 - S_l) \tag{8.4}$$

The evaporable water consists of the liquid water (capillary and adsorbed) and water vapor. Therefore, the evaporable water content, previously defined as the mass of evaporable water per unit volume of the porous medium, can be expressed as[2]

$$w_e = m_l + m_v = \rho_l n_p S_l + \rho_v n_p (1 - S_l) \tag{8.5}$$

It is important to distinguish between the evaporable water content in concrete (taken per unit volume of concrete) and in cement paste (per unit volume of paste). In some experimental reports and models, the amount of evaporable water is character-ized by the *moisture ratio*, u, defined as the mass of evaporable water per unit mass of the dry material. This variable is dimensionless and can be expressed in percent. To emphasize that it represents the mass fraction and not the volume fraction, it is some-times specified in kg/kg or, equivalently, in g/g. Again, it is important to distinguish between mass fractions with respect to the mass of cement, to dry hardened cement paste, or to dry concrete. The mass of cement is understood here as the initial mass of cement in the mix, i.e., as a constant, while the masses of dry hardened cement paste and dry concrete evolve in time, due to the incorporation of chemically bound water into the solid skeleton.

For instance, the water-cement ratio, w/c, is in fact the initial value of the mass fraction of evaporable (or total) water, taken with respect to the mass of cement.

8.2.2   State Equations

The thermodynamic state of a fluid is characterized by temperature and pressure, from which the density can be determined, using an appropriate state equation.[3] For the present purpose, it can be assumed that the local exchange of energy among the phases is sufficiently fast (compared to the time scale of processes that we investigate here), and thermal equilibrium is restored instantaneously; i.e., all the phases are at the same absolute temperature, T. On the other hand, the liquid pressure[4] p_l and the

---

[2]Density ρ_l refers here to the density of liquid water in the capillary pores. Strictly speaking, the density of adsorbed water is slightly different, but for simplicity, we neglect this difference.

[3]Recall that positive pressure corresponds to compressive (i.e., negative) hydrostatic stress. From the point of view of thermodynamics, temperature and pressure represent the state variables, and differentiation of the appropriate thermodynamic potential (*specific Gibbs free energy*, μ, also called *specific free enthalpy*) with respect to these variables provides the state laws that determine the conjugate thermodynamic forces—specific entropy and specific volume (i.e., the reciprocal value of true density); see Sect. 13.5.5. The specific free enthalpy is linked by the Legendre transformation to other thermodynamic potentials, such as the specific Helmholtz free energy or specific enthalpy. Note that quantities denoted as "specific" refer to a unit mass of the substance. The *chemical potential* is usually understood as the free enthalpy per unit amount of the substance, i.e., per mole. In the literature, the chemical potential is often denoted as μ, but here we reserve this symbol for the specific free enthalpy (which is equal to the chemical potential divided by the molar mass).

[4]Pressure p_l refers here to the pressure in the capillary water. The stress state of adsorbed water is different and is not even hydrostatic; see Sect. 8.2.6.2.

gas pressure p_g are in general different, due to the effect of surface tension along curved boundaries of capillary menisci.

The pore gas (wet air) is a binary mixture of vapor and air, both of which can be considered as *ideal gases*. According to the *Boyle–Mariotte state equation* of ideal gases (in its early form postulated by Robert Boyle in 1662 and by Edme Mariotte in 1676), the densities of air and vapor can be expressed as

$$\rho_a = \frac{M_a p_a}{RT} \tag{8.6}$$

$$\rho_v = \frac{M_w p_v}{RT} \tag{8.7}$$

where p_a and p_v are the *partial pressures* of dry air and water vapor, $M_a = 28.96$ g/mol and $M_w = 18.02$ g/mol are the *molar masses* of dry air and water (masses of one mole of the respective substance), and $R = 8.31446$ J/K mol is the *universal gas constant*.

According to *Dalton's law*, formulated by J. Dalton in 1803, the pressure of the gas mixture is the sum of the partial pressures, i.e.,

$$p_g = p_a + p_v \tag{8.8}$$

For a given vapor concentration, described by the dimensionless ratio

$$C_v = \frac{\rho_v}{\rho_g} = \frac{\rho_v}{\rho_v + \rho_a} \tag{8.9}$$

the partial variables can be eliminated and the gas density can be expressed as a function of the gas pressure and temperature. It turns out that the resulting formula

$$\rho_g = \rho_a + \rho_v = \frac{M_g p_g}{RT} \tag{8.10}$$

again corresponds to an ideal gas, with the molar mass of the wet air given by

$$M_g = \frac{M_w M_a}{(1 - C_v)M_w + C_v M_a} = \left(\frac{1 - C_v}{M_a} + \frac{C_v}{M_w} \right)^{-1} \tag{8.11}$$

The state equation of liquid water is much more complicated than that of an ideal gas; see for instance the IAPWS-95 water and steam tables [500], or formula (13.71) in Chap. 13. For the present purpose, it is sufficient to know that the mass density of water at standard atmospheric pressure (average sea-level pressure) $p_0 = p_{atm} = 101.325$ kPa and at temperature $T_0 = 293.15$ K (20 °C) is $\rho_{l0} = 998.2$ kg/m^3, and its bulk modulus is $K_l = 2.2$ GPa. This means that a change of pressure by 1 MPa changes the mass density by less than 0.05%. Consequently, at constant temperature we will consider the mass density of liquid water, for most purposes, as constant, $\rho_l = \rho_{l0}$. Compared to concrete, the compressibility of liquid water is about 5 to 12 times larger (at standard conditions).

8.2.3   Capillary Pressure and Relative Humidity

At a planar interface between two fluids, mechanical equilibrium requires the pressures on both sides to be equal. The interface behaves like a membrane subjected to the surface tension γ_{gl}, which is dependent on temperature and for water–air interface at 20 °C is equal to 0.0728 N/m. If the interface is curved (which is the case for capillary menisci), the equilibrium equation contains a term proportional to the surface tension multiplied by the curvature, and the pressures on both sides are no longer the same. The equilibrium condition leads to the *Young–Laplace equation* (qualitatively described by Thomas Young in 1805 and mathematically formulated by Pierre-Simon Laplace in 1806)

$$p_g = p_l + \frac{2\gamma_{gl}}{r} \tag{8.12}$$

where

$$\frac{1}{r} = \frac{1}{2}\left(\frac{1}{r_1} + \frac{1}{r_2}\right) \tag{8.13}$$

is the mean curvature of the interface, i.e., the average of the principal curvatures $1/r_1$ and $1/r_2$. Variables r_1 and r_2 represent the principal radii of curvature. Of course, for a spherical interface, we have $r_1 = r_2 = r =$ radius of the sphere.

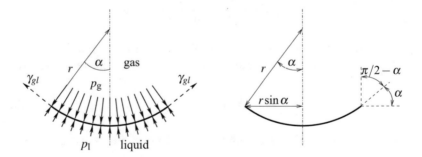

Fig. 8.2 Curved interface between a gas and a liquid, subjected to the gas pressure, p_g, liquid pressure, p_l, and surface tension, γ_{gl}

Equation (8.12) is easiest demonstrated for a part of a spherical meniscus sketched in Fig. 8.2. The vertical resultant of the pressure difference $p_g - p_l$ over the curved interface is given by $(p_g - p_l)\pi(r\sin\alpha)^2$ (it is the same as if $p_g - p_l$ were applied on a flat disk of radius $r\sin\alpha$). The vertical resultant of the surface tension is $\gamma_{gl} \cdot 2\pi r\sin\alpha \cdot \cos(\pi/2-\alpha)$ (because the surface tension acts on a circle of radius $r\sin\alpha$ and is inclined by angle $\pi/2 - \alpha$ with respect to the vertical direction). Setting up the free body equilibrium equation in the vertical direction and taking into account that $\cos(\pi/2-\alpha) = \sin\alpha$, we obtain the condition $(p_g - p_l)r = 2\gamma_{gl}$, which corresponds to (8.12).

The signs in (8.12) have been chosen such that positive curvature corresponds to the case of a concave water meniscus, which is the typical case. The gas pressure is then larger than the pressure in the capillary water. The difference

$$p_c = p_g - p_l \tag{8.14}$$

is called the *capillary pressure* (or the *suction stress*). The Young–Laplace equation (8.12) can thus be rewritten as

$$p_c = \frac{2\gamma_{gl}}{r} \tag{8.15}$$

If the capillary pressure exceeds the gas pressure, the pressure in liquid water, p_l, becomes negative, which means that the water is under tension. Note that some authors define the capillary pressure with the opposite sign, i.e., as $p_c = p_l - p_g$. Here, we adhere to definition (8.14).

High capillary pressures are associated with highly curved gas–liquid interfaces, which typically occur in small pores when the large pores remain empty. At low capillary pressures, capillary menisci are formed in larger pores, while small pores are completely filled by water. It is therefore clear that the capillary pressure is closely related to the saturation degree. Indeed, models for transport of water in soils and other porous geomaterials often express the saturation degree as a function of the suction stress (and potentially of other variables, such as temperature); the graph of this function is then called the *moisture retention curve* (or *capillary pressure curve*). In principle, it seems tempting to deduce the link between the saturation degree and the capillary pressure from purely geometrical considerations. However, this is a very tedious task, even if the detailed morphology of the pore space is known. One reason is that the relation is nonunique, because for one and the same capillary pressure one can find many different configurations of capillary menisci with the mean radius of curvature satisfying the Young–Laplace equation.[5] It is more practical to establish the retention curve experimentally. In soil mechanics, it is often approximated by the analytical expression

$$p_c = p_{\text{entry}} + \pi_0 \left(S_l^{-1/m} - 1 \right)^{1-m} \tag{8.16}$$

proposed by van Genuchten [828]. Here, p_{entry} is the entry pressure, i.e., the value of capillary pressure below which the pores remain fully saturated, and π_0 and m are additional parameters. Baroghel-Bouny et al. [57] fitted their own experimental results by a formula equivalent to (8.16), with p_{entry} set to zero and with parameters $\pi_0 = 18.6$ MPa and $m = 0.44$ for normal-strength concrete and $\pi_0 = 46.9$ MPa and $m = 0.49$ for high-strength concrete; see Fig. 8.3.

---

[5]To obtain a unique relation, one would need to consider the area of the water-air interface per unit volume as an additional variable that characterizes the state of the system; see Gray and Hassanizadeh [436].

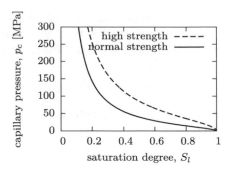

Fig. 8.3 Dependence between capillary pressure and degree of saturation for normal-strength and high-strength concrete, according to Baroghel-Bouny et al. [57]

At the gas–liquid interface, the two phases can exchange not only momentum and energy but also mass. Liquid water can change into vapor by evaporation, and vapor can change into liquid by condensation. This process is driven by the difference in the specific free enthalpies of the two phases. At thermodynamic equilibrium, these potentials have the same value[6] [441, 487]. The specific free enthalpy depends on temperature and pressure, but the function describing this dependence for liquid water, μ_l, is different from that valid for water vapor, μ_v. Therefore, even if both phases are at the same temperature, thermodynamic equilibrium does not imply that their pressures are the same.

Consider first that liquid water is kept at the atmospheric pressure, p_{atm}. If the functions describing the specific free enthalpy of liquid water and vapor, μ_l and μ_v, are known, the vapor pressure at thermodynamic equilibrium, p^*_{sat}, can be computed from the condition

$$\mu_v(p^*_{sat}, T) = \mu_l(p_{atm}, T) \tag{8.17}$$

Pressure p^*_{sat} which satisfies (8.17) corresponds to the partial pressure of saturated vapor just above a flat interface between liquid water and air at atmospheric pressure.

---

[6]For a fundamental proof, first note that a necessary condition of equilibrium of any system can be characterized by the stationarity of a proper thermodynamic potential. The internal energy, enthalpy, and Helmholtz free energy are here inapplicable because they are functions of the extrinsic system characteristics such as volume (or relative displacement) and entropy. We have a multiphase system in which the phases exchange mass, and in that case the potential must depend on the intrinsic characteristics only. Such a potential is the Gibbs free energy (also called free enthalpy), G, of the multiphase system, which is characterized by stresses (or pressures) and temperature. The Gibbs free energy of a system containing a liquid phase of mass m_l at pressure p_l and temperature T and a vapor phase of mass m_v at pressure p_v and temperature T is $G(p_v, p_c, T) = \mu_l(p_l, T)m_l + \mu_v(p_v, T)m_v$, where μ_l and μ_v are the specific (i.e., per unit mass) Gibbs free energies of liquid water and water vapor. Now, consider that mass dm_l is transferred from vapor to liquid at constant pressures and constant temperature. To conserve mass of the entire system, $dm_v = -dm_l$, and so $dG = \mu_l\, dm_l + \mu_v\, dm_v = (\mu_l - \mu_v)\, dm_l$. Noting that thermodynamic equilibrium is maintained if and only if $dG = 0$ (e.g., [441], Sect. 18), we thus conclude that, at equilibrium, $\mu_l = \mu_v$, and for changes maintaining thermodynamic equilibrium, $d\mu_l = d\mu_v$.

The vapor is saturated in the sense that if the actual vapor pressure were smaller than p_{sat}^*, the evaporation rate would exceed the condensation rate and the vapor pressure would increase up to the saturation value (and conversely, if the actual vapor pressure were above p_{sat}^*, the condensation rate would exceed the evaporation rate and the vapor pressure would decrease down to the saturation value). Temperature is present in (8.17) as a parameter, and the resulting value of p_{sat}^* depends on it; thus, we consider p_{sat}^* as a function of T.

Even though Eq. (8.17) properly defines a certain temperature-dependent characteristic value of vapor pressure p_{sat}^* with a clear physical meaning, it could not be used at temperatures that substantially exceed 100 °C, i.e., the boiling point of water at atmospheric pressure. Therefore, it seems preferable to define the *saturated vapor pressure*, p_{sat}, by the condition

$$\mu_v(p_{sat}, T) = \mu_l(p_{sat}, T) \tag{8.18}$$

which describes thermodynamic equilibrium between liquid water and water vapor that are subjected to the same pressure. Physically, this would correspond to the conditions in a closed container in which the "atmosphere" above a flat surface of liquid water contains only water vapor, and so the pressures in both phases of water are the same. Using this definition, p_{sat} can be evaluated for temperatures up to the critical point of water. In fact, the pairs of values p_{sat} and T that satisfy Eq. (8.18) represent the boundary between the domains corresponding to liquid phase and gas phase in the $p - T$ phase diagram of water. This so-called coexistence curve extends from the triple point at $p = 611.657$ Pa and $T = 273.16$ K to the critical point at $p = 22.064$ MPa and $T = 647$ K.

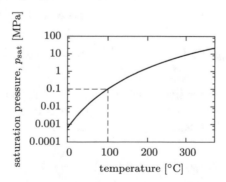

Fig. 8.4 Saturated vapor pressure as a function of temperature, according to Antoine's equation (8.19)

A good approximation of the dependence of saturated vapor pressure on temperature is provided, e.g., by the equation [38]

$$p_{sat}(T) = p_{atm} \exp\left(\frac{11.9515(T - 373.15 \text{ K})}{T - 39.724 \text{ K}}\right) \tag{8.19}$$

If p_{sat} in logarithmic scale is plotted against T in linear scale, the graph of (8.19) is a hyperbola; see Fig. 8.4. At $T = 293.15$ K ($20\,^\circ$C), the saturated vapor pressure is only $0.023\,p_{atm} = 2.33$ kPa. However, for temperatures above $300\,^\circ$C it exceeds 8 MPa. Note that $p_{sat}(373.15$ K$) = p_{atm}$, which is related to the fact that 373.15 K ($100\,^\circ$C), is the boiling point of water at atmospheric pressure.

Formula (8.19) is accurate in the range of temperatures between $0\,^\circ$C and $100\,^\circ$C. For higher temperatures, up to the critical point $374\,^\circ$C (above which there is no difference between the liquid and gaseous phases of water), better accuracy is obtained if coefficients 11.9515 and 39.724 are replaced by 12.1074 and 28.665, but the difference is below 4%.

8.2.4 Kelvin Equation

The saturated vapor pressure p_{sat} corresponds to the case when the liquid and vapor are at the same pressure. In general, thermodynamic equilibrium at temperature T can be attained even for liquid water at pressure p_l and water vapor at a different pressure p_v, provided that

$$\mu_v(p_v, T) = \mu_l(p_l, T) \tag{8.20}$$

Instead of looking for the specific form of potentials μ_v and μ_l and solving the nonlinear equation (8.20), we can exploit the fact that the thermodynamic variable conjugate to the pressure is the specific volume, i.e., volume occupied by a unit mass of the substance, which is the reciprocal value of the mass density (mass of the substance occupying a unit volume). According to the state law (13.70), which will be derived in Sect. 13.5.5, the specific volume is equal to the partial derivative of specific free enthalpy with respect to the pressure p. Therefore, we can write

$$\mu_v(p_v, T) = \mu_v(p_{sat}, T) + \int_{p_{sat}}^{p_v} \frac{\partial \mu_v(p, T)}{\partial p}\, dp = \mu_v(p_{sat}, T) + \int_{p_{sat}}^{p_v} \frac{dp}{\rho_v(p, T)} \tag{8.21}$$

where p is a dummy integration variable that represents the pressure continuously changing from p_{sat} to p_v. Expressing the vapor density ρ_v according to the state equation of ideal gas (8.7) and integrating, we get

$$\mu_v(p_v, T) = \mu_v(p_{sat}, T) + \frac{RT}{M_w} \ln \frac{p_v}{p_{sat}} \tag{8.22}$$

Compared to vapor, liquid water can be considered as incompressible; i.e., ρ_l can be treated as a constant.[7] This leads to

---

[7]Of course, the density of water, ρ_l, depends on temperature, but for simplicity, we do not mark it explicitly. The same comment applies to the saturated vapor pressure, p_{sat}, in (8.21)–(8.24).

$$\mu_l(p_l, T) = \mu_l(p_{sat}, T) + \int_{p_{sat}}^{p_l} \frac{\partial \mu_l(p, T)}{\partial p} \, dp = \mu_l(p_{sat}, T) + \int_{p_{sat}}^{p_l} \frac{dp}{\rho_l} =$$

$$= \mu_l(p_{sat}, T) + \frac{p_l - p_{sat}}{\rho_l} \tag{8.23}$$

Substituting (8.22)–(8.23) into (8.20) and taking into account (8.18), we obtain the *Kelvin equation* (sometimes called the *Kelvin-Laplace equation*)

$$\frac{RT}{M_w} \ln \frac{p_v}{p_{sat}} = \frac{p_l - p_{sat}}{\rho_l} \tag{8.24}$$

Recall that the ratio between the actual partial vapor pressure p_v and the saturated vapor pressure p_{sat} is the relative humidity; see (8.1). Equation (8.24) can thus be rewritten as

$$p_l - p_{sat}(T) = \frac{RT\rho_l(T)}{M_w} \ln h, \quad \text{or} \quad h = \exp\left(\frac{M_w(p_l - p_{sat}(T))}{RT\rho_l(T)}\right) \tag{8.25}$$

This is the "rigorously derived" form of Kelvin equation, of course, with a certain small error caused by the assumptions of (i) incompressibility of liquid water and (ii) behavior of water vapor as an ideal gas.

In the literature, one can often find the Kelvin equation in somewhat different forms. For instance, the left-hand side of the first equation in (8.25) is sometimes replaced by $p_l - p_{atm}$ or by $p_l - p_g \equiv -p_c$. Also, p_{sat} in the fraction on the right-hand side of formula (8.1) defining the pore relative humidity could be understood as p_{sat}^* obtained from (8.17) rather than p_{sat} obtained from (8.18). It is good to realize that all these modified forms of Kelvin equation differ by terms that are comparable to the error induced by the simplifications needed to derive (8.25), as will be shown next. Readers not interested in detail can skip the following discussion and continue reading after Eq. (8.43).

First, let us clarify the relation between p_{sat} and p_{sat}^*. Since (8.24) holds for all pairs of pressures p_v and p_l that satisfy condition (8.20), it is applicable to $p_v = p_{sat}^*$ and $p_l = p_{atm}$, which yields

$$\ln \frac{p_{sat}^*}{p_{sat}} = \frac{M_w(p_{atm} - p_{sat})}{RT\rho_l} \tag{8.26}$$

The right-hand side of (8.26) is typically very small. To see that clearly, let us introduce auxiliary quantities

$$\rho_{atm} = \frac{M_w p_{atm}}{RT}, \qquad \rho_{sat} = \frac{M_w p_{sat}}{RT} \tag{8.27}$$

which represent the density of vapor under the atmospheric pressure and under the saturation pressure, computed from the Boyle–Mariotte state law (8.7). With this notation, Eq. (8.26) can be rewritten as

$$\ln \frac{p_{sat}^*}{p_{sat}} = \frac{p_{atm} - p_{sat}}{\rho_l} \tag{8.28}$$

and it becomes apparent that the right-hand side is very small. Indeed, in the temperature range below $100\,°C$, $0 < p_{sat} < p_{atm}$ and thus $0 < \rho_{sat} < \rho_{atm}$ and $0 < \rho_{atm} - \rho_{sat} < \rho_{atm}$. The value of ρ_{atm} computed for $T = 273$ K $(0\,°C)$ is about $0.79\ \text{kg/m}^3$, and for higher T, it is still smaller. Since the density of liquid water, ρ_l, is much higher, near $1000\ \text{kg/m}^3$, the ratio on the right-hand side of (8.28) remains below $0.8 \cdot 10^{-3}$ and the relative difference between p_{sat} and p_{sat}^* is not greater than 0.08%. The same holds for the relative difference between relative humidities defined as p_v/p_{sat} or as p_v/p_{sat}^*. So the alternative definition of the saturated vapor pressure by condition (8.17) would be acceptable at usual atmospheric temperatures (but not at high temperatures such as those induced by fire).

The derivation leading to (8.24) can be repeated with the reference equilibrium state characterized by $p_l = p_{atm}$ and $p_v = p_{sat}^*$ instead of $p_l = p_v = p_{sat}$. The resulting form of Kelvin equation is then

$$\frac{RT}{M_w} \ln \frac{p_v}{p_{sat}^*} = \frac{p_l - p_{atm}}{\rho_l} \tag{8.29}$$

Note that (8.29) is fully equivalent with (8.24) and could be obtained directly by combining (8.24) with (8.26). Equations analogous to (8.25) can be written as

$$p_l - p_{atm} = \frac{RT\rho_l(T)}{M_w} \ln h^*, \qquad \text{or} \qquad h^* = \exp\left(\frac{M_w(p_l - p_{atm})}{RT\rho_l(T)}\right) \tag{8.30}$$

in which $h^* = p_v/p_{sat}^*$ is a slightly modified pore relative humidity. If h^* is replaced in (8.30) by the standard pore relative humidity, $h = p_v/p_{sat}$, the additional error remains below 0.08% (at temperatures below $100\,°C$).

Yet another admissible approximate form of the Kelvin equation can be constructed by rewriting (8.24) as

$$\frac{RT}{M_w} \ln \frac{p_v}{p_{sat}} = \frac{p_v - p_{sat}}{\rho_l} + \frac{p_l - p_v}{\rho_l} \tag{8.31}$$

and then neglecting the first term on the right-hand side. This approximation is justified by the fact that (8.31) corresponds to the following relation involving integrals:

$$\int_{p_{sat}}^{p_v} \frac{dp}{\rho_v(p)} = \int_{p_{sat}}^{p_v} \frac{dp}{\rho_l} + \int_{p_v}^{p_l} \frac{dp}{\rho_l} \tag{8.32}$$

Since $\rho_v \ll \rho_l$, the first integral on the right-hand side is much smaller than the integral on the left-hand side and thus can be neglected, which leads to

$$\frac{RT}{M_w} \ln \frac{p_v}{p_{sat}} = \frac{p_l - p_v}{\rho_l} \tag{8.33}$$

In the literature, one can also find an interpretation of the Kelvin equation as a relation between the relative pore humidity and the capillary pressure, written as

$$p_c = -\frac{RT\rho_l(T)}{M_w} \ln h, \qquad \text{or} \qquad h = \exp\left(-\frac{M_w p_c}{RT\rho_l(T)}\right) \qquad (8.34)$$

Recall that the capillary pressure is defined as $p_c = p_g - p_l$, where p_g is the pressure of the pore gas, equal to the sum of partial pressures of vapor and dry air. If the pores contain no air, relations (8.34) directly correspond to (8.33), because then $p_g = p_v$ and $p_c = p_v - p_l$. But even if the pore gas is a mixture of dry air and vapor, the partial pressure of dry air is usually negligible compared to the capillary pressure, and (8.34) can still be used, as an approximation.

Historical note:

In the literature on partially saturated porous media, the Kelvin equation is often understood in the sense described above, i.e., as a relation between the pressures of liquid water and water vapor at thermodynamic equilibrium, or as the relation between the capillary pressure and relative humidity. For instance, Schrefler [766] or Gawin, Pesavento and Schrefler [421] presented the Kelvin equation in a form that corresponds to our formula (8.34), while Coussy [328] presented it in a form that corresponds to (8.29).

It is interesting to note though that William Thomson, who became Lord Kelvin in 1892, derived in his short paper [812] a somewhat different and less accurate equation, which would in the present notation read

$$p_v = p_{\text{sat}} - \left(\frac{1}{r_1} + \frac{1}{r_2}\right)\frac{\gamma_{gl}\rho_v}{\rho_l - \rho_v} \qquad (8.35)$$

where the term in parentheses could be replaced by $2/r$; see (8.13). The objective of Thomson [812] was to show that the pressure of vapor near a curved interface with liquid water is different from the vapor pressure near a flat interface. He considered a sufficiently large closed container filled by water and its vapor, separated by a flat interface. If a vertical capillary tube is placed inside the container, the capillary meniscus of mean radius r separating liquid water and vapor in the tube will, at equilibrium, be located at a certain height H above the flat reference surface. The verbal reasoning of Thomson [812] implies that he expressed the vapor pressure p_v and liquid pressure p_l near the meniscus as

$$p_v = p_{\text{sat}} - \rho_v g H \qquad (8.36)$$
$$p_l = p_{\text{sat}} - \rho_l g H \qquad (8.37)$$

where p_{sat} is the pressure of both liquid water and vapor at the flat interface. The difference between the pressures is thus $p_v - p_l = (\rho_l - \rho_v)g H$, and by combining this with the Laplace equation (8.12) in which $p_g = p_v$, we can express the height

$$H = \frac{2\gamma_{gl}}{(\rho_l - \rho_v)gr} \tag{8.38}$$

Finally, substituting this into (8.36) and replacing $2/r$ by $1/r_1 + 1/r_2$, as justified by (8.13), yields (8.35).

A simplifying assumption implicitly contained in the reasoning leading to the "original Kelvin equation" (8.35) was that the vapor density ρ_v was considered as constant even though the vapor pressure must actually vary from p_{sat} to p_v along the path from the flat surface to the curved meniscus. The state equation for gas was not used at all, and no integration was necessary. This is also the reason why no logarithmic term appeared in the resulting equation.

Instead of combining Eqs. (8.36)–(8.37) with the Laplace equation, as done by Thomson [812], one can use (8.37) to express the height

$$H = \frac{p_{sat} - p_l}{\rho_l g} \tag{8.39}$$

and then eliminate H from (8.36) and invoke the state equation of ideal gas for ρ_v. After simple manipulations, the resulting equation can be converted into

$$\frac{RT}{M_w}\left(1 - \frac{p_{sat}}{p_v}\right) = \frac{p_l - p_{sat}}{\rho_l} \tag{8.40}$$

This looks similar to (8.24), except that, on the left-hand side, $\ln(p_v/p_{sat})$ is replaced by $1 - p_{sat}/p_v$. If the fraction p_v/p_{sat} is close to 1, one can write

$$\ln \frac{p_v}{p_{sat}} = -\ln \frac{p_{sat}}{p_v} = -\ln\left(1 + \frac{p_{sat}}{p_v} - 1\right) \approx -\left(\frac{p_{sat}}{p_v} - 1\right) = 1 - \frac{p_{sat}}{p_v} \tag{8.41}$$

Therefore, assumptions used in the derivation of the original Kelvin equation (8.35) combined with the state law of ideal gas (and with no need to exploit the Laplace equation) lead to Eq. (8.40), which can be considered as a good approximation of the "modern Kelvin equation" (8.24), as long as p_v does not substantially differ from p_{sat}. This is exactly the case that Thomson [812] had in mind—he gave an example with $H = 13$ m, leading to $p_v = 0.999\, p_{sat}$.

The approach used by Thomson [812] can easily be adapted to cases in which p_v differs substantially from p_{sat}. Instead of writing the expression for p_v directly in the total form (8.36), one would have to consider the cumulative effect of differential increments $dp_v = -\rho_v g\, dH$ with variable ρ_v related to p_v by the state equation. After the separation of variables and integration, the corrected form of (8.36) would be

$$\ln \frac{p_v}{p_{sat}} = -\frac{M_w}{RT}gH \tag{8.42}$$

and elimination of H based on (8.39) would lead to the modern Kelvin equation (8.24).

To conclude this historical remark, let us mention that von Helmholtz [842] studied the vapor pressure near the surface of a spherical liquid droplet of radius $\hat{r}$ and obtained a relation that would in the present notation read

$$\frac{p_v}{p_{sat}} = \exp\left(\frac{2\gamma_{gl}M_w}{\hat{r}RT\rho_l}\right) \tag{8.43}$$

If the term $2\gamma_{gl}/\hat{r}$ is replaced by $p_l - p_g \equiv -p_c$, as follows from the Laplace equation (8.12) (in which r corresponds to $-\hat{r}$, because the curvatures of a droplet and of a capillary meniscus have opposite signs), then (8.43) becomes equivalent with our Eq. (8.34).

It is useful to note that the term $RT\rho_l/M_w$ has the meaning of a (fictitious) pressure that would compress vapor to the density of liquid water if the ideal gas state law were still applicable at such high pressures. It is therefore not surprising that this term is much larger than the atmospheric pressure. At 20 °C, we get $RT\rho_l/M_w$ = 135 MPa, and Kelvin equation (8.25) leads to $p_l = p_{sat} + 135$ MPa $\times \ln h$. This indicates that the magnitude of liquid pressure is usually much higher than the vapor pressure (except for pore relative humidities close to 1), because the saturated vapor pressure at 20 °C is just 2.33 kPa. To give an example, for $h = 0.9$ one gets $p_l = -14.2$ MPa and $p_v = 0.9 \times 2.33$ kPa $= 2.1$ kPa. Replacing $p_l - p_{sat}$ by p_l or by $-p_c$ then induces a negligible error.

For pore relative humidity $h = 0.5$, the liquid pressure evaluated from the Kelvin equation is $p_l = -93.6$ MPa (tension) and, according to the Young–Laplace equation (8.12), the mean radius of the capillary meniscus is $r = 1.56$ nm. This is approximately at the limit of applicability of bulk thermodynamics. Capillary tension 93.6 MPa is not much smaller than the tensile strength of water, and the radius of the meniscus is about six times the effective diameter of a water molecule (0.27 nm). Therefore, at pore relative humidities below 50%, concrete contains almost no capillary water. For details, see, e.g., Brinker and Scherer [257]. The mean pressure change in hindered adsorbed water from the saturation state is roughly equal to the capillary pressure [78], which means that microstress peaks of the order of 100 MPa are exerted locally by hindered adsorbed water onto the solid skeleton of cement gel (in more detail, see Sect. 8.2.6.2).

For $h = 0.3$, one gets $p_l = -162.5$ MPa, which is comparable to the tensile strength of liquid water on the molecular scale. So, capillary (liquid) water cannot exist at all for pore relative humidities below 30%.

The magnitudes of the capillary pressure and of the vapor pressure are comparable only for pore relative humidities very close to 1. For $h = 1$, the capillary pressure vanishes. Capillary interfaces (menisci) still exist, but they must have a zero mean curvature, which is the case for a plane or for anticlastic (saddle) surfaces with $r_2 = -r_1$. The menisci may exist even for $h > 1$, in which case $1/r_1 + 1/r_2 < 0$; see Fig. I.7 in Appendix I.2.

8.2.5   Sorption Isotherm

As already explained, the degree of liquid saturation S_l of a porous solid can be linked to the suction p_c (capillary pressure) and temperature T by a function that depends on the pore structure and pore size distribution. Kelvin equation (8.34) provides a link between the capillary pressure p_c and the pore relative humidity h (provided that the pore gas pressure is close to the atmospheric pressure, or at least negligible compared to the capillary pressure). Therefore, the degree of saturation, which is related by a certain function Φ_{pT} to the suction and temperature, can alternatively be expressed using a function Φ_{hT} of the pore relative humidity and temperature:

$$S_l = \Phi_{pT}(p_c, T) = \Phi_{pT}\left(-\frac{\rho_l RT}{M_w}\ln h, T\right) \equiv \Phi_{hT}(h, T) \qquad (8.44)$$

At the same time, the saturation degree is related to the evaporable[8] water content w_e (mass of evaporable water per unit volume of the porous material [kg/m^3]) by Eq. (8.5), which also contains densities ρ_l and ρ_v. The mass density of liquid water, ρ_l, can be considered as a constant, and the mass density of water vapor, ρ_v, can be expressed in terms of the pore relative humidity and temperature by combining the state equation for vapor (8.7) with the definition of relative humidity (8.1):

$$\rho_v = \frac{M_w p_v}{RT} = \frac{M_w p_{sat}}{RT}h \qquad (8.45)$$

Combining (8.5) and (8.44)–(8.45), we can proceed to an alternative description of moisture storage in the pore space, with the moisture retention curve (saturation-suction relation) transformed into a relation between the evaporable water content and the pore relative humidity (of course affected by temperature):

$$w_e = n_p\left[\rho_l\Phi_{hT}(h, T) + \frac{M_w p_{sat}}{RT}h\left(1 - \Phi_{hT}(h, T)\right)\right] \qquad (8.46)$$

Equation (8.46) takes into account both liquid water and water vapor. However, at moderate temperatures, the liquid water dominates. This can be demonstrated by evaluating the fraction in the second term at $T = 293$ K (20 °C):

$$\frac{M_w p_{sat}}{RT} = \frac{0.018 \times 2330}{8.31 \times 293}\frac{\text{kg}}{\text{m}^3} = 0.0172\frac{\text{kg}}{\text{m}^3} \qquad (8.47)$$

The value of $M_w p_{sat}/RT$ is seen to be much lower than the liquid water density $\rho_l \approx 1000\,\text{kg/m}^3$, which multiplies the saturation degree in the first term, representing the contribution of liquid water. Consequently, (8.46) can be simplified to

---

[8]We refer here to the **evaporable** water because the chemically bound water is considered as a part of the solid skeleton and thus does not occupy the pore space. But note that the evaporable water includes the adsorbed water, which acts mechanically as part of the solid skeleton.

$$w_e = n_p \rho_l \Phi_{hT}(h, T) \qquad (8.48)$$

At constant temperature, the relation between the evaporable water content w_e and the pore relative humidity h is known as the *sorption isotherm*. More specifically, the *desorption isotherm* describes w_e as a function of h at decreasing water content (drying), while the *adsorption isotherm* (sometimes called just the sorption isotherm) applies to increasing water content (wetting).

As illustrated in Fig. 8.5, the desorption and adsorption isotherms exhibit pronounced hysteresis, which means that they lie significantly lower for sorption than for desorption. On subsequent cycles, the area of the hysteretic loops does not vanish. Sorption–desorption reversals at intermediate h between 0 and 1 follow still different paths. Generally, for $dh < 0$, the minimum possible loss dw_e and, for $dh > 0$, the minimum possible gain dw_e take place. At high relative humidities, the hysteretic phenomena can be attributed to the nonuniqueness of the surfaces of capillary menisci of liquid water in larger pores, caused, e.g., by the ink-bottle effect [274]. At low relative humidities, capillary water does not exist, but the hysteresis is still observed. This has traditionally been explained by pore collapse but, according to Bažant and Bazant [109], such an explanation is not valid.[9] Two different phenomena suffice to explain the hysteresis:

1. nonequilibrium snap-through of water content of nanopores caused by nonuniqueness in the misfit disjoining pressures engendered by a difference between the nanopore width and an integer multiple of the thickness of a monomolecular adsorption layer; and
2. molecular condensation in flow along the nanopores, under the influence of solid surface forces.

These phenomena are discussed in more detail and from other view points in Sect. 8.2.6.3 and Appendix I.2.

Figure 8.5a shows the first desorption isotherm (solid) and the first sorption isotherm (dashed) according to the data reported by Baroghel-Bouny [55] for concrete after 1 year of sealed curing. The concrete mix was composed of 1936 kg of aggregates, 353 kg of cement, and 152 kg of water per cubic meter of concrete (water/cement ratio 0.43), and the mean compressive strength measured on cylinders at 28 days was 49.4 MPa. The isotherm was determined at 23 °C, and the reference "dry" state was considered as the state at 3% pore relative humidity (instead of the state after drying in an oven, which could alter the pore structure). The experiment started by desorption from 100% relative humidity to 3% relative humidity; see the solid curve in Fig. 8.5a. As demonstrated in Fig. 8.5b, the measured desorption isotherm is in the range between 30 and 100% relative humidity very close to a straight line. The lower dashed curve in Fig. 8.5a shows the isotherm for adsorption

---

[9]The experimental data presented, e.g., by Baroghel-Bouny [55] show that the second desorption roughly follows the first, which would be impossible if pore collapsed (how would they get rebuilt?). Also, if one would calculate the amount of shrinkage based on the idea of collapsed pores, its magnitude would be an order of magnitude higher than observed.

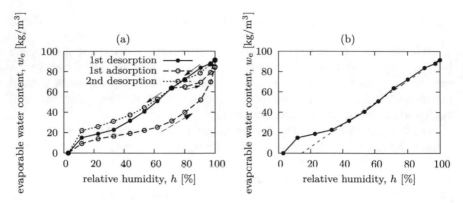

Fig. 8.5 (a) Difference between the sorption and desorption isotherms (experimental data adopted from Baroghel-Bouny [55]), (b) approximation of the desorption isotherm by a straight line

from 3 to 100% relative humidity, which is highly curved and lies below the desorption isotherm. The upper (short) dashed curve illustrates the behavior when the adsorption starts after previous desorption down to 72% relative humidity. Strong hysteresis is observed. For comparison, Fig. 8.5a also contains the second desorption isotherm (plotted by the dotted curve), which closely follows the first desorption isotherm, at least in the range from 100 to 50% pore relative humidity. This proves that "pore collapse" cannot be the cause of sorption hysteresis, since auto-rebuilding of pore walls is unimaginable.

Many specific formulae for approximation of sorption isotherms have been proposed in the literature. Some of them are described in Appendix I.1. They typically contain several parameters that need to be determined by fitting of experimental data. For practical applications, it would be very useful to predict the parameters from basic information such as the composition of the concrete mix. A model of this kind has been developed by Xi, Bažant and Jennings [883], based on the BSB model [277], which is a three-parameter generalization of the BET model [276]. The details are presented in Appendix I.1. For illustration, Fig. I.5 shows the isotherms predicted by the model of Xi et al. [883] for different ages, cement types, water-cement ratios, and temperatures. Note that these isotherms refer to adsorption (increasing humidity), are valid for cement paste only and are plotted in terms of the moisture content (or total water content) by mass.

Approximation of the measured isotherms from Fig. 8.5a by various analytical formulae is shown in Appendix I.1; see Fig. I.1. The appendix also contains another example of isotherms for concrete, based on the data reported by Ahlgren [22]; see Fig. I.2. The desorption isotherm shown there is again close to a straight line. Therefore, in most calculations of concrete drying, we will deal with linear isotherms. Such an approximation is quite good for mature concretes with low water-cement ratio (and thus low porosity).

If we focus on the description of drying, the relation between water content and humidity at constant temperature is formally described by $w_e = \phi(h)$, where ϕ is a function characterizing the desorption isotherm. It is important to bear in mind that this is not a universal function—it depends on the characteristics of the pore space, which evolve in time due to aging. For simplicity,[10] let us consider ϕ as a function of the so-called *equivalent age* t_e, representing the equivalent hydration period (to be defined in Sect. 10.6.1). The rate of growth of t_e depends on the pore relative humidity and temperature and is given by $\dot{t}_e = \beta_{eT}(T)\,\beta_{eh}(h)$; see Eq. (10.28). Dependence of the evaporable water content on pore relative humidity and equivalent age is described by

$$w_e = \phi(h, t_e) \tag{8.49}$$

Later, we will need to express the rate of the total water content, w_t, which enters the water mass balance equation (8.76). In addition to the evaporable water content, w_e, we have to consider the content of chemically bound water, w_n, which increases in time due to the hydration reactions (of course at the expense of the evaporable water) and can be described as a function of t_e. Differentiating the relation $w_t = w_e + w_n$ and taking into account (8.49) and (10.28), we obtain

$$\dot{w}_t = \dot{w}_e + \dot{w}_n = \frac{\partial \phi(h, t_e)}{\partial h}\dot{h} + \frac{\partial \phi(h, t_e)}{\partial t_e}\dot{t}_e + \frac{dw_n(t_e)}{dt_e}\dot{t}_e =$$
$$= \frac{\partial \phi(h, t_e)}{\partial h}\dot{h} + \left(\frac{\partial \phi(h, t_e)}{\partial t_e} + \frac{dw_n(t_e)}{dt_e}\right)\beta_{eT}\,\beta_{eh}(h) \tag{8.50}$$

The derivative $\partial\phi/\partial h$ represents the slope of the desorption isotherm and is called the *moisture capacity* [kg/m^3]. The *inverse slope of the isotherm*,

$$k(h, t_e) = \left[\frac{\partial \phi(h, t_e)}{\partial h}\right]^{-1} \tag{8.51}$$

is also called the *reciprocal moisture capacity* [m^3/kg]. For a linear isotherm, the moisture capacity is independent of h. For example, the straight line in Fig. 8.5b has slope $1/k = 108$ kg/m^3 and the corresponding reciprocal moisture capacity is $k = 9.26 \times 10^{-3}$ m^3/kg.

The second term on the right-hand side of (8.50) is related to the hydration reaction and represents the rate at which water would need to be supplied (per unit volume of concrete) in order to maintain constant relative humidity in the pores. Equation (8.50) can be rewritten in the inverse form

$$\dot{h} = k(h, t_e)\left[\dot{w}_t - \left(\frac{\partial \phi(h, t_e)}{\partial t_e} + \frac{dw_n(t_e)}{dt_e}\right)\beta_{eT}\,\beta_{eh}(h)\right] = k(h, t_e)\dot{w}_t + h_s^*(h, t_e) \tag{8.52}$$

where

---

[10] Since the hydration kinetics depend on humidity and temperature, a complete model should in fact deal with a coupled thermo-hygro-chemical problem.

$$h_s^*(h, t_e) = -k(h, t_e) \left(\frac{\partial \phi(h, t_e)}{\partial t_e} + \frac{dw_n(t_e)}{dt_e} \right) \beta_{eT} \, \beta_{eh}(h) \qquad (8.53)$$

is the humidity rate due to self-desiccation.

8.2.6   Free and Hindered Surface Adsorption, Disjoining Pressure, and Its Continuum Thermodynamics*

8.2.6.1   Adsorption—Free and Hindered*

In strongly hydrophilic solids such as the Portland cement paste, the pore walls are always covered by an adsorption layer and the transition to a capillary meniscus is smooth, with a virtually zero contact angle. The thickness of the adsorption layer, δ_a, depends on h and T and reaches no more than 5 molecular layers (each of them being 0.27 nm thick, which is the the effective diameter of one H_2O molecule). The layer is always in contact with the capillary water or the vapor. The water molecules constantly leave and others reenter the adsorption layer, lingering in it for a certain time called the lingering time. Generally, the adsorbed molecules linger for about 1 ns, which means they move approximately after each 10^5 thermal atomic vibrations.

The mass of free multimolecular adsorbed layer at pore humidity h and temperature T is given by the BET isotherm [276, 345, 479, 888]:

$$\frac{\Gamma_a}{\Gamma_1} = \frac{1}{1-h} - \frac{1}{1 - h(1 - C_0 e^{Q_a/RT})} \qquad (8.54)$$

This equation, statistically derived by E. Teller, is accurate for $h \in (0.05, 0.5)$; here, Γ_a = surface water concentration = mass of adsorbed water per unit surface area [kg/m^2], $\Gamma_1 \approx \rho_l \times 0.27$ nm $= 263 \times 10^{-9}$ kg/m^2 = mass of full monomolecular layer per unit surface area, Q_a = latent heat of adsorption minus latent heat of condensation of vapor (always > 0), and C_0 = constant depending on adsorption entropy ($C_0 \approx 1$). Note that the volumetric mass density of adsorbed water is close to the density of liquid water, ρ_l. The surface water concentration Γ_a is equal to $\rho_l \delta_a$, where δ_a is the equivalent thickness of the adsorbed layer (Fig. 8.1). The dimensionless fraction Γ_a/Γ_1 could be replaced by δ_a/δ_1, where $\delta_1 = 0.27$ nm. A generalization of (8.54) that includes interactions along the adsorbed layer was statistically derived by Bazant and Bažant [216].

Baroghel-Bouny [55] showed that if the experimentally determined adsorption isotherms are plotted in terms of the average thickness of the adsorbed layer (instead of the water content or moisture ratio), a universal "master curve" is obtained for different cementitious materials in the range of low pore relative humidities, up to 63%. This confirms that, in this range, multilayer surface adsorption is the dominant moisture storage mechanism. For higher humidities, capillary water becomes important and the pore size distribution (affected, e.g., by the water/cement ratio) starts playing a role. A linear relationship between w/c and the water content at the relative humidity of 90% was reported by Baroghel-Bouny [54, 55], both for adsorption and for desorption.

In pores of thickness less than 10 water molecules (2.7 nm), which are plentiful in Portland cement paste, the full thickness of free adsorption layers cannot develop. At sufficiently high relative humidities, the adsorption layer is prevented by the pore walls from attaining its full thickness $\delta_{a,\max}$ and the adsorption is hindered; see the right part of Fig. 8.1. Consequently, a transverse pressure p_d, called the *disjoining pressure*, must develop [71, 77, 348, 349]. As will be shown in Chap. 10, it plays a major role in creep of concrete.

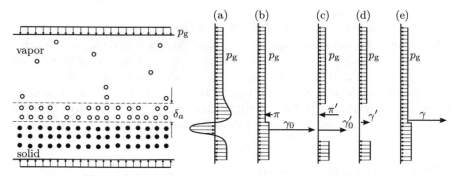

Fig. 8.6 Free adsorption and stresses in the surface phase: (a) actual distribution of longitudinal stress, (b)–(e) equivalent stress and force distributions. Note that the sum of solid surface tension and of the spreading pressure in the adsorption layer is normally the surface tension. For this reason, the tensile resultant γ_0' is shown to be greater than the compression resultant π'. The fact that the stress is tensile causes the solid to shrink. The dashed lines show the adsorption layer of thickness δ_a, which can reach up to five molecular layers

In contrast to bulk liquid water, the adsorbed layers are not in a hydrostatic stress state, which is caused by their interaction with the solid surface. If the surface effects were neglected, the solid particles as well as the adsorbed water layers could be imagined to be in mechanical equilibrium at the same hydrostatic pressure p_g as the pore gas. However, atoms and molecules near an interface such as the surface of a solid particle tend to rearrange into the energetically most favorable configuration, which leads to a stress redistribution. The *transverse stress*, i.e., the normal stress acting in the direction perpendicular to the solid–fluid interface, is typically very small, while the *longitudinal stress*, i.e., the normal stress acting in parallel to the interface, can be very large. As schematically shown in Fig. 8.6, the excess tensile longitudinal stress in the solid, lumped into a force per unit length, represents the *solid surface tension* γ_0, and the excess compressive longitudinal stress in the adsorbed layer, lumped into a force per unit length, represents the *spreading pressure*[11] π, considered as positive in compression. The sum $\gamma = \gamma_0 + (-\pi)$ is then the *surface*

---

[11] Here, the term "pressure" is a historically rooted misnomer; its dimension is not N/m², but force per unit length, N/m.

tension on the solid–liquid interface (recall that the surface tension on the gas–liquid interface, introduced in Sect. 8.2.3, was denoted as γ_{gl}). In a free adsorbed layer, the transverse stress (normal to the solid surface) is virtually unaffected, and the spreading pressure π is denoted more specifically as π_a. In a hindered adsorbed layer, the transverse stress increases by the disjoining pressure, p_d, and the spreading pressure is denoted as π_d. One can also define the *total spreading pressures π'_a* and π'_d, which are the resultants of the total stress (not only of the excess stress) across the thickness of the adsorbed layer; see Figs. 8.1 and 8.6c. Approximately, $\pi'_a = \pi_a + \delta_a p_g$ (where p_g is the pore gas pressure), if we neglect the fact that, up to the distance of 5 molecules from the solid surface, the pore gas pressure p_g is influenced by the solid surface forces.

In contrast to free adsorption, the exchange of hindered adsorbed water molecules with the capillary water and vapor is not almost immediate and may take a longer time. It requires diffusion of water molecules along the hindered adsorbed water layers, driven by the difference in chemical potentials, or spreading pressures. This diffusion was analyzed in Bažant [78] and Bažant and Moschovidis [164] and was proposed as one cause of delayed thermal expansion [72, 78]. It is doubtless also one cause of the relaxation of microprestress (Chap. 10).

8.2.6.2   Disjoining Pressure in Hindered Adsorbed Water in Nanopores*

To estimate the average nanopore width, first note that most of the adsorbate in nanoporous solids is in the form of hindered adsorption layers, i.e., layers confined in the nanopores, which are usually defined as pores less than about 3 nm wide [47]. Further note that 1 cm^3 of hardened Portland cement paste contains the internal surface of about 500 m^2, while the total porosity is typically about 50%, with the large capillary pores occupying about 15% and the nanopores about 35% of the material volume. The capillary pores contribute almost nothing to the internal pore surface, because their surface-to-volume ratio is several orders of magnitude smaller than that of the nanopores. So, if the nanopores were flat, with two opposite planar surfaces each, the average nanopore width would be about 2×0.35 cm^3/500 m^2 = 1.4 nm, which is about 5 water molecule diameters (for cylindrical nanopores, a similar calculation would give the average nanopore diameter 2.8 nm, but flat nanopores are probably closer to reality). The hindered adsorbed layers in such nanopores have no surface directly exposed to vapor and communicate with the vapor in macropores only by diffusion along the adsorption layer. It has been well known that a large transverse stress, called the *disjoining pressure* [348, 349] (aka *solvation pressure* [47]), must develop in these layers.

Development of the theory of hindered adsorption for concrete was stimulated by Powers' general qualitative ideas on the creep mechanism [703]. Its mathematical formulation for calcium silicate hydrates (C-S-H) gradually emerged in Bažant [71, 72, 77, 78] and was reviewed in a broad context in Bažant [80]. The recent advent of molecular dynamics (MD) simulations is advancing the knowledge of nanoporous solids and gels or colloidal systems in a profound way [317–319, 526, 527, 595,

673, 781, 831]. Particularly exciting have been the new results by Rolland Pellenq and co-workers at the Concrete Sustainability Hub at MIT led by Franz-Josef Ulm [246, 260, 261]. These researchers use numerical MD simulations to study sorption and desorption in nanopores of coal and calcium silicate hydrates. MD simulations have been used to study the effect of a molecular interlayer of water on the rate of sliding of two opposite walls of C-S-H [830].

MD simulations (mainly in the group of F.-J. Ulm and R. Pellenq at MIT) have recently shown that water molecules in nanopores too thin to accommodate the full thickness of the water adsorption layers exert on the walls of hydrated cement enormous forces, which show high alternating scatter between negative and positive values, although the mean force is compressive. Because of the scatter, one needs an overall model for the mean behavior, which must be based on continuum thermodynamics and has been derived long ago; see Bažant [78], with preliminary versions in [71, 77]. Continuum thermodynamics is appropriate and necessary for longitudinal mass transfer along the nanopores to and from the large capillary pores because a vast number of molecules is involved in the longitudinal direction.

Consider now an idealized pore with planar rigid adsorbent walls and a width $2l_d$ that is smaller than the combined width $2\delta_a$ of the free adsorption layers at the opposite walls given by the BET equation (8.54), in which the dimensionless ratio Γ_a/Γ_1 is equal to δ_a/δ_1, with $\delta_1 = 0.27$ nm. Then, the adsorbate has no surface in contact with the vapor and full adsorption layers cannot build up freely at the opposite pore walls, i.e., the adsorption is hindered and a disjoining pressure p_d must develop. Since the free water adsorption layer on each surface can be up to 5 molecules thick, the pores less than 10 molecules wide ($2l_d < 2.7$ nm) cannot accommodate the full thickness of opposite free water adsorption layers. This leads to hindered adsorption with disjoining pressure (at high enough relative humidity). The adsorbed water communicates by diffusion of adsorbed water molecules along the pore with the water vapor in an adjacent macropore.

Let us now derive a relation between the disjoining pressure and the relative humidity. The arguments are similar to the derivation of Kelvin's equation in Sect. 8.2.4. In a process in which thermodynamic equilibrium is maintained, the specific free enthalpies (i.e., Gibbs free energies per unit mass) of the vapor and of its adsorbate, μ_v and μ_{ad}, must remain equal, and the same holds for their increments:

$$d\mu_{ad} = d\mu_v \qquad (8.55)$$

A detailed justification has already been provided in footnote 6. Under isothermal conditions, this means that, for a nanopore filled by adsorbed water,

$$\frac{dp_d + 2\,dp_{ad}}{3\rho_{ad}} = \frac{dp_v}{\rho_v} \qquad (8.56)$$

where ρ_{ad} is the average mass density of the adsorbate (which probably is, in the case of water, somewhere between the mass densities of ice and liquid water). Factors 2 and 3 appear in (8.56) because we assume a planar pore filled by water layer under in-plane (longitudinal) biaxial stress and a different transverse stress. The disjoining pressure $p_d = 0$ as long as the nanopore is not filled because p_d is considered as the excess transverse pressure. Symbol $p_{ad} = \pi_d/l_d$ denotes the longitudinal stress in the hindered adsorption layer averaged through the layer thickness (here we assume the gas pressure p_g to be negligible, or else p_{ad} would equal π_d'/l_d); p_{ad} has the dimension of N/m^2 and (in contrast to stress) is taken positive for compression. Recall that π_d is the longitudinal spreading "pressure" [N/m] in the adsorption half-layer of thickness l_d; π_d is superposed on the solid surface tension γ_0 which is generally larger in magnitude, and so the surface tension, $\gamma = \gamma_0 - \pi_d$, is actually tensile; see Fig. 8.6b,c. Thus the decrease of spreading pressure with decreasing h causes an increase of surface tension, which is large at nanoscale globules (of typical size 10 nm) and is one of the causes of shrinkage.

Further note that if p_d and p_{ad} were equal, the left-hand side in (8.56) would be $\mathrm{d}p_d/\rho_{ad}$, which is the standard form for a bulk fluid. Also, in contrast to solid mechanics, the left-hand side in (8.56) cannot be replaced by $\varepsilon_y \, \mathrm{d}p_d + 2\varepsilon_x \, \mathrm{d}p_{ad}$ because strains ε_x and ε_y cannot be defined (since the molecules in adsorption layers migrate and the difference between p_d and p_{ad} is caused by the forces from solid adsorbent wall rather than by strains).

Based on the ideal gas equation for vapor (8.7), the vapor density ρ_v can be expressed in terms of the vapor pressure p_v, and Eq. (8.56) can be rewritten as

$$\frac{\mathrm{d}p_d + 2\,\mathrm{d}p_{ad}}{3\rho_{ad}} = \frac{RT}{M_w}\frac{\mathrm{d}p_v}{p_v} \tag{8.57}$$

Assuming constant density of adsorbed water, ρ_{ad}, integration of (8.57) leads to

$$p_d + 2p_{ad} = \frac{3\rho_{ad}RT}{M_w} \ln p_v + C \tag{8.58}$$

where C is an integration constant, which must determined from a suitable initial condition. Before the nanopore gets filled, the disjoining pressure p_d is zero. The pore relative humidity h_f at which the nanopore just gets filled can be approximately obtained from a quadratic equation representing the inverse of the BET isotherm,[12] and the corresponding vapor pressure is $p_{v,f} = h_f p_{\mathrm{sat}}$. The disjoining pressure at this state is still zero. Denoting the value of the longitudinal pressure when the nanopore just gets filled as $p_{ad,f}$, we can substitute all the values that characterize this state into (8.58) and determine constant C. Equation (8.58) can then be rewritten as

$$p_d + 2p_{ad} - 2p_{ad,f} = \frac{3\rho_{ad}RT}{M_w} \left(\ln p_v - \ln p_{v,f} \right) = \frac{3\rho_{ad}RT}{M_w} \ln \frac{h}{h_f} \tag{8.59}$$

---

[12]The calculation based on the BET isotherm is only approximate since the BET statistics is one-dimensional and inaccurate when the pore is almost full, i.e., when the adsorption layers on the opposite surfaces are almost touching, with almost no vapor between them.

where, as usual, $h = p_v/p_{sat}$ = pore relative humidity. In this form, the equation is easier to interpret, but it still contains an unknown constant, $p_{ad,f}$. The advantage compared to (8.58) is that the constant now has a direct physical meaning (unlike the original constant C).

For pore relative humidities below h_f, the disjoining pressure p_d is identically zero, and (8.59) could be used to evaluate the changes of p_{ad}. But for pore relative humidities above h_f, both p_d and p_{ad} vary, and without some additional information, it is impossible to separate their increments. As a convenient approximation, it can be assumed that the ratio of the increments of in-plane spreading pressure p_{ad} and disjoining pressure p_d, mathematically defined as

$$\kappa = \frac{dp_{ad}}{dp_d} \tag{8.60}$$

remains constant. This new parameter will be called the *disjoining ratio*. If the adsorbate were a fluid, κ would equal 1. Since it is not, $\kappa \neq 1$. The role of κ is somewhat similar to the Poisson ratio ν of elastic solids (or, more precisely, to $\nu/(1 - \nu)$). It can probably be assumed that κ is smaller but not much smaller than 1. A rigorous calculation of κ would require introducing a constitutive equation relating p_{ad} and p_d (this was attempted in [164], but led to a complex hypothetical model with many uncertain parameters). Of course, relation (8.60) can be used only when the nanopore is filled, i.e., for $h \geq h_f$. For an unfilled nanopore, the changes in disjoining pressure are zero.

Assuming that the disjoining ratio κ is a known constant and cumulating the increments of p_d and p_{ad} with respect to the state at which the nanopore just gets filled (and at which $p_d = 0$ and $p_{ad} = p_{ad,f}$), we obtain

$$p_{ad} - p_{ad,f} = \kappa p_d \qquad \text{for } h \geq h_f \tag{8.61}$$

Based on this relation, we can eliminate p_{ad} from (8.59) and construct a formula for the disjoining pressure,

$$p_d = \frac{3\rho_{ad}RT}{(1 + 2\kappa)M_w} \ln \frac{h}{h_f} \qquad \text{for } h \geq h_f \tag{8.62}$$

If $\kappa = 1$ (and if an additional term p'_f is added), Eq. (8.62) coincides with Eq. 29 derived in 1972 by Bažant [78], but $\kappa = 1$ was an oversimplification.

Equation (8.62) gives the disjoining pressure caused by longitudinal interactions along the nanopore with the capillary macropore. The actual transverse pressures in the nanopores are statistically scattered, and formula (8.62) gives the average disjoining pressure. Superposed on this average must be transverse pressures due to accommodating within the nanopore discrete layers of water molecules. Due to this discreteness and small number of layers across the width, such additional pressures are scattered and cannot be calculated by thermodynamics. They require discrete analysis, which is pursued next.

8.2.6.3 Hysteresis Due to Snap-Through Instability in Nanopore Filling*

The BET isotherm, as well as Eq. (8.62) for hindered adsorption, is perfectly reversible. However, even though the adsorbed water represents most of the evaporable pore water for humidities below 80%, the sorption–desorption isotherms are observed to be highly irreversible, and even in the low humidity range [55, 372, 382, 705, 717], MD simulations indicate the same [246, 260, 261]. This irreversibility has been a perplexing feature for more than half a century. Some explanations have been proposed, but they were intuitive and in conflict with other observations. For example, it was speculated that water desorption (or drying) causes pore collapse. However, if formulated mathematically, this would require the shrinkage to be at least an order of magnitude higher than observed, and it would also conflict with the recent observation that water sorption (or wetting) is irreversible upon second drying, too (it is inconceivable that collapsed pores would rebuild themselves during wetting).

The thermodynamic equation (8.62) describes the pressure due to continuum fluid-type interaction along the pore. However, it cannot capture the additional transverse forces due to the discreteness of molecular layers confined in nanopores [109]. This discreteness provides a theoretically consistent explanation of the sorption hysteresis at low humidities. To explain, consider Fig. 8.7, which shows the curve of interatomic pair potential Φ (e.g., the Lennard-Jones potential), and the corresponding curve of interatomic force $F = d\Phi/ds$, as a function of distance s between the neighboring atoms.

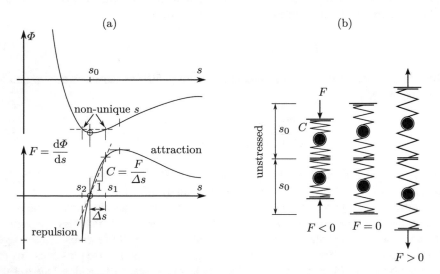

Fig. 8.7 (a) Interatomic pair potential Φ and the corresponding interatomic force F and secant stiffness C, (b) interatomic forces between opposite pore walls visualized by springs

A noteworthy point is that, for a given value of Φ (shown by the dashed horizontal line in Fig. 8.7a), there generally exist two different equilibrium states (s_1, F_1) and (s_2, F_2), as shown in Fig. 8.7a. Since the adsorbed water can behave like a fluid, the transition from one of these states to the other can take place with no supply and no loss of energy (in a solid this is not possible). This property is one of the causes (aside from molecular coalescence along the nanopores [216]) of sorption hysteresis in narrow nanopores.

It is instructive to consider one or two adsorbate molecules (standing for one or two molecular layers) between two walls representing the opposite nanopore walls, as shown in Fig. 8.7b. The interatomic forces can be imagined as the forces in springs connecting the atoms (or molecules) and the pore walls. For the state in Fig. 8.7b–right, the pore is too wide for one molecule, and thus, the springs transmit tension ($F > 0$). For the state in Fig. 8.7b–left, the same pore is too narrow for two molecules, and thus, the springs transmit compression ($F < 0$). The point is that if the energies are for both states equal, a molecule can be added in Fig. 8.7b–right or removed in Fig. 8.7b–left with no energy loss nor gain. This is possible since adsorbed water molecules are a fluid. They linger in the same position for about 10^{-9} s (which means that there is one interatomic bond break per 10^5 thermal vibrations of atoms, since the atomic thermal vibration frequency is about 10^{14}/s). Then, the atoms jump to another position, 10^9 times per second. Thus, the transition from the compressed to the tensile state, which corresponds to the horizontal dashed line in Fig. 8.7a, can occur at no change of energy, i.e., no work done, no energy obtained. So it must occur spontaneously. In fact, the transition must occur dynamically, in a manner that is analogous to snap-through of flat arches or shells.

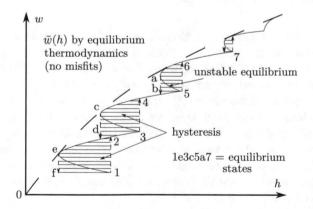

Fig. 8.8 Dynamic snap-throughs of adsorbate content

Based on this observation, one can show [109] that, during the filling of a pore of variable thickness, the transition from one molecular layer to two layers, or from two to three, etc., occurs at no energy change and spontaneously; see Fig. 8.8. The consequence for the adsorption/desorption isotherm of specific evaporable water content w_e versus relative humidity h is that the smooth curve obtained according to thermodynamics is modified by the superposed wavy curve shown in Fig. 8.8. When the nanopores are drying, the path 76ab4cdef must be followed, and when they are wetting, a different path, namely path f1234567, must be followed. The curved segments of these paths represent stable equilibrium states, while the vertical segments represent dynamic jumps, analogous to jumps called "snap-through" in the theory of structural stability. Thus, it is clear that the paths of drying and wetting must be different—hence the hysteresis in the filling of the narrowest nanopores.

What is the relation to continuum thermodynamics? It accounts only for interactions along the pore. The disjoining pressure derived from these interactions, given by Eq. (8.62), is then modified by transverse interactions which are discrete and not describable by continuum thermodynamics and produce forces added to the disjoining pressure obtained from thermodynamics.

This kind of hysteresis gets weaker as the nanopore gets wider. It disappears for nanopores more than about 10 water molecules wide. But there is another mechanism of sorption hysteresis which operates even for much wider nanopores—the molecular condensation with attractive lateral interactions described by Cahn–Hilliard equation for gradient energy. It is similar to the ink-bottle effect [274] on the capillary (or micrometer) scale; for detail, see [216].

8.3 Moisture Transport

8.3.1 Transport Mechanisms

As transpires from the previous section, the state of pore fluids in concrete can be characterized by the temperature, T, liquid water pressure, p_l, pore gas pressure, p_g, and partial pressure of water vapor, p_v. The partial pressure of dry air and the capillary pressure are easily expressed as $p_a = p_g - p_v$ and $p_c = p_g - p_l$, while the spreading pressures and the disjoining pressure may be considered to be uniquely related to p_v because local thermodynamic equilibrium may be assumed to prevail macroscopically at all times. For the same reason, the liquid pressure p_l is linked to the vapor pressure p_v and temperature T by Kelvin equation (8.24). Thus, in general, three variables remain as primary unknowns. In general, they vary in space and in time, and their evolution must be determined from equations that take into account the transport of heat and mass.

In the present chapter, we restrict attention to the transport of pore fluids under isothermal conditions, with the temperature playing the role of a fixed parameter T_0 (uniform in space and constant in time). Extensions to variable temperature will be discussed in Chap. 13. At isothermal conditions, two independent unknown pressure fields, e.g., p_l and p_g, must be determined from two mass balance equations (written for moisture and for dry air) combined with suitable constitutive equations that describe the relevant transport mechanisms. In the present context, the following specific mechanisms can be distinguished:

1. molecular diffusion (ordinary diffusion),
2. effusion (Knudsen diffusion),
3. surface diffusion,
4. advective water flow,
5. advective flow of pore gas (wet air).

The first two mechanisms correspond to transport of water vapor, which is also affected by mechanism 5. Surface diffusion (mechanism 3) is a process of transport of molecules adsorbed at the pore walls, which can be considered either as a special phase, or as a part of the liquid phase (which is the approach followed here). Mechanism 4 corresponds to transport of liquid water.

Molecular diffusion is driven by the gradient of molar concentration of vapor in the pore gas. It dominates in capillary pores with diameters substantially exceeding the mean free path of water vapor molecules (which is 80 nm at 25 °C). In pores smaller than 50 nm and especially in nanopores, the collisions against pore walls provide the main diffusion resistance, and this case is known as the *Knudsen diffusion*. The diffusivity decreases with decreasing pore size. At low humidity, most of the water is adsorbed in one or several layers at the pore walls, and *surface diffusion* with molecules of water climbing along the walls (or moving in the manner of "random walk") can become dominant. On the other hand, at high saturation degrees, the dominant mechanism is the *advective flow of liquid water* driven by the gradient of its pressure (more precisely, by the gradient of the total head, as will be explained in Sect. 8.3.2), similar to water flow in fully saturated porous media. The *advective flow of pore gas* is driven by the gradient of gas pressure.

The following two subsections present separately the main ingredients needed by a simple but complete transport model. Section 8.3.2 introduces the Darcy law that governs advective flow as the dominant transport mechanism in a fully saturated porous medium. Section 8.3.3 is devoted to the mass balance equation for moisture (all phases of water combined). In Sect. 8.3.4, simple models for moisture transport in saturated and partially saturated concrete are constructed by combining the mass balance equation and the transport law (in the Darcy form for the saturated case, or in another form proposed by Bažant and Najjar [166] for the partially saturated case) with the state equation of water for the saturated case or with the desorption isotherm for the partially saturated case. For each case, we obtain a single partial differential equation with one primary unknown field. A broader description of alternative and more general transport models for porous materials can be found in Appendix J.

8.3.2 Darcy's Law

For transport of liquid water in a fully saturated porous medium, it may seem natural to assume that the driving force is the gradient of liquid pressure p_l. However, in some applications, the gravity forces may play an important role and their contribution needs to be subtracted from the liquid pressure before the gradient is computed. Note that the hydrostatic pressure in water contained in a reservoir and subjected to gravity forces increases linearly with the depth below the free surface (gas–liquid interface) but, despite the existence of a pressure gradient, the water is in a state of static equilibrium. The pressure variation is balanced by the body forces due to the self-weight, and there is no reason to expect a flow induced by the pressure gradient. A similar reasoning can be applied to a fully saturated porous medium. Therefore, it can be expected that flow is related to the deviation of the actual pressure from the gravity induced hydrostatic pressure.

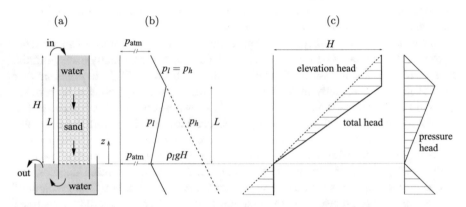

Fig. 8.9 Darcy's filtration experiments: (a) scheme of the experimental setup, (b) distribution of pressures p_h and p_l along the height, (c) distribution of the total head, elevation head, and pressure head along the height

In his classical experiments, Darcy [337] measured the flow of water through vertical pipes filled by sand. The experimental setup is schematically shown in Fig. 8.9a. Darcy found the steady-state flow rate to be proportional to the total height of the water column, $H = h_1 - h_2$, and inversely proportional to the height of the sand column, L. The flow rate can be characterized by the *filtration velocity* (or Darcy velocity), v_l, which represents the magnitude of the volume flux of liquid water and is defined as the volume of liquid passing across the section of a unit area (of the porous material, including the solid skeleton) per unit of time [$m^3/m^2s=m/s$]. The experimentally found relation is then written as

$$v_l = K_h \frac{H}{L} \tag{8.63}$$

where K_h is the *hydraulic conductivity* or *hydraulic permeability* [m/s]. This quantity has a direct physical meaning—it represents the filtration velocity at which a layer of water (bounded by two horizontal free surfaces that are kept at atmospheric pressure) subjected to gravity would pass vertically through the porous medium. For this reason,

K_h is sometimes called the *characteristic filtration velocity* or *filtration coefficient.* Note that the filtration velocity is smaller than the actual average velocity at which the water is passing through the pores, because the flux v_l is taken per unit total area and not per unit area of the pores.

To postulate the general form of Darcy's law, let us analyze the distribution of pressure in the filtration experiment. Figure 8.9b shows the liquid pressure p_l as a function of the vertical coordinate z, which is measured from the bottom surface of the sand sample and is positive above that surface. The dashed line corresponds to the distribution of the hydrostatic pressure that would build up if the bottom section of the pipe were closed. This pressure distribution is given by $p_h(z) = p_{atm} + \rho_l g (H - z)$, where p_{atm} is the atmospheric pressure (pressure of the ambient air, equal to the liquid pressure at the planar free surface), ρ_l is the mass density of the liquid, and g is the gravity acceleration. When the bottom section of the pipe is open and water can freely flow out, the liquid pressure distribution in the reservoir remains the same, but in the pore space inside the pipe, it is redistributed until a linear distribution indicated by the solid line is attained. For $0 \le z \le L$, the liquid pressure during steady-state flow is given by $p_l(z) = p_{atm} + \rho_l g (H - L)z/L$. As already justified, the flow is driven by the gradient of the difference between the actual pressure p_l and the hydrostatic pressure p_h. Since both pressures depend only on the vertical coordinate z, this gradient is a vertically oriented vector, and its magnitude is in the porous sample evaluated as

$$\frac{d[p_l(z) - p_h(z)]}{dz} = \rho_l g \frac{d}{dz} \left[\frac{(H-L)z}{L} - H + z \right] = \rho_l g \frac{H}{L} \tag{8.64}$$

This means that the norm of the pressure difference gradient is the ratio H/L that appears in the "original" Darcy law (8.63) multiplied by a factor $\rho_l g$, which represents the specific weight of water. To be able to identify H/L directly with the gradient of a certain variable, we divide the pressures by $\rho_l g$. This motivates the definition of the *total head*

$$H_t(z) = H + \frac{p_l(z) - p_h(z)}{\rho_l g} = \frac{p_l(z) - p_{atm}}{\rho_l g} + z \tag{8.65}$$

The added constant H has no influence on the gradient and has been included to get a more direct physical meaning of the resulting quantity (Fig. 8.9c). On the right-hand side of (8.65), the second term is simply the height of the point above the reference plane and the first term is the height of the water column that would lead to hydrostatic pressure p_l. Thus, the sum of these two terms, $H_t(z)$, represents the level to which water would rise in a fictitious pipe[13] connected to the current point z, as illustrated in Fig. 8.9. Note that in a water reservoir, we would have $p_l(z) = p_h(z)$ and $H_t(z) = H = $ const., which confirms that no water flow is expected if the gradient of H_t vanishes. The first term on the right-hand side of (8.65) is called the

---

[13]The fictitious pipe that illustrates the physical meaning of the total head is considered as sufficiently wide, so that the capillary pressure induced by the pipe itself be negligible.

pressure head (or *hydraulic head*, or *water potential*), denoted as H_p, and the second term is the *elevation head*; see Fig. 8.9c.

Since the factor H/L in (8.63) corresponds to the (magnitude of the) gradient of the total head, Darcy's law in its more general form can be written as

$$v_1 = -K_h \nabla H_t \tag{8.66}$$

where v_1 is the filtration velocity vector (or volumetric water flux), and ∇ (pronounced "nabla") is a differential operator which, in the present context, provides the gradient of the scalar field on which it is applied (the components of ∇ are partial derivatives with respect to individual spatial coordinates).

Of course, in a general situation (different from the Darcy filtration experiment), the liquid pressure p_l and thus also the pressure head

$$H_p = \frac{p_l - p_{atm}}{\rho_l g} \tag{8.67}$$

depend on all spatial coordinates. The same is true for the total head H_t, and so the filtration velocity vector v_1 can have a general direction. The elevation head still depends on the vertical coordinate only, and its gradient is a unit vector e_z in the direction opposite to the gravity acceleration vector. We can thus write the general form of (8.65) as

$$H_t(x) = H_p(x) + z \tag{8.68}$$

The filtration velocity v_1 represents the volume flux (volume of liquid water passing through unit area per unit of time), and its multiplication by the mass density ρ_l provides the *mass flux* of liquid water, $j_l = \rho_l v_1$ [kg/m²s]. Combining this with (8.65) and (8.66), we obtain an equivalent form of the Darcy law, expressed in terms of the mass flux and pressure gradient instead of the volume flux and total head gradient:

$$j_l = \rho_l v_1 = -\rho_l K_h \nabla \left(\frac{p_l - p_{atm}}{\rho_l g} + z \right) = -\frac{K_h}{g} \nabla p_l - \rho_l K_h e_z \tag{8.69}$$

Note that the density of liquid water, ρ_l, has been treated as a constant.

The second term on the right-hand side of (8.69) is important in geotechnical problems, when the porous body of interest has large dimensions and the hydrostatic pressure caused by the self-weight of the fluid is nonnegligible compared to the actual pore pressure. For concrete members of usual sizes, it can be omitted, but it would need to be kept for large structures such as dams. In general, this term is negligible if the vertical dimension of the investigated structure or structural member is much smaller than the pressure head. To get an idea, the pressure head corresponding to the excess pressure of 1 MPa is about 100 m.

For fully saturated porous media, the Darcy law can be derived by linearizing the fluid momentum balance equation under certain simplifying assumptions

[469]. According to Gray and Hassanizadeh [436], an additional term (accounting for changes in energy due to changes in saturation) would need to be added in the partially saturated case, but such an extended form of the law would be hard to calibrate. A phenomenological generalization of the Darcy law to partially saturated porous media, which is referred to as the Darcy–Buckingham law, is presented in Appendix J.2, see equation (J.11).

The hydraulic permeability K_h introduced in (8.63) has a direct physical meaning and is relatively easy to measure, but it depends not only on the porous material in which the flow takes place but also on the fluid. Darcy's law has been presented here for liquid water flow, because this is the problem for which it was originally postulated. However, the same type of law can be used for advective flow of other fluids, e.g., of air.

Therefore, it is useful to separate the influence of the pore geometry from that of the pore fluid. An incompressible Newtonian fluid can be characterized by its mass density ρ_f and dynamic shear viscosity η_f (the proportionality factor between the shear stress and the shear strain rate). Except for high temperature and some cases of thin bodies, the flow velocity is low enough for the flow to be laminar. For an idealized model of a pore as a cylindrical pipe of radius r, the Stokes equations governing the flow can be solved analytically and the mean velocity (averaged over the cross section of the pipe) is found to be equal to the negative derivative of total head along the pipe multiplied by $r^2 \rho_f g / 8\eta_f$, i.e., to be proportional to the square of the pore radius and to the mass density of the fluid, and inversely proportional to the dynamic viscosity of the fluid [357]. To get the corresponding filtration velocity, the mean velocity in the pipe must be multiplied by the porosity, because the filtration velocity is defined as volume flux per unit total area of the porous medium (including the solid skeleton, in which there is no flow).

It is natural to expect that similar relations will be valid for a general porous medium. Based on dimensional analysis, the hydraulic permeability can be expressed as

$$K_h = \frac{K_0 \rho_f g}{\eta_f} \qquad (8.70)$$

where K_0 is the *intrinsic permeability* [m^2], which is a property of the pore space only, independent of the properties of the pore fluid. With the gravity effect neglected, Eq. (8.69) is then rewritten as

$$\boldsymbol{j}_l = -\frac{K_0 \rho_l}{\eta_l} \nabla p_l \qquad (8.71)$$

where subscripts l indicate that we consider the flow of a liquid. An analogous equation

$$\boldsymbol{j}_g = -\frac{K_0 \rho_g}{\eta_g} \nabla p_g \qquad (8.72)$$

can be set up for the flow of a pore gas (subscript g). The driving force is now the gradient of the gas pressure.

The intrinsic permeability K_0 is proportional to the square of a characteristic pore size. For concrete, its typical values are in the range from 10^{-21} to 10^{-16} m^2 [328]. For instance, Baroghel-Bouny et al. [57] obtained $K_0 = 3 \times 10^{-21}$ m^2 for concrete with water/cement ratio 0.48 and compressive strength 49.4 MPa, and $K_0 = 0.5 \times 10^{-21}$ m^2 for high-strength concrete with water/cement ratio 0.26 and compressive strength 115.5 MPa, while Gawin et al. [420] used $K_0 = 100 \times 10^{-21}$ m^2 for concrete with water/cement ratio 0.45 and compressive strength 49.4 MPa, and $K_0 = 0.5 \times 10^{-21}$ m^2 for cement paste with water/cement ratio 0.37. Powers et al. [707] measured the evolution of hydraulic permeability K_h of cement paste with water/cement ratio 0.7, and they obtained approximately 10^{-10} m/s at the age of 7 days and 10^{-12} m/s at the age of 28 days, which corresponds to intrinsic permeabilities K_0 of 10^{-17} and 10^{-19} m^2, respectively.

The advantage of the separation of geometrical effects from material properties of the fluid is that if the intrinsic permeability is determined from measurements of flow of a specific fluid, the hydraulic permeability for any other fluid (of known density and viscosity) can be calculated without the need for additional experiments. For liquid water, the dynamic viscosity is approximately $\eta_l = 10^{-3}$ kg/m $\cdot$ s, and the density is approximately $\rho_l = 1000$ kg/m^3. The factor $\rho_l g / \eta_l$ is thus 10^7/m $\cdot$ s. For dry air, the dynamic viscosity is approximately $\eta_a = 1.8 \cdot 10^{-5}$ kg/m $\cdot$ s, and the density at atmospheric pressure and room temperature is $\rho_a = 1.2$ kg/m^3, which gives $\rho_a g / \eta_a = 0.667 \times 10^6$/m $\cdot$ s. Consequently, for a given porous medium, the hydraulic permeability to water is 15 times larger than the hydraulic permeability to air (provided that the air is under atmospheric pressure). Of course, here, we are comparing the flow of liquid water in a fully saturated medium with the flow of air in a perfectly dry medium.

The terminology related to transport properties is not fully unified and may partially depend on the specific field of application (hydrology, geotechnical engineering, building physics, concrete structures). The law that describes the dependence of a flux on a gradient can be written in 6 different formats, because (i) the flux can be considered as the volume flux or mass flux and (ii) the gradient operator can be applied on pressure, total head, or relative humidity. For each of these formats, the proportionality coefficient corresponds to a certain kind of "permeability." To avoid confusion, it is always advisable to check the units in which the permeability measure is expressed, because the units uniquely identify the specific meaning of that parameter. All the theoretically possible combinations of flux and gradient are summarized in Table 8.1. In this book, the following three permeability measures are used:

- the *hydraulic permeability* (also called hydraulic conductivity or characteristic filtration velocity), K_h, measured in m/s, which links the volume flux to the gradient of total head;
- the *permeability coefficient*, K_h/g, measured in s, which links the mass flux to the gradient of pressure; and

Table 8.1 Possible forms of the relationship between flux and gradient, with the corresponding units in which the proportionality coefficient is measured

		Gradient of		
		Pressure $\frac{Pa}{m} = kg \cdot m^{-2} \cdot s^{-2}$	Total head $\frac{m}{m} = 1$	Relative humidity $\frac{1}{m} = m^{-1}$
Volume flux	$\frac{m^3}{m^2 s} = m \cdot s^{-1}$	$kg^{-1} \cdot m^3 \cdot s$	$m \cdot s^{-1}$	$m^2 \cdot s^{-1}$
Mass flux	$\frac{kg}{m^2 s} = $ $kg \cdot m^{-2} \cdot s^{-1}$	s	$kg \cdot m^{-2} \cdot s^{-1}$	$kg \cdot m^{-1} \cdot s^{-1}$

- the *moisture permeability*, c_p, measured in $kg/m \cdot s$, which links the mass flux to the gradient of relative humidity.

The permeability coefficient is the factor multiplying ∇p_l in (8.69), which is equal to the hydraulic permeability divided by gravity acceleration. Therefore, no special symbol for this coefficient is needed—it will be denoted as K_h/g.

It has been shown that the hydraulic permeability to liquid water is about 15 times higher than the hydraulic permeability to air (for the same concrete). This may sound surprising, since intuitively one expects that air "passes more easily" through a given porous medium than water does. The clue is in the physical meaning of hydraulic permeability, which is also called the characteristic filtration velocity. It is the filtration velocity at which a layer of the fluid would pass vertically through the medium due to gravity forces, provided that the pressure in the fluid is uniform. The density of air is about 800 times smaller than that of water, and thus, the filtration velocity of air would be lower, despite the fact that water has about 55 times higher dynamic viscosity than air. Here, we are comparing the volume fluxes under a given gradient of total head.

On the other hand, if we compare horizontal volume fluxes under a given pressure gradient, the hydraulic permeability needs to be divided by the weight of fluid per unit volume and the resulting proportionality coefficient K_0/η_a characterizing air flow would be 55 times higher than the coefficient K_0/η_l characterizing water. This means that if a porous wall is subjected at one face to a higher pressure of the surrounding fluid than at the other face, the volume of the fluid passing through the wall is 55 times higher if the fluid is air than if it is water.

8.3.3 Mass Balance Equation

Consider the transport of a certain substance (such as water or air) in a porous medium. Looking at the mass flux across the boundaries of a small control volume inside the medium, we can construct the mass balance equation, which describes mass

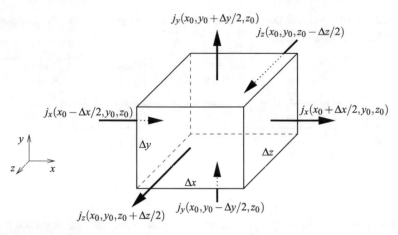

Fig. 8.10 Three-dimensional flux into and out of a prismatic control volume

conservation of the considered species and serves as the fundamental field equation for the transport problem. In the absence of internal mass sources or mass sink, the mass conservation law requires that the mass contained in the volume increases by the net flux into the volume, integrated over the corresponding time interval.

Let us examine a small prismatic control element of sides Δx, Δy, Δz and volume $\Delta V = \Delta x\, \Delta y\, \Delta z$, centered at point (x_0, y_0, z_0), as shown in Fig. 8.10. The mass that **enters** the control volume during time Δt through the left face, which is perpendicular to the x-axis and has area $\Delta y\, \Delta z$, can be expressed as $j_x(x_0 - \Delta x/2, y_0, z_0)\, \Delta y\, \Delta z\, \Delta t$, where j_x is the x-component of the mass flux vector $\boldsymbol{j}$. On the opposite face, the mass that **leaves** the control volume during time Δt is $j_x(x_0 + \Delta x/2, y_0, z_0)\, \Delta y\, \Delta z\, \Delta t$. To get the net mass **gain** in the control volume, we subtract the outward from the inward flux, which gives

$$[j_x(x_0 - \Delta x/2, y_0, z_0) - j_x(x_0 + \Delta x/2, y_0, z_0)]\Delta y\, \Delta z\, \Delta t =$$
$$= -\frac{j_x(x_0 + \Delta x/2, y_0, z_0) - j_x(x_0 - \Delta x/2, y_0, z_0)}{\Delta x}\Delta V\, \Delta t \qquad (8.73)$$

In the limit $\Delta x \to 0$, the fraction in (8.73) becomes the partial derivative $\partial j_x/\partial x$ evaluated at point (x_0, y_0, z_0). Similar expressions can be constructed for the contribution of the outward and inward fluxes in the y and z directions (Fig. 8.10). Summing the net gains and taking the limit for Δx, Δy, Δz, and Δt approaching zero, one finds that the total mass gain per unit volume and unit time is

$$-\left(\frac{\partial j_x}{\partial x} + \frac{\partial j_y}{\partial y} + \frac{\partial j_z}{\partial z}\right) \equiv -\nabla \cdot \boldsymbol{j} \qquad (8.74)$$

where $\nabla \cdot \boldsymbol{j}$ is the *divergence*[14] of the mass flux.

To keep the presentation simple and easy to follow, the derivation of the mass balance equation has been presented here in a somewhat sloppy fashion (from the mathematical point of view). If the dimensions of the control volume are considered as finite, the mass passing the left face per unit of time would not be exactly equal to $j_x(x_0 - \Delta x/2, y_0, z_0)\, \Delta y\, \Delta z$; it would differ from this value by higher-order terms that vanish in the limit of $\Delta y \to 0$ and $\Delta z \to 0$.

An alternative derivation from a broader principle can be based on the divergence theorem (Gauss theorem). Consider a control volume V of an arbitrary shape, bounded by a surface denoted as S. The mass (per unit time) leaving the control volume through an infinitesimal part of the boundary $\mathrm{d}S$ with outward normal $\boldsymbol{n}$ is given by $\boldsymbol{j} \cdot \boldsymbol{n}\, \mathrm{d}S$. Integrating over the entire boundary S, we obtain the net loss per unit time. Based on the divergence theorem, the surface integral can be transformed into a volume integral:

$$\int_S \boldsymbol{j} \cdot \boldsymbol{n}\, \mathrm{d}S = \int_V \nabla \cdot \boldsymbol{j}\, \mathrm{d}V \qquad (8.75)$$

Shrinking the control volume to a point, we find that the net loss of mass per unit volume and unit time is $\nabla \cdot \boldsymbol{j}$. The net gain has of course the opposite sign.

The mass balance equation has a particularly simple form for those substances that are not affected by phase changes or chemical reactions. This is the case for dry air and for the total moisture (which includes all phases of water, both evaporable and chemically bound). In contrast to that, the mass balance equation for water vapor only (or for the pore gas, consisting of dry air and water vapor) needs to contain a source/sink term reflecting evaporation and condensation. The mass balance equation for evaporable water needs to contain a sink term reflecting hydration.

Let us now construct the specific form of the *mass balance equation for moisture*. The mass of moisture per unit volume corresponds to the total water content, w_t. Mass conservation requires that the net gain given by (8.74) be equal to the time derivative of the total water content. If the total mass flux of moisture is denoted as $\boldsymbol{j}_w$, the resulting equation reads

$$\dot{w}_t = -\nabla \cdot \boldsymbol{j}_w \qquad (8.76)$$

For further developments, it is convenient to split w_t into the sum of w_e and w_n and express w_e according to (8.5). The balance equation (8.76) is then presented in the form

$$\frac{\mathrm{d}}{\mathrm{d}t}\left[\rho_l n_p S_l + \rho_v n_p (1 - S_l)\right] + \dot{w}_n = -\nabla \cdot \boldsymbol{j}_w \qquad (8.77)$$

---

[14]The dot $\cdot$ between ∇ and $\boldsymbol{j}$ in (8.74) denotes the contraction (scalar product) of two vectors (first-order tensors), which produces a scalar. Without this dot, $\nabla \boldsymbol{j}$ would be interpreted as the gradient of the flux, which is a second-order tensor. In engineering (or matrix) notation, the divergence is written as $\nabla^T \boldsymbol{j}$ where $\nabla^T = \{\partial/\partial x, \partial/\partial y, \partial/\partial z\}$ is a row matrix of differential operators.

An analogous mass balance equation can be set up for dry air; see equation (J.22) in Appendix J.4.

8.3.4 Differential Equations for Moisture Transport

Having prepared all the necessary ingredients, we can combine the basic equations and construct the governing partial differential equation(s) in terms of the primary unknown field(s).

A great variety of models for moisture transport in porous media have been proposed in the literature. Their complete description and detailed discussion is out of scope of the present treatise. For illustration, Table 8.2 shows a selection of models with indication of the gradients that were considered as driving the moisture transport mechanisms. Of course, this table is by far not complete; it only gives a flavor of the multitude of approaches found in the literature.

Table 8.2 Selected models for moisture transport and the gradients driving transport mechanisms in porous materials (concrete where marked)

Publication	∇w_e	∇p_l	∇p_g	∇p_v	∇p_c	$\nabla\left(\dfrac{p_v}{p_g}\right)$	∇h	∇H_t
Richards [724]								•
Krischer [553]	•			•				
Lykov [587]	•							
Pihlajavaara [693], concrete	•							
Bažant and Najjar [165], concrete							•	
Pedersen [672], Nicolas [656]				•	•			
Künzel [556]				•	•			
Gawin, Majorana and Schrefler [414], concrete		•	•		•	•		
Hagentoft et al. [444]				•	•			
Moonen [636]		•	•	•				
Coussy [328]		•	•			•		
Meftah, Pont and Schrefler [624]		•		•				

8.3.4.1 Water Transport in Fully Saturated Concrete

A simple model for liquid water transport in fully saturated concrete based on the water mass balance equation combined with the Darcy law, can be considered as a special case of the model proposed by Richards [724] and often used in soil mechanics; see Appendix J for more detail.

At full saturation (i.e., for $S_l = 1$ and $\boldsymbol{j}_w = \boldsymbol{j}_l$), the mass balance equation (8.77) reduces to

$$\frac{\mathrm{d}}{\mathrm{d}t}\left(\rho_l n_p\right) + \dot{w}_n = -\nabla \cdot \boldsymbol{j}_l \qquad (8.78)$$

On the right-hand side, the water flux $\boldsymbol{j}_l$ can be expressed using the Darcy law (8.69). Noting that $\nabla \cdot \nabla = \nabla^2 = $ Laplace operator, and expanding the time derivative on the left-hand side, we get

$$\dot{\rho}_l n_p + \rho_l \dot{n}_p + \dot{w}_n = \frac{K_h}{g}\nabla^2 p_l + K_h \frac{\partial \rho_l}{\partial z} \qquad (8.79)$$

The liquid water density ρ_l is sometimes considered as constant (e.g., in the derivation of Kelvin equation). In the present context, it is important to take into account the compressibility of water. The temperature is assumed to be constant, and so the linearized state equation of water can be written as

$$\rho_l(p_l) = \rho_{l0}\left(1 + \frac{p_l - p_{\mathrm{atm}}}{K_l}\right) \qquad (8.80)$$

where K_l is the bulk modulus of water and ρ_{l0} is the water density at atmospheric pressure. Substituting (8.80) into (8.79), we obtain, after a simple rearrangement,

$$\dot{p}_l = C_l \nabla^2 p_l + \frac{K_h}{n_p}\frac{\partial p_l}{\partial z} - \frac{K_l}{\rho_{l0} n_p}\dot{w}_h \qquad (8.81)$$

where

$$C_l = \frac{K_l K_h}{\rho_{l0} n_p g} \qquad (8.82)$$

is the *diffusivity of liquid water in saturated concrete* [m²/s], and

$$\dot{w}_h = \dot{w}_n + \rho_l \dot{n}_p \qquad (8.83)$$

is the rate of water deficiency due to hydration, consisting of two contributions with opposite signs. Note that the porosity rate $\dot{n}_p$ is negative. The loss of water due to chemical reactions (first term, positive) is partially compensated by the reduction of porosity (second term, negative).

An application of Eqs. (8.81)–(8.83) to the analysis of spreading of a hydraulic pressure front in a dam will be presented in Sect. 8.5.

8.3.4.2 Moisture Transport in Partially Saturated Concrete (Bažant–Najjar Model)

Carlson [293] was probably the first to notice that the linear diffusion theory is inapplicable to concrete, grossly overpredicting the long-time moisture loss. Pickett [690] proposed predicting the long-time moisture loss with a linear diffusion theory but greatly reduced diffusivity. The errors with long-time prediction were further documented by Pihlajavaara [693]. Cognizant of these observations, and benefiting from the first measurements of pore humidity distributions enabled by the development of Monfore gauge [8] at Portland Cement Association (Skokie, Illinois), Bažant and Najjar proposed [165] and elaborated [166] a nonlinear model for moisture transport in concrete, with the pore relative humidity h as the primary unknown. Their model, which has been embodied in the fib Model Code [381], directly postulates that, under constant temperature, the total moisture flux is driven by the gradient of pore relative humidity. Mathematically, such a transport law is written as

$$j_w = -c_p \nabla h \tag{8.84}$$

where c_p is the moisture permeability [kg/m·s], to be determined experimentally.

Equation (8.84) relates the moisture flux to the humidity h and its gradient, but the moisture mass balance equation (8.76) is written in terms of the total moisture content w_t. To obtain a differential equation for humidity as the primary unknown, it is necessary to combine these equations with formula (8.52), which provides the link between the rates of w_t and of h. In this way, we obtain the governing equation of the Bažant–Najjar model,

$$\dot{h} = k(h, t_e) \nabla \cdot \left[c_p(h, t_e)\nabla h\right] + h_s^*(h, t_e) \tag{8.85}$$

Recall that k is the reciprocal moisture capacity (inverse slope of the sorption isotherm) and h_s^* is the rate of change of humidity due to self-desiccation. Both are functions of the humidity h and equivalent age t_e. In general, the moisture permeability c_p depends on the equivalent age, too, because the hydration process results into changes of the pore structure. Identification of the specific form of functions k, c_p and h_s^* is quite tedious and requires extensive experimental data. Due to the scarcity of such data, various simplified versions of Eq. (8.85) can be useful.

The changes of pore humidity due to self-desiccation are virtually negligible for normal strength concretes with no admixtures and with water-cement ratio $w/c >$ 0.45, in which self-desiccation typically reduces h to about 0.98 and never below 0.93. In that case, the humidity sink term h_s^* on the right-hand side of (8.85) can be omitted. It can also be omitted for sufficiently old concretes, of any kind, in which the humidity drop due to further progress of self-desiccation is negligible. For such concretes, the evolution of microstructure can be neglected as well, which means that the dependence of the moisture capacity and permeability on the equivalent age can be disregarded. The governing differential equation then takes on the simplified form

$$\dot{h} = k(h) \, \nabla \cdot \left[c_p(h) \nabla h \right] \tag{8.86}$$

The possibility to omit the sink term, which is what motivated Bažant and Najjar [166], is a big modeling advantage of the approach based on the gradient of h. On the other hand, for high-strength concretes, in which the water-cement ratio is very low ($0.18 \sim 0.35$), the self-desiccation may reduce h substantially (even as low as 0.64; see [57]). For modern concretes with admixtures, the self-desiccation can be significant even if $w/c \geq 0.5$. In these cases, the sink term cannot be omitted from the right-hand side of (8.85).

If the moisture capacity is considered as constant (i.e., if the isotherm is approximated by a straight line), Eq. (8.86) can further be simplified to

$$\dot{h} = \nabla \cdot [C(h)\nabla h] \tag{8.87}$$

where

$$C(h) = k c_p(h) \tag{8.88}$$

is the *moisture diffusivity* [m²/s]. For mature good quality concretes, its values are very low—roughly 10 to 20 mm²/day at saturation ($h = 1$), and for high-strength concrete even much less.

For concrete, the dependence of diffusivity on pore relative humidity is very strong (as noticed already by Carlson [293]) and cannot be ignored in computations. A jump in diffusivity to roughly 5% of its value at full saturation (or less) occurs mainly between 85 and 60% humidity. This fact, established by Bažant and Najjar [165] on the basis of experimental data on the evolution of pore humidity distributions in cylinders and slabs, is illustrated in Figs. 8.25a–c and 8.28. The dotted curves, representing optimum fits by the linear diffusion theory, are in blatant disagreement with the data of Abrams and Orals [9]. The solid curves, which fit much better, are the result of optimization of the function $C(h)$, which established that the humidity dependence of concrete diffusivity may be approximately described by the empirical law [165]

$$C(h) = C_1 \left(\alpha_0 + \frac{1 - \alpha_0}{1 + \left(\dfrac{1 - h}{1 - h_c} \right)^r} \right) \tag{8.89}$$

where $C_1 = C(1)$ is the diffusivity at full saturation, α_0 is the ratio $C(0)/C(1)$, h_c is the pore relative humidity at which $C(h_c) = [C(0) + C(1)]/2$, and r is a parameter affecting the shape of the curve; see Fig. 8.11. Typical values identified by Bažant and Najjar [165, 166] based on fitting of experimental data[15] are $\alpha_0 \approx 0.05$ and $h_c \approx 0.75$.

---

[15]Whether the value of $C(h)$ is constant for $h < 0.60$ is an open question. Constancy was assumed because of the lack of data. It might be that $C(h)$ approaches almost 0 as $h \rightarrow 0$ (as suggested by the optimal fit of the steady state profile in Fig. 8.29). Tests are needed.

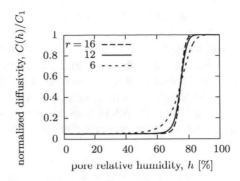

Fig. 8.11 Dependence of normalized diffusivity $C(h)/C_1$ on relative humidity h according to formula (8.89) proposed by Bažant and Najjar [165], for different values of exponent r and for $\alpha_0 = 0.05$ and $h_c = 0.75$

According to many classical data [8, 9, 31, 476, 497] analyzed by Bažant and Najjar [165, 166], the optimum value is $r = 16$, and according to some other classical data [456], the optimum is $r = 6$. In some calculations of structures (in which, of course, many phenomena other than drying intervened), various intermediate values of r seemed optimum. In the absence of more recent precise measurements, it is recommended to use $r = 12$ as a compromise. As demonstrated in Fig. 8.11, high values of r lead to an abrupt drop of diffusivity near $h = h_c$, and the curves for $r = 12$ and $r = 16$ are not too different. Therefore, the precise value of r is not as important as the choice of the other parameters. The values recommended by the *fib* Model Code 2010 [381] are $\alpha_0 = 0.05$, $h_c = 0.8$, $r = 15$ and

$$C_1 = \frac{10^{-8}\,\text{m}^2/\text{s}}{(\bar{f}_c/\text{MPa}) - 8} \tag{8.90}$$

For instance, for concrete with mean compression strength $\bar{f}_c = 40\,\text{MPa}$, this formula gives $C_1 = 3.125 \times 10^{-10}\,\text{m}^2/\text{s} = 27\,\text{mm}^2/\text{day}$.

For a linear desorption isotherm, the reciprocal moisture capacity k is constant and the moisture permeability, calculated as $c_p(h) = C(h)/k$, exhibits the same strong dependence on h as the diffusivity. Even if the isotherm is nonlinear, the change of its slope is relatively small and the observed dramatic reduction of diffusivity at lower humidities must again be due to a strong reduction of moisture permeability.

For constant diffusivity, (8.87) reduces to the *linear diffusion equation*

$$\dot{h} = C \, \nabla^2 h \tag{8.91}$$

where $\nabla^2 = \nabla \cdot \nabla$ denotes the Laplace operator. Equations of this form typically arise in science and engineering when a certain phenomenon is described by (i) a balance equation that contains the time derivative of a certain variable and the divergence of its flux, and (ii) a constitutive equation that links the flux to the gradient of the primary variable. A prominent example is the heat conduction equation, which combines energy balance with Fourier's law postulating proportionality of the heat flux to the temperature gradient. Equations (8.85)–(8.87) are nonlinear forms of the diffusion equation. Note that the term "diffusion" is used here in a broad sense and covers all transport mechanisms, while the "molecular diffusion" as one of the mechanisms has a more specific meaning.

The linear equation (8.91) can easily be solved by Fourier series expansions [297]. A closed-form solution for a half-space will be derived in Sect. 8.4.4.1; see (8.181). A similar one-dimensional formulation is also possible for spherically or axially symmetric problems, for which the linear problem is solved by series expansions in terms of spherical or cylindrical harmonics. Unfortunately, though, the assumption of constant diffusivity is realistic for concrete only in rare situations—when both the initial humidity h_0 and the ambient humidity h_{env} are within the range in which the curve of $C(h)$ in Fig. 8.11 is almost flat, which occurs only if either both are between 85 and 100%, or both below 65%.

8.3.4.3 Boundary and Initial Conditions

In Sects. 8.3.4.1–8.3.4.2, two typical models for moisture transport at constant temperature have been presented:

1. The model describing water flux in fully saturated concrete based on the Darcy law deals with the liquid water pressure p_l as the primary unknown field, and the governing differential equation is given by (8.81).
2. The model of Bažant and Najjar [165, 166] uses the pore relative humidity as the primary unknown, and the governing differential equation is given by (8.85) in the most general case, by (8.86) if the hydration reaction with its effect on the pore structure is neglected, by (8.87) if the sorption isotherm is approximated by a straight line, and by (8.91) if the diffusivity is approximated by a constant.

To make the solution of a differential equation unique, appropriate boundary conditions need to be specified at each point of the boundary. For problems described by a single differential equation with one scalar unknown field, only one scalar boundary condition is used.

At an impervious part of the boundary, the fluxes normal to the boundary vanish. This can be ensured by enforcing zero normal components of the gradients that drive the fluxes. The normal component of the gradient is in fact the directional derivative in the direction perpendicular to the boundary. For the model from Sect. 8.3.4.1, a zero flux is equivalent to a vanishing normal derivative of the total head. This condition can be rewritten in terms of the liquid pressure as

$$\frac{\partial p_l}{\partial n} = -\rho_l g e_z \cdot \boldsymbol{n} \tag{8.92}$$

where n is the outward unit vector normal to the boundary. The scalar product $e_z \cdot n$ vanishes if the boundary segment is vertical. For the Bažant–Najjar model, a zero normal derivative of pore relative humidity must be enforced at an impervious boundary. The same type of boundary condition is applicable at a plane of symmetry.

At a permeable boundary in contact with the surrounding atmosphere, it is reasonable to assume that the pore liquid pressure is equal to the atmospheric pressure. If the permeable boundary is in contact with water in a reservoir (which is the case, e.g., at the upstream face of a dam), the pore liquid pressure is set equal to the hydrostatic pressure at the corresponding depth below the free surface. For the Bažant–Najjar model, the simplest condition at a permeable boundary in contact with the surrounding atmosphere is that the pore relative humidity is equal to the ambient relative humidity, h_{env}. This follows from the assumptions that the vapor pressure in the partially saturated pores is equal to the ambient vapor pressure, $p_{v,env}$, and that the temperature and thus the saturation pressure at the boundary are the same as their environmental values.

More refined boundary conditions take into account that the exchange of vapor between the surface and the environment is not instantaneous. This can be reflected by the condition that the outward normal flux $j_w \cdot n$ is proportional to the difference in pressures:

$$j_w \cdot n = B_v \left(p_v - p_{v,env} \right) \tag{8.93}$$

The proportionality factor B_v is called the *surface vapor transfer coefficient*. In general, it depends on the movement of air, illumination, or partial insulation, and its typical values are 75×10^{-9} s/m for outdoor conditions and 25×10^{-9} s/m for indoor conditions [556].

For the Bažant–Najjar model, condition (8.93) is reformulated as

$$j_w \cdot n = \eta_e \left(\ln h - \ln h_{env} \right) \tag{8.94}$$

with proportionality factor η_e, called the *surface emissivity* [kg/m²s], which corresponds to the water vapor transfer coefficient B_v from (8.93) multiplied by the saturated vapor pressure. At 20°C, we have $p_{sat} = 2.33$ kPa, and the surface emissivity can be estimated as $\eta_e \approx 175 \times 10^{-6}$ kg/m²s ≈ 15 kg/m²day for outdoor conditions and $\eta_e \approx 58 \times 10^{-6}$ kg/m²s ≈ 5 kg/m²day for indoor conditions.

Instead of relating the boundary flux to the jump in logarithm of relative humidity, as done in (8.94), it is possible to consider it as proportional to the jump in relative humidity, which leads to a slightly modified boundary condition

$$j_w \cdot n = \eta_e^* \left(h - h_{env} \right) \tag{8.95}$$

To get a similar behavior for conditions (8.94) and (8.95), the proportionality factor η_e^* used in (8.95) should be somewhat higher than η_e from (8.94). At very early stages of the drying process, when h is close to 1, both conditions would be almost equivalent if $\eta_e^* = \eta_e \ln h_{env} / (h_{env} - 1)$, while for very late stages, when h is close to h_{env}, one would need to set $\eta_e^* = \eta_e / h_{env}$. For instance, for $h_{env} = 0.5$, these relations, respectively, lead to $\eta_e^* = 1.39\eta_e$ and $\eta_e^* = 2\eta_e$.

For ease of computations, Bažant and Najjar [166] suggested that the boundary condition of finite emissivity can be approximately represented by imagining the concrete surface to be covered by an additional concrete layer of some small thickness δ_e, and prescribing $h = h_{env}$ at the surface of this layer. The thickness of such a fictitious layer can be roughly estimated as the ratio between the permeability and the surface emissivity. It appears that generally $\delta_e \leq 1$ mm. The smallness of δ_e documents that the boundary conditions of prescribed h are usually adequate for typical concrete members (though not for extremely thin mortar or hardened cement paste specimens).

The boundary condition usually deals with a fixed (time-independent) value of h_{env}, which represents the mean ambient humidity. This is perfectly justified for laboratory tests in a climatic chamber with controlled humidity (and temperature), but less so for real structures exposed to natural variations. In linear diffusion, the variability of environmental humidity and temperature could be taken easily into account by spectral analysis of diffusion for a spectral environment [92, 192, 193]. But the linear diffusion model is too crude. For nonlinear diffusion, Monte Carlo simulation is a suitable approach. Some insight can also be gained by looking at the effects of periodic daily and annual variations; see Sect. 8.4.6.

In addition to boundary conditions, it is also necessary to prescribe the initial distribution of the primary unknown field (i.e., pore relative humidity for the Bažant–Najjar model) in the form of an initial condition. Most accurately, the simulation should start at casting and the governing equation should include the sink term due to self-desiccation. In this case, the initial state would be fully saturated and the initial condition $h = 1.0$ would be imposed at time $t = 0$. However, often an approximate analysis starts at the moment of exposure to environment, t_0, at which concrete must have self-desiccated to some extent. Then, the initial value for normal concrete could be about 0.97, but for modern concretes of very small water-cement ratios, it could be as low as 0.8.

To give an example, typical simple initial and boundary conditions for moisture transport across a concrete slab of thickness D considered as a one-dimensional problem and described by the Bažant–Najjar model are

$$h(x, t_0) = h_0 \qquad \text{for } 0 \leq x \leq D \qquad (8.96)$$

$$h(0, t) = h_{env} \qquad \text{for } t > t_0 \qquad (8.97)$$

$$h(D, t) = h_{env} \qquad \text{for } t > t_0 \qquad (8.98)$$

where $h_0 =$ initial humidity in concrete pores, and $h_{env} =$ ambient relative humidity, considered here as constant. If the finite emissivity of the surface is taken into account, conditions (8.97)–(8.98) are replaced by

$$j_x(0, t) + \eta_e^* h(0, t) = \eta_e^* h_{env} \qquad \text{for } t > t_0 \qquad (8.99)$$

$$-j_x(D, t) + \eta_e^* h(D, t) = \eta_e^* h_{env} \qquad \text{for } t > t_0 \qquad (8.100)$$

From the mathematical point of view, conditions (8.97)–(8.98) are Dirichlet boundary conditions, because they prescribe the value of the primary unknown h. The zero-flux condition on an impervious boundary could be rewritten in terms of the gradient of h and thus corresponds to a Neumann condition. Conditions (8.99)–(8.100) combine the value of the unknown function and its derivative and are classified as Robin boundary conditions. For zero surface emissivity ($\eta_e^* = 0$), they would be equivalent to Neumann conditions, and for infinite surface emissivity ($\eta_e^* = \infty$) to Dirichlet conditions.

8.3.5 Scaling Properties

The governing equation (8.86) of the Bažant–Najjar model (with the sink term due to self-desiccation neglected) can be considered as an example of a nonlinear diffusion equation. The word "diffusion" should be understood here in the general sense, not just as one specific transport mechanism, but as a general process governed by a parabolic differential equation, same as, e.g., heat conduction. Dimensional analysis of the diffusion equation provides important information on the scaling properties of the solution.

Instead of formal analysis, we will use simple arguments. If the geometric shape of the structure remains the same but all its dimensions are multiplied by a scaling factor, β, we can map each point of the original structure onto a point of the scaled structure by a homothetic transformation. Any given distribution of humidity across the structure is mapped onto the corresponding humidity distribution across the scaled structure. After this transformation, all the first derivatives with respect to the spatial coordinates are reduced β times (because all distances increase β times), and the boundary conditions of prescribed value or zero flux remain the same.

Consequently, the rate at which humidity decreases is reduced β^2 times, because on the right-hand side of (8.86), the components of the humidity gradient are reduced β times; then, they are multiplied by the diffusivity (which, at the same humidity, remains the same) and transformed by the divergence operator that uses spatial derivatives again, so the result is once more reduced β times. Thus, for the same concrete, the times needed to reach the same level of drying are proportional to the square of the structure size.

The foregoing arguments show that if $h_1(\boldsymbol{x}, t)$ is the solution of the diffusion problem (8.86) obtained on a certain domain of characteristic size D_1 (e.g., on a slab of thickness D_1) for initial humidity h_0, constant environmental humidity h_{env}, and neglected self-desiccation, then the solution of the same problem on a geometrically similar domain of characteristic size $D_2 = \beta D_1$ can be written as

$$h_2(\boldsymbol{x}, \hat{t}) = h_1\left(\frac{\boldsymbol{x}}{\beta}, \frac{\hat{t}}{\beta^2}\right) \qquad (8.101)$$

For convenience, we have written the solution as a function of the drying time $\hat{t} = t - t_0$, measured from the onset of drying at concrete age t_0. Otherwise, the scaling relation for the time variable would look more complicated. Since $\hat{t}$ and t differ just by a constant, there is no need to distinguish between partial derivatives with respect to $\hat{t}$ and t, and we will use the simpler notation $\partial/\partial t$.

The scaling rule (8.101) has an important implication: The drying times of geometrically similar structures of different sizes exposed to the same environmental conditions are proportional to the square of the size. This is true independently of the precise definition of "drying time," which can be taken, e.g., as the time after which the maximum humidity in the structure attains a given threshold, or as the time after which the average humidity in the structure attains a given threshold. The scaling rule provides justification of the general form of Eq. (3.17), in which the shrinkage halftime is taken proportional to the square of the effective thickness D.

On the other hand, if we consider the same structure size but a concrete with β times larger diffusivity (i.e., with β times larger permeability c_p or β times smaller moisture capacity $1/k$), the rate of drying increases β times and the time needed to reach the same level of drying is inversely proportional to the diffusivity. This scaling rule justifies why parameter k_t in (3.17) should be inversely proportional to the diffusivity.

8.3.6 Effect of Distributed Cracking on the Rate of Drying

Concrete structures, especially reinforced ones, function with large zones intersected by systems of closely spaced parallel hairline cracks. One might expect these cracks to enhance the overall moisture permeability c_p of concrete in the directions parallel to the cracks. A simple model for the enhancement was set up by Bažant and Raftshol [184] under the assumption that crack walls are perfectly planar, with uniform opening δ. Assuming viscous laminar flow of air with water vapor, they found the additional diffusivity to be

$$\Delta C_v = \alpha_c \frac{\rho_v p_{sat} k}{12\eta_v} \times \frac{\delta^3}{s} \tag{8.102}$$

where $\alpha_c = 1$ if the crack surfaces are perfectly planar, ρ_v and $\eta_v =$ mass density and viscosity of water vapor, $p_{sat} =$ saturated vapor pressure, $k =$ reciprocal moisture capacity, and δ and $s =$ width and spacing of the cracks. To get an idea how strong this influence could be, consider typical values $\rho_v = 0.013$ kg/m^3 (at 75% pore relative humidity), $\eta_v = 18 \times 10^{-6}$ Pa·s, $p_{sat} = 2338$ Pa and $k = 9 \times 10^{-3}$ m^3/kg, from which

$$\frac{\rho_v p_{sat} k}{12\eta_v} = \frac{0.013 \times 2338 \times 0.009}{12 \times 18 \times 10^{-6}} \; \text{s}^{-1} \approx 1.27 \times 10^3 \; \text{s}^{-1} \tag{8.103}$$

For idealized, perfectly planar smooth cracks ($\alpha_c = 1$) having the width of $\delta = 0.1$ mm and spacing of $s = 7$ cm, (8.102) would then give

$$\Delta C_v = 1.27 \times 10^3 \times \frac{(0.1 \times 10^{-3})^3}{0.07}\ \text{m}^2/\text{s} = 18.14 \times 10^{-9}\ \text{m}^2/\text{s} \approx 1570\ \text{mm}^2/\text{day}$$
$$(8.104)$$

which would roughly be a 150-fold increase in diffusivity (compared to typical values in the order of 10 mm^2/day).

(a)   (b)

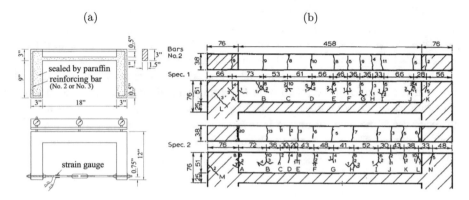

Fig. 8.12 (a) C-shaped specimen with crack-producing tie rod, (b) map of crack shapes and crack widths observed by microscope, with values in 0.01 mm; figure originally published in ACI Materials Journal, Sep-Oct 1987

To check this prediction, Bažant et al. [122] tested C-shaped beams deformed permanently by a tie rod (Fig. 8.12). Cracking localization was prevented (and thus uniform crack spacing achieved) by placing reinforcing steel bars on the tensile side. For the width of 0.1 mm and spacing 7 cm, the measured diffusivity was only 2.25× greater than the diffusivity of uncracked concrete. So, the empirical value of α_c in (8.102) is about $1.25/156 = 0.008$, for this case. The logical explanation is that the crack space must be partly discontinuous or contain narrow necks obstructing the flow.

The laws that govern crack spacing will be discussed in more detail in Sect. 12.2. Empirical expressions that relate the increase of permeability to a damage variable will be presented in Sect. 12.7.

8.4 One-Dimensional Moisture Transport

8.4.1 One-Dimensional Diffusion Equation

In many practical problems, moisture transport can approximately be described in the one-dimensional setting. This is the case in particular for flat slabs or walls with two parallel boundary surfaces exposed to the ambient humidity (not necessarily the same on both sides). If the other parts of the boundary are impermeable or remote (so that one can assume an infinite slab or wall), moisture transport takes place in the direction perpendicular to the midplane of the flat member. All variables of interest become independent of the in-plane coordinates y and z and can be considered as functions of a single spatial coordinate x, measured in the direction perpendicular to the mid-surface. Of course, we must also assume that the material properties depend exclusively on x, e.g., that the material is homogeneous or formed by homogeneous layers parallel to the midplane.

Under the foregoing assumptions, Eq. (8.86) reduces to

$$\dot{h} = k(h) \left[c_p(h) h' \right]'$$

(8.105)

where $h(x, t)$ is the unknown function describing the pore relative humidity, and the prime denotes differentiation with respect to x. For constant moisture capacity, (8.105) simplifies to

$$\dot{h} = \left[C(h) h' \right]'$$

(8.106)

which is the one-dimensional version of (8.87). For a slab of thickness D exposed to the same ambient humidity on both sides, these equations can be combined with initial condition (8.96) and either Dirichlet boundary conditions (8.97)–(8.98) or, if a finite surface emissivity is considered, Robin boundary conditions (8.99)–(8.100). If the material properties are symmetric with respect to the midplane, it is possible to solve the problem on the interval $[0, D/2]$ and replace the second boundary condition by the zero-flux condition

$$j_x(D/2, t) = 0 \qquad \text{for } t > t_0$$

(8.107)

which is equivalent to the homogeneous Neumann condition

$$h'(D/2, t) = 0 \qquad \text{for } t > t_0$$

(8.108)

Another type of problem that can be described using a single spatial variable is radial moisture transport in an axisymmetric body, e.g., in a cylindrical specimen. The corresponding differential equation could be derived by formal manipulations of the general diffusion equation (8.86), based on a transformation from Cartesian to polar coordinates. However, it is more instructive to construct it from the modified form of the mass balance equation.

Consider a small element of an axisymmetric body shown in Fig. 8.13, with moisture flux exclusively in radial direction. The mass flux j_r is thus a function of the
radial coordinate r only. Neglecting the higher-order terms, the net flux (difference
between the influx through the section at r and the outflux through the section at
$r + \Delta r$) per unit volume can be expressed as

$$
\frac{j_r \times r \, \Delta\alpha \, \Delta z - (j_r + \Delta j_r) \times (r + \Delta r) \Delta\alpha \, \Delta z}{r \, \Delta\alpha \, \Delta z \, \Delta r} \approx -\frac{r \, \Delta j_r + j_r \, \Delta r}{r \, \Delta r} = -\frac{\Delta j_r}{\Delta r} - \frac{j_r}{r}
\tag{8.109}
$$

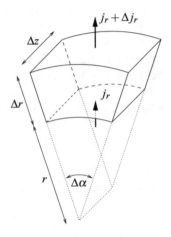

Fig. 8.13 Radial flux into and out of a control volume extracted from an axisymmetric body

Taking the limit of $\Delta r \to 0$, the fraction $\Delta j_r / \Delta r$ becomes the derivative $\mathrm{d} j_r / \mathrm{d} r$.
Note the presence of an additional term, $-j_r/r$, which stems from the difference
between the areas of the sections at r and at $r + \Delta r$. Consequently, the moisture
mass balance equation (8.76) is, in the present case of axisymmetric radial flow,
replaced by

$$
\dot{w}_t = -\frac{\mathrm{d} j_r}{\mathrm{d} r} - \frac{j_r}{r}
\tag{8.110}
$$

According to the Bažant–Najjar transport law (8.84), the radial flux is given by

$$
j_r = -c_p(h)\frac{\mathrm{d} h}{\mathrm{d} r}
\tag{8.111}
$$

If the rate of the water content $\dot{w}_t$ is expressed in terms of the humidity rate $\dot{h}$ using
(8.52) and the effect of self-desiccation is neglected, Eq. (8.110) can be rewritten as

$$
\dot{h} = k(h)\left[\frac{\mathrm{d}}{\mathrm{d} r}\left(c_p(h)\frac{\mathrm{d} h}{\mathrm{d} r}\right) + \frac{c_p(h)}{r}\frac{\mathrm{d} h}{\mathrm{d} r}\right]
\tag{8.112}
$$

This is the one-dimensional nonlinear diffusion equation corresponding to the Bažant–Najjar model for axisymmetric radial moisture transport. For constant moisture capacity k, it could be written as

$$\dot{h} = \frac{\mathrm{d}}{\mathrm{d}r}\left(C(h)\frac{\mathrm{d}h}{\mathrm{d}r}\right) + \frac{C(h)}{r}\frac{\mathrm{d}h}{\mathrm{d}r} \qquad (8.113)$$

where $C(h) = kc_p(h)$ is the moisture diffusivity.

8.4.2   Numerical Solution by Finite Differences

The diffusion equation (8.105) or (8.112) is a partial differential equation for the unknown pore relative humidity, which depends on the temporal variable t and spatial variable x. Approximate numerical solutions of such problems are usually based on a suitable discretization technique that converts the problem into a set of algebraic equations for a finite number of unknowns. Perhaps the most straightforward approach is to replace the partial derivatives (both in space and in time) by finite differences.

8.4.2.1   Forward Euler Method

Let us start from the simplest case of a one-dimensional linear diffusion equation

$$\dot{h}(x, t) = C h''(x, t) \qquad (8.114)$$

which corresponds to (8.106) with constant diffusivity C. The spatial interval $[0, L]$ can be divided into m equally sized subintervals of length $\Delta x = L/m$, and the approximate solution is sought at points $x_i = i\,\Delta x$, $i = 0, 1, \ldots m$, and at time instants t_k, $k = 0, 1, \ldots n$, where t_0 is the initial time at which the initial condition is prescribed and n is the total number of time steps. The (approximately computed) value of $h(x_i, t_k)$ will be denoted as $h_i^{(k)}$.

Suppose that the solution is known at time t_k, and we would like to proceed to time $t_{k+1} = t_k + \Delta t$ (the time step can be variable, but when discussing a single time step, we denote it simply as Δt). Within the time interval from t_k to t_{k+1}, it is natural to approximate the time derivative on the left-hand side of (8.114) evaluated at point x_i by the finite difference expression $\left(h_i^{(k+1)} - h_i^{(k)}\right)/\Delta t$. If the second spatial derivative on the right-hand side of (8.114) is approximated by a finite difference expression based on the currently known values (i.e., on the values at time t_k), the resulting equations

$$\frac{h_i^{(k+1)} - h_i^{(k)}}{\Delta t} = C \frac{h_{i-1}^{(k)} - 2h_i^{(k)} + h_{i+1}^{(k)}}{(\Delta x)^2}, \qquad i = 1, 2, \ldots m - 1 \qquad (8.115)$$

for unknown values $h_i^{(k+1)}$ become decoupled and their solution can be directly expressed as

$$h_i^{(k+1)} = h_i^{(k)} + \frac{C \Delta t}{(\Delta x)^2} \left(h_{i-1}^{(k)} - 2h_i^{(k)} + h_{i+1}^{(k)} \right), \qquad i = 1, 2, \ldots m - 1 \quad (8.116)$$

This procedure is fully explicit in the sense that no set of coupled equations needs to be solved and the solution is obtained in a straightforward way by evaluating the update formula on the right-hand side of (8.116), which can be done recursively for $k = 0, 1, 2, \ldots n - 1$. The approach based on the finite difference scheme (8.115) can be considered as the *forward Euler method* because the right-hand side represents an approximation of the second spatial derivative at time t_k and the approximation of the time derivative on the left-hand side is a forward finite difference from t_k to t_{k+1}.

To get a complete algorithm of the forward Euler method, an additional detail has to be clarified. In Eqs. (8.115)–(8.116), the boundary points with $i = 0$ and $i = m$ have been deliberately omitted. Their treatment depends on the type of boundary conditions. If the pore relative humidity on the boundary is set equal to the prescribed ambient humidity (Dirichlet condition (8.97) or (8.98)), we simply set $h_0^{(k+1)} = h_{\text{env}}(t_{k+1})$ or $h_m^{(k+1)} = h_{\text{env}}(t_{k+1})$. If the boundary condition involves the moisture flux (which depends on the humidity gradient), one can approximate the humidity gradient (i.e., spatial derivative) on the boundary by a finite difference expression. Instead of using a one-sided formula, e.g., $\left(h_m^{(k+1)} - h_{m-1}^{(k+1)} \right) / \Delta x$ on the right boundary, it is better to evaluate the central difference, which uses a (fictitious) value of the solution outside the domain, at a "ghost" point x_{m+1}. However, by combining the boundary condition with Eq. (8.115) written at the boundary point (i.e., for $i = m$), the additional unknown can be eliminated and a closed-form update formula for the value at the boundary is obtained.

For instance, the homogeneous Neumann condition (8.108) imposed on the right boundary, i.e., at $x = L \equiv D/2$, can be approximated by

$$\frac{h_{m+1}^{(k)} - h_{m-1}^{(k)}}{2 \Delta x} = 0 \qquad (8.117)$$

which implies that the ghost-point value $h_{m+1}^{(k)}$ should be set equal to $h_{m-1}^{(k)}$. Using this identity in (8.116) written for $i = m$, we obtain

$$h_m^{(k+1)} = h_m^{(k)} + \frac{2C \Delta t}{(\Delta x)^2} \left(h_{m-1}^{(k)} - h_m^{(k)} \right) \qquad (8.118)$$

On the left boundary, subscripts $m - 1$ and m would be replaced by 1 and 0.

Example 8.1. Accuracy and stability of the forward Euler method

Let us illustrate the performance of the forward Euler method by solving a problem for which a closed-form solution is available. Consider the linear diffusion equation (8.114) with the initial humidity distribution[16]

$$h(x,0) = 0.6 + 0.4 \sin \frac{\pi x}{2L}, \qquad 0 \le x \le L \qquad (8.119)$$

and with boundary conditions

$$h(0,t) = 0.6, \qquad t > 0 \qquad (8.120)$$
$$h'(L,t) = 0, \qquad t > 0 \qquad (8.121)$$

It is easy to check that the exact solution is given by

$$h(x,t) = 0.6 + 0.4 \sin \frac{\pi x}{2L} \exp\left(-\frac{t}{\tau}\right) \qquad (8.122)$$

where

$$\tau = \frac{4L^2}{C\pi^2} \qquad (8.123)$$

is a characteristic time, introduced for convenience.

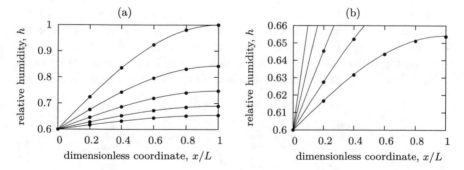

Fig. 8.14 Relative humidity profiles at times $t = 0, 0.5\tau, \tau, 1.5\tau,$ and 2τ: exact solution (solid curves) versus numerical results (isolated points)

The initial values $h_i^{(0)}$ at time $t_0 = 0$ are obtained from the initial condition (8.119), and the values at times t_1, t_2, etc., are then calculated using the boundary condition (8.120) for point $x_0 = 0$, the recursive update formula (8.116) for points x_1 to x_{m-1},

---

[16]For simplicity, in the present section devoted to numerical methods, we set the initial time t_0 to 0, without linking the time scale to the age of concrete. Consequently, there is no need to distinguish here between the "actual time," t, and the duration of drying, $\hat{t} = t - t_0$.

and formula (8.118) for point $x_m = L$. The results can be conveniently presented in terms of the dimensionless spatial coordinate x/L and the dimensionless time t/τ. Figure 8.14a compares the exact solution at times 0, 0.5τ, τ, 1.5τ, and 2τ, plotted by solid curves, with the approximate values computed by the forward Euler method with grid spacing $\Delta x = L/5 = 0.2L$ and time step $\Delta t = \tau/40 = 0.025\tau$. A visual comparison indicates that the solution is sufficiently accurate for engineering purposes, even on such a coarse spatial grid (only 4 internal grid points plus 2 points on the boundary). Some limited numerical error is discernible only if the figure is replotted with a modified scale on the vertical axis; see Fig. 8.14b. Of course, the error is low owing to a high regularity of the problem. If a sharp drying front penetrates into a specimen, steep humidity gradients arise and the error increases, especially for a nonlinear diffusion model.

Let us explore in more detail the dependence of the error on the grid spacing Δx and time step Δt. For the present problem, the exact value of humidity at the right boundary and at time $t = \tau$ (truncated to 6 significant digits) is $h(L, \tau) = 0.747152$ while a numerical solution by the forward Euler method with $\Delta x/L = 0.2$ and $\Delta t/\tau = 0.025$ leads to $h_5^{(40)} = 0.746520$, which corresponds to an absolute error of 6.32×10^{-4}. The choice of numerical parameters Δx and Δt must be made judiciously, because the error is affected by the spatial as well as the temporal discretization. For instance, if $\Delta x/L = 0.2$ is fixed and Δt tends to zero, the numerical result does not converge to the exact value but to 0.748363, and the magnitude of the error becomes larger than for the finite step $\Delta t = 0.025\tau$.

One might think that a good idea would be to keep a constant ratio between the grid spacing and the time step. However, if Δx and Δt are both reduced to one half, i.e., to $0.1L$ and 0.0125τ, the error decreases only slightly, from 6.32×10^{-4} to 6.20×10^{-4}. Even worse, if the reduction of both parameters to one half is repeated, the error blows up. The computed humidity profiles at times 0.1τ and 0.2τ are plotted in Fig. 8.15a and the history of humidity at $x = L$ (right boundary) is shown in Fig. 8.15b. It is clear that the solution becomes polluted by an oscillatory mode,

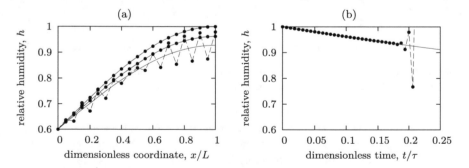

Fig. 8.15 Loss of numerical stability in a computation with $\Delta x = 0.05L$ and $\Delta t = 0.00625\tau$: (a) relative humidity profiles at times $t = 0$, 0.1τ, and 0.2τ, (b) evolution of relative humidity at the right boundary

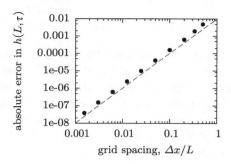

Fig. 8.16 Convergence diagram for the forward Euler method; the dashed line has slope 2 in the logarithmic scale

the magnitude of which increases exponentially. This is a manifestation of numerical instability.

Later, it will be shown that a safe estimate (upper bound) of the critical time step is given by

$$\Delta t^*_{\text{crit}} = \frac{(\Delta x)^2}{2C} = \frac{(\Delta x)^2 \pi^2 \tau}{8L^2} \tag{8.124}$$

Therefore, numerical stability is guaranteed if the time step is selected such that

$$\frac{\Delta t}{\tau} \leq \frac{\pi^2}{8} \left(\frac{\Delta x}{L}\right)^2 \tag{8.125}$$

Since the critical time step is proportional to the square of the grid spacing Δx, it makes sense to vary the numerical parameters such that $\Delta t/(\Delta x)^2$ remains constant. Starting again from $\Delta x/L = 0.2$ and $\Delta t/\tau = 0.025$ (for which condition (8.125) is satisfied), we can proceed to $\Delta x/L = 0.2/2 = 0.1$ and $\Delta t/\tau = 0.025/4 = 0.00625$, then to $\Delta x/L = 0.1/2 = 0.05$ and $\Delta t/\tau = 0.00625/4 = 0.0015625$, etc. The error in the numerically computed relative humidity at the right boundary and at time $t = \tau$ is plotted in Fig. 8.16 for various values of the grid spacing Δx, ranging from $L/2$ to $L/640$. For each computation, the time step is determined as $\Delta t = 0.625\tau(\Delta x/L)^2$, which is at about 51% of the critical time step estimated according to (8.124). The dashed straight line in Fig. 8.16 is an auxiliary line of slope 2 in the logarithmic scale. It is clear that the error is proportional to $(\Delta x)^2$ (and thus proportional to Δt). This means that if the grid spacing is reduced to one half and the time step is reduced to one quarter, the error drops to one quarter, but the computational time increases $2 \times 4 = 8$ times. Reducing the error by two orders of magnitude requires a thousand times higher computational effort. ∎

The forward Euler method is very simple and easy to use, but its disadvantage is that it possesses only conditional stability. If the numerical solution at time t_k is perturbed by $\delta h_i^{(k)}$, $i = 0, 1, 2, \ldots m$, application of formula (8.116) leads to a perturbation at time t_{k+1} given by

$$\delta h_i^{(k+1)} = \theta \, \delta h_{i-1}^{(k)} + (1 - 2\theta)\delta h_i^{(k)} + \theta \, \delta h_{i+1}^{(k)}, \qquad i = 1, 2, \ldots m - 1 \quad (8.126)$$

where, for convenience, the dimensionless fraction $C \, \Delta t / (\Delta x)^2$ is denoted as θ. Numerical instability occurs if the perturbation (which is always induced by finite arithmetics) is magnified by the algorithm, instead of being damped. On an infinite interval (or even on a finite interval with homogeneous Neumann boundary conditions), the most dangerous perturbation is an oscillatory mode with $\delta h_i^{(k)} = (-1)^i a_k$, where a_k is the magnitude of the perturbation at time t_k. Substitution of this assumed perturbation mode into (8.126) leads to

$$\delta h_i^{(k+1)} = \theta \, (-1)^{i-1} a_k + (1 - 2\theta)(-1)^i a_k + \theta(-1)^{i+1} a_k =$$
$$= (1 - 4\theta)(-1)^i a_k = (1 - 4\theta) \, \delta h_i^{(k)} \qquad (8.127)$$

The factor $1 - 4\theta$ is always smaller than 1 and so the perturbation cannot grow monotonically. However, if $1 - 4\theta$ becomes negative and larger than 1 in magnitude, the perturbation at a given point changes sign in each time step and its magnitude grows, which corresponds to numerical instability. The stability condition $1 - 4\theta \geq -1$ (i.e., $\theta \leq 1/2$) can be rewritten in terms of the original parameters as

$$\frac{C \, \Delta t}{(\Delta x)^2} \leq \frac{1}{2} \qquad (8.128)$$

The resulting estimate of the critical time step is thus given by the already used relation (8.124). It is denoted as Δt_{crit}^* (with an asterisk) to emphasize that the actual critical time step can be different (slightly larger), if the prescribed boundary conditions act as constraints that do not allow the development of the most dangerous oscillatory mode in its pure form.

For instance, if the problem from Example 8.1 is solved with $\Delta x / L = 0.1$ and $\Delta t / \tau = 0.01242$, the time step is actually slightly above the estimated critical time step $\Delta t_{\text{crit}}^* = (\pi^2/8)0.1^2 \tau = 0.012337\tau$, but the solution remains numerically stable. However, if the time step is increased to 0.01243τ, numerical stability is lost. For time steps just above the critical one, the growth of numerical perturbations from the level of machine precision to the level of appreciable oscillations takes many time steps. This is why the computation with an unstable time step $\Delta t = 0.0125\tau$ (which was actually mentioned in Example 8.1) leads to quite a small error at time $t = \tau$. Nevertheless, to make sure that the results can be trusted even for long times, computations with unstable time steps should be avoided.

8.4.2.2   Implicit Methods

The time step imposed by the stability condition can become prohibitively short in problems which require a good spatial resolution to capture steep gradients of humidity, e.g., at a propagating drying front. For instance, if the grid spacing near the exposed surface of a drying specimen is set to $\Delta x = 0.1$ mm, a typical diffusivity of concrete at saturation $C = 20$ mm^2/day leads to an estimated critical time step of 2.5×10^{-4} day, and 4000 computational time steps are required to cover one day of drying. In simulations of the drying process over years, dozens of millions of computational steps would be needed. Therefore, it is useful to develop numerical schemes that are unconditionally stable, i.e., stable for arbitrarily large time steps.

Unconditionally stable schemes can be obtained by modifying the approximation of the second spatial derivative on the right-hand side of the diffusion equation. Recall that the approximation on the right-hand side of (8.115) corresponds to time t_k, i.e., to the time at the beginning of the current step, while the approximation of the time derivative on the left-hand side can be expected to provide the highest accuracy in the middle of the time step from t_k to t_{k+1}. A more general family of methods is constructed by expressing the right-hand side as a weighted average of the second spatial derivatives evaluated at t_k and at t_{k+1}. This leads to an approximation of the diffusion equation (8.114) by

$$\frac{h_i^{(k+1)} - h_i^{(k)}}{\Delta t} = C \left((1-\alpha)\frac{h_{i-1}^{(k)} - 2h_i^{(k)} + h_{i+1}^{(k)}}{(\Delta x)^2} + \alpha\frac{h_{i-1}^{(k+1)} - 2h_i^{(k+1)} + h_{i+1}^{(k+1)}}{(\Delta x)^2} \right)$$
(8.129)

where α is a numerical parameter between 0 and 1. The choice of $\alpha = 0$ corresponds to the original forward Euler scheme (8.115), the choice of $\alpha = 1$ gives the *backward Euler method*, and the case of $\alpha = 0.5$, which could be interpreted as the trapezoidal rule, is traditionally called the *Crank–Nicolson method* [329].

For all nonzero values of parameter α, Eq. (8.129) written for $i = 1, 2, \ldots m - 1$ contains the unknown values at time t_{k+1} not only on the left-hand side but also on the right-hand side, and each equation contains not only $h_i^{(k+1)}$ but also $h_{i-1}^{(k+1)}$ and $h_{i+1}^{(k+1)}$. Consequently, the equations become coupled and the method is no longer explicit. Moving all unknown terms to the left-hand side and all known terms to the right-hand side, we can rewrite the equations as

$$-\theta\alpha h_{i-1}^{(k+1)} + (1 + 2\theta\alpha)h_i^{(k+1)} - \theta\alpha h_{i+1}^{(k+1)} =$$
$$= \theta(1-\alpha)h_{i-1}^{(k)} + [1 - 2\theta(1-\alpha)]h_i^{(k)} + \theta(1-\alpha)h_{i+1}^{(k)}$$
(8.130)

where, as before, $\theta = C \Delta t/(\Delta x)^2$. If the unknown values at time t_{k+1} are collected into a column matrix and Eq. (8.130) with $i = 1, 2, \ldots m - 1$ is rewritten in the matrix form, the system matrix multiplying the unknowns on the left-hand side is tridiagonal, with coefficients $1 + 2\theta\alpha$ on the diagonal and coefficients $-\theta\alpha$ just below and just above the diagonal.

For Dirichlet boundary conditions, the value of $h_0^{(k+1)}$ or $h_m^{(k+1)}$ is set to the prescribed ambient humidity at time t_{k+1} and the corresponding terms in the equations that correspond to $i = 1$ or $i = m - 1$ are moved to the right-hand side. For a homogeneous Neumann boundary condition (zero flux, i.e., zero humidity gradient) at the right boundary, the ghost-point values $h_{m+1}^{(k)}$ and $h_{m+1}^{(k+1)}$ are set equal to $h_{m-1}^{(k)}$

and $h_{m-1}^{(k+1)}$, and Eq. (8.130) is rewritten for $i = m$ as

$$- 2\theta\alpha h_{m-1}^{(k+1)} + (1 + 2\theta\alpha)h_m^{(k+1)} = 2\theta(1 - \alpha)h_{m-1}^{(k)} + [1 - 2\theta(1 - \alpha)]h_m^{(k)} \quad (8.131)$$

If the zero-flux condition is imposed on the left boundary, subscripts $m - 1$ and m are replaced by 1 and 0.

For coupled linear equations with tridiagonal matrices, efficient storage schemes and fast solvers are available, with computational complexity proportional to the number of unknowns. Each time step becomes more expensive than for the forward Euler method, but, with a proper choice of parameter α, the method is unconditionally stable and the time step can be in principle arbitrarily large (of course, accuracy may deteriorate for very large time steps, but stability is never lost).

Numerical stability can be assessed by checking the potential growth of oscillatory perturbations. Assuming that $\delta h_i^{(k)} = (-1)^i a_k$ and $\delta h_i^{(k+1)} = (-1)^i a_{k+1}$, $i = 0, 1, 2, \ldots m$, and substituting these expressions into (8.130), one finds that the magnitudes a_k and a_{k+1} are linked by

$$(1 + 4\theta\alpha)a_{k+1} = [1 - 4\theta(1 - \alpha)]a_k \quad (8.132)$$

The factor $1 + 4\theta\alpha$ on the left-hand side is always positive and larger than the factor $1 - 4\theta + 4\theta\alpha$ on the right-hand side (for positive θ and for α between 0 and 1). Therefore, the growth of perturbations is prevented if

$$1 + 4\theta\alpha \geq -(1 - 4\theta + 4\theta\alpha) \quad (8.133)$$

which is equivalent with

$$2\theta(1 - 2\alpha) \leq 1 \quad (8.134)$$

Recall that α is a fixed parameter characterizing the numerical method, while θ is proportional to the time step. If $\alpha \geq 1/2$, condition (8.134) is satisfied for any time step and the method is unconditionally stable. This includes the Crank–Nicolson scheme with $\alpha = 1/2$ as well as the backward Euler scheme with $\alpha = 1$. For $\alpha < 1/2$, the method is conditionally stable and the critical time step can be estimated as

$$\Delta t_{\text{crit}}^* = \frac{(\Delta x)^2}{2C(1 - 2\alpha)} \quad (8.135)$$

As a special case, for $\alpha = 0$, we obtain the critical time step (8.124) of the forward Euler method.

Example 8.2. Accuracy and stability of implicit methods

The same problem as in Example 8.1 is now solved by the backward Euler (BEu) method and by the Crank–Nicolson (CN) method. For the initial choice of grid spacing $\Delta x = 0.2L$ and time step $\Delta t = 0.025\tau$, the error in the relative humidity at the right boundary at time $t = \tau$ is 3.02×10^{-3} for BEu and 1.2×10^{-3} for CN. This is actually higher than the error of the forward Euler (FEu) method evaluated in Example 8.1, which was just 6.32×10^{-4}. However, the short time step used by the FEu method was dictated by stability considerations while the unconditionally stable BEu and CN methods can be run with a much longer time step. A better balance between the spatial and temporal discretization can lead to a higher accuracy at a comparable computational cost. For instance, if the numerical parameters are set to $\Delta x = 0.1L$ and $\Delta t = 0.1\tau$, the error of the CN method is reduced to 1.8×10^{-4} and the total number of unknowns is $m \times n = 10 \times 10 = 100$ instead of $5 \times 40 = 200$ unknowns in the previous case.

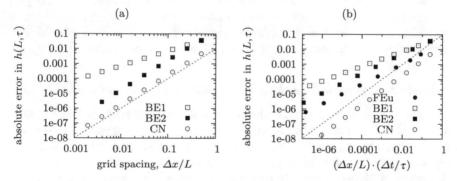

Fig. 8.17 Convergence diagrams: dependence of error (a) on grid spacing, (b) on the product of grid spacing and time step

Since the BEu and CN methods are unconditionally stable, during refinement of the spatial grid, it is not necessary to reduce the time step proportionally to $(\Delta x)^2$. Instead, one can keep the ratio $\Delta t / \Delta x$ fixed. The convergence diagram in Fig. 8.17a shows that, for the CN method represented by hollow circles, the error is then proportional to $(\Delta x)^2$ (the dashed line has slope 2 in the logarithmic scale). This means that if the grid spacing and the time step are simultaneously reduced to one half, the error is reduced to one quarter, while the computational effort increases four times. For the BEu method, the results are less favorable. The hollow squares in Fig. 8.17a correspond to the BEu method with time step proportional to the grid spacing (marked as BE1 in the figure legend). Upon refinement, the error is seen to decrease much more slowly than for the CN method. Faster convergence is obtained if the BEu method is used with time step proportional to the square of the grid spacing (filled squares, marked as BE2). The slope of the diagram in Fig. 8.17a is then the same as for the CN method, but the computational effort grows faster. To obtain a fair comparison, the error is replotted as a function of the product of the dimensionless grid spacing $\Delta x / L$ and the dimensionless time step $\Delta t / \tau$ in Fig. 8.17b. The variable on the horizontal axis thus corresponds to $1/(mn)$ and is inversely proportional to the computational effort. As expected, the fastest convergence rate is achieved for the CN method and the comparison between BE1 and BE2 reveals that reduction of the time step proportionally to the square of the grid spacing is indeed a better choice.[17] For the sake of interest, the results obtained with the explicit FEu method are plotted here as well; see the filled circles. For this particular problem, the FEu method is seen to outperform the BEu method.   ∎

The results obtained so far may seem to indicate that the backward Euler method is not a good choice. However, such a conclusion would be premature. In Examples 8.1 and 8.2, the exact solution was a harmonic function of the spatial coordinate, with a relatively large wavelength (compared to the size of the computational domain) and with an amplitude exponentially decaying in time. In typical simulations of drying specimens or structures, the initial humidity distribution is uniform and a part of the boundary is suddenly exposed to a lower ambient humidity. In such cases, a drying front propagates from the exposed surface into the core. This is accompanied by very high humidity gradients, especially at early stages of the process. The effect of such high gradients on the performance of finite difference methods is illustrated by the next example.

Example 8.3. Numerical simulations of drying from an exposed surface

Consider a slab of thickness D with an initially uniform pore relative humidity $h(x, 0) = 1$, exposed at both boundary surfaces (i.e., at $x = 0$ and $x = D$) to the ambient humidity $h_{env} = 0.6$. Due to symmetry, it is sufficient to solve the diffusion equation (8.114) on the interval $[0, L]$ where $L = D/2$, with the same boundary conditions (8.120)–(8.121) as in Examples 8.1 and 8.2. For easy comparison with the previous results, let us again use the characteristic time $\tau = 4L^2/C\pi^2$.

At early stages of drying, a discernible reduction of humidity is limited to a narrow layer adjacent to the exposed surface. This phenomenon obviously could not be captured properly if the computational grid spacing is too wide. Therefore, let us use a division of the interval $[0, L]$ into $m = 10$ subintervals of length $\Delta x = 0.1L$ as the coarsest grid. If the problem is solved by the forward Euler (FEu) method, the time step must be kept below the critical one. Formula (8.124) leads to an estimated critical time step $\Delta t^*_{crit} = (\pi^2\tau/8)(\Delta x/L)^2 \approx 1.234\tau(\Delta x/L)^2 = 0.01234\tau$, and so the simulation should remain stable with the actual time step set to $\Delta t = 0.01\tau$. Humidity profiles at selected times between $t = 0$ and $t = \tau$ are plotted in Fig. 8.18. The isolated points correspond to the numerical solution obtained by the FEu method with $\Delta x = 0.1L$ and $\Delta t = 0.01\tau$, and the curves represent a highly accurate numerical solution obtained with a much smaller time step on a very fine grid. The accuracy of the solution on the relatively coarse grid is seen to be satisfactory. The BEu and CN methods with the same numerical parameters $\Delta x = 0.1L$ and

---

[17]The convergence rates of the methods examined here are closely related to the accuracy of the underlying finite difference approximations. For sufficiently smooth functions, an approximation of the second spatial derivative by the central difference formula is second-order accurate. The formula for the first temporal derivative on the left-hand side of (8.115) or (8.129) is second-order accurate as an approximation of the value of $\dot{h}$ at $(t_k + t_{k+1})/2$ (because then it corresponds to a central difference in time) but only first-order accurate when used as a forward difference at t_k or as a backward difference at t_{k+1}.

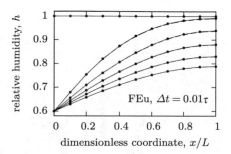

Fig. 8.18 Humidity profiles at times 0, 0.2τ, 0.4τ, 0.6τ, 0.8τ, and τ after exposure of the left boundary to ambient humidity $h_{env} = 0.6$; the isolated points have been computed using the forward Euler method with grid spacing $\Delta x = 0.1L$ and time step $\Delta t = 0.01\tau$, and the solid curves correspond to a highly accurate numerical solution

$\Delta t = 0.01\tau$ would produce very similar plots. The potential advantage of the implicit methods is that they can be run with a longer time step, while the time step of the explicit FEu method is constrained by the stability condition. This advantage of implicit methods becomes important, especially for finer grids, because the critical time step of the FEu method is proportional to the square of the grid spacing. For instance, for $\Delta x = 0.01L$, one would get $\Delta t^*_{crit} = 0.0001234\tau$, and about 10,000 time steps would be needed to cover the time interval from 0 to τ using the FEu method, while the BEu or CN methods could in theory use arbitrarily large steps.

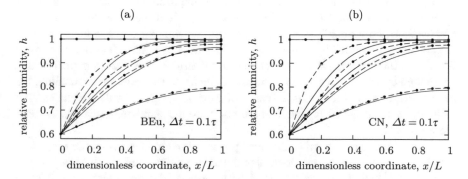

Fig. 8.19 Humidity profiles at times 0, 0.1τ, 0.2τ, 0.3τ, and τ after exposure of the left boundary to ambient humidity $h_{env} = 0.6$, computed with grid spacing $\Delta x = 0.1L$ and time step $\Delta t = 0.1\tau$ using the (a) backward Euler method, (b) Crank–Nicolson method; the solid curves correspond to a highly accurate numerical solution and the dashed lines connect points that represent the coarse-grid solution

Of course, a deterioration of accuracy can be expected if the time step is too large. Let us therefore repeat the simulation on the same grid ($\Delta x = 0.1L$), using the implicit methods with a ten times longer time step, $\Delta t = 0.1\tau$. Figure 8.19 shows the initial uniform humidity profile and the profiles after steps 1, 2, and 3 (i.e., at times $t = 0.1\tau, 0.2\tau$ and 0.3τ) and after step 10 (at time $t = \tau$). For the BEu method

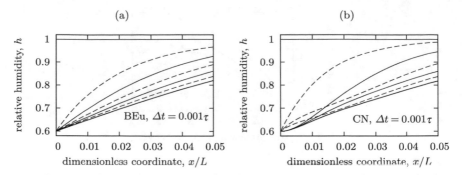

Fig. 8.20 Humidity profiles near the exposed boundary at times $0, 0.001\tau, 0.002\tau, \ldots 0.006\tau$ after exposure to ambient humidity $h_{\text{env}} = 0.6$, computed with grid spacing $\Delta x = 0.001L$ and time step $\Delta t = 0.001\tau$ using the (a) backward Euler method, (b) Crank–Nicolson method; the solid curves correspond to even steps and the dashed curves to odd steps

(Fig. 8.19a), the results at early stages exhibit some error, but the overall evolution of the profiles is reasonable. In contrast to that, the results obtained with the CN method (Fig. 8.19b) exhibit a pathology—the profiles obtained in the second and third computational steps intersect, which is caused by a nonmonotonic evolution of humidity at the first internal grid point, i.e., at $x = 0.1L$. The drying process should lead to a gradual reduction of humidity at each internal point, but the numerically computed values of relative humidity at $x = 0.1L$ after steps 1 to 6 are 0.799, 0.666, 0.681, 0.653, 0.656, and 0.644. Such oscillations are clearly nonphysical.

An oscillatory evolution of humidity near the exposed boundary is found in calculations based on the CN method even on fine grids if the time step is not sufficiently short. Figure 8.20b shows humidity profiles number 0 to 6 in the vicinity of the boundary (note that the scale on the horizontal axis is from 0 to 0.05) obtained for $\Delta x = 0.001L$ and $\Delta t = 0.001\tau$. Since the grid is fine, the numerical results are plotted by curves instead of by isolated points, and for better understanding, the even profiles are shown by solid curves while the odd profiles by dashed curves. No pathologies are detected for the BEu method with the same numerical parameters; see Fig. 8.20a.

To avoid the pathological oscillations produced in simulations with a large time step by the CN method, one could use the BEu method, but the accuracy of the solution would still be compromised. A better remedy can be sought in an adaptive modification of the computational time step. Initial stages of the drying process, when high gradients of humidity lead to fast moisture transport near the exposed boundary, require sufficiently short time steps. Later on, when the humidity profile becomes more flat and the transport slows down, longer steps can be used to increase the efficiency of the numerical procedure. Since the moisture flux is driven by the humidity gradient, a simple adaptive scheme can be based on inverse proportionality of the time step to the humidity gradient at the beginning of the step.

Figure 8.21 shows that if the time step is properly adjusted, the CN method can correctly capture the evolution of the humidity profile, with no pathological oscil-

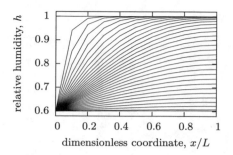

Fig. 8.21 Humidity profiles after individual computational steps, obtained using the Crank–Nicolson method with grid spacing $\Delta x = 0.1L$ and time step adaptively increased from $\Delta t_1 = 0.01\tau$ to $\Delta t_{30} = 0.97\tau$; the solid lines connect points that represent the coarse-grid solution, and the simulation covers the time range up to $t_{max} = 5\tau$

lations, and at the same time provide computational efficiency. The simulation was performed on a coarse grid with $\Delta x = 0.1L$, using an initial time step $\Delta t_1 = 0.01\tau$, which is slightly below the critical time step of the FEu method. At the end of the first step, the maximum slope $h'_{max,1}$ of the humidity profile was determined, and the second step was performed still with the same time increment, $\Delta t_2 = \Delta t_1$. After the second step, the maximum slope $h'_{max,2}$ of the humidity profile was determined, and the time increment for the third step, Δt_3, was evaluated from the condition $\Delta t_3 h'_{max,2} = \Delta t_2 h'_{max,1}$. Using an analogous rule, the subsequent time steps were progressively increased up to $\Delta t_{30} = 0.97\tau$. The simulation was supposed to cover the drying process up to time $t_{max} = 5\tau$, and so the very last time step was set to $\Delta t_{31} = t_{max} - t_{30}$. In this way, the time interval of interest was covered in just 31 steps, while a simulation with the initial time step kept fixed (which would be necessary for the conditionally stable FEu method) would require 500 steps. Of course, such savings of the computational time are irrelevant for problems solved on a coarse one-dimensional spatial grid, but they could be substantial in large-scale multidimensional simulations of structures drying over many years.

Despite the coarseness of the spatial grid, the evolution of relative humidity at the grid points is captured by the CN method with an adaptively increased time step quite well, as shown in Fig. 8.22a. The solid curve corresponds to a highly accurate solution, which used not only a very short (and fixed) time step but also a fine spatial grid. The filled circles show the results of the CN method with the time step adaptively increased from $\Delta t_1 = 0.01\tau$, which required 31 steps over the time period up to time $t_{max} = 5\tau$, while the empty circles correspond to the CN method with a fixed time step $\Delta t = 0.1\tau$ and a total of 50 steps. For easier evaluation of the differences, Fig. 8.22b shows the same results over the initial period up to time $t = \tau$. It is clear that the computation with adaptive time stepping provides a good resolution of the fast initial drying, thanks to the initially short time step 0.01τ. In contrast to that, the computation with the time step fixed to 0.1τ gives a slower reduction of the relative humidity at $x = L$ and an oscillatory evolution at the grid point near the exposed boundary ($x = 0.1L$), which is consistent with the results shown in Fig. 8.19b. ∎

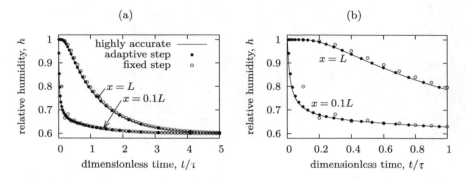

Fig. 8.22 Relative humidity at the impervious boundary ($x = L$) and at the internal grid point closest to the exposed boundary ($x = 0.1L$): (a) evolution over the entire time period up to $t = t_{max} = 5\tau$, (b) initial evolution up to time $t = \tau$

8.4.2.3 Axisymmetric Problems

So far we have considered the linear diffusion equation in the form (8.114), which corresponds to moisture flux across the thickness of an infinite slab. The linear version of Eq. (8.113), which describes an axisymmetric problem (radial moisture flux in an infinite cylinder), can be written as

$$\dot{h}(x, t) = C \left(h''(x, t) + \frac{1}{x} h'(x, t) \right) \tag{8.136}$$

where the diffusivity C has been considered as constant and the radial coordinate r has been renamed to x, to keep the same formalism as for the drying slab.

Extension of finite difference methods to Eq. (8.136) is relatively easy. It suffices to enrich the discretized equation by a suitable approximation of the additional term $Ch'(x, t)/x$. The first spatial derivative is best approximated by a central difference formula. Consequently, approximation (8.129) based on the generalized trapezoidal rule is extended to

$$\frac{h_i^{(k+1)} - h_i^{(k)}}{\Delta t} = C \left((1 - \alpha) \frac{h_{i-1}^{(k)} - 2h_i^{(k)} + h_{i+1}^{(k)}}{(\Delta x)^2} + \alpha \frac{h_{i-1}^{(k+1)} - 2h_i^{(k+1)} + h_{i+1}^{(k+1)}}{(\Delta x)^2} \right)$$

$$+ C \left((1 - \alpha) \frac{h_{i+1}^{(k)} - h_{i-1}^{(k)}}{2x_i \, \Delta x} + \alpha \frac{h_{i+1}^{(k+1)} - h_{i-1}^{(k+1)}}{2x_i \, \Delta x} \right) \tag{8.137}$$

Moving all known terms to the right-hand side and all unknown terms to the left-hand side, we can rewrite (8.137) as

$$-\theta\alpha\left(1 - \frac{\Delta x}{2x_i}\right)h_{i-1}^{(k+1)} + (1 + 2\theta\alpha)h_i^{(k+1)} - \theta\alpha\left(1 + \frac{\Delta x}{2x_i}\right)h_{i+1}^{(k+1)} =$$

$$= \theta(1 - \alpha)\left(1 - \frac{\Delta x}{2x_i}\right)h_{i-1}^{(k)} + [1 - 2\theta(1 - \alpha)]h_i^{(k)} + \theta(1 - \alpha)\left(1 + \frac{\Delta x}{2x_i}\right)h_{i+1}^{(k)}$$

$$(8.138)$$

which is a generalization of (8.130). Recall that $\theta = C \Delta t/(\Delta x)^2$. If the axisymmetric problem is solved on the interval $[0, L]$ where $x = 0$ corresponds to the cylinder axis and L is the cylinder radius (and if the grid is regular, which is an assumption already used in the derivation of all the foregoing finite difference approximations), fractions $\Delta x/x_i$ can be replaced by $1/i$. Equation (8.138) is also applicable to a hollow cylinder, with $x \in [L_1, L_2]$, where L_1 is the inner radius and L_2 is the outer radius.

For $\alpha = 0$, the left-hand side of (8.138) reduces to $h_i^{(k+1)}$ and an explicit update formula

$$h_i^{(k+1)} = \theta\left(1 - \frac{\Delta x}{2x_i}\right)h_{i-1}^{(k)} + (1 - 2\theta)h_i^{(k)} + \theta\left(1 + \frac{\Delta x}{2x_i}\right)h_{i+1}^{(k)} \qquad (8.139)$$

corresponding to the forward Euler method is obtained as a generalization of (8.116). For $\alpha > 0$, the scheme is implicit and a coupled set of equations with a tridiagonal matrix has to be solved.

For a solid cylinder, the left computational boundary is actually at the axis of symmetry and the appropriate boundary condition is $h'(0, t) = 0$. This can be translated into the condition $h_{-1}^{(k)} = h_1^{(k)}$ for the fictitious value of relative humidity at the ghost node located at $x_{-1} = -\Delta x$. However, Eq. (8.138) cannot be directly used for $i = 0$ because of the presence of $x_0 = 0$ in the denominator. Instead of that, one needs to go back to the original multidimensional linear diffusion equation (8.91) and take into account the fact that if the relative humidity depends exclusively on the radial coordinate r, the Laplacian at the axis of symmetry[18] is equal to $2\,\mathrm{d}^2h/\mathrm{d}r^2$, which in the present notation corresponds to $2h''$ evaluated at $x = 0$. Therefore, the finite difference approximation of the diffusion equation for $i = 0$ is written as

$$\frac{h_0^{(k+1)} - h_0^{(k)}}{\Delta t} = 2C\left((1 - \alpha)\frac{h_{-1}^{(k)} - 2h_0^{(k)} + h_1^{(k)}}{(\Delta x)^2} + \alpha\frac{h_{-1}^{(k+1)} - 2h_0^{(k+1)} + h_1^{(k+1)}}{(\Delta x)^2}\right)$$

$$(8.140)$$

and substituting $h_{-1}^{(k)} = h_1^{(k)}$ and $h_{-1}^{(k+1)} = h_1^{(k+1)}$, we obtain

---

[18]Note that if z is the axis of symmetry and if the humidity depends only on the radial coordinate $r = \sqrt{x^2 + y^2}$, the second derivatives at the axis of symmetry are $\partial^2h/\partial x^2 = \partial^2h/\partial y^2 = \mathrm{d}^2h/\mathrm{d}r^2$ and $\partial^2h/\partial z^2 = 0$, and so $\nabla^2h = 2\mathrm{d}^2h/\mathrm{d}r^2$. Another argument leading to the same result is that as $r \to 0$, we have $\mathrm{d}h/\mathrm{d}r \to 0$ and the ratio $(\mathrm{d}h/\mathrm{d}r)/r$ tends to $\mathrm{d}^2h/\mathrm{d}r^2$ based on the L'Hospital rule. Consequently, the second term on the right-hand side of (8.113) tends to $C(h)\mathrm{d}^2h/\mathrm{d}r^2$ and, for constant diffusivity, the right-hand side of (8.113) evaluated at $r = 0$ becomes equal to $2C\,\mathrm{d}^2h/\mathrm{d}r^2$.

$$\frac{h_0^{(k+1)} - h_0^{(k)}}{\Delta t} = \frac{4C}{(\Delta x)^2} \left((1 - \alpha)(h_1^{(k)} - h_0^{(k)}) + \alpha(h_1^{(k+1)} - h_0^{(k+1)}) \right) \qquad (8.141)$$

from which

$$(1 + 4\theta\alpha)h_0^{(k+1)} - 4\theta\alpha h_1^{(k+1)} = [1 - 4\theta(1 - \alpha)]h_0^{(k)} + 4\theta(1 - \alpha)h_1^{(k)} \qquad (8.142)$$

This equation together with Eq. (8.138) written for $i = 1, 2, \ldots m - 1$ forms a set of m linear equations for unknowns $h_i^{(k+1)}$, $i = 0, 1, 2, \ldots m - 1$ (assuming that a Dirichlet boundary condition is prescribed at $x = L$ and thus $h_m^{(k+1)}$ is known).

Example 8.4. Drying of a cylinder

Consider a cylinder of radius $R = L$ with uniform initial relative humidity $h(x, 0) = 1$, exposed at time $t = 0$ to ambient humidity $h_{env} = 0.6$. The cylinder is considered either as very long (theoretically infinite), or as sealed on both flat ends, so that moisture flux occurs at the curved boundary only and the problem can be described by the one-dimensional linear diffusion equation (8.136). Properties of the basic finite difference methods have already been discussed in the previous examples, and for the axisymmetric problem, they would be very similar. Let us, therefore, focus on the differences between the evolution of humidity in a slab and in a cylinder, assuming that the numerical solution has been found with a high level of accuracy.

For easier comparison, for both cases, we denote the length of the spatial interval as L (equal to one half of the thickness D for a slab and to the radius R for a cylinder) and we use a characteristic time $\tau = 4L^2/C\pi^2$ introduced in (8.123). Accurately computed profiles of relative humidity at selected times after exposure ranging from 0 to τ are plotted in Fig. 8.23a, and the evolution of the maximum relative humidity (i.e., of the relative humidity at the cylinder axis) is shown in Fig. 8.23b. For comparison, the evolution of the maximum relative humidity in a slab is plotted by the dashed curve.

Another variable of interest is the average humidity, because its rate is (for a model with constant moisture capacity) proportional to the rate of total water loss. The average is understood here as the volume average over the actual specimen, and thus, it is defined as

$$\bar{h}(t) = \frac{1}{L} \int_0^L h(x, t) \, dx \qquad (8.143)$$

for a slab and as

$$\bar{h}(t) = \frac{1}{\pi L^2} \int_0^L h(x, t) 2\pi x \, dx = \frac{2}{L^2} \int_0^L h(x, t) x \, dx \qquad (8.144)$$

for a cylinder. The evolution of average humidity in a slab and in a cylinder is shown in Fig. 8.24, both in the linear scale and in the semilogarithmic scale. The value of $\bar{h} = 0.8$, which is "halfway" between the initial value of $\bar{h} = 1$ and the asymptotically

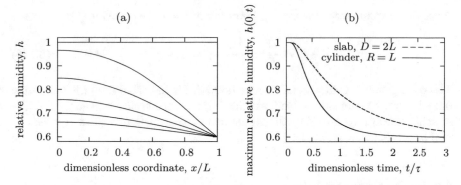

Fig. 8.23 (a) Humidity profiles in a cylinder at times $0, 0.2\tau, 0.4\tau, 0.6\tau, 0.8\tau$, and τ after exposure to drying, (b) evolution of maximum relative humidity in a cylinder of radius $R = L$ and in a slab of thickness $D = 2L$

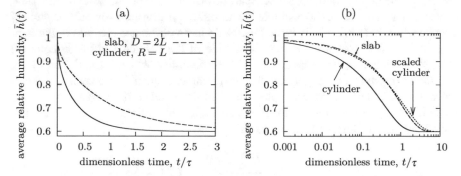

Fig. 8.24 Evolution of average relative humidity in a slab and in a cylinder: (a) linear scale, (b) semilogarithmic scale

approached final value of $\bar{h} = 0.6$, is attained at time $t_{\text{slab}} = 0.4854\tau$ for the slab and at time $t_{\text{cyl}} = 0.1555\tau$ for the cylinder.

It is natural to expect that the drying process is faster for the cylinder, because we have considered a cylinder of a radius equal to one half of the slab thickness. Since the drying time scales with the square of the size (see Sect. 8.3.5), the average humidity of 0.8 would be reached simultaneously (at time 0.4854τ) by a slab of thickness $D = 2L$ and a cylinder of radius

$$R = \sqrt{\frac{t_{\text{slab}}}{t_{\text{cyl}}}} \frac{D}{2} = \sqrt{\frac{0.4854}{0.1555}} L = 1.767L \qquad (8.145)$$

The effective size of a radially drying cylinder, defined as twice the volume divided by the area of the exposed surface, corresponds to the radius. Consequently, the shape factor that should multiply the effective size (radius) of a cylinder to get the equivalent thickness of a slab that would dry approximately at the same rate (in the sense that it would reach the average humidity of 0.8 at the same time) is

$$k_s = \frac{D}{R} = \frac{2L}{1.767L} = 1.132 \tag{8.146}$$

The corresponding evolution of average humidity is indicated in Fig. 8.24b by the dotted curve. A systematic evaluation of the shape factor and an improved criterion for matching the average humidity curves obtained for specimens of different geometries will be presented in Sect. 8.4.5.2.   ∎

8.4.2.4 Nonlinear Diffusion

So far, all numerical simulations of drying have been performed using a linear form of the diffusion equation, with a constant diffusivity C. For concrete, the diffusivity strongly depends on the pore relative humidity (or moisture content), as already explained in Sect. 8.3.4.2. To extend the basic finite difference methods to the nonlinear case, first note that the right-hand side of the nonlinear diffusion equation (8.106) contains the first spatial derivative of humidity, which is multiplied by the humidity-dependent diffusivity and then differentiated again. Using again the generalized trapezoidal rule, we can replace the right-hand side of (8.106) by a weighted average of approximations at times t_k and t_{k+1}, with weights $1 - \alpha$ and α. The resulting finite difference equation reads

$$\frac{h_i^{(k+1)} - h_i^{(k)}}{\Delta t} = (1 - \alpha) \frac{C_{i+1/2}^{(k)} \dfrac{h_{i+1}^{(k)} - h_i^{(k)}}{\Delta x} - C_{i-1/2}^{(k)} \dfrac{h_i^{(k)} - h_{i-1}^{(k)}}{\Delta x}}{\Delta x} +$$
$$+ \alpha \frac{C_{i+1/2}^{(k+1)} \dfrac{h_{i+1}^{(k+1)} - h_i^{(k+1)}}{\Delta x} - C_{i-1/2}^{(k+1)} \dfrac{h_i^{(k+1)} - h_{i-1}^{(k+1)}}{\Delta x}}{\Delta x} \tag{8.147}$$

To simplify the notation, the diffusivity that corresponds to humidity $h_{i+1/2}^{(k)} = \left(h_i^{(k)} + h_{i+1}^{(k)} \right) / 2$ has been denoted as $C_{i+1/2}^{(k)}$, and an analogous meaning has been attributed to $C_{i-1/2}^{(k)}$, $C_{i+1/2}^{(k+1)}$ and $C_{i-1/2}^{(k+1)}$.

Collecting all unknown terms on the left-hand side and all known terms on the right-hand side, we can rewrite (8.147) as

$$-\alpha \theta_{i-1/2}^{(k+1)} h_{i-1}^{(k+1)} + \left[1 + \alpha \left(\theta_{i-1/2}^{(k+1)} + \theta_{i+1/2}^{(k+1)} \right) \right] h_i^{(k+1)} - \alpha \theta_{i+1/2}^{(k+1)} h_{i+1}^{(k+1)} =$$
$$= (1 - \alpha) \theta_{i-1/2}^{(k)} h_{i-1}^{(k)} + \left[1 - (1 - \alpha) \left(\theta_{i-1/2}^{(k)} + \theta_{i+1/2}^{(k)} \right) \right] h_i^{(k)} + (1 - \alpha) \theta_{i+1/2}^{(k)} h_{i+1}^{(k)} \tag{8.148}$$

in which $\theta_{i-1/2}^{(k+1)} = C_{i-1/2}^{(k+1)}\Delta t/(\Delta x)^2$ etc. For $\alpha = 0$, the left-hand side of (8.148) reduces to $h_i^{(k+1)}$ and the formula becomes explicit. For any nonzero value of parameter α, each of the equations obtained by writing Eq. (8.148) for $i = 1, 2, \ldots m - 1$ contains three unknowns and a coupled set of equations needs to be solved. In contrast to (8.130), these equations are now nonlinear because factors $\theta_{i-1/2}^{(k+1)}$ and $\theta_{i+1/2}^{(k+1)}$ depend on the unknowns $h_{i-1}^{(k+1)}$, $h_i^{(k+1)}$, and $h_{i+1}^{(k+1)}$. This calls for an iterative solution, e.g., by the Newton–Raphson method, which is based on a successive linearization of the expression on the left-hand side around the current approximate solution. The ratio between the computational costs of implicit methods and of the explicit forward Euler method is thus higher than in the linear case.

In each time step, the iterative process is initialized by setting the approximation of the humidity values at the end of the step, $h_i^{(k+1,0)}$, equal to the already known (converged) values at the beginning of the step, $h_i^{(k)}$, with $i = 1, 2, \ldots m - 1$ (the second superscript in $h_i^{(k+1,j)}$ refers to iteration number j). The right-hand side of (8.148) is directly evaluated (because it depends exclusively on the already known values), and the left-hand side is replaced by the linear part of the Taylor expansion around the initial approximation. For instance, the first term on the left-hand side of (8.148) is linearized as follows:

$$-\alpha\theta_{i-1/2}^{(k+1)}h_{i-1}^{(k+1)} \approx -\alpha\theta_{i-1/2}^{(k+1,0)}h_{i-1}^{(k+1,0)} - \alpha\theta_{i-1/2}^{(k+1,0)}\delta h_{i-1}$$
$$-\alpha\theta_{i-1/2}'^{(k+1,0)}\frac{1}{2}(\delta h_{i-1} + \delta h_i)h_{i-1}^{(k+1,0)} \qquad (8.149)$$

Here, δh_{i-1} and δh_i are unknown iterative corrections of the approximate values $h_{i-1}^{(k+1,0)}$ and $h_i^{(k+1,0)}$, and coefficients $\theta_{i-1/2}^{(k+1,0)}$ and $\theta_{i-1/2}'^{(k+1,0)}$ are defined as

$$\theta_{i-1/2}^{(k+1,0)} = \frac{\Delta t}{(\Delta x)^2} C\left(\frac{h_{i-1}^{(k+1,0)} + h_i^{(k+1,0)}}{2}\right) \qquad (8.150)$$

$$\theta_{i-1/2}'^{(k+1,0)} = \frac{\Delta t}{(\Delta x)^2} \frac{dC}{dh}\left(\frac{h_{i-1}^{(k+1,0)} + h_i^{(k+1,0)}}{2}\right) \qquad (8.151)$$

After linearization, Eq. (8.148) can be rewritten as

$$-\alpha A_i^{(k+1,0)}\delta h_{i-1} + (1 + \alpha B_i^{(k+1,0)})\delta h_i - \alpha C_i^{(k+1,0)}\delta h_{i-1} = R_i^{(k+1,0)} \qquad (8.152)$$

where

$$A_i^{(k+1,0)} = \theta_{i-1/2}^{(k+1)} + \frac{1}{2}\theta_{i-1/2}'^{(k+1)}\left(h_{i-1}^{(k+1,0)} - h_i^{(k+1,0)}\right) \tag{8.153}$$

$$B_i^{(k+1,0)} = \theta_{i-1/2}^{(k+1)} - \frac{1}{2}\theta_{i-1/2}'^{(k+1)}\left(h_{i-1}^{(k+1,0)} - h_i^{(k+1,0)}\right)$$
$$+\theta_{i+1/2}^{(k+1)} - \frac{1}{2}\theta_{i+1/2}'^{(k+1)}\left(h_{i+1}^{(k+1,0)} - h_i^{(k+1,0)}\right) \tag{8.154}$$

$$C_i^{(k+1,0)} = \theta_{i+1/2}^{(k+1)} + \frac{1}{2}\theta_{i+1/2}'^{(k+1)}\left(h_{i+1}^{(k+1,0)} - h_i^{(k+1,0)}\right) \tag{8.155}$$

are dimensionless coefficients and $R_i^{(k+1,0)}$ is the residual, defined as the difference between the right-hand side and the left-hand side of (8.148).

Equations (8.152) written for $i = 1, 2, \ldots m - 1$ are linear and sparse (the corresponding matrix is tridiagonal) and can be solved using the same efficient algorithm as the set of equations (8.130) corresponding to the linear diffusion model. Once the corrections δh_i are computed, the approximation of humidities at the end of the current step is updated to $h_i^{(k+1,1)} = h_i^{(k+1,0)} + \delta h_i$ and the residual $R_i^{(k+1,1)}$ is evaluated for this improved approximation. If the residual is not small (i.e., if its norm exceeds a prescribed tolerance), the left-hand side of (8.148) is linearized again, but this time around $h_i^{(k+1,1)}$, and additional corrections δh_i are computed from a set of linear equations analogous to (8.152) but with coefficients $A_i^{(k+1,1)}$, $B_i^{(k+1,1)}$, and $C_i^{(k+1,1)}$ and residual $R_i^{(k+1,1)}$ evaluated from $h_i^{(k+1,1)}$ instead of $h_i^{(k+1,0)}$. The iterative process continues until the residual becomes sufficiently small. When the convergence criterion is satisfied, the humidity approximations $h_i^{(k+1,j)}$ computed in the last iteration number j are considered as the converged values $h_i^{(k+1)}$ and the simulation proceeds to the next time step.

The main idea of the Newton–Raphson procedure has been explained for the set of typical finite difference equations (8.148) with $i = 1, 2, \ldots m - 1$. These equations are sufficient if Dirichlet conditions are prescribed on both ends of the interval and thus the boundary values h_0 and h_m are known for an arbitrary time t_k or t_{k+1}. For boundary conditions that involve the humidity gradient, additional equations must be set up, as already exemplified by (8.131) in the linear case. Let us derive the additional equation that corresponds to Robin boundary condition (8.99) prescribed on the left boundary. Since our primary unknown is the pore relative humidity, we express the moisture flux in terms of the humidity gradient as $j_x = -c_p h'$, with c_p = moisture permeability, and then multiply both sides of (8.99) by the reciprocal moisture capacity k. The resulting boundary condition

$$-C\left(h(0, t)\right)\frac{\partial h(0, t)}{\partial x} + k\eta_e^* h(0, t) = k\eta_e^* h_{\text{env}} \tag{8.156}$$

can be replaced by the corresponding finite difference approximation

$$-C_0^{(k)}\frac{h_1^{(k)} - h_{-1}^{(k)}}{2\Delta x} + k\eta_e^* h_0^{(k)} = k\eta_e^* h_{\text{env}} \tag{8.157}$$

from which we get

$$h_{-1}^{(k)} = h_1^{(k)} + \frac{2\Delta x \, k\eta_e^*}{C_0^{(k)}} \left(h_{\mathrm{env}} - h_0^{(k)} \right) \tag{8.158}$$

Note that the factor k that multiplies η_e^* represents the reciprocal moisture capacity and not the time step number. Formula (8.158) is applicable to an arbitrary time step and thus remains valid if the superscript $^{(k)}$ is in all terms replaced by $^{(k+1)}$. Exploiting this formula, the ghost-node humidities $h_{-1}^{(k)}$ and $h_{-1}^{(k+1)}$ can be eliminated from Eq. (8.148) written for $i = 0$.

Examples of humidity profiles obtained by numerical solution of nonlinear diffusion equation (8.106) will be presented in the next section for specific cases with available experimental data that can be used for parameter identification.

Let us conclude the present numerical section by a short remark on alternative spatial discretization techniques. Finite difference methods are convenient for problems solved in one spatial dimension, i.e., on a spatial domain that corresponds to an interval, which can be easily covered by a uniform grid. For multidimensional problems, the finite element method provides much more flexibility regarding the spatial discretization. Arbitrarily shaped domains can be handled, and the computational mesh can be refined in regions with high humidity gradients. Finite element methods for modeling of fluid flow in porous media are extensively treated, e.g., in Lewis and Schrefler [572].

For problems that involve sharp moving interfaces between oversaturated and dried concrete, a numerical method that leads to an exact satisfaction of the mass balance equation is preferable. A spatial discretization that ensures exact mass balance can be constructed by the finite volume method [377, 592, 593], to be discussed in Sect. 13.4.

8.4.3   Drying of a Slab

8.4.3.1   Fitting of Experimental Data

To illustrate the influence of variable diffusivity and of the boundary condition on the evolution of humidity profiles, consider first the experiments of Abrams and Orals [9] and their fits by the Bažant–Najjar [166] model, described in Sect. 8.3.4.2. Abrams and Orals [9] measured the pore relative humidity in 6-inch specimens subjected to one-dimensional drying at three different environmental humidities ($h_{\mathrm{env}} = 0.1, 0.35$ and 0.5). Figure 8.25a-c shows the measured values (isolated points), for each case at four different locations and at two different times of drying. The solid curves represent fits obtained with humidity-dependent diffusivity and Dirichlet boundary conditions ($h = h_{\mathrm{env}}$ on the exposed surface). The parameters of the diffusivity function (8.89) are listed in Table 8.3, in the column denoted as "set A." For comparison, the dashed curves show the results obtained with Robin boundary conditions (8.99)–(8.100) and slightly modified parameters of function (8.89), denoted as "set B." Finally, the dotted curves correspond to the profiles that would be obtained with a linear model (i.e., constant diffusivity) and Dirichlet boundary conditions, with parameters denoted as "set C."

Note that the moisture permeability c_p and the reciprocal moisture capacity k do not need to be specified separately, but only their product $kc_p = C$ (the diffusivity) matters. The moisture capacity would play a role only if we wanted to compute the flux or water loss. Similarly, the surface emissivity η_e^* does not need to be specified directly, because if the boundary condition (8.99) is rewritten in terms of the humidity gradient and both sides are multiplied by k (to convert the moisture permeability into the diffusivity), the resulting Eq. (8.156) contains the product $k\eta_e^*$ but not η_e^* separately. This is why Table 8.3 specifies the product $k\eta_e^*$ as a relevant parameter that characterizes the boundary. Typical values of the moisture capacity are in the order of 100 kg/m^3, and so the actual surface emissivity for set B would be roughly $\eta_e^* \approx 10^{-3}$m/day$\times 100$ kg/m$^3 = 0.1$ kg/m^{2}day. This is much less than the typical values mentioned in Sect. 8.3.4.3, and thus, the illustrative solutions plotted by the dashed curves in fact correspond to specimens with partial protection against drying. For realistic values of surface emissivity, the solutions would be virtually the same as with the Dirichlet boundary condition.

Table 8.3 Parameters used for fitting of the humidity profiles measured by Abrams and Orals (sets A-C), Abrams and Monfore (sets D and E), and Nilsson (sets F and G)

Parameters	set A	set B	set C	set D	set E	set F	set G
α_0	0.05	0.05	1	0.12	1	0.3	0.4
h_c	0.8	0.75	–	0.75	–	0.7	0.9
r	16	16	–	6	–	6	20
C_1 [mm^2/day]	60	70	8	25	8	5	6
$k\eta_e^*$ [mm/day]	–	1	–	0.1	0.1	–	–
Boundary conditions	Dirichlet	Robin	Dirichlet	Dir./Robin	Dir./Robin	Dirichlet	Dirichlet

Figure 8.25d shows the data of Abrams and Monfore [8] and their fits obtained with parameter sets D and E, listed in Table 8.3. The experiments were performed on a 6-inch slab with one surface exposed to the environment of relative humidity $h_{\text{env}} = 0.1$ and the opposite surface sealed. If the sealed boundary had been truly impervious, the test would have been equivalent to symmetric drying of a 12-inch slab. However, the actually recorded pore humidities are not monotonically increasing with increasing distance from the exposed boundary (see the isolated points in Fig. 8.25d). This indicates that the seal was leaking and the homogeneous Neumann condition at the sealed surface is not appropriate. The simulation with the Bažant–Najjar model is thus done with the Dirichlet condition at the exposed boundary ($x = 0$) and Robin condition at the sealed boundary ($x = 152$ mm), using a low value of surface emissivity. The measured shape of the humidity profiles can be

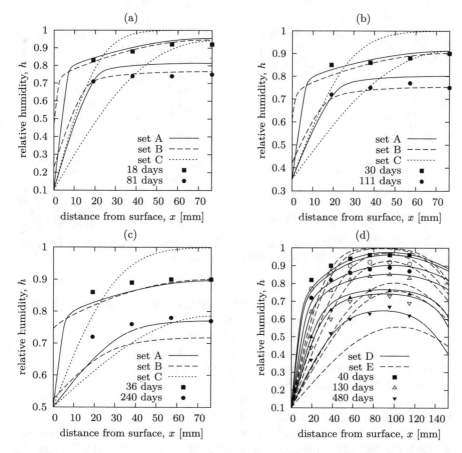

Fig. 8.25 Humidity profiles and their fits by the Bažant–Najjar model: (a) data of Abrams and Orals, $h_{\text{env}} = 0.1$, (b) data of Abrams and Orals, $h_{\text{env}} = 0.35$, (c) data of Abrams and Orals, $h_{\text{env}} = 0.5$, (d) data of Abrams and Monfore, $h_{\text{env}} = 0.1$, drying times (from top to bottom): 40, 50, 90, 130, 230, 270, and 480 days

approximated quite well with the nonlinear diffusion model (parameter set D) if $k\eta_e^*$ is set to 0.1 mm/day; see the solid curves in Fig. 8.25d. For comparison, the dashed curves show the results obtained with a linear diffusion model (parameter set E), which leads to large deviations from the experimental data.

The values of diffusivity at saturation, C_1, determined by fitting of the experimental results of Abrams and Orals [9] and listed in columns A and B of Table 8.3, are between 60 and 70 mm²/day. Much lower values have been found by fitting of two data sets reported by Nilsson [661], who tested slabs with the thickness of 160 mm exposed to 40% environmental humidity after 3 and 28 days of curing. The corresponding parameters are listed, respectively, in columns F and G of Table 8.3, and the computed curves are shown along with the measured values in Fig. 8.26. Only

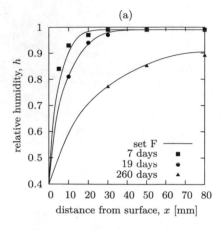
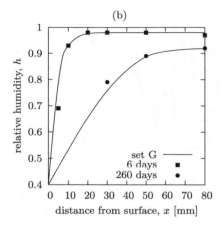

Fig. 8.26 Humidity profiles and their fits by the Bažant–Najjar model, data of Nilsson [661], exposure to $h_{\mathrm{env}} = 0.4$ after (a) 3 days of curing, (b) 28 days of curing

Dirichlet boundary conditions are presented here because the results with Robin boundary conditions are not much different [473].

Good approximations of the flat part of the humidity profiles in the specimen core have been made possible by taking into account the self-desiccation and setting the initial pore relative humidity in the entire specimen to 99% for set F (3 days of curing) and to 98% for set G (28 days of curing). Note that, for the present optimized data sets, the ratio $\alpha_0 \equiv C(0)/C(1)$ is well above the range between 0.025 and 0.1 recommended by Bažant and Najjar [166].

The humidity profiles in Figs. 8.25 and 8.26 have been limited to the drying times at which the measured humidity values were reported in the original papers. An example of the complete evolution of the humidity profile is provided in Fig. 8.27a, based on a simulation with parameter set A. For comparison, Fig. 8.27b shows the profiles that would be obtained with constant diffusivity (parameter set C). It is seen that the shapes of the profiles at late stages of drying are quite similar (but they are attained at different times). This is not surprising, because when the maximum humidity in the specimen core becomes sufficiently smaller than 0.8 (which is the value of parameter h_c), the diffusivity evaluated from (8.89) is almost constant across the whole specimen and the nonlinearity of the governing equation becomes very weak.

The differences between the models with variable and constant diffusivity are also illustrated by Fig. 8.28, which shows the time evolution of relative humidity at the distance of 76 mm from the exposed surface of the slab used by Abrams and Monfore [8]. At humidities close to 1, the nonlinear model with set-D parameters exhibits a higher diffusivity than the linear model with set-E parameters. Therefore, the initial stage of drying is faster for the nonlinear model. After some time, the zone with lower humidities near the boundary gets thicker and its lower diffusivity

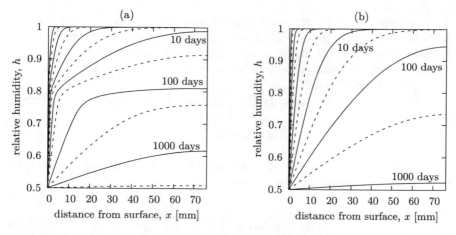

Fig. 8.27 Humidity profiles in a 6-inch slab exposed to 50% ambient relative humidity up to long times, computed with (a) variable diffusivity (parameter set A), (b) constant diffusivity (parameter set C)

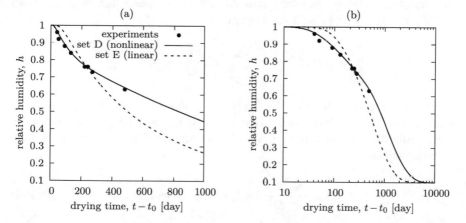

Fig. 8.28 Evolution of relative humidity at a point of a slab exposed to 10% ambient relative humidity: (a) linear time scale, (b) logarithmic time scale

acts as a barrier for the moisture flux. The drying process is thus slowed down, as compared to the linear model. Figure 8.28 clearly shows that a good agreement with experimental results cannot be achieved without accounting for the nonlinearity.

8.4.3.2 Long-Time Asymptotics of Drying

Consider now the idealized case of constant ambient humidity, h_{env}. Near the end of the drying process, the pore humidity $h(x, t)$ throughout the wall is close to h_{env}, and so the reciprocal moisture capacity, k, and the permeability, c_p, may be considered as almost constant. They are also independent of concrete age, because for pore humidity below about 65 to 80% (depending on the type of cement) the hydration reaction stops, and the environmental humidity is often lower. Therefore, the final stage of drying is properly described by the linear diffusion equation (8.91) with constant diffusivity $C_e = C(h_{env})$.

A slab of thickness D is described by a one-dimensional version of the diffusion equation,

$$\dot{h}(x, t) = C_e h''(x, t) \qquad (8.159)$$

The solution can be constructed by expanding the humidity profiles into Fourier series in terms of the spatial coordinate x:

$$h(x, t) = h_{env} + \sum_{n=1}^{\infty} h_n(t) \sin \frac{n\pi x}{D} \qquad (8.160)$$

The coefficients h_n multiplying individual harmonic terms are taken as time-dependent, and their evolution is determined from the diffusion equation. Substituting (8.160) into (8.159), we obtain the relation

$$\sum_{n=1}^{\infty} \dot{h}_n(t) \sin \frac{n\pi x}{D} = -\sum_{n=1}^{\infty} C_e \frac{n^2 \pi^2}{D^2} h_n(t) \sin \frac{n\pi x}{D} \qquad (8.161)$$

which can be satisfied for all $x \in (0, D)$ only if the terms multiplying individual harmonic functions on both sides are the same. Therefore, the original second-order partial differential equation is transformed into a series of first-order ordinary differential equations

$$\dot{h}_n(t) = -\frac{n^2 \pi^2 C_e}{D^2} h_n(t), \qquad n = 1, 2, \ldots \qquad (8.162)$$

Separation of variables leads to the solutions

$$h_n(t) = h_n(\bar{t}) \exp\left(-\frac{n^2 \pi^2 C_e}{D^2}(t - \bar{t})\right), \qquad n = 1, 2, \ldots \qquad (8.163)$$

Time $\bar{t}$ would normally be taken as the initial time t_0, and the corresponding value $h_n(\bar{t})$ would be obtained from the initial distribution of humidity. However, the diffusion problem is in general nonlinear and its linear version applies only to the late stages of the drying process, when the humidity profile is almost uniform. So we cannot precisely specify the initial condition, but formula (8.163) still provides valuable information on the asymptotic behavior at very long times.

As time t tends to infinity, the argument of the exponential function in (8.163) tends to minus infinity, and the exponential term thus tends to zero. As expected, the solution (8.160) approaches a uniform humidity distribution. But the important point is that the exponential terms in (8.163) do not decay at the same rate—those with a higher value of n decay faster, due to the presence of n^2 in the argument of the exponential. If the initial distribution of humidity is symmetric, the even harmonic terms in (8.160) vanish from the very beginning, and the rate of decay of the leading term with $n = 1$ is an order of magnitude lower than the rate of decay of higher-order terms with $n \geq 3$. Therefore, at long times, higher-order harmonics will become negligible compared to the first harmonic term. This is clearly seen in the numerical solutions shown in Fig. 8.27.

We can conclude that the asymptotic solution is given by

$$h(x, t) = h_{\text{env}} + \Delta \bar{h} \sin \frac{\pi x}{D} \exp \left(-\frac{\pi^2 C_e}{D^2} (t - \bar{t}) \right) \qquad (8.164)$$

where $\Delta \bar{h}$ is a constant that corresponds to the difference between the humidity $h(D/2, \bar{t})$ in the middle of the slab at some large time $\bar{t}$ and the environmental humidity h_{env}. The distribution of humidity across the slab is given by a sine half-wave with an amplitude exponentially decreasing in time. A similar distribution of humidity was considered in Examples 8.1 and 8.2, where it was assumed that the diffusion equation is linear and the initial humidity distribution is harmonic, and thus, (8.164) was the exact solution, not just the asymptotic one.

Research under way at Northwestern at the time of proof indicates that, for ambient humidities above cca 0.8, Eq. (8.164) needs a correction by adding a term evolving either logarithmically or as a power function of low exponent such as 0.2. The reason is self-desiccation, which evolves in this way for at least 10 years, probably for a century or more.

8.4.3.3 Steady-State Permeation Through an Unsaturated Wall

Permeability of many materials is measured by steady-state water flux through a wall whose opposite surfaces are kept at different humidities $h(0) = h_0$ and $h(D) = h_D$. However, due to very small permeability, a steady-state permeation in concrete is attainable only for thin concrete slabs—within about 5 years for a concrete slab < 5 cm thick, and within 100 days for a slab of mortar or cement paste < 1 cm thick. No wonder that there exist only scant experimental data for the steady state. The results of Wierig [868], converted by Bažant and Najjar [166] from the water content to relative humidity assuming a linear isotherm, are reproduced in Fig. 8.29. It should be noted that the steady state of the humidity profile is highly nonlinear. This is a direct manifestation of the strong dependence of diffusivity on relative humidity, as given by (8.89).

At steady state, the water content w_t remains constant in time and the water flux j_x is thus uniform across the slab (this follows from the one-dimensional version of the mass balance law (8.76) with a zero left-hand side). The one-dimensional version of the transport law (8.84) can be written as

$$j_x = -c_p(h) \frac{dh}{dx} \tag{8.165}$$

After separation of variables and integration, one gets

$$j_x x = -\int_{h_0}^{h(x)} c_p(h) \, dh \tag{8.166}$$

The yet unknown flux j_x is determined from the condition

$$j_x D = -\int_{h_0}^{h_D} c_p(h) \, dh \tag{8.167}$$

which follows from the boundary condition $h(D) = h_D$.

For a given function $c_p(h)$ describing the dependence of permeability on pore relative humidity, we can evaluate (at least numerically) the function

$$\Gamma_p(h) = -\int_{h_0}^{h} c_p(\tilde{h}) \, d\tilde{h} = \int_{h}^{h_0} c_p(\tilde{h}) \, d\tilde{h} \tag{8.168}$$

and rewrite (8.166) as

$$\frac{x}{D} = \frac{\Gamma_p(h(x))}{\Gamma_p(h_D)} \tag{8.169}$$

By solving this equation, we could get a formula describing the dependence of $h(x)$ on x, based on the inverse function of Γ_p. Instead of that, we can construct the humidity profile by computing the values of x/D corresponding to selected values of $h(x)$ directly from (8.169).

For instance, for boundary values $h_0 = 0.95$ and $h_D = 0$ and for the permeability function $c_p(h)$ in the same form as the Bažant–Najjar diffusivity function (8.89) with somewhat unusual parameter values $\alpha_0 = 0.005$, $h_c = 0.98$ and $r = 1.4$ (see Fig. 8.29b), the steady-state humidity profile plotted in Fig. 8.29a by the solid curve fits the measured values (extracted by Bažant and Najjar [166] from the experimental data of Wierig [868]) very well. Note that parameter c_{p1} (permeability at saturation) and the wall thickness D do not affect the shape of the profile (although they do affect the water flux).

The desorption isotherm affects the conversion of water content into pore relative humidity but, as long as its slope (the reciprocal moisture capacity k) is constant, it has no influence on the shape of the steady-state humidity profile. For a model with constant permeability, a constant flux would imply a constant humidity gradient and the humidity profile would be linear; see the dashed line in Fig. 8.29a. It is obvious that the deviation from linearity is substantial. The "optimal" diffusivity function, graphically shown in Fig. 8.29b, has a similar character to the function proposed by Roncero [732] and described in Appendix I.4.1, see Eq. (I.18) and Fig. I.10.

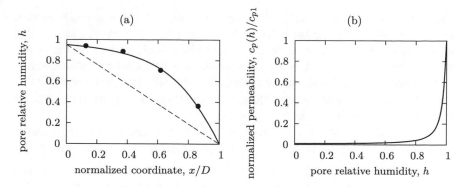

Fig. 8.29 (a) Distribution of pore relative humidity across a wall; dots correspond to experimental data of Wierig [868], solid curve to the theoretical solution based on variable diffusivity according to (8.89), dashed line to linear diffusion theory; (b) dependence of moisture permeability on pore relative humidity providing a good fit of the humidity profile

8.4.4 Initial Drying and Analysis of Infinite Half-Space

8.4.4.1 Penetration Depth of Drying Front After Sudden Exposure: Initial Asymptotics

The depth of penetration of a drying front beneath the body surface may be defined as the distance to the point where a measurable change of humidity occurs. Let us at first ignore aging and consider the initial penetration of drying fronts from the opposite surfaces into a slab or a wall. Before the drying fronts meet, the problem is equivalent to the penetration of a drying front into a half-space. The diffusion equation can be reduced to one spatial dimension, because the solution depends only on the coordinate x defined as the distance from the surface exposed to the environmental humidity. To facilitate the scaling analysis, it is convenient to consider the pore relative humidity $h(x, \hat{t})$ as a function of x and of the drying time $\hat{t} = t - t_0$. Equation (8.105) is then solved with the initial condition $h(x, 0) = 1$ for $x \geq 0$, and the boundary condition $h(0, \hat{t}) = h_{env}$ for $\hat{t} > 0$, in which h_{env} is the given environmental humidity.

As already explained in Sect. 8.3.5, simple scaling arguments show that if $h_1(x, \hat{t})$ is the solution of the diffusion problem on a certain domain of size D_1, then the solution of the diffusion problem on a geometrically similar domain of size $D_2 = \alpha D_1$ is given by $h_2(x, \hat{t}) = h_1(x/\alpha, \hat{t}/\alpha^2)$, provided that the initial and boundary conditions are consistent with the scaling relation. Since the infinite half-space is self-similar (i.e., is mapped onto itself by a homothetic mapping with an arbitrary scaling factor α), both solutions h_1 and h_2 refer to the same domain and we obtain the condition

$$h(x, \hat{t}) = h(\alpha x, \alpha^2 \hat{t}) \qquad \text{for all } \alpha > 0 \qquad (8.170)$$

Consequently, if we know the humidity profile as a function of x at one fixed time instant, we can easily construct the profile at any other time by simple rescaling of the spatial coordinate. Thus, the solution can be described by a function of one variable only, and the partial differential equation (8.105) can be transformed into an ordinary differential equation. To achieve that, we could introduce a new variable $\xi = x/\sqrt{\hat{t}}$ and, taking into account the scaling condition (8.170), write the solution as a function of ξ only, e.g., as $h(x, \hat{t}) = f(\xi)$.

However, it is better to work with dimensionless, normalized variables, because then the transformed problem has the simplest possible form. Therefore, we introduce the dimensionless variable $\xi = x/\sqrt{k_1 c_{p1} \hat{t}}$, where $k_1 = k(1)$ and $c_{p1} = c_p(1)$ are, respectively, the reciprocal moisture capacity at full saturation and the permeability at full saturation. Due to the scaling condition (8.170), the dependence of humidity on the spatial variable and time can be expressed in the form

$$h(x, \hat{t}) = h_{\text{env}} + (1 - h_{\text{env}}) \, f(\xi), \qquad \xi = \frac{x}{\sqrt{k_1 c_{p1} \hat{t}}} \qquad (8.171)$$

Here, f is a new unknown function, which is normalized such that its value is 0 when $h = h_{\text{env}}$ and is 1 when $h = 1$. The advantage is that the boundary conditions for f then do not depend on the environmental humidity h_{env}, as will be shown later.

According to the chain rule, partial derivatives with respect to space and time can be expressed as

$$\frac{\partial \bullet}{\partial x} = \frac{\partial \bullet}{\partial \xi} \frac{\partial \xi}{\partial x} = \frac{1}{\sqrt{k_1 c_{p1} \hat{t}}} \frac{\partial \bullet}{\partial \xi} \qquad (8.172)$$

$$\frac{\partial \bullet}{\partial t} = \frac{\partial \bullet}{\partial \xi} \frac{\partial \xi}{\partial \hat{t}} = -\frac{x}{2\sqrt{k_1 c_{p1} \hat{t}^3}} \frac{\partial \bullet}{\partial \xi} = -\frac{\xi}{2\hat{t}} \frac{\partial \bullet}{\partial \xi} \qquad (8.173)$$

Using (8.171)–(8.173) in (8.105), we get an ordinary differential equation

$$\xi \frac{\partial f}{\partial \xi} + 2\kappa(f) \frac{\partial}{\partial \xi} \left[\gamma(f) \frac{\partial f}{\partial \xi} \right] = 0 \qquad (8.174)$$

in which

$$\kappa(f) = \frac{k(h_{\text{env}} + (1 - h_{\text{env}})f)}{k_1} \qquad (8.175)$$

$$\gamma(f) = \frac{c_p(h_{\text{env}} + (1 - h_{\text{env}})f)}{c_{p1}} \qquad (8.176)$$

are the normalized reciprocal moisture capacity and normalized permeability, considered as functions of the normalized humidity variable f.

Equation (8.174) is a nonlinear ordinary differential equation, to be solved for the unknown function $f(\xi)$ on the interval $[0, \infty)$. Since it is an equation of the second order, it requires two boundary conditions. One of them, $f(0) = 0$, directly follows from the boundary condition $h(0, \hat{t}) = h_{\text{env}}$ of the original partial differential equation. The solution should also reflect the initial condition of the original problem. For fixed $x > 0$ and $\hat{t}$ approaching zero from above, variable $\xi = x/\sqrt{k_1 c_{p1} \hat{t}}$ tends to plus infinity, so the initial condition $h(x, 0) = 1$ will be translated into the condition $f(\xi) \to 1$ as $\xi \to \infty$ (i.e., in shorthand notation, $f(\infty) = 1$).

Example 8.5. Humidity profiles for the linear diffusion model

Equation (8.174) remains nonlinear even if the original partial differential equation (8.105) is linear, i.e., for constant values of k and c_p. Nevertheless, in this special case, the solution can be constructed in closed form. Denoting $g = \partial f/\partial \xi$, we can rewrite (8.174) with constant k and c (i.e., with functions κ and γ defined in (8.175) and (8.176) identically equal to 1) as

$$\xi g + 2\frac{\partial g}{\partial \xi} = 0 \tag{8.177}$$

Integration after separation of variables then gives

$$g(\xi) = g_0 \, \exp\left(-\frac{\xi^2}{4}\right) \tag{8.178}$$

where g_0 is an integration constant. Integrating g and using the boundary conditions $f(0) = 0$ and $f(\infty) = 1$ to determine g_0 and another, newly emerged integration constant, we finally get

$$f(\xi) = \text{erf}\left(\frac{\xi}{2}\right) \tag{8.179}$$

where erf is the well-known error function, closely related to the Gaussian cumulative distribution function and defined by the formula

$$\text{erf}(x) = \frac{2}{\sqrt{\pi}} \int_0^x e^{-s^2} \, ds \tag{8.180}$$

∎

Function f given by (8.179) and plotted in Fig. 8.30 by the dashed line represents the solution of the linear diffusion problem on a half-space, in terms of the original physical variables written as

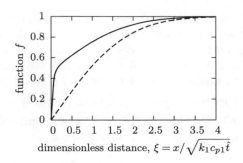

Fig. 8.30 Dimensionless humidity profile $f(\xi)$ for the Bažant–Najjar model with parameters $\alpha_0 = 0.05, h_c = 0.8,$ and $r = 15$ (solid), compared to the profile corresponding to linear diffusion (dashed)

$$h(x, \hat{t}) = h_{\text{env}} + (1 - h_{\text{env}}) \, \text{erf} \left(\frac{x}{2\sqrt{kc_p\hat{t}}} \right) \tag{8.181}$$

For fixed $\hat{t}$, we obtain the profile of humidity through the half-space at the given time, described by a linearly transformed error function of x. For fixed x, we obtain the history of humidity at the given distance from the surface. If we prescribe a fixed value of h between h_{env} and 1, we obtain a relation between x and $\hat{t}$ characterizing the propagation of the plane of given humidity into the half-space.

For instance, if we define the drying front as the set of points at humidity $1 - \Delta h$, where Δh is a small constant, we obtain from the condition $h(x_d, \hat{t}_d) = 1 - \Delta h$ the following description of the depth of penetration x_d as a function of the drying time $\hat{t}_d$:

$$x_d = 2\sqrt{kc_p\hat{t}_d} \, \text{erf}^{-1} \left(1 - \frac{\Delta h}{1 - h_{\text{env}}} \right) = \text{constant} \times \sqrt{\hat{t}_d} \tag{8.182}$$

More specifically, if we consider the drying front as the point where the pore relative humidity differs from 1 by only 1% of the value by which the environmental relative humidity differs from 1, we have $\Delta h/(1 - h_{\text{env}}) = 0.01$ and (8.182) can be rewritten as

$$x_d \approx 3.6\sqrt{kc_p\hat{t}_d} \tag{8.183}$$

This is very close to the formula $x_d = \sqrt{12kc_p\hat{t}_d} \approx 3.46\sqrt{kc_p\hat{t}_d}$, widely used as a simple engineering estimate based on an approximation of the humidity profile by a parabola (to be developed later, see formula (8.195)).

Independently of the choice of Δh, the depth of penetration, x_d, is always proportional to the square root of the drying time. This is a logical consequence of the fact that a given value of humidity uniquely corresponds to a certain value of $\xi = x/\sqrt{kc_p\hat{t}}$ and, to make ξ constant, x must be proportional to $\sqrt{\hat{t}}$ or, put the other way around, $\hat{t}$ proportional to x^2.

We can also characterize the role of the diffusivity, $C = kc_p$. From (8.182), it is easily seen that the drying depth (at a fixed time) is proportional to the square root of diffusivity and also that the drying time (at a fixed distance from the surface) is inversely proportional to the diffusivity.

Interestingly enough, the foregoing results remain qualitatively correct even if the diffusion problem is nonlinear, i.e., if the parameters k and c_p are functions of humidity. In such a case, a closed-form solution of the differential equation (8.174) is in general not available, but the solution can be constructed numerically. No matter how the resulting function $f(\xi)$ looks, a constant humidity implies constant ξ, i.e., a constant ratio $x/\sqrt{k_1 c_{p1} \hat{t}}$. So again, for fixed material properties, we have $x_d \propto \sqrt{\hat{t_d}}$ and $\hat{t_d} \propto x_d^2$.

This explains why, for short drying times, the shrinkage function $S(\hat{t})$ should be proportional to $\sqrt{\hat{t}}$; see Eq. (3.16). If the entire function $k(h)$ is multiplied by a constant β_k and the entire function $c_p(h)$ is multiplied by a constant β_c, then the normalized functions $\kappa(f)$ and $\gamma(f)$ defined in (8.175) and (8.176) do not change and thus also the solution $f(\xi)$ of Eq. (8.174) remains the same. The scaling factors will affect only the relation between x and $\hat{t}$ for given ξ, because constants k_1 and c_{p1} are scaled by β_k and β_c, and so drying depths are proportional to $\sqrt{\beta_k \beta_c}$ and drying times inversely proportional to $\beta_k \beta_c$.

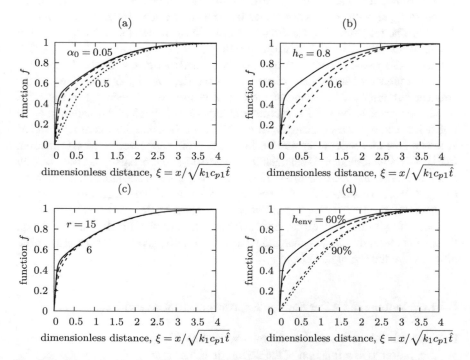

Fig. 8.31 Dimensionless humidity profile $f(\xi)$ for the Bažant–Najjar model: (a) influence of parameter α_0, (b) influence of parameter h_c, (c) influence of parameter r, (d) influence of ambient humidity h_{env}

Example 8.6. Humidity profiles for the Bažant–Najjar model

Consider the Bažant–Najjar model with diffusivity approximated by function (8.89) and with the parameters recommended by the *fib* Model Code: $\alpha_0 = 0.05$, $h_c = 0.8$, and $r = 15$. The normalized reciprocal moisture capacity function κ defined by (8.175) is identically equal to 1. At $h_{env} = 60\%$, the normalized permeability function γ defined by (8.176) can be expressed as

$$
\gamma(f) = \alpha_0 + \frac{1 - \alpha_0}{1 + \left(\frac{1 - h_{env} - (1 - h_{env})f}{1 - h_c} \right)^r} = 0.05 + \frac{0.95}{1 + \left(\frac{1 - 0.6 - 0.4f}{1 - 0.8} \right)^{15}} =
$$
$$
= 0.05 + \frac{0.95}{1 + [2(1 - f)]^{15}} \tag{8.184}
$$

Equation (8.174) simplifies to

$$
\xi \frac{\partial f}{\partial \xi} + 2 \frac{\partial}{\partial \xi} \left[\gamma(f) \frac{\partial f}{\partial \xi} \right] = 0 \tag{8.185}
$$

The numerically constructed solution is plotted by the solid line in Fig. 8.30. On the vertical axis, the values of f between 0 and 1 correspond to the humidities between 60 and 100%. It is clearly seen that the humidity gradient is much higher in the boundary layer with f between 0 and 0.5 (i.e., h between 60 and 80%) than in the core. The reason is that, for the present nonlinear model, the diffusivity at pore humidities below $h_c = 80\%$ is strongly reduced. For comparison, solution (8.179) corresponding to the linear diffusion model is plotted in Fig. 8.30 by the dashed line.

The influence of individual parameters of the Bažant–Najjar model and of the ambient humidity on the shape of the humidity profile is illustrated in Fig. 8.31. In all four graphs, the solid curve corresponds to the standard *fib* parameters. In Fig. 8.31a, parameter α_0 is changed from 0.05 to 0.1, 0.3, and 0.5. As α_0 approaches 1, the nonlinearity fades away and the humidity profile approaches the erf function. In Fig. 8.31b, parameter h_c is changed from 0.8 to 0.7 and 0.6. Again, the nonlinearity fades away, because the transition to low diffusivity takes place at lower humidities. In Fig. 8.31c, the exponent r is changed from 15 to 9 and 6. The effect on the shape of the humidity profile is rather weak. For lower exponents, the transition between the regions with low and high diffusivity becomes more gradual. Finally, in Fig. 8.31d, the ambient humidity h_{env} is changed from 60 to 70, 80 and 90%. This has a similar effect as a reduction of h_c at fixed h_{env}. ∎

8.4.4.2   Water Loss After Sudden Exposure: Initial Asymptotics

It is also interesting to calculate the total water loss W_L (per unit area of the drying surface, [kg/m^2]) as a function of the drying time. The rate of W_L corresponds to the flux j_x on the surface, but it has the opposite sign because the normal to the surface is oriented in the negative direction (for our particular formulation on the interval $[0, \infty)$). So the total water loss at time $\hat{t}$ can be evaluated as

$$W_L(\hat{t}) = -\int_0^{\hat{t}} j_x(0, t') \, dt' \tag{8.186}$$

According to the one-dimensional version of (8.84) combined with (8.171)–(8.172), we have

$$j_x(0, \hat{t}) = -c_p(h(0, \hat{t})) \frac{\partial h(0, \hat{t})}{\partial x} = -c_p(h_{\mathrm{env}}) \frac{1 - h_{\mathrm{env}}}{\sqrt{k_1 c_{p1} \hat{t}}} \frac{df(0)}{d\xi} \tag{8.187}$$

and so

$$W_L(\hat{t}) = \frac{(1 - h_{\mathrm{env}}) c_p(h_{\mathrm{env}}) f_{\xi 0}}{\sqrt{k_1 c_{p1}}} \int_0^{\hat{t}} \frac{dt'}{\sqrt{t'}} = \frac{2(1 - h_{\mathrm{env}}) c_p(h_{\mathrm{env}}) f_{\xi 0}}{\sqrt{k_1 c_{p1}}} \sqrt{\hat{t}} \tag{8.188}$$

where $f_{\xi 0} = df(0)/d\xi$ is the derivative of f with respect to ξ evaluated at $\xi = 0$. As may have been expected, the total water loss is proportional to the square root of the drying time. The proportionality factor depends on the permeability function and on the desorption isotherm. For instance, for the linear diffusion problem, we can use the solution $f(\xi)$ given by (8.179) to evaluate $f_{\xi 0} = 1/\sqrt{\pi}$, and the proportionality factor is obtained explicitly as

$$\frac{2(1 - h_{\mathrm{env}}) c_p f_{\xi 0}}{\sqrt{k c_p}} = 2(1 - h_{\mathrm{env}}) \sqrt{\frac{c_p}{\pi k}} \tag{8.189}$$

Since $k c_p = C = $ diffusivity, the resulting expression for the total water loss after drying time $\hat{t}$ can be presented as

$$W_L(\hat{t}) = \frac{1 - h_{\mathrm{env}}}{k} \frac{2}{\sqrt{\pi}} \sqrt{C \hat{t}} \tag{8.190}$$

Here, the first fraction, $(1 - h_{\mathrm{env}})/k$, represents the mass of water that is removed from a unit volume of concrete when it dries from full saturation to relative humidity h (recall that $1/k$ is the moisture capacity). The second fraction, $2/\sqrt{\pi}$, is a dimensionless factor related to the shape of the humidity profile. The last term, $\sqrt{C \hat{t}}$, reflects the effect of diffusivity and of the drying time. The ratio $W_L(\hat{t})/[(1 - h_{\mathrm{env}})/k]$ defines an equivalent drying depth in the sense that the water mass actually removed from the half-space is equal to the water mass that would be removed by uniformly drying a surface layer up to this depth.

A frequently used simple engineering estimate of the depth of penetration is based on the approximation of the humidity profile by a parabola with apex at the prop-

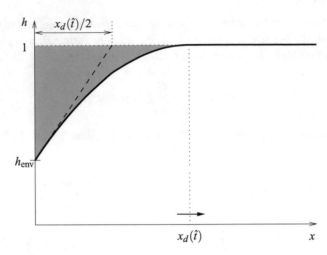

Fig. 8.32 Parabolic humidity profile used by a simple engineering estimate of the drying front position

agating front; see Fig. 8.32. Of course, such a quadratic humidity distribution does not satisfy the diffusion equation pointwise, but the dependence of the penetration depth x_d on time can be estimated using the relation between the flux at the boundary and the water loss. The flux at the boundary can be linked to the humidity gradient, which corresponds to the slope of the parabola (see Fig. 8.32 for a geometrical interpretation):

$$j_x(0, \hat{t}) = -c_p \frac{\partial h(0, \hat{t})}{\partial x} = -c_p \frac{1 - h_{\text{env}}}{x_d(\hat{t})/2} \tag{8.191}$$

The water loss at a given time corresponds to the shaded area above the humidity profile (see Fig. 8.32) multiplied by the moisture capacity:

$$W_L(\hat{t}) = \frac{1}{k} \int_0^{x_d(\hat{t})} [1 - h(x, \hat{t})] \, dx = \frac{1}{k} \times \frac{1}{3} (1 - h_{\text{env}}) x_d(\hat{t}) \tag{8.192}$$

Substituting (8.191)–(8.192) into the differential form of Eq. (8.186),

$$dW_L(\hat{t}) = -j_x(0, \hat{t}) \, d\hat{t} \tag{8.193}$$

we obtain

$$\frac{1 - h_{\text{env}}}{3k} \, dx_d(\hat{t}) = c_p \frac{2(1 - h_{\text{env}})}{x_d(\hat{t})} \, d\hat{t} \tag{8.194}$$

Separation of variables and integration with initial condition $x_d(0) = 0$ then leads to the formula

$$x_d(\hat{t}) = \sqrt{12kc_p\hat{t}} = \sqrt{12C\hat{t}} \tag{8.195}$$

which is often used as an engineering estimate of the penetration depth. Comparison to (8.182) shows that this estimate corresponds to the point at which the exact value of humidity deviates from the initial value by 1.43% of the difference $1 - h_{env}$ between the initial value and the ambient humidity applied at the boundary.

The time-square-root initial evolution of the penetration depth of drying front after a sudden environmental exposure is a salient asymptotic feature of the diffusion equation. It is well known for the linear diffusion equation, but it applies to the nonlinear diffusion equation as well [146], which is an important point for concrete drying and shrinkage. The aging due to hydration merely shortens the period of validity of the time-square-root law but does not kill it because the square root initially always grows infinitely faster than the hydration degree.

8.4.5 Evolution of Total Water Loss from a Specimen

8.4.5.1 Asymptotics of the Water Loss Process

Let us denote the water loss from a flat slab (or wall), per unit area of its plane, as ΔW [kg/m^2]. Two drying fronts propagate into the slab from its two surfaces exposed to the environmental humidity, and initially, they do not influence each other. Therefore, the initial evolution of the water loss as a function of the drying time $\hat{t}$ can be approximated by $2W_L(\hat{t})$, where W_L is the water loss per unit area of the drying surface, computed for a semi-infinite specimen. Recalling formula (8.188), we can write[19]

$$\Delta W(\hat{t}) \approx 2W_L(\hat{t}) = \frac{4(1 - h_{env})c_p(h_{env})f_{\xi 0}}{\sqrt{k_1 c_{p1}}}\sqrt{\hat{t}} \tag{8.196}$$

To characterize the kinetics of drying, it is useful to introduce a dimensionless function

$$S(\hat{t}) = \frac{\Delta W(\hat{t})}{D\,\Delta w_\infty} \tag{8.197}$$

which grows from 0 to 1 and describes the evolution of ΔW from 0 to the final (asymptotic) value $\Delta W_\infty = \lim_{\hat{t}\to\infty} \Delta W(\hat{t}) = D\,\Delta w_\infty$. Here, D is the slab thickness and

---

[19]Recall that factor $f_{\xi 0}$ depends on the specific nonlinear characteristics of the permeability c_p and reciprocal moisture capacity k and on the ambient humidity. The dependence of $f_{\xi 0}$ on h_{env} is not marked in (8.196) because, in contrast to moisture permeability c_p, factor $f_{\xi 0}$ is not directly available as an explicit function of h_{env}, but it is determined by solving differential equation (8.174) on a semi-infinite domain $[0, \infty)$ and evaluating the derivative of the solution at $\xi = 0$.

Δw_∞ is the final water loss per unit volume of concrete, which can be determined from the ambient humidity h_{env} using the desorption isotherm. For sufficiently short times, the approximation of function S based on (8.196) can be presented in the form

$$S(\hat{t}) \approx \frac{2W_L(\hat{t})}{D\,\Delta w_\infty} = \frac{4(1 - h_{env})c_p(h_{env})f_{\xi 0}}{D\,\Delta w_\infty \sqrt{k_1 c_{p1}}}\sqrt{\hat{t}} = \sqrt{\frac{\hat{t}}{\tau_{w0}}}, \qquad \text{for } \hat{t} \ll \tau_{w0} \tag{8.198}$$

where

$$\tau_{w0} = k_1 c_{p1} \left(\frac{\Delta w_\infty}{1 - h_{env}} \frac{D}{4f_{\xi 0}c_p(h_{env})} \right)^2 \tag{8.199}$$

is a certain characteristic time of the drying process.

If the desorption isotherm is approximately linear and its reciprocal slope k is taken as a constant (equal to k_1), then the fraction $\Delta w_\infty/(1 - h_{env})$ can be replaced by $1/k$ and (8.199) simplifies to

$$\tau_{w0} = C_1 \left(\frac{D}{4f_{\xi 0}C(h_{env})} \right)^2 \tag{8.200}$$

where $C_1 = kc_{p1}$ is the diffusivity at full saturation and $C(h_{env}) = kc_p(h_{env})$ is the diffusivity for pore relative humidity equal to the ambient relative humidity h_{env}. For a linear diffusion problem with constant diffusivity $C(h) = C_1$, the value of $f_{\xi 0}$ is $1/\sqrt{\pi}$ (as follows from (8.179)–(8.180)) and we get

$$\tau_{w0} = \frac{\pi D^2}{16C_1} \tag{8.201}$$

Fitting Eq. (8.198) to short-time water loss data is a fast method to measure moisture transport properties of concrete, but one must make sure that the exposure to drying environment is really sudden and that, prior to exposure, the specimen seal has not leaked any moisture (and if the specimen is too thin, a correction for finite surface emissivity may be necessary). It is clear that (8.198) has limited validity, because for drying times $\hat{t} > \tau_{w0}$, it would yield $S > 1$, which is physically impossible.

During the terminal stage of the drying process, the asymptotic solution (8.164) can be used, and the approach of function S, describing the dimensionless water loss, to the final asymptotic value of 1, is found to be exponential:

$$1 - S(\hat{t}) = 1 - \frac{\Delta W(\hat{t})}{\Delta W_\infty} = \frac{\Delta W_\infty - \Delta W(\hat{t})}{\Delta W_\infty} \approx \frac{1}{\Delta W_\infty}\frac{1}{k(h_{env})}\int_0^D (h(x,\hat{t}) - h_{env})\,dx =$$

$$= \frac{\Delta \bar{h}}{D\,\Delta w_\infty k(h_{env})}\int_0^D \sin\frac{\pi x}{D}\,dx\,\exp\left(-\frac{\pi^2 C_e}{D^2}(t_0 + \hat{t} - \bar{t})\right)$$

Here, $\Delta\bar{h}$ is the difference between the pore relative humidity at the slab center and the ambient relative humidity at a fixed (large enough) time $\bar{t}$. To emphasize the structure of the resulting expression, we can rewrite the asymptotic approximation as

$$1 - S(\hat{t}) \approx A e^{-\hat{t}/\tau_{w\infty}}, \qquad \text{for } \hat{t} \to \infty \tag{8.202}$$

where

$$A = \frac{2\Delta\bar{h}}{\pi \Delta w_\infty k(h_{\mathrm{env}})} \exp\left(\frac{\pi^2 C_e(\bar{t} - t_0)}{D^2}\right) \tag{8.203}$$

$$\tau_{w\infty} = \frac{D^2}{\pi^2 C_e} \tag{8.204}$$

An asymptotic matching formula satisfying both (8.198) and (8.202) can be set up, but is not simple. However, only the initial asymptotic behavior described by (8.198) can be calibrated experimentally because the terminal phase of drying arrives so late that it is impossible to run drying experiments long enough (except for specimens < 1 cm in thickness). A simple approximate form of the dimensionless function $S(\hat{t})$ is

$$S(\hat{t}) = \tanh\sqrt{\frac{\hat{t}}{\tau_w}} \tag{8.205}$$

where τ_w is yet another characteristic time, usually referred to as the *halftime of drying*, even though it does not correspond to the time at which $S = 1/2$. If τ_w is set equal to τ_{w0}, formula (8.205) matches the initial asymptotic form in (8.198) because $\tanh s \approx s$ for small s. As for the terminal asymptotic form (8.202), there is a difference from (8.205); indeed, denoting $\sqrt{\hat{t}/\tau_w} = s$, we have, for large s,

$$1 - S(\hat{t}) = 1 - \tanh\sqrt{\frac{\hat{t}}{\tau_w}} = 1 - \frac{e^s - e^{-s}}{e^s + e^{-s}} = \frac{2e^{-2s}}{1 - e^{-2s}} \approx 2e^{-2s} = 2e^{-2\sqrt{\hat{t}/\tau_w}} \tag{8.206}$$

A function S of a similar form, but possibly with a different characteristic time τ_{sh}, can be used for the evolution of shrinkage. The tanh-root function is used by the B3 model. Its predecessor, model BP [175], used the function $S(\hat{t}) = \sqrt{\hat{t}/(\tau_{\mathrm{sh}} + \hat{t})}$ proposed by Bažant, Osman and Thonguthai [174]. This function also begins in proportion to $\sqrt{\hat{t}}$ but approaches the final asymptotic value differently, namely as $1/\hat{t}$, which can be checked as follows:

$$1 - S(\hat{t}) = 1 - \sqrt{\frac{\hat{t}}{\tau_{\mathrm{sh}} + \hat{t}}} = 1 - \frac{1}{\sqrt{1 + \frac{\tau_{\mathrm{sh}}}{\hat{t}}}} \approx 1 - \frac{1}{1 + \frac{\tau_{\mathrm{sh}}}{2\hat{t}}} = \frac{\frac{\tau_{\mathrm{sh}}}{2\hat{t}}}{1 + \frac{\tau_{\mathrm{sh}}}{2\hat{t}}} \approx \frac{\tau_{\mathrm{sh}}}{2\hat{t}} \tag{8.207}$$

To compare the asymptotics, let $f(\hat{t})$ be the function on the right-hand side of (8.202), describing the asymptotic behavior of the "exact" water loss, and $f_1(\hat{t})$ and $f_2(\hat{t})$ be, respectively, the functions on the right-hand sides of (8.206) and (8.207), describing the asymptotic behavior of the B3 formula and of the old BP formula, and let us calculate the limit

$$\lim_{\hat{t}\to\infty} \frac{f_1(\hat{t}) - f(\hat{t})}{f_2(\hat{t}) - f(\hat{t})} = \lim_{s\to\infty} \frac{2e^{-2s} - Ae^{-\tau_w s^2/\tau_{w\infty}}}{\frac{\tau_{sh}}{2\tau_w s^2} - Ae^{-\tau_w s^2/\tau_{w\infty}}} = \lim_{s\to\infty} \frac{2s^2 e^{-2s} - As^2 e^{-\tau_w s^2/\tau_{w\infty}}}{\frac{\tau_{sh}}{2\tau_w} - As^2 e^{-\tau_w s^2/\tau_{w\infty}}} = 0$$

(8.208)

So we see that (8.206) is infinitely closer to the water loss asymptotics (8.202) than is (8.207). However, using a similar procedure, one can show that (8.206) is still "infinitely far" from the correct asymptotics (8.202), because the limit of f_1/f is also zero. Thus, we have in (8.206) and (8.207) two simple formulas with poor asymptotics, but the latter is poorer than the former. This is the reason why the function used in model BP was replaced by the tanh-root function. Nevertheless, Gardner co-opted the BP model function for his second model, denoted as GL [407].

The existing experimental data are not of sufficient duration to verify the long-time asymptotics. They can be fitted equally well by either of the aforementioned functions.

8.4.5.2   Effective Thickness for Drying

The scaling relation (8.101) shows that all one-dimensional drying problems are similar [52], with slab thickness D being merely a parameter in the general solution. According to (8.199)–(8.201), the characteristic time of the one-dimensional water loss process is proportional to the square of the slab thickness and can be expressed as

$$\tau_{w0} = k_{tw0} D^2$$

(8.209)

where k_{tw0} is a certain proportionality factor. For a linear diffusion problem, formula (8.201) implies that the proportionality factor is given by the simple expression

$$k_{tw0} = \frac{\pi}{16 C_1}$$

(8.210)

where C_1 is the (constant) diffusivity.

For a nonlinear diffusion problem with a nonlinear desorption isotherm and variable permeability, factor k_{tw} is not a pure material parameter but is affected by the environmental humidity. According to (8.199), it can be expressed as

$$k_{tw0} = \frac{k_1 c_{p1}}{[4 f_{\xi 0} \bar{k}(h_{\text{env}}) c_p(h_{\text{env}})]^2}$$

(8.211)

where

$$\bar{k}(h_{\text{env}}) = \frac{1 - h_{\text{env}}}{\Delta w_\infty} = \frac{1 - h_{\text{env}}}{w_0 - w_\infty}$$

(8.212)

is the average reciprocal moisture capacity in the range between the environmental humidity and full saturation. Recall that $f_{\xi 0}$ is the derivative of the dimensionless function f, which can be computed by solving Eq. (8.174) with boundary conditions $f(0) = 0$ and $f(\infty) = 1$. Note that the dependence of k_{tw0} on the environmental humidity h_{env} stems not only from the presence of $\bar{k}(h_{\mathrm{env}})$ and $c_p(h_{\mathrm{env}})$ in the denominator of (8.211) but also from the fact that the normalized functions κ and γ that appear in (8.174) depend, according to (8.175)–(8.176), on h_{env}.

If the desorption isotherm can be approximated by a straight line with constant slope $1/k$, formula (8.211) simplifies to

$$k_{tw0} = \frac{C_1}{[4 f_{\xi 0} C(h_{\mathrm{env}})]^2} = \frac{1}{16 C_1 f_{\xi 0}^2 \gamma^2(0)} \tag{8.213}$$

where $C_1 = k_1 c_{p1}$ is the diffusivity at saturation and $C(h_{\mathrm{env}}) = k c_p(h_{\mathrm{env}}) = C_1 \gamma(0)$ is the diffusivity at the environmental humidity.

Example 8.7. Dependence of characteristic time of drying on environmental humidity

Consider again the Bažant–Najjar model with diffusivity approximated by function (8.89) and with the parameters recommended by the fib Model Code, $\alpha_0 = 0.05$, $h_c = 0.8$ and $r = 15$. The graph of function f, evaluated numerically, has been shown in Fig. 8.30. The derivative of function f at $\xi = 0$ is $f_{\xi 0} \approx 6.17$. For $h_{\mathrm{env}} = 0.6$, formulae (8.184) and (8.213) yield

$$\gamma(0) = 0.05 + \frac{0.95}{1 + 2^{15}} \approx 0.05003 \tag{8.214}$$

$$k_{tw0} = \frac{1}{16 C_1 f_{\xi 0}^2 \gamma^2(0)} \approx \frac{0.655}{C_1} \tag{8.215}$$

We have not specified the diffusivity at saturation, C_1, to emphasize that the value of k_{tw0} is inversely proportional to C_1. The proportionality factor of 0.655 is more than three times higher than the factor $\pi/16 \approx 0.196$ which would, according to (8.210), apply to a linear diffusion problem.

To illustrate the effect of environmental humidity on the value of k_{tw0} (and thus on the characteristic time of drying), Fig. 8.33 shows the dimensionless product $k_{tw0} C_1$ as a function of the environmental humidity h_{env}. Note that if h_{env} is close to 1, the diffusion problem is almost linear and $k_{tw0} C_1$ is close to $\pi/16$. This applies to environmental humidities above 0.8, which is the value of parameter h_c. For lower h_{env}, the product $k_{tw0} C_1$ increases with decreasing environmental humidity. This is natural, since the diffusivity at lower humidities is reduced and the drying process slows down, which is reflected by an increase of its characteristic time. ∎

The previous example has confirmed that the characteristic time is inversely proportional to the diffusivity. Recall that the B3 model uses formula (3.17), which is

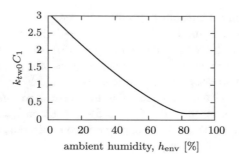

Fig. 8.33 Dependence of dimensionless product $k_{tw0}C_1$ on the ambient humidity, computed using the Bažant–Najjar model with parameters $\alpha_0 = 0.05$, $h_c = 0.8$ and $r = 15$

analogous to (8.209), for evaluation of the shrinkage halftime, with the proportionality factor k_t considered as a material property and estimated using the empirical formula in line 5 of Table C.2 in Appendix C. It should be stressed that factor k_{tw0} evaluated in Example 8.7 and factor k_t used by the B3 model do not have exactly the same meaning. First, the example deals with the water loss process, while the B3 formula refers to the drying shrinkage process, which can have somewhat different kinetics. Second, the characteristic time τ_{w0} studied in the example is related to the initial phase of the process (before the drying fronts meet) while the B3 model uses the halftime τ_{sh} for fitting of the entire evolution of shrinkage strain, and the optimal value may differ from the one that gives the best initial asymptote (this is also the reason why we distinguish between τ_{w0} and τ_{w}). Nevertheless, it is interesting to compare the values for a particular case. For concrete with mean compressive strength $\bar{f}_c = 40$ MPa, exposed to the ambient humidity at $t_0 = 7$ days, the empirical formula from Table C.2 gives

$$k_t = 0.085\, t_0^{-0.08}\, \bar{f}_c^{-1/4} \text{ day/mm}^2 = 0.029 \text{ day/mm}^2 \qquad (8.216)$$

According to the *fib* formula (8.90), the diffusivity of this concrete at saturation would be

$$C_1 = \frac{10^{-8}}{40 - 8} \text{ m}^2/\text{s} = 3.125 \times 10^{-10} \text{ m}^2/\text{s} = 27 \text{ mm}^2/\text{day} \qquad (8.217)$$

and, according to (8.215), the corresponding factor is

$$k_{tw0} = \frac{0.655}{27} \text{ day/mm}^2 \approx 0.024 \text{ day/mm}^2 \qquad (8.218)$$

This means that, in this particular case, the shrinkage halftime τ_{sh} evaluated according to the B3 model is about $1.21\times$ the characteristic time τ_{w0} evaluated by asymptotic fitting of the initial water loss data computed with the Bažant–Najjar model with parameters recommended by *fib*, assuming 60% ambient humidity.

Proportionality of the water loss characteristic time to the square of the structure size is not limited to one-dimensional problems. Similar scaling arguments apply to axisymmetric and spherically symmetric problems, as well as to any set of geometrically similar bodies, such as a set of cubes or spheres of various sizes. Interestingly, if the effective size is defined as $D = 2V/S_e$, where V is the volume of the body and S_e is the area of its surface exposed to the ambient humidity, the characteristic time τ_{w0} that refers to the initial phase of drying is given by the same formula (8.209) as for a slab, independently of the shape of the body. This can be theoretically justified by the fact that the water loss per unit surface area in an infinite half-space is, according to (8.188), proportional to the square root of drying time, with a proportionality factor that depends on the material and on the environmental humidity. In a finite body, such proportionality holds during the initial phase of drying, when the drying front propagates from each boundary point in the direction perpendicular to the boundary and nonnegligible changes of water content occur only in a narrow surface layer. Therefore, if $\hat{t}$ is sufficiently small, the dimensionless function S that characterizes the kinetics of drying (defined as the ratio between the water loss at time $\hat{t}$ and the final water loss at $\hat{t} \to \infty$) can be approximated as

$$S(\hat{t}) \approx \frac{S_e W_L(\hat{t})}{V \, \Delta w_\infty} = \frac{2W_L(\hat{t})}{(2V/S_e) \, \Delta w_\infty} \tag{8.219}$$

This result is completely analogous to formula (8.198) with slab thickness D replaced by effective thickness $2V/S_e$. Consequently, characteristic time τ_{w0} is still given by (8.199), later rewritten as (8.209), with factor k_{tw0} dependent on the material and on the environmental humidity but independent of the shape of the body.

To confirm the expected shape independence of characteristic time τ_{w0}, let us reuse the results of Example 8.4, in which the drying process was simulated using a linear diffusion model for a slab of thickness $D = 2L$ and for a cylinder of radius $R = L$, and the evolution of the average pore relative humidity $\bar{h}$ was plotted in Fig. 8.24a. For a linear isotherm, the loss of water content is proportional to the reduction of pore relative humidity and the dimensionless function

$$S(\hat{t}) = \frac{1 - \bar{h}(\hat{t})}{1 - h_{env}} \tag{8.220}$$

is easily computed from the evolution of the average pore relative humidity, $\bar{h}$, introduced in Example 8.4 and defined by formula (8.143). For matching of the initial asymptotics, it is convenient to transform the time variable such that the short-time approximation (8.198) corresponds to a straight line. This is achieved by introducing a scale proportional to the square root of drying time.

In Fig. 8.34a, the thick solid curves correspond to function S for a cylinder of radius $R = L$ and for a slab of thickness $D = 2L$, and the dashed straight lines are tangents at the origin. Note that the dimensionless variable on the horizontal axis is $\sqrt{\hat{t}/\tau}$, where $\tau = 4L^2/(C\pi^2)$ = auxiliary time-scale parameter. Since the

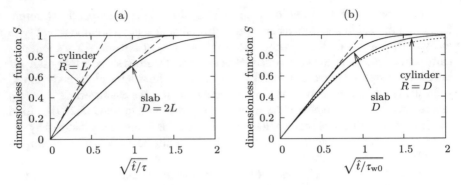

Fig. 8.34 Function S determined using a linear diffusion model: (a) drying time scaled by auxiliary constant $\tau = 4L^2/(C\pi^2)$, (b) drying time scaled by characteristic time $\tau_{w0} = \pi D^2/(16C)$

simulations have been based on a linear diffusion model, the expected value of characteristic time τ_{w0} is $\pi D^2/(16C)$, as follows from (8.201). For the slab, this gives $\tau_{w0} = \pi(2L)^2/(16C) = \pi L^2/(4C) = (\pi^3/16)\tau$, and the short-time asymptotic form of function S is

$$S(\hat{t}) \approx \sqrt{\frac{\hat{t}}{\tau_{w0}}} = \frac{4}{\sqrt{\pi^3}}\sqrt{\frac{\hat{t}}{\tau}} \qquad (8.221)$$

Indeed, it can be verified that the slope of the initial tangent in Fig. 8.34a is $4/\sqrt{\pi^3} \approx 0.718$. For a cylinder, the effective size $2V/S_e$ turns out to be equal to the cylinder radius (assuming that the cylinder is infinitely long or that the flat parts of the surface are sealed). Since the simulation has been done for $R = L$, we must substitute $D = L$ and formula (8.201) gives $\tau_{w0} = \pi L^2/(16C) = (\pi^3/64)\tau$. This is why the slope of the initial tangent in Fig. 8.34a is $8/\sqrt{\pi^3} \approx 1.437$, which is twice as much as for the slab. As noted already in Example 8.4, if the cylinder diameter is equal to the slab thickness, the cylinder dries out faster. Now, we know exactly that the ratio of the characteristic times (of initial drying) for the cylinder and the slab is in this case 1:4 because the characteristic time is proportional to the square of effective size and the ratio of the effective sizes is 1:2.

If the cylinder diameter is doubled (i.e., if the cylinder radius is set equal to the slab thickness), both specimens have the same effective size and thus the same characteristic time τ_{w0}, and the initial evolution of function S has the same short-time asymptote for both specimens. This is documented in Fig. 8.34b, where the curves from Fig. 8.34a are replotted with dimensionless variable $\sqrt{\hat{t}/\tau_{w0}}$ on the horizontal axis. The initial tangents now coincide. However, at later stages the drying process in a slab of thickness D proceeds faster than in a cylinder of radius D. For comparison, the dotted curve shows the graph of the tanh-sqrt function (8.205) for the choice $\tau_w = \tau_{w0}$. It is seen that this analytical expression provides a reasonable approximation of function S for the cylinder but systematically underestimates the values of S computed for the slab. Clearly, identification of the drying halftime τ_w from asymptotic matching of the initial drying is not the best approach if the objective is to obtain a good overall agreement.

A more appropriate procedure is to adjust parameter τ_w such that a suitably defined measure of the deviation of the analytical approximation from the "actual" function S would be minimized. It seems reasonable to minimize the difference between $S(\hat{t})$ and $\tanh \sqrt{\hat{t}/\tau_w}$ in the least-square sense, using the square-root scale for the time of drying on the horizontal axis. The objective function to be minimized is thus defined as

$$\Phi(\tau_w) = \int_0^\infty \left[S(\hat{t}) - \tanh \sqrt{\hat{t}/\tau_w} \right]^2 d\sqrt{\hat{t}} = \frac{1}{2} \int_0^\infty \left[S(\hat{t}) - \tanh \sqrt{\hat{t}/\tau_w} \right]^2 \frac{d\hat{t}}{\sqrt{\hat{t}}} \tag{8.222}$$

and condition $d\Phi(\tau_w)/d\tau_w = 0$ leads to a nonlinear equation

$$\int_0^\infty \frac{S(\hat{t}) - \tanh \sqrt{\hat{t}/\tau_w}}{\cosh^2 \sqrt{\hat{t}/\tau_w}} d\hat{t} = 0 \tag{8.223}$$

which needs to be solved numerically, e.g., by the Newton method.

An alternative, simpler approach is to enforce satisfaction of condition $S(\hat{t}) = \tanh \sqrt{\hat{t}/\tau_w}$ at one single time instant, $\hat{t}^*$, defined e.g., as the time at which $S(\hat{t}^*) = 0.5$. The corresponding halftime is then evaluated as

$$\tau_w = \frac{\hat{t}^*}{\text{atanh}^2(0.5)} \approx 3.314 \, \hat{t}^* \tag{8.224}$$

A similar condition was used by Donmez and Bažant [356] in a somewhat different context (they did not determine the halftime leading to the best analytical approximation but the ratio between the halftimes corresponding to different geometries).

If function S is evaluated using the linear diffusion model with diffusivity C, Eq. (8.223) gives the optimal value of drying halftime $\tau_w = 0.1284 \, D^2/C = 0.6539 \, \tau_{w0}$ for a slab and $\tau_w = 0.1817 \, D^2/C = 0.9256 \, \tau_{w0}$ for a cylinder. The resulting graphs of numerically computed functions S and their analytical approximations by function (8.205) are plotted in Fig. 8.35a. The variable on the horizontal axis is $\sqrt{\hat{t}/\tau_w}$, and so the graph of the analytical function is the same for both geometries. For comparison, Fig. 8.35b shows analogous results for drying halftimes determined from formula (8.224), which leads to $\tau_w = 0.1630 \, D^2/C = 0.8300 \, \tau_{w0}$ for a slab and $\tau_w = 0.2089 \, D^2/C = 1.0637 \, \tau_{w0}$ for a cylinder. It transpires that least-square fitting provides a better overall fit (Fig. 8.35a), while matching of the values at $S = 0.5$ leads to a good agreement in the initial stage but to larger deviations later on (Fig. 8.35b).

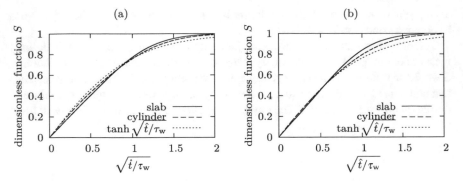

Fig. 8.35 Function S, obtained using a linear diffusion model, and its analytical approximation (8.205) with drying halftime τ_w determined by (a) least-square fitting, (b) matching the times that correspond to $S = 0.5$

The results shown in Figs. 8.34 and 8.35 refer to the linear diffusion model, which is not realistic for concrete. For highly nonlinear models, such as the one based on the Bažant–Najjar formula (8.89) for humidity-dependent diffusivity, the shape of numerically evaluated function S as well as the optimal value of drying halftime τ_w depends on the ambient humidity. For the characteristic time τ_{w0} that refers to the initial asymptotics of the drying process, such dependence has already been demonstrated in Example 8.7 and plotted in Fig. 8.33 in terms of the product $k_{tw0}C_1$, which represents the dimensionless factor that multiplies D^2/C_1 when evaluating τ_{w0}, with C_1 denoting the diffusivity at saturation. In a similar fashion, the drying halftime τ_w can be characterized by the dimensionless factor $k_{tw}C_1$, which corresponds to the ratio between τ_w and D^2/C_1.

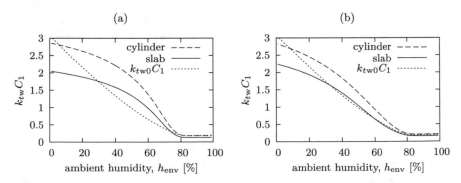

Fig. 8.36 Dependence of dimensionless product $k_{tw}C_1$ on the ambient humidity, computed using the Bažant–Najjar model with parameters $\alpha_0 = 0.05$, $h_c = 0.8$, and $r = 15$ by (a) least-square fitting, (b) matching the times that correspond to $S = 0.5$

Numerically evaluated dependence of factor $k_{tw}C_1$ on the ambient relative humidity is shown in Fig. 8.36, in which solid curves correspond to a slab and dashed curves to a cylinder. For comparison, the graph of factor $k_{tw0}C_1$ from Fig. 8.33 is reproduced here using dotted curves. The results obtained by least-square fitting are shown in Fig. 8.36a, while Fig. 8.36b refers to the halftimes determined by matching the values at $S = 0.5$. It turns out that the latter method gives somewhat lower halftimes in the range of ambient humidities between 40 and 70%. The reason is explained in Fig. 8.37, which shows the numerically computed functions S for 60% ambient humidity and their analytical approximations. Least-square fitting leads to $\tau_w = 0.9790 D^2/C_1$ for a slab and $\tau_w = 1.3806 D^2/C_1$ for a cylinder (i.e., to $k_{tw}C_1 = 0.9790$ and 1.3806), and the analytical approximation is initially below the actual S-curve and later above it (Fig. 8.37a). Matching at $S = 0.5$ leads to $\tau_w = 0.6486 D^2/C_1$ for a slab and $\tau_w = 0.9274 D^2/C_1$ for a cylinder, and the analytical approximation is initially very good but later substantially overestimates the actual values (Fig. 8.37b) because for $h_{env} = 0.6$ the slope of the numerically computed S-curves abruptly decreases when S exceeds 0.5.

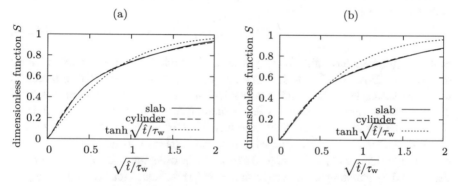

Fig. 8.37 Function S, obtained using a nonlinear diffusion model for ambient humidity $h_{env} = 0.6$, and its analytical approximation (8.205) with drying halftime τ_w determined by (a) least-square fitting, (b) matching the times that correspond to $S = 0.5$

The foregoing examples indicate that the initial asymptotics of the drying process is the same for a slab of thickness D and for a cylinder of radius D (for linear as well as nonlinear diffusion) but later the cylinder dries out more slowly, which is manifested by higher values of the drying halftime determined by fitting of the numerically calculated function S that characterizes the kinetics of drying. Similar trends can be expected for specimens or structural members of other geometries. Even if the size of the body is characterized by the ratio $2V/S_e$, the drying halftime still depends on the specific shape of the body and is somewhat different from the drying halftime of an infinite slab of thickness $D = 2V/S_e$. It is convenient to describe the effect of shape by another dimensionless factor k_s, which transforms $2V/S_e$ into an equivalent thickness of a slab with the same drying halftime as the original body. The complete formula for the characteristic time of drying, analogous to Eq. (3.17), is thus written as

$$\tau_w = k_{tw}(k_s D)^2 \tag{8.225}$$

For an infinite slab, $k_s = 1$ and D is the actual thickness. The B3 model uses fixed shape factors recommended by Bažant et al. [174], namely $k_s = 1.15$ for an infinite cylinder (or a finite cylinder with sealed ends), 1.25 for an infinite square prism, 1.30 for a sphere, and 1.55 for a cube. For bodies of other shapes, it is recommended to interpolate or estimate the k_s value by engineering judgment. However, if nonlinearity of moisture transport is taken into account, it turns out that the values of k_s should also depend on the environmental humidity, h_{env}. Such dependence is ignored by the B3 model.

If the drying halftimes are extracted from numerical simulations of a cylinder of radius D and a slab of thickness D drying at various ambient relative humidities, the corresponding humidity-dependent shape factor can be evaluated based on (8.225). Since the slab has by definition a unit shape factor, we can present the halftime determined for the slab as $\tau_w^{slab} = k_{tw}D^2$, while the halftime determined for the cylinder is $\tau_w^{cyl} = k_{tw}(k_s^{cyl}D)^2 = (k_s^{cyl})^2 \tau_w^{slab}$. The resulting shape factor for a cylinder is thus obtained as

$$k_s^{cyl} = \sqrt{\tau_w^{cyl}/\tau_w^{slab}} \tag{8.226}$$

Similar arguments have already been used in Example 8.4 in the context of linear diffusion; cf. Eqs. (8.145)–(8.146). The analysis can now be extended to nonlinear diffusion, which leads to humidity-dependent shape factors. It turns out that if the drying halftimes are determined by least-square fitting (Fig. 8.36a), the dependence of the shape factor on the ambient humidity is rather weak; see the solid curve in Fig. 8.38. The value of k_s^{cyl} evaluated in this way remains between 1.181 and 1.199 over the entire range of h_{env} between 0 and 1. On the other hand, the drying halftimes determined by matching of the values at $S = 0.5$ (Fig. 8.36b) lead to highly variable k_s^{cyl}, with values between 1.122 and 1.202; see the dashed curve in Fig. 8.38. The dotted horizontal line corresponds to the humidity-independent shape factor used by the B3 model.

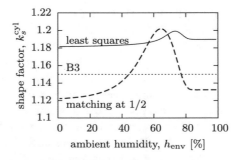

Fig. 8.38 Dependence of shape factor k_s^{cyl} on the ambient humidity, computed using the Bažant–Najjar model with parameters $\alpha_0 = 0.05$, $h_c = 0.8$, and $r = 15$

8.4.6 Effects of Variable Environmental Humidity

Cycling of environmental conditions, typical in nature, may affect creep and shrinkage appreciably, as documented by the tests of Bernhardt [237, 238], Al-Alusi, Bertero and Polivka [24] and Hansen [451], discussed by Bažant and Wang [194]. The magnitude of this effect decreases with increasing thickness of cross section and vanishes for massive structures, except in their surface layer. The influence of humidity variations on creep can be captured by the microprestress-solidification theory, to be presented in Chap. 10. Here, we focus on the evaluation of pore relative humidity in a specimen subjected to variable environmental conditions.

If the evolution of ambient humidity is periodic, it can be expected that, after a transitional period and at negligible rate of aging, a periodic evolution of the pore humidity distribution will be approached. An analytical solution can easily be developed for the one-dimensional linear diffusion problem described by Eq. (8.114) and solved on the semi-infinite interval $[0, \infty)$. Let us first assume that the ambient humidity

$$h_{\text{env}}(\hat{t}) = \bar{h}_{\text{env}} + \hat{h}_{\text{env}} \cos \frac{2\pi \hat{t}}{T_h} \tag{8.227}$$

oscillates around its mean value $\bar{h}_{\text{env}}$ harmonically, with period T_h and peak amplitude $\hat{h}_{\text{env}}$. The asymptotically approached periodic pore humidity evolution can be expected to have the form

$$h(x, \hat{t}) = \bar{h}_{\text{env}} + A_1(x) \cos \frac{2\pi \hat{t}}{T_h} + A_2(x) \sin \frac{2\pi \hat{t}}{T_h} \tag{8.228}$$

where A_1 and A_2 are functions of the spatial variable only. Evaluating the spatial and temporal derivatives and substituting them into the one-dimensional diffusion equation (8.114), we can convert the problem into two coupled ordinary second-order differential equations

$$CA_1''(x) - \frac{2\pi}{T_h} A_2(x) = 0 \tag{8.229}$$

$$CA_2''(x) + \frac{2\pi}{T_h} A_1(x) = 0 \tag{8.230}$$

with boundary conditions $A_1(0) = \hat{h}_{\text{env}}$, $A_2(0) = 0$, $A_1(\infty) = 0$, and $A_2(\infty) = 0$. The standard solution procedures for coupled linear differential equations with constant coefficients lead to

$$A_1(x) = \hat{h}_{\text{env}} e^{-\kappa_h x} \cos \kappa_h x \tag{8.231}$$
$$A_2(x) = \hat{h}_{\text{env}} e^{-\kappa_h x} \sin \kappa_h x \tag{8.232}$$

where

$$\kappa_h = \sqrt{\frac{\pi}{CT_h}} \tag{8.233}$$

is an auxiliary parameter introduced for convenience. The resulting periodic evolution of pore relative humidity is

$$h(x, \hat{t}) = \bar{h}_{\text{env}} + \hat{h}_{\text{env}} e^{-\kappa_h x} \cos\left(\frac{2\pi \hat{t}}{T_h} - \kappa_h x\right) \tag{8.234}$$

At each fixed spatial point, the evolution of humidity is harmonic, with an amplitude that decays exponentially with the distance from the exposed boundary, x, and with a phase shift that varies proportionally to the distance. The spatial distribution of humidity at a fixed time instant has the character of an attenuated wave, shown in Fig. 8.39 by the solid curves (which correspond to four equally spaced stages of the cycle). The dashed exponential curves represent the upper and lower limits between which the humidity varies at each point.

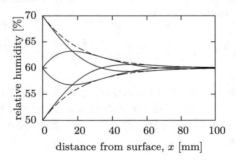

Fig. 8.39 Humidity profiles due to harmonic variation of ambient humidity with mean value 60%, peak amplitude 10%, and period $T_h = 1$ year, plotted for moisture diffusivity $C = 4 \text{ mm}^2/\text{day}$

The depth of the boundary layer affected by the harmonic variation of ambient humidity, δ_c, can be defined, e.g., as the distance at which the amplitude is $100\times$ smaller than the amplitude of ambient humidity cycles. From the condition $\exp(-\kappa_h \delta_c) = 0.01$, we get

$$\delta_c = \frac{\ln 100}{\kappa_h} = \ln 100 \sqrt{\frac{CT_h}{\pi}} \approx 2.6\sqrt{CT_h} \tag{8.235}$$

If the thickness D of a wall is greater than $2\delta_c$, the pores in the core of the wall will not feel the cyclic humidity effect; if it is smaller, the entire wall thickness will be affected. The solution has been derived for linear diffusion, but it can be used as an estimate for nonlinear diffusion if the diffusivity does not vary too much in the humidity range between $\bar{h}_{\text{env}} - \hat{h}_{\text{env}}$ and $\bar{h}_{\text{env}} + \hat{h}_{\text{env}}$. For example, considering the average diffusivity in this range as $C = 4$ mm^2/day, one finds that the daily cycles of environmental humidity will penetrate concrete to the depth of only 5.2 mm, while the annual cycles will penetrate to the depth of 99 mm (as illustrated in Fig. 8.39). The core of any wall thicker than 200 mm will not feel the annual humidity cycles. All the depths are reduced to one half if one considers a tenfold reduction of the amplitude (instead of a reduction by a factor of 100) in the definition of δ_c. The factor of 2.6 in (8.235) is then replaced by 1.3 (because $\ln 10 = (\ln 100)/2$). For instance, if the amplitude of daily ambient humidity variations is 20%, the pore humidity at the distance of 2.6 mm from the exposed surface will vary with amplitude 2% and at the distance of 5.2 mm with amplitude 0.2%.

The penetration depth calculated for a harmonic evolution of ambient humidity represents an upper bound for the penetration depth under an arbitrary periodic evolution with period T_h. The reason is that a general periodic evolution can be expanded into a Fourier series which contains harmonic terms with periods T_h, $T_h/2$, $T_h/3$, etc. Due to the linearity of the problem, solutions constructed for individual harmonic terms can be superposed. Higher-frequency terms attenuate faster than the first term, because T_h in (8.233) is replaced by T_h/n with $n \geq 2$.

The linear diffusion model with no aging, considered so far, permits an analytical treatment and closed-form estimation of the penetration depth under cyclic variations of the environmental humidity. To get a more complete picture, let us look at the effects of variable (humidity-dependent) diffusivity, which can be studied by numerical simulations.

Example 8.8. Concrete wall exposed to variable ambient humidity

Consider a concrete wall of thickness $D = 300$ mm, exposed on both sides to the same environmental humidity h_{env}. To capture seasonal variations, at least in an approximate way, the history of environmental humidity is prescribed as

$$h_{\text{env}}(\hat{t}) = \bar{h}_{\text{env}} + \hat{h}_{\text{env}} \cos \frac{2\pi(\hat{t} - \hat{t}_p)}{T_h} \qquad (8.236)$$

with period $T_h = 365$ days and with $\hat{t}_p = t_p - t_0$ denoting the time shift between the time instant t_p at which the harmonic approximation of the typical annual variation of ambient humidity attains its peak and the time at the onset of drying, t_0. For the real data collected at Fresno, California, during 2000–2014 and plotted in Fig. 6.1a, the optimal fit by function (8.236) is obtained with $\bar{h}_{\text{env}} = 0.61$, $\hat{h}_{\text{env}} = 0.15$ and t_p corresponding to January 12.

To simulate moisture diffusion in concrete, let us use again the Bažant–Najjar model with age-independent diffusivity approximated by function (8.89) and with the parameters recommended by the fib Model Code, i.e., $\alpha_0 = 0.05$, $h_c = 0.8$ and $r = 15$. The initial pore relative humidity is considered as uniform and equal to 1. Due to symmetry, the problem is solved on the interval $[0, L]$, where $L = D/2 = 150$ mm, with boundary conditions $h(0, \hat{t}) = h_{env}(\hat{t})$ and $h'(L, \hat{t}) = 0$. The diffusivity at saturation is set to $C_1 = 20$ mm^2/day.

Long-term effects of periodic ambient humidity described by (8.236) with $\bar{h}_{env} = 0.61$ and $\hat{h}_{env} = 0.15$ are presented in the left column of Fig. 8.40. It is assumed that the first exposure of the wall to the ambient humidity h_{env} occurred on July 12, when h_{env} was at its minimum, 0.46, which means that $\hat{t}_p$ in (8.236) is set to $T_h/2$. Figure 8.40a shows the pore relative humidity profiles in the left half of the wall at selected times of exposure ranging from 5 to 160 years (always referring to the state after an integer number of annual cycles, on July 12). It is seen that the drying process takes many decades. After 40 years of exposure, humidity in the core of the wall is still decreasing. Moisture transport is slowed down by a dramatic reduction of diffusivity at humidities below 0.8. The pore relative humidity at the plane of symmetry (at the distance of 150 mm from both exposed surfaces) is reduced to 0.661 after 40 years, 0.623 after 80 years, and 0.613 after 160 years. The approach to a periodic response takes a very long time, but eventually, the solution gets close to the theoretical one, given by (8.234) in which $\hat{t}$ is replaced by $\hat{t} - \hat{t}_p = \hat{t} - T_h/2$. This modification accounts for the phase shift due to the specific choice of the day on which the wall is first exposed to the environment (July 12). The profile corresponding to the humidity distribution on July 12 according to the periodic solution based on a linear diffusion model is indicated in Fig. 8.40a by the dashed curve. Parameter κ_h has been evaluated from (8.233) using $C = 1.0008$ mm^2/day, which is the diffusivity at $h = \bar{h}_{env}$.

Figure 8.40c depicts pore relative humidity profiles at time instants that correspond to quarters of the annual cycle during the 80th year of exposure, and Fig. 8.40e shows how the pore relative humidity at the exposed surface and at the depths of 10 mm, 20 mm, and 30 mm evolves during the 80th year. The solid curves represent the numerically computed results; they are very close to the dashed curves, which have been obtained from formula (8.234) with $\hat{t}$ replaced by $\hat{t} - \hat{t}_p$.

The results presented so far may seem to confirm that, after a (rather long) transitional phase, the humidity history in a wall exposed to periodic ambient humidity approaches the periodic solution obtained for the linear diffusion model. However, the results plotted in the left column of Fig. 8.40 refer to a particular case in which the periodic solution is confined to a range with a relatively weak dependence of diffusivity on pore relative humidity. Indeed, the ambient relative humidity varies here between $\bar{h}_{env} - \hat{h}_{env} = 0.46$ and $\bar{h}_{env} + \hat{h}_{env} = 0.76$, and the moisture diffusivity evaluated according to the Bažant–Najjar formula (8.89) with the present parameters is almost constant below $h = 0.7$ (it varies from 1 mm^2/day at $h = 0$ to 1.043 mm^2/day at $h = 0.7$) and between $h = 0.7$ and $h = 0.76$ it varies between 1.043

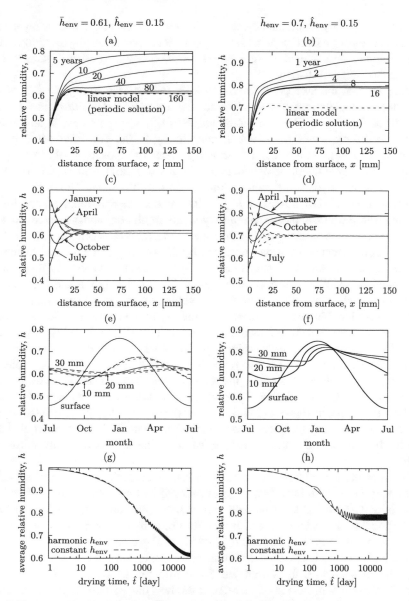

Fig. 8.40 (a) Profiles of relative humidity in July after 5, 10, 20, 40, 80, and 160 years, (b) profiles of relative humidity in July after 1, 2, 4, 8, and 16 years, (c) profiles of relative humidity in July, October, January, and April during the 80th year of exposure, (d) profiles of relative humidity in July, October, January, and April during the 20th year of exposure, (e) variation of relative humidity at the exposed surface and 10, 20, and 30 mm from the surface during the 80th year of exposure, (f) variation of relative humidity at the exposed surface and 10, 20, and 30 mm from the surface during the 20th year of exposure, (g–h) evolution of spatially averaged relative humidity

and 2.158 mm²/day. Consequently, the nonlinearity affects only a narrow layer near the surface, during a fraction of the annual cycle. It is thus interesting to explore what would happen for ambient humidity ranges that involve more dramatic variations of the corresponding diffusivity.

Suppose that the average ambient humidity is increased to $\bar{h}_{env} = 0.7$, while all other parameters remain fixed. The results are graphically shown in the right column of Fig. 8.40. The transitional stage of the drying process is now much shorter than for $\bar{h}_{env} = 0.61$, as documented in Fig. 8.40b. The difference between the humidity profiles in July after 8 years and 16 years of exposure is very small (pore relative humidity at the plane of symmetry slightly decreases from 0.795 to 0.788), and in subsequent years, the profile remains virtually the same. The most striking difference compared to the case of $\bar{h}_{env} = 0.61$ is that the pore relative humidity in the core of the wall oscillates (with a very small amplitude) about a value close to 0.788, which is substantially higher than the average ambient humidity $\bar{h}_{env} = 0.7$. This is a consequence of strong nonlinearity of the transport process. At higher humidities, the moisture transport is faster, which breaks the "symmetry" of positive and negative deviations from average values. Drying during the summer months with a low ambient humidity is slower than wetting during the winter months with a high ambient humidity, and the resulting humidity in the core is thus higher than it would be for a linear diffusion model (or even for a nonlinear model if the ambient humidity remained at its average value and did not vary at all). Predictions obtained with a linear model, plotted in Fig. 8.40b,d by dashed curves, are now completely off. The actual profiles are flatter in the top part (January) and steeper in the bottom part (July). The annual variation of pore relative humidity at a given internal point is far from harmonic, as shown in Fig. 8.40f. For the sake of clarity, the corresponding variations predicted by the linear model are not shown in Fig. 8.40f; they would have exactly the same shape as the dashed curves in Fig. 8.40e, just vertically shifted by 0.09.

To complete the analysis, let us have a closer look at the evolution of spatially averaged pore relative humidity, $\bar{h}(\hat{t})$, introduced in Example 8.4 and defined by formula (8.143). If the sorption isotherm is linear, the difference $\bar{h}(0) - \bar{h}(\hat{t})$ is proportional to the water loss, which is in turn roughly proportional to the average shrinkage strain. The two graphs at the bottom of Fig. 8.40 present the evolution of $\bar{h}(\hat{t})$ in a slab of thickness $D = 300$ mm. Since the transitional phase can be quite long, the logarithmic scale is used here for the time of exposure on the horizontal axis. For comparison, the dashed curve indicates how the spatially averaged humidity would evolve at a constant ambient humidity, $h_{env}(\hat{t}) = \bar{h}_{env}$. In the terminal, almost periodic phase, the spatially averaged pore relative humidity oscillates around a value which is very close to the average ambient humidity $\bar{h}_{env} = 0.61$ in Fig. 8.40g but is substantially higher than the average ambient humidity $\bar{h}_{env} = 0.7$ in Fig. 8.40h. By averaging in time over the period of $T_h = 365$ days (1 year), we find that the value

$$\bar{\bar{h}} = \frac{1}{T_h} \int_{\hat{t}_{max} - T_h}^{\hat{t}_{max}} \bar{h}(\hat{t}) \, d\hat{t} \tag{8.237}$$

around which $\bar{h}(\hat{t})$ oscillates in the terminal phase is 0.613 for $\bar{h}_{env} = 0.61$ and 0.782 for $\bar{h}_{env} = 0.7$. Time $\hat{t}_{max}$ in formula (8.237) corresponds to the end of the simulation ($\hat{t}_{max} = 50,000$ days in the present case, but already after 30,000 days, the values of $\bar{\bar{h}}$ would be, respectively, 0.619 and 0.782). ∎

In the previous example, we have shown that if the moisture transport is governed by a nonlinear diffusion equation, the average pore relative humidity in the structure after a long-term exposure, $\bar{\bar{h}}$, does not need to be equal to (or close to) the average ambient relative humidity, $\bar{h}_{env}$. How much $\bar{\bar{h}}$ deviates from $\bar{h}_{env}$ depends on the range in which the ambient humidity varies and on the sensitivity of moisture diffusivity to the pore humidity in that range. For the range of $h_{env} = 0.61 \pm 0.15$, there is almost no difference between $\bar{\bar{h}}$ and $\bar{h}_{env}$, while for the range of $h_{env} = 0.7 \pm 0.15$ the difference is appreciable.

To see the broad picture, let us perform a series of simulations for various ranges of ambient humidities and evaluate $\bar{\bar{h}}$ in the terminal phase of each simulation, when the response becomes sufficiently close to a periodic one. The results obtained for a wall of thickness $D = 300$ mm with the Bažant–Najjar model using the diffusion parameters from Example 8.8 are shown in Fig. 8.41. The terminal mean pore relative humidity, $\bar{\bar{h}}$, is plotted as a function of the peak amplitude of ambient relative humidity variation, $\hat{h}_{env}$, for different values of the mean ambient relative humidity, $\bar{h}_{env}$. Each curve starts from $\hat{h}_{env} = 0$ and $\bar{\bar{h}} = \bar{h}_{env}$, because for zero amplitude of h_{env}, the transport process corresponds to monotonic drying at a constant ambient humidity. As the peak amplitude $\hat{h}_{env}$ grows, $\bar{\bar{h}}$ increases as well, but the effect becomes strong only if the range $\bar{h}_{env} \pm \hat{h}_{env}$ has a substantial overlap with the range of highly variable diffusivity. Each curve in Fig. 8.41 is terminated at $\hat{h}_{env} = 1 - \bar{h}_{env}$, to make sure that the prescribed ambient relative humidity during the cycle never exceeds 1.

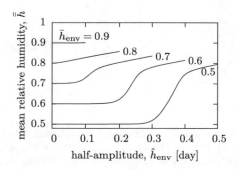

Fig. 8.41 Dependence of mean pore relative humidity in the terminal phase, $\bar{\bar{h}}$, on the peak amplitude of ambient relative humidity, $\hat{h}_{env}$, for various levels of mean ambient relative humidity, $\bar{h}_{env}$

According to Fig. 8.41, if the mean ambient relative humidity is 0.6, the asymptotically approached mean pore relative humidity remains very close to 0.6 as long as the peak amplitude of the ambient relative humidity does not exceed 0.16. On the other hand, if the ambient relative humidity varies around the same mean over a larger range, e.g., with a peak amplitude of 0.25, the resulting mean pore relative humidity increases to 0.73. In Example 8.8, we prescribed $\bar{h}_{env} = 0.61$ and $\hat{h}_{env} = 0.15$ as values that correspond to the best harmonic approximation of the actual variation of humidity recorded in Fresno, California, during the period of 2000–2014. The period in the harmonic approximation was set to 1 year, to capture the annual cycles. As shown in Fig. 6.1a, the actual values recorded in Fresno varied over a wider range, but the deviations from the smooth harmonic approximation can be considered as random fluctuations due to weather changes, with zero mean. Since such fluctuations rapidly change sign (typically after a few days), one might think that their effect on the overall behavior is negligible because, in similarity to the daily cycles discussed at the beginning of the present section, they affect only a narrow layer near the exposed surface. However, such reasoning is correct only if the transport process can be described by a linear model.

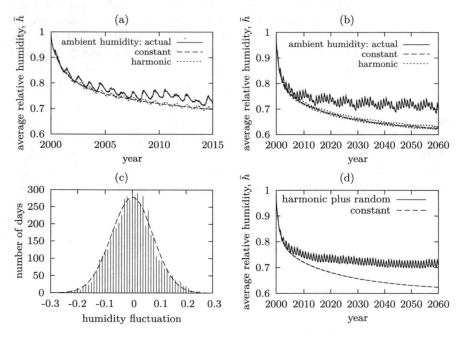

Fig. 8.42 (a–b) Evolution of spatially averaged pore relative humidity for various histories of ambient humidity (constant, given by a harmonic function, specified by actually measured values), (c) histogram of deviations of ambient relative humidity (daily averages) from a harmonic approximation of seasonal variations, (d) evolution of spatially averaged pore relative humidity for ambient humidity given by a harmonic function superposed with random fluctuations

For a highly nonlinear behavior, even short-term fluctuations with zero mean can have an effect on the overall long-term evolution of the pore humidity distribution. A simulation of the same wall as in Example 8.8 subjected to an ambient humidity history directly taken from the actual weather station records reveals that the pore relative humidity in the wall after 15 years of exposure is substantially higher than the results obtained for ambient humidity approximated by the best-fit harmonic function with a period of 1 year; see Fig. 8.42a.

To demonstrate the long-term trends, the simulation has been extended to 60 years, with the measured data from 2000–2015 repeated in the subsequent 15-year periods. Figure 8.42b shows that, after a transitional period of about 12 years, the "actual" spatially averaged humidity $\bar{h}$ fluctuates between 0.69 and 0.74. In contrast to that, the solution computed for a harmonic approximation with period 1 year and peak amplitude 0.15 slowly approaches a periodic solution oscillating around 0.61 (as already shown in Example 8.8; see Fig. 8.40g). A response similar to the "actual" one could be obtained by increasing the amplitude of the harmonic function, but such an adjustment would be rather artificial. Instead of that, one can subtract the smooth harmonic approximation from the measured ambient humidity values and treat the difference as a random variable. From the histogram in Fig. 8.42c, it is clear that this variable can be considered as normally distributed, with zero mean and with a standard deviation of 0.077 (see the dashed curve). The best agreement with the results computed for the measured data is achieved if the harmonic function is combined with normally distributed random fluctuations characterized by zero mean and a standard deviation of 0.09; see Fig. 8.42d.

8.5 Spreading of Hydraulic Pressure Front Into Unsaturated Concrete

In submerged structures, a saturated zone under hydraulic pressure p_l may be propagating into a zone of unsaturated concrete. In that case, as shown by Bažant [79], it is extremely important to take into account the self-desiccation of concrete. It causes most of the concrete to become unsaturated, and thus deficient in water, which means that the permeation of water under hydraulic pressure must supply to the pressure front enough water to replenish the missing water before the pressure front can propagate deeper into the structure. Calculations show that the filling of the pores at pressure front is so slow that it leaves enough time for the profile of hydraulic pressure to become almost linear.

The diffusion equation (8.85) written in terms of the pore relative humidity as the primary unknown field describes moisture transport in partially saturated concrete, in which liquid water coexists with water vapor in the pore gas. To describe the transport of liquid water in the fully saturated region between the upstream boundary of the dam and the saturation front, a model based on the Darcy law and presented in Sect. 8.3.4.1 can be used. Recall that the primary unknown field is then the liquid pressure p_l, which can be computed by solving differential equation

$$\dot{p}_l = C_l \nabla^2 p_l + \frac{K_h}{n_p} \frac{\partial p_l}{\partial z} - \frac{K_l}{\rho_{l0} n_p} \dot{w}_h \tag{8.238}$$

in which

$$C_l = \frac{K_l K_h}{\rho_{l0} n_p g}$$

(8.239)

is the diffusivity of liquid water in saturated concrete [m^2/s], and

$$\dot{w}_h = \dot{w}_n + \rho_l \dot{n}_p$$

(8.240)

is the rate of water deficiency due to hydration (see Fig. 8.43 for a schematic illustration of the physical origin of water deficiency).

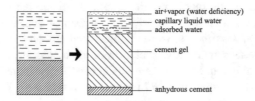

Fig. 8.43 Change of pore space and water deficiency caused by hydration

Powers [701] estimated that the water deficiency amounts to about 28% of all chemically combined water, which in turn represents (at the end of hydration) about 22% of the weight of cement. So the terminal water deficiency $w_{h,\infty}$ is about 28% × 22% = 6.2% of cement weight. For a lean dam concrete (without pozzolans), this gives $w_{h,\infty} = 0.062 \times 112 \approx 7$ kg of water per m^3 of concrete (for recent concretes with pozzolans this would be about 3 kg/m^3). The time variation may be considered as $w_h(t) = w_{h,\infty} f_w(t)$, where $f_w(t)$ is a function monotonically increasing from 0 to 1, roughly in proportion to the strength gain formula (to take into account the heating due to hydration, replace t with the equivalent hydration period t_e; see Sect. 10.6.1). Equation (8.238) is a linear parabolic differential equation for the unknown pore liquid pressure p_l.

For a lean dam concrete, the hydraulic permeability $K_h \approx 50 \times 10^{-12}$ m/s. For instance, for a concrete with 112 kg of cement per m^3 of concrete, Carlson [294] observed the values of 175, 60, and 40 × 10^{-12} m/s at 3, 12, and 24 months of age, respectively. For very lean concretes, with only 43 kg of cement per m^3, 70 to 200 kg/m^3 of pozzolans, and 80 to 150 kg of water per m^3, the hydraulic permeability at the age of 3 months is about 6 × 10^{-12} m/s. These hydraulic permeabilities are several orders of magnitude higher than those of dense structural concretes (but this is not harmful for gravity dams because their leakage depends almost exclusively on joints and cracks).

In the foregoing Eqs. (8.238)–(8.239), the fraction n_p/K_l, further denoted as Λ, represents the compressibility of pore water, taking into account the effect of porosity. According to the detailed theoretical discussion by Bažant [79], the value of Λ should be modified because the problem is complicated by several phenomena:

1. deformation of solid skeleton of cement gel incorporating layers of load-bearing hindered adsorbed water;
2. uncertainty in delineating the part of hindered adsorbed water that should be excluded from n_p (this water phase is load-bearing though evaporable); and
3. increase in accessible nanopore space with increasing pressure.

According to the experiments of Murata [644], Λ should be taken as approximately equal to the bulk compressibility of concrete as a whole, which is about 100×10^{-12}/Pa for a one-year old lean dam concrete (see also [294]).

Consequently, the diffusivity can be estimated as

$$C_l = \frac{K_h}{\rho_{l0} \Lambda g} \approx \frac{50 \times 10^{-12}}{1000 \times 100 \times 10^{-12} \times 10} \text{ m}^2/\text{s} = 50 \times 10^{-6} \text{ m}^2/\text{s} = 4.32 \text{ m}^2/\text{day}$$

(8.241)

which is near the upper bound of the diffusivities identified by Murata [644] from penetration depth measurements. This is about 200,000× greater than the typical diffusivity of unsaturated concrete just below saturation ($C_1 \approx 20 \text{ mm}^2/\text{day}$). The reason for this huge difference is that much water needs to flow into unsaturated pores to change pore humidity significantly, while very little water needs to be forced into saturated pores to change the hydraulic pressure significantly.[20]

Due to the high diffusivity in the saturated region, the gradient of hydraulic head of water diffusing, for example, into a concrete dam from its reservoir, must be almost uniform up to the saturation front at the interface with unsaturated concrete, where the liquid water is at (or near) the atmospheric pressure. The saturation front slowly moves into the unsaturated region, and the speed of this motion is controlled mainly by the rate at which the water supplied to the front can compensate for the water deficiency in the unsaturated region caused by hydration.

To get insight into the physical mechanisms governing this process, let us look at a simplified one-dimensional problem of water flow through a horizontal pipe filled by concrete. All variables depend just on the horizontal spatial coordinate x (and on time). Consequently, the gradient of the elevation head vanishes and the Darcy law (8.69) can be simplified to

$$j_x = -\frac{K_h}{g} \frac{\partial p_l}{\partial x}$$

(8.242)

Consider horizontal water flux from a boundary at $x = 0$, subjected to a given excess pressure p_0 applied at time t_1, to the saturation front located at a certain distance x_s, which varies in time. For high diffusivity C_l, the pressure profile can be expected to be almost linear along the pipe and can be described by

$$p_l(x, t) \approx p_{\text{atm}} + p_0 \left(1 - \frac{x}{x_s(t)}\right), \qquad 0 \leq x \leq x_s(t)$$

(8.243)

---

[20]What is the ratio of permeabilities of saturated and unsaturated concretes? To answer it, it is necessary to consider permeabilities that refer to gradients of the same variable, which is the chemical potential. In this way, it was concluded that the ratio is about 6000 [79].

Differentiating with respect to x and substituting into the Darcy law (8.242), we get a relation between the flux and the position of the saturation front:

$$j_x(t) = \frac{K_h}{g} \frac{p_0}{x_s(t)} \tag{8.244}$$

If the boundary pressure is applied after completion of the hydration process, the water deficiency in the unsaturated region is equal to its terminal value, $w_{h,\infty}$. The amount of water (per unit area of the saturation front) needed to resaturate the concrete and move the saturation front by an increment dx_s is $w_{h,\infty}dx_s$. This water must be supplied by the flux j_x during the time increment dt, which leads to the condition

$$w_{h,\infty}dx_s = j_x(t)dt \tag{8.245}$$

After substitution from (8.244) and separation of variables, we obtain

$$x_s(t)dx_s = \frac{K_h p_0}{g w_{h,\infty}} dt \tag{8.246}$$

Integrating this relation from time t_1 at which the boundary pressure was applied to the current time t, we get

$$\frac{x_s^2(t)}{2} = \frac{K_h p_0}{g w_{h,\infty}} (t - t_1) \tag{8.247}$$

from which

$$x_s(t) = \sqrt{\frac{2 K_h p_0}{g w_{h,\infty}}(t - t_1)} = \sqrt{C_w(t - t_1)}, \qquad t \geq t_1 \tag{8.248}$$

Using the aforementioned typical values $K_h = 50 \times 10^{-12}$ m/s and $w_{h,\infty} = 7$ kg/m^3 and considering, as an example, the excess pressure $p_0 = 0.49$ MPa (which corresponds to the applied pressure head $p_0/g\rho_l = 49$ m, i.e., to the depth of 49 m below the reservoir water level), we can evaluate the factor

$$C_w = \frac{2K_h p_0}{g w_{h,\infty}} = \frac{2 \times 50 \times 10^{-12} \times 0.49 \times 10^6}{10 \times 7} \text{ m}^2/\text{s} = 0.7 \times 10^{-6} \text{ m}^2/\text{s} \tag{8.249}$$

The corresponding penetration depths x_s of the saturation front after $t - t_1 = 1$, 10 and 100 years of exposure to the excess pressure are 4.7 m, 15 m, and 47 m. Since the excess pressure p_0 grows proportionally to the depth d below the water level, factor C_w will be proportional to the d and the distance x_s of the penetration front from the boundary will be proportional to $\sqrt{d}$. Therefore, it can be expected that the shape of the penetration front will be approximately parabolic, as schematically shown in Fig. 8.44. Of course, this is just a simplified solution, valid for a constant water level in the reservoir. Simultaneously, a drying front will propagate from the part of the dam boundary which is in contact with air. The present simplified solution is not applicable to the region near the dam crest where both fronts meet, but it gives a rough idea about the time it takes for the saturated region to reach the core of the dam.

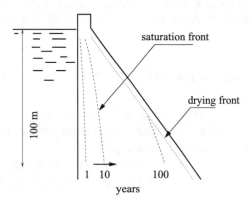

Fig. 8.44 Advance of the fronts of saturation and of drying in an ideal dam having no cracks, no cracked joints, and no drainage (for a typical concrete

If the aggregate is porous and not completely saturated by the time of mixing, or if air bubbles are entrained in concrete, the penetration depths can be easily reduced by a factor of 2 to 4. On the other hand, for modern high-strength concrete, which can have permeability 4 orders of magnitude lower and can undergo much stronger self-desiccation, the penetration depth of saturation front would be several orders of magnitude smaller.

If dam concrete (with no air bubbles) underwent no self-desiccation, we could cancel the sink term with $\dot{w}_h$ and Eq. (8.238) would, for one-dimensional flow in a horizontal pipe, reduce to the standard linear diffusion equation

$$\dot{p}_l = C_l p_l'' \tag{8.250}$$

In analogy to (8.195), the penetration depths can then be estimated as $x_s = \sqrt{12 C_l \hat{t}}$, and for the typical value of $C_l = 50 \times 10^{-6}$ m^2/s calculated in (8.241), the one-year penetration depth would be $x_s \approx 138$ m—much greater than the value of 4.7 m obtained if the effect of self-desiccation is included.

The calculation of hydraulic pressure in concrete is important for the old, yet still discussed, problem of uplift in dams [294, 392], which is further complicated by the question of applicability of Terzaghi's concept of effective stress (for a detailed discussion, see [79]).

To clarify uplift, measurements of water pressure in dam concrete have been made and surprisingly low pressure values have been recorded. For example, at the University of California in Berkeley in the 1930s, a horizontal pipe filled with dam concrete was connected at one end to a vertical pipe running through several stories, filled permanently by water. Pore pressure gauges were installed in the horizontal pipe, but even after 30 years, no pressure was measured even at the first gauge, at the distance of 0.3 m from the wet end under pressure (private communication by R. Carlson and D. Pirtz in March 1969). The only possible explanation is either strong self-desiccation (not likely in dam concrete), or prevalence of anticlastic capillary surfaces leaving significant empty pore space even at 100% humidity.

8.6   Shrinkage and Stresses Due to Nonuniform Drying

Let us now consider the drying shrinkage under the assumption that the autogenous shrinkage is negligible. This hypothesis is acceptable for most normal-strength concretes, but not for high-strength concretes.[21]

It is a widely accepted empirical observation that the drying shrinkage strain ε_{sh} is, at least in a certain range, approximately proportional to the loss of (total) water content, i.e.,

$$\dot{\varepsilon}_{sh} = k_{sh}\dot{w}_t \tag{8.251}$$

where k_{sh} is an empirical constant, which may be called the *shrinkage coefficient*. For mature concrete with a linear desorption isotherm, (8.251) can be rewritten as

$$\dot{\varepsilon}_{sh} = k_{sh}\frac{\dot{h}}{k} = k_{sh}^*\dot{h} \tag{8.252}$$

where k is the reciprocal moisture capacity and $k_{sh}^* = k_{sh}/k$ is another proportionality coefficient, which may be called the *shrinkage ratio*. The B3 model and many researchers postulate the proportionality between the change of pore relative humidity and the shrinkage strain increment directly, independently of the isotherm, with k_{sh}^* considered as the primary parameter. The limitations of simple relations (8.251) and (8.252) with constant proportionality coefficients can be illustrated by two examples of measured data sets.

Granger [430] presented an experimental study of creep and shrinkage of concretes used in six French power plants, with compressive strength ranging from 34.3 MPa (Penly) to 64.5 MPa (Flamanville). Among other tests, he measured shrinkage and water loss on cylinders of 16 cm in diameter, with sealed ends. The specimens were cured in sealed conditions until the age of 28 days and then were exposed to an ambient relative humidity of 50%. The dependence between the water loss (in mass percent) and shrinkage strain is shown in Fig. 8.45a. It is clear that the simple proportionality of the shrinkage strain increment to the change of water content can provide a good approximation of the measured data in the intermediate range

---

[21]Can Eq. (8.252) be used to calculate autogenous shrinkage from the self-desiccation in high-strength concretes? Settled though this question is not, it seems that a calculation based just on (8.252) would underestimate the autogenous shrinkage.

only. The corresponding coefficients k_{sh} for individual concretes are between 9.5 and 18.5×10^{-6} m^3/kg (slopes of the dashed lines in Fig. 8.45a divided by the density of dry material, estimated as $\rho_d = 2300$ kg/m^3). During the early and late stages of the drying process, the evolution of shrinkage lags behind the water loss, and the value of k_{sh} would need to be reduced. The values of this coefficient averaged over the entire range of data are between 6.6 and 15.4×10^{-6} m^3/kg.

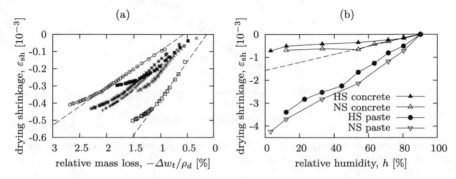

Fig. 8.45 (a) Relation between water mass loss and shrinkage strain measured by Granger [430] for six different types of concrete, (b) relation between relative humidity and shrinkage strain measured by Baroghel-Bouny et al. [57] for normal-strength (NS) and high-strength (HS) concrete and hardened cement paste

Baroghel-Bouny et al. [57] performed water vapor desorption experiments on slices about 3 mm thick, wet-sawn from cylinders after curing with no water exchange for 1 year. The specimens were placed into sealed cells with relative humidity controlled by saturated salt solutions and brought into moisture equilibrium. In addition to the water loss at each level of relative humidity, the change of disk diameter was measured by dial gauges and the results were converted into shrinkage strains, taking the state at 90.4% relative humidity as a reference. The tested materials included normal-strength concrete ($w/c = 0.48$, $\bar{f}_c = 49.4$ MPa, $\rho = 2285$ kg/m^3), high-strength concrete ($w/c = 0.26$, $\bar{f}_c = 115.5$ MPa, $\rho = 2385$ kg/m^3), and also normal-strength and high-strength hardened cement pastes. The results plotted in Fig. 8.45b confirm that cement paste shrinks much more that the concrete as a whole. For cement paste, the graph is close to a straight line over the entire range of humidities, down to extremely low values. For concrete, especially for normal-strength concrete, the graph is close to a straight line for humidities above 55%. At lower humidities, the shrinkage strain remains almost constant. In the linear range, the coefficient of proportionality between the change of humidity and the shrinkage strain (slope of the dashed line in Fig. 8.45b) is, for normal-strength concrete, approximately $k_{sh}^* = 1.76 \times 10^{-3}$. Using moisture capacity $1/k \approx 99$ kg/m^3, estimated from the desorption isotherm reported for this material by Baroghel-Bouny et al. [57], we get $k_{sh} = kk_{sh}^* \approx 17.8 \times 10^{-6}$ m^3/kg.

It would be of course possible to fit the measured data in Fig. 8.45 by more general, nonlinear relations, leading to curves of variable slope. For instance, van Zijl [829]

considered the shrinkage ratio k_{sh}^* as a linear or hyperbolic function of the relative humidity, while Idiart [502] used a linear or quadratic dependence of the shrinkage coefficient k_{sh} on the water mass loss. However, such refined models involve several parameters that are hard to predict based on composition.

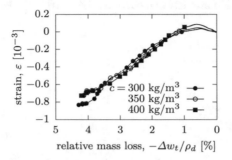

Fig. 8.46 Relation between water mass loss and strain in load-free specimens measured by Kuchar-czyková et al. [554] for three types of concrete

In a recent study, Kucharczyková, Daněk, Kocáb and Misák [554] measured simultaneously the water mass loss and length changes in concrete specimens, starting already 1 hour after casting. Fresh concrete was cast into 1-m-long molds with cross-sectional dimensions 100×60 mm, and the top surface was exposed to the ambient relative humidity of 83%. The molds, equipped with a gauge measuring the displacement of one end (with the other end fixed), were placed on a special weighing table that allowed continuous monitoring of the weight loss. In the initial stage of the test, the recorded length changes were slightly positive, culminating approximately after 1 day, which can be attributed to the thermal expansion caused by hydration heat or swelling from imbibed water before the drying front reached the specimen core, or both. After 3 days, the specimens were demolded and subsequently stored at an ambient relative humidity of 55%. Shrinkage and mass loss were then monitored until the age of 300 days. The curves depicted in Fig. 8.46 show the results for concrete mixes without additives, with Portland cement content ranging from 300 to 400 kg/m^3 and water-cement ratio from 0.61 to 0.5. Remarkably, after an initial stage of limited swelling (partly affected by thermal expansion), the increments of shrinkage strain turned out to be almost perfectly proportional to the mass changes, with shrinkage coefficient $k_{sh} \approx 12 \times 10^{-6}$ m^3/kg.

A pervasive weakness of many comparisons with test data is that Eq. (8.251) or (8.252) is considered and used as if representing material properties (i.e., locally, point-wise), while the measured values usually represent average shrinkage over the cross section, affected by residual stresses, creep due to these stresses, and distributed cracking (with possible localizations); see the analysis in [150]. To measure true material shrinkage, the environmental humidity must be reduced sufficiently slowly [102, 212]. In a realistic analysis of shrinkage data, one must fit them inversely by a finite element program based on experimentally documented strength and creep properties, as attempted by Bažant and Chern [117, 120], though with a shrinkage and creep model inferior to B3 or B4.

The effect of pore humidity variation on creep properties is important but very difficult to measure. Its direct measurement would require varying the pore humidity in time while keeping it almost uniform over the cross section. Conceivably, such a situation can be achieved if the environmental humidity h_{env} is varied continuously at a sufficiently small rate. For cement paste, limited tests of this kind have been performed, e.g., by Bažant et al. [102], but their results were too scattered.

Let us estimate at which rate the ambient humidity would need to be varied in order to keep the pore relative humidity almost uniform, which would permit a direct measurement of free shrinkage and drying creep. Consider a wall of thickness D exposed to slowly decreasing environmental humidity $h_{env}(\hat{t})$. If the distribution of the relative pore humidity across the wall varies between $h_{env}(\hat{t})$ on the boundaries and $h_{env}(\hat{t}) + \Delta h$ at the center of the wall, with Δh sufficiently small, then the reciprocal moisture capacity $k(h)$ and permeability $c_p(h)$ are almost uniform throughout the entire thickness, close to $k(h_{env})$ and $c_p(h_{env})$. The problem can then be accurately described by the linear diffusion equation (8.91) with time-dependent diffusivity $C(\hat{t}) \equiv k(h_{env}(\hat{t}))c_p(h_{env}(\hat{t}))$.

Suppose that, after a transitional period, the difference between the pore humidity and the environmental humidity approaches a certain time-independent function $f(x)$. The solution of the diffusion equation has then the form

$$h(x, \hat{t}) = h_{env}(\hat{t}) + f(x) \tag{8.253}$$

Substituting this into Eq. (8.91), we obtain the condition

$$\dot{h}_{env}(\hat{t}) = C(\hat{t})f''(x) \tag{8.254}$$

which can be satisfied only if $f''(x)$ is a constant independent of x. Therefore, function f must be quadratic, satisfying boundary conditions $f(0) = 0$ and $f(D) = 0$, which follow from $h(0, \hat{t}) = h_{env}(\hat{t})$ and $h(D, \hat{t}) = h_{env}(\hat{t})$. Taking into account that $f(D/2)$ should be equal to Δh, we get

$$f(x) = \frac{4\Delta h}{D^2}(D - x)x \tag{8.255}$$

and substituting this into (8.254) yields the condition

$$\dot{h}_{env}(\hat{t}) = -\frac{8C(\hat{t})\,\Delta h}{D^2} \tag{8.256}$$

linking the rate of environmental humidity, $\dot{h}_{env}$, to the magnitude of the humidity variation across the thickness, Δh. Of course, this relation has been derived under certain simplifying assumptions and is only approximate. The dependence of diffusivity on pore humidity is not known accurately, and in practice, it would be difficult to vary the rate of environmental humidity exactly in proportion to the current diffusivity, as required by condition (8.256). Nevertheless, this equation provides at least a rough estimate of the rate at which the environmental humidity could be changed in order to maintain an almost uniform humidity distribution across the thickness.

To directly observe free shrinkage and drying creep, the variation of stress across the test specimen would have to be negligible, so as to avoid cracking as well as creep due to residual stress (since creep is partly irreversible). This requires that the differences in h within the cross section would not exceed about 3%. Consider a typical value of diffusivity near saturation, $C_1 = 20 \, \text{mm}^2/\text{day}$. For a slab of thickness $D = 100 \, \text{mm}$, the rate of environmental humidity evaluated from (8.256) is

$$\dot{h}_{\text{env}} = -\frac{8 \times 20 \times 0.03 \, \text{mm}^2/\text{day}}{100^2} \frac{}{\text{mm}^2} = -0.48 \times 10^{-3}/\text{day} \qquad (8.257)$$

This means that the environmental humidity can be brought from 100 to 90% in $0.1/\dot{h}_{\text{env}} \approx 208$ days. At lower humidities, the diffusivity is also lower and the admissible rate of environmental humidity would need to be further reduced. For instance, if the diffusivity at 55% relative humidity is 20 times lower than at saturation, the rate computed from (8.256) would be $-0.024 \times 10^{-3}/\text{day}$ and the environmental humidity should decrease from 60 to 50% in about 4160 days (more than 11 years).

Example 8.9. Environmental humidity variation slow enough to observe free shrinkage and drying creep

The foregoing analytical estimates can be confirmed by numerical simulations. For the concrete from Example 3.1, with mean compressive strength $\bar{f}_c = 45.4 \, \text{MPa}$, the *fib* formula (8.90) for diffusivity at saturation gives $C_1 \approx 23 \, \text{mm}^2/\text{day}$. Let us simulate drying of an initially saturated slab of thickness $D = 100 \, \text{mm}$, using the Bažant–Najjar model with parameters recommended by *fib*.

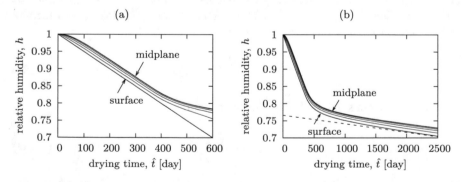

Fig. 8.47 Evolution of pore relative humidity at the surface and at distances 10, 20, 30, 40, and 50 mm from the surface of a slab of 100 mm in thickness, exposed to ambient humidity decreasing (a) at a constant rate, (b) at a variable rate proportional to the diffusivity

If the ambient humidity is decreased by 0.05% per day, formula (8.256) with $C(\hat{t})$ set to C_1 predicts the difference between the maximum and minimum pore relative humidities across the section to be $\Delta h = 2.72\%$. The computed evolution of pore relative humidity at six selected points (at distances 0, 10, 20, 30, 40, and 50 mm from the surface) is plotted in Fig. 8.47a. After an initial transitional period of 100 days, the actual value of Δh is found to be 2.4% and then it slowly increases, reaching 2.7% after 300 days and 3.0% after 400 days of drying. Subsequently, a strong reduction of diffusivity in the range of relative humidities around 80% slows down the transport process and the pore relative humidity at midplane decreases more slowly, which leads to an increase of Δh to 8.4% after 600 days of drying.

To obtain an almost constant Δh during the entire drying process, the rate of ambient humidity must be adjusted according to formula (8.256) in proportion to the current diffusivity. When this is done, the difference between maximum and minimum pore relative humidity is indeed kept roughly constant, even in the low-diffusivity range, as documented in Fig. 8.47b. The terminal rate of humidity decrease (slope of the dashed line in Fig. 8.47b) is 20 times smaller than the initial one. The numerically computed value of Δh initially grows to 2.7%, subsequently drops to 1.8%, and afterward slowly increases up to 2.7% again. For such a drying program, 3360 days would be needed to reach an average relative humidity of 70%, and 400 days would be needed for every additional reduction of 1%, which means that 50% would be reached after 11,360 days.   ∎

Obviously, it is impossible to directly observe free shrinkage on concrete specimens of regular sizes. Its characteristics must be deduced indirectly, by using finite elements to fit the test results on specimens with highly nonuniform pore humidity distribution. This is not easy and has been a great impediment to progress.

As follows from (8.256), the admissible rate of drying is inversely proportional to the square of the specimen thickness D. If a specimen of 1 mm in thickness is used, the admissible drying times are reduced 10,000 times (as compared to $D = 100$ mm), provided that we consider the diffusivity to remain the same.[22] Then, it is possible to dry the specimen from 100 to 50% within 1 or 2 days while keeping an almost uniform distribution of pore relative humidity. This was exploited by Bažant et al. [102] in their tests of cement paste performed on hollow cylinders with a wall thickness of 0.71 mm.

Since the humidity profiles in real structures are nonuniform, the shrinkage strains caused by humidity changes are incompatible, and so they must induce stresses to achieve compatibility. In the central part of a long prism, compatibility requires that (for symmetric drying) the total axial strains ε at all points of the cross section be equal, i.e., that the cross sections remain planar. Figure 8.48 schematically explains why self-equilibrated axial stresses σ must develop to ensure compatibility within the cross section (see also Fig. 3.17).

If the only goal is to avoid cracking, the calculation presented in the previous example must be adjusted. The humidity does not need to be uniform, but the maxi-

---

[22] Of course, specimens 1 mm thick can be made only of cement paste, which has a higher diffusivity. This must lead to a reduction factor less than 10,000, though not too different.

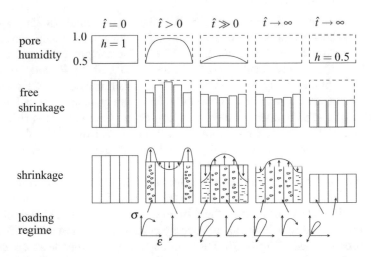

Fig. 8.48 Evolution of humidity, shrinkage, stress, and strain in various layers of a cross section, with induced cracking; reproduced with permission from [94]

mum stress should not exceed the strength limit. How large differences in humidity h within the cross section can be tolerated without causing tensile cracks? The answer requires a numerical calculation of stresses induced by nonuniform shrinkage and relaxed by creep; see Example 8.10. To get a quick impression of magnitudes, suppose that a saturated concrete is suddenly exposed to $h_{env} = 50\%$. Right below the surface the shrinkage is almost instantaneous and creep negligible. For the concrete from Example 3.1, with Young's modulus $E = 32$ GPa (4640 ksi), the free shrinkage can be estimated as $\varepsilon_{sh} = -(1-h_{env}^3)\varepsilon_s^\infty = -(1-0.5^3) \times 702.4 \times 10^{-6} \approx -615 \times 10^{-6}$, and if no cracking occurred, tensile stress $\sigma = -E\varepsilon_{sh} = -32$ GPa $\times (-0.000615) =$ 19.7 MPa $= 2860$ psi would develop. This greatly exceeds the typical value of tensile strength for such a concrete, $\bar{f_t} = 3.4$ MPa (490 psi). So cracking begins immediately after exposure to environmental humidity. All suddenly exposed concrete structures contain hairline shrinkage cracks. On the other hand, if the ambient humidity is reduced gradually and sufficiently slowly, the self-equilibrated stresses induced by nonuniform shrinkage can be kept below the tensile strength and cracking can be avoided.

To illustrate the interplay between shrinkage and creep, let us use a simplified model that takes into account basic creep only. Drying creep is neglected because, in the present context of the material point approach, its inclusion would require advanced models such as the microprestress-solidification theory, to be covered in Chap. 10. Restriction to basic creep greatly simplifies the formulation because it permits to treat the relaxation operator $\mathcal{R}$ as independent of the spatial coordinate. For an infinite slab subjected to drying from its surfaces, the internal stress state is equibiaxial ($\sigma_x = 0, \sigma_y = \sigma_z$) and the values of in-plane normal stresses depend only on the x-coordinate measured across the thickness (and on time, of course). Since the corresponding in-plane normal strains ε_y and ε_z are equal and independent of x, the general biaxial viscoelastic stress–strain law

$$\sigma_y(x, t) = \frac{1}{1 - \nu^2} \mathcal{R} \left\{ \varepsilon_y(x, t) - \varepsilon_{\text{sh}}(x, t - t_0) + \nu \left[\varepsilon_z(x, t) - \varepsilon_{\text{sh}}(x, t - t_0) \right] \right\}$$

(8.258)

can be rewritten as

$$\sigma(x, t) = \frac{1}{1 - \nu} \mathcal{R} \left\{ \varepsilon(t) - \varepsilon_{\text{sh}}(x, t - t_0) \right\}$$

(8.259)

where ν is the Poisson ratio, σ_y and ε_y are denoted simply as σ and ε, and the shrinkage strain is considered as a function of the time of drying, $\hat{t} = t - t_0$. In the absence of external loads and boundary constraints, the stress resultant must vanish, which is described by the condition

$$\int_0^D \sigma(x, t) \, dx = 0$$

(8.260)

In combination with (8.259), this leads to

$$\varepsilon(t) = \frac{1}{D} \int_0^D \varepsilon_{\text{sh}}(x, t - t_0) \, dx$$

(8.261)

Suppose that the shrinkage strain is linked to the pore relative humidity by Eq. (8.252). If the shrinkage ratio k_{sh}^* is constant, the rate equation can easily be integrated in time to yield

$$\varepsilon_{\text{sh}}(x, \hat{t}) = k_{\text{sh}}^* \left[h(x, \hat{t}) - 1 \right]$$

(8.262)

Substituting this into (8.261) and then into (8.259), we obtain

$$\varepsilon(t) = k_{\text{sh}}^* \left[\bar{h}(t - t_0) - 1 \right]$$

(8.263)

$$\sigma(x, t) = \frac{k_{\text{sh}}^*}{1 - \nu} \mathcal{R} \left\{ \bar{h}(t - t_0) - h(x, t - t_0) \right\}$$

(8.264)

where $\bar{h}$ denotes the pore relative humidity averaged across the slab thickness.

A simple analytical estimate can be constructed if creep is neglected and the evolution of humidity is approximately described by (8.253) with function f given by (8.255). The average humidity is then

$$\bar{h}(\hat{t}) = h_{\text{env}}(\hat{t}) + \frac{4 \Delta h}{D^2} \frac{1}{D} \int_0^D (D - x) x \, dx = h_{\text{env}}(\hat{t}) + \frac{2}{3} \Delta h$$

(8.265)

and the stress can be expressed as

$$\sigma(x,t) = \frac{Ek^*_{sh}}{1-\nu}\left(\frac{2}{3}\Delta h - \frac{4\Delta h}{D^2}(D-x)x\right) \tag{8.266}$$

The maximum tensile stress is attained at the surface (i.e., at $x = 0$ or $x = D$) and is given by

$$\sigma_{max} = \frac{2Ek^*_{sh}}{3(1-\nu)}\Delta h \tag{8.267}$$

To satisfy the condition $\sigma_{max} \le \bar{f_t}$, the humidity difference is constrained by

$$\Delta h \le \frac{3(1-\nu)\bar{f_t}}{2Ek^*_{sh}} \tag{8.268}$$

and the corresponding constraint for the rate of ambient humidity, deduced from (8.256), reads

$$|\dot{h}_{env}(\hat{t})| \le \frac{12(1-\nu)\bar{f_t}\,C(\hat{t})}{Ek^*_{sh}D^2} \tag{8.269}$$

Example 8.10. *Environmental humidity variation slow enough to avoid cracking*

Consider again a slab of thickness $D = 100$ mm, made of concrete with properties from Example 3.1. To obtain a rough estimate of the drying rate that will not lead to cracking, we can use formula (8.269) with material parameters $C_1 = 23$ mm^2/day, $E = 32$ GPa, $\nu = 0.2$, $\bar{f_t} = 3.4$ MPa and $k^*_{sh} = 1.76 \times 10^{-3}$. The resulting maximum allowed magnitude of the environmental humidity rate is 0.133%/day near saturation (i.e., at the early stage of the drying process, when $C(\hat{t}) \approx C_1$). If the diffusivity at low humidities is $20\times$ smaller than near saturation, the rate would need to remain below 0.0067%/day at late stages of the drying process. This corresponds to a change of environmental humidity by 10% in 1500 days.

The foregoing rough estimate is based on the elastic stress–strain law, but in reality, the stresses will be reduced by relaxation. Therefore, for numerical simulations, let us set the rate of environmental humidity to a somewhat higher value, $\dot{h}_{env} = -0.2\%$/day, which corresponds to a reduction from 100 to 50% in 250 days. The corresponding evolution of pore relative humidity at selected points, computed using the Bažant–Najjar model with parameters recommended by *fib*, is shown in Fig. 8.49a. Between 50 and 100 days of drying, the lines corresponding to humidities at individual points are almost parallel, which means that the shape of the humidity profile is getting close to the quadratic function that corresponds to (8.255). However, after 100 days of drying, the ambient humidity attains 0.8 and nonlinear effects gain in

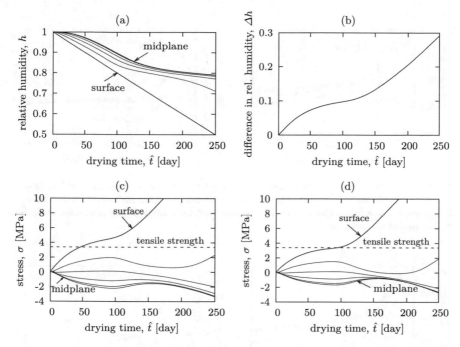

Fig. 8.49 Drying of a slab of 100 mm in thickness at a constant rate of ambient humidity: evolution of (a) pore relative humidity at the surface and at distances 10, 20, 30, 40, and 50 mm from the surface, (b) difference between maximum and minimum pore relative humidity, (c) stresses computed using an elastic stress–strain law, (d) stresses computed using a viscoelastic stress–strain law, with drying creep neglected

importance. To keep a constant shape of the humidity profile, the rate of drying would need to be reduced, but at a constant rate of drying, the difference Δh between maximum and minimum pore relative humidities dramatically increases; see Fig. 8.49b. If the stress–strain law is assumed to be linear elastic, with Young's modulus of 32 GPa, the maximum tensile stress (right at the exposed surface) exceeds the tensile strength already after 47 days; see Fig. 8.49c. For a viscoelastic law with basic creep only, the shrinkage-induced stresses partially relax and the tensile strength is exceeded after 95 days of drying; see Fig. 8.49d. In this calculation, it has been assumed that the onset of drying occurred at $t_0 = 28$ days.

   To avoid cracking, the drying rate needs to be reduced when the humidity enters the range in which the diffusivity decreases. If the rate of ambient humidity is varied in proportion to the current diffusivity, starting from the initial value of -0.2%/day, the pore relative humidity at the exposed surface (equal to the ambient relative humidity) decreases to 0.7 in about 600 days (Fig. 8.50a) and the stresses remain below the tensile strength during this entire period (Fig. 8.50b). The peak value of maximum stress, 3.24 MPa, is attained at 86 days, and after that the stress decreases to levels

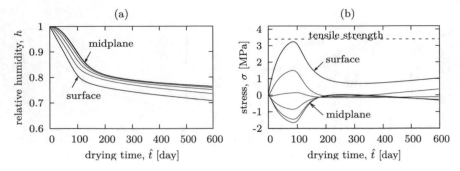

Fig. 8.50 Drying of a slab of 100 mm in thickness at a variable rate of ambient humidity, proportional to the diffusivity: evolution of (a) pore relative humidity at the surface and at distances 10, 20, 30, 40, and 50 mm from the surface, (b) stresses computed using a viscoelastic stress–strain law, with drying creep neglected

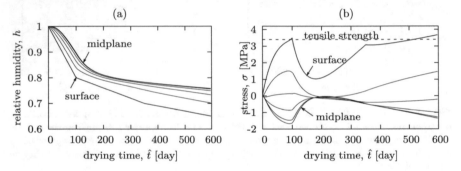

Fig. 8.51 Drying of a slab of 100 mm in thickness at a piecewise constant rate of ambient humidity: evolution of (a) pore relative humidity at the surface and at distances 10, 20, 30, 40, and 50 mm from the surface, (b) stresses computed using a viscoelastic stress–strain law, with drying creep neglected

close to 1 MPa. This indicates that the drying process during this stage could be somewhat faster.

Figure 8.51 shows the evolution of pore relative humidity and stress in a drying process with the ambient humidity rate kept at -0.2%/day during the first 100 days, then reduced to -0.04%/day for the next 250 days, and afterward reduced to -0.02%/day for the last 250 days. In this way, the value of $h_{\text{env}} = 0.65$ is reached in 600 days and the computed stresses only slightly exceed the tensile strength; see Fig. 8.51b. It can be expected that drying creep, which has been neglected in the present simulation, would further reduce the stresses and the strength limit would not be attained. ∎

It should be emphasized that the results presented in the previous example referred to a particular case, with specific values of material parameters. The purpose of the example was to illustrate the basic trends and to show that cracking due to shrinkage-induced stresses can be avoided only if the drying rate is sufficiently low. For concretes with a lower shrinkage ratio or higher diffusivity than in the example, the allowable rates of ambient relative humidity are higher. Also, if the drying specimen is a long prism instead of a slab, the stress state is uniaxial instead of equibiaxial and the factor $1/(1 - \nu)$ in formula (8.259) can be dropped, which means that the stresses are reduced as compared to the case of a slab. On the other hand, if the prism has all faces exposed to the ambient humidity, the one-dimensional diffusion equation is not applicable and the transport problem would need to be solved in 2D or even in 3D.

In simple calculations presented in Examples 8.9 and 8.10, it was assumed that the shrinkage strain is proportional to the change of pore relative humidity. Such a linear law was applied locally, at individual material points. On the other hand, in most experiments, the measured shrinkage strain corresponds to an average over the whole specimen, and the drying process is nonuniform. Consequently, a direct link between the relative humidity and the shrinkage strain is lost. Data obtained from standard shrinkage experiments indicate that the relation between the ambient relative humidity, h_{env}, and the magnitude of the ultimate shrinkage, $\varepsilon_{\text{sh}}^{\infty}$, is strongly nonlinear.[23] Most models determine $\varepsilon_{\text{sh}}^{\infty}$ as the product of a theoretical value of ultimate shrinkage extrapolated to zero humidity environment and a humidity-dependent correction factor. Models B3 and B4, as well as the *fib* model, specify the correction factor k_h by the cubic formula

$$k_h = 1 - h_{\text{env}}^3 \tag{8.270}$$

where h_{env} is the ambient relative humidity; see the solid curve in Fig. 8.52a. The GL model uses a fourth-order polynomial indicated in Fig. 8.52a by the dotted curve, and the ACI model uses a piecewise linear function indicated by the dashed lines. Despite these differences, the overall dependence of k_h (and thus of $\varepsilon_{\text{sh}}^{\infty}$) on h_{env} is approximately the same for all models and is highly nonlinear.

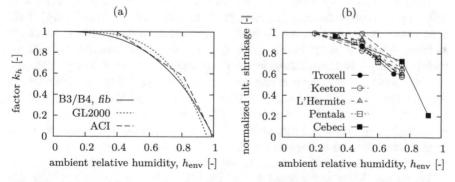

Fig. 8.52 Influence of the ambient relative humidity on drying shrinkage: (a) correction factor k_h reflecting ambient relative humidity h_{env} according to various models, (b) normalized experimental data measured on concrete specimens [301, 538, 576, 577, 674, 817]

---

[23] Besides, ongoing research at Northwestern University suggests that an ultimate value of shrinkage might not exist. The pore humidity decrease causes compressive stress in the solid skeleton. Multiplied by a nonzero factor for the effective porosity, this compression produces creep strain that is superposed on the shrinkage and evolves roughly logarithmically, with no upper bound.

A closer inspection of the concrete creep and shrinkage database [488] reveals that there are only a few experiments that can be directly used to study the relation between the ultimate shrinkage and the ambient humidity. Normalized plots extracted by Havlásek and Jirásek [474] from 12 selected data sets are shown in Fig. 8.52b. The autogenous shrinkage in most of these tests was probably very low and is neglected. To obtain dimensionless values that can be compared for different concretes (which exhibit different levels of ultimate shrinkage in a dry environment), the values of $\varepsilon_{sh}^{\infty}$ measured for each concrete are scaled by a common factor such that the point that corresponds to the lowest tested humidity would lie exactly on the theoretical curve given by (8.270). The shape of the resulting graph in Fig. 8.52b confirms that the theoretical formula describes the actual dependence reasonably well.

As already demonstrated in Fig. 8.45, the drying shrinkage of hardened cement **paste** measured on very thin specimens at a gradually decreasing relative humidity is often found to be an almost linear function of the relative humidity. This means that the origin of the strongly nonlinear dependence of the ultimate drying shrinkage of **concrete** on humidity has to be sought in effects that arise during upscaling from the cement paste level to the concrete level.

Even though the specimens for experimental measurement of drying shrinkage are kept unloaded, internally they are not stress-free. Self-equilibrated stresses develop due to a combination of the self-restraint and the aggregate restraint. Both of these restraints arise due to incompatibility of locally evaluated free shrinkage strains. The self-restraint is related to macroscopic nonuniformity of the drying process (discussed in the previous examples), while the aggregate restraint is caused by differences in shrinkage of concrete constituents (aggregates do not shrink, or shrink considerably less than cement paste). The stresses generated by the restraints are relieved by creep and, if the tensile strength is locally exceeded, also by microcracking.

The role of various phenomena and their effects on the ultimate shrinkage were addressed in a comprehensive numerical study by Havlásek and Jirásek [474]. They used a two-dimensional mesoscale model with inclusions (aggregates) considered as linear elastic disks and matrix (mortar) as a viscoelastic material that exhibits creep, shrinkage, and tensile cracking. A linear relationship between shrinkage and humidity was assumed for the mortar. Basic creep was captured by the B3 model and drying creep by the microprestress-solidification theory (MPS), to be described in Chap. 10. An interfacial transition zone (ITZ) with a reduced strength was inserted between the aggregates and mortar. Tensile cracking was taken into account using an approach based on damage mechanics, with a scalar damage variable driven by the maximum principal effective stress and with an exponential softening law.

The aggregates were supposed to occupy 70% of the total area, and their size distribution was defined by the Fuller curve; see the top part of Fig. 8.53. The maximum aggregate size was set to 20 mm. The position of aggregates in the surrounding matrix was randomly generated, avoiding overlaps. In order to make the analysis computationally feasible, only grains larger than 2 mm were explicitly modeled as circular inclusions in the mesostructure. Consequently, the inclusions actually represented about two thirds of the total aggregates. Smaller grains were combined with the cement paste into the matrix, which thus corresponded to mortar (not to

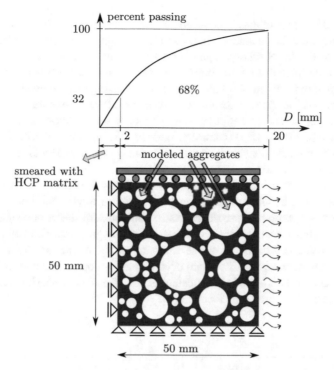

Fig. 8.53 Aggregate grading curve and a simplified scheme of the modeling approach and boundary conditions

pure cement paste). The simulated specimen was a square extracted from a 100 mm thick "infinite wall" drying at $h_{env} = 50\%$. The geometry of the specimen and the boundary conditions are shown in the bottom part of Fig. 8.53.

The dependence of the ultimate average shrinkage strain (evaluated from the total change of length in the vertical direction in Fig. 8.53) on the ambient humidity is shown in Fig. 8.54. The left part of the figure refers to a homogeneous material (with the effective properties of concrete) and the right part to a material with a heterogeneous mesostructure described in the previous paragraph. Dotted lines represent the results obtained if the mechanical response is assumed to be linear elastic (no creep, no cracking). Dashed curves correspond to a material with basic creep only (no drying creep, no cracking), dash-dotted curves to a material with basic and drying creep (no cracking), and finally the solid curves to a material with basic and drying creep and cracking. For the mesoscale model, these characteristics of the material refer to the mortar, while the aggregates are always assumed to be linear elastic (with a higher modulus than the mortar) and nonshrinking.

In the **homogeneous** case, the dependence of $\varepsilon_{sh}^{\infty}$ on h_{env} is perfectly linear not only if the material is assumed to be linear elastic, but also if basic creep is taken into account. The reason is that the relaxation operator is in such cases the same for all material points, and Eqs. (8.258)–(8.261) are applicable. The resulting strain is then equal to the spatial average of the local free shrinkage strains and is not affected by the specific values of elastic constants or creep parameters. This is no longer true when drying creep is considered, because locally computed drying creep evaluated according to the MPS theory depends on the evolution of humidity at the given material point. Since drying is nonuniform, relaxation of stresses is not governed by the same operator at all points of the sample, and $\mathcal{R}$ in Eq. (8.259) must be considered as dependent on the spatial coordinate, x. When (8.259) is substituted into (8.260), the position-dependent relaxation operator cannot be taken out of the spatial integral, and the averaging rule (8.261) cannot be deduced. As shown in Fig. 8.54a, inclusion of drying creep leads to some deviations from linearity, but the additional effect of cracking remains limited.

When a **heterogeneous** mesostructure is considered, the differences between individual approaches become more pronounced, see Fig. 8.54b. Nonlinearity appears already for basic creep because the relaxation operator for the viscoelastic mortar is different from the operator characterizing the elastic aggregates. The ultimate shrinkage at $h_{env} = 0.2$ is reduced by 4% as compared to the linear case (elastic mortar). This reduction increases to 18% when drying creep is considered and to 28% when cracking is added. This already is a substantial deviation from linearity, but still much smaller than 59% predicted by formula (8.270).

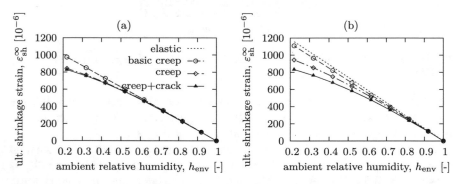

Fig. 8.54 Computed ultimate shrinkage strain of concrete subjected to various levels of ambient relative humidity: (a) homogeneous material, (b) mesoscale approach

The mesoscale model discussed so far neglected the possible nonlinear effects of fine aggregates (smaller than 2 mm), which were "mixed" with cement paste to form the mortar. The shrinkage strain in mortar was assumed to be proportional to the change of pore relative humidity. However, such an assumption was originally postulated for cement paste, and it is logical to expect that the presence of fine aggregates in mortar would lead to deviations from linearity already at this level. In the last part of their study, Havlásek and Jirásek [474] proposed a simplified **multiscale** approach, in which the nonlinear relation between shrinkage strain and pore relative humidity was first extracted from fine-scale simulations of a heterogeneous mortar (viscoelastic matrix with the properties of cement paste and elastic inclusions representing fine aggregates) and subsequently used as input for mesoscale simulations of concrete. When the shrinkage ratio for cement paste was considered as constant over the entire range of humidities ($k_{sh}^* = 0.006$), the ultimate shrinkage strain at $h_{env} = 0.2$ was found to be reduced by 36% as compared to the linear extrapolation from values at high humidity; see the dashed curve in Fig. 8.55b.

The last improvement consisted in assuming a piecewise linear dependence of shrinkage strain on humidity for cement paste, as illustrated in Fig. 8.55a. Even though the proportionality between ε_{sh} and h is a reasonable approximation for cement paste, as already demonstrated in Fig. 8.45b, a closer look at the experimental data of Rougelot, Skoczylas, and Burlion [739] for a paste with $w/c = 0.5$ and of Baroghel-Bouny et al. [57] for a paste with $w/c = 0.35$ (vertically shifted with respect to Fig. 8.45b) reveals certain deviations from linearity at low humidities. A good bilinear fit can be obtained with the shrinkage ratio k_{sh}^* reduced from 0.006 to 0.0036 in the range below $h = 0.5$. With such a nonlinear model for shrinkage of cement paste, the multiscale approach resulted in a more pronounced deviation from linearity for concrete; see the solid curve in Fig. 8.55b. The total reduction of the ultimate shrinkage strain of concrete at $h_{env} = 0.2$ was found to be 46% (as compared to a linear extrapolation from high humidities).

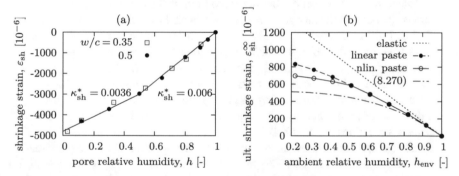

Fig. 8.55 (a) Evaluation of the shrinkage ratio k_{sh}^* for cement paste, (b) dependence of the ultimate shrinkage strain of concrete on ambient humidity, computed using a multiscale approach

It can be concluded that the nonlinearity of the dependence of the ultimate shrinkage strain on the ambient humidity can, to a large extent, be attributed to the effects of differential creep and cracking on the relaxation of self-equilibrated stresses that are induced by nonuniform shrinkage in nonuniformly drying heterogeneous concrete. Nevertheless, the strong nonlinearity shown in Fig. 8.52b and reflected by formula (8.270) is not explained completely. Possibly, one may need to take into account also the huge differences in permeability in the bulk and in cement paste, and the dramatic changes in paste permeability caused by aging due to hydration.

8.7   Effects of Self-Desiccation and Autogenous Shrinkage in Drying or Swelling Specimens—A Problem Requiring Further Research

8.7.1   Recent Paradigm-Changing Observations

Recently, exhaustive literature search related to the work on a new large database led to the following paradigm-changing experimental observations, which will require further research.

1. **Drying and Autogenous Shrinkages Are not Additive:** The drying and autogenous shrinkages have traditionally been considered as separate phenomena, to be superposed. However, after a specimen is exposed to drying, the autogenous shrinkage continues in the core until the arrival of the drying front with pore humidity lower than the self-desiccation humidity, in which case the drying and autogenous shrinkage interact. Should one consider their sum or maximum? It would be the sum if the autogenous shrinkage were of direct chemical origin, unrelated to self-desiccation, but the maximum of the two if not only the drying shrinkage but also the autogenous one were caused by a decrease of pore humidity, i.e., if the autogenous shrinkage were driven by self-desiccation. If the former were true, then, due to high water supply, the autogenous shrinkage in normal concretes with high w/c and only mild self-desiccation would have to be at least as large as it is in modern concretes with strong self-desiccation. But it is far from that. Therefore, both drying and autogenous shrinkages must be driven by a decrease of pore humidity. Consequently, they cannot be additive (which will require a revision of model B4).

2. **Swelling:** In water immersion, most concretes are swelling. Since the pore relative humidity is maintained at 100% in water immersion, the expansion cannot be driven by rising pore humidity. This indicates it must be driven by chemical expansion during hydration. Since the hydration cannot be expansive in water immersion and contractive without immersion, one must infer that the hydration should always cause the hardened cement paste to expand, even in specimens exposed to drying. Since drying specimens shrink, the magnitude of contraction due to a decrease of pore humidity must be greater, even much greater, than the magnitude of expansion due to hydration. And why specimens in fog room, where the relative humidity is 100%, do not swell? Probably because the fog cannot supply sufficient flux of water into the specimens to compensate for self-desiccation (here the boundary condition would need to be written in terms of surface emissivity).

3. **Autogenous Shrinkage:** As established at the dawn of cement research by Le Chatelier and confirmed by Powers and others, the cement hydration reaction is always contractive, i.e., the volume of the cement gel produced by hydration is always smaller than the sum of the original volumes of anhydrous cement and water used for hydration. So how can we explain why the hardened cement paste

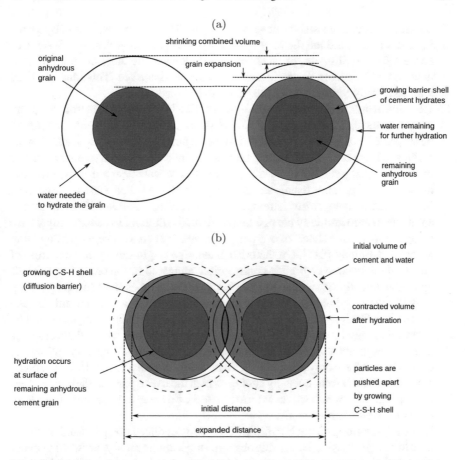

Fig. 8.56 (a) Isolated cement grain surrounded by the exact volume of water needed for hydration (note the shrinkage), (b) interaction of two initially touching grains surrounded by water needed for their hydration (note the increase of separation of grains even though the combined volume of all reactants decreased)

expands? The explanation is probably the development of porosity. The growth of contacting C-S-H shells around the cement grains produces crystallization pressure and thus pushes the adjacent shells apart (see Fig. 8.56).

4. **Long-Term Logarithmic Volume Change:** Misled by an illusion of approach to an asymptotic bound caused by plots in linear time scale, vast majority of experimenters have terminated the self-desiccation, autogenous shrinkage, and swelling tests within just a few months. There exist nevertheless some data showing autogenous shrinkage as well as swelling in water to continue logarithmically for almost ten years [262, 679]. This gives a different picture than the standard calorimetric procedure, which shows the hydration of cement to terminate within about a year. It thus appears that hydration can continue for many years, probably even decades and centuries (in fact, the C_3S grains in some submerged Roman

concrete still contain small unhydrated cores). This can be explained by growing C-S-H shells around the remaining anhydrous cement grains, which serve as
barriers for the diffusion of water molecules toward the anhydrous grain rem-
nant and of ions from the grain surface to the shell surface. This idea has been
mathematically formulated by Bažant et al. [125].

5. **Creep as a Source of Long-Term Logarithmic Shrinkage:** Aside from the delay
caused by the diffusion barriers of C-S-H shells, another phenomenon that might
be suspected as a source of the decade-long logarithmic evolution of autogenous
shrinkage is the creep. Maintaining thermodynamic equilibrium requires equality
of the chemical potential in all the phases of pore water. So a decrease of pore
humidity must produce tensile stress changes in all the water phases, which must
be balanced by compressive stress changes in the solid skeleton. The stresses in
solid due to pore humidity change might in turn be thought to cause compressive
creep in the solid skeleton of cement paste, which is an idea suggested by Ulm,
Maou, and Boulay [827]. On the other hand, one might object that this kind of
creep should make a negligible contribution compared to the creep due to loads
applied externally on the macroscale, because the stresses introduced by a drop
in pore humidity fluctuate between tensile and compressive on the micro- and
nanoscales. For instance, a drop in disjoining pressure in the nanopore could be
balanced mainly by compressive forces in the solid bridges across the nanopores
and in the adjacent C-S-H sheets. These bridges and sheets are nanocrystalline and
thus should deform only elastically, with no creep, because the macroscopic creep
must, for other reasons, occur by sliding, which must take place in the connections
between the sheets, rather than within the sheets. So creep as a source of long-term
shrinkage needs further study.

6. **Logarithmic Long-Term Swelling:** If humidity-induced creep existed, it would
have to be compressive. Hence, the observed logarithmic multiyear swelling under
water immersion cannot be explained by creep. But it can be explained by diffusion of water and ions through the growing barrier shell of C-S-H enveloping
the anhydrous cement grains (typical size 20–100 μm). The diffusion halftimes
increase as the square of the growing thickness of the shell. A thick enough shell
can slow down hydration to take many years or even decades (Fig. 8.56).

8.7.2 Improved Aging Characterization via a Model for Hydration

Calculation of the effects of aging also needs to be improved. The aging is caused
by the chemical process of silicate hydration. It is widely accepted that hydration
terminates within about a year, but that is confirmed only for small specimens with
ample supply of water. The fact that the autogenous shrinkage can continue for many
years implies that the hydration must, too.

The widely accepted idea that the hydration causes volume contraction must also be revised. It is doubtless true that the volume of cement hydration products is slightly smaller than the original volume of cement and reacted water [568, 705], which is what has been called the chemical shrinkage. However, this does not mean that the hydration reaction causes hardened cement paste to contract. In fact, once the C-S-H shells that grow around anhydrous cement grains come into contact, they cannot overlap and thus push the neighbors apart, applying crystallization pressure. This causes the solid framework of cement paste to expand.

In Bažant et al. [125], a new idea was proposed—this expansion always dominates over the contraction, i.e., the hydration is, in the bulk, **always expansive**, while the source of all of the observed shrinkage, whether autogenous or due to external drying, is a decrease of pore humidity. This decrease, in turn, causes a decrease of chemical potential of pore water, which is manifested as an increase of capillary tension in micropores, a decrease of disjoining pressure in nanopores, and an increase of surface tension on the nanoscale C-S-H globules (due to a decrease of spreading pressure in adsorption layers). All these changes produce compressive elastic strain in the solid C-S-H framework, which usually prevails over the expansion, except for specimens immersed in water.

From the aforementioned long-term tests showing that the autogenous shrinkage grows logarithmically in time over many years, it may be inferred that the growing C-S-H shells enveloping the remnants of anhydrous cement grains must act as diffusion barriers, inhibiting the diffusion of water molecules through these shells toward the grain remnants, as well the diffusion of ions in the opposite direction. This must slow down the hydration process and could explain why the hydration proceeds slowly (and logarithmically) for many years and probably even decades and centuries.

This kind of hydration model might provide a more realistic basis for the age effects on creep and shrinkage. It would also require updates in models B3 and B4.

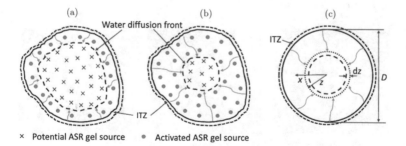

Fig. 8.57 ASR gel formation due to water diffusion into reactive aggregate: (a) early stage of diffusion, (b) late stage of diffusion, (c) idealization with spherical diffusion, after Bažant and Rahimi-Aghdam [185], reproduced with permission from ASCE

8.8   Creep and Diffusion as Processes Controlling Alkali–Silica Reaction

The *alkali–silica reaction* (ASR) is a destructive long-term reaction which afflicts mineral aggregates in concrete if they contain imperfectly crystalline silica (SiO_2), which is a condition hard to predict. Ions of sodium and potassium slowly diffuse into the aggregate pieces and combine with silica, producing the ASR gel. The gel can imbibe an enormous amount of water, if available. Thus, the ASR damage occurs only in sufficiently massive structures such as dams, reactor containments, or large bridges in which high pore humidity persists in the cross-sectional core for many years. Because of the slowness of water and ion diffusion into the aggregate pieces, the ASR typically begins only after several decades, often causing severe strength degradation and fracturing. A good mathematical model is important for making inferences from accelerated laboratory tests and for predicting the remaining lifetime of the structure once the ASR is detected.

The history of production of ASR gel requires calculating the diffusion of water from the pores into an aggregate piece considered, for simplicity, as spherical [186]; see Fig. 8.57. The next step is the formulation of a nonlinear diffusion model for the penetration of gel into the micro- and nanopores in a mineral aggregate grain, into the interface transition zone (ITZ) and into the microcracks very near the ITZ; see Fig. 8.58 (the gel that penetrates farther into the pores and cracks in cement paste or mortar is irrelevant since it calcifies by reacting with CaOH and thus stops expanding). Calculating the effects of expansion requires considering the creep. This was first done by Alnaggar, Cusatis, and Di-Luzio [33], who considered only the gel production but not diffusion and used their lattice discrete particle model (LDPM), assuming, as a simplification necessary to their LDPM approach, that the expansion occurs between, rather than within, the particles.

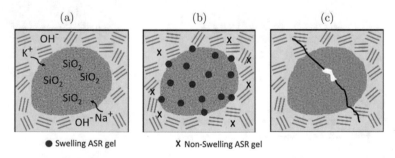

Fig. 8.58 ASR reaction process and schematic illustration of ASR-induced damage: (a) alkali–silica reaction, (b) formation of swelling and nonswelling ASR gel, (c) ASR-induced damage and cracking, after Bažant and Rahimi-Aghdam [185], reproduced with permission from ASCE

The time span of the gel diffusion analysis spans from minutes to decades, which poses problems of numerical stability of standard algorithms. Unconditional numerical stability has been achieved with a novel algorithm [186, Eq. 25], which converts the gel diffusion analysis to calculating the relaxation of the average gel pressure p_{gel} at constant total gel mass and at no material deformation (no creep) during each time step, with sudden pressure jumps between the steps (in a spirit akin to the exponential algorithm for creep, presented in Sects. 5.2.2–5.2.3).

Under these assumptions, the relaxation of the average gel pressure p_{gel} within the time step due to diffusive redistribution of gel within the fixed current volume of accessible pores and cracks is given by the equation

$$\frac{dp_{gel}}{dt} + \frac{p_{gel}}{\tau_p} = 0 \qquad (8.271)$$

where τ_p is a characteristic time dependent on the gel compressibility (about the same as water) and effective Darcy permeability, and on the current total volume fraction of gel.

The gel expansion in the aggregate and the ITZ causes fracturing damage in the surrounding concrete, which is analyzed by microplane model M7 [290, 291]; see Sect. 12.8.1 for more details on this advanced constitutive model. For realistic modeling of the effects of gel expansion, it proved important to incorporate into M7 the aging creep with a broad continuous retardation spectrum and exponential algorithm [185].

The gel and the damaged concrete are macroscopically treated as a two-phase (solid–fluid) medium. The medium is of nonstandard type, because of the nanopore water that is load-bearing but mobile. The condition of equilibrium between the phases is what mathematically introduces the fracture-producing load into the concrete. Depending on the stress tensor in the solid phase, the expansion is directional, producing oriented cracking. The creep is found to have a major mitigating effect on multidecade evolution of ASR damage and is important even for interpreting accelerated laboratory tests lasting a few weeks.

Note that the eigenstrain caused by the ASR is of a different nature than the thermal (or shrinkage) eigenstrain. The latter is simply additive to the stress-induced deformation and is assumed to cause no damage to unrestrained concrete. On the other hand, the ASR gel expansion into the pores and microcracks applies, from within, a tensile hydrostatic stress to the porous solid skeleton of the material. It is actually an eigenstrain on a sub-scale of the material (some investigators modeled this eigenstrain by assigning different thermal expansion coefficients to the aggregate and the mortar matrix and imposing a temperature change; but this approach is correct only for certain values of temperature change and could hardly be combined with diffusion or creep effects).

In Rahimi-Aghdam, Bažant, and Caner [714], the theory for the material and structural damage due to the alkali–silica reaction in concrete was calibrated and validated by finite element fitting of the main test results from the literature, including those of Multon and Toutlemonde [643], Ben Haha [227] and Poyet, Sellier, Capra, Thévenin-Foray, Torrenti, Tournier-Cognon, and Bourdarot [708]. The frac-

(a)   (b)   (a)   (d)

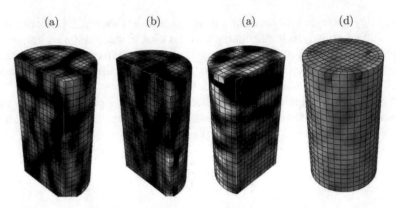

Fig. 8.59 Numerically calculated cracking patterns for different stress states: (a) unconfined load-free, (b) radially confined load-free, (c) unconfined axially loaded, (d) radially confined axially loaded; after Rahimi-Aghdam et al. [714], reproduced with permission from ASCE

ture mechanics aspects were handled by the crack band model; see Sect. 12.2. The theory was shown to capture quite well

1. the effects of various loading conditions and stress states on the ASR-induced expansion, and the orientation of cracking (Fig. 8.59);
2. degradation of the mechanical properties of concrete, particularly its tensile and compressive strength, and elastic modulus;
3. the effect of temperature on ASR-induced expansion; and
4. the ASR dependence on external drying, for cross sections of various sizes, with or without simultaneous temperature effect, and for various cross section thicknesses.

The finite element simulations utilized microplane model M7. The creep (with aging), embedded in M7 using the approach described in Sect. 12.8.2, was found to mitigate the predicted ASR damage significantly. The moisture diffusivity, both the global one for external drying and the local one for mortar near the aggregate, was considered to decrease 20 times as the pore humidity drops (Sect. 8.3.4.2). Close fits have been achieved, and the model appears ready for predicting the ASR effects in large structures.

Chapter 9
Solidification Theory for Aging Effect on Stiffness and Basic Creep

Abstract The rate of concrete creep not only attenuates with the load duration but also decreases, at a decaying rate, with the age at loading. This phenomenon, called aging, complicates the mathematical modeling of creep. Since the phenomenological approach that deals with age-dependent material properties has no physical underpinning, we embark on a more physical approach, based on the analysis of solidification of cement, which is the physical cause of aging (aside from microprestress relaxation, discussed in the next chapter). We show that one can attribute the aging to the growth of volume fraction of a nonaging constituent of hydrating cement, approximately considered as the C-S-H. The fact that this constituent can be considered as nonaging brings about a considerable simplification of the material model. Then, we show how the concept of solidification requires the compliance curves for different ages at loading not to diverge with time from each other, which in turn rules out creep recovery curves with nonmonotonic decay. We also explain the problems with creep compliance models giving a relaxation curve that crosses to the opposite sign. We compare the behavior of a number of models from the literature and design codes, and we show that many of them lead to divergence of compliance curves or to relaxation crossing to the opposite sign, at least under certain specific conditions. We conclude by pointing out the thermodynamic restrictions on rheological Kelvin and Maxwell chains and their implications for the properties of compliance and relaxation functions.

The characteristics of concrete creep evolve at a declining rate over many years and even beyond a century. This aging (or maturing) behavior complicates the mathematical description as well as experimental identification. Traditionally, it has been modeled as a gradual variation of the material parameters in the compliance function for creep [94]. However, this classical modeling approach was purely phenomenological, with no physical underpinning. An approach based on the analysis of solidification of cement, which is (aside from microprestress relaxation, discussed in Chap. 10) the physical cause of aging, is more realistic and, as it turns out, also brings about a simplification of the constitutive model for creep.

© Springer Science+Business Media B.V. 2018

Z.P. Bažant and M. Jirásek, *Creep and Hygrothermal Effects*
in Concrete Structures, Solid Mechanics and Its Applications 225,
https://doi.org/10.1007/978-94-024-1138-6_9

9.1 Growth of Volume Fraction of Calcium Silicate Hydrates and Polymerization Hypothesis

It is logical to assume that the chemical reactions of hydration cause the aging property of creep. In these reactions, the anhydrous cement grains are gradually dissolving in the pore water and then recombining as the calcium silicate hydrate (C-S-H) gel, which exhibits viscoelasticity, and other solid hydration products, e.g., calcium hydroxide, which exhibit elastic behavior. The newly formed C-S-H gel is then being deposited at the surfaces of the pores in hardened cement paste. The volume of hydration products (per unit volume of concrete), and thus also the overall volume of the load-bearing solid skeleton in the cement paste, are growing at the expense of the volumes of free (evaporable) water and anhydrous cement, which are diminishing; see Fig. 8.43.

Thermodynamics of a system described only by properties varying in time is impossible to formulate. It can be formulated only if the system is broken down to time-invariant constituents whose concentrations are varying in time. In hardened Portland cement paste, such a constituent is the C-S-H gel, which is nearly identical to the gel of a natural mineral called tobermorite.

Thus, the aging property of creep must be attributed to volume growth of cement gel, treated as a substance of time-invariant properties. This is the central idea of the *solidification theory* proposed by Bažant [82] and developed in detail by Bažant and Prasannan [179, 180]. The specific volume of cement gel that has solidified up to time t will be denoted as $v(t)$. This function can be normalized such that it approaches 1 as the time tends to infinity.

It is also possible that the C-S-H undergoes polymerization in which additional bonds are gradually formed [179, 180]. Evidently, polymerization can cause an increase of stiffness even if the volume of C-S-H does not grow. What really matters is not the total volume growth of the C-S-H but the volume growth of that part of C-S-H which forms a contiguously interconnected solid skeleton. Loose, unconnected parts of C-S-H do not matter. Thus, strictly speaking, the variable $v(t)$ should better be interpreted as the volume fraction of the interconnected solid skeleton of C-S-H, and the term "volume growth" will henceforth be used, for the sake of brevity, to represent the growth of this fraction. By the time of writing, though, the phenomenon of polymerization of C-S-H has not yet been conclusively demonstrated. It is nevertheless a plausible hypothesis, supported by some leading cement physicists (e.g., H. Jennings).

The aforementioned concept of the solidification process in hydrating Portland cement paste, pictured in Fig. 9.1, governs the viscoelastic strain, ε_v, which is imagined to represent the effective microstrain to which the cement gel is subjected. Consider that, at time t', material of relative volume dv solidifies from the pore solution (dashed hatching in Fig. 9.1) and gets attached to the surface of the previously solidified volume $v(t')$, i.e., to the pore wall. All the previously solidified material volume is subjected to the same strain increments after time t' at which it solidified.

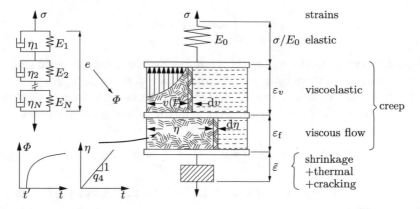

Fig. 9.1 Model for role of solidification in creep

The macroscopic (or continuum) stress σ is the resultant of the microstresses from the entire solidified material volume.

Viscoelastic strain increment $d\varepsilon_v(t')$ introduced at time t' causes at time t the microstress $\Psi(t - t')d\varepsilon_v(t')$, where $\Psi(t - t')$ is the relaxation function of the cement gel. This function depends on the time lag $t - t'$ only (rather than on t and t' separately) because the solidifying constituent is nonaging. The contribution of the microstress to the macroscopic stress is proportional to the effective area on which the microstress acts, which in turn is proportional to the specific volume of the solidified material. Thus, the contribution of the microstress generated by the viscoelastic strain increment $d\varepsilon_v(t')$ introduced at time t' to the macrostress at time t is $v(t')\Psi(t - t')\, d\varepsilon_v(t')$, and the total macroscopic stress at time t generated by all the previously applied viscoelastic strain increments is

$$\sigma(t) = \int_0^t v(t')\Psi(t - t')\, d\varepsilon_v(t') \tag{9.1}$$

It is now convenient to introduce an auxiliary strain-like variable $e(t)$, defined by the rate equation $\dot{e}(t) = v(t)\dot{\varepsilon}_v(t)$ and initial condition $e(0) = 0$. This variable can be interpreted as a certain type of *effective viscoelastic strain* in the cement gel, which is fully recoverable after unloading. On the other hand, ε_v is only partially recoverable, which is caused by the growth of v. Since $de = v\, d\varepsilon_v$, Eq. (9.1) can be rewritten as

$$\sigma(t) = \int_0^t \Psi(t - t')\, de(t') \tag{9.2}$$

This has the same form as the integral stress–strain relation of nonaging viscoelastic material (2.23), with relaxation function $\Psi(t - t')$ and with the strain replaced by e. In other words, the actual stress σ in the solidifying material is linked to the effective viscoelastic strain e by the constitutive relation characterizing the nonaging constituent. It is thus easy to formally invert the stress–strain relation (9.2) and write

$$e(t) = \int_0^t \Phi(t - t') \, d\sigma(t') \tag{9.3}$$

where $\Phi(t - t')$ is the inverse kernel (or resolvent) of the Volterra integral equation (9.2), representing the nonaging compliance function of the cement gel.

To transform the effective strain e into the actual viscoelastic strain ε, we need to proceed to the rate form of (9.3) and divide it by the volume growth function[1]:

$$\dot{\varepsilon}_v(t) = \frac{\dot{e}(t)}{v(t)} = \frac{1}{v(t)} \left[\Phi(0)\dot{\sigma}(t) + \int_0^t \dot{\Phi}(t - t') \, d\sigma(t') \right] \tag{9.4}$$

Comparing this to the general expression (2.17) for the strain rate of an aging viscoelastic material,

$$\dot{\varepsilon}(t) = \frac{\dot{\sigma}(t)}{E(t)} + \int_0^t J(t, t') d\sigma(t') \tag{9.5}$$

we can identify the *instantaneous compliance*

$$\frac{1}{E_v(t)} = \frac{\Phi(0)}{v(t)} \tag{9.6}$$

and the *compliance rate*

$$\dot{J}_v(t, t') = \frac{\dot{\Phi}(t - t')}{v(t)} \tag{9.7}$$

where the dot over J denotes differentiation with respect to the first argument and the subscript "v" at E and J emphasizes that these characteristics refer to the part of total strain in concrete caused by the viscoelastic strain in the gel (and not yet to the total strain in concrete).

Finally, using the instantaneous compliance $J_v(t', t') = 1/E_v(t') = \Phi(0)/v(t')$ as the initial value and integrating (9.7) with respect to t (which is renamed as s, so that t can be used as the upper integration limit), we obtain the macroscopic compliance function of the gel deduced from solidification theory,

$$J_v(t, t') = J_v(t', t') + \int_{t'}^t \dot{J}_v(s, t') ds = \frac{\Phi(0)}{v(t')} + \int_{t'}^t \frac{\dot{\Phi}(s - t')}{v(s)} ds \tag{9.8}$$

This formula provides the link between the microscopic and macroscopic descriptions of the solidifying constituent. On the microscopic level, the material properties are represented by the compliance function of the nonaging constituent Φ and by the solidification function v describing the growth of the specific volume of that constituent in time. Each of these functions depends on a single variable, which is the

---

[1]A slightly longer derivation of (9.4), avoiding the use of relaxation function, was given by Bažant [82], see also Bažant and Prasannan [179]. The present derivation was given by Bažant [96], as an abbreviation of that by Carol and Bažant [296].

duration of loading for Φ and the age for v. On the macroscopic level, we obtain the compliance function J_v of aging viscoelasticity as a function of two variables—the age at which the stress was applied and the current age.

The cement gel is the solid component of the hardened cement paste. This paste also includes the capillary pores (while the nanopores, containing the load-bearing hindered adsorbed water, are considered as part of the cement gel). The foregoing constitutive relation has been formulated with the cement paste in mind. However, as long as the aggregate is elastic, such a constitutive relation may also be used to describe cement mortar and concrete, albeit with different coefficients and different microcompliance function $\Phi(t)$. The reason is that the elastic restraint imposed by the aggregate on the cement paste cannot change the form of the constitutive relation.

9.2 Basic Creep Model for Concrete

Extensive studies of test data showed that the viscoelastic strain $\varepsilon_v(t)$ alone does not suffice to describe concrete creep. It turns out that the so-called *flow strain* $\varepsilon_f(t)$ must be added to $\varepsilon_v(t)$. The need for this additional term is also supported by the fact that the hydration reaction almost stops after one year and the strength of concrete does not increase at later ages (Fig. 9.2a), while the effective modulus $1/J(t' + 365, t')$ grows with age t' for many years (Fig. 9.2b). Note that the effective modulus considered here corresponds to the strain after one year of loading caused by unit stress applied at age t'. Both strength and effective modulus are normalized by their values at age of 28 days. The strength evolution depicted in Fig. 9.2a is based on formula (E.39) recommended by the *fib* Model Code, with parameter $s = 0.25$. The evolution of the effective modulus in Fig. 9.2b is based on model B3 with parameters determined by fitting the creep data for Wylfa Vessel concrete (shown in Fig. 9.6d).

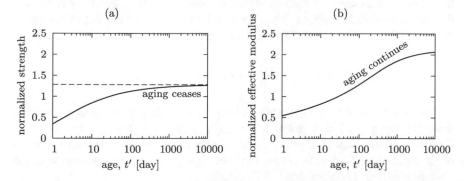

Fig. 9.2 Effective modulus $E_{\text{eff}} = 1/J(t' + 365, t')$ grows long after the strength has ceased to grow

The purely viscous strain (flow strain) ε_f is the completely irrecoverable part of the creep strain, modeled by a viscous dashpot. Its physical origin will be explained in detail in Chap. 10. Here, we simply postulate that the rate of flow

$$\dot{\varepsilon}_f(t) = \frac{\sigma(t)}{\eta(t)} \tag{9.9}$$

depends only on the current stress $\sigma(t)$, although the corresponding viscosity $\eta(t)$ is age-dependent. The fact that the flow strain has no memory (i.e., no heredity, no delayed elastic effect) simplifies the formulation. To be able to incorporate the contribution of the flow strain into the general framework of viscoelasticity, we need to evaluate the compliance function of an aging viscous dashpot. If the dashpot is loaded at time t' by constant stress $\hat{\sigma}$, the evolution of the flow strain from its initial value $\varepsilon_f(t') = 0$ is governed by the rate equation $\dot{\varepsilon}_f(t) = \hat{\sigma}/\eta(t)$. Renaming t as s, integrating from t' to t, and dividing the result by $\hat{\sigma}$, we obtain the flow compliance function

$$J_f(t, t') = \int_{t'}^{t} \frac{ds}{\eta(s)} \tag{9.10}$$

The complete model for basic creep of concrete is illustrated by the rheological scheme in Fig. 9.1. In the absence of cracking, it considers the mechanical strain (i.e., the total strain minus the hygrothermal strain) as the sum of instantaneous elastic strain, viscoelastic strain evaluated by the solidification model, and additional flow strain:

$$\varepsilon(t) = \frac{\sigma(t)}{E_0} + \varepsilon_v(t) + \varepsilon_f(t) \tag{9.11}$$

Here, E_0 = asymptotic elastic modulus representing the material stiffness for extremely fast deformations (extrapolated to a zero duration of load). Combining the compliance function of the nonaging elastic spring, $H(t - t')/E_0$, the viscoelastic compliance function of the solidifying constituent $J_v(t, t')$ given by (9.8), and the flow compliance function $J_f(t, t')$ given by (9.10), we can present the strain evaluation formula in the usual format (2.14) with the total compliance function

$$J(t, t') = \frac{1}{E_0} H(t - t') + J_v(t, t') + J_f(t, t') \tag{9.12}$$

The flow (or aging viscous) strain is important for long-time creep of concrete loaded at young age. This is revealed by comparing Fig. 9.4a, b where the creep data for Canyon Ferry Dam [455, 458] are fitted without and with the flow term. On the other hand, when $t - t'$ is not large, the flow term is unimportant compared to the viscoelastic term (as documented by Fig. 3.18c, d).

Note that the flow term makes no contribution to creep recovery. The recovery is due entirely to the viscoelastic strain. A significant part of the viscoelastic strain is irrecoverable, due to aging.

9.3 Basic Creep Compliance Function of Model B3

To obtain a practically usable form of the compliance function based on the theoretical framework presented in the preceding two sections, we need to select specific formulae for the nonaging microcompliance function of the solidifying constituent, $\Phi(t)$, for the specific volume growth function, $v(t)$, and for the time-dependent flow viscosity, $\eta(t)$, so that the partial compliance functions $J_v(t, t')$ and $J_f(t, t')$ can be evaluated from (9.8) and (9.10).

Experimental data reveal that the compliance curves for basic creep have the form of power curves $(t - t')^n$ for short load durations $t - t'$ and of logarithmic curves for long durations, with a smooth transition in between. Since $\ln(1 + x^n)$ is close to x^n for small x and close to $n \ln x$ for large x, and also $\int dt/\eta(t) \propto \ln t$ if $\eta(t) \propto t$, we may set

$$\Phi(t - t') = q_3 \ln\left[1 + \left(\frac{t - t'}{\lambda_0}\right)^n\right] \tag{9.13}$$

$$\frac{1}{\eta(t)} = \frac{q_4}{t} \tag{9.14}$$

in which n, λ_0, q_3, and q_4 are empirical material parameters (the last two already represent the parameters of model B3). Note that $\Phi(0) = 0$, because experimental data suggest that the asymptotic modulus is approximately age-independent (with the exception of very young ages), and thus, the instantaneous compliance is not attributed to the solidifying constituent but added separately through the nonaging spring in Fig. 9.1. The linear dependence of viscosity on age postulated in (9.14) is designed such that the flow compliance function evaluated from (9.10) has the logarithmic form

$$J_f(t, t') = \int_{t'}^{t} \frac{ds}{\eta(s)} = q_4 \ln\frac{t}{t'} \tag{9.15}$$

For the solidification effect, it is logical to adopt the experimentally justified function from the double-power law and log-double-power law [175]:

$$\frac{1}{v(t)} = 1 + \frac{1}{\alpha}\left(\frac{\lambda_0}{t}\right)^m \tag{9.16}$$

where m and α are further empirical constants. For this specific choice, the compliance function of the solidifying constituent evaluated from (9.8) is

$$J_v(t, t') = \frac{\Phi(0)}{v(t')} + \int_{t'}^{t} \frac{\dot{\Phi}(s - t')}{v(s)}\,ds = \int_{t'}^{t}\left[1 + \frac{1}{\alpha}\left(\frac{\lambda_0}{s}\right)^m\right]\dot{\Phi}(s - t')\,ds =$$

$$= \int_{t'}^{t} \dot{\Phi}(s - t')ds + \int_{t'}^{t} \alpha^{-1}\lambda_0^m s^{-m}\dot{\Phi}(s - t')\,ds =$$

$$= \Phi(t - t') - \Phi(0) + \int_{t'}^{t} \frac{\alpha^{-1} \lambda_0^m s^{-m} n q_3 ds}{s - t' + \lambda_0^n (s - t')^{1-n}} =$$

$$= q_3 \ln \left[1 + \left(\frac{t - t'}{\lambda_0} \right)^n \right] + \frac{q_3}{\alpha} Q(t, t') \tag{9.17}$$

where

$$Q(t, t') = n \lambda_0^m \int_{t'}^{t} \frac{s^{-m} ds}{s - t' + \lambda_0^n (s - t')^{1-n}} \tag{9.18}$$

is a function which cannot be integrated in closed form and needs to be computed by numerical integration or approximated by a suitable formula; see (C.2) in Appendix C.

Substituting the specific forms of compliances (9.17) and (9.15) into the general expression (9.12), we recognize the part of the compliance function of model B3 that corresponds to instantaneous elasticity and basic creep; cf. Eqs. (3.3) and (3.11) in Chap. 3. Data fitting revealed that, for most concretes, one can use the approximate values $n = 0.1$, $m = 0.5$, and $\lambda_0 = 1$ day. This is why parameter λ_0 is not explicitly used in (3.11), but it is implicitly present in the condition that all time variables should be substituted in days. Furthermore, in model B3, the constant q_3/α (which appears in the second term of the final expression in (9.17)) is denoted as q_2, and the inverse of the asymptotic modulus, $1/E_0$, is denoted as q_1. So the final formula is

$$J(t, t') = q_1 H(t - t') + q_2 Q(t, t') + q_3 \ln \left(1 + \frac{(t - t')^n}{\lambda_0^n} \right) + q_4 \ln \frac{t}{t'} \tag{9.19}$$

With the standard values of n, m, and λ_0, only parameters q_1, q_2, q_3, and q_4 need to be identified from test data. An important advantage is that these four free parameters are involved linearly. Hence, they can be obtained from the test data by linear regression (which always gives a unique result, unlike nonlinear regression).

Note that the aging of concrete is described in the stress–strain relation not by one but by two functions, $v(t)$ and $\eta(t)$, and that the aging rate is further modified by the dependence of the increments of the equivalent hydration period t_e on the relative pore humidity h and temperature T; see Eqs. (10.28), (10.31), and (10.34) in Chap. 10. Inspired by the time-temperature shift (or time-temperature superposition principle) for thermorheologically simple materials, widely used for polymers [313, 388, 770], there were many attempts to use a single time transformation to reduce the stress–strain relation for concrete creep to an age-independent form, but only a narrowly selected part of the experimental evidence could be modeled because, as we see, more than one transformation is actually required.

Algorithmic treatment of a creep model with compliance function in the format inspired by the solidification theory is straightforward and can be based on the procedure developed for a solidifying Kelvin chain; see Sect. 5.2.6. For this purpose, the compliance function Φ characterizing the nonaging constituent needs to be approximated by Dirichlet series (5.72) corresponding to a nonaging Kelvin chain. This can be achieved by least-square fitting or better by using the concept of a continuous retardation spectrum, as described in detail in Sect. F.2.

Once the retardation times τ_μ are selected and the corresponding partial moduli E_μ are determined, Algorithm 5.3 can be applied. In its original form, the algorithm covers only the part of response that corresponds to the viscoelastic compliance function J_v. To incorporate the nonaging spring of stiffness $E_0 = 1/q_1$ and the aging dashpot with compliance function J_f, the incremental modulus should be evaluated according to formula (5.83) and the creep strain increment according to formula (5.84). Note that the partial moduli E_μ are determined only once, at the beginning of the analysis, and need not be recomputed after each step, owing to the nonaging character of compliance function Φ.

9.4 Absence of a Characteristic Time as the Reason for Using Power Functions*

Creep is the result of several processes which are not known to possess any distinct characteristic time, i.e., a time at which the behavior would drastically change. Such processes are termed *self-similar*, and the characteristic times are implied by the transitions between these processes. It is illuminating to consider now one such process in isolation.

Because of self-similarity, the ratio of compliances $\Phi(t_2)/\Phi(t_1)$ for any two positive times t_1 and t_2 should depend exclusively on the ratio of times t_2/t_1. We can thus consider a dimensionless compliance function f of dimensionless time $\tau = t/t_{\text{ref}}$ applicable to a time period in which a single self-similar process dominates. The self-similarity is characterized by the condition that

$$f\left(\frac{t}{t_{\text{ref}}}\right) = \frac{\Phi(t)}{\Phi(t_{\text{ref}})} \tag{9.20}$$

where t_{ref} is an arbitrarily selected reference time. Function f is independent of the choice of t_{ref}; otherwise, the process would possess an intrinsic time scale. Equation (9.20) directly implies that $f(1) = 1$. Using the fact that the reference time t_{ref} is arbitrary, it is easy to prove[2] that

$$f(\beta\tau) = f(\beta)f(\tau) \tag{9.21}$$

for any positive numbers β and τ. This is a functional equation for the unknown function f. Differentiating (9.21) with respect to β (while treating τ as a constant), we obtain

---

[2]The proof of (9.21) directly follows from the relations $f(\beta) = \Phi(\beta\tau\, t_{\text{ref}})/\Phi(\tau\, t_{\text{ref}})$ and $f(\tau) = \Phi(\tau\, t_{\text{ref}})/\Phi(t_{\text{ref}})$, which are based on (9.20). Note that the first relation uses $\tau\, t_{\text{ref}}$ instead of t_{ref} as the reference time.

$$\tau f'(\beta\tau) = f'(\beta)f(\tau) \tag{9.22}$$

where f' is the derivative of f. Now, we set $\beta = 1$ and denote $r = f'(1)$. The resulting differential equation

$$\tau f'(\tau) = r\, f(\tau) \tag{9.23}$$

can be integrated by separation of variables, which gives $f(\tau) = C\tau^r$. Condition $f(1) = 1$ implies that the integration variable C is equal to 1, and the final solution is

$$f(\tau) = \tau^r \tag{9.24}$$

Consequently, in the absence of a characteristic time, the dimensionless function f must be a power function. This proof is similar to that used by Bažant [95] for the size effect in structures and by Barenblatt [51] and others in fluid mechanics.

    Going back to (9.20), we can see that the self-similar function Φ must be a power function, too, and it has the form

$$\Phi(t) = \Phi_{\text{ref}} \left(\frac{t}{t_{\text{ref}}} \right)^r \tag{9.25}$$

Note that this formula contains only two (not three) independent model parameters, because t_{ref} can be chosen arbitrarily. Parameter Φ_{ref} corresponds to the value of Φ at $t = t_{\text{ref}}$. If t_{ref} is replaced by βt_{ref}, it is sufficient to replace Φ_{ref} by $\beta^r \Phi_{\text{ref}}$ and function Φ remains exactly the same. Any function different from a power function would require a model parameter with the dimension of time. So functions such as logarithmic, exponential, sinh could be expected to apply only for physical processes possessing some characteristic time or terminal equilibrium state. For the logarithmic function, however, its derivative is a power function (with exponent $r = -1$) and thus possesses no characteristic time. This means that the characteristic time of the logarithmic function is given by the integration constant.

    In model B3, the logarithmic function describes the long-time creep because, for very long creep times, $t - t' \gg t'$, one process, namely the viscous flow corresponding to a dashpot of viscosity $\eta(t)$, should control the rate of creep. The decay of the long-time creep rate is the consequence of aging, i.e., of the increase of $\eta(t)$ with t (which itself should be regarded, for long times, as the consequence of micropre-stress relaxation rather than hydration reaction; see Chap. 10). Since the aging is the only time-dependent process as far as $\eta(t)$ is concerned, and since no characteristic time is known for this process, a power function should again be expected for $\eta(t)$. Empirically, viscosity $\eta(t) \propto t$ is found to represent the rate of long-time creep well. The integration of $\dot{\varepsilon}_f(t) = \hat{\sigma}/\eta(t) \propto \hat{\sigma}/t$ with initial condition $\varepsilon_f(t') = 0$ confirms that the flow strain ε_f should evolve as $\ln(t/t')$, as used in model B3, and the role of the characteristic time is played by the age at loading, t'.

Combination of, or transition between, two or more self-similar processes involves, of course, a characteristic time at which the transition is centered. Thus, e.g., Eq. (9.16) for volume growth involves a characteristic time, which could be defined, e.g., as the halftime of hydration (i.e., the time at which $v = 0.5$, which gives $t_{char} = \lambda_0 \alpha^{-1/m}$). Likewise, the compliance function represents a transition between two power laws.

The foregoing considerations support (though do not unambiguously prove) the use of power functions of time in the expression of the compliance rate based on the solidification theory; see Eq. (3.12) in model B3, which is the simplest combination of power functions satisfying the physical requirements of the solidification theory. Besides, the use of power functions happens to lead to the best agreement with test data.

9.5 Asymptotic Matching Properties of Solidification Theory and Insufficiency of Log-Double-Power Law

The way to combine the power functions for various self-similar processes can be helped by considering the opposite short-time and long-time asymptotic behaviors of these processes. The transition between the opposite asymptotes is called the asymptotic matching. This matching is illuminating and useful for comparing various models, even though the asymptotic regimes are reached only far beyond the practical range of interest.

For sustained stress values in the linearity range (which is the service stress range), extensive experimental evidence [272, 455, 458, 546, 551, 580, 598, 642, 697, 734, 736] reveals some simple asymptotic properties, which should be reflected in the constitutive model. These properties are exhibited by both the simple log-double-power law (3.9) and the solidification theory.

Matching of the asymptotic behavior for short and long times is most conveniently done in terms of the compliance rate $\dot{J}(t, t')$. The complete expressions for the compliance rate are

$$\dot{J}(t, t') = \frac{(\alpha + t'^{-m}) n q_s \psi}{(t - t')(\alpha + t'^{-m}) \psi + (t - t')^{1-n}} \qquad (9.26)$$

for the log-double-power law (3.9) and

$$\dot{J}(t, t') = \dot{J}_v(t, t') + \dot{J}_f(t, t') = \frac{\dot{\Phi}(t - t')}{v(t)} + \frac{q_4}{t} = \frac{(q_3 + q_2 t^{-m}) n}{t - t' + (t - t')^{1-n}} + \frac{q_4}{t} \qquad (9.27)$$

for the compliance function (9.19) of model B3. For simplicity, we have set $\lambda_0 = 1$ day and then omitted this parameter, even though in rigorous writing it should be kept for the sake of dimensionality (but the final conclusions would not be affected by that).

Consider first the case of **extremely short load durations**. For short enough durations (actually for $t - t' < 10^{-10}$ day if $n = 0.1$), the term $t - t'$ is negligible compared to $(t - t')^{1-n}$, and the term $1/t$ is negligible compared to $1/(t - t')^{1-n}$. So, the compliance rates asymptotically behave as

$$\dot{J}(t, t') \approx (\alpha + t'^{-m})nq_s\psi \left(t - t'\right)^{n-1} \tag{9.28}$$

for the log-double-power law and

$$\dot{J}(t, t') \approx (q_3 + q_2 t'^{-m})n(t - t')^{n-1} \tag{9.29}$$

for model B3. Obviously, the short-time asymptotic behavior is identical (for an arbitrary age t') if and only if both models use the same value of parameters n and m, and parameters q_2 and q_3 of model B3 are equal to $q_s\psi$ and $\alpha q_s\psi$, resp., where α, q_s, and ψ are parameters of the log-double-power law.

Now, consider the **long-time limit**, in which $t - t'$ dominates over $(t - t')^{1-n}$ (for $n = 0.1$, this actually occurs for $t - t' > 10^{10}$ days) and is approximately equal to t. The compliance rates then asymptotically behave as

$$\dot{J}(t, t') \approx \frac{nq_s}{t} \tag{9.30}$$

for the log-double-power law and

$$\dot{J}(t, t') \approx \frac{q_3n + q_4}{t} \tag{9.31}$$

for model B3. Matching is achieved if $nq_s = q_3n + q_4$, and since we already know from the short-time matching that $q_3 = \alpha q_s\psi$, we obtain the condition that parameter q_4 of model B3 should be equal to $nq_s(1 - \alpha\psi)$.

We have demonstrated that, with an appropriate choice of model parameters, the asymptotic behavior of model B3 for both very short times and very long times is identical with the asymptotic behavior of the log-double-power law. Since the asymptotic portions of optimized fits by the log-double-power law were shown to match quite well numerous broad-range creep data (see Fig. 2 of [118]), the foregoing comparisons support (though do not prove) the choice of functions $\Phi(t - t')$, $\eta(t)$, and $v(t)$ in (9.13), (9.14), and (9.16) and of their standard parameters $n = 0.1$, $m = 0.5$, and $\lambda_0 \approx 1$ day.

It now seems that the remaining free parameters q_i, $i = 1, 2, 3, 4$, could be deduced from parameters E_0, α, ψ, and q_s of the log-double-power law. However, as is clearly demonstrated in Fig. 9.3, it is actually necessary to determine them directly by fitting of the experimental curves because, aside from the far-out asymptotics, the log-double-power law is distinctly inferior to the solidification theory. One reason is that the transitions between the short-time and the long-time asymptotes differ considerably from the solidification theory and their error shifts the long-time asymptotes vertically, as shown in Fig. 9.3a (to demonstrate that the ultimate slope of both curves is the same, the graph covers an extremely wide range of load durations, reaching far beyond the range of practical interest). The second reason is that the prediction of the log-double-power law coefficients, specified in the short-form B3 model, is fraught with greater errors. The third is a violation of the nondivergence requirement for creep curves, to be discussed in Sect. 9.6.

Finally, it must be noted that no test data suffice to determine the asymptotes alone, disregarding the transition between them. It is for this reason that data fittings by the solidification theory and by the log-double-power law give rather different parameter values. For instance, the parameters of model B3 determined by direct fitting of experimental data (Fig. 9.3a) were $q_1 = 21.9$, $q_2 = 67.5$, $q_3 = 5.8$, and $q_4 = 7.7$ (all in 10^{-6}/MPa), while the parameters determined by asymptotic matching[3] of the LDPL (Fig. 9.3b) would be $q_1 = 0$, $q_2 = 67.5$, $q_3 = 12.8$, and $q_4 = 249$. In the latter case, the agreement of the LDPL with experimental data within the range covered

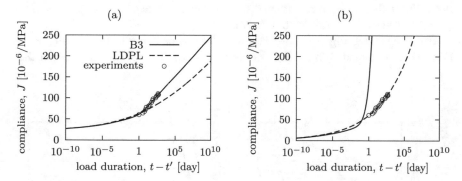

Fig. 9.3 Compliance curves corresponding to (a) model B3 with parameters determined by fitting of experimental data (solid curve) and log-double-power law with parameters determined by asymptotic matching (dashed curve), (b) log-double-power law with parameters determined by fitting of experimental data and model B3 with parameters determined by asymptotic matching

---

[3]The long-time asymptotes of the dashed curve and of the solid curve in Fig. 9.3b seem to be quite different, but the matching is based on their slope (compliance rate).

by the tests seems to be excellent, but it could be achieved only with a zero (or very low) instantaneous compliance and the extrapolation to long-term loading would probably greatly overestimate the actual creep rate.

Example 9.1. Fitting of basic creep data

The basic creep data reported by Hanson [455] and Harboe [458] refer to the Canyon Ferry Dam concrete and contain compliance curves measured for loading at five different ages, ranging from 2 days to 1 year. In Fig. 9.4, the experimental results are represented by isolated points. Figure 9.4a shows that fitting by a pure solidification model (i.e., without the flow term, with $q_4 = 0$) does not lead to satisfactory results. At first glance, the agreement seems to be quite good for old concrete (two bottom curves, ages at loading $t' = 90$ days or 365 days), but it has been achieved with parameter q_1 artificially set to zero (and with $q_2 = 100 \times 10^{-6}$/MPa and $q_3 = 30 \times 10^{-6}$/MPa). For physically realistic values of q_1, the slope of all the compliance curves would be underestimated by the model. With $q_1 = 0$, this underestimation occurs only for loading at younger ages (top curves in Fig. 9.4a). Even for older concrete with seemingly good fits within the range of load durations covered by the experiments (less than 300 days), extrapolation to much longer durations would lead to gross errors. Available long-term creep tests indicate that for old concrete the increase of slope of the creep curve in semilogarithmic scale happens later.

With the flow term included, a substantial improvement can be achieved and extrapolation to shorter or longer load durations becomes much more reliable. The fits in Fig. 9.4b have been constructed with parameter values $q_1 = 21.9$, $q_2 = 67.5$, $q_3 = 5.8$, and $q_4 = 7.7$ (all in 10^{-6}/MPa).

For comparison, Fig. 9.5 shows the fits of the same experimental data (Canyon Ferry Dam) obtained with the log-double-power law. The agreement is better than

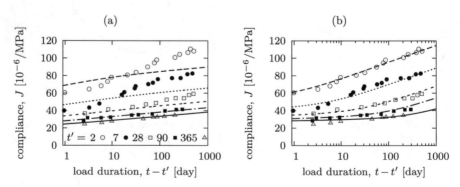

Fig. 9.4 Creep data for Canyon Ferry Dam concrete (for different ages at loading, t', given in the legend in days) fitted by the B3 model (a) without the flow term, (b) with the flow term

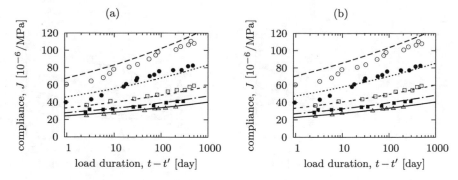

Fig. 9.5 Creep data for Canyon Ferry Dam fitted by the log-double-power law with two different sets of parameters: (a) $E_0 = 125$ GPa, $q_s = 2500 \times 10^{-6}$/MPa, $\psi = 0.027$, $\alpha = 0.19$, (b) $E_0 = 250$ GPa, $q_s = 300 \times 10^{-6}$/MPa, $\psi = 0.28$, $\alpha = 0.18$

that for the pure solidification theory without the flow term (Fig. 9.4a), but not as good as it is for the full B3 model (Fig. 9.4b). The two parts of Fig. 9.5 demonstrate that a comparable quality of fits can be achieved for rather different parameter sets. This is related to the fact that if all the parameters are considered to be free, the parameter identification problem becomes in this case ill-conditioned. ∎

There has been a tendency to make statistical comparisons of creep and shrinkage models only to the database as a whole. However, due to the large scatter of the database, this cannot reveal the incorrectness in the shape of creep curves. To check this shape, comparisons must be made with the individual measured curves. For the B3 model based on the solidification theory, some of such comparisons are shown in Fig. 9.6.

Note that the individual measured creep curves for basic creep demonstrate that the terminal shape should be a logarithmic curve (as proposed already in [114]). The basic information on the data is summarized in Bažant and Panula [175]. The optimal parameter values listed in Table 9.1 have been taken from Bažant and Prasannan [180] and converted to SI units.

Table 9.1 Model parameters (all in 10^{-6}/MPa) obtained by optimum fitting of various sets of experimental data

Test data	Reference	q_1	q_2	q_3	q_4
Canyon Ferry Dam	[455, 458]	20.89	64.40	5.51	6.96
Ross Dam	[455, 458]	23.50	108.78	1.44	6.95
L'Hermite et al.	[576, 580]	6.90	78.61	16.82	2.84
Wylfa Vessel	[272]	20.16	41.05	17.61	12.81

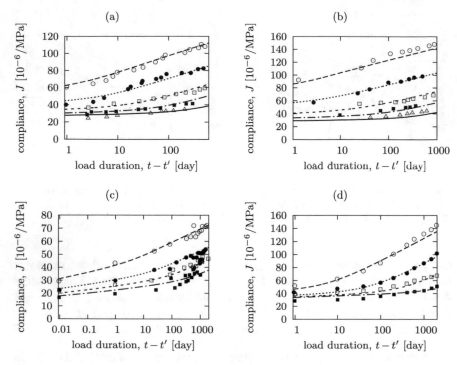

Fig. 9.6 Measured values of basic compliance functions at different ages t' and their fits by model B3: (a) Canyon Ferry Dam (ages $t' = 2, 7, 28, 90$, and 365 days), (b) Ross Dam (ages 2, 7, 28, 90, and 365 days), (c) L'Hermite et al. (ages 7, 28, 90, and 365 days), (d) Wylfa Vessel (ages 7, 60, 400, and 4560 days)

9.6 Nondivergence of Compliance Curves

Apart from its physical justification, the solidification theory, in contrast to the log-double-power law, double-power law, and all the previously proposed concrete creep models used in design recommendations, has the advantage of providing nondivergent creep curves. Geometrically, the *condition of nondivergence* means that the unit creep curves for different loading ages t', plotted as functions of current age t (rather than of the load duration $t - t'$), must converge toward each other, i.e., their vertical distance must be diminishing (Fig. 9.7a).

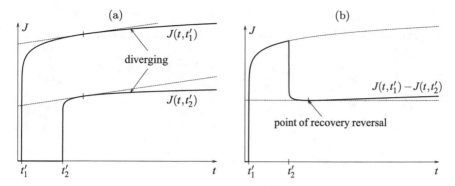

Fig. 9.7 (a) Divergence of creep curves, (b) nonmonotonic recovery curve

Mathematically, the condition of nondivergence reads

$$\frac{\partial \dot{J}(t, t')}{\partial t'} \geq 0 \quad \text{or, equivalently} \quad \frac{\partial^2 J(t, t')}{\partial t \, \partial t'} \geq 0 \quad \text{for all } t \geq t' \tag{9.32}$$

Example 9.2 will demonstrate that violation of this condition does not necessarily mean that the model is thermodynamically inadmissible. On the other hand, there exist no thermodynamic or other fundamental arguments requiring divergence to occur. Although instances of divergence are detected in a few test data, it is not clear whether such instances might merely be random occurrences due to inevitable statistical scatter. The divergence is definitely not a systematic feature of the creep database (Fig.9.8). According to the solidification theory, which is the only available thermodynamically justified theory for aging creep, the divergence is impossible. The Kelvin chain, as a matter of principle, cannot exhibit divergence, provided that all its moduli and viscosities remain positive. Since the few observed instances of divergence are mild enough to be ascribable to inevitable statistical scatter, and since most of the data can be closely fitted by the Kelvin chain, it appears that compliance functions exhibiting divergence ought to be avoided.

Example 9.2. Creep model exhibiting divergence of compliance curves

The mechanism leading to divergence of compliance curves can be illustrated by simple examples. One was given in Bažant and Kim [152], and another one is provided here. Consider a Maxwell chain consisting of two units, one of which is aging (Fig. 9.9a). The relaxation times of both Maxwell units, τ_1 and $\tau_2 = 10\tau_1$, are considered as constant, and the initial values of moduli E_1 and E_2 are equal. The first unit is nonaging, and so modulus E_1 remains constant. Modulus E_2 increases from its initial value E_1 and asymptotically approaches $5E_1$; its growth is described by the function

$$E_2(t) = E_1 \left(5 - 4e^{-3t/\tau_2}\right) \tag{9.33}$$

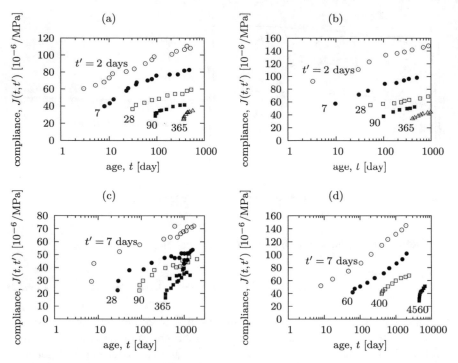

Fig. 9.8 Compliance curves plotted as a function of age: (a) Canyon Ferry Dam, (b) Ross Dam, (c) L'Hermite et al., (d) Wylfa Vessel

The relaxation function of the chain can be expressed analytically as

$$R(t, t') = E_1 e^{-(t-t')/\tau_1} + E_2(t') e^{-(t-t')/\tau_2} =$$
$$= E_1 \left[e^{-(t-t')/\tau_1} + \left(5 - 4e^{-0.3t'/\tau_1}\right) e^{-0.1(t-t')/\tau_1} \right] \quad (9.34)$$

The numerically computed compliance function $J(t, t')$ is plotted in Fig. 9.9c (in the dimensionless form, normalized by $1/E_1$) as a function of the current time t (normalized by τ_1) for different values of t', ranging from $0.05\tau_1$ to τ_1. The curves slightly diverge—for each fixed value of t, the compliance rate (slope of the compliance curve) is lower for compliance tests started at a later age. Consequently, the corresponding strain recovery curves in Fig. 9.9d are increasing (of course after the vertical drop that reflects the instantaneous strain change due to unloading).

To understand the mechanism that leads to divergent compliance curves, let us analyze the redistribution of partial stresses σ_1 and σ_2 carried by individual units of the chain. The behavior of the model is in general described by the relations

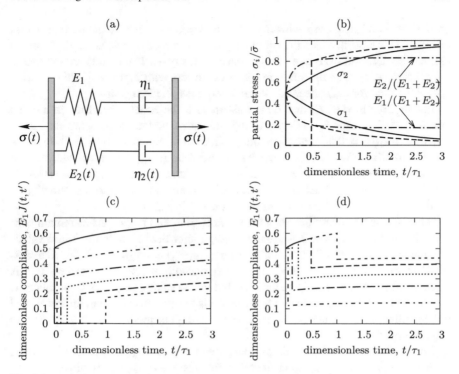

Fig. 9.9 (a) Aging Maxwell chain, (b) evolution of partial stresses, (c) divergence of compliance curves, (d) strain recovery

$$\dot{\varepsilon} = \frac{\dot{\sigma}_1}{E_1} + \frac{\sigma_1}{\eta_1} = \frac{\dot{\sigma}_2}{E_2} + \frac{\sigma_2}{\eta_2} \tag{9.35}$$

$$\sigma = \sigma_1 + \sigma_2 \tag{9.36}$$

where $\eta_1 = \tau_1 E_1$ and $\eta_2 = \tau_2 E_2$ are the viscosities of individual dashpots in the Maxwell chain. In a creep test, the total stress remains constant and its rate vanishes. The rate of partial stress σ_2 can be eliminated from the second equation in (9.35) by substituting $\dot{\sigma}_2 = -\dot{\sigma}_1$. The rate of partial stress σ_1 is then expressed in terms of the current values of both partial stresses:

$$\dot{\sigma}_1 = \frac{E_1 E_2}{E_1 + E_2} \left(\frac{\sigma_2}{\eta_2} - \frac{\sigma_1}{\eta_1} \right) \tag{9.37}$$

Substituting this into the first equation in (9.35), we obtain an expression for the strain rate

$$\dot{\varepsilon} = \frac{1}{E_1 + E_2} \left(\frac{\sigma_1}{\tau_1} + \frac{\sigma_2}{\tau_2} \right) \tag{9.38}$$

Interestingly, the current strain rate in a creep test depends (aside from the current material properties) only on the current values of partial stresses σ_1 and σ_2 but not on the details of the previous history (the history is reflected by the current values of internal variables). Since the sum of σ_1 and σ_2 in a creep test is fixed, the creep rate can be deduced from one of these variables, say σ_1. If the units are numbered such that $\tau_1 < \tau_2$, then the strain rate in a creep test with higher σ_1 is higher. This provides a clue for understanding of the divergence phenomenon.

Initially, the moduli E_1 and E_2 are assumed to be equal. Therefore, in a creep test A started at time $t'_A = 0$, the initial values of partial stresses are the same: $\sigma_{A1}(0) = \sigma_{A2}(0) = 0.5\bar{\sigma}$ where $\bar{\sigma}$ is the applied stress. Since $\eta_2(0) > \eta_1$, the expression in parentheses in (9.37) is negative and partial stress σ_{A1} decreases (while partial stress σ_{A2} simultaneously increases at the same rate). This means that the stress is redistributed from the "fast unit" (with a short relaxation time) to the "slow" unit (with a long relaxation time); see the solid curves in Fig. 9.9b.

Suppose for a moment that the chain is nonaging and its properties remain constant. If another creep test B is started at some later time $t'_B > 0$, the initial values of partial stresses in test B are again the same, $\sigma_{B1}(t'_B) = \sigma_{B2}(t'_B) = 0.5\bar{\sigma}$. In test A, which started earlier, the stresses are already partially redistributed and $\sigma_{A1}(t'_B)$ is below its initial value $\sigma_{A1}(0)$ and thus also below the value of $\sigma_{B1}(t'_B)$. Consequently, the strain rate at time t'_B evaluated from (9.38) is lower for test A than for test B, and the compliance curves are converging. For a nonaging model, this is of course an expected result.

Consider now the effect of aging, in our case manifested by an increase of modulus E_2 (and also of viscosity η_2, since the relaxation time $\tau_2 = \eta_2/E_2$ remains constant). In creep test B which started at time $t'_B > 0$, the initial values of partial stresses are no longer the same. It is easy to show that the stresses are initially distributed in proportion to the moduli of the corresponding units, since the dashpots do not exhibit any instantaneous strain and the model reacts to a suddenly applied stress in the same way as two springs coupled in parallel. Therefore, the initial value of partial stress in unit 1 is $\sigma_{B1}(t'_B) = \bar{\sigma} E_1/[E_1 + E_2(t'_B)]$. For $E_2(t'_B)$ substantially larger than E_1, the fraction $E_1/[E_1 + E_2(t'_B)]$ can be so small that $\sigma_{B1}(t'_B)$ is less than $\sigma_{A1}(t'_B)$. If that happens, the strain rate in test B, evaluated from (9.38), is smaller than in test A, and the compliance curves diverge. This is exactly the case for the model considered here. Figure 9.9b shows the evolution of normalized partial stresses σ_{A1} and σ_{A2} in a creep test A started at time 0 (solid curves) and of the ratios $E_1/[E_1 + E_2(t)]$ and $E_2/[E_1 + E_2(t)]$ for the given evolution of modulus E_2 according to (9.33) (dash-dotted curves). Finally, the dashed curves correspond to the evolution of partial stresses σ_{B1} and σ_{B2} in a creep test B started at time $t'_B = 0.5\,\tau_1$. The fact that the dashed curves lie outside the range bounded by the solid curves implies divergence.

A similar explanation can be provided for the behavior of the model after unloading. If the creep test runs, e.g., from time $t'_A = 0$ till time $t'_B = 0.5\,\tau_1$, the partial stress in the fast unit 1 is reduced with respect to its initial value, and the partial stress in the slow unit 2 is augmented. Upon full unloading at time t'_B, the total stress drops to zero, but the individual units in general carry some nonzero self-equilibrated stresses (i.e., partial stresses in units 1 and 2 have the same magnitude but opposite signs).

1. In a typical (nondivergent) case, the partial stress in unit 1 becomes negative because, before unloading, it was lower than the partial stress in the other unit. At constant strain, the negative stress in unit 1 would relax faster than the positive stress in unit 2, and the total stress would thus become positive. Since the total stress is prescribed as zero, fast relaxation of negative stress must be compensated for by a reduction of the strain (common to both units coupled in parallel), which brings both units into self-equilibrium.
2. The previous arguments may be reversed if unit 2 is influenced by sufficiently fast solidification during the creep test. Its current stiffness at time t'_B becomes very high, and instantaneous unloading leads to a positive residual stress in unit 1 and a negative residual stress in unit 2. Subsequently, fast relaxation of the positive stress must be compensated for by an increase of the strain, which leads to the type of recovery curves plotted in Fig. 9.9d. ∎

In the preceding example, the recovery curves were, after the initial drop, monotonically increasing. As discussed by Bažant and Kim [152], for a chain with 3 or more Maxwell units it may happen that the sign of strain rate during recovery changes from negative to positive, as schematically shown in Fig. 9.7b. This is called nonmonotonic recovery.

Bažant and Kim [152] also examined in detail the conditions of thermodynamic admissibility of an aging Maxwell chain. They showed that appropriate expressions for the free energy and dissipated energy can be constructed if the viscosities of all Maxwell units are nonnegative and if the stiffness moduli are nonnegative and nondecreasing in time. All these conditions are satisfied by the model analyzed in the preceding example.

In conclusion, even though divergence of compliance curves may be thermodynamically admissible, in combination with the principle of superposition it causes a pathological response: The curves of creep recovery after a sudden decrease of stress are not monotonic; i.e., decreasing strain is after some time reversed to increasing strain; see Fig. 9.7b. Another pathology that may arise for models based on a prescribed compliance function is that the relaxation curves (computed using the principle of superposition) eventually cross to the negative sign; i.e., the stress becomes opposite to previous loading (Fig. 9.15b–d). This phenomenon will be examined in Sect. 9.7.

It has been speculated that nonmonotonic recovery might indicate that the principle of superposition fails when the strain magnitude is decreasing, and thus, a nonlinear creep model might be required. There have also been many attempts of nonlinear creep models at nondecreasing strains. However, as it turned out, these nonlinearities are due to separate phenomena such as microcracking and drying, which modify the results obtained with the principle of superposition. Therefore, it seems inappropriate to introduce nonlinear concepts to model creep. Anyway, it would greatly complicate creep analysis, especially in proper triaxial formulation.

Let us now proceed to a deeper analysis of the nondivergence inequality (9.32) for the general form of solidification theory with the compliance rate given by (9.7). Differentiating with respect to t', we get

$$\frac{\partial \dot{J}_v(t,t')}{\partial t'} = -\frac{\ddot{\Phi}(t-t')}{v(t)} \tag{9.39}$$

Since $v(t)$ is always a positive function, condition (9.32) is satisfied if and only if $\ddot{\Phi}(t-t') \leq 0$. Thus, the sufficient and necessary condition for nondivergence of compliance curves generated by the solidification theory is that the compliance function of the nonaging constituent be concave. In other words, the rate of creep must monotonically decrease in time or remain constant, but must never increase (which is called the *principle of fading memory*). This is a very natural condition and easy to satisfy; it is satisfied by the function (9.13) used by the B3 model with positive parameters q_3 and λ_0 and with exponent n between 0 and 1. The compliance function of model B3 is in fact the sum of $1/E_0$, $J_v(t,t')$ and $J_f(t,t')$, but the flow compliance $J_f(t,t')$ defined in (9.10) always satisfies the nondivergence condition, for any age-dependent viscosity $\eta(t)$, because its rate $\dot{J}(t,t') = 1/\eta(t)$ does not depend on t', and thus the derivative in (9.32) vanishes. This is related to the fact that the compliance curves $J_f(t,t')$ plotted as functions of age t for different values t' are just shifted vertically (recall that the starting assumption was that the rate of the flow strain depends only on the current stress).

Note that condition (9.32) would be satisfied by the more general expression

$$\dot{J}(t,t') = \sum_{s=1}^{N_s} \frac{\dot{\Phi}_s(t-t')}{v_s(t)} + \frac{1}{\eta(t)} \tag{9.40}$$

which corresponds to the case of several solidifying constituents with growing volumes $v_s(t)$ and nonaging compliance functions $\Phi_s(t-t')$. For some further interesting aspects, see [152].

So far we have considered only the basic creep compliance. For model B3, the drying creep compliance, too, satisfies the nondivergence condition, as can be proven by differentiation of (3.20):

$$\frac{\partial^2 J_d(t,t')}{\partial t \partial t'} = \frac{q_5 \, \dot{g}(t-t_0) \, \dot{g}(t'-t_0) \, e^{-g(t-t_0)-g(t'-t_0)}}{4 \left(e^{-g(t-t_0)} - e^{-g(t'-t_0)} \right)^{3/2}} \tag{9.41}$$

Recall that g denotes here a positive decreasing function given by (3.23), and so the derivative $\dot{g}$ is always negative. For all $t > t' > t_0$, the numerator and the denominator in (9.41) are both positive, which means that condition (9.32) is satisfied.

We have shown that creep models based on the solidification theory, in particular the B3 model, never exhibit divergence of the compliance curves. On the other hand, for many simple empirical models, divergence does occur, typically when the load duration $t - t'$ exceeds a certain limit that depends on the age t' at the start of the creep test. The compliance functions used in various codes and recommendations (see Appendix E) often have the general form

$$J(t, t') = \frac{1}{E(t')} + c\, \frac{f(t - t')}{g(t')} \tag{9.42}$$

in which $E(t')$ is the aging elastic modulus, c is a positive constant (independent of t and t', but possibly dependent on the concrete properties and environmental conditions), $f(t - t')$ is an increasing function describing the shape of the creep curve, and $g(t')$ is an increasing function that reflects aging (lower creep if the concrete is loaded at a higher age). With appropriate choices of c, f, and g, Eq. (9.42) gives the double-power law (3.8), formula (E.38) recommended by ACI, formula (E.13) used by the CEB Model Code, or formula (F.83) suggested by JSCE. Differentiation of (9.42) yields

$$\frac{\partial J(t, t')}{\partial t} = c\, \frac{\dot{f}(t - t')}{g(t')} \tag{9.43}$$

$$\frac{\partial^2 J(t, t')}{\partial t\, \partial t'} = c\, \frac{-\ddot{f}(t - t')g(t') - \dot{f}(t - t')\dot{g}(t')}{g^2(t')} \tag{9.44}$$

and since $c > 0$, the nondivergence condition $\partial^2 J(t, t')/\partial t\, \partial t' \geq 0$ is violated if

$$\ddot{f}(t - t')g(t') + \dot{f}(t - t')\dot{g}(t') > 0 \tag{9.45}$$

For simplicity, dots over f and g denote derivatives of these functions, no matter whether the argument is denoted as $t - t'$ or t'. Since $\dot{f}$ and g are positive functions, the last condition can be rewritten as

$$\frac{\ddot{f}(t - t')}{\dot{f}(t - t')} > -\frac{\dot{g}(t')}{g(t')} \tag{9.46}$$

The left-hand side of this inequality depends only on the load duration, $t - t'$, and the right-hand side only on the age at loading, t'. Therefore, for each value of t', we can evaluate the range of values of $t - t'$ in which divergence occurs. Typically, this happens when the load duration $t - t'$ exceeds a certain limit, which can be computed from (9.46) if the inequality sign is replaced by equality.

Example 9.3. Divergence of creep curves in design recommendations and codes

The compliance function stipulated by the **CEB Model Code** (see Sect. E.2.1) has the form (9.42) with

$$f(\hat{t}) = \left(\frac{\hat{t}}{\beta_H \beta_T + \hat{t}}\right)^{0.3}, \qquad g(t') = 0.1 + t'^{0.2} \tag{9.47}$$

where β_H and β_T are parameters given by formulae (E.11) and (E.12), and $\hat{t} = t - t'$. Evaluating the derivatives, substituting them into (9.46) and solving for $t - t'$ with t' considered as given, we find that divergence occurs if

$$t - t' > \frac{1}{2}\left[10t' + t'^{0.8} - \beta_H + \sqrt{(10t' + t'^{0.8} - \beta_H)^2 + 1.4\beta_H(10t' + t'^{0.8})}\right]$$
(9.48)

Parameter β_T depends on the temperature and at 20°C is equal to 1. Parameter β_H depends on the compressive strength, environmental humidity, and equivalent thickness of the member. For concrete of mean compressive strength 35 MPa, it must be in the range from 250 to 1500 (days). Figure 9.10a shows the critical load duration $t - t'$ after which divergence occurs as a function of the age at loading t' for the extreme values of β_H and for an intermediate value $\beta_H = 750$. It is clear that the specific value of β_H has only a marginal influence and that in general divergence occurs when the load duration exceeds the age at loading by a factor of 10, for younger concrete even less than that. A pathological consequence of divergent creep curves combined with the principle of superposition is nonmonotonic recovery, illustrated for the CEB Model Code in Fig. 9.11a.

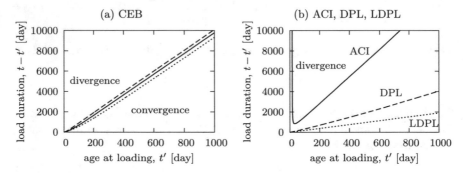

Fig. 9.10 Critical load duration after which divergence of creep curves occurs, plotted as a function of the age at loading for (a) CEB Model Code with different values of parameter β_H, (b) ACI model, double-power law (DPL) and log-double-power law (LDPL)

A similar analysis can be performed for the double-power law (3.8), the log-double-power law (3.9), and the ACI model (described in Sect. E.3); the resulting critical load durations are plotted in Fig. 9.10b. The analysis reveals that the double-power law and especially the log-double-power law exhibit divergence after load durations exceeding the age at loading only a few times (approximately 4 times for the DPL and 2 times for the LDPL). Figure 9.11b presents examples of nonmonotonic recovery for the ACI model and Fig. 9.11c for the log-double-power law. Only the B3 model always leads to monotonic recovery, as confirmed by Fig. 9.11d.

The **double-power law** (3.8) violates the nondivergence condition (9.32) for $t - t' > (1 - n)(1 + \alpha t'^m)t'/m$ [152]. For standard parameter values $n = 0.1$, $m = 1/3$, and $\alpha = 0.05$, this happens if the load duration $t - t'$ exceeds the age at loading t' approximately 3 to 4 times; see the dashed curve in Fig. 9.10b.

For the **log-double-power law** (3.9), the analysis is somewhat more tedious because the compliance function does not have the form (9.42). Nevertheless, it is still possible to analyze the nondivergence inequality and show that it is violated

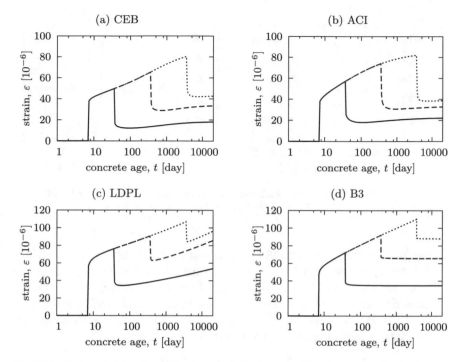

Fig. 9.11 Strain recovery obtained by the principle of superposition for concrete loaded by stress 1 MPa from age of 7 days, with stress removed after 1 month, 1 year, or 10 years, as predicted by the (a) CEB model, (b) ACI model, (c) log-double-power law, (d) B3 model

for $t - t' > (t'^{1+m}/m)(\alpha + t'^{-m})[1 - n + \psi(\alpha + t'^{-m})(t - t')^n]$. This condition is implicit, because t appears also on the right-hand side, but for standard parameter values $n = 0.1$, $m = 0.5$, $\alpha = 0.001$, and $\psi = 0.3$, the term with $(t - t')^n$ is negligible. Except for very early ages, the critical load duration is very close to $t'(1 - n)/m$, which is just $1.8\, t'$; see the dotted curve in Fig. 9.10b.

As shown by Bažant and Kim [152], the **ACI model** with $f(\hat{t}) = \hat{t}^\psi/(d + \hat{t}^\psi)$ and $g(t') = t'^m/\sqrt{b + a/t'}$ violates the nondivergence condition (9.32) for

$$\frac{mb - aT(t - t') - \sqrt{D(t - t')}}{2bT(t - t')} < t' < \frac{mb - aT(t - t') + \sqrt{D(t - t')}}{2bT(t - t')} \qquad (9.49)$$

where

$$T(\hat{t}) = \frac{1}{\hat{t}}\left(1 + \psi\frac{\hat{t}^\psi - d}{\hat{t}^\psi + d}\right) \qquad (9.50)$$

$$D(\hat{t}) = \left[aT(\hat{t}) + mb\right]^2 - 2abT(\hat{t}) \qquad (9.51)$$

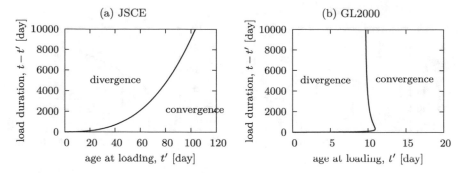

Fig. 9.12 Critical load duration after which divergence of creep curves occurs, plotted as a function of the age at loading for (a) JSCE model, (b) Gardner–Lockman model (GL2000)

For standard parameters $m = 0.118$ (moist curing), $\psi = 0.6$, $a = 4$, $b = 0.85$, and $d = 10$, the results are plotted in Fig. 9.10b by the solid curve. Divergence occurs for those combinations of t' and $\hat{t} \equiv t - t'$ that lie above the curve.

The basic creep compliance function of the **JSCE model** [786] has the form (9.42) with $f(\hat{t}) = 1 - \exp\left(-0.09\,\hat{t}^{0.6}\right)$ and $g(t') = (\ln t')^{0.67}$; see Appendix E.5. The divergence condition (9.46) can be transformed into the inequality

$$\frac{1}{t - t'} + \frac{0.135}{(t - t')^{0.4}} < \frac{1.675}{t' \ln t'} \tag{9.52}$$

The regions of divergence and convergence are graphically presented in Fig. 9.12a. For instance, for loading at age $t' = 28$ days, divergence occurs for load durations $t - t'$ exceeding 280 days. ∎

Would it be possible at all to construct a compliance function in the form (9.42) which does not exhibit divergence and at the same is realistic for concrete? Detailed analysis of the divergence condition (9.46) shows that the answer is negative. The left-hand side maximized over all $t - t' \geq 0$ would need to be no larger than the right-hand side minimized over all $t' \geq t'_{min}$ where t'_{min} is the youngest age at which the concrete could potentially be loaded. Minimization of the right-hand side gives a certain negative constant, say $-G$, and then, it is easy to show that the time derivative of function f would need to satisfy the constraint $\dot{f}(\hat{t}) \leq \dot{f}(\hat{t}_1)\exp[-G(\hat{t} - \hat{t}_1)]$ for all $\hat{t} > \hat{t}_1 \geq 0$. Note that $\dot{f}(\hat{t})$ is proportional to the strain rate in the creep test. To avoid divergence, this rate would need to approach zero exponentially. For commonly used models with compliance functions of the general form (9.42), the creep rate typically approaches zero as a power function (e.g., for the CEB Model Code with function f defined in (9.47), the creep rate is for long load durations proportional to $1/\hat{t}^2$), and so the condition of exponential decay is violated. The only model with a nonpower decay is the JSCE model, for which the long-term asymptotic behavior of the creep rate is governed by an exponential function of $-\hat{t}^{0.6}$ but even that is too slow compared to the exponential of $-\hat{t}$ and the nondivergence condition is violated, same as for models with a power-type decay.

Experiments indicate that the asymptotic decay of the creep rate is proportional to $1/\hat{t}$, because the long-time creep evolution has a logarithmic character. So constructing a model with an exponential decay of the creep rate (which is in fact the property of the simple Kelvin model) just to satisfy the nondivergence condition would not be reasonable. Fortunately, instead of tuning up the description of aging in the format of (9.42), we can use a completely different approach—the solidification theory. This is so far the only practical way of ensuring nondivergence over the entire range of admissible ages at loading and load durations.

Example 9.4. *Divergence of creep curves of the Gardner–Lockman model*

Model GL2000, proposed by Gardner and Lockman [407] and described in Sect. E.4, can be considered as a generalized version of (9.42), given by

$$J(t, t') = \frac{1}{E(t')} + c\left[\frac{f(t - t')}{g(t')} + h(t - t')\right] \tag{9.53}$$

The analysis becomes more tedious, but the condition of divergence can again be established. Divergence occurs if

$$\ddot{h}(t - t')g^2(t') + \ddot{f}(t - t')g(t') + \dot{f}(t - t')\dot{g}(t') > 0 \tag{9.54}$$

Equation 9.53 covers the case of basic creep, with coefficient c_h in (E.45) set to zero. For the specific form of function

$$g(t') = \sqrt{t'} \tag{9.55}$$

used by Gardner and Lockman [407], condition (9.54) can be rewritten as

$$2\ddot{h}(t - t')\,t'^{3/2} + 2\ddot{f}(t - t')\,t' + \dot{f}(t - t') > 0 \tag{9.56}$$

With $t - t'$ considered as given, condition (9.56) is a cubic inequality in terms of $\sqrt{t'}$. For the functions

$$f(t - t') = \sqrt{\frac{7(t - t')}{t - t' + 7}} \tag{9.57}$$

$$h(t - t') = \frac{2(t - t')^{0.3}}{(t - t')^{0.3} + 14} \tag{9.58}$$

used by Gardner and Lockman [407], the results are plotted in Fig. 9.12b. It turns out that condition (9.56) is never satisfied for $t' > 11$ days, and for $t' < 9$ days, it is satisfied for virtually all load durations. This means that the GL model exhibits divergence only for loading at young age, and in that case, divergence occurs almost immediately. This conclusion is supported by Fig. 9.13, which shows the strain recovery curves according to model GL2000 for concrete loaded at age of 1 day (Fig. 9.13a) or 7 days (Fig. 9.13b) and unloaded 1 day, 1 week, 1 month, or 1 year after loading. In the extreme case of very young concrete and load removal after a very short period (solid curve in Fig. 9.13a), nonmonotonic recovery is clearly visible. If the load is removed after a longer period, or if the first loading occurs at a higher age, nonmonotonic recovery is less pronounced or totally disappears. ∎

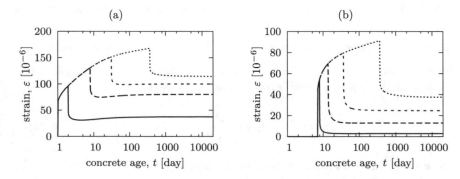

Fig. 9.13 Strain recovery obtained by the principle of superposition for concrete loaded by stress 1 MPa from age (a) 1 day, (b) 7 days, with stress removed after 1 day, 1 week, 1 month, or 1 year, as predicted by the GL2000 model

Example 9.5. Divergence of creep curves of the new fib *model*

The recently approved fib **Model Code 2010** [381] uses the compliance function in the form

$$J(t, t') = \frac{1}{E(t')} + c_1 \ln\left(1 + \left(0.035 + \frac{30}{t'}\right)^2 (t - t')\right) + \frac{c_2}{0.1 + t'^{0.2}} \left(\frac{t - t'}{\beta_H + t - t'}\right)^{\gamma(t')} \tag{9.59}$$

where constants c_1 and c_2 depend on the compression strength and conventional elastic modulus and c_2 also depends on the ambient relative humidity, and notional member size. The second term on the right-hand side of (9.59) represents the basic creep compliance, J_b, and the third term represents the drying creep compliance, J_d.

Due to its logarithmic form, the basic creep compliance function satisfies the nondivergence condition, because its second mixed derivative

$$\frac{\partial^2 J_b}{\partial t \partial t'} = \frac{c_1 (0.035 t' + 30)\left[(0.035 t' + 30)^3 - 60 t'\right]}{\left[t'^2 + (0.035 t' + 30)^2 (t - t')\right]^2} \tag{9.60}$$

is positive for all $t > t' > 0$.

The drying creep compliance function of the *fib* model has a form similar to the previous CEB Model Code 1990, see (9.42) and (9.47), but the fixed exponent 0.3 is replaced by the function

$$\gamma(t') = \frac{1}{2.3 + \dfrac{3.5}{\sqrt{t'}}} \tag{9.61}$$

The second mixed derivative of the drying creep compliance function (i.e., of the last term on the right-hand side of (9.59)) is given by a lengthy formula

$$\frac{\partial^2 J_d}{\partial t \partial t'} = \frac{c_2 \beta_H \gamma(t')}{(0.1 + t'^{0.2})(\beta_H + t - t')^2} \left(\frac{t - t'}{\beta_H + t - t'}\right)^{\gamma(t')} \times \tag{9.62}$$

$$\times \left[\frac{\dot{\gamma}(t')}{\gamma(t')} + \dot{\gamma}(t') \ln \frac{t - t'}{\beta_H + t - t'} - \frac{2}{t'^{0.8} + 10t'} + \frac{2}{\beta_H + t - t'} + \frac{\beta_H (1 - \gamma(t'))}{(t - t')(\beta_H + t - t')}\right]$$

The nondivergence condition is satisfied if the expression in square brackets in the second line of (9.62) is nonnegative. Function γ is increasing, and its values are between 0 and $1/2.3 \approx 0.435$, which means that the term with $1 - \gamma(t')$ in the numerator and $t - t'$ in the denominator tends to plus infinity as $t - t' \to 0^+$. The term with $\ln(t - t')$ tends to minus infinity but more slowly than the term with $1/(t - t')$ tends to plus infinity, and the other terms in square brackets tend to a finite limit. Therefore, for any age t', the nondivergence condition is always satisfied if the load duration $t - t'$ is sufficiently short. On the other hand, it can be shown that for a fixed age t' greater than 6.47 days the expression in the brackets tends to a negative limit as $t \to \infty$, and so the nondivergence condition is violated if the load duration is sufficiently long.

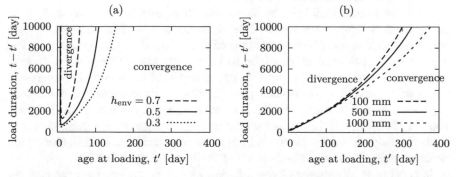

Fig. 9.14 Critical load duration after which divergence of drying creep curves occurs, plotted as a function of the age at loading for the *fib* Model Code 2010: (a) $\bar{f}_c = 45.4$ MPa, $D = 100$ mm, h_{env} ranging from 0.3 to 0.7, (b) $\bar{f}_c = 20$ MPa, $h_{env} = 0.3$, D ranging from 100 to 1000 mm

The foregoing analysis has revealed that the second mixed derivative of function $J_d(t, t')$ that reflects the contribution of drying creep is negative for some combinations of parameters. However, this does not automatically mean that the creep curves diverge, because what matters is the sign of the second mixed derivative of the total creep compliance function, which combines basic creep with drying creep. Whether divergence occurs or not depends on the ratio between factors c_1 and c_2, which control the contributions of basic creep and drying creep in (9.59). Comparing the compact formula (9.59) with the details of the *fib* model, described in Appendix E.2.2, we find that $c_1 = 1.8/(\bar{f}_c^{0.7} E_{28})$ and $c_2 = 412\phi_{RH}/(\bar{f}_c^{1.4} E_{28})$ where $\bar{f}_c$ is the mean compressive strength, E_{28} is the conventional elastic modulus, and ϕ_{RH} is a factor given by formula (E.20), which depends on the ambient relative humidity and notional member size. The ratio

$$\frac{c_2}{c_1} = \frac{412\phi_{RH}}{1.8\bar{f}_c^{0.7}} \approx 228.9\frac{\phi_{RH}}{\bar{f}_c^{0.7}} \tag{9.63}$$

depends on the compressive strength, ambient relative humidity and notional member size. For a given ratio c_2/c_1 and for each fixed age t', one can study the dependence of $\partial^2 J_b/\partial t\partial t' + \partial^2 J_d/\partial t\partial t'$ on $t - t'$ numerically and find the critical value above which the expression becomes negative and the nondivergence condition is violated.

For illustration, let us consider the same concrete and the same conditions as in Example 3.1. For $\bar{f}_c = 45.4$ MPa, $h_{env} = 0.7$, and $D \equiv 2A_c/u = 100$ mm, we obtain $\beta_H = 370$, $\phi_{RH} = 0.6463$, and $c_2/c_1 = 10.2$. The corresponding regions of convergence and divergence are separated in Fig. 9.14a by the dashed curve. Divergence is indeed detected, but for this particular combination of parameters it occurs within the first 10,000 days of loading only for concrete loaded at a relatively young age (between 8 and 59 days). For instance, if the load is applied at 28 days, divergence would occur for load durations that exceed 2166 days.

For comparison, Fig. 9.14a also shows the boundaries between the regions of convergence and divergence for lower levels of ambient humidity, namely 0.5 and 0.3. Since factor ϕ_{RH} in (9.63) is proportional to $1 - h_{env}$, divergence of creep curves is promoted by low ambient humidities and low compressive strengths, which lead to an increase of the ratio c_2/c_1 and thus to a stronger influence of drying creep. For an extreme case with $h_{env} = 0.3$ and $\bar{f}_c = 20$ MPa, the convergence and divergence regions are plotted in Fig. 9.14b. The notional member size is varied here between 100 and 1000 mm, and the results indicate that the divergence phenomenon occurs earlier for larger members. However, even in the most unfavorable case, divergence never occurs for load durations shorter than 20 times the age at loading.

In conclusion, divergence of creep curves cannot be excluded for the new *fib* model, but is much less pronounced than for the previous CEB model. ∎

9.7   Change of Sign of Relaxation Function

Another natural restriction on the behavior of viscoelastic models is that the stress relaxation curves may not cross the horizontal axis into stress values of opposite sign.

Indeed, applying compressive mechanical strain on a specimen cannot produce a tensile stress (note that shrinkage is treated separately). For all the standard recommendations or codes, it is generally accepted that the response to variable stress can be evaluated according to the principle of superposition, and this is the way the relaxation curves are calculated. For **model B3**, crossing of the horizontal axis has never been observed. This is confirmed in Fig. 9.15a, which shows that the relaxation function remains positive up to extremely long times. All the examples of relaxation curves plotted in Figs. 9.15 and 9.16 correspond to a concrete with the same strength and composition as in Example 3.1.

For the **CEB model**, the crossing of the relaxation curves (calculated from the principle of superposition) into values of opposite sign can occur; see Fig. 9.15b, in which the relaxation function for concrete loaded at the age of 1 day becomes negative after about 13 years of loading. For the **ACI model** and the **JSCE model**, a change of sign occurs much earlier, approximately after 100 days of loading; see Fig. 9.15c, d. All these models exhibit such a strange behavior only if the concrete is loaded at a very young age. For the short form of the B3 model, based on the **log-double-power law**, negative values of relaxation function occur even for relaxation started at higher ages; see Fig. 9.15e. Interestingly, for the **GL2000 model** and the *fib* **Model Code 2010**, the relaxation function remains positive, at least for the data used in the present example; see Fig. 9.15f, g.

For models based on the solidification theory, negative values of relaxation function never occur. For more general aging models, it is not clear which condition should be satisfied by the compliance function in order to avoid negative values of the relaxation function. Of course, if a model is defined by its relaxation function, it is easy to check whether this function is nonnegative. However, as already explained, concrete creep models are usually defined by their compliance functions, because (i) much more experimental data are available for creep tests than for relaxation tests, (ii) measuring relaxation requires more expensive equipment, and (iii) stress histories in structures are usually closer to the creep test.

All the curves presented in Fig. 9.15 correspond to sealed specimens (basic creep only). Drying creep aggravates the problem with negative relaxation values for the CEB model and ACI model (Fig. 9.16b, c), and the problem still persists for the JSCE model (Fig. 9.16d). For models B3, GL2000, and *fib*, the computed relaxation curves do not cross the zero axis even when the effect of drying at ambient relative humidity $h_{env} = 50\%$ is taken into account (Fig. 9.16a, e, f).

A detailed parametric study reveals that the *fib* model could lead to negative values of the relaxation function, but only for very unusual parameter combinations that are outside the declared range of applicability of this model. For instance, for a slab of 1 m in thickness, made of concrete with a mean compressive strength of 12 MPa,

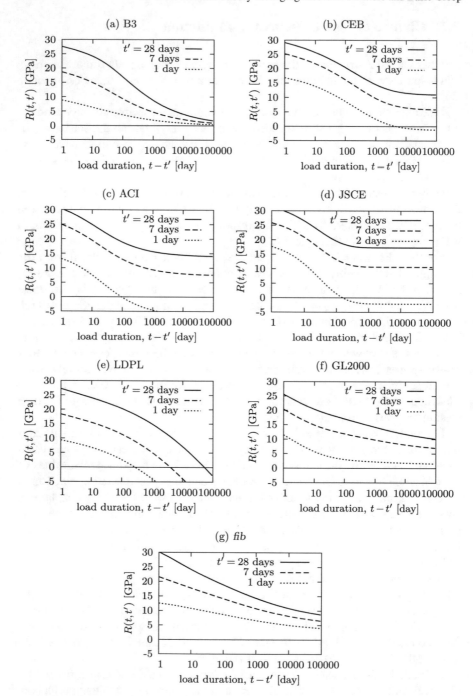

Fig. 9.15 Relaxation curves for saturated concrete, with the onset of relaxation at three different ages: (a) B3 model, (b) CEB model, (c) ACI model, (d) JSCE model, (e) log-double-power law, (f) GL2000 model, (g) *fib* Model Code 2010

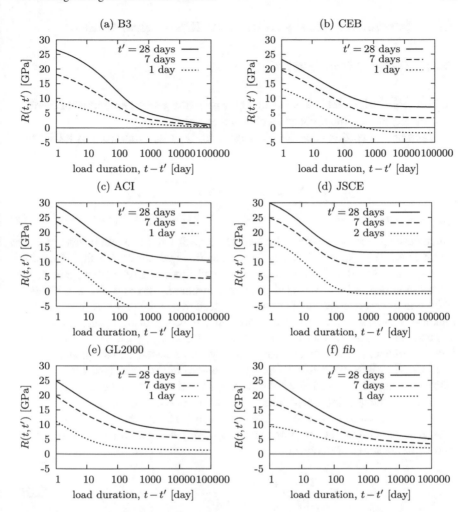

Fig. 9.16 Relaxation curves for a concrete slab of thickness 100 mm, exposed to an environment of 50% relative humidity from age 1 day, with the onset of relaxation at three different ages: (a) B3 model, (b) CEB model, (c) ACI model, (d) JSCE model, (e) GL2000 model, (f) *fib* Model Code 2010

exposed to an environment of 10% relative humidity, and loaded at the age of 7 days, the relaxation function would become negative after about 8500 days. However, the code is applicable to concretes with a mean compressive strength of at least 20 MPa exposed to an environment of at least 40% relative humidity. So, even though the theoretical requirements are not satisfied perfectly, there is no problem in practical applications.

9.8   Thermodynamically Admissible Rheological Chains*

9.8.1   General Properties*

To gain more insight, let us systematically explore the general properties of thermo-dynamically admissible rheological chains with aging units.

A general **relaxation function** can conveniently be approximated by a Dirichlet series of the form

$$R(t, t') = \sum_{\mu=1}^{M} E_\mu(t') \, e^{-(t-t')/\tau_\mu}, \qquad \text{for } t \geq t' \qquad (9.64)$$

which corresponds to an aging Maxwell chain with fixed relaxation times τ_μ; see formula (A.33) in Appendix A.4.1. Such a chain is thermodynamically admissible if all the partial moduli E_μ are nonnegative at all times (the relaxation times τ_μ are always assumed to be preselected positive constants). Maxwell chains with fixed relaxation times are convenient because they lead to a simple form of the relaxation function (9.64) but, in theoretical developments, one could also consider a more general case with partial moduli E_μ and partial viscosities η_μ specified by independent functions of age. Thermodynamic admissibility is guaranteed if all the moduli as well as all the viscosities are positive (springs with zero moduli and dashpots with zero viscosities can be removed without affecting the behavior of the chain). It is good to recall that the age-dependent spring moduli must be inserted into an incremental form of the stress–strain law ($\dot{\sigma} = E\dot{\varepsilon}$) and not into a total form ($\sigma = E\varepsilon$).

A general **compliance function** can conveniently be approximated by a generalized Dirichlet series of the form

$$J(t, t') = \frac{1}{E_0(t')} + \sum_{\mu=1}^{M} \frac{1 - e^{-(t-t')/\tau_\mu}}{D_\mu(t')}, \qquad \text{for } t \geq t' \qquad (9.65)$$

which corresponds to a special type of aging Kelvin chains; see formula (A.40) in Appendix A.4.2. Again, one could consider a general aging Kelvin chain with independent partial moduli E_μ and partial viscosities η_μ, for which the admissibility conditions are simply $E_\mu > 0$ and $\eta_\mu > 0$ (for all μ and at all times). Formula (9.65) is written in terms of time-dependent parameters D_μ which do not have a direct physical meaning and are linked to moduli E_μ by condition $D_\mu - \tau_\mu \dot{D}_\mu = E_\mu$ in which τ_μ are constant (positive) retardation times. The conditions of thermodynamic admissibility written in terms of time-dependent parameters D_μ then read $D_\mu > 0$ (this replaces $\eta_\mu > 0$ because $\eta_\mu = \tau_\mu D_\mu$) and $D_\mu/\tau_\mu \geq \dot{D}_\mu$; the latter condition can be violated even for a positive D_μ if the growth of D_μ is too fast.

General properties of rheological chains with positive moduli and viscosities can be summarized as follows:

- **M-R**: For a thermodynamically admissible aging Maxwell chain, the relaxation function is guaranteed to remain positive.
- **M-D**: There exists a thermodynamically admissible aging Maxwell chain that leads to divergence of creep curves.
- **K-R**: For a thermodynamically admissible aging Kelvin chain, the relaxation function is guaranteed to remain positive.
- **K-ND**: For a thermodynamically admissible aging Kelvin chain, divergence of creep curves cannot occur.

Property M-R, i.e., positiveness of the relaxation function for any aging Maxwell chain with positive partial moduli and viscosities, directly follows from the representation of the relaxation function by formula (A.30), which in the special case of a chain with constant relaxation times reduces to formula (9.64).

Property M-D, i.e., the existence of an aging Maxwell chain with positive partial moduli and viscosities for which the creep curves diverge, is proven by Example 9.2. It transpires that divergence of creep curves and change of sign of relaxation function are two different phenomena, even though the specific models examined in Example 9.3 usually suffer by both problems.

Property K-ND, i.e., nondivergence of creep curves corresponding to a Kelvin chain with positive partial moduli and viscosities, is proven at the end of Appendix A.4.2 based on formula (A.43) for the mixed second derivative of the compliance function.

Finally, property K-R, i.e., positiveness of the relaxation function for any aging Maxwell chain with positive partial moduli and viscosities, can be proven by contradiction. The starting assumption is that the evolution of all variables (stress, partial strains, moduli) is continuously differentiable in time. This is not a major constraint, since all models discussed so far are based on very regular compliance functions, and the strain history in a relaxation test is constant (the initial jump is embedded in the initial conditions). We consider an aging Kelvin chain with a positive strain $\hat{\varepsilon}$ applied abruptly at time t_1 and kept constant afterward. The objective is to show that the stress can never become negative, provided that the partial moduli and viscosities satisfy at all times the conditions $E_\mu \geq 0$ and $\eta_\mu > 0$. Since the proof is relatively complicated, let us split it into two major steps that will be summarized first and subsequently elaborated in detail:

1. If t_3 is the first time instant at which the stress ceases to be positive (i.e., crosses or touches the zero axis), then at least one of the Kelvin units must exhibit a positive strain rate at time t_3.
2. If one of the Kelvin units exhibits a positive strain rate at time t_3, then there must exist a time instant $t_2 \leq t_3$ at which the total stress transmitted by that unit is negative. However, this is in contradiction with the definition of t_3 as the **first** time instant at which the stress ceases to be positive.

Before going into detail, let us recall the basic notation and relations describing a Kelvin chain. Such a chain is a serial arrangement of an elastic spring (unit number 0) and M Kelvin units, each of which consists of an elastic spring and a viscous dashpot, coupled in parallel. The partial strains in individual units are denoted as ε_μ, the partial stresses in individual springs and dashpots are denoted as $\sigma_{e\mu}$ and $\sigma_{v\mu}$, and the basic equations are

$$\varepsilon = \sum_{\mu=0}^{M} \varepsilon_\mu \tag{9.66}$$

$$\sigma = \sigma_{e0} \tag{9.67}$$

$$\sigma = \sigma_{e\mu} + \sigma_{v\mu}, \qquad \mu = 1, 2, \ldots M \tag{9.68}$$

$$\dot{\sigma}_{e\mu} = E_\mu \dot{\varepsilon}_\mu, \qquad \mu = 0, 1, 2, \ldots M \tag{9.69}$$

$$\sigma_{v\mu} = \eta_\mu \dot{\varepsilon}_\mu, \qquad \mu = 1, 2, \ldots M \tag{9.70}$$

The initial response after a sudden application of strain $\varepsilon(t_1) = \hat{\varepsilon}$ is purely elastic, and only the zeroth unit (spring with no dashpot) deforms. The corresponding initial conditions are given by

$$\varepsilon_0(t_1) = \hat{\varepsilon} \tag{9.71}$$

$$\sigma_{e0}(t_1) = E_0(t_1)\hat{\varepsilon} \tag{9.72}$$

$$\sigma(t_1) = E_0(t_1)\hat{\varepsilon} \tag{9.73}$$

$$\varepsilon_\mu(t_1) = 0, \qquad \mu = 1, 2, \ldots M \tag{9.74}$$

$$\sigma_{e\mu}(t_1) = 0, \qquad \mu = 1, 2, \ldots M \tag{9.75}$$

$$\sigma_{v\mu}(t_1) = E_0(t_1)\hat{\varepsilon}, \qquad \mu = 1, 2, \ldots M \tag{9.76}$$

The two steps of the proof that was outlined above can now be elaborated as follows:

1. The initial stress $\sigma(t_1)$ given by (9.73) is positive, and subsequently, the stress evolves in a continuous manner. If σ should ever become negative, there would need to be a time instant t_3 at which $\sigma(t_3) = 0$ for the first time. Since $\sigma(t) > 0$ for all $t < t_3$, the stress rate at t_3 cannot be positive. The cases in which $\dot{\sigma}(t_3) < 0$ and $\dot{\sigma}(t_3) = 0$ are treated separately:

   a. Suppose that $\dot{\sigma}(t_3)$ is strictly negative. Since $\dot{\varepsilon}(t_3) = \dot{\varepsilon}_0(t_3) + \sum_{\mu=1}^{M} \dot{\varepsilon}_\mu(t_3) = 0$ and $\dot{\varepsilon}_0(t_3) = \dot{\sigma}(t_3)/E_0(t_3) < 0$, there must exist at least one Kelvin unit (further referred to by subscript $k \in \{1, 2, \ldots M\}$) for which $\dot{\varepsilon}_k(t_3) > 0$.

   b. The statement that $\dot{\varepsilon}_k(t_3) > 0$ for some k can be extended to the special case when $\dot{\sigma}(t_3) = 0$. In this case, we get $\dot{\varepsilon}_0(t_3) = 0$ and $\sum_{\mu=1}^{M} \dot{\varepsilon}_\mu(t_3) = 0$. If none of the rates $\dot{\varepsilon}_\mu(t_3)$ was positive, all of them would have to be zero. However, this would imply that all the partial stresses in viscous dashpots, $\sigma_{v\mu}(t_3)$, vanish, and by combining this with the condition of zero total stress, $\sigma(t_3) = 0$, we would also get zero partial stresses $\sigma_{e\mu}(t_3)$ in all elastic units. If the model attains such a state of zero stress in all units and the strain remains constant, all the stresses (partial and total) remain at zero forever and negative stress is never reached.

The partial conclusion is that relaxation to negative stress implies the existence of at least one unit with a positive strain rate at the time instant t_3 when the stress for the first time ceases to be positive.

2. Now, we assume that, for a given $k \in \{1, 2, \ldots M\}$, we have $\dot{\varepsilon}_k(t_3) > 0$, and we also know that $\sigma(t_3) = 0$ and $\sigma(t) > 0$ for all t between t_1 and t_3. For partial stresses in the kth unit, we get $\sigma_{vk}(t_3) = \eta_k(t_3)\dot{\varepsilon}_k(t_3) > 0$ and $\sigma_{ek}(t_3) = \sigma(t_3) - \sigma_{vk}(t_3) = 0 - \sigma_{vk}(t_3) < 0$; i.e., the stress in the dashpot is positive and the stress in the spring is negative. According to (9.75), the stresses in all springs (except the zeroth spring) are initially zero. Since the stress in the kth spring evolves continuously from $\sigma_{ek}(t_1) = 0$ to $\sigma_{ek}(t_3) < 0$, there must exist a certain time instant $t_2 \leq t_3$ at which $\sigma_{ek}(t_2) < 0$ and $\dot{\sigma}_{ek}(t_2) < 0$. But then $\dot{\varepsilon}_k(t_2) = \dot{\sigma}_{ek}(t_2)/E_k(t_2) < 0$ and $\sigma_{vk}(t_2) = \eta_k(t_2)\dot{\varepsilon}_k(t_2) < 0$; i.e., the stress in the kth dashpot is negative at t_2. The total stress $\sigma(t_2) = \sigma_{ek}(t_2) + \sigma_{vk}(t_2)$ is thus negative at t_2, which is in contradiction with the facts that $t_2 \leq t_3$ and that t_3 was defined as the **first** time instant at which the stress ceases to be positive.

This completes the proof of the K-R property.

9.8.2 Relation to Retardation Spectrum*

It has been shown that thermodynamically admissible Kelvin chains lead to nondiverging creep curves and nonnegative relaxation functions. At the same time, the conditions of nondivergence and nonnegative relaxation are violated by many models scrutinized in the present section and in the preceding one, in particular by the ACI, CEB, JSCE, and LDPL models. A logical conclusion is that the compliance functions of these models cannot be presented in the form of Dirichlet series (9.65) with parameters satisfying the conditions $D_\mu > 0$ and $\dot{D}_\mu < D_\mu/\tau_\mu$ for all μ and at all times. Violation of such conditions can be demonstrated directly, using the concept of a retardation spectrum, explained in detail in Appendix F.

In analogy to the representation of the compliance function of a nonaging linear viscoelastic material by formula (F.1), the compliance function of an aging material can be expressed as

$$J(t, t') = \frac{1}{E_{as}(t')} + \int_{\tau=0}^{\infty} L(\tau, t')\left[1 - \exp\left(-\frac{t - t'}{\tau}\right)\right] d\ln\tau \qquad (9.77)$$

where $E_{as}(t')$ is the instantaneous (asymptotic) modulus at age t' and $L(\tau, t')$ is the age-dependent continuous retardation spectrum. As shown in Appendix F.1, the retardation spectrum is closely related to the inverse Laplace transform of the compliance function (more precisely, of its continuous part, which is why the initial jump at $t - t' = 0$ is taken into account separately by the asymptotic compliance $1/E_{as}(t')$). For a given compliance function J and a fixed age t', the auxiliary function $\Phi(t - t') = J(t, t')$ can be treated in the same way as a compliance function of a nonaging material and its retardation spectrum $L(\tau)$ can be approximated using the techniques described in Appendix F. The result of course depends on the age t', and thus, the retardation spectrum of an aging material is considered as a function of both τ and t'.

A big advantage of the representation of viscoelastic properties by a retardation spectrum is that, if the spectrum is known, one can directly proceed to an approximation of the compliance function by a Dirichlet series with explicitly evaluated coefficients that correspond to individual Kelvin units (with no need for determination of the coefficients by least-square fitting or similar optimization techniques). For illustration, consider the usual choice of fixed retardation times forming a geometric progression with quotient 10, which means that $\tau_\mu = 10^{\mu-1}\tau_1$, $\mu = 2, 3, \ldots M$. Based on a simple numerical quadrature scheme, the integral in (9.77) can be approximated as follows:

$$\int_{\tau=0}^{\infty} L(\tau, t')\left[1 - \exp\left(-\frac{t-t'}{\tau}\right)\right] d\ln\tau \approx \tag{9.78}$$

$$\approx \int_{\tau=0}^{\tau_1/\sqrt{10}} L(\tau, t')\, d\ln\tau + \sum_{\mu=1}^{M} L(\tau_\mu, t')\left[1 - \exp\left(-\frac{t-t'}{\tau_\mu}\right)\right] (\ln 10)$$

The first term on the right-hand side of (9.78) represents the contribution of the part of the spectrum with very short retardation times. If the first retardation time, τ_1, is sufficiently small, this term could be neglected; otherwise, it is estimated based on an analytical approximation of $L(\tau, t')$ for small τ or evaluated numerically. The result is a function of t' only (i.e., is independent of the load duration, $t - t'$) and corresponds to an additional short-term compliance,

$$\frac{1}{E_0^*(t')} = \int_{\tau=0}^{\tau_1/\sqrt{10}} L(\tau, t')\, d\ln\tau \tag{9.79}$$

which is to be added to the truly instantaneous compliance, $1/E_{as}(t')$. The resulting approximation of the original compliance function by Dirichlet series then reads

$$J(t, t') = \frac{1}{E_0(t')} + \sum_{\mu=1}^{M} \frac{1}{D_\mu(t')}\left[1 - \exp\left(-\frac{t-t'}{\tau_\mu}\right)\right] \tag{9.80}$$

where

$$E_0(t') = \left(\frac{1}{E_{as}(t')} + \frac{1}{E_0^*(t')}\right)^{-1} \tag{9.81}$$

$$D_\mu(t') = \frac{1}{(\ln 10)\, L(\tau_\mu, t')} \tag{9.82}$$

The physical moduli that correspond to the coefficients D_μ given by (9.82) are evaluated as

$$
E_\mu(t') = D_\mu(t') - \tau_\mu \frac{dD_\mu(t')}{dt'} = \frac{1}{(\ln 10)\, L(\tau_\mu, t')} + \tau_\mu \frac{1}{(\ln 10)\, L^2(\tau_\mu, t')} \frac{\partial L(\tau_\mu, t')}{\partial t'} =
$$

$$
= \left(1 + \frac{\tau_\mu}{L(\tau_\mu, t')} \frac{\partial L(\tau_\mu, t')}{\partial t'}\right) D_\mu(t') \tag{9.83}
$$

The model is guaranteed to be thermodynamically admissible if the partial moduli E_μ as well as the partial viscosities $\tau_\mu D_\mu$ of all Kelvin units are positive at all ages $t' > 0$. This is true independently of the specific choice of retardation times τ_μ if the retardation spectrum satisfies the conditions

$$
L(\tau, t') > 0 \tag{9.84}
$$

$$
L(\tau, t') + \tau \frac{\partial L(\tau, t')}{\partial t'} > 0 \tag{9.85}
$$

for all retardation times $\tau > 0$ and at all ages $t' > 0$. Usually, $L(\tau, t')$ is a nonincreasing function of t', and so the second condition represents a more severe restriction than the first one.

It is also interesting to recall that if the mixed second derivative

$$
\frac{\partial^2 J(t, t')}{\partial t \partial t'} = \int_{\tau=0}^{\infty} \frac{1}{\tau^2} \left(L(\tau, t') + \tau \frac{\partial L(\tau, t')}{\partial t'}\right) \exp\left(-\frac{t - t'}{\tau}\right) d\ln \tau \tag{9.86}
$$

obtained by double differentiation of (9.77) is nonnegative, then the creep curves cannot diverge. Condition (9.85) is thus sufficient to guarantee nondivergence.

As shown by Jirásek and Havlásek [522], the spectral values L are positive for virtually all compliance functions proposed for concrete.[4] In the absence of aging, $L(\tau, t')$ does not depend on t' and the derivative $\partial L(\tau, t')/\partial t'$ vanishes. In that case, conditions (9.84) and (9.85) are equivalent, and positivity of the retardation spectrum is sufficient for thermodynamic admissibility as well as for nondivergence of creep curves. On the other hand, for models with aging, the spectral content typically decreases with increasing age t', and so the derivative $\partial L(\tau, t')/\partial t'$ is negative and condition (9.85) is potentially violated, even for a positive spectrum.

---

[4]Slightly negative spectral values were detected, in a narrow range of retardation times, for the drying creep compliance function of the B3 model with some specific combinations of parameters. However, this function represents only one contribution to the total compliance. When it is combined with the basic creep compliance function, the spectral values are expected to become positive, at least in the physically meaningful range of parameters.

Example 9.6. Retardation spectrum and Kelvin moduli for basic creep compliance function of fib *model*

For illustration, let us check conditions (9.84)–(9.85) for the *fib* Model Code 2010. In the absence of drying, the compliance function recommended by this code can be presented in the form

$$J(t, t') = \frac{1}{E(t')} + a \, \ln\left(1 + \frac{t - t'}{g(t')}\right) \tag{9.87}$$

where $a = 1.8/(E_{28} \bar{f}_c^{0.7})$ is a parameter that depends on the specific concrete and

$$g(t') = \left(0.035 + \frac{30}{t'}\right)^{-2} \tag{9.88}$$

is an increasing function that reflects the effect of aging (with t' substituted in days, as usual). For fixed t', the second term on the right-hand side of (9.87) is a special case of the log-power law (F.15) with exponent $n = 1$. As shown in Appendix F.2, the corresponding retardation spectrum

$$L(\tau, t') = a \, e^{-g(t')/\tau} \tag{9.89}$$

is available in a closed form, which is rather an exception among the widely used creep models. Since a is a positive constant, condition (9.84) is easily verified.

To check condition (9.85), let us evaluate

$$L(\tau, t') + \tau \frac{\partial L(\tau, t')}{\partial t'} = a \, e^{-g(t')/\iota} + \tau a \, e^{-g(t')/\iota} \left(-\frac{1}{\tau}\right) \frac{dg(t')}{dt'} =$$

$$= L(\tau, t') \left(1 - \dot{g}(t')\right) \tag{9.90}$$

For the sake of brevity, the derivative of g with respect to its argument t' is denoted as $\dot{g}$. Since we already know that $L(\tau, t') > 0$, the right-hand side of (9.90) is positive if $\dot{g}(t') < 1$ at all ages $t' > 0$. For the specific form of function $g(t')$ given by (9.88), we get

$$\dot{g}(t') = \frac{60 \, t'}{(0.035 \, t' + 30)^3} \tag{9.91}$$

It is easy to check that $60 \, t' < (0.035 \, t' + 30)^3$ for all $t' > 0$. Consequently, condition (9.85) is satisfied, and the approximation of compliance function (9.87) by formula (9.80) with parameters D_μ given by (9.82) always leads to a thermodynamically admissible Kelvin chain, for an arbitrary choice of retardation times τ_μ. In fact, it follows from (9.90) that, for the present model, formula (9.83) can be rewritten as

$$E_\mu(t') = \left(1 - \dot{g}(t')\right) D_\mu(t') \tag{9.92}$$

This means that the partial moduli E_μ are obtained from coefficients D_μ by simple scaling by a dimensionless factor $1 - \dot{g}(t')$. The dependence of this factor on age is presented in Fig. 9.17c.

For a concrete with compressive strength $\bar{f}_c = 45.4$ MPa and conventional modulus $E_{28} = 32$ GPa (the same values as in the comparative example in Appendix E.6), parameter $a = 1.8/(E_{28}\bar{f}_c^{0.7})$ is found to be equal to 3.892×10^{-6}/MPa. In Fig. 9.17a, the continuous retardation spectrum (9.89) is plotted as a function of the retardation time τ, for selected ages t' ranging from 1 day to 10,000 days. Figure 9.17b shows the corresponding function $L(\tau, t') + \tau \, \partial L(\tau, t')/\partial t'$ given by (9.90), which, as expected, remains positive. From these two functions, it is possible to evaluate parameters D_μ given by (9.82) and partial moduli E_μ given by (9.83) or, equivalently, by (9.92). The resulting partial compliances $1/E_\mu$ are shown in Fig. 9.17d for age $t' = 100$ days and for discrete retardation times τ_μ ranging from 1 day to 10^5 days. It is confirmed that all the partial moduli are positive.

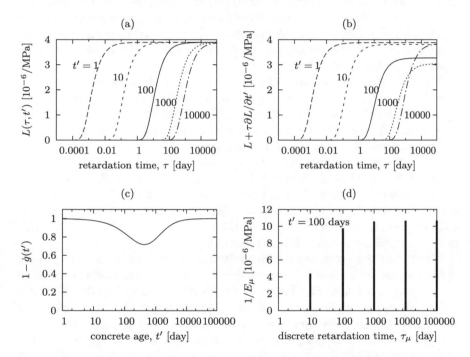

Fig. 9.17 Model for basic creep specified by the *fib* Model Code 2010: (a) continuous retardation spectrum, $L(\tau, t')$, (b) function $L(\tau, t') + \tau \partial L(\tau, t')/\partial t'$, (c) age-dependent reduction factor $1 - \dot{g}(t')$, (d) partial compliances $1/E_\mu$ at age $t' = 100$ days

As shown in Fig. 9.17a, the spectral values are extremely small for retardation times below a certain age-dependent limit, which is about $t'/100$ for ages up to 100 days and not less than $t'/1000$ for ages up to 10,000 days. Parameters D_μ evaluated from formula (9.82) for retardation times $\tau_\mu \ll t'$ are then huge (compared, e.g., to the elastic modulus), and the physical moduli E_μ evaluated from (9.92) are not much smaller (at most by 30%, since $1 - \dot{g}(t') > 0.7$ for all ages t'). This means that the springs in Kelvin units with short retardation times are extremely stiff and their contribution to the total strain of the Kelvin chain is negligible.

For instance, at age $t' = 100$ days, the stiffness of the spring in the unit with retardation time 1 day is 695×10^3 GPa and its reciprocal value (compliance) is 1.44×10^{-9} /MPa, which is several orders of magnitude below the compliances of units with longer retardation times; see Fig. 9.17d. Therefore, the Dirichlet series approximating the compliance function of the *fib* model does not need to include terms with retardation times below a certain limit.

This might look like an advantage from the computational point of view. However, such an extremely low compliance of units with short retardation times is related to the poor ability of the model to capture short-term creep, which is a consequence of the selected form of the compliance function. For $t - t' \ll g(t')$, i.e., for short load durations, the compliance function (9.87) can be approximated by $J(t, t') \approx 1/E(t') + a(t - t')/g(t')$, which means that the creep strain grows proportionally to the duration of loading. As discussed in detail in Sect. 3.2, experimental data indicate proportionality of short-term creep to a power function $(t - t')^n$ with a low exponent, typically $n = 0.1$; see Figs. 3.5 and 3.6.

This feature is properly reflected by the B3 model, for which the basic creep strain after the first minute of loading is roughly one half of the basic creep strain after the first day (because for $n = 0.1$ we get $(60 \times 24)^n \approx 2.07$). For the *fib* model, the basic creep strain after the first minute is by three orders of magnitude smaller than the basic creep strain after the first day, because for $n = 1$ we get $(60 \times 24)^n = 1440$.

For very long retardation times, the spectral value $L(\tau, t')$ is almost constant and tends to $a = 3.892 \times 10^{-6}$ /MPa as $\tau \to \infty$; see Fig. 9.17a. The corresponding coefficients D_μ tend to $1/(a \ln 10) \approx 111.6$ GPa, and compliances $1/E_\mu$ that characterize the springs in Kelvin units with long retardation times get close to $a \ln 10/(1 - \dot{g}(t'))$. At age $t' = 100$ days, this gives $1/E_\mu \to 10.66 \times 10^{-6}$ /MPa as $\tau_\mu \to \infty$; see the values in the right part of Fig. 9.17d.

The fact that there is no upper limit on the retardation times of units that provide significant contributions to the creep strains is related to the logarithmic nature of the compliance function. This feature of the *fib* model is shared by the B3 model, which also uses a compliance function with logarithmic growth; see the retardation spectrum of the log-power law with exponent $n = 0.1$ in Fig. F.3c. For bounded compliance functions, the long-term part of the spectrum is very different, as will be demonstrated in Example 9.7.   ∎

As already mentioned in Sect. 9.6, many of the compliance functions used by creep models have the general form (9.42). The corresponding retardation spectrum is then described by

$$L(\tau, t') = \frac{c\, L_f(\tau)}{g(t')} \tag{9.93}$$

where c is a constant, $g(t')$ is an increasing function that describes aging, and $L_f(\tau)$ is the retardation spectrum obtained if function $f(\hat{t})$ from (9.42) is treated as a nonaging compliance function. For commonly used functions $f(\hat{t})$, the nonaging retardation spectrum $L_f(\tau)$ is positive and condition (9.84) is satisfied. To check condition (9.85), let us evaluate

$$L(\tau, t') + \tau \frac{\partial L(\tau, t')}{\partial t'} = \frac{c\,L_f(\tau)}{g(t')} - \tau \frac{c\,L_f(\tau)\dot{g}(t')}{g^2(t')} = \frac{c\,L_f(\tau)}{g(t')}\left(1 - \tau \frac{\dot{g}(t')}{g(t')}\right)$$
(9.94)

For each age t', the ratio $T_g(t') = g(t')/\dot{g}(t')$ is positive and corresponds to a certain characteristic time of the aging process. The expression in parentheses on the right-hand side of (9.94) becomes negative for all retardation times $\tau > T_g(t')$, and condition (9.85) is violated.

Example 9.7. Retardation spectrum and Kelvin moduli for compliance function of ACI model

For the ACI model, the compliance function has the form (9.42) with $f(\hat{t}) = \hat{t}^\psi/(d + \hat{t}^\psi)$ and $g(t') = t'^m/\sqrt{b + a/t'}$ and with typical parameters $m = 0.118$, $\psi = 0.6$, $a = 4$, $b = 0.85$, and $d = 10$. The characteristic time of the aging process is then given by

$$T_g(t') = \frac{g(t')}{\dot{g}(t')} = \frac{2t'(a + bt')}{a + 2m(a + bt')}$$
(9.95)

and condition (9.85) is violated for all $\tau > T_g(t')$. For large t', the value of $T_g(t')$ is close to $t'/m \approx 8.5\,t'$.

The multiplicative constant c in (9.93) is given by $c = 2.35\,\gamma/E_{28}$ where γ is the product of factors γ_1 to γ_6 that depend on the concrete composition, type of curing, ambient humidity, and member size; see Appendix E.3. For the concrete and conditions considered in the comparative example in Appendix E.6, factor γ is equal to 0.757 and the conventional modulus is $E_{28} = 32$ GPa, which gives $c = 55.6 \times 10^{-6}/$MPa. For this value of c, the continuous retardation spectrum (9.93) of the ACI model is plotted in Fig. 9.18a, based on a high-order approximation of function $L_f(\tau)$ described in Appendix F.3. The effect of age on the spectrum consists in vertical scaling and is totally different from the effect of age for the fib model, which was manifested in Fig. 9.17a by horizontal shifting in the logarithmic scale. This is related to differences in the form of compliance functions. For the fib compliance function (9.87), an age-dependent factor scales the load duration, while for the ACI compliance function (and all other compliance functions that have the form (9.42)) another age-dependent factor scales the creep strain.

Figure 9.18b shows the function $L(\tau, t') + \tau\partial L(\tau, t')/\partial t'$ for different ages t' ranging from 1 day to 10,000 days. It is confirmed that formula (9.94) leads to negative values for $\tau > T_g(t')$ where $T_g(t')$ is approximately one order of magnitude larger than t'. For instance, at age $t' = 100$ days, negative values are obtained for all retardation times that exceed 712 days.

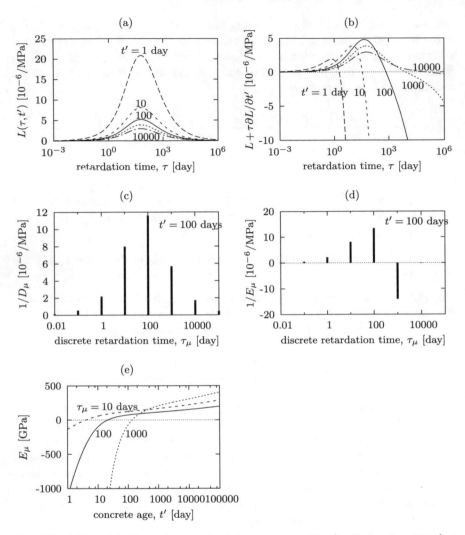

Fig. 9.18 ACI model: (a) continuous retardation spectrum, $L(\tau, t')$, (b) function $L(\tau, t') + \tau \partial L(\tau, t')/\partial t'$, (c) parameters $1/D_\mu$ for $t' = 100$ days, (d) partial compliances $1/E_\mu$ for $t' = 100$ days, (e) dependence of partial moduli E_μ on age

Since the retardation spectrum is positive, parameters D_μ evaluated from (9.82) are also positive, for all retardation times and at all ages. Their reciprocal values $1/D_\mu$ represent the coefficients in Dirichlet series (9.65) and are plotted in Fig. 9.18c for discrete retardation times ranging from 0.01 day to 100,000 days and for a fixed age $t' = 100$ days. Only Kelvin units with retardation times between 1 day and 10,000 days provide an important contribution to the compliance function. For units with retardation times 0.1 day and 100,000 days, the coefficients $1/D_\mu$ are, respectively, 0.523×10^{-6}/MPa and 0.455×10^{-6}/MPa and are more than 20 times smaller than the coefficient that corresponds to the retardation time of 100 days. The contribution of units with very short or very long retardation times (below 0.1 day or above 100,000 days) to the compliance function is negligible.

Even though the compliance function is approximated by Dirichlet series with coefficients $1/D_\mu$ that remain positive at any age, the physical moduli E_μ, which represent age-dependent stiffnesses of individual Kelvin units, are not always positive. At $t' = 100$ days, the compliances $1/E_\mu$ are positive for units with retardation times up to 100 days, but the unit with $\tau_\mu = 1000$ days is characterized by a negative compliance $1/E_\mu = -14 \times 10^{-6}/\text{MPa}$, which has a similar magnitude as the positive compliance of the unit with $\tau_\mu = 100$ days; see Fig. 9.18d. For units with still longer retardation times, the compliances are also negative but very small in magnitude.

To complete the picture, the dependence of moduli E_μ on concrete age t' for three selected Kelvin units is presented in Fig. 9.18e. It turns out that these moduli are increasing functions of age and their initial values are negative. The modulus of the unit with $\tau_\mu = 10$ days becomes positive at age $t' = 3.91$ days, when the condition $T_g(t') = 10$ days is satisfied. The moduli of the units with $\tau_\mu = 100$ days and 1000 days become positive, respectively, at 21 and 135 days. ∎

Chapter 10
Microprestress-Solidification Theory and Creep at Variable Humidity and Temperature

Abstract After an initial period of less than one year, the relative decrease of creep rate with the age at loading is much stronger than the relative decrease of the growth rate of hydration degree. This suggests another source of long-term aging, which is explained by relaxation of the so-called microprestress. We conceive the microprestress as an overall characteristic of the disjoining pressures in nanopores filled by hindered adsorbed water and the counterbalancing tensile stresses in the nanostructure of hydrated cement. We model mathematically how the microprestress gets generated by the volume changes due to hydration as well as the pore water content changes and temperature changes. To anchor the model physically, we discuss the pore structure of hydrated cement and water adsorption on its enormous internal surface. We also present an alternative computational approach in which the microprestress changes are replaced by viscosity variation and show how the microprestress theory is easily incorporated into finite element programs for creep.

The solidification theory explained in the preceding chapter separates viscoelasticity of the solid constituent, the cement gel, from the chemical aging of the hardened cement paste caused by solidification of gel particles and characterized by the growth of volume fraction of hydration products. This permits considering the viscoelastic constituent as nonaging, and the decrease of creep compliance is explained by volume growth of cement gel into the pores. However, this cannot explain the multidecade continuation of significant compliance decrease because the volume growth of cement gel becomes rather slow after about one year. Neither can this explain the drying creep effect (or Pickett effect) and the transitional thermal creep. This chapter will show that all these important phenomena can be explained by one unified concept, the concept of microprestress.[1]

---

[1] The microprestress would perhaps better be called the nanoprestress, but the term microprestress got fixed before everything "nano" became fashionable.

© Springer Science+Business Media B.V. 2018
Z.P. Bažant and M. Jirásek, *Creep and Hygrothermal Effects in Concrete Structures*, Solid Mechanics and Its Applications 225,
https://doi.org/10.1007/978-94-024-1138-6_10

The *microprestress* characterizes self-equilibrated stress fields on the nanoscale level which stretch the bonds and thus facilitate their breakage. The microprestress is independent of the applied load, is initially produced by incompatible volume changes in the microstructure during hydration, and later builds up when changes of moisture content and temperature create thermodynamic imbalance between the chemical potentials of vapor and adsorbed water in the nanopores of cement gel. Further it is shown that the microprestress buildup and relaxation also capture a third effect: the transitional thermal creep, i.e., the transitional creep increase due to temperature change. A reduction in the number of parameters can be achieved by eliminating the microprestress from the formulation and reformulating the governing differential equation in terms of viscosity as the primary internal variable. For computations, an efficient integration algorithm is developed. Numerical simulations of creep tests at variable temperature and humidity are used to identify the constitutive parameters, and a satisfactory agreement with typical test data is documented.

10.1   Overview of Physical Mechanisms

The quest for a realistic physically based creep and shrinkage model for Portland cement concrete has been confounded for decades by three intriguing phenomena:

1. The *aging of concrete*, which causes compliance $J(t, t')$ at constant load duration $t - t'$ to decrease with age t' at loading. The aging is of two kinds:

   a. Short-term *chemical aging*, which becomes weak (at room temperature) at the age of about one year and is explained by the solidification theory presented in Chap. 9; and

   b. long-term *nonchemical aging*, manifested by the fact that the decrease of creep with the age at loading continues unabated even for many years, and probably many decades after the degree of hydration of cement ceased to grow. This phenomenon was explained by the relaxation of a nanoscale microprestress [132], although, as already pointed out in Chap. 9, a long-term increase of bonding, due to "polymerization" in calcium silicate hydrates, might also play a role [179].

2. The *drying creep effect*, also called the Pickett effect (or stress-induced shrinkage). This is a transient effect consisting in the fact that the apparent creep *during* drying is much larger than the basic creep (i.e., creep at moisture saturation) while the creep *after* drying (i.e., after reaching thermodynamic equilibrium with the environmental humidity) is much smaller than the basic creep. The physical source of drying creep was shown [201] to involve two different mechanisms:

   a. One is an *apparent macroscopic mechanism*, arising from the impossibility of measuring the "true" material shrinkage, i.e., the theoretical shrinkage which would occur if the decreasing specific water content during drying could be kept uniform throughout the companion shrinkage specimen. Since concrete is primarily used to resist compressive loads, creep is typically tested under uniaxial compression. The creep (together with the initial elastic deformation) is defined as the deformation difference between the loaded specimen and a load-free companion. The nonuniformity of local drying shrinkage in the load-free companion specimen produces distributed microcracking (or softening damage), initially in the surface layer and later in the core. This causes the observed shrinkage to be less than the "true" shrinkage which would be observed if a uniform water content could be maintained throughout, and the creep identified from the measured deformation difference is larger than the "true" creep [80, 117, 198, 201, 431, 874–876, 879]. Besides, by including the cracking strain from the companion specimen in the drying creep, one gets a false impression of a nonlinear dependence on the mean stress applied on the creep specimen. Obviously, the contribution of microcracking would be different for bending creep, and this is how the portion of drying creep due to microcracking has been experimentally identified [201].

b. The other is a *true nanoscale mechanism*, which resides in the nanostructure and is explained by the hypothesis that the rate of shear (slip) due to breakages and restorations of bonds in the calcium silicate hydrates is reduced (or amplified) by a decrease (or increase) in the magnitude of microprestress that is acting across the slip planes and is controlled by changes in the chemical potential (i.e., the Gibbs free energy per mole) of pore water due to drying [78, 80].

3. The *transitional thermal creep*, which represents a transient increase of creep after a temperature change, either heating or cooling. For similar reasons as the drying creep effect, this effect also has two mechanisms:

a. An *apparent macroscopic mechanism*, due to thermally induced microcracking that is analogous to drying creep; and

b. a *"true" nanoscale mechanism*, due to a change in the level of microprestress caused by a thermally induced change of chemical potential of nanopore water.

While the apparent mechanisms 2a and 3a operate on the macroscopic scale of the whole specimen or structure (on the scale of centimeters and meters), mechanism 1a operates on the scale of capillary pores (which is the micrometer scale), and mechanisms 1b, 2b, and 3b operate on the scale of nanopores in calcium silicate hydrates (which is the nanometer scale).

The solidification theory [179, 180], presented in the preceding chapter, showed that the chemical aging (mechanism 1a) can be separated from the viscoelastic constitutive model if that model is formulated not for concrete as a whole but specifically for its solidifying constituent—the hardened cement gel, and if the chemical aging is interpreted as a growth of the volume fraction of the solidifying constituent.

Mechanisms 1b and 2b were initially modeled separately [179]. In a later study [131, 132], both 1b and 2b were explained by one unified physical theory resting on the idea of relaxation of microprestress—a stress that is created in the solid nanostructure of cement gel either by microscopic chemical volume changes of various chemical species during hydration or by an imbalance of chemical potentials among the four phases of pore water (vapor, capillary, adsorbed, and hindered adsorbed). The microprestress theory can also cover mechanism 3b. The fact that phenomena as diverse as the long-term aging, drying creep and transitional thermal creep can all be explained by one theory lends credence to its validity.

As another possible explanation of the drying creep effect and the transitional thermal creep, it was speculated that the local microdiffusion flux of water molecules between the nanoscale hindered adsorbed layers and the adjacent micrometer-scale capillary pores in the hardened cement paste might accelerate the process of breakage of atomic bonds in the C-S-H, which is the cause of creep [71, 78, 164]. Ongoing molecular dynamics simulations at Northwestern revealed that a movement of water molecules along hindered adsorbed layers must indeed accelerate the sliding of opposite walls of planar nanopores [779]. For more detail, see Sect. 12.4.

10.2   Relevant Aspects of Pore Structure and Water Adsorption in Hardened Cement Gel*

For the subsequent analysis, it is helpful to review first some relevant characteristics of the pore structure and pore water. The hardened Portland cement paste is a strongly hydrophilic porous material whose pores have an enormous internal surface (about $500 \, \text{m}^2$ per cm^3). This implies that pores of width 0.3 to 1 nm must occupy a major portion of pore volume [109], and causes the surface adsorption forces to dominate the stress levels in the microstructure, far exceeding any stresses that can be produced by applied loads, as suggested already by Powers and Brownyard [705, 706] and Powers [704]. The paste consists of unhydrated cement grains, hydration products in the form of hardened cement gel (or xerogel, [874]) and capillary pores. The *capillary pores* are defined here as the pores wider than $0.1 \, \mu\text{m}$, which are those having a well-defined capillary meniscus (obeying the Laplace equation) and contain liquid water unaffected by surface forces of the solid. The pores in the cement gel are subcapillary and range from 0.26 nm (the effective diameter of one water molecule) to about 100 nm.

The capillary water (region 5 in Fig. 10.1a) is in tension, i.e., under negative pressure p_l, proportional to the total curvature of capillary menisci. The magnitude of p_l is very large—when the relative humidity of water vapor in the capillary pore is $h = 50\%$ at $20\,^{\circ}\text{C}$, thermodynamic calculations based on the Kelvin equation (8.25) show that $p_l = -93.5 \, \text{MPa}$, which is not much less than the tensile strength of liquid water. The layers of adsorbed water (region 6) on the pore walls exposed to water vapor (region 4) are subjected to spreading pressure (e.g., Bažant [78]), which reduces the surface tension of the solid, γ_a, as a function of the mass of water

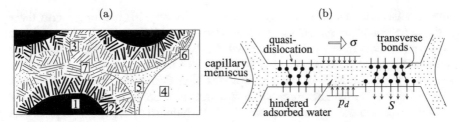

Fig. 10.1 (a) Idealized microstructure of hardened cement paste: 1–anhydrous cement, 2–inner hydration products (interface transition zone), 3–outer hydration products, 4–air with water vapor, 5–capillary water (note that the contact angle of capillary meniscus is virtually zero), 6–free adsorbed water, 7–hindered adsorption (dots = water molecules), (b) nanopore in cement gel, disjoining pressure p_d, microprestress S, and applied macroscopic stress σ; reproduced from [131] with permission from ASCE

molecules adsorbed per unit area of the surface. The effective thickness δ_a of the adsorption layers (as well as the radius of the capillary menisci) is also a function of h. A monomolecular adsorption layer of thickness $\delta_a = 0.263$ nm becomes full at about $h = 12\%$ [702]. From about 40% relative humidity to saturation (100% relative humidity), δ_a has the maximum thickness, which is five water molecules (or 1.32 nm). For $h < 80\%$, the majority of pore water is adsorbed water, free and hindered.

Most pores are less than ten water molecules in width. In such small nanopores[2] (or gel pores), the full thickness of the adsorption layers cannot develop, i.e., the adsorption is hindered. Consequently, a transverse compressive stress p_d, called the *disjoining pressure*,[3] must be exerted by the hindered adsorbed layers on the nanopore walls [71, 78, 164, 347–349, 704, 874–876], as schematically shown in Fig. 10.1b. Calculations of adsorption thermodynamics [78] show the disjoining pressures to be enormous; in a pore two water molecules wide, $p_d = 174$ MPa at saturation ($h = 100\%$) and temperature 25 °C. There are other phenomena, such as the crystal growth pressure, which act similarly.

The solid microstructure of C-S-H also contains water, called the interlayer hydrate water, which is to some extent mobile and can evaporate in a zero-humidity environment [383]. Doubtless this water acts similarly to the hindered adsorbed water and has not been from it mathematically distinguished. Therefore, when speaking of the hindered adsorbed water and its disjoining pressure, we also have in mind the interlayer hydrate water and its analogous pressure.

The disjoining pressure (as well as the crystal growth pressure) must be balanced by tensile forces. These forces are carried partly by the solid framework around the nanopore, and partly by bridges or bonds between the opposite walls of the same nanopore (note that, unlike the schematic picture in Fig. 10.1b, the walls are doubtless very rough and have islands of contact bridging the opposite walls). The opposite walls are also bridged by adjacent C-S-H sheets that are approximately normal to

---

[2]Originally, the gel pores were called "micropores", but since their characteristic size is much below a micron, often just a few nanometers, it seems that "nanopore" is a more appropriate expression.

[3]The disjoining pressure is a concept due to Deryagin [349]; see also Schmidt-Döhl and Rostásy [760].

these walls. As a reaction to the disjoining pressure (or crystal growth pressure), these bridges (or bonds) must transmit very large tensions, which produce on the nanoscale a self-equilibrated stress field. In other words, the solid part of microstructure of the hardened cement gel is permanently in a pretensioned state (which partly explains why the tensile strength is low but improves after drying if cracking is avoided). The buildup of tensile microprestress in the bridges or bonds is also caused by local volume expansions near the nanopore, which are induced by the chemical processes of hydration.

Because the distances of migration of water molecules diffusing between the nanopores and the adjacent capillary pore are very short, the hindered adsorbed water doubtless establishes thermodynamic equilibrium (i.e., the equality of chemical potentials) with the capillary water very fast, probably within seconds [117]. Thus, all the phases of pore water may be assumed to always be locally in thermodynamic equilibrium, and the disjoining pressure to respond almost instantly to the changes in surface tension and in the pressure p_l in capillary water. According to equation 44 in Bažant [78], the surface tension at the interface between the solid particle and free adsorbed water layers is given by an equation of the form

$$\gamma_a = -\frac{C_\gamma RT}{M_w} \ln h + \gamma_{a1} \tag{10.1}$$

where R is the universal gas constant, M_w is the molar mass of water, T is the absolute temperature, h is the pore relative humidity, and C_γ and γ_{a1} are constants. Hence, at constant temperature,

$$\dot{\gamma}_a = -\frac{C_\gamma RT}{M_w} \frac{\dot{h}}{h} \tag{10.2}$$

Comparison with the Kelvin equation (8.25) or (8.34) reveals that the changes in the liquid pressure, capillary pressure, and surface tension are virtually proportional. So are the changes in disjoining pressure, as noted from Eq. (8.62).

10.3   The Concept of Microprestress and Its Relaxation

The challenge is to find a mechanism of aging that does not require volume growth of the solidified constituent having fixed properties in time. The only possible mechanism is an increase of the resistance to slip, i.e., slip viscosity, and that in turn can be attributed to a relaxation of a transverse normal tensile stress in the solid phase (called *microprestress*, although it is a nanoscale feature). This key idea, advanced in Bažant [95], can at the same time explain the Pickett effect and the transitional thermal creep.

Why do we have in mind only the slip and not a crack-like separation of the opposite surfaces of nanopores, which was the main feature of the intuitive cement gel model advanced by Feldman and Sereda [383]? The slips breaking the bond at one atomic site must terminate with bond reformation at the next site and thus, similar to dislocations in crystals, do not change material stiffness and do not cause major volume expansion. In contrast to that, separations between the opposite surfaces of nanopores or interlayers in C-S-H would not lead to bond reformations, and since there are very many parallel nanopores and interlayers, simple calculations show that such separations would have to cause a major volume expansion and stiffness loss, which is not seen in experiments. So the hypothesis of a crack-like separation mechanism is not viable.

But can slips explain volumetric creep, which does occur in concrete? They can. Due to high porosity, a volume change can be produced by slipping only. Because of high porosity and microstructure heterogeneity, a hydrostatic macroscopic stress causes in the microstructure not only normal stresses but also shear stresses. Slipping in the microstructure can lead to changes of the pore volume and thus to macroscopic volumetric strain.

The idea of slip viscosity dependent on transverse microprestress cannot be described by classical rheological models such as the Kelvin chain because they are uniaxial. A multiaxial model is required. The bond breakages leading to shear slips driven by shear stress are obviously influenced by normal stresses across the slip direction. Therefore, the rheological model shown in Fig. 10.2 reflects interactions between forces and deformations of two directions: axial and transverse.

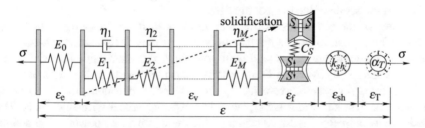

Fig. 10.2 Rheologic scheme of complete microprestress-solidification theory: serial coupling of an elastic spring, solidifying Kelvin chain, flow element affected by microprestress relaxation, and units corresponding to free shrinkage and thermal expansion

The microstructure of hardened cement gel contains dispersed, highly localized sites at which the atomic bonds are under a high tensile stress. Many of such highly stressed bonds may represent bridges across the nanopores containing the hindered adsorbed water, and the layers of hindered adsorbed water are the slip planes giving rise to creep. The creep is the macroscopic result of numerous interatomic bond breaks happening at different times at different overstressed sites (creep sites) in the hindered adsorbed layers. The bond breaks happen with a certain frequency determined by the kinetic energy of random thermal vibrations of atoms and by the magnitude of the activation energy barrier. Because the tensile stress reduces the activation energy barrier for bond rupture, this frequency of breaks increases as the tensile stress across the slip plane increases, as one could easily show by calculations according to the rate process theory. A temperature rise, which increases the kinetic energy of the atomic vibrations, must obviously also increase the frequency of bond breakage, and this can explain the transitional thermal creep.

So the *microprestress theory* postulates that the source of creep (in the linear range) is the shear slip at localized overstressed creep sites, represented by bridges across the nanopores. The shear slip is the result of a dislocation-like series of breakages and restorations of interatomic bonds (Fig. 10.1b). Each bond breakage and restoration at one creep site relaxes the shear stress at that site. As the stress gets redistributed, another creep site may become overstressed, causing again a bond to break, etc. A progressive relaxation of the shear stress at the creep sites and gradual exhaustion of the available overstressed creep sites cause the creep rate under a constant applied macrostress to decline with time.

Bond breakages without restoration are of course possible, too. They cause tensile microcracks to nucleate and grow. Because the crack growth is time-dependent, and because the time dependence of crack growth has been shown to be significant even for rather slow rates of growth [128, 140], the result is also creep. However, because microcracks cause a decrease of stiffness on the macroscale, the result can only be a nonlinear creep, which occurs beyond the service stress range of concrete. So the source of the nonlinear part of creep of concrete, which occurs on approach to the strength limit, is the time dependence of microcrack growth [95]. This phenomenon is beyond the scope of the present treatise.

In view of the roughness of the pore surfaces in hydrated cement, the notion that new bonds can form in a hardened cement gel may at first seem questionable. However, the hardened cement gel with pore water is a metastable material whose surfaces are strongly hydrophilic and readily adapt their shape. That new bonds form quite easily is for example revealed by the classical experiment of Powers [704], in which cement and water were placed in a rotating vessel with steel balls. This produced a completely hydrated cement which was still a fine powder. The powder was then stored in a container under a pressure of only about 0.01 MPa. In a few days, the powder solidified into a body whose strength was about the same as that of a normally cured cement paste. So, the hydrated cement particles, when brought to contact, stick together after a while, and nanocracks heal.

In another experiment revealing the capability to restore bonds [144, 259], concrete specimens confined under enormous pressure (many times the uniaxial compression strength) were subjected to shear deformation in which the shear angle reached $70°$. The uniaxial compressive strength of the cores drilled from such an enormously deformed concrete was not zero but 25 to 30% of the strength of the cores drilled from virgin identically cured specimens.

The microprestress, S, characterizing the magnitude of the locally self-equilibrated tensile stresses acting at the creep sites, i.e., across the bridges between the opposite nanopore walls, is virtually independent of the applied macroscopic stress, σ (Figs. 10.1b and 10.2). The microprestress is generated as a reaction to the disjoining pressure in the nanopore or to the high local shrinkage of small volumes in the material. An increase of microprestress may also be caused by the crystal growth pressures in the nanopore (e.g., in the bridges in Fig. 10.1b) which are induced by the growth of crystalline hydration products precipitated from aqueous solution. Large crystal growth pressures were documented, e.g., by Schmidt-Döhl and Rostásy [760].

The fact that the applied macroscopic stress σ must have a virtually negligible effect on the microprestress (and on the disjoining pressure) is corroborated by the previous inference that the disjoining pressure must respond to the changes in the capillary tension and surface tension almost instantaneously [117] and proportionally; these tensions are, in turn, functions of the relative humidity h in the capillary pores, which has been shown to be independent of the applied macroscopic stress (the other possible contributions to the microprestress, such as the high local volume changes and crystal growth pressure during hydration, are independent, too). This fact was, for example, confirmed by the experiments of Hansen [450, 451], which defeated the consolidation theory of concrete creep. They showed that identical loaded and load-free specimens dry equally fast. The solid framework of concrete and the hardened cement paste is so stiff that the applied load cannot change the pore space, and thus the relative humidity in the pore, by more than a fraction of a percent.

In the aforementioned study of Feldman and Sereda [383], it was surmised that the applied compressive macrostress causes the walls of the nanopores to come into contact and stick by establishing new bonds. However, a quantitative analysis shows this to be impossible, for two reasons:

1. Because the nanopores with hindered adsorbed water and the interlayer spaces are not sparsely distributed but occupy a large fraction of any cross section through the cement gel, the relative change of thickness of the nanopores and interlayer spaces that can be produced by the applied macrostress is of the same order of magnitude as the macroscopic strains, i.e., of the order of 0.1%; but this value is too small for establishing new bonds.
2. Squeezing the opposite walls of nanopores together would cause much of the interlayer and hindered adsorbed water to be expelled from compressed concrete specimens, but this is contradicted by the experiments of Hansen [450, 451].

Since the concrete creeps, it is natural to expect the tensile microprestress S at the creep sites (the bridges across the nanopores) to relax with time. The relaxation reduces the rate of slip at the creep sites and also makes some of the creep sites inactive. Such phenomena reduce the macroscopic creep rate and are manifested as aging. This is the main idea that explains the long term aging. Obviously, the aging due to microprestress relaxation is not associated with the volume growth of the hydration products.

10.4   Generation and Relaxation of Microprestress

The macroscopic creep due to shear slip at the creep sites may be modeled by a viscous flow element serially coupled to the solidifying Kelvin chain, as shown in Fig. 10.2. The bonds across the slip plane in the flow element, representing a nanopore filled by hindered adsorbed water, are subjected to two stresses: the macroscopic applied stress σ causing shear slip, which acts in the figure horizontally, and the tensile microprestress S, which acts in the figure vertically. The rate of strain in the flow element is

$$\dot{\varepsilon}_f = \frac{\sigma}{\eta(S)} \tag{10.3}$$

where the effective viscosity η is a decreasing function of S.

The relaxation of the microprestress S, acting in Fig. 10.2 vertically, is imagined to be the result of another similar series coupling of a flow element and a spring of stiffness C_S, as shown in Fig. 10.2. Because the material is essentially isotropic, and because the phenomena that generate the microprestress have no preferred orientation, the value of S, which is actually the macroscopic characteristic of the average microprestress at all creep sites, must be the same for all directions (the only anisotropic phenomenon is the macroscopic stress and strain tensors, but they can have no significant effect on the disjoining pressure, as already pointed out). Therefore, the slip plane of the flow element governing the relaxation must also be subjected to normal stress S (horizontal in Fig. 10.2), and so the viscosity must also be $\eta(S)$. Hence, the equation governing the relaxation of the microprestress may be written as

$$\frac{\dot{S}}{C_S} + \frac{S}{\eta(S)} = \frac{\dot{s}}{C_S} \tag{10.4}$$

Here $\dot{s}$ denotes the rate of the instantaneous microprestress induced by changes of capillary tension, surface tension, and crystal growth pressure [131, 132].

To complete the model, we need to set up a suitable expression for $\eta(S)$. Because we do not know of any characteristic value of S separating ranges of qualitatively different behavior, the function $\eta(S)$ should be self-similar with respect to changes of the scale by which S is measured. As explained in Sect. 9.4, the only function satisfying this condition is a power function. So we set

$$\frac{1}{\eta(S)} = cpS^{p-1} \tag{10.5}$$

where c and p are positive constants, and $p > 1$ because $\eta(S)$ is a decreasing function.

The disjoining pressure and the microprestress S first develop during the initial hardening of concrete. Initially, the cement hydration is rapid, microprestress builds up mainly as a result of crystal growth pressures and very large localized shrinkage at locations close to the nanopores. Later, after the volume changes due to hydration have almost ceased, the microprestress changes are caused mainly by changes in the disjoining pressure, which respond almost instantly to the changes in the capillary tension and surface tension, and are governed by an equation of the same form as (10.1).

Therefore, in analogy to (10.2), we may assume that the rate of additionally generated microprestress is given by

$$\dot{s} = -c_1 \frac{\dot{h}}{h} \tag{10.6}$$

where c_1 is a constant. Note that (10.2), and thus also (10.6), holds at constant temperature. A generalization to variable temperature will be developed in Sect. 10.6.1. Equations (10.4)–(10.6) lead to the nonlinear differential equation

$$\dot{S} + c_0 S^p = -c_1 \frac{\dot{h}}{h} \tag{10.7}$$

governing the microprestress relaxation. For simplicity, we have denoted $C_S c p$ as c_0. Note that the microprestress S is independent of the applied macroscopic stress σ and strain ε. The microprestress relaxation law (10.7) combined with relation (10.5) replaces the explicit age dependence of viscosity. The long-term aging of flow viscosity is simply a consequence of relaxation of the microprestress.

In the limiting case of no drying nor wetting ($\dot{h} = 0$, sealed specimens, basic creep), we have

$$\dot{S} = -c_0 S^p \tag{10.8}$$

Another argument for this equation is offered in Bažant et al. [132], page 1192.

Equation (10.7) (or Eq. (10.8) as its special case) is a first-order differential equation governing the evolution of microprestress S. To make the solution unique, a suitable initial condition needs to be postulated.[4] It can be deduced from the requirement that, in the absence of drying, the evolution of viscosity evaluated from the microprestress theory according to (10.5) should be the same as in model B3, i.e., should be given by

$$\eta(t) = \frac{t}{q_4} \tag{10.9}$$

where q_4 is the parameter of model B3 that controls long-term creep. According to (10.5), the corresponding microprestress evolution is described by

$$S(t) = \left(\frac{q_4}{cpt} \right)^{1/(p-1)} \tag{10.10}$$

As long as there is no drying, microprestress can be computed directly from (10.10). The value

$$S_0 = \left(\frac{q_4}{cpt_0} \right)^{1/(p-1)} \tag{10.11}$$

---

[4]Strictly speaking, the initial condition for microprestress should be imposed at some fixed time, independent of the times at which drying starts or the load is applied, because the evolution of microprestress is considered as a phenomenon independent of the applied stress. Theoretically, the microprestress starts from a zero value at the set of concrete, then builds up to some maximum, and afterward relaxes. The very early stage of microprestress evolution is hard to describe accurately. Fortunately, for creep calculations we need to know the value of microprestress only from the time of load application.

determined from (10.10) at time t_0 (onset of drying) can be used in the initial condition for microprestress,

$$S(t_0) = S_0 \tag{10.12}$$

supplementing differential equation (10.7).

However, we need to make sure that the "assumed" microprestress evolution (10.10) is consistent with differential equation (10.8), which is the special form of (10.7) in the case of no drying. Evaluating the derivative

$$\dot{S}(t) = -\frac{1}{p-1} \left(\frac{q_4}{cp} \right)^{1/(p-1)} t^{-p/(p-1)} \tag{10.13}$$

and substituting it along with (10.10) into (10.8), it can be easily shown that the governing differential equation is satisfied at all times t if and only if

$$\frac{pc}{(p-1)c_0} = q_4 \tag{10.14}$$

This is a constraint that links parameters p, c, and c_0 of the microprestress theory to parameter q_4 of the B3 model. Since there exists an empirical formula for estimating q_4 from the composition, it is reasonable to consider q_4 as one of the primary model parameters for the microprestress theory as well, and take for instance parameter c as a dependent one, given by

$$c = \frac{p-1}{p} c_0 q_4 \tag{10.15}$$

In this way, the microprestress theory reuses parameter q_4 and introduces additional parameters p and c_0 (and also c_1, which appears in the full form of equation (10.7)). Using (10.15), parameter c can be eliminated, and Eqs. (10.5) and (10.10) can be rewritten as

$$\frac{1}{\eta(S)} = (p-1)c_0 q_4 S^{p-1} \tag{10.16}$$

$$S(t) = \left[\frac{1}{(p-1)c_0 t} \right]^{1/(p-1)} \tag{10.17}$$

Recall that the explicit formula (10.17) for microprestress evolution applies only to the case of no drying. Note that the evolution of viscosity calculated from (10.16) and (10.17) is given by $\eta(t) = t/q_4$ and is independent of parameters p and c_0 (and also of c_1, which is obvious). This indicates that parameters p, c_0, and c_1 have no effect on basic creep; they control the additional compliance due to drying creep.

Fitting of some test data showed that the value $p = 2$ works best [123, 132]. For this special choice, Eqs. (10.16) and (10.17) simplify to

$$\frac{1}{\eta(S)} = c_0 q_4 S \tag{10.18}$$

$$S(t) = \frac{1}{c_0 t} \tag{10.19}$$

It is also worth noting that if the pore pressure changes very fast ($\dot{h} \to \infty$), the term $c_0 S^p$ in (10.7) is negligible compared to the rate terms and the equation reduces to

$$\dot{S} = -c_1 \frac{\dot{h}}{h} \tag{10.20}$$

This integrates to $S = -c_1 \ln h + \text{const.}$, which is (and ought to be) of the same form as the change of chemical potential and surface tension based on thermodynamics (Kelvin and Laplace equations and ideal gas approximation for water vapor).

10.5 Unification of Microprestress and Solidification Models

Aside from the long-term aging due to relaxation of microprestress, there is, of course, the aging due to volume growth of hydration products, which dominates at early ages. From extensive studies of test data [104, 180] and theoretical analysis [295], it appears that this type of aging can be described by a solidifying Kelvin chain (Fig. 10.2), in which the relaxation times are constant and all the spring moduli and dashpot viscosities grow in proportion to the volume growth function $v(t)$. Such variation ensues from the volume growth of cement gel, which itself is characterized by a Kelvin chain model with age-independent properties; see Chap. 9.

In general, the microprestress might affect not only the viscosity of the flow term but also the viscosities of the dashpots in the Kelvin units of the solidifying chain (Fig. 10.2) and, vice versa, the volume growth might affect the viscosity of the flow term and thus complicate the mathematical formulation. Fortunately, analysis of the available test data shows that such a possibility need not be implemented in the model. In other words, the effects of microprestress and of volume growth can be assumed to be separated, acting on different parts of the chain model.

The compliance function of concrete according to the *microprestress-solidification theory* has the same general form

$$J(t, t') = q_1 H(t - t') + J_v(t, t') + J_f(t, t') \tag{10.21}$$

as in the solidification theory, and uses the same expressions for the first two terms, which characterize the asymptotic elastic compliance and the viscoelastic compliance affected by the volume growth of hydration products. The last term, $J_f(t, t')$, representing the contribution of long-term viscous flow, is, instead of the explicit logarithmic formula (9.15), given by

$$J_f(t, t') = \int_{t'}^{t} \frac{dr}{\eta(S(r))} = (p - 1)c_0 q_4 \int_{t'}^{t} S^{p-1}(r) \, dr \tag{10.22}$$

which follows from (10.3) and (10.16). The microprestress S is dependent on the evolution of relative pore humidity and must be solved from the differential equation (10.7).

As mentioned before, in the solidification theory it is more convenient to write the detailed expression for the compliance in terms of its rate [104, 179],

$$\dot{J}(t, t') = \frac{n(q_2 \lambda_0^m t^{-m} + q_3)}{(t - t') + \lambda_0^n (t - t')^{1-n}} + (p - 1)c_0 q_4 S^{p-1}(t) \tag{10.23}$$

in which q_2, q_3, m, n, and λ_0 are empirical constants. According to extensive data fitting, one can always take $\lambda_0 = 1$ day, $m = 0.5$, and $n = 0.1$. On the other hand, parameters q_2, q_3, and q_4 depend on the type of concrete. They can be predicted on the basis of compression strength of concrete and the composition of concrete mix using the formulae of the B3 or B4 models given in Appendices C and D, although it is preferable to determine at least q_2 on the basis of short-time creep tests of the given concrete using the updating procedure specified in Sect. 3.8.1; see also Bažant and Baweja [104].

10.6 Temperature and Humidity Effects

10.6.1 Effects on Creep

The effect of temperature on concrete creep is twofold, generated by two different mechanisms:

1. A temperature increase accelerates the bond breakages and restorations causing creep, and thus increases the creep rate.
2. The higher the temperature, the faster is the chemical process of cement hydration and thus the aging of concrete, which reduces the creep rate.

Usually, the former effect prevails, and then the overall effect of temperature rise is an increase of creep. Nevertheless, for very young concretes in which hydration progresses at elevated temperatures rapidly, heating can have the opposite effect.

At lower pore humidity, both the hydration and creep processes are slowed down, but this is often overpowered by transient creep acceleration due to the rate of change of humidity (or temperature). The decrease of creep of hardened cement pastes with decreasing moisture content was reported by Cilosani [316] and Ruetz [741, 742] and clearly demonstrated by the tests of Wittmann [870], Fig. 10.3. Cylinders of diameter 18 mm and height 60 mm were cured for 28 days at room temperature in sealed conditions and then oven-dried for 2 days at 105 °C. Afterward, they were partially resaturated for at least 3 months (until equilibrium was reached) at various constant humidities, ranging from 0 to 98%. The same environmental humidity was then preserved during the entire creep test. Since the strain measurements during early stages of loading exhibited scatter and were considered as unreliable, the reported values correspond to strain increments with respect to the state at 1h after loading. Therefore, the compliance values plotted in Fig. 10.3 are actually the increments $\Delta J(t, t') = J(t, t') - J(t' + \Delta t, t')$ with $\Delta t = 1$ h.

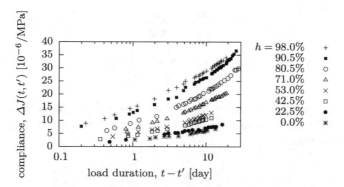

Fig. 10.3 Wittmann's data [870] on creep of hardened cement paste at various constant water contents (compliance at 1 h creep duration is subtracted from all values shown)

The effects of temperature and humidity on processes in the microstructure can be described by introducing three transformed time variables:

- the *equivalent age* t_e (equivalent hydration period, or "maturity"), which indirectly characterizes the degree of hydration,
- the *reduced time* t_r, characterizing the changes in the rate of bond breakages and restorations on the microstructural level, and
- the *reduced microprestress time* t_s.

Under standard conditions, i.e., at room temperature ($T_0 = 293$ K) and for sealed specimens ($h \approx 0.98$), all these times are by definition equal to the actual age of concrete t. Under higher temperatures, all processes are accelerated, which is taken into account by the acceleration of the transformed times with respect to the physical

time. Under lower humidity, all processes are slowed down. Under general temperature and humidity histories, equations describing the evolution of the microstructure
are written in terms of the transformed times.

The hydration process under standard conditions is characterized by a function
$v(t)$ that specifies the relative volume of hydration products at age t. Under general
conditions, t is replaced by the equivalent age t_e, and the first part of Eq. (9.4) is
rewritten as

$$\dot{\varepsilon}_v(t) = \frac{\dot{e}(t)}{v\left[t_e(t)\right]} \tag{10.24}$$

In the integral formula (9.3) describing creep of the solidifying constituent, the load
duration is expressed in terms of the reduced time t_r instead of the physical time t,
and the formula is written as

$$e(t) = \int_0^t \Phi\left[t_r(t) - t_r(t')\right]\, d\sigma(t') \tag{10.25}$$

In a similar spirit, Eqs. (10.3) and (10.4) are generalized to

$$\frac{d\varepsilon_f}{dt_r} = \frac{\sigma}{\eta(S)} \tag{10.26}$$

$$\frac{1}{C_S}\frac{dS}{dt_s} + \frac{S}{\eta(S)} = \frac{1}{C_S}\frac{ds}{dt_s} \tag{10.27}$$

The rates at which the transformed times evolve are defined as products of two
functions, which respectively characterize the effects of temperature and of humidity
[86, 96]. Therefore, we write

$$\frac{dt_e}{dt} = \beta_{eT}(T)\,\beta_{eh}(h) \tag{10.28}$$

$$\frac{dt_r}{dt} = \beta_{rT}(T)\,\beta_{rh}(h) \tag{10.29}$$

$$\frac{dt_s}{dt} = \beta_{sT}(T)\,\beta_{sh}(h) \tag{10.30}$$

Functions describing the influence of temperature have the form motivated by the
rate process theory,

$$\beta_{eT}(T) = \exp\left[\frac{Q_e}{R}\left(\frac{1}{T_0} - \frac{1}{T}\right)\right] \tag{10.31}$$

$$\beta_{rT}(T) = \exp\left[\frac{Q_r}{R}\left(\frac{1}{T_0} - \frac{1}{T}\right)\right] \tag{10.32}$$

$$\beta_{sT}(T) = \exp\left[\frac{Q_s}{R}\left(\frac{1}{T_0} - \frac{1}{T}\right)\right] \tag{10.33}$$

where T is the absolute temperature, R is the universal gas constant,[5] and Q_e, Q_r, and Q_s are *activation energies* for the hydration, viscous processes, and microprestress relaxation, respectively. In previous works, based on the B3 model described in Bažant [96], the ratios of activation energies to the universal gas constant varied in the range from 2700 to 5000 K. Model B4 uses 4000 K for all these ratios.

Functions describing the influence of humidity have been postulated in the form

$$\beta_{eh}(h) = \frac{1}{1 + [\alpha_e(1 - h)]^4} \tag{10.34}$$

$$\beta_{rh}(h) = \alpha_r + (1 - \alpha_r)h^2 \tag{10.35}$$

$$\beta_{sh}(h) = \alpha_s + (1 - \alpha_s)h^2 \tag{10.36}$$

Analysis of experimental test data shows that the value of parameter α_e is in the order of 10, while α_r and α_s are in the order of 0.1 [96]. The graphs of functions β_{eh} and β_{rh} (or β_{sh}) for typical parameter values are plotted in Fig. 10.4. Note that, at room temperature and 100% humidity, all factors β in (10.31)–(10.36) have unit values.

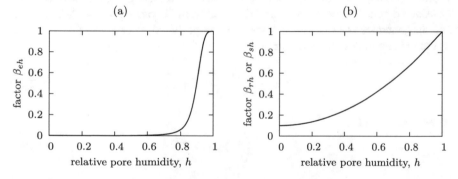

Fig. 10.4 Dependence of factors (a) β_{eh}, (b) β_{rh} or β_{sh} on relative pore humidity

For a given history of temperature and relative pore humidity, we can introduce functions

$$\psi_e(t) = \beta_{eT}(T(t))\,\beta_{eh}(h(t)) \tag{10.37}$$

$$\psi_r(t) = \beta_{rT}(T(t))\,\beta_{rh}(h(t)) \tag{10.38}$$

$$\psi_s(t) = \beta_{sT}(T(t))\,\beta_{sh}(h(t)) \tag{10.39}$$

and rewrite (10.28)–(10.30) as $dt_e = \psi_e(t)dt$, $dt_r = \psi_r(t)dt$, and $dt_s = \psi_s(t)dt$. Equations (10.26) and (10.27) then become

---

[5]In some papers, the Boltzmann constant k_B is used instead of the universal gas constant R. Both approaches are equivalent, and they differ only by the meaning of activation energy Q, which is taken per mole if R is used, or per elementary entity (atom or molecule) if k_B is used.

$$\dot{\varepsilon}_{\mathrm{f}} = \frac{\psi_r \sigma}{\eta(S)} \tag{10.40}$$

$$\frac{\dot{S}}{C_S} + \frac{\psi_s S}{\eta(S)} = \frac{\dot{s}}{C_S} \tag{10.41}$$

Substituting (10.5) into (10.41) and recalling that $C_S cp$ has been denoted as c_0, we obtain

$$\dot{S} + \psi_s c_0\, S^p = \dot{s} \tag{10.42}$$

which is the generalized version of (10.7).

In Eq. (10.42), the right-hand side term $\dot{s}$ depends on capillary tension, surface tension, crystal growth pressure, and disjoining pressure, all of which are sensitive to temperature and humidity changes. As already explained, all the phases of water can be assumed to be in local thermodynamic equilibrium. Under this assumption, it follows from Kelvin's equation (8.25), Eq. (8.62) for disjoining pressure and similar equations that can be derived from the equality of chemical potentials at thermodynamic equilibrium (as summarized in [78] and [80]) that the dependence of all the aforementioned quantities on T and h has the general form (10.1). In Sect. 10.4, we considered constant temperature and deduced from the rate equation (10.2) that the rate of generated microprestress should be proportional to $\dot{h}/h$; see (10.6). Under variable temperature, the rate form of (10.1) is

$$\dot{\gamma}_a = -\frac{C_\gamma R}{M_w} \frac{\mathrm{d}}{\mathrm{d}t} (T \ln h) \tag{10.43}$$

Therefore, we set

$$\dot{s} = -k_1 \left(\dot{T} \, \ln h + T \frac{\dot{h}}{h} \right) \tag{10.44}$$

where k_1 is a model parameter that corresponds to c_1 in Eq. (10.6) divided by the room temperature T_0. Upon substitution of (10.44) into (10.42), the governing equation of the microprestress at simultaneous humidity and temperature variation becomes

$$\dot{S} + \psi_s c_0\, S^p = -k_1 \left(\dot{T} \, \ln h + T \frac{\dot{h}}{h} \right) \tag{10.45}$$

Equation (10.45) gives results that depend on the sign of $\dot{T}$ and $\dot{h}$. In particular, negative increments of h (drying) and positive increments of T (heating) lead to an increase of the microprestress (i.e., of the magnitude of stress peaks in the microstructure), and thus to an increase of the slip rate and the rate of creep flow, counteracting (and usually overpowering) the effect of β_{rh}.

When drying switches to wetting, or heating to cooling, i.e., $\dot{h}$ or $\dot{T}$ changes its sign, the right-hand side of (10.45) becomes negative and the microprestress relaxation is accelerated, which must reduce the creep rate at least temporarily. After a while, this switch must deactivate the current creep sites, i.e., the sites of the highest microprestress at which shear slip must take place. However, new creep sites will form at other locations in the microstructure which were previously unstressed. At the new sites, creep will again be promoted by the change in $\dot{h}$ or $\dot{T}$. To model such effects of the reversal of $\dot{h}$ or $\dot{T}$ in a simplified manner, Cusatis [331] and Bažant et.al. [123] suggested introducing the absolute value sign into (10.45), making the right-hand side always nonnegative:

$$\dot{S} + \psi_s c_0 S^p = k_1 \left| \dot{T} \ln h + T \frac{\dot{h}}{h} \right| \tag{10.46}$$

Although this simple formulation does not capture the expected temporary decrease of creep rate right after a change of sign of $\dot{h}$ or $\dot{T}$, it does reflect in the simplest manner the overall increase of creep rate likely to occur with some delay. Experimental data [403, 690] indeed indicate an increase of the creep rate for both drying and wetting. As for heating and cooling, the evidence is ambiguous; some data [378, 848] suggest creep rate increase under cooling but others [452, 504] suggest the opposite, and so no experimental validation is possible. However, the similarity of cooling with wetting suggests that cooling should also enhance creep.

10.6.2   Hygrometric and Thermal Strains

Changes of the disjoining pressure, capillary tension, and surface tension must also cause elastic deformation of the solid skeleton of cement gel, which represents the free shrinkage, i.e., the shrinkage that occurs in infinitesimal elements of a continuum at no macroscopic stress. By fitting of many experimental data, it transpires that the rate of *free shrinkage strain* is approximately proportional to the rate of humidity, i.e.,

$$\dot{\varepsilon}_{sh} = k_{sh}^* \dot{h} \tag{10.47}$$

where k_{sh}^* is an empirical shrinkage ratio, which can be taken approximately as constant (see also Sect. 8.6)).

The additional terms due to the so-called *stress-induced shrinkage*, which were introduced by Bažant and Chern [117, 120] and are described in Sect. 13.3.3.2, do not belong into the present expression for the shrinkage. They are now taken care of by the dependence of viscosity on the history of humidity. So it transpires that the microprestress theory does not represent a negation but a refinement of the previous formulations [117, 120, 201]. Likewise, the present theory may be regarded as an extension of the theory for the effect of hindered adsorption and disjoining pressure on drying creep [78, 80, 164].

Temperature changes cause the *thermal strain* ε_T, proportional to the change of temperature:

$$\dot{\varepsilon}_T = \alpha_T \dot{T} \tag{10.48}$$

In reality, the *coefficient of thermal expansion* α_T depends on T, but only weakly and, as an approximation, for moderate temperatures can be assumed to be constant. Coefficient α_T of concrete can be estimated from the coefficients of thermal expansion of the constituents, cement paste and aggregate, and from their volume fractions and elastic properties. The coefficient of thermal expansion of cement paste depends on many factors. Although there is no significant difference between α_T for saturated cement paste and α_T for dry cement paste, measurements show α_T to have higher values for partial saturation. The hydration process tends to diminish α_T. Typical values of α_T for hardened cement paste are between 11 and $22 \times 10^{-6}/°C$ [142, 628, 653]. However, since the cement paste occupies a much smaller relative volume than the aggregates, the coefficient of thermal expansion of concrete is governed chiefly by the aggregates, for which α_T depends on their chemical composition and varies from $4 \times 10^{-6}/°C$ for limestone to almost $12 \times 10^{-6}/°C$ for quartzite [247]. The resulting thermal expansion coefficient of concrete is between 6 and $13 \times 10^{-6}/°C$, with $\alpha_T = 10 \times 10^{-6}/°C$ as the typical value; see Table 8.4 in Neville [653] for more details.

10.7   Alternative Computational Approach: Viscosity Evolution Equation

The concept of microprestress appears to be essential for the theoretical and physical justification of evolving viscosity and of the general format of governing equations. On the other hand, the microprestress cannot be directly measured, and a separate calibration of the microprestress relaxation equation (10.46) and of Eq. (10.16) describing the dependence of viscosity on microprestress is difficult, if not impossible. It turns out that the microprestress can be completely eliminated, and the governing equation can be reformulated in terms of viscosity [523]. Although the elimination of microprestress severs the model from its physical basis, the resulting model is still fully equivalent to the original one while its structure is simplified, the fitting of model parameters becomes more transparent and the number of relevant parameters gets reduced.

From (10.16), we can express the microprestress in terms of the viscosity as

$$S = [(p - 1)c_0 q_4 \eta]^{-1/(p-1)} \tag{10.49}$$

and differentiation with respect to time leads to

$$\dot{S} = -(c_0 q_4)^{-1/(p-1)}[(p - 1)\eta]^{-p/(p-1)}\dot{\eta} \tag{10.50}$$

Substituting (10.49) and (10.50) into (10.46), we obtain, after some rearrangements,

$$\dot{\eta} + k_1[c_0 q_4 (p-1)^p]^{1/(p-1)} \left| \dot{T} \ln h + T \frac{\dot{h}}{h} \right| \eta^{p/(p-1)} = \frac{\psi_s}{q_4} \tag{10.51}$$

For the standard value of parameter $p = 2$, the resulting equation is

$$\dot{\eta} + k_1 c_0 q_4 \left| \dot{T} \ln h + T \frac{\dot{h}}{h} \right| \eta^2 = \frac{\psi_s}{q_4} \tag{10.52}$$

This is a nonlinear first-order differential equation for viscosity. It can be solved directly, without resorting to the microprestress, which provides some degree of computational simplification. At constant humidity and temperature, the second term on the left-hand side vanishes and the equation simplifies to

$$\dot{\eta} = \frac{\psi_s}{q_4} \tag{10.53}$$

At variable temperature or humidity, the second term on the right-hand side of (10.52) slows down the growth of viscosity and, if sufficiently large, can even lead to a temporary reduction of viscosity.

Reformulation of the governing differential equation in terms of viscosity clarifies the role of individual parameters. Parameter p fully determines the exponent $p/(p-1)$ at η in the nonlinear term. Parameters k_1 and c_0 appear only in the expression for the multiplicative factor in the nonlinear term, and they can be replaced by one single parameter. Indeed, the values of k_1 and c_0 do not need to be known separately—what matters is only the value of $k_1 c_0^{1/(p-1)}$. For the standard choice $p = 2$, this means that only the product $k_1 c_0$ matters. However, the evolution of microprestress changes if k_1 and c_0 are varied at constant product $k_1 c_0$, which renders the physical basis ambiguous. To avoid this ambiguity, future research should aim to improve the physical basis, which is ultimately rooted in the relaxation of microprestress.

Since the microprestress is not directly measurable, parameters k_1 and c_0 cannot be determined separately. Thus, optimal fitting based on macroscopically measured variables would require choosing (arbitrarily) either k_1 or c_0 [472]. To avoid this choice, k_1 and c_0 may be replaced by a single parameter which can be uniquely determined. For instance, one could use $k_1 c_0^{1/(p-1)}$ as the new single parameter. To avoid units with noninteger exponents (in the general case of p different from 2) and to obtain a parameter with intuitive meaning, it is suggested to introduce a parameter μ_s with the dimension of fluidity (reciprocal value of viscosity) and to cast Eq. (10.51) in the form

$$\dot{\eta} + \frac{1}{\mu_s T_0} \left| \dot{T} \ln h + T \frac{\dot{h}}{h} \right| (\mu_s \eta)^{p/(p-1)} = \frac{\psi_s}{q_4} \tag{10.54}$$

where T_0 is the room temperature.

A comparison of the second terms in (10.51) and (10.54) reveals that

$$\mu_S = c_0 T_0^{p-1} k_1^{p-1} q_4 (p-1)^p \tag{10.55}$$

This relation provides a link between the original parameters k_1 and c_0 and the new parameter μ_S. Note that, at constant room temperature, $k_1 T_0$ in (10.44) corresponds to c_1 in (10.6), and so relation (10.55) can be rewritten as

$$\mu_S = c_0 c_1^{p-1} q_4 (p-1)^p \tag{10.56}$$

For the standard choice $p = 2$, Eqs. (10.54)–(10.56) simplify to

$$\dot{\eta} + \frac{\mu_S}{T_0} \left| \dot{T} \, \ln h + T \, \frac{\dot{h}}{h} \right| \eta^2 = \frac{\psi_s}{q_4} \tag{10.57}$$

$$\mu_S = c_0 T_0 k_1 q_4 = c_0 c_1 q_4 \tag{10.58}$$

The initial condition for viscosity could be "derived" from the initial condition for microprestress combined with relation (10.16). However, recall that initial condition (10.12) with the initial value S_0 given by (10.11) was actually derived from the assumption that, in the absence of drying and at room temperature, the evolution of viscosity should be the same as in models B3 and B4, i.e., should be given by (10.9). Therefore, the initial condition for viscosity can be directly postulated as

$$\eta(t_0) = \frac{t_0}{q_4} \tag{10.59}$$

Recent comparisons with a few selected data, especially those of Bryant and Vadhanavikkit [278], suggested that the microprestress solidification theory might be giving an excessive delay of drying creep behind shrinkage, reversed size effect on drying creep [133], and excessive creep under multiple temperature cycles [523]. Ongoing research aims at explaining or reducing these discrepancies by (i) considering a spectrum of microprestress relaxation times, (ii) separating the shrinkage data into drying and autogenous, (iii) fitting of shrinkage and drying creep data for specimens of different sizes, (iv) considering nanoscale volume expansion due to long-term hydration reaction [125], and (v) considering more test data from the new worldwide NU database [488].

On-going research, partly inspired by MD simulations, indicates that the theory can be improved by distinguishing nano- and macro-viscosities and considering the latter to depend on the microprestress and on the pore humidity rate. This improvement can eliminate the reversed size effect on drying creep, the excessive delay of drying creep after shrinkage and the excessive creep at temperature cycles.

10.8 Numerical Implementation

For the sake of step-by-step finite element analysis, an incremental stress–strain relation needs to be formulated. Recall that the complete microprestress-solidification model can be considered as a serial coupling of an elastic spring, a solidifying Kelvin chain, a dashpot with viscosity depending on the microprestress, and additional units representing the strains due to shrinkage and thermal expansion; see Fig. 10.2. Numerical treatment of a solidifying Kelvin chain coupled in series with an aging dashpot has already been discussed in Sect. 5.2.6. Certain adjustments are needed for the present model, because of the following two differences:

1. Viscosity of the aging dashpot (representing long-term viscous flow) is not an explicit function of age but is governed by the differential evolution equation (10.54).
2. The effect of temperature and humidity on the rate of various processes in the microstructure is reflected by three transformed times, which replace the standard physical time in the governing equations.

10.8.1 Evaluation of Flow Viscosity

In this subsection, the numerical algorithm is formulated on the basis of the viscosity evolution equation (10.54) with initial condition (10.59), as suggested by Jirásek and Havlásek [523]. A similar algorithm can be formulated in terms of microprestress [131]. That approach will be inevitable for the expected refinement of the micropre-stress concept, in which the viscosity would depend on additional variables.

Suppose that the value of viscosity η_k at time t_k is known from the previous time step or from the initial condition, and we want to obtain an approximation of the value $\eta_{k+1} = \eta_k + \Delta\eta_k$ at time $t_{k+1} = t_k + \Delta t_k$. If the temperature and humidity are constant within the current step, the governing equation reduces to (10.53) and the increment

$$\Delta\eta_k = \frac{\psi_s}{q_4}\Delta t_k \tag{10.60}$$

can be evaluated exactly. At variable temperature or humidity, the coefficient multiplying $\eta^{p/(p-1)}$ in the second term of (10.54) can be, within one time step, approximated by a constant. Since T_0 and μ_S are constants, only the time derivative of $T \ln h$ (i.e., the expression in the absolute value) needs to be approximated. Replacing time derivatives by finite differences and approximating η by the average of η_k and η_{k+1}, we rewrite (10.54) as

$$\frac{\Delta\eta_k}{\Delta t_k} + \frac{|\Delta(T \ln h)_k|}{T_0 \Delta t_k}\mu_S^{1/(p-1)}(\eta_k + 0.5\Delta\eta_k)^{p/(p-1)} = \frac{\psi_s}{q_4} \tag{10.61}$$

Here, $\Delta(T \ln h)_k$ denotes the increment of the product $T \ln h$ over the time step number k, which could be written more explicitly as

$$\Delta(T \ln h)_k = T_{k+1} \ln h_{k+1} - T_k \ln h_k \tag{10.62}$$

The evolution of temperature and pore humidity is evaluated separately (from the corresponding equations for heat and moisture transport) and is affected neither by microprestress, nor by viscosity. Therefore, T_{k+1} and h_{k+1} can be considered as known quantities. The only unknown in (10.61) is thus the increment of viscosity, $\Delta\eta_k$. The equation is nonlinear, because of the exponent $p/(p-1)$ in the second term. The solution can be computed iteratively, e.g., by the Newton method.

Interestingly, for the standard value of $p = 2$, Eq. (10.61) is quadratic and thus can be solved in closed form. But an even better approach is to exploit the exact solution of the original differential equation (10.57) with the time derivative of $T \ln h$ replaced by a constant. Indeed, for $p = 2$ and with $d(T \ln h)/dt$ replaced by $\Delta(T \ln h)_k/\Delta t_k$, Eq. (10.57) has the form

$$\dot{\eta} + A^2 \eta^2 = B^2 \tag{10.63}$$

where

$$A = \sqrt{\frac{\mu_S |\Delta(T \ln h)_k|}{\Delta t_k T_0}} \tag{10.64}$$

$$B = \sqrt{\frac{\psi_s}{q_4}} \tag{10.65}$$

are auxiliary constants introduced for convenience. Differential equation (10.63) is still nonlinear, but its analytical solution can be found by separation of variables[6]:

$$\int_{\eta_k}^{\eta_{k+1}} \frac{d\eta}{B^2 - A^2 \eta^2} = \int_{t_k}^{t_{k+1}} dt \tag{10.66}$$

$$\frac{1}{2AB} \left[\ln(B + A\eta) - \ln|B - A\eta| \right]_{\eta_k}^{\eta_{k+1}} = t_{k+1} - t_k \tag{10.67}$$

$$\frac{B + A\eta_{k+1}}{B + A\eta_k} \frac{B - A\eta_k}{B - A\eta_{k+1}} = e^{2AB\Delta t_k} \tag{10.68}$$

Note that in the last step, we have omitted the absolute value signs around $B - A\eta_k$ and $B - A\eta_{k+1}$. The reason is that these terms always have the same sign (both positive, or both negative), and so their ratio is always positive.

Equation (10.68) can be transformed into a linear equation in terms of the viscosity at the end of the step, η_{k+1}, and its solution reads

$$\eta_{k+1} = \frac{B}{A} \frac{B(1 - \tilde{e}) + A\eta_k(1 + \tilde{e})}{B(1 + \tilde{e}) + A\eta_k(1 - \tilde{e})}, \text{ with } \tilde{e} = e^{-2AB\Delta t_k} \tag{10.69}$$

---

[6]Note that if parameter p is different from its standard value 2, the solution cannot be constructed analytically because a closed-form expression for the integral of $1/(B^2 - A^2 \eta^{p/(p-1)})$ is not available for general p.

The advantage compared to the standard finite difference scheme leading to Eq. (10.61) is that if the time derivative of $T \ln h$ is considered constant within the time step, the solution is exact for an arbitrarily large Δt_k. Therefore, one can expect improved accuracy for large time steps even in cases when $T \ln h$ varies in a more general fashion. This idea is an extension of the exponential algorithm to the case of a nonlinear differential equation.

Formula (10.69) gives an explicit expression for updating of viscosity. It is useful to reformulate it such that it would remain applicable in the degenerate case when the second term in (10.54) vanishes, i.e., when the coefficient A in (10.63) is zero. Since A appears in the denominator of the first fraction in (10.69), the formula cannot be applied directly. Of course, one could treat this special case separately, since the exact solution is known to have the simple form (10.60). However, if A is not exactly zero but is very small, the full formula needs to be used and numerical problems may arise due to the truncation error, because the numerator of the second fraction in (10.69) is in such a case small as well. Fortunately, the entire expression on the right-hand side tends to a finite limit corresponding to (10.60). To make it more obvious, we deduce from (10.69) the expression for the viscosity increment (simply by subtracting η_k from both sides) and then transform the resulting formula into

$$\Delta \eta_k = \frac{1 - \tilde{e}}{A} \frac{B^2 - A^2 \eta_k^2}{B(1 + \tilde{e}) + A\eta_k(1 - \tilde{e})} \tag{10.70}$$

If A tends to zero, the first fraction in (10.70) approaches $2B\Delta t_k$, as can be shown by expanding $\tilde{e}$ into Taylor series:

$$\tilde{e} = e^{-2AB\Delta t_k} = 1 - 2AB\Delta t_k + 2A^2 B^2 (\Delta t_k)^2 - \dots \tag{10.71}$$

$$\frac{1 - \tilde{e}}{A} = 2B\Delta t_k - 2AB^2 (\Delta t_k)^2 + \dots \tag{10.72}$$

Therefore, if A is very small (compared to $1/(B\Delta t_k)$), one can replace the first fraction in (10.70) by $2B\Delta t_k(1 - AB\Delta t_k)$, which properly tends to $2B\Delta t_k$ as A approaches zero. The second fraction in (10.70) tends to $B/2$ (because $\tilde{e}$ tends to 1), and so the increment $\Delta \eta_k$ computed from (10.70) for vanishing A is equal to $B^2 \Delta t_k$, which is the correct value implied by (10.60) and (10.65).

Formula (10.70) also properly reflects another special case when η_k happens to be equal to B/A and the rate $\dot{\eta}$ evaluated from (10.63) vanishes. In this case, the viscosity remains constant during the entire step. Indeed, the numerator of the second fraction in (10.70) vanishes for $\eta_k = B/A$, and the increment of viscosity is correctly evaluated as zero.

10.8.2 Evaluation of Flow Strain Increment

Once the increment of viscosity is known, the viscosity can be updated and the increment of viscous strain can be evaluated by numerical integration of Eq. (10.40):

$$\eta_{k+1} = \eta_k + \Delta\eta_k \tag{10.73}$$

$$\Delta\varepsilon_{f,k} = \int_{t_k}^{t_{k+1}} \frac{\psi_r(t)\sigma(t)}{\eta(t)}\, dt \approx \frac{\Delta t_k}{2}\left(\frac{\psi_{r,k}\sigma_k}{\eta_k} + \frac{\psi_{r,k+1}\sigma_{k+1}}{\eta_{k+1}}\right) \tag{10.74}$$

Formula (10.74) is based on numerical integration by the trapezoidal rule, which is exact if the integrand $\psi_r(t)\sigma(t)/\eta(t)$ is a linear function of time. As a minor improvement, one can use a rule that gives the exact result for $\sigma(t)$ and $\eta(t)/\psi_r(t)$ considered separately as two linear functions of time (this covers the case of basic creep or creep at constant temperature and humidity, when $\sigma(t) = \text{const.}$, $\psi_r(t) = \text{const.}$, and $\eta(t) = \eta_k + (\psi_s/q_4)(t - t_k)$). To simplify the notation, we introduce the modified viscosity

$$\tilde{\eta} = \frac{\eta}{\psi_r} \tag{10.75}$$

The flow strain increment is then approximated as

$$\Delta\varepsilon_{f,k} \approx \int_{t_k}^{t_{k+1}} \frac{\sigma_k + (t - t_k)\Delta\sigma_k/\Delta t_k}{\tilde{\eta}_k + (t - t_k)\Delta\tilde{\eta}_k/\Delta t_k}\, dt =$$

$$= \frac{\Delta t_k}{\Delta\tilde{\eta}_k}\left[\left(\sigma_k - \Delta\sigma_k\frac{\tilde{\eta}_k}{\Delta\tilde{\eta}_k}\right)\ln\left(1 + \frac{\Delta\tilde{\eta}_k}{\tilde{\eta}_k}\right) + \Delta\sigma_k\right] \tag{10.76}$$

Due to the presence of $\Delta\tilde{\eta}_k$ in the denominator, the formula fails for constant modified viscosity and may lead to a high error for a very small increment of modified viscosity. Therefore, if $|\Delta\tilde{\eta}_k| \ll \tilde{\eta}_k$, it is preferable to use an alternative formula

$$\Delta\varepsilon_{f,k} = \frac{\Delta t_k}{\tilde{\eta}_k}\left[\sigma_k\left(1 - \frac{\Delta\tilde{\eta}_k}{2\tilde{\eta}_k}\right) + \Delta\sigma_k\left(\frac{1}{2} - \frac{\Delta\tilde{\eta}_k}{3\tilde{\eta}_k}\right)\right] \tag{10.77}$$

constructed from (10.76) by expanding the logarithmic function into truncated Taylor series.

Formulae (10.76) and (10.77) have the general form

$$\Delta\varepsilon_{f,k} = \Delta\varepsilon''_{f,k} + \bar{C}_{f,k}\Delta\sigma_k \tag{10.78}$$

where $\Delta\varepsilon''_{f,k}$ represents the increment of flow strain that would occur under constant stress and $\bar{C}_{f,k}$ is the incremental flow compliance. For instance, according to (10.76) we have

$$\Delta\varepsilon''_{f,k} = \frac{\Delta t_k}{\Delta\tilde{\eta}_k}\sigma_k\ln\left(1 + \frac{\Delta\tilde{\eta}_k}{\tilde{\eta}_k}\right) \tag{10.79}$$

$$\bar{C}_{f,k} = \frac{\Delta t_k}{\Delta\tilde{\eta}_k}\left[1 - \frac{\tilde{\eta}_k}{\Delta\tilde{\eta}_k}\ln\left(1 + \frac{\Delta\tilde{\eta}_k}{\tilde{\eta}_k}\right)\right] \tag{10.80}$$

and analogous expressions following from (10.77) are obvious. Due to serial coupling of the viscous flow unit with the other model units (Fig. 10.2), the flow strain increment $\Delta\varepsilon''_{f,k}$ represents an additive contribution to the overall creep increment $\Delta\varepsilon''_k$, and the incremental flow compliance $\bar{C}_{f,k}$ represents an additive contribution to the overall incremental compliance (which is the reciprocal value of the incremental modulus $\bar{E}_k$). These contributions will be included in formulae (10.113) and (10.114).

10.8.3 Incorporation of Transformed Times

Having covered the numerical treatment of microprestress relaxation (by means of the viscosity evolution equation) and flow strain increment, we can turn attention to the modifications due to temperature and humidity effects, reflected by the three transformed times which were introduced in Sect. 10.6.1.

The reduced microprestress time t_s has already been incorporated through parameter ψ_s into Eq. (10.42) governing the evolution of microprestress, which is reflected in the viscosity evolution equation (10.54) and, through parameter B defined in (10.65), in Eqs. (10.69) and (10.70) for the numerical treatment of viscosity evolution. Recall that ψ_s depends on current temperature and relative pore humidity and is given by (10.33), (10.36), and (10.39).

The reduced time t_r affects the creep rate according to (10.26) and is incorporated through parameter ψ_r into Eq. (10.40) for the rate of the viscous flow, which is reflected in (10.74) and, through the modified viscosity (10.75), in (10.76) and (10.77).

The reduced time t_r also affects the growth of viscoelastic strain in the solidifying material. It is incorporated into the integral formula (10.25) for creep of the solidifying constituent, which is more efficiently handled in the differential format, with the microcompliance function Φ of the solidifying constituent approximated by the Dirichlet series (5.72) corresponding to a nonaging Kelvin chain. Under standard conditions, the differential equation governing a typical Kelvin unit number μ has the same form as (5.44),

$$\frac{\sigma_{v\mu}}{\tau_\mu} + \frac{d\sigma_{v\mu}}{dt} = \frac{d\sigma}{dt} \tag{10.81}$$

where $\sigma_{v\mu}$ is the viscous stress in the unit and τ_μ is the retardation time. Under general conditions, with variable temperature and humidity, the derivatives should be taken with respect to the reduced time, t_r. Recall that $dt_r = \psi_r dt$, where the factor ψ_r depends on the current temperature and relative pore humidity according to (10.32), (10.35), and (10.38). Numerical treatment of the modified form of Eq. (10.81) can follow the procedure valid under standard conditions (see Sect. 5.2.6), in which the increment Δt_k of real physical time is replaced by the reduced time increment

$$\Delta t_{r,k} = \int_{t_k}^{t_{k+1}} \psi_r(t)\, dt \approx \psi_r(t_{k+1/2}) \Delta t_k \tag{10.82}$$

In this approximation, ψ_r is evaluated at the midpoint

$$t_{k+1/2} = t_k + \frac{1}{2}\Delta t_k \tag{10.83}$$

The reduced time increment $\Delta t_{r,k}$ replaces Δt_k in expressions (5.63) for factors $\beta_{k\mu}$ and $\lambda_{k\mu}$.

The last modification of the standard algorithm is related to the equivalent age t_e, which controls the solidification process. The values of equivalent age $t_{e,k} = t_e(t_k)$ are obtained by cumulating the increments

$$\Delta t_{e,k} = \int_{t_k}^{t_{k+1}} \psi_e(t)\, dt \approx \psi_e(t_{k+1/2})\Delta t_k \tag{10.84}$$

The relative volume of solidified material, v, is then computed from the equivalent age instead of the real physical age. The average value of v in step number k is estimated according to the midpoint rule as

$$v_{k+1/2} = v(t_{e,k} + \frac{1}{2}\Delta t_{e,k}) \tag{10.85}$$

10.8.4 Incremental Stress Evaluation Algorithm

The numerical procedure for incremental stress–strain evaluation based on the microprestress-solidification theory, with the effects of temperature and humidity taken fully into account, can be conceived as an extended version of Algorithm 5.3.

Before the actual incremental analysis, parameters E_μ^∞ and τ_μ ($\mu = 1, 2, \ldots M$) of Dirichlet series approximating the nonaging microcompliance function Φ need to be determined. This can be conveniently done using the technique based on continuous retardation spectrum, as explained in Appendix F. Since Φ is typically given by the log-power law (9.13), its retardation spectrum can be approximated by the formulae given in Section F.2. The corresponding volume growth function v is given by (9.16). With this choice, the resulting basic creep compliance function is the same as in model B3 or B4.

To take into account the effects of temperature and humidity, the evolution of T and h must be known. In some cases, it can be prescribed or estimated, but in general it is determined by a separate analysis of heat transfer and moisture diffusion, which is either performed before the mechanical analysis, or runs concurrently. The values of temperature and relative pore humidity at Gauss points of the finite element mesh used by the mechanical analysis can be obtained by standard spatial interpolation. If the analyses use different time steps, temporal interpolation is needed as well. In this way, the temperature and humidity values such as $T_{k+1/2}$, T_{k+1}, $h_{k+1/2}$, and h_{k+1} are obtained; they enter the stress evaluation algorithm as known input values.

The mechanical analysis starts at time t_1, which is taken as the time of first loading in the generalized sense—it can be the time when the self-weight or other forces are applied on the structure, or time at the onset of drying, or time when temperature starts deviating from the room temperature, whichever comes first.[7] At time t_1, we set the initial values

$$\eta_1 = \frac{t_1}{q_4} \tag{10.86}$$

$$\sigma_{v\mu,1} = 0, \; \mu = 1, 2 \ldots M \tag{10.87}$$

$$t_{e,1} = t_1 \tag{10.88}$$

$$\psi_{r,1} = \beta_{rh}(h_1)\,\beta_{rT}(T_1) \tag{10.89}$$

$$\tilde{\eta}_1 = \frac{\eta_1}{\psi_{r,1}} \tag{10.90}$$

$$k = 1 \tag{10.91}$$

where h_1 and T_1 are the relative pore humidity and temperature at time t_1. Typically, $h_1 = 1$ and $T_1 = T_0 = $ room temperature, in which case $\psi_{r,1} = 1$ and $\tilde{\eta}_1 = \eta_1$.

Algorithm 10.1 Incremental stress–strain relation according to the micro-prestress-solidification theory

For a given time step from t_k to $t_{k+1} = t_k + \Delta t_k$ and for given history of temperature and relative pore humidity:

1. Using formulae (10.31)–(10.36), evaluate factors

$$\psi_{e,k+1/2} = \beta_{eh}(h_{k+1/2})\,\beta_{eT}(T_{k+1/2}) \tag{10.92}$$

$$\psi_{r,k+1} = \beta_{rh}(h_{k+1})\,\beta_{rT}(T_{k+1}) \tag{10.93}$$

$$\psi_{s,k+1/2} = \beta_{sh}(h_{k+1/2})\,\beta_{sT}(T_{k+1/2}) \tag{10.94}$$

2. Update the equivalent age and evaluate the volume growth function v given by (9.16) at midpoint:

$$\Delta t_{e,k} = \psi_{e,k+1/2}\Delta t_k \tag{10.95}$$

$$t_{e,k+1} = t_e + \Delta t_{e,k} \tag{10.96}$$

$$v_{k+1/2} = v\left(t_{e,k} + \frac{1}{2}\Delta t_{e,k}\right) \tag{10.97}$$

---

[7] Strictly speaking, the self-weight is applied immediately after casting (on the structure supported by the formwork), and the initial temperature is usually different from the room temperature and very soon starts increasing due to the production of hydration heat. If such effects are studied in detail, the simulation would need to start at a very early age, and its purpose would be to determine the evolution of internal stresses due to nonuniform thermal effects, etc. However, in many practical applications, the simulation starts at the end of curing or at formwork removal, and the previous effects are neglected.

3. Calculate the increment of reduced time and factors for the exponential algorithm:

$$\Delta t_{r,k} = \frac{1}{2} \left(\psi_{r,k} + \psi_{r,k+1} \right) \Delta t_k \tag{10.98}$$

$$\beta_{\mu k} = e^{-\Delta t_{r,k}/\tau_\mu}, \qquad \mu = 1, 2 \ldots M \tag{10.99}$$

$$\lambda_{\mu k} = (1 - \beta_{\mu k}) \frac{\tau_\mu}{\Delta t_{r,k}}, \qquad \mu = 1, 2 \ldots M \tag{10.100}$$

4. Prepare auxiliary factors for viscosity evaluation:

$$A = \sqrt{\frac{\mu_S | T_{k+1} \ln h_{k+1} - T_k \ln h_k |}{\Delta t_k T_0}} \tag{10.101}$$

$$B = \sqrt{\frac{\psi_{s,k+1/2}}{q_4}} \tag{10.102}$$

4. Compute viscosity of the flow unit at the end of the step:
   If $AB\Delta t_k > 10^{-6}$
   then evaluate

$$\tilde{e} = e^{-2AB\Delta t_k} \tag{10.103}$$

$$\eta_{k+1} = \frac{B}{A} \frac{B(1 - \tilde{e}) + A\eta_k(1 + \tilde{e})}{B(1 + \tilde{e}) + A\eta_k(1 - \tilde{e})} \tag{10.104}$$

   else evaluate

$$\eta_{k+1} = \frac{\eta_k + B^2 \Delta t_k}{1 + A^2 \eta_k \Delta t_k} \tag{10.105}$$

6. Evaluate the modified viscosity and its increment:

$$\tilde{\eta}_{k+1} = \frac{\eta_{k+1}}{\psi_{r,k+1}} \tag{10.106}$$

$$\Delta \tilde{\eta}_k = \tilde{\eta}_{k+1} - \tilde{\eta}_k \tag{10.107}$$

7. Compute the viscous flow strain increment at constant stress and the incremental viscous flow compliance.
   If $|\Delta \tilde{\eta}_k| > 10^{-4} \tilde{\eta}_k$
   then evaluate

$$L_k = \ln \left(1 + \frac{\Delta \tilde{\eta}_k}{\tilde{\eta}_k} \right) \tag{10.108}$$

$$\Delta \varepsilon''_{f,k} = \frac{\Delta t_k}{\Delta \tilde{\eta}_k} L_k C_v \sigma_k \tag{10.109}$$

$$\bar{C}_{f,k} = \frac{\Delta t_k}{\Delta \tilde{\eta}_k} \left(1 - \frac{\tilde{\eta}_k}{\Delta \tilde{\eta}_k} L_k \right) \tag{10.110}$$

else evaluate

$$\Delta \varepsilon''_{f,k} = \frac{\Delta t_k}{\tilde{\eta}_k} \left[1 - \frac{\Delta \tilde{\eta}_k}{2\tilde{\eta}_k} + \frac{1}{3} \left(\frac{\Delta \tilde{\eta}_k}{\tilde{\eta}_k} \right)^2 \right] C_v \sigma_k \tag{10.111}$$

$$\bar{C}_{f,k} = \frac{\Delta t_k}{\tilde{\eta}_k} \left(\frac{1}{2} - \frac{\Delta \tilde{\eta}_k}{3\tilde{\eta}_k} \right) \tag{10.112}$$

8. Compute the incremental modulus

$$\bar{E}_k = \left(\frac{1}{E_0} + \frac{1}{v_{k+1/2}} \sum_{\mu=1}^{M} \frac{1 - \lambda_{\mu k}}{E_\mu^\infty} + \bar{C}_{f,k} \right)^{-1} \tag{10.113}$$

9. Evaluate the strain increment due to creep (at constant stress), and the increments of shrinkage and thermal strain.

$$\Delta \varepsilon''_k = \frac{C_v}{v_{k+1/2}} \sum_{\mu=1}^{M} \frac{1 - \beta_{\mu k}}{E_\mu^\infty} \sigma_{v\mu k} + \Delta \varepsilon''_{f,k} \tag{10.114}$$

$$\Delta \varepsilon_{sh,k} = k^*_{sh} (h_{k+1} - h_k) \tag{10.115}$$

$$\Delta \varepsilon_{T,k} = \alpha_T (T_{k+1} - T_k) \tag{10.116}$$

Note that $\varepsilon_{sh,k}$ and $\varepsilon_{T,k}$ are scalars corresponding to normal strain components, which are the same for all directions, provided that the material is isotropic.

10. For a given strain increment $\Delta \varepsilon_k$, compute the stress increment

$$\Delta \sigma_k = \bar{E}_k D_v (\Delta \varepsilon_k - \Delta \varepsilon''_k - \Delta \varepsilon_{sh,k} i - \Delta \varepsilon_{T,k} i) \tag{10.117}$$

Here, $i = (1, 1, 1, 0, 0, 0)^T$ is a column matrix with unit normal components and zero shear components.

11. Update internal variables using the formula

$$\sigma_{v\mu,k+1} = \lambda_{\mu k} \Delta \sigma_k + \beta_{\mu k} \sigma_{v\mu k}, \qquad \mu = 1, 2, \ldots N \tag{10.118}$$

12. Increment the step counter k by 1 and proceed to the next step (go to 1).

The algorithm is used for stress evaluation in the context of incremental structural analysis and at the same time describes the evaluation of the incremental modulus and of the strain increment at constant stress, which are needed for the assembly of the structural stiffness matrix and of the right-hand side of the discretized equilibrium equations. So the algorithm is in fact exploited twice within the same time step. In the first run, the objective is to determine $\bar{E}_k$, $\Delta \varepsilon''_k$, $\Delta \varepsilon_{sh,k}$, and $\Delta \varepsilon_{T,k}$, and so the execution is suspended after step 9. Once the displacement increments are computed from the global equilibrium equations, the strain increment $\Delta\varepsilon_k$ can be evaluated and the execution of the algorithm execution is resumed (provided that the values of $\bar{E}_k$ and $\Delta\varepsilon_k''$ have been stored, otherwise some of the preceding steps of the algorithm would need to be repeated). For the update of internal variables according to (10.118), the factors $\lambda_{\mu k}$ and $\beta_{\mu k}$ need to be stored as well, otherwise they would have to be recomputed. In addition to the viscous stresses $\sigma_{\nu\mu,k+1}$, several other variables such as the viscosity η_{k+1}, equivalent age $t_{e,k+1}$, and factor $\psi_{r,k+1}$ are stored (of course only temporarily, until the end of the next step).

Certain details of Algorithm 10.1 could be modified. For instance, when the increments of times t_e, t_r, and t_s are determined, we have a choice between the midpoint and trapezoidal rules for the integration of factors ψ_e, ψ_r, and ψ_s. Numerical solution of the viscosity evolution equation could be simplified. The advantage of the present algorithm is that creep at constant temperature and humidity is reproduced exactly, for arbitrarily large time steps. However, this is a real advantage only if the time steps are indeed large. For small steps, simpler formulae can be used, with a negligible error. Branching in the code can be eliminated, viscosity can be updated according to (10.105), and formulae (10.108)–(10.112) can be replaced by

$$\Delta\varepsilon''_{f,k} = \frac{\Delta t_k}{\tilde{\eta}_{k+1/2}} C_v \sigma_k \tag{10.119}$$

$$\bar{C}_{f,k} = \frac{\Delta t_k}{2\tilde{\eta}_{k+1/2}} \tag{10.120}$$

where

$$\tilde{\eta}_{k+1/2} = \frac{\eta_k + \eta_{k+1}}{2\psi_{r,k+1/2}} \tag{10.121}$$

is the approximation of the modified viscosity at midstep.

Let us also note that the algorithm assumes the standard value of parameter p of the microprestress theory, namely $p = 2$. For other values of p, the evaluation of $\Delta\eta_k$ would need to be based on an iterative solution of nonlinear algebraic equation (10.61).

10.9 Analysis of Experimental Data on Temperature Effect

To illustrate the fitting capabilities of the model and to provide examples of parameter identification, typical experiments reported in the literature have been numerically simulated. For each tested concrete, parameters q_1, q_2, q_3, and q_4 are obtained by fitting the experimental data for basic creep at room temperature. Parameter μ_S, which controls the evolution of viscosity under general conditions, must be identified from transient analysis of tests at variable temperature or humidity. Unless stated otherwise, activation energies in (10.31)–(10.33) and parameters α used in formulae

(10.34)–(10.36) are taken by their default values recommended by Bažant et al. [123]: $Q_e/R = 2700$ K, $Q_r/R = 5000$ K, $Q_s/R = 3000$ K, $\alpha_e = 10$, $\alpha_r = 0.1$, and $\alpha_s = 0.1$. Parameters q_1 to q_4, which control basic creep, are summarized in Table 10.1. Individual columns refer to the sets of experimental data that will be analyzed in the following sections.

Table 10.1 Basic creep parameters used for fitting of experimental data reported by Nasser and Neville (NN), Komendant et al. (KPPp—predicted from composition, KPPa—adjusted), and Fahmi et al. (FPB)

Parameter	Unit	NN	KPPp	KPPa	FPB
q_1	[10^{-6}/MPa]	15.0	18.9	14.0	19.5
q_2	[10^{-6}/MPa]	80.0	122.9	60.0	160.0
q_3	[10^{-6}/MPa]	24.0	0.751	16.0	5.25
q_4	[10^{-6}/MPa]	5.0	7.27	6.0	12.5

10.9.1 Basic Creep at Constant Elevated Temperature

The effect of temperature on basic creep is analyzed first, using two sets of experimental data.

10.9.1.1 Data of Nasser and Neville

In the tests of Nasser and Neville [645], the specimens were heated to the test temperature immediately after casting and protected against water loss. For this reason, it is appropriate to consider $h \approx 1$ during the analysis ($h = 0.98$ is assumed). As will be shown in Sect. 10.9.1.3, if the temperature increases sufficiently long before the onset of loading and then remains constant, parameter μ_S (or the original parameters k_1 and c_0) has a negligible effect, and so it cannot actually be identified. The thermal dependence of basic creep is then primarily due to the thermal activation of processes such as hydration or viscous flow, which is reflected by the transformed times t_e, t_r, and t_s and controlled by the activation energies Q_e, Q_r, and Q_s.

Nasser and Neville [645] studied the creep of cylindrical concrete specimens subjected to three different levels of temperature. In their experiments, all specimens were sealed in watertight jackets and placed in a water bath in order to guarantee constant temperature, and they were loaded at the age of 14 days. The values of parameters $q_1 = 15$, $q_2 = 80$, $q_3 = 24$, and $q_4 = 5$ (all in 10^{-6}/MPa) determined by Bažant et al. [123] provide a good agreement with experimental data at room temperature; see Fig. 10.5a.

For a higher temperature, $T = 71\,°C$, the agreement is good up to 30 days of load duration, but afterward the computed rate of creep is too low (see the solid curve in

Fig. 10.5b). A remedy can be sought in modifying the activation energy. Reduction of Q_s/R from the default value 3000 K to 2300 K leads to an excellent fit (see the dashed curve in Fig. 10.5b). Unfortunately, the prediction for the highest tested temperature, $T = 96\,°C$, would be quite poor [472].

Changes in activation energy have no influence on the results when the temperature is close to the room temperature. Before loading, the specimens had been subjected to an environment at the given temperature, which accelerated the hydration processes, i.e., the maturity of concrete. In other words, the higher the temperature, the lower the initial compliance. On the other hand, for longer periods of loading, the higher temperature increases the rate of bond breakages and thus accelerates creep. This justifies the shape of the obtained curve for the medium temperature in Fig. 10.5b; it is different from the one published by Bažant et al. [123], who did not consider the effect of elevated temperature before loading and obtained, for the test at the medium temperature, a higher initial compliance than at the room temperature.

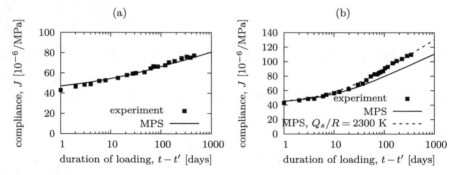

Fig. 10.5 Experimental data of Nasser and Neville [645] and compliance functions for temperatures of (a) $21\,°C$, (b) $71\,°C$

10.9.1.2 Data of Komendant et al.

Komendant et al. [551] performed creep tests under various sustained levels of temperature and for different ages at loading. Two concretes made of very similar mixtures (which differed mainly by the type of aggregate, referred to as York and Berks) were examined. Since their composition and the resulting strength were almost the same, only one set of material parameters is used here for both concretes. The creep experiments were carried out on cylindrical specimens sealed against the moisture loss. All specimens were cured at $23\,°C$, and 5 days prior to loading the temperature started increasing at constant rate of $13.33\,°C/day$ until the target value $43\,°C$ or $71\,°C$ was reached.

Figure 10.6a shows the values measured at room temperature (here taken as $23\,°C$) and the computed compliance functions. Parameters q_i estimated from the composition of the concrete mixture using empirical formulae from Appendix C are $q_1 = 18.9$, $q_2 = 122.9$, $q_3 = 0.751$, and $q_4 = 7.27$ (all in 10^{-6}/MPa). This set of parameters leads to an overestimation of aging; see the dashed curves, corresponding to three different ages at loading. After an adjustment of parameters to $q_1 = 14$, $q_2 = 60$, $q_3 = 16$, and $q_4 = 6$ (all in 10^{-6}/MPa), an excellent fit is obtained; see the solid curves in Fig. 10.6a.

As shown by Havlásek and Jirásek [472], the numerical results published by Bažant et al. [123] are reproduced only when the numerical simulation starts at the age of loading t' and the temperature history is simplified as constant. In such a case, the results are almost insensitive to the choice of parameter μ_S, same as in the previous section. However, this simplification does not correspond to the actual heating history. Heating of the specimen shortly before the onset of loading generates substantial microprestress that does not have sufficient time to fully relax before the creep test starts. The early creep rate then becomes strongly influenced by the elevated microprestress (which depends on age much less than on the time elapsed after heating), and higher values of μ_S make this effect stronger. As shown by the solid curves in Fig. 10.6b, the results obtained with the heating history properly taken

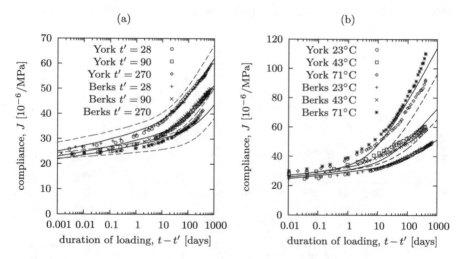

Fig. 10.6 Experimental data of Komendant et al. [551] and computed compliance functions (a) for ages at loading $t' = 28, 90,$ and 270 days and temperature $T = 23\,°C$ (dashed lines represent predictions of the B3 model based on composition, solid lines are the optimum fit), (b) for age at loading $t' = 90$ days and different temperature levels (solid lines computed with the correct temperature history, dashed lines with a simplified approach)

into account agree very well with the experimental data if μ_S is set to $87.5 \times 10^{-6}\,\mathrm{MPa}^{-1}\,\mathrm{day}^{-1}$. For comparison, the dashed curves show the results that would be obtained if the simulation started at the onset of loading and the microprestress generated by the temperature increase was ignored.

10.9.1.3   Temperature Effects on Microprestress, Viscosity, and Compliance

Creep tests presented in Sect. 10.9.1.1 as well as 10.9.1.2 were performed under elevated temperatures, but differed in one important detail. Nasser and Neville [645] stored the specimens at the desired elevated temperature right after casting, i.e., a long time before loading, while Komendant et al. [551] stored them first at the room temperature and then gradually increased the temperature to the prescribed test level only a few days before the onset of loading. In the first case, the microprestress generated by the change of temperature at an early age had enough time to relax before the test, while in the second case it did not, which had an effect on the creep compliance. Since the temperature affects not only the microprestress but also the transformed times, the interplay among various partial effects is quite intricate and it is useful to analyze their role in more detail and illustrate it by a simple example.

Consider again the tests of Komendant et al. [551], in which the specimens were initially kept at 23 °C and the temperature was increased at a constant rate starting at the age of 23 days, up to the level of 71 °C reached at the age of 26.6 days. The creep test started shortly after the heating, at the age of 28 days. The corresponding evolution of the flow viscosity η, computed from Eq. (10.57) with parameters $q_4 =$

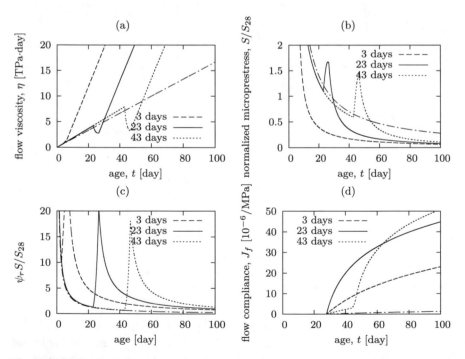

Fig. 10.7 Evolution of various quantities for specimens stored at 23 °C until different ages (3, 23, or 43 days) and then heated in 3.6 days to 71 °C: (a) flow viscosity, (b) normalized microprestress, (c) effective flow viscosity, (d) flow compliance for $t' = 28$ days

$6 \times 10^{-6}\,\text{MPa}^{-1}$ and $\mu_S = 87.5 \times 10^{-6}\,\text{MPa}^{-1}\,\text{day}^{-1}$, is shown by the solid curve in Fig. 10.7a. Up to the age of 23 days, it follows a straight line described by $\eta = \psi_s\, t/q_4$, which corresponds to the evolution of flow viscosity at constant temperature. In this particular test, the room temperature $T = 23\,^\circ\text{C}$ was slightly higher than the reference room temperature $T_0 = 20\,^\circ\text{C}$, and so factor $\psi_s = 1.11$ is slightly larger than 1 (factor β_{sh} is set to 1 because the specimens were sealed[8] and $\beta_{sT} = 1.11$ is computed from (10.33) with $Q_s/R = 3000$ K). During the heating phase, the flow viscosity is reduced from 4.23 TPa·day at 23 days to 2.79 TPa·day at 26 days. After the age of 26.6 days, at constant elevated temperature $T = 71\,^\circ\text{C}$, the flow viscosity grows linearly in time, but with a much higher slope than at the room temperature, because factor ψ_s increases to 4.4. For comparison, the dash-dotted straight line shows the linear evolution of flow viscosity at the reference room temperature $T_0 = 20\,^\circ\text{C}$, which corresponds to basic creep at standard conditions. The dashed and dotted curves illustrate what would happen if the heating phase is shifted by 20 days. If the heating takes place from 3 days to 6.6 days (dashed curve), the evolution of flow viscosity is almost the same as if the specimen was stored at the elevated temperature right after casting. If the heating takes place from 43 days to 46.6 days (dotted curve), the drop in viscosity during the heating phase is more pronounced and the subsequent linear growth is parallel to the other two cases, just shifted in time.

From the physical point of view, it is interesting to examine the evolution of microprestress, shown in Fig. 10.7b. The microprestress could be computed from differential equation (10.45) or, since the flow viscosity is already known, simply from relation (10.49), which for $p = 2$ reduces to

$$S(t) = \frac{1}{c_0 q_4 \eta(t)} \tag{10.122}$$

As already explained, the actual values of microprestress depend on parameters k_0 and c_1, which cannot be calibrated separately based on available "macroscopic" data. Therefore, the values plotted in Fig. 10.7b are normalized by the value of microprestress that would be attained at the age of 28 days at the reference room temperature and sealed conditions, which is given by $S_{28} = 1/(c_0 \times 28\,\text{days})$. The normalized value

$$\frac{S(t)}{S_{28}} = \frac{28\,\text{days}}{q_4 \eta(t)} \tag{10.123}$$

can then be evaluated without specifying parameter c_0. At the reference room temperature, the evolution of the normalized microprestress would follow the hyperbola plotted in Fig. 10.7b by the dash-dotted line. When the specimen is heated, the microprestress increases, and subsequently, at constant elevated temperature, relaxes faster

---

[8]Strictly speaking, β_{sh} computed from (10.36) would be equal to 1 only for $h = 1$ while here we consider the relative pore humidity to be $h = 0.98$ (for normal concrete). But for a sealed specimen at room temperature, we should get basic creep, and so it is reasonable to set $\beta_{sh} = 1$.

than it would at the room temperature. This is caused by the temperature effect on the rate of the reduced microprestress time t_s, described by factor ψ_s.

From the practical point of view, the most important variable is the creep compliance. Its part J_f attributed to the viscous flow is, according to (10.22), obtained by integrating the reciprocal value of flow viscosity. For the standard value of parameter $p = 2$, considered here, the flow viscosity is inversely proportional to the microprestress, and so the flow compliance $J_f(t, t')$ is proportional to the area under the graph of microprestress evolution in time, taken from the onset of loading t' to the current time t. However, one also needs to take into account the effect of temperature on the creep rate, which was not yet considered in (10.22). Replacing the differential of time by the differential of the reduced time t_r (and still assuming $p = 2$), we can write

$$J_f(t, t') = c_0 q_4 \int_{t'}^{t} S(r) \psi_r \, dr = \frac{q_4}{28 \text{ days}} \int_{t'}^{t} \frac{S(r)}{S_{28}} \psi_r \, dr \qquad (10.124)$$

where factor ψ_r depends on temperature and humidity. Therefore, what matters for the flow compliance is the product of the normalized microprestress S/S_{28} with factor ψ_r. This product is plotted in Fig. 10.7c. Since the activation energy Q_r is larger than Q_s, the factor $\psi_r = \beta_{rh} = 12.6$ computed at elevated temperature $71\,^\circ$C is higher than $\psi_s = \beta_{sh} = 4.6$, and the effect of ψ_r on creep acceleration is stronger than the effect of ψ_s on acceleration of microprestress relaxation. This is reflected by the resulting flow compliance J_f, plotted in Fig. 10.7d as function of the current age t for a fixed age at loading $t' = 28$ days. Note that J_f is just the part of compliance attributed to viscous flow, and the full compliance contains additional terms that depend on parameters q_1, q_2, and q_3 and correspond to the elastic spring and solidifying viscoelastic unit.

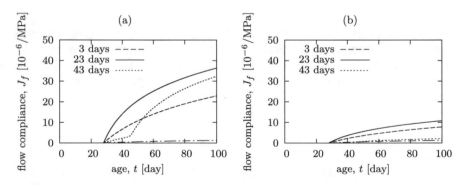

Fig. 10.8 Evolution of the flow compliance $J_f(t, 28)$ of specimens stored at 23 °C until different ages (3, 23, or 43 days) and then heated in 3.6 days to 71 °C: (a) computed with $\mu_S = 0$, (b) computed with $Q_s = 0$ and $Q_r = 0$

To elucidate the role of individual parameters, it is instructive to evaluate the flow compliance that would be obtained if the microprestress generated by temperature increase was neglected (Fig. 10.8a), or if the effect of temperature on reduced times t_r

and t_s was neglected (Fig. 10.8b). In the first case, parameter μ_S is set to zero, which has virtually no effect on the flow compliance of the specimen heated a long time before loading (the dashed curve in Fig. 10.8a is almost the same as in Fig. 10.7d). For specimens heated shortly before loading or after loading, the flow compliance would be reduced if the generated microprestress was neglected. This explains why the simulation of the tests done by Komendant et al. [551] is sensitive to the value of parameter μ_S while the simulation of the tests of Nasser and Neville [645] is not. Finally, Fig. 10.8b shows the compliance curves computed with activation energies Q_r and Q_s set to zero, which means that the influence of temperature on factors ψ_r and ψ_s is artificially suppressed. The reduction of compliance is dramatic, and this confirms that the effect of temperature on the rate of relevant processes is substantial.

10.9.2 Transitional Thermal Creep

Consider now the transitional thermal creep tests of Fahmi et al. [378], in which the temperature was changed during the load application. Mortar specimens had the shape of a hollow cylinder and were cured for 21 days at 100% relative humidity and 23 °C. During the tests, the specimens were either sealed ($h \approx 0.98$) or exposed to an environment of 50% relative humidity. The temperature was varied either in one cycle (two-step heating followed by cooling) or in multiple cycles. Different combinations of sealed/drying conditions and one or multiple temperature cycles resulted into four testing programs, two of which are summarized in Table 10.2.

Table 10.2 Testing programs used by Fahmi et al. [378]

(a) Program #1 – sealed specimen

Duration	T	h	σ
day	[°C]	[%]	[MPa]
21	23	100	0
37	23	98	−6.27
26	47	98	−6.27
82	60	98	−6.27
10	23	98	−6.27
25	23	98	0

(b) Program #2 – drying specimen

Duration	T	h_{env}	σ
day	[°C]	[%]	[MPa]
18	23	100	0
14	23	50	0
37	23	50	−6.27
108	60	50	−6.27
10	23	50	−6.27
25	23	50	0

Due to the nonuniform distribution of humidity resulting from the drying process, shrinkage strains and creep compliance are also nonuniform across the specimen. Therefore, the simulation cannot be limited to one material point (representing the specimen under uniform conditions), and a finite element model is needed. The creep test takes place under constant force but, due to nonuniform creep and shrinkage, the stresses at individual points can vary, giving a constant resultant. Details regarding the finite element mesh and boundary conditions are available in Jirásek and Havlásek [523].

The four parameters of the B3 model describing the basic creep, q_1 to q_4, are determined from the composition of the concrete mixture and from the compressive strength using empirical formulae according to Appendix C. Parameters obtained from this prediction require only minor adjustments to get the optimal fit; see the first stage of the strain evolution in Fig. 10.9. The following values are used: $q_1 = 19.5$, $q_2 = 160$, $q_3 = 5.25$, and $q_4 = 12.5$ (all in 10^{-6}/MPa).

Besides the exponent p with standard value 2, the MPS theory reformulated in terms of viscosity (Sect. 10.7) uses only one additional parameter, μ_S, which is varied until the best agreement with experimental data is obtained. All other parameters are initially set according to standard recommendations. A really good fit of the first experimental data set (sealed specimen, one temperature cycle, Table 10.2a) is obtained for $\mu_S = 875 \times 10^{-6}$ MPa^{-1}day^{-1}; see Fig. 10.9a. The agreement is very satisfactory, except for the last stage, which corresponds to unloading. It is worth noting that the thermally induced part of creep accounts for more than a half of the total creep (compare the experimental data with the solid curve labeled as "basic creep" in Fig. 10.9a).

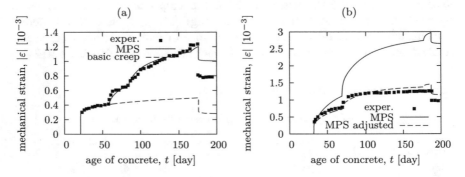

Fig. 10.9 Mechanical strain evolution for specimens loaded by compressive stress 6.27 MPa: (a) sealed specimens (pore relative humidity assumed to be 98%), loaded at time $t' = 21$ days, (b) specimens drying at 50% environmental relative humidity, loaded at time $t' = 32$ days

To obtain an accurate creep evolution for the second loading history (drying and one thermal cycle, Table 10.2b), it is first necessary to calibrate parameters of the Bažant–Najjar model for moisture transport. The distribution of relative humidity across the section was not measured, but the parameters can be identified from the time evolution of shrinkage and thermal strains of the unloaded companion specimen (Fig. 10.10). Parameters $\alpha_0 = 0.05$, $h_c = 0.8$, and $r = 15$ are set to their default values according to *fib* recommendations [381]. A good agreement is reached with maximum diffusivity $C_1 = 25$ mm^2/day, shrinkage ratio $k_{sh}^* = 0.0039$, and coefficient of thermal expansion $\alpha_T = 8 \times 10^{-6}$ K^{-1}.

Fig. 10.10 Evolution of the sum of shrinkage and thermal strains for specimens drying at 50% environmental relative humidity

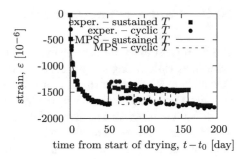

Figure 10.9b shows the comparison between the experimental and numerical results. As usual, the difference between the total strain measured under load and the total strain measured on a load-free companion specimen is reported as a measure of creep strain and plotted against the load duration, although it actually includes also the shrinkage reduction in the companion specimen due to diffuse cracking caused by the nonuniformity of temperature and humidity over the specimen. To analyze this effect, Bažant et al. [123] performed a cracking simulation, with the conclusion that the role of cracking is for these particular tests negligible because the specimen dimensions are very small.

Unfortunately, with default values of the other parameters, the value of μ_S calibrated on data set #1 cannot be used to fit another experimental data set reported by Fahmi et al. [378], because it would lead to a gross overestimation of the creep. These data can be accurately reproduced if the values of parameters α_r and α_s are set to 0.01 and 1.0. But such a dramatic adjustment of the default values would be rather artificial and physically unexplained. Parameters α_s and α_r control the effect of reduced humidity on the rate of microprestress relaxation and the rate of bond breakages, and thus their adjustment would have no effect on the response of sealed specimens, with the good fit in Fig. 10.9a being preserved.

Simulations of testing programs #3 and #4, which involved several temperature cycles, have been reported by Jirásek and Havlásek [523] but are not reproduced here because many questions remain open. It turns out that if the MPS model is calibrated on one temperature cycle, it overestimates the creep caused by subsequent repeated cycles. According to the limited experimental evidence that exists, the actual effect of temperature changes on creep fades away starting from the second cycle, which is not reflected by the original microprestress source term on the right-hand side of (10.46). A tentative remedy has been proposed by Jirásek and Havlásek [523], but a more fundamental extension of the microprestress theory would be needed. This will be no doubt the subject of future research. Recently, a good fit of all tests of Fahmi et al. [378] with one set of parameters has been obtained by Gasch, Malm, and Ansell [410], who considered parameter μ_S as variable, dependent on the pore relative humidity.

For further examples of transitional thermal creep, see Figs. 13.17 and 13.18 in Sect. 13.3.3.

10.10   Comment on Applications and Review of Main Points

The microprestress-solidification theory serves as the theoretical basis for model B3 (as well as B4). Strong simplifications, however, have been made for this purpose. One of them is the replacement of a solution of the diffusion problem of drying by simplified formulae for the mean pore humidity in the cross section. It is nevertheless not difficult to program a numerical integration of the diffusion equation for drying, and its computer solution would not be very demanding—certainly not in comparison to the overall design effort for a large creep-sensitive structure.

The experimental verification of the theory is still limited. Better calibration of its parameters is needed, especially in regard to multidecade predictions. Inverse analysis of the observed long-time deflection of creep-sensitive structures, such as prestressed box girders, would be helpful. Nevertheless, the fact that phenomena as diverse as the long-term aging, drying creep and transitional thermal creep can all be explained by one and the same model lends credence to the validity of the microprestress-solidification theory. Aside from long-time deflections, shortening of tall building columns, and redistribution of internal forces due to creep, the unification provided by the microprestress-solidification theory may be expected to provide a more realistic common basis for the analysis of many practically important problems involving thermal effects—nuclear containments, response to various hypothetical extreme nuclear reactor accidents, long-term effects of radioactive waste disposal [150], ablation of contaminated surface layer concrete by microwave blast [213], behavior of chemical technology vessels, effects of hydration heat in massive structures, and effects of environmental variations on structures.

Since the microprestress-solidification theory is relatively complex, it may be useful to summarize the main points:

1. In the solidification theory for basic creep, formulated in Chap. 9, the aging is explained by volume growth of a nonaging viscoelastic material (the cement gel) into the pores. Although this theory, supplemented by the flow term, agrees with the available test data as well as one might hope in view of inevitable random scatter, it has, from the physical viewpoint, two shortcomings:

   - The effect of age at loading on the creep compliance remains strong even after many years, whereas the volume growth of the hydration products slows down markedly after about one year.
   - The drying creep (Pickett effect) and the transitional thermal creep are not explained by the theory, and their modeling requires separate assumptions.

The microprestress-solidification theory removes these shortcomings and brings about simplification and unification; see the schematic overview in Table 10.3. Model B3 in the context of the sectional approach is described in Chap. 3 and used in the first part of this book. The material approach to model B3 (also applicable to B4), based on the concept of stress-induced shrinkage, is briefly sketched in Appendix C.3. The microprestress-solidification theory (MPS), discussed in the present chapter, uses exclusively the material approach.

Table 10.3 Representation of different components of creep according to models B3/B4 and microprestress-solidification theory (MPS)

Model	Short-term Creep	Long-term Creep	Drying Creep and Shrinkage
B3/B4	solidifying	logarithmic	sectional approach
	Kelvin	flow term	material approach
MPS	chain		microprestress theory

2. The long-term aging, which cannot be fully accounted for by the volume growth of hydration products, can be explained by relaxation of a tensile microprestress in the bonds or bridges across the nanopores in hardened cement gel filled by hindered adsorbed water. The microprestress represents a reaction to the disjoining pressure exerted on the nanopore walls by hindered adsorbed water, and its initial buildup is caused by high local shrinkage and crystal growth pressure at locations close to the nanopore.

3. The long-term creep, deviatoric as well as volumetric, is considered to originate from viscous shear slips between the opposite walls of the nanopores in which the bonds or bridges that cross the nanopores and carry the microprestress break and reform, in a manner similar to the movement of dislocations through a crystal lattice. The shear has the property that bonds can restore and thus the macroscopic stiffness of concrete does not get reduced (tensile breaks do not allow bond restorations; rather they lead to cracks and reduced macroscopic stiffness, which characterizes the nonlinear creep above the service stress range).

4. Due to creep in the direction transverse to the slip plane, the tensile microprestress undergoes relaxation. This relaxation reduces the effective viscosity of the shear slips and thus brings about long-term aging associated with the flow term in the creep model.

5. Since the tensile microprestress S is the reaction to the disjoining pressure, it changes with the disjoining pressure, which in turn changes almost instantaneously with the changes in the relative humidity h in the adjacent capillary pore in the hardened cement paste. This is the cause of the Pickett effect (drying creep or stress-induced shrinkage).

6. Analysis of the available test data confirms that the microprestress relaxation needs to be introduced only for the viscous flow term. The solidifying part of the model need not be considered to be affected by any microprestress. Its viscosity varies only as a consequence of volume growth. On the other hand, the volume growth does not affect the flow term. This separation of solidification and microprestress greatly simplifies the mathematical formulation.

7. The microprestress theory has been formulated in a way that does not require a change in the solidification theory for typical creep and shrinkage tests, for which extensive test data exist. For drying creep, the results of the microprestress-solidification theory are different, but not significantly different, from those of the previous combination of solidification theory with stress-induced shrinkage

[117, 120, 201], described in Sect. 13.3.3.2 and Appendix C.3. This is not surprising, because the microprestress-solidification theory can be regarded as a refinement of these previous formulations.

8. For generalization for the effect of temperature (not exceeding 100 °C), the rates of creep and of volume growth can be characterized by three transformed time variables based on the activation energies.

9. The concept of microprestress achieves a grand unification of theory which simultaneously captures three seemingly disparate basic phenomena:

   - The creep decrease with increasing age at loading after the growth of the volume fraction of hydrated cement has slowed down markedly;
   - the drying creep, i.e., the transient creep increase due to drying (Pickett effect) which overpowers the effect of steady-state moisture content (i.e., less moisture—lesser creep); and
   - the transitional thermal creep, i.e., the transient creep increase due to temperature change.

10. To reduce the number of parameters, the microprestress can be eliminated from the formulation, albeit at the cost of losing contact with the physical theory, which makes sense only in terms of microprestress relaxation. The differential equation for microprestress relaxation is then reformulated in terms of viscosity of the flow unit. Apart from standard parameters q_1, q_2, q_3, and q_4, which control the elastic response and basic creep, only one additional parameter μ_S needs to be introduced in order to capture drying creep.

11. For computations, an efficient integration algorithm is available. A satisfactory agreement with some of the test data can be achieved. However, certain discrepancies between experimental data and predictions of the MPS theory arise. For instance, simulations based on MPS lead to a delay of the drying creep behind shrinkage, to a reversed size effect on drying creep, and to excessive creep under multiple temperature cycles. Besides, the possible role of multiyear autogenous shrinkage and self-desiccation needs to be explored and understood. These phenomena are the subject of new ongoing research (it appears they could be captured by taking into account the viscosity dependence on other variables, as suggested by MD simulations, particularly the humidity rate and the microprestress).

12. The microprestress-solidification theory is a general (pointwise) constitutive model. For continuum modeling of concrete structures exposed to drying and temperature changes, it may be supplemented by a model for distributed tensile damage and its localization.

Chapter 11
Physical and Statistical Justifications of Models B3 and B4 and Comparisons to Other Models

Abstract In structural engineering, it is necessary to design structures with incomplete knowledge of the creep and shrinkage characteristics of the concrete to be used. Therefore, prediction based on concrete strength and composition is required. After summarizing the criteria for a sound prediction model, we discuss in detail the theoretical justification of model B3, including the thermodynamic restrictions, reasons for using power functions, consequences of microprestress relaxation and of activation energy, problems of characterizing aging by strength gain, consequences of diffusion of pore water for size and shape effects on shrinkage and drying creep and their asymptotics, and separation of cracking effects. Then, we focus on unbiased fitting of the existing worldwide database, which is characterized by limited range and complicated by variable data density. We present a statistical evaluation of models B3 and B4 and their statistical comparisons to other prediction models, and we describe the procedure that was used for calibration of the constitutive parameters by fitting a combined database of several thousand laboratory curves of limited time range and of about seventy histories of excessive multidecade deflections of large-span prestressed bridges. Finally, we briefly mention analytical methods for prediction of creep and shrinkage via homogenization.

A good creep prediction model is of paramount importance for realistic assessment of creep and shrinkage effects over the design lifetime of structures. Development of such a model requires: (a) anchoring the creep and shrinkage equations in the physical theory of the mechanism of creep, shrinkage, and pore water diffusion; (b) developing approximate formulae for estimating the parameters of these equations from the concrete strength, composition, environmental conditions, and curing; and (c) calibrating this formulation by extensive test data collected worldwide for various concretes, paying attention not only to laboratory tests but also to creep observations on structures. In its entirety, this is a formidable problem. Considerable progress has nevertheless been achieved and is expounded in this broad chapter.

© Springer Science+Business Media B.V. 2018
Z.P. Bažant and M. Jirásek, *Creep and Hygrothermal Effects in Concrete Structures*, Solid Mechanics and Its Applications 225, https://doi.org/10.1007/978-94-024-1138-6_11

11.1   Main Criteria of Evaluation

The traditional approach in which a creep and shrinkage model is validated and calibrated by comparisons with only a few subjectively selected data sets is not only insufficient but also no longer inevitable. Science progressed and computers have made statistical comparisons to the complete existing database and some of the evidence from structures quite easy. The criteria of optimum model selection are now discussed in detail.

1. The mathematical form of the model should conform to a **physically based theory** and to principles of mechanics.
2. The model should achieve the lowest possible coefficients of variation of the deviations of its predictions (root-mean-square error divided by the mean of data) from a **comprehensive database** that includes all the relevant test data from literature, except those that are suspect for some objective reason. The bias in the distribution of data points, due to the crowding of data into short-time and small-size ranges and to unsystematic sampling of measurement times, needs to be counteracted by proper weighting of data in the least-square evaluation of errors and in calibration of model parameters by optimal multivariate regression of the database.
3. To verify the form of model equations, it is necessary to check how closely they can fit **individual test curves** for one and the same concrete. Such curves have far less scatter than the database in which data for many different concretes and environmental conditions are mixed. Preference should be given to long-term tests, which are more indicative of the general trends.
4. Due to the scarcity of long-term data from laboratory experiments, it is very useful to consider statistics of errors in the predicted deflections and prestress losses compared to **in-situ measurements on creep-sensitive structures** such as large-span bridges or tall buildings. Such comparisons can provide valuable hints on the accuracy of extrapolations from the range of available laboratory data.
5. The model should have a form that allows updating of the model parameters by short-time small-size test data for the given concrete and gives **realistic extrapolations to long times and large sizes**. A form for which the updating can be carried out easily (preferably by linear regression) is preferred.

Criterion 1 will be thoroughly discussed in Sect. 11.2. **Criterion 2**, introduced in Bažant and Panula [175] after the compilation of the first comprehensive creep and shrinkage database, has become almost the only validation method used in the committees developing the standard design recommendations and codes. Nevertheless, **criterion 3**, which has often been ignored, is equally important because the huge scatter in a database involving many different concretes obscures the shape of creep and shrinkage curves.

In view of the acute scarcity of multidecade laboratory tests, **criterion 4**, which has been discussed in Chap. 7, is essential for calibrating and validating the long-time prediction capability. There have been many instances of excessive multidecade deflections and cracking, as well as a few of collapses due to long-time buckling, but precise data are usually not published and are hard to obtain. A salient example, the deflections of the bridge in Palau, has been discussed in Chap. 7. Even if long-time data from measurements on structures are available, the interpretation faces several difficulties:

1. Aside from creep and shrinkage, there may be many other factors that contributed to poor long-time performance, and it might not be easy to sort them out.
2. Often many factors and many of the data needed for evaluation have not been recorded.
3. Data for many similar structures, which would be needed for statistical evaluations, are not available.

Nonetheless, partial multidecade deflection data have recently been obtained for 69 long-span prestressed box-girder bridges, many of them deflecting excessively. The data made it possible to determine the ratio of 20-year deflection to 3-year deflection, which is very useful for calibrating the parameters of model B3 (and B4) that control long-term creep. In the laboratory database, the information on long-term creep is almost nonexistent.

Finally, **criterion 5** is important for practical applicability of the model. For designer's convenience, a low number of influencing factors in the model is considered highly desirable. So is a low number of parameters (or coefficients) in the equations. However, since the computer evaluation of any prediction model nowadays takes only a fraction of a second, it appears inappropriate to omit any significant parameter or factor expressly for this reason.

It is nevertheless important to keep low the number of those free parameters that the user may wish to update by fitting of the limited data he might have. If one tries to identify too many parameters from limited data, the least-square optimization problem becomes ill-conditioned. The updating procedure recommended for model B3 and described in Sect. 3.8 and Appendix H satisfies this limitation—it involves only two free parameters for creep and two for shrinkage.

11.2 Theoretically Based Physical Justifications of Model B3*

Before calibration of any model by fitting available test data is attempted, it should be made sure that the model is theoretically sound and physically justified to the maximum possible extent.

*11.2.1 Overview of Mechanisms and Phenomena**

The current understanding of creep and shrinkage of concrete does not suffice for deriving the constitutive laws mathematically. However, the laboratory test data are quite incomplete and cover only a small part of the range of interest. Particularly, the data for modern concrete compositions, diverse environmental and curing histories, multidecade behavior, and large cross-sectional sizes are scant and hard to obtain. Yet it is useless to initiate a 100-year test of a new type of concrete that we need to use now. This situation is very different from other aspects of concrete design, such as the beam strength dependence on the steel ratio or shear span, the range of which can be fully covered by experiments, thus obviating any need for a theory. To succeed in formulating the constitutive law for creep and shrinkage, it is thus important that various features of the constitutive law be justified as much as possible theoretically, from physical phenomena that are reasonably well understood. Since most of the phenomena have already been discussed in the preceding chapters, we will merely summarize them, to portray the overall picture.

Currently, a number of physical requirements and mechanisms are already understood sufficiently well to base on them the prediction model. There are essentially seven **physical mechanisms** related to creep and shrinkage:

1. Solidification as a mechanism of aging, particularly at early times.
2. Microprestress relaxation as a mechanism of long-time aging.
3. Bond ruptures caused by stress-influenced thermal excitations controlled by activation energy.
4. Transport of pore water in various forms, including liquid and vapor phases and surface diffusion.
5. Surface tension on nanoscale globules of C-S-H, capillarity, hindered adsorption, and disjoining pressure in nanopores.
6. Cracking caused by self-equilibrated stresses and applied load.
7. Expansion of C-S-H during hydration (which can extend over many years due to diffusion barriers) and the counteracting pore humidity changes causing autogenous shrinkage. The importance of this mechanism is made clear in Bažant et al. [125].

These mechanisms allow making **inferences** for the proper mathematical form of the creep and shrinkage prediction model:

1. The drying shrinkage and drying creep are, according to many experiments, approximately proportional to the water loss from the specimen (Sect. 8.6), which also implies boundedness of the drying shrinkage and of the drying part of creep.
2. The curves of water loss, drying shrinkage, and drying creep ought to begin as a square root of time of drying, i.e., as $\sqrt{t - t_0}$, and approach their final value as a decaying exponential of a power function of time. It follows that approximations of these curves can be obtained by asymptotic matching of these asymptotic laws; see Fig. 3.15.

3. The halftimes of water loss, drying shrinkage, and drying creep should be proportional to the square of the cross-sectional size (or effective thickness D); see Sect. 8.3.5.
4. The dependence of effective thickness D on the cross-sectional shape ought to follow the diffusion theory.
5. The penetration depths of cyclic environmental humidity and of cyclic temperature ought to follow the diffusion theory.
6. The difference in cracking of compressed creep specimens and companion shrinkage specimens explains that adsorption phenomena and microprestress cannot be the sole source of drying creep (Pickett effect).
7. The requirement of nonnegative energy dissipation during the creep and shrinkage process (Sect. 9.6 and Appendix G) restricts the form of rate-type creep laws [84, 126, 139, 199].
8. Nonmonotonic creep recovery (i.e., a reversal of recovery, Sect. 9.6) and stress relaxation to opposite sign (Sect. 9.7) should be avoided.
9. The absence of characteristic times for various processes is the reason for using power functions (Sect. 9.4).
10. Microprestress relaxation is one phenomenon which can explain gradual multi-year decrease of creep (Chaps. 3 and 9) even if the hydration becomes within one year very small.
11. The rates of creep, shrinkage, and aging are controlled by Arrhenius-type temperature dependence of the rate processes (Sect. 10.6.1).

11.2.2 Thermodynamic Restrictions*

It suffices to summarize the main points, most of which have been covered by the detailed exposition in Chap. 9:

1. The compliance function $J(t, t')$ must be such that when it is approximated by a Kelvin chain, its moduli $E_\mu(t')$ and viscosities $\eta_\mu(t')$ are positive and nondecreasing functions of time.
2. Because the newly solidified material on the pore walls and the newly formed bonds are stress-free when they form, the elastic moduli $E_\mu(t')$ must relate the rates of the stress and strain, and not their values [84, 199].
3. Divergence of compliance curves for different t', which leads to nonmonotonic recovery, is thermodynamically impossible for the Kelvin chain model. However, an aging Maxwell chain that is thermodynamically admissible yet exhibits a divergence of the compliance curves can be constructed (Example 9.2).
4. The stress relaxation curves should not reach into the opposite sign of stress. For aging Maxwell chain, this is thermodynamically inadmissible.

All these properties are satisfied by the compliance functions of the solidification theory and models B3 and B4, but some of them are violated by those of ACI, CEB, fib, GL, and JSCE. For instance, as shown in Chap. 9, all these models under certain circumstances exhibit divergence of creep curves (which leads to nonmonotonic recovery), and the ACI and CEB models in some cases give stress relaxation into opposite sign.

11.2.3 Microprestress Relaxation and the Question of Characterizing Creep Aging by Strength Gain*

As already explained in Chap. 9, the age effect on the strength of concrete depends on the evolution of hydration, which in turn depends on the global drying process and on self-desiccation. Thus, the strength gain may terminate within one year (Fig. 9.2a), or may proceed for many years, as documented for the concrete in the KB Bridge in Palau. On the other hand, creep at fixed load duration always decreases with age for many years.

Consequently, it is unrealistic to characterize the decrease of creep with age by the strength gain function, as done in the GZ model [408]. The long-time aging cannot be captured in such a manner, which is one basic objection to the GZ model.

A physical explanation of the long-term aging in creep is, at least partly, provided by the relaxation of microprestress produced in the microstructure by unequal volume changes of various constituents during the initial hydration [132] and further buildup of microprestress caused by changes of moisture content; see Chap. 10. However, multiyear hydration (due to reduction of hydration rate by diffusion barriers around cement grains) can also cause multiyear aging. More research is needed.

11.2.4 Activation Energy, Power Laws, and Lack of Bounds*

If a physical process has no characteristic time, then its dependence on time must be described by a power function, as follows from the scaling rule explained in detail in Sect. 9.4. Unlike, e.g., drying, no characteristic times are known for the physical process in concrete creep (slip due to bond breaks and restorations), and so the function $J(t, t')$ should be built from power functions. But any transition from a power function for one process to another one for another process, of course, sets a characteristic time.

The characteristic time must be distinguished from the setting of the unit to time, the choice of which is arbitrary. However, the fact that Eq. (3.12) of model B3 needs no time constant if it is written with times given in days does imply that 1 day happens to be a characteristic time (the same for λ_0 set as 1 day in Eq. (9.18)). Were the times given in years or in hours, a time constant representing a characteristic time would have to be introduced in these equations.

The rate of creep as well as the rates of hydration and microprestress relaxation are governed by the dominant activation energy for the breakage and restoration of interatomic bonds. In this theory, called the rate process theory, the temperature dependence is generally given by the Arrhenius formula of the type $\exp(-Q/RT)$ in which Q is the activation energy, T is the absolute temperature, and R is the universal gas constant. The creep and the hydration are physically different processes, and so they are governed by different activation energies.

The activation energy underlies, in model B3, the definition of the equivalent age, also called the maturity (or equivalent hydration period). However, deeper inferences can be made from the activation energy theory. As Wittmann argued [871, 873, 874], under certain simplifying assumptions the activation energy theory suggests that the short-time creep curves ought to be power curves.

Power laws exhibit no bounded asymptotic value. Many previous creep models assumed that creep curves are bounded, but this is not confirmed by long-term experiments. Particularly in the logarithmic scale, basic creep curves always end with a straight line of significant slope, and there is no physical reason why this should change for still longer load durations. The multidecade deflections of bridges evolve logarithmically, but this could not be so if the compliance curve had a bound. It is therefore safe to assume that basic creep has no bound.[1]

A long held hypothesis has been that a bound must exist for drying shrinkage and drying creep, because they are driven by water loss and the amount of water to be lost is finite. Since the amount of reactants (calcium silicates, calcium aluminates, and calcium aluminoferrites) is finite, the autogenous shrinkage must have a bound, but the bound might be reachable only after decades or centuries (depending on water diffusion through the C-S-H barrier shells surrounding the remnants of unhydrated cement grains and on water supply to the pores).

Remark on Shrinkage Asymptotics: Ongoing research, however, suggests a more complicated picture. The changes in pore water stresses during drying (i.e., in disjoining pressure, solid surface tension and capillary water tension) must be equilibrated by compressive stress changes in the solid skeleton of the hardened cement paste. The drying shrinkage is the elastic volumetric compression of the solid skeleton caused by these stress changes. However, it is inconceivable that these stresses would produce no nanoscale compressive volumetric creep. They must.

Were this creep the same as the creep due to external loading, the shrinkage would be unbounded, terminating with a logarithmic growth in time. But the nanoscale volumetric creep due to pore water stress changes has probably much smaller magnitude and duration than the macroscale creep due to external hydrostatic pressure, for three reasons: 1) The mechanism of creep in concrete, including volumetric creep, must consist of the sliding of adjacent parallel C-S-H sheets, but it is doubtful that volumetric pore water stresses, applied on these nanoscale sheets, could cause much sliding;

---

[1]This conclusion is reinforced by a comparison with rocks, which all creep on the geologic scale. All the creep of concrete treated in this book is the primary creep, which, after a time called the Maxwell time, always gradually transits into the secondary creep, characterized by a constant rate. The Maxwell time could be thousands of years. A nanoscale mechanism explaining why this transition is necessary for shale was proposed in [305]. It may also apply to concrete.

2) the pore water stress changes act over only a portion of the cross section, defined by a certain porosity factor that is probably significantly smaller than 1; and 3) the nanoscale creep, unlike the usual, macroscale, creep, may be short-time and bounded, i.e., have a retardation spectrum of limited breadth.

The last point may explain why the few long-time tests of drying shrinkage that exist nevertheless suggest the existence of a final asymptotic bound, while the presence of at least some nanoscale creep due to pore water stresses can explain why the approach to the final asymptotic value appears to be slower than the exponential approach indicated by the diffusion theory.

The creep effect on shrinkage due to pore water stresses is ignored in this book because it appears to be small, and because it has not been adequately researched.

11.2.5 Diffusion Theory for Pore Water*

Several characteristic features, validated by experiments, can be inferred by applying the diffusion theory to the migration of pore water. This is in spite of the fact that the diffusion of water in concrete is highly nonlinear (because the diffusion coefficient strongly decreases with a decreasing relative humidity in the pores).

11.2.5.1 Final Asymptotic Form and Boundedness of Drying Shrinkage Curve*

The drying shrinkage is caused by the loss of free pore water to the environment. After all the moisture needed to restore thermodynamic equilibrium with the environment has evaporated, the drying shrinkage must stop. So, because the water loss is finite, the drying shrinkage should have a finite asymptotic value (if the creep due to pore water stresses is bounded). This condition has been violated by some models, e.g., by a steep terminal slope in the original GZ model [408]. Tests by Wittmann et al. [878] confirm that drying shrinkage does not terminate with a rising straight line of significant slope in the semilogarithmic scale.

Further note that a part of shrinkage, called the autogenous shrinkage, is due to chemical reactions causing self-desiccation. These reactions must eventually also come to a stop once all the reacting constituents have reacted fully. But recent observations indicate that an asymptotic bound could be reached only after many decades.

Only few careful and statistically significant measurements had a long enough duration to give information on the approach to the final drying shrinkage value; see the data points in Fig. 3.3 [878] which have a high statistical significance because they represent the average of 36 identical, precisely controlled shrinkage tests. On the other hand, the existence of a final drying shrinkage value has been well documented for hardened cement paste; thanks to the fact that the specimens can be made thin enough to dry to an equilibrium water content within a short enough time, e.g., Wittmann [874]. Since the shrinkage of cement paste is what causes the shrinkage of concrete, it follows that the drying shrinkage of concrete, too, cannot be unbounded. When the hardened cement paste in concrete stops shrinking, the concrete stops shrinking.

11.2.5.2 Asymptotics of Shrinkage and Drying Creep Curves*

As shown in Sect. 8.4.5.1, the diffusion theory implies that the shrinkage curve must initially evolve as $\sqrt{t - t_0}$ and must approach its asymptotic value as a decaying exponential of the drying time [146]. These properties were supposed to be disturbed by finiteness of moisture emissivity at the surface, by initial hairline shrinkage cracks, and by possible deviation of local shrinkage rate from proportionality to the pore humidity rate. Since the finiteness of emissivity is roughly equivalent to adding a layer of 1 mm thickness to the specimen surface, it does matter for thin cement paste wafers. But experiments on concrete, especially those of Wittmann [874] and Wittmann et al. [878], verify the initial square-root-time evolution very closely; see Fig. 3.14. Models B3 and B4 satisfy these properties, while the ACI model and some others do not.

According to ongoing research [125], the self-desiccation may significantly alter the long-time asymptotics of shrinkage. For instance, multidecade self-desiccation may prevent total shrinkage from approaching a final horizontal asymptote.

11.2.5.3 Shape Effect on Shrinkage*

The diffusion theory makes it also possible to determine theoretically the factor k_s that gives the correction to the drying shrinkage halftime depending on the shape of the cross section. Its theoretically calculated values (based on the plots in Bažant and Najjar [166]) provide a good agreement with test data. They were adopted for models B3 and B4 (and previously for model BP). A refinement that takes into account the dependence of the shape factor on ambient humidity has recently been proposed by Donmez and Bažant [356].

11.2.5.4 Drying Creep, Flow, and Aging or Nonaging Viscoelasticity*

The diffusion source of drying creep further indicates that the additional creep due to drying should be related to the drying shrinkage function, as formulated in models B3 and B4 (in a manner that satisfies the nondivergence condition; see Sect. 9.6). The initial and final shapes of the drying creep curves, as well as the effect of cross-sectional thickness, should therefore be similar to the drying shrinkage curve shape. This feature is reflected in models B3 and B4.

Furthermore, it is advantageous to separate in the creep formula the additive components of creep having different physical origins and meanings, as will be shown in Eq. (11.3).

11.2.5.5   Effect of Environmental Humidity and Thickness*

Since the environmental humidity represents a boundary condition for the diffusion equation, and the solutions of the diffusion equation scale down if the value of boundary humidity is reduced, the diffusion theory indicates that the influence of environmental humidity should get manifested as a multiplicative factor in the formula of the shrinkage curve (except for possible strong autogenous shrinkage). Thus, a change in the environmental humidity should result in vertical scaling of the shrinkage curve and of the drying creep curve.

Such scaling, which is verified by test results, contrasts with the effect of cross-sectional thickness, which is manifested, in the semilogarithmic plot, by horizontal shifts of the shrinkage curve and of the part of the creep curve attributed to drying.

11.2.6   Effect of Cracking*

The tensile stresses caused by nonuniformity of drying throughout the cross section are known to produce tensile cracking. The cracking causes the observed shrinkage of specimens to be less than the true shrinkage of the material. This difference is one cause of the drying creep effect (Pickett effect).

Although generally the effects of cracking seem hard to quantify by simple formulas, they have to be taken into account in finite element analysis. Since cracking contributes an additional deformation, the drying creep should properly be taken into account as an additive term rather than a multiplicative term. This is what is done in models B3 and B4.

11.3   Statistical Aspects of Model Calibration and Validation

11.3.1   Unbiased Statistical Verification of Model

Criterion 2 postulated in Sect. 11.1 means that comparisons should not be restricted to a limited set of test data. Unless the test data used for comparison are chosen truly randomly (e.g., by casting dice, or by a random number generator), the statistics can get blatantly slanted by using a selected data set. This was demonstrated by examples in Bažant and Panula [177]. They showed that:

- when 25 most favorable data sets among 36 data sets for shrinkage (available in the literature by 1980) were selected, the coefficient of variation of the model errors (defined as the root-mean-square error divided by the mean of all data) was reduced from 31.6 to 21.5%, and when 8 most favorable data sets were selected, it was reduced to 8.7%;

- when 8 most favorable data sets among 12 available data sets for creep were selected, the coefficient of variation was reduced from 52.2 to 20.7%. Yet 8 data sets in a paper would impress most readers as plentiful.

These huge changes in error statistics clearly show the danger of making a subjective selection of the data sets to which a model should be compared. Unfortunately, such bias is often found in the literature. Most often this bias is unintended, but sometimes it is too tempting to say that "something must have gone wrong" with the tests that do not agree with a proposed formula.

For more detail and the choice of statistical indicators, see Sects. 11.4 and 11.5 and Appendix K.

11.3.2 Importance of Validating Model Form by Individual Tests on Many Different Concretes*

Comparisons with the complete database can serve only as a partial validation of a model. The reason is that there are large random differences among data for different concretes from different laboratories. When they are all mixed in one and the same database, the scatter band is inevitably very wide. The scatter due to differences among concretes overwhelms the scatter of creep per se. One consequence is that a reasonable curve such as A in Fig. 11.1a does not give an appreciably lower coefficient of variation of the deviations from the test data than a totally unreasonable curve such as B.

Comparisons with the complete database are insufficient to validate the shape of the curves of the model. It is necessary to check that the shape of the curves of a proposed model can fit closely the experimental curves from various **individual** tests, for which a much narrower scatter band is achievable (Fig. 11.1b).

A physical justification of the mathematical formula of the model is important because its practical use inescapably implies extrapolations far out of the range of the main existing evidence. One serious deficiency of the existing databases is that most of the data points pertain to relatively short creep durations. For the NU 2014 database [488], the number of tests is plotted against the duration they reach or exceed in Fig. 11.2. Only 5.2% of all creep tests exceed 6 years in duration, only 3.5% attain 12 years, and only 1.8% attain 18 years while large bridges and other large structures are supposed to have the lifetimes of at least 100 years. The data of Brooks [267] cover 30 years of creep, but some of his creep and shrinkage curves suddenly change slope or exhibit jumps and oscillations after about 6 years of loading or drying and thus are not totally trustworthy. In his paper, Brooks admits that "after five years the air conditioning began to malfunction and was eventually replaced after ten years." The article of Browne and Bamforth [272] has "12 1/2 years" in the title, but it is only an abstract, with no data.

Consequently, blind statistics based on the comprehensive database imply improper weighting of the data, with a far greater weight put on the data for short times, short ages, and small thicknesses than for long times (Fig. 11.1c), high ages,

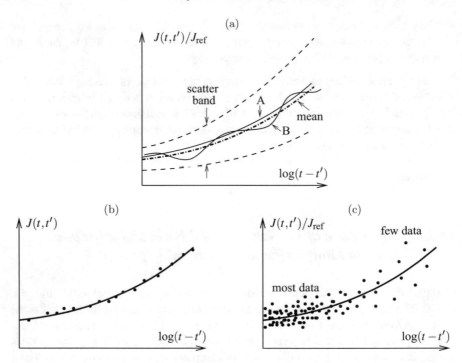

(a)

Fig. 11.1 (a) Due to a very high scatter band width when all the existing data are combined, reasonable (A) and unreasonable (B) curves have about the same coefficient of variation of errors, (b) a narrow scatter band width achievable only for individual tests, (c) the majority of existing test data is concentrated at short times (short creep durations and low ages at loading)

Fig. 11.2 Number of tests in the NU 2014 database plotted against the duration they reach or exceed

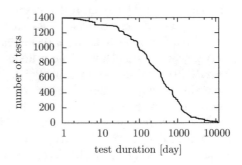

and large thicknesses, while the latter is what is most important. One way to mitigate it is to conduct multivariate regression (linear or nonlinear) with bias-countering weights. But this requires introducing a theoretically well-founded mathematical model, which may be questioned or distrusted by some. Another way, which requires no model, is to consider subsequent decades of time and assign each data point a weight inversely proportional to the number of points in that decade, as presented for several models in Sect. 11.5.2. However, if one evaluates the averages of secondary

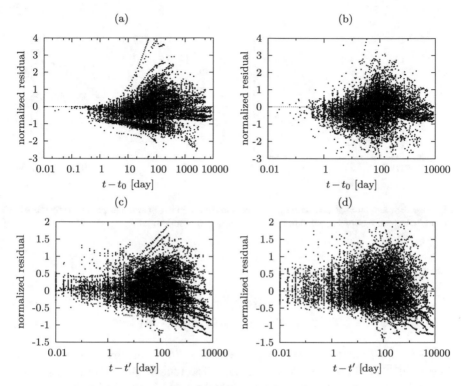

Fig. 11.3 Residuals (errors) of ACI model normalized by weighted mean, compared to all the data in the NU database: (a) shrinkage, original model, (b) shrinkage, perturbed model with $\alpha = 90 \times 10^{-6}$, (c) creep, original model, (d) creep, perturbed model with $\alpha' = 13.0 \times 10^{-6}$/MPa

parameters such as the strength, humidity, thickness, or age at loading in subsequent decades of load duration, one finds that they vary significantly from one decade to the next. Therefore, the database should better be filtered to extract data subbases such that the averages of the secondary parameters be the same in each time decade (this has not yet been done for creep or shrinkage, but has been successful in revealing purely statistically the trend of size effect on structural strength).

To make it blatantly conspicuous how the scatter due to differences among different concretes covers up the underlying statistical trends of creep or shrinkage, Figs. 11.3 and 11.4 present an artificial example (of an obviously erroneous model), used by Bažant, Wendner, Boumakis and Hubler [195] to invalidate one researcher's claim that the 1971 ACI-209 model was still the best. Figure 11.3 compares the residuals (or errors) of the ACI-209 shrinkage curves and compliance curves when these curves either are, or are not, perturbed by large (nonsensical) vertical sinusoidal oscillations with random horizontal shifts (Fig. 11.4). All the residuals (i.e., errors, predictions minus the measurements) are normalized by the weighted mean of the experimental curve. Weights inversely proportional to the number of data points in subsequent constant intervals of $\log(t - t')$ are used, to compensate for the data

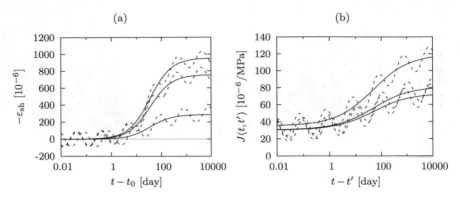

Fig. 11.4 Typical plots of (a) shrinkage strains and (b) creep compliance as predicted by the ACI model, with perturbations (dashed lines) and without them (solid lines)

density bias. The perturbations are

$$\Delta\varepsilon_{sh}(t) = \alpha\,\sin[2\pi(\log t - s)] \tag{11.1}$$

$$\Delta J(t, t') = \alpha'\,\sin[2\pi(\log t - s)] \tag{11.2}$$

Computations have been run using $\alpha = 90 \times 10^{-6}$ for shrinkage and $\alpha' = 13.0 \times 10^{-6}$/MPa for creep, with t in days. The random phase shift s, with uniform distribution in $(-0.5, 0.5)$, prevents bias due to the location of inflexion points.

The errors (or residuals) of the curves compared to the points of the NU 2014 database [488] are computed and then plotted in Fig. 11.3a,c as a function of logarithmic time for all the data sets contained within the database. Ditto for the perturbed curves in Fig. 11.3b,d. The perturbation is found to change the coefficients of variation (C.o.V.) of the errors (root-mean-square error divided by the data mean) only marginally, for shrinkage from 0.63 to 0.66 and for creep from 0.60 to 0.64. Obviously, the overall statistics of the database cannot distinguish a nonsensical model from a realistic one. How could it then be trusted for comparing various models?

The ACI-209 model cannot be used for extrapolation of short-time tests to long times, to larger thicknesses, to higher ages at loading, and to higher temperatures, and especially not for modern high-performance concretes. If used in design for lifetimes exceeding 20 years, it is likely to lead to structural damage.

11.3.3   Need for Short-Time Data Extrapolation by Linear Regression

The need to keep the extrapolation of short-time test data simply means that all or most of the free (adjustable) parameters of the model should be involved linearly, so as to allow linear regression. This is the only foolproof, unambiguous method of data fitting, a method that always works and a method that always gives a unique answer.

Model B3, as well as B4, has been constructed so that all its 5 free (adjustable) parameters governing the basic creep and the drying creep (parameters q_1, q_2, q_3, q_4 and q_5) be involved linearly. Its compliance function has the general form

$$J(t, t') = \underbrace{q_1}_{\text{asymptotic elastic}} + \underbrace{q_2\, Q(t, t')}_{\text{aging viscoelastic}} + \underbrace{q_3\, \ln[1 + (t - t')^n]}_{\text{nonaging viscoelastic}}$$

$$+ \underbrace{q_4\, \ln(t/t')}_{\text{aging flow}} + \underbrace{q_5\, F(t, t'; t_0, h_{\text{env}}, D)}_{\text{drying creep}} \quad (11.3)$$

where the individual terms represent physically well-identifiable distinct components of creep, and the functions multiplying individual parameters q_i are fixed and do not have to be changed in fitting the test data for any concrete, probably not even high-strength or lightweight concrete. Recall that t is the current time, t' is the age at the application of sustained stress, $Q(t, t')$ is a fixed function given by (3.12), $n = 0.1$, and F is a function that depends on factors influencing the drying process, such as the time at the onset of drying, t_0, environmental humidity, h_{env}, and the effective thickness of the concrete part, D.

By contrast, the ACI, CEB, and GL models have the basic form

$$J(t, t') = \underbrace{p_1 E_{28}/E(t')}_{\text{conventional elastic}} + \underbrace{p_2\, G(t, t'; t_0, h_{\text{env}}, D, p_3, p_4...)}_{\text{all creep}} \quad (11.4)$$

in which p_1 and p_2 are parameters of the model, G is a nonlinear function, $E_{28} =$ chosen reference value (for age $t' = 28$ days) of the elastic modulus, and $E(t') =$ assumed static modulus of elasticity, which corresponds (according to ACI formula (3.5) for estimating the elastic modulus from strength) to the loading duration of about 0.01 day and has a given dependence on age t' of concrete; p_1 in this case represents the value of the conventional elastic compliance at the age of 28 days giving the best fit of the given creep data. Specific forms of function G for individual models can be deduced from the description and equations in Appendix E. In the ACI, CEB and other models, there are, of course, further parameters (in (11.4) denoted by the general symbols p_3, p_4) which could be modified, but they would require nonlinear regression. None of these models has five parameters identifiable by linear regression.

An experimentally proven feature of models B3 and B4, which simplifies the formulation but is not exploited to advantage in the ACI, CEB, fib and GL models, is the fact that if the curves of $J(t, t')$ for various ages t' at loading, plotted as functions of $(t - t')^n$, are extrapolated leftward, they all meet approximately at one point corresponding to q_1 (see Fig. 3.5 in Chap. 3). The load duration at which this value gets approached is too short to have any practical meaning. It corresponds to a hypothetical, truly instantaneous elastic compliance, whose inverse is called the asymptotic modulus.

The physical explanation is that the creep process has been found to possess no characteristic time below which the creep would cease to exist (in other words, the retardation spectrum is continuous and nonzero into the shortest durations). By virtue of this fact, the term with q_1 in (11.3) is constant. But the term with p_1 in (11.4) is not, which unnecessarily increases the number of parameters (and also makes it impossible to capture creep for very short times, although this is not of concern to engineers interested in long-life design).

In (11.4), only one elastic parameter, p_1, and only one creep parameter, namely the overall multiplying factor p_2, are involved linearly, while the others are not. This nonlinearity, and the lack of separation of the compliance function into its additive components of different physical meanings, is an impediment to using the ACI, CEB, or GL model for extrapolating given short-time data and for updating the model by fitting it to the given limited data for the given particular concrete. However, in the final version of the *fib* Model Code 2010, the basic and drying creep are split additively, i.e., linearly superposed.

In passing, it may be noted that a linear regression has three potent advantages over a nonlinear one:

1. it is much easier to carry out;
2. a solution is always obtained and is unique, while nonlinear regression might not converge or might converge to a solution corresponding to a local rather than global minimum of the error; and
3. the statistics such as the correlation coefficient or the coefficient of variation of errors are clear and easily obtained.

Figures 3.21 and H.2–H.3 show examples of extrapolation exercises in which it was pretended that only the initial data points from long-range measurements were known. For creep they are successful, but for shrinkage only sometimes. The reason probably is that separation of the drying shrinkage from the autogenous shrinkage in the specimen core before it dries needs to be better understood and would require a more refined experimental approach.

11.4 Statistical Methods Applied to Model Evaluation

Following mostly Bažant and Li [161], this section addresses the problem of selecting the most realistic creep and shrinkage prediction model solely on the basis of a large experimental database. In Sect. 11.5.2, comparisons with individual experimental curves, which have relatively low scatter, are used to validate the shape of the creep and shrinkage curves of models B3 and B4.

The statistical evaluation based on an experimental database [160] should follow a well-known standard method [224, 229, 330, 398], whose outcome should be almost unique. However, much confusion has been caused by the use of various nonstandard statistical methods [14, 407, 408, 641]. The result was that a model rated as superior according to one statistical method was rated as inferior according to another.

Are all the statistical methods used in various creep and shrinkage studies correct? Most of them are not. In the case of creep and shrinkage, in which one deals with central-range statistics of errors (and not with the far-out distribution tail which matters for structural safety), it is actually clear what is the correct statistical approach. It is the *method of least squares*—the standard method which, as shown by Gauss [411], maximizes the likelihood function and is consistent with the central limit theorem of the theory of probability [248, 285, 384]. There are, of course, many debatable points, but they concern only the details such as the sampling and weighting of data, transformation of random variables, or the relevance and admissibility of data, rather than the least-square statistical indicator per se. This chapter will present correct statistical comparisons of the main prediction models for creep and shrinkage of concrete, and explain why various nonstandard statistical indicators have led to dubious conclusions. **Five models** will be considered:

1. Model **B3**, which was approved as the international RILEM Recommendation [104] and later slightly updated [107]. This model is a refinement of model BP [175] and of its improvement known as model BP-KX [148, 151] and is featured in 2008 ACI Guide [14].
2. **ACI** model [11], based on 1960s research [253, 254], which was reapproved in 2008 [14].
3. Model of Comité Euro-International du Béton, labelled **CEB**, which is based on the work of Müller and Hilsdorf [641]. It was adopted in 1990 by CEB [322] and updated in 1999 by *fib* [390], was co-opted in 2002 for Eurocode 2, and is featured in 2008 ACI Guide [14].
4. Gardner and Lockman's model, labelled **GL** [407], which is featured in 2008 ACI Guide [14].
5. Gardner's earlier model, labelled **GZ** [408].

Sakata's model [748, 749] will be omitted from consideration. Also omitted will be the oversimplified old models of Dischinger, Illston, Nielsen, Rüsch and Jungwirth, Mörsch, Maslov, Arutyunyan, Aleksandrovskii, Ulickii, Gvozdev, Prokopovich, and others, conceived during 1935–1970 [66, 80, 87, 126].

The first comprehensive database, comprising about 400 creep tests and about 300 shrinkage tests, was compiled at Northwestern University in 1978 [175], mostly from American and European tests. In collaboration with CEB, begun at the 1980 Rüsch Workshop [323], this database was slightly expanded by an ACI-209 subcommittee chaired by L. Panula. A further slight expansion was undertaken in a subcommittee of RILEM TC-107, chaired by H. Müller. It led to what became known as the *RILEM database* [638, 640, 641], which contained 518 creep tests and 426 shrinkage tests. A reduced database, consisting of 166 creep tests and 106 shrinkage tests extracted from the RILEM database, was used in Gardner's studies [405–407]. Subsequently, an enlarged database, named *NU-ITI database* [160] and consisting of 621 creep tests and 490 shrinkage tests, was assembled in the Infrastructure Technology Institute of Northwestern University by adding some Japanese and Czech data. After an additional major expansion, the new *NU database* contains 1403 creep curves and 1809 shrinkage curves and covers also the effect of various kinds of admixtures [488].

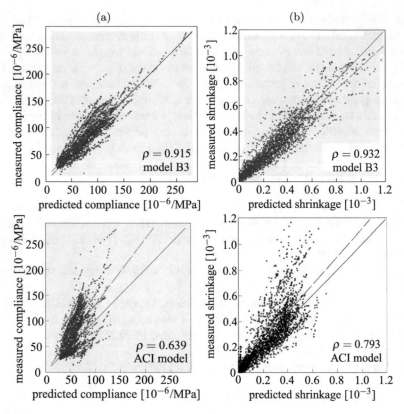

Fig. 11.5 Scatter plots of the measured versus predicted values of (a) creep and (b) shrinkage (dashed lines are regression lines and ρ is the correlation coefficient)

Among concrete researchers, **two methods to validate a model** have gained wide popularity:

1. In one method, the measured values y_k ($k = 1, 2, \ldots n$) from an experimental database are plotted against the corresponding model predictions Y_k (Fig. 11.5).
2. In the other, the errors $e_k = y_k - Y_k$ are plotted against the independent variable, in this case the time [25, 26, 616].

If the model were perfect and the tests scatter-free, the first method would give a straight line of slope 1, and the second a horizontal line of ordinate 0. Figure 11.5 shows examples of plots based on the first method for some of the aforementioned models and the NU-ITI database. One immediately notes that, in this kind of comparison, there is relatively little difference among the prediction models examined, even among those which are known to give very different long-time predictions (e.g.,

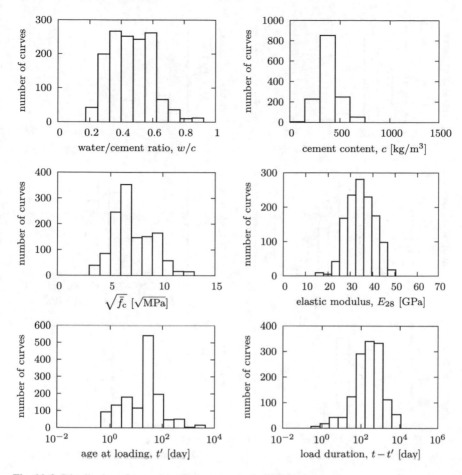

Fig. 11.6 Distribution of creep input parameters in the NU database

ACI and B3).[2] These plots demonstrate the spread of scatter, but are not effective for our purpose. Several **shortcomings** have been pointed out:

1. The statistical trends in time, age, thickness, etc., are not reflected in such plots.
2. The statistics are dominated by the data for short times $t - t'$, low ages t' at loading, and small specimen sizes D, while predictions for long times are of main interest for practice. This is due to highly nonuniform data distributions evident from the

---

[2]The same is true for a variant of the second method, popular with concrete researchers, where the ratio $r_k = y_k / Y_k$ is plotted versus time, for which, if the model were perfect and the tests scatter-free, one would ideally get a horizontal plot $r_k = 1$ (see Fig. 11.9a). For problems with such kind of statistics, see the comments on Eqs. (K.8)–(K.9) in Appendix K. Although the plots of residuals reduced to population statistics (Fig. 11.9a) have been popular in the literature, the plots showing the data trend, as in Fig. 11.9b, are far more informative.

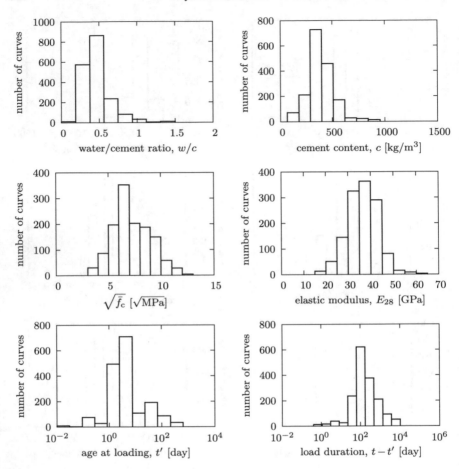

Fig. 11.7 Distribution of shrinkage input parameters in the NU database

histograms in Figs. 11.6 and 11.7. Note that, while in the top left histogram in Fig. 11.6, the actual t of each particular data point is considered; in the bottom left, all the tests running **at least** until t are included. This could make a false impression of having many very short tests, but actually the first column reflects all the tests, including long ones.

3. Because of their longer test durations and high creep and shrinkage, the statistics are also dominated by the data for old low-strength concretes not in use any more. Long-duration tests of modern high-strength concretes, which creep little, are still rare, as documented by Fig. 11.6.

4. The variability of concrete composition and other parameters in the database causes enormous scatter masking the scatter of creep and shrinkage evolution.

If the worldwide testing in the past could have been planned centrally, so as to follow the proper statistical design of experiments, the chosen sampling of the relevant parameters and reading times of creep and shrinkage tests would have been completely different than found in the databases. One must compensate for these deficiencies.

If the time, age, and specimen size are transformed to variables that make the trends uniform and the data set almost homoscedastic [37], and if these variables are subdivided into intervals of equal importance, the number of tests and the number of data points within each interval should ideally be about the same. However, this is far from true for every existing database (Figs. 11.6 and 11.7). There is nonetheless no choice but to extract the best information possible from the imperfect database that exists.

11.4.1   Suppressing Database Bias Due to Nonuniform Sampling of Parameter Ranges

From Figs. 11.6 and 11.7, showing the histograms of the existing data from worldwide testing, it is seen that their distribution in the existing database is highly nonuniform. This nonuniformity is not an objective property but a result of human choice. It thus leads to a bias in data evaluation.

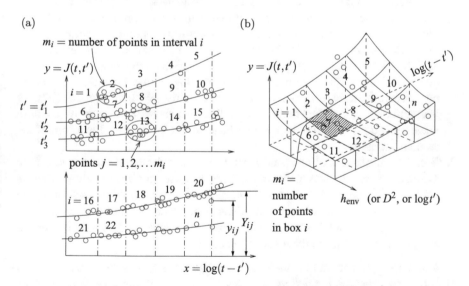

(a)   (b)

Fig. 11.8 Sketches explaining subdivision of database variables into (a) one-dimensional intervals and (b) two-dimensional boxes of equal importance; figure originally published in ACI Materials Journal, Nov-Dec 2008

This bias must be counteracted by proper weighting of the data. To this end, one may first subdivide the load duration $t-t'$, age at loading t', effective specimen thick-

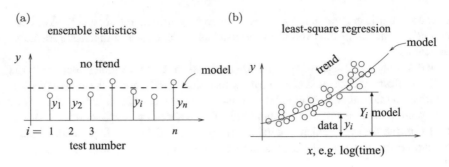

Fig. 11.9 Sketches explaining the difference between (a) ensemble (or population) statistics and (b) regression statistics

ness D, and environmental humidity h_{env} into intervals of roughly equal importance, which ought to have approximately the same weight in the statistical evaluation (in shrinkage tests, the age at loading, t', is replaced by the age at the start of drying, t_0). This is achieved by subdividing $\log(t - t')$ into equal intervals in the logarithmic scale (Fig. 11.8a), which means that the subdivisions of $t - t'$ form a geometric progression. The reason for this kind of subdivision into intervals is twofold:

1. One reason, already invoked, is that the least-square statistical regression gives best results when conducted in variables in which the data appear as approximately *homoscedastic*, i.e., have an approximately uniform conditional variance about the regression line [36, 285]. Plotting the creep or shrinkage data in terms of the load duration $t - t'$ or drying duration $t - t_0$, one finds the data to be markedly *heteroscedastic*, with a variance rapidly decreasing in time. To make them nearly homoscedastic, transformation of the variables is the standard approach [37]. As is generally the case when the relative, rather than absolute, changes of response matter, this transformation happens to be achieved by taking a logarithm of the random variable.

2. Because shrinkage and creep at constant load are decaying processes, a time increment of, say, 10 days makes much difference when the test duration is 10 days but little difference when it is 1000 days. In other words, intervals forming an arithmetic progression cannot have equal importance. By contrast, tripling the duration is about equally important in both cases, and this corresponds to intervals of equal length, $\log 3$, in the logarithmic scale.

A similar argument can be made in regard to the effective thickness (or size) D of the cross section, defined as $D = 2V/S_e = 2 \times$ volume/surface ratio of the specimen. Since a uniform coverage of the shrinkage halftimes is desirable, and since $\tau_{sh} \propto D^2$, the proper coordinate transformation is from D to D^2. This transformation is indicated by the diffusion theory (Chap. 8), which shows that the halftime of drying (or shrinkage) is proportional to D^2. As for the environmental humidity h_{env}, no transformation seems necessary, although small uniform intervals of h_{env} are not possible because of gaps in the distribution of h_{env} in the database.

For creep, there are four independent variables which need to be subdivided into intervals of equal statistical weight: $t - t'$, t', D, and h_{env}; for shrinkage, also four: $t - t_0$, t_0, D, and h_{env}.

Ideally, all these subdivisions should be introduced simultaneously, which creates many multidimensional subdomains (or hypercubes), henceforth called 'boxes'. With four independent variables, the boxes are four-dimensional. Their use, however, runs into a difficulty: For the database that exists, it appears that the number of data points in many of the four-dimensional boxes is 0. No statistics can be taken for such boxes, and so these boxes have to be deleted. But deletion of different boxes from the evaluations of different data sets implies the relative weights of various data sets to be unequal. Since boxes of lesser dimensions have a lesser chance of containing no points or too few points, two-dimensional boxes of $\log(t - t')$ and h_{env} for creep (Fig. 11.8b), and $\log(t - t_0)$ and D^2 for shrinkage appear to be preferable over three- or four-dimensional boxes.

Should differences in weights be also considered for data sets obtained on different concretes and in different laboratories? Maybe they should, but this would be a subjective judgment exposed to criticism. Besides, such differences in weights would certainly be much smaller than an order of magnitude. This makes introduction of such weights unimportant in comparison with the weights w_i for the data boxes, which must differ by more than one order of magnitude in order to compensate for the huge differences in the number of data points in different boxes.

Another debatable point is whether the boxes for long creep or shrinkage durations should not actually receive a greater weight than those for short durations. Probably they should, since accuracy of multidecade prediction is most important.

11.4.2 Reducing Anti-High-Strength Bias

The tests of old types of concretes with high water-cement ratios, lacking modern admixtures, dominate the database. Of little relevance though such concretes are today, these tests cannot be ignored because they supply most of the information on very long creep and shrinkage durations. Besides, these tests are not irrelevant for our purpose because the time curves for low- and high-strength concretes are known to have similar shapes. This is not surprising since, in both, the sole cause of creep is the calcium silicate hydrate. The difference resides mainly in the scaling of creep and shrinkage magnitudes. This scaling depends strongly on the water-cement ratio and admixtures, in a way that is not yet predictable mathematically.

Therefore, the data for old kinds of concrete must be used. But their bias must be counteracted. Since the overall magnitude of creep and shrinkage strains is roughly proportional to the elastic compliance, and since the elastic modulus is roughly proportional to the square root of strength, we can reduce the bias by scaling all the measured compliances and shrinkage strains y in inverse proportion to $\sqrt{\bar{\bar{f}}_c}$ where $\bar{f}_c$ is the cylindrical compressive strength. The measured compliance and shrinkage data y may thus be replaced by the adjusted values

$$\tilde{y} = y \sqrt{\frac{f_c^0}{\bar{f}_c}} \qquad (11.5)$$

where $f_c^0 = 5000$ psi $= 34.5$ MPa is a reference strength introduced to retain convenient dimensions.

11.4.3   Standard Least-Square Regression Statistics of the Database

Based on the subdivision into boxes of equal importance, the weighted standard error s of the prediction model (representing the *standard error of regression*) is defined as [36, 285, 446]

$$s = \sqrt{\frac{N}{N-p} \frac{1}{n} \sum_{i=1}^{n} \frac{1}{m_i} \sum_{j=1}^{m_i} (y_{ij} - Y_{ij})^2} \qquad (11.6)$$

where $n =$ number of boxes; $m_i =$ number of data points in box number i; $N = \sum_{i=1}^{n} m_i =$ number of all the data points in the database; $p =$ number of input parameters of the model ($p \leq 12$ for model B3); $y_{ij} =$ measured creep or shrinkage data of which the database is comprised; $Y_{ij} =$ corresponding model predictions; and $y_{ij} - Y_{ij} = e_{ij} =$ errors of the predictions.

Since $N \gg p$, the multiplier $N/(N-p)$ in (11.6) is very close to 1 and could thus be dropped. This multiplier serves to eliminate a different kind of bias, namely, to prevent the variance of regression errors of the database with a finite number N of data points from being systematically smaller than the variance of a theoretical database with $N \to \infty$ [36, 446]. Another reason why this multiplier is mathematically necessary is that p points can in theory be fitted exactly, so that only the remaining $N - p$ points contribute to s. To counteract the human bias in sampling, the statistical weights $1/m_i$ of the individual data points in each box are chosen as inversely proportional to the number m_i of data points in that box.

To compare various models, it is convenient to use dimensionless statistical indicators of scatter, e.g., the *coefficient of variation* (C.o.V.) of regression errors, defined as

$$\omega = \frac{s}{\bar{y}} \qquad (11.7)$$

where

$$\bar{y} = \frac{1}{n} \sum_{i=1}^{n} \frac{1}{m_i} \sum_{j=1}^{m_i} y_{ij} \qquad (11.8)$$

is the weighted mean of all the measured values y_{ij} in the database. The coefficient of variation, ω, which should be minimized, characterizes the ratio of the scatter band width to the data mean. For a model that gives predictions in perfect agreement with the measured data, it would be equal to zero.

11.4.4 Bias Due to Different Density of Readings

It should be mentioned that another type of bias arises from differences in the density of readings in time. For instance, one over-diligent experimenter might crowd 1000 readings into the time interval from 1000 to 2000 days and another experimenter might be satisfied with only 3 readings in that same interval. Obviously, 3 readings suffice and 1000 readings are superfluous. If not compensated, the data with 1000 readings will dominate the statistical comparisons and the data with 3 readings will be rendered almost irrelevant. To compensate for this kind of bias, Bažant and Panula [175] smoothed (by hand) each experimental time curve and placed on it two data points per decade in the logarithmic scale. The hand smoothing, admittedly, reduced the C.o.V., though only slightly (especially compared to differences among different concretes) and introduced a certain subjective bias (this bias could have been avoided by fitting local groups of points by a polynomial, but the gain in the accuracy of C.o.V. would have been negligible).

Consider that the experimental curves labelled as $j = 1, 2, \ldots N_i$ within interval i contain very different numbers of data points. The corresponding bias may be suppressed by redefining the formulae for weighted averaging as follows:

$$\bar{y} = \frac{1}{n} \sum_{i=1}^{n} \frac{1}{N_i} \sum_{j=1}^{N_i} \frac{1}{l_{ij}} \sum_{k=1}^{l_{ij}} y_{ijk} \tag{11.9}$$

$$s = \sqrt{\frac{N}{N-p} \frac{1}{n} \sum_{i=1}^{n} \frac{1}{N_i} \sum_{j=1}^{N_i} \frac{1}{l_{ij}} \sum_{k=1}^{l_{ij}} (y_{ijk} - Y_{ijk})^2} \tag{11.10}$$

Here, subscripts i refer to the interval or box, j to the test (author), and k to the data point, and l_{ij} is the number of data points on curve number j within interval i. However, in the statistics reported in Bažant and Li [161] and here, this correction was not carried out. The reason is that the differences were not large, and thus, the potential gain in accuracy of comparisons seemed to be quite small.

11.5 Statistical Comparison of Creep and Shrinkage Models

11.5.1 Model Evaluation by Standard Regression Statistics

Although a new greatly expanded database [488] has been compiled by the time of publication, it will suffice to illustrate the procedure using the previous NU-ITI database [160]. Bažant and Li [161] used this database to evaluate the coefficients of variation ω of the five aforementioned prediction models. Their results are summarized in Table 11.1, with all values of ω expressed as percentage.

The comparison in Table 11.1 is based on one-dimensional boxes (intervals), two-dimensional boxes, and three-dimensional boxes (cubes). Four-dimensional boxes, which numbered 1400 for compliance and 1120 for shrinkage, have also been tried but found statistically useless because more than half of them were empty. In addition to the actual compliance and shrinkage, the evaluation has been done for their relative values, defined as $J(t, t')/J_{ref}$ and $\varepsilon_{sh}(t, t_0)/\varepsilon_{sh,ref}$ where J_{ref} is the compliance for $t - t' = 3$ days and $\varepsilon_{sh,ref}$ is the shrinkage for 28 days of drying. The table presents results without the adjustment removing the strength bias. The adjustment turns out to have a relatively weak influence on the results and does not affect the ranking of individual models. For instance, for two-dimensional boxes, the coefficients of variation of the error in compliance prediction before and after strength bias removal would be 27.3% versus 27.0% for B3, 42.6% versus 42.3% for ACI, 31.0% versus 31.8% for CEB, 30.2% versus 28.8% for GL, and 41.9% versus 39.6% for GZ.

Table 11.1 Coefficients of variation of errors (expressed as percentage) of various prediction models

	Compliance					Relative compliance				
	B3	ACI	CEB	GL	GZ	B3	ACI	CEB	GL	GZ
5 intervals, $\log(t - t')$	26.2	41.9	29.7	28.5	43.8	26.4	66.0	33.0	29.8	32.9
4 intervals, $\log t'$	27.4	37.1	29.9	28.8	48.2	26.9	74.3	33.3	30.5	33.0
10 intervals, h_{env}	24.4	44.2	29.0	30.7	44.6	21.0	52.6	28.0	25.4	28.6
50 boxes, $\log(t - t')$, h_{env}	27.3	42.6	31.0	30.2	41.9	23.8	55.0	30.2	27.6	31.8
200 cubes, $\log(t - t')$, $\log t'$, h_{env}	28.3	38.8	30.6	28.5	39.5	24.4	59.0	29.3	27.3	35.7
	Shrinkage					Relative shrinkage				
	B3	ACI	CEB	GL	GZ	B3	ACI	CEB	GL	GZ
4 intervals, $\log(t - t_0)$	29.4	40.8	48.0	37.7	49.3	34.5	49.5	46.0	43.3	54.7
4 intervals, $\log t_0$	42.8	48.6	56.0	53.9	64.2	44.9	52.8	57.6	54.0	64.7
10 intervals, h_{env}	38.4	52.0	46.9	54.4	46.6	41.6	55.6	43.0	41.9	45.6

In all these comparisons, model B3 is seen to be the best, except for one case where it is one of two equal best. As the second best comes out Gardner's newer model GL [407], which modifies his original model GZ [408] by co-opting two key aspects of model BP [175], namely the shrinkage function and the quadratic dependence on the size or volume–surface ratio. Considerably worse but the third best overall is seen to be the CEB model. The current ACI-209 model [11], established on the basis of 1960s research and reapproved in 2008 [14], comes out as the worst in creep and second worst in shrinkage, which is not surprising since it is the oldest.

11.5.2 Statistical Justification of Model B3

The original statistical justification and calibration of the B3 model [105] used the same least-square statistics as just described with one exception in weighting (and so did, to a large extent, the justification of the BP model by Bažant and Panula [176]). The C.o.V. of regression errors was evaluated separately, with proper weights, for the data of each author and was denoted ω_r for author number r. Assuming the distribution of errors to be approximately Gaussian when the concrete type is varied, one should combine the individual ω_r values to get the overall C.o.V. according to the root-mean-square rule

$$\omega_{\text{all}} = \sqrt{\frac{1}{N_r} \sum_{r=1}^{N_r} \omega_r{}^2} \qquad (11.11)$$

where N_r is the total number of authors. This definition of C.o.V. is more egalitarian, since it ascribes the same weight to each author's data set. On the other hand, definition (11.6)–(11.7) used by Bažant and Li [161] ascribes the same weight to each test curve, which means that the author who conducted more tests receives greater weight. But the difference between these two definitions seems to have only a minor effect on the statistics. Another difference of the original statistics of Bažant and

Table 11.2 Example from Bažant and Baweja [105] of coefficients of variation of errors (expressed as percentage) of shrinkage predictions for three classical models, evaluated separately for test data sets of 21 authors, and coefficients of variation implying the same weight for each data set

	B3	ACI	CEB
1. Hummel et al. [498]	27.0	30.0	58.7
2. Rüsch [744]	31.1	35.2	44.8
3. Wesche et al. [862]	38.4	24.0	36.1
4. Rüsch et al. [746]	34.7	13.7	27.8
5. Wischers and Dahms [869]	20.5	27.3	35.9
6. Hansen and Mattock [453]	16.5	52.9	81.5
7. Keeton [538]	28.9	120.6	48.3
8. Troxell et al. [817]	34.1	36.8	47.4
9. Aschl and Stökl [41]	57.2	61.3	44.2
10. Stökl [790]	33.0	19.5	29.6
11. L'Hermite et al. [580]	66.7	123.1	69.4
12. York et al. [887]	30.6	42.8	8.9
13. Hilsdorf [482]	11.7	24.7	29.6
14. L'Hermite and Mamillan [576–579]	46.1	58.7	45.5
15. Wallo et al. [848]	22.0	33.0	55.6
16. Lambotte and Mommens [563]	39.1	30.7	31.3
17. Weigler and Karl [855]	31.3	29.6	21.3
18. Wittmann et al. [878]	23.7	65.4	40.0
19. Ngab et al. [655]	20.4	45.3	64.6
20. McDonald [617]	5.1	68.8	21.4
21. Russell and Burg [747]	38.5	51.0	58.1
$\omega_{\text{all}} = \sqrt{\sum_r \omega_r^2 / N_r}$	34.3	55.3	46.3

Panula [176] and of Bažant and Baweja [105] is that they were conducted on smaller databases.

Bažant and Baweja's [105] statistics of the errors of model B3 for shrinkage compared to the test data sets in the RILEM database are presented in Table 11.2 separately for each author (a similar table with fewer data was published by Bažant and Panula [175] for the BP model). For comparison, the statistics of errors of the ACI model and the CEB model are also included. Analogous tables can be found in Bažant and Baweja [105] for basic creep, creep at drying, and creep at elevated temperature. The overall coefficients of variation are summarized in Table 11.3.

Table 11.3 Overall coefficients of variation of errors (expressed as percentage) of creep predictions evaluated for three classical models by Bažant and Baweja [105]

	BP	ACI	CEB
Basic creep	23.6	58.1	35.0
Creep at drying	23.0	44.5	32.4
Creep at elevated temperature	28.1	–	–

The scatter of shrinkage (and drying creep) data is generally higher than the scatter of basic creep. One cause is the differences in the method of measuring shrinkage, which was not reported for many data. Some measurements were made along the axis of the cylinder, others on the surface, and the gage length sometimes reached close to the ends of the specimen, sometimes not [20]. In many tests, the ends of cylinders or prisms were not sealed, which caused shrinkage warping at the ends, and the specimens were not long enough, and gauge points are not far enough from the specimen ends. Thus, the complex deformation of specimen ends contaminated the results in an undocumented way. Another problem was the microcracking. It is very random and is more pronounced in shrinkage specimens than in compressed creep specimens. Unknown vertical shifts of the shrinkage curves were caused when the experimenters started measuring shrinkage with some undocumented delay after the stripping of the mold.

Table 11.4 Example from Bažant and Baweja [105] of statistics of errors of ACI model for basic creep and creep at drying, calculated separately for subsequent constant intervals of $\log t'$, with t' in days (similar tables were calculated for other classical models and for intervals of $\log(t - t')$)

ACI model	$t' \leq 10$	$10 < t' \leq 100$	$100 < t' \leq 1000$	$t' > 1000$
$t - t' \leq 10$	60.3	30.7	33.3	
$10 < t - t' \leq 100$	45.7	36.7	49.9	97.1
$100 < t - t' \leq 1000$	34.6	39.9	51.7	93.9
$t - t' > 1000$	36.8	39.9	40.9	

According to Table 11.4, the coefficients of variation of model B3 as well as B4 remain low even for the last interval of the age at loading (over 1,000 days), while for some other models they become very large for that range. Correct representation of creep for loading ages over 1,000 days is important for calculating from the principle of superposition the long-time stress relaxation and the shrinkage stresses of thick members, as well as for calculating the stress variation in structures over long periods of time.

It must be emphasized that the largest deviations from the data points seen in the figures are caused by errors in the prediction of model parameters from the composition and strength of the concrete. If the model parameters are adjusted, all these data can be fitted very closely, but then one is not evaluating the prediction capability of the model. The figures showing most of the data from the database were presented in Bažant et al. [151] and Bažant and Kim [148] along with the basic information on the tests.

Some creep and shrinkage models have been justified by a limited selection of the existing data sets. Such a selection can easily be misleading. To give an example, consider that among the 21 data sets in Table 11.2 only the 12 most favorable data sets are used (which might seem like plenty for justifying a model). But the ω_{all} value drops from 34.3 to 23.7%. Likewise, if among the 17 available data sets for basic creep, only the 7 most favorable data sets are selected, the ω_{all} value drops from 23.6 to 10.7%. These observations, and similar ones already mentioned [177], document the deception that could be hidden in a selective use of test data. Justifying some model by 12 or 7 data sets may look like plenty, yet it can be deceptive (with the possible exception of randomly chosen data).

If no weights are used in the regression statistics, a smaller coefficient of variation of errors can be achieved because short-time data are much more plentiful, and most models are better for short times. However, what matters most is long-time predictions.

Another visual perception of the degree of scatter can be obtained from the plots of the measured value y_k versus corresponding predicted values Y_k of creep or shrinkage; see Fig. 11.5. If the models were perfect and no scatter existed, these plots would be straight lines of slope 1 passing through the origin. Thus, deviations from this line represent errors of the model predictions plus inevitable scatter of the measurements. The errors of model B3 are seen to be smaller than those for the previous ACI 209 model and the CEB model, especially for large creep strains (corresponding to long times), which are most important. Generally, however, as emphasized before, this kind of plot is not very useful when the scatter is large.

The *correlation coefficient* ρ (aka coefficient of determination) of the population of the measured values y_i and the corresponding model predictions Y_i, $i = 1, 2, \ldots N$, is given in each figure. It has been calculated according to the standard formula

$$\rho = \frac{\sum_{i=1}^{N}(y_i - \bar{y})(Y_i - \bar{Y})}{\sqrt{\sum_{i=1}^{N}(y_i - \bar{y})^2 \sum_{i=1}^{N}(Y_i - \bar{Y})^2}} \tag{11.12}$$

where $\bar{y}$ and $\bar{Y}$ are the mean values of the measured data and model predictions, resp. Note that ρ characterizes only the grouping of the data about the regression lines of the plots, which are drawn as the dashed lines in Fig. 11.5. The regression lines do not have slope 1 and do not pass through the origin, which represents a measure of how well the data trend is captured by the model and represents another kind of error that is not reflected in the value of ρ. We can see from these plots that, in the case of model B3, the regression line for creep is close to the line of slope 1 through the origin, and also the correlation coefficient ($\rho = 0.915$ for creep and $\rho = 0.932$ for shrinkage) is quite close to its maximum possible value 1.

11.6 Statistical Justification of RILEM Model B4

The new 2015 RILEM Model B4 [136] uses the same general form of the creep compliance function (C.1) and shrinkage function (3.15) as model B3, but provides improved empirical formulae for the estimation of parameters, which are presented in detail in Appendix D and will be often referred to in the present section. Calibration of the model was based on an enlarged creep and shrinkage database, combined with a simplified inverse analysis of data of multidecade deflections of 69 large-span prestressed concrete bridges, which are the only source of information for durations approaching the design lifetimes (often 100 years or more) and for structural members of large thicknesses with very long diffusion halftimes. A simplified variant of the model, denoted as B4s, uses the mean compressive strength $\bar{f}_c$ as the only input parameter.

The method of statistical justification of the B4 model used by Hubler, Wendner, and Bažant [489] and Wendner, Hubler and Bažant [860] was, for the most part, similar as that already described for model B3. Therefore, only the differences and improvements need to be discussed.

11.6.1 Shrinkage

Figure 11.10 presents the B4 predictions of the total shrinkage and the autogenous part of shrinkage for a typical concrete composition as function of size D, ambient humidity h_{env}, temperature T, and cement type.[3] In each diagram of the figure, the parameters that are not varied are set to $c = 400 \, \text{kg/m}^3$, $w/c = 0.35$, $a/c = 4$, $\bar{f}_c = 40 \, \text{MPa}$, $D = 150 \, \text{mm}$, $T = 20 \, ^\circ\text{C}$, $h_{env} = 65\%$, and cement type R.

Because of the huge scatter due to differences among various concretes, the shape of these functions cannot be verified by fitting the entire database. Same as B3, model B4 could be verified by the ability to fit only individual shrinkage curves of a duration long enough to cover the initial phase and the terminal phase of decreasing slope

---

[3]The European classification of cements is selected for model B4 since it is directly related to the reaction rate of the cement instead of the type of application, which is the basis of other classification systems. It should be noted, though, that the class labels used by B4 (RS = rapid hardening, R = normal, and SL = slow hardening) are somewhat different from the class labels used in Eurocode 2 (R = rapid hardening, N = normal, and S = slow hardening) and in CEB Model Code 1990 (RS = rapidly hardening high-strength concrete, R = normal, N = normal, and SL = slow hardening).

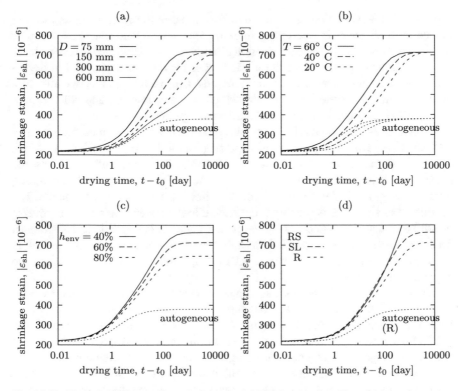

Fig. 11.10 Typical shrinkage curves given by model B4, showing the effect of (a) specimen size, (b) temperature, (c) ambient humidity, and (d) cement type

in log-time. The shrinkage data from Wischers and Dahms [869] and Keeton [538] are the only data that reveal both phases. The initial phase, with ε_{sh} proportional to $\sqrt{t - t_0}$, is dictated by diffusion theory and may be verified by L'Hermite et al. [580], Wittmann et al. [878] and Wallo et al. [848]. To verify the effect of environmental humidity, the most useful data are Troxell, Raphael, and Davis [817], L'Hermite et al. [580], Keeton [538], Mamillan [599, 600], and Pentala and Rautanen[674]; for ambient temperature England and Ross [371], Ayano and Sakata [45] and Jensen and Hansen [515]; for specimen size Wittmann et al. [878] and Shritharan [777]; and for the age at exposure to environment Wischers and Dahms [869], Yang, Sato and Kawai [885] and Wallo et al. [848].

11.6.1.1   Decomposition of Shrinkage into Autogenous and Drying

Unlike B3, the development of B4 required an extensive study of the effect of various *additives* and *admixtures*. While some admixtures such as *silica fume* and *air-entraining agents* primarily affect the microstructure of concrete, and thus can be captured by a recalibration of the basic diffusion-based formulation, others lead to self-desiccation, which is especially pronounced in high-strength concretes. The pore humidity decrease during self-desiccation directly causes shrinkage but also reduces the pore humidity gradient, which retards further the rates of drying and of drying shrinkage. Complications arise from the diverse admixtures and reactive additives used in modern concretes (e.g., superplasticizer, water reducer, and silica fume), which have interdependent opposing or attenuating effects.

B4 also takes into account the aggregate-type effect which, as some studies show [30], can change the 28-day shrinkage by as much as 400×10^{-6}. The density of rock aggregate appears to have the greatest effect on the shrinkage halftime, the elastic modulus, and the final shrinkage, which can be explained by a parallel coupling model. The aggregate density also affects the effective permeability of concrete. The magnitude of shrinkage is roughly proportional to the material compliance (inverse of stiffness) and, in consequence, also to the contributing aggregate compliance. This dependence is captured in B4 through empirically obtained correction factors $k_{\tau a}$ and $k_{\varepsilon a}$, which scale the shrinkage halftime, τ_{sh}, and the final shrinkage, ε_s^∞; see Table D.2 in Appendix D.

The parameters of the B4 shrinkage model were statistically calibrated by optimizing the fit of the full 2014 NU database [488], which contains 1217 shrinkage curves of specimens exposed to drying (more than twice as many curves as the RILEM database used for B3). They represent the total shrinkage, i.e., the sum of drying shrinkage and autogenous shrinkage. The database also contains 417 curves of autogenous shrinkage, observed on sealed specimens. Among these data, there are 177 shrinkage tests in which both the total and autogenous shrinkages were measured. Their difference yields the 'pure' drying shrinkage, but was not used in data fitting because possible cross effects between the drying and autogenous parts were not clear at the time of fitting.

There is a lack of consensus on the effect of many admixtures or additives. This is largely because the market is continuously evolving and no testing standard exists. There are important exceptions, though. One is the tests of Brooks [266, 269, 271], which consistently show the *superplasticizer* to reduce short-term shrinkage, and in particular its autogenous component. Based on Brooks [265, 266] and Wei, Hansen, Biernacki, and Schlangen [853], the *blast furnace slag* increases the long-term shrinkage, especially in the case of a high w/c ratio. *Silica fume* increases both the short-term and long-term shrinkage, and in particular the long-term slope of the shrinkage curve in log-time [242, 266, 457, 503, 514, 677, 804]. Tests by Buil and Acker [283], as well as de Larrard and Bostvironnois [342], indicate the long-term autogenous shrinkage to be twice as large as it is for normal concrete. Low doses of *fly ash* show no effect on shrinkage [265], but a replacement of between 25–50% of the cement results in an increase of autogenous shrinkage, while a replacement above 50% reduces autogenous shrinkage compared to normal concrete [807]. The *viscosity agents* are consistently seen to increase shrinkage [15]. But the *water reducers* and *retarders*, which are the most popular admixtures, show no consistent trends [15, 266].

To systematize the effects of admixtures, a classification system was introduced. Studies of correlation between the admixture contents and the observed shrinkage revealed the magnitude and functional form of these effects, as compared to normal concrete. Subsets were then created relating the admixture dosages or combinations, or both. The number of subsets with low or high admixture dosages were determined by the data availability and the overall fit quality.

After an approximate formula for the composition effects on autogenous shrinkage has been identified from the database, the tests of shrinkage in a drying environment have been fitted assuming a contribution, whether major or minor, of the autogenous shrinkage as predicted by the formula. In this aspect, model B4 differs from B3, in which the autogenous shrinkage contribution in drying shrinkage tests has not been separated.

Formulas for autogenous shrinkage, all of exponential decay form, were proposed by Jonasson and Hedlund [524], Tazawa and Miyazawa [804] and Miyazawa and Tazawa [634] and recommended by RILEM [726] and CEB-FIP [389]. The key similarities among these equations is their dependence on the water-cement or water–binder ratio, cement type, and compressive strength of the concrete. Since the mechanism causing autogenous shrinkage, comprising a multitude of chemical reactions, is a microscale mechanism still not sufficiently clarified, the goal is a simple conservative estimate with the fewest parameters. Because the amount of reactants is finite, autogenous shrinkage must eventually approach a final asymptotic value, but ongoing studies indicate that this value might be approached only after many years of even decades because diffusion barriers can slow down the chemical processes enormously.

Sensitivity studies using the 2014 NU database revealed the strongest effects to be those of the water-cement and aggregate–cement ratios of each mix. Autogenous shrinkage begins right after mixing but what is of interest for mechanics is only the autogenous shrinkage after the moment of set. It is important to realize that the autogenous shrinkage continues in the core of drying specimens until the front of the external drying arrives, which can take even years, depending on specimen size. Optimizing the fits with Eq. (D.17), empirical formulae (D.19)–(D.20) for the parameters of the final autogenous shrinkage and autogenous shrinkage halftime have been identified [489].

Parameters dependent on the cement type are given in Table D.4. Exponents -4.5 in (D.17), -0.75 in (D.19) and 3 in (D.20) were also optimized but since their value turned out to be independent of the cement type, they are represented in the final formulae as fixed numbers. The effect of admixtures and reactive additives is taken into account by correction factors that scale the basic values of parameters and are presented in Table D.6.

When the mold is stripped and drying begins, some autogenous shrinkage and self-desiccation (i.e., a reduction of pore humidity) has already taken place. Thus, the humidity difference between the specimen and the environment is reduced. This tends to reduce the apparent shrinkage of high-strength concretes, compared to normal concretes. Therefore, in the drying shrinkage tests, of high-strength concretes in particular, one must measure the autogenous shrinkage before the drying exposure and, of course, also afterward, on companion specimens for the entire duration of creep and shrinkage tests. Unfortunately, most data in the database miss this information.

In formulating model B4, it was not clear whether the autogenous shrinkage was caused (a) directly by volume changes during the chemical reactions of hydration or (b) indirectly by the pore humidity drop during self-desiccation. Were the former dominant, the total shrinkage would be the sum of drying and autogenous shrinkages, and if the latter were, the total shrinkage would approximately be the maximum of drying and autogenous shrinkages. Regrettably, there are no data to decide (however, theoretical studies under way at the time of proof unequivocally show that (b) is true, and a revised prediction model for autogenous shrinkage is in preparation).

Most tests in the database that were aimed at drying shrinkage actually measured only the total (drying plus autogenous) shrinkage in drying environment and did not include separate tests of the autogenous shrinkage alone. Such tests are indispensable for determining the separate contribution of drying shrinkage, since the autogenous shrinkage continues even after stripping the mold and, in the specimen core, is unaffected by drying for a long time (months, years or decades, depending on thickness), until the moment at which the drying front penetrates to a point at the boundary of the self-desiccation zone in the core. After that, the external drying begins decreasing the pore humidity at that point even more than by self-desiccation alone. The time to reach this point obviously depends on the specimen size.

Since the database does not suffice to distinguish between the aforementioned formulations, the additive formula (D.15) has been adopted for B4, as it is more conservative and simpler to use. The tests of shrinkage in a drying environment have been fitted taking into account a contribution of the autogenous shrinkage as predicted from this formula.

11.6.1.2   Parameter Identification and Optimization

The general optimization algorithm and strategy used to calibrate model B4 are described in Wendner, Hubler, and Bažant [859]. Compared to the calibration of creep, the minimum of the sum of squared errors in shrinkage that is to be minimized is less sharp (or flatter), pushing the optimization problem closer to ill-posedness [203]. This is because most shrinkage data do not satisfy the following three requirements:

1. To be able to identify the shrinkage halftime and the final value, the time range of the shrinkage test must be long enough for the logarithmic scale plot to flatten off and show an approach to the final bound;
2. to avoid the initial offset, i.e., the strain and time shift of the entire shrinkage curve, the first reading must be taken right after the stripping of the mold, preferably within a few seconds; and
3. the prior and concurrent autogenous shrinkage must be measured, too.

A step-by-step approach, based on the fact that the shrinkage curve must initially evolve as the square root of time, was used to make corrections to the reported data for the initial offset. Furthermore, a multidimensional weighting scheme was used to counteract various kinds of bias that exist in the database, as detailed in Wendner et al. [859].

The first step in developing B4 was to study the sensitivity of various aspects of shrinkage to the main composition parameters (w/c, c, a/c, ...). Once the strongest dependencies were identified and introduced into the formulae, parameters were assigned to scale their effects. To obtain the optimum model for the common concrete compositions, first the complete parameter set was optimized using only data sets for normal Portland cement with no admixtures, and for temperature cca 20 °C. At the same time, the data subsets that cover the full time range were used to tune variables characterizing the shrinkage rate. The long-term data that show the full S-shaped curve in log-time were identified and assigned higher weights in the optimization phase dedicated to finding the predictor equations and scaling factors related to the final shrinkage value. Once the final parameter set for the average mix composition was determined, a different set of scaling parameters was introduced and optimized for each deviation from the average composition (e.g., for different cement types or admixtures). Factors whose effect on the quality of fit turned out to be within the inevitable range of scatter for one concrete were considered as constant.

The shrinkage dependence on the aggregate type may serve as an example. Most data in the NU database do not specify the aggregate type. The portion of the tests that does was subdivided into categories with at least 5 tests. The typical values and ranges of Young's modulus, density, porosity, and moisture expansion of each aggregate type were found in the literature [30, 397, 429, 439, 653, 894]. Theoretically, the aggregate stiffness must have a restraining effect on shrinkage and the density may affect the overall permeability. A sensitivity study indicated that the shrinkage halftime was most correlated to the aggregate density and the final shrinkage value to the elastic modulus of aggregate. Factors could then be introduced on these two parameters for optimization. The halftime factor was optimized to fit the S-shaped test curves of full time range, and the final value factor to fit the long-time curves. For the optimized aggregate scaling factors, see Table D.2.

To quantify the effect of various admixtures, by themselves and in combination, the sensitivity of the C.o.V. of errors to the model parameters has been studied. As expected, the admixtures influence primarily the autogenous shrinkage. The only detectable effect is that of cement type factor, τ_{cem}, on the drying shrinkage halftime.

For admixture classes with insufficient data, marked in Table D.6 by asterisk, the correction factors have been obtained by linear interpolation of calibrated neighboring values. Table D.3 lists the cement-type parameters in the drying and autogenous shrinkage models. The rapid hardening (RS) cements reveal a substantial decrease of the shrinkage halftime, explained by accelerated hydration of the material which leads to a decreased gradient of pore humidity. The RS-type cements are seen to lead to higher shrinkage. The slow hardening (SL) cements do not have a significant influence on autogenous shrinkage. However, the SL cements alter the dependence on the composition parameters associated with the shrinkage rate.

A quantification of model parameter uncertainty has been obtained by individually refitting all the shrinkage curves, with parameters τ_{sh}, ε_{sh}^∞, τ_{au} and ε_{au}^∞ considered as freely adjustable. The ratio between the optimal value of a parameter for a given test and the value predicted from an empirical formula of the B4 model is called the correction factor, ψ. As discussed by Wendner et al. [859], such correction factors primarily characterize the intrinsic material uncertainty, biased by random errors associated with the testing. They may serve as input for long-term performance predictions and lifetime analyses [791, 792, 861]. To estimate the 5 and 95% confidence limits, Hubler et al. [489] constructed histograms of the correction factors from the database and found that they are closer to the log-normal than normal distribution; see Fig. 11.11. The confidence intervals based on log-normal distribution were [0.5, 2.5] for the scaling factor of τ_{sh}, [0.5, 3.1] for the scaling factor of ε_{sh}^∞, [0.6, 4.6] for the scaling factor of τ_{au}, and [0.6, 5.7] for the scaling factor of ε_{au}^∞.

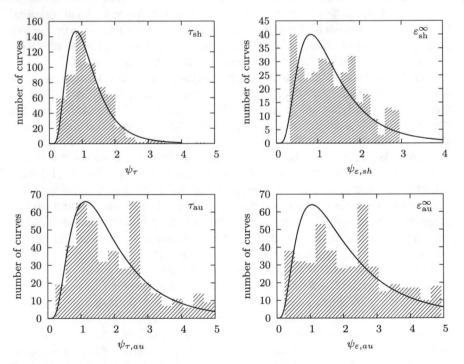

Fig. 11.11 Histograms of correction factors ψ for individual shrinkage parameters and their approximation by log-normal distribution

The scatter plots of predicted versus measured values, popular in the past, were not used because they are dominated by short-time data and the scatter due to composition

Table 11.5 Quality of fit (each shrinkage curve scaled separately) and quality of prediction characterized by the coefficients of variation: (a) fit of 7 selected sets with sufficient data in initial asymptotic part of curve, (b) fit of 32 selected sets with sufficient data in final part of curve, (c) prediction for all concrete compositions in the NU database, (d) prediction for concrete compositions without admixtures

Model	(a) Fit, initial part	(b) Fit, final part	(c) Prediction, all concretes	(d) Prediction, no admixtures
B4	0.104	0.040	0.309	0.316
B4s	0.157	0.035	0.399	0.326
B3	0.188	0.053	0.422	0.410
fib MC10	0.169	0.038	0.510	0.408
CEB MC99	0.169	0.038	0.517	0.415
GL 2000	0.198	0.057	0.363	0.330
ACI92	0.120	0.058	0.435	0.377

variability completely masks the errors in the shape of time evolution curves. More meaningful comparisons are presented in Table 11.5, which shows the C.o.V.s for models B4, B4s, B3, fib MC10, CEB MC99, GL 2000, and ACI92.

The comparison in column (a) of Table 11.5 is based on 7 selected data sets that include the initial phase and reach into the final phase of decreasing slope in log-time, and the comparison in column (b) is based on 32 selected data sets that all give some information on the final phase but only some for the initial phase. For each model, parameters that allow horizontal and vertical scaling (in semilogarithmic scale) were identified, and these parameters were adjusted for each measured curve individually, to get the best possible fit with the given form of the shrinkage evolution function. Table 11.6 summarizes the time functions used by the compared models. It also gives the number of intrinsic parameters and the number of fitted (scaling) parameters for each model. The linear (shape-preserving) transformation of the curve by vertical and horizontal scalings is necessary to isolate the errors in the curve shape from the overall errors. The quality of optimum fit of data achievable by the vertical and horizontal scalings is what reveals how realistic the curve shape is. Figures 11.12 and 11.13 present examples of fits obtained with model B4 and, for comparison, with the ACI92 model.

After verification of the functional form, the overall calibration quality of the models was evaluated by comparing their predictions (based on composition and test conditions, without a priori knowledge of the measured values) with the shrinkage curves in the database. Table 11.5 specifies the corresponding C.o.V.s, computed first for the entirety of all concretes (column (c) in the table) and then only for the concretes without admixtures (column (d)). The graphs in Fig. 11.14 give the evolution of C.o.V. computed separately for each mid-decade in $\log(t - t_0)$. The data sets giving insufficient information on the input parameters of even one model had to be omitted for all the models.

From the statistical comparisons in columns (c) and (d) of Table 11.5, note that both B4 and GL00, the two models that have been calibrated by experiments includ-

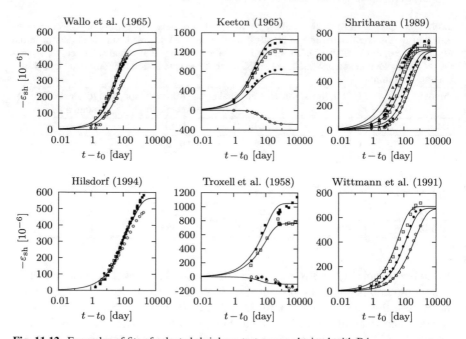

Fig. 11.12 Examples of fits of selected shrinkage test curves obtained with B4

Table 11.6 Summary of time functions and parameters of various shrinkage models and the corresponding number of intrinsic parameters and fitted parameters used for the model comparisons

Model	Time function	Autogenous time function	Intrinsic parameters	Fitted parameters
B4	$\tanh\sqrt{\dfrac{\hat{t}}{\tau_{sh}}}$	$\left[1+\left(\dfrac{\tau_{au}}{t}\right)^{\alpha}\right]^{-r_t}$	4	4
B4s	$\tanh\sqrt{\dfrac{\hat{t}}{\tau_{sh}}}$	–	3	2
B3	$\tanh\sqrt{\dfrac{\hat{t}}{\tau_{sh}}}$	–	4	2
MC10	$\sqrt{\dfrac{\hat{t}}{\tau_{sh}+\hat{t}}}$	$1-\exp\left(-\sqrt{\dfrac{t}{\tau_{au}}}\right)$	2	4
MC99	$\sqrt{\dfrac{\hat{t}}{\tau_{sh}+\hat{t}}}$	$1-\exp\left(-\sqrt{\dfrac{t}{\tau_{au}}}\right)$	2	4
GL00	$\sqrt{\dfrac{\hat{t}}{\tau_{sh}+\hat{t}}}$	–	2	2
ACI92	$\dfrac{\hat{t}}{f+\hat{t}}$	–	2	2

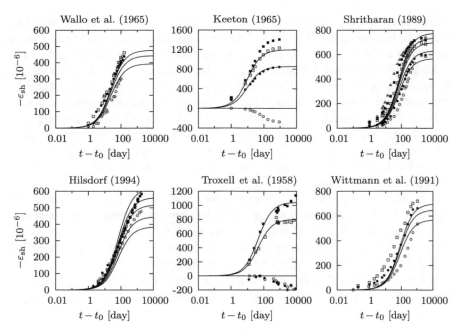

Fig. 11.13 Examples of fits of selected shrinkage test curves obtained with ACI92

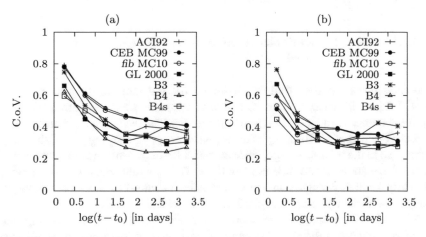

Fig. 11.14 Quality of prediction according to various shrinkage models: development of coefficient of variation with drying time for (a) all concrete compositions, (b) concrete compositions without admixtures

ing modern concretes, give lower C.o.V.'s in comparison with the full database. Model GL00, however, cannot capture well the shrinkage evolution in time, mainly because it lacks a separate function for autogenous shrinkage; see column (b) in Table 11.5. This function introduces important additional parameters, which allow a more accurate representation of the shrinkage curve and, of course, reflect the shrinkage mechanism more realistically. As a result, model B4 shows the lowest C.o.V. of errors in the comparisons with the data covering both long and short times (columns (a) and (b)), and the best overall quality of prediction over the longest time range (columns (c) and (d) and Fig. 11.14). Inclusion of the autogenous shrinkage and the parameters for admixtures and aggregate type is crucial for predicting the shrinkage of high-strength concretes.

The ACI92 model shows the most inconsistent behavior. In the beginning, the shrinkage is underestimated, between 30 and 1000 days it is overestimated, and ultimately tends to underestimations again. The resulting scatter band is very wide. Model B4 is not the best in each statistics but is best overall. It is interesting that model B4s, which considers the compression strength instead of the composition, performs almost as well as B4. However, it cannot predict the influence of composition in presence of admixtures, which is important in selecting the concrete mix.

11.6.2 Creep

11.6.2.1 Overview and Explanations of B4 and Compliance Functions to Be Compared

The compliance function used by model B4 has the same form as in model B3 [104, 179], with two exceptions. One is that minor improvements are made in the equivalent times introducing the temperature effect. The second is that, unlike B3, the drying creep part of the B4 compliance is related only to the drying part of shrinkage, rather than to the total shrinkage, since in model B4 the drying and autogenous parts of shrinkage are split into separate functions. This is a refinement that is important primarily for high-strength concretes for which, in contrast to normal concretes, the autogenous shrinkage is not a negligible part of total shrinkage.

The functions defining the relations of the basic parameters of the compliance function to the composition and strength of concrete and to the environmental conditions have been identified and optimized with the help of a new large laboratory database featuring about 1400 creep tests, and another database featuring multi-decade deflections of 69 bridge spans, both assembled at Northwestern University [138, 488]. Note that all the creep tests used for calibration were conducted under centric uniaxial compression. Therefore, the available models including B4 can have large errors in the case of bending or highly eccentric loads. The reason is that the microcracking distribution and the interaction of stress distribution with pore humidity are different. However, this is not a problem for bridge box girders when the walls are subdivided through the thickness into finite elements because the eccentricity of the compression resultant in each such element is always minor.

11.6.2.2 Effects of Temperature, Cement Type, and Admixtures

Model B4 [136] introduces equivalent times based on Arrhenius-type equations for the temperature effects on the creep rate, aging (or hydration) rate, and drying shrinkage rate. In principle, their activation energies can be different but, because of data ambiguity, the activation energy Q of each is considered the same ($Q/R \approx 4000$ K, with R = gas constant), as formulated in [175] and roughly supported by several experimental studies [698, 849]. This temperature dependence does not apply above $75\,°C$, because of phase changes and because different activation energies dominate in different temperature ranges.

In basic creep, the activation energies of creep rate and of hydration compete with each other, the former accelerating and the latter decelerating the creep as temperature rises. The effect of the latter disappears once hydration is complete (which often occurs after about one year but can take much longer, especially in modern concretes). The drying part of creep also depends on the activation energy of drying (or diffusion process), which leads to an acceleration of the drying creep term when the temperature is raised. These effects are captured in model B4 by a series of scaling parameters.

Admixtures have a smaller effect on creep than on shrinkage. The effects of water-reducers, retarders, superplasticizers, air-entraining agents, accelerators, shrinkage-reducing agents, and mineral admixtures have been studied for creep. Many test data on the effects of cement type and of admixture type and amount exist, but they are so scattered that no systematic trends can be detected.

The differences in the effects on the rate and the magnitude of total creep attributable to admixtures depend on their diverse effects on evolution of microstructure. There is no consensus on the contribution of water-reducers and superplasticizers, as the data lie in the range of experimental uncertainty. While the analysis of the full laboratory database shows that the addition of accelerators and the fly ash replacement exceeding 15% systematically cause some increase of creep, generally the air-entraining agents, shrinkage reducing admixtures, and low amounts of fly ash replacement are found to have no consistent, systematic and statistically verifiable effect on creep.

The high-strength concrete has been shown by various researchers [27, 633, 730] to have a creep coefficient about 1.8–2.4 times smaller (in spite of that, the creep effects on the structural scale can be important because the structures made of high-strength concrete, typically prestressed, are more slender and more flexible). The creep reduction is due to the lower w/c ratio and the addition of silica fume or fly ash. The self-consolidating concrete has similar a creep as the normal concrete [678].

What is clear at present is that the effects of admixtures are highly variable statistically and no unique time functions exist. For the mean behavior, it seems sufficient to introduce empirical coefficients that scale only the creep magnitude. As for the effect on multidecade creep in particular, no data exist. Recalibrations should be performed in the future as new data become available.

Similar studies were made for the effect of cement type on the basic and drying creep. Calibrated parameters capturing the cement type dependence exist in all models for creep. Predictions are complicated by the fact that cement classifications as well as cement products and production standards have changed over time and various cement replacements have been introduced. This engenders a large scatter and uncertainty in the model calibration. The type of cement used shows a strong correlation to the observed basic and drying creep when using the data in the NU database. On the other hand, contrary to shrinkage, there is little correlation to the aggregate type classes. Even though an effect of the aggregate type is perceived to exist [30], there is a lack of consistent and repeated test data. For each type of aggregate, there exist only few curves, in the current NU database at most 6, which is not enough for statistical inferences.

11.6.2.3   Optimization of Fit of Combined Laboratory and Bridge Databases

Large bridges and other creep-sensitive structures are generally designed for service lives of 50–150 years. However, 95% of the laboratory creep tests available in the largest worldwide laboratory database [488] with 1370 creep curves do not exceed 6 years in duration. Only 3% of the data sets, many of them with questionable reliability of long-term measurements, exceed 12 years.

Consequently, the laboratory data used for calibration of a creep model must be supplemented by inverse inference from multidecade structural observations. Most informative for that purpose are the data on deflections of large-span prestressed concrete segmental box girder bridges, provided that the deflections are excessive (if they are not, it means that a large gravity deflection is offset by a large upward deflection due to prestress, which is a small difference of two large random numbers and is too scattered to be useful). Data on multidecade shortening of prestressed bridge girders would be useful even if the deflections are small, but such data are unavailable. Data on multidecade shortening of columns of tall buildings would also be useful but are unavailable as well.

The most useful bridge paradigm is the Koror–Babeldaob bridge in Palau, described in detail in Sect. 7.2. As explained in Sect. 7.7.1, Bažant et al. [210] found that the creep equations in the standard recommendations or design codes of engineering societies severely underestimated the midspan deflections. Their predictions amounted to 31–43% of the measured values.

The new Northwestern University (NU) [488] database, which more than doubles the size of the previous laboratory database [638], includes also the data on relative multidecade deflection histories of 69 large bridge spans from nine countries and four continents [138]. These data are used in statistical inverse analysis and are crucial for calibrating the terminal trend of creep. A complete inverse analysis was unfortunately impossible due to a lack of information on the concrete composition and strength, structural geometry, and prestressing for most of the bridges.

Instead, based on the method formulated by Bažant et al. [138], the mean terminal deflection development was transformed into an approximate terminal compliance evolution based on estimating likely average properties of these bridges and their concretes. These estimated properties included: the required design strength, which was converted to the mean strength of concrete, the average effective cross-sectional thickness, the environmental humidity (based on the bridge location), and the cement composition. Errors stemming from these simplifying assumptions mostly compensate each other in a statistical sense, and so the mean relative compliance development deduced from all the 69 bridge spans is probably roughly correct even though the absolute residuals are, of course, rendered meaningless by these estimations.

The analysis of bridge data showed a systematic underestimation of the terminal trend of creep and led to an adjustment of the compliance function that minimizes the error in matching the terminal deflections of these 69 bridges. In the optimization, the transformed bridge deflection data were considered to have 1/3 of the total weight (and the laboratory database 2/3). The terminal bridge deflections were introduced only for optimizing the parameters that control the terminal slope of the compliance function in the logarithmic time scale. Since the database mostly contains data of much shorter durations (<6 years), only the scaling parameters (and not the formulas for the intrinsic and extrinsic influences) were optimized for the bridges. Thus, the optimization of the effects of concrete composition and environment was not biased by the incompleteness of bridge data.

11.6.2.4 Parameter Identification and Optimization Method

While the initial goal of the update of the creep model was solely a recalibration (keeping the functional form and theoretical foundation of model B3), several assumptions in the model were re-examined before proceeding with the optimization process.

The first is the initial elastic modulus for static load application. As mentioned in Model B3 RILEM Recommendation 107-GCS [104], the inverse of the 28-day elastic modulus given by the ACI empirical equation (3.6) corresponds to the compliance for 5–20 min after load application. However, a better agreement can be reached between standard 28-day modulus and total compliance after roughly 1–2 min ($\Delta t_s = 0.001$ days). This conclusion is the basis of the calibration of model B4 as well as B4s. Figure 11.15 shows the histograms of ratios between the strength-based estimate of conventional modulus, $E_{28}^{(s)}$, and the value of $E_{28}^{(c)} = J(28+\Delta t_s, 28)$ with compliance function J determined using the B4 and B4s predictions.

It is not surprising that the simplified model B4s leads to a better agreement between the values of elastic modulus determined from creep compliance and those estimated from strength. B4s uses the compressive strength $\bar{f}_c$ as the only input parameter, and thus, both $E_{28}^{(c)}$ and $E_{28}^{(s)}$ are unique functions of $\bar{f}_c$. One could even introduce a constraint $E_{28}^{(c)} = E_{28}^{(s)}$ into the calibration procedure and enforce both values to match exactly, as discussed in Appendix D.8.2 and reflected by formula (D.66). However, it remains to be checked whether reduction of the number of free parameters does not lead to a deterioration of other aspects of the model behavior, such as the long-time response.

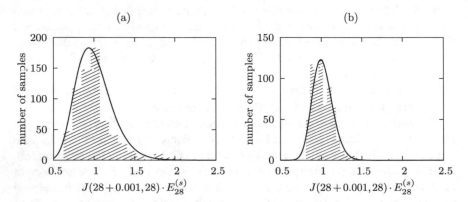

Fig. 11.15 Histograms of the ratio between creep compliance after $\Delta t_s = 0.001$ days and elastic compliance $1/E_{28}^{(s)}$ and their approximation by log-normal distribution for (a) model B4, (b) model B4s

Second, the exponents n and m of the load duration and age were calibrated by short to medium range data from the NU database as well as nanoindentation creep data for cement paste obtained by Vandamme et al. [834]. Only the basic creep tests of normal concrete, unaffected by drying and autogenous shrinkage, were used in this analysis. Unbiased optimizations with different starting points confirmed that, in an average sense, the previously assumed parameters $n = 0.1$ and $m = 0.5$ [105] still provide the best and, more importantly, consistently good, fits. For certain compositions, the prediction quality could be improved by varying n between 0.08 and 0.12. However, no consistent trend or dependency on composition parameters or cement type could be identified.

Third, the calibration of the creep model was in general highly sensitive to the value of the initial elastic strain. So, exponent p_1 in the estimation (D.4) of the instantaneous compliance in terms of the conventional elastic modulus had to be optimized first and then prescribed for all the subsequent optimization steps. Two approaches were pursued and turned out to yield similar results: optimization of the full formulation of model B4 (with fixed average long-term parameters) and a linear fit in power-law scale of the short-term test data with at least 3 measured data points within the first minutes to hours of measurement, depending on the age of concrete at load application (e.g., up to 4 hours for concretes loaded at 7 days). The limit is based on an empirical formulation that is derived for the functional form of model B4 based on sensitivity studies.

Fourth, recent important test data from M.I.T. on nanoindentation creep [834] have also been analyzed, for validation purposes. Since the tests were made on hardened cement paste, the compliance magnitude cannot be compared with the tests on concrete, but the exponent n of the load duration must be about the same. Figure 11.16 shows the measurement data for durations $t - t'$ from 0.1 s to 200 s, compared to the best fits by a logarithmic time function, by a power law with exponent $n = 0.1$, and a power law with optimum exponent. Sampling bias toward later ages with denser point spacing was removed through a weighting scheme with equal weights for each mid-decade in the log-scale.

The overall fit in Fig. 11.16 clearly shows that an exponent $n = 0.10$ is a good approximation. The best fit, with a coefficient of determination greater than 0.99, is attained for $n = 0.089$. The logarithmic function is a fair approximation but by no means an optimum. Figure 11.16b shows the fit to the first measurements for durations <1 s and its extrapolations to longer times. Again $n = 0.10$ works well. The optimum fit within [0.1 s, 1 s] leads to exponent $n = 0.577$, but the reason is that inserting the indenter took much longer than 0.1 s.

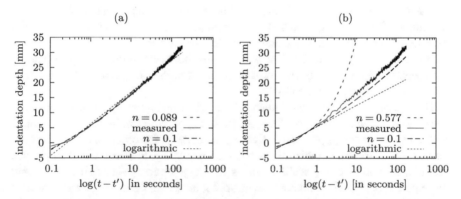

Fig. 11.16 Best fit of nanoindentation test data by Vandamme et al. [834] by logarithmic time function, power law with exponent $n = 0.1$, and power law with optimum exponent: (a) fit of the full data range, optimum $n = 0.089$, (b) fit of the first second only, optimum $n = 0.577$

The next stage required re-evaluating the form of the dependence of material parameters on concrete composition. The existing model (B3) depended on both the mix characteristics (i.e., the water-cement ratio, aggregate–cement ratio, and cement content) and the mean mechanical characteristics (i.e., the 28-day strength and the Young's modulus). It is well known that water-cement ratio, compressive strength, and Young's modulus are highly correlated. With decreasing w/c, both the strength and the elastic modulus increase. As a consequence of this high correlation, a simultaneous use of the strength and w/c brings about little gain and in fact makes the optimization problem ill-conditioned, yielding arbitrary and nonunique results. Furthermore, the compressive strength typically only serves as convenient indicator for other material properties.

Therefore, two sets of predictor equations, for two versions of model B4, have been formulated and calibrated, one using the mix proportions only (named B4), and one using the mean compressive strength only (named B4s). The B4s version for initial design, having fewer parameters, was expected to be inferior but, surprisingly, turned out to be on average almost as good as the B4 version that allows more refined predictions if the mix design is known. Young's modulus is used in both versions since it is the most important characteristic for the instantaneous deformation, but, if it is unknown, it can be estimated from compressive strength using the ACI empirical formula (3.6).

All the effects of composition and strength enter the material parameters in the form of products of power functions. This has the advantage of a linear relation between the logarithms of the input and response and thus helps convergence of the optimization (another reason for power functions is that they are self-similar, which is appropriate when no characteristic value is known). To keep the input values dimensionless, these functions have all been normalized by their typical values. This avoids most dimensional inputs, which also minimizes the chance of user's error in dimensions.

The water-cement ratio was found to be the most important input parameter for the magnitude of all the components of the compliance function. This is consistent with other studies and agrees with the creep mechanisms considered in the microprestress solidification theory (Chap. 10). The second most important is the aggregate–cement ratio, which affects the nonaging viscoelastic creep, the flow, and the drying creep terms of the compliance function.

The individual influencing parameters were identified by a step-by-step procedure using various statistical approaches. At first, the potential influencing parameters were selected as those reported by most experimenters. The objective was to identify the relations of these parameters to the basic parameters q_1 to q_5 of the B4 compliance function, as well as to the scaling factors for temperature, various admixtures and the cement type. For each unknown relation, for example, the effect of water-cement ratio on the scaling factor of the nonaging viscoelastic creep term, one could identify on the creep curve the time range of maximum sensitivity (one or a few decades in the logarithmic time scale).

Subsequently, the relations of model parameters to input material parameters affecting this time range were optimized, so as to minimize the C.o.V. of the differences between the predicted curve and the data points in this time range (relative to the mean of data, not of the differences) [859]. The optimization also yielded an R^2 error measure, a full Jacobian matrix for sensitivity analysis, and the fit of each curve for visual shape analysis. The evaluation of the Jacobian matrix revealed correlations between the model parameters and the input properties, as well as between both groups. This process allowed adjusting the formulation and a converged selection of input material parameters of the creep model (for normal concrete under standard conditions). Further scaling parameters were introduced to capture the effects of temperature, admixtures and cement type. The general optimization algorithm, strategy, and process used to develop the full model B4 are described in Wendner et al. [859]. The exponents p_1 and p_2 of the scaling factors in basic creep, and p_5 in drying creep, showed the strongest dependence on the cement type; see Table D.1. The effects of admixtures were best described by scaling the exponents p_2, p_3, and p_4 for basic creep and p_5 for drying creep; see Table D.5.

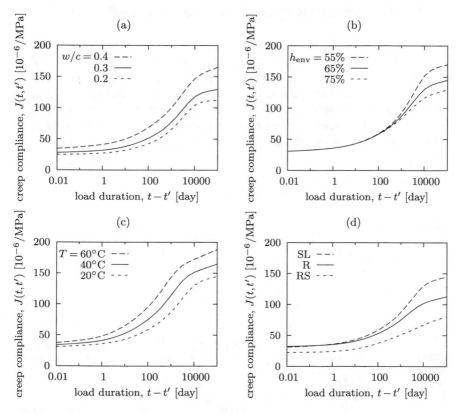

Fig. 11.17 Typical creep curves given by model B4, showing the effect of (a) water-cement ratio, (b) ambient humidity, (c) temperature, and (d) cement type

The changes in functional form of the B4 creep formula are illustrated in Fig. 11.17 by compliance curves that correspond to $D = 150$ mm, $c = 400$ kg/m^3, $a/c = 4$, and $\bar{f}_c = 40$ MPa. In each diagram of the figure, the parameters that are not varied are set to $w/c = 0.35$, $T = 20$ °C, $h_{env} = 65\%$ and cement type R. For standard conditions, an increase in w/c increases the creep rate as well as the vertical scaling factor of the creep curve (Fig. 11.17a). A decrease in the relative humidity of the environment (Fig. 11.17b) increases the vertical scaling factor but has no significant effect on the characteristic time of the creep function, which gives the horizontal scaling in a linear time plot (or a horizontal shift in log-time plot). An increase in temperature (Fig. 11.17c) generally engenders in the database concretes an increased rate and magnitude of creep (except possibly for very young concretes for which the hydration acceleration, which reduces creep, may prevail). The last diagram in the figure shows the change in the creep curve shape due to a change of cement type (Fig. 11.17d).

11.6.2.5   Verification of the Shape of Predicted Individual Curves

As described in Sect. 11.6.1, a separate statistical analysis aimed at verifying the shape of model B4 creep and shrinkage curves was performed at the outset. If the shape of the individual curves of some model is not realistic, it makes no sense to optimize that model by the database. However, by comparisons with the entire database it is impossible to check whether or not the shape of the creep or shrinkage curves is correct because the database scatters due to concrete type, composition, and admixtures dwarfs and obscures any strange features in the curve shape.

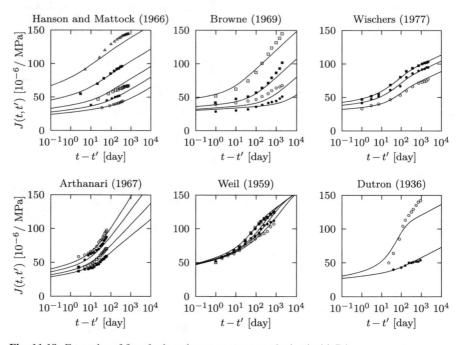

Fig. 11.18 Examples of fits of selected creep test curves obtained with B4

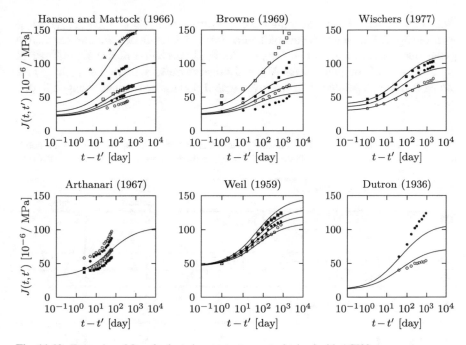

Fig. 11.19 Examples of fits of selected creep test curves obtained with ACI92

Figure 11.18 shows such comparisons of the model B4 curves with individual measured curves, using only the data from the tests whose duration range was long enough for the comparison to be meaningful. Figure 11.19 shows similar individual comparisons for the curves of the ACI92 model. To examine the capability of the general form of the models, the composition-dependent horizontal and vertical scaling parameters have been optimized, consistently for all the curves of each concrete batch.

These graphical comparisons are followed by statistical comparisons in terms of coefficients of variation documenting the capability of the form of each model to capture the shape of the individual creep curves (Table 11.8) as a function of load duration $t - t'$ and their dependence on the age t' at loading. Similar indicators are also used to compare the predictions obtained with the model for the full database (Table 11.9). A detailed study of the development of residuals gives insight into the model calibration.

Example fits in the top row of Fig. 11.18 show the capability of model B4 to fit tests of long durations or a broad range of ages at loading, selected from the NU database. The bottom row shows that test series with broad variations of environmental conditions (temperature and humidity) and specimen size can also be fitted well. The trends in the experiments on the same concrete could be recreated with a C.o.V. of less than 10% even though only the free scaling parameters were adapted, consistently, of course, for all curves of the same series (and thus the same concrete).

None of the parameters influencing the dependence on t', T, $D \equiv 2V/S_e$, and h_{env} were changed. Depending on the particular form of each model, the number of free parameters varied between two (i.e., the initial deformation plus the multiplier of the creep part of compliance) for ACI and other models, and five (i.e., q_1 to q_5) for model B4. The dependence on the investigated parameter was not changed in any case.

Table 11.7 Summary of time functions and parameters of various creep models and the corresponding number of intrinsic parameters and fitted parameters used for the model comparisons (times in days)

Model	Creep time function	Intrinsic parameters	Fitted parameters
B4	$Q(t, t')$, $\ln[1 + (t - t')^{0.1}]$, $\ln\left(\frac{t}{t'}\right)$, $\sqrt{\left(e^{-g(t-t_0)} - e^{-g((t'-t_0))}\right)}$	4	5
B4s	$Q(t, t')$, $\ln[1 + (t - t')^{0.1}]$, $\ln\left(\frac{t}{t'}\right)$, (D.3)	3	5
B3	(D.2), (D.3)	4	5
fib MC10	$\ln(1 + ct)$, $\left(\frac{t}{\beta + t}\right)^{\gamma}$	2	3
CEB MC99	$\left(\frac{t}{\beta + t}\right)^{0.3}$	2	2
GL 2000	$\frac{t^{0.3}}{14 + t^{0.3}}$, $\left(\frac{t}{7 + t}\right)^{0.5}$, $\left(\frac{t}{\gamma + t}\right)^{0.5}$	2	2
ACI92	$\frac{t^{\psi}}{d + t^{\psi}}$	2	2

Table 11.7 defines the creep time function, the number of intrinsic parameters (as a gauge of function flexibility), and the number of fitted parameters used for each model. *Intrinsic parameters* are herein defined as those parameters that describe the concrete composition, such as w/c, a/c, c, but also the strength and elastic modulus. In addition to a visual evaluation of the capability of the model to capture the shape of the creep curves, shape statistics are also calculated using a selection of curves with sufficient data in the initial and final range. The resulting coefficients of variation based on the laboratory data are presented in Table 11.8. A number of inferences can be made from this comparison.

If only data sets with the influence of drying are analyzed (column (a) in Table 11.8), model B4 based on concrete composition outperforms the other models, followed by B4s. The reason is that it can separate the drying shrinkage from the autogenous shrinkage and thus realistically describes the influence of drying creep in the presence of admixtures. Models without this split in autogenous and drying shrinkage (GL00 and ACI92) perform worst for total creep, even though the quality

Table 11.8 Quality of fit (each creep curve scaled separately) characterized by the coefficients of variation of errors for (a) 38 selected sets with data on total creep, (b) 43 selected sets with data on basic creep, and (c) 81 sets combined

Model	(a) Total creep	(b) Basic creep	(c) Combined
B4	0.131	0.106	0.118
B4s	0.155	0.126	0.140
B3	0.183	0.166	0.174
fib MC10	0.187	0.177	0.182
CEB MC99	0.178	0.147	0.162
GL 2000	0.257	0.186	0.222
ACI92	0.237	0.198	0.217

of fit for basic creep (no influence of drying) is only slightly inferior (column (b) in Table 11.8). It is interesting to note that the now replaced MC99 outperforms all other models except B4 and B4s with regard to short-term basic creep. The combined set of comparisons is presented in column (c) of Table 11.8 and follows the ranking governed by the influence of drying creep.

After evaluating, for various models, the functional form of compliance, i.e., the shape of the time curve, the next step is to investigate and compare their capability to predict the dependence on the age at loading. This step is omitted here for the sake of brevity; the findings are detailed in Wendner et al. [859].

The third step is to investigate and compare the overall prediction quality, considering the full NU database. To distinguish the quality of fit in early and later stages of creep, we first separately consider the laboratory data (mostly <6 year in duration) and the multidecade bridge data. Columns (a) and (b) in Table 11.9 show the quality of fit of different models for the laboratory data only, and column (c) shows the same for the multidecade relative bridge deflections. The combination of long-term laboratory data (longer than 1000 days) and relative bridge deflections is given in column (d).

The C.o.V. of residuals for short-term laboratory creep test data is found to be the lowest for the B4 and GL models. Their near equivalence may be due to the similar flexibility of the time function used and the fact that the GL model was empirically based on a carefully handpicked selection of creep tests that showed a clear trend in time rather than the complete data set, as has been done with the B3 and B4 models. In terms of global statistics MC10 outperforms its predecessor MC99 for short-term creep and reaches a close tie for long-term laboratory data even though the individual shape statistics of Table 11.8 show the opposite trend. The reason likely lies in a better overall calibration of the model (note that Table 11.8 illustrates the potential of the formulation, not its calibration). Model B3 suffers from the missing split in autogenous and drying shrinkage. This compromises the long-term prediction, due to the distortion of the drying creep component, in spite of its correct functional form as revealed in column (c) of Table 11.9. A wrong functional form (horizontal asymptote) as formulated for MC99, and ACI92 is clearly revealed

Table 11.9 Quality of fit of the new B4 and B4s creep models as compared to existing creep models using the coefficient of variations of errors as the quantifier: (a) short-term laboratory creep data $(t - t' \leq 1000$ days), (b) long-term laboratory creep data $(t - t' > 1000$ days), (c) relative bridge deflection data, (d) bridge and long-term laboratory creep data combined, with $\omega_{comb} = \sqrt{\omega_{lab}^2 + \omega_{bridge}^2}$

Model	Fitted parameters	(a) Short-term lab	(b) Long-term lab	(c) Bridges	(d) Combined
B4	5	0.150	0.147	0.230	0.273
B4s	5	0.201	0.247	0.199	0.317
B3	5	0.192	0.299	0.211	0.366
fib MC10	3	0.332	0.216	0.264	0.341
CEB MC99	2	0.416	0.204	0.424	0.470
GL 2000	2	0.170	0.300	0.210	0.366
ACI92	2	0.220	0.395	0.403	0.565

in the statistics of multidecade structural evidence. The GL00 model is an exception as its functional form corresponds to MC99 but is calibrated in such a way that it approaches a horizontal asymptote only far beyond the longest measurement times and thus mimics a terminal slope of the creep compliance in logarithmic time.

This fact underscores the need for a separate investigation of the functional form, and in particular its asymptotics, see column (c) in Table 11.9. Clearly all the models that can capture the correct asymptotics (B3, B4, MC10) or that approach it (GL00) outperform models that do not (ACI92, MC99).

If the long-term laboratory creep test data are combined with the bridge deflection information, a more balanced perspective of the long-term prediction quality is obtained. As expected, models B4 and B4s show the lowest C.o.V., followed by B3, MC10, GL00. The MC99 model cannot catch up with the competitors but still exceeds the prediction quality of ACI92 by far.

In future studies, in which not only the relative but also actual deflections of bridges should be used for calibration, it will be important to use a realistic model for steel relaxation as affected by temperature and strain variation; see Sect. 4.3.4.1.

To give a more detailed insight into the development of the prediction quality, Fig. 11.20 shows the residuals of all models plotted against $\log(t - t')$. The residuals (or errors) are defined as the differences $J - \hat{J}$ between the predicted compliances J and the measured compliances $\hat{J}$. Model B4, and also the simplified strength-based model B4s, consistently show a very small mean value of the residuals. The scatter band given by the 5 and 95% percentiles is largely symmetric, which confirms no bias toward over- or underestimation. The ACI92 and GL00, on the other hand, tend to underestimate creep for long times, as seen in the mean value trend and especially the scatter band. The scatter of MC10, interestingly, is symmetric. But it exceeds the scatter of all the other models in the range between 10 and 1000 days while decreasing for long times.

To reduce the scatter for long times, information on the concrete composition must accompany future structural measurements. So must the information on the bridge dimensions, prestress and environment.

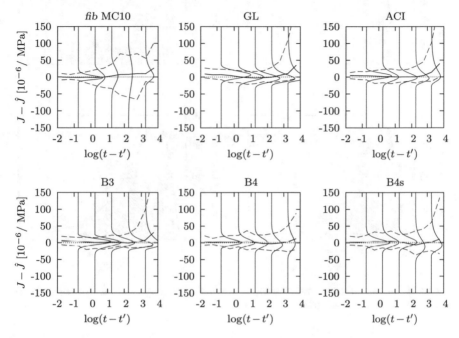

Fig. 11.20 Development of residuals with $\log(t - t')$ according to prediction model (the thick line corresponds to the mean and the dashed lines to 5 and 95% confidence limits)

A quantification of model parameter uncertainty was obtained by individually refitting all the creep curves with parameters q_1 to q_5 considered as uncertain and evaluating the correction factors, denoted as ψ_{qi}. Their distribution is shown in Fig. 11.21, and the corresponding intervals between 5 and 95% confidence limits are [0.6, 1.8] for ψ_{q1}, [0.4, 3.3] for ψ_{q2} and ψ_{q3}, [0.4, 2.7] for ψ_{q4}, and [0.4, 3.1] for ψ_{q5}.

11.7   Analytical Methods for Predicting Concrete Creep from Its Composition*

Although models B3 and B4 rest on many theoretical concepts (Sect. 11.2), the heterogeneity of the microstructure, e.g., the effect of cement-aggregate ratio, is still taken into account in an empirical way and some effects, such as that of grading of the aggregate (i.e., the grain size distribution) are not considered explicitly at

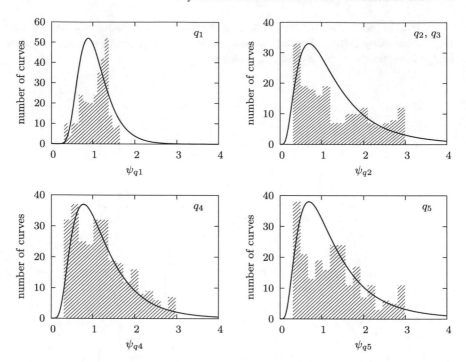

Fig. 11.21 Histograms of correction factors ψ_{qi} for individual creep parameters and their approximation by log-normal distribution

all. By contrast, the *homogenization theories* of fiber composites, other particulate composites, and metals can now provide good predictions of the macrocontinuum properties from the heterogeneous mesostructure (e.g., Hashin [464], Hill [478], Mori and Tanaka [637], Dvorak [362]) and thus facilitate optimization of material composition. But concrete is a much more complex material, for which the effects of heterogeneity must still be handled empirically. Nevertheless, some progress in this direction has been achieved and is briefly reviewed next.

11.7.1   *Predicting Creep and Shrinkage from Heterogenous Microstructure Using Homogenization Theory** *

At the mesoscale of mineral aggregates and cement mortar, the problem is essentially mechanical, while the chemical, thermal, and diffusional aspects can be cast aside. Numerous studies have been devoted to this problem; for elastic modulus, see Hansen [451], Dougill [358] and Counto [327]; for creep see Popovics [699], de Larrard and Roy [343], Nilsen and Monteiro [659], Bernard, Ulm, and Lemarchand [236], Šmilauer and Bittnar [845], and Pichler and Lackner [689]. In recent years, interest in applying homogenization theories to concrete creep surged and various homogenization methods have been extended to creep [260, 361, 565, 752, 753, 756, 773, 774, 809, 815, 832, 833, 844, 891, 892].

Baweja, Dvorak, and Bažant [215] approached the problem within the framework of Dvorak's *transformation field theory* [363, 364], which is an advanced version of the homogenization theory initiated by Hashin [464] and Hill [478]. Baweja et al. [215] simplified concrete as a two-phase composite, consisting of mineral aggregate and cement mortar. They made predictions on the basis of the volume fractions of aggregate and mortar, elastic constants of aggregate, and aging creep properties of mortar. The creep was treated in a simplified way by the age-adjusted effective modulus method. This eliminated time from the creep problem, making it quasi-elastic. The analysis led to explicit expressions for the aging creep properties of concrete. The model was calibrated and validated by test data of Ward, Neville, and Singh [851] and Counto [327] and a close match was achieved.

A simpler model, in which a combination of series and parallel couplings was used instead of the transformation field theory, was presented by Granger and Bažant [432]. The model also fitted the aforementioned test data well but, unlike the transformation field theory, its triaxial behavior did not capture properly the tensorial aspects.

The use of homogenization theory, however, faces multiple obstacles. For example, the mortar itself is a composite, and several levels of homogenization are needed. The grain shape and roughness can have a large effect. The closest gaps separating the aggregate pieces play a bigger role, as suggested by the fact that the prepacked concrete, which is made by infiltrating dry aggregate mass with a cement slurry and thus has aggregate pieces in contact, has a greatly reduced creep. The interface transition zone, which probably creeps much more than the rest of cement paste, should be modeled separately. The porosity of aggregates and ingress of water would need to be taken into account. Various levels of porosity, with disjoining pressure in nanopores, solid surface tension on C-S-H nanoglobules and capillary tension in micropores, would have to be part of the homogenization process for creep and shrinkage, and an analytical model of long-term hydration and self-desiccation (completed at the time of proof) would have to be introduced, etc. It was because of these obstacles that the homogenization theory was eschewed in developing models B3 and B4.

11.7.2 Extracting Creep Properties of C-S-H via Cement Paste Homogenization*

Inverse analysis with homogenization techniques can also be used to identify the properties of components hard to isolate in bulk. To identify the creep properties of calcium silicate hydrates (C-S-H) in the hardened cement paste from its heterogeneous composition, Šmilauer and Bažant [844] formulated a powerful and robust cement paste homogenization method using the spectral approach based on fast Fourier transform (FFT). Their identification was contingent upon the linearity of the creep law of C-S-H. To characterize the cement paste microstructure, they adopted the model developed by Bentz [233] at the National Institute of Standards and Technology (NIST), which has the resolution of $1 \mu m$. The basic form of model B3 (or B4) was assumed to be valid for the creep of C-S-H in cement paste.

The exponential algorithm for creep (Sect. 5.2) was extended to a heterogeneous viscoelastic composite. The numerical homogenization techniques relied on replacing the real microstructure by a *representative volume element* (RVE) of the material. The RVE must be statistically representative, i.e., must be large enough to contain sufficient information about morphology, and the RVE response must be independent of the type of imposed boundary conditions. The periodic boundary conditions were used since they were shown [535] to yield the smallest dependence of the elastic homogenized stiffness on the RVE size.

The RVE of cement paste was approximated by Bentz's [233] *discrete hydration model* (CEMHYD3D), which has the resolution of $1 \mu m$. Therefore, the periodic microstructure of cement paste was simulated on a grid of voxels $1 \times 1 \times 1 \mu m$ in size. Bentz's hydration model captured the main chemical phases and reactions occurring during cement hydration and was previously shown by Šmilauer and Bittnar [845] to yield excellent results for elastic homogenization of hardened cement pastes with periodic boundary conditions.

Experiments were conducted on cement pastes 2 and 30 years old, having the water-cement ratio of 0.5. The viscoelastic properties of C-S-H were identified at the resolution of $1 \mu m$. The inverse homogenization by FFT showed that close fitting of the homogenized properties of cement paste required increasing the exponent of the short-time power-law asymptote of the compliance function $J(t, t')$ from $n \approx 0.10$ to $n \approx 0.35$. The exponent increase means that the attenuation of creep rate with time is slower in cement paste than it is in concrete. This difference could be explained by differences in stress redistribution among the components. In cement pastes, the stress is gradually transferred from the creeping to the noncreeping components, such as the CaOH crystals, while in concrete the stress is also transferred to the aggregate.

Chapter 12
Effect of Cracking and Fracture Mechanics Aspects of Creep and Shrinkage Analysis

Abstract The nonuniformity of drying shrinkage and drying creep, as well as the stress redistributions due to nonuniform creep, typically cause distributed cracking and continuous fractures, which lead to ingress of corrosive agents into concrete and compromise durability. In this chapter, we present an analysis of these phenomena from the viewpoint of fracture mechanics. We explain the crack band model, which is a simple and effective way of avoiding spurious mesh sensitivity in finite element simulations. We analyze the role that cracking plays in drying creep. The creep and rate effects influence the crack propagation, which we describe in terms of the cohesive crack model. We also explain the origin of the rate effect in cohesive fracture, which lies in the fracture kinetics on the atomic scale. Further, we point out that cracking is also a partial cause of the irreversibility of shrinkage and creep, and causes nonlinear stress dependence of drying creep. Finally, we discuss material models combining damage and creep.

Nonuniform shrinkage and thermal strains, as well as redistributions of internal forces caused by creep, produce significant stresses, which typically lead to distributed cracking. If the cracks are large, they may endanger durability, providing conduits for the ingress of various corrosive agents. If the cracking remains distributed, i.e., does not localize, it can be described by stress–strain relations with strain softening, provided that there are sufficient constraints (e.g., by reinforcement, or an adjacent zone under compression) to prevent the localization instability. In some respects, distributed cracking may be beneficial; for example, similar to plasticity, it may cause favorable stress redistribution with a reduction of stress peaks.

If the constraints of the cracking zone are insufficient, the distributed cracking is unstable and will localize into much larger distinct cracks [115, Sects. 12.5 and 13.2]. Proper treatment of such cracks, as well as the problem of crack spacing, calls for *fracture mechanics*, a theory in which the failure is characterized by energy per unit crack surface, called the *fracture energy* G_f (dimension J/m^2 or N/m), usually in conjunction with the tensile strength, $\bar{f}_t$ (dimension N/m^2). A fracture theory in which both G_f and $\bar{f}_t$ play a role is the *cohesive crack model*, introduced by Barenblatt [49], finalized by Rice [723], and in concrete engineering pioneered by Hillerborg, Modéer and Peterson [481], or the *crack band model* [168, 178].

© Springer Science+Business Media B.V. 2018
Z.P. Bažant and M. Jirásek, *Creep and Hygrothermal Effects
in Concrete Structures*, Solid Mechanics and Its Applications 225,
https://doi.org/10.1007/978-94-024-1138-6_12

As shown by Irwin [507], a failure criterion involving both G_f and $\bar{f}_t$ implies the existence of the *material characteristic length* $l_0 = EG_f/\bar{f}_t^2$, which approximately represents the length of the *fracture process zone* (or cohesive zone) in front of a crack (E = Young's modulus of elasticity). If l_0 is much smaller than the characteristic dimension of the cross section, D, the fracture process zone can be treated as a point. The limit case of a vanishing l_0, which theoretically corresponds to $\bar{f}_t \rightarrow \infty$, corresponds to the classical *linear elastic fracture mechanics* (LEFM), whose foundation was laid down by Griffith [438]; see, e.g., Bažant and Planas [178].

A non-negligible ratio l_0/D, which is a typical situation in concrete engineering, inevitably leads to nonstatistical (deterministic) size effect in structural response [100, 178]. This size effect must also intervene in creep response of structures, but in most cases probably to a negligible extent except in the rare case of structural failure resulting from creep (such as creep buckling of concrete columns or shells, or loads much higher than the service loads).

12.1 Limitations of Simplistic Nonlinear Models for Concrete Creep

There have been repeated efforts to generalize linear aging viscoelasticity to nonlinear behavior [28, 29, 731, 754], generally restricted to the uniaxial setting. For instance, $d\sigma(t')$ in the integrand of the superposition integral (2.14) for uniaxial stress was replaced by $df(\sigma(t'))$, or $\sigma(t')$ in the impulse memory integral (2.21) was replaced by $f(\sigma(t'))$, where f is some nonlinear function. Also, the nonlinearity modeled as multiple time integrals was tried. However, such approaches can describe only a limited range of test data for the initial deviations from linearity. They turned out to be unrealistic, mainly for two reasons:

1. the complex triaxial tensorial nature of nonlinear behavior of concrete and
2. the effect of distributed microcracking.

The triaxial aspect of nonlinearity of creep must properly be approached as a time-dependent generalization of nonlinear triaxial models for time-independent behavior of concrete. This is a vast subject, which will be briefly touched in Sect. 12.8.

As it eventually transpired, microcracking can explain a major part of the nonlinearity of creep, or deviations from the principle of superposition (an exception where creep nonlinearity is dominant is the case of high triaxial confining pressure, where the behavior is viscoplastic). As one demonstration, Fig. 12.1 shows the creep isochrones (i.e., lines connecting strains reached at the same time under different stresses) measured by Mamillan [599, 600] and their fits computed by finite elements with a linear creep law and distributed cracking modeled as strain softening under triaxial stress [119].

The microcracking, or fracturing, brings about a second physical source of time dependence, due to breakage of the interatomic bonds. One can distinguish two distinct sources of nonlinearity:

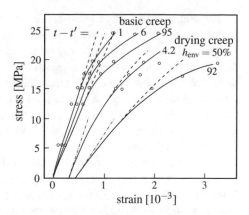

Fig. 12.1 Fits of creep isochrones measured by Mamillan [599, 600] for concrete loaded at age $t' = 28$ days; solid lines – with nonlinearity, dashed lines – without nonlinearity, $t - t'$ specified in days

1. development of microcracks, which is properly described by nonlinear triaxial models, combined with linear aging viscoelastic behavior of the material between the microcracks and
2. the growth of microcracks or microfractures in time due to the separation of interatomic bonds as a rate process governed by the activation energy of interatomic bonds [96, 101, 112, 128, 129, 162].

The second source operates even in materials that do not creep, e.g., granite [103]. It appears, though, that the importance of this source lies almost exclusively in very rapid deformations, and particularly in the extrapolation of static short-time laboratory tests (of 1–60 min. duration) to the loading by impact, shock, explosion, or earthquake [140, 162, 548, 582, 632, 719, 720, 735, 877].

It is self-evident and needs no elaboration that effects of inertia must be included in problems of impact, explosions, shock waves, earthquake, etc. However, the time-dependent forces do not belong into the constitutive equation, except for impact comminution of material when the release of kinetic energy of strain rate field in forming particles exceeds the maximum possible strain energy that can be stored in the material [218].

12.2 Fracture Mechanics Aspects and Crack Band Model

A typical situation in concrete and reinforced concrete is a system of parallel cracks. The cracks may be so thin (e.g., <0.1 mm) that they are undetectable to naked eye, but can also localize into wide cracks (>0.5 mm) that endanger durability. The cracks can be caused by shrinkage due to drying, or by cooling. In massive structures such as nuclear containments, parallel crack systems are caused by thermal stress due to hydration heat. Thin parallel cracks also form on the tensile side of reinforced beams or slabs subject to bending.

For concrete and other quasi-brittle materials, the formation of a macroscopic stress-free crack is preceded by the development of a *fracture process zone*, i.e., of a region characterized by a highly localized strain and by the initiation and growth of microcracks or other defects, which reduce the cohesion of the material and lead to softening. *Cohesive models* lump the inelastic effects and replace the process zone by a discontinuity surface, across which the displacement field has a jump. The displacement jump, which represents the cumulative contributions of all the microdefects summed over the thickness of the process zone, can be physically interpreted as the opening of a *cohesive crack*. The stress transmitted by the cohesive crack depends on the crack opening and vanishes when the opening attains a critical level. Models of this type originated in the pioneering work of Barenblatt [49, 50]. They have been developed for metals and composites under the name of *cohesive zone models* [649, 819] and for concrete under the name of *fictitious crack models* [481], also called *cohesive crack models*. A related early pioneering model of Dugdale [360] considered a plastic zone as a segment on the crack extension line and, instead of a cohesive law, determined the length of plastic segment by solving the elasticity problem of infinite space.

According to the cohesive crack model, the cohesive (crack-bridging) stress σ is linked to the crack opening w_c by the *cohesive stress-crack opening law*

$$\sigma = f_{cr}(w_c) \tag{12.1}$$

Function f_{cr} characterizes the fracture properties of the material (in a uniaxial-stress simplification), and typically has a convex graph with a relatively long tail. It is often taken as bilinear, as shown in Fig. 12.2a, or exponential (a linear softening was also used, but is not realistic). The area under the cohesive stress-crack opening curve corresponds to the *fracture energy*, G_f, i.e., to the energy dissipated per unit area of the final macroscopic stress-free crack. Aging can be taken into account by making the parameters of the cohesive law, such as the tensile strength $\bar{f}_t$ or crack opening at full fracture w_f, depend on the current age, as indicated in Fig. 12.2b.

Since the cohesive crack does not always open monotonically, it is also necessary to define the rules for unloading and reloading. An idealized simple rule used by Bažant et al. [150] is shown graphically in Fig. 12.3. More realistic (but also more complicated) nonlinear unloading–reloading rules were proposed by Reinhardt, Cornelissen and Hordijk [721]. Another more realistic rule was used in the drying creep studies of Bažant and Chern [117, 119].

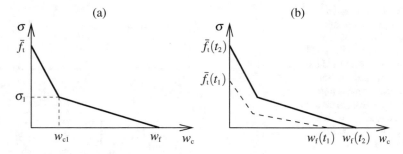

Fig. 12.2 (a) Bilinear softening curve of cohesive stress σ versus crack opening displacement w_c for cohesive crack model, (b) upward shift of this curve due to strength gain with age

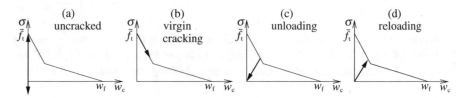

Fig. 12.3 Idealized rules for cohesive crack response—uncracked, virgin cracking (softening), unloading and reloading

A cohesive crack actually represents the cumulative effects of cracking in a fracture process zone of a certain width h_c, and so the spacing s_c between neighboring parallel cohesive cracks must not be smaller than h_c; see Fig. 12.4a. According to the *crack band model* [81, 85, 168], the crack band width h_c is considered to be a material property, best defined as the minimum possible spacing of continuous parallel cracks. As a crude estimate, $h_c \approx 2d_a$ to $3d_a$ where d_a = maximum aggregate size.

Parallel thermal or shrinkage cracks in unreinforced or lightly reinforced concrete grow as the penetration front of cooling or drying advances deeper into the concrete mass. When they become too long, every other crack stops growing and then gradually closes (Fig. 12.4b) while the remaining cracks compensate for the closing of their neighbors by significantly increasing their width. The doubling of spacing gets repeated whenever the ratio of the length a of the dominant cracks to their spacing s_c exceeds a certain critical value. This is a phenomenon of bifurcation of equilibrium path of a system of interacting cracks. Stability analysis of such systems [170, 171, 184, 190, 191], summarized by Bažant and Cedolin [115, Chap. 12], showed that the spacing of the dominant (open) cracks increases roughly as

$$s_c = \phi\, a \tag{12.2}$$

Fig. 12.4 (a) Parallel shrinkage or thermal cracks at their closest possible spacing given by the width of the crack band (width of the fracture process zone, dependent on maximum aggregate size); (b) cracks at doubled spacing, with a closed initial crack in between each pair

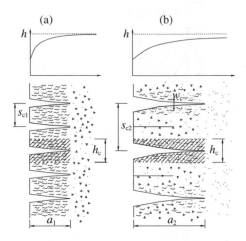

where $\phi \approx 0.69$ is a dimensionless coefficient. However, the crack spacing cannot be less than the effective width of the crack band in the crack band model, h_c. Thus, the general rule for the dominant crack spacing may be approximately written as

$$s_c = \max(h_c, \phi a) \tag{12.3}$$

The value $\phi = 0.69$ applies only for unreinforced or weakly reinforced walls, and it needs to be reduced if the wall has a load-bearing reinforcement. Stability analysis showed that a reduction of ϕ is required when the reinforcement ratio exceeds about 0.2% [191]. This value happens to correspond to the minimum shrinkage and temperature reinforcement empirically introduced in the ACI-318 design code[1] [16].

If a sufficiently heavy and dense three-dimensional reinforcing mesh prevents the cracks from localizing, then $s_c \approx 3d_a =$ constant, for any crack length. This is equivalent to setting $\phi = 0$ and $h_c \approx 3d_a$ in (12.3). If we consider a wall of thickness D, such as a massive nuclear containment wall subjected to significant hydration heat, the parallel cracks propagating symmetrically from both opposite faces cannot get longer than $D/2$. So the maximum possible crack spacing is

$$s_{c,\max} \approx \frac{1}{2}\phi D \tag{12.4}$$

In numerical simulations, the cohesive crack model can be used directly only if it is combined with a spatial discretization technique that represents individual cracks as discontinuities in the displacement field. This can be achieved, e.g., by inserting special cohesive interface elements between neighboring finite elements that discretize the bulk material, or by discontinuous enrichments of the standard finite element shape functions (see, e.g., [517, 793] for overviews of such enrichment

---

[1]In fact, preventing a localization instability of the crack system, and thus an increase of the crack width, is the theoretical reason imposing the requirement for minimum reinforcement.

techniques). In practical simulations, it is often more convenient to stick to standard techniques based on continuous displacement interpolations. In that case, opening of the cohesive crack must be replaced by an equivalent inelastic strain, smeared over a certain distance.

The problem is that the distance h_s over which the crack opening is smeared should in theory correspond to the thickness h_c of the physical cracking band represented by the cohesive crack. On the other hand, in numerical simulations cracking can localize into bands of a different thickness, which is strongly dependent on the size of finite elements and on other factors, e.g., on the angle between the crack band direction and the preferred directions of the finite element mesh. If all parameters of the stress–strain law are considered as fixed material properties, simulations of localized cracking suffer by pathological sensitivity to the size of finite elements. The reason is that the thickness of the computationally resolved band tends to zero as the mesh is refined while the dissipated energy per unit volume (i.e., the area under the stress–strain diagram) remains constant. Consequently, the total dissipated energy tends to zero and the computed structural response becomes extremely brittle as the mesh is refined.

The energy dissipated in the fracture process zone and the global structural response can properly be reproduced by a *smeared crack model* if the cohesive crack opening is smeared over the actual width of the numerically resolved crack band, which is typically formed by one layer of finite elements. Graphically, this idea is illustrated in Fig. 12.5. The cohesive law, i.e., the dependence of the cohesive stress on the crack opening, is considered as the primary characteristic of the inelastic material behavior (Fig. 12.5a). Uniform smearing of the crack opening over the crack band width h_s leads to the *smeared cracking strain* $\varepsilon_{sc} = w_c/h_s$. The cohesive law (12.1) is transformed into a relation between the stress and the cracking strain (Fig. 12.5c) and, in combination with the elastic stress–strain law that describes the

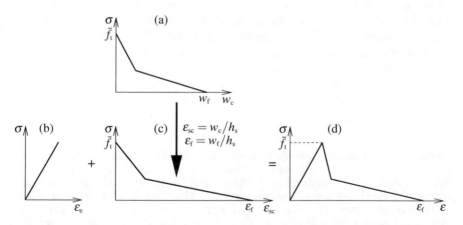

Fig. 12.5 Stress-strain diagram derived from a cohesive stress-crack opening law

bulk material without cracks (Fig. 12.5b), it provides the overall stress–strain law with a softening branch (Fig. 12.5d).

As already explained, the computational band width h_s is closely related to the element size and, therefore, the softening part of the stress–strain curve constructed by the aforementioned transformation of the cohesive law is not a true material property. In fact, the area under the stress–strain curve is equal to G_f/h_s where G_f is the fracture energy (material property) and h_s is the band width, dependent on the finite element mesh. As the mesh is refined, the area under the stress–strain curve increases but its product with the decreasing band width h_s remains constant and the pathological sensitivity of the numerical results to the size of finite elements is thus eliminated. This technique was in 1981 developed for shear softening in plasticity by Pietruszczak and Mroz [692] and in 1982 formulated for mode-I softening under the name of the *crack band model* [85, 168].

The effective value of the numerical crack band width h_s must be deduced from the mesh characteristics. In the one-dimensional setting, cracking localizes into a single element, and so h_s is equal to the length of that element.[2] In multiple dimensions, the cracking strain in general localizes into a band of elements running across the mesh. Usually, this band is the smallest possible pattern that still allows separation of nodes on its opposite sides. The average thickness of the band is affected not only by the sizes of finite elements but also by their shapes and by the inclination of the crack band with respect to the mesh lines. This is illustrated in Fig. 12.6b, which explains why the correct value of h_s for a zigzag band propagating along the diagonals of a regular square mesh is $\sqrt{2}$ times larger than for a straight band aligned with the element sides (Fig. 12.6a).

Based on similar considerations, Bažant [91] proposed the basic rules for estimation of the band width, h_s, and later Rots [738] refined them for a number of special situations on the basis of extensive numerical studies. A sophisticated general approach was developed by Oliver [664]. In practical simulations, it seems to be reasonable to compute h_s as the size of the element projected onto the crack normal; see Fig. 12.6c. Jirásek and Bauer [520] compared a number of techniques that provide estimates of h_s and demonstrated the role of additional factors such as the element shape, order of interpolation, and integration scheme.

The approach described above is based on an adjustment of the softening branch of the stress–strain diagram, motivated by smearing of the cohesive crack opening across the width of the band of cracking finite elements. This technique removes

---

[2]For two-node elements with linear displacement interpolation, the strain is constant over the whole element and localization of strain into shorter intervals is impossible. For higher-order elements, this is no longer true. For instance, for a one-dimensional element with a quadratic displacement interpolation, there exist solutions with cracking localized into 1 out of 2 Gauss integration points, or into 2 out of 3 Gauss integration points; see Jirásek and Bauer [520] for a detailed discussion. Such solutions satisfy equilibrium equations in the weak sense, but the corresponding stress fields are nonuniform and can become quite irregular, especially in multiple dimensions for crack bands inclined with respect to the mesh lines. Therefore, higher-order elements are not suited for simulations of localized failure based on the smeared crack approach.

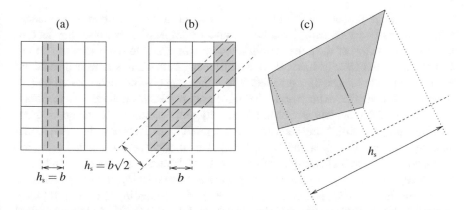

Fig. 12.6 (a) Straight crack band parallel to element sides, (b) zigzag crack band along element diagonals, (c) crack band width estimated by projecting a cracking element of an arbitrary shape onto the direction perpendicular to the expected crack direction (estimated, e.g., based on the principal directions of strain)

pathological sensitivity of the numerical results to the element size only if cracking indeed localizes into narrow bands. On the other hand, if cracking remains distributed in many elements, the transformation from cohesive crack opening to smeared cracking strain should be done using a constant, mesh-independent smearing distance. In some cases, it is known in advance whether diffuse or localized cracking can be expected, and the appropriate technique can be adopted. However, in general a concrete structure can exhibit diffuse cracking in some regions and localized fracture in other regions, or the cracking process can evolve from diffuse to localized. It is then hard to select an approach that would guarantee objective results on arbitrary finite element meshes.

A more fundamental modification of the material model, potentially leading to objective results and proper energy dissipation for both diffuse and localized cracking, can be based on an enhancement of the constitutive equations by a special term that supplies the missing information on the intrinsic material length and acts as a *localization limiter*, i.e., prevents localization of cracking or other inelastic processes into arbitrarily thin bands. In concrete mechanics, such enhancements often incorporate weighted spatial averages of certain internal variables, or gradients of internal variables. Such advanced techniques are described in specialized literature; see, e.g., Bažant and Planas [178, Chap. 13], Jirásek and Bažant [521, Chap. 26] and Bažant and Jirásek [141].

The fracturing of concrete is promoted by a compressive stress parallel to the crack (if this stress is high enough, it alone can cause splitting). This effect cannot be captured by the standard cohesive crack model. It requires a triaxial inelastic constitutive model, such as the microplane model; see Sect. 12.8 in this book, or Chap. 14 in Bažant and Planas [178].

Pure damage models, which reduce stiffness but do not incorporate permanent strains, are not very realistic for concrete under compression because they lead to unloading branches of the stress–strain diagram that return to the origin and do not capture permanent strains observed in experiments. Permanent strains can be added as enhancements to pure damage models, but they can also be incorporated by formulating the constitutive law within the framework of plasticity. To give an example, a model based on multisurface chemo-plasticity and accounting for creep, microcracking, chemical shrinkage, and aging was applied to analysis of a shotcrete tunnel shell by Lackner, Hellmich and Mang [562]. This model was formulated using the framework of thermo-chemo-plasticity [824, 825]. Models for concrete combining the concepts of damage mechanics and plasticity were developed among others by Grassl and Jirásek [433] and Grassl, Xenos, Nystrom, Rempling and Gylltoft [434]. Recently, the damage-plastic model of Grassl and Jirásek [433] has been enriched by aging, creep, and shrinkage and applied to shotcrete by Neuner, Gamnitzer and Hofstetter [651]. A model for concrete at high temperatures, incorporating the effects of mechanical and thermal damage, plasticity, and transitional thermal creep, was proposed by Nechnech, Meftah and Reynouard [648].

12.3 Role of Cracking and Irreversibility in Shrinkage

The development of internal stresses due to shrinkage was schematically explained in Fig. 8.48, which is reproduced in an expanded form as Fig. 12.7; see Bažant and Wittmann [196] and Bažant and Chern [117]. If a prismatic specimen were sliced as shown in the figure, the slices would shrink differently, as shown in the second row (free shrinkage). But in a long enough prism, plane cross sections will be preserved, which requires development of shrinkage stress, typically large enough to cause microcracking, as shown in the third row.

The shrinkage stresses are relaxed by creep. The relaxation is stronger (and cracking milder) in thicker specimens since the drying process is slower and allows for more creep. The change of behavior in the presence of an axial compressive or tensile force and in the presence of a bending moment is sketched in the bottom half of Fig. 12.7.

Figure 12.7 also explains the consequence of irreversibility of microcracking, i.e., of the impossibility of opened microcracks to close fully upon unloading. After the surface layers have dried up, they resist the shrinking of the core because the previously formed microcracks cannot fully close. This causes a reversal of the stress profile. Initially, the surface layers are in tension, balanced by compression in the core. Later, the shrinking core tries to close the surface microcracks, but since it cannot, the surface layers revert into compression while the core gets into tension and suffers microcracking, too.

The effect of creep irreversibility due to aging is similar to the effect of microcracking irreversibility, but is milder. It gets manifested more in thicker specimens since their slower drying allows for more aging. If irreversibility of both the microcracking and creep were absent, the terminal stress profile would be uniform, i.e., the residual stresses would eventually vanish (see the rightmost column in Fig. 12.7).

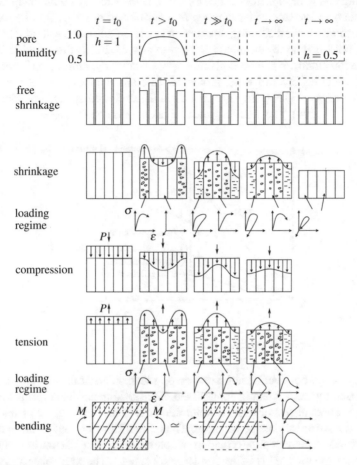

Fig. 12.7 Evolution of humidity, shrinkage, stress and strain in various layers of a cross section, with induced cracking; reproduced with permission from [90]

The schematic sketches in Fig. 12.7 illustrate the origin of cracking due to stresses that are needed to restore compatibility between material "layers" drying at different rates. Additional self-equilibrated stresses are generated by the nonuniformity of drying and shrinkage within the mesostructure of concrete; see, e.g., the analysis performed by Havlásek and Jirásek [474] and its summary in Sect. 8.6.

The cracking and irreversibility also explain *Domone's paradox* [355], which consists in the fact that the sum of swelling at no load in water and of creep without moisture intake (sealed conditions) is significantly less than the creep with moisture intake. The explanation is analogous to that of drying creep; see the next section.

12.4   Role of Cracking in Drying Creep (Pickett Effect)

The drying creep, or Pickett effect [690, 691], is the excess of total creep of drying specimens over the sum of basic creep and shrinkage; see Fig. 12.8. Its nanoscale mechanism involving disjoining pressure in hindered adsorbed water layers can be explained and modeled by the microprestress theory, as already discussed in Chap. 12. More fundamentally, it can be explained by molecular dynamics simulations of the effect of a moving layer of water molecules in a slit nanopore on the rate of shear across the nanopore [779]. However, not all of the drying creep can be explained that way [117, 198, 874]. The cracking must be taken into account, too.

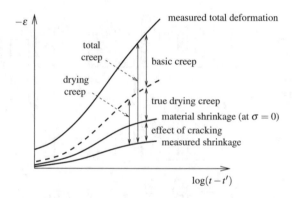

Fig. 12.8 Influencing factors for drying creep

The cracking, however, is not a real creep mechanism. It is an apparent mechanism, caused by the fact that in compressed creep specimens the deformation due to cracking is either absent or much smaller than it is in the load-free companion specimens that measure shrinkage cracking. Thus, the difference, which normally used to be defined as the creep deformation of loaded specimens, actually includes the cracking deformation of load-free shrinkage specimens. The strain observed on these specimens does not give the true drying shrinkage. Rather, the observed shrinkage is smaller in magnitude than the true drying shrinkage, being reduced by the cracking strain [117, 198, 874]. So, the overall increase of creep during drying has two causes:

1. the so-called *true drying creep*, which is a true mechanism residing only in the creep specimen and is caused, on the nanoscale, by increased microprestress (Chap. 10) and by the acceleration of shear slip by flow of water molecules along nanopores, and
2. the additional *apparent contribution due to microcracking* in the companion shrinkage specimen.

Because of these effects, the creep at drying is always larger than the sum of the observed basic creep and shrinkage.

The only way to directly observe true shrinkage, as well as true drying creep, is to place very thin hardened cement paste specimens into a chamber in which the environmental humidity can be lowered at a controlled rate [102]. The rate must be so low that the pore humidity difference between the surface and the specimen core would not exceed about 0.03. For standard concrete specimens, the drying would need to be extremely slow; see the analytical estimates and numerical examples in Sect. 8.6.

The tests of Bažant et al. [102] used hardened cement paste tubes of thickness 0.71 mm, for which the admissible rate of environmental humidity was about 2% per hour. Calculations show [184] that for the standard concrete test cylinders it would be about 2% per month. This makes it virtually impossible to observe true shrinkage of concrete. The true shrinkage characteristics can only be extracted by inverse analysis of shrinkage and creep tests.

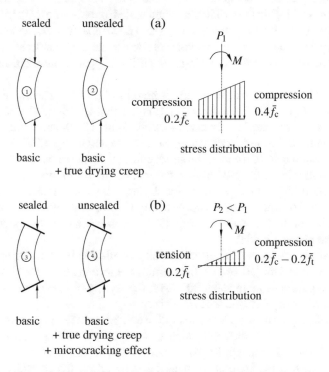

Fig. 12.9 Experimental plan for (a) small and (b) large eccentricity of loading

The contribution of cracking to drying creep can nevertheless be observed directly. It can be done under flexural or torsional loading because no deformation of any companion specimen needs to be subtracted, as drying shrinkage alone causes no curvature and no twisting. Flexural tests were carried out by Bažant and Xi [201]; see also Bažant and Xi [200]. Curvature creep of beams subjected to the same bending moment M was observed under different axial forces that either prevent or allow cracking.

Figure 12.9 shows prismatic specimens subjected to different sustained axial compressive loads P_1 and P_2 with different eccentricities. The loading was chosen so that, for a linear stress distribution, the opposite face stresses would be $\sigma = -0.2\bar{f}_c$ and $-0.4\bar{f}_c$ for load P_1, and $\sigma = 0.2\bar{f}_t$ and $0.2\bar{f}_t - 0.2\bar{f}_c$ for load P_2 (with $\bar{f}_c$ and $\bar{f}_t$ denoting the compressive and tensile strengths of concrete). Of course, the stresses generated by the applied load must be superposed with self-equilibrated stresses induced by nonuniform shrinkage. Load P_1 sufficed to prevent any cracking, while load P_2 allowed cracking to occur at one face.

For each loading case, there were 2 groups of specimens, sealed and unsealed. The growth of curvature $\kappa(t)$ in the fully compressed specimens (under load P_1) represents the basic creep when sealed, and basic creep with the true part of drying creep when unsealed. The growth of curvature $\kappa(t)$ in the partly compressed specimens (under load P_2) represents again the basic creep when sealed, but when unsealed, it represents the basic creep with the total drying creep (true plus apparent). So, by comparisons, one can separate the curvature growths due to (i) the true nanoscale mechanism of drying creep and (ii) microcracking, which is an apparent mechanism of drying creep.

The results of these experiments are plotted in Fig. 12.10. The curves represent the averaged compliance function $\bar{J}(t, t')$, calculated as $I\kappa(t)/M$ where I is the centroidal moment of inertia of the cross section. Figure 12.10a shows the compliance functions obtained for the four measured cases presented in Fig. 12.9. The difference between the top and bottom curves (large eccentricity drying minus large eccentricity sealed) corresponds to the total drying creep, and the difference between the two curves in the middle (small eccentricity drying minus small eccentricity sealed) corresponds to the true drying creep; see Fig. 12.10b. The difference between the two curves in Fig. 12.10b, plotted in Fig. 12.10c, corresponds to the apparent drying creep (microcracking effect).

Note that the microcracking effect initially dominates but later disappears. For the prisms of side 4 in. (10.16 cm), the dominance terminates in about 10 days and microcracking disappears in about 200 days; the scaling of these times should vary roughly quadratically with the specimen size.

The true drying creep was also measured in another set of experiments, performed on specimens made of a different batch of concrete and using small eccentricity only (Fig. 12.10d). Bažant and Xi [201] showed that the observed deformations can be closely matched by finite element analysis based on the solidification theory combined with a model for stress-induced shrinkage and taking into account cracking with its irreversibility.

Fits of the most important data on the drying influence on creep, collected from the literature [270, 455, 601, 602, 690, 817, 850], are reproduced from Bažant and Chern [117]. They include shrinkage (Fig. 12.11), basic creep and drying creep in compression (Figs. 12.12–12.13), tension (Figs. 12.14–12.15), and bending (Fig. 12.16). The comparison in Fig. 12.15a documents that the drying creep effect is stronger in tension than it is in compression, which is explained by a greater contribution from microcracking.

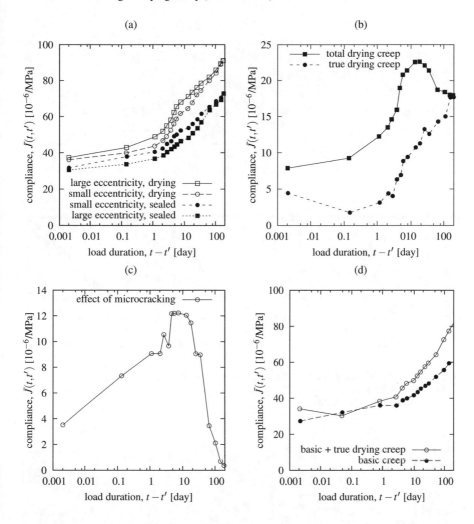

Fig. 12.10 (a) Averaged compliance functions, (b) decomposition of different drying creep mechanisms, (c) effect of microcracking on drying creep, (d) averaged compliance functions in another set of experiments (small eccentricity only)

Figure 12.16 presents the fits of Pickett's bending tests obtained by Bažant and Chern [117], who modeled the true drying creep as stress-induced shrinkage using an early version of the approach described in Sect. 13.3.3.2. Figure 12.16a shows the deflection of a simply supported concrete beam loaded by a concentrated force at midspan. The experiments and simulations cover the cases of loading combined with drying (LD), loading without drying (L) and drying at no load (D). The effect of parameter r, which controls the sensitivity of the shrinkage coefficient to the stress according to formula (13.103), is illustrated in Fig. 12.16b. Parameter r' was set to zero in all simulations. For $r = 0$, the effect of stress on shrinkage would be

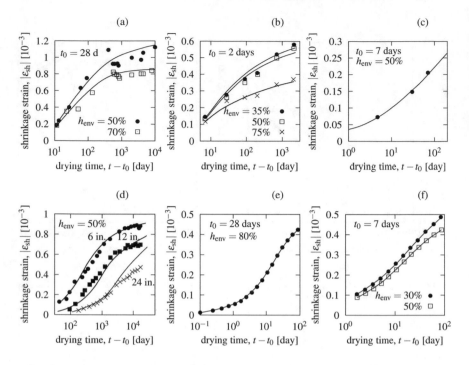

Fig. 12.11 Fits of shrinkage data: (a) Troxell, Raphael and Davis [817], (b) L'Hermite, Mamillan and Lefèvre [580], (c) McDonald [617], (d) Hansen and Mattock [453], (e) Brooks and Neville [270], (f) Ward and Cook [850]

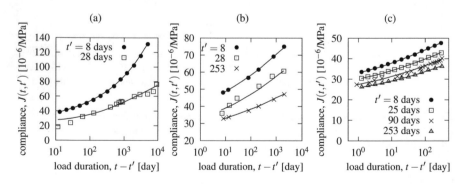

Fig. 12.12 Fits of basic creep data for compression: (a) Troxell, Raphael and Davis [817], (b) L'Hermite and Mamillan [576, 580], (c) McDonald [617]

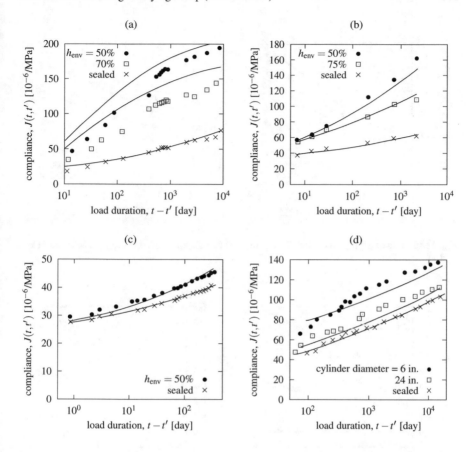

Fig. 12.13 Fits of drying creep data for compression: (a) Troxell, Raphael and Davis [817], (b) L'Hermite and Mamillan [576, 580], (c) McDonald [617], (d) Hansen and Mattock [453]

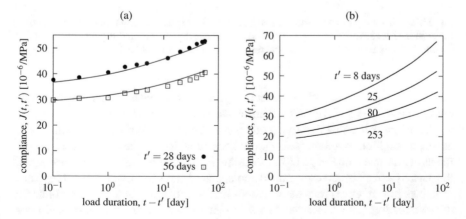

Fig. 12.14 Fits of basic creep data for tension: (a) Brooks and Neville [270], (b) Ward and Cook [850]

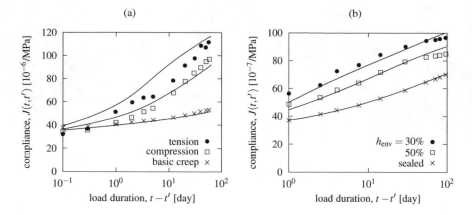

Fig. 12.15 Fits of drying creep data for tension: (a) Brooks and Neville [270], (b) Ward and Cook [850]

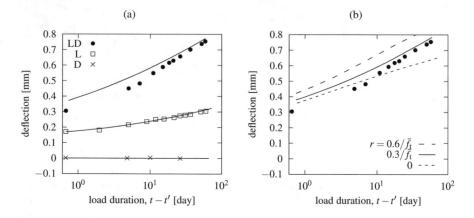

Fig. 12.16 Fits of drying creep data for bending [690]: (a) experimental data and numerical results for combined loading and drying (LD), loading without drying (L), and drying at no load (D), (b) effect of parameter r on the numerical results for the case of combined loading and drying

completely suppressed, and the deflection growth would be underestimated; see the bottom curve in Fig. 12.16b. Good results were obtained with $r = 0.3/\bar{f}_t$.

Further evidence that the drying creep is caused by an intrinsic nanoscale mechanism, in addition to microcracking, has been provided by the tensile creep experiments of Altoubat and Lange [34]. In a recent study, Sinko et al. [779] have analyzed the Pickett effect by coarse-grained molecular dynamics simulations. They have examined how relative creep deformations across a slit pore subjected longitudinally to shear loading get accelerated by the motion of water along the pore due to drying forces. The drying that drives water flow along the nanopores lowers both the activation energy of pore walls sliding past one another and the apparent viscosity of water molecules confined in the pore. This lowering can be captured with an analytical Arrhenius relationship accounting for the role of water flow in overcoming the energy barriers. These findings have established the scaling relationships that explain how the creep driving force, drying force, and fluid properties are related. It has been shown that the movement of the molecular layer within the pore accelerates the relative shear creep displacement across the pore. In this way, Sinko et al. [779] have established the nanoscale origin of the Pickett effect and provided strategies for minimizing the additional displacements arising from this effect. They have also found that the drying creep strain at the nanopore level is not linearly dependent on the applied creep stress.

12.5   Role of Creep in Cohesive Fracture and Size Effect on Structural Strength

If the material is elastic, plastic hardening, viscoelastic, or viscoplastic, and if the material strength is not random, the maximum stress σ_{max} at failure of a structure without notches or macroscopic cracks is independent of the structure size when geometrically similar structures are considered. The size independence then also holds for the *nominal strength*, defined as the maximum load P_{max} divided by a characteristic area of the structure. In the general case of three-dimensional scaling, the characteristic area is proportional to the square of a characteristic size of the structure, D. For instance, for pullout of a headed stud from a massive concrete block (theoretically an infinite half-space), the characteristic size would be the embedment depth. For prismatic beams with a rectangular cross section, it is natural to set the characteristic area equal to the cross-sectional area bD where D is the depth and b is the width of the section. This definition covers not only the full three-dimensional scaling, when all dimensions are scaled proportionally, and thus, bD is proportional to D^2, but also the case of two-dimensional scaling, with b fixed and only the depth and span scaled proportionally.

Of course, if the nominal strength is evaluated simply as $\sigma_N = P_{max}/bD$, it does not have a direct physical meaning, except under uniaxial tension or compression. For a simply supported beam of span L, loaded by a concentrated force P at midspan, the maximum stress according to the elastic beam theory would be $\sigma_{max} = (PL/4)/(bD^2/6) = 1.5PL/(bD^2)$, which equals to P/bD multiplied by the dimensionless factor $1.5L/D$. For geometrically similar beams, this shape factor is constant, and thus, the nominal stress P/bD is proportional to the maximum elastically evaluated stress. If the failure is controlled exclusively by strength, and if the randomness of material properties is neglected, the nominal strength turns out to be independent of the characteristic structure size, D (which can be defined arbitrarily since only relative sizes matter). Any deviation from this behavior is called the *size effect* (on strength).

Since the work of Mariotte in the seventeenth century, it was believed that any observed size effect must be of statistical origin. In 1939, Weibull provided experimental demonstration and mathematical formulation of the statistical size effect, in terms of what became known as the Weibull distribution. However, in the early 1980s it was shown at Northwestern University that, in concrete and other quasibrittle materials, a major size effect is caused by energy release associated with stress redistribution due to a large fracture process zone (type-1 size effect) or to a large cohesive crack with a large fracture process zone (type-2 size effect). In normal-size concrete structures, this *energetic size effect* totally dominates. An exception, rare for concrete structures, occurs if the structure is very large and has a geometry for which the failure occurs right at the initiation of macrofracture from a smooth surface. Only in this case the statistical (or Weibull) size effect takes place [90, 100, 178].

The type-2 size effect is described by Bažant's [90] *size effect law*

$$\sigma_N = \frac{\sigma_0}{\sqrt{1 + \dfrac{D}{D_0}}} \tag{12.5}$$

where D_0 is the *transitional size* and σ_0 is the asymptotic value of nominal strength for very small specimens. These two parameters can be extracted from experimentally determined values of nominal strength for specimens of different sizes, and they can also be deduced from certain characteristics of the material and specimen geometry. The formulae derived by Bažant and Kazemi [143] read

$$D_0 = \frac{c_f g'}{g}, \quad \sigma_0 = \sqrt{\frac{EG_f}{c_f g'}} \tag{12.6}$$

where G_f is the fracture energy, c_f is the material length representing about a half of the length of a fully developed fracture process zone, g is a dimensionless energy release function of linear elastic fracture mechanics, characterizing the geometry of the specimen, and g' is the derivative of g with respect to the relative crack length (the values of g and g' to be substituted into (12.6) are those that correspond to the initial relative crack length). As usual, E denotes the elastic modulus. In linear elastic fracture mechanics, the product EG_f is equal to the square of the *fracture toughness* K_c, which represents the critical value of the *stress intensity factor*, needed for crack propagation.

The size effect law (12.5) is plotted in the logarithmic scale in Fig. 12.17a. With increasing size D, this law represents a smooth transition from quasi-plastic behavior with no size effect to perfectly brittle behavior that corresponds to LEFM. For sizes much smaller than D_0, the size effect curve approaches a horizontal asymptote $\sigma_N = \sigma_0$, while for size much larger than D_0 it approaches the asymptote $\sigma_N = \sigma_0(D_0/D)^{1/2} = (EG_f/gD)^{1/2}$, which characterizes the size effect of linear elastic fracture mechanics and in the logarithmic plot is represented by an inclined straight line of slope $-1/2$. The ratio $\beta = D/D_0$ is called the brittleness number.

For $\beta \gg 1$, the behavior is brittle, with a strong size effect, and for $\beta \ll 1$ it is ductile, with almost no size effect. This is why D_0 is called the transitional size—it is at the center of the transition between ductile and brittle behavior if the size of the specimen or structure is varied.

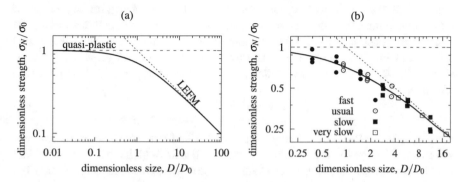

Fig. 12.17 (a) Size effect law (12.5) in terms of dimensionless variables, (b) change in brittleness with loading rate revealed by experimental data of Bažant and Gettu [128]

What is the effect of creep on the energetic size effect? This question was investigated by Bažant and Gettu [128]. It turns out that a decrease of the loading rate causes the corresponding points on the size effect curve (plotted in terms of dimensionless variables D/D_0 and σ_N/σ_0, with D_0 and σ_0 determined for each rate separately) to move to the right, i.e., to higher brittleness or stronger size effect. This is supported by the test data from Bažant and Gettu [128] reproduced in Fig. 12.17b, which gives the nominal strengths of similar notched three-point bend beams of three different sizes scaled in the ratio $1 : 2 : 4$, tested at four different rates for which the average times to peak load were 1.4 s, 8.33 min, 3.79 h, and 2.93 days. Note that the slower the loading, the stronger the size effect, and the higher the brittleness.

At first the increase of brittleness seems counterintuitive, but it may be explained by a decrease of c_f with decreasing loading rate. This means that relaxation of stresses around the fracture process zone, engendered by creep, leads to a shortening of the fracture process zone.

There is, however, another rate effect in fracture that has nothing to do with creep, and is of the same kind as in other materials, even those which do not creep (e.g., granite). As found by Bažant et al. [129], Tandon, Faber, Bažant and Li [800], and Bažant [100], the relation between the crack-bridging stress σ and the crack opening (or separation) w_c can be considered as unique only if the loading rate is approximately constant. If the rate varies significantly, then this relation needs to be generalized. In [129, 162, 800], and [100, Sect. 7.1], an approximate rate-dependent cohesive crack model was formulated and verified by fracture tests with loading rate changes. However, the derivation of this model was oversimplified.

Closely following the exposition by Bažant, Le and Bazant [159] based on atomistic fracture mechanics, the next section presents a more rigorous derivation, which
leads to the relation

$$\sigma = \varphi(\dot{w}_c) = C e^{Q_0/nk_B T} \dot{w}_c{}^{1/n} \tag{12.7}$$

where $\dot{w}_c$ is the crack opening rate, Q_0 is the activation energy of separation of
interatomic bonds, k_B is the Boltzmann constant,[3] T is the absolute temperature, and
C and n are constants that must be considered empirical at present. The value of n is
the same as in the empirical power law for the subcritical crack growth velocity v_c
[178, 183, 307, 375, 376, 814], called the creep growth law or *Charles-Evans law*,
which reads

$$v_c = A e^{-Q_0/k_B T} K^n \tag{12.8}$$

Here, v_c is the velocity of macrocrack growth, A and n are positive empirical constants, and K is the stress intensity factor at macroscale, which must be proportional
to cohesive stresses σ in the macrocrack. Experiments show n to range from 10 to
30.

Formula (12.7) has been derived from the frequency of state jumps over the
activation energy barriers on the potential surface of an atomic lattice block; for a
simplified derivation, see Bažant [96] and Bažant and Li [162]. A more rigorous
derivation [159], outlined in Sect. 12.6, gives a different stress dependence than that
in the previous papers, but the previous fitting of test data remains unaffected since
the tests were conducted at only one stress level.

Equation (12.7) needs to be combined with the softening law $\sigma = f_{cr}(w_c)$ for
very slow opening of cohesive crack ($\dot{w}_c \to 0$). This can be done in a similar spirit
as in viscoplasticity, where the yield stress corresponding to a rate-independent yield
condition is increased by a viscous overstress that depends on the plastic strain rate;
see, e.g., Sluys [780] and Chap. 27 in Jirásek and Bažant [521]. For a rate-dependent
cohesive crack model, the generalized stress-separation law at any point of the crack
is written as

$$\sigma = f_{cr}(w_c) + C e^{Q_0/nk_B T} \dot{w}_c{}^{1/n} \tag{12.9}$$

This can further be improved by considering C as a function of w_c that tends to zero
for very large w_c.

The importance of $\dot{w}_c$ was evidenced by the discovery of a reversal of softening
to hardening when the loading rate of a notched specimen was increased suddenly—
1000-times for the test results in Fig. 12.18a, b [100, 129, 162, 800]. It turned out
that this reversal cannot be modeled merely by considering creep in the material
around the fracture process zone. Creep can cause a decrease of softening slope, but
never a reversal of softening to hardening.

---

[3] Since we work here at the atomistic scale, it is appropriate to use the Boltzmann constant k_B instead
of the universal gas constant R. The activation energy Q_0 then refers to one bond instead of one
mole (as it would if R was used in the denominator).

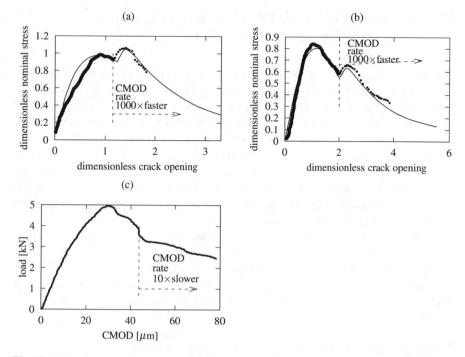

Fig. 12.18 (a) Experimental and predicted load versus crack mouth opening displacement (CMOD) response of a medium-size, high-strength concrete beam for a 1000-fold rate increase, (b) experimental and predicted load versus CMOD response of a large-size, high-strength concrete beam for a 1000-fold rate increase, (c) experimental load versus CMOD response of a large-size, normal-strength concrete beam for a tenfold rate decrease after a load drop to 70%

On the other hand, Eq. (12.7) cannot explain another phenomenon—load relaxation when the rate of a controlled growth of load-point displacement is suddenly decreased (Fig. 12.18c). In the extreme case when the displacement growth is stopped, the load remains constant if there is no creep. This is evidenced by testing notched specimens of granite [103], which does not creep. For concrete, which creeps, the load gradually relaxes.

These phenomena are sure to occur also for microcracks. Hence, to obtain a realistic constitutive model for nonlinear creep of concrete, a smeared continuum version of (12.7) (with strain $= w_c$/microcrack spacing), or its generalization for a broad range of rates, would need to be added to the aging linear viscoelastic model for concrete creep.

It should also be pointed out that the crack growth rate is important for relating the statistics of short-time material strength to the statistics of lifetime under sustained stress [157–159, 569, 750].

12.6   Derivation of Crack Opening Rate Effect from Fracture Kinetics at Atomic Scale*

The fracture begins on the atomic scale, and the atomic scale is where the rate effects originate. For the sake of clarity, imagine fracture in a nanoscale size atomic lattice block (Fig. 12.19a), although generalization to a disordered nanostructure of hardened cement gel poses no problems. The separation w_c between the opposite atoms across the crack gradually increases by small increments as the distance from the crack front grows. The work of the force F_b transmitted across each pair of opposite atoms on their relative displacement w_c defines a certain local potential $\Pi_1(w_c)$, which is a part of the overall potential function Π (or free energy) of the atomic lattice block. The equilibrium states of these atomic pairs (bonds) are marked on the local potential curves $\Pi_1(w_c)$ by circles.

As the fracture separation grows from one atomic pair to the next, the state marked by the circle moves on the curve $\Pi_1(w_c)$ up and right (Fig. 12.19b). The same states are also marked on the corresponding curves of bond force $F_b(w_c) = d\Pi_1/dw_c$ between the opposite atoms in each pair (Fig. 12.19c). The local bond failure begins when the peak point of the curve $F_b(w_c)$ is reached. This point corresponds to the point of maximum slope of the curve $\Pi_1(w_c)$ (state 3 in Fig. 12.19b) and represents the tip of the cohesive crack. The real crack tip is located at the pair where the bond force is reduced to 0 (state 1 in Fig. 12.19b). The cohesive zone (or fracture process zone, FPZ) roughly spans between state 1 and state 3, which lie many atoms apart.

If the lattice were treated as a continuum, the diagram $P(u)$ of load $P = \partial\Pi/\partial u$ applied on the atomic lattice block, versus the associated displacement u caused by elasticity of lattice and by fracture growth, would have the usual shape shown in Fig. 12.19d (note that the curvature of the rising portion is caused by finiteness of the length of growing FPZ). A continuum, though, the nanoscale lattice is not. The fracture advances by random jumps over the activation energy barriers Q on the surface of the state potential Π (free energy) of the atomic lattice block. These barriers superpose undulation on the diagrams of $\Pi(u)$ and $P(u)$; see Fig. 12.19e top.

The jumps of the propagating interatomic crack are equal to the initial atomic spacing δ_a. During each jump, one barrier on the potential Π as a function of u must be overcome (see the undulating potential profile in Fig. 12.19f). After each jump, at each new crack length, there is a small decrease (Fig. 12.19e bottom) of the overall potential Π of the atomic lattice block, corresponding to a small advance along the descending load-deflection curve $P(u)$ (Fig. 12.19e top). A point to note is that the separation of opposite atoms (in their equilibrium positions) increases during each jump by only a small fraction of their initial distance δ_a.

Due to thermal activation, the states of the atomic lattice block fluctuate and can jump over the activation energy barrier in either direction (forward of backward, Fig. 12.19e, f), though not with the same frequency. When crack length a (defined by the location of state 3 in Fig. 12.19a) jumps by one atomic spacing, i.e., from a_i to $a_i + \delta_a$, $i = 1, 2, 3,...$), the activation energy barrier Q changes by only a small

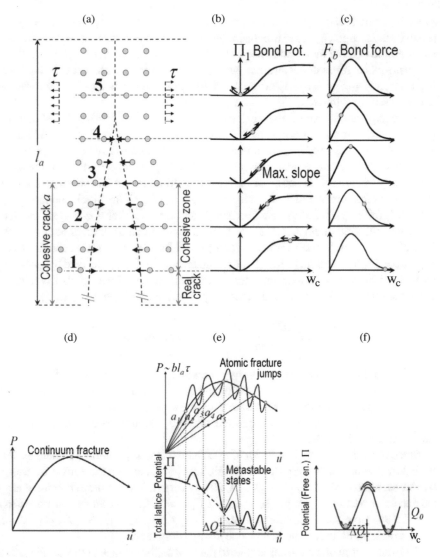

Fig. 12.19 Nanoscale fracture: (a–c) fracture of atomic lattice, (d–f) load-displacement curve of atomic lattice block; reproduced from Bažant, Le and Bazant [569]; in parts (e) and (f), ΔQ marks the vertical distance between the subsequent minima of the wavy curve

amount ΔQ which corresponds to the energy release by fracture (Fig. 12.19e, f) associated with the equilibrium load drop P.

To calculate ΔQ, consider planar three-dimensional cracks that grow in an affine, or self-similar, manner (e.g., expanding concentric circles or squares). The effect of the FPZ may be approximated according to the equivalent linear elastic fracture mechanics (LEFM) in which the tip of an equivalent sharp LEFM crack is considered to lie roughly in the middle of the FPZ. The LEFM stress intensity factor (of mode I, mode II, or mode III) is generally expressed as $K_a = \tau \sqrt{l_a}\, k_a(\alpha)$ where $\alpha = a/l_a =$ relative crack length, $k_a(\alpha) =$ dimensionless stress intensity factor, $\tau = c\sigma =$ remote stress applied on the nanoscale on the atomic lattice block of size l_a (Fig. 12.19a); c is a nano-macrostress concentration factor; and $\sigma =$ macroscale stress in the RVE. For a circular (or penny-shaped) crack of radius a loaded in mode I by a remote stress, $k_a(\alpha) = \sqrt{4\alpha/\pi}$.

The energy release rate of the crack (with respect to crack length rather than time), per unit length of crack perimeter, is $\mathscr{G}_a(\alpha) = K_a^2/E_1 = k_a^2(\alpha) l_a \tau^2 / E_1$. Here, E_1 = Young's (elastic) modulus for a continuum approximation of the lattice (which is larger than the macroscopic Young modulus E); $k_a^2(\alpha)$ represents the dimensionless energy release rate function of LEFM for continuous bodies, characterizing the geometry of fracture and of the atomic lattice block [178] (if the block boundaries are distant, $k_a^2(\alpha) \propto \gamma\alpha$). Let γ_1 be a geometry constant such that $\gamma_1 a = \gamma_1 \alpha l_a =$ crack perimeter, the crack being assumed to grow radially in an affine manner ($\gamma_1 = 2\pi$ for a circular crack of radius $a = \alpha l_a$, and $\gamma_1 = 8$ for a square crack). Similar to the expression in Bažant et al. [158, 159] and Bažant and Le [157] (derived by assuming two- rather than three-dimensional cracks), the increment of energy that is released when the crack advances by δ_a along its entire perimeter of length $\gamma_1 \alpha l_a$ is

$$\Delta Q = \delta_a \frac{\partial \Pi^*(P, a)}{\partial a} = \delta_a \gamma_1 \alpha l_a \mathscr{G}_a = V_a(\alpha) \frac{\tau^2}{E_1} \tag{12.10}$$

Here, Π^* is the complementary energy potential (Gibbs free energy) of the atomic lattice block, and $V_a(\alpha) = \delta_a \gamma_1 \alpha l_a^2 k_a^2(\alpha)$ is an activation volume (note that if the stress tensor is written as τs where $\tau =$ stress parameter, one may write $V_a = s : v_a$ where $v_a =$ activation volume tensor, as in the atomistic theories of phase transformations in crystals [46]).

Since the cohesive crack is much longer than δ_a, the separation w_c changes very little during each crack jump by one atomic spacing δ_a, and so the activation energy barrier for a forward jump, $Q_0 - \Delta Q/2$, differs very little from the activation energy barrier for a backward jump, $Q_0 + \Delta Q/2$ ($Q_0 =$ activation energy at no stress); Fig. 12.19e. Note that multiple activation energy barriers $Q_0 = Q_1, Q_2, \dots$ are always present, however, the lowest one always dominates.[4]

The jumps from one metastable state to the next on the surface of the atomic lattice block potential Π^* must be happening in both forward and backward directions, although at slightly different frequencies. According to the transition rate theory [537, 687], the first-passage time for each transition (in the limit of a large free-energy barrier, $Q_0 \gg k_B T$) is given by Kramer's formula [728]. Thus, the net frequency of crack length jumps is (see Eq. 2 in [158]; also [157, 159])

---

[4]The reason why the lowest barrier Q_1 dominates is that the factor $e^{-Q_1/k_B T}$ is very small, typically 10^{-12}. For instance, if $Q_2/Q_1 = 1.2$ or 2, then $e^{-Q_2/k_B T} = 0.004\, e^{-Q_1/k_B T}$ or $10^{-12} e^{-Q_1/k_B T}$ and thus makes a negligible contribution, and if $Q_2/Q_1 = 1.02$, then Q_1 and Q_2 can be replaced by a single activation energy $Q_0 = 1.01 Q_1$.

$$f_1 = v_T \left[e^{(-Q_0 + \Delta Q/2)/k_B T} - e^{(-Q_0 - \Delta Q/2)/k_B T} \right] = 2 v_T e^{-Q_0/k_B T} \sinh[V_a(\alpha)/V_T]$$
(12.11)

where $V_T = 2 E_1 k_B T / \tau^2$; v_T is a characteristic attempt frequency for the reversible transition, given by $v_T = k_B T / h$ where $h = 6.626 \cdot 10^{-34}$ Js is the Planck constant (i.e., the ratio between the energy of a photon and the frequency of its electromagnetic wave); T is the absolute temperature, and $k_B = 1.381 \cdot 10^{-23}$ J/K is the Boltzmann constant. Since δ_a is of the order of 0.1 nm, $V_a \sim 10^{-26}$ m^3. Volume V_T depends on $\tau = c\sigma$ where c is expected to be >10. For example, in the nanostructure of hardened Portland cement gel, the nanoscale remote stress may perhaps be $\tau \approx 20$ MPa, which gives $V_T \sim 10^{-25}$ m^3, and so $V_a/V_T \ll 1$. Therefore, Eq. (12.11) becomes $f_1 \approx e^{-Q_0/k_B T} [v_T V_a(\alpha)/k_B T] \tau^2/E_1$. The velocity of advance of the FPZ through the atomic lattice may thus be written as

$$v_{FPZ} = v_1 e^{-Q_0/k_B T} \tau^2, \quad v_1 = \delta_a^2 \gamma_1 \alpha l_a / E_1 h$$
(12.12)

Now, we note that this equation is analogous to Eq. (12.8) for the macrocrack growth rate, but exponent n, which equals about 10–30, is much larger than 2. Why?

This discrepancy has been explained [158, 159] using the condition that the energy dissipation power of the macroscale crack a must be equal to the combined energy dissipation power of all the active nanocracks a_i ($i = 1, \ldots N$) in the FPZ of the macroscale crack, i.e., $(\partial \Pi^*/\partial a)\dot{a} = \sum_i (\partial \Pi^*/\partial a_i)\dot{a}_i$, or

$$\mathscr{G} \dot{a} = \sum_{i=1}^{N} \mathscr{G}_i \dot{a}_i$$
(12.13)

where $\mathscr{G}$ and $\mathscr{G}_i$ denote the energy release rates with respect to a and a_i. The increase of the exponent was explained by a power-law increase in the number of cracks when zooming from the macroscales through the subsequent finer scales to the nanoscale [159].

Equation (12.8) gives the velocity of FPZ as a whole. In the cohesive crack model, the rates of separation $\dot{w}_c(x)$ at various points x must, on the average, be proportional to macroscopic crack growth rate v_c. Also, the cohesive stresses $\sigma(x)$ must, on the average, be proportional to K. This leads to the relation

$$\dot{w}_c = e^{-Q_0/k_B T} (\sigma/C)^n$$
(12.14)

where C is a constant. Relation (12.14) has the form of the Charles-Evans law. By inversion, (12.7) ensues.

12.7   Models Combining Damage and Creep

Smeared crack models are well suited for the description of nonlinearities caused by tensile stresses. However, compressive loading may also lead to significant deviations from the principle of superposition. A direct extension of the concept of smeared cracking to such cases would be quite artificial and inaccurate. To capture the non-linear time-dependent behavior of concrete under general triaxial stress states, one needs to develop general tensorial stress–strain laws that combine linear viscoelastic-ity with modeling techniques originally developed for rate-independent inelasticity, for instance those based on continuum damage mechanics or plasticity theory. A detailed coverage of all relevant models of this kind is out of scope of the present book. The following presentation is limited to a few basic observations and comments on the features of selected models.

Traditional approaches formulated within the framework of damage mechanics take into account the degradation of stiffness caused by growing defects in the mate-rial microstructure. For quasi-brittle materials, such defects are typically distributed cracks, and their propagation and coalescence lead to a reduction of the *effective area*, i.e., the area that is still able to transmit stresses. Simple *isotropic damage models* consider a reduction of all material stiffness coefficients by the same factor $1 - \omega$ where ω is a scalar *damage variable* evolving from 0 for the virgin (undamaged) material to 1 for the fully damaged material. Evolution of the damage variable is often linked to the so-called *equivalent strain*, $\tilde{\varepsilon}$, which plays the role of a scalar measure of the strain level. To emphasize the prominent role of tension, the frequently used Mazars definition of equivalent strain [613],

$$\tilde{\varepsilon} = \sqrt{\sum_{I=1}^{3} \langle \varepsilon_I \rangle^2} \tag{12.15}$$

is based on the positive parts of principal strains. Here, ε_I, $I = 1, 2, 3$, are the principal strains, and the positive part $\langle \varepsilon_I \rangle$ is equal to ε_I if $\varepsilon_I > 0$; otherwise, it is set to zero.

The stress–strain law used by rate-independent isotropic damage models with one scalar damage variable reads

$$\sigma = (1 - \omega) D_e \varepsilon \tag{12.16}$$

where σ is the column matrix of stress components, ε is the column matrix of strain components, and D_e is the elastic material stiffness matrix. Equation (12.16) can be interpreted as Hooke's law linking the strain to the effective stress,

$$\bar{\sigma} = \frac{\sigma}{1 - \omega} \tag{12.17}$$

which represents the stress transmitted by the undamaged contiguous solid material between the defects. The presence of defects leads to stress amplification, reflected by the factor $1/(1 - \omega)$, which grows from 1 to ∞ as the damage variable ω grows from 0 to 1.

In a similar spirit, one can construct a stress–strain law that combines damage with viscoelasticity and reads

$$\sigma(t) = (1 - \omega(t))\mathscr{R}\{\varepsilon(t)\} \tag{12.18}$$

or, equivalently,

$$\varepsilon(t) = \mathscr{J}\left\{\frac{\sigma(t)}{1 - \omega(t)}\right\} \tag{12.19}$$

The reason why the relaxation operator in (12.18) is applied to the strain history and the factor $1 - \omega(t)$ is kept outside is that the viscoelastic law is supposed to link the strain history, $\varepsilon(t)$, to the history of the effective stress, $\overline{\sigma}(t) = \sigma(t)/(1 - \omega(t))$, as becomes clear from (12.19).

In the one-dimensional setting, a rate-independent damage model can be motivated by the idea of a bundle of parallel elastic-brittle fibers that are all subjected to the same strain (see, e.g., [518, 519] for a detailed discussion). The fibers break when they attain a critical strain level, which is different for each infinitesimal fiber. Consequently, the overall growth of damage is continuous and the macroscopic material stiffness is reduced gradually. The yet unbroken fibers carry the effective stress, which is linked to the strain by Hooke's law. For a model combining damage with creep, each fiber is considered as viscoelastic, and multiplication of the current strain by the elastic modulus is replaced by application of the relaxation operator to the strain history. The output is then interpreted as the effective stress history, and its multiplication by $1 - \omega(t)$ leads to the nominal stress history.

For loading at low levels, before the damage threshold is attained, the damage variable remains equal to zero and the response described by (12.18)–(12.19) is purely viscoelastic. On the other hand, if the damage threshold is exceeded and the loading process takes place on the time scale of conventional short-term tests, the delayed viscous strain has no time to develop. The stress–strain response then corresponds to a pure damage model, provided that the damage variable is linked to the strain in the same way as in the original rate-independent damage formulation. The key question is which variable should drive the evolution of damage in a general case of long-term loading at moderate and high stress levels, when the response is affected simultaneously by damage growth and by viscous processes.

The simplest choice would be to make the damage depend again on the total (equivalent) strain. However, such assumption would lead to a response that does not match experimental observations. This is illustrated in Fig. 12.20a, which shows the strain histories corresponding to basic creeps tests performed under uniaxial compression at three stress levels, namely $|\hat{\sigma}| = 14.5, 22.0$ and $29.3\,\text{MPa}$, for concrete with standard compression strength $f_c = 45.4\,\text{MPa}$ loaded at the age of 90 days. The isolated points represent real data measured by Komendant et al. [551], and the continuous curves are obtained from the strain history measured at the lowest stress level (i.e., $14.5\,\text{MPa}$) by vertical scaling using the stress ratios $22.0/14.5$

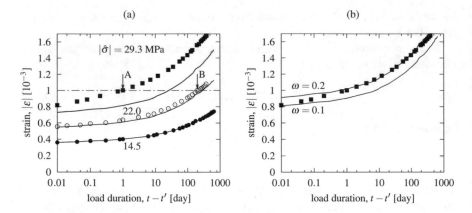

Fig. 12.20 Strain history in basic creep tests at various stress levels [551]: (a) measured strains at three stress levels (points) and proportional scalings of the curve corresponding to the lowest stress, (b) measured strains at the highest stress (points) and proportional scalings of the curve corresponding to the lowest stress amplified by factors $1/(1 - \omega)$ with $\omega = 0.1$ and 0.2

and 29.3/14.5 as scaling factors. The curves characterize the behavior according to linear viscoelasticity, with the compliance function determined from the test at the lowest stress level. For the medium stress level, the deviation from linearity is almost negligible, which confirms that the test performed at the lowest stress level can be considered as unaffected by damage. On the other hand, the test at the highest stress level exhibits a significant deviation from the prediction based on linear viscoelasticity. After about 1 day of loading, the strain magnitude in this test attains 10^{-3}; see point A in Fig. 12.20a. The deviation from linearity at this point is well pronounced. In the test at the medium stress level, the strain magnitude of 10^{-3} is attained after about 200 days of loading; see point B in Fig. 12.20a. In this case, the deviation from linearity is still very small. Consequently, it is obvious that damage cannot be uniquely linked to the total strain.

States A and B in Fig. 12.20a correspond to the same values of total strain, but the contributions of the instantaneous (elastic) strain and the delayed (creep) strain are different. In state A, the instantaneous strain is larger and the delayed strain is smaller than in state B. Since the damage in state A is seen to be higher, it appears logical to relate damage to the sum of the full value of instantaneous strain and the value of delayed strain reduced by a factor smaller than 1. Such an approach was used, e.g., by Mazzotti and Savoia [614] and adopted by other authors. A rather low value of the reduction factor was recommended, typically around 0.1 or 0.2. Even if this factor is kept small, the approach predicts a continuous growth of damage during a creep test at elevated constant stress, because the delayed strain keeps growing.

It is not straightforward to compute the strain history in a creep test at variable damage because the effective stress becomes variable, too, even at constant nominal (applied) stress. However, one can easily obtain strain histories that would correspond to constant nonzero damage (thought of, e.g., as the damage induced at the beginning of the test by the instantaneous strain). In the case of constant damage ω and constant applied stress $\hat{\sigma}$, formula (12.19) gives

$$\varepsilon(t) = \mathscr{J} \left\{ \frac{\hat{\sigma} H(t - t')}{1 - \omega} \right\} = \frac{\hat{\sigma}}{1 - \omega} J(t, t') \qquad (12.20)$$

This means that the corresponding strain curve is obtained from the strain curve that corresponds to linear viscoelasticity by an additional vertical scaling, using a constant amplification factor $1/(1 - \omega)$. In Fig. 12.20b, such scaled curves are plotted for two fixed values of damage, $\omega = 0.1$ and 0.2, along with the points that represents the measured response at the highest stress level. It is seen that with $\omega = 0.1$ one would obtain a good prediction of the "instantaneous" strain (load duration of 0.01 day) but, during the first day of loading, the measured strain grows somewhat faster than it would at constant damage. On the other hand, for load durations longer than 1 day the measured evolution of strain follows the curve that would be obtained with damage set to 0.2 and kept constant during the whole test. If damage kept growing after 1 day of loading, the points would have to climb above the curve obtained by vertical scaling of the linear viscoelastic response.

The foregoing analysis of one particular series of experiments reported by Komendant et al. [551] indicates that damage may remain constant even during a test with growing strain, far beyond the range of linear viscoelasticity. In theory, this could be an exception, but similar trends are found for other creep experiments performed at elevated stress levels. Figure 12.21 presents the dimensionless compliance amplification factors constructed as the ratios between the actual strain measured in the creep test and the strain that would be predicted by linear viscoelasticity, with the compliance function deduced from the creep test for the same concrete at the lowest reported stress level. At constant damage ω, such amplification factors correspond to $1/(1 - \omega)$ because (12.20) can be rewritten as

$$\frac{1}{1 - \omega} = \frac{\varepsilon(t)}{\hat{\sigma} J(t, t')} = \frac{\varepsilon(t)}{\varepsilon_{lve}(t)} \qquad (12.21)$$

where ε is the actual strain and ε_{lve} denotes the strain predicted by linear viscoelasticity. In Fig. 12.21a–c, the fraction on the right-hand side of (12.21) is plotted as a function of the load duration. In Fig. 12.21d, a similar fraction is used, with the total strain values replaced by the delayed part of strain (e.g., with the instantaneous strains subtracted).

The experimental data presented in Fig. 12.21 have been extracted from Komendant et al. [551], Weil [856] and Roll [731]. For each case, the ratio between the applied stress $\hat{\sigma}$ and the standard compression strength $\bar{f}_c$ is reported. The test performed at the lowest stress level always serves as a reference for determination of the compliance function, and so the corresponding ratio $\varepsilon(t)/\varepsilon_{lve}(t)$ is identically equal to 1. For higher stress levels, these ratios are typically larger than 1 and, in most cases, are found to fluctuate without a clear indication of a growth in time. The only notable exception is the test at the highest stress level, $|\hat{\sigma}| = 0.64 \bar{f}_c$, in

Fig. 12.21b, but here the values for load durations between 1 and 10 days could be questioned (because they are too close to the values for stress $|\hat{\sigma}| = 0.57\,\bar{f}_c$) and, if they are omitted, the growth of the amplification factor appears to be relatively mild. The data presented in Fig. 12.21d have been adapted based on Fig. 10 from Mazzotti and Savoia [614]. Only the creep (delayed) strains were reported in that source, and so the amplification factor is evaluated as $\varepsilon_c(t)/\varepsilon_{c,lve}(t)$ where ε_c is the measured creep strain and $\varepsilon_{c,lve}$ is its prediction based on linear viscoelasticity. The results corresponding to $|\hat{\sigma}| = 0.35\,\bar{f}_c$ have been included only for completeness, but they cannot be trusted since three out of five measurements lead to $\varepsilon_c(t) < \varepsilon_{c,lve}(t)$.

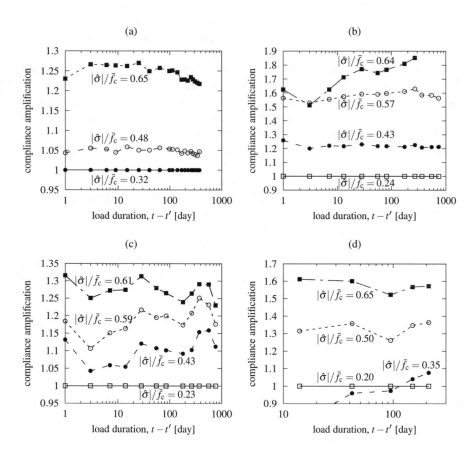

Fig. 12.21 Dimensionless compliance amplification factors for creep tests at various stress levels, based on the data of (a) Komendant et al. [551], $\bar{f}_c = 45.4\,\text{MPa}$, basic creep, (b) Weil [856], $\bar{f}_c = 31.2\,\text{MPa}$, drying creep at $h_{\text{env}} = 65\%$, (c) Weil [856], $\bar{f}_c = 50.8\,\text{MPa}$, drying creep at $h_{\text{env}} = 65\%$, (d) Roll [731], $\bar{f}_c = 42\,\text{MPa}$

The graphs in Fig. 12.21 clearly show that, at least in the range of load durations longer than 1 day, reasonable fits of the measured strain by a model combining viscoelasticity and creep could be obtained with the damage variable considered as constant during each creep test. Since, in each test, the total (mechanical) strain increases and the creep strain (after subtraction of the instantaneous strain) increases as well, neither of these strains can be used as the damage-driving variable. On the other hand, each test was performed at constant stress and also at a constant value of the instantaneous strain. It is therefore reasonable to consider the damage variable as dependent either on the instantaneous strain, or on the (nominal) stress. The first option is consistent with standard rate-independent damage models that link damage to the equivalent strain, and the second option would correspond to stress-based damage models, which are slightly less convenient from the computational point of view but can have other advantages.

A simple stress-based damage model can be described by *loading–unloading conditions* in the form

$$\dot{\omega} \geq 0, \qquad f_\omega(\sigma, \omega) \leq 0, \qquad \dot{\omega} f_\omega(\sigma, \omega) = 0 \qquad (12.22)$$

in which f_ω is a suitable function constructed such that condition $f_\omega(\sigma, \omega) < 0$ characterizes those states in which damage does not grow, condition $f_\omega(\sigma, \omega) = 0$ corresponds to those states in which damage does grow, and condition $f_\omega(\sigma, \omega) > 0$ characterizes inadmissible states that cannot occur. Damage plays here the role of a hardening-softening variable. For a fixed value of ω, condition $f_\omega(\sigma, \omega) \leq 0$ describes the set of stress states that can be attained without inducing further damage growth. For instance, for $\omega = 0$, condition $f_\omega(\sigma, 0) \leq 0$ describes the initial elastic domain in which the material keeps its virgin stiffness. With growing ω, the admissible domain in the stress space can expand, which corresponds to hardening (an ascending branch of the stress–strain diagram), but it may also shrink, which corresponds to softening (a descending branch of the stress–strain diagram). In the hardening range, function f_ω can be presented in the form

$$f_\omega(\sigma, \omega) = g_\omega(\sigma) - \omega \qquad (12.23)$$

where function g_ω assigns to each (admissible) stress state the value of damage that would be induced by monotonic loading up to the given stress level.

To provide an illustrative example, consider again the tests of Komendant et al. [551], performed on sealed concrete with $\bar{f}_c = 45.4\,\text{MPa}$, loaded at the age of 90 days. Since the reported value of $\bar{f}_c$ refers to the standard 28-day strength, it is good to estimate the current strength at loading, $\tilde{f}_c$, and relate the stress levels to this modified strength value. Based on the *fib* formula

$$\tilde{f}_c(t) = \bar{f}_c \exp\left(s[1 - \sqrt{28/t}]\right) \tag{12.24}$$

with parameter $s = 0.25$, we obtain $\tilde{f}_c(90) \approx 50.7\,\text{MPa}$. The tests were performed at stress levels $|\hat{\sigma}| = 14.5$, 22.0 and $29.3\,\text{MPa}$, which correspond to ratios $|\hat{\sigma}|/\tilde{f}_c(90)$ equal to 0.286, 0.434 and 0.578, respectively. From Fig. 12.21a, we can estimate the amplification factors $1/(1-\omega)$ to be around 1.05 at the medium stress level $(0.434\,\tilde{f}_c)$ and around 1.25 at the high stress level $(0.578\,\tilde{f}_c)$. This corresponds to damage values $\omega = 0.048$ and 0.2, respectively. As a rough approximation, one can consider the amplification factor to be a linear function[5] of the stress level (of course, in the range above the initial damage threshold, which is assumed to be located between the low and the medium stress levels). Such a linear function is described by

$$\frac{1}{1 - \omega} = 1.388\frac{|\sigma|}{\tilde{f}_c} + 0.448, \tag{12.25}$$

and thus, as a rough approximation, one can consider function g_ω from (12.23) in the form

$$g_\omega(\sigma) = \left\langle 1 - \frac{1}{1.388|\sigma|/\tilde{f}_c + 0.448} \right\rangle \approx \frac{\langle|\sigma| - 0.40\,\tilde{f}_c\rangle}{|\sigma| + 0.32\,\tilde{f}_c} \tag{12.26}$$

The positive-part brackets (Macauley brackets) have been added to make sure that damage is set to zero (and not no negative values) for states below the damage threshold, i.e., for $|\sigma| < 0.4\tilde{f}_c$. The lowest stress at which the creep test was performed, $|\hat{\sigma}| = 0.286\,\tilde{f}_c$, was indeed in this range.

The basic creep compliance function can be determined from the strain history measured in the test at the lowest stress level, $|\hat{\sigma}| = 14.5\,\text{MPa} = 0.286\,\tilde{f}_c$. The B3 model with parameters estimated in Example 3.1 from the mix composition and concrete strength would, in this case, provide a very good "blind" prediction, with slightly overestimated compliance values for short load durations. An almost perfect fit is obtained with adjusted parameters $q_1 = 16$, $q_2 = 165$, $q_3 = 1$, and $q_4 = 7.7$, all in $10^{-6}/\text{MPa}$; see the lowest solid curve in Fig. 12.22a.

---

[5]Based on the tests of Komendant et al. [551], it is not possible to determine whether the assumed linear dependence of the amplification factor on the stress level is reasonable because the experimental data cover only two stress levels exceeding the damage threshold. The tests of Weil [856] reported in Fig. 12.21b cover three stress levels exceeding the damage threshold and the linear interpolation is found to be acceptable, too. Of course, for stress levels above the range covered by experiments, the law would no doubt require appropriate modifications based on additional data.

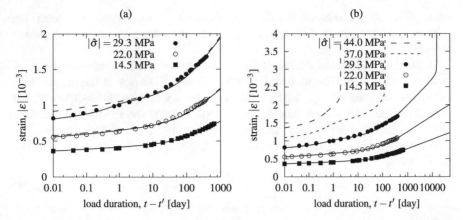

Fig. 12.22 Strain history in creep tests at various stress levels: data by Komendant et al. [551] and theoretical curves (a) without tertiary creep, (b) with tertiary creep (and with additional hypothetical tests at higher stress levels)

The dashed curves in Fig. 12.22a represent the strain histories calculated for the medium and high stress levels using a damage model based on loading–unloading conditions (12.22), with function f_ω in the form given by (12.23) and function g_ω specified in (12.26). In this simplest version of the model, damage is assumed to directly depend on the stress level and, therefore, it increases instantaneously at load application and afterward remains constant during the creep test. The resulting strain histories are obtained from (12.20), with $\omega = g_\omega(\hat{\sigma})$ being constant in time.

The agreement with experimental data is good, except for the early response at the high stress level, $|\hat{\sigma}| = 29.3$ MPa. For load durations below 1 day, the computed strains are higher than the actually measured ones. In terms of the present modeling approach, this indicates that damage does not attain the value corresponding to the given stress level instantaneously. The damage value after the conventional delay of 0.01 day is found to be about 50% of the "terminal" value that is closely approached for load durations above 1 day. If this short-term damage growth is considered as important, the model can be refined by assuming that damage needs some time to develop, which motivates a modified damage law in the rate form

$$\dot{\omega} = \frac{\langle \tilde{g}_\omega(\sigma/\tilde{f}_c) - \omega \rangle}{\tau_\omega} \tag{12.27}$$

where τ_ω is a characteristic time of the damage development process and $\tilde{g}_\omega$ is a function that corresponds to the original function g_ω from (12.26) but is considered as function of the dimensionless stress $\sigma/\tilde{f}_c$; its formal definition reads

$$\tilde{g}_\omega(s) = \frac{\langle \tilde{\sigma} - 0.40 \rangle}{\tilde{\sigma} + 0.32} \tag{12.28}$$

where $\tilde{\sigma} = \sigma/\tilde{f}_c$ represents the dimensionless stress-to-strength ratio. Since the current strength $\tilde{f}_c$ grows in time (due to aging), the ratio $\sigma/\tilde{f}_c$ is variable even at constant stress.

In general, one should deal with the whole spectrum of characteristic times of damage evolution, similar to what is done for creep. However, in the present case there is no visible damage growth for load durations above 5 days, and so it is sufficient to consider one characteristic time $\tau_{\omega,1}$ in the order of 1 day and another characteristic time $\tau_{\omega,0}$ which is much shorter than the conventional delay (0.01 day), and thus, the corresponding damage growth is almost instantaneous. The refined damage law then has the form

$$\dot{\omega} = \frac{\langle \tilde{g}_{\omega,0}(\sigma/\tilde{f}_c) - \omega \rangle}{\tau_{\omega,0}} + \frac{\langle \tilde{g}_\omega(\sigma/\tilde{f}_c) - \omega \rangle}{\tau_{\omega,1}} \tag{12.29}$$

The present example deals with a creep test in which the stress is applied abruptly at $t_1 = 90$ days. The corresponding damage quickly grows to $\omega_1 = \tilde{g}_{\omega,0}(\hat{\sigma}/\tilde{f}_c(t_1))$. There is no need to integrate the damage law during the extremely short initial period numerically. Instead, one can directly set $\omega(t_1) = \omega_1$ as the initial value. Afterward, the first term on the right-hand side of (12.29) becomes inactive (because the value inside Macauley brackets ceases to be positive) and the damage law can be rewritten in the form of the differential equation

$$\tau_{\omega,1}\dot{\omega}(t) + \omega(t) = \tilde{g}_\omega\left(\frac{\hat{\sigma}}{\tilde{f}_c(t)}\right) \tag{12.30}$$

If the aging effect on strength is neglected (i.e., $\tilde{f}_c$ is treated as a constant), Eq. (12.30) with initial condition $\omega(t_1) = \omega_1$ has the analytical solution

$$\omega(t) = \omega_\infty - (\omega_\infty - \omega_1)\exp\left(-\frac{t - t_1}{\tau_{\omega,1}}\right) \tag{12.31}$$

where

$$\omega_\infty = \tilde{g}_\omega\left(\frac{\hat{\sigma}}{\tilde{f}_c}\right) \tag{12.32}$$

is the asymptotically approached terminal value of damage. If the aging effect is included, Eq. (12.30) can be solved numerically, e.g., using the exponential algorithm; see Sects. 5.2.2–5.2.3. Simultaneously, one can determine the strain evolution by a numerical application of the compliance operator to the effective stress $\sigma(t)/(1 - \omega(t))$ (see the right-hand side of (12.19)), or by a step-by-step solution of the corresponding rate-form equations, resulting from an approximation of the compliance function by Dirichlet series.

The numerically computed strain histories for a model based on (12.29) combined with (12.19) are plotted as solid curves in Fig. 12.22a. The difference compared to the simpler model with instantaneous damage is clearly seen for the high stress level.

Function $\tilde{g}_{\omega,0}$ is taken here simply as $\tilde{g}_{\omega,0}(\tilde{\sigma}) = 0.5\tilde{g}_{\omega}(\tilde{\sigma})$, the precise value of $\tau_{\omega,0}$ is irrelevant (as long as it is much smaller than 0.01 day) and $\tau_{\omega,1} = 0.5$ day.

It is clear that a direct extrapolation of the approach based on (12.29) to very high stress levels would not lead to realistic results, especially because such a simple model does not take into account the effects of *tertiary creep* (i.e., accelerated creep leading to material failure). It is well known that failure can occur even at sustained stresses smaller than the short-term strength if the load duration is sufficiently long. Experiments indicate that the dependence of the time to failure on the sustained load level can be approximated by inverse proportionality to a high power, with an exponent in the range of 20 to 30 [632, 893]. This can be taken into account by a modification inspired by the original damage model of Kachanov [530]. A damage law in the form

$$\dot{\omega} = \frac{1}{\tau_f} \left(\frac{|\sigma|}{(1-\omega)\tilde{f}_c} \right)^{n_f} \tag{12.33}$$

with parameters τ_f and n_f leads to an acceleration of damage growth when the state of failure, characterized by $\omega = 1$, is approached. For σ and $\tilde{f}_c$ considered as constant and for initial condition $\omega(t_1) = \omega_1$, Eq. (12.33) can be solved analytically and the resulting damage evolution can be described by

$$\omega(t) = 1 - \left[(1-\omega_1)^{n_f+1} - (n_f+1) \left(\frac{|\sigma|}{\tilde{f}_c} \right)^{n_f} \frac{t-t_1}{\tau_f} \right]^{1/(n_f+1)} \tag{12.34}$$

The resulting *time to failure*, determined from condition $\omega(t_f) = 1$, is then given by

$$t_f - t_1 = \frac{(1-\omega_1)^{n_f+1}\tau_f}{n_f+1} \left(\frac{\tilde{f}_c}{|\sigma|} \right)^{n_f} \tag{12.35}$$

and is inversely proportional to $|\sigma|^{n_f}$.

Since the exponent n_f is typically around 20, the damage rates computed from (12.33) at low damage and stress levels are negligible. Therefore, a combined model can easily be constructed by summing the right-hand sides of (12.33) and (12.29). The damage rate is then given by

$$\dot{\omega} = \frac{\langle \tilde{g}_{\omega,0}(\sigma/\tilde{f}_c) - \omega \rangle}{\tau_{\omega,0}} + \frac{\langle \tilde{g}_{\omega}(\sigma/\tilde{f}_c) - \omega \rangle}{\tau_{\omega,1}} + \frac{1}{\tau_f} \left(\frac{|\sigma|}{(1-\omega)\tilde{f}_c} \right)^{n_f} \tag{12.36}$$

With an appropriate choice of parameters, the first term on the right-hand side of (12.36) controls instantaneous damage, the second term controls short-term damage, and the third term controls tertiary creep leading to failure under sustained load. For concrete, it is natural to consider $\tilde{f}_c$ at the age-dependent current strength.

For illustration, Fig. 12.22b shows the evolution of strain in creep tests, computed with parameters $\tau_f = 10,000$ days, $n_f = 20$, and all other parameters set to the same values as before: $\tau_{\omega,0} \ll 0.01$ day, $\tau_{\omega,1} = 0.5$ day, $t_1 = 90$ days, $\tilde{g}_{\omega}$ given by (12.28),

$\tilde{g}_{\omega,0}$ set to $0.5\tilde{g}_{\omega}$, and $\tilde{f}_c$ determined from (12.24) with $\bar{f}_c = 45.4$ MPa and $s = 0.25$. In addition to three stress levels actually used in the experiments of Komendant et al. [551], two hypothetical tests at stress levels $|\hat{\sigma}| = 37$ and 44 MPa are presented. As shown in the figure, the added term that reflects tertiary creep has almost no influence on the strain growth for the two lowest stress levels, even when the simulated test duration is extended to 50,000 days. For $|\hat{\sigma}| = 29.3$ MPa, the strain blows up after approximately 15,300 days of loading. For the hypothetical tests at still higher stress levels, the times to failure would be about 130 and 2 days, respectively. It should be emphasized that parameters τ_f and n_f, which control tertiary creep, have not been calibrated by real experimental data for the same concrete. The calculated curves merely illustrate the trend that can be obtained with the present type of model and should not be considered as realistic predictions.

Note that the foregoing depiction of tertiary creep is only a crude approximation. In reality, the tertiary creep is a manifestation of structural instability and depends on the rate of release of strain energy from the entire structure. In finite element programs with a rate-dependent progressive damage law and a proper localization limiter, the tertiary creep develops spontaneously as part of structural analysis. This analysis, however, demands a realistic triaxial damage constitutive law, such as the microplane model, to be discussed in Sect. 12.8.

Let us conclude this section by a brief comment on the effects of damage on moisture transport, extending the discussion started already in Sect. 8.3.6. It is clear that cracking leads not only to a reduction of material stiffness, but also to an increase of the space accessible to pore fluids and to formation of additional paths for the flow of water and vapor or wet air. This is reflected by an increase of permeability, which can become quite dramatic if the loading induces severe damage. For instance, Picandet, Khelidj and Bastian [688] found that the intrinsic permeability[6] of concrete loaded to 90% of uniaxial compression strength and then unloaded was by an order of magnitude higher than in the original intact state. Choinska, Khelidj, Chatzigeorgiou and Pijaudier-Cabot [312] measured the intrinsic permeability of concrete subjected to compressive loading at various temperatures. They found that, at room temperature, low and moderate levels of the applied compressive stress lead to a slight decrease of permeability, due to the compaction effect. However, at peak stress, the permeability increased five times at room temperature and by an order of magnitude at 105 and 150 °C. In these conditions, the permeability measured during loading was found to be larger than after unloading. Extensive experimental data acquired within the HITECO project at temperatures up to 700 °C were presented and analyzed by Gawin, Alonso, Andrade, Majorana and Pesavento [413].

Refined theoretical models formulated within the framework of damage mechanics describe the effect of cracking on moisture transport by explicit formulae relating the intrinsic permeability to the damage variable, ω, or, in applications to coupled hygro-thermo-mechanical problems, to temperature and to pore pressure. The sim-

---

[6]Recall that the intrinsic permeability, K_0, is supposed to characterize the transport properties of the pore space independently of the type of pore fluid and is linked to the hydraulic permeability by formula (8.70).

plest version of such a law has traditionally been written in the form

$$K_0 = K_{0,\text{ref}} 10^{A_\omega \omega} \tag{12.37}$$

where $K_{0,\text{ref}}$ is the intrinsic permeability in a reference undamaged state and A_ω is a parameter, but the same relation could equivalently be written as

$$K_0 = K_{0,\text{ref}} \exp(C_\omega \omega) \tag{12.38}$$

where $C_\omega = A_\omega \ln 10$ is a conveniently transformed version of the original parameter A_ω. Equation (12.37) or (12.38) is often attributed to Bary [58], even though Bary himself referred in his dissertation to the original idea of Bourdarot [250]. Typically it is used with $A_\omega = 4$, which means that complete damage increases the intrinsic permeability by four orders of magnitude. A slightly generalized expression

$$K_0 = \begin{cases} K_{0,\text{ref}} & \text{if } \omega \le \omega_0 \\ K_{0,\text{ref}} 10^{A_\omega(\omega-\omega_0)} & \text{if } \omega > \omega_0 \end{cases} \tag{12.39}$$

was used for granite by Souley, Homand, Pepa and Hoxha [784] and for concrete by Jason, Pijaudier-Cabot, Ghavamian and Huerta [511], with parameters $A_\omega = 8.67$ and $\omega_0 = 0.035$. Another generalized expression

$$K_0 = K_{0,\text{ref}} \exp(C_\omega \omega^{\gamma_\omega}) \tag{12.40}$$

was adopted, e.g., by Picandet et al. [688] with parameters $C_\omega = 11.3$ and $\gamma_\omega = 1.64$. A general expression that covers (12.39) and (12.40) as special cases can be found in Bary [58]. It is worth noting that Bary emphasized the anisotropic character of crack-induced permeability increase and adopted a tensorial description of permeability.

In some studies, the intrinsic permeability was expressed as a function of temperature and pore gas pressure. For instance, Gawin et al. [414] suggested to use

$$K_0 = K_{0,\text{ref}} 10^{A_T(T-T_0)} \left(\frac{p_g}{p_{\text{atm}}}\right)^{A_p} \tag{12.41}$$

where A_T and A_p are parameters, T_0 is the reference (room) temperature, p_g is the gas pressure, and p_{atm} is the atmospheric pressure. Later, Gawin, Pesavento and Schrefler [416] added the effect of load-induced damage and generalized (12.41) to

$$K_0 = K_{0,\text{ref}} 10^{A_T(T-T_0)} \left(\frac{p_g}{p_{\text{atm}}}\right)^{A_p} 10^{A_\omega \omega} \tag{12.42}$$

which includes (12.37) as a special case. Formula (12.41) was used with parameters $A_T = 0.005$ and $A_p = 0.368$ while the extended formula (12.42) was used with parameters $A_T = 0.0025$, $A_p = 0.368$, and $A_\omega = 2$.

Clearly, all these formulae are empirical in nature and the parameter values have been derived by fitting of experimental data for a specific concrete. The proposed expressions must be used with care and, ideally, they should be recalibrated for each new application. A comparative study and critical assessment was published by Davie, Pearce and Bićanić [340]. Interestingly, they obtained a very good agreement with the experimental data of Kalifa, Menneteau and Quenard [533] using the intrinsic permeability given by the simplest formula (12.37).

12.8   Microplane Modeling of Cracking Damage with Creep

12.8.1   Basic Ideas of Microplane Modeling

Models that combine tensorial description of inelastic behavior and cracking with the aging and moisture-dependent creep represent the traditional approach, which has a limited modeling potential. For such models, it is difficult to realistically reproduce the damage due to cracking, always combined with frictional microslips, which has an oriented character. The induced anisotropy can be best captured by the vector-based *microplane modeling* concept.

Its history is long. It began in 1938 with G I. Taylor's idea to formulate the constitutive relation in terms of the stress and strain vectors acting on a generic plane within the material [801]. Initially, the stress vector was assumed to be the projection of the stress tensor on this plane. This was a *static constraint*, which led to Batdorf and Budianski's slip theory of plasticity [60] and culminated with the success of the Taylor models for plastic hardening of polycrystals (e.g. [39, 619]).

In 1984, it was shown that for quasi-brittle materials, which exhibit softening damage, the static constraint must be replaced, for reasons of stability (as well as explicitness of computations), by a *kinematic constraint* [89, 127]. In that constraint, the strain (rather than stress) vector on a generic plane in the material microstructure (for which the term "microplane" was coined) is a projection of the continuum strain tensor, while the stress vector is calculated from the strain vector by the microplane constitutive law.

Furthermore, it was shown that, in the case of softening damage, the simple superposition of the plastic strain vectors used in Taylor models must be replaced by virtual work (variational) equivalence between the stress tensor and the microplane stress vectors, and that, in contrast to Taylor model, the elasticity, too, must be included in the microplane constitutive law rather than on the tensorial macrolevel. For isotropic randomly heterogeneous materials, the microplanes may be regarded as the tangent planes of a unit sphere surrounding every material point. By tracing the history of internal variables on individual microplanes, the anisotropy induced by inelastic processes such as cracking or frictional slip can be taken into account in a simple and elegant manner.

The microplane model has been progressively improved for concrete through versions M1, M2, ...M7 [113, 169, 181, 205, 217, 289, 290] and has been widely applied in finite element analysis (e.g. [100, 293]). It has also been adapted to other isotropic randomly heterogeneous quasi-brittle materials, particularly rocks [214, 308] and clays [182, 220], including a generalization to orthotropic gas shale [581]. The thermodynamic restrictions of microplane model M7 have been elucidated in Bažant and Caner [219]. The microplane model for concrete is now embedded in various commercial softwares (e.g., ATENA, DIANA), open-source codes (e.g., OOFEM), and large wavecodes (e.g., EPIC, PRONTO, MARS). For slip in jointed rock mass [309], it is featured as a user subroutine in ANSYS. Initially, the microplane model was too demanding computationally. But the inexorable rise of computer power removed this obstacle after the advent of the twenty-first century.

To avoid model instability in post-peak softening and facilitate explicit step-by-step integration, a kinematic constraint must be used instead of a static one [89, 169]. Thus, the strain vector on each microplane is the projection of the macroscopic strain tensor, i.e.,[7]

$$\varepsilon_N = \varepsilon_{ij} N_{ij}, \quad \varepsilon_M = \varepsilon_{ij} M_{ij}, \quad \varepsilon_L = \varepsilon_{ij} L_{ij} \qquad (12.43)$$

where ε_N, ε_M, and ε_L are the normal and tangential components of the strain vector corresponding to the microplane characterized by a unit normal vector $\boldsymbol{n}$ with components n_i, and $N_{ij} = n_i n_j$, $M_{ij} = (n_i m_j + m_i n_j)/2$, and $L_{ij} = (n_i l_j + l_i n_j)/2$ $(i, j = 1, 2, 3)$ are the components of auxiliary second-order tensors. Here, m_i and l_i are the components of two mutually orthogonal vectors $\boldsymbol{m}$ and $\boldsymbol{l}$ that are tangential to the microplane. This means that each microplane is equipped with three vectors $\boldsymbol{n}$, $\boldsymbol{m}$ and $\boldsymbol{l}$ that form a local orthonormal basis.

The kinematic constraint combined with the principle of virtual work provides a tool for conversion of the stresses on the microplanes (σ_N, σ_M, and σ_L) into the components of the macroscopic stress tensor σ_{ij}. The *virtual work equality* can be understood as a statement of equivalence between the microscopic and macroscopic stresses. The resulting stress evaluation formula reads [113]

$$\sigma_{ij} = \frac{3}{2\pi} \int_{\Omega} (N_{ij}\sigma_N + M_{ij}\sigma_M + L_{ij}\sigma_L) \, d\Omega \qquad (12.44)$$

where Ω is the unit hemisphere.

In practical computations, the integration is carried out numerically, by special integration rules designed for the hemisphere. The microplane stresses are tracked at a finite number of selected microplanes, and formula (12.44) is approximately replaced by

$$\sigma_{ij} = 6 \sum_{\mu=1}^{N_m} w_\mu \left(N_{ij}\sigma_N + M_{ij}\sigma_M + L_{ij}\sigma_L \right)_\mu \qquad (12.45)$$

---

[7]In the present section, tensorial notation is used and the summation convention is adopted. In product-like terms, a sum over all indices that appear twice is implied. For instance, $\varepsilon_{ij} N_{ij}$ actually corresponds to a double sum over i and j running from 1 to 3 (number of spatial dimensions).

The sum is taken over all microplanes that correspond to the integration points of a quadrature formula for integration over the unit hemisphere. N_m is the number of such microplanes, and subscript μ refers to microplane number μ, with $\mu = 1, 2, \ldots N_m$. The integration weight associated with microplane number μ is denoted as w_μ and the subscript μ after the expression in parentheses in (12.45) means that the expression is evaluated on microplane number μ. The integration weights are normalized so that [169]

$$\sum_{\mu-1}^{N_m} w_\mu = \frac{1}{2} \tag{12.46}$$

The kinematic constraint (12.43) and the stress evaluation formula (12.44) link the macroscopic quantities (second-order tensors) to their microplane counterparts (vectors). The core component of a microplane model is a microplane-level constitutive law that relates, on each microplane, the strain vector with components ε_N, ε_M, and ε_L to the stress vector with components σ_N, σ_M and σ_L. *Microplane model M7* [290] uses an explicit procedure that incrementally calculates the microplane stresses from the microplane strains. To model inelastic behavior, M7 uses the so-called *stress–strain boundaries* (or strain-dependent strength limits) within which the behavior is elastic or damaged-elastic.

The main idea is schematically illustrated in Fig. 12.23, in which ε represents a certain microplane strain component and σ is the corresponding microplane stress component. Point A corresponds to the state at the beginning of a computational step number k, and point B to the trial state at the end of the step, with the trial stress increment computed from the given strain increment using an elastic law (taking into account possible degradation of the elastic stiffness). If the trial state is beyond the stress–strain boundary (thick curve in Fig. 12.23), the stress value is reduced and the state point drops at constant strain to the boundary (point C), with the drop interpreted as an inelastic stress increment, $\Delta\sigma''$.

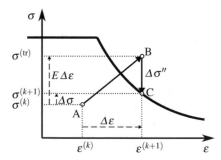

Fig. 12.23 Stress evaluation based on the concept of stress–strain boundaries

Microplane stress–strain laws based on the concept of stress–strain boundaries have traditionally been written in terms of finite computational increments, but they can be reformulated in terms of stress and strain rates, in order to separate the actual constitutive model from its numerical approximation. To illustrate the basic formalism, consider the relation between the normal microplane strain ε_N and normal microplane stress σ_N as a typical example. Suppose that the stress–strain boundary is described by the relation $\sigma_N = \sigma_{Nb}(\varepsilon_N)$ where σ_{Nb} is a given function. In analogy with plasticity and damage mechanics, it is useful to introduce a loading function

$$f_N(\sigma_N, \varepsilon_N) = \sigma_N - \sigma_{Nb}(\varepsilon_N) \tag{12.47}$$

All admissible combinations of stress and strain must satisfy the condition $\sigma_N \leq \sigma_{Nb}(\varepsilon_N)$, which is equivalent to $f_N(\sigma_N, \varepsilon_N) \leq 0$. From this point of view, function f_N plays a similar role to the yield function in plasticity; however, the current "yield stress" $\sigma_{Nb}(\varepsilon_N)$ is evaluated from the total strain and not from an internal variable such as the accumulated plastic strain.

If the current state satisfies condition $f_N(\sigma_N, \varepsilon_N) < 0$, the corresponding point is below the bounding curve and the rate form of the stress–strain law reads $\dot{\sigma}_N = E_N \dot{\varepsilon}_N$ where E_N is the microplane-level normal stiffness, to be specified later. On the other hand, if the current state satisfies condition $f_N(\sigma_N, \varepsilon_N) = 0$ and the stress rate evaluated as $\dot{\sigma}_N = E_N \dot{\varepsilon}_N$ would lead to states that are located above the bounding curve, the actual response must follow that curve and the stress–strain law is described simply by $\sigma_N = \sigma_{Nb}(\varepsilon_N)$. To cover both above-mentioned cases by one unified set of conditions, we can write the rate form of the stress–strain law in the general form

$$\dot{\sigma}_N = E_N \dot{\varepsilon}_N - \dot{\sigma}_N'' \tag{12.48}$$

and impose conditions

$$\dot{\sigma}_N'' \geq 0, \qquad f_N(\sigma_N, \varepsilon_N) \leq 0, \qquad \dot{\sigma}_N'' f_N(\sigma_N, \varepsilon_N) = 0 \tag{12.49}$$

Here, $\dot{\sigma}_N''$ is the *inelastic stress rate*, defined as the difference between the elastically computed stress rate $E_N \dot{\varepsilon}_N$ and the actual stress rate $\dot{\sigma}_N$. Note that conditions (12.49) have the same formal structure as conditions (12.22) used by the simple damage model presented in Sect. 12.7, and also as the standard loading–unloading conditions used in the flow theory of plasticity (see, e.g., Sect. 15.2.2 in [521]). From the mathematical point of view, these are the famous *Karush–Kuhn–Tucker conditions* [536, 555].

The microplane stiffness E_N is in general variable and reflects the degradation of material stiffness due to damage. Its initial value, valid for an intact material, is given by

$$E_{N0} = \frac{E}{1 - 2\nu} \tag{12.50}$$

where E is the macroscopic elastic modulus and ν is the Poisson ratio of the material. Note that E_{N0} is in fact equal to three times the macroscopic bulk modulus. The evolution of the microplane stiffness must be described by an additional law that depends on the particular version of microplane model. Model M7 uses a relatively complicated formula dependent on σ_N, σ_V, ε_N, and the sign of $\dot{\varepsilon}_N$, and also on three internal variables, which are denoted as ζ, $\varepsilon_{N\,max}$ and $\varepsilon_{N\,min}$. All these internal variables can be evaluated directly from the strain history. Variable ζ is obtained by integration of the positive parts of volumetric strain increments, and variables $\varepsilon_{N\,max}$ and $\varepsilon_{N\,min}$ (in the original paper denoted as ε_N^{0+} and ε_N^{0-}) represent the maximum and minimum normal strains reached so far on the given microplane. For future use, it is convenient to write the formula for normal microplane stiffness in the general form

$$E_N = E_{N0}\,\beta_N\,(\sigma_N, \sigma_V, \varepsilon_N, \text{sgn}(\dot{\varepsilon}_N), \zeta, \varepsilon_{N\,max}, \varepsilon_{N\,min}) \qquad (12.51)$$

where β_N is a dimensionless function. The specific expression for β_N can be deduced from the rules given by Caner and Bažant [290].

To keep the presentation simple, we have so far discussed only one bounding curve, considered as a curve that limits the tensile normal microplane stresses. To capture the complex behavior of concrete under general triaxial loading, model M7 uses a combination of bounding curves that limit the components of the microplane stress vector in various ways. In addition, the bounding stresses are not defined as unique functions of one component of microplane strain but they are affected by additional variables that characterize the state of the material on the macroscopic scale. These variables include the volumetric strain[8] ε_V, volumetric (mean) stress σ_V, and maximum and minimum principal strains, ε_1 and ε_3. The following stress bounding functions are used:

1. The **tensile normal** bounding function $\sigma_{Nb}(\varepsilon_N, \sigma_V)$, which captures progressive tensile fracturing.
2. The **compressive volumetric** bounding function $\sigma_{Vb}(\varepsilon_V, \sigma_V, \varepsilon_1, \varepsilon_3)$, which simulates pore collapse under extreme pressures.
3. The **compressive deviatoric** bounding function $\sigma_{Db}(\varepsilon_N, \varepsilon_V)$, which simulates compression softening at low confinement. Primarily, σ_{Db} was designed as a function of the deviatoric microplane stress, defined as $\varepsilon_D = \varepsilon_N - \varepsilon_V$, with an additional dependence on ε_V. Formally, it can be rewritten as a function of ε_N and ε_V.
4. The **shear** bounding function $\sigma_{Tb}(\sigma_N, \varepsilon_V)$, which simulates frictional shear slip and contributes to the description of compression softening at low confinement.

The specific form of functions σ_{Nb}, σ_{Vb}, σ_{Db}, and σ_{Tb} is rather complicated; it can be found in Caner and Bažant [290].

The normal microplane stress, σ_N, is bounded from above by σ_{Nb} and from below, i.e., in compression, by $\sigma_{Vb} + \sigma_{Db}$ (this bounding stress is always negative). To describe such rules formally, we need to consider $\dot{\sigma}_N''$ from (12.48)–(12.49) as the

---

[8]In microplane theory, it is customary to consider the volumetric strain ε_V as the mean normal strain, i.e., one-third of the sum of principal strains. This definition is respected in the present section and differs from the definition used in Sect. 13.2.3, where ε_V denotes the relative change of volume, i.e., the sum of principal strains.

tensile part of the inelastic normal stress rate, further denoted as $\dot{\sigma}_N''^+$, and introduce another, compressive part of the inelastic normal stress rate, denoted as $\dot{\sigma}_N''^-$. Equations (12.48)–(12.49) are then extended to

$$\dot{\sigma}_N = E_N \dot{\varepsilon}_N - \dot{\sigma}_N''^+ + \dot{\sigma}_N''^- \tag{12.52}$$

$$\dot{\sigma}_N''^+ \geq 0, \quad f_N(\sigma_N, \varepsilon_N, \sigma_V) \leq 0, \quad \dot{\sigma}_N''^+ f_N(\sigma_N, \varepsilon_N, \sigma_V) = 0 \tag{12.53}$$

$$\dot{\sigma}_N''^- \geq 0, \quad f_{VD}(\sigma_N, \varepsilon_N, \varepsilon_V, \sigma_V, \varepsilon_1, \varepsilon_3) \leq 0, \quad \dot{\sigma}_N''^- f_{VD}(\sigma_N, \varepsilon_N, \varepsilon_V, \sigma_V, \varepsilon_1, \varepsilon_3) = 0 \tag{12.54}$$

in which

$$f_N(\sigma_N, \varepsilon_N, \sigma_V) = \sigma_N - \sigma_{Nb}(\varepsilon_N, \sigma_V) \tag{12.55}$$

$$f_{VD}(\sigma_N, \varepsilon_N, \varepsilon_V, \sigma_V, \varepsilon_1, \varepsilon_3) = -\sigma_N + \sigma_{Vb}(\varepsilon_V, \sigma_V, \varepsilon_1, \varepsilon_3) + \sigma_{Db}(\varepsilon_N, \varepsilon_V) \tag{12.56}$$

The volumetric stress, σ_V, corresponds to the mean macroscopic stress (one-third of the sum of principal macroscopic stresses) and can be obtained by averaging of normal stresses σ_N over all microplanes. At the same time, σ_V is needed for evaluation of σ_N because it is one of the arguments of functions f_N and f_{VD}. To keep the stress evaluation procedure explicit, the numerical algorithm uses the already known value of σ_V at the beginning of the step for evaluation of σ_N at the end of the step.

Once the normal microplane stresses have been computed, the bounding value of the shear microplane stress is determined by evaluating function $\sigma_{Tb}(\sigma_N, \varepsilon_V)$. This value is then imposed as a limit on the norm of the microplane shear stresses, defined as $\sigma_T = \sqrt{\sigma_M^2 + \sigma_L^2}$. If the trial value of this norm exceeds the limit, the trial values of shear microplane stresses are scaled proportionally, to satisfy the condition $\sigma_T = \sigma_{Tb}$. The corresponding rate form of the constitutive equation can be written as

$$\dot{\sigma}_M = E_T \dot{\varepsilon}_M - \dot{\lambda} \frac{\partial f_T(\sigma_M, \sigma_L, \sigma_N, \varepsilon_V)}{\partial \sigma_M} = E_T \dot{\varepsilon}_M - \frac{\dot{\lambda} \sigma_M}{\sqrt{\sigma_M^2 + \sigma_L^2}} \tag{12.57}$$

$$\dot{\sigma}_L = E_T \dot{\varepsilon}_L - \dot{\lambda} \frac{\partial f_T(\sigma_M, \sigma_L, \sigma_N, \varepsilon_V)}{\partial \sigma_L} = E_T \dot{\varepsilon}_L - \frac{\dot{\lambda} \sigma_L}{\sqrt{\sigma_M^2 + \sigma_L^2}} \tag{12.58}$$

$$\dot{\lambda} \geq 0, \quad f_T(\sigma_M, \sigma_L, \sigma_N, \varepsilon_V) \leq 0, \quad \dot{\lambda} f_T(\sigma_M, \sigma_L, \sigma_N, \varepsilon_V) = 0 \tag{12.59}$$

in which $\dot{\lambda}$ is an auxiliary multiplier,

$$E_T = \frac{(1 - 4v)E_{N0}}{1 + v} = \frac{(1 - 4v)E}{(1 + v)(1 - 2v)} \tag{12.60}$$

is the microplane shear stiffness, and

$$f_T(\sigma_M, \sigma_L, \sigma_N, \varepsilon_V) = \sqrt{\sigma_M^2 + \sigma_L^2} - \sigma_{Tb}(\sigma_N, \varepsilon_V) \qquad (12.61)$$

is the loading function for shear, similar to the yield function in plasticity of cohesive-frictional materials. Note that, unlike the microplane normal stiffness, the microplane shear stiffness E_T is considered as constant, unaffected by damage.

In combination with suitable localization limiters taking into account the material characteristic length, microplane model M7 has been proven to give rather realistic predictions of the constitutive and fracturing behavior of quasi-brittle materials over a broad range of loading scenarios, including uniaxial, biaxial, and triaxial loadings with post-peak softening, compression-tension load cycles, opening and mixed mode fractures, compression-shear failure and axial compression followed by torsion (or vertex effect) [290, 291]. The model has also been extended to fatigue loading of concrete up to several thousands of cycles [547].

12.8.2 Incorporation of Creep with Aging and Shrinkage into Microplane Constitutive Laws

The combination of creep with microplane inelasticity has recently been formulated in detail for the purpose of simulating the damage due to expansive alkali–silica reaction (ASR) [185, 714], as an extension of rate-dependent microplane damage constitutive model M7 [290, 291] and rate-dependent microplane model for dynamics [101]. In this model, linear viscoelastic aging creep is considered to occur in the undamaged concrete, which is between the cracks, while a rate-dependent fracturing (or damage) strain represents smeared cracks. Therefore, the creep strain is additive to the other types of strain.

The main idea of the approach used in 2017 by Bažant and Rahimi-Aghdam [185] is that the creep strain is subtracted from the mechanical strain at the macroscopic level. Recall that the mechanical strain is obtained by subtracting the thermal and shrinkage strains from the total strain. The difference between the mechanical strain and the creep strain, considered as the elastic strain, is then projected on individual microplanes and used as the microplane strain vector that drives the evolution of microplane stresses. On the other hand, the bounding stress and the reduced microplane stiffness are still computed from the mechanical strain, which includes the creep strain. The stress–strain boundary can thus be attained and inelastic effects can arise even during creep at constant stress, because the growth of mechanical strain may lead to a reduction of the bounding stress.

To incorporate the above idea into the microplane framework presented in the previous section, we first need to split the mechanical strain rate into the elastic and creep parts. In standard linear viscoelasticity, the relation between the rates of macroscopic strain and macroscopic stress is described by Eq. (2.26), which can be rewritten as[9]

$$\frac{\dot{\sigma}(t)}{E(t)} = \dot{\varepsilon}_\sigma(t) + \frac{1}{E(t)} \int_0^t \dot{R}(t, t') \, d\varepsilon_\sigma(t') \tag{12.62}$$

Realizing that the expression on the left-hand side corresponds to the elastic strain rate, $\dot{\varepsilon}_e(t)$, we can identify the creep strain rate

$$\dot{\varepsilon}_c(t) = \dot{\varepsilon}_\sigma(t) - \dot{\varepsilon}_e(t) = -\frac{1}{E(t)} \int_0^t \dot{R}(t, t') \, d\varepsilon_\sigma(t') \tag{12.63}$$

An alternative expression for the creep strain rate, based on Eq. (2.17), would be

$$\dot{\varepsilon}_c(t) = \int_0^t \dot{J}(t, t') \, d\sigma(t') \tag{12.64}$$

Both foregoing definitions of $\dot{\varepsilon}_c(t)$ are equivalent. Formula (12.63) clearly shows that the creep strain rate is uniquely determined by the history of the mechanical strain, and formula (12.64) links the creep strain rate to the stress history.

In numerical simulations, rates are replaced by increments over finite time steps. In a generic step number k, Eq. (12.62) is replaced by

$$\frac{\Delta\sigma_k}{\bar{E}_k} = \Delta\varepsilon_{\sigma,k} - \Delta\varepsilon_k'' \tag{12.65}$$

where $\bar{E}_k$ is the incremental modulus and $\Delta\varepsilon_k''$ is the creep strain increment. An efficient implementation, based on an approximation of the compliance function by Dirichlet series, makes it possible to determine $\bar{E}_k$ and $\Delta\varepsilon_k''$ from a limited number of internal variables that are updated in each step. The corresponding formulae were derived in Sect. 5.2.

The foregoing equations describe pure viscoelasticity and, after a straightforward generalization to multiaxial states, permit evaluation of the tensorial components of the elastic strain rate, $\dot{\varepsilon}_{e,ij}$. As already mentioned, Bažant and Rahimi-Aghdam [185] substituted the microplane counterparts of the elastic strain rate into the modified microplane laws, as a replacement of the total strain rates that were used by the original rate-independent microplane model. This idea leads to a modified form of Eqs. (12.52) and (12.57)–(12.58), now written as

$$\dot{\sigma}_N = E_N \dot{\varepsilon}_{Ne} - \dot{\sigma}_N''^+ + \dot{\sigma}_N''^- \tag{12.66}$$

$$\dot{\sigma}_M = E_T \dot{\varepsilon}_{Me} - \frac{\lambda\sigma_M}{\sqrt{\sigma_M^2 + \sigma_L^2}} \tag{12.67}$$

---

[9]Note that ε in (2.26) was actually meant to be the mechanical strain. To emphasize that, we denote this strain in (12.62) as ε_σ.

$$\dot{\sigma}_L = E_T \dot{\varepsilon}_{Le} - \frac{\dot{\lambda}\sigma_L}{\sqrt{\sigma_M^2 + \sigma_L^2}} \tag{12.68}$$

where

$$\dot{\varepsilon}_{Ne} = \dot{\varepsilon}_{e,ij}\, N_{ij} \tag{12.69}$$

$$\dot{\varepsilon}_{Me} = \dot{\varepsilon}_{e,ij}\, M_{ij} \tag{12.70}$$

$$\dot{\varepsilon}_{Le} = \dot{\varepsilon}_{e,ij}\, L_{ij} \tag{12.71}$$

are the normal and shear components of the elastic microplane strain rate vector. The microplane stiffnesses must also be adjusted as compared to the original model M7, because they should reflect not only degradation by damage but also the effects of aging. Therefore, the normal and shear microplane stiffnesses are expressed as

$$E_N(t) = \frac{E(t)}{1 - 2\nu}\, \beta_N \left(\sigma_N, \sigma_V, \varepsilon_{N\sigma}, \operatorname{sgn}(\dot{\varepsilon}_{N\sigma}), \zeta_\sigma, \varepsilon_{N\sigma\,\max}, \varepsilon_{N\sigma\,\min}\right) \tag{12.72}$$

$$E_T(t) = \frac{(1 - 4\nu)E(t)}{(1 + \nu)(1 - 2\nu)} \tag{12.73}$$

where $E(t)$ represents the age-dependent elastic modulus and β_N is a dimensionless factor evaluated in the same way as in the rate-independent version of the microplane model, but with all strain variables determined from the mechanical strain (this is indicated in (12.72) by additional subscripts σ). Loading–unloading conditions (12.53)–(12.54) and (12.59) are also kept in the same form as in the original model M7, again with all strain variables derived from the mechanical strain and denoted as $\varepsilon_{N\sigma}$, $\varepsilon_{V\sigma}$, $\varepsilon_{\sigma,1}$, and $\varepsilon_{\sigma,3}$.

Based on the postulated constitutive equations of microplane model M7 extended by creep, shrinkage, and thermal expansion, computational procedures dealing with finite increments can be developed. The main steps of the stress evaluation algorithm that processes a generic time step number k are summarized below. The purpose of the algorithm is to compute the stresses $\sigma_{ij}^{(k+1)}$ at the end of a time step during which the strains increase from $\varepsilon_{ij}^{(k)}$ to $\varepsilon_{ij}^{(k+1)} = \varepsilon_{ij}^{(k)} + \Delta\varepsilon_{ij}$. Superscripts k and $k + 1$ are used to distinguish between values at the beginning of the current step and at its end (i.e., at the beginning of the next step). For simplicity, we do not use such superscripts if no confusion can arise (e.g., for increments of various quantities).

Algorithm 12.1

1. Subtract the increments of shrinkage and thermal strains from the normal components of total strain increments, to get the increments of mechanical strain

$$\Delta\varepsilon_{\sigma,ij} = \begin{cases} \Delta\varepsilon_{ij} - \alpha_T \Delta T - \Delta\varepsilon_{sh} & \text{for } i = j \\ \Delta\varepsilon_{ij} & \text{for } i \neq j \end{cases} \tag{12.74}$$

2. Using a standard algorithm for rate-type creep models, evaluate the incremental effective modulus $\bar{E}$ and the creep strain increments $\Delta\varepsilon_{ij}''$. In general, the compliance function of an aging viscoelastic model can be approximated by Dirichlet series corresponding to the compliance function of an aging Kelvin chain, and then formulae (5.92) and (5.92) can be used for evaluation of $\bar{E}$ and $\Delta\varepsilon_{ij}''$. Of course, this means that certain creep history variables must be stored and updated in each step; see Algorithm 5.4 for details.

3. Subtract the creep strain increments from the mechanical strain increments to get the elastic strain increments

$$\Delta\varepsilon_{e,ij} = \Delta\varepsilon_{\sigma,ij} - \Delta\varepsilon_{ij}''. \tag{12.75}$$

4. Based on the mechanical strain at the end of the step, $\varepsilon_{\sigma,ij}^{(k+1)}$, evaluate variables that will later be used on all microplanes. These include the volumetric strain

$$\varepsilon_{V\sigma}^{(k+1)} = \frac{1}{3}\left(\varepsilon_{\sigma,11}^{(k+1)} + \varepsilon_{\sigma,22}^{(k+1)} + \varepsilon_{\sigma,33}^{(k+1)}\right) \tag{12.76}$$

the history variable

$$\zeta_\sigma^{(k+1)} = \zeta_\sigma^{(k)} + \frac{1}{3}\left(\Delta\varepsilon_{\sigma,11}^{(k+1)} + \Delta\varepsilon_{\sigma,22}^{(k+1)} + \Delta\varepsilon_{\sigma,33}^{(k+1)}\right) \tag{12.77}$$

and the maximum and minimum principal mechanical strains, $\varepsilon_{\sigma,1}^{(k+1)}$ and $\varepsilon_{\sigma,3}^{(k+1)}$.

5. In a loop over all microplanes (used as integration points), do the following:

a. Project the mechanical strains at the end of the step, $\varepsilon_{\sigma,ij}^{(k+1)}$, and the elastic strain increments, $\Delta\varepsilon_{e,ij}$, on the given microplane, to get their microplane counterparts

$$\varepsilon_{N\sigma}^{(k+1)} = \varepsilon_{\sigma,ij}^{(k+1)} N_{ij} \tag{12.78}$$

$$\Delta\varepsilon_{Ne} = \Delta\varepsilon_{e,ij} N_{ij} \tag{12.79}$$

$$\Delta\varepsilon_{Me} = \Delta\varepsilon_{e,ij} M_{ij} \tag{12.80}$$

$$\Delta\varepsilon_{Le} = \Delta\varepsilon_{e,ij} L_{ij} \tag{12.81}$$

b. Update the maximum and minimum microplane normal strains reached so far:

$$\varepsilon_{N\sigma\,\text{max}}^{(k+1)} = \max\left(\varepsilon_{N\sigma}^{(k+1)}, \varepsilon_{N\sigma\,\text{max}}^{(k)}\right) \tag{12.82}$$

$$\varepsilon_{N\sigma\,\text{min}}^{(k+1)} = \min\left(\varepsilon_{N\sigma}^{(k+1)}, \varepsilon_{N\sigma\,\text{min}}^{(k)}\right) \tag{12.83}$$

c. Determine the stiffness modification factor

$$\beta = \beta_N \left(\sigma_N^{(k)}, \sigma_V^{(k)}, \varepsilon_{Ne}^{(k+1)}, \operatorname{sgn}(\Delta\varepsilon_{Ne}), \zeta_\sigma^{(k+1)}, \varepsilon_{N\sigma\,\max}^{(k+1)}, \varepsilon_{N\sigma\,\min}^{(k+1)} \right) \quad (12.84)$$

and the incremental normal microplane stiffness

$$E_N = \frac{\beta \bar{E}}{1 - 2v} \quad (12.85)$$

and evaluate the trial normal microplane stress

$$\sigma_N^{(tr)} = \sigma_N^{(k)} + E_N \, \Delta\varepsilon_{Ne} \quad (12.86)$$

d. Evaluate the bounding stresses

$$\sigma_{Nb}^{(k+1)} = \sigma_{Nb}\left(\varepsilon_{N\sigma}^{(k+1)}, \sigma_V^{(k)} \right) \quad (12.87)$$

$$\sigma_{Vb}^{(k+1)} = \sigma_{Vb}\left(\varepsilon_{N\sigma}^{(k+1)}, \sigma_V^{(k)}, \varepsilon_{\sigma,1}^{(k+1)}, \varepsilon_{\sigma,3}^{(k+1)} \right) \quad (12.88)$$

$$\sigma_{Db}^{(k+1)} = \sigma_{Db}\left(\varepsilon_{N\sigma}^{(k+1)}, \varepsilon_{V\sigma}^{(k+1)} \right) \quad (12.89)$$

e. Check whether the trial normal microplane stress exceeds the bounding values and determine the final normal microplane stress:

$$\text{if } \sigma_N^{(tr)} > \sigma_{Nb}^{(k+1)} \text{ then } \sigma_N^{(k+1)} = \sigma_{Nb}^{(k+1)} \quad (12.90)$$

$$\text{else if } \sigma_N^{(tr)} < \sigma_{Vb}^{(k+1)} + \sigma_{Db}^{(k+1)} \text{ then } \sigma_N^{(k+1)} = \sigma_{Vb}^{(k+1)} + \sigma_{Db}^{(k+1)} \quad (12.91)$$

$$\text{else } \sigma_N^{(k+1)} = \sigma_N^{(tr)} \quad (12.92)$$

f. Determine the shear microplane stiffness

$$E_T = \frac{(1 - 4v)\bar{E}}{(1 + v)(1 - 2v)} \quad (12.93)$$

and evaluate the trial shear microplane stresses

$$\sigma_M^{(tr)} = \sigma_M^{(k)} + E_T \, \Delta\varepsilon_{Me} \quad (12.94)$$

$$\sigma_L^{(tr)} = \sigma_L^{(k)} + E_T \, \Delta\varepsilon_{Le} \quad (12.95)$$

and their norm

$$\sigma_T^{(tr)} = \sqrt{\left(\sigma_M^{(tr)}\right)^2 + \left(\sigma_L^{(tr)}\right)^2} \quad (12.96)$$

g. Evaluate the bounding shear stress

$$\sigma_{Tb}^{(k+1)} = \sigma_{Tb}\left(\sigma_N^{(k+1)}, \varepsilon_{V\sigma}^{(k+1)} \right) \quad (12.97)$$

h. Check whether the norm of trial shear microplane stresses exceeds the bounding value and determine the final shear microplane stresses:

$$\text{if } \sigma_T^{(tr)} \leq \sigma_{Tb}^{(k+1)} \text{ then } \sigma_M^{(k+1)} = \sigma_M^{(tr)}, \ \sigma_L^{(k+1)} = \sigma_L^{(tr)} \tag{12.98}$$

$$\text{else } \sigma_M^{(k+1)} = \frac{\sigma_{Tb}^{(k+1)}}{\sigma_T^{(tr)}} \sigma_M^{(tr)}, \ \sigma_L^{(k+1)} = \frac{\sigma_{Tb}^{(k+1)}}{\sigma_T^{(tr)}} \sigma_L^{(tr)} \tag{12.99}$$

6. Evaluate the final tensorial stress components by summing the contributions of individual microplanes based on the numerical quadrature formula (12.45):

$$\sigma_{ij}^{(k+1)} = 6 \sum_{\mu=1}^{N_m} w_\mu \left(N_{ij}\sigma_N^{(k+1)} + M_{ij}\sigma_M^{(k+1)} + L_{ij}\sigma_L^{(k+1)} \right)_\mu \tag{12.100}$$

7. Evaluate the volumetric stress

$$\sigma_V^{(k+1)} = \frac{1}{3} \left(\sigma_{11}^{(k+1)} + \sigma_{22}^{(k+1)} + \sigma_{33}^{(k+1)} \right) \tag{12.101}$$

8. Store the volumetric stress $\sigma_V^{(k+1)}$, the macroscopic history variable $\zeta_\sigma^{(k+1)}$ and, for all microplanes, microplane stresses $\sigma_N^{(k+1)}$, $\sigma_M^{(k+1)}$, and $\sigma_L^{(k+1)}$, and microplane history variables $\varepsilon_{N\sigma\,max}^{(k+1)}$ and $\varepsilon_{N\sigma\,min}^{(k+1)}$, all of which will be needed in the next incremental step.

As an example of microplane creep analysis of the cracking caused by alkali–silica reaction, Fig. 8.59 shows some of the results obtained in 2017 by Rahimi-Aghdam et al. [714]. The microplane model made it possible to show the development of cracking of different orientations for an unconfined load-free specimen, radially confined load-free specimen, unconfined axially loaded specimen, and radially confined axially loaded specimen.

As seen from this example, the microplane approach can simulate the oriented character of damage depending on the triaxial stress state. Comparison with the solution for no creep shows that creep has a major effect. It relaxes the self-equilibrated internal stresses caused by the ASR. Another major effect in these simulations of ASR (Fig. 8.59) is the stress relaxation due to diffusion of expanding silica gel into the pores of cement and concrete. The gel diffusion occurs slowly, over many years, and has a strong delayed stress mitigating effect, similar to creep.

Microplane-based models for concrete with incorporated creep and shrinkage effects were also developed by other authors. For instance, Di Luzio and Cusatis [351] combined an updated version of the M4 microplane model [350] with the microprestress-solidification theory and proposed a model that can capture the early-age behavior of concrete. Their approach is somewhat different from the one used by Bažant and Rahimi-Aghdam [185]. It is based on an additive split of the macroscopic strain into parts that correspond to instantaneous elasticity, creep, damage (cracking), shrinkage, and thermal expansion. Elasticity and creep are described within the framework of the microprestress-solidification theory, in the spirit of the B3 model.

The additional damage strain is computed using a microplane model subjected to the same stress as the other units of the rheological chain. Since the elastic strain is already accounted for by the creep model, the microplane model uses a fictitious elastic stiffness and the corresponding elastic strain is subtracted on the macroscopic level. This makes sure that, as long as the response on all microplanes remains elastic, the resulting damage strain vanishes.

Yet another approach was used by Ožbolt and Reinhardt [666, 667], who incorporated creep and rate effects into the specific version of microplane model previously developed by Ožbolt, Li and Kožar [665].

12.8.3 The Lattice Discrete Particle Model Generalized for Creep

The lattice discrete particle model (LDPM), progressively developed by Cusatis et al. in a series of papers [332, 333], is able to resolve the ASR effect locally, in the aggregate mesostructure. The gel diffusion has not been included in an earlier model of ASR by Alnaggar et al. [33] based on lattice particle model, although the creep was. In this model, the inelastic behavior, creep, and shrinkage are introduced in the contact plates of adjacent simulated particles, using a vectorial constitutive law on these planes that is analogous to the constitutive law on the microplanes. The LDPM is very realistic in reproducing the local small-scale behavior concrete, including creep, but its computational demands are prohibitive for real concrete structures.

The LDPM has been extended to ASR [33] by including the ASR gel expansion on the contact planes of adjacent particles, although the ASR gel diffusion into the nanopores has not been included.

12.9 Why Creep Rate at Low Stress Depends on Stress Linearly

At low stress, no microcracks form (or else E-modulus would decrease). So, $[\partial \Pi / \partial w_c]_{w_c=0} = 0$, and the rate of energy release from the structure is zero. But imagine w_c replaced by x in Fig. 12.19f. Because the applied local shear stress $\tau = \partial \Pi / \partial x$, the valleys of potential Π decrease with distance x of the interatomic slip that causes creep. Their decrease, $\Delta \Pi = \Delta Q$, per atomic spacing Δx is very small. So, similar to Eq. (12.11), we must substitute $\Delta Q \propto \tau \Delta x$ into the left part of Eq. (12.11). Hence, the frequency of interatomic bond breaks and restorations, determining the creep rate, is $f_1 \propto \sinh(C \tau \Delta x) \propto \tau \propto \sigma$ (where C = stress-independent coefficient). This explains why, at low stress, the stress dependence of concrete creep rate must be (and is) **linear**, even though the interatomic bond breakages and restorations are a highly nonlinear phenomenon. This fact provides one refutation of recent claims that the transition from compressive to tensile creep is nonlinear (the apparent nonlinearity must be explained differently—probably by forgetting to measure and subtract non-negligible autogenous shrinkage).

Chapter 13
Temperature Effect on Water Diffusion, Hydration Rate, Creep and Shrinkage

Abstract Temperature and heat transfer have a large effect on concrete creep as well as on the moisture transport. Great advances in this regard have been made in research motivated by concrete nuclear power plant structures, and more recently by tall building fires and tunnel fires. First we review the mathematical modeling of heat transfer in concrete, including the effect of heat on cement hydration and the heating caused by early-age hydration. Then we turn attention to combined heat and moisture transfer and hygrothermal effects in heated concrete. To model the strains and stresses at high temperature, we discuss the combined thermal and hygral volume changes, the extension of creep models to high temperatures, and explain the phenomena of explosive thermal spalling in fire and surface layer ablation by means of microwave heating. We emphasize the efficacy and necessity of using the finite volume method for the moisture and heat transfer in concrete in the case of a moving dry–wet interface. Finally, we discuss the mass, momentum, and energy balance laws, which provide the theoretical basis for the modeling of hygrothermal processes in multiphase media.

In regular service, concrete structures are normally subjected to temperatures below 50 °C. In fire, however, concrete is exposed to temperatures of at least several hundred degree Celsius and often up to 1000 °C. Similar temperatures can be produced in concrete by a microwave blast used to ablate a thin surface layer of contaminated concrete. Still higher temperatures would occur in nuclear containments and prestressed concrete pressure vessels as a consequence of hypothetical nuclear accidents. It was this application of concrete that stimulated, during 1965–1985, the greatest advance in the modeling of high temperature exposure. Further, significant progress was made in response to a series of *tunnel fires*, which occured in the Channel Tunnel in 1996, Mont Blanc Tunnel in 1999, Gotthard Tunnel in 2001, or Storebaelt Tunnel in 2006; see Ulm, Coussy and Bažant [826], Ulm, Acker and Lévy [823], Vuilleumier, Weatherill and Crausaz [847], Abraham and Dérobert [7], Voeltzel and Dix [837].

Special refractory concretes, used in coal gasification and liquefaction vessels and other chemical technology vessels, have to serve at temperatures from about 500 °C to almost 2000 °C. Moderately elevated temperatures can also be produced in concrete by various heat machines, pipes, and chemical process vessels.

© Springer Science+Business Media B.V. 2018
Z.P. Bažant and M. Jirásek, *Creep and Hygrothermal Effects in Concrete Structures*, Solid Mechanics and Its Applications 225,
https://doi.org/10.1007/978-94-024-1138-6_13

Since creep and delayed volume changes get intensified by high temperatures, they are an important consideration in all these engineering applications, even for short time periods of a few hours or minutes.

A detailed exposition of the thermal effects on concrete is found in Bažant and Kaplan's (1996) book. This chapter presents a brief summary, with updates for later advances. A definite mathematical model, however, is impossible at present since the existing experimental information on the multitude of phenomena at high temperature is still very limited. Instead of the current, theoretically unguided, phenomenological approach to testing, special experiments aimed to prove or disprove various conceivable theories are urgently needed.

13.1 Heat Transfer in Concrete

At temperatures much below 100 °C, the heat gets transferred solely by conduction. The diffusivity of heat in concrete at **room temperature** is about 3–4 orders of magnitude greater than the diffusivity of moisture (see the quantitative comparison in Sect. 13.1.2). Consequently, compared to the moisture diffusion, the halftimes of the heat conduction process are also 3–4 orders of magnitude shorter, and so, at comparable times, the penetration depths of heat are 30–100 times larger. For computations, this has the pleasant consequence that the initial-boundary value problems of heat and moisture diffusion are essentially decoupled, and the heat transfer problem can be solved prior to that of moisture diffusion (here it is of course assumed that the cracking is so fine that it can increase the effective permeability of moisture only by less than an order of magnitude).

With **increasing temperature**, the moisture transfer gets accelerated. The major acceleration comes after exceeding 100 °C. On passing from about 95–105 °C, the moisture permeability jumps up by two orders of magnitude. Aside from conduction, the heat may also be transferred by convection in diffusing pore water, and the transfers of heat and water through concrete become two-way coupled.

13.1.1 Heat Equation

In problems with simultaneous mass and heat transfer, the mass conservation equation already discussed in Chap. 8 must be supplemented by an appropriate energy conservation equation. In its primary form, energy conservation corresponds to the first law of thermodynamics, according to which the increase in total energy is equal to the sum of supplied work and heat. The total energy is the sum of kinetic energy and internal energy. Since a consistent derivation of the energy balance equation for a multiphase medium is relatively lengthy and complicated, it is postponed to Sect. 13.5, in which a theoretically inclined reader can find all the details, with an in-depth discussion of the origin and meaning of individual terms within the frame-work of continuum thermodynamics.

For practical calculations, it is sufficient to know that, after certain transformations and simplifications described in Sects. 13.5.4–13.5.5, the energy balance equation can be converted into the *heat equation*. A rather general form of the heat equation for concrete considered as a multiphase medium consisting of a solid skeleton with pores filled by liquid water and gas (wet air) reads

$$\rho C_p \dot{T} + (C_{ps}\boldsymbol{j}_s + C_{pl}\boldsymbol{j}_l + C_{pg}\boldsymbol{j}_g)\cdot\nabla T = \rho r - \nabla\cdot\boldsymbol{q} - \dot{m}_{\text{deh}}\Delta h^{\text{w}}_{\text{s,l}} - \dot{m}_{\text{ev}}\Delta h^{\text{w}}_{\text{l,g}} \quad (13.1)$$

where ρ is the mass density and C_p is the effective specific heat capacity of concrete (per unit mass); T is the temperature and ∇T is its gradient; C_{ps}, C_{pl}, and C_{pg} are the specific heat capacities of the solid skeleton, liquid water and pore gas; $\boldsymbol{j}_s$, $\boldsymbol{j}_l$, and $\boldsymbol{j}_g$ are the mass fluxes (per unit total area) of the solid skeleton, liquid water, and pore gas; r is the effective distributed heat source (per unit mass); $\boldsymbol{q}$ is the conductive heat flux (per unit total area) and $\nabla\cdot\boldsymbol{q}$ is its divergence; m_{deh} and m_{ev} are the contents of water released by dehydration and of evaporated water (masses per unit volume); and $\Delta h^{\text{w}}_{\text{s,l}}$ and $\Delta h^{\text{w}}_{\text{l,g}}$ are the specific enthalpies of dehydration and of vaporization. A detailed derivation of Eq. (13.1), along with an in-depth discussion of related theoretical aspects, is provided in Sect. 13.5; see Eq. (13.186).

Let us now elucidate the physical meaning of individual terms in the heat equation. The whole equation can be interpreted as expressing the balance of supplied and con-sumed heat per unit time, written for an infinitesimal control volume fixed in space. The **first term on the left-hand side** of (13.1) corresponds to the heat consumed by temperature increase, per unit volume and unit time. As will be shown in Sect. 13.5.5, the specific heat capacity of a pure substance (e.g., of water) is formally defined as the partial derivative of specific enthalpy with respect to temperature, taken at constant pressure; see formula (13.173). The specific enthalpy is a thermodynamic potential obtained from the specific internal energy by Legendre transformation (13.166), and it is linked to the specific Gibbs free energy by another Legendre transformation (13.169). Enthalpy can be thought of as the energy stored in the material (i.e., the internal energy) plus the work of pressure on the volume occupied by the material.

The effective specific heat capacity of concrete,

$$C_p = \frac{1}{\rho}\left[(1-n_p)\rho_s C_{ps} + n_p S_l \rho_l C_{pl} + n_p(1-S_l)\rho_g C_{pg}\right] \quad (13.2)$$

is obtained by weighted averaging of the specific heat capacities of the solid skeleton, liquid water, and pore gas, with mass fractions playing the role of weight coefficients. Recall that n_p denotes the porosity, S_l is the saturation degree, and ρ_s, ρ_l, and ρ_g are the intrinsic mass densities of the solid skeleton, liquid water, and pore gas. The product $n_p S_l$ is thus the volume fraction occupied by liquid water, and $n_p S_l \rho_l$ is the mass of liquid water per unit volume of concrete. The mass density of concrete can be expressed as

$$\rho = (1-n_p)\rho_s + n_p S_l \rho_l + n_p(1-S_l)\rho_g \quad (13.3)$$

and $n_p S_l \rho_l / \rho$ is the mass fraction of liquid water.

The **second term on the left-hand side** of the heat equation (13.1) reflects the effect of heat convection. When liquid water flows through an infinitesimal control volume and the temperature is not uniform, water that flows into the control volume through a part of its boundary has a slightly different temperature from the water that flows out through the complementary part of the boundary. If cold water flows in and warm water flows out, some heat must be spent to keep the temperature of the material in the control volume constant. Mathematically, this heat (per unit time and unit volume) is described by the product $C_{pl} \boldsymbol{j}_l \cdot \nabla T$ where C_{pl} is the specific heat capacity of liquid water, $\boldsymbol{j}_l$ is the flux of liquid water, and ∇T is the temperature gradient. An analogous effect arises due to the flow of pore gas and is reflected by the product $C_{pg} \boldsymbol{j}_g \cdot \nabla T$. In theory, one should also include the product $C_{ps} \boldsymbol{j}_s \cdot \nabla T$ that reflects the effect of the "flow" of the solid skeleton. Of course, the skeleton does not flow in the same sense as pore fluids, but in general it deforms and since the control volume is fixed in space, solid particles also cross the boundary of the control volume. The solid mass flux, $\boldsymbol{j}_s$, is generally very small compared to the flux of liquid water, $\boldsymbol{j}_l$, because the motion of material particles of the solid skeleton is much slower than the motion of liquid water particles. Gas particles can move fast but the gas mass flux is usually also small, because of the low mass density of the pore gas. Nevertheless, Eq. (13.1) displays all the convective terms, for completeness.

The **first term on the right-hand side** of (13.1), ρr, represents the heat supplied by a distributed heat source to a unit volume of concrete per unit time, and the **second term**, $-\nabla \cdot \boldsymbol{q}$, is the net rate of heat supplied by conduction. Note the analogy with a similar term, $-\nabla \cdot \boldsymbol{j}_w$, which appears in the mass balance equation (8.76) and represents the net influx of mass. The **last two terms on the right-hand side**, without the minus signs, correspond to the rates of heat consumed by the processes of dehydration and vaporization. Vaporization is a typical example of a phase change. When liquid water attains the boiling point, it can be converted into vapor if a sufficient amount of energy is supplied. This energy is the *latent heat* of vaporization (or enthalpy of vaporization). When taken per unit mass, it corresponds to the difference

$$\Delta h_{l,g}^w = h_g^w - h_l^w \tag{13.4}$$

between the specific enthalpy of water vapor, h_g^w, and the specific enthalpy of liquid water, h_l^w. In (13.1), the specific enthalpy of vaporization $\Delta h_{l,g}^w$ is multiplied by the time derivative of the mass of evaporated water per unit volume. The process of evaporation "consumes heat," which is why the product $\dot{m}_{ev} \Delta h_{l,g}^w$ appears on the right-hand side of (13.1) with a negative sign.

The processes of hydration and dehydration are not pure phase changes because they involve chemical reactions. Despite that, the heat consumed by dehydration at temperatures above $100\,°C$ is conveniently included in the heat equation in a manner analogous to the heat of vaporization, using the specific enthalpy (or specific latent heat) of dehydration,

$$\Delta h_{s,l}^w = h_l^w - h_s^w \tag{13.5}$$

where h_s^w is the specific enthalpy of water in hydrates that form a part of the solid skeleton. In (13.1), the specific enthalpy of dehydration $\Delta h_{s,l}^w$ is multiplied by the time derivative of the mass of dehydrated water per unit volume. On the other hand, when hydration takes place at temperatures below $100\,°C$, the heat released by hydration reactions is usually considered as part of the distributed heat source, ρr, even though it could be reflected by a term analogous to the dehydration term, with a negative value of latent heat.

In theory, the conductive heat flux q in concrete could be composed of fluxes in the solid skeleton, liquid water, and pore gas; see Eq. (13.189). However, it is more convenient to deal directly with the total heat flux and simply adjust the heat transport properties depending, e.g., on the degree of hydration or on the moisture content.

It is important to realize that (13.1) represents just one possible form of the heat equation for concrete, derived from certain assumptions. Concrete is considered as a multiphase medium consisting of a solid skeleton with pores filled by two fluid phases, namely liquid water and gas (wet air). Two types of mass exchanges between the phases are accounted for, namely evaporation (also covering condensation as the reverse process) and dehydration. Depending on the intended application (and also on the choices made by particular researchers), certain terms in the heat equation can be neglected and other terms can be added. In fact, in Sect. 13.5.5, it will be shown that a fully consistent heat equation would contain two kinds of additional terms, reflecting the contribution of mechanical dissipation and the effect of variable pressure; see Eq. (13.164), derived for a one-phase material. However, it will also be demonstrated that, in the context of hygrothermal modeling of concrete, such terms are negligible.

13.1.2   Characteristic Times of Heating and Drying

In general, heat transfer in concrete is coupled with moisture transport, and the heat equation (13.1) needs to be combined with one or more mass balance equations, depending on the modeling assumptions and the choice of primary unknown variables. Such coupled hygrothermal models will be treated in Sect. 13.2. They play an important role in analyses of extreme loading scenarios, such as an exposure of a concrete structure to fire. On the other hand, under normal conditions, heat transfer and moisture transport can usually be treated separately because they take place on different time scales. This statement can be supported by an analysis of the dominant terms in the heat equation.

Let us start from the simplest case, in which convective terms, distributed heat sources, phase changes, and chemical reactions are neglected. Heat conduction is then the only mechanism of energy transfer, and Eq. (13.1) is reduced to

$$\rho C_p \dot{T} = -\nabla \cdot q \tag{13.6}$$

Conductive heat flux is driven by the temperature gradient and, for isotropic materials, it is natural to describe their relation by the isotropic Fourier law

$$q = -k_T \nabla T \tag{13.7}$$

where k_T is the *thermal conductivity*. Substituting (13.7) into (13.6), we obtain the classical heat conduction equation

$$\rho C_p \dot{T} = \nabla \cdot (k_T \nabla T) \tag{13.8}$$

which is a partial differential equation of a parabolic type, with temperature T as the unknown field. The thermal conductivity k_T as well as the volumetric heat capacity ρC_p may depend on temperature, but if these coefficients are considered as constant, equation (13.8) can further be simplified and rewritten as

$$\dot{T} = D_T \nabla^2 T \tag{13.9}$$

where ∇^2 is the Laplace operator and

$$D_T = \frac{k_T}{\rho C_p} \tag{13.10}$$

is the *thermal diffusivity* [m/s^2].

For the purpose of a comparative analysis, it is useful to note the formal equivalence between the equations describing heat transfer and moisture transport (in their simplest form). Equations (13.8)–(13.9) have mathematically exactly the same structure as equations (8.86) and (8.91), with relative humidity h replaced by temperature T, moisture capacity $1/k$ replaced by volumetric heat capacity ρC_p, moisture permeability c_p replaced by thermal conductivity k_T, and moisture diffusivity C replaced by thermal diffusivity D_T. Therefore, all the results previously obtained for the linear diffusion equation (8.91) apply to the linear heat equation (13.9) as well. The thermal diffusivity has the same units [m^2/s] as the moisture diffusivity, and so their values can be directly compared.

As explained in Chap. 8, the typical value of moisture diffusivity C in saturated mature concrete is about $20\,\text{mm}^2/\text{day}$, which corresponds to $2.3 \cdot 10^{-10}$ m^2/s. In concrete only 1 week old, it can be roughly $10 \times$ larger, and in dry concrete it can be $10 \times$ smaller. It also decreases significantly with age of concrete.

The thermal conductivity k_T can range from 1.4 to 3.6 W/(m·K). Its temperature dependence is only mild and may be neglected. The mass density ρ is around $2300\,\text{kg/m}^3$, and the specific heat capacity C_p can range from 840 to 1170 J/(kg·K). For more details, see Neville [653], p. 491, or Bažant and Kaplan [142], pages 60, 65 and 212. For the typical values $k_T = 2.3$ W/(m·K) and $C_p = 1000$ J/(kg·K), one gets the thermal diffusivity $D_T = k_T/\rho C_p = 1$ mm^2/s $= 10^{-6}$ m^2/s, which is 4 orders of magnitude larger than the typical moisture diffusivity.

For constant diffusivities, the problem of linear heat conduction in a wall with a prescribed temperature evolution on its surfaces is mathematically equivalent to the problem of linear moisture diffusion with a prescribed evolution of relative humidity on the surfaces (provided that the boundary conditions are, for both problems, of the same type, e.g., of the Dirichlet type). As discussed in Chap. 8, the time for a front of drying to penetrate from the surface to a given depth is inversely proportional to the diffusivity; see formula (8.195). For a typical reactor containment of thickness $D = 0.9$ m, the time needed to reach the core of the wall can be estimated as[1]

$$t_{ch,w} = \frac{(D/2)^2}{12\,C} = \frac{0.45^2}{12 \times 2.3 \cdot 10^{-10}} \frac{m^2}{m^2/s} \approx 7.34 \cdot 10^7 \text{ s} \approx 850 \text{ days} \quad (13.11)$$

An analogous formula is valid for heat conduction, and the time for a heat front to propagate from a heated face to the core is estimated as

$$t_{ch,T} = \frac{(D/2)^2}{12\,D_T} = \frac{0.45^2}{12 \times 1 \cdot 10^{-6}} \frac{m^2}{m^2/s} = 16875 \text{ s} \approx 4.7 \text{ h} \quad (13.12)$$

In young concrete heated by hydration, elevated temperatures may persist for several weeks.

Now note that 4.7 h, as well as the period of several weeks, is much less than 850 days. Therefore, heat conduction and moisture diffusion are essentially decoupled. For much thicker bodies, though, e.g., gravity dams, the elevated temperatures persist far longer and the heat transfer is coupled with moisture diffusion because heat produced by hydration causes water migration. In fire problems, the coupling between heat conduction and water and vapor diffusion is always important [188, 189].

Changes in moisture content produce shrinkage and drying creep, but creep has no effect on moisture diffusion or heat conduction. Therefore, the evolution of temperature and humidity can be solved first, and the mechanical analysis can follow. However, cracking could accelerate overall moisture diffusion enough to influence the heating of a containment wall. According to Bažant et al. [122, 150], the acceleration is less than an order of magnitude for normal crack width. This is not enough to make the moisture and heat transfer problems coupled.

13.1.3 Boundary Conditions for Heat Transfer

Heat gets transferred at the surface to or from the surrounding environment. Physically accurate modeling of the conditions near the surface, which would call for

---

[1] The actual time at which the drying front reaches the core would be even much longer than 850 days because the estimate in (13.11) is valid for a linear diffusion problem while in reality the moisture diffusivity strongly decreases with decreasing pore relative humidity; see (8.11).

nonlinear hydrodynamics, is not necessary. For heat transfer, it suffices to use *Newton's law of cooling* (e.g., Chapman [306])

$$\mathbf{q} \cdot \mathbf{n} = B_T \, (T - T_{\text{env}}) \tag{13.13}$$

where B_T is the *surface heat transfer coefficient*, $\mathbf{n}$ is the unit outward normal of the boundary surface, T_{env} is the ambient temperature, and T is the surface temperature. In the case of free convection of air near the surface, B_T is in the range of 5–25 W/(m²·K) (e.g., Chapman [306]). A perfectly insulated surface with a *Neumann-type boundary condition*

$$\mathbf{q} \cdot \mathbf{n} = 0 \tag{13.14}$$

is a limiting case of (13.13) for $B_T = 0$, and perfect heat transmission characterized by a *Dirichlet-type boundary condition*

$$T = T_{\text{env}} \tag{13.15}$$

is a limiting case of (13.13) for $B_T \to \infty$.

When a heated solid is placed in vacuum, it still releases heat. It does so by radiation. This is described by *Stefan's radiation law* [245, 789]

$$\mathbf{q}_r \cdot \mathbf{n} = \gamma_e \sigma_{SB}(T^4 - T_{\text{env}}^4) \tag{13.16}$$

where $\mathbf{q}_r$ is the radiation heat flux, $\sigma_{SB} = 5.67 \times 10^{-8}$ W/(m²·K⁴) is the *Stefan-Boltzmann constant*, and γ_e is the *surface heat emissivity*, which varies in the range [0, 1]. For a perfectly black surface, $\gamma_e = 1$; for brick, $\gamma_e = 0.9$ [525]. The same emissivity as for brick is assumed for concrete. In absence of vacuum, both the surface heat transfer and the radiation take place simultaneously, and the total heat flux at the boundary is equal to the sum of the right-hand sides of (13.13) and (13.16).

13.1.4   Role of Heat Convection*

The linear diffusion equation formulated for a half-space (i.e., an "infinitely thick" wall) can be reduced to one spatial dimension and solved analytically; see formula (8.181) derived in Chap. 8. An analogous analytical solution of the linear heat equation (13.9) can be exploited for an assessment of the relative importance of the convective terms that were initially present in (13.1) but later were neglected.

Contribution of the solid skeleton to heat convection is certainly negligible because the solid mass flux is much smaller than the mass fluxes of pore fluids. In the spirit of the Bažant–Najjar model (see Sect. 8.3.4.2), let us describe the transport of all phases of water by one single moisture flux, j_w, which combines the fluxes of liquid water (including adsorbed water) and of water vapor because these fluxes cannot be distinguished as the capillaries are usually discontinuous. The linear heat equation enhanced by the convective term would read

$$\rho C_p \dot{T} = k_T \nabla^2 T - C_{pw} \boldsymbol{j}_w \cdot \nabla T \tag{13.17}$$

where C_{pw} is the specific heat of moisture. The convective term $C_{pw} \boldsymbol{j}_w \cdot \nabla T$ replaces the sum of $C_{pl} \boldsymbol{j}_l \cdot \nabla T$ and $C_{pg} \boldsymbol{j}_g \cdot \nabla T$ from (13.1), and C_{pw} should in theory depend on the relative contributions of the liquid water flux and vapor flux (the contribution of the dry air flux to heat convection being negligible, due to the low density of air). For the present purpose, it is sufficient to know that C_{pw} is a weighted average of the specific heats of liquid water, C_{pl}, and of water vapor, C_{pv}. At temperatures between 0 and 100 °C, C_{pl} ranges from 4178 to 4219 J/(kg·K) while C_{pv} ranges from 1859 to 1890 J/(kg·K). The maximum possible value of C_{pw} is thus 4219 J/(kg·K).

Consider a massive concrete wall at initial temperature T_{in} and initial pore relative humidity $h_{in} = 1$, exposed at time t_0 to an environment of constant temperature T_{env} and constant relative humidity h_{env}. For a half-space, the one-dimensional form of the linear moisture diffusion equation (8.91) with the Dirichlet boundary condition (8.97) has an analytical solution given by

$$h(x, \hat{t}) = h_{env} + (1 - h_{env}) \, \mathrm{erf} \left(\frac{x}{2\sqrt{C\hat{t}}} \right) \tag{13.18}$$

where $C = k c_p$ is the moisture diffusivity and $\hat{t} = t - t_0$ is the time of drying. When the linear heat equation is considered in the simple form (13.9), i.e., without the convective term, and is combined with the Dirichlet boundary condition (13.15), it has a formally similar analytical solution

$$T(x, \hat{t}) = T_{env} + (T_{in} - T_{env}) \, \mathrm{erf} \left(\frac{x}{2\sqrt{D_T \hat{t}}} \right) \tag{13.19}$$

Recall that erf is the so-called error function, defined by the formula

$$\mathrm{erf}(x) = \frac{2}{\sqrt{\pi}} \int_0^x e^{-z^2} \, dz \tag{13.20}$$

Now we would like to estimate whether the solution of (13.9) changes substantially if the equation is enhanced by the convective term. For this purpose, let us substitute the analytical solutions (13.18) and (13.19) into the right-hand side of the one-dimensional form of Eq. (13.17) and compare the relative importance of the conductive term, $k_T T''$, and the convective term, $-C_{pw} j_w T'$ (primes denote here derivatives with respect to the spatial coordinate x). It would not be a good idea to compare these terms pointwise, because if one of them is zero at a certain point, the other appears to be "infinitely more important" even if it is very small. Therefore, let us first integrate the terms with respect to the spatial coordinate, x. Physically, this makes very good sense because the first term corresponds to minus the spatial derivative of the conductive heat flux (see (13.6)). By integrating the first term and taking into account that the heat flux q tends to zero as $x \to \infty$, we obtain the heat flux on the boundary:

$$\int_0^\infty k_T T''(x, \hat{t}) \, dx = -\int_0^\infty q'(x, \hat{t}) \, dx = -\left[q(x, \hat{t})\right]_{x=0}^{x=\infty} = q(0, \hat{t}) \quad (13.21)$$

The result corresponds to the energy supplied to the wall by heat conduction, per unit time and unit area of the boundary.

The second (convective) term on the right-hand side of (13.17) could be thought of as a fictitious heat source or sink due to moisture transport in the presence of a temperature gradient. For instance, if $T_{env} > T_{in}$, the temperature is a decreasing function of x and the moisture is transported in the direction of growing temperature ("to the left"). Consider a small control volume, fixed in space. Cold water enters this volume through its right boundary, heats up (by an exchange of energy with the skeleton), and then leaves in a warmer state through the left boundary. This process has a cooling effect on the skeleton and is equivalent to a distributed heat sink. By integrating the convective term with respect to the spatial variable, we obtain the energy extracted from the wall, per unit time and unit area of the wall boundary. This energy can be expressed as

$$q_{conv}(\hat{t}) = -\int_0^\infty C_{pw} j_w(x, \hat{t}) T'(x, \hat{t}) \, dx = \int_0^\infty C_{pw} c_p h'(x, \hat{t}) T'(x, \hat{t}) \, dx$$

$$(13.22)$$

Note that the flux j_w has been replaced by $-c_p h'$, based on the one-dimensional version of the Bažant–Najjar transport law (8.84).

For the humidity and temperature histories given by (13.18)–(13.19), we obtain

$$h'(x, \hat{t}) = \frac{1 - h_{env}}{\sqrt{\pi C \hat{t}}} \exp\left(-\frac{x^2}{4C\hat{t}}\right) \quad (13.23)$$

$$T'(x, \hat{t}) = \frac{T_{in} - T_{env}}{\sqrt{\pi D_T \hat{t}}} \exp\left(-\frac{x^2}{4D_T \hat{t}}\right) \quad (13.24)$$

$$q(0, \hat{t}) = -k_T T'(0, \hat{t}) = \frac{k_T(T_{env} - T_{in})}{\sqrt{\pi D_T \hat{t}}} \quad (13.25)$$

$$q_{conv}(\hat{t}) = \int_0^\infty C_{pw} c_p h'(x, \hat{t}) T'(x, \hat{t}) \, dx =$$

$$= C_{pw} c_p \frac{1 - h_{env}}{\sqrt{\pi C \hat{t}}} \frac{T_{in} - T_{env}}{\sqrt{\pi D_T \hat{t}}} \int_0^\infty \exp\left(-\frac{x^2}{4C\hat{t}}\right) \exp\left(-\frac{x^2}{4D_T \hat{t}}\right) \, dx =$$

$$= \frac{C_{pw} c_p}{\sqrt{\pi (C + D_T)\hat{t}}} (1 - h_{env})(T_{in} - T_{env}) \quad (13.26)$$

The relative importance of the convective heat flux is given by the ratio between the absolute values of $q_{conv}(\hat{t})$ and $q(0, \hat{t})$. This dimensionless ratio,

$$\varepsilon_q = \frac{|q_{\mathrm{conv}}(\hat{t})|}{|q(0,\hat{t})|} = \frac{C_{pw}c_p}{k_T}\sqrt{\frac{D_T}{C+D_T}}(1-h_{\mathrm{env}}) \tag{13.27}$$

turns out to be independent of the duration of exposure and can be bounded from above by $C_{pw}c_p/k_T$, because the fraction under the square root in (13.27) is always smaller than 1, and the same holds for the factor $1-h_{\mathrm{env}}$. Substituting typical values $C_{pw} = 4200\,\mathrm{J/(kg\cdot K)}$, $c_p = 2.3\cdot 10^{-8}\,\mathrm{kg/(m\cdot s)}$, and $k_T = 2.3\,\mathrm{W/(m\cdot K)}$, we get

$$\varepsilon_q < \frac{C_{pw}c_p}{k_T} \approx \frac{4200\times 2.3\cdot 10^{-8}}{2.3} = 4.2\cdot 10^{-5} \tag{13.28}$$

This rough estimate clearly shows that the effect of heat convection is at least by four orders of magnitude less important than the effect of heat conduction and thus can be safely neglected.

The foregoing estimate has been obtained for a half-space, but the analytical solution characterizes the initial asymptotics of the drying and heating processes for a finite body of an arbitrary shape. It has also been assumed that the temperature remains below 100 °C. A comparative numerical study of Gawin, Pesavento, and Schrefler [422] revealed that, in concrete heated to temperatures above 100 °C, the convective flux can grow to the same order of magnitude as the conductive flux. Despite that, the effect of convective flux on the evolution of temperature and pore pressure in representative examples that covered slow and fast heating turned out to be negligible [423].

13.1.5 Hydration Heat

Hydration of cement is an exothermic chemical reaction (or rather a group of reactions), which means that it releases heat. This effect can be reflected by a distributed heat source on the right-hand side of the heat equation. Such a term was present in the general heat equation (13.1) but later neglected in Sects. 13.1.2–13.1.4, devoted to the analysis of a wall heated from the surface. In the presence of a heat source, Eq. (13.8) must be rewritten in a generalized form

$$\rho C_p \dot{T} = \nabla \cdot (k_T \nabla T) + \rho r \tag{13.29}$$

where r is the specific power of the heat source [W/kg]. In the present context, the term ρr corresponds to the mass of reacted cement per unit time and unit volume of concrete, multiplied by the difference between specific enthalpies (of cement plus the corresponding amount of water) before and after the reaction.

If the hydration process is dominated by a single reaction (which is a reasonable assumption for ordinary Portland cement, typically containing about 70% of alite), the mass of reacted material is proportional to the degree of hydration and can be expressed as a unique function of the equivalent age. The concept of equivalent age was already introduced in Sect. 10.6.1. If the hydration process takes place at room temperature and near saturation, the equivalent age is equal to the time elapsed from the onset of hydration. In a test, the temperature can be kept constant if all the heat generated by chemical reactions is almost instantaneously extracted from the hydrating specimen. Such procedure is used by isothermal calorimetry, which provides the dependence of hydration heat on equivalent time. Another extreme case is an adiabatic test, in which the temperature growth is measured in a thermally insulated sample. At elevated temperature, the hydration process is accelerated, which is reflected by a faster evolution of the equivalent age. In real concrete members and structures, especially massive ones, the heat released by hydration is evacuated by conduction, with a certain delay. The temperature distribution becomes nonuniform, and the temperature rise is smaller than in an adiabatic test.

Table 13.1 Typical composition of silicate clinker and potential hydration heat of its chemical components

Chemical component	Mass fraction [%] (min – average – max)	H_{max} [kJ/kg]
C_3S (tricalcium silicate)	45–63–80	517
C_2S (dicalcium silicate)	5–20–32	262
C_3A (tricalcium aluminate)	4–8–16	1144
C_4AF (tetracalcium aluminoferrite)	3–7–12	725
Free CaO (calcium oxide)	0.1–1–3	1150
Free MgO (magnesium oxide)	0.5–1.5–4.5	840

Table 13.2 Potential hydration heat of various types of cement

Cement type	H_{max} [kJ/kg]
Generic Portland cement	375–525
Blast furnace slag cement	355–440
Sulfate-resistant cement	350–440
Pozzolanic cement	315–420
High alumina cement	545–585

The total amount of heat released by complete hydration of a unit mass of cement, called the *potential hydration heat* and denoted as H_{max}, depends on composition. Typical values of H_{max} for the main chemical components found in cement clinker are given in Table 13.1, and the ranges of H_{max} for various types of cement are specified in Table 13.2. The data are taken from Czernin [334], Taylor [802], and Bentz [235].

Hydration kinetics can be described by the dependence of the hydration heat released up to the current time, H_c, on the equivalent age, t_e. Equivalently, one can use the dimensionless ratio $\xi = H_c/H_{max}$, called the *hydration degree*.[2] The dependence of the hydration degree or hydration heat on the equivalent age can be described either in the total form, by specifying ξ or H_c directly as functions of t_e, or in the rate form, by linking the rate of hydration to the current hydration degree.

The total form was used, e.g., by Bažant et al. [150], who suggested to approximate the hydration heat (per unit mass of cement) by a function that can be presented in the form

$$H_c(t_e) = \begin{cases} 0 & \text{for } t_e \le t_d \\ \dfrac{H_\infty}{\left(1+\left(\dfrac{t_c}{t_e - t_d}\right)\right)^n} & \text{for } t_e > t_d \end{cases} \tag{13.30}$$

where H_∞ is the *ultimate hydration heat* (per unit mass of cement), t_d is the delay time for the onset of hydration, t_c is a certain characteristic time of hydration, and n is a dimensionless exponent. The ultimate hydration heat H_∞ is equal to the potential hydration heat H_{max} only if the hydration process asymptotically approaches the state of complete hydration. In general, $H_\infty = \xi_\infty H_{max}$ where ξ_∞ is the ultimate hydration degree. If H_c is replaced by ξ and H_∞ by ξ_∞, formula (13.30) can be used for evaluation of the hydration degree. For illustration, Fig. 13.1a shows experimental data presented by Bažant et al. [150] for cement pastes with water-cement ratio $w/c = 0.4$ at a constant temperature of $20\,°C$, and their fits by formula (13.30) with parameters $H_\infty = 455$ kJ/kg, $t_d = 0.16$ day, $t_c = 1.18$ day and $n = 0.78$ for type-I cement, and $H_\infty = 435$ kJ/kg, $t_d = 0.35$ day, $t_c = 1.27$ day, and $n = 0.81$ for type-V cement.

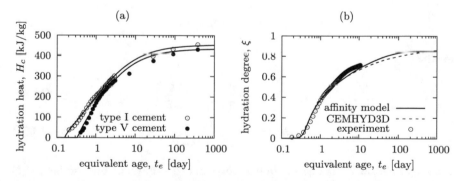

Fig. 13.1 (a) Test data on evolution of hydration heat for cements of type I and V measured under isothermal conditions and their fit by formula (13.30), (b) test data on evolution of hydration degree for cement CEM I measured under isothermal conditions and their fits by the affinity hydration model (13.31)–(13.32) and by the CEMHYD3D model

---

[2] According to its primary definition, the hydration degree is the mass of already hydrated cement divided by the total mass of cement. If the hydration process is dominated by one chemical reaction, the released hydration heat is proportional to the hydration degree, and one can use an alternative definition, $\xi = H_c/H_{max}$.

In the rate form, evolution of the hydration degree is described implicitly, as the solution of the differential equation

$$\frac{d\xi}{dt_e} = A(\xi) \tag{13.31}$$

where A is a generalized thermodynamic force that drives the hydration process, called the *affinity*. Jendele, Šmilauer and Červenka [512] proposed to approximate the dependence of affinity on the current hydration degree by a function of the form

$$A(\xi) = B_1 \left(\frac{B_2}{\xi_\infty} + \xi \right) (\xi_\infty - \xi) \exp\left(-\frac{\eta\xi}{\xi_\infty} \right) \tag{13.32}$$

where ξ_∞ is the ultimate hydration degree and B_1, B_2, and η are additional parameters. The dependence of ξ on t_e can be established by numerical integration. Using parameters $B_1 = 30.4/\text{day}$, $B_2 = 8 \cdot 10^{-6}$, $\eta = 7.4$, and $\xi_\infty = 0.85$, Jendele et al. [512] obtained a good agreement with calorimetric experimental data, and also with the development of hydration heat obtained by a sophisticated discrete hydration model called CEMHYD3D [233]. Their fit is plotted in Fig. 13.1b.

Equations (13.30) and (13.31) are formulated in terms of the equivalent age, t_e. Acceleration of the hydration reaction under increased temperature can be taken into account by an Arrhenius-type factor. On the other hand, low relative humidity reduces the rate of hydration. As already discussed in Sect. 10.6.1, these considerations motivate the definition of the equivalent age by the rate equation

$$\frac{dt_e}{dt} = \beta_{eT}(T)\beta_{eh}(h) \tag{13.33}$$

in which

$$\beta_{eT}(T) = \exp\left(\frac{Q_e}{RT_0} - \frac{Q_e}{RT} \right) \tag{13.34}$$

$$\beta_{eh}(h) = \frac{1}{1 + [\alpha_e(1 - h)]^4} \tag{13.35}$$

Here, Q_e is the activation energy of hydration, R is the universal gas constant, T_0 is the room temperature, T is the current temperature, and h is the current pore relative humidity. For concrete with ordinary Portland cement classified as CEM I 42.5 R (according to the European standards), Kada-Benameur, Wirquin, and Duthoit [532] performed isothermal tests at different temperatures and obtained optimal values of Q_e/R ranging between 3850 and 5500 K. The B4 model sets $Q_e/R = 4000$ K. Parameter α_e used in (13.35) is typically in the order of 10.

The heat source term ρr, included on the right-hand side of (13.29), describes the rate of heat released by hydration, per unit volume of concrete. Since the hydration heat H_c represents the cumulative value (heat released up to the current time) and is taken per unit mass of cement, the corresponding heat source term is evaluated as

$$\rho r = c\frac{\mathrm{d}H_c}{\mathrm{d}t} = c\frac{\mathrm{d}H_c}{\mathrm{d}t_e}\beta_{eT}(T)\beta_{eh}(h) \tag{13.36}$$

where c is the mass of cement per unit volume of concrete.

For a model working with the hydration degree, ξ, it suffices to replace H_c by $\xi H_{\max}$ where $H_{\max}$ is the potential hydration heat for the given type of cement (heat that would be obtained by complete hydration of a unit mass of cement). If the hydration process is described by the rate equation (13.31), relation (13.36) can be rewritten as

$$\rho r = c H_{\max}\frac{\mathrm{d}\xi}{\mathrm{d}t_e}\beta_{eT}(T)\beta_{eh}(h) = c H_{\max} A(\xi)\beta_{eT}(T)\beta_{eh}(h) \tag{13.37}$$

The models for cement hydration developed prior to finishing this book (2017) have some significant limitations. They do not take into account the complete range of variation of pore relative humidity and temperature, the best (such as CEMHYD3D) are computationally too intensive for use in finite element programs, and apply over durations limited from up to a few months to about a year. However, recent tests of autogenous shrinkage and swelling in water (reported in Bažant et al. [125]) imply that the hydration degree may grow, roughly logarithmically, for decades, even centuries, provided that a not too low relative humidity (above cca 0.65) persists in the pores for a long time, as expected for the cores of thick concrete structural members. Therefore, and because design lifetimes of over hundred years are required for large concrete structures, a new hydration model for a hundred year lifespan has been developed [715], by extending and refining a preliminary version from ConCreep-10 conference [125].

This new model considers that, after the first day of hydration, the remnants of anhydrous cement grains, gradually consumed by hydration, are enveloped by contiguous, gradually thickening, spherical barrier shells of calcium silicate hydrate (C-S-H). The hydration progress is controlled by diffusion of water from capillary pores through the barrier shells toward the interface with anhydrous cement. The diffusion is driven by a difference of humidity, defined by equivalence with the difference in chemical potential of water. Although, during the first 4–24 h, the C-S-H forms discontinuous nanoglobules around each cement grain, an equivalent barrier shell control was formulated for ease and effectiveness of calculation. The entire model was calibrated and validated by published test data on the evolution of hydration degree and hydration heat for various cement types, particle size distributions, water-cement ratios and temperatures. Computationally, this model is sufficiently effective for calculating the evolution of hydration degree (or aging) at every integration point of every finite element in a large structure.

13.1.6 Temperature Increase Induced by Hydration

Example 13.1. Hydration of a thick wall (simple model)

For illustration, let us analyze the development of temperature caused by hydration in a wall of thickness $D = 1$ m. Concrete properties are the same as in the analysis of a segment of Opárno Bridge by Jendele et al. [512]. The content of Portland cement CEM I 42.5 R is set to $c = 410$ kg/m^3, and the potential hydration heat is considered as $H_{max} = 510$ kJ/kg, which gives $cH_{max} = 209.1$ MJ/m^3. Hydration is decribed by the affinity model (13.31) and (13.32) with parameters $B_1 = 23.4$/day, $B_2 = 7 \cdot 10^{-4}$, $\eta = 6.7$, and $\xi_\infty = 0.85$. The effect of temperature on the rate of hydration is reflected by factor β_{eT} evaluated from (13.34) with $Q_e/R = 4600$ K, and the effect of pore humidity is neglected.

Table 13.3 Composition of concrete mix and evaluation of effective heat capacity

	$\eta_\alpha \rho_\alpha$ [kg/m^3]	$C_{p\alpha}$ [kJ/(kg·K)]	$\eta_\alpha \rho_\alpha C_{p\alpha}$ [kJ/(m^{3}K]
Water	178	4.18	744
Cement	410	0.79	324
Aggregates	1785	0.84	1499
Filler	86	1.00	86
Concrete	2459		2653

In the first simple analysis, the volumetric heat capacity and the thermal conductivity $k_T = 1.44$ W/(m·K) are considered as constant. The effective value of heat capacity, $\rho C_p = 2.653$ MJ/(m^3·K), is determined for the given composition of concrete mix by summing the contributions of cement, water, aggregates, and filler; see Table 13.3. Note that the volumetric heat capacity of the concrete mix is calculated as a weighted average of the volumetric heat capacity of individual constituents, which can be formally written as $\rho C_p = \sum_\alpha \eta_\alpha \rho_\alpha C_{p\alpha}$. The products $\eta_\alpha \rho_\alpha$ correspond to masses of individual constituents per unit volume of concrete, i.e., to the water content, w, cement content, c, aggregate content, a, and filler content, f. The effective value of thermal conductivity, $k_T = 1.44$ W/(m·K), is obtained using an analytical multilevel homogenization procedure, following the procedure described in Jendele et al. [512].

Evolution of temperature T and hydration degree ξ is computed by numerically solving the one-dimensional heat equation

$$\rho C_p \dot{T} = k_T T'' + c H_{max} \dot{\xi} \tag{13.38}$$

combined with the hydration equation

$$\dot{\xi} = A(\xi)\beta_{eT}(T) \tag{13.39}$$

Initial conditions are set to $\xi(x, 0) = 0$ and $T(x, 0) = T_0 = 293$ K (20°C), and boundary condition (13.13) is rewritten as

$$k_T T'(0, t) = B_T (T(0, t) - T_{\text{env}}) \tag{13.40}$$

where the surface heat transfer coefficient is given by $B_T = 4$ W/(m^{2}K) and the ambient temperature T_{env} is set to T_0. The selected value of B_T corresponds to a 20-mm-thick layer of plywood used as formwork, combined with the effect of air convection.[3] Owing to symmetry, the problem can be solved on the interval $[0, D/2]$, with a zero-flux boundary condition

$$T'(D/2, t) = 0 \tag{13.41}$$

imposed at the plane of symmetry.

Differential equations (13.38)–(13.39) can be approximated by finite differences using similar numerical schemes as for the moisture diffusion equation in Sect. 8.4.2. For instance, the forward Euler approach leads to the explicit update formulae

$$\Delta \xi_i = \Lambda \left(\xi_i^{(k)} \right) \beta_{eT} \left(T_i^{(k)} \right) \Delta t \tag{13.42}$$

$$\Delta T_i = \frac{k_T}{\rho C_p} \frac{\Delta t}{(\Delta x)^2} \left(T_{i+1}^{(k)} - 2T_i^{(k)} + T_{i-1}^{(k)} \right) + \frac{c H_{\max}}{\rho C_p} \Delta \xi_i \tag{13.43}$$

$$\xi_i^{(k+1)} = \xi_i^{(k)} + \Delta \xi_i \tag{13.44}$$

$$T_i^{(k+1)} = T_i^{(k)} + \Delta T_i \tag{13.45}$$

where Δt is the time step, Δx is the spacing of grid points, and $\xi_i^{(k)}$ and $T_i^{(k)}$ are numerical approximations of the hydration degree and temperature at point $x = x_i = i \, \Delta x$ and time $t = t_k \equiv k \, \Delta x$.

It can be shown that, in analogy to formula (8.124) derived in Sect. 8.4.2 for the moisture diffusion equation, a safe estimate of the critical time step is given by

$$\Delta t_{\text{crit}}^* = \frac{(\Delta x)^2}{2 \, D_T} \tag{13.46}$$

where $D_T = k_T / (\rho C_p)$ is the thermal diffusivity. For the present set of parameters, $D_T = 0.543 \cdot 10^{-6}$ m^2/s. If the interval $[0, D/2]$ is divided into 100 equally sized segments, we have $\Delta x = 5 \cdot 10^{-3}$ m and the estimated critical time step is $\Delta t_{\text{crit}}^* \approx$

---

[3] A standard value of the surface heat transfer coefficient for a boundary in direct contact with the atmosphere would be $B_{T1} = 10$ W/(m^{2}K). This value already includes, in an approximate way, the effect of radiation. A layer of plywood, characterized by thermal conductivity 0.13 W/(m·K) and thickness 20 mm, would alone lead to $B_{T2} = (0.13/0.02)$ W/(m^{2}K)= 6.5 W/(m^{2}K). Due to serial coupling (fluxes are equal, temperature differences are summed), the reciprocal values of B_T are additive, which leads to $B_T = 1/(1/B_{T1} + 1/B_{T2}) = 3.94$ W/(m^{2}K) ≈ 4 W/(m^{2}K).

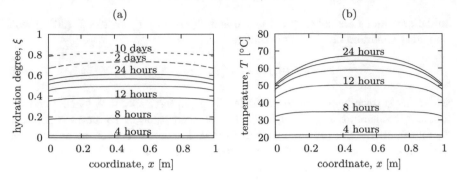

Fig. 13.2 Simple analysis of a wall heated by hydration: profiles of (a) hydration degree and (b) temperature at selected times

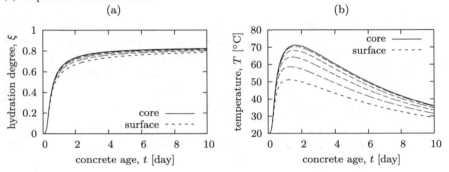

Fig. 13.3 Simple analysis of a wall heated by hydration: evolution of (a) hydration degree and (b) temperature at selected points

23 s. To guarantee numerical stability, the actual time step must be shorter, e.g., $\Delta t = 10$ s.

The computed profiles of hydration degree and temperature at selected times are plotted in Fig. 13.2. In an initial period of several hours, hydration is slow and the temperature increases only slightly. Afterward, the hydration reaction speeds up, which results into a fast temperature growth. In boundary layers, the temperature is partially reduced by heat conduction, and the temperature profiles become nonuniform. Due to thermal activation, the hydration process develops faster in the core than near the boudaries, and the distribution of hydration degree becomes nonuniform, too. After about 1 day, the hydration degree in the core reaches 0.61 and the reaction slows down. The temperature attains its maximum and starts decreasing. For the sake of clarity, the profiles of temperature are plotted in Fig. 13.2b only up to 24 h. The overall evolution of temperature is documented in Fig. 13.3b. In this plot, we can identify the maximum temperature of 71°C, attained in the core after 40 h of hydration. At the surface, the maximum temperature is only 51°C and is reached already after 31 h.

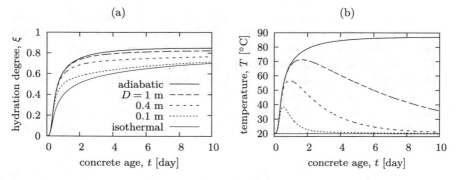

Fig. 13.4 Influence of wall thickness on the evolution of (a) hydration degree and (b) temperature in the core

Figure 13.3a shows the evolution of the hydration degree at selected points. Asymptotically, the hydration degree approaches the same ultimate value, $\xi_\infty = 0.85$, at all points.

Figure 13.4 illustrates the influence of wall thickness on the evolution of the hydration degree and temperature in the core (i.e., at the plane of symmetry). For comparison, the graphs also contain curves that correspond to adiabatic conditions and to isothermal conditions, which can be considered as the limit cases of an infinitely thick wall and a zero-thickness wall. Under adiabatic conditions, the temperature would increase from the initial value of 20°C up to 87°C. This is in agreement with the theoretical ultimate temperature increase under adiabatic conditions,

$$\Delta T_{max} = \frac{\xi_\infty c H_{max}}{\rho C_p} = \frac{0.85 \times 410 \times 510 \cdot 10^3}{2.653 \cdot 10^6} \, \text{K} = 67 \, \text{K} \qquad (13.47)$$

evaluated as the ultimate hydration heat per unit volume, $\xi_\infty c H_{max}$, divided by the volumetric heat capacity, ρC_p.   ■

In Example 13.1, we have considered the heat capacity and thermal conductivity as constant. However, these properties are affected by the current temperature and moisture content (or pore relative humidity), and they also vary in the course of hydration. The dependence on temperature and humidity will be discussed later in the context of complex models for heat and moisture transfer under extreme conditions such as fire; see Sects. 13.2.5–13.2.6. For young hydrating concrete, it is more important to take into account the variation due to changes in microstructure.

Some experimental measurements indicate that hydration leads to a gradual reduction of thermal conductivity. According to Ruiz, Schindler, Rasmussen, Kim, and Chang [743], the dependence on the degree of hydration can be described by the linear function

$$k_T(\xi) = (1 - 0.248\xi) \, k_{T0} \qquad (13.48)$$

where k_{T0} is the thermal conductivity of fresh concrete. On the other hand, Bentz [234] found the thermal conductivity of hydrating cement pastes to exhibit no systematic dependence on the degree of hydration, within the experimental error.

Heat capacity of concrete can be evaluated by combining the contributions of individual constituents. Formula (13.2) is applicable to mature concrete, considered as a multiphase material consisting of a solid skeleton and pore fluids. In a similar spirit, for fresh concrete, one should combine the contributions of water, cement, and aggregates (plus other additives and admixtures, if present). The corresponding formula reads

$$\rho C_{p0} = a\, C_{pag} + w\, C_{pl} + c\, C_{pc} \tag{13.49}$$

where C_{p0}, C_{pag}, C_{pl}, and C_{pc} are the specific heat capacities of fresh concrete, aggregates, liquid water, and cement, and a, w, and c are the masses of aggregates, water and cement per unit volume of concrete. Of course, if the coarse and fine aggregates have different specific heat capacities, the term $a\, C_{pag}$ should be replaced by $a_c C_{pag,c} + a_f C_{pag,f}$.

At room temperature, the specific heat capacities of water and cement are $C_{pl} = 4180$ J/(kg·K) and $C_{pc} = 790$ J/(kg·K). Hydration converts cement and water into hardened cement paste, and the heat capacity of the hydration products is smaller than the combined heat capacity of the reactants. Based on the results of Bentz [234], this effect can be taken into account by setting

$$\rho C_p(\xi) = a\, C_{pag} + \left(w\, C_{pl} + c\, C_{pc} \right) \left[1 - 0.26 \left(1 - e^{-2.9\xi} \right) \right] \tag{13.50}$$

The reduction factor in the square brackets was obtained from experiments on cement pastes in sealed conditions. The effect of water loss due to drying (or of water gain due to imbibition into concrete cured under water) needs to be taken into account separately.

Example 13.2. Hydration of a thick wall (refined model)

Let us analyze the same problem as in Example 13.1, but this time accounting for the effects of hydration on heat capacity and thermal conductivity. The values $k_{T0} = 1.44$ W/(m·K) and $\rho C_{p0} = 2.653$ MJ/(m³·K) are now considered as characterizing fresh concrete, and the dependence of thermal properties on the hydration degree is described by Eqs. (13.48) and (13.50). Based on the values from Table 13.3, the contribution of aggregates (and filler) to the volumetric heat capacity is taken as $a\, C_{pag} + f\, C_{pf} = 1.499 + 0.086 = 1.585$ [MJ/(m³·K)], and the contribution of water and cement as $w\, C_{pl} + c\, C_{pc} = 0.744 + 0.324 = 1.068$ [MJ/(m³·K)].

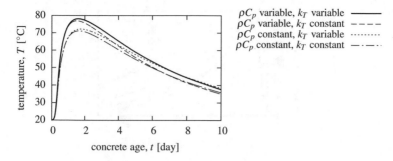

Fig. 13.5 Influence of model choice on the computed evolution of the core temperature (heat capacity and thermal conductivity are considered either as constant, or as decreasing functions of the hydration degree)

The dependence of k_T and ρC_p on ξ is easily incorporated into the computational procedure, e.g., into formula (13.43) of the explicit scheme. The computed evolution of the core temperature is plotted in Fig. 13.5. The dash-dotted curve corresponds to the results from Example 13.1, where the effects of hydration on thermal properties (capacity and conductivity) were neglected and the maximum temperature was found to be 71 °C. The solid curve presents the results of a refined analysis with the effects of hydration included. The hydration-induced reduction of both capacity and conductivity leads to higher temperatures, with a peak value of 78 °C. To provide more insight, the graph also contains curves that correspond to one property considered as variable and the other as constant. The dashed curve shows that the reduction of heat capacity is responsible for an increase in the peak temperature, and the dotted curve shows that the reduction of thermal conductivity is responsible for a slower decay of temperature after the peak.

To complete the picture, let us also investigate the sensitivity of the results to the boundary condition, in particular to the value of the surface heat transfer coefficient, B_T. The choice of $B_T = 4$ W/(m²·K) in Example 13.1 was supposed to correspond to a boundary covered by a 20-mm-thick layer of plywood. In the absence of such a layer, a realistic choice could be $B_T = 10$ W/(m²·K), or even $B_T = 25$ W/(m²·K) if the concrete surface is ventilated. Figure 13.6 shows the temperature evolution in the core that corresponds to different values of the surface heat transfer coefficient. In all the cases, the refined model with variable heat capacity and thermal conductivity is used. With increasing B_T, heat flows through the boundary "more easily," and the peak temperature in the core is reduced. The lowest peak is obtained with the Dirichlet boundary condition (13.15), which corresponds to the limit of B_T approaching infinity. ∎

It is worth noting that da Silva and Šmilauer [335] developed an easy-to-use nomogram for a quick prediction of the maximum temperature attained during hydration of mass concrete. The nomogram calculations have been embedded in a mobile application, called "Mass Concrete App," which is available for free download from the App Store. The temperature nomogram accounts for the cement type, total binder content,

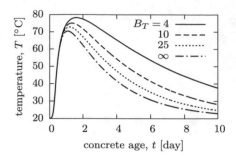

Fig. 13.6 Influence of the surface heat transfer coefficient, B_T, on the computed evolution of the core temperature

effective percentage of supplementary cementitious materials, member thickness, initial temperature, and average ambient temperature. Field validations indicated a prediction error of less than $4\,°C$.

13.2   Heat and Moisture Transfer, and Hygrothermal Effects in Heated Concrete

The previous section was devoted to analyses of heat transfer, uncoupled from the transport of moisture. Such modeling is appropriate, e.g., for evaluation of the effects of hydration heat on a sealed specimen or a structure protected against moisture loss. Also, moderate heating or cooling of mature concrete caused by climatic variations of ambient temperature can be treated separately from long-term drying processes because of the huge difference in characteristic times of heat conduction and moisture transport, as discussed in Sect. 13.1.2. On the other hand, rapid heating of concrete caused by fire or microwave irradiation calls for coupled hygrothermal modeling.

Many models of hygrothermal effects in porous materials have been proposed in the literature; they range from rather general models developed for applications in building physics [444, 508, 553, 556, 587, 672] to specialized models focusing explicitly on the behavior of structural concrete. An excellent overview of the latter class of approaches can be found in the comparative study of Gawin et al. [423], who summarized the main features of the models proposed by Abdel-Rahman and Ahmed [5], Bažant and Thonguthai [188, 189], Ichikawa and England [501], Tenchev, Li and Purkiss [806], Tenchev and Purnell [805], Davie, Pearce and Bićanić [339], Consolazio, McVay and Rish III [324], Chung, Consolazio and McVay [315], Dwaikat and Kodur [365], Ulm et al. [823, 826], and of the family of multiphase models developed over the years by Gawin, Schrefler, Pesavento and coworkers [414, 415, 417, 418, 421, 767].

A detailed presentation and discussion of all competing formulations from the literature is out of scope of the present book. We will focus our attention on the model proposed by Bažant and Thonguthai [188, 189] and later slightly modified by Bažant and Zi [213], which will be referred to as the BT model. For comparison, the main ideas of a family of multiphase models developed by Gawin, Schrefler, Pesavento, and coworkers will be outlined in Sect. 13.6.

13.2.1 Structure of Bažant–Thonguthai Model

Models of hygrothermal effects in porous materials usually start from suitably selected mass and energy balance equations. The specific choice of such equations is closely related to the choice of primary unknown variables. As will be explained in Sect. 13.5.6, the balance equations can be written separately for individual phases, or even for individual components of a phase (such as dry air and water vapor as components of the pore gas). But they can also be combined such that certain interaction terms are canceled. This combination is equivalent and simpler, making the phase separation an unnecessary complication.

Bažant and Thonguthai [188, 189] argued that for concrete it is not realistic to consider separate flows of liquid (capillary) water, water vapor, and adsorbed water because the capillaries in hardened cement paste are not continuous [325]. To get from one capillary pore to the next, water molecules must pass through nanopores while they are in the adsorbed state, held by surface forces of the C-S-H. The resistance to flow in the capillary portions of the flow channel is orders of magnitude smaller than it is in the nanopore portions, and thus irrelevant. So the flow of adsorbed water is what controls the moisture transport. The evaporation–condensation of water molecules does not affect the overall moisture transport. Vaporized water molecules cannot pass through nanopores because their mean free path is an order of magnitude greater than the nanopore width (80 nm vs. 1–20 nm at 25 °C).

In their analysis of concrete exposed to elevated temperatures, Bažant and Thonguthai [188, 189] neglected the contribution of dry air, and they considered vapor combined with liquid water as moisture, characterized by the evaporable water content w_e, and by the moisture flux, $\boldsymbol{j}_w = \boldsymbol{j}_l + \boldsymbol{j}_g$. Consequently, the governing equations of the BT model are based on the mass balance of moisture and on a simplified form of the heat equation (which originates from the energy balance equation).

The moisture mass balance equation (8.76) was derived by intuitive arguments in Sect. 8.3.3, and for the present purpose, it is written as

$$\dot{w}_e + \dot{w}_n = -\nabla \cdot \boldsymbol{j}_w \tag{13.51}$$

where $\dot{w}_e$ and $\dot{w}_n$ are the (spatial) rates of the evaporable and nonevaporable water contents, and $\nabla \cdot \boldsymbol{j}_w$ is the divergence of the moisture mass flux. Equation (13.51) can also be interpreted as the sum of the balance equations written separately for liquid water and for water vapor, i.e., of Eqs. (13.179) and (13.180) with the pore gas corresponding exclusively to water vapor (note that $n_p S_l \rho_l + n_p (1 - S_l) \rho_g = w_e$ and $\dot{m}_{\mathrm{deh}} = -\dot{w}_n$).

The energy balance equation, transformed into the heat equation (13.1), was initially postulated by Bažant and Thonguthai [188] in a somewhat modified form. However, in the actual computations, many terms were neglected [142, 189] and the heat equation was reduced to its simple form

$$\rho C_p \dot{T} = \rho r - \nabla \cdot \boldsymbol{q} \tag{13.52}$$

In fact, Bažant and Thonguthai did not use the distributed heat source term, ρr, but later Bažant and Zi [213] applied the model to concrete heated by microwaves and added this term.

The mass and energy balance equations (13.51) and (13.52) represent the starting point for the development of a complete moisture and heat transport model. The fluxes must now be described by appropriate equations that reflect various transport mechanisms, and the equations must be converted into a set of two partial differential equations with two primary unknown fields. One of them will obviously be the temperature, T, but the choice of the other primary variable is ambiguous. This has already been amply discussed in Chap. 8, in the context of moisture transport modeling at constant temperature. Bažant and Thonguthai [188] decided to select the pore vapor pressure, p_v, as the second primary variable. Recall that the vapor pressure is closely related to the pore relative humidity,

$$h = \frac{p_v}{p_{sat}} \tag{13.53}$$

which was used as the primary variable of the Bažant–Najjar model in Sect. 8.3.4.2.

The formulation presented here can be considered as an extension of the Bažant–Najjar model to the saturated case and to high temperatures. In Chap. 8, we treated only the cases when $h \leq 1$, i.e., $p_v \leq p_{sat}$, considering the case of $h = 1$ as fully saturated concrete. However, this was just a simplification. Perfect saturation is probably never achieved in practice. Several attempts were made to measure the water pressure in sealed or massive concrete specimens heated to several hundred °C. But the pressure measured was always several orders of magnitude lower than that predicted by the steam and water tables for a rigid container filled before heating by liquid water [895].

When the vapor pressure p_v reaches the temperature-dependent saturation pressure p_{sat}, it does not mean that all vapor condenses into liquid water. At thermodynamic equilibrium with $p_v = p_{sat}$, the capillary pressure p_c evaluated from the Kelvin equation (8.34) for $h = 1$ vanishes and, according to the Laplace equation (8.12), the mean curvature of the liquid–gas interfaces is zero. But even for vapor pressures exceeding p_{sat}, when the mean curvature of the interfaces becomes negative, vapor can still coexist in equilibrium with liquid water. If, at saturation, all pores were perfectly filled with liquid water and their volume was fixed, the properties of liquid water (e.g., from ASTM tables) would predict an imperceptibly small increase of specific water content beyond the saturation point, even if the liquid water is brought to high pressures. However, the actually observed increase of w_e is several orders of magnitude higher. The logical explanation is that a nonnegligible part of pore volume still contains vapor, even for $p_v > p_{sat}$. This implies the existence of menisci whose total (Gaussian) curvature is zero at saturation and negative beyond, the menisci surfaces being anticlastic in connected vapor-filled pore space; see Appendix I.1.

Since vapor can indeed exist above the saturation point, it is possible to extend the definition of relative humidity to that range and admit values of $h > 1$. In a more fundamental approach, the chemical potential of water could be selected as the primary variable, because it can be defined for any phase of water. But since its physical meaning is not so straightforward (which is also true of relative humidity above saturation), the choice here is to base the general formulation on the vapor pressure, p_v.

The balance equations (13.51)–(13.52) now have to be rewritten in terms of the primary variables, T and p_v. All other variables must be expressed as functions of the primary variables, or at least linked to their rates or gradients. This is done as follows:

- The **evaporable water content**,

$$w_e = \phi(h, T) \tag{13.54}$$

is expressed as a function of the pore relative humidity and temperature, based on the concept of sorption isotherms (extended to the saturated range, as will be discussed in Sect. 13.2.3). Note that, in Sect. 8.2.5, function ϕ was considered as dependent on the pore relative humidity and equivalent age, because the temperature was assumed to remain constant. In the present section we aim at describing mature concrete subjected to variable temperature, and the effect of aging (i.e., the dependence of the pore structure on equivalent age) is neglected. This is fully justified in applications to extreme events of a short duration, such as fire.

- The **nonevaporable water content**, w_n, does not need to be specified by its value— it suffices to express the rate of its change, $\dot{w}_n$. Here, the nonevaporable water content is considered to be reduced by dehydration at high temperatures. Approximately, one can consider the mass of water released by dehydration (per unit volume of concrete) to be a unique function of temperature, denoted as $w_d(T)$.[4] The rate of nonevaporable water content is thus expressed as

$$\dot{w}_n = -\dot{m}_{deh} = -\frac{dw_d}{dT}\dot{T} \tag{13.55}$$

where the negative sign means that an increase in the amount of water released by dehydration corresponds to a decrease in the nonevaporable (chemically bound) water content. The specific form of function w_d will be provided in Sect. 13.2.2.

---

[4] For consistency with the notation used by Bažant and Thonguthai in their original papers, we denote here the content of water released by dehydration as w_d, but it is essentially the same quantity as m_{deh} in Sects. 13.5–13.6 and Eq. (13.1).

- The **saturated vapor pressure**, p_{sat}, is a unique function of temperature, described, e.g., by the Antoine equation (8.19).
- The **moisture flux**,

$$\boldsymbol{j}_w = -\frac{a}{g} \nabla p_v \tag{13.56}$$

is assumed to be proportional to the gradient of vapor pressure. This relation, resembling the Darcy law, will be discussed in detail in Sect. 13.2.4. Coefficient a is the *permeability*,[5] strongly dependent on pore humidity and temperature, and g is the gravity acceleration.
- Finally, the **heat flux** is governed by the isotropic Fourier law (13.7), and the distributed heat source ρr reflects external heating effects that are specified depending on the given application.

Based on (13.53)–(13.54), the rates of pore relative humidity and of evaporable water content can be linked to the rates of primary variables:

$$\dot{h} = \frac{\partial}{\partial t} \left(\frac{p_v}{p_{sat}} \right) = \frac{\dot{p}_v}{p_{sat}} - \frac{p_v}{p_{sat}^2} \frac{\mathrm{d} p_{sat}}{\mathrm{d} T} \dot{T} \tag{13.57}$$

$$\dot{w}_e = \frac{\partial \phi}{\partial h} \dot{h} + \frac{\partial \phi}{\partial T} \dot{T} = \frac{1}{p_{sat}} \frac{\partial \phi}{\partial h} \dot{p}_v + \left(\frac{\partial \phi}{\partial T} - \frac{p_v}{p_{sat}^2} \frac{\mathrm{d} p_{sat}}{\mathrm{d} T} \frac{\partial \phi}{\partial h} \right) \dot{T} \tag{13.58}$$

Substituting (13.58), (13.55)–(13.56), and (13.7) in the balance equations (13.51)–(13.52), we obtain the final set of governing equations

$$\frac{1}{p_{sat}} \frac{\partial \phi}{\partial h} \dot{p}_v + \left(\frac{\partial \phi}{\partial T} - \frac{p_v}{p_{sat}^2} \frac{\mathrm{d} p_{sat}}{\mathrm{d} T} \frac{\partial \phi}{\partial h} - \frac{\mathrm{d} w_d}{\mathrm{d} T} \right) \dot{T} = \frac{1}{g} \nabla \cdot (a \nabla p_v) \tag{13.59}$$

$$\rho C_p \dot{T} = \rho r + \nabla \cdot (k_T \nabla T) \tag{13.60}$$

This is a parabolic system of two partial differential equations, which can be solved, e.g., by the finite element method (or, on simple domains, by the finite difference method). The equations are of the first order in time and of the second order in space. Initial conditions specify the initial distribution of temperature and vapor pressure (which can be deduced from the temperature and pore relative humidity by using (13.53)). Boundary conditions for temperature have the same structure as for the uncoupled heat equation discussed in Sect. 13.1.6. Boundary conditions for the vapor pressure can be obtained from boundary conditions (8.94) or (8.95), again by exploiting (13.53).

Having clarified the choice of primary variables and the structure of governing equations, we can proceed to details specific to the BT model. The dependence of

---

[5]Coefficient a used in (13.56) has the same physical dimension [m/s] as the hydraulic permeability K_h, introduced in Sect. 8.3.2 in connection with the Darcy law. Here it is called simply the "permeability", to be consistent with the nomenclature used by Bažant and Thonguthai in their original papers. Permeability a should not be confused with the moisture permeability, c_p, which is used by the Bažant–Najjar model and has a different physical dimension [kg/(m · s)]; see Sect. 8.3.4.2.

the saturation vapor pressure on the temperature is given by the Antoine equation
(8.19), but the form of functions $\phi(h, T)$ and $w_d(T)$, as well as the dependence of
permeability a, volumetric heat capacity C_p, and thermal conductivity k_T on the
temperature and vapor pressure (or pore relative humidity), still need to be specified.
This will be done next.

13.2.2   Distributed Source of Water

Hydration reactions consume free water and convert it into chemically bound water.
This effect, which can be described by a positive rate $\dot{w}_n$, is important mainly at early
ages, although it probably evolves logarithmically and proceeds at a decaying rate
for many years and even for a century [125, 715]. On the other hand, when mature
concrete is heated high enough, the chemically bound water is freed and gets released
into the pores. Such dehydration is reflected by a negative term, $\dot{w}_n = -\dot{w}_d < 0$, in
the mass balance equation (13.51). Recall that w_n denotes the content of chemically
bound water, which increases during hydration and decreases during dehydration.

The dehydration of cement paste is strong and causes a significant strength reduction above cca 400°C [121, 142, p. 282]. Below 400°C, the strength loss due to
dehydration is only mild (apparently, the first dehydration sites are those not important for strength). If temperatures fall below 100°C, the dehydration may get reversed
by rehydration of cement.

The amount of dehydrated water is obtained experimentally, by weight loss measurements. However, the escape of gas from some aggregates (e.g., CO_2 from limestone) should be subtracted from the measured mass loss.

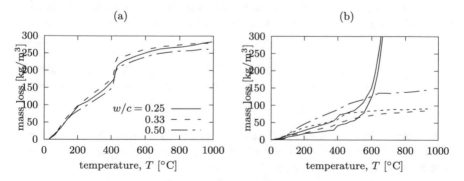

Fig. 13.7 Examples of mass loss measured at elevated temperatures for (a) cement paste, (b)
concrete (gases other than water vapor included)

Graphs of the mass loss at elevated temperatures, determined for various materials by thermogravimetric tests, were summarized in the report of Harmathy [462].
Figure 13.7a shows such plots for Portland cement pastes, and Fig. 13.7b for several

types of concrete. The mass loss at temperatures below $100\,°C$ must be solely the loss of evaporable water, and this could also be a significant part of the mass loss seen below $200\,°C$. Up to about $400\,°C$ for cement paste and $500\,°C$ for concrete, the mass loss (not necessarily of water only) is seen to be approximately proportional to the change of temperature. The behavior of concrete at higher temperatures is strongly affected by the type of aggregates. The solid curves in Fig. 13.7b correspond to concretes with dolomitic aggregates; for the other concretes, the type of aggregate was not specified. As discussed by Harmathy and Allen [463], aggregates with predominantly siliceous constituents are, under elevated temperatures, in general more stable than calcareous aggregates (such as dolomite). The highest degree of stability is exhibited by lightweight aggregates, expanded slag, shale, and clay.

Dwaikat and Kodur [365] suggested to estimate the mass of water released by dehydration by a piecewise linear function of temperature, e.g., of the form

$$w_d(T_C) = \begin{cases} 0 & \text{for } T_C \le 100\,°C \\ 4 \cdot 10^{-4}\, c\,(T_C - 100) & \text{for } 100\,°C < T_C \le 700\,°C \\ 0.24\, c & \text{for } 700\,°C < T_C \end{cases} \quad (13.61)$$

where c is the mass of cement per unit volume of concrete and T_C is the temperature in $°C$. To illustrate the correspondence with experimental data for certain concretes, the curves from Fig. 13.7b are replotted in terms of the dimensionless ratio w_d/c. The thick dash-dotted line in Fig. 13.8 corresponds to formula (13.61). However, it must be emphasized that equations such as (13.61) may be quite different for concretes with different aggregates, different w/c and different types of cement, and that the data may be distorted by unrecorded escape of gases other than water vapor (e.g., of CO_2). For accurate analysis, it is important to make measurements on the given concrete.

Gawin et al. [414] considered w_d as a linear function of T, with dw_d/dT between 0.04 and 0.08 kg/(m^{3}K). Tenchev et al. [806] used a piecewise linear function

$$w_d(T_C) = \begin{cases} 0 & \text{for } T_C \le 200\,°C \\ 7 \cdot 10^{-4}\, c\,(T_C - 200) & \text{for } 200\,°C < T_C \le 300\,°C \\ 0.07\, c + 0.4 \cdot 10^{-4}\, c\,(T_C - 300) & \text{for } 300\,°C < T_C \le 800\,°C \\ 0.09\, c & \text{for } 800\,°C < T_C \end{cases}$$

$$(13.62)$$

Referring to the experimental data of Schneider and Herbst [765], Gawin et al. [422] suggested to describe the mass of dehydrated water per unit volume by a cubic function of temperature, given by

$$w_d(T_C) = f_s \xi c \left[a_1 \langle T_C - 105 \rangle + a_2 \langle T_C - 105 \rangle^2 + a_3 \langle T_C - 105 \rangle^3 \right] \quad (13.63)$$

where T_C is again the temperature substituted in $°C$, $f_s = 0.32$ is a stoichiometric coefficient, ξ is the hydration degree, and c is the cement content. In comparative calculations presented in Gawin et al. [423], the coefficients of the cubic polynomial were set to $a_1 = 1.715 \cdot 10^{-3}$, $a_2 = -4 \cdot 10^{-7}$, and $a_3 = -2.95 \cdot 10^{-10}$, the

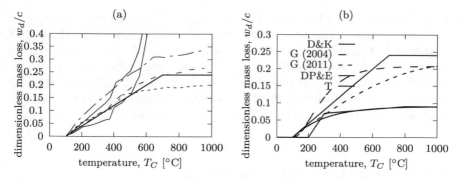

Fig. 13.8 Dimensionless mass loss for concrete at elevated temperatures (mass loss divided by mass of cement): (a) experimental data possibly containing contributions of other gases than water vapor, (b) theoretical functions used by various authors

hydration degree was $\xi = 0.65$ and the cement content was $c = 510 \, \text{kg/m}^3$. For these parameters, the initial value of dw_d/dT was equal to $f_s \xi c a_1 = 0.182 \, \text{kg/(m}^3\text{K)}$.

In a private communication, Pesavento [681] stated that the calculations in Gawin et al. [418] were based on the formula

$$w_d(T_C) = \frac{w_{d\infty}}{2} \left[1 + \sin \left(\frac{\pi}{2} \left(1 - 2 e^{-0.004 \langle T_C - 105 \rangle} \right) \right) \right] \qquad (13.64)$$

where $w_{d\infty}$ is the asymptotically approached final value of dehydrated water content and T_C has the same meaning as above.

In detailed models, the dehydration process is sometimes described by a separate differential equation which links the rate of dehydration to the difference between the current hydration degree and the equilibrium value of hydration degree corresponding to the current temperature. At fixed temperature, the hydration degree asymptotically approaches the equilibrium value. Ulm et al. [826] estimated the characteristic time of the transition to equilibrium as roughly 5 min at 20 °C, 0.1 second at 300 °C, and 0.001 s at 700 °C. This means that the dehydration process is very fast with respect to the typical time scale of the heating process (e.g., in fire), and so one can simplify the description and assume that the hydration degree always corresponds to the equilibrium value and is a unique function of temperature. In contrast to that, Dal Pont and Ehrlacher [336] used a characteristic time of 10,800 s, and described the dehydrated water content at equilibrium by an exponential function of temperature:

$$w_d(T) = 0.075 \, m_{\text{eq}} \left(1 - e^{-0.005 \langle T_C - 105 \rangle} \right) \qquad (13.65)$$

The same formula was also used by Feraille-Fresnet, Tamagny, Ehrlacher, and Sercombe [386], who referred to the experimental data of Feraille-Fresnet [385]. In Feraille-Fresnet et al. [386], m_{eq} was described as the "equilibrium mass" at 378 K (105 °C). From the context, one can infer that what is meant is the mass of cement paste. Therefore, in applications to concrete, m_{eq} should be set equal to $c + f_s \xi c$ where the first term is the mass of cement (per unit volume of concrete) and the second represents the mass of water in hydrates. For instance, for $f_s = 0.32$ and $\xi = 0.65$, one would get $m_{eq} = 1.208\,c$, and the asymptotically approached limit value of w_d would be $0.0906\,c$, which is very close to the limit value of $0.09\,c$ considered by Tenchev et al. [806] but much lower than the values attained at high temperatures by the expressions in (13.61) and (13.63)–(13.64); see Fig. 13.8b.

13.2.3 Isotherms at High Temperatures

Except for temperatures above the critical point of water (374 °C), one must distinguish the vapor from the liquid water in the pores of concrete. These two phases of water can be assumed to be locally (within the micropores) always in thermodynamic equilibrium. The BT model based on this hypothesis was shown to give acceptable match of the few pertinent test data that existed [188, 189]. One important component of the model is the specific form of function ϕ that links the evaporable water content, w_e, to the pore relative humidity, h, and temperature, T. At fixed temperature, the graph of w_e as function of h is called the isotherm.

In theory, one might prefer to deduce the isotherms completely from the morphology of the pore space and from the known thermodynamic properties of water, as defined in the ASME steam tables [626] or in the IAPWS-95 water and steam tables [500]. However, the complexity of the pore system, and especially the role of water adsorbed in the nanopores of hydrated cement paste, prevent that. Therefore, the formulation of the isotherms must be semiempirical.

An abrupt transition between an unsaturated region and a saturated region implies a jump in the sorption isotherms, which causes computational difficulties. To circumvent this problem, a linear transition in the sorption isotherm was introduced between relative humidities $h = 0.96$ and 1.04 [188]. Such gradual transition is, in fact, logical because of the necessary existence of anticlastic surfaces, as discussed in Appendix I.1.

Based on certain thermodynamic considerations for capillary water, Bažant and Thonguthai [188] proposed to describe the sorption isotherms of concrete **below saturation** by the power law

$$w_e = \phi(h, T) = c \left(\frac{w_1}{c} h \right)^{1/m(T)} \qquad \text{for } h \le 0.96 \qquad (13.66)$$

in which c is the (initial) mass of cement per unit volume of concrete [kg/m^3], w_1 is a parameter representing the evaporable water content [kg/m^3] in saturated concrete at 25 °C, and the reciprocal value of the exponent is given by the empirical function[6]

$$m(T) = 0.04 + \frac{1}{1 + (T - 263)^2/27370} \qquad (13.67)$$

with the absolute temperature T substituted in K. Note that $1/m = 1$ at $T = 297$ K (24 °C) and $1/m = 2.12$ at $T = 453$ K (180 °C). At temperatures close to the room temperature, the shape of the isotherm in the humidity range below 96% is almost linear; see the top curve in the left part of the diagram in Fig. 13.9. With increasing temperature, parameter m decreases, exponent $1/m$ increases, and the isotherms become convex.

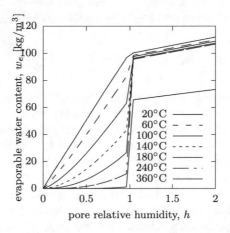

Fig. 13.9 Isotherms for a wide range of temperatures according to a simplified version of the BT model

For **saturated concrete**, Bažant and Thonguthai [188] proposed to describe the isotherms by

$$w_e = \phi(h, T) = (1 + \varepsilon_V)\, \rho_l(h, T)\, n_p(h, T) \qquad \text{for } h \geq 1.04 \qquad (13.68)$$

where $\varepsilon_V = \varepsilon_x + \varepsilon_y + \varepsilon_z$ is the relative volume change, n_p is the porosity accessible to water, and ρ_l is the specific mass of liquid water, evaluated for the given relative humidity and temperature based on the equation of state for water.

The volumetric strain ε_V is typically much smaller than one and is not so easy to evaluate. In a rigorous approach, one would need to take into account not only

---

[6]The original formula proposed by [188] was $m = 1.04 - T'/(22.34 + T')$, in which $T' = [(T_C + 10)/35]^2$ and T_C = temperature substituted in °C. Formula (13.67) is an equivalent expression in terms of the absolute temperature.

the dependence of ε_V on temperature but also on the stress arising in the material, which would introduce a coupling between the mechanical problem and the transport problem. The error introduced by omitting the factor $1 + \varepsilon_V$ from formula (13.68) is certainly smaller than the error associated with the approximate dependence of porosity n_p on the relative humidity and temperature. Davie, Pearce, and Bićanić [338] used (13.68) in the simplified form

$$w_e = \phi(h, T) = \rho_l(h, T) n_p(h, T) \qquad \text{for } h \geq 1 \qquad (13.69)$$

over an extended range of relative humidities ≥ 1 (instead of ≥ 1.04).

The density of liquid water, ρ_l, is linked to the temperature, T, and pressure (of liquid water), p_l, by the state equation. For sufficiently small changes, the state equation can be linearized around a reference point and written as

$$\rho_l = \rho_{l0} \left[1 - \alpha_{vT}(T - T_0) + \frac{p_l - p_{l0}}{K_l} \right] \qquad (13.70)$$

where ρ_{l0} is the density at the reference temperature T_0 and reference pressure p_{l0}, α_{vT} is the volumetric coefficient of thermal expansion (corresponding to three times the linear coefficient of thermal expansion, which is typically used for solids) and K_l is the bulk modulus of water (about 2.2 GPa at standard conditions). Due to the low compressibility of liquid water, the effect of pressure on density is often negligible and water can be considered as incompressible; see, e.g., the derivation of Kelvin's equation in Sect. 8.2.4. The effect of temperature is stronger and, for temperature variations of several hundred degrees, it becomes markedly nonlinear. Therefore, to cover a wider range of temperatures, up to the critical point, it is better to use a nonlinear form of the empirical equation of state.

Furbish [401] fitted experimental data by a fifth-order polynomial function of temperature, with coefficients considered as linear functions of pressure. The resulting equation can be presented in the form

$$\rho_l = \sum_{k=0}^{5} (A_k + a_k p_l) T_C^k \qquad (13.71)$$

where T_C is the temperature substituted in °C. Coefficients A_k and a_k, obtained by optimum fitting to experimental data [401, Section 4.10.2], are summarized in Table 13.4. Their units are adjusted such that formula (13.71) gives the density in kg/m^3 if T_C is substituted in °C and p_l in Pa.

Table 13.4 Coefficients used in Furbish's empirical state equation for liquid water, Eq. (13.71), and in its simplified version, Eq. (13.72)

k	A_k	a_k	C_k
0	1011.5	$4.8863 \cdot 10^{-7}$	1016.4
1	-0.74071	$-1.6528 \cdot 10^{-9}$	-0.75724
2	0.0087324	$1.8621 \cdot 10^{-12}$	0.0087510
3	$-9.6971 \cdot 10^{-5}$	$2.4266 \cdot 10^{-13}$	$-9.4545 \cdot 10^{-5}$
4	$3.6733 \cdot 10^{-7}$	$-1.5996 \cdot 10^{-15}$	$3.5134 \cdot 10^{-7}$
5	$-5.0775 \cdot 10^{-10}$	$3.3703 \cdot 10^{-18}$	$-4.7404 \cdot 10^{-10}$

(a)   (b)

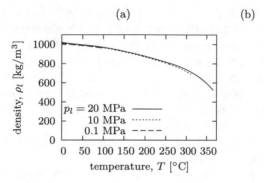

Fig. 13.10 Dependence of liquid water density on temperature at various pressures, according to Furbish's empirical state equation (13.71); each curve terminates at the boiling point corresponding to the given pressure

Formula (13.71) provides a good approximation for pressures between the atmospheric pressure and the critical pressure, and for temperatures between the room temperature and the critical temperature. As demonstrated in Fig. 13.10, the sensitivity to pressure is, in this range, relatively weak. Gawin et al. [415] neglected the dependence on pressure and expressed the density as a polynomial function of temperature,

$$\rho_l = \sum_{k=0}^{5} C_k T_C^k \tag{13.72}$$

where coefficients $C_k = A_k + a_k p_1$ are obtained from (13.71) with p_l set to $p_1 = 10^7$ Pa, which is near the middle of the range from 0.1 to 20 MPa. For convenience, the values of coefficients C_k are provided in the last column of Table 13.4.

The actual slope of the isotherm is significantly higher than the slope that would be predicted by (13.68) if the porosity n_p were assumed constant. The original partial explanation was that nanopores inaccessible to water at small pressures (see the discussion of ink-bottle pore geometries in Appendix I.1) become gradually accessible at increasing pressures. But the main reason probably is the existence of reverse menisci with pressurized liquid water that does not fill the pores completely. In view of this phenomenon, sometimes called the "pore space inflation," the porosity (relative pore volume accessible to water), estimated from considerations of weight loss, may be expressed as [188]

$$n_p(h, T) = \left(n_{p0} + \frac{w_d(T)}{\rho_{l0}}\right)[1 + 0.12(h - 1.04)] \quad \text{for} \ h \geq 1.04 \qquad (13.73)$$

where n_{p0} is the capillary porosity at the reference temperature, ρ_{l0} is the water density at the reference temperature and atmospheric pressure, and $w_d(T)$ is the amount of free water released into the pores by dehydration if the temperature is increased from the reference temperature T_0 to T. The dependence of w_d on T can be obtained from thermogravimetric measurements [463]; see Sect. 13.2.2.

Isotherms corresponding to formulae (13.66) for $h \leq 0.96$ and (13.69) for $h \geq$ 1.04, with linear interpolation in between, are plotted in Fig. 13.9 for temperatures ranging from 20 to 360 °C. Porosity n_p is evaluated from (13.73), with mass of dehydrated water approximated by (13.61), and liquid water density is considered as a function of temperature given by (13.72). Parameters $c = 300$ kg/m^3 and $w_1 =$ 100 kg/m^3 are the same as in Bažant and Thonguthai [188], and the corresponding initial porosity is $n_{p0} = w_1/\rho_{l0} = 0.1$.

It must be emphasized that equations (13.69), (13.73), and (13.61) leading to the right part of the diagram in Fig. 13.9 are merely very crude estimates. Theoretical inferences from the pore structure and its chemical change run into extreme (and unwarranted) complexity. More, and better controlled, experiments are needed.

Certain authors evaluate porosity using even simpler expressions than (13.73). For instance, based on the experimental data of Schneider and Herbst [765], Gawin et al. [414] considered porosity as a linear function of temperature, i.e., they set

$$n_p(T) = n_{p0} + A_n(T - T_0) \qquad (13.74)$$

where n_{p0} is the porosity at reference temperature T_0, and A_n is a parameter dependent on the type of concrete. For concretes tested by Schneider and Herbst [765], the optimal values of parameter A_n were 195×10^{-6}/K for silicate concrete, 163×10^{-6}/K for limestone concrete and 170×10^{-6}/K for basalt concrete. The porosity at room temperature, n_{p0}, varied between 0.06 and 0.087.

Exploiting the results of Feraille–Fresnet [385], Feraille-Fresnet et al. [386] postulated a linear relation between porosity and dehydrated water content, which would lead to

$$n_p(T) = n_{p0} + (0.72 \cdot 10^{-3} \text{m}^3/\text{kg}) w_d(T) \qquad (13.75)$$

where the dependence of w_d on T is given by (13.65).

13.2.4   Permeability at High Temperatures

In some transport models for multiphase pore water, the water transport is assumed to be controlled by the flow of liquid (or capillary) water and to involve evaporation and condensation at gas-liquid interfaces. However, such models are not realistic for concrete, because the capillaries in good quality concretes are discontinuous, and water molecules must pass through the nanopores in the hardened cement paste.

As already mentioned in Sect. 13.2.1, the width of nanopores (from 0.3 to about 20 nm) is, at room temperature, much smaller than the mean free path of water molecules in vapor (about 80 nm at 25 °C). Therefore, the molecules of vapor in the nanopores bounce most of the time randomly from the rough pore walls, or, more likely, linger at the surfaces in the form of adsorbed water. Since the molecules rarely impact each other, they cannot transmit vapor pressure along the nanopore. Consequently, the water molecules cannot pass through the nanopores in a vaporized state. Rather, they must become adsorbed on the pore walls and migrate along adsorption layers. This migration is very slow, far slower than the water flow through the capillary pores [78, 80], which are approximately those larger than 0.1–1 μm (the transition is not sharp). Physically, the main driving force for moisture diffusion in concrete is probably the spreading pressure, which is proportional to the logarithm of pore relative humidity. This argument supports the Bažant–Najjar transport law (8.84), widely used in Chap. 8 for analysis of moisture diffusion at isothermal conditions. Since, at constant temperature, the vapor pressure p_v is proportional to the relative humidity h, a fully equivalent formulation would be obtained with a transport law based on the gradient of p_v.

The gradient of h is problematic as a driving force when the temperature changes, especially if it increases above 100 °C. At high temperatures, the capillaries become continuous (i.e., percolating), and then it is appropriate to use the gradient of p_v as the driving force.[7] It was for this reason that Bažant and Thonguthai [188] wrote the equation governing the moisture flux as[8]

$$j_w = -\frac{a}{g}\nabla p_v \qquad (13.78)$$

---

[7]Here it should be noted that, because of the existence of anticlastic meniscus surfaces of negative total curvature, vapor still exists in the pores even if $p_v > p_{\text{sat}}(T)$.

[8]Some models for water and heat transfer through porous media use the moisture content w_e and the temperature T as primary variables. The fluxes are then described by the coupled equations

$$j_w = -a_{ww}\nabla w_e - a_{wT}\nabla T \qquad (13.76)$$

$$q = -a_{Tw}\nabla w_e - a_{TT}\nabla T \qquad (13.77)$$

in which coefficients a_{ww}, a_{wT}, a_{Tw} and a_{TT} depend on w_e and T. Note that $a_{wT} \neq a_{Tw}$ because ∇w_e and ∇T are not the thermodynamic driving forces associated with j_w and q [341, 584, 585, 858]. In (13.76), the first and second terms are sometimes called the Fick flux and Soret flux, and in (13.77), the first and second terms are called the Dufour flux and Fourier flux.

where a is called simply the *permeability* [m/s]. If the pore space was really filled exclusively by vapor, a would be the hydraulic permeability of concrete to vapor (in Chap. 8 denoted as K_h), and (13.78) would be the Darcy law (8.72) written for vapor instead of for gas. At constant temperature, one can compare (13.78) with the Bažant–Najjar equation (8.84), where h corresponds to p_v/p_{sat}. This leads to the relation $a = c_p g/p_{sat}$, which makes it possible to estimate the order of magnitude of a. A typical value of moisture permeability can be obtained from the relation $c_p = C/k$ where C is the moisture diffusivity and k is the inverse moisture capacity; see formula (8.88). Substituting typical values of $C \approx 20$ mm^2/day $= 2.3 \cdot 10^{-10}$ m^2/s and $1/k \approx 100$ kg/m^3, we get $c_p \approx 2.3 \times 10^{-8}$ kg/(m·s). For gravity acceleration $g \approx 10$ m/s^2 and saturated vapor pressure $p_{sat} = 2.33$ kPa (at 20 °C), we then obtain $a = c_p g/p_{sat} \approx 10^{-10}$ m/s. This correspondence would be exact only under constant (or at least uniform) temperature, otherwise the expressions differ by a term containing the temperature gradient. Also, the substituted value of moisture diffusivity refers to a state near saturation. At low relative humidities, it is by an order of magnitude smaller. For comparison, the value of permeability a used by Bažant and Thonguthai [188] for cement paste of age 40 days was 10^{-12} m/s, and they stated that the permeability of mature cement pastes is between 10^{-10} and 10^{-14} m/s. For concrete, it is not necessarily smaller, because concrete (unlike cement paste) contains interfacial transition zones around aggregates, and these zones are probably much more permeable than the rest of cement paste.

The question now is whether the transport law based on the gradient of vapor pressure can be applied above the saturation point. The most logical answer is that it can. The permeability may, of course, continuously change beyond the saturation point. The alternatives would be to assume the water flux to be proportional to the gradient of pressure p_l in the liquid water, or to the gradient of evaporable water content w_e. But at all temperatures below the critical one (374 °C), there exists a one-to-one relation between the liquid and vapor pressures. So this is really a moot point.

From the thermodynamic viewpoint, a fundamental approach would be to use the gradient of chemical potential of water (at thermodynamic equilibrium, the chemical potential of liquid water has the same value as the chemical potential of vapor, so it is uniquely defined). This would work even for the perfectly saturated concrete in which all the evaporable water would be liquid or adsorbed.

The fact that permeability is completely controlled by migration of adsorbed water molecules along the nanopores explains the extremely low values of concrete permeability at normal temperatures. However, when the temperature is increased above 100 °C, the permeability jumps sharply up. This has been explained by heat-induced changes in the structure of the smallest pores. Two kinds of changes were hypothesized:

1. Initially, Bažant and Thonguthai [188] suggested that, because of the tendency to minimize the free energy of pore surfaces (or maximize their entropy), the pore walls reduce their surface and get smoother, which eliminates the narrowest necks of nanometer dimensions that control water migration along the passages through

the hardened cement gel. It is not clear, however, whether this mechanism plays a significant role.
2. The calcium silicate hydrates (C-S-H) produced by hydration are of two kinds: high-density gel, which surrounds the unhydrous Portland cement grains, and low-density gel. As shown by DeJong and Ulm [346] and Jennings, Thomas, Gevrenov, Constantinides and Ulm [513], high temperature converts the low-density C-S-H gel to high-density gel. It was suggested that this conversion produces new capillary channels, which can explain the permeability jump.

No information apparently exists to show whether concrete cooled down after heating to 100 °C is much more permeable than concrete never heated. If it is not, then the foregoing two phenomena would become questionable since they ought to be irreversible on cooling. Further research is needed.

Based on the fitting of older test data, the permeability of concrete heated above 100 °C was inferred to jump up about 200 times [188]. Such a dramatic increase of permeability by two orders of magnitude was confirmed for high-performance concrete by Schneider [764]. These observations are supported by the fact that a 15-cm-diameter cylinder of concrete takes at room temperature more than ten years to dry to a nearly uniform pore humidity while, in an oven of 105°, the near uniform drying can be achieved in a few days. A suitable permeability function, used in computations by Bažant and Thonguthai [188], has the form

$$a(h, T) = \begin{cases} a_0 f_1(h, T) f_2(T) & \text{for } T \le T_{\text{tr}} \\ a_0 f_2(T_{\text{tr}}) f_3(T) & \text{for } T > T_{\text{tr}} \end{cases} \tag{13.79}$$

where a_0 is the reference value of permeability at saturation and room temperature (here considered as 25 °C), and T_{tr} is the temperature at the beginning of the transition to high permeability. Function

$$f_1(h, T) = \alpha(T) + \frac{1 - \alpha(T)}{1 + 256\langle 1 - h \rangle^4} \tag{13.80}$$

has the same form as the Bažant–Najjar formula (8.89) for moisture diffusivity, with $h_c = 0.75$, and with a temperature-dependent parameter α. At room temperature, α is quite low, while at T_{tr} it is set to 1 and then function f_1 is equal to 1 for arbitrary h, which assures a continuous transition to the high-temperature range.

The temperature dependence of permeability below T_{tr} is given by an Arrhenius-type equation,

$$f_2(T) = \exp\left[\frac{Q_w}{R}\left(\frac{1}{T_0} - \frac{1}{T}\right)\right] \tag{13.81}$$

where T and $T_0 = 298$ K are substituted in the absolute scale [K], Q_w is the activation energy for water migration at low temperature, and R is the universal gas constant. Based on data fitting, the value $Q_w/R = 2700$ K was recommended [166].

Function $f_3(T)$, which describes the abrupt increase of permeability near 100 °C, was initially proposed in the form

$$f_3(T) = \exp\left(\frac{T - T_{tr}}{0.881 + 0.214(T - T_{tr})}\right) \tag{13.82}$$

where $T_{tr} = 368$ K ($95°$C) was chosen to represent the beginning of the transition [188]. This was combined with a linear dependence of parameter α on temperature below T_{tr}, described by

$$\alpha(T) = 0.05 + 0.95\frac{T - T_0}{T_{tr} - T_0} \tag{13.83}$$

The resulting dependence of permeability on temperature is shown in Fig. 13.11a.

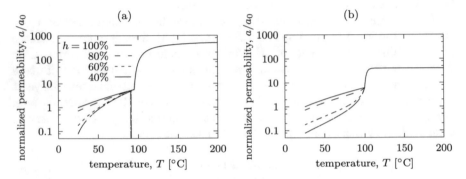

Fig. 13.11 Dependence of permeability (normalized by permeability at $25°$C and 100% humidity) on temperature for various relative humidities, according to (a) Bažant and Thonguthai [188], (b) Bažant and Zi [213]

Later, Bažant, and Zi [213] noticed that the fit of their data would improve by considering that the permeability may jump up, upon passing $100°$C, only about 5–10 times. It is unclear whether the difference in jump magnitude could be attributed to differences among different concretes. Bažant and Zi [213] decided to start the transition at $T_{tr} = 373$ K ($100°$C). They revised function f_3 as

$$f_3(T) = \frac{11}{1 + \exp\left[-0.455(T - T_{tr})\right]} - 4.5 \tag{13.84}$$

and combined it with

$$\alpha(T) = \frac{1}{1 + 0.253(T_{tr} - T)} \tag{13.85}$$

The resulting dependence of permeability on temperature is shown in Fig. 13.11b. Based on this updated formulation, reasonable fits of experimental data were obtained, see Fig. 13.12.

It should be noted that the reference permeability, a_0, is strongly influenced by the hydration degree (or age). Bažant and Thonguthai [188] fitted the experimental data

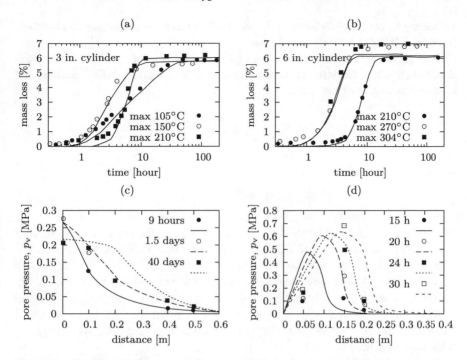

Fig. 13.12 Fits of (a–b) Bažant and Thonguthai's [188] experiments, (c) England and Ross' (1970) experiments, and (d) Zhukov and Shevchenko's [895] experiments by the model proposed in Bažant and Zi [213]; symbols represent experimental data and curves correspond to simulations

of Powers and Brownyard [705], valid for a type-I cement paste with $w/c = 0.7$, by the function

$$a_0(t_e) = 10^{\sqrt{40/t_e}} \times 10^{-13} \text{ m/s} \tag{13.86}$$

where t_e is the equivalent age in days.

13.2.5 Thermal Conductivity

The thermal conductivity of cement paste varies with the changes of temperature and moisture content appreciably, though far less than the permeability (discussed in the previous section). For concrete, the thermal conductivity is much less sensitive to temperature and moisture changes than it is for the hardened cement paste. The reason for this difference is that the mineral aggregates, which represent most of the volume of concrete and usually are chemically stable [461], conduct heat no less than the cement paste but do not transfer moisture significantly. Thus, the thermal conductivity depends on the volume fraction of aggregate and its type [653]. Typical values are in the range between 0.75 and 1.75 W/(m·K) for concrete and between 0.25 and 1.0 W/(m·K) for cement paste.

According to Perre and Degiovanni [675] and Gawin et al. [414], the dependence of the effective thermal conductivity of concrete on pore relative humidity and temperature can be expressed as

$$k_T(h, T) = k_T^{(\mathrm{dry})}(T) \left(1 + \frac{4 n_p(T) \rho_l(T) S_l(h, T)}{(1 - n_p(T)) \rho_s(T)} \right)$$ (13.87)

where

$$k_T^{(\mathrm{dry})}(T) = k_T^{(\mathrm{dry},0)}[1 + A_\lambda(T - T_0)]$$ (13.88)

is the effective thermal conductivity of dry concrete, S_l is the saturation degree, ρ_l is the density of liquid water, and ρ_s is the density of the solid skeleton. The dependence of porosity n_p on temperature is given by the linear function (13.74). Densities ρ_l and ρ_s are treated as temperature-dependent but their dependence on relative humidity (or pore pressure) is considered as negligible. In (13.88), parameter $k_T^{(\mathrm{dry},0)}$ is the effective thermal conductivity of dry concrete at room temperature T_0, and a typical value of coefficient A_λ is $-5 \cdot 10^{-4} \, \mathrm{K}^{-1}$ (note that the thermal conductivity decreases with increasing temperature).

In a somewhat different context (simulation of early-age behavior), Jendele et al. [512] demonstrated that the effective thermal conductivity of concrete can be estimated from the thermal conductivities of individual constituents (unreacted cement, water, filler, air, and aggregates) by a four-step homogenization procedure exploiting the Mori-Tanaka scheme and Hashin–Shtrikman bounds. This approach was exploited (without detailing its partial steps) for determination of k_T in Example 13.1.

13.2.6 Heat Capacity

In general, the effective specific heat capacity of a composite material is obtained by weighted averaging of the specific heat capacities of individual constituents, with weight factors that correspond to mass fractions. For concrete, considered as consisting of a solid skeleton, liquid water and pore gas, this rule of mixture leads to formula (13.2). In theory, one could even further decompose the pore gas into dry air and water vapor, as done, e.g., by Gawin et al. [414]. However, the mass fraction of pore gas in concrete is always negligible, because of very low mass densities of both air and vapor. The mass fraction of water in concrete rarely exceeds 6%, even at saturation. The effect of water content on the heat capacity of concrete is stronger than that, because the specific heat of water at room temperature is more than four times higher than the specific heat of the skeleton and, at high temperatures, the specific heats of liquid water and of water vapor increase dramatically. Still, the specific heat capacity of oven-dried concrete can be used as a crude approximation for C_p, as suggested by Bažant and Thonguthai [188]. For structural concrete, typical values of C_p are between 800 and 1170 J/(kg·K) [380].

The heat capacity can be considered as constant within a limited range of temperatures, but in general it is variable. According to BRITE Euram III BRPR-CT95 HITECO [258], the specific heat capacity of the solid skeleton (i.e., of dry concrete) can be approximated by a linear function of temperature,

$$C_{ps}(T) = C_{ps0}[1 + A_c(T - T_0)] \tag{13.89}$$

where C_{ps0} is the value of C_{ps} at the reference temperature T_0, and A_c is a parameter. Typical values for concrete C60 are $C_{ps0} = 855$ J/(kg·K) at $T_0 = 293$ K, and $A_c = 0.226$ K^{-1} [681].

Davie et al. [339] described the specific heat capacity of the solid skeleton by a nonlinear function

$$C_{ps}(T) = 900 + 80 \, \frac{T - 273.15}{120} - 4 \left(\frac{T - 273.15}{120} \right)^2 \tag{13.90}$$

in which T is substituted in K and C_{ps} is obtained in J/(kg·K). For liquid water, the same authors suggested to use

$$C_{pl}(T) = 2.4768 \, T + 3368.2 + \left(\frac{1.0854 \, T}{513.15} \right)^{31.44}, \qquad \text{for } T < T_{cr} \tag{13.91}$$

again with T substituted in K and C_{pl} obtained in J/(kg·K). Recall that $T_{cr} = 647.3$ K is the temperature at the critical point of water.

For specific heat capacities of water vapor and of dry air, Davie et al. [339] recommended expressions

$$C_{pv}(T) = \begin{cases} 7.14 \, T - 443 + \left(\dfrac{1.1377 \, T}{513.15} \right)^{29.4435} & \text{for } T < T_{cr} \\ 45800 & \text{for } T \geq T_{cr} \end{cases} \tag{13.92}$$

$$C_{pa}(T) = 1012.5 - 0.1216 \, T + 3.564 \cdot 10^{-4} \, T^2 - 9.9 \cdot 10^{-8} \, T^3 \tag{13.93}$$

with T substituted in K, and C_{pv} and C_{pa} obtained in J/(kg·K). At $T_0 = 293$ K, the capacities are $C_{pv}(T_0) = 1649$ J/(kg·K) and $C_{pa}(T_0) = 1005$ J/(kg·K). The specific heat capacity of vapor increases dramatically as the critical temperature is approached, but the specific heat capacity of dry air increases only slightly, to $C_{pa}(T_{cr}) = 1056$ J/(kg·K). Consequently, the contribution of dry air to the heat capacity of concrete is totally negligible, because of the extremely low mass of air contained in concrete. Even for fully dry concrete with porosity as high as 15%, the mass of dry air contained in the pores at atmospheric pressure and room temperature would be as low as 0.18 kg, and the contribution of dry air to the volumetric heat capacity would be $180 \, \text{J}/(\text{m}^3\text{K})$ while the typical volumetric heat capacity of concrete exceeds $2 \, \text{MJ}/(\text{m}^3\text{K})$; see Example 13.1.

In contrast to the contribution of dry air, which is always negligible, the contribution of water vapor to the heat capacity of concrete may play some role as the temperature is increased. For instance, let us assume that the pores are filled by vapor at temperature $T_{cr} = 647.3$ K and pressure $p_v = 2$ MPa, which can indeed happen, as will be illustrated by an example in Sect. 13.3.4. The corresponding mass density calculated from the state equation of ideal gas is $\rho_v = 6.7 \, \text{kg/m}^3$ (the actual mass density being even somewhat higher), and the porosity at this temperature can easily reach 15%, which gives the mass content of vapor around 1 kg per cubic meter of concrete. The contribution of vapor to the volumetric heat capacity is then $45.8 \, \text{kJ}/(\text{m}^3\text{K})$.

For young concrete, the specific heat capacity is affected by the hardening process. As already mentioned in Sect. 13.1.6, experimental data of Bentz [234] indicate that the specific heat capacity of hardening cement paste can be estimated as a function of the hydration degree using formula (13.50). Weighted averaging with the specific heat capacity of aggregates then provides the effective specific heat capacity of concrete.

13.2.7   Latent Heat

According to Gawin et al. [414], the latent heat of vaporization can be evaluated using the Watson [852] formula

$$\Delta h_{1,g}^w(T) = 2.672 \cdot 10^5 (T_{cr} - T)^{0.38}, \qquad \text{for } T < T_{cr} \qquad (13.94)$$

with T substituted in K and $\Delta h_{1,g}^w$ obtained in J/kg. Above the critical temperature of water, $T_{cr} = 647.3$ K, the latent heat is zero (there is no liquid–gas phase transition in this range). At $T = 373.15$ K ($100\,^\circ$C), formula (13.94) gives $\Delta h_{1,g}^w = 2.256$ MJ/kg.

The latent heat of dehydration, $\Delta h_{s,1}^w$, can be considered as constant, equal to 2.4 MJ/kg [339]. Since the specific heat of concrete is about $1 \, \text{kJ}/(\text{kg·K})$, dehydration that releases 1 kg of hydrate water consumes an amount of energy that could heat 2400 kg of concrete by 1 degree. In fact, if the mass of water released by dehydration is considered as a unique function of temperature, we can express the third term on the right-hand side of the heat equation (13.1) as $\dot{m}_{deh} \Delta h_{s,1}^w = (\text{d}w_d/\text{d}T)\dot{T} \, \Delta h_{s,1}^w$ and combine it with the first term on the left-hand side, which is also proportional to the temperature rate. This means that (13.1) can be rewritten as

$$\left(\rho C_p + \frac{\text{d}w_d}{\text{d}T} \Delta h_{s,1}^w\right) \dot{T} + (C_{ps}\boldsymbol{j}_s + C_{pl}\boldsymbol{j}_l + C_{pg}\boldsymbol{j}_g) \cdot \nabla T = \rho r - \nabla \cdot \boldsymbol{q} - \dot{m}_{ev} \Delta h_{1,g}^w$$

$$(13.95)$$

Thus, instead of including the heat consumed by dehydration as a separate term in the heat equation, we can simply increase the effective specific heat capacity by $(dw_d/dT)\Delta h_{s,1}^w$. Of course, this adjustment should be used only in the range of temperatures in which dehydration takes place (and, strictly speaking, only as long as the temperature keeps increasing).

To get an idea about the relative importance of the corrective term $(dw_d/dT)\Delta h_{s,1}^w$, let us estimate the derivative dw_d/dT according to (13.61) as $4 \cdot 10^{-4}c$. For cement content $c = 350$ kg/m^3, we get

$$\frac{dw_d}{dT}\Delta h_{s,1}^w \approx 4 \cdot 10^{-4} \times 350 \times 2.4 \cdot 10^6 \frac{J}{m^3} = 336 \frac{kJ}{m^3} \qquad (13.96)$$

This corresponds to 14% of a typical value of ρC_p, taken as 2400 J/m^3.

13.3   Strains and Stresses at High Temperature

In the preceding sections, we analyzed coupled heat and mass transport, without paying attention to mechanical aspects of the problem. Now we will complete the picture by adding the laws that govern the evolution of strains and stresses at high temperature.

13.3.1   Thermal and Hygral Volume Changes

The normal strain rates due to **thermal expansion** are expressed by the standard formula

$$\dot{\varepsilon}_{Tx} = \dot{\varepsilon}_{Ty} = \dot{\varepsilon}_{Tz} = \alpha_T \dot{T} \qquad (13.97)$$

where α_T is the (linear[9]) thermal expansion coefficient. For cement mortar, α_T changes appreciably with temperature, which means that the relation between thermal strain and temperature is nonlinear. This is caused by moisture effects. For concrete, however, the change of α_T with temperature is much less pronounced, which is explained by the restraining effect of aggregates on the cement paste [142, 461, 653]. As an approximation, the thermal expansion coefficient α_T of concrete may simply be taken as constant.

---

[9]Note that the thermal expansion coefficient α_T used for solids is the "linear" expansion coefficient, in the sense that it describes the relative change of "linear dimensions" (length). In contrast to that, the volumetric thermal expansion coefficient α_{vT} used for fluids describes the relative change of volume; see (13.70) or (13.172).

The thermal expansion coefficient of hardened cement paste (measured over intervals not allowing significant water escape) varies significantly with the water content and averages about $\alpha_T \approx 25 \cdot 10^{-6}/\text{K}$. For most aggregates, $\alpha_T \approx 5\text{–}8 \cdot 10^{-6}/\text{K}$. For concretes, depending on the water content and the mineralogical type of aggregate, $\alpha_T \approx 5\text{–}16 \cdot 10^{-6}/\text{K}$, with the average value of about $10 \cdot 10^{-6}/\text{K}$ (for more information, see Bažant and Kaplan [142]).

A linear thermal expansion law with a constant value of the thermal expansion coefficient α_T can be used as an approximation only within a limited range. Over a wide range of temperatures, the law becomes nonlinear, which means that α_T must be considered as temperature-dependent. For instance, according to Sect. 3.3.1 of Part 1-2 of Eurocode 2 [374], the thermal expansion coefficient can be evaluated as

$$\alpha_T = \begin{cases} 9 \cdot 10^{-6} + 6.9 \cdot 10^{-11} T^2 & \text{for concrete with siliceous aggregates} \\ 6 \cdot 10^{-6} + 4.2 \cdot 10^{-11} T^2 & \text{for concrete with calcareous aggregates} \end{cases}$$

$$(13.98)$$

where T is the temperature in Kelvin and α_T is obtained in K^{-1}. At room temperature (20 °C), the Eurocode formula gives $14.9 \cdot 10^{-6} \, \text{K}^{-1}$ and $9.6 \cdot 10^{-6} \, \text{K}^{-1}$ (concrete with siliceous or calcareous aggregates, respectively). At 100 °C, these values increase to $18.6 \cdot 10^{-6} \, \text{K}^{-1}$ and $11.8 \cdot 10^{-6} \, \text{K}^{-1}$, and at 400 °C to $40.3 \cdot 10^{-6} \, \text{K}^{-1}$ and $25.0 \cdot 10^{-6} \, \text{K}^{-1}$.

A part of the thermal strains is irreversible. Gawin et al. [418] used the thermal expansion law (13.97) for the description of reversible thermal strains, which are to be added to the *thermo-chemical strains*, dependent on an irreversible thermo-chemical damage variable.

The modeling of **shrinkage** has already been discussed in detail in Chap. 8, in particular in Sect. 8.6. The ultimate shrinkage strain of hardened cement paste varies with environmental humidity and cement type, and averages roughly $\varepsilon_{sh}^\infty \approx 4 \cdot 10^{-3}$. For concrete, it is about an order of magnitude less, due to the restraining effect of the aggregates, which do not shrink and are relatively stiffer.

The behavior becomes quite complex when the environmental humidity and temperature vary simultaneously. As shown in Fig. 13.13, the heating drives out evaporable water and the resulting shrinkage of cement paste overpowers the thermal expansion and produces large contraction [763]. Also shown is the effect of water loss during heating of concrete, which is much less, because of the aggregate restraint.

Generally, the hygrothermal deformations cannot be predicted realistically without a finite element program for heat and water transfer with creep and cracking or damage.

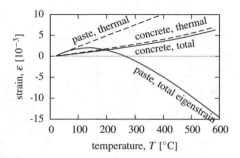

Fig. 13.13 Temperature dependence of strain due to thermal expansion and of total eigenstrain due to thermal expansion and drying shrinkage in cement paste and concrete (at no load)

13.3.2 Mechanical Properties at High Temperature

The dehydration processes at high temperatures have a strong effect on the microstructure of concrete and, consequently, also on the mechanical properties such as stiffness or strength. To give a picture of the experimentally observed behavior, Fig. 13.14a-c shows the aggregated data of Anderberg and Thelandersson [35], Purkiss and Dougill [711], Schneider [762] and Furamura [400] used for empirical input to the computations, including the dependence of the conventional elastic modulus (Fig. 13.14a), uniaxial compressive strength (Fig. 13.14b) and ultimate strain (Fig. 13.14c) on the temperature. This is complemented by Thelandersson's data [810] on the split-cylinder (Brazilian) tensile strength in Fig. 13.14d.

The effect of high temperatures on compression strength and elastic modulus was experimentally investigated also by Phan and Carino [684] and Neuenschwander, Knobloch and Fontana [650] for normal-strength concrete, Castillo and Durrani [299] for high-strength concrete, and Khaliq and Kodur [540], Bamonte and Gambarova [48] and Persson [680] for self-compacting concrete. Behnood and Ghandehari [225] measured the compression strength and tensile splitting strength of high-strength concretes after exposure to high temperatures. Envelopes of the dependencies of strength and elastic modulus on temperature based on experimental data were constructed by Phan and Carino [686].

The compressive stress–strain curves at various temperatures and constant strain rate measured by Schneider [762] are shown in Fig. 13.15a. Their fitting [120] revealed that the post-peak softening was caused by both invisible microcracking and stress relaxation.

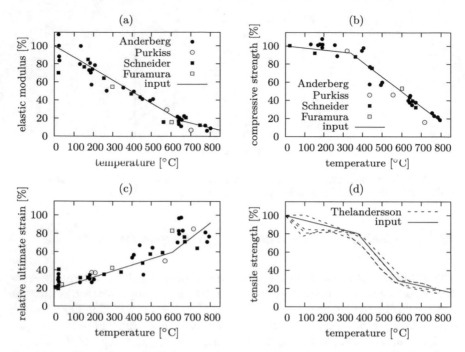

Fig. 13.14 Effects of temperature on (a) static elastic modulus, (b) compressive strength, (c) ultimate strain, measured on heated unstressed specimens, (d) split-cylinder tensile strength in hot state and after cooling (normal concrete)

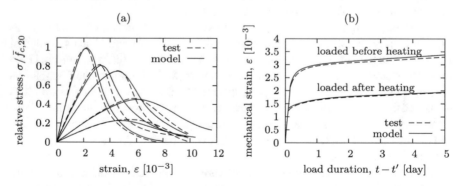

Fig. 13.15 Schneider's [762] tests of basalt concrete and their fits by Bažant and Chern [120]: (a) stress–strain curves at constant strain rate and various temperatures, (b) creep curves before and after heating to 300 °C for loading by stress $\sigma = 0.3\bar{f}_c$ (the label "loaded before heating" means that right after loading the temperature started rising; it did so at the rate of 2 °C/min up to 300 °C, which must have induced transitional thermal creep)

13.3.3 Extension of Creep Models to High Temperature

High-temperature creep depends strongly on the water content. Unlike at room temperature, concrete above 100 °C dries quickly and, like at room temperature, this reduces its creep by about one order of magnitude (see the dashed curves labeled as "wet" and "dry" in Fig. 13.16). The rapid escape of water after 100 °C is exceeded causes the creep rate curve to cross over from the curve for wet concrete to the curve for dry concrete; see the solid curves in Fig. 13.16.

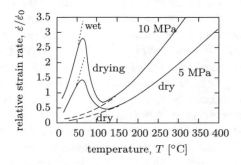

Fig. 13.16 Strain rates (at fixed load duration) of wet and dry concrete and of initially wet concrete that dries out, at two uniaxial stress levels (the strain rates have been normalized by the reference strain rate, $\dot{\varepsilon}_0$, considered as the rate in dry concrete at approximately 200 °C)

The Poisson ratio, v, which is essential for multiaxial generalization, seems to increase with temperature, which suggests that the deviatoric creep is accelerated by temperature rise more than the volumetric creep. This effect appears to be strong for the hardened cement paste, for which v rises up to 0.46 at 200 °C [154]. Due to the aggregates, the Poisson ratio for concrete remains nearly constant (see Eq. (13) in Bažant [88]).

The graphs in Fig. 13.15b, showing the creep of 8-cm-diameter cylinders loaded before or after heating, demonstrate a phenomenon that is called the *transitional thermal creep*. Note that the graphs show only the mechanical strain, with the thermal strain (measured at zero load) subtracted. The specimen loaded before heating (i.e., heated after the application of the load) exhibits much larger strain than the specimen that was first heated and then loaded. This means that the combined effects of stress and variable temperature cannot be expressed simply as the sum of the effects of (i) variable temperature under zero stress and (ii) stress under constant temperature. This is analogous to the well-known drying creep, resulting from combined effects of stress and variable humidity.

The fact that a change in temperature of concrete during its creep under load causes a significant transient increase of creep was documented by the experimental studies of Wallo et al. [848], Hansen and Eriksson [452], Fahmi et al. [378], Illston and Sanders [504], and Schneider [761]. Recall that the drying creep can alternatively be considered as stress-induced shrinkage (Sect. 10.1). In a similar spirit, the transitional thermal creep is often interpreted as the *load-induced thermal strain*, or LITS [543]. These two possible points of view are also reflected by two alternative modeling approaches, discussed next.

13.3.3.1 Microprestress Solidification Theory

The effects of variable temperature and humidity on creep have already been described in the context of the microprestress solidification (MPS) theory in Sect. 10.6.1. Recall that MPS theory takes these effects into account in differential equation (10.45) that describes the evolution of microprestress, which in turn affects the viscosity that controls the creep flow rate; see (10.3) and (10.5). This modeling framework incorporates the concept of transformed times that reflect the acceleration or deceleration of physical processes such as hydration, creep, or microprestress relaxation. All these processes are considered as thermally activated and the effect of temperature is introduced through Arrhenius-type expressions (10.31)–(10.33), each of which contains a certain activation energy.

In a more refined formulation, one might need to consider that there are multiple activation energies and different ranges of temperature are usually dominated by different activation energies. Above $100\,^\circ$C, water evaporates from normal test specimens ($h \to 0$) and then the hydration stops, which is captured by β_{eh} approaching zero[10] and makes β_{eT} from (10.31) irrelevant. From $200\,^\circ$C to about $400\,^\circ$C, the evaporable water is absent and there is a mild dehydration of concrete, with almost no loss of strength but appreciable loss of weakly held hydrate water. Above $400\,^\circ$C, the dehydration of normal Portland cement concrete becomes more pronounced and a significant strength reduction begins.

13.3.3.2 Stress-Induced Thermal Expansion and Shrinkage

The numerical results presented in Figs. 13.15, 13.17 and 13.18 were obtained by Bažant and Chern [120] with an older creep model, which included the transitional thermal creep and the drying creep in the form of mathematically equivalent stress-induced thermal expansion and stress-induced shrinkage. In this approach, the effect of stress on thermal strains is taken into account by making the thermal expansion coefficient depend on the current stress. Since the stress state is in general not isotropic (i.e., not purely hydrostatic), the scalar thermal expansion coefficient α_T is generalized to a tensor with components[11]

---

[10]More precisely, as h tends to zero, β_{eh} evaluated from (10.31) tends to $1/(1+\alpha_e^4)$. Since parameter α_e is typically in the order of 10, the rate of hydration is reduced by four orders of magnitude and becomes negligible.

[11]Unlike (13.99), Eq. (14) in Bažant and Chern [120] contained a negative sign in front of the stress-dependent term, but this misprint was corrected in Eq. (2.58) in Bažant [94].

$$\alpha_{T,ij} = \alpha_T[\delta_{ij} + (\rho\sigma_{ij} + \rho'\sigma_m\delta_{ij})\,\mathrm{sgn}\,(\dot{h} + a_T\dot{T})] \tag{13.99}$$

which is then used in the rate form of the thermal expansion law,

$$\dot{\varepsilon}_{T,ij} = \alpha_{T,ij}\dot{T} \tag{13.100}$$

Note that this is a generalized form of the isotropic thermal expansion law (13.97), which could be written in the tensorial notation as

$$\dot{\varepsilon}_{T,ij} = \alpha_T\delta_{ij}\dot{T} \tag{13.101}$$

In equation (13.99), α_T, ρ, ρ', and a_T are constant parameters, δ_{ij} is the Kronecker delta (equal to 1 if $i = j$ and to 0 otherwise), and σ_m is the mean stress. Parameter α_T corresponds to the standard thermal expansion coefficient, which is valid for thermal expansion at zero stress (and potentially depends on the pore relative humidity). Parameters ρ and ρ' control the influence of the total stress and its hydrostatic part on the stress-induced thermal strains. Based on their calculations, Bažant and Chern [120] decided to set $\rho' = 0$, which simplifies the model. The factor $\mathrm{sgn}\,(\dot{h} + a_T\dot{T})$ controls the sign of the stress-induced thermal strains. At constant pore relative humidity ($\dot{h} = 0$) and with $\rho' = 0$, Eqs. (13.99) and (13.100) can be combined into

$$\dot{\varepsilon}_{T,ij} = \alpha_T\delta_{ij}\dot{T} + \rho\alpha_T\sigma_{ij}|\dot{T}| \tag{13.102}$$

Thus, according to the present model, both heating and cooling generate additional stress-induced thermal strains that have the same signs as the corresponding stresses and therefore increase the creep effects; see Fig. 13.18b for experimental confirmation.

Simultaneously with stress-induced thermal strains, the model of Bažant and Chern [120] treats stress-induced shrinkage by making the shrinkage ratio k_{sh}^* depend on the stress. Again, the scalar coefficient is replaced by a tensor with components

$$k_{sh,ij}^* = -\varepsilon_s^\infty g_{sh}(t_e)\frac{dk_h(h)}{dh}[\delta_{ij} + (r\sigma_{ij} + r'\sigma_m\delta_{ij})\,\mathrm{sgn}\,(\dot{h} + a_T\dot{T})] \tag{13.103}$$

This tensor is then used in the rate form of the shrinkage law,

$$\dot{\varepsilon}_{sh,ij} = k_{sh,ij}^*\dot{h} \tag{13.104}$$

Parameter ε_s^∞ is positive and corresponds to the theoretical value of free ultimate shrinkage at zero stress extrapolated to zero humidity. Parameter r needs to be determined from experiments, and parameter r' is usually set to zero. Function g_{sh} usually defined as

$$g_{sh}(t_e) = \frac{E(t_0)}{E(t_e)} \tag{13.105}$$

characterizes the reduction in shrinkage due to hardening and depends on the equivalent age, t_e. Function k_h describes the dependence of the normalized shrinkage strain at zero stress on the pore relative humidity and is usually defined as $k_h(h) = 1 - h^3$. The derivative of k_h with respect to h is negative and, because of the negative sign on the right-hand side of (13.103), the resulting value of $k_{\mathrm{sh},ij}^*$ is positive.

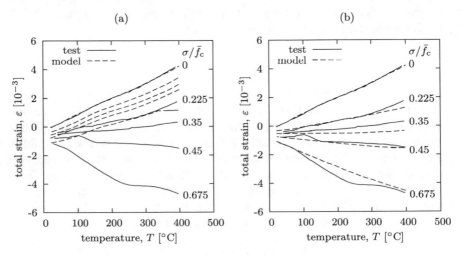

Fig. 13.17 Measured total strain at variable temperature and several compressive stress levels (solid curves), and dependences calculated by Bažant and Chern [120] (dashed curves) with the stress-induced thermal strain (a) ignored, (b) taken into account

The major importance of transitional thermal creep (or delayed stress-induced thermal expansion) is demonstrated in Fig. 13.17, in which Anderberg and Thelandersson's [35] data on total strain at constant rate of loading at various compressive stress-strength ratios $\sigma/\bar{f}_c$ and increasing temperature applied before loading are fitted with this phenomenon ignored (Fig. 13.17a) and taken into account (Fig. 13.17b).

Figure 13.18a shows Hansen and Ericksson's [452] data on deflections of beams immersed in water ($<100°$C) and loaded before or after heating at different heating

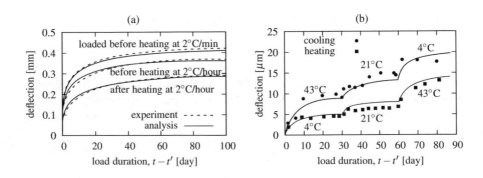

Fig. 13.18 Fits of test data by (a) Hansen and Ericksson [452], (b) Wallo et al. [848]

rates (2 °C per minute or per hour). The responses for loading before and after heating, respectively, include or exclude the transitional thermal creep. The data of Wallo et al. [848] in Fig. 13.18b demonstrate that not only heating but also cooling produces additional transitional thermal creep.

The additional strains resulting from a combination of loading and heating are often referred to as the *load-induced thermal strains* (LITS). Their existence was experimentally discovered by Hansen and Eriksson [452] and confirmed by many others. An extensive experimental campaign by Khoury, Grainger, and Sullivan [544] indicated that the dependence of LITS on the load level is approximately linear in the range between 10 and 60% of the maximum load. This is reflected by simple models that approximate LITS by a product of the normalized stress level and a suitable function of temperature. For instance, the model of Thelandersson [811] is based on the relation

$$\varepsilon_{LITS}(T, \sigma) = \frac{\sigma}{\bar{f}_c} \beta \varepsilon_{T0}(T) \tag{13.106}$$

where ε_{LITS} is the load-induced thermal strain at stress level σ and temperature T, $\bar{f}_c$ is the compression strength, β is a model parameter, and $\varepsilon_{T0}(T)$ is the usual thermal strain at temperature T measured in the absence of loading. Equation (13.106) can be obtained as a special case from Eqs. (13.99)–(13.100) that describe the model used by Bažant and Chern [120]. Indeed, at constant humidity, monotonically increasing temperature and uniaxial compression, Eq. (13.99) written for $i = j = 1$ reduces to

$$\alpha_{T,11} = \alpha_T \left[1 + \left(\rho + \frac{\rho'}{3} \right) \sigma_{11} \right] \tag{13.107}$$

where σ_{11} is the axial stress. Substituting (13.107) into (13.100) and integrating from the initial state at reference temperature to the current state at temperature T, we obtain

$$\varepsilon_{T,11}(T, \sigma_{11}) = \varepsilon_{T0}(T) + \varepsilon_{T0}(T) \left(\rho + \frac{\rho'}{3} \right) \sigma_{11} \tag{13.108}$$

Note that $\varepsilon_{T,11}$ represents here the total thermally induced axial strain, which consists of (i) the standard thermal strain ε_{T0} (at zero stress), obtained by integrating the rate equation $\dot{\varepsilon}_T = \alpha_T \dot{T}$, and (ii) the additional stress-induced thermal strain, analogous to ε_{LITS} from (13.106). If we set the parameter of the Thelandersson model to

$$\beta = \left(\rho + \frac{\rho'}{3} \right) \bar{f}_c \tag{13.109}$$

then the right-hand side of (13.106) coincides with the second term on the right-hand side of (13.108).

The Thelandersson model was implemented by Khennane and Baker [542] and extended to tensorial form by Nechnech, Meftah, and Reynouard [648]. Slight modifications were proposed by Terro [808] and Nielsen, Pearce, and Bicanic [657], who approximated the dependence of LITS on temperature by a function independent of the function $\varepsilon_{T0}(T)$ that describes the standard thermal strain. Terro [808] used a fourth- or fifth-order polynomial while Nielsen et al. [657] adopted a piecewise quadratic function. Furthermore, Terro [808] replaced the proportionality of LITS to the stress level by proportionality to $k + \sigma/\bar{f}_c$ where k is an additional parameter. This improves the fitting of experimental data by Khoury et al. [544] in the medium stress range but the model would give inconsistent results in the limit of zero stress. A comparative study by Robson, Davie, and Gosling [729] revealed that all LITS models based on a universal temperature function become inaccurate for temperatures exceeding 500 °C.

13.3.4   Application Example: Explosive Thermal Spalling Due to Microwave Heating

A fundamental, still debated, problem is the explosive thermal spalling of concrete walls or slabs subjected to intense rapid heating. Fires have produced such spalling in tunnels, and some spalling observed in tall buildings after fire is attributable to rapid heating. The results of simulations of microwave blast effects presented by Bažant and Zi [213] illuminate this problem.

In nuclear facilities, concrete gets slightly contaminated by radionuclides in thin surface layers, only 1–10 mm thick. To guarantee a safe long-time work environment, the contaminated layers need to be removed and properly disposed of as nuclear waste. One of possible decontamination techniques is based on rapid heating generated by microwaves emitted from a powerful applicator. Such treatment allows a fast removal of the contaminated layer, within only about 10 s [864].

The model used by Bažant and Zi [213] is a slightly modified version of the BT model, with temperature and vapor pressure as two primary unknowns. The governing equations are based on the moisture mass balance equation (13.51) and heat equation (13.52).

In the moisture mass balance equation, the rate of nonevaporable water content is expressed as $\dot{w}_n = \dot{w}_h - \dot{w}_d$, which is an extended version of (13.55). The additional term $\dot{w}_h$ reflects hydration at temperatures below 100 °C; the hydrate water content w_h is considered as a function of equivalent hydration period t_e, given by $w_h(t_e) = 0.21c[t_e/(\tau_e + t_e)]^{1/3}$, with parameter $\tau_e = 23$ days. The moisture flux is assumed to be controlled by the transport law (13.78), with permeability given by (13.79)–(13.81) and (13.84) and (13.85). The isotherm linking the evaporable water content to the pore relative humidity and temperature is described by equations (13.66)–(13.68) and (13.73).

In the heat equation, the heat capacity ρC_p is considered as constant. The heat flux is governed by the Fourier law (13.7) with constant thermal conductivity. The distributed heat source is calculated on the basis of the standing wave normally incident to the concrete wall. Since the microwave time period is much shorter than the time a heating front takes to propagate over the length of a microwave, and since concrete is heterogeneous, the ohmic power dissipation rate is averaged over both the time period and the wavelength.

The hygrothermal analysis is followed by a mechanical analysis, taking into account the strains induced by changes of temperature and pore humidity. Thermal strains are estimated using a simple linear law (13.97) with a constant thermal expansion coefficient, and shrinkage strains are obtained from a similar linear law (8.252) with a constant shrinkage ratio. The three-dimensional stress–strain relation is described by the M4 microplane model [113, 288]. To prevent spurious mesh sensitivity, the crack band approach (see Sect. 12.2) is used to analyze tensile softening due to distributed cracking in an axisymmetric situation.

Figure 13.19 shows the profiles of pore vapor pressure and temperature after 5, 10 and 15 s of heating of a concrete wall by blasts of microwaves having the same initial power of $1.1\,\mathrm{MW/m^2}$ but different frequencies ($f = 2.45, 10.6$ and $18.0\,\mathrm{GHz}$), which produce different rates of heating, because the dissipation of microwave energy increases with its frequency. Maximum temperature is typically found on the surface while the peaks of pore pressure are located somewhat deeper inside. The maximum pore pressure found in Fig. 13.19 is about 3 MPa. If this pore pressure acted on an unrestrained element of concrete, it would have to be balanced in the solid framework of concrete by the tensile stress of approximately $0.1\,\mathrm{MPa} \times 3.0 = 0.3\,\mathrm{MPa}$ where the value 0.1 is adopted for the typical porosity of concrete. This stress is only about 10% of the tensile strength of ordinary concrete, typically $\bar{f}_\mathrm{t} \approx 3\text{–}4\,\mathrm{MPa}$. The concrete subjected to the peak pressure is at temperatures near $250\,^{\circ}\mathrm{C}$, and so

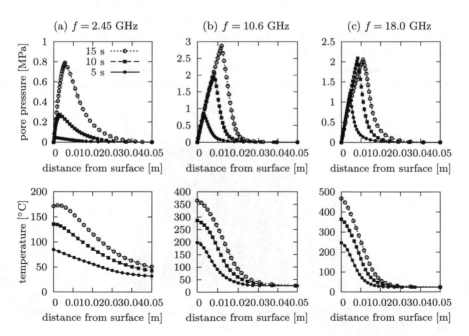

Fig. 13.19 Effect of different microwave frequencies on pore pressure and temperature profiles; after Bažant and Zi [213]

its tensile strength is reduced, but not substantially; see Fig. 13.14d. So, although
the effect of pore pressure is not entirely negligible, it cannot be the main cause of
spalling.

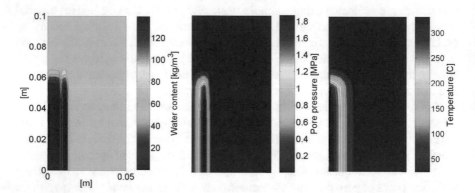

Fig. 13.20 Contour plots of water content [kg/m³], pore pressure [MPa], and temperature [°C]
after 10 s of microwave heating; after Bažant and Zi [213], reproduced with permission from ASCE

Figure 13.20 shows the contour plots of water content, pore pressure, and
temperature that develop after 10 seconds of microwave heating at frequency
$f = 18.0$ GHz and power input 1.1 MW/m². The heated zone is seen to be very
localized and located near the heated surface. Figure 13.21 depicts the deformed
finite element mesh (displacements exaggerated 100 times) and the contour plots of
the computed strain field; it shows the radial mechanical strain, i.e., the total strain
minus the hygrothermal strain (produced by changes of temperature and water content). It is found that the maximum principal mechanical strain in the surface layer
exceeds $5 \cdot 10^{-3}$ in tension and the stress state is essentially biaxial. This strain value

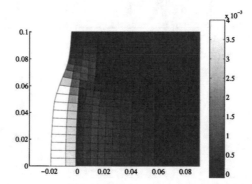

Fig. 13.21 Deformed finite element mesh and mechanical strain after 10 s of microwave heating;
after Bažant and Zi [213], reproduced with permission from ASCE

is much higher than the typical strain at peak in uniaxial tension (which is about $0.2 \cdot 10^{-3}$). Therefore, it was concluded that the concrete must suffer disintegration by cracking [213].

Is the explosive spalling triggered by pore pressure, or by compressive stresses that develop in directions parallel to the surface due to confined thermal expansion? This question has been debated since the 1960s. One school of thought, initiated by Harmathy [460, 461], holds that because the pore water cannot escape fast enough (which is a phenomenon called the "moisture clog"), the vapor pressures that must develop in the pores are so high that they cannot be balanced by the tensile strength of concrete [304, 509, 534, 685]. Another school of thought holds that the thermal expansion of the saturated heated zone, resisted by the cold concrete mass that surrounds the heated zone, leads in the surface layer to very high compressive stresses parallel to surface which either crush concrete or cause the compressed surface layer to split off and buckle.

The relative significance of these two mechanisms must of course depend on the type of problem, and can be different for microwave heating in the bulk of concrete and for conductive heating by fire. In the microwave heating problem, the highest pore pressure calculated has the value of 3 MPa, which causes in concrete a tension of about 0.3 MPa. This does not suffice to spall concrete. Besides, as soon as a crack starts to open, the volume available to pore water vapor increases by several orders of magnitude. This must cause a sudden drop of pore pressure since further water can be transferred into the open crack from the surrounding pores only with a significant delay [97, 213].

So it appears that the main cause of explosive thermal spalling must be the compressive stresses parallel to the surface, caused by restrained thermal expansion, rather than the pore pressure. These stresses, engendered by the resistance of cold concrete to the thermal expansion of the heated zone, reach values as high as about 50 MPa in the microwave blast problem.

The fact that explosive thermal spalling is predominantly caused by biaxial compressive thermal stresses parallel to surface was also proven by Ulm et al. [823], who presented a sophisticated finite element analysis of chemoplastic softening due to dehydration in the Channel Tunnel fire. Experimentally, the role of these stresses was documented by Hertz [477]. For completeness, let us also mention proposals of alternative explanations of the spaling phenomenon based on the boiling liquid expanding vapor explosion theory [4] or on high hydraulic pressures due to pore saturation [256].

13.4 Finite Volume Method for Problems with Moving Interfaces*

In the case of rapid heating, for example in the case of a wall exposed to fire (or a containment exposed to a hypothetical nuclear accident), a sharp moving interface between oversaturated and dried concrete typically develops. Such a situation cannot be handled by the finite element method in its normal form, but it can be handled by the finite volume method [377, 592, 593]. The reason is that the finite element method normally does not ensure exact mass balance while the finite volume method always does.

In the *finite volume method*, the domain is divided into discrete control volumes (Fig. 13.22). The interfaces (or boundaries) of a control volume are placed midway between adjacent representative points, which is generally accomplished by Voronoi tessellation, although such tessellation is not needed when regular meshes can be used.

Due to severe nonlinearity of the problem, modified Picard's iteration [303], in which the coefficients of the fluxes are taken as constants during each iteration, is often adopted. To simplify the calculation of the fluxes at the control volume interfaces, linear distributions of the state variables between the representative points are assumed. Since the flux through the common boundary of two adjacent control volumes is computed in the same way for both control volumes, the condition of mass balance is satisfied for each control volume exactly (even if the fluxes are only approximate). This is a fundamental advantage of the finite volume method, important for avoiding spurious wild oscillations in highly nonlinear problems with high local gradients and sharp fronts. For detailed formulation, see Bažant and Zi [213].

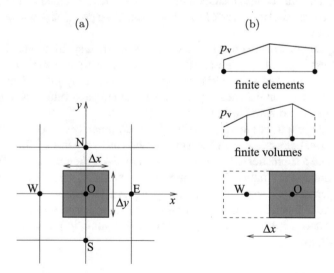

Fig. 13.22 (a) Two-dimensional discretization of finite volume where W, E, S, and N are labels for West, East, South, and North; shaded rectangle represents control volume; and (b) difference in pressure approximations implied by finite element and finite volume methods

The advantage of the finite volume method for the analysis of moisture transfer in concrete was recognized already by Eymard et al. [377] and was explored by Mainguy and Coussy [592] in the problem of calcium leaching from concrete, although in the absence of heat transfer. Mainguy et al. [593] showed an effective application of the finite volume method to drying of porous materials.

In the standard *finite element method*, by contrast, the local mass balance cannot be satisfied exactly. For example, if finite elements with linear shape functions are used (Fig. 13.22b), then the flux at the west boundary of a control volume surrounding a node of the FE mesh is generally different from the flux at the east boundary of the adjacent control volume lying to the west; note the different slopes adjoining the element interface in Fig. 13.22b. Identical results with the finite element method can nevertheless be achieved with the lowest-order finite elements and implicit time integration, which leads to pressure interpolation imitating the finite volume method [594].

13.5 Mass, Momentum and Energy Balance Laws*

As shown in the previous sections and illustrated by a number of examples, modeling of heat transfer is based on the so-called heat equation, which originates from the energy balance law. The heat equation can be presented in various forms, depending on the specific problem and on the level of simplification. In Sect. 13.1.1, it was simply postulated as Eq. (13.1), and the physical meaning of individual terms was briefly described. The present section provides a rigorous derivation of the heat equation and a detailed discussion of related theoretical aspects.

Before we proceed to the form of the heat equation applicable to concrete and similar multiphase materials, it is useful to revisit the mass conservation equation and develop a systematic procedure that can then be extended to other balance laws. Also, the momentum balance equation needs to be derived, because it will be exploited when converting the original form of the energy balance equation into a simpler form that serves as the governing equation in the description of heat, possibly coupled with moisture transport.

13.5.1 Mass Conservation*

Consider first the motion of a body of single-phase material (solid or fluid) in free space. If $V(t)$ is the spatial domain occupied by the body at time t and $\rho(x, t)$ is the density as a function of spatial coordinates x and time t, the total mass of the body is expressed as

$$M(t) = \int_{V(t)} \rho(x, t) \, dV \tag{13.110}$$

In the absence of mass sources or sinks, mass conservation requires that $M(t) = $ const. Since the integral in (13.110) is taken over a domain that varies in time, the time derivative of the integral is evaluated as

$$\dot{M}(t) = \frac{d}{dt} \int_{V(t)} \rho(x, t) \, dV = \int_{V(t)} \frac{\partial \rho(x, t)}{\partial t} \, dV + \int_{S(t)} \rho(x, t) v_n(x, t) \, dS$$

$$(13.111)$$

where $S(t)$ is the boundary of $V(t)$ and v_n is the velocity of boundary points in the direction normal to the boundary, given by $v_n = v \cdot n$, in which v is the velocity vector and n is the unit vector normal to the boundary. Using the Gauss theorem, the boundary integral can be converted to a volume integral:

$$\int_{S(t)} \rho(x, t) v_n(x, t) \, dS = \int_{V(t)} \nabla \cdot (\rho(x, t) v(x, t)) \, dV = \int_{V(t)} \nabla \cdot j(x, t) \, dV$$

$$(13.112)$$

Here, $j = \rho v$ is the mass flux and $\nabla \cdot j$ is its divergence (evaluated based on partial derivatives with respect to the spatial coordinates x).

Based on (13.111) and (13.112), the mass conservation condition $\dot{M}(t) = 0$ can be written as

$$\int_{V(t)} \left(\frac{\partial \rho(x, t)}{\partial t} + \nabla \cdot j(x, t) \right) \, dV = 0 \qquad (13.113)$$

Conservation of mass must hold not only for the whole body of interest but also for its arbitrarily small part, which means that the integral in (13.113) vanishes even if the integration domain $V(t)$ is replaced by its arbitrarily small subdomain. Consequently, the integrand must vanish (mathematically speaking, almost everywhere) and we obtain the local form of the *mass conservation equation*,

$$\frac{\partial \rho(x, t)}{\partial t} = -\nabla \cdot j(x, t) \qquad (13.114)$$

This is in principle the same equation as (8.76), which was derived (in a slightly different notation) by tracing the mass that enters an elementary control volume through its boundaries.

On the left-hand side of (13.114), ρ is the mass per unit current volume (at position x and time t) and the derivative with respect to t is taken at constant x, i.e., at a fixed spatial location. The "mass-carrying" material particles flow through the space and the derivative $\partial \rho(x, t)/\partial t$ corresponds to the rate at which the mass density changes at a fixed location in space (given by x). Alternatively, we could imagine an elementary control volume that follows the motion of a selected material particle, characterized by its initial[12] position X. The motion is in general described by $x = \xi(X, t)$ where ξ is a vector-valued function that specifies the spatial position x occupied at time t by the material particle that was initially at position X.

Mathematically speaking, function $\tilde{\rho}$ that describes the dependence of density on X and t is different from function ρ that describes the dependence of density on x and t, but both functions refer to the same physical field (current mass density) and

---

[12]By "initial" position we mean the position in a reference state, which can be selected arbitrarily. For solids, it is usually considered as the initial "undeformed" state of the body.

are directly linked by the relation $\tilde{\rho}(X, t) = \rho(\xi(X, t), t)$. If the objective is to find the rate at which the density changes if the observer follows the motion of a particle that was initially at X, function $\tilde{\rho}$ must be differentiated with respect to t at fixed X. Making use of the chain rule of differentiation, we obtain

$$\frac{\partial \tilde{\rho}(X, t)}{\partial t} = \frac{d}{dt}\rho(\xi(X, t), t) = \frac{\partial \rho(\xi(X, t), t)}{\partial x} \frac{\partial \xi(X, t)}{\partial t} + \frac{\partial \rho(\xi(X, t), t)}{\partial t} \quad (13.115)$$

The first "fraction" on the right-hand side of (13.115) is the spatial gradient of density, $\nabla \rho = \partial \rho / \partial x$, and the second "fraction" is the time derivative of the position of the particle that was initially at X, i.e., the velocity considered as a function of X and, in a rigorous notation, denoted as $\tilde{v}$. The last term on the right-hand side of (13.115) is the rate of density at a fixed point in space, $\partial \rho(x, t)/\partial t$, evaluated at $x = \xi(X, t)$.

From (13.115), it is clear that the time derivative computed at fixed x is in general different from the time derivative computed at fixed X. The derivative with respect to t at fixed x is called the *spatial time derivative* and, for simplicity, we will futher denote it by a superposed dot. The derivative with respect to t at fixed X is called the *material time derivative*, and we will denote it by D_t preceding the symbol of the differentiated function. To further simplify the notation, we will usually omit the arguments (such as X, x and t) and also leave out the superposed tilde when referring to a function that describes a certain physical quantity as a function of X and t. With all these conventions in mind, we can rewrite (13.115) as

$$D_t \rho = \dot{\rho} + v \cdot \nabla \rho \quad (13.116)$$

Material and spatial time derivatives are seen to differ by the term $v \cdot \nabla \rho$, which represents the convective flow of mass. Note again that the operator ∇ refers here to differentiation with respect to the spatial coordinates x and in a more precise notation it would be denoted as ∇_x, to avoid confusion with the operator ∇_X that corresponds to differentiation with respect to the material coordinates, X.

Relation (13.116) between the material time derivative and the spatial time derivative has a general validity and applies not only to the density but to any other field, because it was obtained in (13.115) by applying mathematical rules of differentiation, with no reference to the physical meaning of function ρ. It is therefore possible to replace ρ by any other symbol denoting a scalar field, or even a tensorial field, and the relation remains valid. The convective term, i.e., the difference between the material time derivative and the spatial time derivative, is given by the dot product (contraction) of the velocity field and the gradient of the field for which the time derivatives are computed. Physically, $\dot{\rho}$ describes the change that occurs at a fixed position in space. If we want to evaluate the rate by which the density varies as we follow the trajectory of a moving material particle, we must consider that in a time interval dt the particle is displaced by $v \, dt$ and such a change of spatial location results into an additional change in ρ given by the scalar product of $v \, dt$ with the spatial gradient of ρ.

Formula (13.116) permits transformations between material time derivatives and spatial time derivatives. For instance, expression (13.114) for the spatial time derivative of density, rewritten as

$$\dot{\rho} = -\nabla \cdot \boldsymbol{j} = -\nabla \cdot (\rho \boldsymbol{v}) = -(\nabla \rho) \cdot \boldsymbol{v} - \rho \nabla \cdot \boldsymbol{v} \qquad (13.117)$$

can be transformed into the corresponding material time derivative of density,

$$D_t \rho = \dot{\rho} + \boldsymbol{v} \cdot \nabla \rho = -\rho \nabla \cdot \boldsymbol{v} \qquad (13.118)$$

The physical interpretation of this result becomes apparent if the density is replaced by the specific volume, $v = 1/\rho$. The material time derivative obeys the standard rules of differentiation, and so

$$D_t v = D_t \left(\frac{1}{\rho} \right) = -\frac{D_t \rho}{\rho^2} = \frac{\nabla \cdot \boldsymbol{v}}{\rho} = v \nabla \cdot \boldsymbol{v} \qquad (13.119)$$

The resulting relation $(D_t v)/v = \nabla \cdot \boldsymbol{v}$ means that when we follow a material particle, the relative change of volume per unit time is equal to the divergence of the velocity field. This is the rate form of the standard relation between the volumetric strain and the divergence of the displacement field.

Derivation of the mass conservation equation (13.114) based on differentiation of an integral over a domain that evolves in time may seem to be unnecessarily complicated because essentially the same equation was derived in Sect. 8.3.3 in a simpler way, by looking at the mass flux that crosses the boundary of an elementary control volume. The main advantage of the approach used here is that it is relatively easily extensible to more complicated conservation or balance laws.

13.5.2 Momentum Balance*

The balance of momentum represents the continuum version of Newton's second law, roughly stated as "the time derivative of the momentum of a mass point is equal to the resultant force acting on that point." The momentum is the product of mass and velocity and, for a continuous body with distributed mass, can be obtained by integrating the mass density multiplied by velocity. The external force acting on the body is the resultant of body forces with intensity $\boldsymbol{b}$ (per unit mass) and surface forces with intensity t (per unit current area of the boundary). The momentum balance equation thus reads

$$\frac{d}{dt} \int_{V(t)} \rho \boldsymbol{v} \, dV = \int_{V(t)} \rho \boldsymbol{b} \, dV + \int_{S(t)} t \, dS \qquad (13.120)$$

Expanding the time derivative of the integral over a time-dependent domain on the left-hand side and making use of the relation $t = n \cdot \sigma$ between the surface forces t and the Cauchy stress tensor σ in the last term on the right-hand side, we can rewrite (13.120) as

$$\int_{V(t)} (\dot{\rho} v + \rho \dot{v})\, dV + \int_{S(t)} \rho v (v \cdot n)\, dS = \int_{V(t)} \rho b\, dV + \int_{S(t)} n \cdot \sigma\, dS \quad (13.121)$$

Using the Gauss theorem, we then convert both surface integrals to volume integrals, and by the same reasoning as before (namely that $V(t)$ can be replaced by its arbitrarily small subdomain) we arrive at the local form of momentum balance,

$$\dot{\rho} v + \rho \dot{v} + \nabla \cdot (\rho v \otimes v) = \rho b + \nabla \cdot \sigma \quad (13.122)$$

On the right-hand side of (13.122), we recognize the terms known from the standard static equilibrium equations (Cauchy equations), $\nabla \cdot \sigma + \rho b = 0$. For a body in motion, we expect an additional term that corresponds to inertia forces. At first one may wonder why the left-hand side of (13.122) contains not only the term $\rho \dot{v}$ (density times "acceleration") but also two other terms. The reason is that $\dot{v}$ is not really the acceleration, because the dot denotes here the spatial time derivative while the true acceleration of a mass point corresponds to the material time derivative of velocity. Indeed, taking into account (13.117), the third term on the left-hand side of (13.122) can be worked out as follows:

$$\nabla \cdot (\rho v \otimes v) = [\nabla \cdot (\rho v)] v + \rho v \cdot \nabla v = -\dot{\rho} v + \rho v \cdot \nabla v \quad (13.123)$$

Consequently, the left-hand side of (13.122) can be rewritten as $\rho \dot{v} + \rho v \cdot \nabla v$, which corresponds to $\rho D_t v$, i.e., to the product of the current density and the material time derivative of velocity. This product is the true inertia force (per unit current volume). One of possible forms of the *momentum balance equation* thus is

$$\rho D_t v = \rho b + \nabla \cdot \sigma \quad (13.124)$$

To facilitate future manipulations with time derivatives of integrals over time-dependent domains, it is useful to develop an alternative approach. The main idea is that the integral is transformed to a fixed (time-independent) domain, the differentiation with respect to time is then simply applied to the transformed integrand, and the result is transformed back to the original time-dependent domain. The time-independent reference domain V_0 corresponds to the domain occupied by the body in its reference state at time t_0, when the spatial positions of individual particles, x, coincide with their "material coordinates", X, i.e., when $\xi(X, t_0) = X$. Transformation from $V(t)$ to V_0 requires a replacement of the differential volume dV by the expression $J\, dV_0$ where dV_0 is a differential volume in the reference domain and $J = dV / dV_0$ is the Jacobian of the mapping ξ, i.e., the determinant of the deformation gradient $\partial \xi / \partial X$. The product $J\rho$ is equal to the mass density in the reference state, ρ_0, i.e., to the mass per unit volume in the reference state (this statement follows from the identity $\rho\,dV = \rho_0\,dV_0 = $ mass of the material that initially occupies volume dV_0 and, at time t, occupies volume dV). Consequently, we can write

$$\frac{d}{dt}\int_{V(t)} \rho v\,dV = \frac{d}{dt}\int_{V_0} \rho v J\,dV_0 = \frac{d}{dt}\int_{V_0} \rho_0 v\,dV_0 \qquad (13.125)$$

Now, since the reference domain V_0 and the initial density ρ_0 are time-independent, differentiation with respect to time applies exclusively to the velocity, v. In the integral over V_0, the velocity is considered as a function of X and t and the derivative is taken at constant X, which means that it is the material time derivative, denoted as $D_t v$. Finally, the integral is transformed back to $V(t)$, using identities $dV_0 = dV/J$ and $\rho_0/J = \rho$:

$$\frac{d}{dt}\int_{V_0} \rho_0 v\,dV_0 = \int_{V_0} \rho_0 D_t v\,dV_0 = \int_{V(t)} \rho_0 D_t v J^{-1}\,dV = \int_{V(t)} \rho D_t v\,dV \quad (13.126)$$

Combining (13.125) and (13.126), we obtain a very simple rule for differentiation of time-dependent integrals. For the present case, it is written as

$$\frac{d}{dt}\int_{V(t)} \rho v\,dV = \int_{V(t)} \rho D_t v\,dV \qquad (13.127)$$

but the velocity vector field v can be replaced by any other field, even a tensorial one. An important point is that the integrand must contain the considered field multiplied by the mass density, in other words, that the field must represent a specific quantity (i.e., a quantity taken per unit mass). In fact, the velocity can be interpreted as the specific momentum, because it corresponds to the momentum per unit mass.

13.5.3 Energy Balance*

Now we can proceed to the energy balance equation, which, in a modified form, will serve as the governing equation for heat transfer (possibly coupled with mass transfer). According to the first law of thermodynamics, the increase in the total energy of a body is equal to the sum of the work done on the body by external forces and the heat supplied to the body. In the rate form, this statement can be written as

$$\frac{d}{dt}\left(\int_{V(t)} \rho u\,dV + \int_{V(t)} \tfrac{1}{2}\rho v\cdot v\,dV\right) = \int_{V(t)} \rho b\cdot v\,dV + \int_{S(t)} t\cdot v\,dS +$$
$$+ \int_{V(t)} \rho r\,dV - \int_{S(t)} q_n\,dS \quad (13.128)$$

where u is the specific internal energy (per unit mass), r is the specific heat source (heat supplied by sources distributed in the body, per unit time and unit mass), and q_n is the (conductive) heat flux in the direction normal to the boundary, which can be expressed as $n \cdot q$ where q is the heat flux vector. The surface integral of q_n is preceded by a negative sign because n is the outward unit normal and if $q_n = n \cdot q$ is positive, heat is transferred from the body to its environment and the heat supplied to the body is thus negative. The distributed heat source term on the right-hand side of (13.128) can originate, e.g., from microwave heating, which is of interest for curing of concrete, and also for the technique of ablating a contaminated surface layer by a microwave blast [213]; see Sect. 13.3.4.

Equation (13.128) can be converted to the local form using the procedure that has already been explained in detail in the context of mass conservation and momentum balance. Based on formula (13.127), which remains valid if v is replaced on both sides by $u + v \cdot v/2$, the left-hand side of (13.128) can be rewritten as the integral of $\rho D_t (u + v \cdot v/2)$. On the right-hand side, we use the relations $t = n \cdot \sigma$ and $q_n = n \cdot q$ and transform the surface integrals into volume integrals using the Gauss theorem. The resulting local form of the energy balance equation is then

$$\rho D_t u + \tfrac{1}{2} \rho D_t (v \cdot v) = \rho b \cdot v + \nabla \cdot (\sigma \cdot v) + \rho r - \nabla \cdot q \qquad (13.129)$$

Making use of the rules of differentiation and of formula (13.124), we can express the second term on the left-hand side as

$$\tfrac{1}{2} \rho D_t (v \cdot v) = \rho v \cdot D_t v = (\rho D_t v) \cdot v = \rho b \cdot v + (\nabla \cdot \sigma) \cdot v \qquad (13.130)$$

After substitution into (13.129) we find that the term $\rho b \cdot v$ cancels with the first term on the right-hand side, and the term $(\nabla \cdot \sigma) \cdot v$ cancels with a part of the second term on the right-hand side, because

$$\nabla \cdot (\sigma \cdot v) = (\nabla \cdot \sigma) \cdot v + \sigma : \nabla v \qquad (13.131)$$

where the ":" operator stands for the double contraction of tensors, and ∇v is the spatial gradient of velocity. The described adjustments lead to a shorter and more convenient form of the *energy balance equation,*

$$\rho D_t u = \sigma : \nabla v + \rho r - \nabla \cdot q \qquad (13.132)$$

The first term on the right-hand side of (13.132) represents the stress power density (work done by stress, per unit time and unit current volume). In the small-strain theory, it would reduce to $\sigma : \dot{\varepsilon}$, i.e., to the double contraction of the stress tensor and the rate of the small-strain tensor. In the general case, σ is the Cauchy stress (true stress, force per unit current area) and ∇v is the spatial velocity gradient. Since σ is symmetric, ∇v could be replaced by its symmetric part, which is also called the rate-of-deformation tensor. The second term on the right-hand side of (13.132) corresponds to the heat supplied to the control volume by a distributed heat source and the third term to the heat supplied by conductive heat flux through the boundary of the control volume.

If the material time derivative of specific internal energy, $D_t u$, is transformed into the spatial time derivative, $\dot{u} = D_t u - v \cdot \nabla u$, then the energy balance equation (13.132) reads

$$\rho \dot{u} = \sigma : \nabla v + \rho r - \nabla \cdot q - \rho v \cdot \nabla u \qquad (13.133)$$

Note the additional term on the right-hand side, in which $\rho v = j$ corresponds to the mass flux and the whole term, $-j \cdot \nabla u$, represents the net internal energy that is brought into the control volume (fixed in space) by the flux of matter through the boundary of the control volume, i.e., by convection.

13.5.4 Entropy Balance*

The concept of entropy is closely related to the second law of thermodynamics, which states that there exists a certain state variable, called *entropy*, with the following property: In a closed system (i.e., in a system that does not exchange mass with its environment), the rate of entropy increase is never smaller than the rate of externally supplied entropy, defined as the rate of heat supply divided by the absolute temperature. For a thermally insulated system, this means that the total entropy can never decrease.

The difference between the rate of entropy increase and the rate of external entropy supply is the rate of internal entropy production, and this quantity must always be nonnegative. Processes that result into no internal entropy production are reversible, because when the signs of all rates are reverted, the internal entropy production is still zero and the reverse process is admissible. Processes that result into a positive internal entropy production are irreversible, because when the signs of all rates are reverted, the internal entropy production becomes negative, which is not admissible.

For a continuous medium, one can define the *specific entropy*, s, as the entropy per unit mass, and the total entropy of a body is obtained by integrating ρs over the corresponding spatial domain. Entropy can be supplied to the body (or extracted from it) by exchanges of heat. For a body with a uniform temperature distribution, the rate of supplied entropy is defined as the rate of heat supply divided by the absolute temperature. Under nonuniform temperature, one needs to apply this definition locally at each elementary volume and elementary surface segment and integrate. The second law of thermodynamics can then be written as the inequality

$$\frac{d}{dt} \int_{V(t)} \rho s \, dV \geq \int_{V(t)} \frac{\rho r}{T} \, dV - \int_{S(t)} \frac{q_n}{T} \, dS \qquad (13.134)$$

in which the left-hand side represents the rate of total entropy and the right-hand side is the rate of supplied entropy.

As usual, the surface integral in (13.134) can be converted to a volume integral, making use of the fact that $q_n = \boldsymbol{n} \cdot \boldsymbol{q}$. The inequality must hold for an arbitrarily small domain and thus can be converted into the local condition

$$\rho D_t s \geq \frac{\rho r}{T} - \nabla \cdot \left(\frac{\boldsymbol{q}}{T} \right) \tag{13.135}$$

which is sometimes called the "entropy imbalance," because of the inequality sign. However, one can also introduce the auxiliary notion of *specific entropy production*, s^*, defined by the relation

$$\rho s^* = \rho D_t s - \frac{\rho r}{T} + \nabla \cdot \left(\frac{\boldsymbol{q}}{T} \right) \tag{13.136}$$

which is equivalent to

$$\rho D_t s = \frac{\rho r}{T} - \nabla \cdot \left(\frac{\boldsymbol{q}}{T} \right) + \rho s^* \tag{13.137}$$

Inequality (13.135), which originates from the second law of thermodynamics, can then be rewritten simply as the condition of nonnegative entropy production,

$$s^* \geq 0 \tag{13.138}$$

and Eq. (13.137) can be considered as the *entropy balance equation*.

Expanding the divergence of $\boldsymbol{q}/T$ as $\nabla \cdot (\boldsymbol{q}/T) = (\nabla \cdot \boldsymbol{q})/T - \boldsymbol{q} \cdot (\nabla T)/T^2$ and multiplying both sides of (13.137) by T, we obtain

$$\rho T D_t s = \rho r - \nabla \cdot \boldsymbol{q} + \frac{1}{T} \boldsymbol{q} \cdot \nabla T + \rho T s^* \tag{13.139}$$

The last term on the right-hand side is called the *dissipation* (more precisely, the dissipation density rate, i.e., the dissipated energy per unit volume and unit time).

Equation (13.139) contains the terms $\rho r - \nabla \cdot \boldsymbol{q}$, which correspond to the rate of heat supply per unit volume, and the same terms are found in the energy balance Eq. (13.132). Combining both equations such that the heat supply be eliminated, we obtain

$$\rho D_t u = \boldsymbol{\sigma} : \nabla \boldsymbol{v} + \rho T D_t s - \frac{1}{T} \boldsymbol{q} \cdot \nabla T - \rho T s^* \tag{13.140}$$

13.5.4.1   Fluids*

For fluids, the internal state can be uniquely characterized by two state variables, e.g., by the pressure and temperature. In the present context, it is better to select the specific volume, v, and the specific entropy, s, as the independent state variables, and consider all the other state variables (e.g., the pressure, temperature, and specific internal energy) as functions of v and s. Furthermore, the stress state in a fluid at equilibrium is hydrostatic, given by $\sigma = -p\mathbf{1}$ where p is the pressure and $\mathbf{1}$ is the unit second-order tensor (spherical tensor). The negative sign reflects the sign convention: a positive normal stress means tension while a positive pressure means compression. For a fluid in motion, an additional deviatoric stress σ_D can arise due to viscous effects. This part of stress is related to the deviatoric part of the rate-of-deformation tensor (i.e., of the symmetric part of the velocity gradient, ∇v) and vanishes if $\nabla v = \mathbf{0}$.

Setting $\sigma = -p\mathbf{1} + \sigma_D$, we can rewrite the first term on the right-hand side of (13.140) as $\sigma : \nabla v = -p\mathbf{1} : \nabla v + \sigma_D : \nabla v = -p\nabla \cdot v + \sigma_D : \nabla v$. Substituting this expression into (13.140), multiplying the whole equation by $v = 1/\rho$ and recognizing the term $v\nabla \cdot v$ as the material time derivative of the specific volume (as stated in Eq. (13.119)), we obtain

$$D_t u = -pD_t v + v\sigma_D : \nabla v + TD_t s - \frac{v}{T}q \cdot \nabla T - Ts^* \qquad (13.141)$$

If the internal energy is considered as a function of specific volume and specific enthalpy, its rate on the left-hand side of (13.141) can be expressed in terms of the rates of v and s, which leads to[13]

$$\frac{\partial u}{\partial v}\bigg|_s D_t v + \frac{\partial u}{\partial s}\bigg|_v D_t s = -pD_t v + v\sigma_D : \nabla v + TD_t s - \frac{v}{T}q \cdot \nabla T - Ts^* \quad (13.142)$$

This equality must hold for an arbitrary combination of rates $D_t v$ and $D_t s$, which is possible only if

$$p = -\frac{\partial u}{\partial v}\bigg|_s \qquad (13.143)$$

$$T = \frac{\partial u}{\partial s}\bigg|_v \qquad (13.144)$$

$$\rho Ts^* = \sigma_D : \nabla v - \frac{1}{T}q \cdot \nabla T \qquad (13.145)$$

Equations (13.143)–(13.144) are state laws that link the pressure and temperature to the derivatives of the specific internal energy with respect to the specific volume and specific entropy. Equation (13.145) provides a recipe for evaluation of the dissipation, which is extremely useful, because the dissipation (or the closely related entropy production rate) is now linked to the actual processes in the material. Such a link can be exploited in two ways.

---

[13]A vertical bar with a subscript which follows a partial derivative specifies which variable is kept constant when the derivative is taken. For instance, $(\partial u/\partial v)|_s$ means the partial derivative of u with respect to v at constant s. This notation removes possible ambiguity, because in principle we can consider the internal energy density as a function of various pairs of internal variables, not necessarily of v and s. If u was considered as a function of v and T, its partial derivative with respect to v (at constant T) would have a different meaning.

First, inequality (13.138), which reflects the second law of thermodynamics, can be rewritten as

$$\sigma_D : \nabla v - \frac{1}{T} q \cdot \nabla T \geq 0 \qquad (13.146)$$

The left-hand side should be nonnegative for all admissible processes. The first term, $\sigma_D : \nabla v$, represents the *mechanical part of dissipation*, and the second term, $-q \cdot (\nabla T)/T$, is the *thermal part of dissipation*. The second law states only that the sum of these terms must be nonnegative. This must hold for all admissible processes, including (i) those in which the temperature is uniform, i.e., in which $\nabla T = 0$ and the second term vanishes, and (ii) those in which the material particles are at rest, i.e., in which $v = 0$ and the first term vanishes. It is therefore reasonable to stipulate that each term be nonnegative even when considered separately, which leads to conditions

$$\sigma_D : \nabla v \geq 0 \qquad (13.147)$$
$$q \cdot \nabla T \leq 0 \qquad (13.148)$$

These conditions play the role of constraints on the constitutive equations that complement the balance laws and describe the behavior of the given material. In the simplest case, the constitutive equations could be postulated as

$$\sigma_D = \eta_f (\nabla_s v)_D \qquad (13.149)$$
$$q = -k_T \nabla T \qquad (13.150)$$

where η_f is the dynamic viscosity of the fluid, $(\nabla_s v)_D$ is the deviatoric part of the symmetric velocity gradient, and k_T is the *thermal conductivity*. For such linear isotropic laws, conditions (13.147)–(13.148) are satisfied if the dynamic viscosity and thermal conductivity are both nonnegative, which is a very natural requirement. Equation (13.150) is the isotropic form of the *Fourier law*, widely used by models of heat conduction.

Formula (13.145) can also be exploited for elimination of the entropy production rate from equations (13.139) and (13.141). The resulting equations

$$\rho T D_t s = \rho r - \nabla \cdot q + \sigma_D : \nabla v \qquad (13.151)$$
$$D_t u = -p D_t v + T D_t s \qquad (13.152)$$

are much simpler than the original ones. In fact, Eq. (13.152) directly follows from the rules of differentiation of u as a function of v and s, combined with the state laws (13.143) and (13.144).

13.5.4.2   Solids*

For solids, it is not sufficient to characterize the internal state by two scalar variables. Even for an elastic solid, the specific internal energy is not a unique function of the specific volume and specific entropy. The specific volume is related to the volumetric part of strain but, for solids, the deviatoric part of strain stores energy, too. Moreover, if the solid is inelastic, its behavior may depend not only on the current state but also on the previous history. This dependence is conveniently incorporated through internal state variables.

Since the focus here is on concrete, it is sufficient to restrict attention to a simplified formulation based on the assumption of small displacements and small displacement gradients (which implies small strains and small rotations). Within this theory, it is not necessary to distinguish between the material and spatial coordinates, between the material and spatial rates, and between the initial and current density. A detailed discussion of the thermodynamic approach to constitutive modeling of inelastic solids is available in Chap. 23 of Jirásek and Bažant [521]. Here we just summarize the main equations analogous to those derived in the previous section for fluids.

Equation (13.140) is replaced by

$$\rho \dot{u} = \boldsymbol{\sigma} : \dot{\varepsilon} + \rho T \dot{s} - \frac{1}{T} \boldsymbol{q} \cdot \nabla T - \rho T s^* \tag{13.153}$$

and the specific energy u is considered as a function of the strain, ε, specific entropy, s, and a set of internal variables that depend on the choice of material model (viscoelasticity, plasticity, damage mechanics, etc.). In the general setting, the internal variables are collectively denoted by the symbol α. If the time derivative on the left-hand side of (13.153) is expanded using the chain rule, we obtain an equation analogous to (13.142) multiplied by ρ:

$$\rho \frac{\partial u}{\partial \varepsilon}\bigg|_{s,\alpha} : \dot{\varepsilon} + \rho \frac{\partial u}{\partial s}\bigg|_{\varepsilon,\alpha} \dot{s} + \rho \frac{\partial u}{\partial \alpha}\bigg|_{\varepsilon,s} \cdot \dot{\alpha} = \boldsymbol{\sigma} : \dot{\varepsilon} + \rho T \dot{s} - \frac{1}{T} \boldsymbol{q} \cdot \nabla T - \rho T s^* \tag{13.154}$$

The state laws analogous to (13.143) and (13.144) now read

$$\boldsymbol{\sigma}_Q = \rho \frac{\partial u}{\partial \varepsilon}\bigg|_{s,\alpha} \tag{13.155}$$

$$T = \frac{\partial u}{\partial s}\bigg|_{\varepsilon,\alpha} \tag{13.156}$$

$$\beta_Q = \rho \frac{\partial u}{\partial \alpha}\bigg|_{\varepsilon,s} \tag{13.157}$$

where $\boldsymbol{\sigma}_Q$ is the so-called *quasi-conservative stress* [896], and β_Q are the *quasi-conservative thermodynamic forces* conjugate to the internal variables α.

As a generalization of (13.145), the expression for the dissipation rate that follows from (13.154)–(13.157) is given by

$$\rho T s^* = \left(\sigma - \sigma_Q\right) : \dot{\varepsilon} - \beta_Q \cdot \dot{\alpha} - \frac{1}{T} q \cdot \nabla T \tag{13.158}$$

The last term on the right-hand side of (13.158) is the thermal dissipation, given by the same expression as for fluids. The first two terms constitute the mechanical part of dissipation, which vanishes if the solid is elastic. For an inelastic solid, the difference $\sigma - \sigma_Q = \sigma_D$ is identified as the *dissipative stress*,[14] and $-\beta_Q = \beta_D$ can be considered as the *dissipative thermodynamic forces*. The dissipative stress and dissipative thermodynamic forces are linked to the state variables and their rates by special laws that describe the inelastic part of the model and must be constructed such that the mechanical dissipation would not become negative in any admissible process. Finally, substituting (13.158) into (13.139) and making use of the definitions of σ_D and β_D, we obtain

$$\rho T \dot{s} = \rho r - \nabla \cdot q + \sigma_D : \dot{\varepsilon} + \beta_D \cdot \dot{\alpha} \tag{13.159}$$

which is a generalized version of (13.151).

Extension of the description of inelastic solids to the case of large strains is not needed for the present purpose; interested readers can find the main ideas in Chap. 24 of Jirásek and Bažant [521].

13.5.5 Heat Equation*

Equation (13.151) (or its modified form (13.159), valid for solids under small strains) forms the basis of the governing equation that is used in thermal analysis, possibly coupled with stress analysis and mass transport analysis. In thermal analysis, the fundamental unknown field is typically the temperature, and the other state variable needed to characterize the internal state of a fluid is usually taken as the pressure. It is therefore useful to rewrite (13.151) in terms of the pressure and temperature rates. If we consider the specific entropy s as a function of pressure, p, and temperature, T, the material time derivative of s can be expressed as

$$\mathrm{D}_t s = \left.\frac{\partial s}{\partial p}\right|_T \mathrm{D}_t p + \left.\frac{\partial s}{\partial T}\right|_p \mathrm{D}_t T \tag{13.160}$$

Later it will be shown that the partial derivatives of specific entropy with respect to pressure and temperature can be transformed into equivalent expressions with a

---

[14]In Sect. 13.5.4.1, devoted to fluids, symbol σ_D was used for the deviatoric part of stress. Since deviatoric stresses in fluids do not have any quasi-conservative part, the interpretation of σ_D as the dissipative stress applies to the case of fluids, too.

direct physical meaning, which will explain the motivation behind the definition of the *isobaric specific heat capacity*[15] (or simply the *specific heat*)

$$C_p = T \frac{\partial s}{\partial T}\bigg|_p \tag{13.161}$$

and of the *isobaric volumetric coefficient of thermal expansion*

$$\alpha_{vT} - -\rho \frac{\partial s}{\partial p}\bigg|_T \tag{13.162}$$

The left-hand side of (13.151) can now be written as

$$\rho T D_t s = -T \alpha_{vT} D_t p + \rho C_p D_t T \tag{13.163}$$

and (13.151) is replaced by

$$\rho C_p D_t T = \rho r - \nabla \cdot \boldsymbol{q} + \boldsymbol{\sigma}_D : \nabla \boldsymbol{v} + T \alpha_{vT} D_t p \tag{13.164}$$

This is the *heat equation* in its general form (sometimes also called the enthalpy balance equation).

If a material sample is kept at constant pressure and its temperature is slowly varied such that it expands or shrinks uniformly, with no changes in shape, the last two terms on the right-hand side of (13.164) vanish (because $D_t p = 0$ and $\boldsymbol{\sigma}_D = \boldsymbol{0}$) and the equation simplifies to

$$\rho C_p D_t T = \rho r - \nabla \cdot \boldsymbol{q} \tag{13.165}$$

In this reduced form, the heat equation permits an easy interpretation of coefficient C_p, which represents the heat per unit mass that needs to be supplied to the material at constant pressure in order to increase its temperature by 1 degree (Kelvin or Celsius). This explains the term "isobaric specific heat". Multiplication by the mass density, ρ, converts C_p into the heat per unit volume (and per unit temperature increase).

The interpretation of α_{vT} defined in (13.162) as a thermal expansion coefficient is less straightforward. It requires a conversion of (13.162) into an alternative formula, based on the notion of enthalpy, which is one of the modified thermodynamic potentials that can be obtained from the internal energy u by Legendre transformations.

---

[15]The **specific** heat capacity is defined as the heat capacity per unit mass. When multiplied by the density, it is converted into the **volumetric** heat capacity, taken per unit volume. The adjective **isobaric** means that the derivative in (13.161) is taken at constant pressure. One could also define the **isochoric** heat capacity, at constant volume. For solids, the difference between the isobaric and isochoric heat capacities is negligible, for liquids it is small, but for gases it is substantial.

For fluids, the *specific enthalpy* (i.e., enthalpy per unit mass) is defined as[16]

$$h^* = u + pv \tag{13.166}$$

where p is the pressure and v is the specific volume. The specific enthalpy plays the role of a potential when it is considered as a function of pressure and specific entropy. Taking the derivatives and making use of state equations (13.143)–(13.144), we obtain

$$\left.\frac{\partial h^*}{\partial p}\right|_s = \left.\frac{\partial u}{\partial v}\right|_s \left.\frac{\partial v}{\partial p}\right|_s + v + p\left.\frac{\partial v}{\partial p}\right|_s = v \tag{13.167}$$

$$\left.\frac{\partial h^*}{\partial s}\right|_p = \left.\frac{\partial u}{\partial v}\right|_s \left.\frac{\partial v}{\partial s}\right|_p + \left.\frac{\partial u}{\partial s}\right|_v + p\left.\frac{\partial v}{\partial s}\right|_p = T \tag{13.168}$$

In a similar fashion, we can define the *specific Gibbs free energy*

$$\mu = h^* - Ts \tag{13.169}$$

which serves as a potential when considered as a function of pressure and temperature, because

$$\left.\frac{\partial \mu}{\partial p}\right|_T = \left.\frac{\partial h^*}{\partial p}\right|_s + \left.\frac{\partial h^*}{\partial s}\right|_p \left.\frac{\partial s}{\partial p}\right|_T - T\left.\frac{\partial s}{\partial p}\right|_T = v \tag{13.170}$$

$$\left.\frac{\partial \mu}{\partial T}\right|_p = \left.\frac{\partial h^*}{\partial s}\right|_p \left.\frac{\partial s}{\partial T}\right|_p - s - T\left.\frac{\partial s}{\partial T}\right|_p = -s \tag{13.171}$$

With the state law (13.171) for specific entropy at hand, definition (13.162) can be rewritten as

$$\alpha_{vT} = -\rho\left.\frac{\partial s}{\partial p}\right|_T = \rho\frac{\partial^2 \mu}{\partial p \partial T} = \rho\left.\frac{\partial v}{\partial T}\right|_p = \frac{1}{v}\left.\frac{\partial v}{\partial T}\right|_p \tag{13.172}$$

The last expression presents α_{vT} as the relative change of volume per unit temperature increase at constant pressure, which explains the meaning of α_{vT} as the isobaric coefficient of thermal expansion, considered in the volumetric sense (not in the usual linear sense as for solids). One could also exploit (13.171) and (13.169) to convert (13.161) into an alternative definition of the isobaric specific heat:

$$C_p = T\left.\frac{\partial s}{\partial T}\right|_p = T\left.\frac{\partial s}{\partial T}\right|_p + s + \left.\frac{\partial \mu}{\partial T}\right|_p = \frac{\partial (Ts + \mu)}{\partial T}\bigg|_p = \left.\frac{\partial h^*}{\partial T}\right|_p \tag{13.173}$$

---

[16]The standard symbol for specific enthalpy would be h but here we use h^*, to avoid confusion with the pore relative humidity. The asterisk is omitted when we refer to the specific enthalpy of a given substance, e.g., to the specific enthalpy of liquid water, denoted as h_1^w.

The last expression corresponds to the change of specific enthalpy per unit temperature increase at constant pressure.

Strictly speaking, the simplified form (13.165) of the heat equation is valid only under constant pressure and in the absence of mechanical dissipation (represented by the third term on the right-hand side of (13.164)). However, the simplified form can be used as a convenient approximation even under more general conditions, because the last two terms on the right-hand side of (13.164) are usually negligible. To get an idea about their relative importance, let us assess these terms separately.

For liquid water at atmospheric pressure and room temperature (20 °C), the iso-baric specific heat is $C_p = 4181.8$ J/(kg·K), the isobaric volumetric coefficient of thermal expansion is $\alpha_{vT} = 2.07 \cdot 10^{-4}$/K, and the mass density is $\rho = 998.2$ kg/m^3. At constant pressure and under adiabatic conditions (no heat supply), the only nonzero term on the right-hand side of (13.164) is the third one, $\sigma_D : \nabla v$. Suppose that a layer of water is subjected to shearing at a constant velocity, leading to a constant shear strain rate $\dot{\gamma}_{xy}$. According to the linear law (13.149), the generated stress is proportional to the deviatoric part of the strain rate, with proportionality coefficient $\eta_1 = 1.002 \cdot 10^{-3}$ Pa·s = dynamic viscosity of water at room temperature. In this case, the product $\sigma_D : \nabla v$ reduces to $\tau_{xy}\dot{\gamma}_{xy} = \eta_1\dot{\gamma}_{xy}^2$, and the temperature rate evaluated from (13.164) is

$$D_t T = \frac{\eta_1}{\rho C_p} \dot{\gamma}_{xy}^2 = 2.4 \cdot 10^{-10} \, \text{K} \cdot \text{s} \, \dot{\gamma}_{xy}^2 \tag{13.174}$$

At a shear strain rate of 1/s (which is high), the temperature would increase by 1 °C in about $4.2 \cdot 10^9$ s, i.e., in 132 years (provided that the water layer is perfectly thermally insulated, so that no heat is lost by conduction through the boundary). It is obvious that the effect of the third term on the right-hand side of (13.164) is, for water at room temperature and atmospheric pressure, totally negligible. It might play a role for highly viscous fluids under extreme deviatoric strain rates.

If the water layer is uniformly compressed under adiabatic conditions, the first three terms on the right-hand side of (13.164) vanish and the relation between the rates of temperature and pressure is given by

$$D_t T = \frac{T\alpha_{vT}}{\rho C_p} D_t p = 1.45 \cdot 10^{-8} \, \frac{\text{K}}{\text{Pa}} \, D_t p \tag{13.175}$$

If the pressure is increased from the atmospheric pressure to its double (i.e., is increased by 0.1 MPa), the temperature rises only by 0.00145 K. Again, this effect is totally negligible.

Of course, coefficients C_p and α_{vT} are not constants—they strongly depend on the internal state. So there can be certain extreme conditions under which the terms discussed above play an important role, and then the full form of the heat equation (13.164) should be preferred to the reduced form (13.165).

For completeness, let us also derive an expression for the rate of specific enthalpy. Making use of the state laws (13.167)–(13.168) and of formula (13.151) for the rate of specific entropy, we get

$$\rho D_t h^* = \rho \left.\frac{\partial h^*}{\partial p}\right|_s D_t p + \rho \left.\frac{\partial h^*}{\partial s}\right|_p D_t s = \rho v D_t p + \rho T D_t s = D_t p + \rho r - \nabla \cdot \boldsymbol{q} + \boldsymbol{\sigma}_D : \nabla \boldsymbol{v}$$

(13.176)

This relation is sometimes called the *enthalpy balance equation*, even though it is not really an independent balance law—it has been deduced by combining the laws of energy conservation and entropy balance.

13.5.6 Balance Laws for Multiphase Media*

So far we have been dealing with the motion of a "material body" consisting of one single phase of a material (solid, liquid, gas) that completely fills the available space. The main balance equations are summarized in Table 13.5. The next step is to proceed to multiphase systems, such as a porous solid body with pores filled by one or more fluids.

Table 13.5 Overview of balance laws and related equations for a single-phase medium (operator ∇ refers to differentiation with respect to the spatial coordinates $\boldsymbol{x}$)

	Material time derivative	Spatial time derivative
Mass	$D_t \rho = -\rho \nabla \cdot \boldsymbol{v}$	$\dot{\rho} = -\nabla \cdot (\rho \boldsymbol{v})$
	$D_t v = v \nabla \cdot \boldsymbol{v}$	$\dot{v} = v^2 \nabla \cdot (\boldsymbol{v}/v)$
Momentum	$\rho D_t \boldsymbol{v} = \rho \boldsymbol{b} + \nabla \cdot \boldsymbol{\sigma}$	$\rho \dot{\boldsymbol{v}} = \rho \boldsymbol{b} + \nabla \cdot \boldsymbol{\sigma} - \rho \boldsymbol{v} \cdot \nabla \boldsymbol{v}$
Energy	$\rho D_t u = \boldsymbol{\sigma} : \nabla \boldsymbol{v} + \rho r - \nabla \cdot \boldsymbol{q}$	$\rho \dot{u} = \boldsymbol{\sigma} : \nabla \boldsymbol{v} + \rho r - \nabla \cdot \boldsymbol{q} - \rho \boldsymbol{v} \cdot \nabla u$
Entropy[†]	$\rho T D_t s = \rho r \quad \nabla \cdot \boldsymbol{q} + \boldsymbol{\sigma}_D : \nabla \boldsymbol{v}$	$\rho T \dot{s} - \rho r - \nabla \cdot \boldsymbol{q} + \boldsymbol{\sigma}_D : \nabla \boldsymbol{v} - \rho T \boldsymbol{v} \cdot \nabla s$
Enthalpy[†]	$\rho D_t h^* = D_t p + \rho r - \nabla \cdot \boldsymbol{q} + \boldsymbol{\sigma}_D : \nabla \boldsymbol{v}$	$\rho h^* = D_t p + \rho r - \nabla \cdot \boldsymbol{q} + \boldsymbol{\sigma}_D : \nabla \boldsymbol{v}$ $-\rho \boldsymbol{v} \cdot \nabla h^*$

[†]... written in a form valid for fluids

A fully general and rigorous derivation of balance laws for multiphase media is quite a tedious and cumbersome procedure, for multiple reasons. Equations derived for a single phase apply at the microscopic scale, with the part of space occupied by each phase characterized by the specific shape of the spatial domain, resolved in detail. On the other hand, structural problems are solved at the macroscopic scale, at which individual phases are treated as "overlaid" and the description of the part of space they occupy is reduced to the volume fraction. To obtain the quantities that characterize the state of each phase at the macroscopic scale, one needs to apply suitable averaging procedures to the microscopic quantities, and such procedures give rise to additional terms in the balance equations. Also, one needs to take into account interactions among the phases, i.e., internal exchanges of mass, momentum, energy, etc. The description is further complicated by the fact that individual phases can consist of several components, e.g., the gas phase consists of dry air and water vapor. Additional difficulties arise when the contributions of several components or phases have to be combined, because the material time derivatives in individual equations do not have the same meaning—they refer to motions of individual components or phases, characterized by different velocity vectors. It is thus necessary to convert them to spatial time derivatives, but then the sum of convection terms containing different velocity vectors is not always equivalent to one simple convection term for an effective medium.

A systematic treatment of the complex problem mentioned in the previous paragraph can be found, e.g., in the works of Hassanizadeh and Gray [467, 468], Gray and Hassanizadeh [435], and Lewis and Schrefler [572]. Another level of complexity is added if one considers not only the thermodynamic variables associated with the bulk material (distributed in volume) but also additional terms associated with interfaces, such as the surface energy [10, 437, 465]. Such developments have a high theoretical value and provide insight into the structure of the fundamental equations. However, they lead to models with extremely high numbers of parameters, which are often hard to identify. For practical applications, simplified models with a limited number of parameters are needed. In fact, many terms that appear in a detailed description accounting for a multitude of phenomena either turn out to be negligible, or can be combined with other terms in order to reduce the number of parameters.

For the present purpose, it is sufficient to use the simplified form of the heat equation (13.165), in which the contributions of mechanical dissipation and pressure changes have already been neglected. When the equation is written for a specific phase within a multiphase medium, an extra term that represents the energy received by the given phase from all other phases must be added. Also, when phase changes are accounted for, another term that represents the enthalpy contained in the "newly received material" must be added to the heat equation, and the mass balance equation must also be modified accordingly.

To be specific, let us describe a model that represents concrete as a solid skeleton with pores filled by water as the liquid phase and by a mixture of dry air and water vapor as the gas phase. Phase changes that will be taken into account are evaporation and dehydration (strictly speaking, dehydration is a chemical reaction; considering it as a phase change is a convenient simplification). Evaporation is a transition from liquid water to water vapor, and dehydration is a transition from chemically bound water in the solid skeleton to evaporable liquid water. The rates of evaporation and dehydration are, respectively, denoted as $\dot{m}_{ev}$ and $\dot{m}_{deh}$ and defined as the masses of evaporated and dehydrated water per unit volume of concrete[17] and unit time. A negative rate of evaporation corresponds to condensation, and a negative rate of dehydration corresponds to hydration.

---

[17]By "concrete" we mean here the multiphase material, in contrast to the "solid skeleton," which is just the solid part, without pore fluids. Therefore, masses "per unit volume of concrete" are taken per unit total volume, which also includes the pore volume.

All balance equations will be written in terms of spatial time derivatives, because such derivatives are taken at constant spatial coordinates x, which are the same for all phases. In contrast to that, material time derivatives reflect the motion of particles, which is different for different phases. When working with such derivatives, one would need to define special rate symbols, such as D_t^l for the liquid and D_t^g for the gas. On the other hand, if the balance laws are written in terms of spatial time derivatives, the velocity of each particular phase is reflected in the convective terms, and the time derivatives have the same meaning for all phases and can be summed, if needed.

At the macroscopic level,[18] the *mass balance equation* for a generic phase denoted by subscript α can be written as

$$\frac{\partial}{\partial t}\left(\eta_\alpha^* \rho_\alpha\right) = -\nabla \cdot (\rho_\alpha v_\alpha) + \sum_\beta \dot{m}_{\beta,\alpha} \tag{13.177}$$

where the derivative $\partial/\partial t$ is taken at constant x, i.e., it is the spatial time derivative, η_α^* is the volume fraction of phase α and $\dot{m}_{\beta,\alpha}$ is the rate of phase change from phase β to phase α, expressed as mass per unit volume of concrete and unit time. The sum in the last term on the right-hand side of (13.177) represents the net increase of mass of phase α per unit volume of concrete and unit time. Since Eq. (13.177) contains such a mass "source" term, it is more appropriate to call it the mass **balance** equation rather than the mass **conservation** equation. Conservation of mass is reflected by the fact that $\dot{m}_{\beta,\alpha} = -\dot{m}_{\alpha,\beta}$ for all α and β.

Setting subscript α, respectively, to s, l and g, we obtain from (13.177) the mass balance equations for the solid, liquid, and gas phases. However, for an easier interpretation, it is better to adjust the notation and rename certain variables. For instance, the volume fraction of the solid skeleton, η_s^*, corresponds to $1 - n_p$ where n_p is the porosity. The volume fraction of the liquid phase, η_l^*, can be expressed as $n_p S_l$ where S_l is the liquid saturation degree, and the volume fraction of the gas, η_g^*, is then given by $n_p(1 - S_l)$. The product $\rho_\alpha v_\alpha$ corresponds to the mass flux of phase α (per unit total area), denoted as j_α. Finally, the mass source terms corresponding to evaporation/condensation and dehydration/hydration are respectively given by $\dot{m}_{l,g} = -\dot{m}_{g,l} = \dot{m}_{ev}$ and $\dot{m}_{s,l} = -\dot{m}_{l,s} = \dot{m}_{deh}$ while the other $\dot{m}_{\beta,\alpha}$ terms are zero. After these formal changes, the resulting mass balance equations read

---

[18]It is important to note that equations written at the macroscopic level deal with quantities taken per unit total volume and per unit total area, and not per unit volume or area of the respective phase. For instance, the velocity of the liquid phase, v_l, is not the "true" velocity at which the liquid particles move through the pores, but it is the volume flux (also called the filtration velocity), which is equal to the volume of liquid per unit total area and unit time; see the discussion of the Darcy law in Sect. 8.3.2. In the notation used in this book, ρ_α has the meaning of intrinsic density of phase α, i.e., of mass per unit volume occupied by that phase. Multiplication by the volume fraction η_α^* converts ρ_α into the partial (or apparent) density, defined as the mass of phase α per unit total volume.

$$\frac{\partial}{\partial t}\left[(1-n_p)\rho_s\right] = -\nabla \cdot \boldsymbol{j}_s - \dot{m}_{\text{deh}} \tag{13.178}$$

$$\frac{\partial}{\partial t}\left[n_p S_l \rho_l\right] = -\nabla \cdot \boldsymbol{j}_l + \dot{m}_{\text{deh}} - \dot{m}_{\text{ev}} \tag{13.179}$$

$$\frac{\partial}{\partial t}\left[n_p(1-S_l)\rho_g\right] = -\nabla \cdot \boldsymbol{j}_g + \dot{m}_{\text{ev}} \tag{13.180}$$

The solid mass flux $\boldsymbol{j}_s$ (due to deformation of the skeleton) is usually neglected but we include it for completeness.

Taking the sum of Eqs. (13.178)–(13.180), we cancel the internal mass exchange terms and arrive at the total mass conservation equation,

$$\dot{\rho} = -\nabla \cdot (\boldsymbol{j}_s + \boldsymbol{j}_l + \boldsymbol{j}_g) \tag{13.181}$$

in which

$$\rho = (1-n_p)\rho_s + n_p S_l \rho_l + n_p(1-S_l)\rho_g \tag{13.182}$$

is the density of (partially wet) concrete and $\boldsymbol{j}_s + \boldsymbol{j}_l + \boldsymbol{j}_g$ is the total mass flux. The first term on the right-hand side of (13.182) is dominant—it represents the density of dry concrete, $\rho_d = (1-n_p)\rho_s$. The second term corresponds to the increase of density due to the presence of liquid water in the pores. The last term is negligible because the density of pore gas, ρ_g, is by three orders of magnitude smaller than the densities of liquid water and solid skeleton, ρ_l and ρ_s.

Equation (13.181) is not used directly but its derivation illustrates how the internal exchange terms cancel and how the effective density of the multiphase medium is assembled from the partial densities of individual phases. In a similar spirit, one can construct the heat equation for concrete as a multiphase material. The assumption of local thermal equilibrium [466, 469] implies that the temperature of each phase is the same (at the same macroscopic point) and thus can be denoted simply as T. In analogy to (13.165), the heat equations for individual phases read[19]

---

[19] At a first glance, it may seem strange why the terms containing the rates of evaporation and dehydration in equations (13.183)–(13.185) have signs that are opposite to the signs of analogous terms in (13.178)–(13.180). The reason is that these terms describe just the effects of "pure mass exchange" while energy exchange is described separately by terms $e_{\alpha,\beta}$. For instance, the negative term $-\dot{m}_{\text{ev}}$ in (13.179) corresponds to the mass of liquid water lost in evaporation (per unit volume and unit time), while the positive term $\dot{m}_{\text{ev}} h_1^{\text{w}}$ in (13.184) corresponds to the enthalpy "liberated" from this mass and contributing to the enthalpy of the remaining liquid water. Of course, in reality this enthalpy does not really stay in the liquid phase, but the transfer of energy from the liquid phase to the gas phase is described by other terms, namely $e_{g,l}$ in (13.184) and $e_{l,g}$ in (13.185).

$$(1 - n_p)\rho_s C_{ps}\dot{T} + C_{ps}\boldsymbol{j}_s \cdot \nabla T = (1 - n_p)\rho_s r_s - \nabla \cdot \boldsymbol{q}_s + e_{l,s} + e_{g,s} + \dot{m}_{\text{deh}}h_s^{\text{w}}$$
(13.183)

$$n_p S_l \rho_l C_{pl}\dot{T} + C_{pl}\boldsymbol{j}_l \cdot \nabla T = n_p S_l \rho_l r_l - \nabla \cdot \boldsymbol{q}_l + e_{s,l} + e_{g,l} + (\dot{m}_{\text{ev}} - \dot{m}_{\text{deh}})h_l^{\text{w}}$$
(13.184)

$$n_p(1 - S_l)\rho_g C_{pg}\dot{T} + C_{pg}\boldsymbol{j}_g \cdot \nabla T = n_p(1 - S_l)\rho_g r_g - \nabla \cdot \boldsymbol{q}_g + e_{s,g} + e_{l,g} - \dot{m}_{\text{ev}}h_g^{\text{w}}$$
(13.185)

Here, C_{ps}, C_{pl}, and C_{pg} are the isobaric specific heat capacities of the solid skeleton, liquid water, and gas (wet air); r_s, r_l, and r_g are the distributed heat sources for these three phases (per unit mass of the respective phase); $\boldsymbol{q}_s$, $\boldsymbol{q}_l$, and $\boldsymbol{q}_g$ are the conductive heat fluxes in the three phases (per unit total area); $e_{\beta,\alpha}$ is the energy transferred from phase β to phase α (per unit volume of concrete and unit time), including mechanical interaction and excluding enthalpy exchange associated with mass exchange (phase changes); and h_s^{w}, h_l^{w}, and h_g^{w} are the specific enthalpies of water chemically bound in the solid skeleton, of liquid water and of water vapor. The second terms on the left-hand sides of equations (13.183)–(13.185) correspond to heat convection. The contribution of convection due to deformation of the solid skeleton, i.e., the term $C_{ps}\boldsymbol{j}_s \cdot \nabla T$ in (13.183), is often neglected.

The terms $e_{\beta,\alpha}$ that describe energy exchange between the phases cancel when Eqs. (13.183)–(13.185) are summed, because $e_{\beta,\alpha} = -e_{\alpha,\beta}$ for all α and β. On the other hand, the terms that reflect the phase changes (i.e., the last terms on the right-hand side of these equations) do not cancel because the enthalpies of water in different forms (hydrates in the skeleton, liquid phase, vapor) are not the same. Their differences correspond to the latent heat associated with phase changes. Summing Eqs. (13.183)–(13.185), we obtain

$$\rho C_p \dot{T} + (C_{ps}\boldsymbol{j}_s + C_{pl}\boldsymbol{j}_l + C_{pg}\boldsymbol{j}_g) \cdot \nabla T = \rho r - \nabla \cdot \boldsymbol{q} - \dot{m}_{\text{deh}}\Delta h_{s,l}^{\text{w}} - \dot{m}_{\text{ev}}\Delta h_{l,g}^{\text{w}} \quad (13.186)$$

where

$$C_p = \frac{1}{\rho}\left[(1 - n_p)\rho_s C_{ps} + n_p S_l \rho_l C_{pl} + n_p(1 - S_l)\rho_g C_{pg}\right] \quad (13.187)$$

is the effective specific heat capacity of concrete (per unit mass of partially wet concrete),

$$r = \frac{1}{\rho}\left[(1 - n_p)\rho_s r_s + n_p S_l \rho_l r_l + n_p(1 - S_l)\rho_g r_g\right] \quad (13.188)$$

is the effective distributed heat source (per unit mass of concrete),

$$\boldsymbol{q} = \boldsymbol{q}_s + \boldsymbol{q}_l + \boldsymbol{q}_g \quad (13.189)$$

is the total conductive heat flux (per unit total area),

$$\Delta h_{s,l}^{w} = h_l^{w} - h_s^{w} \tag{13.190}$$

is the specific enthalpy (specific latent heat) of dehydration, and

$$\Delta h_{l,g}^{w} = h_g^{w} - h_l^{w} \tag{13.191}$$

is the specific enthalpy (specific latent heat) of vaporization. For water at 100 °C, the specific latent heat of vaporization is 2.257 MJ/kg, and at 20 °C it is 2.454 MJ/kg. The latent heat of vaporization decreases with increasing temperature and tends to zero as the temperature approaches the critical point.

Note that the effective specific heat capacity C_p defined in (13.187) is obtained from the specific heat capacities of individual phases by weighted averaging, with weights proportional to the relative mass of each phase. An analogous rule applies to the effective distributed heat source r defined in (13.188). On the other hand, the total fluxes are obtained by simple summation, because each partial flux is already expressed per unit total area (and not per unit area of the respective phase). This applies to mass fluxes as well as heat fluxes; see Eqs. (13.181) and (13.189).

13.6 Comments on Multiphase Modeling of Hygrothermal Processes and Creep

Gawin, Schrefler, Pesavento, and coworkers have developed a family of rather general and sophisticated models for creep and shrinkage in a porous hygroscopic reactive solid, such as concrete, containing a multiphase system of pore fluids and reacting solids [414, 415, 417–421, 424, 682, 766]. These models are based on the hybrid mixture theory [467, 468, 572, 766] and take into account a large number of phenomena, including various mass and energy transport mechanisms, phase changes and chemical reactions, load-free and load-induced thermal strains, shrinkage, creep, cracking and thermo-chemical degradation of concrete. Their scope has been extended to temperatures above the critical point of water [415], to early-age behavior [419, 420], and to high-strength concrete [418, 767].

One version of the model developed in previous papers was summarized by Gawin et al. [422] and will be briefly described here. Four primary unknown fields (displacement, gas pressure, capillary pressure, temperature) are calculated by solving four governing equations, which originate from the balance of linear momentum, mass balances of water and air, and enthalpy (energy) balance. In addition to the primary state variables, three internal variables (the degree of dehydration, chemical damage variable, and mechanical damage variable) characterize the state of the material. Their growth is determined by three evolution equations. Of course, the model also exploits a number of auxiliary variables that are functions of the primary and internal variables. For instance, the degree of saturation by liquid water is linked to the capillary pressure and temperature by the sorption isotherm.

An overview of the complete set of equations is available in Gawin et al. [422] and their detailed derivation was provided in Gawin et al. [417]. It makes little sense to reproduce the complete model here. Nevertheless, to conclude the present chapter, it might be useful to describe the part of the model related to thermal effects. Yet another multiphase model for the hygrothermal behavior of concrete at elevated temperatures, proposed by Beneš and Štefan [228], is presented in Appendix J.7.

Leaving aside formal differences in notation, it can be stated that the heat equation (or enthalpy balance equation) used in Gawin et al. [422] has the same form as (13.186), with the convective term originating from the "solid flux" neglected and with the external distributed heat source excluded. In the present notation, the equation would read

$$\rho C_p \dot{T} + (C_{pl} \boldsymbol{j}_l + C_{pg} \boldsymbol{j}_g) \cdot \nabla T = -\nabla \cdot \boldsymbol{q} - \dot{m}_{\text{deh}} \Delta h_{\text{s,l}}^{\text{w}} - \dot{m}_{\text{ev}} \Delta h_{\text{l,g}}^{\text{w}} \qquad (13.192)$$

For clarity, the processing of individual terms in (13.192) will be discussed separately and their physical meaning will be specified:

- Heat accumulation, $\rho C_p \dot{T}$.
  The effective heat capacity of concrete, ρC_p, is determined using a refined version of formula (13.187), with $\rho_g C_{pg}$ replaced by $\rho_a C_{pa} + \rho_v C_{pv}$ where ρ_a and ρ_v are the mass densities of dry air and water vapor and C_{pa} and C_{pv} are the specific heat capacities of these two components of the pore gas. The heat capacity of the solid skeleton, C_{ps}, to be substituted into the modified form of (13.187), is taken as a linear function of temperature, given by (13.89).
- Heat loss by convective flow, $(C_{pl} \boldsymbol{j}_l + C_{pg} \boldsymbol{j}_g) \cdot \nabla T$.
  The mass fluxes, $\boldsymbol{j}_l$ and $\boldsymbol{j}_g$, are evaluated from appropriate transport laws, which reflect the diffusive and advective flow of air and vapor and the advective flow of liquid water.
- Heat loss by conductive flow, $\nabla \cdot \boldsymbol{q}$.
  Heat conduction is described by the isotropic Fourier law (13.150), with the effective thermal conductivity, k_T, evaluated from (13.87)–(13.88).
- Heat needed for dehydration, $\dot{m}_{\text{deh}} \Delta h_{\text{s,l}}^{\text{w}}$.
  The water content released by dehydration, m_{deh}, is treated as proportional to the dehydration degree, which is a dimensionless variable dependent on the maximum temperature reached so far and described by a cubic function.
- Heat needed for vaporization, $\dot{m}_{\text{ev}} \Delta h_{\text{l,g}}^{\text{w}}$.
  The rate of vaporization, $\dot{m}_{\text{ev}}$, is determined from the water mass conservation equation; this approach leads to a rather lengthy expression consisting of seven terms. The latent heat of vaporization, $\Delta h_{\text{l,g}}^{\text{w}}$, is a function of temperature given by the Watson formula (13.94).

The models mentioned above are certainly highly sophisticated and theoretically appealing. Their practical applicability still remains limited, because of a high complexity of the model equations, as well as because of difficulties associated with proper calibration of numerous parameters and potential problems with numerical fragility of large-scale simulations of coupled multiphysics systems. Also, even though these models cover a multitude of phenomena, they neglect certain other effects that may play an equally important role. For instance, it may seem necessary to also take into account the evolution of the capillary menisci, geometry of the liquid-vapor interface, hysteresis and irreversibility of sorption–desorption isotherms, nanoscale changes in pore surface morphology caused by drying and wetting, transport of various solute ions with their associated electro-chemical effects and electric potential gradients, crystallization and dissolution of calcium hydroxide and other species [83].

It should be noted that the path dependence of the sorption–desorption process is not mathematically represented by the models considered here. This appears to be acceptable as long as only desorption (decreasing pore humidity) occurs in the entire pore system. However, omission of the path dependence of the isotherms would be a major drawback of refined general models intended for arbitrary variation of humidity and temperature if they were applied to problems with nonmonotonic changes of humidity. The benefit of introducing various secondary refinements while ignoring some other, possibly more important, effects is questionable.

The complexity of moisture–temperature interactions in a hydrophilic nanoporous material such as concrete could probably be alleviated by nanoscale modeling. The recent modeling studies at the Cement Sustainability Hub at MIT (led by F.-J. Ulm) and in the CEE Department at Northwestern University might eventually lead to progress in this direction. So far, these studies, based on molecular dynamics simulations of C-S-H or simplified atomic arrangements [610, 611, 778, 779, 830], were confined to constant temperature. However, these models include activation energy barriers and thermal transition rates, which is what controls the temperature effects. In this way, they open an avenue to more fundamental analysis of combined moisture–temperature effects.

Appendix A
Viscoelastic Rheologic Models

Development of constitutive equations for inelastic materials can be inspired by rheologic models, representing idealized schemes of deformation processes in the material microstructure. For viscoelasticity, such models combine two basic types of rheologic units—elastic springs and viscous dashpots, as shown in Fig. 2.4. The stress transmitted by a linear elastic spring, σ_e, is related to the strain in that spring, ε_e, by Hooke's law

$$\sigma_e = E\,\varepsilon_e \tag{A.1}$$

where E is the elastic modulus. The stress transmitted by a linear viscous dashpot,

$$\sigma_v = \eta\,\dot{\varepsilon}_v \tag{A.2}$$

is proportional to the strain rate in that dashpot, $\dot{\varepsilon}_v$, and the proportionality factor η is called the viscosity

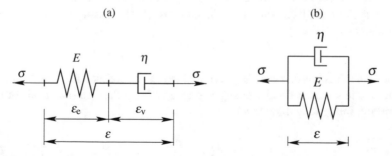

Fig. A.1 Basic rheologic models: (a) Maxwell model, (b) Kelvin model

© Springer Science+Business Media B.V. 2018
Z.P. Bažant and M. Jirásek, *Creep and Hygrothermal Effects in Concrete Structures*, Solid Mechanics and Its Applications 225,
https://doi.org/10.1007/978-94-024-1138-6

A.1   Maxwell Model

The *Maxwell model* [612] is obtained by serial coupling of a spring and a dashpot
(Fig. A.1a). Two units coupled in series transmit the same stress, and the total strain is
obtained by summing the partial strains in individual units, which can be interpreted
as the elastic strain and the viscous strain:

$$\sigma = \sigma_v = \sigma_e \tag{A.3}$$

$$\varepsilon - \varepsilon_e + \varepsilon_v \tag{A.4}$$

In addition to the total stress σ and total strain ε, which are of main interest,
Eqs. (A.1)–(A.4) contain certain partial stresses and strains in individual units, which
play the role of internal variables. By a suitable manipulation of the basic equations,
the internal variables can be eliminated and the direct stress–strain relation can be
deduced. To this end, we differentiate (A.4) with respect to time and express the
partial strain rates $\dot{\varepsilon}_e$ and $\dot{\varepsilon}_v$ using (A.1) and (A.2), with partial stresses σ_e and σ_v
replaced by σ according to (A.3). This leads to the differential equation

$$\dot{\varepsilon}(t) = \frac{\dot{\sigma}(t)}{E} + \frac{\sigma(t)}{\eta} \tag{A.5}$$

in which we have explicitly indicated that ε and σ are functions of time, while E and
η are at first considered as fixed model parameters.

A.1.1   Compliance Function

For a given stress history, the corresponding strain history is obtained easily, by
integrating the right-hand side of (A.5). In a creep test starting at time $t = 0$, the
stress history is given by

$$\sigma(t) = \hat{\sigma} H(t) \tag{A.6}$$

where H is the Heaviside step function, defined in (2.3). Since $H(t) = 1$ for $t \geq 0$,
the stress rate vanishes for all nonnegative times t,[1] and the right-hand side of (A.5)
is constant. Integrating the relation

$$\dot{\varepsilon}(t) = \frac{\hat{\sigma}}{\eta} \tag{A.7}$$

---

[1] Strictly speaking, the stress is not differentiable at $t = 0$, but it has at least the derivative from the
right, which is sufficient for the present purpose. The jump in stress at $t = 0$ will be reflected by
the initial condition.

we obtain the strain history

$$\varepsilon(t) = \varepsilon_0 + \frac{\hat{\sigma}}{\eta} t \qquad (A.8)$$

where ε_0 is an integration constant that has the meaning of the initial strain at $t = 0$ and must be determined from an initial condition. That condition reflects the state of the model at time $t = 0$, just after the application of stress $\hat{\sigma}$. It is easy to see that the strain in the dashpot cannot change by a jump (as long as the stress remains finite), and so $\varepsilon_v(0) = 0$, while the spring responds instantaneously and its strain jumps to $\varepsilon_e(0) = \hat{\sigma}/E$. Since $\varepsilon = \varepsilon_e + \varepsilon_v$, the appropriate initial condition is $\varepsilon(0) = \hat{\sigma}/E$, from which $\varepsilon_0 = \hat{\sigma}/E$ and

$$\varepsilon(t) = \frac{\hat{\sigma}}{E} + \frac{\hat{\sigma}}{\eta} t \qquad (A.9)$$

Formula (A.9) describes the strain history in a creep test at stress level $\hat{\sigma}$ started at $t' = 0$. According to the definition of the compliance function of a nonaging material (Sect. 2.1), we get

$$J_0(t) = \frac{\varepsilon(t)}{\hat{\sigma}} = \frac{1}{E} + \frac{t}{\eta} = \frac{1}{E} \left(1 + \frac{t}{\tau} \right) \qquad (A.10)$$

where $\tau = \eta/E$ is a parameter introduced for convenience and called the *relaxation time*. The foregoing derivation is valid for all $t \geq 0$, while for $t < 0$, the value of J_0 is by definition zero. To obtain an expression valid at all times, including negative ones, we can write

$$J_0(t) = \frac{1}{E} \left(1 + \frac{t}{\tau} \right) H(t) \qquad (A.11)$$

The graph of the compliance function is plotted in Fig. A.2a, which shows the meaning of parameters E and η. Parameter E is the reciprocal value of the instantaneous compliance $J_0(0)$, and parameter η is the inverse slope of the straight line representing the compliance function of this simple model.

The compliance function (A.11) can be interpreted as the sum of the compliance function of the elastic spring, $(1/E)H(t)$, and the compliance function of the viscous dashpot, $(t/\eta)H(t)$. For units coupled in series, the total strain is the sum of partial strains, and so the total compliance is the sum of partial compliances (this general rule will also be applied to the Kelvin chain in Sect. A.3.1). The relaxation time $\tau = \eta/E$ is the duration of loading after which the contributions of the elastic spring and of the viscous dashpot to the total compliance are the same. It sets a certain intrinsic time scale of the model. Loadings applied over time intervals much shorter than τ lead to viscous strain that is negligible compared to the elastic strain, and the model response is close to an elastic spring. Under loadings applied over time intervals much longer than τ, the viscous strain dominates and the model response is close to a viscous dashpot. A truly viscoelastic response is obtained only at time scales comparable to τ.

Due to the nonaging character of the model, the compliance function has been considered as dependent on one argument—the duration of loading. The compliance function with two arguments, which represent the current time, t, and the time at load application, t', is easily constructed as $J(t, t') = J_0(t - t')$, where the time lag $t - t'$ corresponds to the load duration.

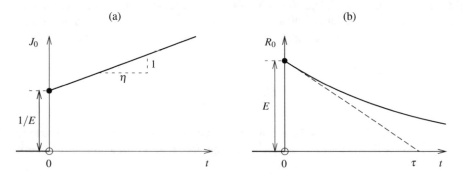

Fig. A.2 Maxwell model: (a) compliance function, (b) relaxation function

A.1.2   Relaxation Function

Recall that the differential stress–strain equation (A.5) is valid for the Maxwell model under general loading. In a relaxation test starting at time $t = 0$, the strain history is given by

$$\varepsilon(t) = \hat{\varepsilon} H(t) \tag{A.12}$$

and, for $t \geq 0$, the strain rate $\dot{\varepsilon}$ vanishes. From (A.5), we obtain the differential equation

$$\frac{\dot{\sigma}(t)}{E} + \frac{\sigma(t)}{\eta} = 0 \tag{A.13}$$

which can be rewritten as

$$\tau \dot{\sigma}(t) + \sigma(t) = 0 \tag{A.14}$$

where, as before, $\tau = \eta/E$ is the relaxation time. Since τ is constant, the general solution of (A.14) is

$$\sigma(t) = C e^{-t/\tau} \tag{A.15}$$

and the integration constant C can be determined from the initial condition $\sigma(0) = E\hat{\varepsilon}$, which follows from an analysis of the instantaneous response after sudden application of the strain. Substituting $C = E\hat{\varepsilon}$ into (A.15), we obtain the particular solution

$$\sigma(t) = E\hat{\varepsilon}e^{-t/\tau} \tag{A.16}$$

and the corresponding relaxation function

$$R_0(t) = Ee^{-t/\tau}H(t) \tag{A.17}$$

The graph of this function is plotted in Fig. A.2b. Relaxation of the stress transmitted by the Maxwell model is an exponentially decaying process, and its time scale is set by the relaxation time τ. After load duration τ, the stress relaxes to the $1/e$ multiple of its initial value. A more direct interpretation is that if the initial rate of relaxation remained constant, the stress would relax to zero after time τ. Again, at time scales much shorter than τ, the model responds almost as an elastic spring (relaxation is negligible), and at time scales much longer than τ, it responds almost as a viscous dashpot (stress is completely relaxed). It is clear that such a simple model could approximate the complex viscoelastic behavior of concrete only roughly, over a very limited range of times.

A.2　Kelvin Model

The *Kelvin model* (also called the *Voigt model*) is obtained by parallel coupling of a spring and a dashpot (Fig. A.1b). Two units coupled in parallel share the same strain, and the total stress is obtained by summing the partial stresses transmitted by individual units, which can be interpreted as the elastic stress and the viscous stress:

$$\varepsilon = \varepsilon_e = \varepsilon_v \tag{A.18}$$

$$\sigma = \sigma_e + \sigma_v \tag{A.19}$$

Combining this with Eqs. (A.1) and (A.2) and eliminating the partial strains and stresses, we obtain the differential stress–strain equation

$$\sigma(t) = E\varepsilon(t) + \eta\dot{\varepsilon}(t) \tag{A.20}$$

first proposed by Meyer [627] and reintroduced by Voigt [838]. Nevertheless, the model will be referred to as the Kelvin model, after Lord Kelvin, formerly W. Thomson, who contributed significantly to the understanding of viscous phenomena [813].

A.2.1　Compliance Function

For the stress history (A.6) prescribed in a creep test, the corresponding strain history for $t \geq 0$ is obtained by solving the differential equation

$$E\varepsilon(t) + \eta\dot{\varepsilon}(t) = \hat{\sigma} \tag{A.21}$$

with initial condition $\varepsilon(0) = 0$. The resulting particular solution is

$$\varepsilon(t) = \frac{\hat{\sigma}}{E}\left(1 - e^{-t/\tau}\right) \tag{A.22}$$

where parameter $\tau = \eta/E$ is in this context called the *retardation time*. The compliance function of the Kelvin model is thus given by

$$J_0(t) = \frac{1}{E}\left(1 - e^{-t/\tau}\right) H(t) \tag{A.23}$$

and its graph is shown in Fig. A.3a. Since the strain in the dashpot must evolve continuously and is equal to the total strain, the model cannot capture any instantaneous effects and the compliance grows continuously from zero to the asymptotic value $1/E$. The retardation time sets the time scale for the delayed response of the model—if the initial strain rate remained constant, the final strain would be attained after time τ. For load durations much shorter than τ, the model behaves almost like a lone dashpot (the elastic stress is negligible because the strain does not have enough time to develop), while for load durations much longer than τ, it behaves almost like an elastic spring (the viscous stress is negligible because the strain remains almost constant). This is exactly the opposite of the behavior of the Maxwell model.

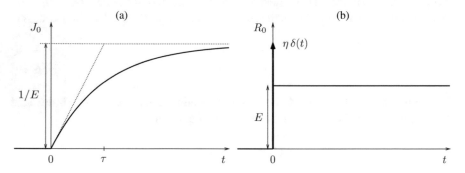

Fig. A.3 Kelvin model: (a) compliance function, (b) relaxation function

A.2.2 Relaxation Function

At a first glance, determination of the relaxation function for the Kelvin model seems to be an easy task, because for a prescribed strain history, the corresponding stress history should be obtained simply by evaluating the right-hand side of (A.20). However, the problem is that the strain history in a relaxation test is not differentiable at the time instant $t' = 0$ when the strain is suddenly applied. If we ignore this fact and simply consider $\varepsilon(t) = \hat{\varepsilon} =$ constant, substitution into (A.20) leads to the stress history given by $\sigma(t) = E\hat{\varepsilon}$, which represents the response of the elastic spring, without any dashpot attached to it. It would obviously be wrong to set $R_0(t) = EH(t)$, because the presence of the dashpot would be totally ignored.

At finite stress, the evolution of strain in the Kelvin model must be continuous, because the strain rate in the viscous dashpot (equal to the total strain rate) must remain finite. A jump in the strain generates an infinite stress, acting over an infinitesimal time interval. Formal mathematical description of such singular phenomena can be based on the concept of distributions (generalized functions); cf. Schwartz [769]. In this generalized sense, the derivative of the unit step function $H(t)$ is the so-called Dirac distribution, $\delta(t)$, which physically represents a unit impulse at time $t = 0$. Substituting $\varepsilon(t) = \hat{\varepsilon}H(t)$ into the right-hand side of (A.20), we obtain $\sigma(t) = E\hat{\varepsilon}H(t) + \eta\hat{\varepsilon}\delta(t)$, and so the relaxation "function" (more rigorously, distribution) of Kelvin's model is

$$R_0(t) = EH(t) + \eta\delta(t) \tag{A.24}$$

This is schematically shown in Fig. A.3b, where the singular part of the distribution is represented by a vertical arrow.

The concept of distributions is handy because it permits a unified treatment of the basic relations describing the mechanical behavior of simple rheologic models. For instance, due to the parallel coupling of two units in the Kelvin model, the resulting relaxation function can be expected to be the sum of the relaxation functions of individual units. Indeed, $EH(t)$ is the relaxation function of the elastic spring and $\eta\delta(t)$ can be interpreted as the relaxation distribution of the viscous dashpot. Nevertheless, in this book, we do not really need to exploit such formalism, because all rheologic models that are actually used for concrete must be capable of reflecting instantaneous elasticity. Therefore, the Kelvin model is never used directly in its elementary form. It is rather exploited as a building block of a chain arranged such that the jump in strain could be accommodated at finite stress levels and thus no singularity would arise in the relaxation function.

A.3 Rheologic Chains

Rheologic models obtained by serial or parallel coupling of an elastic spring and a viscous dashpot are too simple to approximate viscoelastic behavior of real materials in a realistic manner. The deficiencies of such simple models are obvious: Maxwell model leads to a constant creep rate if the stress is kept constant, while the actual creep process slows down in time. Kelvin model cannot capture any instantaneous elastic strain and does not exhibit gradual relaxation of stress under constant strain. Each of these models has a certain intrinsic time scale and on much shorter or much longer time scales does not behave as a truly viscoelastic model but degenerates

(a)   (b)

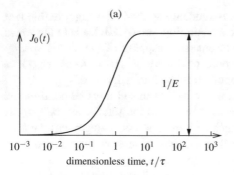
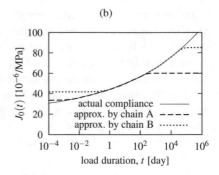

Fig. A.4 (a) Compliance function of Kelvin model (in semilogarithmic scale), (b) actual compliance function of concrete (for a fixed age at the onset of loading) approximated by two different Kelvin chains

either into a spring or into a dashpot, as already discussed before. This becomes clear if we plot the compliance function in the semilogarithmic scale (compliance against the logarithm of load duration). The real compliance function of concrete is shown by the solid curve in Fig. A.4b. The creep process obviously takes place over many orders of magnitude in the time domain. The Kelvin model can approximate the real behavior only over a limited range of times, comparable to its retardation time τ. As shown in Fig. A.4a, the value $J_0(t)$ of the compliance function of Kelvin's model is close to zero for $t \ll \tau$ and close to the asymptotic limit $1/E$ for $t \gg \tau$. An appreciable variation of J_0 takes place over less than two orders of magnitude of the load durations t, roughly from 0.05τ to 3τ. Within this range, J_0 grows from 5% to 95% of its final value.

A.3.1 Kelvin Chain

Since serial coupling of rheologic models corresponds to summing the strains of individual models and thus also their compliances, it is a good idea to link in series several Kelvin models (furthermore called Kelvin units) and construct the so-called *Kelvin chain*. Each unit in the chain can describe the creep process over a specific range of times, and by combining units with different retardation times, we can extend the range of times over which the rheologic chain approximates the real compliance function. Moreover, if one of the units in the chain is a lone spring without a parallel dashpot, the chain can also capture instantaneous strain and the relaxation function becomes regular.

A general scheme of the Kelvin chain is shown in Fig. A.5. The chain consists of an elastic spring characterized by stiffness E_0 and of M Kelvin units characterized by stiffnesses E_μ and viscosities η_μ, $\mu = 1, 2, \ldots, M$. Instead of the viscosities, we can use the retardation times $\tau_\mu = \eta_\mu/E_\mu$ as the primary parameters. The zeroth unit has a vanishing retardation time, which means that its response is instantaneous. The compliance function of the Kelvin chain,

$$J_0(t) = \left[\frac{1}{E_0} + \sum_{\mu=1}^{M} \frac{1}{E_\mu} \left(1 - e^{-t/\tau_\mu} \right) \right] H(t) \tag{A.25}$$

is easily obtained as the sum of compliance functions of individual units. In mathematical literature, an infinite series of the form $f(t) = \sum_{j=1}^{\infty} a_j e^{-\lambda_j t}$ is called the *Dirichlet series* (or sometimes *Prony series*). The expression in the square brackets in (A.25) is thus a special case of a finite Dirichlet series.

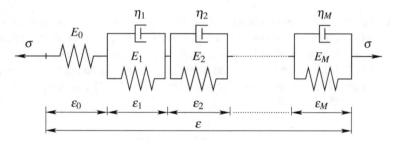

Fig. A.5 Kelvin chain

Good approximation properties of the Dirichlet series (A.25) are obtained if the retardation times τ_μ are sufficiently separated but not too far apart in the logarithmic scale. It is recommended to select them in a geometric progression with quotient 10. A Kelvin chain with M Kelvin units and a spring can then approximate the real compliance function over M orders of magnitude in the space of load durations $t - t'$. This is documented in Fig. A.4b for two different chains with parameters given in Table A.1.

Table A.1 Parameters of Dirichlet series approximating the compliance function of concrete on two different intervals

| | Chain A | | Chain B | |
| | t_{min} = 5 min | t_{max} = 6 months | t_{min} = 1 day | t_{max} = 100 years |
μ	E_μ [GPa]	τ_μ [day]	E_μ [GPa]	τ_μ [day]
0	29.956	0	24.078	0
1	312.99	10^{-2}	188.38	$2 \cdot 10^0$
2	264.92	10^{-1}	160.37	$2 \cdot 10^1$
3	216.44	10^0	131.89	$2 \cdot 10^2$
4	173.42	10^1	106.41	$2 \cdot 10^3$
5	107.12	10^2	66.610	$2 \cdot 10^4$

Each chain consists of a spring and $M = 5$ Kelvin units. Chain A, with retardation times ranging from $\tau_1 = 10^{-2}$ days to $\tau_5 = 100$ days, provides a good approximation of the real compliance function for load durations between 5 min and 6 months (in this range, the dashed curve in Fig. A.4b overlaps with the solid one). Chain B, with retardation times from $\tau_1 = 2$ days to $\tau_5 = 20{,}000$ days, provides a good approximation for load durations between 1 day and 100 years. In general, it is advised to select the retardation times such that $\tau_1 \leq 3\,t_{min}$ and $\tau_M \geq 0.5\,t_{max}$, where t_{min} is the shortest and t_{max} the longest load duration of interest.

A.3.2   Maxwell Chain

In a similar spirit, one can construct a *Maxwell chain*, consisting of several Maxwell units coupled in parallel; see Fig. A.6. This time, the total stress (not strain) is the sum of the stresses in individual units, and thus, the relaxation function (not compliance function) is the sum of the relaxation functions of individual units. Each unit is a Maxwell model with relaxation function of the form (A.17), and so the relaxation function of the chain is given by

$$R_0(t) = \left(\sum_{\mu=1}^{M} E_\mu\, e^{-t/\tau_\mu} \right) H(t) \tag{A.26}$$

where E_μ and τ_μ, $\mu = 1, 2, \ldots, M$, are the stiffnesses and relaxation times of individual units. The expression in the parentheses on the right-hand side of (A.26) is again a special case of Dirichlet series. If all the relaxation times are finite, the relaxation function tends to zero as t approaches infinity. To obtain a positive limit, one could add a constant term corresponding to a spring without a dashpot. From the practical point of view, it is sufficient to make sure that the relaxation times of individual Maxwell units cover a sufficiently wide range, with the maximum relaxation time exceeding the maximum time of interest (e.g., expected lifetime of the structure).

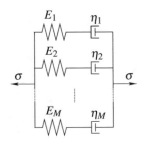

Fig. A.6 Maxwell chain

For nonaging chains, it is possible to prove that every Maxwell chain is exactly equivalent to a certain conjugate Kelvin chain and vice versa [32]. Since the viscoelastic properties of concrete are usually determined by creep tests (and not by relaxation tests), the compliance function is considered as the primary characteristic and Kelvin chains are much more widely used in concrete creep models than Maxwell chains.

A.4 Aging Rheologic Chains

With the exception of very old concrete or short-term loading, aging is an important phenomenon that needs to be taken into account in practical applications. The rheologic models covered so far do not change their properties in time, and their compliance and relaxation functions depend only on the duration of loading. To capture the aging effects, it is necessary to use a generalized form of rheologic chains, with properties of individual units varying in time.

If the properties of basic rheologic units evolve in time, parameters E and η in (A.1) and (A.2) can no longer be considered as constants. Equation (A.2) describing the dashpot can keep its original form, however, with viscosity η dependent on the current time. To emphasize that, we rewrite (A.2) as

$$\sigma_v(t) = \eta(t)\,\dot\varepsilon_v(t) \tag{A.27}$$

Generalization of the elastic law (A.1) is more tricky. It would be wrong to replace the constant modulus E by a function of time and otherwise keep the law in the same form, linking the stress to the strain. This would mean that if, e.g., the strain remains constant and the modulus increases, the stress increases as well. It can be shown that such behavior is inadmissible and violates the fundamental laws of thermodynamics [82, 84]. The increase of stiffness due to hydration can be reflected only by the **increments** of stress and strain and cannot affect their values accumulated in the past. Therefore, the stress–strain law for an aging elastic spring must be written in the rate (incremental) form,

$$\dot\sigma_e(t) = E(t)\,\dot\varepsilon_e(t) \tag{A.28}$$

Note that, on the other hand, for a disintegrating or melting material, such as dehydrating concrete at high temperature, the thermodynamically correct relation is $\sigma_e(t) = E(t)\varepsilon_e(t)$.

A.4.1 Aging Maxwell Chain

The modification of the equations for basic units must be reflected in the differential equations describing the composite rheologic models. For the Maxwell model, it is sufficient to consider E and η in (A.5) as functions of time. The aging Maxwell model is thus described by the first-order differential equation

$$\dot{\varepsilon}(t) = \frac{\dot{\sigma}(t)}{E(t)} + \frac{\sigma(t)}{\eta(t)} \tag{A.29}$$

Note that the first fraction on the right-hand side represents the elastic strain rate, which is consistent with (A.28).

Since the Maxwell model will later be used as a building block of a Maxwell chain, the main interest is in the evaluation of the relaxation function (for the compliance function, there is no simple rule that could be used for a chain with units coupled in parallel). So we need to solve Eq. (A.29) with $\dot{\varepsilon}(t) = 0$ and $\sigma(t)$ considered as unknown. For a general evolution of stiffness $E(t)$ and viscosity $\eta(t)$, this differential equation has variable coefficients and its solution can be constructed by separation of variables.[2] In the special case when both parameters evolve proportionally and their ratio $\eta(t)/E(t) = \tau$ remains constant, the relaxation problem is described by the same differential equation (A.14) as in the nonaging case, and the influence of aging enters only through the initial condition. In a relaxation test started at time t', the initial condition is $\sigma(t') = E(t')\hat{\varepsilon}$, and the stress evolution for $t \geq t'$ is given by

$$\sigma(t) = E(t')\hat{\varepsilon} e^{-(t-t')/\tau} \tag{A.31}$$

The resulting relaxation function of an aging Maxwell model is then

$$R(t, t') = E(t')e^{-(t-t')/\tau} H(t - t') \tag{A.32}$$

where multiplication by the Heaviside function of the load duration ensures that $R(t, t') = 0$ for $t < t'$. Note that, in contrast to the case of a nonaging model, the right-hand side of (A.32) cannot be written as a function of a single variable, the load duration $t - t'$.

Summing the contributions of individual units, it is now easy to construct the relaxation function of an aging Maxwell chain,

$$R(t, t') = \left(\sum_{\mu=1}^{M} E_\mu(t') e^{-(t-t')/\tau_\mu} \right) H(t - t') \tag{A.33}$$

which represents a generalized form of the Dirichlet series (A.26).

---

[2]The resulting formula for the relaxation function of an aging Maxwell model with a general evolution of parameters E and η reads

$$R(t, t') = E(t') \exp\left(-\int_{t'}^{t} \frac{E(s)}{\eta(s)} ds \right) H(t - t') \tag{A.30}$$

In the special case when $E(s)/\eta(s) = 1/\tau = $ const., the integral can be evaluated analytically and (A.30) simplifies to (A.32).

A.4.2 Aging Kelvin Chain

For the Kelvin model, generalization to the case of aging is less straightforward, but still possible. Differential equation (A.20) for the nonaging Kelvin model originates from the stress equivalence relation, $\sigma = \sigma_e + \sigma_v$. As already explained, for an aging spring, we cannot link the elastic stress to the current value of the elastic strain, and we must deal with their rates; cf. (A.28). Therefore, the stress equivalence relation must also be written in the rate form, $\dot{\sigma} = \dot{\sigma}_e + \dot{\sigma}_v$, so that the partial stress rates could be expressed in terms of the strain rate. For $\dot{\sigma}_e$, we directly substitute from (A.28), with $\dot{\varepsilon}_e = \dot{\varepsilon}$, but the rate of the viscous stress must be obtained by differentiation of (A.27), with $\dot{\varepsilon}_v = \dot{\varepsilon}$. The resulting differential equation

$$\dot{\sigma}(t) = E(t)\dot{\varepsilon}(t) + \frac{d}{dt}[\eta(t)\dot{\varepsilon}(t)] \qquad (A.34)$$

is then of the second order. It can be rewritten as

$$\dot{\sigma}(t) = D(t)\dot{\varepsilon}(t) + \eta(t)\ddot{\varepsilon}(t) \qquad (A.35)$$

where $D(t) = E(t) + \dot{\eta}(t)$ is a modified age-dependent modulus, introduced for convenience.

For a creep test with $\dot{\sigma}(t) = 0$, (A.35) becomes a homogeneous second-order differential equation with variable coefficients. The solution is easy if the ratio $\eta(t)/D(t) = \tau$ is assumed to remain constant. In that case, (A.35) reduces to an equation with constant coefficients,

$$\dot{\varepsilon}(t) + \tau\ddot{\varepsilon}(t) = 0 \qquad (A.36)$$

which has the general solution

$$\varepsilon(t) = C_1 + C_2 e^{-t/\tau} \qquad (A.37)$$

The integration constants C_1 and C_2 must be determined from two initial conditions. For a Kelvin model, the strain cannot change by a jump, even if the stress suddenly rises from 0 to $\hat{\sigma}$ at time t'. One initial condition thus is $\varepsilon(t') = 0$. The other conditions follow from the fact that the initial stress in the spring is zero, and the applied stress $\hat{\sigma}$ must be, after its application, fully equilibrated by the viscous stress in the dashpot, $\sigma_v(t')$. Since $\sigma_v(t') = \eta(t')\dot{\varepsilon}(t') = \tau D(t')\dot{\varepsilon}(t')$, the second initial condition is $\dot{\varepsilon}(t') = \hat{\sigma}/[\tau D(t')]$. After the evaluation of the integration constants, we obtain the particular solution

$$\varepsilon(t) = \frac{\hat{\sigma}}{D(t')}\left[1 - e^{-(t-t')/\tau}\right] \qquad (A.38)$$

and the compliance function of the aging Kelvin model,

$$J(t, t') = \frac{1 - e^{-(t-t')/\tau}}{D(t')} H(t - t') \tag{A.39}$$

Finally, the compliance function of an aging Kelvin chain (with an added aging elastic spring),

$$J(t, t') = \left[\frac{1}{E_0(t')} + \sum_{\mu=1}^{M} \frac{1 - e^{-(t-t')/\tau_\mu}}{D_\mu(t')} \right] H(t - t') \tag{A.40}$$

is a generalized form of the Dirichlet series (A.25). The added spring reflects the instantaneous compliance, and its stiffness is generally considered to be a function of t'.

Recall that formula (A.39) has been derived for the special case when the ratio between the time-dependent coefficients $D(t)$ and $\eta(t)$ in (A.35) is constant. In a general case, the compliance function of an aging Kelvin model derived directly from (A.34) can be presented in the integral form

$$J(t, t') = \int_{t'}^{t} \exp\left(-\int_{t'}^{s} \frac{E(r)}{\eta(r)} dr \right) \frac{ds}{\eta(s)} H(t - t') \tag{A.41}$$

For the discussion of the nondivergence condition (9.32) in Sect. 9.6, it is useful to work out the expression for the mixed second derivative of the compliance function. Differentiation of (A.41) leads to the following formulae valid for $t > t'$:

$$\frac{\partial J(t, t')}{\partial t} = \frac{1}{\eta(t)} \exp\left(-\int_{t'}^{t} \frac{E(r)}{\eta(r)} dr \right) \tag{A.42}$$

$$\frac{\partial^2 J(t, t')}{\partial t \, \partial t'} = \frac{E(t')}{\eta(t)\eta(t')} \exp\left(-\int_{t'}^{t} \frac{E(r)}{\eta(r)} dr \right) \tag{A.43}$$

Note that if the modulus E is nonnegative and the viscosity η is positive, the mixed second derivative (A.43) cannot be negative. This means that an aging Kelvin chain with positive moduli and viscosities always satisfies the nondivergence condition (9.32).

A.5 Solidifying Rheologic Chains

A.5.1 Solidifying Kelvin Chain

To obtain the closed-form expression (A.39) for the compliance function of the aging Kelvin model, we had to make the assumption that the ratio $D(t)/\eta(t)$ remains constant. However, $D(t)$ is not equal to the incremental stiffness of the aging spring,

$E(t)$; it is also affected by the evolution of the viscosity, according to the formula $D(t) = E(t) + \dot{\eta}(t)$. So the assumption that $D(t)$ grows proportionally to $\eta(t)$ seems to be quite artificial. Indeed, it is hard to imagine a physical reason why the evolution of stiffness and viscosity should satisfy the relation $E(t) = \eta(t)/\tau - \dot{\eta}(t)$ with $\tau = $ const. On the other hand, the assumption used in the derivation of relaxation function for an aging Maxwell model in Sect. A.4.1, namely that the stiffness grows proportionally to the viscosity, is much more natural. It can be justified by the concept of solidification, thoroughly discussed in Chap. 9.

According to the solidification theory, the changes of apparent rheological properties are due to solidification of a nonaging constituent, e.g., due to the hydration reaction in concrete, which leads to the formation of cement gel. If $v(t)$ is a function describing the growth of the relative volume of the solidified material, starting from $v(0) = 0$ and approaching a limit value $v^{(\infty)}$ as $t \to \infty$, it is logical to expect that the elastic stiffness and the viscosity are both proportional to $v(t)$, and thus, their ratio remains constant.

It is interesting to check whether the compliance function of the aging Kelvin model can be evaluated under the assumption that $E(t) = E^{(\infty)}v(t)$ and $\eta(t) = \eta^{(\infty)}v(t)$, where $E^{(\infty)}v^{(\infty)}$ and $\eta^{(\infty)}v^{(\infty)}$ represent the final values of elastic stiffness and viscosity at full solidification. Substituting these expressions into (A.34) and setting $\dot{\sigma}(t) = 0$ (since we consider the creep test at constant stress level $\hat{\sigma}$), we obtain

$$E^{(\infty)}v(t)\dot{\varepsilon}(t) + \frac{\mathrm{d}}{\mathrm{d}t}\left[\eta^{(\infty)}v(t)\dot{\varepsilon}(t)\right] = 0 \qquad (A.44)$$

In the standard derivation, originally presented by Bažant and Prasannan [179], the product $v(t)\dot{\varepsilon}(t)$ was denoted as a new variable, $\gamma(t)$, which can be interpreted as the rate of strain in the nonaging constituent. Here, we will use a somewhat different notation, with $\eta^{(\infty)}v(t)\dot{\varepsilon}(t)$ denoted as $\sigma_{\mathrm{v}}(t)$ and interpreted as the stress transmitted by the viscous dashpot (the difference is only formal, since $\sigma_{\mathrm{v}}(t) = \eta^{(\infty)}\gamma(t)$, where $\eta^{(\infty)}$ is a constant). The second term in (A.44) can simply be written as $\dot{\sigma}_{\mathrm{v}}(t)$. The first term, $E^{(\infty)}v(t)\dot{\varepsilon}(t)$, corresponds to the rate of elastic stress and can be expressed as $(E^{(\infty)}/\eta^{(\infty)})\sigma_{\mathrm{v}}(t)$. Recognizing $E^{(\infty)}/\eta^{(\infty)}$ as the reciprocal value of the retardation time τ, we can rewrite Eq. (A.44) as

$$\frac{\sigma_{\mathrm{v}}(t)}{\tau} + \dot{\sigma}_{\mathrm{v}}(t) = 0 \qquad (A.45)$$

This is a first-order differential equation with constant coefficients, which is easy to solve. Imposing the initial condition $\sigma_{\mathrm{v}}(t') = \hat{\sigma}$ justified by the fact that immediately after application of stress $\hat{\sigma}$ at time t', the elastic spring is unstretched and the entire stress must be transmitted by the viscous dashpot, we obtain the particular solution

$$\sigma_{\mathrm{v}}(t) = \hat{\sigma}\mathrm{e}^{-(t-t')/\tau} \qquad (A.46)$$

So the evolution of viscous stress in the creep test can be expressed in a closed form, but we still need to evaluate the strain, by integrating the strain rate $\dot{\varepsilon}(t) = \sigma_v(t)/\eta(t) = \sigma_v(t)/(\tau E^{(\infty)}v(t))$ while taking into account the initial condition $\varepsilon(t') = 0$. Since $v(t)$ is in the present general setting an arbitrary function, the integration cannot be performed analytically, but we can formally write the result as

$$\varepsilon(t) = \int_{t'}^{t} \frac{\sigma_v(s)}{\eta(s)}\, ds = \frac{\hat{\sigma}}{\tau E^{(\infty)}} \int_{t'}^{t} \frac{e^{-(s-t')/\tau}}{v(s)}\, ds \qquad (A.47)$$

The last integral scaled by $1/\tau E^{(\infty)}$ obviously represents the compliance function of the solidifying Kelvin model.[3] Instead of specifying the compliance function by such an integral, it is simpler to provide a formula for the rate of compliance, i.e., the derivative of the compliance function with respect to its first argument (current time). The rate of compliance is equal to the strain rate in the creep test divided by the stress level $\hat{\sigma}$, and so it is given by

$$\dot{J}(t, t') = \frac{e^{-(t-t')/\tau}}{\eta(t)} = \frac{e^{-(t-t')/\tau}}{\tau E^{(\infty)}v(t)}, \qquad t \geq t' \qquad (A.48)$$

As usual, the compliance of a solidifying Kelvin chain is obtained by summing the compliances of individual solidifying Kelvin units. It is reasonable to assume that these units have different retardation times τ_μ and final moduli $E_\mu^{(\infty)}$ but that the evolution of the moduli and viscosities is governed by a single function $v(t)$ that characterizes the solidification process. The resulting expression for the rate of compliance is

$$\dot{J}(t, t') = \frac{1}{v(t)} \sum_{\mu=1}^{M} \frac{e^{-(t-t')/\tau_\mu}}{\tau_\mu E_\mu^{(\infty)}}, \qquad t \geq t' \qquad (A.49)$$

If $v(t) = 1$ for all $t \geq t'$, formula (A.49) must reduce to the rate of compliance of a nonaging Kelvin chain with partial moduli $E_\mu^{(\infty)}$ and retardation times τ_μ. Indeed, the sum in (A.49) represents the time derivative of a function given by

$$\Phi(t - t') = \sum_{\mu=1}^{M} \frac{1 - e^{-(t-t')/\tau_\mu}}{E_\mu^{(\infty)}} H(t - t') \qquad (A.50)$$

which can be interpreted as the compliance function of a Kelvin chain representing the nonaging constituent. Consequently, the rate of compliance of the solidifying Kelvin chain can be written as

---

[3]The expression for the compliance function that follows from (A.47) can also be derived from the general formula (A.41) for an aging Kelvin unit if $E(r)/\eta(r)$ is replaced by $1/\tau$ and $\eta(s)$ by $\tau E^{(\infty)}v(s)$.

$$\dot{J}(t, t') = \frac{\dot{\Phi}(t - t')}{v(t)} \tag{A.51}$$

and the compliance function

$$J(t, t') = \frac{1}{E_0(t')} + \int_{t'}^{t} \frac{\dot{\Phi}(s - t')}{v(s)} \, ds, \qquad t \geq t' \tag{A.52}$$

is obtained by integration. Note that since the integration is performed with respect to the first argument, t, the integration "constant" can be a general function of the second argument, t'. It is represented by the first term in (A.52), denoted as $1/E_0(t')$, where $E_0(t')$ has the meaning of (asymptotic) elastic modulus at age t'. The specific version of solidification theory incorporated into model B3 considers this term as a constant, $1/E_0$, corresponding to a load duration much shorter than any of practical interest. The constancy of E_0 is motivated by the fitting of experimental data but is not a physical requirement. In general, the term $1/E_0(t')$ corresponds to the compliance function of an aging spring (e.g., of a Kelvin unit with a dashpot of zero viscosity).

A.5.2 Solidifying Maxwell Chain

Expression (A.33) for the relaxation function of an aging Maxwell chain was derived in Sect. A.4.1 under the assumption that, for each unit, the ratio between the age-dependent viscosity and the age-dependent modulus, $\eta_\mu(t)/E_\mu(t) = \tau_\mu$, remains constant in time. This assumption is fully consistent with the solidification theory, which postulates that $\eta_\mu(t) = \eta_\mu^{(\infty)} v(t)$ and $E_\mu(t) = E_\mu^{(\infty)} v(t)$, where $\eta_\mu^{(\infty)}$ and $E_\mu^{(\infty)}$ are the time-independent properties of the solidifying constituent and $v(t)$ is a monotonic function evolving between 0 and $v^{(\infty)}$, which characterizes the solidification process and is the same for all units of the chain. Replacing $E_\mu(t')$ in (A.33) by $E_\mu^{(\infty)} v(t')$, we realize that the relaxation function of a solidifying Maxwell chain can be written as

$$R(t, t') = v(t') \Psi(t - t') \tag{A.53}$$

where

$$\Psi(t - t') = \left(\sum_{\mu=1}^{M} E_\mu^{(\infty)} e^{-(t-t')/\tau_\mu} \right) H(t - t') \tag{A.54}$$

is the relaxation function of the nonaging constituent.

It turns out that the solidification theory leads to a simpler expression for the relaxation function than for the compliance function; cf. Eqs. (A.53) and (A.52). The reason is that relaxation takes place at constant strain and the newly deposited solidified material remains unstrained and thus also unstressed. Therefore, in a test starting at time t', relaxation takes place in the material of relative volume $v(t')$ and follows the curve corresponding to the nonaging constituent, only scaled by $v(t')$. On the other hand, in a creep test, the strain increases and the solidifying material deposited after the start of the test get gradually activated. An increment of strain at time t activates stress in the already solidified material occupying relative volume $v(t)$, and this is why the expression for the compliance rate has the simple form (A.51). The compliance must be obtained by integration, which complicates the resulting expression (A.52). However, the values of compliance function are not really needed if the numerical solution is based on the rate-type approach, which leads to efficient numerical schemes; see Sect. 5.2.

Appendix B
Historical Note on Old Creep Models

Up to the 1970s, the quest for simplicity led to the formulation of several competing models based on various simplifications of the compliance function $J(t, t')$. The errors of these simplifications became gradually apparent as data from creep tests of longer durations became available. Here, we briefly review several old models, which were all rendered obsolete by the discovery of the age-adjusted effective modulus method, which is much more accurate than any one of them, yet (with the exception of the effective modulus method) allows simpler structural creep analysis. Knowing these methods is nevertheless useful for understanding the older literature on concrete creep. Besides, these methods might still be instructive for clarifying some aspects of the creep effects, if not their magnitudes.

1. Effective Modulus Method. Proposed by McMillan [622, 623], this was the first method to calculate creep and shrinkage effect (already mentioned in Chap. 4). This method uses the elastic analysis with effective modulus $E_{ef} = 1/J(t, t_1)$, where t_1 = age at the first loading, which is an approach widely used in practice for viscoelasticity of polymers. However, it gives good estimates only for stresses that change little after the first loading, or for durations less than the age at loading. In the plot of compliance J versus load duration $t - t'$, this methods implies the curves for various ages t' at loading to be identical, which is far from reality and overestimates the creep for high t'. The drop of stress in relaxation problems and the reduction of shrinkage stresses by creep are significantly underestimated.

2. Rate-of-Creep Method (Dischinger and Glanville). To capture the effects of nonsteady loads, imposed displacements, and changes of structural system introduced at later ages, a simplified creep law in the form of an easily tractable differential equation was desired. This goal was facilitated by Whitney's [865] hypothesis that the curves of compliance J versus current age t for various ages at loading t' can be considered to be approximately parallel and vertically shifted relative to each other (Fig. B.1). This simplifying hypothesis, which is acceptable only for short load durations and inevitably implies far too little creep for high t', means that

© Springer Science+Business Media B.V. 2018
Z.P. Bažant and M. Jirásek, *Creep and Hygrothermal Effects in Concrete Structures*, Solid Mechanics and Its Applications 225,
https://doi.org/10.1007/978-94-024-1138-6

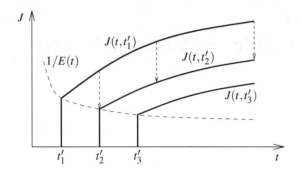

Fig. B.1 Vertically shifted compliance curves

$$J(t, t') = \frac{1}{E(t')} + \frac{\psi(t) - \psi(t')}{E_{\text{ref}}} \tag{B.1}$$

where E_{ref} is a chosen reference value of the elastic modulus, usually taken as E_{28} (i.e., as $E(t')$ for $t' = 28$ days), and ψ is an empirical dimensionless function, playing the role of a generalized creep coefficient. Substituting (B.1) into the integral expression (2.17) for the strain rate, one finds that

$$\dot{\varepsilon}(t) = \frac{\dot{\sigma}(t)}{E(t)} + \int_0^t \frac{\dot{\psi}(t)}{E_{\text{ref}}} \, d\sigma(t') = \frac{\dot{\sigma}(t)}{E(t)} + \frac{\dot{\psi}(t)}{E_{\text{ref}}} \sigma(t) \tag{B.2}$$

which may be rewritten as

$$\dot{\varepsilon} = \frac{1}{E}\dot{\sigma} + \frac{\sigma}{E_{\text{ref}}}\dot{\psi} \tag{B.3}$$

This ordinary differential equation, known as the *theory of aging* or *Dischinger's method*, was first proposed by Glanville [426, 427] and was extensively used in Europe until the 1960s to analyze various structural creep problems, particularly by Dischinger [353, 354], Glanville and Thomas [428], Sattler [755], Ulickii et al. [821], Bažant [61–66] and many other European engineers; cf. [66].

The following expressions were used: $\psi(t) = \psi_\infty(1 - e^{-at})$ [353] and $\psi_\infty\sqrt{1 - e^{-\sqrt{at}}}$, where $a = $ constant. Parameter ψ_∞ was expressed by Ullickii (and also in Soviet design standard) as a product of several coefficients taking empirically into account the type of concrete, the age at loading, the average environmental humidity, and the thickness of cross section, in a way similar as later introduced by Branson and Christiason [253] and adopted for the ACI-209 recommendation (although without a recommendation of the rate-of-creep method).

Note that (B.3) corresponds to modeling creep by a single aging Maxwell unit with spring constant $E(t)$ and dashpot viscosity $\eta(t) = E_{\text{ref}}/\dot{\psi}(t)$. Obviously, such a model is incapable of representing creep recovery. The creep due to stress increments at higher ages is severely underestimated. The loss of stress in relaxation tests is overestimated, and the shrinkage stress or buckling deflection are underestimated.

3. Compromise 1: Using Both Rate-of-Creep Method and Effective Modulus. The errors of the aforementioned two methods are usually of opposite signs, mutually compensating. Thus, a better estimate is obtained by averaging the results for both methods. A relatively safe approach to design is to solve the structure by both the effective modulus and the rate-of-creep method and make the design satisfy both solutions.

4. Compromise 2: Rate-of-Flow Method (Improved Dischinger). Another compromise was proposed by England and Illston [369], was further refined by Nielsen [658], Rüsch et al. [746], and Rüsch and Jungwirth [745], cf. Bažant and Thonguthai [187], and was also used in the CEB Model Code. The creep strain was subdivided into a reversible part, called delayed elasticity, and an irreversible part, called flow. The compliance function was assumed in the form

$$J(t, t') = \frac{f(t - t')}{E(t')} + \frac{\psi(t) - \psi(t')}{E_{28}} \tag{B.4}$$

The first term, representing delayed elasticity, was assumed to be short-lived, reaching within an initial period of a few months the final value of $1/E_{ef}^*(t') = f(\infty)/E(t')$, taken about 40% larger than the initial elastic compliance. It was also assumed that only the creep effects after the initial period needed to be calculated. Thus, for long-term analysis of structural creep effects, (B.4) was replaced by

$$J(t, t') = \frac{1}{E_{ef}^*(t')} + \frac{\psi(t) - \psi(t')}{E_{28}} \tag{B.5}$$

This equation is formally identical to Eq. (B.1) underlying the rate-of-creep method, and so the structural effects of creep and shrinkage may again be calculated according to a first-order differential equation analogous to (B.3). Although the results for creep over a few years are more realistic than with the rate-of-creep method, the creep for a few months as well as for more than a few years is not captured. There are significant deviations from the principle of superposition, much greater than with the age-adjusted effective modulus method [167]. Optimizations of the fits of various experimental compliance functions $J(t, t')$ according to (B.4) showed that good fits of test data are unattainable [187]. The errors of the approach were discussed in Bažant and Osman [172] and three subsequent discussions and replies.

5. Maslov–Arutyunyan's Method. Proposed by Maslov [609], developed by Arutyunian [39] for solving structural problems and promulgated by Gvozdev [442, 443], this method is based on assuming the compliance function to be of the form

$$J(t, t') = \frac{1}{E(t')} + \left(A + \frac{C}{t'}\right)\left(1 - e^{-(t-t')/\tau_1}\right) \tag{B.6}$$

where A, C, and τ_1 are empirical constants. In contrast to the rate-of-creep method, this compliance function can exhibit a nonzero creep for loads applied at very high ages and exhibits nonzero recovery. In contrast to the effective modulus method, it includes the effect of the age at loading t'. Writing the first and second derivatives of the superposition integral, one finds the same integral to be repeated in both. Eliminating it, one finds that (B.6) is equivalent to the following linear ordinary second-order differential equation with variable coefficients [66, 67]:

$$E(t)\ddot{\varepsilon}(t) + a(t)\dot{\varepsilon}(t) = \ddot{\sigma}(t) + b(t)\dot{\sigma}(t) \tag{B.7}$$

An attractive feature of this stress–strain relation was that various simple structural creep problems with a single unknown variable could be solved analytically in terms of the incomplete gamma function [39], though not as easily as with the rate-of-creep method. However, comparisons with long-term creep tests showed poor agreement when plotted in the logarithmic, rather than linear, time scale. This limitation is due to the fact that the exponential in (B.6) and the function $A + C/t'$ approach a constant too rapidly. If this function is replaced by another that decays more slowly, analytical solutions become impossible. This limitation is so severe that, for load durations over 15 years, the Maslov–Arutyunyan method is even worse than the classical effective modulus method and the rate-of-creep method [167, 187].

6. Trost Method. A major departure from all the classical simplified models, which went in the right direction and can be regarded as a predecessor of the age-adjusted effective modulus method, was the Trost method [816]. Like the effective modulus method, it made it possible to use the compliance function $J(t, t')$ as measured, with no simplifications. The Trost method used a so-called relaxation coefficient, derived from $J(t, t')$ semiempirically. It did not take into account the aging of elastic modulus E and expressed the incremental Young modulus for the time period from t' to t simply by $E'' = E_{28}/[1 + \rho\varphi(t, t')]$, where $\varphi(t, t') = E_{28}J(t, t') - 1$ is the creep coefficient and ρ is Trost's empirical relaxation coefficient (typically fixed as 0.8). A simple replacement of E_{28} by $E(t')$ and of ρ by the aging coefficient χ calculated from $J(t, t')$, as proposed by Bažant [76] and presented in (4.55), ensures exact results according to the principle of superposition for a broad range of strain histories and provides simple yet accurate approximations for many practical problems. The reason for changing the name "relaxation" coefficient to "aging" coefficient was the finding that, in absence of aging, the value of this coefficient should be close to 1, including the case of stress relaxation.

Appendix C
Estimates of Parameters
Used By RILEM Model B3

Model B3 [104, 107] covers creep and shrinkage of concrete, including their coupling. The effects of drying (i.e., drying shrinkage and drying creep) can be taken into account at two different levels of accuracy and sophistication:

1. The *"sectional" approach* only takes into account the mean effects of drying averaged over the cross section of a beam or slab. The environmental humidity must be given, but the time evolution of the pore humidity distribution throughout the structure is not solved. The influence of the cross-sectional size and shape on the kinetics of drying is incorporated approximately, through semiempirical coefficients. The self-equilibrated stresses caused by nonuniform drying cannot be captured. The sectional approach was calibrated by tests under centric axial loading and works for such loading better than it does for flexural loading. In the presence of a large bending moment, this approach is useful only for simple engineering applications. Note that the flexural moments are always small in the elements of walls of box girders because the longitudinal stress resultant in these elements is always nearly centric.

2. The *"material" approach* takes into account the time evolution of the distribution of pore humidity throughout the structure, which must be obtained by solving a nonlinear diffusion equation, with the environmental humidity providing the boundary conditions. Once the evolution of pore humidity distribution is known, its effect on the constitutive behavior of the material can be evaluated from a physical law that links the increments of shrinkage strains to the changes of pore humidity. This approach is useful for more accurate (and more demanding) analyses of drying-sensitive structures.

C.1 Sectional Approach

As explained in Chap. 3, the sectional approach defines the compliance function of concrete in the form

© Springer Science+Business Media B.V. 2018
Z.P. Bažant and M. Jirásek, *Creep and Hygrothermal Effects in Concrete Structures*, Solid Mechanics and Its Applications 225,
https://doi.org/10.1007/978-94-024-1138-6

$$J(t, t') = q_1 + q_2 Q(t, t') + q_3 \ln[1 + (t - t')^n] + q_4 \ln\left(\frac{t}{t'}\right) + J_d(t, t') \quad (C.1)$$

where $q_1 = 1/E_0$ is the inverse of the asymptotic elastic modulus (cf. Sect. 3.2), the terms containing q_2, q_3, and q_4 represent the aging viscoelastic compliance, nonaging viscoelastic compliance, and flow compliance, respectively (Sect. 3.3), and $J_d(t, t')$ is the additional compliance due to drying, which is also influenced by the time t_0 at the beginning of drying (Sect. 3.6).

In the definition of the compliance function (C.1), times t and t' must be in days, q_1, q_2, q_3, and q_4 are empirical constitutive parameters that depend on the specific type of concrete, and function $Q(t, t')$ is defined by expression (3.12) in which $m = 0.5$ and $n = 0.1$ are empirical parameters whose values can be taken the same for all normal concretes. The values of $Q(t, t')$ can be calculated by numerical evaluation of the integral in (3.12) or by interpolation from a table given in Bažant and Baweja [104], or evaluated (for $n = 0.1$ and $m = 0.5$ with an error under 1%) using the approximate explicit formula

$$Q(t, t') = Q_f(t') \left[1 + \left(\frac{Q_f(t')}{Z(t, t')}\right)^{r(t')}\right]^{-1/r(t')} \quad (C.2)$$

in which

$$r(t') = 1.7(t')^{0.12} + 8 \quad (C.3)$$

$$Z(t, t') = (t')^{-m} \ln[1 + (t - t')^n] \quad (C.4)$$

$$Q_f(t') = [0.086(t')^{2/9} + 1.21(t')^{4/9}]^{-1} \quad (C.5)$$

If the sectional approach is used, the average drying shrinkage strain in a cross section, ε_{sh}, is estimated using formulae (3.15)–(3.19) given in Sect. 3.5, and the drying creep compliance J_d is estimated using formulae (3.20) and (3.23)–(3.24) given in Sect. 3.6. For the reader's convenience, we reproduce here the main equations in a slightly more compact format:

$$\varepsilon_{sh}(t) = -\varepsilon_{sh}^\infty \left(1 - h_{env}^3\right) S(t - t_0) \quad (C.6)$$

$$J_d(t, t') = q_5 \sqrt{e^{-g(t-t_0)} - e^{-g(t'-t_0)}} \quad (C.7)$$

with the auxiliary functions given by

$$S(\hat{t}) = \tanh\sqrt{\frac{\hat{t}}{\tau_{sh}}}, \quad \tau_{sh} = k_t (k_s D)^2 \quad (C.8)$$

$$g(\hat{t}) = 8\left[1 - (1 - h_{env})S(\hat{t})\right] \quad (C.9)$$

In the above, h_{env} is the environmental relative humidity, t_0 is the age of concrete at the end of curing, D is the equivalent thickness of the concrete member, and k_s is a correction factor that takes into account the shape of the member (see Table 3.1). Parameters $\varepsilon_{sh}^{\infty}$, k_t, and q_5 need to be determined by fitting of experimental data or estimated from composition.

C.2   Prediction of Model Parameters

Ideally, the material parameters should be determined from long-term creep and shrinkage tests performed on the specific concrete that will be used in the particular structure. In practical applications, it is impossible to wait with the analysis until such tests are finished. The parameters can be crudely estimated from empirical formulae that have been established by fitting a large experimental database for various types of concrete and correlating the obtained model parameters to the concrete strength and composition. The high uncertainties in predicting creep and shrinkage from the basic characteristics of concrete can be drastically reduced by statistical updating of the model based on short-time tests of creep and shrinkage [104, 107], as explained in Sect. 3.8 and Chap. 6. The shrinkage tests must be accompanied by the measurements of water loss due to drying; see Sect. H.1.

Table C.1 Concrete properties serving as input data

Property	Symbol	in-lb units	SI units
Mean compression strength	$\bar{f}_c$	psi	MPa
Water content	w	lb/ft^3	kg/m^3
Cement content	c	lb/ft^3	kg/m^3
Aggregate content	a	lb/ft^3	kg/m^3

Since the coefficients that appear in the approximate formulae are not dimensionless, their values depend on the choice of the system of units. We present here alternative formulae valid in inch-pound (in-lb) system units and in SI (metric) units. The entire parameter evaluation must be done in one selected system of units. The age of concrete at the onset of drying, t_0, is always given in days.

The required input data specifying the strength and composition of concrete are summarized in Table C.1. The compression strength $\bar{f}_c$ should be determined as the statistical average of test results on cylinders of diameter 15 cm (6 in.) and length 30 cm (12 in.) at age 28 days.[4]

---

[4]Note that design codes deal with a certain "safely estimated" strength value, which is significantly lower than the mean. The CEB code uses the so-called characteristic strength, f_{ck}, while the ACI code uses the specified design strength, f_c'; see Appendix E.1 for details.

The content of water, cement, and aggregates is the mass of the component per unit volume of concrete mix. If the water and cement contents to be used have not yet been decided, they may be estimated from the empirical correlation of the water/cement ratio to the required design strength (mean compressive cylinder strength after 28 days) of concrete [653],

$$\frac{w}{c} = \left(\frac{\bar{f}_c}{f_{ref}} + 0.535 \right)^{-1} \tag{C.10}$$

where the reference strength $f_{ref} = 22.8\,\text{MPa} = 3307\,\text{psi}$ corresponds to $w/c = 0.65$.

The formulae for predicting the creep and shrinkage parameters are presented in Table C.2. Knowing that the formulae are empirical and the scatter of the experimental results is quite high, it may seem strange that some coefficients in the table are specified with a five-digit accuracy. From the practical point of view, truncation of these coefficients to three or even two valid digits is fully acceptable. Nevertheless, using the "exact" values might be useful for comparison of the results during benchmarking of a certain computational tool and for detection of potential implementation errors. For instance, the formula for q_1 has its origin in the simple relation between the asymptotic modulus and conventional modulus, $E_0 = E_{28}/0.6$. Since the conventional modulus is estimated from the compression strength using the ACI formula (3.5), the resulting relation between parameter q_1 (inverse of the asymptotic modulus) and the compression strength is

$$q_1 = \frac{1}{E_0} = \frac{0.6}{E_{28}} = \frac{0.6}{57\,\text{ksi}} \sqrt{\frac{1\text{psi}}{\bar{f}_c}} \tag{C.11}$$

If $\bar{f}_c$ is substituted in psi and q_1 is expected to be evaluated in $10^{-6}/\text{psi}$, the final formula is written as

$$q_1 = 10^3 \frac{0.6}{57} \bar{f}_c^{-0.5} = 10.526 \bar{f}_c^{-0.5} \tag{C.12}$$

After accurate conversion to SI units, we obtain the coefficient 126.77. This can be truncated to 127, but if 10.526 is first truncated to 10.5 and then converted to SI units, it gives 126.45 and after truncation 126. For any practical purpose, the difference below 1% is of course negligible, but when one compares two design aids or computational programs for creep and finds a difference in the order of 1% even though the material parameters (in this case compression strength) are exactly the same, one may suspect that there is a hidden bug that could lead to larger deviations in other cases. The "accurate" coefficients permit a reliable comparison even in cases when the programs work with different systems of units.

Table C.2 Prediction formulae for creep parameters according to the B3 model

Line	Phenomenon	Parameter	Inch-pound formula	Units	SI formula	Units
1	Basic creep	q_1	$10.526\,\bar{f}_c^{-0.5}$	$10^{-6}/\text{psi}$	$126.77\,\bar{f}_c^{-0.5}$	$10^{-6}/\text{MPa}$
2		q_2	$451.1\,c^{0.5}\,\bar{f}_c^{-0.9}$	$10^{-6}/\text{psi}$	$185.4\,c^{0.5}\,\bar{f}_c^{-0.9}$	$10^{-6}/\text{MPa}$
3		q_3	$0.29(w/c)^4 q_2$	$10^{-6}/\text{psi}$	$0.29(w/c)^4 q_2$	$10^{-6}/\text{MPa}$
4		q_4	$0.14(a/c)^{-0.7}$	$10^{-6}/\text{psi}$	$20.3(a/c)^{-0.7}$	$10^{-6}/\text{MPa}$
5		k_t	$190.8\,t_0^{-0.08\,p_g}\,\bar{f}_c^{-1/4}$	days/in^2	$0.085\,t_0^{-0.08}\,\bar{f}_c^{-1/4}$	days/mm^2
6	Shrinkage	ε_s^∞	$\alpha_1\alpha_2\left(26\,w^{2.1}\,\bar{f}_c^{-0.23}+270\right)$	10^{-6}	$\alpha_1\alpha_2\left(0.019\,w^{2.1}\,\bar{f}_c^{-0.28}+270\right)$	10^{-6}
7		$\varepsilon_{\text{sh}}^\infty$	$\varepsilon_s^\infty \times 0.57514\sqrt{3+14/(t_0+\tau_{\text{sh}})}$	10^{-6}	$\varepsilon_s^\infty \times 0.57514\sqrt{3+14/(t_0+\tau_{\text{sh}})}$	10^{-6}
8	Drying creep	q_5	$7.57\times10^5\,\bar{f}_c^{-1}\left(\varepsilon_{\text{sh}}^\infty\right)^{-0.6}$	$10^{-6}/\text{psi}$	$7.57\times10^5\,\bar{f}_c^{-1}\left(\varepsilon_{\text{sh}}^\infty\right)^{-0.6}$	$10^{-6}/\text{MPa}$

The formula in line 7 of Table C.2 originally comes from

$$\varepsilon_{sh}^{\infty} = \varepsilon_{s}^{\infty} \frac{E(7 + 600)}{E(t_0 + \tau_{sh})} \tag{C.13}$$

The fraction on the right-hand side is a correction for aging, because the final value of shrinkage depends not only on the material properties, but also on the interplay between hardening and drying. If the drying process starts later and takes more time, concrete becomes stiffer and its shrinkage is reduced. To take that into account, a reference value ε_{s}^{∞} is defined as the final shrinkage strain exhibited by the given concrete if drying starts at 7 days and shrinkage halftime is 600 days. This value is considered as dependent only on material properties and type of curing and is estimated using the formula in line 6. Under general conditions, it is modified by the above-mentioned factor, defined as the ratio between the elastic modulus at 607 days (i.e., somewhere "in the middle" of the reference drying process) and the elastic modulus at $t_0 + \tau_{sh}$ (i.e., somewhere "in the middle" of the actual drying process). The dependence of elastic modulus on age is then estimated using the adjusted ACI formula (E.29),

$$E(t) = E_{28} \sqrt{\frac{7t}{28 + 6t}} \tag{C.14}$$

which leads to

$$\frac{E(7 + 600)}{E(t_0 + \tau_{sh})} = \sqrt{\frac{7 \times 607}{28 + 6 \times 607}} \sqrt{\frac{28 + 6(t_0 + \tau_{sh})}{7(t_0 + \tau_{sh})}} = 0.57514 \sqrt{\frac{14}{t_0 + \tau_{sh}} + 3} \tag{C.15}$$

Recall that parameter k_t is used in the second part of formula (C.8), which specifies the shrinkage halftime, τ_{sh}. Coefficients α_1 and α_2 that appear in the formulae for the reference shrinkage strain ε_{s}^{∞} in line 6 of Table C.2 are defined as

$$\alpha_1 = \begin{cases} 1.0 & \text{for type I cement} \\ 0.85 & \text{for type II cement} \\ 1.1 & \text{for type III cement} \end{cases} \tag{C.16}$$

$$\alpha_2 = \begin{cases} 0.75 & \text{for steam curing} \\ 1.2 & \text{for sealed or normal curing in air with initial protection} \\ & \quad \text{against drying} \\ 1.0 & \text{for curing in water or at 100\% relative humidity} \end{cases} \tag{C.17}$$

If the specific information is not available, the following reasonable default values can be used: type I cement ($\alpha_1 = 1.0$), curing in air with initial protection ($\alpha_2 = 1.2$). The types of cement are understood here according to the American classification (ASTM C 150-07: Specification for Portland Cement). Type I is ordinary Portland cement, type II is modified cement, type III is rapid-hardening Portland cement, type IV is low-heat Portland cement, and type V is sulfate-resisting Portland cement.

C.3 Material Approach

Creep-sensitive structures are analyzed by layered beam finite element programs or by two- and three-dimensional finite element programs. In such programs, the material properties used at each material point must be the constitutive properties, independent of the cross-sectional dimensions and shape, as well as of the environmental conditions, which represent the boundary conditions of the partial differential equations for drying and heat conduction (see Chaps. 8 and 13). At drying, the constitutive properties cannot be measured directly, but they have been identified by fitting with a finite element program the overall deformation measurements on test specimens [126, 130]. For this kind of analysis, the compliance function of concrete has the form

$$J(t,t') = q_1 + q_2 Q(t,t') + q_3 \ln[1 + (t - t')^n] + q_4 \ln\left(\frac{t}{t'}\right) \qquad (C.18)$$

which means that it only includes the instantaneous response and the basic creep. The additional compliance $J_d(t, t')$ that appears in Eq. (C.1) used by the sectional approach is now deleted, and the effects of drying creep are incorporated either in the viscous flow strain in the spirit of the microprestress theory [131, 132], or in the shrinkage strain, ε_s [117, 120].

The former approach is discussed in detail in Chap. 10. According to the latter approach, presented in Sect. 13.3.3.2, the evolution of the pointwise defined shrinkage strain (which is now different from the average drying shrinkage strain in a cross section, ε_{sh}) is governed by the following three-dimensional rate-type constitutive relation:

$$\dot{\varepsilon}_{s,ij} = -\varepsilon_s^\infty \frac{E(t_0)}{E(t_e)} \frac{dk_h(h)}{dh} [\delta_{ij} + (r\sigma_{ij} + r'\sigma_m \delta_{ij}) \operatorname{sgn}(\dot{h} + a_T \dot{T})]\dot{h} \qquad (C.19)$$

Here, ε_s^∞ is a positive parameter that corresponds to the theoretical value of free ultimate shrinkage at zero stress extrapolated to zero humidity, t_e is the equivalent age, δ_{ij} is Kronecker delta, σ_{ij} are the stress components, and σ_m is the mean stress. Parameters r and a_T need to be determined from experiments, and parameter r' is usually set to zero. Function $k_h(h) = 1 - h^3$ describes the dependence of the normalized shrinkage strain at zero stress on the pore relative humidity, h. The evolution of pore relative humidity, h, and temperature, T, must be obtained by solving the moisture and heat transport equations. When (C.19) is used, the cracking or fracture must also be included in the analysis.

The constitutive relation (C.19) is simpler than Eqs. (3.15)–(3.20) which it replaces. However, at present, the prediction of the values of parameters from the composition and strength of the concrete is rather uncertain, because of limited data and multitude of influencing factors. So, these parameters must be identified by fitting the measured data for drying creep and shrinkage.

Finally, it should be pointed out that the research under way at the time of writing [125] indicates the need for a significant further improvement of the materials approach, particularly of Eq. (C.19). It will be necessary to include further equations to distinguish humidity changes due to self-desiccation, resulting from the chemical process of hydration, and due to moisture transport, caused, e.g., by external drying. In treating shrinkage, one will need to account separately for intrinsic expansion of the porous skeleton of cement paste due to hydration (which causes swelling under water) from the contraction caused by stress changes in capillary water, free adsorbed layers, and hindered adsorbed layers in nanopores. Such an approach will make possible a rational separation of the autogenous shrinkage from the observed total shrinkage, basic creep and drying creep [124, A5].

Appendix D
Estimates of Parameters Used By RILEM Model B4

Model B4 [136] is an updated and improved version of model B3, which uses the same general form of the creep compliance function (C.1) and shrinkage function (3.15) but different empirical formulae for the estimation of parameters based on the concrete mix composition, member size and shape, curing and environmental conditions. The main improvement is that model B4 takes into account the autogenous shrinkage, which was not separated from drying shrinkage in model B3 and was neglected in the case of no moisture exchange. Another improvement is that the applicability range of model B4 is much broader, covering modern concretes with admixtures, high-strength concretes, and durations of many decades, expectably even a century (which is the desired lifetime of large structures). The extension of applicability range has been made possible by calibration with multidecade bridge data and with a broader range of compositions and concrete strengths; see Hubler et al. [488].

The ranges of various parameters for which model B4 has been calibrated are typical for practice and are as follows:

$$0.22 \leq w/c \leq 0.87, \quad 1.0 \leq a/c \leq 13.2$$
$$2{,}070 \text{ psi} \leq \bar{f}_c \leq 10{,}000 \text{ psi}, \quad 12.5 \text{ lb/ft}^3 \leq c \leq 93.6 \text{ lb/ft}^3 \quad (inch\text{-}pound \; system)$$
$$15 \text{ MPa} \leq \bar{f}_c \leq 70 \text{ MPa}, \quad 200 \text{ kg/m}^3 \leq c \leq 1{,}500 \text{ kg/m}^3 \quad (SI \; system)$$
$$-25\,^{\circ}\text{C} \leq T \leq 75\,^{\circ}\text{C}$$
$$12 \text{ mm} \leq V/S_e \leq 120 \text{ mm}$$

As usual, a, c, and w are the mass of aggregates, cement, and water per unit volume of concrete (see Table C.1), $\bar{f}_c$ is the mean 28-day compression strength measured on cylinders, T is the average temperature of the environment, V is the volume of the concrete member, and S_e is its surface exposed to the ambient humidity h_{env}. The actual applicability range might be even broader than indicated above. For instance, the ratio V/S_e corresponds to one half of slab or wall thickness. The limit of 240 mm in thickness might seem too small, but the data of Hansen and Mattock [453], L'Hermite and Mamillan [577], and Wittmann et al. [878] indicate that the effect of size predicted by diffusion theory and incorporated into model B4 is correct even for much larger thicknesses.

© Springer Science+Business Media B.V. 2018
Z.P. Bažant and M. Jirásek, *Creep and Hygrothermal Effects in Concrete Structures*, Solid Mechanics and Its Applications 225,
https://doi.org/10.1007/978-94-024-1138-6

D.1 Creep

Model B4 uses the same type of compliance function as model B3 but estimates the values of parameters in a different way. The general form of the compliance function is

$$J(t, t') = q_1 + J_b(t, t') + J_d(t, t') \tag{D.1}$$

in which q_1 is the asymptotic compliance (reciprocal value of the asymptotic modulus),

$$J_b(t, t') = q_2 Q(t, t') + q_3 \ln[1 + (t - t')^n] + q_4 \ln \left(\frac{t}{t'} \right) \tag{D.2}$$

accounts for the basic creep, and

$$J_d(t, t') = q_5 \sqrt{\langle e^{-g(t-t_0)} - e^{-g((t'-t_0))} \rangle} \tag{D.3}$$

accounts for the additional creep due to drying.

Recall that the current age t and the age at loading t' should be substituted in days, and the standard value of exponent n is 0.1. The values of function $Q(t, t')$ can be calculated by numerical evaluation of the integral in (3.12), or from the approximate explicit formulae (C.2)–(C.5), with $m = 0.5$ and $n = 0.1$. The angular brackets $\langle \ldots \rangle$ in (D.3) denote the positive part, as explained in the discussion below Eq. (3.21). The form of function g will be specified later; see Eq. (D.8) in Sect. D.1.2.

D.1.1 Basic Creep Compliance

According to model B4, the parameters that describe the asymptotic compliance and basic creep are estimated using the following expressions:

$$q_1 = \frac{p_1}{E_{28}} \tag{D.4}$$

$$q_2 = \frac{p_2}{1 \, \text{GPa}} \left(\frac{w/c}{0.38} \right)^3 \tag{D.5}$$

$$q_3 = p_3 q_2 \left(\frac{a/c}{6} \right)^{-1.1} \left(\frac{w/c}{0.38} \right)^{0.4} \tag{D.6}$$

$$q_4 = \frac{p_4}{1 \, \text{GPa}} \left(\frac{a/c}{6} \right)^{-0.9} \left(\frac{w/c}{0.38} \right)^{2.45} \tag{D.7}$$

The conventional elastic modulus E_{28} can be estimated from the compressive strength $\bar{f}_c$ using standard formulae (3.5) or (3.6). Factors p_i depend on the cement type; their values are given in Table D.1 for three types of cement:

- R = normal,
- RS = rapid hardening, and
- SL = slow hardening.

The European classification of cements is selected for model B4 since it is directly related to the reaction rate of the cement instead of the type of application, which is the basis of other classification systems. It should be noted, though, that the class labels used by B4 are somewhat different from the class labels used in Eurocode 2 (R = rapid hardening, N = normal, and S = slow hardening) and in CEB Model Code 1990 (RS = rapidly hardening high-strength concrete, R = rapid hardening, N = normal, and SL = slow hardening). For cements classified according to ASTM, an approximate correspondence can be made: ASTM Type I general-purpose Portland cement may be assumed as type R reactivity. ASTM Type II is a low-heat cement and may be considered as SL. Type III, high-early-heat cements can be assumed as RS. Types IV, V, Ia, IIa, and IIIa should be mapped by their reactivity to the present classification table, and any admixtures which are part of their composition should be considered based on their proportions. All additives such as fly ash are taken into account by model B4 separately.

Table D.1 Creep parameters depending on cement type according to model B4 (* denotes an assumed value, due to lacking data)

Parameter	R	RS	SL
p_1	0.70	0.60	0.80
p_2	58.6×10^{-3}	17.4×10^{-3}	40.5×10^{-3}
p_3	39.3×10^{-3}	39.3×10^{-3}	39.3×10^{-3}
p_4	3.4×10^{-3}	3.4×10^{-3}	3.4×10^{-3}
p_5	777×10^{-6}	94.6×10^{-6}	496×10^{-6}
p_{5H}	8.00	1.00	8.00*

D.1.2 Additional Compliance Due to Drying

The additional compliance due to drying is given by function J_d specified in (D.3), which is the same as in model B3, cf. Eq. (C.7). However, the auxiliary function

$$g(t - t_0) = p_{5H} \left[1 - (1 - h_{env}) \tanh \sqrt{\frac{t - t_0}{\tau_{sh}}} \right] \tag{D.8}$$

is the same as the B3 function defined in (C.8) and (C.9) only for cement types R and SL, for which $p_{5H} = 8$. For cement type RS (rapidly hardening), parameter p_{5H} is reduced to 1; see the last row of Table D.1. As usual, h_{env} is the environmental relative humidity and t_0 is the age of concrete at the end of curing and at the beginning of drying exposure.

Another difference compared to the B3 model is that the shrinkage halftime is given by an adjusted formula

$$\tau_{\text{sh}} = \tau_0 \, k_{\tau a} \left(k_s \frac{D}{1 \text{ mm}} \right)^2 \tag{D.9}$$

in which $D = 2V/S_e$ is the equivalent thickness of the concrete member. The shape factor k_s is the same as in model B3; see Table 3.1. Factor $k_{\tau a}$ is taken from Table D.2, depending on the aggregate type. Its default value, to be used if the type of aggregate is not known, is $k_{\tau a} = 1$. Factor τ_0 takes into account the composition and is given by

$$\tau_0 = \tau_{cem} \left(\frac{a/c}{6} \right)^{-0.33} \left(\frac{w/c}{0.38} \right)^{p_{\tau w}} \left(\frac{6.5 \, c}{\rho} \right)^{p_{\tau c}} \tag{D.10}$$

where factor τ_{cem} and exponents $p_{\tau w}$ and $p_{\tau c}$ depend on the cement type and are taken from Table D.3. The concrete density ρ (total mass of concrete per unit volume) can be considered by its default value $\rho = 2{,}350 \text{ kg/m}^3$.

Parameter q_5 controlling the magnitude of drying creep is estimated as

$$q_5 = \frac{p_5}{1 \text{ GPa}} \left(\frac{a/c}{6} \right)^{-1} \left(\frac{w/c}{0.38} \right)^{0.78} |k_h \varepsilon_{\text{sh}}^\infty (t_0)|^{-0.85} \tag{D.11}$$

where factor p_5 depends on the type of cement and is taken from Table D.1. The effect of the ambient relative humidity h_{env} is reflected in the factor

$$k_h = \begin{cases} 1 - h_{\text{env}}^3 & \text{if } h_{\text{env}} \leq 0.98 \\ 12.94 \, (1 - h_{\text{env}}) - 0.2 & \text{if } 0.98 \leq h_{\text{env}} \leq 1 \end{cases} \tag{D.12}$$

For environmental conditions with 100% relative humidity, the amount of potential water supply must be taken into account. Underwater conditions should be captured by $k_h = -0.2$, which approximates swelling and is obtained from (D.12) for $h_{\text{env}} = 1$. For concrete exposed to fog (having also 100% humidity), it is better to use $h_{\text{env}} = 0.98$ since normally fog cannot supply enough water to advance the hydration that leads to swelling.

The theoretical magnitude of final shrinkage for drying at zero ambient humidity is estimated as

$$\varepsilon_{\text{sh}}^\infty (t_0) = \varepsilon_s^\infty \frac{E \, (7 + 600)}{E \, (t_0 + \tau_{sh})} = \varepsilon_s^\infty \times 0.57514 \sqrt{\frac{14}{t_0 + \tau_{sh}} + 3} \tag{D.13}$$

where

$$\varepsilon_s^\infty = k_{\varepsilon a} \, \varepsilon_{cem} \left(\frac{a/c}{6} \right)^{-0.8} \left(\frac{w/c}{0.38} \right)^{p_{\varepsilon w}} \left(\frac{6.5c}{\rho} \right)^{0.11} \tag{D.14}$$

is the magnitude of the final drying shrinkage for reference conditions, which are defined as drying from age $t_0 = 7$ days with shrinkage halftime $\tau_{sh} = 600$ days. Factor $k_{\varepsilon a}$ with default value 1 depends on the type of aggregate and is taken from Table D.2, while factor ε_{cem} and exponent $p_{\varepsilon w}$ depend on the type of cement and are taken from Table D.3. The fraction in (D.13) is the shrinkage correction for the effect of aging on elastic stiffness, and its form is co-opted from model B3, cf. formula (C.13). Recall that function $E(t)$ describing the dependence of elastic modulus on age was taken according to the adjusted ACI formula (C.14).

Table D.2 Aggregate-dependent parameter scaling factors for shrinkage for model B4 (* denotes uncertain fitted parameters)

Aggregate type	$k_{\tau a}$	$k_{\varepsilon a}$
Diabase	0.06*	0.76*
Quartzite	0.59	0.71
Limestone	1.80	0.95
Sandstone	2.30	1.60
Granite	4.00	1.05
Quartz Diorite	15.0*	2.20*

Table D.3 Shrinkage parameters depending on cement type for model B4

Parameter	R	RS	SL
τ_{cem} [day]	0.016	0.080	0.010
$p_{\tau w}$	-0.06	-2.40	3.55
$p_{\tau c}$	-0.10	-2.70	3.80
ε_{cem}	360×10^{-6}	860×10^{-6}	410×10^{-6}
$p_{\varepsilon w}$	1.10	-0.27	1.00

D.2 Shrinkage

A novel feature of model B4 (as compared to its predecessor, model B3) is that it takes into account not only the shrinkage due to drying but also the autogenous shrinkage. For the sake of consistency with the B3 model, we keep using the symbol ε_{sh} for the drying shrinkage and we introduce new symbols $\varepsilon_{sh,tot}$ for the total shrinkage and ε_{au} for autogenous shrinkage. Model B4 assumes that both parts of shrinkage are additive:[5].

---

[5]By the time of proof it transpired that the hypothesis of additivity is too conservative. A future update of model B4 will consider that all shrinkage is caused by the decrease of pore humidity, and that the autogenous shrinkage is driven solely by self-desiccation (as suggested in [125] and elaborated in more detail in [111]; see also Sec. D.8.2).

$$\varepsilon_{sh,tot}(t, t_0) = \varepsilon_{sh}(t, t_0) + \varepsilon_{au}(t) \tag{D.15}$$

Here, t is the current age at which the shrinkage is evaluated and t_0 is the age at the onset of drying (end of curing).

D.2.1 Drying Shrinkage

The average drying shrinkage strain in a cross section is estimated as

$$\varepsilon_{sh}(t, t_0) = -k_h \, \varepsilon_{sh}^{\infty}(t_0) \, \tanh \sqrt{\frac{t - t_0}{\tau_{sh}}} \tag{D.16}$$

where k_h is a factor reflecting the effect of ambient humidity according to (D.12), $\varepsilon_{sh}^{\infty}(t_0)$ is the theoretical final shrinkage for drying at zero ambient humidity, evaluated from (D.13) and (D.14), and τ_{sh} is the shrinkage halftime, given by (D.9) and (D.10).

D.2.2 Autogenous Shrinkage

The autogenous shrinkage is the strain in a stress-free element at constant total water content and constant temperature. The empirical function describing the autogenous shrinkage is

$$\varepsilon_{au}(t) = -\varepsilon_{au}^{\infty} \left[1 + \left(\frac{\tau_{au}}{t} \right)^{\alpha} \right]^{-4.5} \tag{D.17}$$

This function approximates the result of a large number of chemical reactions among the constituents of the mix. It gives a good estimate of the magnitude and evolution of the autogenous shrinkage contribution to the total shrinkage. Note that its definition does not include the volume change of fresh concrete within the first few hours before the set, which are not relevant to structural analysis.

Exponent α, the magnitude of the final autogenous shrinkage, and the autogenous shrinkage halftime are estimated as

$$\alpha = r_{\alpha} \left(\frac{w/c}{0.38} \right) \tag{D.18}$$

$$\varepsilon_{au}^{\infty} = \varepsilon_{au,cem} \left(\frac{a/c}{6} \right)^{-0.75} \left(\frac{w/c}{0.38} \right)^{r_{\varepsilon w}} \tag{D.19}$$

$$\tau_{au} = \tau_{au,cem} \left(\frac{w/c}{0.38} \right)^{3} \tag{D.20}$$

Parameters r_{α}, $r_{\varepsilon w}$, $\varepsilon_{au,cem}$, and $\tau_{au,cem}$ used in formulae (D.18)–(D.20) depend on the type of cement and are taken from Table D.4.

Table D.4 Autogenous shrinkage parameters depending on cement type for model B4

Parameter	R	RS	SL
r_α	1.00	1.40	1.00
$r_{\varepsilon w}$	-3.50	-3.50	-3.50
$\varepsilon_{au,cem}$	210×10^{-6}	-84.0×10^{-6}	0
$\tau_{au,cem}$ [day]	1.00	41.0	1.00

D.3 Effect of Admixtures

If the concrete contains admixtures, the predictions of creep and shrinkage can be improved by an appropriate modification of certain parameters. Table D.5 provides dimensionless correction factors that should multiply creep-related parameters p_2 to p_5, and Table D.6 provides dimensionless correction factors that should multiply shrinkage-related parameters τ_{cem}, $\varepsilon_{au,cem}$, $r_{\varepsilon w}$, and r_α. Their values have been obtained by statistical optimization of the fit of the new NU database. Note that not all the trends of the commercially available admixtures and additives could have been investigated, because of insufficient data.

Table D.5 Admixture-dependent parameter scaling factors for creep for model B4: *Retarder (Re), Fly Ash (Fly), Superplasticizer (Super), Silica Fume (Silica), Air Entraining Agent (AEA), Water reducer (WR)*

Admixture class (% of c)	$\times p_2$	$\times p_3$	$\times p_4$	$\times p_5$
Re (≤ 0.5), Fly (≤ 15)	0.31	7.14	1.35	0.48
Re (>0.5), Fly (≤ 15)	1.43	0.58	0.90	0.46
Fly (>15)	0.37	2.33	0.63	1.60
Super (>0)	0.72	2.19	1.72	0.48
Silica (>0)	1.12	3.11	0.51	0.61
AEA (>0)	0.90	3.17	1.00	0.10
WR (≤ 2)	1.00	2.10	1.68	0.45
WR ($>2, \leq 3$)	1.41	0.72	1.76	0.60
WR (>3)	1.28	2.58	0.73	1.10

D.4 Simplified Strength-Based Model B4s

Even if concrete composition for a given structure has not yet been decided, it is usually known what the typical concrete composition in a given geographical area is. Nevertheless, engineers wish to estimate creep and shrinkage solely from the chosen required (or characteristic) strength f'_c of concrete to be used in the structure. Most of the existing creep and shrinkage recommendations of engineering societies are formulated that way. Therefore, by means of statistical optimization of the fit of the new NU database, a simplified variant of model B4 using only the mean compressive strength $\bar{f}_c$ has been developed. It should be noted that the mean strength, $\bar{f}_c$, is significantly higher than f'_c. Typically, $\bar{f}_c \approx f'_c + 8$ MPa [639]; see formulae (E.1)–(E.4) in Appendix E, which use a somewhat different notation.

Table D.6 Admixture-dependent parameter scaling factors for shrinkage for model B4; *Retarder (Re), Fly Ash (Fly), Superplasticizer (Super), Silica Fume (Silica), Air Entraining Agent (AEA), Water reducer (WR)*

Admixture class (% of c)	$\times \tau_{cem}$	$\times \varepsilon_{au,cem}$	$\times r_{\varepsilon w}$	$\times r_\alpha$
Re ($\leq$0.5), Fly ($\leq$15)	6.00	0.58	0.50	2.60
Re ($>$0.5, $\leq$0.6), Fly ($\leq$15)	2.00	0.43	0.59	3.10
Re ($>$0.5, $\leq$0.6), Fly ($>$15, $\leq$30)	2.10	0.72	0.88	3.40
Re ($>$0.5, $\leq$0.6), Fly ($>$30)	2.80	0.87	1.60	5.00
Re ($>$0.6), Fly ($\leq$15)	2.00	0.26	0.22	0.95
Re ($>$0.6), Fly ($>$15, $\leq$30)	2.10	1.10	1.10	3.30
Re ($>$0.6), Fly ($>$30)	2.10*	1.10	0.97	4.00
Fly ($\leq$15), Super ($\leq$5)	0.32	0.71	0.55	1.71
Fly ($\leq$15), Super ($>$5)	0.32*	0.55	0.92	2.30
Fly ($>$15, $\leq$30), Super ($\leq$5)	0.50	0.90	0.82	1.25
Fly ($>$15, $\leq$30), Super ($>$5)	0.50*	0.80	0.80	2.81
Fly ($>$30), Super ($\leq$5)	0.63	1.38	0.00	1.20
Fly ($>$30), Super ($>$5)	0.63*	0.95	0.76	3.11
Super ($\leq$5), Silica ($\leq$8)	6.00	2.80	0.29	0.21
Super ($\leq$5), Silica ($>$8)	3.00	0.96	0.26	0.71
Super ($>$5), Silica ($\leq$8)	8.00	1.95	0.00	1.00
Silica ($\leq$8)	1.90	0.47	0.00	1.20
Silica ($>$8)	2.60	0.82	0.00	1.20
Silica ($>$18)	1.00	1.50	5.00	1.00
AEA ($\leq$0.05)	2.30	1.10	0.28	0.35
AEA ($>$0.05)	0.44	4.28	0.00	0.36
WR ($\leq$2)	0.50	0.38	0.00	1.90
WR ($>$2, $\leq$3)	6.00	0.45	1.51	0.30
WR ($>$3)	2.40	0.40	0.68	1.40

* ... lacking data, assumed.

According to the simplified model, denoted as B4s, the creep compliance function is given by (D.1) and (D.2) with parameters

$$q_2 = \frac{s_2}{1 \text{ GPa}} \left(\frac{\bar{f}_c}{40 \text{ MPa}} \right)^{-1.58} \tag{D.21}$$

$$q_3 = 0.976 \, q_2 \left(\frac{\bar{f}_c}{40 \text{ MPa}} \right)^{-1.61} = 0.976 \, \frac{s_2}{1 \text{ GPa}} \left(\frac{\bar{f}_c}{40 \text{ MPa}} \right)^{-3.19} \tag{D.22}$$

$$q_4 = \frac{4 \times 10^{-3}}{1 \text{ GPa}} \left(\frac{\bar{f}_c}{40 \text{ MPa}} \right)^{-1.16} \tag{D.23}$$

$$q_5 = \frac{s_5}{1 \text{ GPa}} \left(\frac{\bar{f}_c}{40 \text{ MPa}} \right)^{-0.45} |k_h \varepsilon_{sh}^\infty|^{-0.85} \tag{D.24}$$

Factors s_2 and s_5 dependent on the cement type are taken from Table D.7, factor k_h dependent on the ambient humidity is evaluated according to (D.12), and the final shrinkage ε_{sh}^∞ is evaluated according to (D.13), but with the auxiliary formula (D.14) replaced by

$$\varepsilon_s^\infty = \varepsilon_{s,cem} \left(\frac{\bar{f}_c}{40 \text{ MPa}} \right)^{s_{\varepsilon f}} \tag{D.25}$$

where factor $\varepsilon_{s,cem}$ and exponent $s_{\varepsilon f}$ are taken from Table D.7. The shrinkage halftime τ_{sh} is evaluated from (D.9) with $k_{\tau a} = 1$, but with the auxiliary formula (D.10) replaced by

$$\tau_0 = \tau_{s,cem} \left(\frac{\bar{f}_c}{40 \text{ MPa}} \right)^{s_{\tau f}} \tag{D.26}$$

where factor $\tau_{s,cem}$ and exponent $s_{\tau f}$ are taken from Table D.7.

Table D.7 Parameters depending on cement type for the B4s model

Parameter	R	RS	SL
s_2	14.2×10^{-3}	29.9×10^{-3}	11.2×10^{-3}
s_5	1.54×10^{-3}	41.8×10^{-6}	150×10^{-6}
$\varepsilon_{s,cem}$	590×10^{-6}	830×10^{-6}	640×10^{-6}
$s_{\varepsilon f}$	-0.51	-0.84	-0.69
$\tau_{s,cem}$ [day]	0.027	0.027	0.032
$s_{\tau f}$	0.21	1.55	-1.84

The autogenous shrinkage is given by

$$\varepsilon_{au}(t) = -\varepsilon_{au}^\infty \left[1 + \left(\frac{\tau_{au}}{t} \right)^{1.73} \right]^{-1.73} \tag{D.27}$$

in which

$$\varepsilon_{au}^\infty = 78.2 \times 10^{-6} \left(\frac{\bar{f}_c}{40 \text{ MPa}} \right)^{1.03} \tag{D.28}$$

$$\tau_{au} = 2.26 \text{ days} \times \left(\frac{\bar{f}_c}{40 \text{ MPa}} \right)^{0.27} \tag{D.29}$$

D.5 Effect of Temperature

The basic version of the model has been calibrated for reference temperature $T_0 = 293$ K (i.e., $20\,^{\circ}$C). If the actual temperature is higher (or lower), various physical and chemical processes are accelerated (or decelerated). The effect of temperature on creep and shrinkage is accounted for by an acceleration of time (horizontal scaling of the creep and shrinkage curves) combined with vertical scaling of selected terms (those that describe basic creep).

The concept of time acceleration at higher temperatures has been introduced in the context of the microprestress-solidification theory in Chap. 10. In model B4, acceleration factors similar to (10.31)–(10.33) are used. The official RILEM recommendation [136] defines three different acceleration factors, denoted as β_{Th}, β_{Ts}, and β_{Tc}, which correspond to the processes of hydration, moisture diffusion (drying), and creep. Since the values of activation energies recommended by Bažant et al. [136] are the same for all three processes, for practical applications it is fully sufficient to consider only one common factor, denoted as β_T. In analogy to (10.31)–(10.33), its dependence on temperature is described by an Arrhenius-type law

$$\beta_T(T) = \exp\left[\frac{Q}{R}\left(\frac{1}{T_0} - \frac{1}{T}\right)\right] \tag{D.30}$$

in which T is the ambient temperature (in absolute scale), $T_0 = 293$ K is the reference temperature, Q is the activation energy, and R is the universal gas constant. Only the ratio Q/R is what matters, and, in the absence of data for the given concrete, it is recommended to use $Q/R = 4000$ K. The equivalent time t_e (for simplicity denoted by the same symbol as the equivalent age or equivalent hydration period in Chap. 10) is then defined by its rate

$$\frac{dt_e}{dt} = \beta_T(T) \tag{D.31}$$

For a general temperature history, the equivalent time can be obtained by integration of (D.31), which is formally represented by the formula

$$t_e(t) = \int_0^t \beta_T(T(t')) \, dt' \tag{D.32}$$

For simplicity, model B4 considers the ambient temperature during curing, T_{cur}, and the average temperature during drying and loading, T_{dl}, to be constant. Then, the integration leads to

$$t_e(t) = t_{e0} + \beta_T^{dl}(t - t_0), \quad \text{for } t \geq t_0 \tag{D.33}$$

where

$$t_{e0} = t_e(t_0) = \beta_T^{cur} t_0 \tag{D.34}$$

is the equivalent time at the end of curing (onset of drying), and factors β_T^{cur} and β_T^{dl} are evaluated from (D.30) for $T = T_{cur}$ and $T = T_{dl}$. Note that the values of t_e for $t < t_0$ are usually not needed at all.

If the activation energies associated with hydration, moisture diffusion, and creep are considered by the same value, Q, incorporation of the temperature effect becomes very easy because it is sufficient to evaluate the creep compliance function and the shrinkage function from the already presented formulae with the actual times t, t', and t_0 replaced by their equivalent values $t_e(t)$, $t_e(t')$, and $t_e(t_0) \equiv t_{e0}$. But, in addition to that, model B4 recommends to increase the basic creep compliance by a multiplicative factor R_T. This factor is computed again from formula (D.30), with the same default value of activation energy $Q/R = 4000$ K as for the evaluation of β_T (but potentially with another value, if it can be determined for the given concrete). If a concrete member is exposed to drying from age t_0 and is loaded by constant stress σ from age t', the strain at age t is evaluated as[6]

$$\varepsilon(t) = \sigma \left[q_1 + R_T J_b(t_e(t), t_e(t')) + J_d(t_e(t), t_e(t')) \right] + \varepsilon_{sh,tot}(t_e(t), t_{e0}) \quad (D.35)$$

where

$$J_d(t_e, t_e') = q_5\sqrt{\left\langle e^{-g(t_e - t_{e0})} - e^{-g(\langle t_e' - t_{e0}\rangle)} \right\rangle} \quad (D.36)$$

Generalization to variable stress leads to the integral formula

$$\varepsilon(t) = \sigma_1 J(t_e(t), t_{e1}) + \int_{t_1}^{t} J(t_e(t), t_e(t')) \dot{\sigma}(t') \, dt' + \varepsilon_{sh,tot}(t_e(t), t_{e0}) \quad (D.37)$$

where $t_{e1} = t_e(t_1)$ and

$$J(t_e, t_e') = q_1 + R_T J_b(t_e, t_e') + J_d(t_e, t_e') \quad (D.38)$$

Note that the lower and upper limits of the integral in (D.37), t_1 and t, are not affected by the transformation to equivalent times, because $\dot{\sigma}$ is still the derivative of stress with respect to the actual (untransformed) time and $\dot{\sigma}(t') \, dt'$ is the infinitesimal increment of stress during an actual time increment dt'. This is why the integration variable, t', still sweeps from t_1 to t.

D.6 Examples of Compliance Curves

Example D.1. Compliance curves predicted by model B4

Application of the empirical formulae presented in Sect. D.1 is illustrated using the same input data as in Example 3.1 for the B3 model. The concrete mix is characterized by water content $w = 170 \, \text{kg/m}^3$, type-I cement content $c = 450 \, \text{kg/m}^3$ and aggregate content $a = 1800 \, \text{kg/m}^3$. In calculations based on model B4, the cement type is considered as R (normal), which corresponds to type-I cement according to the ASTM classification. The standard compression strength is taken as $\bar{f}_c = 45.4$ MPa. The compliance curve is constructed for a concrete slab of thickness $D = 200$ mm,

---

[6]For simplicity, only the mechanical strain and the shrinkage strain are considered in Eq. (D.35). Of course, the thermal strain would need to be added, but this part of strain is directly computed from temperature and is not affected by the acceleration of time discussed here.

cured in air with initial protection against drying until the age $t_0 = 7$ days and subsequently exposed to an average environmental humidity $h_{\text{env}} = 70\%$.

Model parameters that control basic creep are evaluated using formulae (3.6) and (D.4)–(D.7):

$$E_{28} = 4.733 \text{ GPa} \times \sqrt{\frac{\bar{f_c}}{1 \text{ MPa}}} = 4.733 \text{ GPa} \times \sqrt{45.4} = 31.89 \text{ GPa} \qquad \text{(D.39)}$$

$$q_1 = \frac{p_1}{E_{28}} = \frac{0.7}{31.89 \text{ GPa}} = 21.95 \times 10^{-6}/\text{MPa} \qquad \text{(D.40)}$$

$$q_2 = \frac{p_2}{1 \text{ GPa}} \left(\frac{w/c}{0.38}\right)^3 = \frac{58.6 \times 10^{-3}}{1 \text{ GPa}} \left(\frac{170/450}{0.38}\right)^3 = 57.58 \times 10^{-6}/\text{MPa} \qquad \text{(D.41)}$$

$$q_3 = p_3 q_2 \left(\frac{a/c}{6}\right)^{-1.1} \left(\frac{w/c}{0.38}\right)^{0.4} =$$

$$= 0.0393 \times \frac{57.58 \times 10^{-6}}{1 \text{ MPa}} \times \left(\frac{1800/450}{6}\right)^{-1.1} \left(\frac{170/450}{0.38}\right)^{0.4} =$$

$$= 3.527 \times 10^{-6}/\text{MPa} \qquad \text{(D.42)}$$

$$q_4 = \frac{p_4}{1 \text{ GPa}} \left(\frac{a/c}{6}\right)^{-0.9} \left(\frac{w/c}{0.38}\right)^{2.45} =$$

$$= \frac{3.4 \times 10^{-3}}{1 \text{ GPa}} \left(\frac{1800/450}{6}\right)^{-0.9} \left(\frac{170/450}{0.38}\right)^{2.45} =$$

$$= 4.827 \times 10^{-6}/\text{MPa} \qquad \text{(D.43)}$$

To estimate the additional compliance due to drying creep, we first need to determine the shrinkage halftime and the maximum shrinkage strain, using (D.10), (D.9), (D.14), and (D.13):

$$\tau_0 = \tau_{cem} \left(\frac{a/c}{6}\right)^{-0.33} \left(\frac{w/c}{0.38}\right)^{p_{\tau w}} \left(\frac{6.5 \, c}{\rho}\right)^{p_{\tau c}} = \qquad \text{(D.44)}$$

$$= (0.016 \text{ day}) \times \left(\frac{1800/450}{6}\right)^{-0.33} \left(\frac{170/450}{0.38}\right)^{-0.06} \left(\frac{6.5 \times 450}{2350}\right)^{-0.1} =$$

$$= 0.01790 \text{ day}$$

$$\tau_{sh} = \tau_0 \, k_{\tau a} \left(k_s \frac{D}{1 \text{ mm}}\right)^2 = \qquad \text{(D.45)}$$

$$= 0.01790 \text{ day} \times 1 \times \left(1 \times \frac{200 \text{ mm}}{1 \text{ mm}}\right)^2 = 716 \text{ day}$$

$$\varepsilon_s^\infty = k_{\varepsilon a} \, \varepsilon_{cem} \left(\frac{a/c}{6}\right)^{-0.8} \left(\frac{w/c}{0.38}\right)^{p_{\varepsilon w}} \left(\frac{6.5c}{\rho}\right)^{0.11} = \qquad \text{(D.46)}$$

$$= 1 \times 360 \times 10^{-6} \left(\frac{1800/450}{6}\right)^{-0.8} \left(\frac{170/450}{0.38}\right)^{1.1} \left(\frac{6.5 \times 450}{2350}\right)^{0.11} =$$

$$= 506.79 \times 10^{-6}$$

$$\varepsilon_{sh}^{\infty} = \varepsilon_s^{\infty} \frac{E\,(7+600)}{E\,(t_0 + \tau_{sh})} = \varepsilon_s^{\infty} \times 0.57514 \sqrt{\frac{14}{t_0 + \tau_{sh}}} + 3 =$$

$$= 506.79 \times 10^{-6} \times 0.57514 \sqrt{\frac{14}{7+716}} + 3 = 506.48 \times 10^{-6} \tag{D.47}$$

We also need the humidity-dependent factor k_h, which is according to (D.12) given by

$$k_h = 1 - h_{env}^3 = 1 - 0.7^3 = 0.657 \tag{D.48}$$

Based on the above auxiliary values, parameter q_5 is estimated from (D.11) as

$$q_5 = \frac{p_5}{1\,\text{GPa}} \left(\frac{a/c}{6}\right)^{-1} \left(\frac{w/c}{0.38}\right)^{0.78} |k_h \varepsilon_{sh}^{\infty}|^{-0.85} =$$

$$= \frac{777 \times 10^{-6}}{1\,\text{GPa}} \left(\frac{1800/450}{6}\right)^{-1} \left(\frac{170/450}{0.38}\right)^{0.78} |0.657 \varepsilon_{sh}^{\infty}|^{-0.85} =$$

$$= 1049 \times 10^{-6}/\,\text{MPa} \tag{D.49}$$

The additional compliance due to drying is then given by (D.3)–(D.8) with $q_5 = 1049 \times 10^{-6}/\text{MPa}$, $p_{5H} = 8$, $h_{env} = 0.7$, and $\tau_{sh} = 716$ days.

The total compliance function $J(t, t')$ defined in (D.1) is plotted in Fig. D.1 for four different ages at loading t', first in linear scale as a function of the current age t and then in semilogarithmic scale as a function of the load duration $t - t'$. ∎

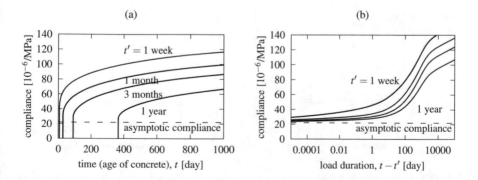

Fig. D.1 Compliance functions predicted by model B4 in Example D.1

Example D.2. Compliance curves predicted by model B4s

As explained in Sect. D.4, the simplified version of model B4, labeled as B4s, takes the mean compression strength $\bar{f}_c$ as the only parameter characterizing the concrete properties. For concrete grades C20 to C60 as defined in the *fib* Model Code, the mean strength values estimated using (E.1) range from 28 to 68 MPa. The corresponding model parameters evaluated from (3.6), (D.4) and (D.21)–(D.23) for cement of type R are listed in Table D.8. Parameter τ_0 is specified in a somewhat unusual unit, 10^{-4} day, because then the value in the table represents the shrinkage halftime of an infinite slab of thickness 100 mm, expressed in days. The resulting creep compliance functions for loading at age $t' = 28$ days are plotted in Fig. D.2.   ■

Table D.8 Creep parameters according to the B4s model for different concrete grades (assuming cement of type R)

Parameter	Unit	C20	C30	C40	C50	C60
$\bar{f}_c$	[MPa]	28	38	48	58	68
E_{28}	[GPa]	25.04	29.18	32.79	36.05	39.03
q_1	[10^{-6}/MPa]	27.95	23.99	21.35	19.42	17.94
q_2	[10^{-6}/MPa]	24.95	15.40	10.65	7.895	6.140
q_3	[10^{-6}/MPa]	43.24	16.32	7.747	4.236	2.550
q_4	[10^{-6}/MPa]	6.050	4.245	3.238	2.599	2.161
ε_s^∞	[10^{-6}]	707.7	605.6	537.6	488.2	450.1
τ_0	[10^{-4} day]	250.5	267.1	280.5	291.9	301.8

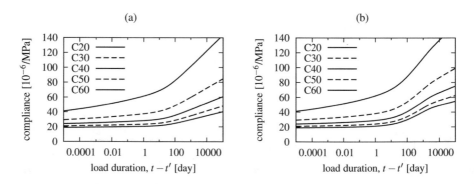

Fig. D.2 Compliance functions for different concrete grades loaded at $t' = 28$ days, as predicted by model B4s in Example D.2: (a) at sealed conditions, (b) at 70% ambient humidity

D.7   Aging Effects

D.7.1   Aging of Elastic Modulus

According to the testing standards [43], the conventional elastic modulus of concrete should be evaluated from the unloading slope of the stress–strain diagram after several cycles. Based on such experiments, empirical formulae for the estimation of elastic modulus from compression strength have been developed. One example is the ACI empirical formula (3.5) or (3.6). If the strength is substituted in MPa and the elastic modulus is evaluated in GPa, the ACI formula can be written as

$$E_{28}^{(s,ACI)} = 4.733\sqrt{\bar{f}_c} \tag{D.50}$$

The elastic modulus can also be expected to be closely related to the "almost instantaneous" deformation in a creep test. The correlation between the predictions of conventional elastic modulus from strength and the reciprocal value of the compliance function was studied by Bažant and Baweja [105]. Using the RILEM creep database available at that time, they determined moduli $E_{28}^{(s,ACI)}$ estimated from (D.50) based on the measured compression strength and moduli $E_{28}^{(c)} = 1/J(28 + \Delta t_s, 28)$ evaluated as the reciprocal values of measured creep compliances, for different values of time delay Δt_s. As documented in Fig. D.3b, the ratios $E_{28}^{(c)}/E_{28}^{(s,ACI)}$ were found to be around 1 for $\Delta t_s = 0.01$ day, while for $\Delta t_s = 0.001$ day, they were systematically larger than 1, though not much larger (Fig. D.3a), and for $\Delta t_s = 0.1$ day systematically smaller than 1 (Fig. D.3c).

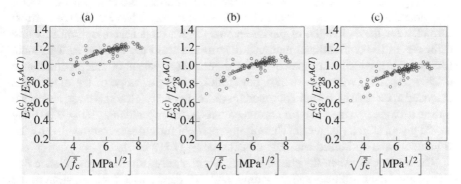

Fig. D.3 Ratio of elastic modulus determined from measured compliance, $E_{28}^{(c)} = 1/J(28 + \Delta t_s, 28)$, to value $E_{28}^{(s,ACI)}$ predicted from compression strength according to the ACI formula (D.50), plotted against the square root of compression strength, for (a) $\Delta t_s = 0.001$ day, (b) $\Delta t_s = 0.01$ day, (c) $\Delta t_s = 0.1$ day

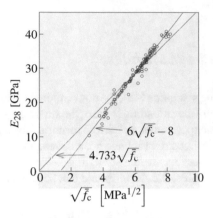

Fig. D.4 Elastic modulus determined from measured compliance versus square root of compression strength (isolated points correspond to experimental data and the straight lines to formulae (D.50) and (D.51))

It is apparent from Fig. D.3 that the ratios $E_{28}^{(c)}/E_{28}^{(s,ACI)}$ are increasing with increasing strength. Therefore, if the conventional elastic modulus to be used for calibration of a creep model is estimated from the compression strength, a closer fit of the reciprocal compliance $E_{28}^{(c)} = 1/J(28.01, 28)$ could be achieved with a modified formula

$$E_{28}^{(s,mod)} = 6\sqrt{\bar{f}_c} - 8 \tag{D.51}$$

where $\bar{f}_c$ is substituted in MPa and E_{28} is obtained in GPa. This formula corresponds to the regression line in Fig. D.4, where the points correspond to measured moduli E_{28} plotted against the square root of compression strength. Nevertheless, empirical formulae for the estimation of parameters of the B3 model were optimized using estimates of the conventional modulus determined from the original ACI formula (3.5) or (3.6), which were matched by the reciprocal values of creep compliance $J(28 + \Delta t_s, 28)$, with $\Delta t_s = 0.01$ day. Later, when the scope of the model was extended to modern high-strength concretes and concretes with admixtures, and when more data were obtained also for normal concretes, it appeared that $\Delta t_s = 0.001$ day would give a slightly better fit. Therefore, empirical formulae recommended by the B4 model were optimized with $E_{28}^{(c)}$ considered as $1/J(28.001, 28)$.

The relation between the elastic modulus and short-term creep compliance can be extended to an arbitrary age at loading, and so it can be used for evaluation of the elastic modulus at age t',

$$E(t') = \frac{1}{J(t' + \Delta t_s, t')} \tag{D.52}$$

with $\Delta t_s = 0.01$ day for the B3 model and $\Delta t_s = 0.001$ day for models B4 and B4s. Recall that the basic creep compliance of models B3 and B4 is given by (3.10). Since the conventional delay Δt_s is always much smaller than the age at loading,

t', one may replace t in the numerator of the first fraction on the right-hand side of (3.10) by t', and then, integration becomes easy and the basic creep compliance can be evaluated in closed form. Substituting the result into (3.3) with $1/E_0 = q_1 =$ age-independent asymptotic compliance, $m = 0.5$ and $J_d(t, t') = 0$ (basic creep only) yields

$$J(t, t') = q_1 + \left(\frac{q_2}{\sqrt{t'}} + q_3 \right) \ln[1 + (t - t')^n] + q_4 \ln \frac{t}{t'} \qquad (D.53)$$

where the last term is negligible for $t - t' \ll t'$ (i.e., for $t \approx t'$). This simplification and substitution into (D.52) lead to a simple expression for the time dependence of elastic modulus in the form

$$E(t') = \frac{1}{A_0 + A_1/\sqrt{t'}} \qquad (D.54)$$

where, for the B3 model,

$$A_0 = q_1 + q_3 \ln(1 + \Delta t_s^n) = q_1 + q_3 \ln(1 + 0.01^{0.1}) = q_1 + 0.489q_3 \qquad (D.55)$$
$$A_1 = q_2 \ln(1 + \Delta t_s^n) = q_2 \ln(1 + 0.01^{0.1}) = 0.489q_2 \qquad (D.56)$$

while for the B4 model,

$$A_0 = q_1 + q_3 \ln(1 + \Delta t_s^n) = q_1 + q_3 \ln(1 + 0.001^{0.1}) = q_1 + 0.406q_3 \qquad (D.57)$$
$$A_1 = q_2 \ln(1 + \Delta t_s^n) = q_2 \ln(1 + 0.001^{0.1}) = 0.406q_2 \qquad (D.58)$$

As a special case, the conventional modulus (elastic modulus at 28 days) is, for model B3, given by

$$E_{28}^{(c,B3)} = E(28) = \frac{1}{A_0 + A_1/\sqrt{28}} = \frac{1}{q_1 + 0.0924q_2 + 0.489q_3} \qquad (D.59)$$

and the limit value of elastic modulus approached asymptotically for high ages is

$$E_{\infty}^{(c,B3)} = \lim_{t' \to \infty} E(t') = \frac{1}{A_0} = \frac{1}{q_1 + 0.489q_3} \qquad (D.60)$$

For the set of parameters used in Example 3.1, we obtain $E_{28} = 32.36$ GPa and $E_{\infty} = 52.15$ GPa, i.e., $E_{\infty}/E_{28} = 1.61$. It is not clear whether such a dramatic increase of elastic modulus due to aging is realistic. In general, the ratio is given by

$$\frac{E_{\infty}^{(c,B3)}}{E_{28}^{(c,B3)}} \approx \frac{q_1 + 0.0924q_2 + 0.489q_3}{q_1 + 0.489q_3} = 1 + \frac{0.0924q_2}{q_1 + 0.489q_3} \qquad (D.61)$$

To obtain a less pronounced aging effect, q_2 would need to be reduced and q_3 increased.

For model B4, the expressions for A_0 and A_1 are given by (D.57) and (D.58), and formulae (D.59) and (D.60) must be replaced by

$$E_{28}^{(c,B4)} = \frac{1}{A_0 + A_1/\sqrt{28}} = \frac{1}{q_1 + 0.0768q_2 + 0.406q_3} \tag{D.62}$$

$$E_{\infty}^{(c,B4)} = \frac{1}{A_0} = \frac{1}{q_1 + 0.406q_3} \tag{D.63}$$

For the set of parameters used in Example D.1 (in which the concrete was assumed to have the same properties as in Example 3.1), we obtain $E_{28} = 35.97$ GPa and $E_{\infty} = 42.77$ GPa, and the resulting ratio $E_{\infty}/E_{28} = 1.19$ is much lower than for the B3 model. This is caused by a different combination of parameters: Empirical formulae used by B4 lead to a lower value of q_2 and a higher value of q_3 than those used by B3 (at least for this particular concrete); see Table D.9.

For model B4s, the delay $\Delta t_s = 0.001$ day is supposed to be the same as for model B4, and thus, Eqs. (D.62) and (D.63) are applicable. For this simplified version of model B4, parameters q_i depend exclusively on the strength and on the cement type. Substituting expressions (D.4) and (D.21)–(D.22) into (D.62) and (D.63), we obtain

$$E_{28}^{(c,B4s)} = \frac{1}{\dfrac{p_1}{4.733\sqrt{\bar{f_c}}} + 0.0768s_2 \left(\dfrac{\bar{f_c}}{40}\right)^{-1.58} + 0.406 \times 0.976s_2 \left(\dfrac{\bar{f_c}}{40}\right)^{-3.19}} \tag{D.64}$$

$$E_{\infty}^{(c,B4s)} = \frac{1}{\dfrac{p_1}{4.733\sqrt{\bar{f_c}}} + 0.406 \times 0.976s_2 \left(\dfrac{\bar{f_c}}{40}\right)^{-3.19}} \tag{D.65}$$

with parameters p_1 and s_2 dependent on the cement type, as shown in Tables D.1 and D.7.

The dimensionless ratio $E_{28}^{(s,ACI)}/E_{28}^{(c,B4s)}$ is plotted as a function of compression strength $\bar{f_c}$ in Fig. D.5a, separately for the three cement types. It turns out that the agreement between $E_{28}^{(s,ACI)}$ and $E_{28}^{(c,B4s)}$ is acceptable for concretes with strength between 30 and 40 MPa for R and SL cements, and between 35 and 45 MPa for RS cement. For lower strengths, the compliance derived from model B4s is too high, especially for RS cement, and for higher strengths, it is too low. Of course, this conclusion is valid under the assumption that the ACI formula (D.50) is sufficiently accurate for high-strength concrete, too.

As explained before (see Fig. D.4), somewhat better estimates of the actual (measured) elastic moduli can be obtained using a modified formula (D.51). The curves in Fig. D.5b show the ratio $E_{28}^{(s,mod)}/E_{28}^{(c,B4s)}$ as a function of $\bar{f_c}$. The agreement is slightly improved as compared to Fig. D.5a, but still, important deviations of the creep compliance $J(28.001, 28)$ from $1/E_{28}^{(s,mod)}$ are found for concretes of low or high strengths.

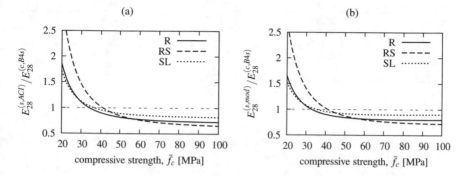

Fig. D.5 Ratio between the short-term creep compliance of model B4s and the inverse of elastic modulus estimated from strength: (a) elastic modulus estimated from the ACI formula, $E_{28}^{(s,ACI)} = 4.733\sqrt{\bar{f_c}}$, (b) elastic modulus estimated from a modified formula, $E_{28}^{(s,mod)} = 6\sqrt{\bar{f_c}} - 8$

A possible improvement of model B4s would be achieved by calibrating the parameters with a constraint that enforces equality of $E_{28}^{(c,B4s)}$ and $E_{28}^{(s,mod)}$. For instance, optimization of the formulae for basic creep could be done just for parameters q_2, q_3, and q_4, with q_1 determined as

$$q_1 = \frac{1000}{6\sqrt{\bar{f_c}} - 8} - 0.0768q_2 - 0.406q_3 \qquad (D.66)$$

where $\bar{f_c}$ is substituted in GPa and q_i in $10^{-6}/\text{MPa}$.

Let us now turn attention to the effect of aging on elastic modulus predicted by model B4s. Based on (D.64) and (D.65), the ratio between the old-age elastic modulus and the conventional one is expressed as

$$\frac{E_\infty^{(c,B4s)}}{E_{28}^{(c,B4s)}} = 1 + \frac{1}{0.4351\dfrac{p_1}{s_2}\left(\dfrac{\bar{f_c}}{40}\right)^{1.08} + 5.1646\left(\dfrac{\bar{f_c}}{40}\right)^{-1.61}} \qquad (D.67)$$

For compressive strength in the range from 20 to 100 MPa, this ratio is between 1.017 and 1.039 for cement of type R, between 1.041 and 1.050 for cement of type RS, and between 1.012 and 1.033 for cement of type SL. Thus, the relative increase of elastic modulus from the age of 28 days to infinity predicted by B4s is very small, in the order of a few percent. For the concrete from Examples 3.1 and D.1 with compressive strength $\bar{f_c} = 45.4$ MPa and cement of type R, model B4s gives an increase of 3.5%, while model B3 predicts 61% and model B4 predicts 19%. This large discrepancy is, of course, related to the values of parameters q_1, q_2, and q_3, which are summarized in Table D.9.

Table D.9 Basic creep parameters for the concrete from Examples 3.1 and D.1 according to models B3, B4, and B4s (all in 10^{-6}/MPa)

	B3	B4	B4s
q_1	18.81	21.95	21.95
q_2	126.9	57.58	11.63
q_3	0.7494	3.527	9.253
q_4	7.692	4.827	3.454
$nq_3 + q_4$	7.767	5.180	4.379

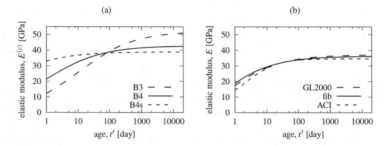

Fig. D.6 Dependence of elastic modulus on age, according to (a) models B3, B4, and B4s, (b) formulae used by model GL and by ACI and *fib* recommendations

The foregoing comparison of ratios $E_\infty^{(c)}/E_{28}^{(c)}$ for the particular concrete from Examples 3.1 and D.1 indicates that the effect of aging on elastic modulus is very strong for model B3, moderate for model B4, and very weak for model B4s. This is confirmed by Fig. D.6a, which shows the overall dependence of elastic modulus $E^{(c)}(t')$ on age t'. For the sake of interest, Fig. D.6b shows similar graphs based on formulae incorporated in design codes. According to the ACI, *fib* and GL models, to be presented in detail in Appendix E, the age dependence of the (static) elastic modulus is, respectively, predicted by empirical relations (E.29), (E.5)–(E.6), and (E.39)–(E.40), which describe the age dependence of elastic modulus $E(t')$ directly, without reference to the values of creep compliance. These code formulae predict a ratio E_∞/E_{28} equal to $\sqrt{7/6} \approx 1.08$ (ACI) or to $\sqrt{e^s}$ with cement-type-dependent parameter s between 0.2 and 0.38 (*fib*), which gives E_∞/E_{28} between 1.11 and 1.21. For ordinary Portland cement and $\bar{f}_c = 45.4$ MPa, the GL model predicts $E_{28} = 33.82$ GPa and $E_\infty = 39.35$ GPa, i.e., $E_\infty/E_{28} = 1.16$. The code formulae have probably been obtained by fitting experimental data, but the range of ages at testing might not have been wide enough, because values of E_∞/E_{28} near 1.1 seem to be too low.

The comparison in Fig. D.6a was constructed only after the publication of model B4. Model B3 had been optimized so as to obtain good agreement with the elastic moduli estimated from strength. On the other hand, model B4 was optimized strictly from creep test data, with no constraint linking the short-term behavior to the strength for tests which did not report the measured value of elastic modulus or short-term deformations. In the future, simultaneous optimization of the databases for creep and for elastic modulus would be appropriate.

In the creep database, it is not so easy to find reliable information on the values of $J(t' + 0.01, t')$ (or of $J(t' + 0.001, t')$) for the same concrete but different ages t'. Dramatic changes of elastic moduli were reported by Brooks [267] in his summary of 30-year tests. The evaluation of Brooks' data is not straightforward, because the elastic modulus at age 14 days was measured at loading, while the elastic modulus at age 15 years or 30 years was measured at unloading (also, there was a lapse of environmental control at 7 years). The moduli measured at high ages were substantially larger for specimens stored under wet conditions than for those stored under dry conditions. For specimens with Stourton aggregates and $w/c = 0.40$, the measured ratios[7] $E_\infty / E(14)$ were 1.74 (dry) and 3.23 (wet). For specimens with North Notts aggregates and $w/c = 0.50$, these ratios were 1.47 (dry) and 1.83 (wet). For comparison, the ACI model predicts $E_\infty / E(14)$ equal to 1.15, the *fib* Model Code predicts at most 1.31 (for the largest value of parameter $s = 0.38$), and the B3 model predicts 1.86. So it cannot be excluded that the actual growth of elastic modulus due to aging is much more pronounced than according to the codes. Doubtless it depends on the long-term supply of water to the pores, which can allow hydration to proceed for decades. However, as already mentioned, the moduli measured by Brooks do not have the same meaning at 14 days (loading modulus) and at 15 or 30 years (unloading modulus, which is always significantly larger than the modulus at first loading). It is also unclear how much "instantaneous" the measurement was.

Looking at the strength evolution reported by Brooks [267], one finds that the ratio between the compression strengths at 30 years and at 14 days was closer to the code formulae than was the case for elastic moduli. For Stourton aggregates and $w/c = 0.40$, the strength after 30 years was almost the same for measurements done on loaded and load-free control specimens and on wet and dry specimens (for some other cases there were large discrepancies). The measured strength increased from 14 days to 30 years by a factor of 1.51, and so the strength increase was actually smaller than the increase in elastic modulus. This observation contradicts the usual assumption that the modulus grows proportionally to the square root of strength. For strength, the ACI model would predict an increase by a factor of 1.33 and the *fib* code by a factor between 1.33 (for $s = 0.2$) and 1.71 (for $s = 0.38$).

Valuable information can also be extracted from the study of Shideler [776], who focused on lightweight concrete but also acquired some data for normal-weight concrete. The measured ratios between elastic moduli at the age of 1 year and at the age of 28 days were ranging from 1.34 to 1.55 for wet conditions (continuously moist-cured specimens) and from 1.09 to 1.17 for dry conditions (7 days of moist curing followed by storage at 50% relative humidity). Interestingly, the relative increase in strength was in all cases smaller than the relative increase in elastic modulus (1.19 to 1.27 for wet conditions and 1.05 to 1.11 for dry conditions). Again, this contradicts the assumed proportionality of the modulus to the square root of strength stipulated by the ACI code.

According to the ongoing research [125, 715], cement hydration may or might not continue for years and even decades or centuries, depending on the cement fineness

---

[7]The value measured at the high age of 15 or 30 years is taken here as an approximation of the limit value, E_∞.

and water supply to the micropores in concrete. The growth of $E(t')$ as well as $\bar{f}_c(t')$ will have to be considered as a function of the degree of hydration, whose evolution will have to be calculated from differential equations proposed by Bažant et al. [125] and refined in Eqs. (13)–(23b) in Rahimi-Aghdam et al. [715]. In these equations, the hydration degree growth from days to centuries is controlled by diffusion of water toward the anhydrous cement grain remnants through accreting barrier shells of C-S-H that surround the remnants.

D.7.2 Effect of Aging on Creep

In the previous section, aging effects on elastic modulus predicted by the simplified model B4s were found to be much weaker than those obtained with the full B4 model. Even though the comparison was done just for one typical concrete, the conclusion has a general validity and can also be extended to the effects of aging on creep. In models of the B3/B4 family, short-term creep is controlled by parameters q_2 and q_3, with q_3 affecting the age-independent part of creep compliance. It is thus natural that higher values of q_2 in general lead to more pronounced aging. For the particular concrete from Examples 3.1 and D.1, the value of q_2 predicted by the empirical formula used by B4 was 5 times higher than the value estimated from strength according to B4s; see Table D.9.

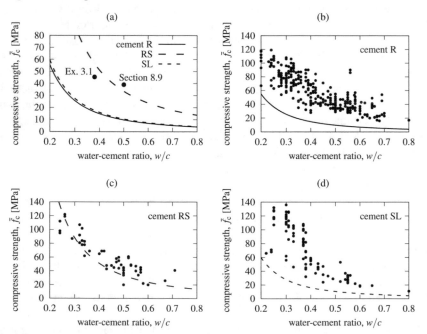

Fig. D.7 Combinations of water-cement ratio and compressive strength that would give the same value of q_2 according to models B4 and B4s (curves), and isolated points corresponding to (a) the values used in Example 3.1 and in Sect. 7.9, (b)–(d) real values from the database

Parameter q_2, which controls aging, is evaluated from the water-cement ratio in model B4 and from the compressive strength in model B4s. The B4 formula (D.5) and the B4s formula (D.21) would give the same result if the water-cement ratio w/c and the compressive strength $\bar{f}_c$ were linked by the relation

$$p_2 \left(\frac{w/c}{0.38} \right)^3 = s_2 \left(\frac{\bar{f}_c}{40} \right)^{-1.58} \tag{D.68}$$

with $\bar{f}_c$ substituted in MPa and with parameters p_2 and s_2 dependent on the cement type.

For three types of cement (R, RS, and SL), Fig. D.7a shows the combinations of w/c and $\bar{f}_c$ that satisfy condition (D.68) (smooth curves) and an isolated point that correspond to the values used in Examples 3.1 and D.1 ($w/c = 0.378$ and $\bar{f}_c = 45.4$ MPa). The point is way above the curve that corresponds to cement type R, used in concrete from the example. For $w/c = 0.378$, model B4s would give the same value of q_2 as model B4 if the compressive strength was $\bar{f}_c = 16.5$ MPa. Since the actual compressive strength is much higher, q_2 evaluated from the expression on the right-hand side of (D.68) (which corresponds to the B4s formula (D.21)) is much lower, due to the negative exponent. This explains why, in this particular case, models B4 and B4s give so different results in terms of aging.

To check whether the same trend can be expected in general, let us plot the combinations of w/c and $\bar{f}_c$ reported for the samples from the NU creep database (of course, leaving out cases in which one of these properties, typically the strength, is not reported). This is done in Fig. D.7b–d, separately for the three types of cement. It turns out that, with the exception of cement RS (Fig. D.7c), all the points that represent real data are above the curve that corresponds to (D.68). This proves that model B4s gives systematically much less aging than model B4, for all creep tests from the database with cements R and SL.

We have shown that higher values of parameter q_2 lead to more pronounced aging of elastic modulus and proven that model B4 always predicts a higher value of this parameter than model B4s does. As given in Table D.9, the value of q_2 predicted by model B3 for the concrete from Example 3.1 is even higher than the B4 prediction. For a general comparison of aging effects according to models B3 and B4, it is convenient to work with the ratio q_3/q_2, which depends only on the water-cement and aggregate–cement ratios. Recall that q_2 controls the aging part and q_3 controls the nonaging part of the viscoelastic compliance modeled by the solidification theory. The formulae in line 3 in Table C.2 and in Eq. (D.6) imply that models B3 and B4 give the same ratio q_3/q_2 if

$$0.29 \, (w/c)^4 = 0.0393 \left(\frac{a/c}{6} \right)^{-1.1} \left(\frac{w/c}{0.38} \right)^{0.4} \tag{D.69}$$

which is true if

$$\frac{a}{c} = 1.3863 \left(\frac{w}{c} \right)^{-3.27} \tag{D.70}$$

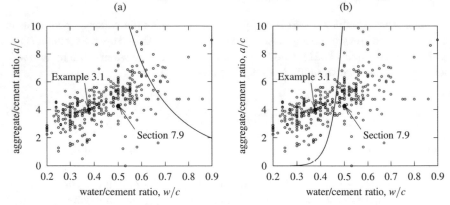

Fig. D.8 Combinations of water-cement and aggregate–cement ratios corresponding to Example 3.1 and to Sect. 7.9 (large filled circles) and to real tests from the creep database (small hollow circles); the solid curves indicate for which combinations models B3 and B4 predict (a) the same ratios between parameters q_3 and q_2, (b) the same values of parameter q_4

If the actual aggregate–cement ratio is lower than the right-hand side of (D.70), then model B3 gives a lower ratio q_3/q_2 than model B4 does, and stronger aging effects can be expected for B3 than for B4. Figure D.8a indicates that this is the case for most (but not all) concrete compositions from the NU creep database and also for the compositions used in Example 3.1 and in Sect. 7.9. Thus, with the exception of concretes with high w/c and a/c ratios (i.e., concretes with a low cement content and low strength), model B4 can be expected to predict less pronounced aging effects than model B3. This is true not only for the elastic modulus (calculated based on creep compliance for very short loading) but also for the evolution of creep compliance, as will be documented in the next example.

Example D.3. Compliance curves predicted by models B3, B4, and B4s for concrete from Example 3.1

The basic creep compliance curves predicted by models B3 (solid), B4 (dashed) and B4s (dotted) for the concrete from Example 3.1 are shown in Fig. D.9a. For each model, three curves corresponding to the ages of 7, 28, and 365 days are plotted. The distances between curves that correspond to the same model at different ages are the largest for B3, smaller for B4, and the smallest for B4s, which confirms what has already been said about the aging effects.

It is somewhat surprising that even though the calibration of B4 and B4s has been partially based on the bridge deflection data, which are expected to enhance long-term creep as compared to B3, the value of the combined parameter $nq_3 + q_4$ that controls the terminal slope of the creep curve in the log-time scale is, for the present typical concrete, the largest for B3, smaller for B4, and the smallest for B4s; see the corresponding line in Table D.9. This is reflected by the terminal slope of the basic creep curves in Fig. D.9a, which is the same as the terminal slope of the total creep curves in Fig. D.9b.

Model B3 gives much higher long-time values of basic creep compliance than B4, and B4 gives higher values than B4s, except for the case of loading at the age of

1 year (the bottom curves in each triplet). On the other hand, as shown in Fig. D.9c, model B4 gives a much higher contribution of drying creep than models B3 and B4s do. Because of that, the long-term total creep compliances of models B3 and B4, plotted in Fig. D.9b, are comparable for concrete loaded at 7 days, even though B4 gives substantially less basic creep. For concrete loaded at 28 days or 1 year, the total creep predicted by B4 is even larger than that predicted by B3, due to the large contribution of drying creep. But this contribution grows mainly between 10 days and 2000 days of load duration and virtually stops after 3000 days, which is why for longer load durations the creep rate predicted by B3 is higher. Model B4s gives rather low values of creep compliance compared to the other models, and the weakest effect of aging.   ∎

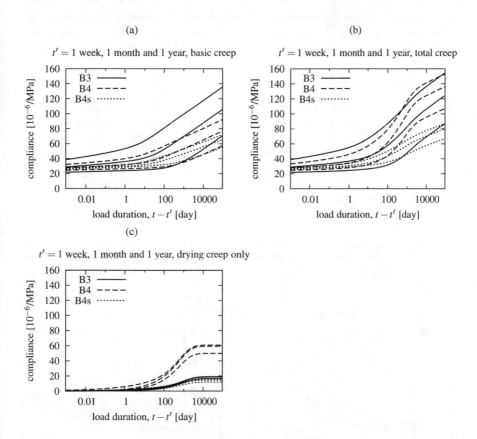

Fig. D.9 Comparison of creep compliance function at ages 7 days, 28 days, and 365 days for models B3, B4, and B4s: (a) basic creep, (b) total creep at drying, (c) drying part of creep

The foregoing comparison has been done for one specific concrete mix from Example 3.1, and not all conclusions are generally valid. Example 3.1 used concrete composition and strength very close to a real concrete tested by Komendant et al. [551]. To complete the picture, let us also compare the predictions for another concrete with properties that were assumed in Sect. 7.9 in the context of extrapolation of bridge deflections (see, e.g., Fig. 7.16).

Table D.10 Creep parameters for the concrete from Sect. 7.9 according to models B3, B4, and B4s (q_i in 10^{-6}/MPa, τ_{sh} in days)

	B3	B4	B4s
q_1	20.3	23.68	23.68
q_2	137.1	133.5	14.78
q_3	2.486	8.556	15.02
q_4	7.373	9.084	4.119
$nq_3 + q_4$	7.62	9.94	5.62
q_5	332.8	916.4	246.9
τ_{sh}	1819	1091	1679

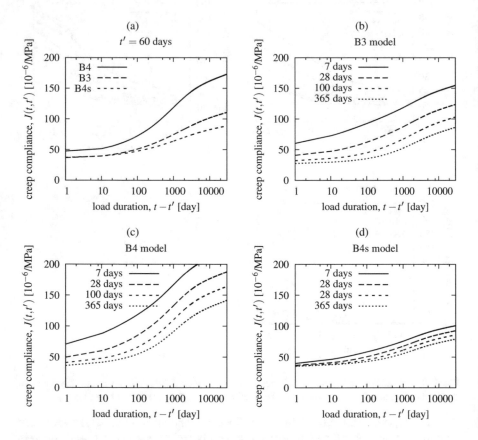

Fig. D.10 Total creep compliance functions for concrete from Sect. 7.9: (a) comparison of models, loaded at 60 days, (b)–(d) aging effect for individual models

Example D.4. Compliance curves predicted by models B3, B4, and B4s for concrete from Sect. 7.9

The concrete used in Sect. 7.9 was defined by the following characteristics: mean strength $\bar{f}_c = 39$ MPa, water-cement ratio $w/c = 0.5$, cement content $c = 400$ kg/m^3, and density $\rho = 2300$ kg/m^3. The effective cross-sectional thickness was set to $D = 0.25$ m and the environmental humidity to $h_{env} = 65\%$. The corresponding estimates of parameters predicted by models B3, B4, and B4s are given in Table D.10, and the creep compliance curves (total creep, including drying creep) for the age of 60 days are shown in Fig. D.10a.

Same as for the concrete from Example 3.1, model B4s gives less creep than B3 and B4 do, and also less aging (Fig. D.10d), because parameter q_2 is by an order of magnitude smaller than for model B3. B4s also gives a less steep terminal slope (controlled by $nq_3 + q_4$). On the other hand, unlike the previous example, model B4 exhibits aging effects comparable to B3 because it predicts almost the same value of parameter q_2 as model B3 does (Fig. D.10b, c). The ratio $E_\infty^{(c)}/E_{28}^{(c)}$ would be 1.59 for B3, 1.38 for B4, and only 1.038 for B4s.

Drying creep is again much higher for B4 (since parameter q_5 is almost three times higher than for B3, see Table D.10), and so the total creep is now very high for B4, and even the terminal slope of B4 is higher than for B3.   ∎

Finally, to provide a more general picture, Fig. D.11 shows the ratios between selected parameters predicted by models B4 and B3 for all concretes from the NU creep database which used cements of type R and for which the basic parameters $\bar{f}_c$, w/c, a/c, and c were reported. The parameters have been evaluated using the basic form of B4 predictions (D.5)–(D.7), with no admixture-dependent adjustments. The following trends can be identified: In comparison with model B3, model B4 gives lower values of parameters q_2 and q_4 for high-strength concretes (Fig. D.11a,c) and higher values of parameter q_3 for all concretes (Fig. D.11b). Furthermore, for high-strength concretes, model B4 predicts higher ratios q_3/q_2 (which means less pronounced aging effects) and lower values of the combined parameter $0.1\,q_3 + q_4$ (which means a less steep terminal slope of creep compliance curves).

D.8   Improvements of Model B4

D.8.1   Better Prediction of Drying Creep

According to the original version of the B4 model [136], parameter q_5 that controls the magnitude of drying creep compliance is estimated using formula (D.11), which can be rewritten as

$$q_5(h_{env}) = |k_h(h_{env})|^{-0.85} q_5^* \tag{D.71}$$

where

$$q_5^* = \frac{p_5}{1\text{ GPa}} \left(\frac{a/c}{6}\right)^{-1} \left(\frac{w/c}{0.38}\right)^{0.78} |\varepsilon_{sh}^\infty(t_0)|^{-0.85} \tag{D.72}$$

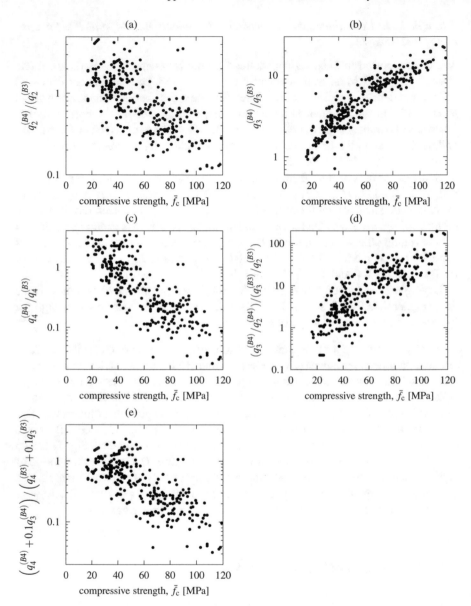

Fig. D.11 Ratios between predictions of models B4 and B3 evaluated for: (a) parameter q_2, (b) parameter q_3, (c) parameter q_4, (d) ratio q_3/q_2, (e) parameter combination $0.1\,q_3 + q_4$

is a reference value that would be obtained at zero ambient humidity. The effect of ambient humidity is in (D.71) incorporated via an auxiliary factor

$$k_h(h_{\text{env}}) = \begin{cases} 1 - h_{\text{env}}^3 & \text{if } h_{\text{env}} \leq 0.98 \\ 12.94\,(1 - h_{\text{env}}) - 0.2 & \text{if } 0.98 \leq h_{\text{env}} \leq 1 \end{cases} \tag{D.73}$$

Recall that model B3 uses a humidity-independent value of q_5, given by formula (3.24), see also line 8 of Table C.2. The dependence of q_5 on ambient humidity is a new feature of model B4. Unfortunately, for h_{env} close to 1, k_h is close to 0 and formula (D.71) with $|k_h|$ raised to a negative power leads to unrealistically high values of q_5. This deficiency results into nonphysical features in the dependence of drying creep compliance on ambient humidity, as illustrated in Fig. D.12 for the concrete from Example 3.1.

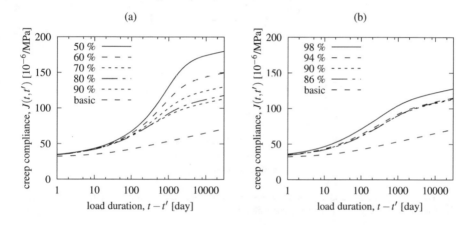

Fig. D.12 Original B4 model: influence of ambient relative humidity on creep compliance resulting from formula (D.71)

Normally, in concretes without admixtures and strong self-desiccation, one should get zero (or negligible) drying creep for $h_{\text{env}} = 0.98$ because this value is supposed to correspond to the sealed conditions, at which all the creep is basic. Figure D.12a shows that, in the range of h_{env} between 0.5 and 0.9, the additional creep compliance due to drying decreases with increasing humidity, in agreement with the expected trend. However, it is somewhat surprising that an increase of h_{env} from 0.8 to 0.9 results only into a mild reduction of drying creep and the total creep compliance curve at $h_{\text{env}} = 0.9$ is still quite far from the basic creep compliance curve. A detailed investigation reveals that when h_{env} is increased from 0.9 to 0.94 and further to 0.98, the natural trend gets reversed and drying creep increases, as documented in Fig. D.12b. This is clearly a nonphysical phenomenon.

The origin of the problem can be detected by examining the dependence of the final drying creep on ambient humidity, for a fixed concrete composition and fixed times at the onset of drying and at the onset of loading. The drying creep compliance function of model B4, given by (D.3), contains an auxiliary function $g(\hat{t})$, defined in (D.8). As the drying time $\hat{t} = t - t_0$ tends to infinity, the value of $g(t - t_0)$ approaches a finite limit, $g_\infty = p_{5H} h_{env}$, while the value of $g(t' - t_0)$ remains constant and can be expressed as $p_{5H} [1 - (1 - h_{env}) S_0]$, where $S_0 = \tanh \sqrt{(t' - t_0)/\tau_{sh}}$ (we consider a typical case with $t' \geq t_0$, i.e., loading after the onset of drying). The asymptotically approached final value of drying creep compliance is then given by

$$J_d^\infty (h_{env}) = |k_h (h_{env})|^{-0.85} q_5^* \sqrt{\exp (-p_{5H} h_{env}) - \exp (-p_{5H} [1 - (1 - h_{env}) S_0])}$$
(D.74)

Since the objective is to analyze the dependence of drying creep on ambient humidity, we have explicitly marked the dependence of J_d^∞ and k_h on h_{env}.

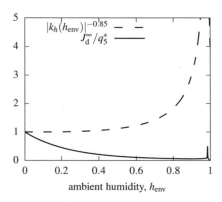

Fig. D.13 Original B4 model: dependence of factor $|k_h (h_{env})|^{-0.85}$ and of the final drying creep J_d^∞ (normalized by q_5^*) on ambient relative humidity

As already mentioned, the origin of the problem is in the negative exponent -0.85. The square-root expression in (D.74) decreases with increasing humidity, as expected, and tends to zero as h_{env} tends to 1. But $|k_h (h_{env})|^{-0.85}$, due to the negative exponent, increases with humidity increasing up to 0.9845; see the dashed curve in Fig. D.13. As an unpleasant consequence, the resulting value of J_d^∞, which is proportional to the product of $|k_h (h_{env})|^{-0.85}$ and the square-root term, does not decrease monotonically with increasing h_{env} over the entire range of humidities. The correct trend is obtained for sufficiently low humidities, for which $|k_h (h_{env})|^{-0.85}$ increases slowly and the decreasing square-root term dominates. On the other hand, in the range of humidities between 0.85 and 0.95, the two mutually counteracting influences are roughly balanced and the dependence of J_d^∞ on h_{env} is weak, with a local minimum. For humidities approaching 0.98 from below, the final drying creep increases, and for still higher humidities, it blows up. There is a singularity at $h_{env} = 0.9845$, when k_h evaluated according to the second line in (D.73) becomes equal to zero. Finally, for h_{env} between 0.9845 and 1, the final drying creep decreases fast from infinity to zero. The graph in Fig. D.13 has been constructed as an example for $S_0 = 0.1$, which corresponds to $t' - t_0 \approx 0.01\tau_{\text{sh}}$. Qualitatively, the same behavior is found for all meaningful values of S_0 between 0 and 1, as demonstrated in Fig. D.14.

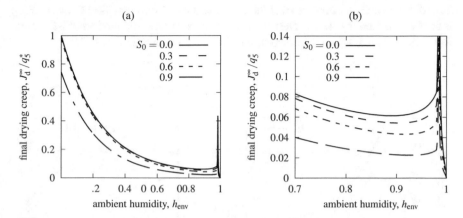

Fig. D.14 Original B4 model: dependence of the final drying creep J_d^∞ (normalized by q_5^*) on ambient relative humidity, plotted for various values of parameter $S_0 = \tanh\sqrt{(t' - t_0)/\tau_{\text{sh}}}$

Formula (D.11) was calibrated on an extensive set of experimental data, but it turns out that it does not possess the correct asymptotic properties for $h_{\text{env}} \rightarrow 1$. The reason why the problem was not detected during calibration is that the database contains only a limited number of drying creep tests performed at relatively elevated humidity levels around 0.8 or 0.9. Most experiments focused either on basic creep, or on drying creep at lower humidities, at which the increase of compliance as compared to sealed conditions is well pronounced. An inspection of the NU database reveals that 620 drying creep tests were run at humidities between 0.5 and 0.7, but only 8 tests at humidities between 0.71 and 0.9.

It is therefore reasonable to suppress the dependence of parameter q_5 on the ambient humidity and replace the humidity-dependent factor $|k_h(h_{\text{env}})|^{-0.85}$ in the originally suggested formula (D.71) by a constant[8] obtained by averaging over the range $0.5 \leq h_{\text{env}} \leq 0.7$. That constant turns out to be equal to 1.26, and so the B4 formula (D.11) for parameter q_5 is better replaced by

$$q_5 = \frac{1.26 p_5}{1\,\text{GPa}} \left(\frac{a/c}{6}\right)^{-1} \left(\frac{w/c}{0.38}\right)^{0.78} |\varepsilon_{\text{sh}}^\infty(t_0)|^{-0.85} \qquad (\text{D.75})$$

Note that $\varepsilon_{\text{sh}}^\infty(t_0)$ is no longer multiplied by k_h. Of course, the factor 1.26 could be absorbed by parameter p_5, but for compatibility with the parameter tables presented

---

[8]Alternatively, one could keep using the original formula for humidities up to 0.7 and replace $|k_h(h_{\text{env}})|^{-0.85}$ by a constant only in the range above 0.7. This modification would work reasonably well for cement types R and SL but not for cement type RS, for which the original formula gives a minimum of J_b^∞ already near $h_{\text{env}} = 0.7$ (due to a much lower value of parameter p_{5H}).

in the official RILEM recommendation, it is preferable to keep the value of p_5 unchanged and explicitly consider that it should be multiplied by 1.26. When q_5 is evaluated from (D.75), the dependence of the final drying creep J_b^∞ on the ambient humidity becomes monotonic over the entire range and the transition to basic creep is continuous; see Fig. D.15. The creep compliance curves obtained for the concrete from Example 3.1 at various ambient humidities are shown in Fig. D.16.

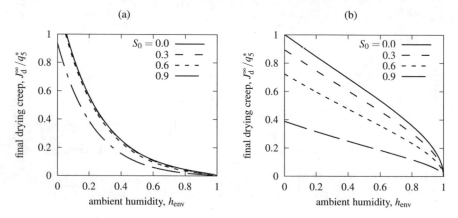

Fig. D.15 Modified formulation with humidity-independent parameter q_5 given by (D.75): dependence of the final drying creep J_d^∞ (normalized by q_5^*) on ambient relative humidity for (a) cements R and SL, (b) cement RS

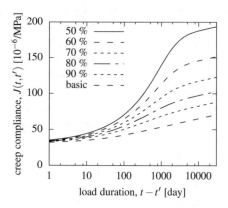

Fig. D.16 Modified formulation with humidity-independent parameter q_5 given by (D.75): influence of ambient relative humidity on creep compliance for the concrete from Example 3.1

The general lesson that can be learned from the detected discrepancies is that one cannot rely on mathematical optimization too much. The development of a sound model must still involve engineering insight, to fill the gaps where data are missing or limited.

D.8.2 Anticipated Future Improvements of Model B4

The analysis presented in Sect. 8.7, based on new experimental evidence for the long-term processes of hydration, self-desiccation, autogenous shrinkage, and swelling in water immersion, will doubtless allow major improvements of model B4 and the microprestress-solidification theory (MPS), with recalibration by a database with thousands of curves. They will probably take a few years and thus cannot be covered in this book (they are planned for the second edition). Nevertheless, let us at least list some of the anticipated developments [124, A5].

1. It has become clear that the drying shrinkage includes autogenous shrinkage in the core of test specimens and that drying creep tests are also affected by the autogeneous shrinkage in the core. The thicker the specimen, the longer the autogenous shrinkage goes on in the core. In the core of thick specimens, it may proceed for years, even decades. Since autogenous shrinkage has no size effect, it weakens the size effect on the shrinkage and on the drying part of creep, as observed on test specimens.
2. When autogenous shrinkage is not negligible, every basic creep test, i.e., creep of a sealed specimen, would have to be accompanied by a test of autogenous shrinkage (on a sealed specimen), which would have to be subtracted from the total strains of the loaded sealed specimen. The companion autogenous shrinkage test would have to be run for the entire duration of the creep test.
3. The thousands of tests of drying shrinkage and drying creep tests in the NU database would have to be reanalyzed, splitting the drying effect from the self-desiccation effect in the specimen core producing the autogenous part of shrinkage. Then, the thousands of data on the drying part and autogenous part of shrinkage tests and drying creep tests will have to be refitted with an update of model B4 (while simultaneously fitting the bridge database).
4. Function tanh for the B4 shrinkage curve will have to be updated. It is based on diffusion theory, which predicts the moisture content and shrinkage to approach a finite asymptotic value, i.e., a horizontal asymptote. Since (depending on the fineness of cement) the autogenous shrinkage may proceed logarithmically for years and probably even decades, it may cause the shrinkage to approach an inclined straight line in the logarithmic scale (some test data for small specimens hint at such an asymptote). The self-desiccation and autogenous shrinkage will also alter the aging effects in the drying shrinkage functions and drying creep functions. The total autogenous shrinkage must nevertheless be bounded, since the amount of reactants is finite. However, it may take decades, perhaps even centuries, to reach the bound.
5. When the autogenous shrinkage is strong, it also represents part of the strain measured in creep tests of sealed specimens. This means that not all of the time-dependent deformation of loaded sealed specimens represents creep. So the autogenous shrinkage would have to be separated from what was previously regarded as the basic creep data.

6. Since the autogenous shrinkage does not depend on the stress level and the true creep is proportional to it, the inclusion of autogenous shrinkage must cause the time-dependent deformation to depend on stress nonproportionally and thus cause a false impression of nonlinear dependence of the basic creep on stress in the low stress range. This would have to be taken into account in evaluating test data.

Appendix E
Creep Models Recommended by Design Codes

E.1 General Structure of Creep Design Formulae

This appendix summarizes the main equations and parameters recommended for creep design calculations by the International Federation for Structural Concrete (known under the French acronym *fib*) and its predecessor CEB-FIP, by the American Concrete Institute (ACI), and by the Japan Society of Civil Engineers (JSCE)

The International Federation for Structural Concrete (in French "fédération internationale du béton, *fib*") was created in 1998 by merging the Euro-International Committee for Concrete (Comité Euro-International du Béton, CEB) and the International Federation for Prestressing (Fédération internationale de la précontrainte, FIP). The *fib* Model Code 2010 is a successor of CEB Model Codes 1990 and 1999. The creep and shrinkage model incorporated into this code will be referred to as the *fib* model. The final version published in 2013 substantially differs from the earlier CEB model.[9]

The American Concrete Institute (ACI) has had a permanent committee TC 209 Creep and Shrinkage in Concrete. The creep model adopted by this committee in 1971 [11] was developed in the late 1960s by Branson and coworkers [252–254]. In 1976, based on Bažant et al. [114], Bažant (as member of ACI TC 209) proposed a model with a much higher multidecade creep, terminating with a logarithmic asymptote, which was not adopted. The ACI 1971 model was reapproved in the 1982 recommendations [12], which were expanded by the AAEM method (see Sect. 4.2). The most recent version, labeled as 209R-92, was published in 1992 [13] and again reapproved (with the sole dissenting vote of the first writer) in 2008. After the recent publication of bridge deflection data [138, 209–211], a debate on a revision has begun in ACI TC

---

[9]Note that the extensive criticism of the bounded nonlogarithmic long-term creep in the Model Code 2010 offered in Bažant et al. [208] applied only to the 2010 draft and not to the final version released in 2012.

© Springer Science+Business Media B.V. 2018

Z.P. Bažant and M. Jirásek, *Creep and Hygrothermal Effects in Concrete Structures*, Solid Mechanics and Its Applications 225, https://doi.org/10.1007/978-94-024-1138-6

209 by the time of writing. The ACI Guide for Modeling and Calculating Shrinkage and Creep in Hardened Concrete [14] for the first time presents not only the ACI model, but also the B3 model, the CEB model (predecessor of the *fib* model), and the Canadian GL2000 model proposed by Gardner and Lockman [407] and slightly modified by Gardner [406]. Model B3 was approved in 1995 as an international RILEM Recommendation [104], as the sole model recommended by RILEM.

Concrete design codes and recommendations usually classify concrete according to its strength. The fundamental property is the uniaxial compressive strength measured on cylinders of diameter 150 mm (6 in.) and height 300 mm (12 in.) at age 28 days. Due to inevitable scatter, the strength does not have a unique value (even for concrete produced in the same plant from the same mix and the same batch using the same technological procedure of mixing, casting, and curing). So, the strength needs to be considered as a random variable. The strength value used in design procedures for the assessment of the ultimate load-carrying capacity of a structure is not the mean value but a certain "safe lower bound."

In creep design, the structure is analyzed in the service state, and the mean value of deflection is of most interest. If the elastic modulus and creep properties of concrete are estimated from strength, the mean value of strength (rather than the lower bound) must be used. The codes specify an approximate formula that links the mean strength to the safe lower bound, called the "characteristic strength" by *fib* and the "specified strength" by ACI.

The *fib* Model Code (as well as the previous CEB Model Codes) uses the *characteristic* compressive strength, f_{ck}, understood as the value below which only 5% of all possible strength measurements may be expected to fall. The *mean* compressive strength, f_{cm}, is estimated as

$$f_{cm} = f_{ck} + 8 \text{ MPa} \tag{E.1}$$

For instance, concrete of grade C30 has the characteristic cylinder strength of 30 MPa and the mean strength of 38 MPa. [407] proposed another formula,

$$f_{cm} = 1.1 f_{ck} + 5 \text{ MPa} \tag{E.2}$$

which gives exactly the same difference between f_{cm} and f_{ck} for concrete of grade C30 but takes into account an increase of the standard deviation for higher grades.

The ACI recommendations [16] are based on the *specified* compressive strength, f'_c, which should satisfy the following two conditions:

- only 1 out of 100 test results is below a minimum value, defined as $f'_c - 500$ psi for low-strength concrete with $f'_c \leq 5000$ psi and as $0.9 f'_c$ for normal-strength concrete with $f'_c > 5000$ psi,
- only 1 out of 100 averages of 3 consecutive test results is below f'_c.

Here, a "test result" is understood as the average of values measured on two samples (cylinders). The required *average* (mean) compressive strength f'_{cr} is estimated from the specified compressive strength f'_c using the formula

$$f'_{cr} = \begin{cases} \max\left(f'_c + 1.34S, \; f'_c + 2.33S - 500 \text{ psi}\right) & \text{if } f'_c \leq 5000 \text{ psi} \\ \max\left(f'_c + 1.34S, \; 0.9f'_c + 2.33S\right) & \text{if } f'_c \leq 5000 \text{ psi} \end{cases} \qquad \text{(E.3)}$$

where S is the standard deviation of tests (for the specific concrete plant, determined from the previous tests). If the standard deviation is not available, the required mean compressive strength is evaluated as

$$f'_{cr} = \begin{cases} f'_c + 1000 \text{ psi} & \text{if } f'_c < 3000 \text{ psi} \\ f'_c + 1200 \text{ psi} & \text{if } 3000 \text{ psi} \leq f'_c \leq 5000 \text{ psi} \\ 1.1f'_c + 700 \text{ psi} & \text{if } f'_c \geq 5000 \text{ psi} \end{cases} \qquad \text{(E.4)}$$

Note that 3000 psi = 20.7 MPa and 5000 psi = 34.5 MPa.

To facilitate the comparison among various codes and models, we use a unified notation for the fundamental constants and variables: t = current time, t' = age at loading, t_0 = age at the end of curing (onset of drying), f'_c = characteristic or specified strength, $\tilde{f}_c$ = mean compressive strength (dependent on age), $\bar{f}_c$ = mean compressive strength at age 28 days, E = elastic modulus (dependent on age), E_{28} = conventional modulus, i.e., elastic modulus at age 28 days, φ = creep coefficient, φ_{28} = creep coefficient in its alternative definition,[10] To avoid ambiguity in structural analysis, the codes should specify the compliance function, J, which includes the initial elastic deformation and is the only characteristic that matters. h_{env} = relative humidity of the environment, V/S_e = ratio between the volume V of the concrete member and surface area S_e exposed to drying. These symbols replace the original ones used in the code specifications (e.g., f_{cm} and E_{cm} for the mean strength and modulus, or RH for the relative humidity). On the other hand, the original notation is kept for parameters that are specific to each individual model. In all the equations to follow, time and age should be substituted in days.

The design code specifications related to creep assessment usually combine the following components:

1. Time evolution of the (mean) elastic modulus, often derived from the time dependence of (mean) compressive strength.
2. Dependence of the creep coefficient on the current time and the age at loading.
3. Definition of the compliance function in terms of the elastic modulus and creep coefficient.

---

[10]The standard definition considers the creep coefficient as the ratio of the creep strain to the elastic strain induced by the same stress (as the creep strain) at the actual age of loading; see Eq. (3.14). This standard creep coefficient is used by the ACI model and will be denoted as φ. However, the *fib* model considers the creep coefficient as the ratio of the (actual) creep strain to the (fictitious) elastic strain that would be caused by the same stress applied at age 28 days. This alternative type of creep coefficient will be denoted as φ_{28}. It is used by the *fib* model and also by GL2000. No matter which specific definition of creep coefficient is used, it should be noted that the very concept of elastic strain is highly ambiguous, as it depends on the duration within which the load is applied, which can vary from 0.1 s to 1 h. To avoid ambiguity in structural analysis, the codes should specify the compliance function, J, which includes the initial elastic deformation and is the only characteristic that matters.

4. Set of rules for the estimation of model parameters based on the concrete mix properties, type of curing, size and shape of concrete member, ambient relative humidity, and similar factors.

E.2 CEB and *fib* Model Codes

A design code developed by CEB and FIP was first approved and published in 1991 under the name of "CEB-FIP Model Code: Design Code" and republished in 1993 in CEB Bulletins No. 213 and 214. The creep description embodied in this code was based on the work of a task group coordinated by Müller and Hilsdorf [641]. CEB and FIP merged in 1998 into *fib*, and an updated version of the code appeared in 1999 in *fib* Bulletins No. 1–3 and was co-opted in 2002 for Eurocode 2. A new update is referred to as the "*fib* Model Code 2010," and its final version was published in 2013 [381]. In June 2016, the *fib* Technical Council approved the start of activities on the development of a new code, under the working title of Model Code 2020.

The CEB and *fib* codes consider the total strain in concrete as the sum of the initial strain (in the sense of the instantaneous elastic strain), creep strain, shrinkage strain, and thermal strain. The cross-sectional approach is used; i.e., the variation of internal stresses and humidity across the cross section is neglected. Equations given below refer to the mean sectional behavior of a concrete member moist-cured at normal temperatures for not more than 14 days. They are intended for ordinary structural concrete with the mean compressive strength between 20 and 130 MPa, subjected to compressive stress not exceeding 40% of the mean strength (at loading) and exposed to an environment with the mean relative humidity between 40% and 100% and mean temperature between 5 and 30 °C. The age at loading should be at least 1 day.

E.2.1 CEB Model

The creep description according to the original CEB model can be summarized by the following equations:

$$E_{28} = 21.5 \text{ GPa} \times \alpha_E \times \left(\frac{\bar{f}_c}{10 \text{ MPa}} \right)^{1/3} \tag{E.5}$$

$$E(t) = E_{28} \sqrt{\exp\left(s[1 - \sqrt{28/t}] \right)} \tag{E.6}$$

$$\varphi_{28}(t, t') = \frac{\phi_{RH}\beta_f}{0.1 + t'^{0.2}} \left(\frac{t - t'}{\beta_H + t - t'} \right)^{0.3} \tag{E.7}$$

$$J(t, t') = \frac{1}{E(t')} + \frac{\varphi_{28}(t, t')}{E_{28}} \tag{E.8}$$

Equation (E.5) estimates the conventional modulus of elasticity from the mean compressive strength at age 28 days. Parameter α_E reflects the influence of aggregate type and is set to 1.0 for quartzite aggregates, 1.2 for basalt and dense limestone aggregates, 0.9 for limestone aggregates, and 0.7 for sandstone aggregates. The modulus of elasticity is understood as the unloading modulus in a static uniaxial compressive test, after previous loading to 40% of the compressive strength. The modulus obtained from (E.5) is expected to correspond to the reciprocal value of the creep compliance for a load duration of approximately 1 s, while the modulus used by ACI and given by (E.28) corresponds to a load duration of about 0.01 day.

Development of the elastic modulus with time (due to aging) is described by Eq. (E.6), in which s is a parameter equal to 0.38, 0.25, or 0.20, depending on the strength class of cement and hardening characteristics (e.g., $s = 0.25$ for normally hardening cement of strength class 42.5 or for rapidly hardening cement of strength class 32.5).

According to (E.7), the creep curves (after subtraction of the initial strain) have the same shape, but their amplitude depends on the age at loading. Parameters ϕ_{RH} and β_f express the influence of relative environmental humidity, h_{env}, and mean compressive strength at age 28 days, $\bar{f}_c$, and are given by the following expressions:

$$\phi_{RH} = \left[1 + 10\frac{1 - h_{\mathrm{env}}}{\sqrt[3]{2A_c/u}}\left(\frac{35}{\bar{f}_c}\right)^{0.7}\right]\left(\frac{35}{\bar{f}_c}\right)^{0.2} \tag{E.9}$$

$$\beta_f = \frac{16.8}{\sqrt{\bar{f}_c}} \tag{E.10}$$

The ratio $2A_c/u$ in (E.9) is the notional size of the member (to be substituted in mm), evaluated from the area of the cross section, A_c, and the perimeter of the member in contact with the atmosphere, u. The compressive strength, $\bar{f}_c$, has to be substituted in MPa. Finally, parameter β_H in (E.7) again depends on the strength, relative environmental humidity, and notional member size and is given by

$$\beta_H = 1.5\frac{2A_c}{u}\left[1 + (1.2h_{\mathrm{env}})^{18}\right] + 250\sqrt{\frac{35}{\bar{f}_c}} \tag{E.11}$$

If the right-hand side of (E.11) exceeds $1500\sqrt{35/\bar{f}_c}$, then β_H is set to this value as a limit value.

The effect of temperature on creep can be incorporated through the factor

$$\beta_T = \exp\left(\frac{1500}{T} - 5.12\right) \tag{E.12}$$

where T is the absolute temperature in K. Factor β_T multiplies β_H in (E.7). It is equal to 1 for $T = 293$ K, which corresponds to the reference temperature 20°C.

In summary, the compliance function of the CEB model, obtained by combining Eqs. (E.6)–(E.8), has the form

$$J(t, t') = \frac{1}{E_{28}} \exp\left(-\frac{s}{2}\left[1 - \sqrt{\frac{28}{t'}}\right]\right) + \frac{\phi_{RH}\beta_f}{E_{28}} \frac{1}{0.1 + t'^{0.2}} \left(\frac{t - t'}{\beta_H \beta_T + t - t'}\right)^{0.3}$$

(E.13)

E.2.2 *fib* Model

The updated version known as the *fib* Model Code 2010 reuses Eqs. (E.5), (E.6), and (E.8) of the original CEB model, but Eq. (E.7) is revised by adopting the separation of basic and drying creep introduced in models BP [175], BPKX [151], and B3.

$$\varphi_{28}(t, t') = \varphi_{bc}(t, t') + \varphi_{dc}(t, t')$$

(E.14)

The basic creep is represented by the unbounded logarithmic function[11]

$$\varphi_{bc}(t, t') = \frac{1.8}{\bar{f}_c^{0.7}} \ln\left[1 + \left(0.035 + \frac{30}{t'_{adj}}\right)^2 (t - t')\right]$$

(E.15)

and drying creep by the bounded function

$$\varphi_{dc}(t, t') = \frac{412}{\bar{f}_c^{1.4}} \frac{\phi_{RH}}{0.1 + (t'_{adj})^{0.2}} \left(\frac{t - t'}{\beta_H + t - t'}\right)^{\gamma(t'_{adj})}$$

(E.16)

with the mean compressive strength $\bar{f}_c$ substituted in MPa.

In formulae (E.15) and (E.16), t'_{adj} is the adjusted age at loading, which reflects the effect of elevated or reduced temperatures and of the cement type on the maturity of concrete. For cement strength classes 32.5 R and 42.5 N, the adjusted age t'_{adj} is equal to the temperature-adjusted age t'_T, defined by the Arrhenius-type rate equation

$$\frac{dt'_T}{dt} = \exp\left(\frac{Q_T}{RT_0} - \frac{Q_T}{RT}\right)$$

(E.17)

in which $Q_T/R = 4000$ K, $T_0 = 293$ K is the reference temperature, and T is the concrete temperature (mean value over the section), substituted in K. For other cement strength classes, the temperature-adjusted age t'_T is transformed into the adjusted age t'_{adj} according to the formula

$$t'_{adj} = t'_T \left(1 + \frac{9}{2 + (t'_T)^{1.2}}\right)^{\alpha}$$

(E.18)

---

[11]The decomposition into basic and drying creep and the asymptotic logarithmic character of the basic creep compliance bring the *fib* model closer to models B3 and B4. These features were not present in the first draft of the *fib* code, published in 2010, which was criticized by Bažant et al. [208]. They were incorporated into the final version, published in 2013.

with exponent $\alpha = -1$ for strength class 32.5 N, $\alpha = 0$ (leading to $t'_{adj} = t'_T$) for strength classes 32.5 R and 42.5 N, and $\alpha = 1$ for strength classes 42.5 R, 52.5 N, and 52.5 R. The code stipulates that $t'_{adj} \geq 0.5$ days.

Note that the load duration $t - t'$ in (E.15) and (E.16) is based on the actual physical time, not on the adjusted one. The influence of the member size on drying creep is in (E.16) reflected by parameter

$$\beta_H = 1.5 \frac{2A_c}{u} + 250 \sqrt{\frac{35}{f_c}} \tag{E.19}$$

Same as for the CEB model, the ratio $2A_c/u$ is the notional size of the member (to be substituted in mm), and if the right-hand side of (E.19) exceeds $1500\sqrt{35/f_c}$, then β_H is set to this value as a limit value. The influence of environmental humidity is taken into account by parameter

$$\phi_{RH} = 10 \frac{1 - h_{env}}{\sqrt[3]{2A_c/u}} \tag{E.20}$$

and the exponent in (E.16) is given by

$$\gamma(t'_{adj}) = \frac{1}{2.3 + 3.5/\sqrt{t'_{adj}}} \tag{E.21}$$

with t'_{adj} substituted in days, as usual.

The temperature-adjusted age at loading t'_{adj} takes into account the effect of temperature prior to loading. The effect of temperature on the development of creep can be incorporated in the same way as for the CEB model, i.e., using factor β_T given by (E.12), which multiplies parameter β_H in (E.16). In addition to that, the basic creep coefficient φ_{bc} should be multiplied by ϕ_T and the drying creep coefficient φ_{dc} should be multiplied by $\phi_T^{1.2}$, with

$$\phi_T = \exp[0.015(T - T_0)] \tag{E.22}$$

where $T_0 = 293$ K is the reference temperature. The code also specified a transient creep coefficient, to be used if the increase of temperature occurs while the structural member is under load.

E.3 ACI Model

The empirical model developed by Branson and Christiason [253] was incorporated into the recommendations of the American Concrete Institute [11] and reapproved in the later versions [12–14]. The question of a replacement is currently under discussion. The version presented here corresponds to the guide No. 209.2R-08, published by ACI Committee 209 in 2008.

The ACI model uses the following equations:

$$\tilde{f}_c(t) = \frac{t}{a + bt} \tilde{f}_c \tag{E.23}$$

$$E(t) = 0.043\sqrt{\rho^3 \tilde{f}_c(t)} \tag{E.24}$$

$$\varphi(t, t') = \frac{(t - t')^{\psi}}{d + (t - t')^{\psi}} \varphi_u(t') \tag{E.25}$$

$$J(t, t') = \frac{1 + \varphi(t, t')}{E(t')} \tag{E.26}$$

Times t and t' have to be substituted in days. Parameters a and b in (E.23) depend on the type of cement and type of curing. For moist-cured concrete and cement of type I (ordinary Portland cement), their recommended values are $a = 4$ and $b = 0.85$. In (E.24), ρ denotes the mass density of concrete in kg/m^3, the strength $\tilde{f}_c$ should be substituted in MPa, and the resulting modulus E is also in MPa. Combining (E.23) with (E.24), we can describe the evolution of the elastic modulus by

$$E(t) = E_{28}\sqrt{\frac{t}{a + bt}} \tag{E.27}$$

where

$$E_{28} = 0.043\sqrt{\rho^3 \tilde{f}_c} \tag{E.28}$$

is the elastic modulus at age 28 days. For normal-weight concrete, formula (E.28) can be replaced by (3.6). If the modulus is measured directly, it is preferable to use in (E.27) the actually measured (mean) value instead of the estimate (E.28). It could be somewhat disturbing that if the typical parameters $a = 4$ and $b = 0.85$ are used, the fraction in (E.27) is not exactly equal to 1 for $t = 28$ days. This is caused by a truncation error. To get the exact coincidence between $E(28)$ and E_{28}, one should use $b = 6/7$ instead of $b = 0.85$, and Eq. (E.27) then reads

$$E(t) = E_{28}\sqrt{\frac{7t}{28 + 6t}} \tag{E.29}$$

The creep coefficient specified in (E.25) has the standard meaning according to the definition (3.14). The recommended values of the parameters in (E.25) are $d = 10$ and $\psi = 0.6$. The ultimate creep coefficient for standard conditions is $\varphi_u = 2.35$. The standard conditions in the sense of the code are described by a number of parameters specifying the concrete composition and curing, member geometry and environment, and loading. For instance, it is assumed that the load is applied at the end of curing, at age 7 days for moist-cured concrete or 1–3 days for steam-cured concrete, and that the ambient relative humidity is $h_{\text{env}} = 40\%$ and the volume–surface ratio is $V/S_e = 38$ mm. The volume-to-surface ratio of the concrete member V/S_e is equivalent to a half of the notional size $2A_c/u$ used by the *fib* Model Code if the ends of prismatic specimens or members are sealed and thus not counted in S_e. For instance, for an infinite slab, V/S_e is one half of the notional size.

For other than standard conditions, φ_u is corrected by the product of six factors that depend on t', h_{env}, V/S_e, slump, ratio of fine aggregate to total aggregate by weight, and air content:

$$\varphi_u(t') = 2.35 \, \gamma_1 \gamma_2 \gamma_3 \gamma_4 \gamma_5 \gamma_6 \, t'^{-m} \tag{E.30}$$

where

$$\gamma_1 = \begin{cases} 1.25 & \text{for moist curing} \\ 1.13 & \text{for steam curing} \end{cases} \tag{E.31}$$

$$\gamma_2 = 1.27 - 0.67 \, h_{env} \tag{E.32}$$

$$\gamma_3 = \frac{2}{3} \left(1 + 1.13 \, e^{-0.0213 \, V/S_e} \right) \tag{E.33}$$

$$\gamma_4 = 0.82 + 0.00264 \, s_{sl} \tag{E.34}$$

$$\gamma_5 = 0.88 + 0.0024 \, a_f/a_t \tag{E.35}$$

$$\gamma_6 = \max(1, \ 0.46 + 0.09 \, \alpha_{air}) \tag{E.36}$$

$$m = \begin{cases} 0.118 & \text{for moist curing} \\ 0.094 & \text{for steam curing} \end{cases} \tag{E.37}$$

In these equations, t' is the age at loading in days, h_{env} is the ambient relative humidity, V/S_e is the volume–surface ratio in mm, s_{sl} is the slump in mm, a_f is the mass fraction of fine aggregate, a_t is the mass fraction of total aggregate, and α_{air} is the air content in percent.

In summary, the compliance function of the ACI model, obtained by combining Eqs. (E.25)–(E.27) and (E.30) and substituting the recommended values of parameters $d = 10$ and $\psi = 0.6$, has the form

$$J(t, t') = \frac{1}{E_{28}} \sqrt{b + \frac{a}{t'}} \left[1 + \frac{2.35\gamma}{(t')^m} \frac{(t - t')^{0.6}}{10 + (t - t')^{0.6}} \right] \tag{E.38}$$

where γ denotes the product of factors γ_1 to γ_6, given by (E.31)–(E.36).

E.4 GL2000 Model

The model proposed by Gardner and Lockman [407] and denoted as the GL2000 model is a major modification of the earlier Atlanta97 model (or GZ model) of Gardner and Zhao [408]. The modification co-opts significant aspects of the 1978 BP model [175], particularly the mathematical form of shrinkage dependence on the drying time and thickness (or the volume/surface ratio), and the additive separation of drying creep from basic creep. Minor adjustments are incorporated in the final version [406], reproduced in the ACI Guide 209.2R-08 [14]. The model is applicable to concretes with characteristic strength up to 82 MPa that do not experience self-desiccation.

The GL2000 model in the form described in the ACI Guide 209.2R-08 can be summarized by the following equations:

$$\tilde{f}_{\mathrm{c}}(t) = \bar{f}_{\mathrm{c}} \exp\left(s(1 - \sqrt{28/t})\right) \tag{E.39}$$

$$E(t) = 3.5 + 4.3\sqrt{\tilde{f}_{\mathrm{c}}(t)} \tag{E.40}$$

$$\varphi_{28}(t, t') = \varPhi\left[\frac{2(t - t')^{0.3}}{(t - t')^{0.3} + 14} + \sqrt{\frac{7(t - t')}{t'(t - t' + 7)}} + c_h\sqrt{\frac{t - t'}{t - t' + 0.12(V/S_{\mathrm{e}})^2}}\right] \tag{E.41}$$

$$J(t, t') = \frac{1}{E(t')} + \frac{\varphi_{28}(t, t')}{E(28)} \tag{E.42}$$

The strength evolution factor in (E.39) has the same form as in the *fib* Model Code, but the relation between the elastic modulus and the strength is different. The recommended values of the strength development parameter s are 0.335 for cement of type I (ordinary Portland cement), 0.4 for type II (modified cement), and 0.13 for type III (rapid-hardening cement). Equation (E.40) is valid in the SI units, with E in GPa and $\tilde{f}_{\mathrm{c}}$ in MPa. Combining (E.39) and (E.40), we can express the evolution of elastic modulus more directly as

$$E(t) = 3.5 + (E_{28} - 3.5)\sqrt{\exp\left(s(1 - \sqrt{28/t})\right)} \tag{E.43}$$

where

$$E_{28} = E(28) = 3.5 + 4.3\sqrt{\tilde{f}_{\mathrm{c}}} \tag{E.44}$$

is the conventional modulus.

In expression (E.41) for the creep coefficient, the volume–surface ratio V/S_{e} needs to be substituted in mm. Parameter

$$c_h = 2.5(1 - 1.086h_{\mathrm{env}}^2) \tag{E.45}$$

depends on the ambient humidity and vanishes for $h_{\mathrm{env}} = 0.96$, which is the value approximately corresponding to the relative pore humidity under sealed conditions. Therefore, the last term in the brackets in (E.41) (which contains c_h) corresponds to drying creep, and the first two terms correspond to basic creep. The model also takes into account the effect of drying before loading. If the member is loaded at the same time as it is exposed to drying, parameter $\varPhi$ is equal to 1 (and thus can be omitted from (E.41)). However, if the first loading occurs at age t_1 larger than the age t_0 at the onset of drying, the correction factor is evaluated as

$$\Phi = \sqrt{1 - \sqrt{\frac{t_1 - t_0}{t_1 - t_0 + 0.12(V/S_e)^2}}} \qquad \text{(E.46)}$$

Note that Φ depends on the age at first loading, t_1, which does not need to coincide with t' in the compliance function. Of course, in a standard creep test, the load is applied at once and $t_1 = t'$. However, if the material response is evaluated at variable stress according to the principle of superposition, t' in the integral stress–strain equation sweeps the interval from t_1 to t, but the factor Φ to be used in (E.41) is evaluated from t_1 and thus remains constant.

In the original version of GL2000 [407], the constant multiplying $(V/S_e)^2$ in (E.41) and (E.46) was 0.15 instead of 0.12, and the evolution of strength was described by

$$\tilde{f}_c(t) = \bar{f}_c \frac{t^{0.75}}{a + bt^{0.75}} \qquad \text{(E.47)}$$

instead of (E.39).

In summary, the compliance function of the GL2000 model in its final version, obtained by combining Eqs. (E.41)–(E.43), has the form

$$J(t, t') = \frac{1}{3.5 + (E_{28} - 3.5) \exp\left(\frac{s}{2}\left[1 - \sqrt{\frac{28}{t'}}\right]\right)} + \qquad \text{(E.48)}$$
$$+ \frac{\Phi}{E_{28}}\left[\frac{2(t - t')^{0.3}}{(t - t')^{0.3} + 14} + \sqrt{\frac{7(t - t')}{t'(t - t' + 7)}} + c_h\sqrt{\frac{t - t'}{t - t' + 0.12(V/S_e)^2}}\right]$$

E.5 JSCE Model

The creep model recommended by the Japan Society of Civil Engineers (JSCE) as a part of the Standard Specifications for Concrete Structures [786] is applicable to concrete with water/cement ratio between 0.4 and 0.65 and strength up to 55 MPa (or up to 70 MPa if the water/cement ratio is reduced to increase the strength), loaded by stresses not exceeding 40% of the strength, at ambient relative humidities between 45 and 80%, and for volume/surface ratios between 100 and 300 mm.

The JSCE guidelines specify the additional strain due to creep separately from the elastic strain. The corresponding compliance function can be written as

$$J(t, t') = \frac{1}{E(t')} + \Delta J(t')\left[1 - \exp\left(-0.09(t - t')^{0.6}\right)\right] \qquad \text{(E.49)}$$

where $E(t')$ is the elastic modulus at age t' and

$$\Delta J(t') = 1.5(c + w)^2 (w/c)^{2.4} (\ln t')^{-0.67} \times 10^{-9} / \text{MPa} + \tag{E.50}$$
$$+ 450(c + w)^{1.4} (w/c)^{4.2} \left(\ln \frac{V}{10 \, S_e} \right)^{-2.2} (1 - h_{\text{env}})^{0.36} t_0^{-0.3} \times 10^{-9} / \text{MPa}$$

is the final increase of compliance due to creep for concrete loaded at age t'. The first term in (E.50) represents the contribution of basic creep, and the second is the additional compliance due to drying creep. The cement and water contents c and w should be substituted in kg/m^3, volume V in mm^3, surface in contact with outside air S_e in mm^2, and times t, t', and t_0 in days.

To take into account the effect of temperature different from 20 °C, the actual ages t, t', and t_0 should be replaced by temperature-adjusted ages computed in the same way as according to the *fib* model; see (E.17).

For high-strength concrete with compressive strength exceeding 55 MPa, a different formula for the compliance function is recommended:

$$J(t, t') = \frac{1}{E(t')} + \frac{4w(1 - h_{\text{env}}) + 350}{12 + f_c(t')} \ln(1 + t - t') \times 10^{-6} / \text{MPa} \tag{E.51}$$

where $f_c(t')$ is the compressive strength at the age of loading, substituted in MPa, w is the water content in kg/m^3, and t and t' are substituted in days. It is interesting that compliance function (E.51) for high-strength concrete is logarithmic, in contrast to the bounded compliance function for normal-strength concrete specified in (E.49).

The JSCE code also approves creep predictions based on the B3 model, ACI model, or CEB model.

E.6 Comparison of Compliance Functions

For illustration, the graphs of compliance functions will be plotted for the five models described in this appendix and for the B3 model. The purpose of the figures is merely to show the shape of the compliance graphs and their main features; a systematic comparison and evaluation are not attempted here. In addition to creep in an environment of 70% relative humidity, basic creep will be considered as a special case.

Parameters of individual models are determined or estimated for a concrete with the same composition and under the same conditions as in Example 3.1. From the mean strength $\bar{f_c} = 45.4$ MPa, the conventional modulus is estimated according to the ACI formula (3.6) as $E_{28} = 31.9$ GPa, while the GL formula (E.44) gives $E_{28} = 32.5$ GPa. Assuming limestone aggregates ($\alpha_E = 0.9$), the *fib* formula (3.7) gives 32.0 GPa. Since these values are quite close, we set $E_{28} = 32$ GPa for all the models.

The values of parameters of model B3 are taken from Example 3.1: $q_1 = 18.81$, $q_2 = 126.9$, $q_3 = 0.7494$, $q_4 = 7.692$, $q_5 = 327.0$, all in 10^{-6}/MPa; $\tau_{sh} = 1121$ days, and $\varepsilon_{sh}^\infty = 701.1 \times 10^{-6}$.

For the CEB model, we consider $s = 0.25$ (normally hardening cement of strength class 42.5), parameter $\beta_f = 2.49$ follows from (E.10), and parameters $\phi_{RH} = 1.46$ and $\beta_H = 376$ follow from (E.9) and (E.11) with $h_{env} = 0.7$ and $2A_c/u = 100$ mm. For basic creep, ϕ_{RH} is set to 1, and parameter β_H evaluated from (E.11) would exceed its maximum allowed value, $1500\sqrt{35/\bar{f_c}} = 1317$, so it is set to that value.

For the *fib* model, parameters $\phi_{RH} = 0.6463$ and $\beta_H = 370$ follow from (E.20) and (E.19) with $\bar{f_c} = 45.4$ MPa, $h_{env} = 0.7$, and $2A_c/u = 100$ mm. The temperature-adjusted age t'_{adj} is considered as equal to the actual age, t'.

For the ACI model, we consider parameters $a = 4$, $b = 6/7$, and $m = 0.118$ (moist-cured concrete and cement of type I). If we knew the mass density ρ, we could estimate the conventional modulus from (E.28). For instance, for $\rho = 2400$ kg/m^3, we would get 34.1 GPa, which is not that far from the value of 31.9 GPa obtained from (3.6). Since the exact mass density is not specified, $E_{28} = 32$ GPa will be used, same as for the other models. It remains to determine parameters $\gamma_i, i = 1, 2, \ldots, 6$, and their product, γ. We set $\gamma_1 = 1.25$ (moist curing), and for $h_{env} = 0.7$ and $V/S_e = 100$ mm, we get $\gamma_2 = 0.801$ and $\gamma_3 = 0.756$. Since the slump, fraction of fine aggregate, and air content are not known, parameters γ_4, γ_5, and γ_6 are set to 1 (default value). The resulting parameter γ is thus $\gamma = 1.25 \times 0.801 \times 0.756 = 0.757$. For basic creep, we take $\gamma_2 = 0.627$ corresponding to $h_{env} = 0.96$, and the resulting value of γ is 0.593. It is somewhat disturbing that parameter γ and thus also the compliance function is affected by the size of the member even in the absence of drying (through γ_3, which depends on V/S_e). Note that the effect of the member size is incorrectly introduced by vertical scaling of the compliance function rather than by its horizontal shift in the log-scale.

For the GL2000 model, V/S_e is set to 100 (mm) and, for $h_{env} = 0.7$, formula (E.45) gives $c_h = 1.17$. According to (E.46), parameter Φ depends on the age at first loading, t_1, and on $t_0 = 7$ (days) and $V/S_e = 100$ (mm). In the present example, we consider loading at ages $t_1 = 7$, 28, and 365 days, and the corresponding values of Φ are 1.0, 0.932, and 0.722. In the case of basic creep, c_h is set to zero and Φ is considered as 1, because drying in fact never starts and thus it does not precede loading.

For the JSCE model, the compliance increase due to creep is determined from the cement content $c = 450$ kg/m^3, water content $w = 170$ kg/m^3, volume/surface ratio $V/S_e = 100$ mm, and onset of drying at $t_0 = 7$ days. Formula (E.50) gives $\Delta J(7) = 35.7 \times 10^{-6}$/MPa, $\Delta J(28) = 24.9 \times 10^{-6}$/MPa, and $\Delta J(365) = 17.0 \times 10^{-6}$/MPa for basic creep, and $\Delta J(7) = 39.2 \times 10^{-6}$/MPa, $\Delta J(28) = 28.4 \times 10^{-6}$/MPa, and $\Delta J(365) = 20.5 \times 10^{-6}$/MPa for creep at drying. The evolution of elastic modulus is estimated using the ACI formula (E.27).

Figure E.1 displays the compliance curves for the six models considered here. The left column corresponds to basic creep and the right one to drying creep, in each case for three different ages at loading (from top to bottom, $t' = 7, 28$, and 365 days).

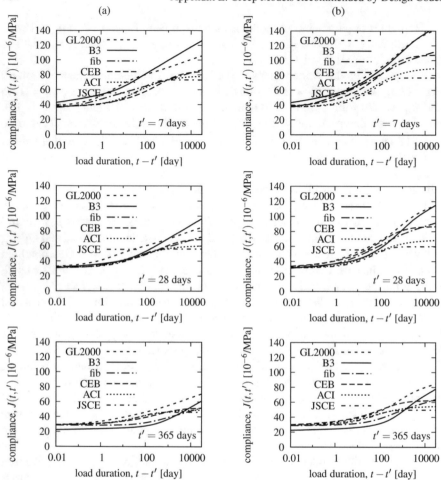

Fig. E.1 Compliance functions predicted by various creep models: (a) basic creep, (b) creep at ambient humidity $h_{\text{env}} = 70\%$

One drawback of the code formulae according to CEB, ACI, and JSCE, which is best shown in Fig. E.1b, is that they lead to bounded creep curves. Long-time experiments indicate that the creep curves approach straight lines in the semilogarithmic scale, which means that, for long load durations, the compliance grows as a logarithmic function and thus is unbounded. This feature is directly incorporated into models B3 and B4 and into the new *fib* model (see the solid curves in Fig. E.1) and is also reasonably well captured by the GL2000 model, which uses a bounded compliance function, but the finite limit would be approached for load durations by several orders of magnitude larger than any durations of practical interest. To illustrate that, the compliance functions for age at loading 28 days (including the effect of drying creep) are replotted in Fig. E.2 for hypothetical load durations up to 10^9 days.

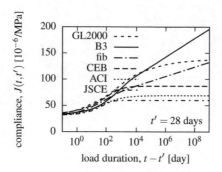

Fig. E.2 Compliance functions at age 28 days and ambient humidity $h_{env} = 70\%$ predicted by various creep models and plotted up to extremely long load durations

Note that the slope of the line approached by the new *fib* model is substantially lower than for the B3 model. This is true in general, because the coefficient multiplying the logarithmic term is $q_4 + 0.1q_3$ for the B3 model and $1.8/(E_{28}\bar{f}_c^{0.7})$ for the *fib* model. In the present example, these coefficients amount to $7.77 \times 10^{-6}/MPa$ and $3.89 \times 10^{-6}/MPa$, resp. In general, $q_4 = 20.3(a/c)^{-0.7} \times 10^{-6}/MPa$, which is for aggregate/cement ratios between 2 and 6 roughly in the range between 6 and $12 \times 10^{-6}/MPa$, while the coefficient for the *fib* model is evaluated as

$$\frac{1.8}{E_{28}} \times \left(\frac{1MPa}{\bar{f}_c}\right)^{0.7} = \frac{1}{\alpha_E}\left(\frac{1MPa}{\bar{f}_c}\right)^{1.03} \times 185 \times 10^{-6}/MPa \qquad (E.52)$$

which is for limestone aggregates ($\alpha_E = 0.9$) and for strengths between 20 and 45 MPa roughly in the range between 3 and $8 \times 10^{-6}/MPa$.

For instantaneous visualization of the creep and shrinkage curves predicted by Model B3 for various input values, K.-T. Kim developed a program described in [545] and downloadable as "Model B3 Creep Design Aid Program" from the first author's website (http://www.civil.northwestern.edu/people/bazant).

Appendix F
Continuous Retardation Spectrum

The concept of a *retardation spectrum*, to be presented in this section, is very effective computationally and is also useful for deeper understanding of the behavior of a viscoelastic material, especially with regard to the time scale at which the viscous processes take place. One Kelvin unit has a well-defined characteristic time, the retardation time $\tau = \eta/E$, which sets the intrinsic time scale of the model and determines which loading rates are considered as "slow" and which ones as "fast." The retardation time τ and the compliance $1/E$ are parameters of the compliance function (A.23) and uniquely characterize the model. A Kelvin chain can be described by the retardation times of the individual units, τ_μ, and the corresponding compliances, $1/E_\mu$, $\mu = 1, 2, \ldots, M$. It turns out that a general viscoelastic model can be characterized by a continuous spectrum of retardation times and compliances. From the mathematical point of view, such a spectrum is related to the inverse Laplace transform of the compliance function.

In a similar spirit, one could define the *relaxation spectrum*, which would be related to the inverse Laplace transform of the relaxation function, and for a Maxwell chain would be discrete. As already mentioned, models for creep of concrete usually specify the compliance function and not the relaxation function, and so we will restrict our attention to the retardation spectrum.

F.1 Relation Between Compliance Function and Retardation Spectrum

Consider the Dirichlet series in (A.25), representing the compliance function of a nonaging Kelvin chain. The diagram of the compliances $1/E_\mu$ versus $\ln \tau_\mu$ is called the retardation spectrum of the material. For a Kelvin chain model with a finite number M of Kelvin units, the spectrum is discrete, consisting of a set of vertical lines (Fig. F.1a). However, it is advantageous to conceive a generalization of equation (A.25) in which the spectrum is continuous (Fig. F.1b); that is, the chain consists of

© Springer Science+Business Media B.V. 2018
Z.P. Bažant and M. Jirásek, *Creep and Hygrothermal Effects in Concrete Structures*, Solid Mechanics and Its Applications 225,
https://doi.org/10.1007/978-94-024-1138-6

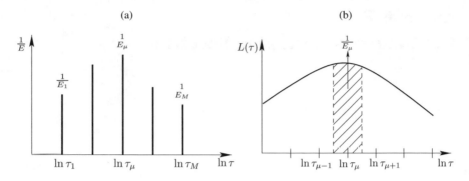

Fig. F.1 (a) Discrete and (b) continuous retardation spectrum

infinitely many Kelvin units with infinitely small compliances $1/E_\mu$ and with the retardation times τ_μ distributed infinitely closely. According to this generalization, well known from classical viscoelasticity (e.g., [818]), one has, as the limit case of (A.25),

$$\Phi(t) = \int_{\tau=0}^{\infty} L(\tau) \left(1 - e^{-t/\tau}\right) d(\ln \tau), \qquad t \geq 0 \tag{F.1}$$

in which function $L(\tau)$ characterizes the continuous spectrum, and the compliance function of the nonaging material is now denoted as Φ instead of J_0, just for formal reasons. It should be noted that if L is a regular function (i.e., with no Dirac-delta-like terms), $\Phi(t)$ is continuous and vanishes at $t = 0$. Therefore, the instantaneous (elastic) compliance is not reflected in the spectrum and must be added separately, which is straightforward.

The continuous spectrum is very useful when a given compliance function (e.g., defined by an analytical formula containing power functions and logarithms) is to be approximated by a Dirichlet series, which is needed for an efficient rate-type numerical approach (Sect. 5.2). Of course, one could try to construct the approximation directly, by minimizing a suitable measure of the difference between the "exact" compliance function and the Dirichlet series. This is a somewhat tedious procedure, which can be circumvented by considering the Dirichlet series (A.25) as a numerical approximation of the integral in (F.1). If function $L(\tau)$ is known and the discrete retardation times τ_μ are selected, determination of the compliances $1/E_\mu$ based on a suitable numerical quadrature scheme leads to explicit formulae.

An important point is that a good approximation of the continuous spectrum can be obtained analytically, exploiting the Post-Widder formula [866, 867] for the inversion of Laplace transform. It can be shown that the sequence of approximations

$$L_k(\tau) = -\frac{(-k\tau)^k}{(k-1)!} \, \Phi^{(k)}(k\tau), \qquad k = 1, 2, \ldots \tag{F.2}$$

converges to the continuous spectrum, i.e.,

$$L(\tau) = \lim_{k \to \infty} L_k(\tau) \tag{F.3}$$

Here, k is the desired order of approximation, and $\Phi^{(k)}$ denotes the kth derivative of the compliance function. This approach can be used only if the compliance function $\Phi(t)$ is sufficiently smooth. Therefore, scattered test data cannot be used directly but must be fitted by a smooth compliance function before the spectrum can be approximated by (F.2). Readers who are not interested in the mathematical background of the Post-Widder formula can skip the remaining part of this section and proceed directly to Sect. F.2.

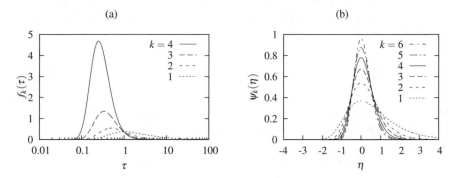

Fig. F.2 (a) Functions $f_k(\tau) = \tau^{-k} e^{-t/\tau}$ for fixed $t = 1$, (b) functions $\psi_k(\eta)$

Formula (F.2) can be derived by the differentiation of Eq. (F.1). Taking the kth derivative with respect to t, we obtain

$$\Phi^{(k)}(t) = (-1)^{k-1} \int_{\tau=0}^{\infty} L(\tau) \tau^{-k} e^{-t/\tau} \, d(\ln \tau) \tag{F.4}$$

Now consider the function

$$f_k(\tau) = \tau^{-k} e^{-t/\tau} \tag{F.5}$$

that multiplies $L(\tau)$ in the above integral. For a fixed value of t, this function of τ tends to zero as $\tau \to 0^+$ and as $\tau \to \infty$ and attains its maximum at $\tau = \bar{\tau}_k \equiv t/k$; see Fig. F.2a. Therefore, the integral is mainly affected by the value of L near this point. If we replace $L(\tau)$ by $L(t/k)$, we can take it out of the integral and construct the approximation

$$\Phi^{(k)}(t) \approx (-1)^{k-1} L\left(\frac{t}{k}\right) \int_{\tau=0}^{\infty} \tau^{-k} e^{-t/\tau} d(\ln \tau) = (-1)^{k-1} L\left(\frac{t}{k}\right) \frac{(k-1)!}{t^k} \tag{F.6}$$

Formal replacement of t by $k\tau$ then leads to the Post-Widder formula (F.2).

On purpose, we have presented the derivation in a somewhat sloppy way, to keep it simple and to emphasize the main idea, namely that the spiky function $f_k(\tau)$ multiplying $L(\tau)$ in the integral acts as a filter that samples the value of L near $\bar{\tau}_k \equiv t/k$. A rigorous mathematical proof can be based on a series of functions ψ_k that are related to f_k and converge to the Dirac distribution, for which the value of the integral indeed depends exclusively on the value of L at $\bar{\tau}_k$. To make sure that all functions in this series are centered around the same point, we first write them as functions of a dimensionless time variable $\xi = \tau/\bar{\tau}_k$. Setting

$$\phi_k(\xi) = f_k(\bar{\tau}_k \xi) \tag{F.7}$$

we obtain functions that all attain their maximum at $\xi = 1$. In the next step, we normalize these functions such that the area under their graphs becomes equal to unity. Since the integral in (F.4) uses a differential of $\ln \tau$, we need to scale the function by factors

$$I_k = \int_{\xi=0}^{\infty} \phi_k(\xi)\, \mathrm{d}(\ln \xi) = \int_{\tau=0}^{\infty} f_k(\tau)\, \mathrm{d}(\ln \tau) = \int_0^{\infty} \tau^{-k-1} \mathrm{e}^{-t/\tau}\, \mathrm{d}\tau = \frac{(k-1)!}{t^k} \tag{F.8}$$

Finally, to facilitate the interpretation of the limit as the Dirac distribution, it is useful to consider $\ln \xi$ as another dimensionless variable, η, which varies from minus to plus infinity.

Combining all these considerations, we define functions

$$\psi_k(\eta) = \frac{1}{I_k} \phi_k(\mathrm{e}^{\eta}) = \frac{1}{I_k} f_k(\bar{\tau}_k \mathrm{e}^{\eta}) \tag{F.9}$$

Their graphs are plotted in Fig. F.2b. Each graph has its peak at $\eta = 0$ (which corresponds to $\xi = 1$ and to $\tau = \bar{\tau}_k$) and the area under it is equal to unity. It can be shown that functions ψ_k tend to the Dirac distribution δ as $k \to \infty$. In terms of these functions, Eq. (F.4) can be rewritten as

$$\Phi^{(k)}(t) = (-1)^{k-1} \int_{\tau=0}^{\infty} L(\tau) f_k(\tau)\, \mathrm{d}(\ln \tau) = (-1)^{k-1} I_k \int_{-\infty}^{\infty} L(\bar{\tau}_k \mathrm{e}^{\eta}) \psi_k(\eta)\, \mathrm{d}\eta \tag{F.10}$$

from which

$$\int_{-\infty}^{\infty} L(\bar{\tau}_k \mathrm{e}^{\eta}) \psi_k(\eta)\mathrm{d}\eta = (-1)^{k-1} \frac{\Phi^{(k)}(t)}{I_k} = (-1)^{k-1} \frac{t^k \Phi^{(k)}(t)}{(k-1)!} = L_k\left(\frac{t}{k}\right) \tag{F.11}$$

where L_k is the function defined in (F.2) and representing the kth order approximation of the retardation spectrum L. As k approaches infinity, the integral on the left-hand side of (F.11) tends to

$$L(\bar{\tau}_k \mathrm{e}^0) = L(\bar{\tau}_k) = L\left(\frac{t}{k}\right) \tag{F.12}$$

which proves the Post-Widder formula (F.3).

Finally, it is worth noting that the original Post-Widder formula applies to the approximate inversion of Laplace transform, and its extension to the approximation of retardation spectrum is, under certain assumptions, justified by splitting the integral in (F.1) into two parts:

$$\Phi(t) = \int_{\tau=0}^{\infty} L(\tau) \, d(\ln \tau) - \int_{\tau=0}^{\infty} L(\tau) e^{-t/\tau} \, d(\ln \tau) \tag{F.13}$$

Using substitution $z = 1/\tau$, the last integral can be rewritten as

$$\int_{\tau=0}^{\infty} L(\tau) e^{-t/\tau} \, d(\ln \tau) = \int_{0}^{\infty} \frac{1}{z} L\left(\frac{1}{z}\right) e^{-tz} \, dz \tag{F.14}$$

which corresponds to the *Laplace transform* of function L^* defined by the formula $L^*(z) = L(1/z)/z$. Since the first integral on the right-hand side of (F.13) is independent of t, it can be concluded that the compliance function Φ differs from the Laplace transform of $-L^*$ only by a constant and therefore has the same derivatives. Application of the original Post-Widder formula to $-L^*$ then leads to (F.2) and (F.3). However, this reasoning is correct only if the first integral on the right-hand side of (F.13) is finite. This is the case for the retardation spectra of bounded compliance functions, such as the functions used by ACI or CEB (to be discussed in Sect. F.3), or the drying creep compliance function of models B3/B4 (Sect. F.4), but not for the retardation spectra of unbounded compliance functions, such as the basic creep function of model B3 (Sect. F.2). For unbounded compliance functions, the integral in (F.1) is still finite, but it cannot be split as suggested in (F.13) because the integrals on the right-hand side would both be infinite. The derivation presented here in (F.7)–(F.12), which follows [522], does not make any direct reference to the Laplace transform and shows that formulae (F.2)–(F.3) are applicable even to models such as B3 or B4, with a logarithmic growth of the compliance function.

Special treatment would be needed for models with an asymptotically linear compliance growth. For instance, for the simple Maxwell model and for $t > 0$, the compliance function (A.11) is linear and its second and higher derivatives vanish, which means that approximations (F.2) of all orders higher than 1 would vanish as well. Since models with creep rate asymptotically approaching a nonzero limit are not realistic for concrete, this case does not need to be considered here.

F.2 Spectrum of Log-Power Law

To demonstrate the convergence properties of the Post-Widder formula and to give an example of a continuous retardation spectrum, we consider the compliance function given by the log-power law

$$\Phi_0(t) = q_3 \ln[1 + (t/\lambda_0)^n]$$ (F.15)

in which $\lambda_0 = 1$ day, and n and q_3 are empirical constants. Note that this function is actually used by the B3 and B4 models (Chap. 3) as the compliance function of the nonaging constituent, in the context of the solidification theory (Chap. 9). Since q_3 is just a scaling factor and the standard value of parameter λ_0 is 1 day, we focus attention on the function

$$\Phi(t) = \ln(1 + t^n) = \ln(f(t))$$ (F.16)

where

$$f(t) = 1 + t^n$$ (F.17)

is an auxiliary function, introduced for convenience.

F.2.1 Straightforward Application of Post-Widder Formula

In terms of the auxiliary function f given by (F.17), the derivatives of Φ can be expressed in a relatively simple form:

$$\Phi' = \frac{f'}{f}$$ (F.18)

$$\Phi'' = \frac{f''}{f} - \frac{f'^2}{f^2}$$ (F.19)

$$\Phi''' = \frac{f'''}{f} - \frac{3f'f''}{f^2} + \frac{2f'^3}{f^3}$$ (F.20)

$$\Phi^{IV} = \frac{f^{IV}}{f} - \frac{4f'f''' + 3f''^2}{f^2} + \frac{12f''f'^2}{f^3} - \frac{6f'^4}{f^4}$$ (F.21)

For convenience, we have omitted the argument t, and we have denoted the derivatives with respect to t by primes instead of dots. For the fourth and higher derivatives, we use Roman superscripts IV, V, etc. A general derivative of order k will be denoted by a superscript in parentheses. The derivatives of function f are given by

$$f^{(k)} = n_k t^{n-k}$$ (F.22)

where[12]

$$n_k = \prod_{i=0}^{k-1}(n - i)$$ (F.23)

---

[12]Formula (F.23) means that $n_1 = n$, $n_2 = n(n - 1)$, $n_3 = n(n - 1)(n - 2)$, etc.

are auxiliary constants introduced for the sake of brevity. Substituting all this into (F.2), we can construct the first four approximations of $L(\tau)$. Only the first three are presented here explicitly:

$$L_1(\tau) = \tau \Phi'(\tau) = \frac{n\tau^n}{1+\tau^n} \tag{F.24}$$

$$L_2(\tau) = -4\tau^2 \Phi''(2\tau) = \frac{n(2\tau)^n[1-n+(2\tau)^n]}{[1+(2\tau)^n]^2} \tag{F.25}$$

$$L_3(\tau) = \frac{27}{2}\tau^3 \Phi'''(3\tau) = \frac{n(n-1)(n-2)(3\tau)^n}{2[1+(3\tau)^n]} - \frac{3n^2(n-1)(3\tau)^{2n}}{2[1+(3\tau)^n]^2} + \frac{n^3(3\tau)^{3n}}{[1+(3\tau)^n]^3} \tag{F.26}$$

For higher orders, the complexity of the formulae increases, but they are still manageable.

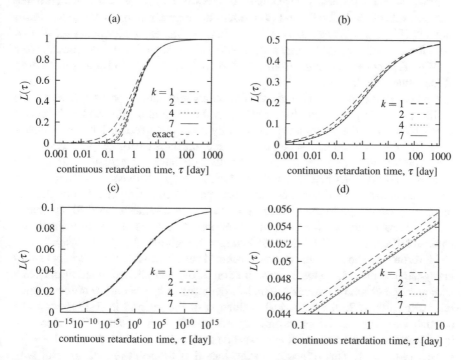

Fig. F.3 Approximations of the continuous retardation spectrum of the log-power law by the Post-Widder formula of various orders for (a) exponent $n = 1$, (b) exponent $n = 0.5$, (c)–(d) exponent $n = 0.1$

Convergence of the approximations $L_k(\tau)$ to the continuous retardation spectrum $L(\tau)$ is first examined for the simplest case with exponent $n = 1$. In this case, the formulae are greatly simplified (because the derivatives of f of second and higher orders vanish). It is even possible to give an explicit formula for the approximations of an arbitrary order,

$$L_k(\tau) = \left(\frac{k\tau}{1+k\tau}\right)^k, \qquad k = 1, 2, \ldots \tag{F.27}$$

and find the limit

$$L(\tau) = \lim_{k \to \infty} \left(\frac{k\tau}{1 + k\tau} \right)^k = e^{-1/\tau} \tag{F.28}$$

Figure F.3a shows, in semilogarithmic scale, selected approximations up to order 7 and their limit (F.28). The value of L grows monotonically from 0 to 1 as τ varies from 0 to infinity. The extreme values are captured correctly by all approximations, but in the intermediate range (τ between 0.1 and 1), the error is considerable even for $k = 7$.

Fortunately, the quality of the approximations is much better for smaller values of exponent n, as shown in Fig. F.3b, c. Here, the exact expression for $L(\tau)$ is not available, but it is clear that for $n = 0.5$, the difference between the approximations of orders 4 and 7 is already very small (Fig. F.3b), and for $n = 0.1$, which is the standard value used in the B3 and B4 models, all approximations seem to be almost identical (Fig. F.3c). We can also see that the range of values τ over which a substantial variation of L takes place becomes much wider if the exponent n is decreased. Note that Fig. F.3c covers the range from 10^{-15} to 10^{15} while Fig. F.3d shows a close-up of the range from 0.1 to 10.

Once a good approximation of the continuous retardation spectrum has become available, we can proceed to discretization of the integral in (F.1) and construct an approximation of the compliance function by a Dirichlet series. This procedure is somewhat complicated by the fact that the integration is performed with respect to $\ln \tau$, which spans an infinite domain. The discrete spectrum contains only a finite number of retardation times τ_μ, but the part of the integral between $\tau = 0$ and $\tau = \tau_1$ corresponds to a semi-infinite interval in the space of $\ln \tau$. One might think that the integration can be performed with respect to τ over a finite interval $[0, \tau_1]$, but if $d(\ln \tau)$ is replaced by $d\tau/\tau$, then the function to be integrated with respect to τ tends to infinity as $\tau \to 0^+$ (provided that $n < 1$, which is always the case).

Nevertheless, the problem can be overcome if one recalls that the Dirichlet series is expected to provide a close approximation of the compliance function only in a certain range of load durations t, which is determined by the choice of the discrete retardation times. So the numerical evaluation of the integral in (F.1) needs to be accurate only for t of the same order of magnitude as τ_1 or larger. Therefore, for those load durations t that are of interest and for all values $\tau \ll \tau_1$, the term $1 - e^{-t/\tau}$ is very close to 1. This brings us to the idea that the semi-infinite interval with $\ln \tau$ ranging from $-\infty$ to $\ln \tau_0$ (where τ_0 will be specified later) can be processed separately, with $1 - e^{-t/\tau}$ set to 1, and the integral of $L(\tau)$ over that interval becomes a constant, which in fact represents the compliance $1/E_0$ of the spring considered as the zeroth unit in the Kelvin chain. The next part of the integral, from τ_0 to τ_M, is discretized using a simple midpoint rule.

To achieve a good compromise between efficiency and accuracy, the discrete retardation times are often selected such that they form a geometric progression with quotient 10, i.e., $\tau_\mu = 10^{\mu-1}\tau_1, \mu = 2, 3 \ldots M$. Each of these times is representative of one order of magnitude, covering the interval from $\tau_\mu/\sqrt{10}$ to $\tau_\mu\sqrt{10}$. It is thus natural to set $\tau_0 = \tau_1/\sqrt{10}$ and to assign equal integration weights to all times τ_μ (this choice corresponds to the midpoint rule). The numerical approximation of (F.1) is then written as

$$\Phi(t) = \int_{\tau=0}^{\infty} L(\tau)\left(1 - e^{-t/\tau}\right)\,d(\ln\tau) \approx$$

$$\approx \int_{\tau=0}^{\tau_1/\sqrt{10}} L_k(\tau)\,d(\ln\tau) + (\ln 10)\sum_{\mu=1}^{M} L_k(\tau_\mu)(1 - e^{-t/\tau_\mu}) \equiv \Phi_k(t) \quad \text{(F.29)}$$

This approximation corresponds to the Dirichlet series with

$$\frac{1}{E_0} = \int_{\tau=0}^{\tau_1/\sqrt{10}} L(\tau)\,d(\ln\tau) \quad \text{(F.30)}$$

$$\frac{1}{E_\mu} = (\ln 10)\,L(\tau_\mu), \qquad \mu = 1, 2, \ldots, M \quad \text{(F.31)}$$

The part of the integral from τ_M to infinity is neglected. The approximation is then inaccurate for $t \gg \tau_M$ but still acceptable for $t \leq \tau_M$. Note that for t in the order of τ_M or smaller and for $\tau \gg \tau_M$, the value of $1 - e^{-t/\tau}$ is very small and the contribution to the integral can be neglected.

For $k = 1$ and $k = 2$, the integral of $L_k(\tau)$ can be evaluated analytically and the resulting formulae for the zeroth term of the approximation (compliance of the elastic spring) read

$$\frac{1}{E_0} = \ln(1 + \tau_0^n) \qquad \text{for } k = 1 \quad \text{(F.32)}$$

$$\frac{1}{E_0} = \ln(1 + (2\tau_0)^n) - \frac{n(2\tau_0)^n}{1 + (2\tau_0)^n} \qquad \text{for } k = 2 \quad \text{(F.33)}$$

where $\tau_0 = \tau_1/\sqrt{10}$.

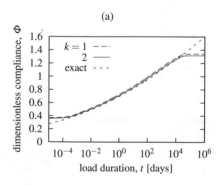

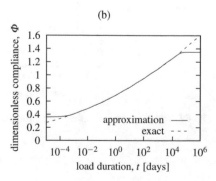

Fig. F.4 Approximations of the compliance function (log-power law) by a Dirichlet series based on (a) Post-Widder formulae of orders 1 and 2, (b) Post-Widder formula of order 2 with additional corrections

Accuracy of the approximation in (F.29) is illustrated by a specific example with $\tau_1 = 10^{-3}$ day and $M = 8$. For the Post-Widder formulae of orders $k = 1$ and $k = 2$, approximation (F.29) is graphically shown in Fig. F.4a, along with the exact compliance function (F.16).

The agreement within the relevant range of times is not bad but, upon closer examination, the approximation by Dirichlet series is found to be systematically above the exact values, as if the whole curve were shifted. This is partially due to the difference between L_k and L, but raising the order of the Post-Widder formula to $k = 3$ or 4 brings only a partial improvement. Since the compliance curves are very close to the exact one, the accuracy is better assessed by looking at the relative error, defined as $[\Phi_k(t) - \Phi(t)]/\Phi(t)$, where Φ_k is the approximation defined in (F.29).

As shown in Fig. F.5, the relative error exhibits oscillations, caused by the finite spacing of the retardation times, but stays (within the range of interest covered by the selected retardation times, i.e., from 10^{-3} day to 10^4 day) below 6% for the first-order and below 3% for the second-order approximation. For higher-order approximations, some effort is required to compute $1/E_0$ based on the integral of $L_k(\tau)$, because analytical integration would be quite tedious. Therefore, $1/E_0$ is evaluated by integrating L analytically from $\tau = 0$ to $\tau = 10^{-20}$ using the exact integral of L_2, and then, the integration of L_k from $\tau = 10^{-20}$ to $\tau = \tau_1/\sqrt{10} = 10^{-3.5}$ day is performed numerically, using the Simpson rule. The error further decreases as compared to $k = 2$, but not substantially, and the compliance is still systematically overestimated; see the curve for $k = 4$ in Fig. F.5, with relative error between 1% and 1.5%. For the third-order approximation (F.26), adopted by Bažant and Xi [202], the relative error is around 2%, which can be acceptable for practical applications. However, the accuracy can be dramatically increased by a modification described in the next subsection, and then, the simpler second-order formula (F.25) turns out to be fully sufficient.

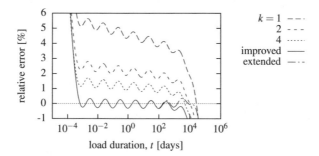

Fig. F.5 Relative error of various approximations of the compliance function (log-power law)

F.2.2 Improved Technique Based on Shifted Retardation Times

As shown in Fig. F.5, improving the accuracy by an increase of the approximation order k is too tedious and not really worth the effort. Fortunately, Jirásek and Havlásek [522] proposed another technique, much easier, albeit somewhat heuristic. If all retardation times are multiplied by the same correction factor (while leaving the associated compliances unchanged), the graph of the Dirichlet series approximation is shifted to the right and the accuracy can be substantially increased. The optimum value of the correction factor is easily found by a trial-and-error procedure. A more rigorous approach could be based on nonlinear optimization techniques, but the result would anyway depend mainly on the choice of the range in which the error (in the least-square sense) should be minimized. In the present example, if the compliances $1/E_\mu$ are determined based on the Post-Widder formula with $k = 2$ and then all retardation times are multiplied by 1.35, the error oscillates around a mean value close to zero (see the solid curve labeled as "improved" in Fig. F.5), and the approximation becomes almost optimal within the range of interest, with relative error below 0.3%.

Interestingly, it is possible to introduce yet another improvement, which expands the range of times over which the approximation is acceptable. The compliance associated with the longest relaxation time, τ_M, can be increased such that the times just above τ_M are better represented. In fact, since the integral is numerically evaluated only up to $\tau = \tau_M \sqrt{10}$ and the remaining part up to $\tau = \infty$ is neglected, it makes sense to increase the upper limit at which the integral is truncated, which corresponds to an increase of the integration weight of the last point and thus of the compliance $1/E_M$. How much to increase the upper limit is hard to determine theoretically, but it can be done again by trial and error, looking at the resulting effect on the compliance function. For the present example, an increase of $1/E_\mu$ by a factor 1.2 has been found to work best. The final approximation of the compliance function is plotted in Fig. F.4b, and the corresponding error is shown by the curve labeled as "extended" in Fig. F.5 (which coincides for times t up to 100 days with the solid curve labeled as "improved").

Let us emphasize that the correction factors 1.35 and 1.2, applied to the retardation times and to the compliance associated with the largest retardation time, were determined empirically by Jirásek and Havlásek [522], but it can be verified that they are applicable independently of the choice of the first retardation time τ_1 and of the number M of terms in the Dirichlet series. The relative error after the correction oscillates around zero and does not exceed ±0.3%. However, all this refers to the log-power law (F.15), which forms the basis of the solidification part of model B3. For other types of compliance functions, such factors need to be re-evaluated.

It may seem strange why we apply correction factors to an approximation that is based on a mathematical formula with proven convergence. The explanation is that while the standard approximations do converge to the exact solution, they do not necessarily do so in the optimal way. The reason becomes apparent if one looks at the graphs in Fig. F.2. Functions ψ_k shown in Fig. F.2b converge to the Dirac distribution but are not symmetric. They have been constructed by a certain transformation of the original functions f_k in Fig. F.2a, based on the condition that the peak value should always be attained at $\eta = 0$. But then the values for positive η are somewhat larger than for negative η with the same magnitude, and since function L is increasing, the integral in (F.10) is necessarily larger than the value of L at the sampling point $\bar{\tau}_k$. It is thus logical that the approximations L_k converge to L monotonically from above and that the corresponding compliances are overestimated. In principle, one could use another criterion for centering functions ψ_k around the origin. For instance, using the center of gravity instead of the maximum point could be regarded as a reasonable choice. In that case, $\bar{\tau}_k = t/k$ would be replaced by $\bar{\tau}_k^* = t/c_k$, where

$$c_k = \exp\left(-\frac{\int_0^\infty \zeta^{-k} e^{-1/\zeta} \ln \zeta \, d(\ln \zeta)}{\int_0^\infty \zeta^{-k} e^{-1/\zeta} \, d(\ln \zeta)}\right) \tag{F.34}$$

The integral in the denominator is equal to $(k-1)!$, but the integral in the numerator must be evaluated numerically. If a modified approximation L_k^* is constructed using the same approach as in (F.7)–(F.12) but with $\bar{\tau}_k$ replaced by $\bar{\tau}_k^*$, the last part of formula (F.11) needs to be replaced by

$$L_k^*\left(\frac{t}{c_k}\right) = (-1)^{k-1} \frac{t^k \Phi^{(k)}(t)}{(k-1)!} \tag{F.35}$$

Comparing this expression with the original form of (F.11), we obtain the relation $L_k^*(t/c_k) = L_k(t/k)$, which can also be written as

$$L_k(\tau) = L_k^*\left(\frac{k}{c_k}\tau\right) \tag{F.36}$$

This means that the compliance evaluated from the standard formula (based on L_k) for a certain retardation time τ_μ should, according to the modified approach (based on L_k^*), actually refer to retardation time $(k/c_k)\tau_\mu$, where k/c_k is a certain correction factor.

For $k = 2$, the value of the integral in the numerator of (F.34) is close to -0.4228, and constant c_2 is thus equal to $e^{0.4228} = 1.526$. The correction factor $2/c_2 = 2/1.526 \approx 1.31$ justified by the foregoing considerations is quite close to the optimal scaling factor 1.35 found by numerical calculations. For higher-order approximations, similar factors could be derived; their values would be lower and would tend to 1 with increasing approximation order k. In this sense, the modified approach converges to the same limit as the standard one, but the convergence is faster.

A good agreement between the theoretical correction factor 1.31 and the empirical optimal factor 1.35 is related to the fact that the actual spectrum L of the log-power law is in a wide range of retardation times an almost linear function of $\ln \tau$; see Fig. F.3d. Since $\ln \tau$ differs from $\ln \xi$ only by a constant, the dependence of L on $\eta = \ln \xi$ is also almost linear. In such a case, the choice of function ψ_2^* with the center of gravity (instead of the maximum point) placed at the origin of the η-space makes the integral of $\eta \psi_2^*(\eta)$ vanish, and the error of the approximation is thus substantially reduced. However, this is not a universal rule. For compliance functions with a nonnegligible curvature of the corresponding spectrum, application of a constant correction factor to all retardation times leads to a partial improvement only, as will be shown in the next sections.

To summarize the results of our analysis and to provide hints for practical applications, we can give the following recommendations:

- For approximation of the log-power law (F.15) with $n = 0.1$ and $\lambda_0 = 1$ day by Dirichlet series, the Post-Widder formula with $k = 2$ is considered as the best compromise between simplicity and accuracy. If an error in the order of 3% is deemed negligible from the engineering point of view, the formula can be used directly. The compliances of the Kelvin chain are then given by simple expressions

$$\frac{1}{E_0} = q_3 \ln(1 + \tilde{\tau}_0) - \frac{q_3 \tilde{\tau}_0}{10(1 + \tilde{\tau}_0)}, \quad \tilde{\tau}_0 = \left(\frac{2\tau_1}{\sqrt{10}}\right)^{0.1} \tag{F.37}$$

$$\frac{1}{E_\mu} = (\ln 10) \frac{q_3 \tilde{\tau}_\mu (0.9 + \tilde{\tau}_\mu)}{10(1 + \tilde{\tau}_\mu)^2}, \quad \tilde{\tau}_\mu = \left(2\tau_\mu\right)^{0.1}, \quad \mu = 1, 2, \dots, M \tag{F.38}$$

where τ_1 is the shortest retardation time and $\tau_\mu = 10^{\mu-1}\tau_1, \mu = 2, 3, \dots, M$.
- If a high accuracy is desired, the retardation times can be multiplied by the empirical factor 1.35 and the compliance $1/E_M$ associated with the longest retardation time τ_M can be multiplied by another empirical factor 1.2. This procedure is much more efficient than increasing the order of Post-Widder formula, k. It is advisable to check the accuracy by comparing the approximation to the exact compliance function and to adjust the empirical factors, if needed.

It is appropriate to stress again the purpose of the developed procedure, since it may seem that we have started from a compliance function given by a very simple analytical formula and after a series of steps we end up with its approximation that has a complicated form and is valid only for a certain range of load durations. True, but the point is that if the compliance function is converted to the Dirichlet series that corresponds to a Kelvin chain, the stress–strain relation can be described by a set of differential equations (governing individual units of the chain) instead of an integral expression. This is inevitable when pore humidity or temperature vary, and has tremendous advantages for numerical solutions covering general loading conditions, as explained in detail in Sect. 5.2.

F.3   Spectra of ACI and CEB Models

Let us now explore the retardation spectra of the creep models recommended by ACI or CEB. These models take into account aging, and their compliance functions depend on two arguments, t and t'. According to both codes, the compliance functions have the general form

$$J(t, t') = \frac{1}{E(t')} + c\,\frac{\Phi(t - t')}{g(t')} \tag{F.39}$$

in which $E(t')$ is the aging elastic modulus, c is a positive constant (independent of t and t', but possibly dependent on environmental humidity, concrete composition, etc.), $\Phi(t - t')$ is an increasing function describing the shape of the creep curve, and $g(t')$ is an increasing function that reflects aging (creep is reduced if the concrete is loaded at a higher age). For simplicity, we omit here multiplication by $H(t - t')$, and so the expression in (F.39) is valid for $t \geq t'$ and should be complemented by $J(t, t') = 0$ for $t < t'$. If we approximate function Φ by a Dirichlet series with constant partial moduli $E_\mu^{(\Phi)}$, $\mu = 0, 1, 2, \ldots, M$, the complete compliance function J can easily be approximated by

$$J(t, t') = \frac{1}{E(t')} + \frac{c}{E_0^{(\Phi)} g(t')} + \sum_{\mu=1}^{M} \frac{c}{E_\mu^{(\Phi)} g(t')} \left(1 - e^{-(t-t')/\tau_\mu}\right) \tag{F.40}$$

This is a Dirichlet series for aging materials, with age-dependent partial moduli.

The compliance function of the ACI model is based on formula (E.38) and can be presented in the form (F.39) with function Φ given by

$$\Phi_A(t) = \frac{t^n}{a + t^n} = 1 - \frac{a}{a + t^n} = 1 - \frac{a}{f(t)} \tag{F.41}$$

where $f(t) = a + t^n$ is an auxiliary function that differs from the function defined in (F.17) only by a constant and thus has the same derivatives, given in (F.22). Parameters n and a correspond to ψ and d in the ACI notation, and their standard values are $n = \psi = 0.6$ and $a = d = 10$ (with t expressed in days). In terms of the derivatives of function f, the derivatives of Φ_A are conveniently expressed as

$$\Phi_A' = a\,\frac{f'}{f^2} \tag{F.42}$$

$$\Phi_A'' = a\left(\frac{f''}{f^2} - \frac{2f'^2}{f^3}\right) \tag{F.43}$$

$$\Phi_A''' = a\left(\frac{f'''}{f^2} - \frac{6f'f''}{f^3} + \frac{6f'^3}{f^4}\right) \tag{F.44}$$

$$\Phi_A^{IV} = a\left(\frac{f^{IV}}{f^2} - \frac{8f'f''' + 6f''^2}{f^3} + \frac{36f'^2 f''}{f^4} - \frac{24f'^4}{f^5}\right) \tag{F.45}$$

$$\Phi_A^V = a\left(\frac{f^V}{f^2} - \frac{10f'f^{IV} + 20f''f'''}{f^3} + \frac{60f'^2 f''' + 90f'f''^2}{f^4} - \frac{240f'^3 f''}{f^5} + \frac{120f'^5}{f^6}\right) \tag{F.46}$$

For the CEB model, the compliance function (E.13) can also be expressed in the form (F.39) with function Φ given by

$$\Phi_C(t) = \left(\frac{t}{\beta + t}\right)^n = [f(t)]^n \tag{F.47}$$

where $n = 0.3$ and β (equal to the product $\beta_H \beta_T$ according to the CEB notation) is a parameter dependent on the compressive strength, environmental humidity, temperature, and equivalent thickness of the concrete member (typical values of β range between 250 and 1500 days). The auxiliary function is now defined as $f(t) = t/(\beta + t)$, and its kth derivative is

$$f^{(k)}(t) = \frac{(-1)^{k-1} k! \beta}{(\beta + t)^{k+1}} \tag{F.48}$$

The derivatives of Φ_C are then expressed as

$$\Phi_C' = n_1 f^{n-1} f' \tag{F.49}$$
$$\Phi_C'' = n_2 f^{n-2} f'^2 + n_1 f^{n-1} f'' \tag{F.50}$$
$$\Phi_C''' = n_3 f^{n-3} f'^3 + 3n_2 f^{n-2} f' f'' + n_1 f^{n-1} f''' \tag{F.51}$$
$$\Phi_C^{IV} = n_4 f^{n-4} f'^4 + 6n_3 f^{n-3} f'^2 f'' + n_2 f^{n-2}(4f' f''' + 3f''^2) + n_1 f^{n-1} f^{IV} \tag{F.52}$$
$$\Phi_C^V = n_5 f^{n-5} f'^5 + n_4 f^{n-4} f'^3 f'' + n_3 f^{n-3}(15 f' f''^2 + 10 f'^2 f''')$$
$$+ n_2 f^{n-2}(10 f'' f''' + 5 f' f^{IV}) + n_1 f^{n-1} f^V \tag{F.53}$$

where n_k are constants related to n and defined in (F.23).

Approximations of the continuous retardation spectrum by the Post-Widder formula of selected orders up to $k = 7$ are plotted in Fig. F.6. The spectra of both models have a similar shape, quite different from the spectrum of the log-power law in Fig. F.3. The most striking difference is that at large retardation times the value of L tends to zero for the ACI and CEB models, while for the log-power law, it

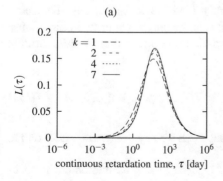
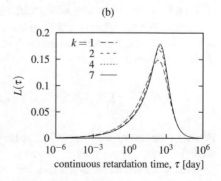

Fig. F.6 Approximations of the continuous retardation spectrum by the Post-Widder formula of various orders for (a) ACI model with $n = 0.6$ and $a = 10$, (b) CEB model with $n = 0.3$ and $\beta = 750$

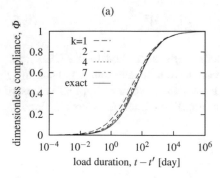
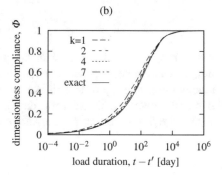

Fig. F.7 Approximations of the compliance function by Dirichlet series with $\tau_1 = 10^{-3}$ day, $M = 10$ and with parameters determined from the Post-Widder formula of various orders for (a) ACI model with $n = 0.6$ and $a = 10$, (b) CEB model with $n = 0.3$ and $\beta = 750$

asymptotically approaches a positive limit, equal to the parameter n. This is related to the fact that the ACI and CEB models assume that creep approaches a finite limit, while for the log-power law (and the B3 and B4 models based on it), the creep is unbounded.

It also appears that the approximation based on the Post-Widder formula converges quite fast for the log-power law and more slowly for the ACI and CEB models. This has a detrimental effect on the quality of the approximation of the compliance function by Dirichlet series, as shown in Fig. F.7. The series consists of $M = 10$ terms for the ACI model and $M = 9$ terms for the CEB model, with the shortest retardation time $\tau_1 = 10^{-3}$ day and with an additional constant term corresponding to the compliance of a spring (zeroth unit of the Kelvin chain). In contrast to the log-power law, it is not sufficient to determine the compliance coefficients simply from the values of the approximated spectrum at individual retardation times τ_μ, according to formula (F.31). This approach corresponds to numerical integration using the midpoint rule, which is sufficiently accurate for the log-power law with almost linear dependence of L on $\ln \tau$ but induces an additional error for models with highly curved spectra. Therefore, the compliance coefficients $1/E_\mu$ are evaluated using two-point Gauss integration in the interval around each retardation time τ_μ. In other words, formula (F.31) is replaced by

$$\frac{1}{E_\mu} = \frac{\ln 10}{2} \left(L(\tau_\mu 10^{-\sqrt{3}/6}) + L(\tau_\mu 10^{\sqrt{3}/6}) \right), \qquad \mu = 1, 2, \ldots, M \qquad \text{(F.54)}$$

Further increase of the integration order does not have any noticeable effect on the results.

Even with accurate integration, the error induced by approximation of the spectrum remains quite high. For the CEB model at load durations up to 10 days, the relative error oscillates around 15% for the second-order approximation and around 5% for the seventh-order approximation; see Fig. F.8b. Only for longer load durations, the error decreases and eventually becomes negligible, independently of the

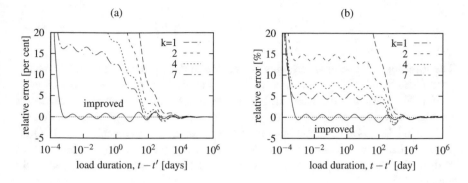

Fig. F.8 Relative error of approximations of the compliance function by Dirichlet series with $\tau_1 = 10^{-3}$ day, $M = 10$ and with parameters determined from the Post-Widder formula of various orders for (a) ACI model, (b) CEB model

approximation order.[13] For the ACI model, the results are still worse, especially at short load durations up to 1 day; see Fig. F.8a. For instance, the fourth-order approximation would overestimate the creep strain at 1 day by 18% and at 10 days by more than 9%.

The quality of the approximation can be dramatically improved by an additional adjustment, similar to the shift (in logarithmic scale) of retardation times used for the log-power law. However, this time, it is not sufficient to multiply the retardation times of all Kelvin units by the same correction factor. This would reduce the positive error for short times but induce a negative error at long times.

As shown in Fig. F.8b, the error of the second-order approximation of the CEB model is quite low for load durations longer than $\beta = 750$ days, and the same holds if another value of β within the reasonable range is used. Therefore, it makes sense to shift the retardation times only in the lower part of the spectrum. Sensitivity to the specific choice of the retardation times can be reduced by a smooth transition from (almost) constant shift for $\tau \ll \beta$ to (almost) no shift for $\tau > \beta$. The solid curve in Fig. F.8b, labeled as "improved," refers to the adjusted Dirichlet series with compliances $1/E_\mu$ based on the second-order approximation but with retardation times τ_μ multiplied by the factor [522]

---

[13]Interestingly, the final value of compliance is captured correctly by all approximations, independently of their order. This is not by chance. The limit value of the approximated compliance function Φ_k at $t \to \infty$ is

$$\Phi_k^\infty = \int_{\tau=0}^\infty L_k(\tau)\, \mathrm{d}(\ln \tau) = -\frac{(-k)^k}{(k-1)!} \int_0^\infty \tau^{k-1} \Phi^{(k)}(k\tau)\, \mathrm{d}\tau \qquad (\text{F.55})$$

For $k = 1$, this gives

$$\Phi_1^\infty = \int_0^\infty \Phi'(\tau)\, \mathrm{d}\tau = \lim_{t \to \infty} \Phi(t) \qquad (\text{F.56})$$

because $\Phi(0) = 0$. The result can be extended to higher values of k by mathematical induction.

$$\alpha_\mu^C = 1 + 0.555 \exp\left(-\frac{4\tau_\mu^2}{\beta^2}\right) \tag{F.57}$$

The resulting relative error remains below 1%, not only for the present choice of parameters $\beta = 750$ days and $\tau_1 = 10^{-3}$ day, but also for other reasonable choices.

For the ACI model, the error exhibits less regular behavior, but the construction of an improved approximation is facilitated by the fact that the compliance function Φ_A defined in (F.39) is always used with the same parameters $n = 0.6$ and $a = 10$. Quite good results, with maximum relative error only slightly above 1%, can be achieved with the fourth-order approximation, if the retardation times τ_μ are multiplied by the factor [522]

$$\alpha_\mu^A = 0.95 + 0.48 \exp\left(-\frac{\tau_\mu^{0.75}}{23}\right) \tag{F.58}$$

The corresponding error is plotted in Fig. F.8a by the solid curve.

F.4 Spectrum of Drying Creep Compliance Function of B3 Model

According to models B3 and B4, the basic creep is described by the log-power law with aging incorporated through the solidification theory and by an additional logarithmic term that reflects viscous flow and corresponds to a dashpot with age-dependent viscosity; see Fig. 9.1. Spectrum of the log-power law was analyzed in Sect. F.2, and the viscous dashpot can be treated directly in the rate form, without the need to construct a Dirichlet series approximating its compliance function. In the presence of drying modeled by the sectional approach, an additional compliance function J_d is used by model B3; its spectrum will be established in the present section.

Recall that the drying creep compliance function J_d given by formula (3.20) depends not only on the current age t and the age t' at loading, but also on the age t_0 at the end of curing (i.e., at the onset of exposure to the environmental humidity). Parameters affecting the drying creep compliance include the shrinkage halftime, τ_{sh}, and the environmental relative humidity, h_{env}. In order to reduce the number of variables and parameters, it is useful to rewrite (3.20) in terms of dimensionless elapsed times $\xi = (t - t')/\tau_{sh}$ and $\xi_0 = (t_0 - t')/\tau_{sh}$. Note that ξ_0 is usually negative, because $t' > t_0$. In the exceptional case when the structure is loaded before the onset of drying, ξ_0 should be set to zero.

Taking into account (3.23) and (3.16), we can write (3.20) as

$$J_d(t, t') = \frac{q_5}{e^4} \Phi_D\left(\frac{t - t'}{\tau_{sh}}, \frac{t_0 - t'}{\tau_{sh}}\right) \tag{F.59}$$

where

$$\Phi_D(\xi, \xi_0) = \sqrt{e^{b \tanh \sqrt{\xi - \xi_0}} - e^{b \tanh \sqrt{-\xi_0}}} \qquad \text{(F.60)}$$

is a dimensionless compliance function of dimensionless times ξ and ξ_0, and $b = 8(1 - h_{env})$ is a parameter introduced for convenience. For a fixed difference $t' - t_0$, the compliance function (F.59) can be considered as a function of the elapsed time $t - t'$ and its spectrum can be approximated according to the Post-Widder formula (F.2). For this purpose, it is sufficient to construct an approximation of the dimensionless compliance function Φ_D, with ξ as the time variable and ξ_0 as a fixed parameter. The spectrum of Φ_D is affected by the environmental humidity (through parameter b) but is independent of the shrinkage halftime.

To approximate the spectrum using formula (F.2), we need to differentiate Φ_D with respect to ξ. For convenience, we will denote (partial) derivatives with respect to ξ by primes, and we will introduce an auxiliary variable $\eta = \sqrt{\xi}$ and auxiliary functions

$$T(\eta) = \tanh \eta \qquad \text{(F.61)}$$
$$S(\xi) = T(\sqrt{\xi}) \qquad \text{(F.62)}$$
$$f(\xi) = \exp(bS(\xi)) \qquad \text{(F.63)}$$

The dimensionless compliance function (F.60) can now be written in the form

$$\Phi_D(\xi, \xi_0) = \sqrt{f(\xi - \xi_0) - f(-\xi_0)} \qquad \text{(F.64)}$$

and its derivatives with respect to ξ are expressed as

$$\Phi'_D = \frac{f'}{2\Phi_D} \qquad \text{(F.65)}$$

$$\Phi''_D = \frac{f''}{2\Phi_D} - \frac{f'\Phi'_D}{2\Phi_D^2} \qquad \text{(F.66)}$$

$$\Phi'''_D = \frac{f'''}{2\Phi_D} - \frac{2f''\Phi'_D + f'\Phi''_D}{2\Phi_D^2} + \frac{f'\Phi_D'^2}{\Phi_D^3} \qquad \text{(F.67)}$$

$$\Phi_D^{IV} = \frac{f^{IV}}{2\Phi_D} - \frac{3f'''\Phi'_D + 3f''\Phi''_D + f'\Phi'''_D}{2\Phi_D^2} + \frac{3f''\Phi_D'^2 + 3f'\Phi'_D\Phi''_D}{\Phi_D^3} - \frac{3f'\Phi_D'^3}{\Phi_D^4} \qquad \text{(F.68)}$$

where

$$f' = bfS' \qquad \text{(F.69)}$$
$$f'' = b(fS'' + f'S') \qquad \text{(F.70)}$$
$$f''' = b(fS''' + 2f'S'' + f''S') \qquad \text{(F.71)}$$
$$f^{IV} = b(fS^{IV} + 3f'S''' + 3f''S'' + f'''S') \qquad \text{(F.72)}$$

To express the derivatives of function S in a manageable form, we denote the derivatives of function T with respect to its argument η by subscripts $1, 2, \ldots$:

$$T_1(\eta) = \frac{1}{\cosh^2 \eta} \tag{F.73}$$

$$T_2(\eta) = -\frac{2 \sinh \eta}{\cosh^3 \eta} \tag{F.74}$$

$$T_3(\eta) = \frac{2(2 \sinh^2 \eta - 1)}{\cosh^4 \eta} \tag{F.75}$$

$$T_4(\eta) = \frac{8 \sinh \eta (2 - \sinh^2 \eta)}{\cosh^5 \eta} \tag{F.76}$$

The derivatives of S with respect to ξ are now given by

$$S' = \frac{1}{2} T_1 \xi^{-1/2} \tag{F.77}$$

$$S'' = \frac{1}{4} \left(T_2 \xi^{-1} - T_1 \xi^{-3/2} \right) \tag{F.78}$$

$$S''' = \frac{1}{8} \left(T_3 \xi^{-3/2} - 3 T_2 \xi^{-2} + 3 T_1 \xi^{-5/2} \right) \tag{F.79}$$

$$S^{IV} = \frac{1}{16} \left(T_4 \xi^{-2} - 6 T_3 \xi^{-5/2} + 15 T_2 \xi^{-3} - 15 T_1 \xi^{-7/2} \right) \tag{F.80}$$

Combining the above relations, approximations of the spectrum up to order 4 can be constructed. Of course, one needs to substitute the appropriate arguments. To illustrate the procedure and to clarify the meaning of individual symbols, we present the full formula for the first-order approximation:

$$
\begin{aligned}
L_1(\tau) = \tau \Phi_D'(\tau, \xi_0) &= \frac{\tau f'(\tau - \xi_0)}{2 \Phi_D(\tau, \xi_0)} = \frac{\tau b f(\tau - \xi_0) S'(\tau - \xi_0)}{2 \sqrt{f(\tau - \xi_0) - f(-\xi_0)}} = \\
&= \frac{\tau b \exp\left(b S(\tau - \xi_0)\right) \frac{1}{2} T_1(\sqrt{\tau - \xi_0})(\tau - \xi_0)^{-1/2}}{2 \sqrt{\exp\left(b S(\tau - \xi_0)\right) - \exp\left(-a + b S(-\xi_0)\right)}} = \\
&= \frac{\tau b \exp\left(b \tanh \sqrt{\tau - \xi_0}\right)}{4 \sqrt{(\tau - \xi_0)} \left[\exp\left(b \tanh \sqrt{\tau - \xi_0}\right) - \exp\left(b \tanh \sqrt{-\xi_0}\right)\right] \cosh^2 \sqrt{\tau - \xi_0}}
\end{aligned} \tag{F.81}
$$

It is clear that such complete formulae for higher-order approximations would be too complicated, but a recursive evaluation of individual terms is still feasible.

Selected approximations up to order 7 are graphically shown in Fig. F.9 for $\xi_0 = -0.1$ and $h_{env} = 60\%$ (i.e., $b = 3.2$). Since Φ_D has been expressed in terms of dimensionless time variables, the retardation time τ on the horizontal axis is also dimensionless and in fact represents the ratio of the actual retardation time and the shrinkage halftime, τ_{sh}. The most important part of the spectrum is concentrated in a narrow interval near $\tau = 1$ (i.e., actually near the shrinkage halftime), and the spectral values at $\tau > 10$ are negligible, which is related to the fact that for such times, the drying process is almost complete and the drying creep compliance function is almost constant. Convergence to the exact spectral function with increasing approximation order k is quite slow, which is documented by the large difference between the fourth-order and seventh-order approximations. It is also worth noting that, for low environmental humidities and onset of loading early after exposure to drying, the spectral values may become negative for a certain limited range of retardation times, as shown in Fig. F.9b.

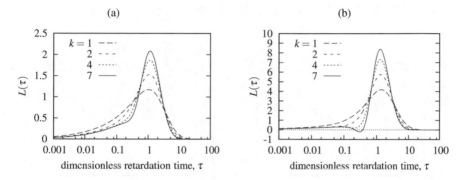

(a)   (b)

Fig. F.9 Approximations of the continuous retardation spectrum by the Post-Widder formula of various orders for the drying creep compliance function of model B3 with parameters (a) $h_{\text{env}} = 60\%$ and $\xi_0 = -0.1$, (b) $h_{\text{env}} = 35\%$ and $\xi_0 = -0.001$

In view of the slow convergence to the exact spectrum, rather poor accuracy of the resulting approximation of the compliance function can be expected. Indeed, the error remains quite large even for high-order approximations, as shown in Fig. F.10. Moreover, since the core part of the spectrum is concentrated in an interval spanning not more than two orders of magnitude of retardation times, the Dirichlet series must be constructed with great care. The standard approach, with neighboring retardation times in ratio 1:10 and with compliances $1/E_\mu$ determined simply from the values of $L_k(\tau_\mu)$, would result into a dramatic sensitivity to the specific choice of the retardation times and to strange shapes of the compliance curve approximations. This is documented in Fig. F.10a, which shows such approximations for $h_{\text{env}} = 60\%$, $\xi_0 = -0.1$, $k = 4$, $M = 6$, and τ_1 set, respectively, to 10^{-4}, 2×10^{-4} and 5×10^{-4}. If the compliances are determined by numerical integration of $L_k(\tau)$ over the entire interval represented by τ_μ using two-point Gauss quadrature, the final value of the compliance function is captured correctly (Fig. F.10b). The sensitivity to the specific choice of retardation times is almost removed if the number of terms in Dirichlet series is increased to $M = 11$ and a denser set of retardation times is used, with neighboring retardation times in ratio $1:\sqrt{10}$ (Fig. F.10c). Nevertheless, the intermediate values exhibit a relative error ranging in this case between 12 and -1.2%; see the dashed curve in Fig. F.10d. The compliance values for short loading times are overestimated and for long times are underestimated. Multiplication of each retardation time by a correction factor [522]

$$\alpha_\mu^D = 0.9 + 0.37 \exp\left(-\frac{\tau_\mu^2}{4}\right) \tag{F.82}$$

reduces the overestimation to 1.2% and slightly increases the underestimation to -1.5%; see the solid curve in Fig. F.10d. This figure refers to the specific case under consideration, with parameters $h_{env} = 60\%$ and $\xi_0 = -0.1$, but it can be verified that the error remains below 2% for all higher humidities and for arbitrary (of course negative) values of ξ_0, while for humidity $h_{env} = 50\%$, it increases to 3%. Such error is still acceptable from the practical point of view.

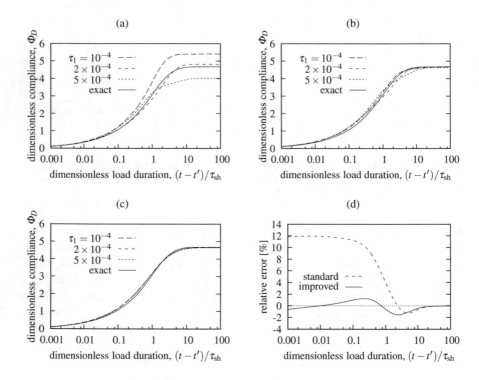

Fig. F.10 (a)–(c) Approximations of the drying creep compliance function of model B3 and (d) their relative error

F.5 Spectrum of JSCE Model

The creep model recommended by the Japan Society of Civil Engineers (JSCE) and described in Appendix E.5 uses a compliance function in the form (F.39) with function Φ given by

$$\Phi_J(t) = 1 - \exp\left(-0.09\, t^{0.6}\right) \tag{F.83}$$

Introducing an auxiliary function

$$f(t) = ct^n \tag{F.84}$$

with $c = -0.09$ and $n = 0.6$, we can write the derivatives of Φ_J as

$$\Phi_J' = -e^f f' = (\Phi_J - 1) f' \tag{F.85}$$
$$\Phi_J'' = (\Phi_J - 1) f'' + \Phi_J' f' \tag{F.86}$$
$$\Phi_J''' = (\Phi_J - 1) f''' + 2\Phi_J' f'' + \Phi_J'' f' \tag{F.87}$$
$$\Phi_J^{IV} = (\Phi_J - 1) f^{IV} + 3\Phi_J' f''' + 3\Phi_J'' f'' + \Phi_J''' f' \tag{F.88}$$
$$\Phi_J^{V} = (\Phi_J - 1) f^{V} + 4\Phi_J' f^{IV} + 6\Phi_J'' f''' + 4\Phi_J''' f'' + \Phi_J^{IV} f' \tag{F.89}$$

where

$$f^{(k)} = cn_k t^{n-k}, \qquad k = 1, 2, \ldots \tag{F.90}$$

and coefficients n_k are given by (F.23).

As shown in Fig. F.11a, the retardation spectrum of the JSCE model has a similar character to the spectra of the ACI and CEB models, but it is concentrated in an even more narrow band, with almost zero values for retardation times exceeding 10^3 days. This is related to the form of function (F.83), which asymptotically approaches 1 and attains the value of 0.999 at time $t \approx 1386$ days; its subsequent growth is negligible. It is also clear that the approximations based on the Post-Widder formula converge quite slowly, and the same holds for the corresponding approximations of the compliance function, plotted in Fig. F.11b. It is disturbing that the asymptotic value for large times, which should be exactly 1, is not captured correctly, not even with the approximation of order 7. Moreover, it can be shown that the asymptotic value as well as the entire approximated compliance function are quite sensitive to the choice of the discrete retardation times.

The relative error of the compliance approximation for different orders k is indicated in Fig. F.11c, which refers to the standard choice of discrete retardation times in a geometric progression with quotient 10. As already discussed for the drying creep compliance function of the B3 model in Sect. F.4, more densely spaced retardation times and higher-order integration are needed for compliance functions with a narrow spectrum. Indeed, the error substantially decreases if the neighboring retardation times are selected in ratio $1:\sqrt{10}$ and the two-point Gauss quadrature formula is used; see Fig. F.11d. Still, the relative error of the seventh-order approximation is up to 10%. It can be substantially reduced by using a correction factor

$$\alpha_\mu^J = 1.15 + 0.5\exp\left(-\frac{\tau_\mu^2}{1000}\right) \tag{F.91}$$

applied to the (relatively simple) second-order approximation $L_2(\tau)$. The resulting relative error remains below 2% over the entire range of times; see the solid curve labeled as "improved" in Fig. F.11d.

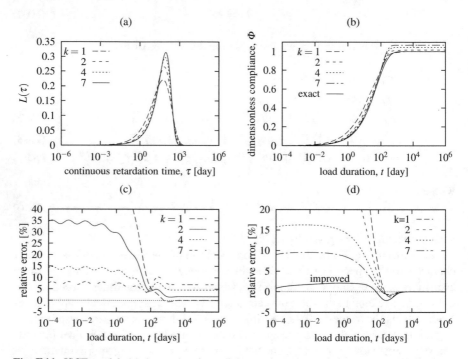

Fig. F.11 JSCE model: (a) Approximations of the continuous retardation spectrum by the Post-Widder formula of various orders, (b) approximations of the compliance function by Dirichlet series, (c)–(d) relative error of approximations of the compliance function by Dirichlet series

F.6   Spectrum of *fib* Model

The *fib* Model Code 2010 uses an additive decomposition of the compliance function into the basic creep compliance and the drying creep compliance. The basic creep compliance function is unbounded and has the logarithmic form

$$J(t, t') = \frac{1}{E(t')} + a \ln\left(1 + \frac{t - t'}{g(t')}\right) \tag{F.92}$$

in which

$$g(t') = \left(0.035 + \frac{30}{t'}\right)^{-2} \tag{F.93}$$

For $t' \geq 1$ day, g is in the range between 1.1×10^{-3} and 816 (in days). For a fixed value of t', the dependence on the load duration $t - t'$ is described by the function

$$\Phi_f(t) = \ln\left(1 + \frac{t}{g}\right) \tag{F.94}$$

which is similar to the log-power law (F.15) with exponent $n = 1$. For this case, the limit of the sequence of approximations $L_k(\tau)$ can be evaluated analytically, in the same fashion as in (F.28). The retardation spectrum of (F.94) can thus be presented in the closed form as

$$L(\tau) = e^{-g/\tau} \tag{F.95}$$

For $g = 1$ day, the spectrum is shown in Fig. F.3a. For a general g, it would be just shifted horizontally. The growth of the value of $L(\tau)$ from 0 to 1 takes place essentially in the interval between $(0.1\,g, 100\,g)$. For such a narrow range, the discrete retardation times need to be spaced more densely; otherwise, the resulting approximation of the compliance function would exhibit oscillations and would be sensitive to the specific choice of times τ_μ. For good accuracy, the two-point Gauss quadrature is used and the discrete retardation times are selected in geometric progression with quotient $\sqrt{10}$. However, this denser spacing is actually needed only in the range in which $L(\tau)$ varies substantially. For larger retardation times, $L(\tau)$ is almost constant (equal to 1) and the standard spacing can be used.

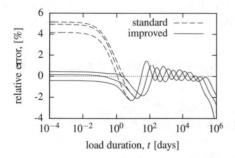

Fig. F.12 Relative error of approximations of the basic creep compliance function of the *fib* model for $g(t') = 1$ day (i.e., for age $t' = 31.1$ days) by Dirichlet series with $M = 10$ and $\tau_1 = 0.05$ day, 0.1 day, and 0.2 day

Assuming that $g = 1$ day, a reasonable choice is to set $\tau_1 = 0.1$ day and use the reduced spacing up to $\tau_5 = 10^2\tau_1 = 10$ days, followed by $\tau_6 = 10^{0.75}\tau_5 = 56.23$ days and afterward increasing the times by a factor of 10 up to $\tau_{10} = 10^4\tau_6 = 562,300$ days. The error of the resulting approximation of the compliance function (F.94) by the Dirichlet series is shown by the dashed curve in Fig. F.12. For comparison, two additional dashed curves corresponding to $\tau_1 = 0.05$ day and $\tau_1 = 0.2$ day have been plotted. The same relative errors are obtained for a general value of g if all the discrete retardation times are divided by g.

Note that in this case, the spectrum $L(\tau)$ is known exactly and the error is exclusively due to the approximation of the integral in (F.1) by a finite sum. The relative error is below 2%, with the exception of very short or very long times. For long times, the accuracy could be improved by adding more terms to the Dirichlet series. For short times, an improvement is achieved at no extra cost if the retardation times are multiplied by the factor

$$\alpha_\mu^f = 1 + 0.15 \exp(-\tau_\mu^2) \tag{F.96}$$

The corresponding relative error is plotted in Fig. F.12 by the solid curves, again corresponding to $g = 1$ day and $\tau_1 = 0.05$ day, 0.1 day, and 0.2 day, respectively.

The drying creep compliance function of the *fib* model has the same general form as the CEB compliance function, with the only difference that the exponent n in (F.47) is not a fixed constant (equal to 0.3) but depends on the age at loading, t'. It is given by

$$n(t') = \frac{1}{2.3 + 3.5/\sqrt{t'}} \tag{F.97}$$

and for $t' \geq 1$ day, which is a restriction imposed by the code, it is in the range between 0.172 and 0.435 (note that $n(t')$ is denoted as $\gamma(t')$ in the actual code notation presented in Appendix E.2.2). For the evaluation of the corresponding retardation spectrum, we can reuse formulae (F.48)–(F.53) and a generalized version of the corrective formula (F.57), with a coefficient dependent on exponent n:

$$\alpha_\mu^C = 1 + (0.255 + n) \exp\left(-\frac{4\tau_\mu^2}{\beta^2}\right) \tag{F.98}$$

The resulting error for extreme values of parameter n within the range of interest is shown in Fig. F.13. The solution marked as "improved" is based on an approximation of order $k = 2$ and corrective formula (F.98). The relative error remains below 1% over the entire range of interest.

F.7 Summary

The continuous retardation spectrum is a useful tool for the description of viscoelastic materials. It can be exploited for a straightforward and efficient evaluation of compliance coefficients in Dirichlet series approximating the compliance function. However, for most of the compliance functions used by various creep codes and recommendations, the spectrum cannot be expressed in a closed form. Application of the Post-Widder formula for the inversion of Laplace transform leads to approximations

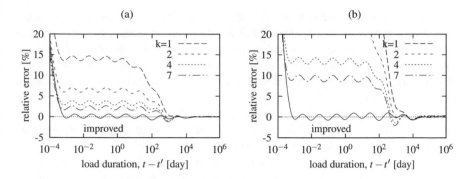

Fig. F.13 Relative error of approximations of the drying creep compliance function of the *fib* model by Dirichlet series with $\tau_1 = 10^{-3}$ day, $M = 10$ and with parameters determined from the Post-Widder formula of various orders, evaluated for exponent (a) $n = 0.172$, (b) $n = 0.435$

of various orders, which converge to the exact spectrum as the order tends to infinity. Since the complexity of the analytical expressions increases with increasing order, low-order approximations are more convenient for practical applications. However, one has to be careful about the accuracy.

In many cases of practical interest, the convergence is rather slow and even approximations of order 7 lead to relative errors of several percent. Fortunately, the accuracy can be dramatically increased by adjustments of the discrete retardation times, applied after the evaluation of the compliance coefficients. Such adjustments are theoretically justified, but their optimal form was determined by Jirásek and Havlásek [522] in a rather heuristic manner. Interestingly, the resulting approximations of the compliance function are sufficiently accurate even for simple expressions derived from a low-order Post-Widder formula (typically, second-order formulae are sufficient for this purpose).

The retardation spectra have quite a different character for models with bounded and unbounded compliance functions. In the former case, represented by the ACI, CEB, and JSCE models and by the drying creep compliance function of the B3 and *fib* models, the spectral values are negligible for both very short and very long retardation times. The approximation by Dirichlet series then does not need to include terms with very long retardation times, for which the creep has almost stopped. On the other hand, the basic creep compliance functions of the B3 and *fib* models are unbounded and have a logarithmic character. The corresponding spectrum approaches a constant nonzero value for long retardation times, and the contribution to the Dirichlet series is essential. The choice of the longest discrete retardation time is then related to the limits of applicability of the approximate compliance function.

For models with a wide and smoothly varying spectrum, it is sufficient to select the discrete retardation times in the Dirichlet series in a geometric progression with ratio 10, which is the standard recommendation from the literature, and the compliance coefficients can be determined simply as the spectral values at individual retardation times multiplied by the size of the covered interval in logarithmic scale, which corresponds to a one-point integration scheme. Special attention is needed if the retardation spectrum is relatively narrow. For the ACI and CEB models, nonnegligible spectral values are found for retardation times over about 8 orders of magnitude, and the graph of the spectrum exhibits in certain ranges a high curvature. In such a case, it is preferable to evaluate the compliance coefficients using a two-point Gaussian quadrature formula.

The drying creep compliance function of the B3 and B4 models and the total compliance function of the JSCE model have spectra concentrated to about 4 orders of magnitude, and the standard choice of discrete retardation times one order of magnitude apart does not provide good approximations of the compliance functions. It is then preferable to use a denser set of retardation times in a geometric progression with ratio $\sqrt{10}$, combined with a two-point Gauss quadrature scheme. Such a denser set is also useful over a certain range of retardation times for the basic creep compliance function of the *fib* model, to provide a better coverage of the highly curved part of the spectrum.

In general, the optimized approximations exhibit relative errors below 2%, which is certainly sufficient from the practical point of view. At the same time, they are based on low-order formulae, which lead to simple expressions and are thus easy to handle. Therefore, results reported in this appendix provide a basis for efficient and accurate computational algorithms.

Appendix G
Free-Energy Potentials for Aging Linear Viscoelasticity

The free energy per unit volume of a viscoelastic material may be expressed as a sum of the strain energies in all the springs in Kelvin or Maxwell chain model. Based on this fact, expressions for the Helmholtz and Gibbs free energy per unit volume of nonaging viscoelastic materials, formulated in terms of the relaxation function, have been derived by Staverman and Schwarzl [787, 788]. For aging viscoelastic materials, the energy of the springs must be modified by the growing volume $v(t)$ of the solidified material.

The following symmetric expression was deduced by Bažant and Huet [139] to express, by means of the compliance function and stress history, the negative of the Gibbs free energy per unit volume of an aging viscoelastic material (in mechanics called the isothermal complementary energy density):

$$\mathscr{F}^*(t) = \frac{1}{2} \int_{r=0}^{t} \int_{s=0}^{t} \max[J(2t - r, s), J(2t - s, r)] \, d\sigma(s) \, d\sigma(r) \qquad (G.1)$$

The maximum operator is used here only because it permits writing the formula in a symmetric way (with respect to the integration variables, r and s). Compliances $J(2t - r, s)$ and $J(2t - s, r)$ correspond to the same load duration, $2t - r - s$, but different ages, s and r. Due to aging, compliance obtained after the same load duration is higher for younger concrete, which means that the maximum is equal to $J(2t - r, s)$ if $s < r$ and to $J(2t - s, r)$ if $r > s$. Therefore, the double integral presented in (G.1), originally taken over a square in the rs plane (Fig. G.1a), can be split into a sum of two double integrals over triangular domains (Fig. G.1b), and the formula can equivalently be written as

$$\mathscr{F}^*(t) = \frac{1}{2} \int_{r=0}^{t} \int_{s=0}^{r} J(2t - r, s) \, d\sigma(s) \, d\sigma(r) + \frac{1}{2} \int_{s=0}^{t} \int_{r=0}^{s} J(2t - s, r) \, d\sigma(r) \, d\sigma(s)$$

$$(G.2)$$

© Springer Science+Business Media B.V. 2018
Z.P. Bažant and M. Jirásek, *Creep and Hygrothermal Effects in Concrete Structures*, Solid Mechanics and Its Applications 225, https://doi.org/10.1007/978-94-024-1138-6

Due to symmetry with respect to r and s, both integrals in (G.2) have the same value. Therefore, the same value of complementary energy density is obtained by integrating over one triangular domain if the factor $1/2$ is dropped:

$$\mathscr{F}^*(t) = \int_{r=0}^{t} \int_{s=0}^{r} J(2t - r, s) \, d\sigma(s) \, d\sigma(r) \tag{G.3}$$

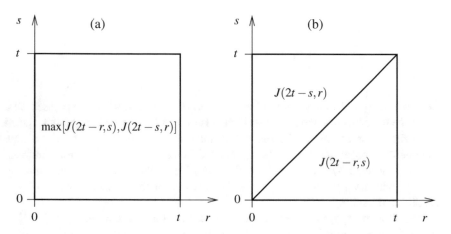

Fig. G.1 Integration domains in the rs plane: (a) square domain used in formula (G.1), (b) decomposition into two triangular domains used in formula (G.2)

To prove that (G.1) (and thus also (G.2) and (G.3)) plays the role of a viscoelastic potential from which the stress–strain law can be deduced, let us differentiate the complementary energy density with respect to the stress value at time t. Since $\mathscr{F}^*$ is not just a function of $\sigma(t)$ but a functional of the stress history, it is good to explain in detail what kind of differentiation we mean.

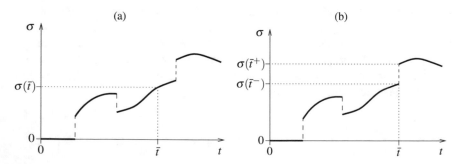

Fig. G.2 Piecewise continuous stress history, with time instant $\bar{t}$ at which the stress is (a) continuous, (b) discontinuous

Consider a given stress history $\sigma(t)$, $t \geq 0$, which is supposed to be piecewise continuous and differentiable, with finite stress jumps occurring at a finite number of isolated time instants (Fig. G.2). The integral in (G.1) is meant to be a Stieltjes integral, and so it is properly defined even if the stress history exhibits discontinuities (leading to singularities in the stress rate). For a fixed time instant $\bar{t}$, we can construct a function $\Delta \mathscr{F}^*$ describing the change of complementary energy density $\mathscr{F}^*(\bar{t})$ that would result from an instantaneous modification of the stress value at time $\bar{t}$.

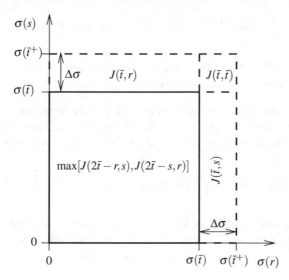

Fig. G.3 Increment of the double integral from (G.1) caused by an instantaneous stress increment $\Delta \sigma$ at time $\bar{t}$

For simplicity, we start from the regular case in which the actual stress history is continuous at $\bar{t}$, as indicated in Fig. G.2a. The actual complementary energy density $\mathscr{F}^*(\bar{t})$ is obtained by evaluating the integral in (G.1) with both upper limits set to $\bar{t}$. In the stress space with axes $\sigma(r)$ and $\sigma(s)$, the corresponding integration domain is represented by the solid square in Fig. G.3. Now suppose that a fictitious change $\Delta \sigma$ is added to the stress $\sigma(\bar{t})$, with the time considered as frozen. This corresponds to an extension of the integration domain in the stress space to a larger square, as indicated in Fig. G.3 by dashed lines. The additional contribution to the integral in (G.1) consists of three terms that are represented in Fig. G.3 by two elongated rectangles and a small square. Formally, they can be derived by considering $\Delta \sigma$ as a jump in stress that occurs at time $\bar{t}$ and leads to the stress level $\sigma(\bar{t}^+) = \sigma(\bar{t}) + \Delta \sigma$ reached "just after" time $\bar{t}$.

Let us adopt a convention that integrals with $\bar{t}$ as the upper limit do not take into account the jump while integrals with $\bar{t}^+$ as the upper limit do. With this notation at hand, the derivation can proceed as follows[14]:

---

[14]In the first line of (G.4), we should actually integrate the maximum of $J(2\bar{t}^+ - r, s)$ and $J(2\bar{t}^+ - s, r)$. Since the compliance function is in general continuous, it is sufficient to write simply

$$\int_{r=0}^{\bar{t}+} \int_{s=0}^{\bar{t}+} \max[J(2\bar{t}-r,s), J(2\bar{t}-s,r)] \, d\sigma(s) \, d\sigma(r) =$$

$$= \int_{r=0}^{\bar{t}} \int_{s=0}^{\bar{t}+} \max[J(2\bar{t}-r,s), J(2\bar{t}-s,r)] \, d\sigma(s) \, d\sigma(r) +$$

$$+ \int_{0}^{\bar{t}+} \max[J(2\bar{t}-\bar{t},s), J(2\bar{t}-s,\bar{t})] \, d\sigma(s) \Delta\sigma =$$

$$= \int_{r=0}^{\bar{t}} \int_{s=0}^{\bar{t}} \max[J(2\bar{t}-r,s), J(2\bar{t}-s,r)] \, d\sigma(s) \, d\sigma(r) +$$

$$+ \int_{0}^{\bar{t}} J(\bar{t},r) \Delta\sigma \, d\sigma(r) + \left(\int_{0}^{\bar{t}} J(\bar{t},s) \, d\sigma(s) + J(\bar{t},\bar{t}) \Delta\sigma \right) \Delta\sigma \quad \text{(G.4)}$$

Taking into account that the definition of $\mathscr{F}^*$ in (G.1) contains a factor of $1/2$ in front of the double integral, we find that the virtual increment of complementary energy density caused by an instantaneous virtual change of stress at frozen time $\bar{t}$ is given by

$$\Delta\mathscr{F}_{\bar{t}}^*(\Delta\sigma) = \int_{0}^{\bar{t}} J(\bar{t},s) \, d\sigma(s) \, \Delta\sigma + \frac{1}{2} J(\bar{t},\bar{t})(\Delta\sigma)^2 \quad \text{(G.5)}$$

Since the time $\bar{t}$ and the previous stress history up to time $\bar{t}$ are considered as fixed, the virtual increment $\Delta\mathscr{F}_{\bar{t}}^*$ is a function of $\Delta\sigma$. The derivative of this function, evaluated at $\Delta\sigma = 0$, provides the constitutive equation

$$\varepsilon(\bar{t}) = \frac{d\Delta\mathscr{F}_{\bar{t}}^*}{d\Delta\sigma} \bigg|_{\Delta\sigma=0} = \int_{0}^{\bar{t}} J(\bar{t},s) \, d\sigma(s) \quad \text{(G.6)}$$

which, leaving aside formal differences in notation, agrees with (2.14). This result confirms that (G.1) can be used as a viscoelastic potential.

The foregoing derivation can be extended to the exceptional case of a time instant at which the actual stress history has a jump. Here, we have to distinguish between the stress "just before the jump," $\sigma(\bar{t}^-) = \lim_{t \to \bar{t}^-} \sigma(t)$, and the stress "just after the jump," $\sigma(\bar{t}^+) = \lim_{t \to \bar{t}^+} \sigma(t)$; see Fig. G.2b. The strain can also be expected to exhibit a jump, and we can use an analogous notation. For the state just **before** the jump, all results derived for the continuous case are directly applicable, with $\sigma(\bar{t})$ understood as $\sigma(\bar{t}^-)$ and the upper integration limit $\bar{t}$ set to $\bar{t}^-$, to emphasize that the integrals do not take into account the jump that occurs at $\bar{t}$. The change of complementary energy resulting from a modification of $\sigma(\bar{t}^-)$ by $\Delta\sigma$ is now given by

---

(Footnote 14 continued)
$J(2\bar{t}-r,s)$ and $J(2\bar{t}-s,r)$. However, we need to be careful with the interpretation of $J(\bar{t},\bar{t})$ in the last line of (G.4), because the compliance function $J(t,t')$ is discontinuous at points, where $t = t'$ (it jumps from zero to the elastic compliance $1/E(t')$). The proper interpretation of $J(\bar{t},\bar{t})$ in (G.4) is that it corresponds to $J(\bar{t}^+,\bar{t})$, i.e., to the reciprocal value of elastic modulus at age $\bar{t}$.

$$\Delta \mathscr{F}_{\bar{t}^-}^*(\Delta\sigma) = \int_0^{\bar{t}^-} J(\bar{t}, s)\, \mathrm{d}\sigma(s)\, \Delta\sigma + \frac{1}{2} J(\bar{t}, \bar{t})(\Delta\sigma)^2 \qquad (G.7)$$

and Eq. (G.6) is rewritten as

$$\varepsilon(\bar{t}^-) = \left.\frac{\mathrm{d}\Delta\mathscr{F}_{\bar{t}^-}^*}{\mathrm{d}\Delta\sigma}\right|_{\Delta\sigma=0} = \int_0^{\bar{t}^-} J(\bar{t}, s)\, \mathrm{d}\sigma(s) \qquad (G.8)$$

An appropriate expression for the strain just **after** the jump can be derived using again function (G.7) but differentiating it at a different point. The reason is that if the fictitious stress state for which we evaluate the complementary energy differs from $\sigma(\bar{t}^-)$ by $\Delta\sigma$, then the virtual change of $\sigma(\bar{t}^+)$ is just $\Delta\sigma - \overline{\Delta\sigma}$, where $\overline{\Delta\sigma} \equiv \sigma(\bar{t}^+) - \sigma(\bar{t}^-)$ corresponds to the actual stress jump. Therefore, to get the strain after the jump, the derivative of $\Delta\mathscr{F}_{\bar{t}^-}^*$ must be evaluated at $\Delta\sigma = \overline{\Delta\sigma}$. Even though these arguments may sound somewhat complicated, the final form of the resulting constitutive equation is fully analogous to the already derived ones:

$$\varepsilon(\bar{t}^+) = \left.\frac{\mathrm{d}\Delta\mathscr{F}_{\bar{t}^-}^*}{\mathrm{d}\Delta\sigma}\right|_{\Delta\sigma=\sigma(\bar{t}^+)-\sigma(\bar{t}^-)} = \int_0^{\bar{t}^-} J(\bar{t}, s)\, \mathrm{d}\sigma(s) + J(\bar{t}, \bar{t})(\sigma(t^+) - \sigma(t^-)) =$$

$$= \int_0^{\bar{t}^+} J(\bar{t}, s)\, \mathrm{d}\sigma(s) \qquad (G.9)$$

In fact, all the cases discussed so far can be covered by the constitutive law

$$\varepsilon(t) = \int_0^t J(t, s)\, \mathrm{d}\sigma(s) \qquad (G.10)$$

in which the interpretation of the upper integration limit determines which value of strain is obtained. In all cases, the expression for evaluation of strain at a given time has been constructed by differentiating the complementary energy density with respect to the stress value at that time. This confirms that the mutually equivalent expressions (G.1)–(G.3) for the complementary energy density can be used as viscoelastic potentials written in terms of the stress history and of the compliance function.

In a completely analogous manner, one can define the Helmholtz free energy density (in mechanics called the isothermal strain energy density) using a symmetric expression [139]

$$\mathscr{F}(t) = \frac{1}{2} \int_{r=0}^t \int_{s=0}^t \min[R(2t - r, s), R(2t - s, r)]\, \mathrm{d}\varepsilon(s)\, \mathrm{d}\varepsilon(r) \qquad (G.11)$$

or using an equivalent (simpler but nonsymmetric) expression

$$\mathscr{F}(t) = \int_{r=0}^{t} \int_{s=0}^{r} R(2t - r, s) \, d\varepsilon(s) \, d\varepsilon(r) \tag{G.12}$$

The change of free-energy density at time $\bar{t}$ caused by an instantaneous fictitious strain change $\Delta\varepsilon$ can be expressed as

$$\Delta\mathscr{F}_{\bar{t}}(\Delta\varepsilon) = \int_{0}^{\bar{t}} R(\bar{t}, s) \, d\varepsilon(s) \, \Delta\varepsilon + \frac{1}{2} R(\bar{t}, \bar{t})(\Delta\varepsilon)^2 \tag{G.13}$$

and its differentiation leads to the stress–strain law

$$\sigma(\bar{t}) = \frac{d\Delta\mathscr{F}_{\bar{t}}}{d\Delta\varepsilon}\bigg|_{\Delta\varepsilon=0} = \int_{0}^{\bar{t}} R(\bar{t}, s) \, d\varepsilon(s) \tag{G.14}$$

It is instructive to link potentials $\mathscr{F}$ and $\mathscr{F}^*$ to the thermodynamic quantities discussed in Sect. 13.5. Recall that $\mathscr{F}^*$ corresponds to minus the Gibbs free energy (i.e., minus the free enthalpy) per unit volume and thus can be expressed as minus the specific Gibbs free energy μ multiplied by the density ρ. In a similar spirit, the strain energy density $\mathscr{F} = \rho\psi$ where ψ is the specific Helmholtz free energy. Since we work here in the context of the small-strain theory, no difference is made between the initial and current volume and between the initial and current density. Consequently, the rates are expressed simply as $\dot{\mathscr{F}} = \rho\dot{\psi}$ and $\dot{\mathscr{F}}^* = -\rho\dot{\mu}$, with no need for additional terms containing $\dot{\rho}$.

The specific Helmholtz free energy ψ is linked to the specific internal energy u by the partial Legendre transformation

$$\psi = u - Ts \tag{G.15}$$

in which T is the absolute temperature and s is the specific entropy. Differentiating (G.15) with respect to time, substituting into $\dot{\mathscr{F}} = \rho\dot{\psi}$ and exploiting Eq. (13.153), we obtain

$$\dot{\mathscr{F}} = \rho\dot{\psi} = \rho\dot{u} - \rho T\dot{s} - \rho s\dot{T} = \sigma : \dot{\varepsilon} - \frac{1}{T}q \cdot \nabla T - \rho Ts^* - \rho s\dot{T} \tag{G.16}$$

The first term on the right-hand side of (G.16) is the stress power density (i.e., the mechanical work supplied to a unit volume of the material per unit time), the second term including the negative sign represents the thermal dissipation (again per unit volume and unit time), and the third term is minus the overall dissipation (sum of mechanical and thermal dissipations). Under isothermal conditions,[15] we have $\dot{T} = 0$ and the last term on the right-hand side of (G.16) vanishes. Realizing that

---

[15] By isothermal conditions, we mean that the temperature remains constant in time at each material point, but we admit its spatial variability (e.g., a linear temperature distribution in space and constant heat flux). Of course, if the temperature is uniform in space, the thermal dissipation vanishes and there is no difference between the mechanical and overall dissipation.

the difference between the overall dissipation and the thermal dissipation, $\mathscr{D}_M = \rho T s^* + (\mathbf{q} \cdot \nabla T)/T$, is the mechanical dissipation, we can rewrite (G.16) under isothermal conditions as

$$\dot{\mathscr{F}} = \sigma : \dot{\varepsilon} - \mathscr{D}_M \tag{G.17}$$

The complementary energy density is linked to the strain energy density $\mathscr{F}$ by the Legendre transformation

$$\mathscr{F}^* = \sigma : \varepsilon - \mathscr{F} \tag{G.18}$$

and its rate can be expressed as

$$\dot{\mathscr{F}}^* = \dot{\sigma} : \varepsilon + \sigma : \dot{\varepsilon} - \dot{\mathscr{F}} = \dot{\sigma} : \varepsilon + \mathscr{D}_M \tag{G.19}$$

where $\dot{\sigma} : \varepsilon$ is the complementary power density.

Equations (G.17) and (G.19) can now be used to construct two alternative expressions for the mechanical dissipation,

$$\mathscr{D}_M = \sigma : \dot{\varepsilon} - \dot{\mathscr{F}} = \dot{\mathscr{F}}^* - \dot{\sigma} : \varepsilon \tag{G.20}$$

For the present choice of viscoelastic potentials, differentiation of (G.12) with respect to time yields

$$\dot{\mathscr{F}}(t) = \int_0^t R(t, s) \, d\varepsilon(s) \dot{\varepsilon}(t) + 2 \int_{r=0}^t \int_{s=0}^r \dot{R}(2t - r, s) \, d\varepsilon(s) \, d\varepsilon(r) \tag{G.21}$$

where $\dot{R}$ denotes the derivative of the relaxation function with respect to its first argument. The first integral in (G.21) is recognized as the stress at time t evaluated from the constitutive law (G.14), and so the first term on the right-hand side of (G.21) corresponds to the stress power density, in the one-dimensional setting written as $\sigma(t)\dot{\varepsilon}(t)$. The second term on the right-hand side of (G.21) thus represents minus the mechanical dissipation (per unit volume and unit time). The negative sign is compensated by the fact that $\dot{R} \leq 0$, which makes the dissipation nonnegative. Since the inequality $\dot{R}(2t - r, s) > \dot{R}(2t - s, r)$ holds if and only if $s < r$, the derived formula for the mechanical dissipation

$$\mathscr{D}_M(t) = -2 \int_{r=0}^t \int_{s=0}^r \dot{R}(2t - r, s) \, d\varepsilon(s) \, d\varepsilon(r) \tag{G.22}$$

can be rewritten in a symmetric form as

$$\mathscr{D}_M(t) = -\int_{r=0}^t \int_{s=0}^r \dot{R}(2t - r, s) \, d\varepsilon(s) \, d\varepsilon(r) - \int_{s=0}^t \int_{r=0}^s \dot{R}(2t - s, r) \, d\varepsilon(r) \, d\varepsilon(s) =$$
$$= -\int_{r=0}^t \int_{s=0}^t \max[\dot{R}(2t - r, s), \dot{R}(2t - s, r)] \, d\varepsilon(s) \, d\varepsilon(r) \tag{G.23}$$

An alternative expression for the mechanical dissipation can be derived from (G.19), based on formula (G.3) for the complementary energy density. Differentiation of (G.3) leads to

$$\dot{\mathscr{F}}^*(t) = \int_0^t J(t, s) \, d\sigma(s) \, \dot{\sigma}(t) + 2 \int_{r=0}^t \int_{s=0}^r \dot{J}(2t - r, s) \, d\sigma(s) \, d\sigma(r) \quad \text{(G.24)}$$

where $\dot{J}$ denotes the derivative of the compliance function with respect to its first argument. The first integral in (G.24) is recognized as the strain at time t evaluated from the constitutive law (G.10), and so the first term on the right-hand side corresponds to the complementary power density, $\varepsilon(t)\dot{\sigma}(t)$. Therefore, the second term on the right-hand side of (G.24) represents the mechanical dissipation

$$\mathscr{D}_M(t) = 2 \int_{r=0}^t \int_{s=0}^r \dot{J}(2t - r, s) \, d\sigma(s) \, d\sigma(r) \quad \text{(G.25)}$$

which can also be presented in a symmetric form as

$$\mathscr{D}_M(t) = \int_{r=0}^t \int_{s=0}^r \dot{J}(2t - r, s) \, d\sigma(s) \, d\sigma(r) + \int_{s=0}^t \int_{r=0}^s \dot{J}(2t - s, r) \, d\sigma(r) \, d\sigma(s) =$$

$$= \int_{r=0}^t \int_{s=0}^t \max[\dot{J}(2t - r, s), \dot{J}(2t - s, r)] \, d\sigma(s) \, d\sigma(r) \quad \text{(G.26)}$$

Example G.1. Maxwell model

To gain more insight, let us consider the specific case of a solidifying Maxwell model with a relaxation function in the form

$$R(t, t') = E^\infty v(t') e^{-(t-t')/\tau} H(t - t') \quad \text{(G.27)}$$

where E^∞ is the final value of elastic modulus, $v(t')$ is an increasing dimensionless function that describes solidification and tends to 1 as the age t' tends to infinity, τ is the relaxation time, and H is the Heaviside function. For this model, the stress–strain law (G.14) reads

$$\sigma(t) = \int_0^t E^\infty v(s) e^{-(t-s)/\tau} \, d\varepsilon(s) \quad \text{(G.28)}$$

and the free-energy density (G.12) is given by

$$\mathscr{F}(t) = \int_{r=0}^t \int_{s=0}^r E^\infty v(s) e^{-(2t-r-s)/\tau} \, d\varepsilon(s) \, d\varepsilon(r) =$$

$$= \int_{r=0}^t e^{-2(t-r)/\tau} \int_{s=0}^r E^\infty v(s) e^{-(r-s)/\tau} \, d\varepsilon(s) \, d\varepsilon(r) =$$

$$= \int_0^t e^{-2(t-r)/\tau} \sigma(r) \, d\varepsilon(r) \quad \text{(G.29)}$$

Substituting

$$\dot{R}(t, t') = \frac{\partial R(t, t')}{\partial t} = -\frac{E^\infty}{\tau} v(t') e^{-(t-t')/\tau} \tag{G.30}$$

into (G.22), we get the mechanical dissipation

$$\mathcal{D}_M(t) = 2 \int_{r=0}^{t} \int_{s=0}^{r} \frac{E^\infty}{\tau} v(s) e^{-(2t-r-s)/\tau} \, d\varepsilon(s) \, d\varepsilon(r) =$$

$$= \frac{2}{\tau} \int_0^t e^{-2(t-r)/\tau} \sigma(r) \, d\varepsilon(r) = \frac{2}{\tau} \mathcal{F}(t) \tag{G.31}$$

In the last integrals in (G.29) and (G.31), the product $\sigma(r) \, d\varepsilon(r)$ represents the elementary work (per unit volume) supplied to the material during an infinitesimal time interval dr. The newly supplied work is first fully converted into an increment of strain energy and afterward gradually dissipated at a decreasing rate, which is described by the decaying exponential function $e^{-2(t-r)/\tau}$. However, one needs to be careful with the interpretation of $\sigma(r)$ at those isolated time instants at which the strain history is discontinuous. If, at a certain time instant $\bar{t}$, the strain increases by a jump from $\varepsilon(\bar{t}^-)$ to $\varepsilon(\bar{t}^+) = \varepsilon(\bar{t}^-) + \Delta\varepsilon$, then the stress evaluated according to (G.28) also jumps from

$$\sigma(\bar{t}^-) = \int_0^{\bar{t}^-} E^\infty v(s) e^{-(\bar{t}-s)/\tau} d\varepsilon(s) \tag{G.32}$$

to

$$\sigma(\bar{t}^+) = \int_0^{\bar{t}^+} E^\infty v(s) e^{-(\bar{t}-s)/\tau} d\varepsilon(s) = \sigma(\bar{t}^-) + E^\infty v(\bar{t}) \Delta\varepsilon \tag{G.33}$$

Due to the jump, the contribution of time instant $r = \bar{t}$ to the integrals in (G.29) and (G.31) is finite and equal to $e^{-2(t-\bar{t})/\tau} \sigma(\bar{t}) \Delta\varepsilon$. But which value of $\sigma(\bar{t})$ should be used here, $\sigma(\bar{t}^-)$ or $\sigma(\bar{t}^+)$? It turns out that it should be the average of the limits from the left and from the right, $(\sigma(\bar{t}^-) + \sigma(\bar{t}^+))/2$.

For instance, in a relaxation test that starts at age t_1, the strain history is given by $\varepsilon(t) = \hat{\varepsilon} H(t - t_1)$ and the corresponding stress history is

$$\sigma(t) = \int_0^t R(t, s) \, d\varepsilon(s) = \hat{\varepsilon} R(t, t_1) = E^\infty v(t_1) \hat{\varepsilon} e^{-(t-t_1)/\tau} H(t - t_1) \tag{G.34}$$

The free-energy density is according to (G.29) evaluated as

$$\mathcal{F}(t) = \int_0^t e^{-2(t-r)/\tau} \sigma(r) \, d\varepsilon(r) = e^{-2(t-t_1)/\tau} \sigma(t_1) \hat{\varepsilon} H(t - t_1) \tag{G.35}$$

where $\sigma(t_1)$ needs to be understood as the average between $\sigma(t_1^-) = 0$ and $\sigma(t_1^+) = E^\infty v(t_1) \hat{\varepsilon}$, i.e., $\sigma(t_1) = E^\infty v(t_1) \hat{\varepsilon}/2$. The product $\sigma(t_1) \hat{\varepsilon}$ is then equal to $E^\infty v(t_1)\hat{\varepsilon}^2/2$, which is the correct value of the strain energy stored in a spring of stiffness $E^\infty v(t_1)$ subjected to strain $\hat{\varepsilon}$. Consequently, (G.35) can be rewritten as

$$\mathscr{F}(t) = \frac{1}{2}E^\infty v(t_1)\hat{\varepsilon}^2 e^{-2(t-t_1)/\tau} H(t - t_1) \tag{G.36}$$

The mechanical dissipation is according to (G.31) evaluated as

$$\mathscr{D}_M(t) = \frac{2}{\tau}e^{-2(t-t_1)/\tau}\sigma(t_1)\hat{\varepsilon}H(t - t_1) = \frac{1}{\tau}E^\infty v(t_1)\hat{\varepsilon}^2 e^{-2(t-t_1)/\tau} H(t - t_1) \tag{G.37}$$

Recall that $\mathscr{D}_M(t)$ is the rate at which energy is dissipated (per unit volume). The cumulative energy dissipation up to time $t_2 > t_1$, denoted as $D(t_2)$, is obtained by integrating $\mathscr{D}_M(t)$ in time:

$$D(t_2) = \int_0^{t_2} \mathscr{D}_M(t)\, dt = \frac{E^\infty v(t_1)\hat{\varepsilon}^2}{\tau}\int_{t_1}^{t_2} e^{-2(t-t_1)/\tau}\, dt = \frac{E^\infty v(t_1)\hat{\varepsilon}^2}{2}\left(1 - e^{-2(t_2-t_1)/\tau}\right) \tag{G.38}$$

It is interesting to note that as t_2 increases, the dissipated energy asymptotically tends to $E^\infty v(t_1)\hat{\varepsilon}^2/2$, which is the energy supplied as mechanical work and initially stored in the form of elastic strain energy. The dissipated energy approaches its asymptotic limit exponentially, with characteristic time $\tau/2$, while the stress relaxes to zero also exponentially but with characteristic time τ.

The reason for the difference between the characteristic times of the stress relaxation and energy dissipation becomes clear if one realizes that the rate of dissipation in a linear viscous dashpot is given by $\sigma_v \dot{\varepsilon}_v = \sigma_v^2/\eta$, where $\sigma_v = \eta\dot{\varepsilon}_v$ is the stress in the dashpot, ε_v is the strain in the dashpot, and η is the viscosity. In a Maxwell model, the dashpot is subjected to the overall stress σ. If the stress relaxes to one half of its initial value, the rate of dissipation is reduced to one quarter of its initial value. This reasoning is valid not only for a nonaging Maxwell model with constant viscosity η, but also for a solidifying Maxwell model relaxing at constant strain. No stress is generated in the material that is newly deposited after time t_1 at which the strain jump is imposed, and so we have $\sigma(t) = \sigma_v(t) = \eta(t_1)\dot{\varepsilon}_v(t)$ where $\eta(t_1) = \tau E^\infty v(t_1)$.

During a relaxation test at constant strain, no additional energy is supplied to or extracted from the sample, and so the sum of the strain energy and dissipated energy remains constant. This directly follows from (G.17) with $\dot{\varepsilon}$ set to zero. Combining the condition $\dot{\mathscr{F}}(t) + \mathscr{D}_M(t) = 0$ with relation $\mathscr{D}_M(t) = (2/\tau)\mathscr{F}(t)$ derived in (G.31), we obtain a differential equation for the evolution of strain energy in the form

$$\dot{\mathscr{F}}(t) + \frac{2}{\tau}\mathscr{F}(t) = 0 \tag{G.39}$$

Equation (G.39) clearly demonstrates that the characteristic time of the conversion from stored to dissipated energy is $\tau/2$. The particular solution of (G.39) satisfying the initial condition $\mathscr{F}(t_1) = E^\infty v(t_1)\hat{\varepsilon}^2/2$ is of course given by the already derived formula (G.36).

For illustration, the evolution of stress σ, strain energy $\mathscr{F}$, and dissipated energy D is plotted in Fig. G.4a. For easier comparison, the stress is normalized by its initial value $\sigma(t_1) = E^\infty v(t_1)\hat{\varepsilon}$, energies are normalized by the supplied work $\mathscr{F}(t_1) = E^\infty v(t_1)\hat{\varepsilon}^2/2$, and elapsed time $t - t_1$ by relaxation time τ_1. ■

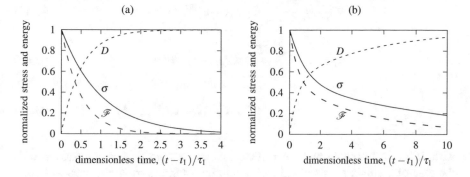

Fig. G.4 Evolution of normalized stress $\sigma(t)/\sigma(t_1)$, normalized free energy $\mathscr{F}(t)/\mathscr{F}(t_1)$, and normalized dissipated energy $D(t)/\mathscr{F}(t_1)$ in a relaxation test of (a) one Maxwell unit, (b) a Maxwell chain consisting of two units with relaxation times τ_1 and $\tau_2 = 10\tau_1$

Example G.2. Maxwell chain

The equations from the previous example can be extended to a solidifying Maxwell chain model consisting of M Maxwell units, which may differ not only by their elastic moduli E^∞_μ and relaxation times τ_μ but also by their volume growth functions $v_\mu(t)$. Stresses and energies corresponding to individual Maxwell units are additive, and the strain is the same for all units. Based on these rules, it is easy to develop generalized version of (G.27)–(G.29) and (G.31):

$$R(t, t') = \sum_{\mu=1}^{M} E^\infty_\mu v_\mu(t') e^{-(t-t')/\tau_\mu} H(t - t') \tag{G.40}$$

$$\sigma(t) = \int_0^t \sum_{\mu=1}^{M} E^\infty_\mu v_\mu(s) e^{-(t-s)/\tau_\mu} \, d\varepsilon(s) \tag{G.41}$$

$$\mathscr{F}(t) = \int_0^t \sum_{\mu=1}^{M} e^{-2(t-r)/\tau_\mu} \sigma(r) \, d\varepsilon(r) \tag{G.42}$$

$$\mathscr{D}_M(t) = 2 \int_{r=0}^t \int_{s=0}^r \sum_{\mu=1}^{M} \frac{E^\infty_\mu}{\tau_\mu} v_\mu(s) e^{-(2t-r-s)/\tau_\mu} \, d\varepsilon(s) \, d\varepsilon(r) =$$

$$= \int_0^t \sum_{\mu=1}^{M} \frac{2}{\tau_\mu} e^{-2(t-r)/\tau_\mu} \sigma(r) \, d\varepsilon(r) \tag{G.43}$$

In the particular case of a relaxation test started at age t_1, Eqs. (G.34) and (G.36)–(G.38) are generalized to

$$\sigma(t) = \hat{\varepsilon} \sum_{\mu=1}^{M} E_{\mu}^{\infty} v_{\mu}(t_1) e^{-(t-t_1)/\tau_{\mu}} \tag{G.44}$$

$$\mathscr{F}(t) = \frac{\hat{\varepsilon}^2}{2} \sum_{\mu=1}^{M} E_{\mu}^{\infty} v_{\mu}(t_1) e^{-2(t-t_1)/\tau_{\mu}} \tag{G.45}$$

$$\mathscr{D}_M(t) = \hat{\varepsilon}^2 \sum_{\mu=1}^{M} \frac{E_{\mu}^{\infty}}{\tau_{\mu}} v_{\mu}(t_1) e^{-2(t-t_1)/\tau_{\mu}} \tag{G.46}$$

$$D(t) = \frac{\hat{\varepsilon}^2}{2} \sum_{\mu=1}^{M} E_{\mu}^{\infty} v_{\mu}(t_1) \left(1 - e^{-2(t-t_1)/\tau_{\mu}}\right) \tag{G.47}$$

with $H(t - t_1)$ omitted and t considered as larger than t_1. For comparison with the case of a single Maxwell unit, the evolution of stress σ, strain energy $\mathscr{F}$, and dissipated energy D is plotted in Fig. G.4b for a Maxwell chain consisting of $M = 2$ units with parameters E_1^{∞}, τ_1, $E_2^{\infty} = E_1^{\infty}$, and $\tau_2 = 10\tau_1$ and with the same volume growth functions $v_1(t) = v_2(t) = v(t)$. The stress is normalized by its initial value $\sigma(t_1) = (E_1^{\infty} + E_2^{\infty})v(t_1)\hat{\varepsilon}$, energies are normalized by the supplied work $\mathscr{F}(t_1) = (E_1^{\infty} + E_2^{\infty})v(t_1)\hat{\varepsilon}^2/2$, and elapsed time by relaxation time τ_1. ∎

For the sake of simplicity, the viscoelastic potentials have been presented here in the one-dimensional setting, but generalizations to triaxial stress and strain are straightforward (at least if the Poisson ratio is considered as fixed). It is also possible to develop expressions for $\mathscr{F}(t)$ in terms of $J(t, t')$ and for $\mathscr{F}^*(t)$ in terms of $R(t, t')$ or $J(t, t')$.

In principle, these potentials could be useful for establishing error bounds, assessing numerical stability of computational algorithm, proving convergence, determining approximate bounds on viscoelastic heterogeneous materials, etc., similar to the theory of elasticity. However, it seems that no such applications have yet been made. For background developments, see Boltzmann [244], Volterra [839, 841], Roscoe [733], Biot [241], Coleman [321], Brun [273], Mandel [603, 605, 606], and a series of papers by Huet [490–496].

Appendix H
Updating Long-Time Shrinkage Predictions from Short-Time Measurements

H.1 Measuring Water Loss to Update Shrinkage Prediction

The reason why direct fitting of measured short-time shrinkage data and their extrapolation to long times does not provide reliable results has already been explained in Sect. 3.8.2. Let us now discuss how to estimate the *shrinkage halftime* τ_{sh}, which is imperative to avoid the ill-posedness of the shrinkage updating problem (Fig. 3.23). The basic idea suggested by Bažant and Baweja [104] is that while the final shrinkage cannot be predicted, the final water loss can.

It has long been known that shrinkage strains (averaged over the cross section) are approximately proportional to the water loss, denoted as $\Delta w(h_{env}, \hat{t})$. For a specimen of a given shape and size, the water loss depends on the environmental humidity h_{env} and on the drying duration $\hat{t} = t - t_0$. At constant environmental humidity h_{env}, the water loss approaches a limit $\Delta w_\infty(h_{env})$ as $\hat{t} \to \infty$. The key point is that, by contrast to shrinkage, the final value of water loss $\Delta w_\infty(0)$ at zero environmental humidity can be estimated easily—by heating the test specimen in an oven to $105\,°C$ right after the short-time test is terminated (the total initial evaporable water content is then the sum of weight losses during shrinkage and in the oven). From that, one can figure out the final value of water loss $\Delta w_\infty(h_{env})$ when hygral equilibrium at the given environmental humidity is reached. So it suffices to correlate the water-loss history to shrinkage history. This can be accomplished by simultaneous weighing of the test specimens during their shrinkage.

To estimate the final water loss $\Delta w_\infty(h_{env})$ for the given environmental humidity h_{env}, we need an approximation for the *desorption isotherm*, i.e., for the curve of water content w versus decreasing relative humidity h in concrete pores, at constant temperature (for further details see Sect. 8.2.5 and Appendix I.1). The problem is that the shapes of the desorption isotherms of concrete vary considerably [23, 55–57, 226, 382, 454, 705, 706, 883]. As discussed in Sect. 8.2.5, many experiments show that the desorption isotherm is virtually linear, from 10% to about 95% humidity. However, the behavior from 95 to 100% is highly varied. This is probably explained

© Springer Science+Business Media B.V. 2018
Z.P. Bažant and M. Jirásek, *Creep and Hygrothermal Effects in Concrete Structures*, Solid Mechanics and Its Applications 225,
https://doi.org/10.1007/978-94-024-1138-6

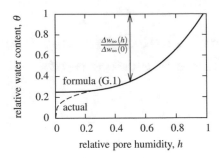

Fig. H.1 Normalized desorption isotherm used by Bažant and Baweja [104], plotted in terms of the relative water content, $\theta(h) = 1 - \Delta w_\infty(h)/\Delta w_\infty(0)$, versus relative pore humidity, h

by the nonuniqueness of water content at saturation, which may be quite pronounced when the biggest capillary pores occupy a large volume fraction. Depending on the extent of chemical self-desiccation, these pores can, at $h = 100\%$, be either filled by liquid water or most of them can still be empty.

Based on an analogy with formula (3.19) for the coefficient k_h that represents the influence of humidity on the final shrinkage, Bažant and Baweja [104] proposed to estimate the final water loss as

$$\Delta w_\infty(h_{\mathrm{env}}) \approx 0.75 \left[1 - \left(\frac{h_{\mathrm{env}}}{0.98} \right)^3 \right] \Delta w_\infty(0) \tag{H.1}$$

This cubic formula, intended for the range $0.25 \leq h \leq 0.98$, is graphically shown in Fig. H.1. Note that the actual isotherm describes the dependence of specific water content on the relative humidity in concrete pores, h, which is in general variable throughout the specimen. Here, we replace h by the relative humidity of the environment, h_{env}, because we are interested in the final water loss, Δw_∞, which corresponds to the equilibrium state with $h = h_{\mathrm{env}}$. Expression (H.1) satisfies the condition that there is no water loss for $h \approx 0.98$, i.e., for a sealed specimen. Recall that for $h_{\mathrm{env}} = 1.0$, i.e., in water immersion, there is water gain and the concrete is swelling. For $h_{\mathrm{env}} < 0.25$, expression (H.1) is invalid (Fig. H.1), but sustained environmental humidities below 25% normally do not occur in practice.

Due to the assumption of proportionality between shrinkage and water loss, the kinetics of water loss should be approximately described by the same function (3.16) as the evolution of shrinkage strain, that is,

$$\frac{\Delta w(h_{\mathrm{env}}, \hat{t})}{\Delta w_\infty(h_{\mathrm{env}})} = \tanh \sqrt{\frac{\hat{t}}{\tau_{\mathrm{w}}}} \tag{H.2}$$

where τ_{w} is the *water-loss halftime*. Based on some limited tests [430, 454], it is assumed that the shrinkage halftime τ_{sh} is about 25% longer than the water-loss halftime τ_{w}, approximately $\tau_{\mathrm{sh}} = 1.25\tau_{\mathrm{w}}$. The reason for $\tau_{\mathrm{sh}} > \tau_{\mathrm{w}}$ might be explained by the fact that the microcracking in the surface layer of drying specimens accelerates water loss but decreases average axial shrinkage in the cross section. Another reason could be the existence of a certain time lag caused by the local microdiffusion of water from gel nanopores to capillary pores.

Equation (H.2) can easily be rearranged to a linear form:

$$\hat{t} = \tau_w \psi \quad \text{with} \quad \psi = \left(\operatorname{atanh} \frac{\Delta w(h_{env}, \hat{t})}{\Delta w_\infty(h_{env})} \right)^2 \tag{H.3}$$

Now consider that, at times $t_i = t_0 + \hat{t}_i$ $(i = 1, 2 \ldots n)$ of shrinkage measurements, the values of water loss Δw_i up to times t_i have simultaneously been measured and the corresponding values of ψ_i have been calculated. The optimum value of τ_w must minimize the sum of squared deviations, i.e.

$$Z_w(\tau_w) = \sum_{i=1}^{n} \left(\tau_w \psi_i - \hat{t}_i \right)^2 = \min \tag{H.4}$$

Since Z_w is a convex function, a necessary and sufficient condition of a minimum is that $dZ_w/d\tau_w = 0$. Thus, one gets the linear equation $\sum_i \left(\tau_w \psi_i - \hat{t}_i \right) \psi_i = 0$, from which

$$\tau_w = \frac{\sum_{i=1}^{n} \hat{t}_i \psi_i}{\sum_{i=1}^{n} \psi_i^2} \tag{H.5}$$

The corresponding value of shrinkage halftime $\bar{\tau}_{sh} = 1.25\tau_w$ may then be used to obtain the update of the final shrinkage value.

The updated shrinkage prediction is obtained simply by scaling the model B3 prediction vertically:

$$\varepsilon_{sh}^*(t, t_0) = p_3 \bar{\varepsilon}_{sh}(t, t_0) \tag{H.6}$$

Here, $\bar{\varepsilon}_{sh}(t, t_0)$ are the values predicted from model B3 based on $\bar{\tau}_{sh}$ (i.e., ignoring the empirical estimate for τ_{sh} based on (3.17), with k_t determined from the formula in line 5 of Table C.2), and p_3 is the scaling parameter to be calculated. Consider that values $\varepsilon'_{sh,i}$ at times t_i $(i = 1, 2, \ldots, n)$ have been measured. The optimum update should again minimize the sum of squared deviations of the updated model from the measured data, i.e.

$$Z_{sh}(p_3) = \sum_i (p_3 \bar{\varepsilon}_{sh,i} - \varepsilon'_{sh,i})^2 = \min \tag{H.7}$$

where $\bar{\varepsilon}_{sh,i} = \bar{\varepsilon}_{sh}(t_i, t_0)$. A necessary (and sufficient) condition of a minimum is that $dZ_{sh}/dp_3 = 0$. This yields the condition $\sum_i (p_3 \bar{\varepsilon}_{sh,i} - \varepsilon'_{sh,i}) \bar{\varepsilon}_{sh,i} = 0$, from which the optimum update parameter is

$$
p_3 = \frac{\sum_{i=1}^{n} \varepsilon'_{\text{sh},i} \bar{\varepsilon}_{\text{sh},i}}{\sum_{i=1}^{n} \bar{\varepsilon}^2_{\text{sh},i}}
\tag{H.8}
$$

The shrinkage halftime $\bar{\tau}_{\text{sh}}$ obtained from the water-loss data should also be used to improve the drying creep compliance function $J_{\text{d}}(t, t')$ (which is a part of function $F(t, t')$ in (3.35)), before the creep update procedure described in Sect. 3.8.1 is invoked.

A nonnegligible error in shrinkage extrapolation may arise if the first reading is not taken immediately after the stripping of the mold [151, 580, 878]. The initial stage of the shrinkage process is very fast, because the average shrinkage strain is proportional to the square root of the drying time, as predicted by the diffusion theory (there is an initial deviation from the square root due to finite moisture emissivity at surface, but this is usually important only for specimens thinner than 1 cm).

Therefore, if the first reading is taken 15 min, or even just 1 min, after the onset of drying, a nonnegligible part of shrinkage must have been missed. The true values of shrinkage strains may then be determined by optimally fitting to the initial data points the theoretical relation $\varepsilon'_{\text{sh},i} + \Delta\varepsilon_{\text{sh}} = k\sqrt{\hat{t}_i}$, in which $\varepsilon'_{\text{sh},i}$ are the measured values of shrinkage after drying times $\hat{t}_i$, and $\Delta\varepsilon_{\text{sh}}$ and k are constants to be determined by optimum fitting. $\Delta\varepsilon_{\text{sh}}$ has the meaning of the missed part of shrinkage strain and should be added to the measured values $\varepsilon'_{\text{sh},i}$ before they are inserted into (H.8).

An irreparable error occurs if the specimen seals leak moisture during their curing before shrinkage test.

The water loss should preferably be measured directly on the shrinkage specimens themselves. If it is measured on companion specimens, they must have the same environmental exposure all the time and must be identical (although random differences cannot be avoided).

H.2   Procedure of Updating Shrinkage Prediction from Short-Time Tests

The procedure developed in Bažant and Baweja [104] and explained in the preceding section can be summarized as follows:

Algorithm H.1

1. Determine the final water loss that would occur at zero relative environmental humidity, $\Delta w_\infty(0)$, obtained when the average final water loss measured upon heating the specimens (after the drying test) to about 105 °C is added to the weight

loss during drying at the specified ambient humidity. If such measurements could not, or have not, been done, the final water loss can be estimated as the total water content of the concrete mix minus the amount of water consumed by the hydration reaction (chemically bound water), whose terminal weight is roughly 20% of the weight of cement.

2. Calculate the estimate of the final relative water loss for drying at given relative environmental humidity $h_{env} \in (0.25, 0.98)$ and room temperature,

$$\Delta w_\infty(h_{env}) \approx 0.75 \left[1 - (h_{env}/0.98)^3\right] \Delta w_\infty(0) \tag{H.9}$$

3. On the specimens sealed until time t_0 and then stored at ambient relative humidity h_{env}, measure the values of relative water loss Δw_i at times $t_i = t_0 + \hat{t}_i$ ($i = 1, 2, \ldots, n$), which should be spaced approximately uniformly in the scale of $\log \hat{t}$.

4. Calculate the auxiliary values

$$\psi_i = \left(\text{atanh}\frac{\Delta w_i}{\Delta w_\infty(h_{env})}\right)^2, \quad i = 1, 2, \ldots, n \tag{H.10}$$

and evaluate the estimate of water-loss halftime

$$\tau_w = \frac{\sum_i \hat{t}_i \psi_i}{\sum_i \psi_i^2} \tag{H.11}$$

Use the improved estimate of shrinkage halftime,

$$\bar{\tau}_{sh} = 1.25\tau_w \tag{H.12}$$

instead of the rough estimate obtained from (3.17).

5. Denote $\varepsilon'_{sh,i} = \varepsilon'_{sh}(t_i) =$ measured short-time values of shrinkage at times t_i. Also, denote

$$\bar{\varepsilon}_{sh}(t) = -\varepsilon_{sh}^\infty k_h \tanh \sqrt{\frac{t - t_0}{\bar{\tau}_{sh}}} \tag{H.13}$$

the shrinkage function of model B3 using $\bar{\tau}_{sh}$ as the shrinkage halftime in formula (3.16). Calculate the values $\bar{\varepsilon}_{sh,i} = \bar{\varepsilon}_{sh}(t_i)$ predicted for the times t_i, and evaluate the scaling parameter

$$p_3 = \frac{\sum_i \varepsilon'_{sh,i}\bar{\varepsilon}_{sh,i}}{\sum_i \bar{\varepsilon}_{sh,i}^2} \tag{H.14}$$

6. The updated values of shrinkage prediction for any time t are

$$\varepsilon_{sh}^*(t) = p_3\, \bar{\varepsilon}_{sh}(t) \tag{H.15}$$

The times of shrinkage measurements and of water-loss measurements should preferably coincide, but the procedure is still applicable even if they do not. The sums in (H.11) and in (H.14) are then taken over two different sets of measurement times, one for water loss and the other for drying.

Equation (H.15) combined with (H.13) is equivalent to formulae (3.15)–(3.16) for the shrinkage of the test specimen if $\varepsilon_{sh}^{\infty}$ is replaced by $\varepsilon_{sh}^{\infty*} = p_3 \varepsilon_{sh}^{\infty}$ and τ_{sh} is replaced by $\bar{\tau}_{sh}$. To obtain the updated values of $\varepsilon_{sh}(t)$ for the real structure for which t_0, D and h_{env} may be different, the formulae for k_t and ε_s^{∞} in lines 5 and 6 of Table C.2 are disregarded. Using the updated value $\bar{\tau}_{sh}$ (instead of the original predicted value τ_{sh}) for the short-time test specimen, Eq. (3.17) is solved for k_t. This yields the updated value, $k_t^* = \bar{\tau}_{sh}/(k_s D)^2$, replacing k_t.

Next, the expression in line 7 of Table C.2 is used with the updated value $\varepsilon_{sh}^{\infty*}$ for the specimen (instead of the original value $\varepsilon_{sh}^{\infty}$) to solve for the updated value

$$\varepsilon_s^{\infty*} = \frac{\varepsilon_{sh}^{\infty*}}{0.57514\sqrt{3 + 14/(t_0 + \bar{\tau}_{sh})}} \tag{H.16}$$

that should replace ε_s^{∞}. With the updated values, k_t^* and $\varepsilon_s^{\infty*}$, and the values of t_0, D, and h_{env} for the structure, formulae (3.15)–(3.17) and line 7 of Table C.2 are used to obtain the updated values of $\varepsilon_{sh}(t)$ for the structure.

The updated $\bar{\tau}_{sh}$ and $\varepsilon_s^{\infty*}$ should also be used in the calculation of the drying creep compliance J_d, which modifies the function $F(t, t')$ in (3.35). This improves the procedure for creep updating described in Sect. 3.8.1. Recall that J_d is given by (3.20), with auxiliary functions $\hat{g}$ and S defined in (3.23) and (3.16), and with parameter q_5 evaluated from (3.24), in which $\varepsilon_{sh}^{\infty*}$ is obtained from $\varepsilon_s^{\infty*}$ using the formula in line 7 of Table C.2.

H.3 Example of Shrinkage Updating

Example H.1. Updating shrinkage prediction

To illustrate the shrinkage updating procedure proposed in Bažant and Baweja [104], consider the shrinkage and water-loss data for French nuclear containments obtained by Granger [430]. The shrinkage was measured over a gauge length of 50 cm on cylinders with a 16 cm diameter and 100 cm length, exposed to an environmental humidity of 50% at age 28 days. The relative weight loss was measured on cylinders with a 16 cm diameter and 15 cm length. The ends of both the shrinkage and the weight-loss specimens were sealed to ensure radial axisymmetric drying.

Granger [430] tested six types of concrete from five sites. The shrinkage updating procedure will be illustrated for two selected concretes, denoted as Civaux B11 and Paluel. Their composition is shown in Table H.1; the cylindrical compression strength was 40.2 MPa for Civaux B11 and 43 MPa for Paluel.

Initial prediction: Based on the given composition, strength, specimen shape, dimensions, and environmental humidity, the parameters affecting shrinkage can be

Table H.1 Composition of two concretes used by Granger [430]; all quantities given in kg/m³

Concrete denomination	Civaux B11	Paluel
Cement, c	350	375
Water, w	195	180
Fine aggregates	629	709
Coarse aggregates	1100	1048
Filler, f	143	63
Total aggregates, a (incl. filler)	1872	1820
Admixtures	1	2
Air entrainer	3	2
Total	2421	2379

Table H.2 Example of shrinkage calculations: (a) predicting model parameters based on data available before the test, (b) updating model parameters based on short-time water-loss and shrinkage measurements on a cylindrical specimen, (c) predicting long-time shrinkage of a real structure

	Shrinkage specimen – initial prediction					
	(cylinder, $D = 2V/S = 8$ cm, $t_0 = 28$ days, $h_{env} = 50\%$)					
	Parameter	Equation	Civaux B11	Paluel	Unit	
(a)	$\bar{f}_c$	Experiment	40.2	43.0	MPa	
	k_t	Line 5 in Table C.2	0.02586	0.02543	day/mm²	
	τ_{sh}	(3.17)	218.9	215.2	day	
	ε_s^∞	Line 6 in Table C.2	846.2	757.1	10^{-6}	
	ε_{sh}^∞	Line 7 in Table C.2	850.9	761.5	10^{-6}	
	k_h	(3.19)	0.875	0.875	–	
	Shrinkage specimen – update					
	(cylinder, $D = 2V/S = 8$ cm, $t_0 = 28$ days, $h_{env} = 50\%$)					
	Parameter	Equation	Civaux B11	Paluel	Unit	
(b)	$\Delta w_\infty(0)$	$w - 0.2(c + f)$	96.4	92.4	kg/m³	
	$\Delta w_\infty(0.5)$	(H.9)	62.7	60.1	kg/m³	
	$\Delta w_\infty(0.5)/\rho$		2.59	2.53	%	
	τ_w	(H.11)	243.0	304.2	day	
	$\bar{\tau}_{sh}$	$1.25 \times \tau_w$	303.8	380.2	day	
	p_3	(H.14)	0.6458	0.7567	–	
	$\varepsilon_{sh}^{\infty*}$	$p_3 \varepsilon_{sh}^\infty$	549.5	576.2	10^{-6}	
	k_t^*	$\bar{\tau}_{sh}/(k_s D)^2$	0.03589	0.04492	day/mm²	
	$\varepsilon_s^{\infty*}$	(H.16)	547.8	571.5	10^{-6}	
	Real structure					
	(Civaux concrete B11, slab, $D = 25$ cm, $t_0 = 10$ days, $h_{env} = 65\%$)					
	Parameter	Equation	Value	Unit		
(c)	τ_{sh}	(3.17) with k_t^*	2243	day		
	ε_{sh}^∞	line 7 in Table C.2 with $\varepsilon_s^{\infty*}$	546.3	10^{-6}		
	k_h	(3.19)	0.725	–		
	$\varepsilon_{sh}(1010, 10)$	(3.15)	−231.1	10^{-6}		

estimated using the empirical formulae given in Appendix C. In addition to the mix
composition and compression strength, the following input parameters are needed:
effective thickness $D = 2V/S = 80$ mm (cylinder radius), shape factor $k_s = 1.15$
(infinite cylinder), parameters $\alpha_1 = 1$ (type-I cement) and $\alpha_2 = 1.2$ (normal curing),
end of curing at $t_0 = 28$ days, and ambient relative humidity $h_{env} = 50\% = 0.5$.
The resulting estimates are summarized in Table H.2a, and the corresponding initial
predictions of the history of shrinkage strain are shown by the dashed curves in
Figs. H.2b and H.3b. The actually measured values, indicated by solid and hollow
circles, are for both concretes substantially overestimated by the model.

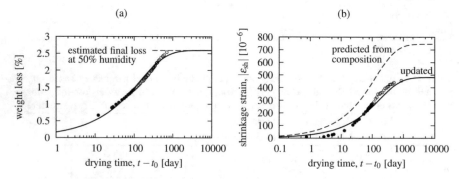

Fig. H.2 Updating shrinkage prediction for Civaux concrete B11 using short-time data: (a) history
of water loss, (b) history of shrinkage strain

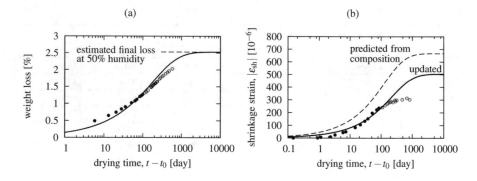

Fig. H.3 Updating shrinkage prediction for Paluel concrete using short-time data: (a) history of
water loss, (b) history of shrinkage strain

Update after 100 days: Pretend now that we know the measured water-loss
data and shrinkage strains only up to 100 days of drying; see the solid circles in
Figs. H.2 and H.3. Table H.2b illustrates the procedure for updating model parameters
on the basis of such short-time tests, described in Algorithm H.1. Unfortunately,
Granger [430] did not measure the final water loss of his samples in a completely dry
environment, $\Delta w_\infty(0)$. Therefore, we need to use a rough estimate, e.g., to assume
that the water used up in the hydration reaction is about 20% (by weight) of the

cement and filler (which is a simplifying assumption since the amount of water combined with the filler is not well known at present). This simple calculation gives the values reported in the first line of Table H.2b. Then, the final water loss at relative environmental humidity 50% is calculated from (H.9) and is divided by the mass density of concrete to get the relative values (by weight) of 2.59% for Civaux B11 and 2.53% for Paluel. Using these values and the points of the weight-loss data up to 100 days of drying, we determine the water-loss halftime τ_w from (H.10) and (H.11) and then estimate the shrinkage halftime as $\bar{\tau}_{sh} = 1.25\tau_w$ from (H.12); see Table H.3a. The updated values of shrinkage halftime and the drying shrinkage data up to 100 days duration are used in (H.13)–(H.15) to calculate the updated parameters $\varepsilon_{sh}^{\infty*}$; see Table H.3b. The updated predictions of shrinkage strain evolution, based on $\varepsilon_{sh}^{\infty*}$ and $\bar{\tau}_{sh}$ instead of $\varepsilon_{sh}^{\infty}$ and τ_{sh}, are shown in Figs. H.2b and H.3b by the solid curves. A significant improvement is achieved for Civaux B11, but only a partial improvement for Paluel.

Table H.3 Example of shrinkage calculations for Civaux B11: (a) evaluation of the water-loss halftime τ_w according to (H.10)–(H.11), (b) evaluation of scaling parameter p_3 according to (H.13)–(H.15)

(a)

i	$\hat{t}_i$	Δw_i	ψ_i	$\hat{t}_i\psi_i$	ψ_i^2
	[day]	[%]	[1]	[day]	[1]
1	12.0	0.666	0.0690	0.8300	0.0048
2	26.6	0.892	0.1285	3.4200	0.0165
3	28.6	0.917	0.1365	3.9061	0.0186
4	34.1	0.989	0.1612	5.4998	0.0260
5	40.4	1.046	0.1828	7.3879	0.0334
6	48.4	1.122	0.2144	10.3773	0.0460
7	55.5	1.171	0.2367	13.1268	0.0560
8	62.1	1.219	0.2601	16.1625	0.0676
9	69.7	1.272	0.2879	20.0529	0.0829
10	76.6	1.308	0.3080	23.6047	0.0949
11	84.0	1.353	0.3347	28.0994	0.1120
12	91.6	1.397	0.3625	33.2122	0.1314
13	98.0	1.425	0.3812	37.3484	0.1453
Σ				203.03	0.8354

$\tau_w = 203.03/0.8354 = 243.0$ [day]

(b)

i	$\hat{t}_i$	$-\varepsilon_{sh,i}$	$-\bar{\varepsilon}_{sh,i}$	$\varepsilon_{sh,i}\bar{\varepsilon}_{sh,i}$	$\bar{\varepsilon}_{sh,i}^2$
	[day]	[10^{-6}]	[10^{-6}]	[10^{-12}]	[10^{-12}]
1	0.0	1.4	7.1	10	51
2	0.1	3.7	11.4	42	130
3	0.8	9.0	37.0	332	1368
4	2.7	11.2	70.5	792	4973
5	3.8	20.8	83.1	1726	6900
6	5.1	27.1	96.1	2608	9235
7	5.8	31.1	101.9	3170	10373
8	12.7	60.6	149.9	9085	22484
9	23.6	105.3	202.4	21305	40968
10	27.2	122.8	216.4	26574	46816
11	32.1	138.0	233.7	32251	54631
12	37.8	150.7	252.3	38029	63643
13	53.2	191.5	294.5	56376	86711
14	61.2	209.8	313.4	65758	98232
15	69.7	220.2	331.6	73002	109930
16	73.9	235.4	340.0	80020	115600
17	81.1	246.5	353.8	87229	125195
18	90.2	258.5	369.8	95590	136745
19	94.5	268.1	377.0	101066	142128
Σ				694966	1076113

$p_3 = 694966/1076113 = 0.6458$

Application to a structure: Now consider, for example, that the designer needs the value of $\varepsilon_{sh}(t, t_0)$ for a slab of thickness $D = 25$ cm, exposed to environmental humidity $h_{env} = 65\%$ from age $t_0 = 10$ days. Table H.2c shows the evaluation of the shrinkage halftime τ_{sh} and final shrinkage ε_{sh}^∞ for such a structure, based on the usual formulae but with parameters k_t and ε_s^∞ replaced by the updated values k_t^* and $\varepsilon_s^{\infty*}$ determined in the previous step (update). Note that these parameters are the same for both the specimen and the structure. They are used along with the given values of D, t_0, and h_{env} for the real structure to obtain the updated shrinkage predictions for the structure. After determining $\tau_{sh} = 2243$ days, $\varepsilon_{sh}^\infty = 546.3 \times 10^{-6}$, and $k_h = 0.725$, we can predict the shrinkage strain after 1000 days of drying (i.e., at age $t = 1010$ days) according to (3.15) as

$$\varepsilon_{sh}(1010) = -546.3 \times 10^{-6} \times 0.725 \times \tanh\sqrt{\frac{1000}{2243}} = -231.1 \times 10^{-6} \quad \text{(H.17)}$$

∎

Remark: Updating Based on Diffusion Size Effect and Role of Autogenous Shrinkage (added in proof).

In view of the aforementioned limitations, it has recently been tried to exploit the diffusion size effect [124], as described in Sect. 3.8.1. However, similar discrepancies as with the weight loss method have been found. The inevitable conclusion is that (1) one must take into account the autogenous shrinkage, which proceeds for a long time in the core of the standard specimen but quickly becomes negligible in the small companion specimen, and (2) the compressive volumetric creep of the solid skeleton of the hardened cement paste, loaded by the stress changes in the liquid and adsorbed pore water. Both these phenomena prolong the final stage of total shrinkage of drying specimens. Further studies are in progress.

Appendix I
Moisture Transport Characteristics

I.1 Sorption Isotherms

The classical Langmuir isotherm [564] describes adsorption in a single molecular layer and has the form

$$u = \frac{abh}{1 + bh} \tag{I.1}$$

where u is the *moisture ratio*, i.e., the ratio between the mass of evaporable water and the mass of the dry sample (also called the *water content in mass percent*), a is the moisture ratio corresponding to a complete monolayer, b is a parameter dependent on temperature, and h is the pore relative humidity,

The isotherm of the BET theory [276] is composed of a series of Langmuir isotherms representing individual molecular layers. For n molecular layers, the result is

$$u = \frac{abh\left[1 - (n+1)h^n + nh^{n+1}\right]}{(1-h)\left[1 + (b-1)h - bh^{n+1}\right]} \tag{I.2}$$

For $n \to \infty$ this tends to

$$u = \frac{abh}{(1-h)\left[1 + (b-1)h\right]} \tag{I.3}$$

This equation is suitable for the description of the isotherm at lower humidities, up to about 50%. Note that it can be deduced from (8.54) by setting $u = \Gamma_a S_{spec} \rho_d$, where S_{spec} is the specific area (surface of pores per unit volume) and ρ_d is the specific mass of dry concrete. Parameter a then corresponds to $\Gamma_1 S_{spec} \rho_d$ and parameter b to $C_0 \exp(Q_a/RT)$ (see Sect. 8.2.6.1 for the meaning of Γ_a, Γ_1, C_0 and Q_a).

A simple empirical relation proposed by Lykow [588] for the range of relative humidities between 10 and 90% reads

© Springer Science+Business Media B.V. 2018
Z.P. Bažant and M. Jirásek, *Creep and Hygrothermal Effects in Concrete Structures*, Solid Mechanics and Its Applications 225,
https://doi.org/10.1007/978-94-024-1138-6

$$u = \frac{c_1 h}{h - c_2} \tag{I.4}$$

where c_1 and c_2 are temperature-dependent constants. To obtain a convex isotherm, parameter c_1 must be negative and parameter c_2 must be larger than 1. For the range between 30 and 100%, Lykow recommended the Posnow equation

$$u = \frac{u_h}{1 - \dfrac{\ln h}{d}} \tag{I.5}$$

derived from experiments with wood. Parameter u_h is the maximum hygroscopic moisture ratio and d is another temperature-dependent parameter. A generalization of (I.5) suggested by Hansen [449] reads

$$u = \frac{u_h}{\left(1 - \dfrac{\ln h}{A}\right)^{1/n}} \tag{I.6}$$

with parameters u_h, A, and n. Künzel [556] proposed a simpler formula,

$$u = \frac{(b - 1)u_h h}{b - h} \tag{I.7}$$

with only two parameters, u_h and b. Some of the isotherms used in building physics are unbounded (u tends to infinity as h approaches 1), and so they cannot be used near saturation [499, 549, 725].

An isotherm can also be constructed from the capillary pressure curve (or moisture retention curve) linking the capillary pressure p_c to the saturation degree S_l, e.g., from the van Genuchten formula (8.16). Equation (8.48) can be rewritten in terms of the moisture ratio as

$$u = \frac{w_e}{\rho_d} = \frac{n_p \rho_l S_l}{\rho_d} \tag{I.8}$$

From the van Genuchten equation (8.16) with $p_{\text{entry}} = 0$, the saturation degree can be expressed as

$$S_l = \left[\left(\frac{p_c}{\pi_0}\right)^{1/(1-m)} + 1 \right]^{-m} \tag{I.9}$$

Combining this with the Kelvin equation in its simplified form (8.34) and substituting into (I.8), we obtain

$$u = \frac{n_p \rho_l}{\rho_d} \left[\left(\frac{p_c}{\pi_0}\right)^{1/(1-m)} + 1 \right]^{-m} = \frac{n_p \rho_l}{\rho_d} \left[\left(-\frac{\rho_l R T}{M_w \pi_0} \ln h\right)^{1/(1-m)} + 1 \right]^{-m} \tag{I.10}$$

The resulting isotherm has the form

$$u = u_1 \left[(-b \ln h)^{1/(1-m)} + 1 \right]^{-m} \tag{I.11}$$

with parameters u_1, b, and m.

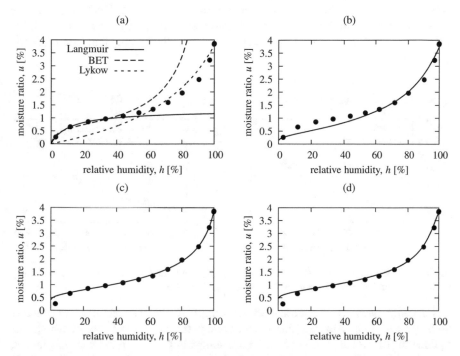

Fig. I.1 Adsorption isotherms measured by Baroghel-Bouny [55] and fitted by (a) Langmuir, BET, and Lykow formulae, (b) Posnow formula (I.5), (c) Freiesleben–Hansen formula (I.6), (d) van Genuchten formula (I.11)

Most of the above-mentioned isotherms were originally proposed for materials other than concrete. We will illustrate their applicability by fitting experimental data reported by Baroghel-Bouny [55] for 1-year-old concrete with $w/c = 0.43$, which were presented in the form of desorption and adsorption isotherms in Fig. 8.5. These data cover the complete range of relative humidities, down to 3%. In the original paper, the state at $h = 3\%$ was considered as the reference "dry" state. Since the theoretical formulae take zero humidity as the dry state with zero evaporable water content, the measured data have been shifted vertically by 0.26%, which is the value determined by fitting the low-humidity data by the Langmuir isotherm. In contrast to Fig. 8.5, here we plot the values of the moisture ratio u instead of the evaporable water content w_e. The fits of the adsorption isotherm are shown in Fig. I.1. They have been obtained with parameters $a = 1.3$ and $b = 8.7$ for the Langmuir isotherm (I.1), $a = 0.68$ and $b = 28$ for the BET isotherm (I.3), $c_1 = -1.9$ and $c_2 = 1.5$ for the Lykow equation (I.4), $u_h = 3.74\%$ and $d = 0.26$ for the Posnow equation (I.5),

$u_h = 3.86\%$, $A = 0.072$, and $n = 2$ for the Freiesleben-Hansen equation (I.6), and $u_1 = 3.8\%$, $b = 20.6$ and $m = 0.3$ for the van Genuchten equation (I.11).

For fitting by the Langmuir and BET isotherms (Fig. I.1a), only the measured data up to 45% humidity have been considered, because these isotherms are not meant to describe capillary condensation. The Lykow formula is seen to give a fair fit only for humidities above 40% (Fig. I.1a), and the Posnow formula is slightly better (Fig. I.1b). A very good fit over the range from 10 to 100%, reproducing the characteristic sigmoidal shape of the adsorption isotherm, is obtained with the Freiesleben–Hansen formula (Fig. I.1c) as well as with the van Genuchten formula (Fig. I.1d).

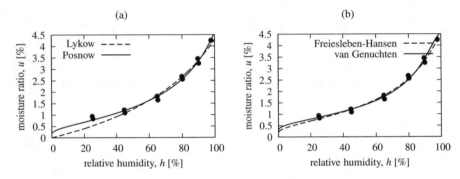

Fig. I.2 Adsorption isotherm measured by Ahlgren [22] and fitted by the (a) Lykow and Posnow formulae, (b) Freiesleben–Hansen and van Genuchten formulae

For comparison, we also present fits of the experimental data recorded for concrete with $w/c = 0.48$ by Ahlgren [22] and reported in the database of Hansen [448]. The optimal fits of the adsorption isotherm are shown in Fig. I.2. As already explained, the Langmuir isotherm and the BET isotherm are appropriate for low humidities only and thus are not presented here. The Lykow formula (I.4) with parameters $c_1 = -3$ and $c_2 = 1.7$ is seen to give quite a good fit for humidities above 40%, and the Posnow formula (I.5) with parameters $u_h = 4.6\%$ and $d = 0.28$ gives a good fit over the entire range covered by experiments (Fig. I.2a). For this concrete, the Freiesleben–Hansen isotherm (I.6) with parameters $u_h = 4.6\%$, $A = 0.26$, and $n = 1.05$ does not give any visible improvement compared to the simpler Posnow formula, because the optimal value of n turns out to be very close to 1. The van Genuchten isotherm with parameters $u_1 = 4.3\%$, $b = 8.3$, and $m = 0.4$ provides a very good fit, too (Fig. I.2b).

The desorption isotherms are often close to a straight line. For the data of Baroghel-Bouny [55] shown in Fig. I.3a, this is true in the range above 40% relative humidity, while for the data of shown in Fig. I.3b, linearity holds over the entire range covered by experiments (from 20% relative humidity). As illustrated in Fig. I.3a, the van Genuchten formula with parameters $u_1 = 4.1\%$, $b = 3.1$, and $m = 0.48$ could provide a good fit over the entire range of humidities, but it is more convenient to use a simple linear relation, $u = 4.6\% \times h - 0.4\%$, which provides the same level of accuracy in the range above 30% relative humidity. For the data in Fig. I.3b, the linear relation $u = 4.54\% \times h + 0.052\%$ is even better than the van Genuchten three-parameter formula. The dry concrete density is in this case $\rho_d = 2300$ kg/m^3, and the evaporable water content can be expressed as $w_e = \rho_d u$. The moisture capacity is thus $1/k = \rho_d \times 4.5\% = 104.4$ kg/m^3, and its reciprocal value is $k = 0.00958$ m^3/kg.

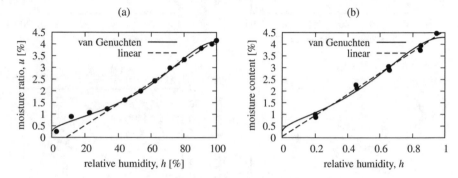

(a)   (b)

Fig. I.3 Desorption isotherms measured by (a) Baroghel-Bouny [55] and (b) Ahlgren [22], approximated by the van Genuchten formula and by a straight line

A host of attempts have been made to modify the BET equation (I.3) in order to obtain better agreement with experimental data. Such modifications include the BDDT model [275], the FHH model [447], Hillerborg's formula [480], and the BSB model [277], which served as a starting point for the development of prediction formulae for the adsorption isotherm of cement paste proposed by Xi et al. [883]. The general form of the BSB model is

$$u(h, T) = \frac{V_m C_T(T) k_T(T) h}{[1 - k_T(T)h][1 + (C_T(T) - 1)k_T(T)h]} \tag{I.12}$$

where T is the absolute temperature, k and V_m are parameters, and

$$C_T(T) = \exp\left(\frac{Q_a}{RT}\right) \tag{I.13}$$

with R = universal gas constant and Q_a = net heat of adsorption per mole. Note that, for $k_T = 1$, formula (I.12) would become equivalent to the BET formula (I.3), with $a \equiv V_m$ and $b \equiv C_T$.

Xi et al. [883] developed empirical formulae for the prediction of parameters V_m and k_T based on the composition of the paste and its age. They took into account the type of cement, water-cement ratio w/c, and equivalent age t_e (in days). Parameter V_m has the meaning of monolayer capacity and can be estimated as

$$V_m = \left(0.068 - \frac{0.22}{t_e}\right)\left(0.85 + 0.45\frac{w}{c}\right) V_{ct} \tag{I.14}$$

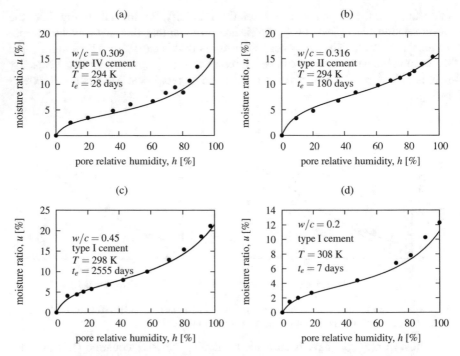

Fig. I.4 Comparison of predicted adsorption isotherms for various cement pastes with data measured by (a)–(b) Powers and Brownyard [706], (c) Hagymassy et al. [445], (d) Mikhail et al. [629]

where $V_{ct} = 0.9$ for cement type I (ordinary Portland cement), 1.0 for type II (modified cement), 0.85 for type III (rapid-hardening Portland cement), and 0.6 for type IV (low-heat Portland cement).

Parameter k is linked to the number n of adsorbed layers at the saturation state and can be expressed as

$$k_T(T) = \frac{(1 - 1/n)\, C_T(T) - 1}{C_T(T) - 1} \qquad (I.15)$$

where

$$n = \left(2.5 + \frac{15}{t_e}\right)\left(0.33 + 2.2\frac{w}{c}\right) N_{ct} \qquad (I.16)$$

and $N_{ct} = 1.1$ for cement type I, 1.0 for type II, 1.15 for type III and 1.5 for type IV.

Parameter Q_a is the net heat of adsorption and could be determined from heat of immersion experiments. For moderate temperatures, Xi et al. [883] suggested to set $Q_a/R = 855$ K. This simple approach is not applicable at high temperatures. For temperatures much higher than the room temperature, and especially above 100 °C, further phenomena come into play and a more complex model is needed [188].

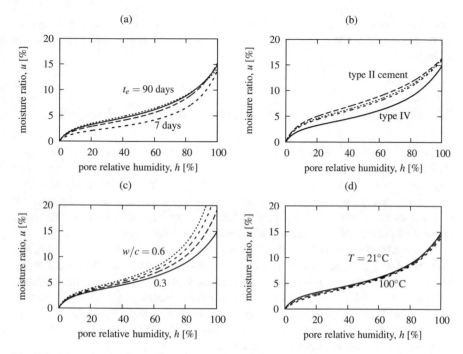

Fig. I.5 Adsorption isotherms for cement paste and their dependence of on (a) age, (b) cement type, (c) water-cement ratio, and (d) temperature

Formulae (I.14) and (I.16) are valid in the range of $t_e > 5$ days and $0.3 \leq w/c \leq 0.7$. Outside this range, t_e or w/c should be replaced by the value of the closest limit (e.g., if $t_e < 5$ days, the formulae should be evaluated for $t_e = 5$ days).

Xi et al. [883] demonstrated a very good agreement of the predicted isotherms with the experimental data of Powers and Brownyard [706], Hagymassy et al. [445], Odler et al. [663], and Mikhail et al. [629]. Selected examples are shown in Fig. I.4. Note that these are the adsorption isotherms, valid at increasing humidity.

Based on this prediction model, it is possible to illustrate the effect of age, cement type, water-cement ratio, and temperature on the adsorption isotherms; see Fig. I.5. The reference curve, plotted in all parts of the figure by a solid line, corresponds to the equivalent age $t_e = 28$ days, cement type IV, water/cement ratio 0.3, and temperature $T = 294$ K (21 °C). In Fig. I.5a, the equivalent age is set to 7 days, 14 days, 28 days, and 90 days (from bottom to top). In Fig. I.5b, the cement types are IV, III, I, and II (from bottom to top). In Fig. I.5c, the water/cement ratios are 0.3, 0.4, 0.5, and 0.6 (from bottom to top). As shown in Fig. I.5d, the effect of temperature within the range from 20 to 100 °C is very weak, which is consistent with the findings of Radjy et al. [713] and Monlouis-Bonnaire [635].

Upscaling of the moisture capacity to the level of concrete was treated by Xi [880, 881], who developed a thermodynamically based model for evaluation of the effective moisture capacity of a composite material and applied it to concrete, considered as a two-phase composite consisting of cement paste as the matrix and aggregates as inclusions.

I.2   Sorption Hysteresis Due to Nonuniqueness of Menisci

For pore relative humidity $h \geq 0.5$, the pores contain the vapor (of negligible mass), the liquid (or capillary) water, and the adsorbed water (free or hindered). The adsorbed water dominates for $h < 0.75$. For $h < 0.5$, the pores contain virtually no liquid water because the capillary tension would be close to the nanoscale tensile strength of water and also because the radius of the capillary menisci would not be much larger than the size of a few molecules of water; see Sect. 8.2.4.

As already mentioned in Sect. 8.2.5 and documented in Figs. 8.5a and I.6, the isotherms lie significantly lower for adsorption than for desorption. For pore humidities $h > 0.75$, at which capillary water dominates, such a hysteresis may be explained by pore geometries idealized in Figs. I.7 and I.8. For instance, the configurations A (empty) and B (filled) in Fig. I.7a can both exist at the same h. Figure I.8a, b shows the so-called *bottle neck effect*, or *ink bottle effect* [274, 320], which allows the large pore to be either full or empty at the same h.

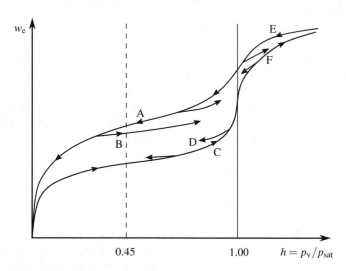

Fig. I.6 Illustrative sketch of irreversibility of desorption and adsorption isotherms, and of sorption reversals

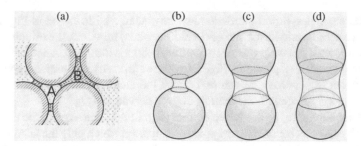

Fig. I.7 Schematic examples of different equilibrium shapes of capillary menisci between liquid water (dashed) and vapor (dotted): (a) convex menisci for $h < 1$, (b) anticlastic meniscus with $r > 0$ at $h < 1$, (c) with $r = 0$ at $h = 1$, and (d) with $r < 0$ at $h > 1$

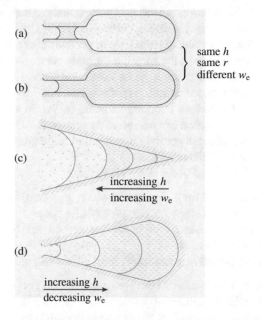

Fig. I.8 Schematic examples of (a)–(b) various degrees of pore filling (in 2D) for the same vapor pressure, (c)–(d) pore (in 2D) whose filling by liquid with increasing h increases or decreases

Figure I.7b–d shows the evolution of the anticlastic surface of a capillary meniscus between two spherical particles. For the meniscus shown in Fig. I.7b, the concave curvature $1/r_1$ is smaller in magnitude than the convex curvature $1/r_2$ (the curvature radius is considered positive, and the arc convex, if the center of curvature is in the vapor). Such shapes of cyclohexane menisci at the periphery of liquid bridges trapped between two crossed cylindrical mica surfaces were observed in the experimental study of Fisher and Israelachvili [395], who verified validity of the Kelvin equation for mean radii of curvature between 4 and 19 nm. At vanishing capillary pressure, the meniscus must have a zero mean curvature, $1/r = 0$. But this condition does not mean that the vapor–liquid interface must be planar. As illustrated in Fig. I.7c, the meniscus can be an anticlastic surface (saddle surface) for which the principal radii of curvature, r_1 and r_2, have the same magnitude but opposite signs. Another anticlastic meniscus geometry, depicted in Fig. I.7d, shows that supersaturated vapor can be present when the concave curvature $1/r_1$ of the meniscus between two spherical particles is larger in magnitude than its convex curvature $1/r_2$.

Behavior similar to the bottle neck effect (Fig. I.8a, b) can explain the nonuniqueness and hysteresis observed at low pore humidities at which only the adsorbed water layers are present. Such phenomena are included in the concept of molecular condensation mentioned in Sect. 8.2.6.3. That section also describes other phenomena that play a role in hysteresis in nanopores.

As h is decreased, the pore water system tries to accommodate so that the water loss be minimized. Thus, the desorption isotherm corresponds to the maximum possible volume of pore water for a given mean curvature $1/r$ of the meniscus (Fig. I.6). As h is increased, the pore water system tries to accommodate so as to achieve the minimum water intake, and thus, the sorption isotherm corresponds to the minimum possible volume of pore water for a given mean curvature $1/r$ of the meniscus. When desorption is reversed to resorption (path A–B in Fig. I.6), again the minimum possible water intake for a given decrease of mean curvature $1/r$ is followed, and the inverse occurs when sorption is reversed to desorption (C–D in Fig. I.6). Similar phenomena occur at variable temperature because the temperature changes also cause changes of the curvature of the meniscus.

The possibility of nonfilled pores at supersaturation, $p_v > p_{sat}$, indicates that:

1. the first sorption isotherm for p_v increasing above p_{sat} should have a much higher slope than the theoretical slope for completely filled pores;
2. the isotherms cannot be unique even at supersaturation, and hysteresis must exist as depicted at E–F in Fig. I.6;
3. the isotherm slope should be continuous through $h = 1$.

As an example, at Northwestern University in 1984 (in connection with nuclear accident research), a thick hermetically sealed steel vessel filled with concrete was placed for 1 month in an oven of $350\,°C$. But a piezoelectric gauge in a small cavity in the concrete filling never registered any pressure. Likewise, many pore pressure measurements on heated concrete showed surprisingly low pressures of only a few atmospheres, orders of magnitude less than indicated by ASTM steam table for water at constant volume. One reason probably is that the nanopores were initially empty, despite saturation of capillary space, and could accommodate water from capillary pores. Another reason probably is the self-desiccation of cement paste.

Another example (private communication by Roy W. Carlson, 1969) is a 30-year experiment at UC Berkeley in which a 2-m-long section of a three-story pipe was filled by concrete. A pressure gauge about 60 cm into the concrete could never register any pore pressure. The disputed experiments with uplift pressure by Paul Fillunger at TU Vienna [392–394] probably suffered from the same problem.

I.3 Permeability

A prediction model for the moisture permeability c_p of cement paste was proposed by Xi et al. [884]. In fact, they worked with a coefficient denoted as D_h, which is, in the present notation, equal to the ratio c_p/ρ_{cp}, where ρ_{cp} is the mass density of the paste. In terms of the moisture permeability (as defined in Sect. 8.3.2), the empirical formula developed by Xi et al. [884] can be rewritten as

$$c_p(h) = \rho_{cp}\alpha_h + \rho_{cp}\beta_h \left[1 - 2^{-10^{\gamma_h(h-1)}}\right] \tag{I.17}$$

where α_h, β_h and γ_h are parameters.

To give a specific example, the values obtained by fitting experimental results for a cement paste with $w/c = 0.5$ cured for 3 days were $\alpha_h = 0.0423$ cm^2/day, $\beta_h = 0.432$ cm^2/day, and $\gamma_h = 3.54$. The corresponding graph of function (I.17) normalized by $c_p(1)$ is shown in Fig. I.9. At low humidities, the moisture permeability is close to $c_p(0) \approx \rho_{cp}\alpha_h$, and at saturation, we have $c_p(1) = \rho_{cp}(\alpha_h + \beta_h/2)$. Xi et al. [884] even proposed empirical formulae for the estimation of parameter values from the water/cement ratio. However, those formulae were calibrated using data for pastes with w/c between 0.5 and 0.75, and extrapolation to lower w/c would be very unreliable (and might even give nonphysical results).

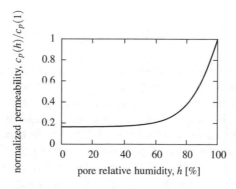

Fig. I.9 Dependence of normalized moisture permeability on pore relative humidity for concrete with $w/c = 0.5$, according to Xi et al. [884]

Studying various kinds of nanoporous gels, Scherer [757–759] proposed an interesting novel method for measuring their permeability, and also applied it to hardened Portland cement paste. He cast cylinders of 5 mm in diameter, and after 2.75 days of curing, he submerged them in water and suddenly subjected them to three-point loading in flexure. He then measured the relaxation of the applied force while holding the deflection constant. His analysis rested on the hypothesis that the relaxation was caused not only by viscoelasticity, but also, and mainly, by squeezing the pore water out of the specimens. To identify the permeability, he used optimum fitting by Biot's

[240] linear theory of consolidation of viscoelastic materials, expressed as a series of error functions.

It is doubtful, however, that this approach could be used for older cement pastes or concretes. It may have worked for a very young paste because the capillaries were still continuous, permeability high, and all the pores filled mostly by liquid water. In mature concrete, this approach cannot work, for two reasons: (1) Self-desiccation causes that the pores in concrete always contain some vapor, even if the specimen is submerged in water, and (2) specimens loaded in compression lose moisture at the same rate as the companion load-free shrinkage specimens (as shown by Hansen [451] and Maney [607]). These findings are what defeated the original consolidation theory of concrete creep mechanism.

I.4 Moisture Diffusivity

I.4.1 Dependence of Diffusivity on Humidity

The standard moisture transport model used in this book (and also adopted by the *fib* Model Code 2010) describes the dependence of moisture diffusivity C on the pore relative humidity h by the Bažant–Najjar formula (8.89). Alternative expressions proposed in the literature include the formula of Roncero [732],

$$C(h) = C_0 + (C_1 - C_0)\frac{h}{h + e^\beta(1 - h)} \tag{I.18}$$

where C_0 is the diffusivity at zero humidity, C_1 is the diffusivity at saturation, and $\beta \geq 0$ is a dimensionless shape factor. For $\beta = 0$, a linear dependence is obtained. Positive values of β give a convex curve with a steep slope near saturation; see Fig. I.10. Note that the shape of this curve is very similar to that obtained with formula (I.17) for moisture permeability; cf. Fig. I.9. In contrast to the Bažant–Najjar formula (8.89), the value of diffusivity according to (I.18) decreases with decreasing humidity already near saturation, and there is no range of high humidities in which the value would be almost constant. With a diffusivity function of this kind, one could fit very well the data of Wierig [868] for steady-state permeation in a wall, discussed in Sect. 8.4.3.3, because the optimal diffusivity function derived from the experiments and shown in Fig. 8.29b has a similar shape to those in Fig. I.10.

I.4.2 Aging of Diffusivity

Even when exposed to drying environment, the pore humidity in the cores of thick cross sections remains, for a long time, very high (albeit < 100%, due to self-desiccation). This causes that hydration continues in thick members for a long time, even for centuries [715]. Then, the effect of hydration, or aging, on diffusivity C (or

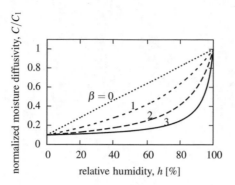

Fig. I.10 Dependence of normalized moisture diffusivity on pore relative humidity according to Roncero [732]

permeability c_p) and on inverse moisture capacity k (demonstrated by the isotherms in Fig. I.5a) needs to be taken into account.

Data on the age dependence of diffusivity were reported by Wierig [868]. They were used by Bažant and Wang [192] in spectral analysis of random environmental effects on a nuclear containment. As seen in Fig. 2 in [192], these data can be approximately fitted by replacing constant C_1 in (8.89) by the function $C_1(t) = C_{1,28}(t/t_{ref})^{-0.21}$ with $t_{ref} = 28$ days and $C_{1,28} =$ diffusivity of saturated concrete at the age of 28 days. Using this simple age dependence, Bažant and Wang [192] reduced the diffusion equation with variable diffusivity $C_1(t)$ to one with constant diffusivity $C_{1,28}$ by replacing the real age t with a new independent variable, the transformed time

$$t^* = \int_{t_0}^{t} \frac{C_1(t')}{C_{1,28}} \, dt' = \frac{t_{ref}^{0.21}}{0.79} \left(t^{0.79} - t_0^{0.79} \right) \tag{I.19}$$

The age dependence shown by Wierig's data is relatively mild (diffusivity at 1 year of age reduced to 58% of its value at 28 days). This is doubtless due to the fact that a decrease of pore humidity arrested the progress of aging in Wierig's laboratory specimens because they were too small and thus dried too fast. For real structures, which can be much larger than Wierig's specimens, the core of cross section will remain at high humidity much longer, even for decades. This will doubtless cause a much greater decrease of diffusivity with age than indicated by the foregoing equation. In any case, a more realistic time transformation should depend on pore humidity (as well as temperature).

Generally, the aging effect of hydration may be captured by replacing the actual time t (age) by the so-called equivalent hydration period t_e (also called the maturity); t_e represents a period of hydration at 25 °C of a saturated material element needed to achieve the same hydration degree as that achieved at variable humidity and temperature during the actual time period t. Functions $C(h, t)$, $k(h, t)$, and $c_p(h, t)$ are then replaced by functions $C(h, t_e)$, $k(h, t_e)$, $c_p(h, t_e)$. For the precise definition of t_e and a more detailed discussion, see Sect. 10.6.1.

Appendix J
Moisture Transport in Porous Materials

The fundamentals of moisture transport modeling with emphasis on approaches applicable to concrete are covered in Chap. 8. This appendix contains supplementary material that provides a broader context, elaborates certain details, and discusses the links among various models and theories. For comparison and clarification of conceptual differences, attention is extended to theories applicable to other porous media, such as soils or building materials, in which the diffusion of water vapor in the pore gas can play an important role. Some aspects of these theories are relevant to concrete.

The Fick law suitable for the description of such a transport mechanism is presented in Sect. J.1. The Darcy law in it original form described in Sect. 8.3.2 is limited to fluid flow in a saturated medium, and in Sect. J.2, it is extended to the Darcy–Buckingham law, which covers the partially saturated case. The Richards model discussed in Sect. J.3 combines the Darcy–Buckingham law with the moisture mass balance equation and represents the most popular transport model for soils. Section J.4 is devoted to the Coussy model as an example of a sophisticated transport model, which uses two primary unknown fields and two mass balance equations and considers multiple transport mechanisms. Potential links between these alternative models and the Bažant–Najjar model from Sect. 8.3.4.2 are briefly discussed in Sect. J.5. Section J.6 discusses the Künzel model, which was conceived as a general model for coupled heat and moisture transport in various building materials. Finally, Sect. J.7 presents the model of Beneš and Štefan [228] for coupled heat and moisture transport in concrete, with a numerically solved example showing the evolution of temperature, vapor pressure, and other relevant quantities in a heated concrete specimen.

J.1 Fick Law

According to the kinetic gas theory, diffusion in a gas mixture is driven by gradients of molar concentrations [328]. Fick's law [391] postulates proportionality between the flux of molar concentration and the negative gradient of molar concentration. For

© Springer Science+Business Media B.V. 2018
Z.P. Bažant and M. Jirásek, *Creep and Hygrothermal Effects*
in Concrete Structures, Solid Mechanics and Its Applications 225,
https://doi.org/10.1007/978-94-024-1138-6

a binary mixture of air and vapor, the *molar concentration* of vapor is given by

$$C_v = \frac{\rho_v/M_w}{\rho_v/M_w + \rho_a/M_a} \tag{J.1}$$

which is a dimensionless quantity. The ratio ρ_v/M_w corresponds to the molar density of vapor and ρ_a/M_a to the molar density of air [mol/m^3]. Fick's law is then written as

$$C_v \boldsymbol{v}_v = -D_{av}\nabla C_v \tag{J.2}$$

where $\boldsymbol{v}_v$ is the *volume flux of vapor* (volume of vapor per unit area and unit time), measured in [m^3/m^{2}s $=$ m/s], and D_{av} is the *free air-vapor diffusion coefficient* [m^2/s]. In a more explicit notation, Eq. (J.2) could be written as

$$C_v v_{vi} = -D_{av}\frac{\partial C_v}{x_i}, \qquad i = 1, 2, 3 \tag{J.3}$$

where v_{vi} is the ith component of vector $\boldsymbol{v}_v$ and x_i is the ith spatial coordinate.

For a mixture of ideal gases, the molar density of each component is proportional to its partial pressure. This follows from the state equation of ideal gas, according to which $\rho_j/M_j = p_j/RT$, with subscript $j = v, a$ referring to individual components of the mixture (vapor, air). Consequently, Eq. (J.1) can be rewritten as

$$C_v = \frac{p_v}{p_v + p_a} = \frac{p_v}{p_g} \tag{J.4}$$

which means that the molar concentration can be expressed in terms of pressures. Substituting (J.4) into (J.2), we obtain

$$p_v \boldsymbol{v}_v = -D_{av} p_g \nabla\left(\frac{p_v}{p_g}\right) \tag{J.5}$$

To get a law governing the *mass flux of vapor*, defined as $\boldsymbol{j}_v = \rho_v \boldsymbol{v}_v$ [kg/m^{2}s], it is sufficient to multiply both sides of (J.5) by M_w/RT and exploit the state equation of vapor (8.7):

$$\boldsymbol{j}_v = \rho_v \boldsymbol{v}_v = \frac{M_w p_v}{RT}\boldsymbol{v}_v = -\frac{M_w}{RT}D_{av}p_g\nabla\left(\frac{p_v}{p_g}\right) \tag{J.6}$$

This equation describes free vapor diffusion in air. The diffusion coefficient D_{av} depends on temperature T and gas pressure p_g and, according to de Vries and Kruger [344], can be approximated as

$$D_{av} = \frac{p_{atm}}{p_g}\left(\frac{T}{T_0}\right)^{1.88} D_0 \tag{J.7}$$

where $D_0 = 2.48 \times 10^{-5}$ m^2/s is the value of the diffusion coefficient at reference temperature $T_0 = 293$ K and reference gas pressure $p_g = p_{atm}$.

For diffusion in a constrained pore space, the diffusion coefficient has to be reduced by the product of two factors—porosity n_p and tortuosity τ. Multiplication by porosity reflects the reduction of the area across which the diffusing molecules can travel (note that the flux is always taken per unit total cross-sectional area of the porous medium, not per unit area of the pores). Multiplication by tortuosity reflects the constraining effect of the pore walls. In the special case of uniform gas pressure p_g equal to the atmospheric pressure p_{atm} and uniform temperature T equal to T_0, the resulting equation for vapor diffusion mass flux reads

$$j_v = -n_p \tau \frac{M_w D_0}{R T_0} \nabla p_v \tag{J.8}$$

The coefficient $n_p \tau M_w D_0 / R T_0$ is called the *permeability coefficient*[16] [s]. For diffusion in free space (i.e., for porosity $n_p = 1$ and tortuosity $\tau = 1$), its value would be

$$\frac{M_w D_0}{R T_0} = \frac{18.02 \times 10^{-3} \times 2.17 \times 10^{-5}}{8.31446 \times 273} \text{s} = 0.17227 \times 10^{-9} \text{ s} \tag{J.9}$$

Porosity n_p has a clear meaning, but the tortuosity τ can be deduced only indirectly, by measuring the actual permeability coefficient and dividing it by $n_p M_w D_0 / R T_0$. In the presence of liquid pore water, the diffusion of vapor is further constrained. The volume fraction of the pore space occupied by gas is reduced to $(1 - S_l) n_p$, where S_l is the liquid saturation degree. Also, the tortuosity factor τ becomes a function of the saturation degree, which is usually nonlinear. Both effects lead to a reduction of the permeability coefficient. The tortuosity factor at partial saturation is sometimes written as the product of the (constant) tortuosity factor at full saturation and a saturation-dependent reduction factor called the partial tortuosity, which is equal to 1 at full saturation. For simplicity, we deal directly with the resulting saturation-dependent tortuosity $\tau(S_l)$.

J.2   Darcy–Buckingham Law

The original Darcy law presented in Sect. 8.3.2 applies to the flow of a fluid that completely fills the pore space. Under partially saturated conditions, the flow can still be considered as driven by the gradient of the total head, but the hydraulic permeability needs to be reduced. This was taken into account by Buckingham [279], who extended Darcy's approach to the partially saturated case, with hydraulic permeability considered as a function of the water content. It is convenient to describe

---

[16]The notion of permeability is primarily linked to the Darcy law, see Sect. 8.3.2, but it can be used in connection with the Fick law, too.

the reduction of permeability by a dimensionless factor k_r, called the *relative permeability*, which depends on the water content and varies from 1 at full saturation to 0 at a certain minimum degree of saturation, below which the flow stops. The *Darcy–Buckingham law* generalizing Eq. (8.66) is then written as

$$v_1 = -k_r K_h \nabla H_t \qquad (J.10)$$

It can also be rewritten in an equivalent form, which generalizes Eq. (8.69) and is formulated in terms of the mass flux and pressure gradient instead of the volume flux and total head gradient:

$$\boldsymbol{j}_l = \rho_l \boldsymbol{v}_1 = -\rho_l k_r K_h \nabla \left(\frac{p_l - p_{\text{atm}}}{\rho_l g} + z \right) = -\frac{k_r K_h}{g} \nabla p_l - \rho_l k_r K_h \boldsymbol{e}_z \qquad (J.11)$$

If the gravity effect is neglected and the hydraulic permeability is expressed in terms of the intrinsic permeability and fluid properties using formula (8.70), Eq. (J.11) can be written as

$$\boldsymbol{j}_l = -\frac{k_{r,l} K_0 \rho_l}{\eta_l} \nabla p_l \qquad (J.12)$$

for the flow of a liquid (which is indicated by subscripts l). Recall that η_l is the dynamic shear viscosity of the liquid. An analogous equation

$$\boldsymbol{j}_g = -\frac{k_{r,g} K_0 \rho_g}{\eta_g} \nabla p_g \qquad (J.13)$$

can be set up for the flow of pore gas (subscript g). The driving force is now the gradient of the gas pressure.

The relative permeability to liquid water (or, in general, to the wetting fluid) depends on the saturation degree, S_l. According to Luckner et al. [583], this dependence can be approximated by a function of the form

$$k_{r,l} = \sqrt{S_l} \left[1 - \left(1 - S_l^{1/m} \right)^m \right]^2 \qquad (J.14)$$

where m is a dimensionless parameter (the same as parameter m used by the van Genuchten retention curve (8.16)). For the flow of gas (or, in general, of the nonwetting fluid), the relative permeability could be approximated by

$$k_{r,g} = \sqrt{1 - S_l} \left(1 - S_l^{1/m} \right)^{2m} \qquad (J.15)$$

Note that as the liquid saturation S_l increases from 0 to 1, $k_{r,l}$ increases from 0 to 1, while $k_{r,g}$ decreases from 1 to 0. In fact, more refined models consider that the flow of liquid water stops at some minimum saturation degree, larger than zero. For instance, formulae

$$k_{r,l} = \left(\frac{S_l - S_{ir}}{1 - S_{ir}} \right)^{A_w} \tag{J.16}$$

$$k_{r,g} = 1 - \left(\frac{S_l}{S_{cr}} \right)^{A_g} \tag{J.17}$$

have been recommended in the literature [221, 396]. Here, S_{ir} is the irreducible saturation value at which the liquid flow stops, S_{cr} is the critical saturation value above which there is no gas flow, and exponents A_w and A_g are usually in the range from 1 to 3.

It must be pointed, however, that Buckingham's extension of Darcy law to a separate flow of gas is physically justified only if the pore space occupied by gas percolates through pores of width > 200 nm (the reason is that this is sufficiently wider than the mean free path of water molecules in vapor, which is 80 nm at 25 °C). In good-quality concrete at temperatures below 100°C, a percolating pore space of pores wider than about 200 nm does not exist; in more detail see the remark at the end of Sect. J.7.

J.3   Richards Model

The classical model for liquid water transport in partially saturated soils, proposed by Richards [724], is based on the water mass balance equation combined with the Darcy–Buckingham equation. No chemical reactions are considered, and so there is no need to distinguish between the total and evaporable water contents. On the left-hand side of the water mass balance equation (8.77), $\dot{w}_n$ is set to zero and the contribution of vapor to the water mass (i.e., the term $\rho_v n_p (1 - S_l)$) is neglected. Traditionally, the Richards equation has been written in terms of the volumetric water content, $\widetilde{w}$, defined as the volume of water per unit volume of the medium [$m^3/m^3 = 1$], which can be interpreted as the relative percentage of volume occupied by water and varies between 0 and n_p (porosity). In fact, $\widetilde{w} = n_p S_l$, where S_l is the saturation degree. As usual, the liquid water density ρ_l is considered as constant, and so the left-hand side of (8.77) can be rewritten as $\rho_l \dot{\widetilde{w}}$. Substituting $j_w = j_l = \rho_l v_l$ into the right-hand side and dividing both sides by ρ_l, we obtain

$$\dot{\widetilde{w}} = -\nabla \cdot v_l \tag{J.18}$$

This is an alternative form of the water balance equation, valid under the present assumptions and written in terms of the volumetric water content $\widetilde{w}$ and volumetric water flux v_l.

The *Richards equation*

$$\dot{\widetilde{w}} = \nabla \cdot (k_r K_h \nabla H_t) \tag{J.19}$$

follows from (J.18) combined with the Darcy–Buckingham equation (J.10). In this form, it is a second-order partial differential equation with two unknown fields, $\widetilde{w}$ and H_t (total head). However, these fields are not independent and one of them can be expressed in terms of the other, using the moisture retention relation. One also needs to take into account the relative permeability k_r as a function of the saturation degree S_l, which can be linked to the pressure head H_p (rather than to the total head H_t). Recall that, according to (8.68), the total head is the sum of the pressure head and the vertical coordinate z. Consequently, Eq. (J.19) can be converted into an equation with a single unknown field, either the pressure head, or the volumetric water content.

In its primary form, the moisture retention relation links the capillary pressure p_c (i.e., suction) to the liquid saturation degree S_l; see Sect. 8.2.3. As already mentioned, the saturation degree multiplied by porosity gives the volumetric water content. The capillary pressure is the difference between the gas pressure and the liquid pressure, with the gas pressure usually close to the atmospheric pressure. This means that, according to (8.67), the capillary pressure p_c is proportional to the pressure head H_p. Therefore, the moisture retention relation can be transformed by simple scaling into a relation between the volumetric water content and the pressure head, described by a certain function $\widetilde{w}(H_p)$. Differentiating this relation, we can rewrite the left-hand side of (J.19) in terms of the time derivative of the pressure head. The right-hand side of (J.19) is also easily expressed in terms of H_p, since $H_t = H_p + z$ and $\nabla H_t = \nabla H_p + e_z$. In this way, the Richards equation is converted into a partial differential equation with the pressure head as the only unknown field:

$$\frac{\mathrm{d}\widetilde{w}(H_p)}{\mathrm{d}H_p}\dot{H}_p = \nabla \cdot \left[k_r(H_p)K_h\nabla H_p\right] + \frac{\mathrm{d}k_r(H_p)}{\mathrm{d}H_p}K_h\frac{\partial H_p}{\partial z} \qquad (\text{J.20})$$

Of course, instead of the pressure head, one could use the liquid water pressure or the capillary pressure as the primary unknown.

Alternatively, the Richards equation can be rewritten in terms of the volumetric water content. The moisture retention relation is invertible, and so one can consider H_p as a function of $\widetilde{w}$ and write $H_t(\widetilde{w}) = H_p(\widetilde{w}) + z$. Elimination of H_t from (J.19) leads to the equation first derived by Klute [550]:

$$\dot{\widetilde{w}} = \nabla \cdot \left[k_r(\widetilde{w})K_h\frac{\mathrm{d}H_p(\widetilde{w})}{\mathrm{d}\widetilde{w}}\nabla\widetilde{w}\right] + \frac{\mathrm{d}k_r(\widetilde{w})}{\mathrm{d}\widetilde{w}}K_h\frac{\partial\widetilde{w}}{\partial z} \qquad (\text{J.21})$$

Note that the relative permeability k_r has originally been introduced as a function of the saturation degree S_l. Using the links between the saturation degree and the pressure head or the volumetric moisture content, the dependence of k_r on S_l can be transformed into a dependence on H_p or $\widetilde{w}$, which is then conveniently used in (J.20) or (J.21).

J.4 Coussy Model

As the first example of a multifield model, we present here the approach summarized by Coussy [328], based on the previous work of Coussy and coworkers. For simplicity, we restrict attention to the mass transport at constant temperature. The fundamental equations are the moisture mass balance equation (8.76) and the dry air mass balance equation, which has a similar format to (8.76), with the following modifications: The moisture flux j_w must be replaced by the air flux j_a, and the water content w_t must be replaced by the mass of dry air per unit volume of the porous medium, m_a, which is equal to the product $(1 - S_l)n_p\rho_a$, where S_l is the saturation degree, n_p is the porosity, and ρ_a is the air density (per unit volume occupied by the pore gas). The resulting equation reads

$$\frac{d}{dt}\left[\rho_a n_p (1 - S_l)\right] = -\nabla \cdot j_a \tag{J.22}$$

The mass balance equations have to be rewritten in terms of two primary unknowns, e.g., of the vapor pressure p_v and the gas pressure p_g. The transport mechanisms include the advective flow of liquid water, advective flow of pore gas, and diffusion of water vapor in the pore gas.

The mass flux of liquid water, j_l, is linked to the gradient of liquid water pressure by the Darcy–Buckingham law (J.12). For convenience, we rewrite this relation as

$$j_l = -a_l^* \rho_l \nabla p_l \tag{J.23}$$

where

$$a_l^* = \frac{k_{r,l} K_0}{\eta_l} \tag{J.24}$$

is an auxiliary coefficient, dependent on the saturation degree (it is in fact the permeability coefficient for liquid water flow divided by the liquid water density). Note, however, that application of the Darcy–Buckingham law to the fluxes of various phases of the same substance (water) has some physically questionable aspects (see the remark at the end of Sect. J.7).

The fluxes of water vapor, j_v, and of dry air, j_a, are obtained by summing the contributions of two simultaneous transport mechanisms—advective flow of the gas mixture and diffusion of individual gases in the mixture. The *advective mass flux of pore gas*, j_g^A, is linked to the gradient of gas pressure by the Darcy–Buckingam law for gas, see Eq. (J.13). In analogy to (J.23) and (J.24), we rewrite this relation as

$$j_g^A = -a_g^* \rho_g \nabla p_g \tag{J.25}$$

where

$$a_g^* = \frac{k_{r,g} K_0}{\eta_g} \tag{J.26}$$

During advective flow, all the molecules move at the same velocity, and so the volumetric fluxes of water vapor and dry air are the same, equal to $v_g = j_g^A/\rho_g$. The advective mass fluxes of vapor and air are proportional to the respective densities:

$$j_v^A = \rho_v v_g = \frac{\rho_v}{\rho_g} j_g^A = -a_g^* \rho_v \nabla p_g \tag{J.27}$$

$$j_a^A = \rho_a v_g = \frac{\rho_a}{\rho_g} j_g^A = -a_g^* \rho_a \nabla p_g \tag{J.28}$$

The diffusion is governed by Fick's law (J.6), with the diffusion coefficient (J.7) further multiplied by $(1 - S_l) n_p \tau$, to take into account the fact that the diffusion takes place in a constrained pore space under partially saturated conditions. To make the resulting relations easier to handle, let us introduce an auxiliary variable

$$B = (1 - S_l) n_p \tau \frac{D_0 p_{atm}}{R T_0} \tag{J.29}$$

which depends on the saturation degree. Fick's law for the *diffusive vapor mass flux* at constant temperature $T = T_0$ can then be presented as

$$j_v^D = -B M_w \nabla \left(\frac{p_v}{p_g} \right) \tag{J.30}$$

At the same time, the dry air diffuses in gas in the direction opposite to the vapor diffusion, and its diffusive flux j_a^D is given by a formula similar to (J.30), but with p_v replaced by p_a (partial pressure of dry air) and M_w replaced by M_a (molar mass of air). Since $\nabla(p_a/p_g) = \nabla((p_g - p_v)/p_g) = -\nabla(p_v/p_g)$, the resulting *diffusive mass flux of air* is

$$j_a^D = -B M_a \nabla \left(\frac{p_a}{p_g} \right) = B M_a \nabla \left(\frac{p_v}{p_g} \right) \tag{J.31}$$

Note that the net diffusive mass flux of gas, $j_v^D + j_a^D$, is not zero. Pure molecular diffusion takes place at uniform gas pressure, and thus at uniform molar density, but not at uniform mass density. Therefore, what vanishes is not the mass flux but the molar flux (number of moles per unit area and unit time), given by $j_v^D/M_w + j_a^D/M_a$.

Summing the contributions of advective and diffusive transport, we obtain the resulting mass fluxes of vapor and air:

$$j_v = j_v^A + j_v^D = -a_g^* \rho_v \nabla p_g - B M_w \nabla \left(\frac{p_v}{p_g} \right) \tag{J.32}$$

$$j_a = j_a^A + j_a^D = -a_g^* \rho_a \nabla p_g + B M_a \nabla \left(\frac{p_v}{p_g} \right) \tag{J.33}$$

The total moisture flux $j_w = j_l + j_v$, with the *liquid water flux* j_l given by (J.12) and the vapor flux j_v given by (J.32), is substituted into the moisture balance equation (8.77). The air flux j_a given by (J.33) is substituted into the air balance equation (J.22). The resulting set of two nonlinear partial differential equations is

$$\frac{d}{dt}\left[\rho_l n_p S_l + \rho_v n_p(1 - S_l)\right] + \dot{w}_n = \nabla \cdot \left[a_l^* \rho_l \nabla p_l + a_g^* \rho_v \nabla p_g + B M_w \nabla \left(\frac{p_v}{p_g}\right)\right] \quad \text{(J.34)}$$

$$\frac{d}{dt}\left[\rho_a n_p(1 - S_l)\right] = \nabla \cdot \left[a_g^* \rho_a \nabla p_g - B M_a \nabla \left(\frac{p_v}{p_g}\right)\right] \quad \text{(J.35)}$$

It contains six unknown fields (S_l, ρ_a, ρ_v, p_v, p_l, p_g), which are linked by four equations: the state laws (8.6) and (8.7), the Kelvin equation (8.24), and the moisture retention relation. Therefore, four variables can be eliminated, and the remaining two (e.g., the vapor pressure p_v and the gas pressure p_g) then play the role of primary unknowns.

In the absence of chemical reactions, the rate of the nonevaporable water content w_n vanishes. Otherwise, it would have to be evaluated from an appropriate hydration model (which is affected by the pore relative humidity, causing the problem to become coupled). The liquid pressure p_l is related to the vapor pressure p_v by the Kelvin equation (8.24). The air density ρ_a is linked to the air pressure $p_a = p_g - p_v$ by the state equation (8.6). Combining the Kelvin equation and the moisture retention relation, we can express the saturation degree as a function of the vapor pressure and the gas pressure, denoted as $S_l(p_v, p_g)$. The liquid density ρ_l is constant, and the vapor density ρ_v is related to p_v by the state equation (8.7).

For simplicity, we will now disregard the hydration reaction and the corresponding changes of pore structure, which means that $\dot{w}_n$ is set to zero and the porosity n_p and tortuosity τ are treated as constants. Substituting the aforementioned relations into (J.34) multiplied by RT_0/M_w and into (J.35) multiplied by RT_0/M_a, we obtain the governing differential equations written explicitly in terms of the primary unknowns, p_v and p_g:

$$b_1\dot{p}_v + b_2\dot{p}_g = \nabla \cdot \left[a_l^* p_0^2 \frac{\nabla p_v}{p_v} + a_g^* p_v \nabla p_g + RT_0 B \nabla \left(\frac{p_v}{p_g}\right)\right] \quad \text{(J.36)}$$

$$b_3\dot{p}_v + b_4\dot{p}_g = \nabla \cdot \left[a_g^*(p_g - p_v)\nabla p_g - RT_0 B \nabla \left(\frac{p_v}{p_g}\right)\right] \quad \text{(J.37)}$$

where

$$b_1(p_v, p_g) = n_p \left[1 - S_l(p_v, p_g) + (p_0 - p_v)\frac{\partial S_l(p_v, p_g)}{\partial p_v}\right] \quad \text{(J.38)}$$

$$b_2(p_v, p_g) = n_p (p_0 - p_v)\frac{\partial S_l(p_v, p_g)}{\partial p_g} \quad \text{(J.39)}$$

$$b_3(p_v, p_g) = n_p \left[-1 + S_l(p_v, p_g) + (p_v - p_g)\frac{\partial S_l(p_v, p_g)}{\partial p_v}\right] \quad \text{(J.40)}$$

$$b_4(p_v, p_g) = n_p \left[1 - S_l(p_v, p_g) + (p_v - p_g)\frac{\partial S_l(p_v, p_g)}{\partial p_g}\right] \quad \text{(J.41)}$$

$$RT_0 B(p_v, p_g) = n_p D_0 p_{atm} \left[1 - S_l(p_v, p_g)\right] \tau \left(S_l(p_v, p_g)\right) \quad \text{(J.42)}$$

The set of two second-order differential equations (J.36) and (J.37) must be supplemented by two boundary conditions written in terms of the primary unknowns, p_v and p_g. On the impervious part of the boundary, zero normal derivatives of the vapor pressure and gas pressure are enforced. Note that, in this last case, conditions $\partial p_v / \partial n = 0$ and $\partial p_g / \partial n = 0$ imply a vanishing normal derivative of p_v / p_g because

$$\frac{\partial}{\partial n}\left(\frac{p_v}{p_g}\right) = \frac{1}{p_g}\frac{\partial p_v}{\partial n} - \frac{p_v}{p_g^2}\frac{\partial p_g}{\partial n} \tag{J.43}$$

Consequently, the normal components of fluxes j_l, j_v and j_a vanish on the impervious boundary. At the permeable part of the boundary in contact with the surrounding atmosphere, the simplest assumption is that the pore gas pressure is equal to the atmospheric pressure and the vapor pressure in the pores is equal to the ambient vapor pressure, $p_{v,env}$.

Let us return to Eqs. (J.34) and (J.35) and discuss their structure and the physical meaning of individual terms. Equation (J.34) represents the mass balance of moisture, and the three terms in brackets on the right-hand side correspond to the contributions of liquid flow, advective gas flow, and vapor diffusion. Equation (J.35) represents the mass balance of dry air, and the two terms in brackets on the right-hand side correspond to the contributions of the advective gas flow and air diffusion. It is possible to eliminate the diffusive terms by creating a weighted sum of these two balance equations, with weights $1/M_w$ and $1/M_a$. Physically, the resulting equation represents the molar balance law for all the fluids combined. If both sides are multiplied by the constant RT_0 and the state laws of vapor and air are taken into account, the equation can be presented in the form

$$\frac{d}{dt}\left[n_p p_0 S_l + n_p p_g (1 - S_l)\right] + \frac{p_0}{\rho_l}\dot{w}_n = \nabla \cdot \left[a_l^* p_0 \nabla p_l + a_g^* p_g \nabla p_g\right] \tag{J.44}$$

where

$$p_0 = \frac{\rho_l R T_0}{M_w} = \frac{998 \text{ kg/m}^3 \times 8.314 \text{ J/(K} \cdot \text{mol)} \times 293 \text{ K}}{18.02 \times 10^{-3} \text{ kg/mol}} = 135 \text{ MPa} \approx 1330 \, p_{atm} \tag{J.45}$$

is a constant that corresponds to the (fictitious) pressure that would be generated in vapor compressed to the density of liquid water (if it were still governed by the state law of vapor). Note that p_0 is three orders of magnitude larger than the atmospheric pressure.

Coussy [328] converted Eqs. (J.34) and (J.44) to a dimensionless form and identified three characteristic times corresponding to individual transport mechanisms. He showed that if the intrinsic permeability K_0 is sufficiently high compared to $n_p \tau \eta_g D_0 / p_{atm}$, the characteristic time of advective gas flow is much shorter than the characteristic times of molecular diffusion and of liquid water flow.

For this class of materials, called by Coussy [328] "quite permeable," the advective gas flow quickly reduces any excess gas pressure, and the gas pressure can be considered as constant and equal to the atmospheric pressure (as dictated by the boundary condition on the permeable part of the boundary). Substituting $p_g = p_{atm}$ into the moisture mass balance equation (J.34), we obtain its simplified form

$$\frac{d}{dt}\left[\rho_l n_p S_l + \rho_v n_p (1 - S_l)\right] + \dot{w}_n = \nabla \cdot \left[a_l^* \rho_l \nabla p_l + \frac{BM_w}{p_{atm}}\nabla p_v\right] \qquad (J.46)$$

The expression in the brackets on the left-hand side is, of course, the evaporable water content, w_e. The expression in the brackets on the right-hand side is the mass flux of moisture, expressed as a sum of the contributions of liquid water flow and vapor diffusion.

Equation (J.46) contains four unknown fields: the liquid pressure p_l, the vapor pressure p_v, the vapor density ρ_v, and the saturation degree S_l. These fields are linked by three relations: the Kelvin equation, the state equation of vapor, and the moisture retention relation (assuming that $p_g = p_{atm}$ and thus $p_c = p_{atm} - p_l$). Consequently, only one of the fields is an independent primary unknown, and the others can be eliminated. Which field is selected as the primary unknown is a matter of convenience and taste (as well as considerations related to the boundary conditions, numerical discretization, etc.). From the mathematical point of view, the resulting formulations are fully equivalent.

J.5 Relation to the Bažant–Najjar Model

The moisture transport model proposed by Bažant and Najjar [165] and described in Sect. 8.3.4.2 directly postulates that, under constant temperature, the total moisture flux is driven by the gradient of pore relative humidity:

$$j_w = -c_p \nabla h \qquad (J.47)$$

Under the assumptions that the pore gas pressure remains equal to the atmospheric pressure, $p_g = p_{atm}$ and that the temperature remains constant, $T = T_0$, equation (J.47) can be derived by summing the liquid flux and the vapor flux, respectively given by Eqs. (J.12) and (J.8) adjusted to the partially saturated case:

$$j_w = j_l + j_v = -\frac{k_{r,l} K_0 \rho_l}{\eta_l}\nabla p_l - n_p \tau (1 - S_l)\frac{M_w D_0}{RT_0}\nabla p_v \qquad (J.48)$$

Note that the contribution of advective gas flow to the vapor transport is not considered, as a consequence of the assumption of constant gas pressure. The vapor pressure p_v can be expressed in terms of the pore relative humidity based on definition (8.1). The liquid pressure p_l is related to the pore relative humidity by the Kelvin equation (8.25). Taking all this into account, (J.48) can be rewritten as

$$j_w = -\frac{k_{r,l}K_0\rho_l}{\eta_l}\frac{\rho_l RT_0}{M_w}\frac{\nabla h}{h} - n_p\tau(1-S_l)\frac{M_w D_0}{RT_0}p_{\text{sat}}\nabla h \qquad (J.49)$$

This corresponds to (J.47), with the moisture permeability given by

$$c_p(h) = \frac{k_{r,l}(h)K_0\rho_l^2 RT_0}{\eta_l M_w h} + \frac{n_p\tau(h)[1-S_l(h)]M_w D_0 p_{\text{sat}}}{RT_0} \qquad (J.50)$$

To emphasize that the moisture permeability is variable, we have marked it explicitly as a function of the pore relative humidity, h. Recall that the tortuosity, τ, and relative permeability, k_r, depend on the saturation degree, S_l. For a given structure of the pore space, the saturation degree can be expressed in terms of the pore relative humidity (and temperature); see (8.44). Therefore, τ and k_r can be considered as functions of the pore relative humidity, h, which plays here the role of the primary unknown field.

Formula (J.50) reflects the complex dependence of permeability on various constants and variables. It consists of two additive terms, respectively, related to vapor diffusion and liquid water flow. In theory, one could perform separate measurements of vapor and liquid water transport, determine the dependence of tortuosity and relative permeability on the saturation degree, and then evaluate the moisture permeability from (J.50). In practice, it is better to determine the moisture permeability directly, by measurements of the total moisture transport. In that case, formula (J.50) is in fact not needed. Nevertheless, separate evaluation of vapor and liquid transport could be useful for more general models that take into account variable temperature.

J.6 Künzel Model

Another model closely related to the Bažant–Najjar model was proposed by Künzel [556], who simulated coupled heat and moisture transport in building materials. At constant temperature, Künzel's model is based on the water mass balance equation combined with the following assumptions regarding moisture transport mechanisms: The moisture flux is expressed as the sum of the vapor flux, driven by the gradient of vapor pressure, and the liquid flux, driven by the gradient of capillary pressure. This can be described by equations

$$j_w = j_v + j_l \qquad (J.51)$$
$$j_v = -\delta_p \nabla p_v \qquad (J.52)$$
$$j_l = K_1 \nabla p_c \qquad (J.53)$$

In Künzel's work, K_1 was called the *permeability coefficient* (which is in accordance with the terminology introduced in Sect. 8.3.2; see Table 8.1), and δ_p was called the *water vapor permeability*, both expressed in [kg/m·s·Pa = s].

Künzel [556] started from the assumption that the gradient of capillary pressure is the fundamental "force" driving the liquid flux, but then he expressed p_c in terms of temperature T and relative humidity h using the approximate Kelvin equation (8.34) and rewrote (J.53) as

$$j_l = K_1 \nabla \left(-\frac{\rho_l RT}{M_w} \ln h \right) = -K_1 \frac{\rho_l R}{M_w} \ln h \, \nabla T - K_1 \frac{\rho_l RT}{M_w} \frac{\nabla h}{h} \tag{J.54}$$

Subsequently, he simplified the model by neglecting the term proportional to the gradient of temperature and presented the resulting equation in the form

$$j_l = -D_h \nabla h \tag{J.55}$$

where

$$D_h = K_1 \frac{\rho_l RT}{M_w h} \tag{J.56}$$

Künzel [556] called D_h the *liquid conduction coefficient* [kg/m·s] but, according to the terminology introduced in the discussion related to Table 8.1, it would be called the *moisture permeability*. More precisely, it is just the part of moisture permeability related to the flux of liquid water, because the total moisture flux given by (J.51) contains an additional term j_v attributed to the flux of vapor. At uniform temperature (vanishing temperature gradient), the corresponding transport law (J.52) can also be rewritten in terms of the humidity gradient (recall that the vapor pressure p_v is equal to the product $h p_{sat}$, where p_{sat} is the temperature-dependent saturation pressure):

$$j_v = -\delta_p \nabla p_v = -\delta_p \nabla (p_{sat} h) = -\delta_p p_{sat} \nabla h \tag{J.57}$$

Substituting (J.55) and (J.57) into (J.51), we obtain

$$j_w = -(\delta_p p_{sat} + D_h) \nabla h \tag{J.58}$$

which exactly corresponds to the Bažant–Najjar law (8.84) with moisture permeability given by

$$c_p = \delta_p p_{sat} + D_h \tag{J.59}$$

It should be emphasized again that a full equivalence is valid only at constant and uniform temperature. In such a case, the Künzel model can be seen as one specific member of the wide class of models considered by Bažant and Najjar. In fact, the actual model used in Chap. 8 in Examples 8.6–8.9 and referred to as the Bažant–Najjar model is another specific member of that class, because it is based on Eq. (8.87) and deals with the moisture diffusivity described by formula (8.89). It is therefore applicable under the assumption that the desorption isotherm is linear (i.e., that the moisture capacity is constant). In contrast to that, the more general Eq. (8.86) deals with the moisture permeability and inverse moisture capacity separately and thus allows for a nonlinear isotherm. It is, however, not very clear how the dependence of moisture permeability on humidity should look if the isotherm is indeed nonlinear. It may thus be useful if the Künzel model can provide some guidelines regarding the potential dependence of parameters D_h and δ_p on relative humidity or another related variable, such that formula (J.59) can be used to determine the moisture permeability.

The water vapor permeability δ_p is, according to Künzel [556], given by the fraction

$$\delta_p = \frac{\delta}{\mu} \tag{J.60}$$

in which δ is the *water vapor diffusion coefficient*, approximately expressed as

$$\delta = 2 \times 10^{-7} \frac{T^{0.81}}{p_{\text{atm}}} \tag{J.61}$$

and μ is a dimensionless parameter μ called the *water vapor diffusion resistance factor*. In formula (J.61), the absolute temperature T should be substituted in Kelvin and the atmospheric pressure p_{atm} in Pa; the resulting coefficient δ is then in seconds. For the standard atmospheric pressure and the room temperature $T_0 = 293$ K, we get $\delta = 1.97 \times 10^{-10}$ s.

Künzel's formula (J.61) may be linked to Fick's law presented in Sect. J.1 by considering that $\delta_p = \delta/\mu$ should correspond to the coefficient of proportionality between the mass vapor flux and the gradient of vapor pressure implied by (J.6)–(J.8), which means that

$$\delta = \frac{M_w D_{av}}{RT} = \frac{M_w}{RT} \frac{p_{\text{atm}}}{p_g} \left(\frac{T}{T_0}\right)^{1.88} 2.48 \times 10^{-5} \text{m}^2/\text{s} =$$

$$= \frac{M_w}{RT_0} \frac{p_{\text{atm}}}{p_g} \left(\frac{T}{T_0}\right)^{0.88} 2.48 \times 10^{-5} \text{m}^2/\text{s} = \frac{p_{\text{atm}}}{p_g} \left(\frac{T}{T_0}\right)^{0.88} 1.83 \times 10^{-10} \text{s} \quad (J.62)$$

while μ should correspond to $1/(n_p\tau)$. This formula gives, at standard conditions, $\delta = 1.83 \times 10^{-10}$ s. The fact that this differs from the aforementioned value of $\delta = 1.97 \times 10^{-10}$ s is probably caused by one-digit truncation of the factor in Künzel's formula (J.61). Note that the exponent of 0.88 is also slightly different from 0.81 used in (J.61).

Parameter μ is the ratio between the diffusion coefficients of water vapor in air and in the given porous material. Its typical values for concrete are between 210 and 260, and the corresponding water vapor diffusion coefficient δ_p is thus in the range from 7.5 to 9.4×10^{-13} s. Künzel [556] admitted that factor μ may depend on the water content, especially near saturation, but due to the lack of reliable data, he recommended to ignore this dependence. The saturation pressure p_{sat} can be evaluated from the temperature using, e.g., the Antoine equation (8.19), even though Künzel [556] recommended a relation having a similar form but different parameters:

$$p_{\text{sat}}(T) = 611 \text{ Pa } \exp\left(\frac{17.08(T - 273.15 \text{ K})}{T + 234.18 - 273.15 \text{ K}}\right) \tag{J.63}$$

While the first term on the right-hand side of (J.59), $\delta_p p_{sat}$, is approximately treated as constant, the second term, D_h, may vary by several orders of magnitude as a function of the relative humidity. Künzel [556] linked the liquid conduction coefficient D_h to the *capillary transport coefficient* D_w [m^2/s]. In the present context, the formula may be written as

$$D_h(h) = \frac{D_w(\phi(h))}{k(h)} \tag{J.64}$$

where ϕ is the function describing the isotherm and k is its inverse slope (reciprocal moisture capacity). Evaluation of D_w may depend on the specific transport process, but the general recommendation of Künzel [556] is to describe it by an exponential function of the evaporable water content w_e, e.g., in the form

$$D_w(w_e) = 3.8 \left(\frac{A}{w_f}\right)^2 1000^{w_e/w_f - 1} \tag{J.65}$$

where w_f is the evaporable water content at free saturation. In the case of sorption, parameter A is called the *water absorption coefficient*, for concrete typically in the range from 0.1 to 1 m^{-2}day$^{-0.5}$.

Künzel's model was developed for general moisture and heat transport in building materials and in its direct application to drying of concrete might be questionable. However, it is interesting to observe the general structure of the resulting equation. Substituting (J.58) into the water mass balance equation (8.76) and using relation (8.52) with the effects of self-desiccation neglected and relation (J.64), one gets

$$\dot{h} = k(h)\nabla \cdot \left[\delta_p p_{sat} + \frac{D_w(\phi(h))}{k(h)}\right]\nabla h \tag{J.66}$$

For simulations in which the isotherm is considered as nonlinear (typically with a steep slope near saturation), it is of advantage to define moisture permeability (i.e., the term in brackets) in this form, with a monotonically increasing function D_w.

J.7 Heat and Moisture Transport—Model of Beneš and Štefan

As an example of a simplified multifield model for coupled transport of heat and moisture in concrete at elevated temperatures, the approach used by Beneš and Štefan [228] will be briefly described, and the results of a simulation of a heated specimen will be presented. Beneš and Štefan [228] started from a rather general framework but then examined the relative importance of various terms (aiming at applications to concrete subjected to fire) and simplified the description. In particular, they neglected the contribution of diffusion to the mass flux of water vapor and the effects of variations of dry air pressure, and they combined the diffusion of adsorbed water with liquid water flux into one single flux term. The resulting model uses the temperature, T, and pore vapor pressure, p_v, as two primary unknown fields. The governing equations are based on moisture mass balance and enthalpy balance in properly adjusted forms.

The moisture mass balance equation is obtained by combining the mass balance of liquid water (13.179) with the mass balance of vapor,

$$\frac{\partial}{\partial t}\left[n_p(1 - S_l)\rho_v\right] = -\nabla \cdot \boldsymbol{j}_v + \dot{m}_{ev} \tag{J.67}$$

which has the same form as the mass balance of pore gas (13.180), but ρ_g is replaced by ρ_v and $\boldsymbol{j}_g$ by $\boldsymbol{j}_v$. Summing (13.179) with (J.67) yields

$$\dot{w}_e = -\nabla \cdot (\boldsymbol{j}_l + \boldsymbol{j}_v) + \dot{m}_{deh} \tag{J.68}$$

where

$$w_e = n_p S_l \rho_l + n_p(1 - S_l)\rho_v \tag{J.69}$$

is the evaporable water content. Equation (J.68) can also be interpreted as a modified form of the moisture mass balance equation (13.51), with the moisture mass flux $\boldsymbol{j}_w$ split into the mass fluxes of liquid and adsorbed water, $\boldsymbol{j}_l$, and of vapor, $\boldsymbol{j}_v$, and with the rate of nonevaporable water content $\dot{w}_n$ replaced by $-\dot{m}_{deh}$.

In the heat equation (13.192), the convective term related to gas flow is limited to vapor; i.e., $C_{pg}\boldsymbol{j}_g$ is replaced by $C_{pv}\boldsymbol{j}_v$. Furthermore, the rate of evaporation $\dot{m}_{ev}$ is expressed using (J.68) and subsequently eliminated. The resulting form of the modified heat equation reads

$$\rho C_p \dot{T} + (C_{pl}\boldsymbol{j}_l + C_{pv}\boldsymbol{j}_v) \cdot \nabla T = -\nabla \cdot \boldsymbol{q} - \dot{m}_{deh}\Delta h^w_{s,1} - \left(\dot{m}_v + \nabla \cdot \boldsymbol{j}_v\right)\Delta h^w_{1,g} \tag{J.70}$$

in which

$$m_v = n_p(1 - S_l)\rho_v \tag{J.71}$$

denotes the mass of vapor per unit volume of concrete.

The transport equations for high temperatures used by Beneš and Štefan [228] are based on the Darcy–Buckingham law, which was previously used mainly for soils. This law represents an extension of the Darcy law to unsaturated porous media and was described in Appendix J.2; see Eqs. (J.12)–(J.13). The water mass flux (incorporating also the flux of adsorbed water) is assumed to be driven by the gradient of liquid water pressure, and the flux corresponding to vapor is included in the advection driven by the gradient of vapor pressure (see the remark at the end of the section).

In the spirit of the Darcy–Buckingham law, separate fluxes of liquid water and vapor are postulated:

$$\boldsymbol{j}_l = -\frac{k_{r,l}K_0\rho_l}{\eta_l}\nabla p_l \tag{J.72}$$

$$j_v = -\frac{k_{r,g} K_0 \rho_v}{\eta_g} \nabla p_v \qquad (J.73)$$

where K_0 is the intrinsic permeability [m^2], $k_{r,l}$ and $k_{r,g}$ are the dimensionless relative permeabilities to liquid and to gas (dependent on the saturation degree), and η_l and η_g are the dynamic viscosities of liquid water and of the pore gas. As usual, ρ_l and ρ_v denote the densities of liquid water and vapor.

The intrinsic permeability K_0 depends on the geometry of the pore space and, neglecting the adsorbed water transport, is considered independent of the specific type of fluid transported in that space. Since the paths in the pore space available for fluid transport are affected by cracking and other microstructural changes caused by mechanical and thermal effects, refined models consider the intrinsic permeability as variable; see a brief overview at the end of Sect. 12.7. Beneš and Štefan [228] described the increase of intrinsic permeability by the Bary formula [340]

$$K_0 = K_{0,\mathrm{ref}} 10^{4\omega} \qquad (J.74)$$

in which $K_{0,\mathrm{ref}}$ is the initial intrinsic permeability of undamaged material and ω is a damage variable with initial value 0 and maximum possible value 1. To capture both the mechanical and thermal effects, the total damage ω is linked to two partial damage variables ω_m and ω_T by the relation $1 - \omega = (1 - \omega_m)(1 - \omega_T)$, from which

$$\omega = \omega_m + \omega_T - \omega_m \omega_T \qquad (J.75)$$

The mechanical damage ω_m is evaluated from the stress induced by the pore pressure and constrained thermal expansion, and the thermal damage ω_T is a function of temperature; see Beneš and Štefan [228] for details.

Since the liquid and fluid phases of water at high temperature are assumed to be in thermodynamic equilibrium, the pressures p_l and p_v are linked by the Kelvin equation (8.25), and so ∇p_l in (J.72) can be expressed in terms of ∇p_v and ∇T. Afterward, the expressions for fluxes are substituted into the governing equations (J.68) and (J.70).

The heat conduction is described in a standard way, using the Fourier law (13.7) with variable thermal conductivity. The dehydration process could be described simply by making the mass of water released by dehydration depend directly on the maximum temperature reached so far. In this case, $\dot{m}_{\mathrm{deh}}$ would be linked to the temperature rate by (13.55), same as for the Bažant–Thonguthai model. However, Beneš and Štefan [228] decided to take into account a possible delay of the dehydration process at fast heating and to describe dehydration by the differential equation suggested by Feraille-Fresnet et al. [386] and Dal Pont and Ehrlacher [336],

$$\dot{m}_{\mathrm{deh}} = \frac{\langle w_{\mathrm{d}}(T) - m_{\mathrm{deh}} \rangle}{\tau_{\mathrm{deh}}} \qquad (J.76)$$

where w_{d} is a given function of temperature and τ_{deh} is a characteristic time of the dehydration process. Consequently, m_{deh} needs to be solved numerically in each incremental step. However, it does not play the role of a primary unknown field because Eq. (J.76) does not contain spatial derivatives and can be solved at each material point separately. Note that $w_d(T)$ now represents the "equilibrium" value of dehydrated water content at temperature T, and that m_{deh} remains close to $w_d(T)$ if the heating process is slow on the time scale determined by τ_{deh}.

The remaining, crucial, component that requires special attention is the link between the primary unknowns, p_v and T, and the liquid water content, $w_l \equiv n_p S_l \rho_l$. In the approach of Beneš and Štefan [228], the Bažant–Thonguthai isotherm is used in a modified sense. The expression proposed by Bažant and Thonguthai [188] for the moisture content is now understood as valid for the liquid water only, and the contribution of vapor is added separately, for the sake of simplicity. Therefore, the relation $w_e = \phi(h, T)$ with function ϕ specified in (13.66) and (13.68) is approximately replaced by $n_p S_l \rho_l = \phi(h, T)$, which can be rephrased as

$$S_l(h, T) = \frac{\phi(h, T)}{n_p(h, T)\rho_l(h, T)} \tag{J.77}$$

This is one basic simplifying assumption, which means that the liquid saturation degree is approximated as a function of the primary unknown variables p_v and T (because $h = p_v / p_{\text{sat}}(T)$), borrowing an isotherm originally meant to describe the total water content. The mass of vapor per unit volume, m_v, can then be expressed by substituting (J.77) into (J.71), leading to

$$m_v(h, T) = \left(n_p(h, T)\rho_l(h, T) - \phi(h, T)\right) \frac{\rho_v(h p_{\text{sat}}(T), T)}{\rho_l(h, T)} \tag{J.78}$$

where the dependence of ρ_v on p_v and T corresponds to the state law of vapor. The contribution of vapor to the mass of moisture is, at low temperatures, very small, but m_v is also needed for evaluation of the rate of evaporation, which may have a significant effect in the heat equation; see (J.68) and the right-hand side of (J.70).

In their simulations, Beneš and Štefan [228] adopted the Bažant–Thonguthai isotherm (13.66) for pore relative humidities below 0.96 and the simplified isotherm (13.69) for pore relative humidities above 1, with a cubic spline bridging the transitional range (a smooth transition provided by the cubic spline improves numerical performance as compared to a linear transition with abrupt changes of slope). In theory, it would be possible to express the evaporable water content

$$w_e = \phi(p_v / p_{\text{sat}}(T), T) + m_v(p_v / p_{\text{sat}}(T), T) \tag{J.79}$$

as a function of p_v and T and then substitute the time derivative of w_e into the left-hand side of the moisture mass balance equation (J.68). For numerical reasons, Beneš and Štefan [228] preferred to include w_e in the set of primary unknowns and determine it along with T and p_v by solving the governing partial differential equations combined with algebraic equation (J.79).

Regarding the specific forms of various empirical functions, the following choices were made by Beneš and Štefan [228]:

- Mass density of liquid water: $\rho_l(T)$ given by the fifth-order polynomial (13.72), with coefficients C_k taken from Table 13.4.
- Mass density of water vapor: $\rho_v(p_v, T)$ determined from the state equation of ideal gas (8.7).
- Mass density of solid skeleton: considered as constant, even though the effect of dehydration could be taken into account.
- Porosity of concrete: $n_p(T)$ given by the linear function (13.74).
- Specific enthalpy of evaporation: $\Delta h_{l,g}^w(T)$ given by Watson formula (13.94).
- Specific enthalpy of dehydration: constant value, $\Delta h_{s,l}^w = 2.4\,\text{MJ/kg}$.
- Specific heat capacity of solid skeleton: $C_{ps}(T)$ given by (13.90).
- Specific heat capacity of liquid water: $C_{pl}(T)$ given by (13.91).
- Specific heat capacity of water vapor: $C_{pv}(T)$ given by (13.92).
- Effective volumetric heat capacity of concrete: ρC_p given by the expression in square brackets on the right-hand side of (13.2), with $\rho_g C_{pg}$ replaced by $\rho_v C_{pv}$.
- Mass of dehydrated water at equilibrium (per unit volume of concrete): $w_d(T)$ given by (13.65), with additional terms in the range above 400 °C (not relevant to the example to be presented here). The characteristic time in (J.76) was set to $\tau_{\text{deh}} = 10,800$ s, which is the value identified by Feraille-Fresnet [385].
- Thermal conductivity of concrete: $k_T(h, T)$ given by (13.87) and (13.88), which can easily be rewritten as $k_T(p_v, T)$.
- Relative permeability to gas: $k_{r,g}(S_l, n_p) = 10^{S_l \psi(n_p)} - S_l 10^{\psi(n_p)}$, where $\psi(n_p) = 0.05 - 22.5 n_p$; formula experimentally determined by Chung and Consolazio [314].
- Relative permeability to liquid water: $k_{r,l}(S_l, n_p) = 10^{(1-S_l)\psi(n_p)} - (1-S_l) 10^{\psi(n_p)}$ where $\psi(n_p) = 0.05 - 22.5 n_p$; formula experimentally determined by Chung and Consolazio [314].
- Dynamic viscosity of liquid water: $\eta_l(T) = 0.6612(T - 229)^{-1.532}$, with T substituted in K; formula recommended by Gawin et al. [414].
- Dynamic viscosity of pore gas:

$$\eta_g(p_v, T) = \eta_v(T) + (\eta_a(T) - \eta_v(T))\left(\frac{p_a}{p_v + p_a}\right)^{0.608} \tag{J.80}$$

where the air pressure p_a is set to p_{atm} and the dynamic viscosities of vapor and dry air are evaluated as

$$\eta_v(T) = 8.85 \cdot 10^{-6} + 3.53 \cdot 10^{-8}(T - T_0) \tag{J.81}$$
$$\eta_a(T) = 17.17 \cdot 10^{-6} + 4.73 \cdot 10^{-8}(T - T_0) + 2.22 \cdot 10^{-11}(T - T_0)^2 \tag{J.82}$$

with T substituted in K and η_g obtained in Pa·s. These formulae are taken from Gawin [412] and Gawin et al. [414].

Using the model described above, Beneš and Štefan [228] simulated tests of heated concrete prisms reported by Kalifa et al. [533] and Mindeguia [630], as well as spalling tests of larger blocks reported by Mindeguia et al. [631]. For illustration, let us present the results obtained for concrete C60 under unidirectional heating by a radiant heater at 600 °C.

The specimens were prisms of size $300 \times 300 \times 120$ mm, made of concrete characterized by standard compression strength $\bar{f}_c = 61$ MPa. They were subjected to heating by a radiant heater placed near the top face for 5 h. All vertical faces were insulated by ceramic blocks, so that the transport can be modeled as a one-dimensional problem. No spalling was observed in the experiments.

The material parameters are summarized in Table J.1. All of them were determined from independent basic data measured by Mindeguia [630], except for the intrinsic permeability, which was calibrated by optimizing the agreement with the simulated experiments.

Table J.1 Material properties and model parameters used by Beneš and Štefan [228]

Parameter	Value	Unit
$\bar{f}_c$	61.0	MPa
$\bar{f}_t$	3.76	MPa
c	550	kg/m^3
n_{p0}	0.1027	–
A_n	$106 \cdot 10^{-6}$	$1/K$
ρ_s	2660	kg/m^3
$k_T^{(\text{dry},0)}$	2.015	$W/(m \cdot K)$
A_λ	$-985 \cdot 10^{-6}$	$1/K$
$K_{0,\text{ref}}$	$4 \cdot 10^{-20}$	m^2

The initial temperature was set to the room temperature, $T_0 = 293.15$ K $(20\,^\circ\text{C})$, which also served as the reference temperature in equations specifying the dependence of porosity and thermal conductivity on temperature. On the other hand, parameter w_1 used by the Bažant–Thonguthai isotherm (13.66) is supposed to represent the water content at full saturation and $25\,^\circ\text{C}$. It can be determined as the product of porosity and liquid water density. Based on Eq. (13.74) and parameters n_{p0} and A_n from Table J.1, porosity at $25\,^\circ\text{C}$ is estimated as $n_{p1} = 0.10323$, and the density of water at $25\,^\circ\text{C}$ is evaluated from (13.72) as $\rho_{l1} = 1001.6$ kg/m^3. Consequently, $w_1 = n_{p1}\rho_{l1} = 103.4$ kg/m^3. The value of exponent m at $20\,^\circ\text{C}$ is obtained from (13.67) as $m_0 = 1.0082$. The initial saturation degree reported by Mindeguia [630] was $S_{l0} = 0.78$, and the corresponding vapor pressure p_{v0} can be determined from the equation

$$c \left(\frac{w_1 p_{v0}}{c p_{\text{sat}}(T_0)} \right)^{1/m_0} = S_{l0} n_{p0} \rho_{l0} \tag{J.83}$$

which is constructed from (13.66) by substituting $p_{v0}/p_{\text{sat}}(T_0)$ for relative humidity and $S_{l0} n_{p0} \rho_{l0}$ for the water content, with $\rho_{l0} = 1004.1$ kg/m^3 denoting the mass density of water at $20\,^\circ\text{C}$ calculated from (13.72), and with $c = 550$ kg/m^3 and $p_{\text{sat}}(T_0) = 2.33$ kPa calculated from (8.19). The resulting initial vapor pressure is given by

$$p_{v0} = \frac{c p_{\text{sat}}(T_0)}{w_1} \left(\frac{S_{l0} n_{p0} \rho_{l0}}{c} \right)^{m_0} = 1.784 \text{ kPa} \tag{J.84}$$

The boundary conditions for heat transport include the effect of radiation and are written in the form

$$\left(\Delta h^{w}_{l,g} j_v - k_T T'\right) n_x = B_T (T - T_{env}) + \gamma_e \sigma_{SB}(T^4 - T^4_{env}) \qquad \text{(J.85)}$$

where $-k_T T'$ corresponds to the conductive heat flux expressed using the Fourier law (13.7); n_x is equal to 1 or -1 and specifies orientation of the outward normal to the boundary in the one-dimensional setting. The surface heat transfer coefficient B_T is set to 20 W/(m²·K) on the exposed face and to 4 W/(m²·K) on the unexposed face, $\sigma_{SB} = 5.67 \times 10^{-8}$ W/(m²·K⁴) is the Stefan–Boltzmann constant, and the value of the surface heat emissivity is $\gamma_e = 0.7$. The ambient temperature, T_{env}, is set to T_0 on the unexposed side, and its evolution on the exposed side is described by

$$T_{env}(t) = \begin{cases} 293.15 + \dfrac{380}{300}t & \text{for } 0 \le t \le 300 \\[2mm] 673.15 + \dfrac{50}{17,700}(t - 300) & \text{for } 300 \le t \le 18,000 \end{cases} \qquad \text{(J.86)}$$

where t is the time (since the onset of heating) in seconds and the temperature is obtained in K. The above equation means that the ambient temperature rises from 20 to 400 °C during the first 5 min and then slowly increases up to the value of 450 °C, reached after 5 h since the onset of heating.

Boundary conditions for moisture transport are written in the form

$$(j_l + j_v)n_x = \beta_c(\rho_v - \rho_{v,env}) \qquad \text{(J.87)}$$

where the convective mass transfer coefficient β_c is set to 0.019 m/s on the exposed face and 0.009 m/s on the unexposed face, and the ambient vapor pressure $\rho_{v0} = 0.0132$ kg/m³ (on both sides) is computed from T_0 and p_{v0} via the state equation of vapor.

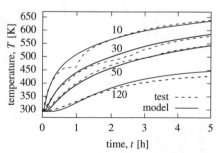

Fig. J.1 Temperature evolution in a heated specimen at 10, 30, 50, and 120 mm from the exposed face; dashed curves correspond to measurements of Mindeguia [630] and solid curves to numerical results of Beneš and Štefan [228]

Using their hygrothermal model with parameters and initial and boundary conditions as described above, Beneš and Štefan [228] simulated the transport of heat and moisture in a heated specimen as a one-dimensional problem. The calculated evolution of temperature and vapor pressure at selected locations (10, 30, 50, and 120 mm from the exposed face) is compared to experimental data in Figs. J.1 and J.2. The agreement between the computed and measured temperatures is very good; see Fig. J.1. The overall evolution of vapor pressures is captured properly, with some deviations at the point closest to the exposed surface (Fig. J.2a). The precise value of pore pressure is difficult to measure without disturbing the original conditions in the pores, and the reproducibility of the test is not perfect, as documented by the differences between pressures measured at the same location in two tests; see Fig. J.2. Taking this into account, the results of simulations can be considered as satisfactory.

To gain more insight into the processes taking place during heating, let us examine the spatial distribution of various important quantities at times ranging from 12 min to 5 h of heating. The results provided by Štefan [846] are plotted in Fig. J.3. The origin

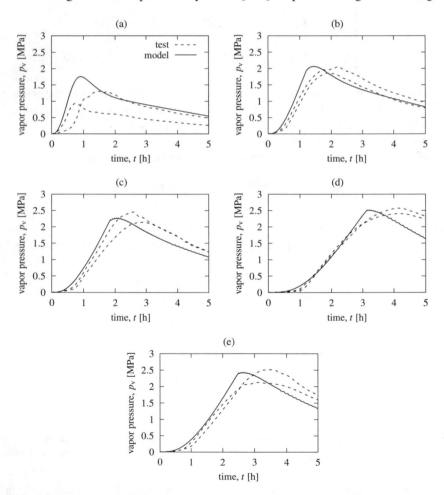

Fig. J.2 Vapor pressure evolution in a heated specimen at (a) 10 mm, (b) 20 mm, (c) 30 mm, (d) 40 mm, and (e) 50 mm from the exposed face; dashed curves correspond to measurements of Mindeguia [630] and solid curves to numerical results of Beneš and Štefan [228]

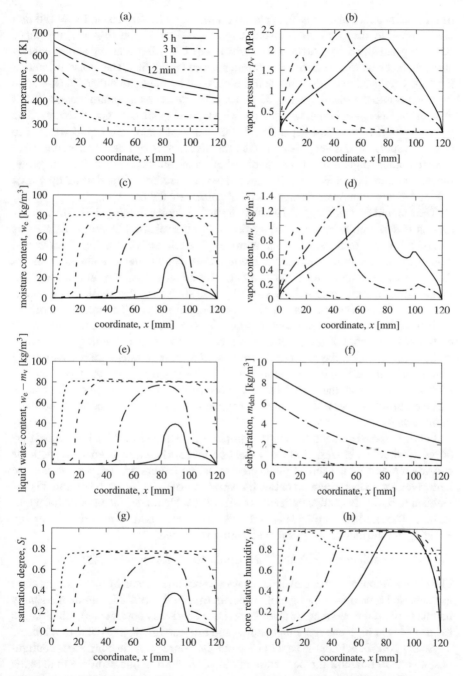

Fig. J.3 Heated specimen: calculated distribution of (a) temperature, (b) vapor pressure, (c) moisture content, (d) vapor content, (e) liquid water content, (f) water released by dehydration per unit volume, (g) liquid saturation degree, (h) pore relative humidity after 12 min (short dashes), 1 h (long dashes), 3 h (dash-dotted curves), and 5 h (solid curves) of heating of the spatial coordinate, $x = 0$, is placed on the exposed surface, and $x = 120$ mm corresponds to the opposite (unexposed) surface. During the first 3 h, the vapor pressure gradually rises up to the maximum value of 2.5 MPa, attained approximately at the distance of 45 mm from the exposed surface (Fig. J.3b). The moisture content is quickly reduced in a thin layer near the exposed boundary and then this dry zone propagates into the specimen (Fig. J.3c). Later, when the temperature increase is felt even at the unexposed boundary, another dry zone appears near that boundary. The graphs in Fig. J.3d,e show separately the contents of vapor and liquid water. The mass of vapor remains very small, and so the distribution of liquid water (Fig. J.3e) is almost undistinguishable from the distribution of moisture (Fig. J.3c). Therefore, the assumption that the Bažant–Thonguthai isotherm refers to liquid water only does not lead to any significant changes of the resulting model behavior.

The amount of water released by dehydration is depicted in Fig. J.3f. Since dehydration is assumed to be governed by Eq. (J.76), the release of water is delayed as compared to the increase of temperature. Figure J.3g shows the liquid saturation degree, which is initially uniform and equal to 0.78 and later is reduced in almost the same way as the liquid water content. The effect of porosity increase at high temperatures is only marginal. Finally, Fig. J.3h shows the distribution of pore relative humidity, which is initially uniform and equal to 0.77. The relative humidity is seen to decrease near the exposed boundary (and later also near the unexposed one), but in the core of the specimen, it increases to levels exceeding 0.96, which is the value at which the isotherms corresponding to high temperatures start rising steeply. It is interesting to observe that, after 5 h of heating and in the zone between 80 and 100 mm from the exposed surface, the saturation degree is below 0.4 but the relative humidity is near 0.96, which is related to the high values of vapor pressure in that zone.

Let us emphasize that the present example has been included for illustrative purposes, to show what kind of choices must be made when a hygrothermal model is constructed and which model parameters must be calibrated. The model adopted here gives a reasonable agreement with several sets of experimental data, but it is not claimed to be a universally applicable and fully reliable model for all potential applications. The field of coupled heat and mass transfer in porous materials is rapidly evolving and further development and testing is needed.

Remark: Percolation Limits and Their Implications for Beneš-Štefan Model

The Beneš-Štefan model postulates a separate vapor flux, driven by the vapor pressure gradient and governed by a separate vapor permeability. But this can be physically justified only if the vapor phase percolates (i.e., is perfectly contiguous), which could happen only at high enough temperatures and low enough degrees of saturation.

Below 100 °C, neither the vapor phase nor the liquid phase in concrete is contiguous, no matter how low the pore humidity is. To pass from one capillary pore to the next, an H_2O molecule in vapor must: (1) enter the liquid phase, (2) next enter the adsorbed phase, (3) then pass through a nanopore, creeping at solid surface along the hindered adsorbed water layer to the next capillary pore, (4) then exit into the liquid phase, and (5) finally enter the vapor phase again. Obviously, what matters for transport of this water molecule is only one permeability, the permeability of the nanopore. Even at a very low degree of saturation, at which the nanopore is almost empty, the vapor pressure cannot be transmitted through the near-empty nanopore and a water molecule must pass through it in an adsorbed state, because the mean free path of H_2O molecule in vapor is longer than the pore width.

A big upward, cca 200-fold, jump in permeability occurs upon exceeding $100\,°C$. It may be explained by transformation of the low density C-S-H into a high density C-S-H. This transformation can make the liquid capillary phase percolate. However, except at low degrees of pore saturation, the vapor will form separate bubbles within a contiguous liquid phase. In that case, an H_2O molecule must still enter and exit the liquid phase in order to pass from one pore to the next, and then only the permeability to liquid water matters.

So the hypothesis of separate fluxes and permeabilities of vapor and liquid cannot be physically valid in general. However, the Beneš-Štefan model circumvents this limitation by setting the vapor permeability for low temperatures (especially below $100\,°C$) virtually to zero, and the liquid permeability so small that it approximately corresponds to flow of the adsorbed phase through the nanopores. Thus, at low temperatures, this model is almost equivalent to the single phase flow and single permeability, same as the Bažant-Thonguthai model.

Appendix K
Nonstandard Statistics Used in Support of Some Creep and Shrinkage Models

K.1 Linear Coefficient of Variation (L.C.o.V.)

In Gardner [406], the logarithmic scales of load duration $t - t'$ and drying duration $t - t_0$ are subdivided into decade-long intervals, labeled by subscripts $i = 1, 2, \ldots, n$, and the individual data points in interval number i are labeled by subscripts $j = 1, 2, \ldots, m_i$. The weighted mean of data is obtained using the standard formula (11.8), giving equal weight to each decade of time. However, the calculation of the overall coefficient of variation of prediction errors, ω_G, is nonstandard:

$$\omega_G = \frac{s_G}{\bar{y}}, \quad s_G = \frac{1}{n} \sum_{i=1}^{n} s_i \tag{K.1}$$

where

$$s_i = \sqrt{\frac{1}{m_i - 1} \sum_{j=1}^{m_i} \left(y_{ij} - Y_{ij}\right)^2} \tag{K.2}$$

The bias due to variation of the means of other variables throughout the intervals is ignored. The bias due to having different numbers m_i of points in different intervals is here compensated by evaluating separately the standard deviation for each interval according to (K.2), which correctly gives to each time interval the same weight. However, the expression in (K.1) for the overall standard deviation $\bar{s}$ of the data from the model predictions is not statistically correct because, instead of averaging the variances s_i^2 (squared standard deviations), what is averaged are the standard deviations s_i. Correctly, the averaging must be applied to the squared errors; i.e., one must take the root mean square (RMS).

The linear averaging of standard deviations s_i is tantamount to denying the validity of the central limit theorem of the theory of probability, underpinning the Gaussian distribution (see Sect. K.6). This implicit denial is untenable (it is true that linear

© Springer Science+Business Media B.V. 2018
Z.P. Bažant and M. Jirásek, *Creep and Hygrothermal Effects in Concrete Structures*, Solid Mechanics and Its Applications 225, https://doi.org/10.1007/978-94-024-1138-6

averaging of errors has been used for some special purposes in financial statistics [248], but that was in problems of extreme value statistics, to which the central limit theorem of the theory of probability does not apply).

For the overall error definition used in comparisons of prediction models to be correct, minimization of the overall error must yield the optimum data fit. In the special case of a linear model, the statistical method must reduce to linear regression statistics. This is a simple but fundamental check on the soundness of the statistical approach to the comparison of prediction models.

In the case of error defined by (K.1) and (K.2), one would have to minimize the expression

$$s_G^2 = \frac{1}{n^2} \left(\sum_{i=1}^{n} \sqrt{\frac{1}{m_i - 1} \sum_{j=1}^{m_i} (y_{ij} - Y_{ij})^2} \right)^2 \tag{K.3}$$

In the special case of a linear model, we have $Y_{ij} = a + bX_{ij}$, where X_{ij} are the coordinates (e.g., the values of $\log(t - t')$) of data points Y_{ij}. The minimizing conditions $\partial s_G^2 / \partial a = 0$ and $\partial s_G^2 / \partial b = 0$ then yield two equations for free parameters a and b. It is easy to see that these equations are nonlinear and thus might not guarantee a unique solution, despite linearity of the model. The nonlinearity of these equations confirms that definition (K.1) is not appropriate.

On the other hand, in the case of the standard error expression (11.6), substitution of $Y_{ij} = a + bX_{ij}$ yields

$$s^2 = \frac{N}{N - p} \frac{1}{n} \sum_{i=1}^{n} \frac{1}{m_i} \sum_{j=1}^{m_i} \left[y_{ij} - (a + bX_{ij}) \right]^2 = \min \tag{K.4}$$

Here, the minimizing conditions $\partial s^2 / \partial a = 0$ and $\partial s^2 / \partial b = 0$ yield linear equations, whose solution gives the well known expressions for slope b and intercept a of the regression line.

Another debatable aspect of Gardner's L.C.o.V. is the expression $m_i - 1$ in the denominator of (K.2). In population statistics, a similar expression is used to obtain an unbiased estimate of the variance. However, this is done in a different context. If we deal with the differences between individual values and their mean, a data set consisting of n points has only $n - 1$ degrees of freedom because the remaining degree of freedom has been removed by subtracting the average computed from the original values. It is clear that if the data set contains only 1 point, no estimate of the variance can be constructed. On the other hand, in the present context, we deal with differences between measured values, y_{ij}, and values predicted by a model, Y_{ij}. The total number of data points is $N = \sum_{i=1}^{n} m_i$, and a model with p parameters can, in principle, be adjusted such that p values are fitted exactly. So the number of remaining degrees of freedom is $N - p$ and this is reflected by the expression in the denominator of the correct error definition (11.6). Gardner's definition (K.1) and (K.2) has the peculiar property that intervals that contain only 1 point need to be excluded, to avoid division by zero, and intervals containing a low number of points but at least two have a stronger influence than intervals with a high number of points. Even if $m_i - 1$ in the denominator of (K.2) were replaced by m_i, the definition of s_G would still be dubious because of the second part of formula (K.1). Consider a data set with intervals containing only 1 point each. Then, the modified definition of s_G with m_i in the denominator of (K.2) leads to averaging of the absolute values of individual errors, $|y_{ij} - Y_{ij}|$, and the parameter optimization problem ceases to have a unique solution.

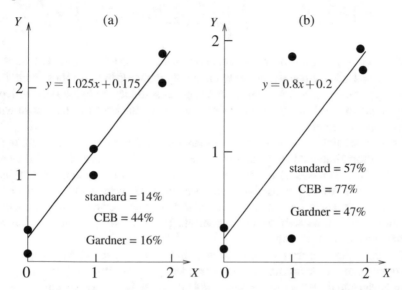

Fig. K.1 (a) Differences in coefficients of variation of errors between standard and nonstandard statistical methods for examples of linear regression

However, can the difference between the statistical indicators s and s_G in (11.6) and (K.1)–(K.2) be significant? Indeed, it can. To document it, consider again the fundamental case of a linear model $Y = a + bX$, for which we know that the optimum data fit generally accepted as correct is obtained if and only if the least-square regression is used. Let us consider 2 sets of 3 pairs of data points shown in 2 diagrams in Fig. K.1. For set 1, the data are $Y = 0.1$ and 0.3 for $X = 0$, $Y = 1.0$ and 1.3 for $X = 1$, and $Y = 2.1$ and 2.4 for $X = 2$; and for set 2, the data are $Y = 0.1$ and 0.3 for $X = 0$, $Y = 0.2$ and 1.8 for $X = 1$, and $Y = 1.7$ and 1.9 for $X = 2$. In each diagram, the regression line is drawn and the values of the coefficient of variation obtained according to the least-square linear regression and according to (K.1) are indicated. For set 1 (left diagram), the correct coefficient of variation (based on linear regression) is $\omega = 14\%$ while (K.1) gives $\omega_G = 16\%$. This is not a great discrepancy. However, for set 2 (right diagram), the correct coefficient of variation is $\omega = 57\%$; this is much larger than the value given by (K.1), which is $\omega_G = 47\%$. Such a discrepancy is not negligible. Note also that for set 1, we get $\omega < \omega_G$, while for set 2, the opposite inequality holds. It is then no surprise that the ranking of models based on Gardner's inappropriate error measure is different from that obtained by standard methods.

K.2   CEB Coefficient of Variation

In Müller and Hilsdorf [641], Al-Manaseer and Lakshmikanthan [25] and Al-Manaseer and Lam [26], the coefficient of variation of prediction model errors was defined as

$$
\omega_{CEB} = \sqrt{\frac{1}{n}\sum_{i=1}^{n}\omega_i^2}, \quad \omega_i = \frac{1}{\bar{y}_i}\sqrt{\frac{1}{m_i-1}\sum_{j=1}^{m_i}(Y_{ij}-y_{ij})^2}, \quad \bar{y}_i = \frac{1}{m_i}\sum_{j=1}^{m_i}y_{ij}
$$

(K.5)

Note that $m_i - 1$ again appears in the denominator and is 0 for a box with only one point, $m_i = 1$. Consequently, not only the empty intervals but also those containing a single point have to be deleted in calculating this statistic, while intervals with only a few points (but at least 2) have a stronger influence on the error than those with a high number of points.

Another debatable aspect is that, compared to the least-square statistical regression, the short-time data get overemphasized and the long-time data get underemphasized. This is caused by the appearance of $\bar{y}_i$ (rather than $\bar{y}$) in the denominator of (K.5) before all ω_i are combined into one coefficient of variation. An interval with a very small $\bar{y}_i$ gives a very large ω_i and thus, incorrectly, dominates the entire statistics. This is especially serious for shrinkage, which starts from zero, but what matters most is the final value.

Can the difference from the correct statistical indicator in (11.6) be significant? It certainly can. To demonstrate it, we consider again the limiting special case of a linear model and the example of 2 sets of data in Fig. K.1. The coefficient of variation for set 1 (the left diagram) is found to be 44%, which is more than twice the correct value of 14% from linear regression. The coefficient of variation for set 2 (the right diagram) is found to be 77%, which is much larger than the correct value of 57%.

K.3   CEB Mean-Square Relative Error

In Müller and Hilsdorf [641], Al-Manaseer and Lakshmikanthan [25] and Al-Manaseer and Lam [26], another comparison is made on the basis of the relative error defined as

$$
S_{CEB} = \sqrt{\frac{1}{n}\sum_{i=1}^{n}S_i^2}, \quad S_i^2 = \frac{1}{m_i-1}\sum_{j=1}^{m_i}\left(\frac{Y_{ij}}{y_{ij}}-1\right)^2 = \frac{1}{m_i-1}\sum_{j=1}^{m_i}w_{ij}\left(Y_{ij}-y_{ij}\right)^2
$$

(K.6)

where $w_{ij} = 1/y_{ij}^2$. Unlike the previous case, this definition of error is consistent with the least-square regression. However, it implies unrealistic weighting of the data. As shown by the last expression, it means that the weights w_{ij} are inversely proportional to y_{ij}^2. This causes the errors in the small compliance or shrinkage values to get greatly overemphasized, and the errors in the large values to get greatly underemphasized. Yet, the long-time predictions are the most important, while the short-time ones are the least.

K.4   CEB Mean Relative Deviation

Still another indicator, called the mean deviation, was employed to compare models [25, 26, 641]:

$$M_{CEB} = \frac{1}{n} \sum_{i=1}^{n} M_i, \quad M_i = \frac{1}{m_i} \sum_{j=1}^{m_i} \frac{Y_{ij}}{y_{ij}} \tag{K.7}$$

This indicator does not correspond to the method of least squares, and for the special case of a linear model, the minimization of $(M_{CEB} - 1)^2$ does not reduce to linear regression. So this approach is afflicted by all the previously described problems that arise for such nonstandard statistics.

K.5   Coefficient of Variation of the Data/Prediction Ratios

Noting that, in a perfect model, the ratios $r_{ij} = y_{ij}/Y_{ij}$ should be as close to 1 as possible, some studies calculate the coefficient of variation of r_{ij} and use it to compare the prediction models. But this approach to statistics, endemic in concrete research, is incorrect. To show the problem, let us replace, for the sake of brevity, the double indices ij by a single index $k = 1, 2, \ldots, K$, where $K = \sum_{i=1}^{n} m_i$. The variance of the population of all $r_k = y_k/Y_k$ is

$$s_R^2 = \sum_{k=1}^{K} w_k \left(\frac{y_k}{Y_k} - \bar{r} \right)^2, \quad \bar{r} = \sum_{k=1}^{K} w_k \frac{y_k}{Y_k} \tag{K.8}$$

where w_k are the weights such that $\sum_{k=1}^{K} w_k = 1$, and $\bar{r}$ is the weighted mean of all r_k. Consider now that the prediction formula giving Y_k is multiplied by any constant factor c, i.e., consider the replacement $Y_k \leftarrow cY_k$. Then, the variance changes from s_R^2 to

$$\tilde{s}_R^2 = \sum_{k=1}^{K} w_k \left(\frac{y_k}{cY_k} - \sum_{m=1}^{K} w_m \frac{y_m}{cY_m} \right)^2 = \frac{1}{c^2} s_R^2 \tag{K.9}$$

So, as we see, the variance of the prediction-data ratios can be made arbitrarily small by multiplying the prediction formula by a sufficiently large factor c. Since the mean $\bar{r}$ is replaced by $\bar{r}/c$, the coefficient of variation $\omega_R = s_R/\bar{r}$ is found to be independent of factor c [99].

Therefore, minimization of s_R^2 cannot be used for the purpose of data fitting. Another problem is that the differences $1 - r_{ij}$ tend to be the greatest for short times, which thus dominate the statistics although the long times are of main interest. It follows that the use of the coefficient of variation ω_R in some studies, intended for statistical comparison of different prediction models, was incorrect and misleading. Further, it follows that the plots of data/prediction ratios r_k versus time (or versus k) should not be used for visual comparison of the goodness of data fits by various creep prediction models.

K.6   Why Is the Method of Least Squares the Only Correct Approach to Central Range Statistics?

The method of least squares was first published by Legendre [571], but its rigorous derivation is due to Gauss [411], who is known to have used it already before 1806. For brevity, let us again replace the double indices ij by a single index $k = 1, 2, \ldots, K$ where $K = \sum_{i=1}^{n} m_i$. The errors are defined as $e_k = y_k - Y_k$, where $X_k =$ coordinates of data points (i.e., influencing parameters such as the load duration, age at loading, thickness, humidity), $Y_k = F(X_k) =$ predicted values, and function F defines the prediction model. The joint probability density distribution of all the measured data, also called the likelihood function $\mathscr{L}$ [285], is

$$\mathscr{L} = f(y_1, y_2, \ldots, y_K) = [\phi_1(y_1)]^{W_1} [\phi_2(y_2)]^{W_2} \ldots [\phi_K(y_K)]^{W_K} \tag{K.10}$$

where $\phi_k(y_k) =$ probability density distribution of measurement y_k alone, and exponent W_k means that we imagine a W_k-fold repetition of the kth measurement, which is equivalent to applying weight W_k to data point k. Let us first assume the errors to be approximately normally distributed; then,

$$\phi_k(y_k) = \frac{1}{s_k \sqrt{2\pi}} \, e^{-(y_k - Y_k)^2 / 2s_k^2}, \qquad k = 1, 2, \ldots, K \tag{K.11}$$

where $s_k^2 =$ (conditional) variance of y_k, which is a constant known (or knowable) a priori.

The objective of optimal data fitting is to maximize the likelihood function $\mathscr{L}$ [285]. This is equivalent to minimizing $- \ln \mathscr{L}$, i.e.,

$$- \ln \mathscr{L} = - \ln \left[\exp \left(- \sum_{k=1}^{K} \frac{W_k (y_k - Y_k)^2}{2s_k^2} \right) \prod_{k=1}^{K} \left(s_k \sqrt{2\pi} \right)^{-W_k} \right] =$$

$$= \sum_{k=1}^{K} w_k (y_k - Y_k)^2 + C = \min \tag{K.12}$$

where $w_k = W_k/2s_k^2$ = modified weights and $C = \sum_{k=1}^{K} w_k \ln(s_k\sqrt{2\pi})$ = constant. Equation (K.12) represents minimization of the sum of weighted squared errors and thus proves validity of the method of least squares.

The histograms of data plotted on the normal probability paper demonstrate that the distributions or errors e_{ij} in creep and shrinkage are approximately normal, although small deviations from normality exist.

What if the distributions $\phi_k(y_k)$ of data y_k are not normal? In that case, the database may be subdivided into data groups labeled as $r = 1, 2, \ldots, N_g$, such that each group r contains a sufficient but not excessive number n_r of adjacent data points located so closely that the statistical trends within each group are negligible ($n_r \approx 6$ appears suitable). The mean of each data group is a scaled sum of random variables,[17] and according to the Central Limit Theorem [248, 285], the distribution of this sum, and thus the group mean, converges to the normal distribution, albeit one with a scaled standard deviation. We may now logically expect that the best fit of y_k can be obtained as the best fit of all the group means, each of which has a Gaussian distribution. The remaining derivation up to (K.12) is the same and leads to the same conclusion.

To be rigorous, it must be admitted that there exist special problems where the least-square regression is insufficient or even inappropriate. One example is the extreme value statistics, leading to Weibull distribution of strength of brittle structures [248]. Another is the extension of the least-square approach to Bayesian optimization, in which the posterior data are supplemented by some sort of prior information [37]. A third example is the robust regression [740], used to emphasize the role of numerous outliers of heavily tailed non-Gaussian distributions. But these special problems do not arise for the typical regression problems of concrete design equations discussed here.

K.7 Comparison of Models by Standard and Nonstandard Statistical Indicators

The standard and nonstandard statistical indicators have been calculated for five prediction models using the NU-ITI database [160], as well as the RILEM database and Gardner's limited database. From the last two databases, it was necessary to delete a few data sets for which the parameters required to evaluate some of the prediction models were not known.

The results are shown in Table K.1. According to the standard indicator (11.7), model B3 appears as the best, while the classical ACI-209 model and the GZ model are by far the worst.

The five creep and shrinkage prediction models considered here were statistically compared in a committee report and ACI 2008 Guide, based on the results reported

---

[17]The errors at closely spaced times must actually be correlated, but the only way to take this aspect into account is to treat creep and shrinkage as random processes; this has been done, but is far more complicated; see, e.g., Çinlar et al. [300] and Sect. K.8.

Table K.1 Comparison of standard and nonstandard statistical indicators for various prediction models, based on NU-ITI database, with 50 boxes of $\log(t - t')$ and H for creep and 28 boxes of $\log(t - t_0)$ and $\sqrt{D}$ for shrinkage

	Creep compliance					Shrinkage				
	B3	ACI	CEB	GL	GZ	B3	ACI	CEB	GL	GZ
Standard indicator [%]	27.3	42.6	31.0	30.2	41.9	28.5	42.3	47.4	31.0	44.4
Gardner's linear C.o.V. [%]	21.2	36.0	26.0	25.2	34.7	24.5	35.8	41.1	25.5	35.4
CEB C.o.V. [%]	23.0	37.9	28.6	28.4	37.9	36.5	46.2	46.8	37.9	45.7
CEB mean-square error [%]	23.5	36.6	29.2	28.4	37.7	36.9	45.8	45.9	38.1	44.5
CEB mean deviation	0.95	0.74	0.94	0.88	0.83	1.06	1.03	0.70	1.12	1.10

by Bažant and Baweja [107], Gardner [406], and Al-Manaseer and Lam [26]. Unfortunately, the nonstandard statistical indicators were considered as equally relevant, and thus, it is no surprise that each different statistical indicator placed a different prediction model on top or bottom. This is documented by Table K.2, extracted from Tables 4.2 and 4.3 in ACI Committee 209 [14], which were reproduced as Table 2 in Bažant and Li [161]. The differences among various models are seen to be minor, and interestingly, it was not even questioned that in some comparisons the 1972 ACI model came on top. These comparisons are in gross disagreement with those obtained in Bažant and Li [161] by standard weighted regression; see Sect. 11.5. Even greater is the disagreement with the model rankings in Chap. 7, which were based on the analysis of observed long-time bridge deflections.

Table K.2 Comparison of standard and nonstandard statistical indicators of errors used by various authors to compare and rank four prediction models, for (a) compliance and (b) shrinkage; extracted from Tables 4.2 and 4.3 in ACI Guide 2008

	(a) Compliance [%]						(b) Shrinkage [%]				
	Indicator	ACI	B3	CEB	GL		Indicator	ACI	B3	CEB	GL
Bažant and Baweja basic creep	ω	58	**24**	35	–	Bažant and Baweja	ω	55	**34**	46	–
Bažant and Baweja drying creep	ω	45	**23**	32	–	Al-Manaseer and Lam	ω_{CEB}	46	41	52	**37**
Al-Manaseer and Lam	ω_{CEB}	48	36	36	**35**		S_{CEB}	83	84	**60**	84
	S_{CEB}	32	35	**31**	34		$M_{CEB}^{\dagger}$	122	**107**	75	126
	$M_{CEB}^{\dagger}$	86	**93**	92	92						
						Gardner	ω_G	41	20	–	**19**
Gardner	ω_G	30	27	–	**22**	Gardner recalculated[††]	ω_G	41	**20**	44	22

[†]Note that the ideal value of indicator M_{CEB} is 100%, while for all the other indicators the ideal value is 0%. The best result according to each criterion is emphasized by bold face.
[††]The values published by Gardner [406] were recalculated by Bažant and Li [161], using Gardner's database and definition of error.

K.8 Stochastic Process for Extrapolating Concrete Creep

The least-square regression is predicated on the hypothesis of statistically independent errors (see Sect. K.6). This hypothesis is doubtless realistic for a large database
in which the errors are dominated by the differences among many different concretes tested in many different laboratories. Not, however, for a single test curve of
one given concrete. If, for instance, the error at 1000 days of creep is positive and
large, it will almost certainly be positive and large at 1001 days. In other words, the
errors at close enough times are, in one and the same test, highly correlated.

To take this correlation properly into account, it is necessary to treat creep (or
shrinkage) curve as a random process [230]. In Çinlar et al. [300], it was shown
that the creep may be realistically modeled as a nonstationary pure jump increasing
stochastic process with statistically independent (or uncorrelated) random increments
of locally gamma distribution. This process is nonstationary, because of gradual
deceleration of the creep, but it can be transformed to a stationary gamma process.

The gamma distribution is justified by its infinite divisibility, which is a property
required for two reasons: (1) a homogeneously stressed specimen can be split into
smaller ones whose responses are summed, and (2) the responses to a sum of stresses
are additive. There are several infinitely divisible distributions which are statistically
tractable: the Gaussian, gamma, Poisson, and stable distributions (Weibull, Gumbel
and Fréchet). The Gaussian, Gumbel, and Fréchet may be questionable because they
could include negative increments, and the others except gamma for various physical
reasons [300]. It was shown that the gamma process of creep may be characterized
by its Laplace transform as follows:

$$E\{\exp[\lambda(J_s - J_r)]\} = \exp\left[-\int_{x=t_r}^{s}\int_{y=0^+}^{\infty}(1 - e^{-\lambda y})m(\mathrm{d}x, \mathrm{d}y)\right] \qquad (K.13)$$

for any constant $\lambda > 0$. Here, E is the expectation, $J_r = J(t_r, t')$, t' is fixed since the
process simulates an individual creep test, and $m(\mathrm{d}x, \mathrm{d}y)$ is a measure defined as

$$m(\mathrm{d}x, \mathrm{d}y) = a'(x)\,\mathrm{d}x\,e^{-b(x)y}\mathrm{d}y/y \qquad (K.14)$$

where $a(x)$, $b(x)$ = scale and shape functions of the gamma process for the given t';
and $a'(x) = \mathrm{d}a(x)/\mathrm{d}x > 0$.

Monte Carlo simulations were used to demonstrate this process for various long-
time creep tests on the concretes for Dworshak Dam, Canyon Ferry Dam, Ross
Dam, and Shasta Dam, and the concrete of York et al. [887]. An initial group of
data points was assumed to be known and used for calibration. The process was
then simulated from the last point of this group both forward and backward, as
shown for the Dworshak Dam concrete; see Fig. 1 in Çinlar et al. [300]. The band of
random simulations enveloped quite realistically the series of subsequent data points
pretended to be unknown.

The gamma process represents a more realistic, but more complicated, alternative to the data extrapolation procedure described in Sect. 3.8.1. In the existing form, however, this extrapolation does not capture the random variations among different concretes and among different batches of the same concrete, and the random effects of environment, curing and microcracking. Neither it includes Bayesian updating (Sect. 6.4). For further details, see Çinlar et al. [300].

Appendix L
Method of Measurement of Creep and Shrinkage

The basic testing of creep and shrinkage deals with two simple cases: the case without moisture exchange with the environment, in which the specimen is sealed, and the case of drying in a stable environment typical of practical situations. The former case is relevant to mass concrete and also to the core of thicker cross sections, as in nuclear containments, large bridges, supertall building columns, and, of course, dams. The latter case typifies the behavior of thin cross sections, but it is also relevant to the behavior of a surface layer in thick cross sections.

Most of the present exposition is based on the RILEM Recommendation [20]; see also [19, 105, 197, 249, 878]. Some updates are mentioned based on current research on modern concretes with high autogenous shrinkage.

L.1 Testing Apparatus

The creep test device must apply a centric compressive force and keep it constant with time, with the accuracy of $\pm 1\%$. This may be achieved by different systems:

1. the helical spring [44, 239], see Fig. L.1a, b,
2. the hydropneumatic accumulator [573], see Fig. L.1c, or
3. electronic control.

The apparatus must allow measuring the applied force with a known precision, whose errors should be reported ($\pm 1\%$ is reasonable). A calibration process ought to be established, applied in the tests and information on it also reported.

The strain must be measured with the same system, both in the creep test and in the companion shrinkage test. It must be measured in the middle portion of the specimen length, along at least three longitudinal lines spaced evenly around the circumference, and on a base not shorter than one diameter. The distance between the strain measurement base and the ends of the specimen must not be smaller than

© Springer Science+Business Media B.V. 2018
Z.P. Bažant and M. Jirásek, *Creep and Hygrothermal Effects in Concrete Structures*, Solid Mechanics and Its Applications 225, https://doi.org/10.1007/978-94-024-1138-6

1.5 diameters (or 1 diameter if the end faces of the specimens are protected against drying). To determine the creep Poisson's ratio, it is useful to measure also the transverse deformation across the central portion of the specimen.

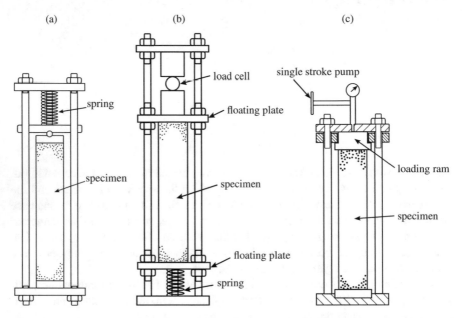

Fig. L.1 (a)–(b) Helical spring loading device (classical method), (c) loading device with hydro-pneumatic accumulator (preferable), after Neville [652]

L.2 Specimens

L.2.1 Form and Dimensions

All the creep and shrinkage specimens should normally be cylinders. A sufficient number of specimens (at least six) should have the same diameter, with a slenderness (length/diameter ratio) not less than 4. The specimens are normally cast in molds. Alternatively, the specimens can be cored from concrete blocks that have lost no water before coring, which is, for example, suitable for the testing of fiber-reinforced concrete. Cores are also suitable for the testing of those parts of massive in-situ structures that have lost no water before the coring, and for diagnostics of a structure in service (note that the surface layer in cast specimens contains more mortar than in cored or saw-cut specimens; this "wall effect" causes some differences, which can be compensated for by calculations).

The molds should be metallic. Alternatively, they can be lined internally with a metallic foil or with a sheet of polymer, provided that the total water absorption of the lining remains, after 24 h, less than 2 grams per liter of capacity of the mold. The diameter should be at least 5 times the size of the largest aggregate, but for cores, 4 (or even 3) times is acceptable.

L.2.2   Specimen Production

The specimens must be manufactured according to the requirements for compression tests. Casting in a horizontal position is unsuitable as it produces properties with a transverse gradient and might be risky for long specimens as it promotes buckling. The age of the specimen is customarily measured from the moment of the filling of the mold with concrete (although from the moment of set is better for short-time data).

L.2.3   Curing

Except in special circumstances (which may occur when the aim is to study the curing effects or to imitate the specific conditions of a construction site), all the specimens must be kept after casting in their mold and in a temperature-controlled room at 20 $\pm$ 1°C or at 25 $\pm$ 1°C (this kind of standardization is needed for comparability of tests from different labs). When the age at exposure to drying is no more than 3 h lower than the lowest age at load application, the molds are stripped, all at about the same time, in the testing room (which must be temperature and moisture controlled). Immediately after stripping, all the specimens intended for autogenous shrinkage and basic creep must be protected by a closely adhering jacket consisting of a metallic (e.g., copper or aluminum) foil. To facilitate evaluation, the end faces of the drying specimens should be protected against desiccation, immediately after stripping, by the same procedure. When the age at exposure to drying is planned to coincide with the first age at loading, all the specimens of the series must be stripped no more than 2 h before applying the load on the first creep specimen, preferably in a wet room of temperature controlled at 20 $\pm$ 1°C or at 25 $\pm$ 1°C, and protected immediately after stripping against desiccation by an adhering protective jacket consisting of a metallic (e.g., copper or aluminum) foil.

L.2.4   Environmental Conditions

Except when the goal is to study the effects of drying or to emulate the drying conditions of a specific practical application, four types of test, requiring four series of specimens, are carried out:

870 Appendix L: Method of Measurement of Creep and Shrinkage

1. Basic creep plus autogenous shrinkage of sealed loaded specimens.
2. Autogenous shrinkage of sealed load-free specimens, starting right after casting and running for the whole duration of basic creep tests (item 1).
3. Drying shrinkage of exposed load-free specimens, which inevitably includes autogenous shrinkage continuing for some time in specimen core (that has not yet been penetrated by the drying front spreading from the surface).
4. Drying creep of exposed loaded specimens, which inevitably includes drying shrinkage and autogenous shrinkage in the core.

The autogenous shrinkage and basic creep tests are carried out at $20 \pm 1°C$ or at $25 \pm 1°C$, on specimens confined by an adhering protective jacket consisting of a metallic (e.g., copper or aluminum) foil; the specimens are kept in a temperature controlled room at $20 \pm 1°C$ or at $25 \pm 1°C$, preferably in a wet room. They must be stripped in the same room 2 h before the start of the shrinkage measurements (which must begin before the first creep test). The surface protection (e.g., coating with a resin or a copper or aluminum foil, or direct application of a self-adhesive aluminum foil) must be applied immediately after the stripping or coring of each specimen.

The drying shrinkage and drying creep tests must be carried out in a room of controlled temperature and humidity, preferably at $20 \pm 1°C$ and at $50 \pm 5\%$ relative humidity, or at $25 \pm 1°C$ and at $65 \pm 3\%$ relative humidity. The shrinkage strain measurements must start immediately (within 3 min, but better 1 min) after stripping of the moisture seal. In any case, the specimens must be kept (before, during and after stripping) in a room in which the temperature is kept constant to within $\pm 1°C$. Except for special cases, the specimens cored from a structure are always kept sealed according to the conditions already specified.

In old concretes of high w/c, the self-desiccation and autogenous shrinkage was small and could be ignored. But in modern concrete, it is big, often even if w/c is high (because of various admixtures). Thus, for modern concretes, the autogenous shrinkage must be measured for the entire duration of the creep test and then used in calculating the actual basic creep.

L.2.5 Companion (or Control) Specimens

These are load-free specimens required to accompany any creep test. They are made and kept under the same conditions as those for the creep tests and for Young's modulus tests.

L.3 Testing Procedure

L.3.1 Preparation of Specimens

The specimens intended for shrinkage measurements must be placed in the creep test room at least two hours before the start of the shrinkage measurements.

Measurements of drying shrinkage must begin as quickly as possible after stripping of the mold or the moisture seal (in any case within ≤ 3 min). Since the placing of a wet surface in contact with a dry atmosphere causes a drop in temperature of the surface [552], it is useful to measure temperature both in the core and near the surface of the specimen, in order to assess this thermal effect.

Immediately after the stripping of specimens intended for creep tests, the two end faces of each of the creep specimens should be ground flat, in order to have plane faces perpendicular to the axis of the cylinder, with a precision of $50\,\mu$. In cored specimens, this should be done during the coring.

Immediately after the stripping and, for the creep specimens, after the grinding of the end faces, the two end sides of all the specimens must be protected against desiccation. This protection must include a metallic (e.g., copper or aluminum) foil and must ensure that there are no air pockets between the concrete and the foil. Immediately after that, the specimen must be weighed (same as in Sect. L.3.4).

All the creep specimens must be weighed just before they are loaded, as well as just after the end of the test. The specimens intended for measuring the autogenous shrinkage or basic creep must be protected immediately after the stripping or grinding, over their entire surface, and must be weighed, too (same as in Sect. L.3.4).

The first measurement of drying shrinkage is carried out within at least 3 min (better 1 min) after exposure. It should be emphasized that a part of shrinkage (the autogenous shrinkage, as well as a part of the drying shrinkage itself, which begins immediately after stripping, with a maximal rate) inevitably remains unrecorded. To determine the total shrinkage, special gauges embedded in concrete need to be used. If the age at loading of the first creep test coincides with the age at exposure to drying, a procedure in which the temperature is measured both in the core and near the surface of the specimen should be used, to make it possible to take into account the surface cooling due to evaporation [552].

L.3.2   Measurements Prior to Loading

At least three specimens should be used for determining the compressive strength and the conventional (static) elastic modulus of concrete at the age at which the creep test begins.

L.3.3   Measurements of Total Strain Under Load

The loading must be applied as quickly as possible. Preference should be given to reducing the time during which the load is raised (to approach a Heaviside step function) and to attaining quickly the required load value (a continuous recording, graphic or digital, of the applied load allows taking into account in calculations the actual loading history, in particular, its difference from the step function). The strain readings ought to be taken at intervals that are spaced uniformly in the logarithmic time scale (e.g., 0.5, 1, 2, 4 min., ..., 1, 2, 4, 8 days, ..., 1, 2, 4 years, ...), i.e., in a geometric progression of reading times.

When a hydropneumatic accumulator is used, the loading must be applied as fast as possible, so that precise measurements of almost instantaneous strain and of the initial rapid creep be obtained. The recording of strain ought to begin no later than one second after load application and proceed at constant intervals in the logarithmic time scale (e.g., $t_i = 2^{i-18}$ day, $i = 1, 2, 3, \ldots$). For at least 1 h after load application, the pressure in the accumulator, or in the cylinder, must be checked and adjusted (as the nitrogen cooled by adiabatic expansion returns to ambient temperature).

L.3.4 Measurements of Water Loss

As mentioned in Sect. 3.8.2, simultaneous measurements of water loss by weighing may be useful for evaluating the tests of shrinkage and creep at drying. At the end of such tests, the total evaporable water content should be determined by measuring the water loss at 110 °C in an oven or on crushed specimens.

L.3.5 Recommended Test Parameters

The purpose of the recommended values that follow is to make the measurements from different laboratories easier to compare.

- Cylinder diameters: $d = 7.5$, 15 and, if possible, 30 cm.
- Cylinder length: $L = 4d$.
- Ages at exposure to drying: $t_0 = 1, 7, 28$ and, if possible, 150 days (t_0 must not be greater than the age at loading of the corresponding creep test).
- Ages at loading: $t' = 1, 28, 150$ days, 2 years.
- Compressive stress: $\sigma = k \tilde{f}_c(t')$, where $\tilde{f}_c(t')$ is the strength at the age t' at loading, and $k = 0.3$ as the reference case, and if possible also 0.5 and 0.7.
- Test duration: As long as possible.

For the purpose of shrinkage extrapolation, it helps to measure also shrinkage on square prisms of side about 2 cm cut by a saw; see Bažant and Donmez [124].

Remark: Shrinkage Extrapolation via Diffusion Size Effect and on Role of Autogenous Shrinkage (added in proof).

In view of the aforementioned limitations of the extrapolation based on water loss, an investigation of an alternative method relying on the diffusion size effect in shrinkage has recently been completed [124]. In a 5-times thinner companion specimen, the drying part of shrinkage is accelerated about 125-times, and so the concave part of the drying shrinkage curve on approach to the alleged bound should become evident. Fitting the 3-month shrinkage data for both specimens by the same diffusion-based model, one can identify their shrinkage halftimes, and thus extrapolate to infinity. However, even though this method is no worse than the weight loss method and clearly better than the traditional extrapolation "by eye," there is again a significant underestimation of the terminal shrinkage curve, similar as that for the weight loss method. The inevitable conclusion is that a successful extrapolation of short-time shrinkage tests will require considering: (1) the autogenous shrinkage, which proceeds in the cores of thick specimens for much longer than in the thin ones, and (2) the compressive volumetric creep of the solid skeleton of cement paste, loaded by the stress changes in the pore water (both adsorbed and capillary). Simultaneous tests of autogenous shrinkage will, of course, be required for both methods.

L.3.6   Reporting of Results

The results should be presented in numerical tables indicating, for each specimen, the age at each reading counted from the end of the casting of concrete into the mold, and the total strain measured (shrinkage plus elastic deformation plus creep). In the case of drying, water-loss data should also be tabulated. If, additionally, the $J(t, t')$ values are calculated and reported, the shrinkage strains that need to be subtracted must be obtained by linear interpolation between the readings in the logarithmic time scale.

The test report must include: the specimen preparation and geometry, mix composition, all environmental conditions, type of sealing, age at loading, applied stress, stress history during load application, compressive strength at the time of loading, standard 28-day compressive strength; strain measurement method, the position, and length of the measurement bases; measured strains in tabular form, weight loss data, and preferably also Young's modulus and Poisson's ratio.

L.4   Ring Test of Restrained Shrinkage and Cracking and Its Limitations

To check the prediction of shrinkage cracking, a useful tool is the *ring test* (e.g., Grzybowski and Shah [440], Shah et al. [772]). In this test, an annular layer of concrete is cast around a stiff steel ring, and the subsequent development of radial cracks in drying environment is observed (Fig. L.2). Up to the start of cracking, the circumferential strain in concrete is zero, while in the radial and axial directions, the stress is nearly zero. After radial cracks develop, the circumferential strain in concrete between the cracks remains nonzero and the stress and strain fields become complicated. Concrete deforms elastically and creeps.

Fig. L.2 Ring test; after Grzybowski and Shah [440], reproduced with permission from ACI

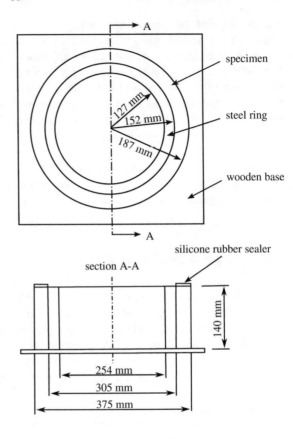

section A-A

The test is a good indicator of shrinkage cracking potential in restrained concrete layers of the same dimensions and has been used to demonstrate the suppression of wide cracks in fiber-reinforced concrete. But it is very difficult to extrapolate this test to layers of different thicknesses or different restraints (e.g., biaxial).

The ring test can also be used as a check for a comprehensive computer model for creep and shrinkage, which solves the diffusion equation, the evolution of the stress field before and after cracking, the formation of cohesive fractures, and damage localization.

References

1. AASHTO. (2002). *Standard specifications for highway bridges* (17th ed.). Washington, DC: American Association of State Highway and Transportation Officials.
2. AASHTO. (2004). *AASHTO LRFD bridge design specification.* Washington, DC: American Association of State Highway and Transportation Officials.
3. Engineers, A. B. A. M. (1993). *Basis for design.* Koror-Babeldaob bridge repairs: Technical report, ABAM Engineers Inc.
4. Abbasi, T., & Abbasi, S. A. (2007). The boiling liquid expanding vapor explosion (BLEVE): Mechanism, consequence assessment, management. *Journal of Hazardous Materials, 141,* 489–519.
5. Abdel-Rahman, A. K., & Ahmed, G. N. (1996). Computational heat and mass transport in concrete walls exposed to fire. *Numerical Heat Transfer, 29,* 373–395.
6. Abdel-Samad, S. R., Wright, R. N., & Robinson, A.-R. (1968). Analysis of box girders with diaphragms. *Journal of the Structural Division, ASCE, 94,* 2231–2255.
7. Abraham, O., & Dérobert, X. (2003). Non-destructive testing of fired tunnel walls: The Mont-Blanc tunnel case study. *NDT&E International, 36,* 411–418.
8. Abrams, M. S., & Monfore, G. E. (1965). Application of a small probe-type relative humidity gauge to research on fire resistance of concrete. *Journal of the Portland Cement Association Research and Development Laboratories, 7,* 2–12.
9. Abrams, M. S., & Orals, D. L. (1965). Concrete drying methods and their effect on fire resistance. *Moisture of materials in relation to fire, STP* (Vol. 385, pp. 52–73). Philadelphia: American Society for Testing Materials (PCA Bulletin 181).
10. Achanta, S., Cushman, J., & Okos, M. (1994). On multicomponent, multiphase thermomechanics with interfaces. *International Journal of Engineering Science, 32*(11), 1717–1738. http://80.apps.webofknowledge.com.dialog.cvut.cz/full_record.do?product=WOS&search_mode=GeneralSearch&qid=5&SID=Q2pWw4ZPkVQTctwemBU&page=1&doc=3.
11. ACI Committee 209. (1971). Prediction of creep, shrinkage and temperature effects in concrete structures. *Designing for the effects of creep, shrinkage and temperature* (pp. 51–93). Farmington Hills, Michigan: American Concrete Institute (ACI SP-27).
12. ACI Committee 209. (1982). Prediction of creep, shrinkage and temperature effects in concrete structures. *Designing for creep and shrinkage in concrete structures* (pp. 193–300). Farmington Hills, Michigan: American Concrete Institute (ACI SP-76).
13. Committee, A. C. I., & 209., (1992). *Prediction of creep, shrinkage and temperature effects in concrete structures* (p. 47). Michigan: Farmington Hills.

© Springer Science+Business Media B.V. 2018
Z.P. Bažant and M. Jirásek, *Creep and Hygrothermal Effects
in Concrete Structures*, Solid Mechanics and Its Applications 225,
https://doi.org/10.1007/978-94-024-1138-6

14. ACI Committee 209. (2008). Guide for modeling and calculating shrinkage and creep in hardened concrete, Technical report 209.2R-08, American Concrete Institute, Farmington Hills, Michigan.

15. Committee, A. C. I., & 212., (1986). *Admixtures for concrete, Technical report ACI 212–1R-81, ACI 212–2R-81*. Farmington Hills, Michigan: American Concrete Institute.

16. Committee, A. C. I., & 318., (2005). *Building code requirements for structural concrete, Technical report ACI 318–05*. Farmington Hills, Michigan: American Concrete Institute.

17. Committee, A. C. I., & 318., (2011). *Building code requirements for structural concrete and commentary, Technical report ACI 318–11*. Farmington Hills, Michigan: American Concrete Institute.

18. Committee, A. C. I., & 318., (2014). *Building code requirements for structural concrete and commentary, Technical report ACI 318–14*. Farmington Hills, Michigan: American Concrete Institute.

19. Acker, P. (1993). *Creep tests of concrete: Why and how? Creep and shrinkage of concrete*. London: Chapman & Hall.

20. Acker, P., Bažant, Z. P., Chern, J. C., Huet, C., & Wittmann, F. H. (1998). RILEM recommendation on "Measurement of time-dependent strains of concrete". *Materials and Structures, 31*, 507–512 (prepared by Subcommittee 4 of RILEM Committee TC107-CSP).

21. Acker, P., & Ulm, F.-J. (2001). Creep and shrinkage of concrete: Physical origins and practical measurements. *Nuclear Engineering and Design, 203*(2–3), 143–158. http://www.sciencedirect.com/science/article/pii/S0029549300003046.

22. Ahlgren, L. (1972). *Moisture fixation in porous building materials, Technical report 36*. Lund Institute of Technology, Lund, Sweden: Division of Building Technology.

23. Ahs, M. S. (2008). Sorption scanning curves for hardened cementitious materials. *Construction and Building Materials, 22*, 2228–2234.

24. Al-Alusi, H. R., Bertero, V. V., & Polivka, M. (1972). *Effects of humidity on the time-dependent behavior of concrete under sustained loading, Technical report UC SESM 72–2*. Structural Engineering Laboratory: University of California at Berkeley, Berkeley, California.

25. Al-Manaseer, A., & Lakshmikanthan, S. (1999). Comparison between current and future design code models for creep and shrinkage. *Revue Française de Génie Civil, 3*, 39–40.

26. Al-Manaseer, A., & Lam, J.-P. (2005). Statistical evaluation of creep and shrinkage models. *ACI Materials Journal, 102*, 170–176.

27. Al-Omaishi, N., Tadros, M. K., & Seguirant, S. J. (2009). Elasticity modulus, shrinkage, and creep of high-strength concrete as adopted by AASHTO. *PCI Journal, 54*, 44–63.

28. Aleksandrovskii, S. V., & Kolesnikov, N. S. (1971). Nonlinear creep of concrete at stepwise varying stress (in Russian). *Beton i Zhelezobeton, 17*, 24–27.

29. Aleksandrovskii, S. V., & Popkova, O. M. (1970). Nonlinear creep strains of concrete at complex load histories (in Russian). *Beton i Zhelezobeton, 16*, 27–32.

30. Alexander, M. G. (1996). Aggregates and the deformation properties of concrete. *ACI Materials Journal, 93*, 569–577.

31. Alexandrovskii, S. V. (1959). On thermal and hygrometric properties of concrete related to heat and moisture exchange (in Russian), Trudy Inst. 4, Akad. Stroit. i Architektury USSR, Nauchno-Issled. Inst. Betona i Zhelezobetona (NIIZhB), Moscow (pp. 184–214).

32. Alfrey, T., & Doty, P. (1945). The methods of specifying the properties of viscoelastic materials. *Journal of Applied Physics, 16*, 700.

33. Alnaggar, M., Cusatis, G., & Di-Luzio, G. (2013). Lattice discrete particle modeling (LDPM) of alkali silica reaction (ASR) deterioration of concrete structures. *Cement and Concrete Composites, 41*, 45–59.

34. Altoubat, S. A., & Lange, D. A. (2002). The Pickett effect at early age and experiment separating its mechanism in tension. *Materials and Structures, 35*, 211–218.

35. Anderberg, Y., & Thelandersson, S. (1976). Stress and deformation characteristic of concrete at high temperatures, 2. Experimental investigation and material behavior model. *Bulletin* (Vol. 54). Lund Institute of Technology, Lund, Sweden.

36. Ang, A. H.-S., & Tang, W. H. (1975). *Probability concepts in engineering planning and design: Basic principles* (Vol. 1). New York: Wiley.
37. Ang, A. H.-S., & Tang, W. H. (1984). *Probability concepts in engineering planning and design: Decision, risk and reliability* (Vol. 2). New York: Wiley.
38. Antoine, C. (1888). Tensions des vapeurs; nouvelle relation entre les tensions et les températures. *Comptes Rendus des Séances de l'Académie des Sciences, 107,* 681–684, 778–780, 836–837.
39. Arutyunian, N. K. (1952). *Some problems in the theory of creep (in Russian), Techteorizdat* (p. 1966). Moscow. Engl. transl: Pergamon Press.
40. Asaro, R. J., & Rice, J. R. (1977). Strain localization in ductile single crystals. *Journal of the Mechanics and Physics of Solids, 33,* 309–338.
41. Aschl, H., & Stöckl, S. (1981). Wärmedehnung, E-Modul, Schwinden, Kriechen und Restfestigkeit von Reaktorbeton unter einachsiger Belastung und erhöhten Temperaturen, Heft 324, Deutscher Ausschuss für Stahlbeton.
42. Ashby, M. F., & Hallam, S. D. (1986). The failure of brittle solids containing small cracks under compressive stress states. *Acta Metallurgica et Materialia, 34,* 497–510.
43. ASTM C469/C469M-14. (2014). *Standard test method for static modulus of elasticity and Poisson's ratio of concrete in compression.* West Conshohocken, PA: ASTM International.
44. ASTM C512-87. (1994). *Standard test method for creep of concrete in compression.* West Conshohocken, PA: ASTM International.
45. Ayano, T., & Sakata, K. (1997). Concrete shrinkage strain under the actual atmosphere. *Proceedings of Japan Concrete Institute, 19,* 709–714.
46. Aziz, M. J., Sabin, P. C., & Lu, G. Q. (1991). The activation strain tensor: Nonhydrostatic stress effects on crystal growth kinetics. *Physical Reviews B, 41,* 9812–9816.
47. Balbuena, P. B., Berry, D., & Gubbins, K. E. (1993). Solvation pressures for simple fluids in micropores. *Journal of Physical Chemistry, 97,* 937–943.
48. Bamonte, P., & Gambarova, P. G. (2012). A study on the mechanical properties of self-compacting concrete at high temperature and after cooling. *Materials and Structures, 45,* 1375–1387.
49. Barenblatt, G. I. (1959). The formation of equilibrium cracks during brittle fracture. General ideas and hypothesis, axially symmetric cracks. *Prikladnaja Matematika i Mechanika, 23,* 434–444.
50. Barenblatt, G. I. (1962). The mathematical theory of equilibrium of cracks in brittle fracture. *Advances in Applied Mechanics, 7,* 55–129.
51. Barenblatt, G. I. (1979). *Similarity, self-similarity and intermediate asymptotics.* New York: Consultants Bureau.
52. Barenblatt, G. I. (1996). *Scaling, self-similarity and intermediate asymptotics.* Cambridge: Cambridge University Press.
53. Barenblatt, G. I. (2003). *Scaling.* Cambridge: Cambridge University Press.
54. Baroghel-Bouny, V. (1994). *Characterization of cement pastes and concretes–methods, analysis, interpretations, in French.* Paris: LCPC.
55. Baroghel-Bouny, V. (2007). Water vapour sorption experiments on hardened cementitious materials. Part I. : Essential tool for analysis of hygral behaviour and its relation to pore structure. *Cement and Concrete Research, 37,* 414–437.
56. Baroghel-Bouny, V. (2007). Water vapour sorption experiments on hardened cementitious materials. Part II.: Essential tool for assessment of transport properties and for durability prediction. *Cement and Concrete Research, 37,* 438–454.
57. Baroghel-Bouny, V., Mainguy, M., Lassabatere, T., & Coussy, O. (1999). Characterization and identification of equilibrium and transfer moisture properties for ordinary and high-performance cementitious materials. *Cement and Concrete Research, 29,* 1225–1238.
58. Bary, B. (1996). Etude du couplage hydraulique-mécanique dans le béton endommagé, Ph.D. thesis, Laboratoire de Mécanique et Technologie, E.N.S. de Cachan, C.N.R.S., Université Paris 6.

59. Baston, G., Ball, C., Bailey, L., Lenders, E., & Hooks, J. (1972). Flexural fatigue strength of steel fibre reinforced concrete beams. *ACI Journal*, *69*, 673–677.
60. Batdorf, S. B., & Budianski, B. (1949). *A mathematical theory of plasticity based on the concept of slip, Technical note 1871*. Washington, D.C.: National Advisory Committee for Aeronautics.
61. Bažant, Z. P. (1961). Effect of creep and shrinkage in statically indeterminate structures with concrete of nonuniform age (in Czech). *Inženýrské Stavby*, *9*, 462–532.
62. Bažant, Z. P. (1962). Theory of creep and shrinkage of concrete in nonhomogeneous structures and cross sections (in Czech with English summary). *Stavebnícky Časopis* (SAV, Bratislava), *10*, 552–576.
63. Bažant, Z. P. (1964). Approximate methods of analysis of creep and shrinkage of complex nonhomogeneous structures and use of computers (in Czech with English summary). *Stavebnícky Časopis* (SAV, Bratislava), *12*, 414–431.
64. Bažant, Z. P. (1964). Time-interaction of statically indeterminate structures and subsoil (in Czech with English summary). *Stavebnícky Časopis* (SAV, Bratislava), *12*, 542–558.
65. Bažant, Z. P. (1964). Die Berechnung des Kriechens und Schwindens nicht-homogener Betonkonstruktionen. In *Proceedings of the 7th Congress, International Association for Bridge and Structural Engineers, Rio de Janeiro* (pp. 887–897).
66. Bažant, Z. P. (1966). *Creep of concrete in structural analysis (in Czech)*. Prague: State Publishers of Technical Literature (SNTL).
67. Bažant, Z. P. (1966). Phenomenological theories for creep of concrete based on rheological models. *Acta Technica ČSAV*, *11*, 82–109.
68. Bažant, Z. P. (1967). Linear creep problems solved by a succession of generalized thermoelasticity problems. *Acta Technica ČSAV*, *12*, 581–594.
69. Bažant, Z. P. (1968). Langzeitige Durchbiegungen von Spannbetonbrücken infolge des Schwingkriechens unter Verkehrslasten. *Beton und Stahlbetonbau*, *63*, 282–285.
70. Bažant, Z. P. (1968). On causes of excessive long-time deflections of prestressed concrete bridges. Creep under repeated live load (in Czech). *Inženýrské Stavby*, *16*, 317–320.
71. Bažant, Z. P. (1970). Constitutive equation for concrete creep and shrinkage based on thermodynamics of multi-phase systems. *Materials and Structures*, *3*, 3–36.
72. Bažant, Z. P. (1970). Delayed thermal dilatations of cement paste and concrete due to mass transport. *Nuclear Engineering and Design*, *24*, 308–318.
73. Bažant, Z. P. (1970). Numerical analysis of creep of an indeterminate composite beam. *Journal of Applied Mechanics, ASME*, *37*, 1161–1164.
74. Bažant, Z. P. (1971). Numerically stable algorithm with increasing time steps for integral-type ageing creep. In *Proceedings of the 1st International Conference on Structural Mechanics in Reactor Technology* (Vol. 3, p. H2/3). Berlin.
75. Bažant, Z. P. (1972). Numerical determination of long-range stress history from strain history in concrete. *Materials and Structures*, *5*, 135–141.
76. Bažant, Z. P. (1972). Prediction of concrete creep effects using age-adjusted effective modulus method. *ACI Journal*, *69*, 212–217.
77. Bažant, Z. P. (1972). Thermodynamics of hindered adsorption with application to cement paste and concrete. *Cement and Concrete Research*, *2*, 1–16.
78. Bažant, Z. P. (1972). Thermodynamics of interacting continua with surfaces and creep analysis of concrete structures. *Nuclear Engineering and Design*, *20*, 477–505.
79. Bažant, Z. P. (1975). Pore pressure, uplift and failure analysis of concrete dams. In D. J. Naylor, K. G. Stagg, & O. C. Zienkiewicz (Eds.), *Criteria and assumptions for numerical analysis of dams* (pp. 781–808). Swansea: Department of Civil Engineering, University of Wales.
80. Bažant, Z. P. (1975). Theory of creep and shrinkage in concrete structures: A précis of recent developments. In S. Nemat-Nasser (Ed.), *Mechanics today* (Vol. 2, pp. 1–93). Oxford: Pergamon Press.
81. Bažant, Z. P. (1976). Instability, ductility, and size effect in strain-softening solids. *Journal of the Engineering Mechanics Division, ASCE*, *102*, 331–344.

82. Bažant, Z. P. (1977). Viscoelasticity of porous solidifying material–concrete. *Journal of the Engineering Mechanics Division, ASCE, 103*, 1049–1067.

83. Bažant, Z. P. (1979). Physical model for steel corrosion in concrete sea structures–theory. *Journal of the Engineering Mechanics Division, ASCE, 105*, 1137–1153.

84. Bažant, Z. P. (1979). Thermodynamics of solidifying or melting viscoelastic material. *Journal of the Engineering Mechanics Division, ASCE, 105*, 933–952.

85. Bažant, Z. P. (1982). Crack band model for fracture of geomaterials. In Z. Eisenstein (Ed.), *Proceedings of the 4th International Conference on Numerical Methods in Geomechanics* (Vol. 3, pp. 1137–1152). Edmonton: University of Alberta.

86. Bažant, Z. P. (1982). Mathematical models for creep and shrinkage of concrete. In Z. P. Bažant & F. H. Wittmann (Eds.), *Creep and shrinkage in concrete structures* (pp. 163–256). New York: Wiley.

87. Bažant, Z. P. (1982). Mathematical models of nonlinear behavior and fracture of concrete. In L. E. Schwer (Ed.), *Nonlinear numerical analysis of reinforced concrete* (pp. 1–25). New York: American Society of Mechanical Engineers.

88. Bažant, Z. P. (1983). Mathematical model for creep and thermal shrinkage of concrete at high temperature. *Nuclear Engineering and Design, 76*, 183–191.

89. Bažant, Z. P. (1984). Microplane model for strain controlled inelastic behavior. In C. S. Desai & R. H. Gallagher (Eds.), *Mechanics of engineering materials* (pp. 45–59). London: Wiley.

90. Bažant, Z. P. (1984). Size effect in blunt fracture: Concrete, rock, metal. *Journal of Engineering Mechanics, ASCE, 110*, 518–535.

91. Bažant, Z. P. (1985). Mechanics of fracture and progressive cracking in concrete structures. In G. C. Sih & A. DiTommaso (Eds.), *Fracture mechanics of concrete: Structural application and numerical calculation* (pp. 1–94). Dordrecht and Boston: Martinus Nijhoff.

92. Bažant, Z. P. (1986). Response of aging linear systems to ergodic random input. *Journal of Engineering Mechanics, ASCE, 112*, 322–342.

93. Bažant, Z. P. (1987). Matrix force-displacement relations in aging viscoelasticity. *Journal of Engineering Mechanics, ASCE, 113*, 1235–1243.

94. Bažant, Z. P. (1988). Material models for structural creep analysis. In Z. P. Bažant (Ed.), *Mathematical modeling of creep and shrinkage of concrete* (pp. 99–215). Chichester and New York: Wiley (RILEM Committee TC-69 2).

95. Bažant, Z. P. (1993). Current status and advances in the theory of creep and interaction with fracture. In Z. P. Bažant & I. Carol (Eds.), *Creep and shrinkage of concrete, Proceedings of ConCreep-5, Barcelona* (pp. 291–307). London: E & FN Spon.

96. Bažant, Z. P. (1995). Creep and damage in concrete. In J. Skalny & S. Mindess (Eds.), *Materials science of concrete IV* (pp. 355–389). Westerville, OH: The American Ceramic Society.

97. Bažant, Z. P. (1997). Analysis of pore pressure, thermal stress and fracture in rapidly heated concrete. In L. Phan (Ed.), *Proceedings of the International Workshop on Fire Performance of High-Strength Concrete, NIST Special Publication* (Vol. 919, pp. 155–164), National Institute of Standards and Technology, Gaithersburg, Maryland.

98. Bažant, Z. P. (2000). Criteria for rational prediction of creep and shrinkage of concrete. In A. Al-Manaseer (Ed.), *Adam Neville symposium: Creep and shrinkage—structural design effects* (pp. 237–260). Farmington Hills, Michigan: American Concrete Institute (ACI SP-194).

99. Bažant, Z. P. (2004). Discussion of "Shear database for reinforced concrete members without shear reinforcement," by K.-H. Reineck, D.A. Kuchma, K.S. Kim and S. Marx. *ACI Structural Journal, 101*, 139–140.

100. Bažant, Z. P. (2005). *Scaling of structural strength* (2nd ed.). Amsterdam: Elsevier.

101. Bažant, Z. P., Adley, M. D., Carol, I., Jirásek, M., Akers, S. A., Rohani, B., et al. (2000). Large-strain generalization of microplane model for concrete and applications. *Journal of Engineering Mechanics, ASCE, 126*, 971–980.

102. Bažant, Z. P., Asghari, A. A., & Schmidt, J. (1976). Experimental study of creep of hardened Portland cement paste at variable water content. *Materials and Structures, 9*, 279–290.

103. Bažant, Z. P., Bai, S.-P., & Gettu, R. (1993). Fracture of rock: Effect of loading rate. *Engineering Fracture Mechanics, 45*, 393–398.

104. Bažant, Z. P., & Baweja, S. (1995). Creep and shrinkage prediction model for analysis and design of concrete structures – model B3. *Materials and Structures, 28*, 357–365. RILEM recommendation, in collaboration with RILEM Committee TC 107-GCS, with Errata, Vol. 29 (March 1996), p. 126.

105. Bažant, Z. P., & Baweja, S. (1995). Justification and refinements of model B3 for concrete creep and shrinkage. 1. *Statistics and sensitivity. Materials and Structures, 28*, 415–430.

106. Bažant, Z. P., & Baweja, S. (1996). Short form of creep and shrinkage prediction model B3 for structures of medium sensitivity. *Materials and Structures, 29*, 587–593 (Addendum to RILEM recommendation TC 107-GCS).

107. Bažant, Z. P., & Baweja, S. (2000). Creep and shrinkage prediction model for analysis and design of concrete structures: Model B3. In A. Al-Manaseer (Ed.), *Adam Neville symposium: Creep and shrinkage—structural design effects* (pp. 1–83). Farmington Hills, Michigan: American Concrete Institute (ACI SP-194).

108. Bažant, Z. P., & Baweja, S. (2000). Creep and shrinkage prediction model for analysis and design of concrete structures: Model B3–short form. In A. Al-Manaseer (Ed.), *Adam Neville symposium: Creep and shrinkage—structural design effects* (pp. 85–100). Farmington Hills, Michigan: American Concrete Institute (ACI SP-194).

109. Bažant, Z. P., & Bazant, M. Z. (2012). Theory of sorption hysteresis in nanoporous solids: Part I. Snap-through instabilities. *Journal of the Mechanics and Physics of Solids, 60*, 1644–1659.

110. Bažant, Z. P., & Buyukozturk, O. (1988). Creep analysis of structures. In Z. P. Bažant (Ed.), *Mathematical modeling of creep and shrinkage of concrete* (pp. 217–273). Chichester and New York: Wiley (RILEM Committee TC-69 3).

111. Bažant, Z. P., & Rahimi-Aghdam, S. (2018). Century-long durability of concrete structures: Expansiveness of hydration and chemo-mechanics of autogenous shrinkage and swelling. Proc. EURO-C 2018, held in Bad Hofgastein, publ. by ASCE, in press.

112. Bažant, Z. P., Caner, F., Adley, M. D., & Akers, S. (2000). Fracturing rate effect and creep in microplane model for dynamics. *Journal of Engineering Mechanics, ASCE, 126*, 962–970.

113. Bažant, Z. P., Caner, F. C., Carol, I., Adley, M. D., & Akers, S. (2000). Microplane model M4 for concrete: I. Formulation with work-conjugate deviatoric stress. *Journal of Engineering Mechanics, ASCE, 126*, 944–953.

114. Bažant, Z. P., Carreira, D., & Walser, A. (1975). Creep and shrinkage in reactor containment shells. *Journal of the Structural Division, ASCE, 101*, 2117–2131.

115. Bažant, Z. P., & Cedolin, L. (1991). *Stability of structures*. New York and Oxford: Oxford University Press.

116. Bažant, Z. P., & Chern, J.-C. (1984). Bayesian statistical prediction of concrete creep and shrinkage. *ACI Journal, 81*, 319–330.

117. Bažant, Z. P., & Chern, J. C. (1985). Concrete creep at variable humidity: Constitutive law and mechanism. *Materials and Structures, 18*, 1–20.

118. Bažant, Z. P., & Chern, J. C. (1985). Log-double power law for concrete creep. *Journal of the American Concrete Institute, 82*, 665–675.

119. Bažant, Z. P., & Chern, J. C. (1985). Strain-softening with creep and exponential algorithm. *Journal of Engineering Mechanics, ASCE, 111*, 391–415.

120. Bažant, Z. P., & Chern, J. C. (1987). Stress-induced thermal and shrinkage strains in concrete. *Journal of Engineering Mechanics, ASCE, 113*, 1493–1511.

121. Bažant, Z. P., Chern, J. C., Abrams, M. S., & Gillen, M. P. (1992). *Normal and refractory concretes for LMFBR applications, Technical report EPRI-NP-2437*. Palo Alto, California: Electric Power Research Institute.

122. Bažant, Z. P., Şener, S., & Kim, J.-K. (1987). Effect of cracking on drying permeability and diffusivity of concrete. *ACI Materials Journal, 84*, 351–357.

123. Bažant, Z. P., Cusatis, G., & Cedolin, L. (2004). Temperature effect on concrete creep modeled by microprestress-solidification theory. *Journal of Engineering Mechanics, ASCE, 130*, 691–699.

124. Bažant, Z. P., & Donmez, A. (2016). Extrapolation of short-time drying shrinkage tests based on measured diffusion size effect: Concept and reality. *Materials and Structures, 49*, 411–420.

125. Bažant, Z. P., Donmez, A., Masoero, E., & Rahimi Aghdam, S. (2015). Interaction of concrete creep, shrinkage and swelling with water, hydration and damage: Nano-macro-chemo. *ConCreep-10: Mechanics and physics of creep, shrinkage, and durability of concrete and concrete structures.* Vienna.

126. Bažant, Z. P. (Ed.). (1988). *Mathematical modeling of creep and shrinkage of concrete.* Chichester and New York: Wiley.

127. Bažant, Z. P., & Gambarova, P. (1984). Crack shear in concrete: Crack band microplane model. *Journal of Structural Engineering, ASCE, 110*, 2015–2035.

128. Bažant, Z. P., & Gettu, R. (1992). Rate effects and load relaxation: Static fracture of concrete. *ACI Materials Journal, 89*, 456–468.

129. Bažant, Z. P., Gu, W. H., & Faber, K. T. (1995). Softening reversal and other effects of a change in loading rate on fracture of concrete. *ACI Materials Journal, 92*, 3–9.

130. Bažant, Z. P., Hauggaard, A. B., & Baweja, S. (1996). Microprestress solidification theory for aging and drying creep of concrete. In A. Gerdes (Ed.), *Advances in building and materials science* (pp. 111–130). Freiburg, Germany: Aedificatio Publishers.

131. Bažant, Z. P., Hauggaard, A. P., & Baweja, S. (1997). Microprestress solidification theory for concrete creep. II: Algorithm and verification. *Journal of Engineering Mechanics, ASCE, 123*, 1195–1201.

132. Bažant, Z. P., Hauggaard, A. P., Baweja, S., & Ulm, F. J. (1997). Microprestress solidification theory for concrete creep. I: Aging and drying effects. *Journal of Engineering Mechanics, ASCE, 123*, 1188–1194.

133. Bažant, Z. P., Havlásek, P., & Jirásek, M. (2014). Microprestress-solidification theory: Modeling of size effect on drying creep. *Computational modelling of concrete structures – proceedings of EURO-C 2014.* Leiden: CRC Press/Balkema.

134. Bažant, Z. P., & Hubler, M. H. (2014). Theory of cyclic creep of concrete based on Paris law for fatigue growth of subcritical microcracks. *Journal of the Mechanics and Physics of Solids, 63*, 187–200.

135. Bažant, Z. P., Hubler, M. H., & Jirásek, M. (2013). Improved estimation of long-term relaxation function from compliance function of aging concrete. *Journal of Engineering Mechanics, ASCE, 139*, 146–152.

136. Bažant, Z. P., Hubler, M. H., & Wendner, R. (2015). Model B4 for creep, drying shrinkage and autogenous shrinkage of normal and high-strength concretes with multi-decade applicability. *Materials and Structures, 48*, 753–770 (RILEM Technical Committee TC-242-MDC).

137. Bažant, Z. P., Hubler, M. H., & Yu, Q. (2011). Excessive creep deflections: An awakening. *ACI Concrete International, 33*, 44–46.

138. Bažant, Z. P., Hubler, M., & Yu, Q. (2011). Pervasiveness of excessive deflections of segmental bridges: Wake-up call for creep. *ACI Structural Journal, 108*, 766–774.

139. Bažant, Z. P., & Huet, C. (1999). Thermodynamic functions for ageing viscoelasticity: Integral form without internal variables. *International Journal of Solids and Structures, 36*, 3993–4016.

140. Bažant, Z. P., & Jirásek, M. (1993). R-curve modeling of rate and size effects in quasibrittle fracture. *International Journal of Fracture, 62*, 355–373.

141. Bažant, Z. P., & Jirásek, M. (2002). Nonlocal integral formulations of plasticity and damage: Survey of progress. *Journal of Engineering Mechanics, ASCE, 128*, 1119–1149.

142. Bažant, Z. P., & Kaplan, M. F. (1996). *Concrete at high temperatures: Material properties and mathematical models.* London: Longman (Addison-Wesley).

143. Bažant, Z. P., & Kazemi, M. T. (1990). Determination of fracture energy, process zone length and brittleness number from size effect, with application to rock and concrete. *International Journal of Fracture, 44*, 111–131.

144. Bažant, Z. P., Kim, J.-J. H., & Brocca, M. (1999). Finite strain tube-squash test of concrete at high pressures and shear angles up to 70 degrees. *ACI Materials Journal, 96*, 580–592.

145. Bažant, Z. P., & Kim, J.-K. (1989). Segmental box girder: Deflection probability and Bayesian updating. *Journal of Structural Engineering, ASCE, 115*, 2528–2547.

146. Bažant, Z. P., & Kim, J.-K. (1991). Consequences of diffusion theory for shrinkage of concrete. *Materials and Structures, 24*, 323–326.

147. Bažant, Z. P., & Kim, J.-K. (1991). Segmental box girder: Effect of spatial random variability of material on deflections. *Journal of Structural Engineering, ASCE, 117*, 2542–2547.

148. Bažant, Z. P., & Kim, J.-K. (1992). Improved prediction model for time dependent deformations of concrete: III. Creep at drying, IV. Temperature effects. *Materials and Structures, 25*(21–28), 84–94.

149. Bažant, Z. P., & Kim, J.-K. (1992). Improved prediction model for time-dependent deformations of concrete: V. Cyclic load and cyclic humidity. *Materials and Structures, 25*, 163–169.

150. Bažant, Z. P., Kim, J.-K., & Jeon, S.-E. (2003). Cohesive fracturing and stresses caused by hydration heat in massive concrete wall. *Journal of Engineering Mechanics, ASCE, 129*, 21–30.

151. Bažant, Z. P., Kim, J.-K., & Panula, L. (1991). Improved prediction model for time dependent deformations of concrete: I. Shrinkage, II. Basic creep. *Materials and Structures, 24*(327–345), 409–442.

152. Bažant, Z. P., & Kim, S.-S. (1978). Can the creep curves for different loading ages diverge? *Cement and Concrete Research, 8*, 601–612.

153. Bažant, Z. P., & Kim, S. S. (1979). Approximate relaxation function for concrete. *Journal of the Structural Division, ASCE, 105*, 2695–2705.

154. Bažant, Z. P., Kim, S. S., & Meiri, S. (1979). Triaxial moisture-controlled creep tests of hardened cement paste at high temperature. *Materials and Structures, 12*, 447–456.

155. Bažant, Z. P., Křístek, V., & Vítek, J. L. (1992). Drying and cracking effects in box-girder bridge segment. *Journal of Structural Engineering, ASCE, 118*, 305–321.

156. Bažant, Z. P., & Le, J.-L. (2009). Nano-mechanics based modeling of lifetime distribution of quasibrittle structures. *Engineering Failure Analysis, 16*, 2521–2529.

157. Bažant, Z. P., & Le, J.-L. (2017). *Probabilistic mechanics of quasibrittle structures: Strength lifetime and size effect*. Cambridge: Cambridge University Press.

158. Bažant, Z. P., Le, J.-L., & Bazant, M. Z. (2008). Size effect on strength and lifetime distributions of quasibrittle structures implied by interatomic bond break activation. In J. Pokluda (Ed.), *Proceedings of the 17th European Conference on Fracture (ECF-17)* (pp. 78–92), Technical University of Brno, Brno, Czech Republic.

159. Bažant, Z. P., Le, J.-L., & Bazant, M. Z. (2009). Scaling of strength and lifetime probability distributions of quasibrittle structures based on atomistic fracture mechanics. *Proceedings of the National Academy of Sciences of the United States of America, 106*, 11484–11489.

160. Bažant, Z. P., & Li, G.-H. (2008). Comprehensive database on concrete creep and shrinkage. *ACI Materials Journal, 106*, 635–638.

161. Bažant, Z. P., & Li, G.-H. (2008). Unbiased statistical comparison of creep and shrinkage prediction models. *ACI Materials Journal*, 610–621.

162. Bažant, Z. P., & Li, Y.-N. (1997). Cohesive crack model with rate-dependent crack opening and viscoelasticity: I. Mathematical model and scaling. *International Journal of Fracture, 86*, 247–265.

163. Bažant, Z. P., & Liu, K.-L. (1985). Random creep and shrinkage in structures: Sampling. *Journal of Structural Engineering, ASCE, 111*, 1113–1134.

164. Bažant, Z. P., & Moschovidis, Z. (1973). Surface diffusion theory for the drying creep effect in Portland cement paste and concrete. *Journal of the American Ceramic Society, 56*, 235–241.

165. Bažant, Z. P., & Najjar, L. J. (1971). Drying of concrete as a nonlinear diffusion problem. *Cement and Concrete Research, 1*, 461–473.

166. Bažant, Z. P., & Najjar, L. J. (1972). Nonlinear water diffusion in nonsaturated concrete. *Materials and Structures, 5*, 3–20.

167. Bažant, Z. P., & Najjar, L. J. (1973). Comparison of approximate linear methods for concrete creep. *Journal of the Structural Division, ASCE, 99*, 1851–1874.

168. Bažant, Z. P., & Oh, B.-H. (1983). Crack band theory for fracture of concrete. *Materials and Structures*, *16*, 155–177.
169. Bažant, Z. P., & Oh, B.-H. (1985). Microplane model for progressive fracture of concrete and rock. *Journal of Engineering Mechanics, ASCE*, *111*, 559–582.
170. Bažant, Z. P., & Ohtsubo, H. (1977). Stability conditions for propagation of a system of cracks in a brittle solid. *Mechanics Research Communications*, *4*, 353–366.
171. Bažant, Z. P., Ohtsubo, H., & Aoh, K. (1979). Stability and post-critical growth of a system of cooling and shrinkage cracks. *International Journal of Fracture*, *15*, 443–456.
172. Bažant, Z. P., & Osman, E. (1975). On the choice of creep function for standard recommendations on practical analysis of structures. *Cement and Concrete Research*, *5*, 129–137. Discussion & reply 5, 631–641; 6 (1976) 149–157; 7 (1977) 119–130; 8 (1978) 129–130.
173. Bažant, Z. P., & Osman, E. (1976). Double power law for basic creep of concrete. *Materials and Structures*, *9*, 3–11.
174. Bažant, Z. P., Osman, E., & Thonguthai, W. (1976). Practical formulation of shrinkage and creep in concrete. *Materials and Structures*, *9*, 395–406.
175. Bažant, Z. P., & Panula, L. (1978). Practical prediction of time dependent deformations of concrete: I. Shrinkage, II. Basic creep, III. Drying creep, IV. Temperature effect on basic creep. *Materials and Structures*, *11*, 307–316, 317–328, 415–424, 425–434.
176. Bažant, Z. P., & Panula, L. (1979). Practical prediction of time dependent deformations of concrete: V. Temperature effect on drying creep, VI. Cyclic creep, nonlinearity and statistical scatter. *Materials and Structures*, *12*(169–174), 175–183.
177. Bažant, Z. P., & Panula, L. (1980). Creep and shrinkage characterization for prestressed concrete structures. *Journal of the Prestressed Concrete Institute*, *25*, 86–122.
178. Bažant, Z. P., & Planas, J. (1998). *Fracture and size effect in concrete and other quasibrittle materials*. Boca Raton: CRC Press.
179. Bažant, Z. P., & Prasannan, S. (1989). Solidification theory for concrete creep: I. Formulation. *Journal of Engineering Mechanics, ASCE*, *115*, 1691–1703.
180. Bažant, Z. P., & Prasannan, S. (1989). Solidification theory for concrete creep: II. Verification and application. *Journal of Engineering Mechanics, ASCE*, *115*, 1704–1725.
181. Bažant, Z. P., & Prat, P. (1988). Microplane model for brittle plastic materials. I: Theory, II: Verification. *Journal of Engineering Mechanics, ASCE*, *114*, 1672–1702.
182. Bažant, Z. P., & Prat, P. C. (1987). Creep of anisotropic clay: New microplane model. *Journal of Engineering Mechanics, ASCE*, *113*, 1000–1064.
183. Bažant, Z. P., & Prat, P. C. (1988). Effect of temperature and humidity on fracture energy of concrete. *ACI Materials Journal*, *85*, 262–271.
184. Bažant, Z. P., & Raftshol, W. J. (1982). Effect of cracking in drying and shrinkage specimens. *Cement and Concrete Research*, *12*, 209–226.
185. Bažant, Z. P., & Rahimi-Aghdam, S. (2017). Diffusion-controlled and creep-mitigated ASR damage via microplane model: I. Mass concrete. *Journal of Engineering Mechanics, ASCE*, *143*, 04016108-1–04016108-10.
186. Bažant, Z. P., & Steffens, A. (2000). Mathematical model for kinetics of alkali-silica reaction in concrete. *Cement and Concrete Research*, *30*, 419–428.
187. Bažant, Z. P., & Thonguthai, W. (1976). Optimization check of certain practical formulations for concrete creep. *Materials and Structures*, *9*, 91–98.
188. Bažant, Z. P., & Thonguthai, W. (1978). Pore pressure and drying of concrete at high temperature. *Journal of the Engineering Mechanics Division, ASCE*, *104*, 1058–1080.
189. Bažant, Z. P., & Thonguthai, W. (1979). Pore pressure in heated concrete walls–theoretical prediction. *Magazine of Concrete Research*, *31*, 67–76.
190. Bažant, Z. P., & Wahab, A. B. (1979). Instability and spacing of cooling or shrinkage cracks. *Journal of the Engineering Mechanics Division, ASCE*, *105*, 873–889.
191. Bažant, Z. P., & Wahab, A. B. (1980). Stability of parallel cracks in solids reinforced by bars. *International Journal of Solids and Structures*, *16*, 97–106.
192. Bažant, Z. P., & Wang, T.-S. (1984). Spectral analysis of random shrinkage stresses in concrete. *Journal of Engineering Mechanics, ASCE*, *110*, 173–186.

193. Bažant, Z. P., & Wang, T.-S. (1984). Spectral finite element analysis of random shrinkage in concrete. *Journal of Structural Engineering, ASCE, 110*, 2196–2211.

194. Bažant, Z. P., & Wang, T.-S. (1985). Practical prediction of cyclic humidity effect in creep and shrinkage of concrete. *Materials and Structures, 18*, 247–252.

195. Bažant, Z. P., Wendner, R., Boumakis, G., & Hubler, M. H. (2015). Discussion of "Statistical comparisons of creep and shrinkage prediction models using RILEM and NU-ITI databases" by A. Al-Manaseer and A. Prado (ACI Materials Journal, 111(1–6), 1–4, Jan.-Dec. 2014, MS No. M-2013-147.R1). *ACI Materials Journal, 112*, 829–831.

196. Bažant, Z. P., & Wittmann, F. H. (Eds.). (1982). *Creep and shrinkage of concrete structures*. London: Wiley.

197. Bažant, Z. P., Wittmann, F. H., Kim, J.-K., & Alou, F. (1987). Statistical extrapolation of shrinkage data–Part I: Regression. *ACI Materials Journal, 84*, 20–34.

198. Bažant, Z. P., & Wu, S. T. (1974). Creep and shrinkage law of concrete at variable humidity. *Journal of the Engineering Mechanics Division, ASCE, 100*, 1183–1209.

199. Bažant, Z. P., & Wu, S. T. (1974). Rate-type creep law of aging concrete based on Maxwell chain. *Materials and Structures, 7*, 45–60.

200. Bažant, Z. P., & Xi, Y. (1993). New test method to separate microcracking from drying creep: Curvature creep at equal bending moments and various axial forces. In Z. P. Bažant & I. Carol (Eds.), *Creep and shrinkage of concrete* (pp. 77–82). London: E & FN Spon.

201. Bažant, Z. P., & Xi, Y. (1994). Drying creep of concrete: Constitutive model and new experiments separating its mechanisms. *Materials and Structures, 27*, 3–14.

202. Bažant, Z. P., & Xi, Y. (1995). Continuous retardation spectrum for solidification theory of concrete creep. *Journal of Engineering Mechanics, ASCE, 121*, 281–288.

203. Bažant, Z. P., Xi, Y., & Baweja, S. (1993). Improved prediction model for time-dependent deformations of concrete: Part 7: Short form of BP-KX model, statistics and extrapolation of short-time data. *Materials and Structures, 26*, 567–574.

204. Bažant, Z. P., & Xiang, Y. (1997). Size effect in compression fracture: Splitting crack band propagation. *Journal of Engineering Mechanics, ASCE, 123*, 162–172.

205. Bažant, Z. P., Xiang, Y., & Prat, P. C. (1996). Microplane model for concrete. I: Stress-strain boundaries and finite strain. *Journal of Engineering Mechanics, ASCE, 122*, 245–254.

206. Bažant, Z. P., & Xu, K. (1991). Size effect in fatigue fracture of concrete. *ACI Materials Journal, 88*, 390–399.

207. Bažant, Z. P., & Yu, Q. (2013). Relaxation of prestressing steel at varying strain and temperature: Viscoplastic constitutive relation. *Journal of Engineering Mechanics, ASCE, 139*, 814–823.

208. Bažant, Z. P., Yu, Q., Hubler, M., Křístek, V., & Bittnar, Z. (2011). Wake-up call for creep, myth about size effect and black holes in safety: What to improve in *fib* model code draft. *Concrete Engineering for Excellence and Efficiency, Proceedings of the* fib *Symposium, Prague*. (pp. 731–746).

209. Bažant, Z. P., Yu, Q., & Li, G.-H. (2012). Excessive long-time deflections of prestressed box girders: I. Record-span bridge in Palau and other paradigms. *Journal of Structural Engineering, ASCE, 138*, 676–686.

210. Bažant, Z. P., Yu, Q., & Li, G.-H. (2012). Excessive long-time deflections of prestressed box girders: II. Numerical analysis and lessons learned. *Journal of Structural Engineering, ASCE, 138*, 687–696.

211. Bažant, Z. P., Yu, Q., Li, G.-H., Klein, G., & Křístek, V. (2010). Excessive deflections of record-span prestressed box girder: Lessons learned from the collapse of the Koror-Babeldaob bridge in Palau. *ACI Concrete International, 32*, 44–52.

212. Bažant, Z. P., & Zebich, S. (1983). Statistical linear regression analysis of prediction models for creep and shrinkage. *Cement and Concrete Research, 13*, 869–876.

213. Bažant, Z. P., & Zi, G. (2003). Decontamination of radionuclides from concrete by microwave heating: I. Theory, II. Computations. *Journal of the Engineering Mechanics Division, ASCE, 129*, 777–784, 785–792.

214. Bažant, Z. P., & Zi, G. (2003). Microplane constitutive model for porous isotropic rocks. *International Journal for Numerical and Analytical Methods in Geomechanics, 27*(1), 25–47. https://doi.org/10.1002/nag.261.

215. Baweja, S., Dvorak, G. J., & Bažant, Z. P. (1998). Triaxial composite model for basic creep of concrete. *Journal of Engineering Mechanics, ASCE, 124*, 959–966.

216. Bazant, M. Z., & Bažant, Z. P. (2012). Theory of sorption hysteresis in nanoporous solids: Part II. Molecular condensation. *Journal of the Mechanics and Physics of Solids, 60*, 1660–1675.

217. Bažant, Z. P., & Caner, F. C. (2005). Microplane model M5 with kinematic and static constraints for concrete fracture and anelasticity. I: Theory. *Journal of Engineering Mechanics, 131*(1), 31–40. http://ascelibrary.org/doi/10.1061/%28ASCE%290733-9399%282005% 29131%3A1%2831%29.

218. Bažant, Z. P., & Caner, F. C. (2013). Comminution of solids caused by kinetic energy of high shear strain rate, with implications for impact, shock, and shale fracturing. *Proceedings of the National Academy of Sciences of the United States of America, 110*(48), 19291–19294. http://www.pubmedcentral.nih.gov/articlerender.fcgi?artid=3845168&tool= pmcentrez&rendertype=abstract.

219. Bažant, Z. P., & Caner, F. C. (2014). Impact comminution of solids due to local kinetic energy of high shear strain rate: I. Continuum theory and turbulence analogy. *Journal of the Mechanics and Physics of Solids, 64*, 223–235. http://www.sciencedirect.com/science/article/ pii/S0022509613002421.

220. Bažant, Z. P., & Kim, J. (1986). Creep of anisotropic clay: Microplane model. *Journal of Geotechnical Engineering, 112*(4), 458–475. https://ascelibrary.org/doi/10.1061/% 28ASCE%290733-9410%281986%29112%3A4%28458%29.

221. Bear, J. (1979). *Dynamics in fluids in porous media*. New York: Dover.

222. Bear, J., & Bachmat, Y. (1990). *Introduction to modeling of transport in porous media*. Dordrecht: Kluwer.

223. Bechyně, S. (1959). *Concrete construction. I. Concrete technology (in Czech)* (pp. 122–136). Prague: SNTL.

224. Beck, J. V., & Arnold, K. J. (1977). *Parameter estimation in engineering science*. New York: Wiley.

225. Behnood, A., & Ghandehari, M. (2009). Comparison of compressive and splitting tensile strength of high-strength concrete with and without polypropylene fibers heated to high temperatures. *Fire Safety Journal, 44*(8), 1015–1022. http://www.sciencedirect.com/science/ article/pii/S0379711209000940.

226. Belie, N. D., Kratky, J., & Vlierberghe, S. V. (2010). Influence of pozzolans and slag on the microstructure of partially carbonated cement paste by means of water vapour and nitrogen sorption experiments and BET calculations. *Cement and Concrete Research, 40*, 1723–1733.

227. Ben Haha, M. (2006). Mechanical effects of alkali silica reaction in concrete studied by SEM-image analysis. Ph.D. thesis, EPFL, Lausanne, Switzerland.

228. Beneš, M., & Štefan, R. (2015). Hygro-thermo-mechanical analysis of spalling in concrete walls at high temperatures as a moving boundary problem. *International Journal of Heat and Mass Transfer, 85*, 110–134.

229. Benjamin, J. R., & Cornell, C. A. (1970). *Probability, statistics and decision for civil engineers* (pp. 522–641). New York: McGraw Hill.

230. Benjamin, J. R., Cornell, C. A., & Gabrielson, B. I. (1965). A stochastic model for creep deflection of reinforced concrete beams. In *Proceedings of the International Symposium on the Flexural Mechanics of Reinforced Concrete* (pp. 557–580). Detroit, Michigan: American Concrete Institute (ACI SP-12).

231. Bennett, E. W., & Muir, S. E. S. J. (1967). Some fatigue tests of high-strength concrete in axial compression. *Magazine of Concrete Research, 19*, 113–117.

232. Benscoter, S. U. (1954). A theory of torsion bending for multi-cell beams. *Journal of Applied Mechanics, ASME, 21*, 25–34.

233. Bentz, D. P. (2005). CEMHYD3D: A three-dimensional cement hydration and microstructure development modeling package. Version 3.0, Technical report, NIST Building and Fire Research Laboratory, Gaithersburg, Maryland.

234. Bentz, D. P. (2007). Transient plane source measurements of the thermal properties of hydrating cement pastes. *Materials and Structures*, *40*(10), 1073–1080. http://www.springerlink.com/index/10.1617/s11527-006-9206-9.

235. Bentz, D. P. (2007). Verification, validation, and variability of virtual standards. In J. J. Beaudoin, J. M. Makar, & L. Raki (Eds.), *12th International Congress on the Chemistry of Cement*.

236. Bernard, O., Ulm, F.-J., & Lemarchand, E. (2003). A multiscale micromechanics-hydration model for the early-age elastic properties of cement-based material. *Cement and Concrete Research*, *33*, 1293–1309.

237. Bernhardt, C. J. (1967). Krypning og Svinn av Betong ved Forskjellige Ytre Forhold. *Nordisk Betong*, *1*, 9–26.

238. Bernhardt, C. J. (1969). Creep and shrinkage of concrete. *Materials and Structures*, *2*, 145–148.

239. Best, C. H., Pirtz, D., & Polivka, M. (1957). A loading system for creep studies of concrete. *ASTM Bulletin* (Vol. 224, pp. 44–47). American Society of Testing Materials.

240. Biot, M. A. (1941). General theory of three-dimensional consolidation. *Journal of Applied Physics*, *12*, 155–164.

241. Biot, M. A. (1954). Theory of stress-strain relations in anisotropic viscoelasticity and relaxation phenomena. *Journal of Applied Physics*, *25*, 1385–1391.

242. Bloom, R., & Bentur, A. (1995). Free and restrained shrinkage of normal and high-strength concretes. *ACI Materials Journal*, *92*, 211–217.

243. Boltzmann, L. (1874). Zur Theorie der elastischen Nachwirkung. Sitzber. Akad. Wiss. Wiener Bericht 70. *Wiss. Abt.*, *1*, 279–306.

244. Boltzmann, L. (1878). Zur Theorie der elastischen Nachwirkung. *Annalen der Physik*, *241*, 430–432.

245. Boltzmann, L. (1884). Ableitung des Stefan'schen Gesetzes, betreffend die Abhängigkeit der Wärmestrahlung von der Temperatur aus der elektromagnetischen Lichttheorie. *Annalen der Physik und Chemie*, *22*, 291–294.

246. Bonnaud, P. A., Coasne, B., & Pellenq, R. J.-M. (2010). Molecular simulation of water confined in nanoporous silica. *Journal of Physics-Condensed Matter*, *22*, 284110.

247. Bonnell, D. G. R., & Harper, F. C. (1951). *The thermal expansion of concrete* (Vol. 7). National building studies. London: HMSO.

248. Bouchaud, J.-P., & Potters, M. (2000). *Theory of financial risks: From statistical physics to risk management*. Cambridge: Cambridge University Press.

249. Boulay, C., & Patiès, C. (1993). Mesure des déformations du béton au jeune âge. *ACI Materials Journal*, *26*, 308–314.

250. Bourdarot, E. (1991). *Application of a porodamage model to analysis of concrete dams*. EDF/CNEH: Technical report.

251. Bourdet, O., & Muttoni, A. (2006). *Evaluation of existing measurement systems for the long-term monitoring of bridge deflections*. Confédération Suisse: Technical report.

252. Branson, D. E. (1977). *Deformations of concrete structures*. New York: McGraw-Hill.

253. Branson, D. E., & Christiason, M. L. (1971). Time-dependent concrete properties related to design-strength and elastic properties, creep and shrinkage. *Designing for the effects of creep, shrinkage and temperature* (Vol. SP-27, pp. 257–277). Farmington Hills, Michigan: American Concrete Institute (reapproved in 1982 and 1992).

254. Branson, D. E., Meyers, B. L., & Kripanarayanan, K. M. (1970). Time-dependent deformation of non-composite and composite prestressed concrete structures. *Highway Research Record*, *324*, 15–43.

255. Bresler, B., & Selna, L. (1964). Analysis of time-dependent behavior of reinforced concrete structures. *Symposium on creep of concrete* (pp. 115–128). American Concrete Institute (ACI SP-9).

256. Breunese, A. J., & Fellinger, J. H. H. (2004). Spalling of concrete – and overview of ongoing research in The Netherlands. In *Proceedings of the 3rd International Workshop on Structures in Fire*, Ottawa, Canada.

257. Brinker, C. J., & Scherer, G. W. (1990). *Sol-gel science*. Boston: Academic Press.

258. BRITE Euram III BRPR-CT95 HITECO. (1999). Understanding and industrial application of high performance concrete in high temperature environment, Final report.
259. Brocca, M., & Bažant, Z. P. (2001). Microplane finite element analysis of tube-squash test of concrete with shear angles up to 70°. *International Journal for Numerical Methods in Engineering, 52*, 1165–1188.
260. Brochard, L., Vandamme, M., & Pellenq, R. J.-M. (2012). Poromechanics of microporous media. *Journal of the Mechanics and Physics of Solids, 60*, 606–622.
261. Brochard, L., Vandamme, M., Pellenq, R. J.-M., & Teddy, F.-C. (2011). Adsorption-induced deformation of microporous materials: Coal swelling induced by CO2/CH4 competitive adsorption. *Langmuir, 28*, 2659–2670.
262. Brooks, J. J. (1984). Accuracy of estimating long-term strains in concrete. *Magazine of Concrete Research, 36*, 131–145.
263. Brooks, J. J. (1989). Influence of mix proportions, plasticizers and superplasticizers on creep and drying shrinkage of concrete. *Magazine of Concrete Research, 41*, 145–153.
264. Brooks, J. J. (1992). *Preliminary state of the art report: Elasticity, creep and shrinkage of concretes containing admixtures, slag, fly-ash and silica-fume* (p. 209). ACI committee: Preliminary report.
265. Brooks, J. J. (1999). How admixtures affect shrinkage and creep. *Concrete International, 21*, 35–38.
266. Brooks, J. J. (2000). Elasticity, creep, and shrinkage of concrete containing admixtures. In A. Al-Manaseer (Ed.), *Adam Neville symposium: Creep and shrinkage—structural design effects* (Vol. ACI SP-194, pp. 283–360). Farmington Hills, Michigan: American Concrete Institute.
267. Brooks, J. J. (2005). 30-year creep and shrinkage of concrete. *Magazine of Concrete Research, 57*, 545–556.
268. Brooks, J. J., & Forsyth, P. (1986). Influence of frequency of cyclic load on creep of concrete. *Magazine of Concrete Research, 38*, 139–150.
269. Brooks, J. J., & Johari, M. A. M. (2001). Effect of metakaolin on creep and shrinkage of concrete. *Cement and Concrete Composites, 23*, 495–502.
270. Brooks, J. J., & Neville, A. M. (1977). A comparison of creep, elasticity and strength of concrete in tension and in compression. *Magazine of Concrete Research, 29*, 475–486.
271. Brooks, J. J., Wainwright, P. J., & Neville, A. M. (1979). *Time-dependent properties of concrete containing a superplasticizing admixture* (pp. 293–314). Farmington Hills: American Concrete Institute (ACI SP-62).
272. Browne, R. D., & Bamforth, P. P. (1975). The long-term creep of the Wylfa P. V. concrete for loading ages up to 12 1/2 years. In *3rd International Conference on Structural Mechanics in Reactor Technology, London, England*. H1/8.
273. Brun, L. (1969). Méthodes énergétiques dans les systèmes évolutifs linéaires. *Journal de Mécanique, 8*, 125–166.
274. Brunauer, S. (1943). *The adsorption of gases and vapors* (p. 398). Princeton: Princeton University Press.
275. Brunauer, S., Deming, L. S., Deming, W. E., & Teller, E. J. (1940). On a theory of the van der Waals adsorption of gases. *Journal of the American Chemical Society, 62*, 1723–1732.
276. Brunauer, S., Emmett, P. T., & Teller, E. (1938). Adsorption of gases in multimolecular layers. *Journal of the American Chemical Society, 60*, 309–319.
277. Brunauer, S., Skalny, J., & Bodor, E. E. (1969). Adsorption on nonporous solids. *Journal of Colloid and Interface Science, 30*, 546–552.
278. Bryant, A. H., & Vadhanavikkit, C. (1987). Creep, shrinkage-size, and age at loading effects. *ACI Materials Journal, 84*, 117–123.
279. Buckingham, E. (1907). Studies on the movement of soil moisture. *Bulletin* (Vol. 38), U.S. Department of Agriculture Bureau of Soils, Washington.
280. Buckingham, E. (1914). On physically similar systems: Illustrations of the use of dimensional equations. *Physical Review, 4*, 345–376.
281. Buckler, J. D., & Scribner, C. F. (1985). Relaxation characteristics of prestressing strand. Engineering studies UILU-ENG-85-2011, University of Illinois.

282. Buil, M. (1990). Comportement physico-chimique du système ciment-fs. *Annales de l'ITBTP*, (483 (271)), 19–29.
283. Buil, M., & Acker, P. (1985). Creep of a silica fume concrete. *Cement and Concrete Research*, *15*, 463–466.
284. Bulmer, M. G. (1967). *Principles of statistics*. New York: Dover.
285. Bulmer, M. G. (1979). *Principles of statistics*. New York: Dover Publications.
286. Burgoyne, C., & Scantlebury, R. (2006). Why did Palau bridge collapse? *The Structural Engineer*, *84*, 30–37.
287. Camus, B. L. (1946). Recherches expérimentales sur la déformation du béton et du béton armé. *Cahiers de la recherche des lab. du bât. et des travaux publ. Circ. ITB* (27, Ser. F).
288. Caner, F. C., & Bažant, Z. P. (2000). Microplane model M4 for concrete: II. Algorithm and calibration. *Journal of Engineering Mechanics, ASCE*, *126*, 954–961.
289. Caner, F. C., & Bažant, Z. P. (2011). Microplane model M6f for fiber reinforced concrete. In *Proceedings of the XI International Conference on Computational Plasticity Fundamentals and Applications, COMPLAS 2011, Barcelona, Spain* (pp. 796–807).
290. Caner, F. C., & Bažant, Z. P. (2013). Microplane model M7 for plain concrete. I: Formulation. *Journal of Engineering Mechanics, ASCE*, *139*, 1714–1723.
291. Caner, F. C., & Bažant, Z. P. (2013). Microplane model M7 for plain concrete. II: Calibration and verification. *Journal of Engineering Mechanics, ASCE*, *139*, 1724–1735.
292. Caner, F. C., & Bažant, Z. P. (2014). Impact comminution of solids due to local kinetic energy of high shear strain rate: II. Microplane model and verification. *Journal of the Mechanics and Physics of Solids*, *64*, 236–248.
293. Carlson, R. (1937). Drying shrinkage of large concrete members. *ACI Journal*, *33*, 327–336.
294. Carlson, R. W. (1957). Permeability, pore pressure and uplift in gravity dams. *Transactions ASME*, *122*, 587–613.
295. Carol, I., & Bažant, Z. P. (1992). Viscoelasticity with aging caused by solidification of nonaging constituent. *Journal of Engineering Mechanics, ASCE*, *119*, 2252–2269.
296. Carol, I., & Bažant, Z. P. (1993). Viscoelasticity with aging caused by solidification of nonaging constituent. *Journal of Engineering Mechanics, ASCE*, *119*, 2252–2269.
297. Carslaw, H. S., & Jaeger, J. C. (1959). *Conduction of heat in solids* (2nd ed.). Oxford: Clarendon Press.
298. Castigliano, A. (1873). Théorie de l'équilibre des systèmes élastiques, Engineering degree thesis, Politecnico di Torino.
299. Castillo, C., & Durrani, A. J. (1990). Effect of transient high temperature on high-strength concrete. *ACI Materials Journal*, *87*, 47–53.
300. Çinlar, E., Bažant, Z. P., & Osman, E. (1977). Stochastic process for extrapolating concrete creep. *Journal of the Engineering Mechanics Division, ASCE*, *103*, 1069–1088.
301. Cebeci, O., Al-Noury, S., & Mirza, W. (1989). Strength and drying shrinkage of masonry mortars in various temperature-humidity environments. *Cement and Concrete Research*, *19*, 53–62.
302. Cederberg, H., & David, M. (1969). Computation of creep effects in prestressed concrete pressure vessels using dynamic relaxation. *Nuclear Engineering and Design*, *9*, 439–448.
303. Celia, M. A., Bouloutas, E. T., & Zarba, R. L. (1990). A general mass-conservative numerical solution for the unsaturated flow equation. *Water Resources Research*, *26*, 1483–1496.
304. Chan, S. Y. N., Peng, G.-F., & Anson, M. (1999). Fire behaviour of high-performance concrete made with silica fume at various moisture contents. *ACI Materials Journal*, *96*, 405–409.
305. Chau, V. T., Li, C., Rahimi-Aghdam, S., & Bažant, Z. P. (2017). The enigma of large-scale permeability of gas shale: Pre-existing or frac-induced? *Journal of Applied Mechanics ASME*, *84*, 061008-1–061008-11.
306. Chapman, A. J. (1987). *Fundamentals of heat transfer*. New York: Macmillan.
307. Charles, R. J. (1957). Energy-size reduction relationships in comminution. *Transactions of the American Institute of Mining and Metallurgical Engineers*, *208*, 80–88.
308. Chen, X., & Bažant, Z. P. (2014). Microplane damage model for jointed rock masses. *International Journal for Numerical and Analytical Methods in Geomechanics*, *38*(14), 1431–1452. https://doi.org/10.1002/nag.2257 (NAG-13-0200.R1).

309. Chen, X., & Bažant, Z. P. (2014). Microplane damage model for jointed rock masses. *International Journal for Numerical and Analytical Methods in Geomechanics, 38*, 1431–1452.
310. Chiorino, M. A. (2005). A rational approach to the analysis of creep structural effects. In J. Gardner & J. Weiss (Eds.), *Shrinkage and creep of concrete, ACI SP* (Vol. 227, pp. 107–141). Farmington Hills: American Concrete Institute.
311. Chiorino, M. A., Napoli, P., Mola, F., & Koprna, M. (1980). CEB design manual on structural effects of time-dependent behavior of concrete. *CEB Bulletin d'Information* (Vol. 136). Comité Euro-International du Béton.
312. Choinska, M., Khelidj, A., Chatzigeorgiou, G., & Pijaudier-Cabot, G. (2007). Effects and interactions of temperature and stress-level related damage on permeability of concrete. *Cement and Concrete Research, 37*(1), 79–88.
313. Christensen, R. M. (2003). *Theory of viscoelasticity* (2nd ed.). New York: Dover Publications.
314. Chung, J. H., & Consolazio, G. R. (2005). Numerical modeling of transport phenomena in reinforced concrete exposed to elevated temperatures. *Cement and Concrete Research, 35*, 597–608.
315. Chung, J. H., Consolazio, G. R., & McVay, M. C. (2006). Finite element stress analysis of a reinforced high-strength concrete column in severe fires. *Computers and Structures, 84*, 1338–1352.
316. Cilosani, Z. N. (1964). On the probable mechanism of creep of concrete. *Beton i Zhelezobeton, 2*, 75–78.
317. Coasne, B., Galarneau, A., Renzo, F. D., & Pellenq, R. J.-M. (2008). Molecular simulation of adsorption and intrusion in nanopores. *Adsorption, 14*, 215–221.
318. Coasne, B., Galarneau, A., Renzo, F. D., & Pellenq, R. J.-M. (2009). Intrusion and retraction of fluids in nanopores: Effect of morphological heterogeneity. *Journal of Physical Chemistry C, 113*, 1953–1962.
319. Coasne, B., Renzo, F. D., Galarneau, A., & Pellenq, R. J.-M. (2008). Adsorption of simple fluid on silica surface and nanopore: Effect of surface chemistry and pore shape. *Langmuir, 24*, 7285–7293.
320. Cohan, L. J. (1938). Sorption hysteresis and the vapor pressure of concave surfaces. *Journal of the American Chemical Society, 60*, 430–435.
321. Coleman, B. D. (1964). Thermodynamics of material with memory. *Archive for Rational Mechanics and Analysis, 17*, 1–46.
322. Comité Euro-International du Béton. (1991). CEB-FIP model code 1990: Design code, Thomas Telford, London. Also published in 1993 by Comité Euro-International du Béton (CEB), Bulletins d'Information No. 213 and 214, Lausanne, Switzerland.
323. Conclusions of the Hubert Rüsch Workshop on Creep of Concrete. (1980). (Vol. 2 (11), p. 77). ACI Concrete International.
324. Consolazio, G. R., McVay, M. C., Rish, J. W., & I. I. I., (1998). Measurement and prediction of pore pressures in saturated cement mortar subjected to radiant heating. *ACI Materials Journal, 95*, 526–536.
325. Copeland, L. E., Kantro, D. I., & Verbeck, G. (1960). Chemistry of hydration of Portland cement. In *4th International Symposium on the Chemistry of Cement* (Vol. 43, pp. 429–465). National Bureau of Standards monograph. Washington, DC.
326. Cottrell, A. (1964). *The mechanical properties of matter*. New York: Wiley.
327. Counto, U. J. (1964). The effect of the elastic modulus of the aggregate on the elastic modulus, creep and creep recovery of concrete. *Magazine of Concrete Research, 16*, 129–138.
328. Coussy, O. (2010). *Mechanics and physics of porous solids*. Chichester: Wiley.
329. Crank, J., & Nicolson, P. (1947). A practical method for numerical evaluation of solutions of partial differential equations of the heat conduction type. *Proceedings of the Cambridge Philosophical Society, 43*, 50–67.
330. Crow, E. L., Davis, F. A., & Maxfield, M. W. (1960). *Statistics manual* (pp. 152–167). New York: Dover Publications.
331. Cusatis, G. (1998). Modellazione della viscosita del calcestruzzo in regime di umidita e temperatura variabili mediante la teoria dei microsforzi, Master's thesis, Politecnico di Milano, Milan, Italy.

332. Cusatis, G., Bažant, Z. P., & Cedolin, L. (2003). Confinement-shear lattice model for concrete damage in tension and compression: I. Theory. *Journal of Engineering Mechanics, ASCE, 129*, 1439–1448.

333. Cusatis, G., Pelessone, D., & Mencarelli, A. (2011). Lattice discrete particle model (LDPM) for failure behavior of concrete. I: Theory. *Cement and Concrete Composites, 33*, 881–890.

334. Czernin, W. (1980). *Cement chemistry and physics for civil engineers* (2nd ed.). Wiesbaden and Berlin: Bauverlag GmbH.

335. da Silva, W. R. L., & Šmilauer, V. (2015). Nomogram for maximum temperature of mass concrete. *Concrete International, 37*, 30–36.

336. Dal Pont, S., & Ehrlacher, A. (2004). Numerical and experimental analysis of chemical dehydration, heat and mass transfers in a concrete hollow cylinder submitted to high temperatures. *International Journal of Heat and Mass Transfer, 47*(1), 135–147.

337. Darcy, H. (1856). *Les Fontaines Publiques de la Ville de Dijon*. Paris: Dalmont.

338. Davie, C., Pearce, C., & Bićanić, N. (2010). A fully generalised, coupled, multi-phase, hygro-thermo-mechanical model for concrete. *Materials and Structures, 43*, 13–33.

339. Davie, C. T., Pearce, C. J., & Bićanić, N. (2006). Coupled heat and moisture transport in concrete at elevated temperatures–effects of capillary pressure and adsorbed water. *Numerical Heat Transfer, 49*, 733–763.

340. Davie, C. T., Pearce, C. J., & Bićanić, N. (2012). Aspects of permeability in modelling of concrete exposed to high temperatures. *Transport in Porous Media, 95*(3), 627–646.

341. de Groot, S. R., & Mazur, P. (Eds.). (1962). *Nonequilibrium thermodynamics*. Amsterdam etc: North-Holland.

342. de Larrard, F., & Bostvironnois, J. L. (1991). On the long term strength losses of silica fume high strength concretes. *Magazine of Concrete Research, 43*, 109–119.

343. de Larrard, F., & Roy, R. L. (1992). Relation entre formulation and quelques propriétés mécaniques des bétons a hautes performances. *Materials and Structures, 25*, 464–475.

344. de Vries, D. A., & Kruger, A. J. (1966). On the value of the diffusion coefficient of water vapour in air. *Phénomènes de transport avec changement de phase dans les milieux poreux ou colloïdaux* (pp. 561–572). CNRS.

345. DeBoer, J. H. (1968). *The dynamical character of adsorption*. New York - Oxford: Oxford University Press.

346. DeJong, M. J., & Ulm, F. J. (2007). The nanogranular behavior of C-S-H at elevated temperatures (up to 700°C). *Cement and Concrete Research, 37*(1), 1–12.

347. Deryagin, B. V. (1933). Elastic form of thin water layers. *Zeitschrift für Physik, 84*, 657–670.

348. Deryagin, B. V. (1940). On the repulsive forces between charged colloid particles and the theory of slow coagulation and stability of lyophole sols. *Transactions of the Faraday Society, 36*(203), 730.

349. Deryagin, B. V. (Ed.). (1963). *Research in surface forces*. New York: Consultants Bureau.

350. Di Luzio, G. (2007). A symmetric over-nonlocal microplane model M4 for fracture in concrete. *International Journal of Solids and Structures, 44*, 4418–4441.

351. Di Luzio, G., & Cusatis, G. (2013). Solidification-microprestress-microplane (SMM) theory for concrete at early age: Theory, validation and application. *International Journal of Solids and Structures, 50*(6), 957–975.

352. DIN. (1973). *Richtlinien für die Bemessung and Ausführung massiver Brücken, Technical report* (p. 1075). Substitute for DIN: German Standards.

353. Dischinger, F. (1937). Untersuchungen über die Kriechsicherheit, die elastische Verformung und das Kriechen des Betons bei Bogenbrücken. *Bauingenieur, 18*, 487–520, 539–552, 595–621.

354. Dischinger, F. (1939). Elastische und plastische Verformungen bei Eisenbetontragwerken. *Bauingenieur, 20*, 53–63, 286–294, 426–437, 563–572.

355. Domone, P. L. (1974). Uniaxial tensile creep and failure of concrete. *Magazine of Concrete Research, 26*, 144–152.

356. Donmez, A., & Bažant, Z. P. (2016). Shape factors for concrete shrinkage and drying creep in model B4 refined by nonlinear diffusion analysis. *Materials and Structures, 49*, 4779–4784.

357. Dormieux, L., Kondo, D., & Ulm, F.-J. (2006). *Microporomechanics*. Chichester: Wiley.
358. Dougill, J. W. (1962). Discussion of 'Modulus of elasticity of concrete affected by moduli of cement paste and aggregate' by T.J. *Hirsch. Proceedings of American Concrete, 59,* 1363–1365.
359. DRC Consultants, I. (1996). Koror-Babelthuap bridge: Force distribution in bar tendons, Technical report, DRC Consultants, Inc.
360. Dugdale, D. S. (1960). Yielding of steel sheets containing slits. *Journal of the Mechanics and Physics of Solids, 8,* 100–108.
361. Dutra, V. F. P., Maghous, S., Campos Filho, A., & Pacheco, A. R. (2010). A micromechanical approach to elastic and viscoelastic properties of fiber reinforced concrete. *Cement and Concrete Research, 40,* 460–472.
362. Dvorak, G. (2013). *Micromechanics of composite materials*. Netherlands: Springer.
363. Dvorak, G. J. (1992). Transformation field analysis of inelastic composite materials. *Proceedings of the Royal Society A, 437,* 311–327.
364. Dvorak, G. J., & Benveniste, Y. (1992). On transformation strains and uniform fields in multiphase elastic media. *Proceedings of the Royal Society A, 437,* 291–310.
365. Dwaikat, M. B., & Kodur, V. K. R. (2009). Hydrothermal model for predicting fire-induced spalling in concrete structural systems. *Fire Safety Journal, 44,* 425–434.
366. Eliáš, J., & Le, J.-L. (2012). Modeling of mode-I fatigue crack growth in quasibrittle structures under cyclic compression. *Engineering Fracture Mechanics, 96,* 26–36.
367. Engineers, B. (1995). Koror-Babeldaob bridge modifications and repairs, Technical report, Berger/ABAM Engineers, Inc.
368. Engineers, B. (1995). Koror-Babeldaob bridge repair project report on evaluation of VECP, Technical report, Berger/ABAM Engineers, Inc. presented by Black Construction Corporation.
369. England, G. L., & Illston, J. M. (1965). Methods of computing stress in concrete from a history of measured strain. *Civil Engineering and Public Works Review, 513–517*(692–694), 845–847.
370. England, G. L., & Ross, A. D. (Eds.), Shrinkage, moisture and pore pressure in heated concrete. In *Proceedings of the ACI International Seminar on Concrete for Nuclear Reactors, West Berlin, Germany* (pp. 883–907) (ACI SP-34).
371. England, G., & Ross, A. D. (1962). Reinforced concrete under thermal gradients. *Magazine of Concrete Research, 14,* 5–12.
372. Espinosa, R. M., & Franke, L. (2006). Influence of the age and drying process on pore structure and sorption isotherms of hardened cement paste. *Cement and Concrete Research, 36,* 1969–1984.
373. Eurocode 2: Design of concrete structures, part 1-1: General rules and rules for buildings (2004). Brussels. BS EN 1992-1-1:2004: E.
374. Eurocode 2: Design of concrete structures, part 1-2: General rules—structural fire design (2004). Brussels. BS EN 1992-1-2:2004: E.
375. Evans, A. G. (1972). A method for evaluating the time-dependent failure characteristics of brittle materials–and its application to alumina. *Journal of Materials Science, 7,* 1137–1146.
376. Evans, A. G., & Fu, Y. (1984). The mechanical behavior of alumina. *Fracture in ceramic materials* (pp. 56–88). Park Ridge: Noyes Publications.
377. Eymard, R., Gallouët, T., Hilhorst, D., & Slimane, Y. N. (1998). Finite volumes and nonlinear diffusion equations. *RAIRO-Mathematical Modelling and Numerical Analysis, 32,* 747–761.
378. Fahmi, H. M., Polivka, M., & Bresler, B. (1972). Effect of sustained and cyclic elevated temperature on creep concrete. *Cement and Concrete Research, 2,* 591–606.
379. Fairhurst, C., & Comet, F. (1981). Rock fracture and fragmentation. In H. H. Einstein (Ed.), *Rock mechanics: From research to application, Proceedings of the 22nd U.S. Symposium on Rock Mechanism* (pp. 21–46). Cambridge: MIT Press.
380. Faria, R., Azenha, M., & Figueiras, J. A. (2006). Modelling of concrete at early ages: application to an externally restrained slab. *Cement and Concrete Composites, 28,* 572–578.

381. Fédération internationale du béton. (2013). *fib model code for concrete structures 2010*. Berlin: Ernst & Sohn.
382. Feldman, R. F., & Sereda, P. J. (1964). Sorption of water on compacts of bottle hydrated cement. I: The sorption and length-change isotherms. *Journal of Applied Chemistry, 14*, 87–93.
383. Feldman, R. F., & Sereda, P. J. (1968). A model for hydrated Portland cement paste as deduced from sorption-length change and mechanical properties. *Materials and Structures, 1*, 509–520.
384. Feller, W. (1957). *An introduction of probability theory and its applications* (2nd ed., Vol. 1). New York: Wiley.
385. Feraille-Fresnet, A. (2000). Le role de l'eau dans le comportement à haute température des bétons. Ph.D. thesis, Ecole Nationale des Ponts et Chaussées, Paris.
386. Feraille-Fresnet, A., Tamagny, P., Ehrlacher, A., & Sercombe, J. (2003). Thermo-hydro-chemical modelling of a porous medium submitted to high temperature: An application to an axisymmetrical structure. *Mathematical and Computer Modelling, 37*(03), 641–650. www.elsevier.nl/locate/mcm
387. Fernie, G. N., & Leslie, J. A. (1975). Vertical and longitudinal deflections of major prestressed concrete bridges. *Symposium of serviceability of concrete* (Vol. n7516). Australia, Melbourne: Institution of Engineers.
388. Ferry, J. D. (1980). *Viscoelastic properties of polymers* (3rd ed.). New York: Wiley.
389. *fib*. (1999). Structural concrete: Textbook on behaviour, design and performance, Fédération Internationale du Béton (fib), Lausanne.
390. *fib*. (1999). Structural concrete: Textbook on behaviour, design and performance, basis of design. fib *Bulletin* (Vol. 2), pp. 35–52
391. Fick, A. (1855). Über diffusion. *Poggendorff's Annalen, 94*, 59.
392. Fillunger, P. (1915). Versuche über die Zugfestigkeit bei allseitigem Wasserdruck. *Österreichische Wochenschrift für den öffentlichen Baudienst, 29*, 443.
393. Fillunger, P. (1929). Auftrieb und Unterdruck in Talsperren. *Die Wasserwirtschaft*, (18, 20, 21), 334–336, 371–377, 388–390.
394. Fillunger, P. (1934). Nochmals der Auftrieb in Talsperren. *Zeitschrift des Österreichischen Ingenieur- und Architekten-Vereines*, (5/6), 28–30.
395. Fisher, L. R., & Israelachvili, J. N. (1979). Direct experimental verification of the Kelvin equation for capillary condensation. *Nature, 277*, 548–549.
396. Forsyth, P. A., & Simpson, R. B. (1991). A two phase, two component model for natural convection in a porous medium. *International Journal for Numerical Mehods in Fluids, 12*, 655–682.
397. Fossen, H. (2010). *Structural geology*. Cambridge: Cambridge University Press.
398. Fox, J. (1997). *Applied regression analysis, linear models and related methods*. Los Angeles: Sage Publications.
399. Freund, J. E. (1962). *Mathematical statistics*. Englewood Cliffs: Prentice Hall.
400. Furamura, F. (1970). Stress-strain relationship in compression for concrete at high temperatures. *Transactions of the Architectural Institute, 174* (Tokyo, Japan).
401. Furbish, D. J. (1997). *Fluid physics in geology*. New York and Oxford: Oxford University Press.
402. Gaede, K. (1962). Über die Festigkeit und die Verformung von Beton bei Druck-Schwellbeanspruchung. *Deutscher Ausschuss für Stahlbeton* (Vol. 144).
403. Gamble, B. R., & Parrot, L. J. (1978). Creep of concrete in compression during drying and wetting. *Magazine of Concrete Research, 30*, 129–138.
404. Gao, L., & Hsu, C.-T. T. (1998). Fatigue of concrete under compression cyclic loading. *ACI Materials Journal, 95*, 575–579.
405. Gardner, N. J. (2000). Design provisions for shrinkage and creep of concrete. In A. Al-Manaseer (Ed.), *Adam Neville symposium: Creep and shrinkage—structural design effects* (Vol. ACI SP-194, pp. 101–104). Farmington Hills, Michigan: American Concrete Institute.
406. Gardner, N. J. (2004). Comparison of prediction provisions for drying shrinkage and creep of normal strength concretes. *Canadian Journal of Civil Engineering, 31*, 767–775.

407. Gardner, N. J., & Lockman, M. J. (2001). Design provisions for drying shrinkage and creep of normal strength. *ACI Materials Journal, 98*, 159–167.
408. Gardner, N. J., & Zhao, J. W. (1993). Creep and shrinkage revisited. *ACI Materials Journal, 90*, 236–246.
409. Garrett, G. G., Jennings, H. M., & Tait, R. B. (1979). The fatigue hardening behavior of cement-based materials. *Journal of Materials Science, 14*, 296–306.
410. Gasch, T., Malm, R., & Ansell, A. (2016). A coupled hygro-thermo-mechanical model for concrete subjected to variable environmental conditions. *International Journal of Solids and Structures, 91*, 143–156.
411. Gauss, K. F. (1809). *Theoria Motus Corporum Caelestium*, Hamburg, reprinted by Dover Publications, New York, 1963.
412. Gawin, D. (1996). A model of hygro-thermic behaviour of unsaturated concrete at high temperatures. In *Proceedings of the 2nd International Scientific Conference on Analytical Models and New Concepts in Mechanics of Concrete Structures, Lodz, Poland* (pp. 65–71).
413. Gawin, D., Alonso, C., Andrade, C., Majorana, C. E., & Pesavento, F. (2005). Effect of damage on permeability and hygro-thermal behaviour of HPCs at elevated temperatures: Part 1. *Experimental results. Computers and Concrete, 2*(3), 189–202.
414. Gawin, D., Majorana, C. E., & Schrefler, B. (1999). Numerical analysis of hygro-thermal behaviour and damage of concrete at high temperature. *Mechanics of Cohesive-Frictional Materials, 4*, 37–74.
415. Gawin, D., Pesavento, F., & Schrefler, B. A. (2002). Modeling of hygro-thermal behavior and damage of concrete at temperature above the critical point of water. *International Journal of Numerical and Analytical Methods in Geomechanics, 26*, 537–562.
416. Gawin, D., Pesavento, F., & Schrefler, B. A. (2002). Simulation of damage-permeability coupling in hygro-thermo-mechanical analysis of concrete at high temperature. *Communications in Numerical Methods in Engineering, 18*, 113–119.
417. Gawin, D., Pesavento, F., & Schrefler, B. A. (2003). Modelling of hygro-thermal behaviour of concrete at high temperature with thermo-chemical and mechanical material degradation. *Computer Methods in Applied Mechanics and Engineering, 192*(13–14), 1731–1771.
418. Gawin, D., Pesavento, F., & Schrefler, B. A. (2004). Modelling of deformations of high strength concrete at elevated temperatures. *Concrete Science and Engineering/Materials and Structures, 37*, 218–236.
419. Gawin, D., Pesavento, F., & Schrefler, B. A. (2006). Hygro-thermo-chemo-mechanical modelling of concrete at early ages and beyond. Part I. Hydration and hygro-thermal phenomena. *International Journal for Numerical Methods in Engineering, 67*, 299–331.
420. Gawin, D., Pesavento, F., & Schrefler, B. A. (2006). Hygro-thermo-chemo-mechanical modelling of concrete at early ages and beyond. Part II. Shrinkage and creep of concrete. *International Journal for Numerical Methods in Engineering, 67*, 332–363.
421. Gawin, D., Pesavento, F., & Schrefler, B. A. (2006). Towards prediction of the thermal spalling risk through a multi-phase porous media model of concrete. *Computer Methods in Applied Mechanics and Engineering, 195*, 5707–5729.
422. Gawin, D., Pesavento, F., & Schrefler, B. A. (2011). What physical phenomena can be neglected when modelling concrete at high temperature? A comparative study. Part 1: Physical phenomena and mathematical model. *International Journal of Solids and Structures, 48*(13), 1927–1944. https://doi.org/10.1016/j.ijsolstr.2011.03.004.
423. Gawin, D., Pesavento, F., & Schrefler, B. A. (2011). What physical phenomena can be neglected when modelling concrete at high temperature? A comparative study. Part 2: Comparison between models. *International Journal of Solids and Structures, 48*(13), 1945–1961. https://doi.org/10.1016/j.ijsolstr.2011.03.003.
424. Gawin, D., & Schrefler, B. A. (1996). Thermo-hydro-mechanical analysis of partially saturated porous materials. *Engineering Computations, 13*, 113–143.
425. Ghali, A., Neville, A. M., & Jha, P. C. (1967). Effect of elastic and creep recoveries of concrete on loss of prestress. *Journal of the American Concrete Institute, 64*, 802–810.

426. Glanville, W. H. (1930). *Studies in reinforced concrete III - Creep or flow of concrete under load, Building Research Technical Paper 12*. London: Department of Scientific and Industrial Research.
427. Glanville, W. H. (1933). Creep of concrete under load. *The Structural Engineer, 11*, 54–73.
428. Glanville, W. H., & Thomas, F. G. (1930). *Studies in reinforced concrete IV - Further investigations on the creep flow of concrete under load, Building Research Technical Paper 21*. London: Department of Scientific and Industrial Research.
429. Goodman, R. E. (1989). *Introduction to rock mechanics* (2nd ed.). New York: Wiley.
430. Granger, L. (1995). Comportement différé du béton dans les enceintes de centrales nucléaires: analyse et modélisation, Technical report, Laboratoire Central des Ponts et Chaussées, Paris. Ph.D. thesis of ENPC.
431. Granger, L., Acker, P., & Torrenti, J.-M. (1994). Discussion of drying creep of concrete: Constitutive model and new experiments separating its mechanisms. *Materials and Structures, 27*, 616–619.
432. Granger, L. P., & Bažant, Z. P. (1995). Effect of composition on basic creep of concrete and cement paste. *Journal of Engineering Mechanics, ASCE, 121*, 1261–1270.
433. Grassl, P., & Jirásek, M. (2006). Damage-plastic model for concrete failure. *International Journal of Solids and Structures, 43*(22–23), 7166–7196.
434. Grassl, P., Xenos, D., Nystrom, U., Rempling, R., & Gylltoft, K. (2013). CDPM2: A damage-plasticity approach to modelling the failure of concrete. *International Journal of Solids and Structures, 50*(24), 3805–3816. https://doi.org/10.1016/j.ijsolstr.2013.07.008.
435. Gray, W. G., & Hassanizadeh, S. M. (1980). General conservation equations for multiphase systems: 3. Constitutive theory for porous media. *Advances in Water Resources, 3*, 25–40.
436. Gray, W. G., & Hassanizadeh, S. M. (1991). Paradoxes and realities in unsaturated flow theory. *Water Resources Research, 27*, 1847–1854.
437. Gray, W. G., & Hassanizadeh, S. M. (1991). Unsaturated flow theory including interfacial phenomena. *Water Resources Research, 27*(8), 1855–1863. http://80.apps.webofknowledge.com.dialog.cvut.cz/full_record.do?product=WOS&search_mode=GeneralSearch&qid=9&SID=Q2pWw4ZPkVQTctwemBU&page=1&doc=9.
438. Griffith, A. A. (1921). The phenomena of rupture and flow in solids. *Philosophical Transactions of the Royal Society of London, Series A, 221*, 163–197.
439. Grübl, P., Weigler, H., & Karl, S. (2002). Beton: Arten, Herstellung und Eigenschaften, Handbuch für Beton-, Stahlbeton- und Spannbetonbau (pp. 68, 226). Ernst & Sohn.
440. Grzybowski, M., & Shah, S. P. (1990). Shrinkage cracking of fiber reinforced concrete. *ACI Materials Journal, 87*, 138–148.
441. Guggenheim, E. A. (1959). Thermodynamics, classical and statistical. In S. Flügge (Ed.), *Encyclopedia of physics* (Vol. III/2). Berlin: Springer.
442. Gvozdev, A. A. (1953). *Thermal-shrinkage deformations in massive concrete blocks (in Russian), Technical report 17(4)*. SSSR, OTN: Izvestia A.N.
443. Gvozdev, A. A. (1966). Creep of concrete (in Russian), *Mekhanika Tverdogo Tela, Proceedings of the 2nd National Conference on Theory and Applied Mechanics Academy of Sciences USSR, Nauka, Moscow* (pp. 137–152).
444. Hagentoft, C. E., Kalagasidis, A. S., Adl-Zarrabi, B., Roels, S., Carmeliet, C., Hens, H., et al. (2004). Assessment method of numerical prediction models for combined heat, air and moisture transfer in building components: Benchmarks for one-dimensional cases. *Journal of Thermal Envelope and Building Science, 27*, 327–352.
445. Hagymassy, J., Odler, I., Yudenfreund, M., Skalny, J., & Brunauer, S. (1972). Pore structure analysis by water vapor adsorption. III. Analysis of hydrated calcium silicates and Portland cements. *Journal of Colloid and Interface Science, 38*, 20–34.
446. Haldar, A., & Mahadevan, S. (1999). *Probability, reliability, and statistical methods in engineering design*. New York: Wiley.
447. Halsey, G. J. (1948). Physical adsorption on nonuniform surfaces. *Journal of Chemical Physics, 16*, 931–937.

448. Hansen, K. K. (1986). *Sorption isotherms: A catalogue, Technical report 162/86*. Building Materials Laboratory: The Technical University of Denmark.
449. Hansen, P. F. (1985). *Coupled moisture/heat transport in cross sections of structures (in Danish)*. Beton og Konstruktionsinstituttet (BKI): Technical report.
450. Hansen, T. C. (1960). Creep and stress relaxation in concrete. In *Proceedings of the Swedish Cement and Concrete Institute (CBI)* (Vol. 31), Royal Institute of Technology, Stockholm.
451. Hansen, T. C. (1960). Creep of concrete: The influence of variations in the humidity of ambient atmosphere. In *6th Congress of the International Association of Bridge and Structural Engineering (IBASE)* (pp. 57–65). Stockholm.
452. Hansen, T. C., & Eriksson, L. (1966). Temperature change effect on behavior of cement paste, mortar, and concrete under load. *ACI Journal, 63*, 489–504.
453. Hansen, T. C., & Mattock, A. H. (1966). Influence of size and shape of member on the shrinkage and creep of concrete. *Journal of the American Concrete Institute, 63*, 267–290.
454. Hansen, W. (1987). Drying shrinkage mechanisms in Portland cement pastes. *Journal of the American Ceramic Society, 70*, 323–328.
455. Hanson, J. A. (1953). *A ten-year study of creep properties of concrete, Concrete Laboratory Report Sp-38*. Bureau of Reclamation, Denver, Colorado: US Department of the Interior.
456. Hanson, J. A. (1968). Effects of curing and drying environments on splitting tensile strength. *ACI Journal, 65*, 535–543 (PCA Bulletin D141).
457. Haque, M. N. (1995). Strength development and drying shrinkage of high-strength concretes. *Cement and Concrete Composites, 18*, 333–342.
458. Harboe, E. M. (1958). *A comparison of the instantaneous and the sustained modulus of elasticity of concrete, Concrete Laboratory Report C-854*. US Department of the Interior, Bureau of Reclamation, Denver, Colorado: Division of Engineering Laboratories.
459. Hardy, G. H., & Riesz, M. (1915). *The general theory of Dirichlet's series* (Vol. 18). Cambridge tracts in mathematics. Cambridge: Cambridge University Press.
460. Harmathy, T. Z. (1965). Effect of moisture on the fire endurance of building materials. *Moisture in materials in relation to fire tests, ASTM special technical publication* (Vol. 385, pp. 74–95). Philadelphia: American Society of Testing Materials.
461. Harmathy, T. Z. (1970). Thermal properties of concrete of elevated temperature. *Journal of Materials, JMLSA, 5*, 47–75.
462. Harmathy, T. Z. (1983). *Properties of building materials at elevated temperatures, Technical report DBR Paper No. 1080*, Division of Building Research, Ottawa.
463. Harmathy, T. Z., & Allen, L. W. (1973). Thermal properties of selected masonry unit concretes. *Journal of the American Concrete Institute, 70*, 132–142.
464. Hashin, Z. (1962). The elastic modulus of heterogeneous materials. *Journal of Applied Mechanics, ASME, 29*, 143–150.
465. Hassanizadeh, S., & Gray, W. G. (1990). Mechanics and thermodynamics of multiphase flow in porous media including interphase boundaries. *Advances in Water Resources, 13*(4), 169–186. http://www.sciencedirect.com/science/article/pii/030917089090040B.
466. Hassanizadeh, S. M. (1986). Derivation of basic equations of mass transport in porous media, Part 1. Macroscopic balance laws. *Advances in Water Resources, 9*, 196–206.
467. Hassanizadeh, S. M., & Gray, W. G. (1979). General conservation equations for multiphase systems: 1 Averaging procedure. *Advances in Water Resources, 2*, 131–144.
468. Hassanizadeh, S. M., & Gray, W. G. (1979). General conservation equations for multiphase systems: 2 Mass, momenta, energy and entropy equations. *Advances in Water Resources, 2*, 191–203.
469. Hassanizadeh, S. M., & Gray, W. G. (1980). General conservation equations for multiphase systems: 3 Constitutive theory for porous media flow. *Advances in Water Resources, 3*, 25–40.
470. Hatt, W. K. (1907). Notes on the effect of time element in loading reinforced concrete beams. *Proceedings of ASTM, 7*, 421–433.
471. Havlásek, P. (2014). Creep and shrinkage of concrete subjected to variable environmental conditions, Ph.D. thesis, Czech Technical University in Prague, Prague, Czech Republic.

472. Havlásek, P., & Jirásek, M. (2012). Modeling of concrete creep based on microprestress-solidification theory. *Acta Polytechnica, 52*, 34–42.

473. Havlásek, P., & Jirásek, M. (2012). *Modeling of nonlinear moisture transport in concrete. Nano and macro mechanics*. Prague: Czech Technical University.

474. Havlásek, P., & Jirásek, M. (2016). Multiscale modeling of drying shrinkage and creep of concrete. *Cement and Concrete Research, 85*, 55–74.

475. Hellesland, J., & Green, R. (1971). Sustained and cyclic loading of concrete columns. *Proceedings of the American Society of Civil Engineers, 97*, 1113–1128.

476. Helmuth, R. A., & Turk, D. H. (1967). The reversible and irreversible drying shrinkage of hardened Portland cement and tricalcium silicate paste. *Journal of the Portland Cement Association Research and Development Laboratories, 9*, 8–21 (PCA Bulletin 215).

477. Hertz, K. D. (2003). Limits of spalling of fire-exposed concrete. *Fire Safety Journal, 38*, 103–116.

478. Hill, R. (1965). A self consistent mechanics of composite materials. *Journal of the Mechanics and Physics of Solids, 13*, 213–222.

479. Hill, T. L. (1960). *Statistical thermodynamics*. Reading: Addison-Wesley.

480. Hillerborg, A. (1985). A modified absorption theory. *Cement and Concrete Research, 15*, 809–816.

481. Hillerborg, A., Modéer, M., & Peterson, P. E. (1976). Analysis of crack propagation and crack growth in concrete by means of fracture mechanics and finite elements. *Cement and Concrete Research, 6*, 773–782.

482. Hilsdorf, H. K. (1980). Unveröffentlichte Versuche an der MPA München, private communication.

483. Hirst, G. A., & Neville, A. M. (1977). Activation energy of creep of concrete under short-term static and cyclic stresses. *Magazine of Concrete Research, 29*, 13–18.

484. Holt, E. (2005). Contribution of mixture design to chemical and autogenous shrinkage of concrete at early ages. *Cement and Concrete Research, 35*, 464–472.

485. Horii, H., & Nemat-Nasser, S. (1982). Compression-induced nonplanar crack extension with application to splitting, exfoliation and rockburst. *Journal of Geophysical Research, 87*, 6806–6821.

486. Hsu, T. C. (1981). *Fatigue of plain concrete. ACI Journal, 78*, 292–305.

487. Huang, K. (1963). *Statistical mechanics*. New York: Wiley.

488. Hubler, M. H., Wendner, R., & Bažant, Z. P. (2015). Comprehensive database for concrete creep and shrinkage: Analysis and recommendations for testing and recording. *ACI Materials Journal, 112*, 547–558.

489. Hubler, M. H., Wendner, R., & Bažant, Z. P. (2015). Statistical justification of model B4 for drying and autogenous shrinkage of concrete and comparisons to other models. *Materials and Structures, 48*, 797–814.

490. Huet, C. (1970). Sur l'évolution des contraintes et déformations dans les systèmes multicouches constitués de matériaux viscoélastiques présentant du vieillissement. *Comptes Rendus de l'Académie des Sciences Paris, 270*, 213–216.

491. Huet, C. (1973). Application à la viscoélasticité non linéaire du calcul symbolique à plusieurs variables. *Rheologica Acta, 12*, 279–288.

492. Huet, C. (1974). Opérateurs matriciels en viscoélasticité linéaire avec vieillissement et application aux structures viscoélastiques hétérogènes. In J. Hult (Ed.), *Mechanics of visco-elastic media and bodies* (pp. 131–146). Berlin: Springer.

493. Huet, C. (1980). Adaptation d'un algorithme de Bazant au calcul des multilames viscoélastiques vieillissants. *Materials and Structures, 74*, 91–98.

494. Huet, C. (1985). Relations between creep and relaxation functions in nonlinear viscoelasticity with or without aging. *Journal of Rheology, 29*, 245–257.

495. Huet, C. (1992). Minimum theorems for viscoelasticity. *European Journal of Mechanics - A/Solids, 11*, 653–684.

496. Huet, C. (1995). Bounds for the overall properties of viscoelastic heterogeneous and composite materials. *Archives of Mechanics, 47*, 1125–1155.

497. Hughes, B. P., Lowe, I. R. G., & Walker, J. (1966). The diffusion of water in concrete at temperatures between 50 and 95°C. *British Journal of Applied Physics, 17*, 1545–1552.
498. Hummel, A., Wesche, K., & Brand, W. (1962). Der Einfluss der Zementart, des Wasser-Zement-Verhältnisses und des Belastungsalters auf das Kriechen von Beton. *Deutscher Ausschuss für Stahlbeton* (Vol. 146). Berlin: W. Ernst & Sohn.
499. Husseini, F. (1982). Feuchtverteilung in porösen Baustoffen aufgrund instationärer Wasserdampfdiffusion, Ph.D. thesis, Universität Dortmund, Dortmund, Germany.
500. IAP. (2009). Revised release on the IAPWS formulation 1995 for the thermodynamic properties of ordinary water substance for general and scientific use. http://www.iapws.org.
501. Ichikawa, Y., & England, G. L. (2004). Prediction of moisture migration and pore pressure build-up in concrete at high temperatures. *Nuclear Engineering and Design, 228*, 245–259.
502. Idiart, A. E. (2009). Coupled analysis of degradation processes in concrete specimens at the meso-level, Ph.D. thesis, Technical University of Catalonia, Barcelona, Spain.
503. Igarashi, S. I., Bentur, A., & Kovler, K. (2000). Autogenous shrinkage and induced restraining stresses in high-strength concretes. *Cement and Concrete Research, 30*, 1701–1707.
504. Illston, J. M., & Sanders, P. D. (1973). The effect of temperature change upon the creep of mortar under torsional loading. *Magazine of Concrete Research, 25*, 136–144.
505. Iman, R. L., Helton, J. C., & Campbell, J. E. (1981). An approach to sensitivity analysis of computer models, Part 1. Introduction, input variable selection and preliminary variable assessment. *Journal of Quality Technology, 13*, 174–183.
506. Ingraffea, A. R. (1977). Discrete fracture propagation in rock: Laboratory tests and finite element analysis, Ph.D. thesis, University of Colorado.
507. Irwin, G. R. (1958). Fracture. In S. Flügge (Ed.), *Handbuch der Physik* (Vol. VI, pp. 551–590). Berlin: Springer.
508. Janssen, H., Blocken, B., & Carmeliet, J. (2007). Conservative modelling of the moisture and heat transfer in building components under atmospheric excitation. *International Journal of Heat and Mass Transfer, 50*, 1128–1140.
509. Jansson, R. (2013). Fire spalling of concrete, Ph.D. thesis, Royal Institute of Technology, Stockholm, Sweden.
510. Japan International Cooperation Agency. (1990). *Present condition survey of the Koror-Babelthuap bridge*. Japan International Cooperation Agency: Technical report.
511. Jason, L., Pijaudier Cabot, G., Ghavamian, S., & Huerta, A. (2007). Hydraulic behaviour of a representative structural volume for containment buildings. *Nuclear Engineering and Design, 237*, 1259–1274.
512. Jendele, L., Šmilauer, V., & Červenka, J. (2013). Multiscale hydro-thermo-mechanical model for early-age and mature concrete structures. *Advances in Engineering Software, 72*, 134–146.
513. Jennings, H. M., Thomas, J. J., Gevrenov, J. S., Constantinides, G., & Ulm, F. J. (2007). A multi-technique investigation of the nanoporosity of cement paste. *Cement and Concrete Research, 37*(3), 329–336.
514. Jensen, O. M., & Hansen, P. F. (1996). Autogenous deformation and change of the relative humidity in silica fume-modified cement paste. *ACI Materials Journal, 93*, 539–543.
515. Jensen, O. M., & Hansen, P. F. (1999). Influence of temperature on autogenous deformation and relative humidity change in hardening cement paste. *Cement and Concrete Research, 29*, 567–575.
516. Jensen, O. M., & Hansen, P. F. (2001). Autogenous deformation and RH-change in perspective. *Cement and Concrete Research*, 1859–1865.
517. Jirásek, M. (2000). Comparative study on finite elements with embedded cracks. *Computer Methods in Applied Mechanics and Engineering, 188*, 307–330.
518. Jirásek, M. (2007). Nonlocal damage mechanics. *Revue européenne de génie civil, 11*, 993–1021.
519. Jirásek, M. (2011). Damage and smeared crack models. In G. Hofstetter & G. Meschke (Eds.), *Numerical modeling of concrete cracking* (pp. 1–50). Berlin: Springer.
520. Jirásek, M., & Bauer, M. (2012). Numerical aspects of the crack band approach. *Computers and Structures, 110–111*, 60–78.

521. Jirásek, M., & Bažant, Z. P. (2002). *Inelastic analysis of structures*. Chichester: Wiley.
522. Jirásek, M., & Havlásek, P. (2014). Accurate approximations of concrete creep compliance functions based on continuous retardation spectra. *Computers and Structures, 135*, 155–168.
523. Jirásek, M., & Havlásek, P. (2014). Microprestress-solidification theory of concrete creep: Reformulation and improvement. *Cement and Concrete Research, 60*, 51–62.
524. Jonasson, J. E., & Hedlund, H. (2000). An engineering model for creep and shrinkage in high performance concrete. *Shrinkage of concrete, shrinkage*, RILEM Proceedings (pp. 507–529).
525. Jones, H. R. N. (2000). *Radiation heat transfer*. Oxford: Oxford University Press.
526. Jönson, B., Nonat, A., Labbez, C., Cabane, B., & Wennerström, H. (2005). Controlling the cohesion of cement paste. *Langmuir, 21*, 9211–9221.
527. Jönson, B., Wennerström, H., Nonat, A., & Cabane, B. (2004). Onset of cohesion in cement paste. *Langmuir, 20*, 6702–6709.
528. JRA. (2002). *Specifications for highway bridges with commentary*. Concrete (in Japanese), Technical report, Japan Road Association (JRA): Part III.
529. JSCE. (1991). Standard specification for design and construction of concrete structures (in Japanese), Technical report, Japan Society of Civil Engineers (JSCE).
530. Kachanov, L. M. (1958). Time of the rupture process under creep conditions. *Izvestija Akademii Nauk SSSR, Otdelenie Techniceskich Nauk, 8*, 26–31.
531. Kachanov, M. (1982). A microcrack model of rock inelasticity-part I. Frictional sliding on microcracks. *Mechanics of Materials, 1*, 19–41.
532. Kada-Benameur, H., Wirquin, E., & Duthoit, B. (2000). Determination of apparent activation energy of concrete by isothermal calorimetry. *Cement and Concrete Research, 30*(2), 301–305.
533. Kalifa, P., Menneteau, F., & Quenard, D. (2000). Spalling and pore pressure in HPC at high temperatures. *Cement and Concrete Research, 30*, 1915–1927.
534. Kanema, M., de Morais, M. V. G., Noumowé, A., Gallias, J. L., & Cabrillac, R. (2007). Experimental and numerical studies of thermo-hydrous transfers in concrete exposed to high temperature. *Heat and Mass Transfer, 44*, 149–164.
535. Kanit, T., Forest, S., Galliet, I., Mounoury, V., & Jeulin, D. (2003). Determination of the size of the representative volume element for random composites: Statistical and numerical approach. *International Journal of Solids and Structures, 40*, 3647–3679.
536. Karush, W. (1939). Minima of functions of several variables with inequalities as side conditions, Master's thesis, Department of Mathematics, University of Chicago.
537. Kaxiras, E. (2003). *Atomic and electronic structure of solids*. New York: Cambridge University Press.
538. Keeton, J. R. (1965). *Study of creep in concrete, Technical report R333-I, R333-II, R333-III*. Naval Civil Engineering Laboratory, Port Hueneme, California: U.S.
539. Kemeny, I. M., & Cook, N. G. W. (1987). Crack models for the failure of rock under compression. In C. S. Desai (Ed.), *Proceedings of the 2nd International Conference on Constitutive Laws for Engineering Materials* (Vol. 2, pp. 879–887). New York: Elsevier Science Publishing Co.
540. Khaliq, W., & Kodur, V. (2011). Thermal and mechanical properties of fibre reinforced high performance self-consolidating concrete at elevated temperatures. *Cement and Concrete Research, 41*, 1112–1122.
541. Khazanovich, L. (1998). Age-adjusted effective modulus method for time-dependent loads. *Journal of Engineering Mechanics, ASCE, 116*, 2784–2789.
542. Khennane, A., & Baker, G. (1992). Thermo-plasticity models for concrete under varying temperature and biaxial stress. *Proceedings of the Royal Society A, 439*, 59–80.
543. Khoury, G. (1995). Strain component of nuclear-reactor-type concretes during first heating cycle. *Nuclear Engineering and Design, 156*, 313–321.
544. Khoury, G. A., Grainger, B. N., & Sullivan, P. J. A. (1985). Transient thermal strain of concrete: Literature review, conditions within specimen and behaviour of individual constituents. *Magazine of Concrete Research, 37*, 131–144.

545. Kim, K.-T., & Bažant, Z. P. (2014). Creep design aid: Open-source website program for concrete creep and shrinkage prediction. *ACI Materials Journal, 111*(4), 423–432.

546. Kimishima, H., & Kitahara, H. (1964). *Creep and creep recovery of mass concrete, Technical report C-64001*. Tokyo, Japan: Central Research Institute of Electric Power Industry.

547. Kirane, K., & Bažant, Z. P. (2015). Microplane damage model for fatigue of quasibrittle materials: Sub-critical crack growth, lifetime and residual strength. *International Journal of Fatigue, 70,* 93–105.

548. Kirane, K., Su, Y., & Bažant, Z. P. (2015). Strain-rate dependent microplane model for high-rate comminution of concrete under impact based on kinetic energy release theory. *Proceedings of the Royal Society A, 20150535,* 1–8.

549. Klopfer, H. (1974). *Wassertransport durch Diffusion in Feststoffen*. Wiesbaden, Germany: Bauverlag.

550. Klute, A. (1952). A numerical method for solving the flow equation for water in unsaturated material. *Soil Science, 73,* 105–116.

551. Komendant, G. J., Polivka, M., & Pirtz, D. (1976). Study of concrete properties for prestressed concrete reactor vessels. Part 2: Creep and strength characteristics of concrete at elevated temperatures, Technical report UC SESM 76-3, Department of Civil Engineering, University of California, Berkeley, California.

552. Kovler, K. (1995). Shock of evaporative cooling of concrete in hot dry climates. *ACI Concrete International, 17,* 65–69.

553. Krischer, O. (1942). Der Wärme und Stoffaustausch im Trocknungsgut. *VDI - Forschungsheft, 415,* 1–22.

554. Kucharczyková, B., Daněk, P., Kocáb, D., & Misák, P. (2017). Experimental analysis on shrinkage and swelling in ordinary concrete. *Advances in Materials Science and Engineering* (3027301).

555. Kuhn, H. W., & Tucker, A. W. (1951). Nonlinear programming. In J. Neyman (Ed.), *Proceedings of the Second Berkeley Symposium on Mathematical Statistics and Probability* (pp. 481–492). University of California Press, Berkeley and Los Angeles.

556. Künzel, H. M. (1995). Simultaneous heat and moisture transport in building components, Technical report, Fraunhofer Institute of Building Physics, IRB Verlag, Stuttgart.

557. Křístek, V., & Bažant, Z. P. (1987). Shear lag effect and uncertainty in concrete box girder creep. *Journal of Structural Engineering, ASCE, 113,* 557–574.

558. Křístek, V., Bažant, Z. P., Zich, M., & Kohoutková, A. (2006). Box girder deflections: Why is the initial trend deceptive? *ACI Concrete International, 28,* 55–63.

559. Křístek, V., & Kohoutková, A. (2002). Serviceability limit state of prestressed concrete bridges. In *Concrete Structures in the 21st Century, Proceedings of the 1st* fib *Congress 2002, Osaka* (Vol. 2, pp. 47–48).

560. Křístek, V., & Vítek, J. L. (1999). Deformations of prestressed concrete structures—measurement and analysis. In *Proceedings of* fib *Symposium "Structural Concrete - The Bridge Between People"*, fib *and ČBS, Prague* (pp. 463–469).

561. Křístek, V., Vráblík, L., Bažant, Z. P., Li, G.-H., & Yu, Q. (2008). Misprediction of long-time deflections of prestressed box girders: Causes, remedies and tendon lay-out effect. In T. Tanabe, K. Sakata, H. Mihashi, R. Sato, K. Maekawa, & H. Nakamura (Eds.), *Creep, shrinkage and durability mechanics of concrete and concrete structures* (pp. 1291–1295). Boca Raton-London: CRC Press/Balkema, Taylor & Francis Group.

562. Lackner, R., Hellmich, C., & Mang, H. A. (2002). Constitutive modeling of cementitious materials in the framework of chemoplasticity. *International Journal for Numerical Methods in Engineering, 53*(10), 2357–2388.

563. Lambotte, H., & Mommens, A. (1976). *L'évolution du fluage du béton en fonction de sa composition, du taux de contrainte et de l'âge, Groupe de travail GT 22*. Brussels: Centre national de recherches scientifiques pour l'industrie cimentière.

564. Langmuir, I. (1918). The adsorption of gases on plane surfaces of glass, mica and platinum. *Journal of the American Chemical Society, 40,* 1361–1403.

565. Lavergne, F., Sab, K., Sanahuja, J., Bornert, M., & Toulemonde, C. (2016). Homogenization schemes for aging linear viscoelastic matrix-inclusion composite materials with elongated inclusions. *International Journal of Solids and Structures*, *80*, 545–560.

566. Lazić, J. D., & Lazić, V. B. (1984). Generalized age-adjusted effective modulus method for creep in composite beam structures. *Part I-Theory. Cement and Concrete Research*, *14*, 819–932.

567. Lazić, J. D., & Lazić, V. B. (1985). Generalized age-adjusted effective modulus method for creep in composite beam structures. *Part II-Applications. Cement and Concrete Research*, *15*, 1–12.

568. Le Chatelier, H. L. (1887). Recherches expérimentales sur la constitution des mortiers hydrauliques, Dissertation of doctor of physical and chemical science, École des Mines, Paris.

569. Le, J.-L., & Bažant, Z. P. (2011). Unified nano-mechanics based probabilistic theory of quasibrittle and brittle structures: II. Fatigue crack growth, lifetime and scaling. *Journal of the Mechanics and Physics of Solids*, *59*, 1322–1337.

570. Lee, K. M., Lee, H. K., Lee, S. H., & Kim, G. Y. (2006). Autogeneous shrinkage of concrete containing granulated blast-furnace slag. *Cement and Concrete Research*, *36*, 1279–1285.

571. Legendre, A. M. (1806). Nouvelles méthodes pour la détermination des orbites des comètes, Paris.

572. Lewis, R. W., & Schrefler, B. A. (1998). *The finite element method in the static and dynamic deformation and consolidation of porous media* (2nd ed.). Chichester: Wiley.

573. L'Hermite, R. (1959). What do we know about the plastic deformation and creep of concrete? *RILEM Bulletin*, *1*, 21–51.

574. L'Hermite, R. (1961). Les déformations du béton. *Cahiers de la recherche I.T.B.T.P.*, *IX*(12).

575. L'Hermite, R. (1961). Que savons nous de la déformation plastique et du fluage du béton? *Annales de l'ITBTP*, *IX*(117).

576. L'Hermite, R. G., & Mamillan, M. (1968). Further results of shrinkage and creep tests. In *Proceedings of the International Conference on the Structure of Concrete* (pp. 423–433). London: Cement and Concrete Association.

577. L'Hermite, R. G., & Mamillan, M. (1968). Retrait et fluage des bétons. *Annales de l'Inst. Techn. du Bâtiment et des Travaux Publiques (Supplément)*, *21*(249), 1334.

578. L'Hermite, R. G., & Mamillan, M. (1969). Nouveaux résultats et récentes études sur le fluage du béton. *Materials and Structures*, *2*, 35–41.

579. L'Hermite, R. G., & Mamillan, M. (1970). Influence de la dimension des éprouvettes sur le retrait. *Annales de l'Inst. Techn. du Bâtiment et des Travaux Publiques (Supplément)*, *23*(270), 5–6.

580. L'Hermite, R. G., Mamillan, M., & Lefèvre, C. (1965). Nouveaux résultats de recherches sur la déformation et la rupture du béton. *Annales de l'Institut Techn. du Bâtiment et des Travaux Publiques*, *18*(207–208), 323–360.

581. Li, C., Caner, F. C., Chau, V. T., & Bažant, Z. P. (2017). Spherocylindrical microplane constitutive model for shale and other anisotropic rocks. *Journal of the Mechanics and Physics of Solids*, *103*, 155–178.

582. Liu, Z.-G., Swartz, S. E., Hu, K. K., & Kan, Y.-C. (1989). Time-dependent response and fracture of plain concrete beams. In S. P. Shah, S. E. Swartz, & B. Barr (Eds.), *Fracture of concrete and rock: Recent developments* (pp. 577–586). London: Elsevier Applied Science.

583. Luckner, L., van Genuchten, M. T., & Nielsen, D. R. (1989). A consistent set of parametric models for the two-phase flow of immiscible fluids in the subsurface. *Water Resources Research*, *25*, 2187–2193.

584. Luikov, A. V. (1966). *Heat and mass transfer in capillary-porous bodies*. Oxford: Pergamon Press.

585. Luikov, A. V., & Mikhailov, Y. A. (1966). *Theory of energy and mass transfer*. Englewood Cliffs: Prentice Hall.

586. Lura, P., Jensen, O. M., & van Breugel, K. (2003). Autogenous shrinkage in high-performance cement paste: An evaluation of basic mechanisms. *Cement and Concrete Research*, *33*, 223–232.

587. Lykov, A. V. (1954). *Javlenija perenosa v kapilljarnoporistych telach*. Moscow: Gosenergoizdat.
588. Lykow, A. W. (1958). *Transporterscheinungen in kapillarporösen Körpern*. Berlin: Akademie-Verlag.
589. Madsen, H., & Bažant, Z. P. (1983). Uncertainty analysis of creep and shrinkage effects in concrete structures. *ACI Journal, 80*, 116–127.
590. Maekawa, K., & El-Kashif, K. F. (2004). Cyclic cumulative damaging of reinforced concrete in post-peak regions. *Journal of Advanced Concrete Technology, 2*, 257–271.
591. Magura, D. D., Sozen, M. A., & Siess, C. P. (1964). A study of stress relaxation in prestressing reinforcement. *PCIJ, 9*, 13–57.
592. Mainguy, M., & Coussy, O. (2000). Propagation fronts during calcium leaching and chloride penetration. *Journal of Engineering Mechanics, ASCE, 126*, 250–257.
593. Mainguy, M., Coussy, O., & Baroghel-Bouny, V. (2001). Role of air pressure in drying of weakly permeable materials. *Journal of Engineering Mechanics, ASCE, 127*, 582–592.
594. Mainguy, M., Ulm, F.-J., & Heukamp, F. H. (2001). Similarity properties of demineralization and degradation of cracked porous materials. *International Journal of Solids and Structures, 38*, 7079–7100.
595. Malani, A., Ayappa, K. G., & Murad, S. (2009). Influence of hydrophillic surface specificity on the structural properties of confined water. *Journal of Physical Chemistry, 113*, 13825–13839.
596. Malcolm, D. J., & Redwood, R. G. (1970). Shear lag in stiffened box girders. *Journal of the Structural Division, ASCE, 96*, 1403–1419.
597. Mal'mejster, A. K. (1957). *Uprugost' i neuprugost' betona (Elasticity and inelasticity of concrete)*. Nauk Latvian SSR, Riga: Izdatel'stvo Akad.
598. Mamillan, M. (1959). A study of the creep of concrete. *RILEM Bulletin, 3*, 15–33.
599. Mamillan, M. (1960). Évolution du fluage et des propriétés du bétons. *Annales de l'Inst. Techn. du Bâtiment et des Travaux Publiques, 13*, 1017–1052.
600. Mamillan, M. (1969). Évolution du fluage et des propriétés du bétons. *Annales de l'Inst. Techn. du Bâtiment et des Travaux Publiques, 21*, 1033.
601. Mamillan, M., & Lelan, M. (1968). Le fluage du béton. *Annales de l'Inst. Techn. du Bâtiment et des Travaux Publiques (Supplément), 21*, 847–850.
602. Mamillan, M., & Lelan, M. (1970). Le fluage du béton. *Annales de l'Inst. Techn. du Bâtiment et des Travaux Publiques (Supplément), 23*, 7–13.
603. Mandel, J. (1957). Sur les corps viscoélastiques dont les propriétés dépendent de l'âge. *Comptes Rendus de l'Académie des Sciences Paris, 247*, 175–178.
604. Mandel, J. (1964). *The statistical analysis of experimental data*. New York: Dover.
605. Mandel, J. (1967). Application de la thermodynamique aux milieux viscoélastiques à élasticité nulle ou restreinte. *Comptes Rendus de l'Académie des Sciences Paris, 264*, 133–134.
606. Mandel, J. (1974). Un principe de correspondance pour les corps viscoélastiques linéaires vieillissants. In J. Hult (Ed.), *Mechanics of visco-elastic media and bodies*. Berlin: Springer.
607. Maney, G. A. (1941). Concrete under sustained working loads: Evidence that shrinkage dominates time yield. *Proceedings of the American Society for Testing and Materials, 41*, 1021–1030.
608. Manjure, P. Y. (2001). Rehabilitation/strengthening of Zuari bridge on NH-15 in Goa. *Indian Roads Congress, 490*, 471.
609. Maslov, G. N. (1940). Thermal stress states in concrete masses, with consideration of concrete creep (in Russian). *Izvestia Nauchno-Issledovatelskogo Instituta Gidrotekhniki, Gosenergoizdat, 28*, 175–188.
610. Masoero, E., Gado, E. D., Pellenq, R. J.-M., Ulm, F.-J., & Yip, S. (2012). Nanostructure and nanomechanics of cement: Polydisperse colloidal packing. *Physical Review Letters, 109*, 155503-1–155503-4.
611. Masoero, E., Gado, E. D., Pellenq, R. J.-M., Yip, S., & Ulm, F.-J. (2014). Nano-scale mechanics of colloidal C-S-H gels. *Soft Matter, 10*, 491–499.
612. Maxwell, J. C. (1867). On the dynamical theory of gases. *Philosophical Transactions of the Royal Society London, 157*, 49–88.

613. Mazars, J. (1984). *Application de la mécanique de l'endommagement au comportement non linéaire et à la rupture du béton de structure.* Thèse de Doctorat d'Etat: Université Paris VI, France.

614. Mazzotti, C., & Savoia, M. (2003). Nonlinear creep damage model for concrete under uniaxial compression. *Journal of Engineering Mechanics, ASCE, 129,* 1065–1075.

615. McDonald, B., Saraf, V., & Ross, B. (2003). A spectacular collapse: The Koror-Babeldaob (Palau) balanced cantilever prestressed, post-tensioned bridge. *The Indian Concrete Journal, 77,* 955–962.

616. McDonald, D. B., & Roper, H. (1993). Accuracy of prediction models for shrinkage of concrete. *ACI Materials Journal, 90,* 265–271.

617. McDonald, J. E. (1975). *Time-dependent deformation of concrete under multiaxial stress conditions, Technical report C-75-4.* Army Engineer Waterways Experimental Station, Vicksburg, Miss: U.S.

618. McDowell, D. L. (2008). Viscoplasticity of heterogeneous metallic materials. *Materials Science and Engineering: R: Reports, 62,* 67–123.

619. McHenry, D. (1943). A new aspect of creep in concrete and its application to design. *Proceedings of the American Society for Testing and Materials, 43,* 1069–1086.

620. McKay, M. D., Beckman, R. J., & Conover, W. J. (1979). A comparison of three methods for selecting values of input variables in the analysis of output from a computer code. *Technometrics, 21,* 239–245.

621. McKay, M. D., Conover, W. J., & Beckman, R. J. (1976). *Report on the application of statistical techniques to the analysis of computer code, Technical report LA-NUREG-6526-MS, NRC-4.* New Mexico: Los Alamos Scientific Laboratory.

622. McMillan, F. R. (1915). Shrinkage and time effects in reinforced concrete. *Studies in Engineering. Bulletin* (Vol. 3), University of Minnesota, 41 pp.

623. McMillan, F. R. (1916). Method of designing reinforced concrete slabs, discussion of A.C. Janni's paper. *Transactions ASCE, 80,* 1738.

624. Meftah, F., Pont, S. D., & Schrefler, B. A. (2012). A three-dimensional staggered finite element approach for random parametric modeling of thermo-hygral coupled phenomena in porous media. *International Journal of Numerical and Analytical Methods in Geomechanics, 36,* 574–596.

625. Mehmel, A., & Kern, E. (1962). Elastische und plastische Stauchungen von Beton infolge Druckschwell- und Standbelastung. *Deutscher Ausschuss für Stahlbeton* (Vol. 153). W. Ernst: Berlin.

626. Meyer, C. A., McClintock, R. B., Silvestri, C. J., & Spencer, R. C. (Eds.). (1993). *ASME Steam Tables: Thermodynamic and Transport Properties of Steam.* New York: ASME Press.

627. Meyer, O. (1874). Zur Theorie der inneren Reibung. *Zeitschrift für reine und angewandte Mathematik, 78,* 130–135.

628. Meyers, S. L. (1951). How temperature and moisture changes may affect the durability of concrete. *Rock Products,* 153–157.

629. Mikhail, R. S., Abo-El-Enein, S. A., & Abd-El-Khalik, M. (1975). Hardened slagcement pastes of various porosities. II. Water and nitrogen adsorption. *Journal of Applied Chemistry and Biotechnology, 25,* 835–847.

630. Mindeguia, J.-C. (2009). Contribution expérimentale à la compréhension des risques d'instabilité thermique des bétons, Ph.D. thesis, UPPA.

631. Mindeguia, J.-C., Pimienta, P., Carré, H., & Borderie, C. L. (2013). Experimental analysis of concrete spalling due to fire exposure. *European Journal of Environmental and Civil Engineering, 17,* 453–466.

632. Mindess, S. (1985). Rate of loading effects on the fracture of cementitious materials. In S. P. Shah (Ed.), *Application of fracture mechanics to cementitious composites* (pp. 617–636). Dordrecht: Martinus Nijhoff Publishers.

633. Mindess, S., Young, J. F., & Darwin, D. (2003). *Concrete* (2nd ed.). Englewood Cliffs: Prentice Hall.

634. Miyazawa, S., & Tazawa, E. (2001). Prediction model for shrinkage of concrete including autogenous shrinkage. In F.-J. Ulm, Z. P. Bažant, & F. H. Wittmann (Eds.), *Creep, shrinkage and durability mechanics of concrete and other quasi-brittle materials* (pp. 735–740). Amsterdam: Elsevier.

635. Monlouis-Bonnaire, J. P. (2003). Numerical modeling of coupled air-water-salt transfers in cementitious materials and terracotta (in French), Toulouse.

636. Moonen, P. (2009). Continuous-discontinuous modelling of hygrothermal damage processes in porous media, Ph.D. thesis, Delft University of Technology.

637. Mori, J., & Tanaka, K. (1973). Average stress in matrix and average elastic energy of materials with misfitting inclusions. *Acta Metallurgica, 21*, 571.

638. Müller, H. S. (1993). Considerations on the development of a database on creep and shrinkage tests. In Z. P. Bažant & I. Carol (Eds.), *Creep and shrinkage of concrete* (pp. 859–872). London: E & FN Spon.

639. Müller, H. S., Anders, I., Breiner, R., & Vogel, M. (2013). Concrete: Treatment of types and properties in fib model code 2010. *Structural Concrete, 14*, 320–334.

640. Müller, H. S., Bažant, Z. P., & Kuttner, C. H. (1999). *Data base on creep and shrinkage tests, RILEM Subcommittee 5 report RILEM TC 107-CSP*. Paris: RILEM.

641. Müller, H. S., & Hilsdorf, H. K. (1990). Evaluation of the time-dependent behaviour of concrete. Summary report on the work of the General Task Force Group 9, bulletin d'information no. 199, Comité Euro-International du Béton (CEB), Lausanne.

642. Mullick, A. K. (1972). Effect of stress history on the microstructure and creep properties of maturing concrete, Ph.D. thesis, University of Calgary, Alberta, Canada.

643. Multon, S., & Toutlemonde, F. (2006). Effect of applied stresses on alkali-silica reaction-induced expansions. *Cement and Concrete Research, 36*, 912–920.

644. Murata, J. (1965). Studies of the permeability of concrete. *Bulletin RILEM, 29*, 47–54.

645. Nasser, K. W., & Neville, A. M. (1965). Creep of concrete at elevated temperatures. *ACI Journal, 62*, 1567–1579.

646. Navrátil, J. (1998). The use of model B3 extension for the analysis of bridge structures (in Czech). *Stavební obzor, 4*, 110–116.

647. Nawy, E. G. (2006). *Prestressed concrete: A fundamental approach* (5th ed.). Upper Saddle River, New Jersey: Pearson Prentice Hall.

648. Nechnech, W., Meftah, F., & Reynouard, J. M. (2002). An elasto-plastic damage model for plain concrete subjected to high temperatures. *Engineering Structures, 24*(5), 597–611. http://linkinghub.elsevier.com/retrieve/pii/S0141029601001250.

649. Needleman, A. (1987). A continuum model for void nucleation by inclusion debonding. *Journal of Applied Mechanics, ASME, 54*, 525–531.

650. Neuenschwander, M., Knobloch, M., & Fontana, M. (2016). Suitability of the damage-plasticity modelling concept for concrete at elevated temperatures: Experimental validation with uniaxial cyclic compression tests. *Cement and Concrete Research, 79*, 57–75.

651. Neuner, M., Gamnitzer, P., & Hofstetter, G. (2017). An extended damage plasticity model for shotcrete: Formulation and comparison with other shotcrete models. *Materials, 10*(1), 82. http://www.mdpi.com/1996-1944/10/1/82.

652. Neville, A. M. (1970). *Creep of concrete: Plain, reinforced and prestressed*. Amsterdam: North-Holland.

653. Neville, A. M. (1997). *Properties of concrete* (4th ed.). New York: Wiley.

654. Neville, A. M., & Hirst, G. A. (1978). Mechanism of cyclic creep of concrete. In *Douglas McHenry International Symposium on Concrete and Concrete Structures* (pp. 83–101). American Concrete Institute (ACI SP-55).

655. Ngab, A. S., Nilson, A. H., & Slate, F. O. (1981). Shrinkage and creep of high strength concrete. *Journal of the American Concrete Institute, 78*, 255–261.

656. Nicolas, P. (1992). Modélisation mathématique et numérique des transfers d'humidité en milieux poreux, Ph.D. thesis, Université Paris VI.

657. Nielsen, C. V., Pearce, C. J., & Bicanic, N. (2004). Improved phenomenological modeling of transient thermal strains for concrete at high temperatures. *Computers and Concrete, 1*, 189–204.

658. Nielsen, L. F. (1970). Kriechen und Relaxation des Betons. *Beton- und Stahlbetonbau, 65*, 272–275.

659. Nilsen, A. U., & Monteiro, P. J. M. (1993). Concrete: A three phase material. *Cement and Concrete Research, 23*, 147–151.

660. Nilson, A. H. (1987). *Design of prestressed concrete* (2nd ed.). New York: Wiley.

661. Nilsson, L. O. (1980). Hygroscopic moisture in concrete – drying, measurements related material properties, Ph.D. thesis, Lund University.

662. Nordby, G. M. (1967). Fatigue of concrete–a review of research. *Journal of the American Concrete Institute, 30*, 191–219.

663. Odler, I., Hagymassy, J., Yudenfreund, M., Hanna, K. M., & Brunauer, S. (1972). Pore structure analysis by water vapor adsorption. IV. Analysis of hydrated Portland cement pastes of low porosity. *Journal of Colloid and Interface Science, 38*, 265–276.

664. Oliver, J. (1989). A consistent characteristic length for smeared cracking models. *International Journal for Numerical Methods in Engineering, 28*, 461–474.

665. Ožbolt, J., Li, Y.-J., & Kožar, I. (2001). Microplane model for concrete with relaxed kinematic constraint. *International Journal of Solids and Structures, 38*, 2683–2711.

666. Ožbolt, J., & Reinhardt, H.-W. (2001). Sustained loading strength of concrete modelled by creep-cracking interaction. *Otto Graf Journal, Annual Journal on Research and Testing of Materials, 12*, 9–20.

667. Ožbolt, J., & Reinhardt, H.-W. (2005). Rate dependent fracture of notched plain concrete beams. In G. Pijaudier-Cabot, B. Gérard, & P. Acker (Eds.), *Proceedings of the 7th International Conference CONCREEP-7, Nantes* (pp. 57–62).

668. Pandolfi, A., & Taliercio, A. (1998). Bounding surface models applied to fatigue of plain concrete. *Journal of Engineering Mechanics, ASCE, 124*, 556–564.

669. Paris, P. C., & Erdogan, F. (1963). Critical analysis of propagation laws. *Journal of Basic Engineering, 85*, 528–534.

670. Parker, D. (1996). Tropical overload. *New Civil Engineer*.

671. Pauw, A. (1960). Static modulus of elasticity of concrete as affected by density. *Journal of the American Concrete Institute, 32*, 679–687.

672. Pedersen, C. R. (1990). Combined heat and moisture transfer in building constructions, Ph.D. thesis, Technical University of Denmark.

673. Pellenq, R. J.-M., Kushima, A., Shashavari, R., Vliet, K. J. V., Buehler, M. J., Yip, S., et al. (2010). A realistic molecular model of cement hydrates. *Proceedings of the National Academy of Sciences of the United States of America, 106*, 16102–17107.

674. Pentala, V., & Rautanen, T. (1990). Microporosity, creep and shrinkage of high-strength concrete. In W. T. Heston (Ed.), *2nd International Symposium on High-Strength Concrete* (ACI SP 121-21, pp. 409–432).

675. Perre, P., & Degiovanni, A. (1990). Simulation par volumes finis des transferts couplés en milieux poreux anisotropes: séchage du bois a basse et a haute température. *International Journal of Heat and Mass Transfer, 33*(11), 2463–2478. http://linkinghub.elsevier.com/retrieve/pii/001793109090004E.

676. Persson, B. (1996). Hydration and strength of high performance concrete. *Advanced Cement Based Materials, 3*, 107–123.

677. Persson, B. (1997). Long-term effect of silica fume on the principal properties of low-temperature-cured ceramics. *Cement and Concrete Research, 27*, 1667–1680.

678. Persson, B. (2001). A comparison between mechanical properties of self-compacting concrete and the corresponding properties of normal concrete. *Cement and Concrete Research, 31*, 193–198.

679. Persson, B. (2002). Eight-year exploration of shrinkage in high-performance concrete. *Cement and Concrete Research, 32*, 1229–1237.

680. Persson, B. (2004). Fire resistance of self-compacting concrete. *Materials and Structures, 37*, 575–584.

681. Pesavento, F. (2017). Private communication by email with M. Jirásek, 5 June 2017.

682. Pesavento, F., Gawin, D., & Schrefler, B. A. (2008). Modelling cementitious materials as multiphase porous media: Theoretical framework and applications. *Acta Mechanica, 201,* 313–339.

683. Pfeil, W. (1981). Twelve years monitoring of long span prestressed concrete bridge. *Concrete International, 3,* 79–84.

684. Phan, L., & Carino, N. (2002). Effects of test conditions and mixture proportions on behavior of high-strength concrete exposed to high temperatures. *ACI Materials Journal, 99*(1), 54–66. http://80.apps.webofknowledge.com.dialog.cvut.cz/full_record.do?product=WOS& search_mode=GeneralSearch&qid=9&SID=Z1xJYslL39jJihgPGew&page=1&doc=3.

685. Phan, L. T. (2008). Pore pressure and explosive spalling in concrete. *Materials and Structures, 41,* 1623–1632.

686. Phan, L. T., & Carino, N. J. (1998). Review of mechanical properties of HSC at elevated temperature. *Journal of Materials in Civil Engineering, 10,* 58–64.

687. Philips, R. (2001). *Crystals, defects and microstructures: Modeling across scales.* New York: Cambridge University Press.

688. Picandet, V., Khelidj, A., & Bastian, G. (2001). Effect of axial compressive damage on gas permeability of ordinary and high-performance concrete. *Cement and Concrete Research, 31*(11), 1525–1532.

689. Pichler, C., & Lackner, R. (2008). A multiscale creep model as basis for simulation of early-age concrete behavior. *Computers and Concrete, 5,* 295–328.

690. Pickett, G. (1942). The effect of change in moisture content on the creep of concrete under a sustained load. *Journal of the American Concrete Institute, 38,* 333–355.

691. Pickett, G. (1946). Shrinkage stresses in concrete. *Journal of the American Concrete Institute, 47*(165–204), 361–397.

692. Pietruszczak, S., & Mróz, Z. (1981). Finite element analysis of deformation of strain-softening materials. *International Journal for Numerical Methods in Engineering, 17,* 327–334.

693. Pihlajavaara, S. E. (1965). A review of the research on drying of concrete. *Bulletin RILEM (Paris), 27,* 61–63.

694. Pilz, M. (1997). The collapse of the KB bridge in 1996, Ph.D. thesis, Imperial College London.

695. Pilz, M. (1999). Untersuchungen zum Einsturz der KB Brücke in Palau. *Beton- und Stahlbetonbau,* (94/5).

696. Pirtz, D. (1968). Creep characteristics of mass concrete for Dworshak Dam. *Structural Engineering Laboratory 65-2,* University of California, Berkeley.

697. Polivka, M., Pirtz, D., & Adams, R. F. (1964). Studies of creep in mass concrete. In *Symposium on creep of concrete* (pp. 257–285). American Concrete Institute (ACI SP-9).

698. Poole, J. L., Riding, K. A., Folliard, K. J., Juenger, M. C. G., & Schindler, A. K. (2007). Methods for calculating activation energy for Portland cement. *ACI Materials Journal, 104,* 86–94.

699. Popovics, S. (1986). A model for deformations of two-phase composites under load. In Z. P. Bažant (Ed.), *Proceedings of the International Symposium on Creep and Shrinkage of Concrete: Mathematical Modeling, RILEM, Northwestern University, Evanston, Ill* (pp. 733–742).

700. Powers, T. C. (1947). A discussion of cement hydration in relation to the curing of concrete. *Proceedings of the Highway Research Board, 27,* 178–188.

701. Powers, T. C. (1955). Hydraulic pressure in concrete. *Proceedings of the American Society of Civil Engineers,81.* Paper 742.

702. Powers, T. C. (1965). Mechanism of shrinkage and reversible creep of hardened cement paste. In *Proceedings of the International Conference on the Structure of Concrete* (pp. 319–344). London: Cement and Concrete Association.

703. Powers, T. C. (1966). Some observations on the interpretation of creep data. *Bulletin RILEM (Paris),* 381.

704. Powers, T. C. (1968). The thermodynamics of volume change and creep. *Materials and Structures, 1,* 487–507.

705. Powers, T. C., & Brownyard, T. L. (1946). Studies of the physical properties of hardened Portland cement paste. *ACI Journal, Proceedings*, *42*, 101–132, 249–366, 469–504.

706. Powers, T. C., & Brownyard, T. L. (1947). Studies of the physical properties of hardened Portland cement paste. *ACI Journal, Proceedings*, *43*, 549–602, 669–712, 854–880, 933–992.

707. Powers, T. C., Copeland, L. E., Hayes, J. C., & Mann, H. M. (1954). Permeability of Portland cement paste. *Journal of the American Concrete Institute*, *51*, 285–298.

708. Poyet, S., Sellier, A., Capra, B., Thèvenin-Foray, G., Torrenti, J.-M., Tournier-Cognon, H., et al. (2006). Influence of water on alkali-silica reaction: Experimental study and numerical simulations. *Journal of Materials in Civil Engineering*, *18*, 588–596.

709. Prokopovich, I. E. (1963). *Effects of long-term processes on stress and deformation state in structures*. Moscow: Gosstroyizdat. (in Russian).

710. Prokopovich, I. E. (1969). Consideration of creep and shrinkage in analysis of reinforced concrete structures (in Russian). *Beton i Zhelezobeton*, *15*.

711. Purkiss, J. A., & Dougill, J. W. (1973). Apparatus for compression tests on concrete at high temperatures. *Magazine of Concrete Research*, *25*, 102–108.

712. Radjy, F., & Richards, C. W. (1973). Effect of curing and heat treatment history on the dynamic response of cement paste. *Cement and Concrete Research*, *3*, 7–21.

713. Radjy, F., Sellevold, E. J., & Hansen, K. (2003). Isosteric vapor pressure—temperature data for water sorption in hardened cement paste: Enthalpy, entropy and sorption isotherms at different temperatures, Technical report BYG DTU R-057, Technical University of Denmark.

714. Rahimi-Aghdam, S., Bažant, Z. P., & Caner, F. C. (2017). Diffusion-controlled and creep-mitigated ASR damage via microplane model: II. Material degradation, drying and verification. *Journal of Engineering Mechanics, ASCE*, *143*, 04016109-1–04016109-10.

715. Rahimi-Aghdam, S., Bažant, Z. P., & Qomi, M. J. A. (2017). Cement hydration from hours to centuries controlled by diffusion through barrier shells of C-S-H. *Journal of the Mechanics and Physics of Solids*, *99*, 211–224.

716. Raiffa, H., & Schlaifer, R. (1961). *Applied statistical decision theory*. Cambridge: Harvard University Press.

717. Rarick, R. L., Bhatty, J. W., & Jennings, H. M. (1995). Surface area measurement using gas sorption: Application to cement paste. In J. Skalny & S. Mindess (Eds.), *Materials science of concrete IV* (pp. 1–41). Westerville: The American Ceramic Society.

718. Rashid, Y. R. (1972). Nonlinear analysis of two-dimensional problems in concrete creep. *Journal of Applied Mechanics, ASME*, *39*, 475–482.

719. Reinhardt, H. W. (1985). Tensile fracture of concrete at high rates of loading. In S. P. Shah (Ed.), *Application of fracture mechanics to cementitious composites* (pp. 559–592). Dordrecht: Martinus Nijhoff Publishers.

720. Reinhardt, H. W. (1986). Strain rate effects on the tensile strength of concrete as predicted by thermodynamic and fracture mechanics models. In S. Mindess & S. P. Shah (Eds.), *Cement-based composites: Strain rate effects on fracture* (pp. 1–14).

721. Reinhardt, H. W., Cornelissen, H. A. W., & Hordijk, D. A. (1986). Tensile tests and failure analysis of concrete. *Journal of Structural Engineering, ASCE*, *112*, 2462–2477.

722. Reissner, E. (1946). Analysis of shear lag in box beams by the principle of minimum potential energy. *Quarterly of Applied Mathematics*, *4*, 268–278.

723. Rice, J. R. (1968). Mathematical analysis in the mechanics of fracture. In H. Liebowitz (Ed.), *Fracture–an advanced treatise* (Vol. 2, pp. 191–308). New York: Academic Press.

724. Richards, L. A. (1931). Capillary conduction of liquids through porous mediums. *Journal of Applied Physics*, *1*, 318–333.

725. Ricken, D. (1989). Ein einfaches Berechnungsverfahren für die eindimensionale, instationäre Wasserdampfdiffusion in mehrschichtigen Bauteilen, Ph.D. thesis, Universität Dortmund, Dortmund, Germany.

726. Rilem, T. C., & 119-TCE., (1997). Avoidance of thermal cracking in concrete at early ages. *Materials and Structures*, *30*, 451–464.

727. Technical Committee, R. I. L. E. M., & 107., (1995). Guidelines for characterizing concrete creep and shrinkage in structural design codes or recommendations. *Materials and Structures*, *28*, 52–55.

728. Risken, H. (1989). *The Fokker-Planck equation* (2nd ed.). New York: Springer.

729. Robson, C. J., Davie, C. T., & Gosling, P. D. (2010). Investigation into the form of the load-induced thermal strain model. *Computational modelling of concrete structures – proceedings of EURO-C 2010*. Leiden: CRC Press/Balkema.

730. Rodway, L. E., & Fedirko, W. M. (1989). Superplasticized high volume fly ash concrete. In *Proceedings of the Third International Conference on Fly Ash, Silica Fume, Slag and Natural Pozzolans, Trondheim, Norway* (pp. 98–112).

731. Roll, F. (1964). Long-time recovery of highly stressed concrete cylinders. *Symposium on creep of concrete* (pp. 95–114). American Concrete Institute (ACI SP-9).

732. Roncero, J. (1999). Effect of superplasticizers on the behavior of concrete in the fresh and hardened states: Implications for high performance concrete, Ph.D. thesis, UPC Barcelona.

733. Roscoe, R. (1950). Mechanical models for the representation of viscoelastic properties. *British Journal of Applied Physics*, *1*, 171–173.

734. Ross, A. D. (1958). Creep of concrete under variable stress. *ACI Journal, Proceedings*, *54*, 739–758.

735. Ross, C. A., & Kuennen, S. T. (1989). Fracture of concrete at high strain-rates. In S. P. Shah, S. E. Swartz, & B. Barr (Eds.), *Fracture of concrete and rock: Recent developments* (pp. 152–161). London: Elsevier Applied Science.

736. Rostásy, F. S., Teichen, K.-T., & Engelke, H. (1972). *Beitrag zur Klärung des Zusammenhanges von Kriechen und Relaxation bei Normal-beton, Technical report 139*. Otto-Graf Institute: University of Stuttgart, Stuttgart, Germany.

737. Rostásy, F. S., Thienel, K.-C., & Schütt, K. (1991). On prediction of relaxation of colddrawn prestressing wire under constant and variable elevated temperature. *Nuclear Engineering and Design*, *130*, 221–227.

738. Rots, J. G. (1988). Computational modeling of concrete fracture, Ph.D. thesis, Delft University of Technology, Delft, The Netherlands.

739. Rougelot, T., Skoczylas, F., & Burlion, N. (2009). Water desorption and shrinkage in mortars and cement pastes: Experimental study and poromechanical model. *Cement and Concrete Research*, *39*, 36–44.

740. Rousseeuw, R. J., & Leroy, A. M. (1987). *Robust regression and outlier detection*. New York: Wiley.

741. Ruetz, W. (1966). Das Kriechen des Zementsteins im Beton und seine Beeinflussung durch gleichzeitiges Schwinden. *Deutscher Ausschuss für Stahlbeton* (Vol. 183). Berlin: W. Ernst.

742. Ruetz, W. (1968). A hypothesis for the creep of hardened cement paste and the influence of simultaneous shrinkage. In *Proceedings of the International Conference on the Structure of Concrete* (pp. 365–387). London: Cement and Concrete Association.

743. Ruiz, J., Schindler, A., Rasmussen, R., Kim, P., & Chang, G. (2001). Concrete temperature modeling and strength prediction using maturity concepts in the FHWA HIPERPAV software. In *7th International Conference on Concrete Pavements, Orlando, Florida*.

744. Rüsch, H. (1968). Festigkeit und Verformung von unbewehrten Beton unter konstanter Dauerlast. *Deutscher Ausschuss für Stahlbeton* (Vol. 198). Berlin: W. Ernst & Sohn (See also *ACI Journal*, *57*, 1–58 (1968)).

745. Rüsch, H., & Jungwirth, D. (1976). *Berücksichtigung der Einflüsse von Kriechen und Schwinden auf das Verhalten der Tragwerke*. Düsseldorf: Werner-Verlag.

746. Rüsch, H., Jungwirth, D., & Hilsdorf, H. (1973). Kritische Sichtung der Einflüsse von Kriechen und Schwinden des Betons auf das Verhalten der Tragwerke. *Beton- und Stahlbetonbau*, *68*, 49–60, 76–86, 152–158.

747. Russell, H., & Burg, R. (1996). Test data on Water Tower Place concrete. In *International Workshop on High Performance Concrete*. Farmington Hills, Michigan: American Concrete Institute (ACI SP-159).

748. Sakata, K. (1993). Prediction of concrete creep and shrinkage. In Z. P. Bažant & I. Carol (Eds.), *Creep and shrinkage of concrete* (pp. 649–654). London: E & FN Spon.
749. Sakata, K., Tsubaki, T., Inoue, S., & Ayano, T. (2001). Prediction equations of creep and drying shrinkage for wide-ranged strength concrete. In F.-J. Ulm, Z. P. Bažant, & F. H. Wittmann (Eds.), *Creep, shrinkage and durability mechanics of concrete and other quasibrittle materials* (pp. 753–758). Amsterdam: Elsevier.
750. Salviato, M., Kirane, K., & Bažant, Z. P. (2014). Statistical distribution and size effect of residual strength of quasibrittle materials after a period of constant load. *Journal of the Mechanics and Physics of Solids, 64*, 440–454.
751. Sammis, C. G., & Ashby, M. F. (1986). The failure of brittle porous solids under compressive stress state. *Acta Metallurgica et Materialia, 34*, 511–526.
752. Sanahuja, J. (2013). Effective behavior of ageing linear viscoelastic composites: Homogenization approach. *International Journal of Solids and Structures, 50*, 2846–2856.
753. Sanahuja, J., & Dormieux, L. (2010). Creep of a C-S-H gel: Micromechanical approach. *International Journal for Multiscale Computational Engineering, 8*, 357–368.
754. Sanzharovskii, R. S., & Manchenko, M. M. (2015). Creep of concrete and its instantaneous nonlinearity of deformation in the structural calculations. *Structural mechanics and constructions of buildings* (Vol. 2). St. Petersburg.
755. Sattler, K. (1953). *Theorie der Verbundkonstruktionen*. Berlin: Verlag W. Ernst.
756. Scheiner, S., & Hellmich, C. (2009). Continuum microviscoelasticity model for aging basic creep of early-age concrete. *Journal of Engineering Mechanics, ASCE, 135*, 307–323.
757. Scherer, G. W. (1992). Bending of gel beams: Method of characterizing mechanical properties and permeability. *Journal of Non-Crystalline Solids, 142*, 18–35.
758. Scherer, G. W. (1994). Relaxation of a viscoelastic gel bar: I. Theory. *Journal of Sol-Gel Science and Technology, 1*, 169–175.
759. Scherer, G. W. (2000). Measuring permeability of rigid materials by a beam-bending method: I. Theory. *Journal of the American Ceramic Society, 83*, 2231–2239.
760. Schmidt-Döhl, F., & Rostásy, F. S. (1995). Crystallization and hydration pressure or formation pressure of solid phases. *Cement and Concrete Research, 25*, 255–256.
761. Schneider, U. (1976). Behavior of concrete under thermal steady state and non-steady state conditions. *Fire and Structures, 1*, 103–115.
762. Schneider, U. (1982). Behavior of concrete at high temperatures. *Deutscher Ausschuss für Stahlbeton* (Vol. 337). Berlin: W. Ernst & Sohn.
763. Schneider, U. (1988). Concrete at high temperatures–a general review. *Fire Safety Journal, 13*, 55–68.
764. Schneider, U. (2010). Permeability of high performance concrete at elevated temperatures. In *Proceedings of the 6th International Conference on Structures in Fire, East Lansing, MI, USA*.
765. Schneider, U., & Herbst, H. J. (1989). Permeabilität und Porosität von Beton bei hohen Temperaturen. (*Heft* 403, pp. 23–52). Deutscher Ausschuss für Stahlbeton.
766. Schrefler, B. A. (2002). Mechanics and thermodynamics of saturated-unsaturated porous materials and quantitative solutions. *Applied Mechanics Reviews, ASME, 55*, 351–388.
767. Schrefler, B. A., Khoury, G. A., Gawin, D., & Majorana, C. E. (2002). Thermo-hydro-mechanical modeling of high performance concrete at high temperature. *Engineering Computations, 19*, 787–819.
768. Schulson, E. M., & Nickolayev, O. Y. (1995). Failure of columnar saline ice under biaxial compression: Failure envelopes and the brittle-to-ductile transition. *Journal of Geophysical Research, 100*, 22383–22400.
769. Schwartz, L. (1950). *Théorie des distributions*. Paris: Hermann.
770. Schwarzl, F., & Staverman, A. J. (1952). Time-temperature dependence of linear viscoelastic behavior. *Journal of Applied Physics, 23*, 838.
771. Selna, L. G. (1969). A concrete creep, shrinkage and cracking law for frame structures. *ACI Journal, 66*, 847–848.

772. Shah, S. P., Weiss, W. J., & Yang, W. (2000). Influence of specimen size/geometry on shrinkage cracking of rings. *Journal of Engineering Mechanics, ASCE, 126*, 93–101.

773. Shahidi, M., Pichler, B., & Hellmich, C. (2014). Viscous interfaces as source for material creep: A continuum micromechanics approach. *European Journal of Mechanics/A: Solids, 45*, 41–58.

774. Shahidi, M., Pichler, B., & Hellmich, C. (2016). Interfacial micromechanics assessment of classical rheological models. *Journal of Engineering Mechanics, ASCE, 142*(04015092), 04015093.

775. Shawwaf, K. (2008). Private communication. Dir., Dywidag Systems International USA, Bollingbrook, Illinois; former structural analyst on KB bridge design team.

776. Shideler, J. J. (1957). Lightweight aggregate concrete for structural use. *ACI Journal, 54*, 299–328.

777. Shritharan, S. (1989). *Structural effects of creep and shrinkage on concrete structures*. Civil Engineer: University Auckland.

778. Sinko, R., Bažant, Z. P., & Keten, S. (2017). A nanoscale perspective on the effects of transverse microprestress on drying creep of nanoporous solids. *Proceedings of the Royal Society A*. In press.

779. Sinko, R., Vandamme, M., Bažant, Z. P., & Keten, S. (2016). Transient effects of drying creep in nanoporous solids: Understanding the effects of nanoscale energy barriers. *Proceedings of the Royal Society A, 472*, 20160490.

780. Sluys, L. J. (1992). Wave propagation, localisation and dispersion in softening solids, Ph.D. thesis, Delft University of Technology, Delft, The Netherlands.

781. Smith, D. E., Wang, Y., Chaturvedi, A., & Whitley, H. D. (2006). Molecular simulations of the pressure, temperature, and chemical potential dependencies of clay swelling. *Journal of Physical Chemistry B, 110*, 20046–20054.

782. SOFiSTiK AG. (2010). AQB design of cross sections and of prestressed concrete and composite cross sections v13.64. Software Manual.

783. Soong, T. T. (2004). *Fundamentals of probability and statistics for engineers*. Chichester: Wiley.

784. Souley, M., Homand, F., Pepa, S., & Hoxha, D. (2001). Damage-induced permeability changes in granite: A case example at the URL in Canada. *International Journal of Rock Mechanics and Mining Sciences, 38*, 297–310.

785. SSFM Engineers (1996). Preliminary assessment of Koror-Babeldaob bridge failure, Technical report, SSFM Engineers, Inc., prepared for US Army Corps of Engineers.

786. Specifications, Standard, & for Concrete Structures - 2007 "Design"., (2010). *Tokyo* (p. 15). JSCE Guidelines for Concrete No: Japan.

787. Staverman, A. J., & Schwarzl, P. (1952). Non-equilibrium thermodynamics of visco-elastic behaviour. *Proceedings of the Academy of Sciences The Netherlands, 55*, 486–492.

788. Staverman, A. J., & Schwarzl, P. (1952). Thermodynamics of viscoelastic behaviour (model theory). *Proceedings of the Academy of Sciences The Netherlands, 55*, 474–485.

789. Stefan, J. (1879). Über die Beziehung zwischen der Wärmestrahlung und der Temperatur. *Sitzungsberichte* (pp. 391–428). Wien: Akademie der Wissenschaften.

790. Stöckl, S. (1978). Lastversuche über den Einfluss von vorangegangenen Dauerlasten auf die Kurzfestigkeit des Betons. *Deutscher Ausschuss für Stahlbeton* (Vol. 196). Berlin: W. Ernst & Sohn.

791. Strauss, A., Hoffmann, S., Wendner, R., & Bergmeister, K. (2009). Structural assessment and reliability analysis for existing engineering structures, applications for real structures. *Structure and Infrastructure Engineering, 5*, 277–286.

792. Strauss, A., Wendner, R., Bergmeister, K., & Costa, C. (2013). Numerically and experimentally based reliability assessment of a concrete bridge subjected to chloride induced deterioration. *Journal of Infrastructure Systems, 19*, 166–175.

793. Sukumar, N., Dolbow, J. E., & Moës, N. (2015). Extended finite element method in computational fracture mechanics: A retrospective examination. *International Journal of Fracture, 196*, 189–206.

794. Suresh, S. (1998). *Fatigue of materials*. Cambridge: Cambridge University Press.
795. Suresh, S., Tschegg, E. K., & Brockenbrough, J. R. (1989). Crack growth in cementitious composites under cyclic compressive loads. *Cement and Concrete Research, 19*, 827–833.
796. T.Y. Lin International. (1996). *Collapse of the Koror-Babelthuap bridge, Technical report*. Lin International: T.Y.
797. Tada, H., Paris, P. C., & Irwin, G. R. (1973). *The stress analysis of cracks handbook*. Hellerton: Del Research Corporation.
798. Tadros, M. K., Ghali, A., & Dilger, W. H. (1975). Time-dependent prestress loss and deflection in prestressed concrete members. *Prestressed Concrete Institute Journal, 20*, 86–98.
799. Tadros, M. K., Ghali, A., & Dilger, W. H. (1979). Long-term stresses and deformations of segmental bridges. *Prestressed Concrete Institute Journal, 24*, 66–87.
800. Tandon, S., Faber, K. T., Bažant, Z. P., & Li, Y.-N. (1995). Cohesive crack modeling of influence of sudden changes in loading rate on concrete fracture. *Engineering Fracture Mechanics, 52*, 987–997.
801. Taylor, G. I. (1938). Plastic strain in metals. *Journal of the Institute of Metals, 62*, 307–324.
802. Taylor, H. F. W. (1990). *Cement chemistry*. New York: Academic Press.
803. Taylor, R. L., Pister, K. S., & Goudreau, G. L. (1970). Thermomechanical analysis of viscoelastic solids. *International Journal for Numerical Methods in Engineering, 2*, 45–60.
804. Tazawa, E. I., & Miyazawa, S. (1995). Influence of cement and admixture on autogenous shrinkage of cement paste. *Cement and Concrete Research, 25*, 281–287.
805. Tenchev, R., & Purnell, P. (2005). An application of a damage constitutive model to concrete at high temperature and prediction of spalling. *International Journal of Solids and Structures, 42*, 6550–6565.
806. Tenchev, R. T., Li, L. Y., & Purkiss, J. A. (2001). Finite element analysis of coupled heat and moisture transfer in concrete subjected to fire. *Numerical Heat Transfer, Part A: Applications, 39*, 685–710.
807. Termkhajornkit, P., Nawa, T., Nakai, M., & Saito, T. (2005). Effect of fly ash on autogenous shrinkage. *Cement and Concrete Research, 35*, 473–482.
808. Terro, M. J. (1998). Numerical modeling of the behavior of concrete structures in fire. *ACI Structural Journal, 95*, 183–193.
809. Thai, M.-Q., Bary, B., & He, Q.-C. (2014). A homogenization-enriched viscodamage model for cement-based material creep. *Engineering Fracture Mechanics, 126*, 54–72.
810. Thelandersson, S. (1971). Effect of high temperatures on tensile strength of concrete. Report, Division of Structural, Mechanical and Concrete Construction, Lund Institute of Technology, Lund, Sweden.
811. Thelandersson, S. (1987). Modeling of combined thermal and mechanical action in concrete. *Journal of Engineering Mechanics, ASCE, 113*, 893–906.
812. Thomson, W. (1871). On the equilibrium of vapour at a curved surface of liquid. *Philosophical Magazine, 42*, 448–452.
813. Thomson, W. (1878). Elasticity. *Encyclopaedia britannica* (9th ed., Vol. VII, p. 803d). New York: Charles Scribner's.
814. Thouless, M. D., Hsueh, C. H., & Evans, A. G. (1983). A damage model for creep crack growth in polycrystals. *Acta Metallica, 31*, 1675–1687.
815. Tran, A. B., Yvonnet, J., He, Q.-C., Toulemonde, C., & Sanahuja, J. (2011). A simple computational homogenization method for structures made of linear heterogeneous viscoelastic materials. *Computer Methods in Applied Mechanics and Engineering, 200*, 2956–2970.
816. Trost, H. (1967). Auswirkungen des Superpositionsprinzip auf Kriech- un Relaxations-Probleme bei Beton und Spannbeton. *Beton- und Stahlbetonbau, 62*(230–238), 261–269.
817. Troxell, G. E., Raphael, J. E., & Davis, R. W. (1958). Long-time creep and shrinkage tests of plain and reinforced concrete. *Proceedings of the American Society for Testing and Materials, 58*, 1101–1120.
818. Tschoegl, N. W. (1989). *The phenomenological theory of linear viscoelastic behavior*. Berlin: Springer.

819. Tvergaard, V., & Hutchinson, J. W. (1992). The relation between crack growth resistance and fracture process parameters in elastic-plastic solids. *Journal of the Mechanics and Physics of Solids, 40*, 1377–1397.

820. Ulickii, I. I. (1963). *Determination of the magnitude of creep and shrinkage deformations of concrete.* Kiev: Stroyizdat. (in Russian).

821. Ulickii, I. I., Chan, U. Y., & Bolyshev, A. V. (1960). *Analysis of reinforced concrete structures considering long-time processes.* Kiev: Gosstroyizdat. (in Russian).

822. Ulm, F.-J., & Acker, P. (1998). *Le point sur le fluage et la recouvrance des bétons* (pp. 73–82). Spécial XX: Bulletin de liaison des Laboratoires des Ponts et Chaussées.

823. Ulm, F.-J., Acker, P., & Lévy, M. (1999). The "Chunnel" fire. II. Analysis of concrete damage. *Journal of Engineering Mechanics, ASCE, 125*, 283–289.

824. Ulm, F.-J., & Coussy, O. (1995). Modeling of thermochemomechanical couplings of concrete at early ages. *Journal of Engineering Mechanics, 121*(7), 785–794.

825. Ulm, F.-J., & Coussy, O. (1996). Strength growth as chemo-plastic hardening in early age concrete. *Journal of Engineering Mechanics, 122*(12), 1123–1132.

826. Ulm, F.-J., Coussy, O., & Bažant, Z. P. (1999). The "Chunnel" fire. I. Chemoplastic softening in rapidly heated concrete. *Journal of Engineering Mechanics, ASCE, 125*, 272–282.

827. Ulm, F., Maou, F. L., & Boulay, C. (2000). *Creep and shrinkage of concrete—kinetics approach* (pp. 134–153). American Concrete Institute (ACI SP-194).

828. van Genuchten, M. T. (1980). A closed-form equation for predicting the hydraulic conductivity of unsaturated soils. *Soil Science Society of America Journal, 44*, 892–898.

829. van Zijl, G. P. (1999). Computational modelling of masonry creep and shrinkage, Ph.D. thesis, Delft University of Technology, Delft, The Netherlands.

830. Vandamme, M., Bažant, Z. P., & Keten, S. (2015). Creep of lubricated layered nano-porous solids and application to cementitious materials. *ASCE Journal of Nanomechanics and Micromechanics, 5*, 04015002-1–04015002-8.

831. Vandamme, M., Brochard, L., Lecampion, B., & Coussy, O. (2010). Adsorption and strain: The CO2-induced swelling of coal. *Journal of the Mechanics and Physics of Solids, 58*, 1489–1505.

832. Vandamme, M., & Ulm, F.-J. (2009). Nanogranular origin of concrete creep. *Proceedings of the National Academy of Sciences of the United States of America, 106*, 10552–10557.

833. Vandamme, M., & Ulm, F.-J. (2013). Nanoindentation investigation of creep properties of calcium silicate hydrates. *Cement and Concrete Research, 52*, 38–52.

834. Vandamme, M., Zhang, Q., Carrier, B., Ulm, F.-J., Pellenq, R., van Damme, H., et al. (2013). *Creep properties of cementitious materials from indentation testing: signification, influence of relative humidity, and analogy between C-S-H and clays, CONCREEP 9.* Boston.

835. Vashy, A. (1892). Sur les lois de similitude en physique. *Annales télégraphiques, 19*, 25–28.

836. Vítek, J. L. (1997). Long-term deflections of large prestressed concrete bridges. *Serviceability models - behaviour and modelling in serviceability limit states including repeated and sustained load, CEB Bulletin d'Information, CEB, Lausanne* (Vol. 235, pp. 215–227 and 245–265).

837. Voeltzel, A., & Dix, A. (2004). A comparative analysis of the Mont Blanc, Tauern and Gotthard tunnel fires. *Routes/Roads, 324*, 18–34.

838. Voigt, W. (1892). Ueber innere Reibung fester Körper, insbesondere der Metalle. *Annalen der Physik, 283*, 671–693.

839. Volterra, V. (1887). Sopra le funzioni che dipendono da altre funzioni. *R. C. Accad. Lincei, 4*, 97–105.

840. Volterra, V. (1909). Sulle equazioni integrodifferenziali della teoria dell'elasticità. *Atti Reale Accad. Lincei, 5*, 296–301.

841. Volterra, V. (1959). *Theory of functionals and of integral and integro-differential equations (1925 Madrid lectures).* New York: Dover.

842. von Helmholtz, R. (1886). Untersuchungen über Dämpfe und Nebel, besonders über solche von Lösungen. *Annalen der Physik, 263*, 508–543.

843. Vořechovský, M., & Novák, D. (2005). Simulation of random fields for stochastic finite element analysis. *Safety and reliability of engineering systems and structures* (pp. 436–443). Rotterdam: Millpress.

844. Šmilauer, V., & Bažant, Z. P. (2010). Identification of viscoelastic C-S-H behavior in mature cement paste by FFT-based homogenization method. *Cement and Concrete Research, 40,* 197–207.

845. Šmilauer, V., & Bittnar, Z. (2006). Microstructure-based micromechanical prediction of elastic properties in hydrating cement paste. *Cement and Concrete Research, 36,* 1708–1718.

846. Štefan, R. (2017). Private communication at the Czech Technical University with M. Jirásek, June 2017.

847. Vuilleumier, F., Weatherill, A., & Crausaz, B. (2002). Safety aspects of railway and road tunnel: Example of the Lötschberg railway tunnel and Mont-Blanc road tunnel. *Tunnelling and Underground Space Technology, 17,* 153–158.

848. Wallo, E. M., Yuan, R. L., Lott, J. L., & Kesler, C. E. (1965). *Sixth progress report: Prediction of creep in structural concrete from short time tests, T&AM report 658.* Urbana, Illinois: University of Illinois.

849. Wang, J., Yan, P., & Yu, H. (2007). Apparent activation energy of concrete in early age determined by adiabatic test. *Journal of Wuhan University of Technology-Materials Science, 22,* 537–541.

850. Ward, M. A., & Cook, D. J. (1969). The mechanism of tensile creep in concrete. *Magazine of Concrete Research, 21,* 151–158.

851. Ward, M. A., Neville, A. M., & Singh, S. P. (1969). Creep of air entrained concrete. *Magazine of Concrete Research, 21,* 205–210.

852. Watson, K. M. (1943). Thermodynamics of the liquid states, generalized prediction of properties. *Industrial and Engineering Chemistry, 35,* 398–406.

853. Wei, Y., Hansen, W., Biernacki, J. J., & Schlangen, E. (2011). Unified shrinkage model for concrete from autogenous shrinkage test on paste with and without ground-granulated blast-furnace slag. *ACI Materials Journal, 108,* 13–20.

854. Weibull, W. (1939). The phenomenon of rupture in solids. *Proceedings of the Royal Swedish Institute for Engineering Research, 153,* 1–55.

855. Weigler, H., & Karl, S. (1981). Kriechen des Betons bei frühzeitiger Belastung. *Betonwerk und Fertigteiltechnik,* (H9/81).

856. Weil, G. (1959). Influence des dimensions et des tensions sur le retrait et le fluage du béton. *RILEM Bulletin, 3,* 4–14.

857. Weiss, W. J., Schiessl, A., Yang, W., & Shah, S. (1998). Shrinkage cracking potential, permeability, and strength for HPC: Influence of W/C, silica fume, latex, and shrinkage reducing admixtures. In *Proceedings of the International Symposium on High-Performance and Reactive Powder Concretes, Sherbrooke, Canada* (Vol. 1, pp. 349–364).

858. Welty, J. R., Wicks, C. E., & Wilson, R. E. (1969). *Fundamentals of heat and mass transfer.* New York: Wiley.

859. Wendner, R., Hubler, M. H., & Bažant, Z. P. (2015). Optimization method, choice of form and uncertainty quantification of model B4 using laboratory and multi-decade bridge databases. *Materials and Structures, 48,* 771–796.

860. Wendner, R., Hubler, M. H., & Bažant, Z. P. (2015). Statistical justification of model B4 for multi-decade concrete creep using laboratory and bridge databases and comparisons to other models. *Materials and Structures, 48,* 815–833.

861. Wendner, R., Strauss, A., Guggenberger, T., Bergmeister, K., & Teply, B. (2010). Approach for the assessment of concrete structures subjected to chloride induced deterioration. *Beton-und Stahlbetonbau, 105,* 778–786.

862. Wesche, K., Schrage, I., & von Berg, W. (1978). Versuche zum Einfluss des Belastungsalters auf das Kriechen von Beton. *Deutscher Ausschuss für Stahlbeton* (Vol. 295). Berlin: W. Ernst & Sohn.

863. Whaley, C. P., & Neville, A. M. (1973). Non-elastic deformation of concrete under cyclic compression. *Magazine of Concrete Research, 25,* 145–154.

864. White, T. L., Foster, D., Wilson, C. T., & Schaich, C. R. (1995). *Phase II microwave concrete decontamination results, Technical report DE-AC05-84OR21400*. Oak Ridge, Tenn: Oak Ridge National Laboratory.
865. Whitney, G. S. (1932). Plain and reinforced concrete arches. *Journal of the American Concrete Institute, 28*, 479–519; *Discussion, 29*, 87–100.
866. Widder, D. V. (1941). *The Laplace transform*. Princeton: Princeton University Press.
867. Widder, D. V. (1971). *An introduction to transform theory*. New York: Academic Press.
868. Wierig, H. J. (1965). Die Wasserdampfdurchlässigkeit von Zementmörtel und Beton. *Zement-Kalk-Gips*, 471–482.
869. Wischers, G., & Dahms, G. (1977). Kriechen von frühbelasteten Beton mit hoher Anfangsfestigkeit. *Beton, H2*, 69–74; *H3*, 104–108.
870. Wittmann, F. (1970). Einfluss des Feuchtigkeitsgehaltes auf das Kriechen des Zementsteines. *Rheologica Acta, 9*, 282–287.
871. Wittmann, F. H. (1971). Bestimmung physikalischer Eigenschaften des Zementsteins. *Rheologica Acta, 20*, 422–428.
872. Wittmann, F. H. (1971). Kriechverformung des Betons unter statischer und unter dynamischer Belastung. *Rheologica Acta, 10*, 422–428.
873. Wittmann, F. H. (1971). Vergleich einiger Kriechfunktionen mit Versuchsergebnissen. *Cement and Concrete Research, 1*, 679–690.
874. Wittmann, F. H. (1974). Bestimmung physikalischer Eigeschaften des Zementsteins. *Deutscher Ausschuss für Stahlbeton* (Vol. 232, pp. 1–63). Berlin: W. Ernst & Sohn.
875. Wittmann, F. H. (1980). Properties of hardened cement paste. In *Proceedings of the International Congress on Chemistry of Cement* (Vol. I). Paris. Subtheme VI-2.
876. Wittmann, F. H. (1982). Creep and shrinkage mechanisms. In Z. P. Bažant & F. H. Wittmann (Eds.), *Creep and shrinkage of concrete structures* (pp. 129–161). New York: Wiley.
877. Wittmann, F. H. (1985). Influence of time on crack formation and failure of concrete. In S. P. Shah (Ed.), *Application of fracture mechanics to cementitious composites* (pp. 593–616). Dordrecht: Martinus Nijhoff Publishers.
878. Wittmann, F. H., Bažant, Z. P., Alou, F., & Kim, J. K. (1987). Statistics of shrinkage test data. *Cement, Concrete and Aggregates, 9*, 129–153.
879. Wittmann, F. H., & Roelfstra, P. E. (1980). Total deformation of loaded drying concrete. *Cement and Concrete Research, 10*, 211–222.
880. Xi, Y. (1995). A model for moisture capacities of composite materials. *Part I: Formulation. Computational Materials Science, 4*(1), 65–77.
881. Xi, Y. (1995). A model for moisture capacities of composite materials, Part II: Application to concrete. *Computational Materials Science, 4*(1), 78–92.
882. Xi, Y., & Bažant, Z. P. (1989). Sampling analysis of concrete structures for creep and shrinkage with correlated random material parameters. *Probabilistic Engineering Mechanics, 4*, 174–186.
883. Xi, Y., Bažant, Z. P., & Jennings, H. M. (1994). Moisture diffusion in cementitious materials: Adsorption isotherms. *Journal of Advanced Cement Based Materials, 1*, 248–257.
884. Xi, Y., Bažant, Z. P., Molina, L., & Jennings, H. M. (1994). Moisture diffusion in cementitious materials: Moisture capacity and diffusivity. *Journal of Advanced Cement Based Materials, 1*, 258–266.
885. Yang, Y., Sato, R., & Kawai, K. (2005). Autogeneous shrinkage of high-strength concrete containing silica fume under drying at early ages. *Cement and Concrete Research, 35*, 449–456.
886. Yee, A. A. (1979). Record span box girder bridge connects Pacific Islands. *Concrete International, 1*, 22–25.
887. York, G. P., Kennedy, T. W., & Perry, E. S. (1970). *Experimental investigation of creep in concrete subjected to multiaxial compressive stresses and elevated temperatures, Research report to Oak Ridge National Laboratory 2864-2*. Department of Civil Engineering: University of Texas, Austin.

888. Young, D. M., & Crowell, A. D. (1962). *Physical adsorption of gases*. Washington, D.C.: Butterworth.

889. Yu, Q., Bažant, Z. P., & Wendner, R. (2012). Improved algorithm for efficient and realistic creep analysis of large creep-sensitive concrete structures. *ACI Structural Journal, 109*, 665–675.

890. Zelinski, R. (2010). *Private communication*. Caltrans, California: Former Bridge Engineer.

891. Zhang, Q., Roy, R. L., Vandamme, M., & Zuber, B. (2014). Long-term creep properties of cementitious materials: Comparing microindentation testing with macroscopic uniaxial compressive testing. *Cement and Concrete Research, 58*, 89–98.

892. Zhang, Y. M., Pichler, C., Yuan, Y., Zeiml, M., & Lackner, R. (2013). Micromechanics-based multifield framework for early-age concrete. *Engineering Structures, 47*, 16–24.

893. Zhou, F. P. (1992). *Time-dependent crack growth and fracture in concrete, Report TVBM-1011*. Lund, Sweden: Lund Institute of Technology.

894. Zhou, F. P., Lydon, F. D., & Barr, B. I. G. (1995). Effect of coarse aggregate on elastic modulus and compressive strength of high performance concrete. *Cement and Concrete Research, 25*, 177–186.

895. Zhukov, V. V., & Shevchenko, V. I. (1974). Investigation of causes of possible spalling and failure of heat-resistant concrete at drying, first healing and cooling. In K. D. Nekrasov (Ed.), *Zharostoikie Betony (Heat-resistant concretes)* (pp. 32–45). Moscow: Stroiizdat.

896. Ziegler, H. (1983). *An introduction to thermomechanics*. Amsterdam etc: North-Holland.

897. Zienkiewicz, O. C., Watson, M., & King, I. P. (1968). A numerical method of viscoelastic stress analysis. *International Journal of Mechanical Sciences, 10*, 807–827.

Index

© Springer Science+Business Media B.V. 2018
Z.P. Bažant and M. Jirásek, *Creep and Hygrothermal Effects
in Concrete Structures*, Solid Mechanics and Its Applications 225,
https://doi.org/10.1007/978-94-024-1138-6

Printed in the United States
By Bookmasters